Connie Church

BUSINESS LAW
Text and Cases

SECOND EDITION

BURT A. LEETE

College of Business and Management, University of Maryland
Member of the Bar, District of Columbia and State of Maryland

MACMILLAN PUBLISHING CO., INC.
NEW YORK
Collier Macmillan Publishers
London

To Jennifer

MACMILLAN PUBLISHING CO., INC.
866 Third Avenue, New York, New York 10022

COLLIER MACMILLAN CANADA, LTD.

Library of Congress Cataloging in Publication Data

Leete, Burt A.
Business law.

Includes index.
1. Commercial law—United States. I. Title.
KF889.L44 1982 346.73'07 81-8472
ISBN 0-02-369360-6 347.3067 AACR2

Printing: 1 2 3 4 5 6 7 8 Year: 2 3 4 5 6 7 8 9

PREFACE

The purpose of the second edition of this book remains the same as the first, to provide a useful tool for the study of legal problems facing not only the student who intends to enter business, but also any other individual. Both legal cases and textual comment are employed. An attempt has been made to mix the classic landmark cases with more recent cases having a bearing on present-day problems. Although most of the cases are of recent vintage, cases that clearly illustrate significant legal concepts have not been eliminated merely because of their age. In addition to the text and cases, discussion questions are presented after a number of cases to challenge students to think beyond the material in the text and to help them determine whether they have mastered the concepts that the cases illustrated. For ease of discussion, the cases are placed in the sections of the text that they are intended to illustrate.

The cases have been edited in such a fashion as to include enough of the facts for the student to get a flavor of the actual situation that led to the lawsuit. Too often cases are condensed to a point where nothing is left but naked statements of law. Of course, more cases could be included at the cost of eliminating much of the flavor and interest from them. I hope a satisfactory compromise has been reached.

Many of the cases found in the first edition have been replaced with newer ones. In some instances a new case has been selected because it illustrates a particular point better. In addition, some cases have been added where none appeared in the original edition.

One problem that constantly faces the business law professor is how to integrate the appropriate Uniform Commercial Code material into the chapters that discuss contracts. Everyone has his or her own approach to teaching this material. In order to facilitate the integration of the Uniform Commercial Code material I have added a section in most of the chapters dealing with contracts that summarize the changes the code has made in the general law of contracts. The purpose is to afford the professor the basis for further discussion of this material if he or she finds it desireable at this point. The material is only a summary and is not intended to replace the complete discussion of this code material that is found in the section on Sales.

In addition to the Uniform Commercial Code, Uniform Partnership Act, and Uniform Limited Partnership Act, The Model Business Corporation Act has been added to the Appendix. Many of the state corporation statutes are based upon this Act. It is also useful for purposes of discussion where a provision of a particular state statute differs from the Model Act.

At the end of each chapter are a number of problem cases. In many instances these are designed to give a contrasting or different point of view from the cases presented in the main body of the text. These problems are based on actual cases that are appropriately identified. A number of new problems have been included in this edition.

If a text in business law is to be useful to the student, it must cover a wide variety of subject areas. The omission of significant topics, such as the Uniform Commercial Code or a proper treatment of corporations, would result in a loss of usefulness. On the other hand, it is not possible to cover all subjects in one text. I have therefore attempted to cover the traditional areas of business law and, in addition, a selection of several others less frequently covered that may be of interest and significance to the student. The areas of torts, antitrust law, and consumer protection are examples. These areas obviously affect us all, whether we are business students or not. The Magnuson–Moss Warranty Act is also covered because of its impact on the Uniform Commercial Code. This edition also includes two chapters on the new Bankruptcy Code.

The text is broken into a number of sections for flexibility and because business law is frequently taught as a two- or three-term course. For example, during the first term one may wish to cover the introductory material in the first section together with contracts, sales, and public policy and business. The Uniform Commercial Code, agency, and business organizations may be covered during the second semester. If necessary, certain sections may be eliminated.

Section Two, on contracts, deals with the traditional legal concepts. Where the Uniform Commercial Code has changed significant traditional contract principles, the change has been mentioned in this section. However, because the code affects only contracts for the sale of goods, it is necessary to discuss general contract law and the Uniform Commercial Code separately. A summary of Uniform Commercial Code changes is found in most of the contracts chapters.

To further aid the student in understanding the full impact of the code on the law of contracts, I have placed the discussion of sales in Section Three, immediately following the general contract section. Many schools treat the Uniform Commercial Code in a separate course. Therefore the sections are independent, and the text is designed so that Section Three may be treated in some other order. Where they are relevant, the appropriate sections of the code are cited.

In the discussion of warranties, I have chosen to discuss not only the code approach, but also the Magnuson–Moss Warranty Act; the act has had great impact on product warranties. Also included in this chapter is a discussion of products-liability problems in general, a matter of relevance to the code warranty material.

Perhaps the most difficult material for students of business law to master is that on secured transactions and negotiable instruments. I believe this is partly because texts frequently do not relate many necessary concepts on the subject to factual situations. I have attempted to be as specific and factual as possible. I have also attempted to explain the individual code provisions in as many instances as possible. My experience has been that merely telling the student what the code says is, in many cases, not sufficient.

The relationship of government and business and consumer protection are areas so broad as to be the subjects of one or more courses separate from the basic course in business law in some schools. It is nevertheless important for the business law student to be aware of some of the more significant concepts in these fields. Two chapters in Section Nine are designed to serve this purpose for the instructor who wishes to give a substantial amount of time to two such important subjects. Cases have been added to the chapter on Consumer Protection. As mentioned earlier, two chapters on the new Bankruptcy Code have been added to this section.

As with any work of this magnitude, thanks are owed to a number of people for their help in preparing this edition. I am indebted to my research assistant, Gail Azaroff. Her comments and suggestions in reviewing the manuscript were most helpful. Her assistance in research for the text material and preparation of test questions is also very much appreciated.

I also thank the editors at Macmillan for their comments and suggestions in the preparation of the final version of this work, as well as the many professors of business law who reviewed the manuscript at various stages. Many of their suggestions were incorporated in the final draft of this text. I am particularly indebted to Professor Frank Land whose thorough review of the manuscript and suggestions were of invaluable help in the preparation of this edition.

Once again I would like to thank Karen for her support of this project. She edited and reviewed this work and prepared a number of the materials found in it. Without her spiritual and moral support the job would have been immeasurably more difficult.

In the final analysis an author is responsible for his own work. I hope that this edition continues the well received philosophy of the first and that it is an improvement upon it.

B.A.L.

TABLE OF CASES

CONTENTS

SECTION I

Introduction

SECTION II

Contracts

SECTION III

Sales

SECTION IV
Commercial Paper

SECTION V
Property

SECTION VI
Agency

SECTION VII

Partnership

SECTION VIII

Corporations

SECTION IX

Government Protection of Business and the Consumer

Appendices

SECTION I

Introduction

CHAPTER 1

Nature, History, and Sources of Law

1:1. Why Study Law? One question that the student inevitably asks himself at some point in a course is, "Why am I studying this subject?" The answer, when the subject matter is law, is relatively easy. Because it affects almost everything that one does, like it or not. This is particularly true in our highly industrialized and complicated society. It is important that the individual have some basic understanding of his rights and responsibilities under the system that governs him. He should also have some knowledge of how that system works. There are other reasons to study law, of course, but this is one of the most practical ones.

Almost any activity that one may engage in is affected in some way by the law. Buying a house or car, charging goods on a credit card, driving a car, and even marriage and dying are examples of some common activities that may bring one into contact with the legal system. It is virtually impossible to escape the impact of the law. It affects us either directly or indirectly almost every day. Pick up any newspaper and you will find it replete with accounts of court decisions and the actions of legislative bodies that may alter your daily life. One ignores the legal system at his peril. He simply cannot escape its impact. Therefore, a basic knowledge of the subject matter is necessary if one is to have knowledge of his responsibilities as well as his rights.

You should find studying the law interesting in addition to being extremely useful. Your study will entail the reading of a number of cases in order to help you grasp the impact of the legal principles involved. Don't forget that the names in those cases represent real people. In many instances, their day in court was one of the most important of their lives. The decision of the court may have resulted in rights being vindicated, hopes dashed, businesses saved, or lives being altered. It is easy to lose sight of this fact when one is reading the unemotional analysis of the legal problem by an appellate court.

In short, a study of the law is both interesting and relevant in some way to everyone. One problem in the study of law is to determine where one should start because the various topics are so interrelated. Of necessity, they must be treated separately. For instance, if one runs a red light and injures another individual, he will be brought into contact with the criminal law with all its various responsibilities, rights, and protections. The civil law of torts will be involved as a result of the injuries to the other party, and contract law may come into play if there is an attempt to settle the case. In addition, there may be constitutional problems in the criminal case, as well as agency problems in the civil one. In short, a half dozen legal topics may be involved in just this one incident. Perhaps the best place to start a study of the law is to examine the purposes the law seeks to fulfill and then examine briefly the particular legal system under which we operate. Finally in this chapter we will look at the sources of the law by which we are governed.

1:2. The Nature and Purposes of Law: A Process. Before going any further, perhaps we should examine very briefly the nature and purposes of this subject called the law. What is it and what function does it serve in our society? To define "the law" precisely is difficult at best. Over the years, literally hundreds of definitions have been given and all are to a certain extent accurate. They range from Mr. Bumble's definition in Charles Dickens's *Oliver Twist* that ". . . the law is a ass, a idiot," to the definition found in Blackstone, *Commentaries* (1847, p. 44), that law is ". . . a rule of civil conduct prescribed by the supreme power in a state, commanding what is right, and prohibiting what is wrong." Oliver Wendell Holmes, one of America's greatest jurists, stated that, "The prophecies of what the courts will do in fact, and nothing more pretentious, are what I mean by the law."[1] In other words if one knows the rules, he can attempt to predict how a court will resolve a particular dispute.

None of these definitions is all-encompassing, however. For instance, the word *law* implies more than a set of rules or the use of those rules to attempt to predict court actions, although this is certainly part of the law. The law is, broadly speaking, a continuous process of administering the conduct of society. Lawyers may advise clients by advising them of the rules and attempting to predict court actions; courts may be settling disputes, police may be tracking down a criminal, and the legislature may be creating new rules and deleting old ones. In other words, the law is a process by which people govern and order their conduct. This process results in constant change in the law. This is one reason Holmes talked about prophesying what courts will do in a given case. Often one cannot state with absolute certainty how a court will treat a given set of facts because the law is in constant change, and each case is influenced by its unique characteristics.

This inability to predict with total accuracy is frustrating to students, not to mention to lawyer's clients. However, once one understands the nature of the process—how it works, how the rules are developed and changed—and gains a knowledge of some of the basic rules or laws, then he should have some idea of the difficulty of making the prophecies Holmes talked about. In order to understand more fully the legal system under which our society operates, one must have some understanding of how it developed.

[1] Holmes, "The Path of the Law," 10 *Harvard Law Review* (1897), p. 461.

1:3. LEGAL HISTORY: THE COMMON-LAW BACKGROUND. The basis of the American legal system was transferred from England with the colonists. That system, in turn, began its formation after the Norman Conquest in 1066. This is not a text on English history, but a brief explanation of the development of this system is necessary to a more complete understanding of our system.

William the Conqueror is credited with centralizing the English government and court system. William set up the Curia Regis, or Kings Court, which had great judicial authority as well as other duties, some ambassadorial and some purely administrative. At first this court traveled with the king, but eventually court was held in each county at regular intervals. There were few, if any, written statutes as we know them today, so the judges decided cases based upon their knowledge of the customs and traditions of the people. One problem with this system, of course, is the absence of certainty that two cases with similar facts will be decided in a similar manner.

In the reign of Edward I (1272–1307), the judges began to publish their decisions in Year Books that could be read by other judges. Thus, when a judge was presented with a case for decision, he could refer to similar cases for guidance in his decision. With the passage of time, this practice became ingrained in the system and eventually evolved into the doctrine known as *stare decisis* (to stand on decided cases). The doctrine of stare decisis required judges to follow decisions of earlier courts when presented with a similar set of circumstances. No law or statute of any kind required the application of this doctrine and procedure; rather, the practice was developed by the courts themselves and has been followed in England down to the present time. The highest court of a given jurisdiction must either follow its prior decision in a similar case or overrule it. A decision might be overruled if it is not in the spirit of the times. This procedure of referring to earlier cases added some degree of certainty to the law and resulted in a body of what is, in effect, judge-made law, known now as the *common law.* This body of law developed over hundreds of years as a result of the courts' following the decisions of earlier courts and adding to the scope of those decisions as the peculiar facts of a case required. The body of law known as the common law did not appear intact overnight but rather grew and developed over a long period of time. The courts developed the basic legal categories of tort, contract, and criminal actions as well as court procedures. For example, the requirements for a valid contract or the definition of a particular crime, such as murder, were developed by the common-law courts and did not result from statutes passed by Parliament. The common law was therefore a process of continuous growth and not a static thing. This is not to say that legislation was not enacted but rather that the basic law was the common law, to which the statutes were added.

There are some drawbacks to this system as well as some common misconceptions. In the first place, if a court decision resulted from a mistaken interpretation of the law, there was, and is, the chance that the mistake might become perpetuated because of the doctrine of stare decisis. Some legal scholars, for example, feel that this process accounts in part for the requirement of consideration for a valid legal contract, which is unique to our legal system and which we will examine in detail later. There is also the notion that once the courts decide a case, that law remains with us forever because of the doctrine of stare decisis. This idea ignores the fact that the common law is constantly changing. One way the common law changes is that the courts may distinguish between a case with one set of facts and the facts and circumstances of another case previously decided. In this way, the common law has grown and changed—slowly to be sure, but it does grow. One should also not ignore the fact that the legislature has had the ability to pass statutes to alter or change the common law. Also courts today can and do overrule prior decisions on occasion.

Another criticism of the common law was that the system eventually became extremely complicated and formalistic, to the point where it was difficult to understand and use as an instrument for justice. One of the parts of the system that became extremely complicated and technical was the use of writs in order to obtain court action.

(1) Common-Law Writs and Actions. In order to get the common-law court to hear one's case, one had to file a writ, which was, as a practical matter, the form of action under which the court would proceed. The writ selected determined the remedy applied by the court. At first there were only a few writs available, but with the passage of years more were added to fill in various gaps until perhaps thirty or forty existed. Today this system has been greatly simplified or eliminated, but many of the concepts and much of the terminology remain.

The common-law writs made three remedies available: money damages, recovery of personal property, and recovery of real property. Eventually the primary remedy of the law courts came to be an award of money damages. Real property is, of course, real estate, that is, land and the things permanently attached to it, such as houses. All other property is personal property. The principal writ for the recovery of the possession of real property was *ejectment.*

The recovery of possession of personal property was affected by the use of the writs of *replevin* for the recovery of property wrongfully taken and retained and *detinue* for property lawfully taken but wrongfully retained. These writs were of little good, however, where the property was damaged in the hands of the wrongdoer.

The writs used for the recovery of money damages can be collected into two groups, writs *ex delicto* (from a tort) and *ex contractu* (from a contract). The writ of *trover* was used for the wrongful interference with one's right to possession of his personal property. The remedy was money damages calculated at the time of the interference. The writ of *trespass* was used to recover money damages for injuries as the result of direct intentional force used against a person, personal property, or real property: *trespass vi et armis* (by force and arms) for injury to the person, *trespass de bonis asportatis* (for goods carried away) for injury to personal property, and *trespass quare clausum fregit* (he who invades the close) for injury to real property. The problem with the writ of trespass, of course, was that injuries resulting from indirect or unintentional force could not be compensated. Later the courts developed the action of *trespass on the case,* sometimes called *case.* This was used for the recovery of money damages for injuries to one's person, personal property, or real property as the result of indirect or unintentional acts. Slander, libel, fraud, and negligence are the type of actions that required the use of this writ.

Actions *ex contractu* were *debt, covenant, assumpsit,* and *account.* The action of debt was used to recover a specific sum of money as the result of a contract, a judgment, or the like. The writ of covenant was used to recover for damages resulting from the breach of a contract under seal. Assumpsit was used to recover for damages resulting from a breach of a simple contract (one not under seal) in which the amount of money was not specific. Account was used to recover money from a person who should render an account to another because of some fiduciary relationship such as a trusteeship or a guardianship.

This system was extremely complicated. It also required that different allegations be made in order to satisfy the requirements of a writ, which in turn provided for a specific remedy. Thus, if one wanted to recover the value of the property at the time it was taken, he would use the writ of trover. If he wanted to recover the property itself, he would use the writ of replevin. In order to fashion the described remedy, the courts came to allow the use of "legal fictions" in statements of the facts necessary to gain the issuance of the writ. These legal fictions did not have to square with the actual facts of the case; they were merely a method used to obtain the proper form of action so that the desired remedy would be applied.

The common-law courts also adopted the use of the jury system. The jury system was used in both common-law and criminal cases, but it really had its earliest antecedents in the criminal law. The use of the jury can be traced historically back to the French inquest nearly a thousand years ago. The Normans adopted the inquest and modified it shortly after they arrived in England. In 1166, the procedure of using a jury to accuse one of a crime was established by statute. This use of the jury is the predecessor of our modern *grand jury.* Once the person was accused, he was tried by another jury, later called the *petit jury.* Originally it functioned quite differently from our modern jury. For instance, it was often composed in part of those acquainted with the facts of the offense, and even the accusers were included so that their opinions as to guilt might be asked. Today, of course, we attempt to exclude from the jury anyone with knowledge of the facts of the case. This procedure did not make the use of a jury overwhelmingly popular. The use of the jury was "encouraged" by a statute passed in 1275, the Statute of Westminster. Under this statute, if one accused of a crime rejected the use of a jury, as was permitted, he was laid on the ground and his chest loaded with successively heavier weights until he either changed his mind or expired. The use of the jury in civil cases developed from these beginnings, and its use was adopted by the common-law courts.

If all this seems confusing, it is only because it was. The technical forms of action became so cumbersome that most jurisdictions in this country have abolished or greatly simplified their use. Many of the concepts, principles, and terms remain in use today, however. You will run across many of these in the cases presented later in the text.

1:4. Equity. There came a time when the law courts grew stagnant. The development of new writs to accommodate changing conditions virtually ceased. Thus there were many occasions when an obvious injustice was suffered, but the law courts could not act because there was no writ to cover the situation. When this happened, the only recourse was to turn to the king, who, since he was sovereign, could effect a remedy.

Eventually the king, no doubt tired of handling these matters himself, appointed chancellors to hear these problems. This system came to be known as the *equity courts.* Because the chancellors in effect represented the king, they were not bound to follow the writ system developed by the law courts and could fashion their own remedies, developing their own separate procedures and requirements. In the first place, equity courts would not hear a case if there was an adequate remedy at law, the most usual remedy being money damages. Thus if one's problem could not be solved or one could not be compensated by an award of money damages, the proper place to go was the equity court. In addition, because the chancellor represented the king, the procedure of using a jury never developed in the equity courts. Although equity was similar to the common law in that it was based on case law, the equity courts were much more flexible in fashioning remedies to meet a particular situation.

(1) Equitable Remedies. As a rule, the equity courts did not grant money damages, although they might in an appropriate case. Their remedies most often required people to do things. Because the law courts primarily gave money damages, it became known that law acted in rem, or on the thing, and equity acted in personam, or on the person. These remedies will be discussed more fully in Chapter 14, but a brief example of each will make their use clear when they arise in the cases that follow.

(a) *Accounting.* When there is a fiduciary relationship between two people and one of the parties is responsible for collecting money that belongs to the other, in a proper case the courts will order an accounting. This often occurs when the party collecting the money refuses to account to the other party for it, or when there is a suspicion that the figures are incorrect for one reason or another. An equity court has the power to order an accounting in such a case. The remedy is often granted in disputes between partners or between beneficiaries and trustees.

(b) *Specific Performance.* An equity court can order a contract to be specifically performed by an individual when the contract is breached but when money damages will not adequately compensate the innocent party. One frequent situation in which money damages are not adequate is when the subject matter of the contract is unique, as, for example, in the case of land or an ancient Chinese vase.

(c) *Reformation and Rescission.* If a written contract or deed does not reflect the actual intent of the parties, an equity court can reform it in order to reflect that intent. The court can also decree a rescission of a contract or deed—that is, call it off—if it resulted from fraud, duress, undue influence, or the like.

(d) *Partition.* Often property is owned by more than one person. Each co-owner may own a percentage of the entire property but not any specific part of the property (an undivided interest). If one co-owner wishes to sell but the other does not, he may ask the equity court to divide the property so that it may be sold and the proceeds divided. Whether this procedure is necessary depends on the type of co-ownership and the nature of the property.

(e) *Quieting Title.* Often there may be some outstanding claim, real or apparent, on a title that must be removed before property can be sold. A suit in equity may be instituted to clear up such a claim. This procedure is rather quaintly known as removing "a cloud" on the title or quieting title.

(f) *Foreclosure of Mortgage.* A mortgage is given on property to secure a debt or obligation to a creditor. If the debt is not paid, the creditor may take the property subject to the debtor's right to reclaim or redeem his property. In order to terminate this right of the debtor so that he might freely use or sell the property, the creditor may institute a proceeding in equity to foreclose the debtor's equity of redemption.

(g) *Injunction.* Perhaps the most well-known equitable remedy is the injunction. The remedy is most often used to prevent the infringement of patents, trademarks, and copyrights; to limit picketing; prevent interference with water rights; and in conjunction with labor disputes. It is essentially an order or decree from a court restricting or requiring some sort of action or behavior.

(2) Maxims of Equity. Because the equity courts were more flexible in their administration of justice, a number of principles developed that governed the courts in deciding whether justice required that relief should be given to an individual. These principles were reduced to specific phrases. These phrases are often stated rather glibly, but some reflection on them should reveal the essential justice behind each one of them. The following are some of these principles or maxims: He who comes to equity must come with clean hands; he who seeks equity must do equity; equity will not suffer a wrong without a remedy; equity abhors a forfeiture; equity considers that done which ought to be done; where there are equal equities, the first in time will prevail; equity looks to intent and not to form; and equity aids the vigilant and not those who slumber on their rights.

This last maxim is often known as the *doctrine of laches.* The right to bring an equitable action is not limited by a statute of limitations as the law actions are. Nonetheless, if one does not act within a reasonable time, the equity court will not hear the case, particularly if the party is aware of these rights.

1:5. Law and Equity Courts. The common-law and equity courts developed separately in England and remained that way. Not only were their procedures and remedies different, but their judges were different and remained separated. This system was transferred to America by the colonists. Today, in most of those states where the distinction between law and equity remains, the same judges administer a single court. If an action is one "at law," then the remedies and procedures of that form of action are used. On the other hand, if the action is an equitable one, then the appropriate procedures and remedies will be applied in that case. Some states such as New York, as well as the federal judicial system, have abolished the distinction between law and equity altogether. However, even in those jurisdictions, vestiges of the distinction still remain in the procedures and remedies that are used in cases that are based on equitable principles.

1:6. The Law Merchant and Ecclesiastical Law. Two other factors have influenced the development of our present-day legal system: the law of the Middle Age merchants and the law of the church. The law merchant or mercantile law developed in the Middle Ages to help settle disputes among merchants particularly those engaged in commerce between countries on the European continent and England. The common-law courts were not really concerned with the practices of the trader, as England was primarily an agrarian society at that time. As a result, merchant guilds were formed, which in turn formed merchants' courts. These courts developed a private commercial law and gave the merchants the prompt relief that commerce demanded. This "law merchant" gradually began to be accepted by the common-law courts in the seventeenth and eighteenth centuries and is the basis for our negotiable instruments law, partnership law, insurance law, and surety and guaranty law.

The ecclesiastical law of England was administered by the church courts. After some struggle with the state, these courts handled cases that arose out of the marriage contract and out of wills as well as strictly church-related problems. The Church of England maintained jurisdiction over these private matters until around the middle of the nineteenth century, when it was divested of all authority except for matters affecting the Church of England.

1:7. The Civil Law. The other great body of law that has governed Western civilization is known as the civil law as distinguished from the common law. It is the legal system that governs the countries of the continent of Europe. The civil law has its antecedents in the old Roman Codes, which were promulgated by the Emperor Justinian in 529 A.D. The civil law is basically code law, as opposed to the case system of the common law. The civil law is statutory, and although the civil-law judges do not ignore previous decisions, there is not the same emphasis on precedent that is used in the common-law system. In addition to the countries on the European continent, the countries that were colonized by them also operate under the civil-law system. These, of course, would be primarily the Latin-American countries.

1:8. THE LAW OF THE UNITED STATES. In addition to England, those countries colonized by the English also operate under the common-law system. Thus the colonists brought the system of common law and equity courts to the territory that is now the United States. There are also some civil-law influences in the United States, most notably in Louisiana, which was influenced by the French, and the far western and southwestern states, which were influenced by the Spanish. Primarily, however, the background of the legal system in this country comes from the English.

The legal system of the United States is, of course, not identical to or restricted to the common law. The common-law influence can perhaps be best visualized as the background of a painting on which other parts of the system are impressed. Thus we also have statutes, constitutions, and treaties that govern our behavior. On the other hand, if there is not a statute to cover some form of our behavior, some principle of the common law may be used. In the criminal field, for example, some states do not have statutes making simple assault and battery a crime. Nonetheless a person may be charged with this offense because it is a crime under the common law.

(1) Sources of Law. The previous discussion has given us an idea as to the origins of our legal system. Let us look now at some of the sources of our law.

(a) *The Constitutions.* When the individual colonies entered the federal union, they each gave up some of their sovereignty. The sovereign rights that they gave up to the federal government are stated in the federal Constitution. All the authority not given up by the states in the federal Constitution they retained. We then have a dual system of government: the federal government with the powers given to it under the federal Constitution and the state governments, which have the residual powers of a sovereign. These powers are, of course, limited by certain prohibitions in the Constitution for the protection of certain rights of the people of the United States.

The Constitution of the United States provides that it, as well as the laws of the United States made pursuant to it, are the supreme law of the land. The states also have their own constitutions, which often are similar to the United States Constitution. Of course, no state constitution may violate the provisions of the federal Constitution, and a state law may not violate a provision of the federal or state constitution.

(b) *Statutes and Ordinances.* Under our legal system the U.S. Congress and state legislatures may enact statutes that regulate various areas of activity in our society. The statutes passed by the Congress must not violate the U.S. Constitution. The Congress can only pass legislation pursuant to a power granted to it under the U.S. Constitution. State legislatures may also enact legislation as long as they do not violate either the U.S. Constitution or the state constitution or conflict with an area already preempted by congressional legislation.

Ordinances are generally enacted by local governmental bodies, such as counties, townships, cities, and towns. Most ordinances are enacted to regulate the health, safety, morals, and welfare of the community. Obviously these ordinances cannot violate either the state or federal constitutions. Ordinances typically regulate such varied things as billboards, the use of sidewalks and streets, building construction, and the use of fires and other potential nuisances.

(c) *Administrative-Agency Regulations.* The Congress and the states are increasingly making use of agencies to regulate various aspects of our society. Many of the statutes creating the agencies empower them to promulgate regulations, to hear and adjudicate controversies, and to issue orders for the purpose of carrying out the objectives of the legislature. For example, Congress has created the Federal Trade Commission to regulate various aspects of business. That agency is empowered to pass regulations governing the use of warranties and advertising and in other areas. One may appeal a final agency decision to the courts, but nonetheless the use of agencies to regulate various aspects of our society seems to be increasing at both the federal and the state levels.

(d) *Cases.* Obviously the legislatures have not passed laws concerning every aspect of our lives. Under our system, however, if there is not a statute covering a particular matter, a court may look to the common-law principles in rendering a decision. To that extent, court decisions have the force of law. In addition, in our legal system the courts interpret the laws passed by the legislatures. Statutes are often ambiguous and it is up to the courts to determine how they should be applied. It is up to the courts to determine whether the laws passed by the legislatures violate either the federal or the state constitution. Not all legal systems provide for the courts to perform this function. Many of the civil-law countries, for instance, do not provide for judicial review of the statutes themselves. In the United States the federal courts decide whether federal or state laws conform to the federal Constitution, and the state courts determine when state statutes violate either the state or the federal constitution.

(e) *Uniform Laws, Treaties, and Other Sources.* There are other sources of law that do not receive as much weight as those previously discussed but that are considered by courts when they make decisions and by legislatures when they enact laws. One result of our system is that there are a great many different laws governing the same subject because of the number of jurisdictions involved. There are many instances in which the law on one specific point may vary from jurisdiction to jurisdiction. This variation may be very inconvenient, particularly with regard to our extensive interstate commerce. There has been an effort to eliminate this diversity among the states by an organization known as the National Conference of Commissioners on Uniform State Laws, which is made up of commissioners from all the states, the District of Columbia, and Puerto Rico. One function of the commission is to draft a statute covering a particular subject and then encourage the states to adopt the proposed legislation. The conference has been successful in getting uniform legislation passed in a number of areas, such as the Uniform Partnership Act. Its most notable success has been in the drafting and general acceptance of the Uniform Commercial Code with the cooperation of the American Law Institute. The code is an attempt to draft a comprehensive statute in the field of commercial transactions. It covers such areas as sales, negotiable instruments, bank deposits and collections, letters of credit, bulk transfers, secured transactions, and documents of title. Each of these subject areas is preceded by a separate uniform statute, which was revised and incorporated into the code. All of the states, the District of Columbia, and the Virgin Islands have adopted the code with the exception of Louisiana, which has adopted parts of it. The code has had a great effect upon our commercial law and will be covered in much detail later in the text.

In deciding cases, the courts also look to the statements of scholars as to what the law on a particular subject is or ought to be. For example, the American Law Institute is an organization composed of some prominent judges, lawyers, and law teachers. One of the major contributions of this organization is its publication of various works known as the *Restatement of the Law,* which are similar to an encyclopedia on a particular subject. Thus the Institute has published treatises or "Restatements" on contracts, torts, property, trusts, and other subjects. These works are not the law and do not have the force of law unless adopted by the courts. They are often considered by the courts when they are making their decisions, however. Other encyclopedic treatises that purport to cover much of the various subject areas of the law are *Corpus Juris Secundum, American Jurisprudence,* and the *American Law Reports,* which are written and published by private companies.

Next we turn to an analysis of our own court system and how it works.

SUMMARY

Law is a process. The basis for our legal system was transferred from England. The English system in turn began after the Norman Conquest of 1066 when William the Conqueror centralized the English government and court system. By the reign of Edward I (1272–1307) the judges began to publish their decisions in year books. This enabled the doctrine of stare decisis to be developed. The result was the development of a body of, in effect, judge-made law, known as the common law.

Under the common law, the system of using writs was developed. The primary writ used to recover real property was the writ of ejectment. Recovery of personal property was accomplished by the writ of replevin or detinue.

The writs for the recovery of money damages were divided into two groups, writs ex delicto and writs ex contractu. The writs of trover, trespass vi et armis, trespass de bonis asportatis, trespass quare clausum fregit and trespass on the case were among the writs ex delicto. The writs ex contractu were debt, covenant, assumpsit and account.

The common law court also adopted the use of the jury system. The petit jury was used to try a person who was accused by the grand jury.

The equity courts were established in addition to the law courts. The chancellors, who represented the king, were not bound by the writ system. Separate trial procedures were also developed and a jury was not used. One could only have his case heard by an equity court if he had an inadequate remedy at law. The equitable remedies were accounting, specific performance, reformation, rescission, partition, quieting title, foreclosure and injunction.

The equity courts were not subject to statutes of limitations as were the law courts. However, if an equity action was not brought within a reasonable period of time it would be subject to the doctrine of laches. Today many, but not all, state court systems distinguish between law and equity. The federal court system has eliminated the distinction.

The law merchant and ecclesiastical law of England also have affected the development of our legal system. The civil law as it developed on the continent of Europe has also had an impact on our system. It is statutory and does not place the same emphasis on precedent as the common-law system.

The sources of law in the United States are the federal and state constitutions, statutes, ordinances, administrative agency regulations, court decisions, uniform laws, and treaties.

PROBLEMS

1. What function does law serve in our society?
2. Describe what is meant by the statement that "Law is a process."
3. What is the essential difference between the civil law and the common law?
4. Describe what is meant by the doctrine of stare decisis.
5. Describe how and why the law merchant was developed.
6. May the legislature pass a statute that is contrary to the common law?
7. If Ira Innocent loaned William Wrongdoer his horse and Wrongdoer refused to return the horse on demand, what old common-law action would Innocent use?
8. How did our use of the jury system develop?
9. Distinguish between the procedures and remedies used by the equity and the law courts.
10. Distinguish between the doctrine of laches and a statute of limitations.

CHAPTER 2

Legal Procedure and the Judicial System

The first chapter focused on the history and sources of the law used in the United States. At this point we should examine the structure of the system that applies that law in order to have some understanding of how it operates.

2:1. Trial Courts and Appellate Courts. First, it is necessary to distinguish between two different types of courts: trial courts and appellate courts. Most people are familiar with the trial court, if not from firsthand experience, then at least from the simulations that appear on television. It is the function of the trial court to determine the facts of the case and then return a verdict based upon the applicable law. There may be a jury involved in hearing the case, or in some instances a judge may hear a case without a jury. It is in the trial court that the partics and witnesses testify and are cross-examined by attorneys. The purpose of this procedure, of course, is to determine the facts of the case. The primary function of the trial court, then, is to determine the facts.

If a party feels that an error was made during the trial, he may appeal to a higher court, the appellate court. The appellate court operates in a very different manner from the trial court. The primary function of the appellate court is to see that legal mistakes are not made in the trial that could influence its outcome. As a general rule the parties appeal questions of law to the appellate court. An appellate court does not relitigate the facts but rather answers questions of law pertaining to the case. An exception would be if there is no evidence to support the decision of the lower court. Because the appellate court hears questions of law rather than fact, its procedure is quite different from the trial court's. We shall discuss this procedure later, but basically there is no jury nor is any testimony taken from witnesses at the appellate level. There are just arguments of legal issues in the case, and the court then hands down a written decision. The arguments usually involve legal theories and technicalities, something not nearly as interesting to the layman as a trial, which probably accounts for the scarcity of television programs concerning the life, loves, and tribulations of the lawyer practicing in an appellate court.

2:2. Jurisdiction and Venue. In order for a court to decide a case and enforce its judgment against a party, it must have the power to do so. This power is known as *jurisdiction.* The jurisdiction of a court is typically thought of in two contexts: jurisdiction over the person, often known as *in personam jurisdiction,* and jurisdiction over the subject matter, or *in rem jurisdiction.* Jurisdiction over the subject matter means that the court has the subject matter within the limits of the court. For instance, a court in one state would have no jurisdiction over land located in another state and would have no power to determine who has the title to that land.

Not all cases involve a specific subject matter, such as land, but in any case a court may have to have jurisdiction over the parties to a lawsuit. For instance, suppose Danny Driver, a resident of state Y, is involved in an auto accident in State X with Caren Careful, who also lives in State X. A court in State Y would not have jurisdiction in this case because Caren is not within the limits of the court and the accident also occurred elsewhere. One may, however, voluntarily submit himself to the jurisdiction of the court.

The court may gain personal jurisdiction over a defendant by serving him with a *summons.* The summons may be served personally by a sheriff or other suitable person within the geographical limits of the state. In a few cases personal jurisdiction may be gained by the publishing of a notice of the suit in a newspaper or in some states even by the tacking of a notice on the courthouse door. Obtaining jurisdiction in this latter manner is generally limited to those cases in which the subject matter of the lawsuit is present in the state. Examples are suits involving land and divorce cases. Most statutes also require some actual attempt to notify the defendant of the suit in some manner, as, for example, by mail. The majority of cases require actual service of the suit on the defendant.

Historically the jurisdiction of a court was limited to cases in which the subject matter, such as land, was within the geographical boundaries of the state or in which the defendant was actually served in the state. As our society grew more mobile, this limitation created problems, particularly with the increased use of the automobile. If one were involved in an accident with an out-of-state motorist, one could not get relief in court unless the out-of-state motorist were somehow served before he left the state. The only alternative would be to travel to the state where the motorist lived and file suit there, which of course was not practical in many cases, so the states found it necessary to develop methods for extending their jurisdiction in these situations.

The states have constantly expanded personal jurisdiction over nonresidents by the use of so-called long-arm statutes. In the case of a motorist the laws of most states provide that by driving on the state's highways, the motorist automatically appoints some designated state official as his agent to accept service of process for any suit arising out of an accident occurring while the motorist is operating a vehicle in the state. The official is most often the secretary of state or the head of the motor-vehicle administration.

More recently long-arm statutes have been passed to cover other situations.

For example, an individual or corporation that does business in a state and causes an injury may be sued in that state even though it may not physically be served with the summons in the state. The only requirement is that the defendant have certain "minimal contacts" with the state so that the suit does not violate constitutional requirements for "fair play and substantial justice."

A concept related to jurisdiction is venue. *Venue* refers to the geographical area within the state or district where a suit may be brought. Statutes regulate the venue of a case. Many state statutes provide, for instance, that a civil suit may be maintained in the county where the plaintiff or the defendant resides. The proper venue for a case involving title to land is the county where the land is situated.

2:3. THE STATE COURT SYSTEMS. Although the court systems of the various states differ in detail, they do follow a general pattern. There is usually a trial court of general jurisdiction, most often called the *circuit court* or *superior court.* Historically the name *circuit court* arose because one justice would ride the circuit from one court to another, hearing cases in each court on the circuit. This is the court where most of the significant lawsuits are instituted and jury and court trials are held. It hears both civil and criminal cases. Many states also have inferior courts of limited jurisdiction that hear smaller civil suits, often of $1,000 or less, although some hear cases up to $10,000. These courts may also hear the less serious criminal cases. Sometimes these courts are called *small-claims courts* and provide for inexpensive litigation under simplified procedures. Some states also have separate traffic courts to hear motor-vehicle cases and related matters. A number of states also allow a justice of the peace to hear minor cases. There are also special courts in some states such as the probate court, the family court, and the juvenile court. These most often are branches of the circuit court and are not necessarily inferior to it.

All of the states have at least one appellate court, and the more populated states—such as New York, Illinois, Pennsylvania, and California—have more than one level of appellate court. The highest court of a state is most often called the *supreme court,* although some states name it differently. If the state has an intermediate appellate court, it is most often called the *court of appeals* or *court of special appeals.* Those states with an intermediate appellate court most often provide for an automatic right of appeal to that court from the trial court of general jurisdiction. Whether the highest appellate court will hear an appeal from the intermediate court is frequently a matter within the discretion of the higher court in many states. In other words, an appeal to the highest court is not a matter of right. In those states where there is only one appellate court, it must hear all the appeals.

2:4. THE FEDERAL COURT SYSTEM. The federal courts are courts of limited jurisdiction because, with the exception of the Supreme Court, they are created by Congress under the power granted to it by the U.S. Constitution. The Constitution provides for a Supreme Court and such inferior courts as Congress shall authorize. Pursuant to this power, Congress has established at least one trial-level court for each state and territory. These are known as *district courts.* Eleven appellate courts known as the *U.S. courts of appeals* have also been created, as well as a number of special courts, such as the Court of Claims, the U.S. Court of Custom and Patent Appeals, the Tax Court, and others.

The basic trial-level court is the U.S. District Court. As mentioned earlier, there is at least one for each state, and the more densely populated states—such as New York, Pennsylvania, Illinois, and California—have more than one district. There are also U.S. magistrates, who issue warrants and hear minor criminal cases, such as misdemeanor and traffic cases, that arise from the use of federally owned roads, as well as other minor cases. Because the federal courts are courts of limited jurisdiction, the district courts may hear only certain types of cases. There are basically two types of cases these courts may hear: (1) those based upon *diversity jurisdiction* and (2) those based upon *federal-question jurisdiction.*

When we speak of diversity, we are talking about diversity of citizenship. Thus the district courts may hear cases in which the parties are (1) citizens of different states; (2) a state and citizens of another state; and (3) a foreign country and citizens of a state. An additional qualification for cases heard based upon diversity is that the amount in controversy must exceed $10,000 exclusive of costs and interest. Congress, of course, has the power to change this amount and has raised it from time to time in an attempt to keep the cases brought to federal court to a manageable number. In order to qualify for diversity jurisdiction, all the parties on one side of a case must be citizens of a different jurisdiction than all parties on the other. For purposes of diversity jurisdiction, a corporation is a "citizen" of the state where it is incorporated and maintains its principal place of business, hence the fact that a corporation merely does business in the same state as an adverse party will not deprive the court of diversity jurisdiction.

Obviously cases that involve citizens of different states may be tried in a state court, but where the requisite diversity exists and the amount in controversy exceeds $10,000, a defendant sued in state court may have the case removed to federal court. The defendant may do this on the notion that because the federal courts draw their juries from a wider area, the party might not be subject to any local prejudice that could exist, or because the party is more familiar with the federal-court procedure than with the rules of the local courts.

Cases that primarily involve a federal question form the other basis for federal jurisdiction. As a practical matter, cases based on federal-question jurisdiction are heard without regard to the amount in controversy. Technically, the $10,000 limitation applies but is rarely mentioned by the courts in cases involving a federal question. A federal questison may involve the violation of a federal statute, the Constitution, a treaty, or proceedings under federal laws. For example, matters relating to bankruptcies, federal tax questions, patents, antitrust laws, federal criminal laws, constitutional rights such as freedom of speech, and tort cases involving actions by agents of the federal government are all tried in the federal courts on the basis of federal-question jurisdiction. Remember that where there is federal-question jurisdiction, it is not necessary that the parties be citizens of different states.

Decisions of U.S. district courts are appealable to the U.S. courts of appeals. There are a few exceptions to this rule, such as when an act of Congress has been held unconstitutional, in which event the appeal may be sent directly to the U.S. Supreme Court. The U.S. courts of appeals were established by Congress in 1891 to relieve the U.S. Supreme Court of the burden of hearing all appeals. The country is divided geographically into ten circuits plus the District of Columbia, with a U.S. court of appeals for each circuit. A justice of the U.S. Supreme Court is assigned to each circuit, largely for supervisory purposes. One has the *right of appeal* to this court from a final decision of the U.S. district court. In addition, in certain cases Congress has provided for appeals of decisions of administrative agencies to the U.S. courts of appeals, usually for the District of Columbia circuit.

The U.S. Supreme Court is the highest court in the federal system and is basically an appellate court. There are some instances provided in the Constitution in which the Supreme Court may act as a trial court of original jurisdiction, as, for example, in the case of a controversy between the United States and a state. The Court is currently composed of a Chief Justice and eight Associate Justices, but this number may be changed by Congress.

It is a surprise to some people to learn that for the typical case there is no right of appeal to the U.S. Supreme Court; rather one must petition the Court to issue a *writ of certiorari* to the lower court, requiring it to send up the record for review. Whether the Court issues the writ is within its discretion. It usually will not hear a case unless at least four of the Justices vote to issue the writ. If the petition for writ of certiorari is denied, it does not mean that the Court has rendered a decision on the case, merely that it has decided not to hear it. A relatively small percentage of the petitions for certiorari are granted, and these usually involve cases that center on important constitutional questions or have a significant impact on the administration of justice or in which there are different interpretations of law by different circuits of the U.S. courts of appeals. There are some cases in which one has a right of appeal to the U.S. Supreme Court, such as when the highest court of a state has ruled that a federal statute is unconstitutional.

2:5. The Anatomy of a Civil Trial. At this stage it may be helpful to take an imaginary case from its inception through its trial in order to explain how the cases that appear later in the text got there and to answer procedural questions that are bound to crop up as you read these cases. In reading this section, keep several things in mind. First, this is an example of a civil as distinguished from a criminal trial. Second, of necessity the explanation is to some extent oversimplified and certainly not all-inclusive. Third, because procedures differ widely from jurisdiction to jurisdiction, there may be some variance from one state to another in the steps outlined.

(1) The Facts of the Case. Assume that Dangerous Dan and Reckless Ronald collide in their automobiles at an intersection that is controlled by a traffic light. Each has a passenger in his car. Each claims that the light was green in his direction of travel, and each suffers some physical injuries as well as property damage to the car. As they cannot agree about who should pay for the damages, the case goes to court. Perhaps we should also assume that each is uninsured.

(2) The Pleadings. Because he is unable to collect from Ronald, Dan files suit. Dan does this by filing a *complaint* (or declaration) with the court that has jurisdiction of the case. Since Dan filed the suit, he is known as the *plaintiff.* The complaint states the facts of the case as Dan alleges them; the legal theory of the suit, in this case the tort of negligence; and the amount that Dan is seeking as damages. The purpose of the complaint is to notify the person being sued, the *defendant,* that a claim is being made against him and what its basis is. The court attaches a *summons* to the complaint and serves it on Ronald, usually by sheriff. The summons commands Ronald to answer the suit within a certain period of time. Ronald does this by filing an *answer* (or plea). In most jurisdictions Ronald would state in the answer the reasons that he feels he is not liable. For instance, he may state that the light was not red but green in his direction. In some states Ronald might have the option of filing a *demurrer* to Dan's complaint. The effect of a demurrer is to ask the court to assume that the facts stated in the complaint are true but that it states no legal cause of action, that is, no basis for lawsuit. Essentially Ronald is replying to Dan's complaint by saying, "So what?" The court may *sustain* Ronald's demurrer or *overrule* it. It could be sustained, for instance, if Dan forgot to allege that Ronald either owned or drove the car. If the court feels that the complaint does state a cause of action, it will overrule the demurrer. Under modern rules of procedure, if the demurrer is overruled, Ronald could proceed to plead a defense, or if the demurrer is sustained, the court will frequently grant *leave to amend* a defective complaint when this is possible. Under the old common law the decision on the demurrer was a final disposition of the case. Today not all state procedures allow for a demurrer, although the same result is achieved by the filing of a *motion to dismiss* on the grounds stated previously.

Since Ronald feels that he had the green light, he may wish to do some suing of his own and could file a *counterclaim* in this case. It serves the same function as the complaint, but it reverses the parties and places the defendant in the plaintiff's posture. He would then be a defendant and counterplaintiff. Ronald decides against filing a counterclaim.

There may be other pleadings. Most common are various types of *motions.* A motion is nothing more than a request that the court do something. It may be made in writing or orally in open court. Among the most frequently filed motions are motions to dismiss for failure to state a cause of action and a motion for *summary judgment* on the grounds that there is no dispute between the parties as to any material fact and therefore the court should decide the case purely upon the applicable law.

(3) **Discovery.** Among the most important recent developments in legal procedure are the more liberalized rules of discovery. These are simply methods whereby each side may determine what the other side's case is about. In the real world today it is rare in an important case for the lawyer's secretary to come into court just before the trial concludes with the secret document or witness that wins the case as happens weekly on television, usually five minutes before the last commercial.

There are several primary discovery techniques.

1) *Interrogatories.* Interrogatories are written questions to a party to be answered in writing under oath. They are usually limited to a specific number of questions or to a reasonable number. In this case Ronald may send interrogatories to Dan asking him to state his version of the accident and the extent of his injuries and to list the witnesses to the accident. One has to answer these questions, usually under oath, or the court will enter judgment against him.

2) *Depositions.* An oral deposition occurs when one conducts an oral examination of another party or witness. The testimony is usually recorded and is given under oath. Usually the lawyer for each side is present, as well as a stenographer and the notary public, and the questions may involve anything relating to the lawsuit. In this case, for example, Ronald may take the deposition of Dan's doctor in order to determine the doctor's opinion of the extent of the injuries and their cause and whether they are permanent. This information is obviously useful in the preparation of a defense. One has a right to take a deposition and submit interrogatories to a party and does not have to ask the permission of the court.

3) *Motion to Produce or Examine.* Because some requests for discovery may place a burden on the other party, the reason for the request must be demonstrated to the court. If one wants to examine the financial records of the other side, for example, he may have to get the court's permission by filing a motion to produce, setting forth the reason why he needs to examine those records. In our case Ronald may wish to have Dan examined by his own doctor so that his doctor may testify as to the extent of Dan's injuries and their permanency. Ronald would file a motion for a physical examination of Dan. The request is usually granted when physical injuries are involved.

(4) **The Trial.** Next comes the most exciting stage of the lawsuit, although many lawyers state that a case is usually won or lost by the activities that precede it. In many civil cases the parties have a choice of whether they want a jury to hear the case or whether the judge alone will decide it. In our case Dan wants a jury trial.

(a) *Impaneling a Jury.* The first step is the selection of a jury. Most often a jury consists of twelve persons who are selected from a panel of people who have been chosen for jury duty during that term of court. The judge or the lawyers question the prospective jurors in order to determine if there is any reason that an individual might not decide the case without bias or prejudice. If there is a showing that an individual might be prejudiced, he is *challenged for cause* and stricken from the panel. In this case Dan's mother is on the panel, and the judge determines that she could be biased and strikes her for cause. The parties may also eliminate a limited number of individuals on a purely arbitrary basis by the use of *peremptory challenges* or strikes. In this case Dan's lawyer does not want a police officer who is on the panel to be on the jury, so he strikes him. This process of examining the jury is known as the *voir dire* in many jurisdictions.

(b) *Opening Statements and Presentation of Evidence.* Once the jury is selected, the parties give their opening statements. This is when the jury first learns about the nature of the case. Because Dan is the plaintiff, his lawyer makes his opening statement first, setting forth the facts as Dan alleges them to be and explaining what Dan expects to prove during the course of the trial. Dan's lawyer confidently predicts that the jury will find for his client. Ronald's lawyer briefly tells what his client's evidence will show and just as confidently predicts victory.

Dan is the plaintiff and therefore has the *burden of proof.* This means that he must prove those facts that he has alleged in his case and that are necessary elements of the cause of action under which he seeks to recover. If he fails to introduce enough evidence from which a jury could reasonably conclude that the facts are actually as alleged, then he has not established a *prima-facie* case and the court may terminate the proceedings at this point. One asks the court to do this by making a motion for a *directed verdict.* Because Dan has the burden of proof, he presents his case first and has an opportunity at a later time to rebut evidence that the defendant, Ronald, puts on. Dan's evidence consists of his testimony and a number of medical bills, as well as the testimony of his doctor and the passenger who was in the car. Ronald has the opportunity to cross-examine Dan's witnesses, but their testimony remains unchanged.

Ronald, the defendant, then puts on his evidence, but he has been unable to locate the passenger of his car since the accident and is without the benefit of that testimony. Although he attempts to tell what the witness told him about the red light, the judge, upon Dan's objection to this testimony, will not permit it under the rules of evidence. The judge decides that this constitutes hearsay evidence, which he will not allow the jury to hear because Dan will not have an opportunity to cross-examine the missing witness as to the truth of his statement. Ronald, of course, testifies that he thought he had the green light. At the end of Ronald's case Dan is entitled to call witnesses in rebuttal, because he has the burden of proof. He does not call anyone. The court determines, after both sides once again request a directed verdict, that there is a jury question as to the facts of the case and that both sides have introduced evidence from which the jury could conclude that the facts were as they contended.

(c) *Instructions and Final Argument.* In many jurisdictions the court next instructs the jury as to the law that should be applied in the case. (Sometimes this is done after the final arguments by the attorneys.) The length of the instructions and their detail depends to a large extent upon the complexity of the case. Each party may request the judge to instruct the jury on certain points of law that they consider applicable. If the judge refuses to give a requested instruction, such a refusal may be the subject of a later appeal. For example, in this case the judge instructs that "a statute prohibits one from going through a red light, and if you find that Ronald went through a red light, that is evidence of negligence." The suit between Dan and Ronald is rather uncomplicated from a legal point of view, so it takes only a brief period for the judge to give his instructions.

Dan's attorney then argues his case to the jury, recalling the statements of Dan's passenger concerning the red light and the testimony of the doctor as to the seriousness and pain of Dan's injuries and finally noting that no witness appeared in support of Ronald's allegations concerning the color of the light. Ronald's lawyer points out the weaknesses in Dan's case and explains the absence of Ronald's witness. Dan's lawyer has an opportunity to rebut this argument. The jury then retires to consider its verdict.

(d) *The Verdict.* The jury does not take long to consider the case and returns to render its verdict within an hour. In a civil case the decision of the jury must usually be unanimous in favor of one party or the other. (A few states permit a verdict by less than a unanimous jury in a civil case.) If the jurors cannot agree, then there is said to be a *hung jury* and the case will be tried again. In deciding who is entitled to their verdict, the jury in a civil case must be convinced by a *preponderance of the evidence.* This means that the evidence submitted by one party outweighs that submitted by the other. Note that this standard is different from that for a criminal case, in which the jury must be convinced of the defendant's guilt "beyond a reasonable doubt." This concept will be discussed in Chapter 3.

The jury steps into the box and announces its verdict: "We find for the plaintiff and determine the damages to be $2,500." Unlike on television nobody rushes up to hug Dan, nor does he cry for joy. He thanks his attorney and leaves, glad the trial is over. Ronald and his attorney, on the other hand, are convinced that the judge made some mistakes in instructing the jury and in refusing to let him testify concerning what his passenger said about the traffic light. Ronald decides to appeal.

(5) The Appeal. As discussed earlier, the appellate courts deal with questions of law rather than issues of fact, which are decided in the trial courts. Because Ronald files the appeal, he is known as the *appellant.* Dan is the *appellee* and is placed in the position of defending the decision of the trial court. It is possible for one who wins in the trial court to appeal. Dan does not do so in this case but might if, for example, he were not satisfied with the amount of damages awarded him.

The first step in the appeal is for the appellant to file a written *brief*, which is a document setting forth the legal issues involved in the appeal. The brief also discusses the legal arguments. One issue, for instance, might be whether the trial court erred in excluding the testimony of Ronald as to what his passenger said concerning the traffic light.

After both parties submit written briefs, the court usually allows the attorneys to argue the case orally. A month or so later, the appellate court hands down a written opinion analyzing the points raised in the appeal. It decides that the trial court made no error that would require them to *reverse* the decision and *remand* it for a new trial. They therefore *affirm* the decision of the trial court.

Sometimes an appellate court will find a mistake made by the trial court that would result in the opposite party's winning the case. In that instance it simply *reverses* the lower court's decision. If the mistake is one that might change the outcome of the trial but it is not certain in what respect, the appellate court *reverses and remands* the case for a new trial.

This is the only appellate court in this hypothetical case. Under the doctrine of *res judicata* ("the thing is decided"), Ronald may not relitigate the matter. That doctrine states that once the matter is finally decided by the highest court of a jurisdiction, the same parties may not relitigate the same issues.

2:6. The Reporting System. The written decision of the appellate court discusses points of law that are applicable to its jurisdiction. For this reason these written decisions are published and distributed, primarily to lawyers, judges, and libraries. It is from these decisions that the cases in this book are taken. Most states print and publish the decisions of their own courts. In addition, a private company, the West Publishing Company, collects and publishes the state and federal court decisions. Since the private company publishes its reports quicker and distributes them more widely than the states, most lawyers and libraries have the West reports. Some states have eliminated their own reports altogether. Let us examine the citation of a typical case and see what the various parts mean:

LUCY v. ZEHMER
196 Va. 453, 84 S.E.2d 516 (1954)

The names of the parties are Lucy and Zehmer. Because some states always list the appellant's name first, whereas others always keep the plaintiff first, one cannot tell who is the plaintiff in a case merely by looking at the title. The first citation in the second line states where the case may be found in the state reporting system. The first number is the volume, the last is the page number, and the middle letters stand for the reporting system. In this example the case may be found in Volume 196 of the Virginia reports at page 453.

The West Publishing Company has divided the states into geographical areas, Northeast (N.E.), Atlantic (Atl. or A.), Southeast (S.E.), South (So.), Southwest (S.W.), Northwest (N.W.), and Pacific (Pac. or P.). It publishes the written decisions of U.S. district courts in the *Federal Supplement* (F.Supp.) and the decisions of the U.S. Supreme Court in the *Supreme Court Reporter* (S.Ct.). Decisions of the U.S. courts of appeals are found in the *Federal Reporter* (F.). After a large number of volumes are accumulated in one reporter system, the numbering system is repeated in a second series. Thus in our example *Lucy* v. *Zehmer* may be found in the West reporting system in the eighty-fourth volume of the *Southeastern Reporter,* second series, at page 516. The case was decided in 1954.

2:7. Nonjudicial Methods of Dispute Settlement. Although this text concerns itself with cases drawn from the judicial system, there are other methods of dealing with disputes. Two of the most significant are the administrative agencies and arbitration.

(1) Administrative Agencies. Our society has grown so complex that state and federal legislatures have found it necessary to establish various agencies to deal with and resolve problems that are prevalent in the various parts of our society. The legislature usually creates the agency to deal with a specific problem area and then establishes guidelines for the agency to follow. Some agencies are given the power to make rules and regulations covering various aspects of our behavior. For instance, under the recently enacted Magnuson–Moss Warranty Act, Congress has given the Federal Trade Commission the power to promulgate detailed regulations concerning product warranties created by suppliers of goods. The agencies thus serve a quasi-judicial function in that they resolve and settle disputes between parties. An example would be the National Labor Relations Board, which hears complaints of unfair labor practices filed by employees or employers.

The creation of rules and regulations and the resolution of disputes are both done after hearings are held by the agencies. The procedure used during these hearings may vary from one agency to another but often involves taking testimony and receiving evidence gathered from other sources. The procedure is usually considerably less formal than that followed in a trial by a court. One may appeal from the decision of an administrative agency to the courts, and the courts may set aside a rule or regulation that is made outside the authority of the agency or that is arbitrary, unreasonable, and capricious. The court may not substitute its judgment for the agency's, however, just because it does not agree with the rule.

Administrative agencies are becoming a more and more important part of our system of government. Such agencies as the Federal Trade Commission, the Federal Communications Commission, the Federal Power Commission, the National Labor Relations Board, and the many state and local agencies, such as consumer-protection agencies, touch many activities in our daily lives.

An alternative to using the courts for the resolution of disputes is the use of *arbitration.* Arbitration is the process of submitting a dispute to a nonjudicial third party for resolution. The decision may or may not be binding upon the parties to the dispute.

Arbitration is usually used by the mutual agreement of the parties, as, for example, when a union or an employee and an employer agree in the employment contract that certain types of disputes should be submitted to an arbitrator for decision. Sometimes a statute requires an arbitrator. For instance, many state workman's compensation statutes require that a dispute between an employer and an employee be made on behalf of the administrative agency by an arbitrator.

There are a number of advantages in using arbitration for the resolution of disputes, not the least of which is that it is less costly than going to court. In addition, if the area of dispute requires special knowledge, the arbitrator can be an expert in the field and thus more knowledgeable than a judge. An additional advantage is that arbitration is much faster than the court system in resolving disputes. The American Arbitration Association is a group established to provide experienced arbitrators for this purpose.

Arbitration should be distinguished from *mediation.* Mediation involves a third party, as does arbitration, but the third party in the mediation procedure does not render a decision in the dispute; he works with the parties so that they may compromise and settle the dispute on their own.

SUMMARY

There are two types of courts, trial courts and appellate courts. Trial courts are the determiners of fact whereas the appellate courts are primarily concerned with questions of law. In order to hear a case a court must have jurisdiction. Jurisdiction may be either in personam or in rem.

In order to obtain jurisdiction over a person, that person must be served with a summons. In some situations, such as those involving the use of an automobile, jurisdiction may be obtained over nonresidents of a state by use of the long-arm statute. The concept of venue refers to the geographical area within the state or district where a suit may be brought.

Each state has its own court system. The federal government also has a court system. The trial court of a state is usually called the circuit court or superior court. Many states also have small claims courts to handle small cases. The highest court of a state is frequently called the supreme court or court of appeals. Some states have an intermediate level of appellate court.

The trial court in the federal system is known as the district court. There are eleven U.S. courts of appeals, one for each circuit. The highest federal court is the U.S. Supreme Court. There are also special courts such as the Tax Court.

The federal courts are courts of limited jurisdiction and have jurisdiction over two types of cases; those involving parties with different citizenship known as diversity jurisdiction and those involving federal questions known as federal-question jurisdiction.

There are two ways cases get to the U.S. Supreme Court. In a few situations a party may have a right of appeal to the Supreme Court. The other method is to petition the Court to issue a writ of certiorari. The issuance of such a writ is up to the discretion of the court.

A civil trial has a number of typical stages in most jurisdictions. The first state may be called the pleading stage. A complaint (or declaration) filed by the plaintiff and an answer (or plea) filed by the defendant are the pleadings necessary to start the case. A party may also file a demurrer or any number of motions such as a motion for summary judgment. In addition, the defendant may file a counterclaim against the plaintiff.

Another stage in the trial may be called the discovery stage. Among the methods for discovery are written interrogatories, oral depositions, and motions to produce or for physical examination. One of the steps in the actual trial involves impaneling a jury, if one is used. The process of examining the prospective jurors is known as the voir dire. As a result of this process, some prospective jurors may be dismissed for cause or eliminated by the use of peremptory challenges. Another step is for each party to make an opening statement and present evidence. The plaintiff has the burden of proof and must establish a prima facie case. If he does not, the judge may direct that a verdict be entered against him. After the evidence is presented the judge may instruct the jury as to the law to be applied in the case. The jury then will hear each side's final argument and retire to arrive at its verdict. The plaintiff must prove his case by a preponderance of the evidence. The parties may then appeal the decision to the appellate court which may affirm, reverse, or remand the decision of the trial court.

The decisions of appellate cases are published. Many states have their own reporting system. A private company also publishes the state and federal decisions. The states are divided into geographic areas, and the decisions for that area are reported for that area.

There are alternatives that may be used instead of going to court. The parties may decide to use arbitration in order to resolve their dispute. In some cases the parties may seek the services of a mediator in order to help them settle their disputes.

PROBLEMS

1. Distinguish between a trial court and an appellate court.
2. What is the difference between having "a right of appeal" and certiorari?
3. Distinguish between the jurisdictional basis for the state courts and the federal courts. What types of cases may be taken to federal court?
4. What are depositions and interrogatories? What purpose do they serve?
5. Define *res judicata*. What reasons can you give for the doctrine?
6. What is meant by *jurisdiction* and *venue?*
7. May one serve a nonresident motorist with process? If so, how?
8. Distinguish between mediation and arbitration.

CHAPTER 3

An Introduction to Torts and Crimes

This chapter concerns a brief introduction to the law of torts and crimes. Both subjects are complex to the point where a detailed analysis is beyond the scope of this text. However, it is necessary for any student of business law to know something of the basic principles of these subject areas in order to avoid some of the common pitfalls that confront any person in business or society in general.

3:1. Torts. It is difficult to define torts in a manner that is useful. The word *tort* connotes a wrong, so perhaps it can best be defined as a civil wrong—as distinguished from a crime—that is not a breach of contract and for which the law provides a remedy. It might also be helpful to compare torts with two other broad categories of legal action: crimes and contracts.

A tort consists of the violation of a standard of behavior established by society. The same is true of a crime. A breach of contract, on the other hand, involves the violation of a standard of behavior established by the parties themselves. Once the standard of behavior is set, however, a violation of that standard in the field of tort or contract results in injury or damages being suffered by an individual for which he may recover. The theory of crimes, however, is that it is the state—originally the king—rather than an individual that is offended by the illegal behavior. Thus, if one is convicted of the crime of robbery, the victim is not compensated in any way as a result of the criminal action. In order to recover his money (assuming the police have not recovered what is identified as actually being his), the victim must file a civil suit in tort.

Crimes, torts, and contracts may be distinguished, therefore, on the basis of who determines the standard of behavior and the party who is the victim of the violation of this standard. The same act, of course, may result in both a crime and a tort. For instance, if an individual goes through a red light and strikes another car, he may be charged with a violation of the traffic laws, a minor criminal charge, and he may be sued by the victim of the incident for damages sustained by the victim under the theory of the tort of negligence. The result of the perpetration of a crime, of course, is that the criminal is punished.

The question arises as to what types of behavior that cause injury should result in the compensation of the victim for his loss. One useful method of classifying these various types of behavior is to visualize them as covering a spectrum of activity ranging from *intentional acts* at one extreme to *unavoidable accidents* at the other. In between are those types of acts that are not intentional but that result from behaving in an unreasonable manner characterized as *negligent* acts.

The law has long imposed liability on those who injure others as a result of their intentional acts. Thus, if one individual strikes another intentionally, he could be sued for the tort of battery by the victim. On the other hand, the law in most cases does not impose liability on those involved in an unavoidable accident. There is no requirement that liability be imposed on someone simply because an accident occurs. Society has decreed that the parties are not responsible for any damages other than their own when the accident is unavoidable. Of course, in most cases accidents are avoidable. At an early time in the development of the law, however, even though an injury resulted from an individual's behaving in an unreasonable manner, no liability was imposed unless there was an intent to injure. Later the tort of negligence was developed, which allows one to recover for those injuries resulting from acts that may not be perpetrated with the intent to cause injury but that are performed in a negligent manner. More recently the law has imposed liability without regard to fault on individuals engaged in dangerous activities as well as in what have become known as *products-liability* cases. This concept is known as *strict liability.* Thus, if one engages in blasting rock and takes all reasonable precautions when doing so, he is still liable in most states for damages that occur as a result of any concussion from the blasting. In another type of case discussed in Chapter 17, a manufacturer of a product may be liable for injuries received from a product manufactured by him when it is inherently dangerous if defective. This liability is imposed without regard to any negligence in the production of the good by the manufacturer. Now let us examine very briefly some of the more common torts for purposes of illustration.

(1) Intentional Torts. (a) *Assault and Battery.* These are two separate torts, the tort of assault and the tort of battery. *Battery* may be defined very simply as an unpermitted touching. The obvious question raised in many cases involves whether touching is permitted or not. One proceeding through a narrow store door in a crowd of holiday shoppers, for instance, may be touched, but the person gives permission to the touching by implication, hence there is no battery. On the other hand, the store detective who pushes a customer without cause may be guilty of battery.

The tort of assault occurs when one's action creates in another a reasonable apprehension of bodily harm. Assault consists of an act that causes an individual to apprehend a battery. For instance, pointing a gun at an individual under circumstances giving rise to a reasonable apprehension of bodily harm would constitute an assault. A battery may never occur, yet it is not required for the tort of assault. On the other hand, it is not necessary that there be any assault preceding a battery. For instance, one who is struck from behind with a lead pipe has been the victim of a battery, but he has not been the victim of an assault since the tort requires an apprehension of harm. In this case the victim would not be aware of the impending blow until it occurred.

(b) *Conversion.* The tort of conversion consists of the wrongful interference with one's right to possession of personal property. The intent required is for the defendant to exercise control over goods that is inconsistent with the plaintiff's right to possession, but the act does not have to be a conscious wrongdoing. For instance, a refusal to deliver property to one who is entitled to it constitutes a conversion. Transferring property from one place to another may also constitute conversion. If goods are left with another for safekeeping or for transfer to another, delivery to the wrong person may constitute a conversion by the one to whom the goods are entrusted.

(c) *Defamation.* Defamation consists of a false communication that tends to injure one's reputation by lessening his esteem, respect, or goodwill in the community. Most often the statement tends to hold one up to ridicule or contempt. A requirement of defamation is that the statement must be published in the technical sense. All that is required for publication is that the statement be made to a third party.

Defamation consists of the torts of *libel* and *slander,* libel being the written defamation and slander the oral defamation. A frequent problem that arises in defamation cases is whether the plaintiff must prove actual or special damages. When the communication is libelous, the existence of damages is presumed from the publication and need not be proved. Defamation by slander is more complicated. The courts long ago determined that some types of slanderous statements were so likely to cause pecuniary loss that it was unnecessary actually to prove special damages. These statements are lumped into a category known as *slander per se.* They consist of the imputation of a crime or of a loathsome disease and of those statements affecting the plaintiff in his business, trade, profession, office, or calling. Later the courts or legislatures added the imputation of unchastity to a woman. All other slanderous statements require proof of special damages regardless of how derogatory the statement may be. This latter category is often called *slander per quod.*

There are circumstances in which an individual is protected against liability for published false statements because of *privilege.* Statements made in judicial proceedings, legislative proceedings, and certain executive communications, for instance, are protected by *absolute privilege.* The purpose of these exceptions is so that these functions may proceed unincumbered by the fear that any statement made may later be submitted to the scrutiny of a jury. This protection is necessary in a free society if our government institutions are to operate freely and openly.

Other statements are made with a *qualified privilege.* The qualification is that the statement must be made in a reasonable manner and for a proper purpose. It is a communication that is "fairly made by a person in the discharge of some public or private duty, whether legal or moral, or in the conduct of his own affairs, in matters where his interest is concerned." Thus any statement reasonably necessary to protect one's reputation against defamatory statements or to protect the mismanagement of a business in which one has an interest is protected by a qualified privilege. A statement by a client to his attorney that may be false is protected.

DEMERS v. MEURET
512 P.2d 1348 (Or. 1973)

Leo John Demers, Sr., sued Sally Meuret for slander as the result of statements made by Meuret at a meeting of the City of Madras Airport Commission. Defendant filed a demurrer to the plaintiff's complaint, which was sustained by the trial court. The plaintiff, Leo Demers, Sr., appealed.

DENECKE, J.

Defendant urged as one ground for her demurrer that the complaint did not state a cause of action because no special damages were alleged. Plaintiff counters that he alleged slander per se and, therefore, no allegation of special damages is necessary. The issue is whether plaintiff alleged a cause of action for slander per se.

The complaint alleged that the plaintiff was the president of a concern which operates an airport under a lease from the City of Madras; that at a meeting of the Airport Commission, which represents the city in regard to plaintiff's lease, the defendant stated to the commission:

> What kind of protection can you give us from this terrible, mean, demented old man and what kind of protection do we have, . . . he might come out and chop our airplanes up with an axe.

Words actionable per se include words "falsely spoken of a party which prejudice such party in his or her profession or trade." Such words can be interpreted to mean that the defendant stated that the plaintiff was insane and might damage airplanes at the airport. Such words could be found by the trier of fact to prejudice the plaintiff in his business of operating an airport.

> A, falsely and without a privilege, says to B that C, a merchant, is insane. A is liable to C. 3 Restatement 180, Torts, § 573, Illustration 6.

The other ground for the demurrer was that the statement was absolutely privileged. The trial court sustained the demurrer because, "it is the court's view that the Airport Commission is quasi-judicial in character. . . ."

In her brief defendant suggests that a defamatory communication to an administrative body is absolutely privileged although that body is acting in a legislative or administrative capacity, rather than a judicial capacity. The general law is not to this effect.

> 3 Restatement of Torts, § 598, provides: An occasion is conditionally (the Restatement term for "qualified") privileged when the circumstances induce a correct or reasonable belief that
> (a) facts exist which affect a sufficiently important public interest, and
> (b) the public interest requires the communication of the defamatory matter to a public officer or private citizen and that such person is authorized or privileged to act if the defamatory matter is true.

Prosser states the principle, "communications made to those who may be expected to take official action of some kind for the protection of some interest, of the public" are protected by a qualified, not an absolute, privilege. Prosser, Torts (4th ed) 791.

The complaint alleges that plaintiff maliciously spoke the defamatory words; therefore, the complaint states a cause of action although the words may have been qualifiedly privileged.

The trial court erred in sustaining the demurrer to the complaint.

(Reversed and remanded.)

Questions for Discussion

1. *The court ruled that the words spoken in this case constituted slander per se. What effect did this ruling have on the plaintiff's case when it was retried?*
2. *What is the difference between an absolute and a qualified privilege? On retrial what would the defendant have had to prove in order to make a successful defense of privilege?*

(d) *Slander of Title and Disparagement of Goods.* Clearly related to defamation but distinct from it are a series of injuries that may result from false statements that may be made about one's product or business or title to property. Any type of property interest that may be legally protected may be the subject of disparagement. The disparagement may consist of a falsehood about land, leases, patents or literary property or statements that reflect on the character of a business or a product. It is necessary that the plaintiff prove actual damages as a result of the disparagement, generally that a third person has refrained from dealing with the plaintiff as a result of the disparagement.

(e) *Interference with Contractual Relations.* Most of the important law concerning the tortious interference with contractual relations has arisen since the start of the twentieth century, although the tort originated earlier. The tort was the basis of a well-known case, *Lumley* v. *Gye,* 118 Eng. Rep. 749 (1853), in which Miss Johanna Wagner, a famous opera singer, was under contract to sing exclusively to Lumley for a definite term. Gye, knowing of this contract, "enticed and procured" Miss Wagner to refuse to carry out her agreement and sing for Gye instead. The court held that it constituted a tort to entice Miss Wagner to break her contract with Lumley.

The courts have found defendants guilty of this tort even when the contract is terminable at will by the third party. It is, however, necessary that the plaintiff prove that the defendant interfered with the contractual relationship, not that he merely reaped the benefits of a broken contract. It is also often stated that an additional requirement of this tort is malice. This may be confusing unless one understands that the term *malice* as used here may mean anything from intending an act, knowing of the plaintiff's contractual interests that will interfere with those interests, to interference for the express purpose of doing harm to the plaintiff for its own sake.

(f) *False Imprisonment.* False imprisonment consists of the intentional unprivileged detention of a person without his consent. It is not limited to the concept of an unlawful arrest without a warrant. False imprisonment may consist of a store manager's detaining a customer on grounds of shoplifting when there is no probable cause for believing that the customer engaged in the shoplifting, or it may consist of an official arrest by a police officer without legal justification.

The essence of the tort is the unlawful detention, but this does not mean that the plaintiff must be locked up by physical force; it may be accomplished by threats of force that result in the plaintiff's detention against his will. On the other hand, if the plaintiff consents to the detention, there is no false imprisonment. Thus there is no liability when a customer agrees to remain in a store to clear himself of suspicion of a theft. Of course, a shopkeeper may detain a suspect when he reasonably believes that the person detained is engaged in shoplifting. Many cases often revolve around the question of whether the shopkeeper was reasonable in his belief.

(g) *The Deceit.* The tort of deceit, otherwise known as *fraud,* consists of a false representation when it is made with knowledge of the falsity, when it is made with the intention of inducing the plaintiff to rely on the statement, and when the plaintiff justifiably relies on the statement and an injury results. Fraud may also be used as a defense in a breach-of-contract action and is more fully discussed in Chapter 7.

(h) *Nuisance.* Nuisance results from a substantial interference with a right or an interest of another person, most often the right to enjoy and use his property. The classes of wrongs that are lumped under the category of *nuisance* are innumerable. The standard—as stated in the *Restatement of Torts,* which is used to determine what sort of behavior constitutes a nuisance—is that of offensiveness, inconvenience, or annoyance to the normal person in the community. Obviously what constitutes a nuisance today may not be considered a nuisance in other times. The use of one's property to create excessive noise, pollution, or smoke may constitute a nuisance although it may otherwise be legal. On the other hand, every use of one's property that may cause others an inconvenience does not necessarily constitute a nuisance. It often depends upon whether one's use of his property is reasonable or not, and the courts are frequently placed in the position of balancing one party's interests against another's. In reality, nuisance is not so much a tort of its own as a way of stating a type of damage to one's property interest.

JEWETT v. DEERHORN ENTERPRISES, INC.
575 P.2d 164 (Or. 1977)

Howard Jewett and 18 other plaintiffs sued Deerhorn Enterprises, Inc. to enjoin the operation of a pig farm on Deerhorn's property. The plaintiffs contended the pig farm constituted a nuisance.

Jewett's property as well as the other plaintiffs' is located in an area that may be characterized as "rural residential." Their houses were built and occupied prior to the time that the defendant began operating the pig farm.

Deerhorn's pig farm had about 400 pigs in residence at any one time. A sewer lagoon was constructed in order to dispose of animal waste from the pig pens. The pigs were fed whey and milk byproducts including raw bread dough and stale bread.

RICHARDSON, J.

In approaching these issues we take note of the proposition that an injunction is an extraordinary remedy and will be granted only upon clear and convincing proof. A nuisance, claimed to be an interference with the use and enjoyment of land, is not actionable unless that interference is both substantial and unreasonable. Whether a particular use of property constitutes an actionable nuisance cannot be determined by fixed general rules but depends on the individual facts of a particular case. We have, however, used a number of guidelines in assessing each fact situation.

Comprehensively stated these guidelines are the location of the claimed nuisance, the character of the neighborhood, the nature of the thing complained of, the frequency of the intrusion, and the effect upon the enjoyment of life, health and property.

Plaintiffs complained of essentially three conditions resulting from the operation of the pig farm: a noxious unpleasant odor from the pens, the lagoon and in some measure from the fermenting whey and bakery products; noise in the form of high pitched squealing made by the pigs while fighting, mating and nursing their young; and finally, a substantial increase in flies after the piggery became operational.

The pig farm is located on approximately 25 acres. However, the 400 animals are confined in two converted greenhouses with an approximately one quarter acre area in which the pigs may roam. The confinement of approximately 400 pigs in a relatively small space results in a concentration of odor and noise. The cleaning procedure followed by defendant is incomplete and results in a residue of animal waste in the pens which gives off an unpleasant odor. The lagoon, covering slightly more than two acres, and the settling tank also emit odors.

The character of the neighborhood surrounding the pig farm is rural residential. All of the plaintiffs' homes are within one quarter mile of the defendant's property. The plaintiffs all use their property as a residence and not for commercial farming. Some years prior to the commencement of defendant's farming operation the area changed from a commercial farm community to a residential community. This fact, coupled with the existence of the nearby subdivision, foreshadows future residential development in the area. If the pig farm is to continue residential development in the area may be curtailed.

The plaintiffs all owned or occupied their property prior to the time defendant purchased the property and began raising pigs. This is not then a situation where plaintiffs came to the nuisance and then registered their complaint. In *Kramer v. Sweet,* 179 Or. 324, 169 P.2d 892 (1946), we said under those circumstances "the court must take a less favorable view of defendant's case than it might have been disposed to take if this business had been maintained in the neighborhood for a long period of time." Indeed, the evident character of the neighborhood should have served as a warning to defendant of the risk they assumed in locating the pig farm in this area.

As indicated plaintiffs complained of odors, noise and flies emanating from the defendant's farm. The odor from the animal waste in the pens, the settling tank and the lagoon was variously described as "nauseating," "vile," "rotten," "very unpleasant" and "suffocating" in its intensity. This odor was at times combined with the odor of fermenting pig food and rancid raw bread dough. To a limited extent there was an odor from the burning plastic bread wrappers.

Whether a particular condition is sufficient to constitute a nuisance depends upon its effect on an ordinarily reasonable man, a normal person of ordinary habits and sensibilities. There was no evidence that any of the plaintiffs possessed other than the sensibilities of normal human beings. Nor was there evidence any plaintiff had an affliction or followed peculiar living habits that made them more sensitive to odor, noise or insects. The odor was experienced in varying degrees on all of plaintiffs' property and was uniformly described as unpleasant.

The noise from the farm is the least of plaintiffs' complaints and if only the noise were proven we would find it difficult to hold the defendant's farm constituted a nuisance. This noise, described as "high pitched squeals" and an "unnerving sound" did exist and can be considered as part of the cumulative effect of the pig farm on the neighborhood.

The substantial increase in the number of large blowflies came after the pig farm was in full operation. None of the plaintiffs had noted a problem with the flies prior to this time. Although there is no direct evidence the insects came from the defendant's farm the conclusion is inescapable that they resulted from the operation of the piggery. The flies existed in large numbers around those plaintiffs' homes which were nearer to the defendant's farm. At times they covered the windows of the houses and the windshields of the cars and rose in large swarms when persons walked across the lawn. The insects were present in significant numbers inside as well as outside plaintiffs' houses. One of the plaintiffs sprayed insecticide in the area around his house on a number of occasions without significantly abating the insect problem. The flies were present the year around even, to a limited extent, during the winter when they are ordinarily dormant.

A critical inquiry in determining if a nuisance exists and if so whether some relief is mandated is the effect of the intrusions on the plaintiffs and whether the interferences with the use of their property is substantial and unreasonable. One primary reason many of the plaintiffs purchased their property was to have ready access to outdoor activities such as picnics, gardening, outdoor barbeques and recreation. Because of the odor and the insects plaintiffs have had to substantially curtail their outdoor activities. During the summer months they are unable to open the doors and windows of the houses because of the pervasive unpleasant odor and the large number of flies. In light of the uses the plaintiffs would normally make of their premises these interferences are substantial. It is unreasonable to require them to adjust their normal living habits in deference to the defendants' use of its property. We conclude, as did the trial court, defendant's pig farm is a nuisance which interferes substantially and unreasonably with plaintiffs' use and enjoyment of their property. Plaintiffs are entitled to relief.

QUESTIONS FOR DISCUSSION

1. *What standard did the Court apply in determining that the defendants' use of their property constituted a nuisance?*
2. *Do you think it should or would make any difference if the defendants had used their property in the manner stated in the case for some time prior to the time when the plaintiffs purchased their property?*

---

(2) Negligence. The tort of negligence covers a large number of types of behavior that result in injuries for which the perpetrator of the act may be liable. It is behavior that is not intended to injure anyone but that involves a failure to exercise due care. Typical negligence cases involve accidents, medical and legal malpractice, and many other types of incidents. There are a number of elements that must be proved in order to establish the tort of negligence.

(a) *Duty.* In order to recover for the tort of negligence, the plaintiff must prove that the tortfeasor owed a duty of care to the plaintiff. For instance, one has a duty of care to fellow motorists to drive on the highways in a prudent and reasonable manner. On the other hand, if one is sitting on the beach, he has no duty to rescue a fellow swimmer who is in trouble, even though he has the means to do so.

Perhaps a better example would be to examine the duty owed to persons who come on one's property. Let us assume the case of a storekeeper. Three classes of persons may come into the store: *invitees, licensees,* and *trespassers.* The storekeeper's duty of care to each of these three classes varies. An invitee is one who comes on the premises for the benefit of the property owner. In this example, a customer would be an invitee. The storekeeper's duty toward an invitee is to make the premises safe not only against those dangers of which he is aware but also against those dangers that he might discover with reasonable care.

A licensee is one who, for his own benefit, comes on one's premises with permission from the owner. In our example, a salesman from a wholesaler would be a licensee. The storekeeper's duty to a licensee is to refrain from any active negligence, but he has no duty to make the premises safe, as for example, by inspection. A trespasser, of course, is one who is on the premises without permission, and the owner of land—our shopkeeper, in this instance—merely must refrain from willfully and intentionally injuring him. There are some exceptions to this rule, the best known being that condition known as *attractive nuisance,* which applies to children. If one maintains a condition on his property that might be dangerous to and reasonably expected to be a lure to children—a swimming pool, for example—he has a duty to exercise reasonable care to protect them, even though technically they may be trespassers. This doctrine has been adopted by a majority but not all states.

(b) *Breach of Duty: The Reasonable Man.* The second element that must be proved is that the duty of care was breached. In the determination of whether an act is negligent an *objective standard*—the "reasonable man"—is used. This concept is more fully developed in Chapter 5, but the basic question is "How would a reasonable man have behaved under the circumstances?" The issue is not how a juror would behave or how the judge or defendant would behave, but how the standard, the reasonable man would behave under the circumstances. For instance, although the speed limit on a road is 55 miles per hour, the reasonable man might not drive that fast because of circumstances affecting driving conditions, such as fog or ice. If the reasonable man would not drive 55 miles per hour, then the defendant has breached his duty of care by driving that fast regardless of whether he thought that speed was safe or not.

(c) *Causation and Injury.* It is not enough to show that the defendant behaved in a negligent manner. It must also be shown that the negligent behavior caused the injury in question. The question of causation is not difficult in rather simple cases, as, for example, when a pedestrian is struck by a negligently driven car. Causation is more difficult to establish in other cases, however. For instance, it would be negligence for a physician to remove a patient's right leg instead of the left one. But suppose that the patient dies while on the operating table as the operation is being completed. The important question is whether the physician's negligence caused the patient's death or whether he died from some other cause.

PERMINAS v. MONTGOMERY WARD & CO.
328 N.E.2d 290 (Ill. 1975)

Adam Perminas slipped and fell while shopping in a store owned by Montgomery Ward, hurting his back. He sued to recover for his injuries and was awarded $85,000 by the trial court. The appellate court reversed on the ground that Perminas failed to prove a prima facie case of negligence. The Supreme Court of Illinois granted plaintiff's petition for leave to appeal.

UNDERWOOD, C.J.

It is undisputed that at the time of the accident the plaintiff was a business invitee on defendant's premises and that defendant owed him a duty to exercise ordinary care in maintaining the premises in a reasonably safe condition.

We turn now to an analysis of the evidence which we believe sufficient to support the trial court's judgment. A detailed statement of the testimony is unnecessary since it is adequately set forth in the appellate court opinion and clearly supports findings that: (1) the plaintiff fell when he stepped on a small triangular shaped object with wheels on the bottom while walking in an aisle in defendant's store; (2) that object was probably an attachment to a floor polisher sold by defendant and displayed on low shelves in the area where plaintiff fell; and (3) the plaintiff suffered a serious back injury as the result of his fall. There was no evidence specifically indicating how this particular object got on the floor, how long it had been there or whether defendant or its servants knew of the presence of the object on the floor.

The crucial testimony was that of Anna Stecyna, who had been a clerk in the department which handled the floor-polisher attachments until May 1, 1966, when she was transferred to another department. As the result of injuries suffered in an automobile accident, she temporarily left work on May 26, 1966, and thus was unaware of actual conditions in the store on May 31, 1966, the date of plaintiff's accident. She testified that prior to May 1, 1966, while working in the department which handled the attachments, she personally observed on a number of occasions, both children and adults (primarily store employees on their "breaks") using these attachments as skateboards to skate up and down the aisles. She also stated that she had often found these attachments on the floor and had picked them up and replaced them on the low shelves where they were displayed with the floor polishers. (T)he evidence was, in our judgment, admissible for the purpose of showing that the defendant, through its servant Anna Stecyna, had noticed at least 30 days prior to the accident that these floor polisher attachments were creating a dangerous condition in that area of the store. Such notice can be inferred through evidence of previous, similar accidents, so direct evidence of notice to defendant through its servant is clearly admissible.

When Anna Stecyna's testimony is considered for the additional purpose of proving notice and combined with the other evidence in the case, we believe a sufficient showing of negligence is established. She was a clerk in the same department where these floor-polisher attachments were sold and as such had the responsibility to correct unsafe conditions or at least report their existence to her superiors. Through her personal observation, she had notice that these attachments were creating a dangerous situation because people used them as skateboards and then left them on the floor. Her testimony does not reveal whether her superiors or other employees knew of this situation or whether she picked up the attachments on order of her supervisors or just on her own initiative. That makes no difference, however, since notice to her is sufficient to give notice to the defendant. Once defendant had notice of the dangerous condition, its obligation was to either correct the condition or give its customers sufficient warning to enable them to avoid harm. . . . On this testimony, the trial judge could have concluded that the store should have either put up a warning of the danger, displayed the attachments in such a manner that they could not be used as skateboards or simply removed them from display altogether. In failing to take any of these steps, defendant could be considered to have breached its duty to exercise ordinary care in maintaining its premises in a reasonably safe condition.

We wish to emphasize that we "firmly adhere to the rule that a storekeeper is not the insurer of his customer's safety." We do not hold, by this decision, that a storekeeper is negligent simply because he displays small objects or objects with wheels which could possibly be dropped on the floor and cause someone to fall. Liability must be founded on fault; and that fault is found here in the evidence that defendant, after receiving notice that a particular product was creating a dangerous situation, failed to return its premises to a safe condition or warn its customers of the existence of the danger.

(Reversed and remanded.)

QUESTIONS FOR DISCUSSION

1. *Why was the testimony of Anna Stecyna so critical in this case?*
2. *Do you think the outcome would have been the same if the person injured had been a trespasser? A licensee?*

(d) *Common Defenses.* The most common defense in a negligence action, other than its absence, is *contributory negligence.* Contributory negligence occurs when the plaintiff is guilty himself of some negligence that contributes to his injury. Thus a pedestrian crossing against a red light will not be able to recover for the injuries received when struck by a speeding car. In most states, contributory negligence is a complete defense. Occasionally a case may arise in which a plaintiff may still recover although guilty of contributory negligence. This occurs when the defendant, once negligent, has a chance to resume exercising due care and does not. This most often is known as the doctrine of *last clear chance.*

Another frequently used defense in a negligence suit is *assumption of risk.* If one knowingly enters an area of danger, he cannot recover as the result of another's negligence. Thus if one sits along the third-base foul line in a baseball park, he assumes the risk of being hit by a foul ball.

HODGES v. NOFSINGER

183 So.2d 14 (Fla. App. 1966)

SWANN, J.

The defendant, Gary C. Hodges, appeals from a final judgment entered for the plaintiff, Mary Nofsinger, after a jury trial, in the sum of $7,500.00. The sole question on appeal is whether the plaintiff was guilty of contributory negligence, as a matter of law, thereby precluding her from any recovery from the defendant in this cause.

The parties have referred to this as the "kissing case." The defendant's version of the facts on appeal are as follows. The parties had seen each other many times prior to the accident. The plaintiff was a single woman, about twenty-five years of age, and the defendant was a member of the United States Air Force, stationed at Homestead, Florida at the time the accident occurred. On that day, the defendant and a friend went to the plaintiff's house. The friend had to return to the base early, but the defendant wanted to stay and the plaintiff agreed to take him back to the base in her automobile later in the evening.

At about 8:00 P.M. they departed for Homestead with the defendant driving the plaintiff's car and the plaintiff sitting close to him on the front seat, "about the middle of the car." She testified that the defendant drove normally, and made the following answers to questions propounded at trial:

. . .

Q. From the time you got onto Allapattah Drive up to the time of the accident, describe what happened.

A. Well, we were just driving along Allapattah, and Gary kissed me, and we went off the road into the canal.

. . .

Q. You didn't protest or object, or push Gary away at all during the kissing, did you?

A. No, I didn't.

. . .

A. The kissing occurred for a number of seconds, and we hit right then. I mean there was no pause in between.

. . .

Q. (By Counsel) Would you please tell me, please, isn't it a fact that you did kiss fully on the mouth?

A. The kiss was fully on the mouth.

. . .

Q. And the kiss continued for a number of seconds, didn't it?

A. I felt at the time that it did.

. . .

Q. And it endured up to the time of the accident?

A. Yes.

. . .

Q. This wasn't the first time you kissed, was it?

A. No.

. . .

The defendant contends that on these facts and circumstances the plaintiff cooperated in the kissing, with a reckless disregard for her safety, and was therefore guilty of contributory negligence as a matter of law.

The plaintiff sets forth the facts in a different light than those of the defendant. The plaintiff contends, and submitted to the jury, the same preliminary factual situation as the defendant. The essential difference in the evidence of the parties is summarized as follows. The defendant was driving in a normal manner, looking straight ahead, and suddenly, without any prior conversation or warning, the defendant kissed the plaintiff. This surprised her and she did not react or cooperate. She did not have an opportunity to protest or object to the defendant's kissing her before the car veered across the road through a guard rail and into a canal, which resulted in her injuries.

It is apparent from the testimony that there are conflicts in the evidence as to the issue of contributory negligence; that is, whether the plaintiff cooperated in the kiss, or whether she was so surprised that she did not have time in which to protest or object to the actions of the defendant. The conflicting evidence on this issue was properly submitted to the jury to be resolved by it.

For the reasons stated, the judgment appealed from is therefore

(Affirmed.)

3:2. CRIMINAL LAW. The field of criminal law is so vast and complicated as to really be beyond the scope of this text. It is appropriate, however, to mention a few general principles by way of contrast and comparison with those areas of the civil law to be discussed later. There are many differences. For instance, historically the person offended in the legal sense by the commission of a crime was the king rather than the victim. Today it is the state against whom the crime is committed, although the state may not be the actual victim. As stated earlier, a criminal act will usually also provide grounds for a civil action, however, and it is by bringing a civil suit that the victim of a crime may be compensated. The result of the criminal action is that the criminal is punished. Thus one who uses force to steal another's goods has committed the crime of robbery, but the same act may also constitute the torts of assault and conversion.

(1) Classification and Procedure. Crimes are classified as *felonies* and *misdemeanors.* Broadly speaking, felonies are the more serious crimes, which usually can result in a sentence to the penitentiary or heavy fine, whereas misdemeanors are lesser crimes, which result in shorter jail terms and/or fines. This is really a great oversimplification, however, because the legislature may designate a crime either a felony or a misdemeanor. Some misdemeanors, therefore, may have maximum sentences far in excess of some felonies. The significance of the difference may be found in the consequences of a conviction for a felony versus a misdemeanor. For instance, in some states a convicted felon cannot vote and the conviction may be grounds for divorce. These consequences do not attach to a conviction for a misdemeanor. As you may have concluded by now, the distinction between a felony and a misdemeanor may be somewhat arbitrarily drawn by the legislature. Typically traffic offenses are misdemeanors, and some states provide that larceny of goods over $100 is a felony while larceny under $100 is often classified as a misdemeanor. Typically robbery, larceny, murder, and arson are felonies, while drunk and disorderly charges, traffic offenses, and, in some states, possession of marihuana are misdemeanors.

In order for one to be charged with a felony in most jurisdictions, he must be *indicted* by the *grand jury.* The grand jury is a body of citizens, usually numbering twenty-three, to which the state presents evidence in order to establish probable cause to believe that the defendant committed the crime in question. If a majority of the grand jurors agree with the state, they will hand down an indictment, which is the charging document setting forth the crime charged and naming the person against whom it was committed, as well as other relevant matters.

The reason for the requirement of an indictment is so that some branch of government other than the executive is involved in the rather serious decision to prosecute a person for the commission of a crime. Because a misdemeanor is a less serious crime, one may be charged by a document prepared by the district or state's attorney known as an *information.* This sets forth basically the same material as that contained in the indictment but does not require action by the grand jury.

The concept of one branch of government's operating as a check on another is found in other areas of criminal procedure. For instance, before one may be arrested, the state must usually obtain an *arrest warrant.* An arrest warrant is issued by a member of the judicial branch of government, usually a judge or a magistrate, upon a showing by the state that there is probable cause to believe that the defendant committed the crime. There are exceptions to this requirement. For example, a police officer may arrest without a warrant when he reasonably believes that a crime was committed and has probable cause to believe that the person being arrested committed it. The state may later have to show a judge that at the time of the warrantless arrest, the officer had probable cause. For similar reasons it is also necessary for the state to obtain a *search warrant* prior to searching a person's home.

After one has been arrested, he or she may be brought to trial. An individual has a constitutional right to a trial by jury. The jury in this case is called the *petit jury,* which is usually composed of twelve persons. The standard for the determination of guilt in a criminal case is different than the standard applied in the civil trial we examined earlier. Here each juror must be convinced that the defendant is guilty beyond a *reasonable doubt,* a test that is much more stringent than that applied in our civil case. It is possible, therefore, for a defendant to be acquitted of a criminal charge and still be liable in a civil suit for damages.

(2) **Business Crimes.** There are a number of crimes that typically seem to involve business and business people. These are sometimes referred to as "business crimes" although this is really a misnomer since there is no requirement that a business person or business itself participate in the crime. Another term frequently used is "white collar crime." This refers to the fact that many of the participants in these crimes hold jobs working in an office. At this point, it might be helpful to examine some of these crimes.

(a) *Antitrust Crimes.* The antitrust laws are discussed in Chapter 46. Many of the laws provide for civil liability if they are violated. Some of the antitrust laws provide for criminal liability as well. They apply both to the corporations as well as to the officers that engage in the criminal activity.

One of the most important antitrust laws is the Sherman Antitrust Act which was passed in 1890. It contains a section which makes it illegal to engage in any contract or agreement to restrain trade. Such an act may result in both civil and criminal liability. An example of a contract to restrain trade would be an agreement between two companies to fix the price of a product at a certain level or to divide the markets in which they would sell their products. Both the corporations and officials involved in the price fixing agreement would be liable.

A 1974 amendment to the Sherman Act provides for the classification of the crime as a felony. Corporations may be fined up to $1 million and an individual $100,000 or imprisoned for up to three years, or both.

(b) *Use of Mails to Defraud.* Many businesses make extensive use of the mails. Unfortunately some unscrupulous persons have taken advantage of the public acceptance of the use of the mails as a method of doing business. As a result, Congress has made it a criminal offense to use the mails in order to further any scheme or artifice to defraud or to obtain money or property by false representations or pretenses. The crime consists of the use of the mails to defraud, it is not the fraud or scheme itself. The elements of the offense are (1) the scheme must be devised, or intended to be devised, in order to defraud or to obtain money by false pretenses and (2) to place or cause to be placed any letter, circular, or similar matter, in any United States post office to be delivered, for the purpose of carrying out such scheme.

(c) *Embezzlement and Larceny.* Embezzlement is a crime created by statute in most states. It is frequently associated with the conduct of a business because it involves a breach of trust that deprives the victim of his property. Larceny is very similar in nature but in many states lacks the special confidential relationship between the criminal and his victim that is required for embezzlement.

Embezzlement by employees is often difficult to prove. The crime may be as simple as the overstatement of expense accounts or overcharging customers and keeping the average. It may amount to the outright theft of equipment from an employer. On the other hand, the scheme may be very complex. The result is the same however, each year employers lose millions of dollars because of the crime which indirectly increases the cost of goods to the consumer.

(d) *Forgery.* Another crime that may be connected with a scheme to defraud the business is forgery. The crime of forgery consists of a false writing or alteration of a written document that is apparently capable of defrauding. The writing or alteration must be done with an intent to defraud.

Frequently the crime consists of falsely writing another's signature to a document such as a check or purchase order. Forgery may also consist of the alteration of the terms of an instrument so that the legal effect of it is changed. On the other hand, altering an instrument in an immaterial way, such as by adding a person's middle initial, would not constitute forgery.

There are many other crimes that may be used by businesses or perpetrated against them. The use of fraud in the sale of goods is an example. Another example is fraudulent home improvement schemes which will be touched upon in our discussion of the concept of holder in due course in Chapter 24.

It is impossible to cover all the crimes adequately in this text. However, you should be aware as a business person of the need to guard against a number of these crimes. Establishing proper accounting systems and audit procedures is one way. The proper training of management is another.

UNITED STATES v. AZZARELLI CONST. CO.

612 F.2d 292 (1979)

The state of Illinois mailed invitations to a number of contractors to bid on numerous highway projects in Illinois. The day before the bidding was to be held, Lawrence Loitz, Larry Boettcher, and John Azzarelli met in a hotel in Springfield in order to discuss the anticipated bidding.

At the bidding on July 29, 1975, item 25 on the bidding list went to Loitz, who was the only bidder. Azzarelli had agreed to let Loitz have that item in reciprocity for a prior contract award to Azzarelli several months before. Loitz agreed that Azzarelli (who turned out to be the only bidder) was to get item 26. Loitz also got item 27. Azzarelli and Boettcher also submitted bids but they were "out of the ball park." Boettcher increased his bid by $1,000,000, and received "payola" of $7,500 from Loitz in return for letting the item go to Loitz.

John Azzarelli was vice president of Azzarelli Construction Company. The corporation and John Azzarelli were convicted of violating Section 1 of the Sherman Act and of twelve counts of mail fraud. The corporation was fined $200,000 on the antitrust count and $12,000 on the mail fraud counts. John Azzarelli was fined $25,000 and sentenced to a suspended two-year prison term with three years probation after serving the first ninety days of the sentence in jail. Azzarelli and the corporation appealed. The other parties were not involved in this appeal.

DUMBAULD, J.

I—THE SHERMAN ACT

The Sherman Antitrust Act of July 2, 1890, 26 Stat. 209, 15 U.S.C. 1, has been likened to a charter of economic liberty, expressing a national policy akin to constitutional principles in importance and impact upon the general welfare.

As well stated by the late Mr. Justice Hugo Black:

The Sherman Act was designed to be a comprehensive charter of economic liberty aimed at preserving free unfettered competition as the rule of trade. It rests on the premise that the unrestrained interaction of competitive forces will yield the best allocation of our economic resources, the lowest prices, the highest quality and the greatest material progress, while at the same time providing an environment conducive to the preservation of our democratic political and social institutions. But even were that premise open to question, the policy unequivocally laid down by the Act is competition. And to this end it prohibits "Every contract, combination . . . or conspiracy, in restraint of trade or commerce among the several States." Although this prohibition is literally all-encompassing, the courts have construed it as precluding only those contracts or combinations which "unreasonably" restrain competition.

However, there are certain agreements or practices which because of their pernicious effect on competition and lack of any redeeming virtue are conclusively presumed to be unreasonable and therefore illegal without elaborate inquiry as to the precise harm they have caused or the business excuse for their use.

Among the practices which the courts have heretofore deemed to be unlawful in and of themselves are price fixing, *United States v. Socony-Vacuum Oil Co.*, 310 U.S. 150, 210, 60 S.Ct. 811, 838, 84 L.Ed. 1129; division of markets, *United States v. Addyston Pipe & Steel Co.*, 6 Cir., 85 F. 271, 46 L.R.A. 122, affirmed 175 U.S. 211 [, 20 S.Ct. 96, 44 L.Ed. 136;] group boycotts, *Fashion Originators' Guild v. Federal Trade Comm.*,

312 U.S. 457 [, 468, 61 S.Ct. 703, 85 L.Ed. 949;] and tying arrangements, *International Salt Co. v. United States,* 332 U.S. 392 [, 68 S.Ct. 12, 92 L.Ed. 20].

The case at bar is a typical or classical example of division of markets or allocation of business by bid-rigging which has been recognized as a *per se* violation since the Taft opinion in *Addyston Pipe.*

It is undeniable that appellants' conspiracy by increasing the cost of highway projects using asphalt concrete made from crude oil and asphaltic cement moving in interstate commerce substantially and adversely affected such commerce in those commodities. These highway construction projects obviously had more than a *de minimis* effect on interstate commerce. The total of the lowest bids on the three items involved was $3,192,085.54.

III—MAIL FRAUD COUNTS

The counts in the indictment dealing with mail fraud also resulted in valid convictions. To prove this charge the Government must establish (1) a scheme to defraud, and (2) use of the mails for the purpose of executing the scheme. *Pereira v. U.S.,* 347 U.S. 1, 9, 74 S.Ct. 358, 98 L.Ed. 435 (1954).[1]

Appellants argue that an antitrust conviction constitutes an exclusive remedy for anticompetitive conduct and precludes prosecution under the mail fraud statute. While under many circumstances a violation of the antitrust laws might occur without involving any fraudulent conduct, there may also be circumstances where fraud is an ingredient of the anticompetitive scheme.

The circumstances of the case at bar, involving collusion to defeat a statutory scheme of competitive bidding prescribed in the public interest by the federal Government and the State of Illinois in order to obtain for taxpayers the lowest possible costs for public works being undertaken, plainly manifest a species of fraud or false representation. This is particularly true where, as here, the bidders were required by law to sign (and did sign) statements that no collusion had been practiced. Under these circumstances appellants' conduct was not only anticompetitive but also fraudulent. Since the two crimes required proof of different facts, conviction of one did not preclude simultaneous conviction for the other.

Accordingly, the judgment of the District Court is affirmed.

QUESTIONS FOR DISCUSSION

1. *What must the government prove in order to prove mail fraud?*
2. *Was it necessary for the government to prove that the defendant's behavior was unreasonable in order to obtain a conviction on the charge of violating the Sherman Act?*

SUMMARY

A tort consists of a violation of a standard of behavior established by society. Some torts may be classified as intentional. Examples are assault, battery, conversion, and defamation. Defamation may be broken down into slander, which is the oral statement, and libel, which is the written. Slander may be either per se or per quod. An individual may be protected against suits for libel or slander because of privilege. The privilege may either be qualified or absolute.

Among the other intentional torts are slander of title, disparagement of goods, interference with contractual relationships, false imprisonment, deceit or fraud, and nuisance.

Another type of tort is negligence. The elements of negligence are a duty

[1] It is not necessary that a defendant mail anything himself; it is sufficient if he caused it to be done. "Where one does an act with knowledge that the use of the mails will follow in the ordinary course of business, or where such use can reasonably be foreseen, even though not actually intended, then he 'causes' the mails to be used." 347 U.S. at 8–9, 74 S.Ct. at 363.

of care, a breach of the duty, causation and injury. Those who own real property owe a duty to three classes of persons, invitees, licensees, and trespassers. A duty may be owed to trespassing children because of the doctrine of attractive nuisance. An objective standard, the reasonable man, is used to determine if one has breached his duty of care.

One common defense to a charge of negligence is the negligence of the plaintiff known as contributory negligence. Sometimes the plaintiff may overcome this defense by the doctrine of last clear chance. Another defense to a negligence charge is that the plaintiff assumed the risk by knowingly placing himself in a position of danger.

Crimes may be classified as felonies or misdemeanors. Felonies are typically more serious crimes then misdemeanors. In order to be charged with a felony one must be indicted by the grand jury. A grand jury action is not necessary to charge a person for a misdemeanor. An information prepared by the district or state's attorney will suffice.

Often, in order to arrest a person for committing a crime, an arrest warrant must be obtained. In order to search a person's house a search warrant first must be obtained from the judicial branch of government.

For one to be convicted of a crime he or she must be proven guilty beyond a reasonable doubt. It is possible to be acquitted of a criminal charge and still be liable in a civil suit for damages arising from the same act.

A number of crimes are frequently committed in connection with business. Among these are crimes involving the antitrust laws, use of the mails to defraud, embezzlement, larceny, and forgery.

PROBLEMS

1. What is a tort? How does it differ from a crime or a breach of contract?
2. Distinguish between assault and battery.
3. When is it not necessary to prove special damages for the tort of slander?
4. Under what circumstances may one speak without fear of liability for slander or libel?
5. Is one liable absolutely if he uses his real property in such a manner as to disturb his neighbor?
6. What are the elements that must be proved in order to establish the tort of negligence?
7. What standard is used in order to determine whether one's behavior amounts to negligence?
8. Distinguish between contributory negligence and assumption of risk.
9. What is the purpose of requiring an indictment to charge a person with a felony and of requiring an arrest warrant?
10. What is the difference between a misdemeanor and a felony?

SECTION II

Contracts

CHAPTER 4

Introduction to Contracts

At one time in the development of our legal system, agreements could not be enforced in court. If one party to the agreement breached it, the other party's only remedy was to rely on what has been quaintly called *self-help* in order to enforce the promised performance. This situation would obviously be intolerable in a society heavily dependent upon commerce and business relations, so the common-law courts began to enforce certain types of agreements that we know today as *contracts*.

Contracts perform some very important functions in our relationships with other people. They may be used to allocate risk between parties and to provide certainty in dealing with people. For example, suppose Carl Clothier contracts to purchase 100 pairs of slacks, at $10 per pair, from William Wholesaler on August 1. Carl now knows that he has a supplier of slacks for his store at a set price per pair. If the price of pants increases prior to August 1, the risk of the price increase is on Wholesaler. If Wholesaler does not perform, Carl would be able to sue Wholesaler for damages. Without a contract, Carl would not be able to plan on the availability of pants at the price stated in the contract. Likewise, Wholesaler knows that he has a buyer for 100 pairs of slacks at a set price per pair. Carl has the risk that the market price will fall prior to August 1.

Contracts enter into everyday life in a multitude of ways. Renting on apartment, purchasing goods at the store, buying a house or a car, and borrowing money are all very common activities that involve the use of contracts. In this section we will examine general principles of contract law, which have been developed by the courts. Where appropriate, certain statutes affecting contracts will also be covered.

One of the most important statutes is the Uniform Commercial Code. The Uniform Commercial Code (or U.C.C.) was drafted in the 1940s and early 1950s for the purpose of standardizing the numerous laws passed by the individual states that affected commerce. The offical text was finished in 1952. There have been several amended versions since that time.

The U.C.C. covers such topics as contracts for the sale of goods, negotiable instruments, secured transactions, and bulk sales as well as a number of other subjects. To date all of the states have adopted all or part of the U.C.C. Some of the states have modified parts of the code, but by and large the goal of standardizing the law in these areas has been achieved.

The part of the U.C.C. that has the most impact upon the general law of contracts (the subject matter of this section) is that known as "Sales." This part of the code deals only with one type of contract, contracts for the sale of goods. The definition of goods will be developed more fully in Chapter 15. At this point, you should know however, that the code states that goods means all things moveable at the time of identification to the contract. This includes specially manufactured goods, growing crops, the unborn young of animals, as well as certain things to be served from land by the seller. It does not include investment securities, things in action, and money in which the price is to be paid. Obviously, contracts for the sale of goods do not include labor contracts and contracts for the sale of land.

Relevant sections of the code will be discussed in this section because the code has had a heavy impact upon certain aspects of general contract law concerning contracts for the sale of goods. Be careful at this point to remember that the U.C.C. only deals with contracts for the sale of goods and therefore would not affect contracts that deal with other subjects.

At the end of each chapter in this section the changes that the U.C.C. has made in the general contract law will be summarized. The U.C.C. coverage of contracts for the sale of goods is covered extensively in Section Three of this text. You should remember that the purpose of the summaries is to make you aware of some of the changes the code has made in the law of contracts where it is applicable. It is not intended as a complete discussion of the U.C.C. as that is accomplished later in the text.

4:1. Definition. There are a number of definitions of a contract, but the one most frequently quoted is from the *Restatement of the Law of Contracts* § 1: "A contract is a promise or a set of promises for the breach of which the law gives a remedy, or the performance of which the law in some way recognizes as a duty." Examination of this definition reveals that the courts will enforce only certain types of promises. Thus, if Jones promises to give Smith $50 and later changes his mind, Smith could not enforce this promise in court because it lacks the elements required for a valid contract. It is nothing more than a promise to make a gift.

A contract is an agreement that is entered into freely and voluntarily between the parties. They are not required, in the legal sense, to enter into the contract. However, once the requirements for a valid contract are met, the law will enforce the duties set forth in the contract or assess damages against the breaching party for failing to live up to its terms.

4:2. Elements of a Contract. A valid contract requires the following elements: (1) an agreement made up of an offer and acceptance; (2) contractual capacity of the parties; (3) a legal purpose and object; and (4) consideration. Each one of these elements will be examined closely in the following chapters. If any one of these elements is missing, the promise or agreement is not a contract and, with certain exceptions, will not be enforced by the court.

4:3. CLASSIFICATION OF CONTRACTS. There are many different types of contracts, which may take numerous forms. As you read the following discussion, bear in mind that one contract may fit a number of the classifications. For instance, a contract may be oral, bilateral, executory, voidable, and expressed.

(1) **Oral or Written.** Contracts may be oral or written. Although it is preferable to have a contract in writing, an oral contract, no matter how complicated, is just as enforceable as a written contract. There are some exceptions to this general rule, which are covered in detail in Chapter 11. In addition, there are some obvious practical reasons for preferring written contracts over oral contracts, particularly in complicated transactions. Nonetheless oral contracts are made for very important matters.

(2) **Formal or Simple.** Under the old common law, a formal contract was one "under seal." A formal contract had to be in writing and the parties literally placed their waxed seal on the document. Today the same result is obtained by the placement of the word *seal* beside one's signature. All other contracts are considered simple contracts. One function of the sealed contract under the old common law was to eliminate the necessity of consideration. Today the use of the seal has been greatly restricted, but it still serves a limited function in some jurisdictions, such as extending the statute of limitations.

(3) **Bilateral or Unilateral.** In discussing bilateral and unilateral contracts, we are concerned with the promises of the parties. In a bilateral contractual situation, both parties to the contract have promised to do something. For example, Barry Buyer promises to buy Sam Seller's car for $400 one month from the date of the contract. Sam Seller promises to sell Buyer his car one month from the date of contract. Buyer and Seller have both made promises. Buyer has promised to give Seller $400 one month from the date of the contract and Seller has agreed to give Buyer his car at that time. This exchange of promises results in a bilateral contract.

Some contractual situations involve a promise in exchange for some requested performance. Only one party to the contract makes a promise, hence it is a unilateral contractual situation. For instance, if Sad Loser promises to pay Any Citizen $25 for the return of his lost watch, Any Citizen need not make a return promise in order for there to be a binding contract. We will see later that all Citizen need do is perform the requested act by returning the lost watch and the contract comes into being. Because a promise is made by only one party, it is called a *unilateral contract.*

(4) **Void, Voidable, and Unenforceable.** A void agreement is one that lacks one of the elements of a contract. A contract never comes into existence because of the absence of one of these elements, and thus to speak in terms of a "void contract" is technically a misnomer. The contract has never existed because of the absence of one of the elements.

A voidable contract is an agreement that contains all the elements of a contract, but for one reason or another one of the parties may be entitled to rescind the contract. One of the individuals may lack full contractual capacity, or there may be circumstances surrounding the execution of the agreement that would allow one of the parties to avoid the agreement at his option. Thus, if the contract was signed under duress, the person who was the subject of the duress would have the option to avoid the contract. Be sure to distinguish between void and voidable agreements.

An unenforceable contract is an agreement that satisfies all the requirements of a contract but may not be enforced by a court. For instance, the Statute of Frauds requires certain types of contracts to be in writing in order to be enforceable. If there is an oral contract that falls under the Statute of Frauds, it may not be enforced in the courts despite the fact that it contains all the requirements of a contract. Passage of the statute of limitations is another common situation in which a valid contract may not be enforced. All states have statutes limiting the time within which an action may be brought to enforce contractual rights. If one fails to file suit within this time period, he may not enforce the contract in court. The contract is not void or voidable but unenforceable.

(5) Executed or Executory. An executory contract is one where the parties have not performed. An executed contract, as the term is used here, is one where the parties have performed the contract. A contract which is partially executed is one where there has been partial performance. The term executed is also sometimes used in contract law to mean that a contract has been signed.

(6) Express or Implied in Fact. An express contract is one in which the parties have manifested their intent to enter into a contract either orally or in writing. They have expressed their intent by words.

It is not necessary that a contractual intent be manifested by words. People often convey their intent by their acts and conduct. When an intent to contract is manifested in a manner other than words, a contract is said to be implied in fact. For example, Harry Homeowner has a sign on his front lawn that states "House for Sale by Owner." He admits a real-estate salesman with a prospective purchaser. Homeowner knows the salesman's occupation, but no words are spoken between the salesman and Harry concerning the salesman's commission. If the prospect buys the house, the salesman is entitled to a commission because the parties are operating under facts and circumstances that indicate that a contract between them has been implied. Their contractual intent would be implied from these acts and circumstances, and an actual contract results.

4:4. Quasi Contract or Contract Implied in Law. There are times when circumstances place people in such a position that the court will create a contract or imply a contract in law in order to avoid the *unjust enrichment* of one party at the expense of the other. Because the parties have no contractual intent and the court creates a contract for equitable reasons, it is not an actual contract at all and thus is called a *quasi contract* or a *contract implied in law.* One who recovers under this theory is often said to *recover in quantum meruit.*

Because the creation of a quasi contract is an equitable action by the court, several limitations should be noted. In the first place, a court will not imply a contract in law when the parties have a contractual intent and have created an actual contract. Additionally the courts require some circumstance other than an intentional act of one of the parties that creates the need for implying a contract in law. One may not volunteer his services and then expect the court to create a contract in law in order to compensate him. Suppose Frank Farmer is away on vacation and Perry Painter, needing work, comes and paints Farmer's barn. Because there had been no prior contract between the parties and Perry has volunteered his services, the court will not grant relief even though Farmer does derive some benefit from Painter's act. Farmer is not unjustly enriched because Painter has volunteered his services. Circumstances did not throw the parties together; rather it was Painter's voluntary act. The law treats Painter's act as a gift.

It is necessary that the party requesting relief in quasi contract should have done something resulting in a benefit to the other party for which he should be compensated. If the other party did not benefit, he was not unjustly enriched. Furthermore you should not conclude that the court will apply this doctrine to correct every injustice or to rescue a party who has failed to prevail in an action based on an alleged actual contract.

DYER CONSTRUCTION CO., INC. v. ELLAS CONSTRUCTION CO., INC.

287 N.E.2d 262 (Ind. App. 1972)

Dyer Construction Company filed suit against Ellas Construction Co. claiming that it had an oral contract with Ellas whereby Ellas agreed to pay at the rate of $1.10 per cubic yard for certain materials and services performed by Dyer. Ellas denied the existence of the contract and counterclaimed, alleging that it had performed certain work and supplied material on other jobs to Dyer for which it had not been paid, plus making cash payments.

The trial court found for Dyer on the theory of quasi contract in the amount of $210,981.94 and for Ellas on its counterclaim in the amount of $230,599.36. Dyer appealed.

HOFFMAN, C.J.

The first issue to be considered is whether the trial court erred in finding that there was no actual contract in existence between the parties. The specific finding of the trial court in the record before us reads as follows:

> The Court is convinced that there was never a meeting of the minds between plaintiff and defendant sufficient to form a contract which could be the basis for legal recovery.

Dyer contends that the above finding is contrary to the evidence and that the evidence before the trial court leads to but one conclusion—"that there was in fact an implied contract existing between the parties." This contention is of no avail to Dyer because an implied contract, or a contract implied in fact, is a true contract and as such requires mutual assent.

In *Retter* v. *Retter* (1942), 110 Ind.App. 659, at 663, . . . it was stated:

> An implied contract, that is, one wherein an agreement is arrived at by the acts and conduct of the parties, is equally as binding as an express contract, wherein the agreement is arrived at by their words, spoken or written. In either case it grows out of the intention of the parties to the transaction. If there has been a meeting of minds and the clear intent of the parties to the transaction is evidenced by their acts and conduct viewed in the light of the surrounding circumstances, then the resultant implied contract differs from an express contract only in the mode of proof.

In the instant case, the record before us contains approximately 550 pages of testimony and exhibits. We will not attempt to summarize the evidence presented to the trial court. The evidence and reasonable inferences flowing therefrom are conflicting. The trial court received the evidence, observed the witnesses and heard their testimony. There is probative evidence to support the conclusion that there was no meeting of the minds between the plaintiff and defendant. We will, therefore, not disturb the conclusion of the trial court that no contract existed between the parties.

While we may not disturb the trial court's finding that there was no meeting of the minds thus preventing an express or implied in fact contract, the uncontroverted evidence and pretrial stipulations of the parties show that appellant was entitled to recover under the doctrine of quasi or constructive contract.

Contracts implied in law, more properly termed constructive or quasi contracts, are not contracts in the true sense. They rest on a legal fiction imposed by law without regard to assent of the parties. They arise from reason, law, and natural equity, and are clothed with the semblance of contract for the purpose of remedy. No action can lie in quasi contract unless one party is wrongfully enriched at the expense of another.

In the instant case, the evidence most favorable to the appellees and the pre-trial stipulations of the parties, shows that Dyer delivered certain materials and performed certain services for Ellas. Ellas, in the course of a construction project, retained the benefits from Dyer's performance. Therefore, even accepting the trial court's finding that there was no meeting of the minds preventing an implied in fact contract, equity will prevent Ellas from being unjustly enriched by imposing on Ellas a duty to pay Dyer for the reasonable value of its services and materials.

(Judgment affirmed.)

4:5. Conflict of Laws. There are times when the parties to a contract reside in two different states or jurisdictions and the laws of these jurisdictions differ. The question arises as to which law governs. This is known as *conflict of laws.* The parties may specify in the contract what state law should be applied if a dispute arises concerning the contract. If they do so, the court will enforce their intention. When the litigation concerns matters of contract validity or interpretation, the courts generally will apply the law of the state where the contract came into being in the absence of an expressed intent. A contract generally comes into being at the location where the final act necessary to the creation of the contract takes place. Disputes concerning contractual performance and damages are usually resolved under the law of the state where the performance of the contract takes place.

4:6. Interpretation of Contracts. Often the terms of a contract are not clearly expressed, or there may be a conflict in the contractual language that requires an interpretation by the court. Some general rules of contractual interpretation have been established by the courts in order to resolve disputes over contract language and conflicts as well as to determine whether the parties have complied with the terms of the contract. These are not hard-and-fast rules that are uniformly applied by the courts in all cases; rather they are guidelines developed by precedent over the years to aid the court in contractual interpretation. Examples of these rules are:

1. If possible, the court will give effect to the common intent of the parties to the contract.

2. The contract is construed as a whole, with effect being given to all its parts where possible.
3. Words and phrases will be given their ordinary meaning unless the parties show that they are used in a different sense. Technical words are given their technical meanings unless the parties intend otherwise.
4. Conflicting and inconsistent terms are construed to carry out the intention of the parties as determined from the entire agreement.
5. When the contract is a printed form, any conflicting words or phrases deliberately added by the parties will prevail. Handwriting will prevail over typewritten words and phrases.
6. Words prevail over figures where they are in conflict.
7. Words and phrases will be given the interpretation that will uphold the contract rather than an interpretation that would render the agreement invalid.
8. Ambiguous contract language is construed against the party who prepares the agreement.

INNES v. WEBB

538 S.W.2d 237 (Tex. 1976)

Neil F. Innes was a real estate agent who prepared an earnest money contract between Hulon Webb and Herbert Huller for the purchase of a residence to be constructed by Huller upon a certain lot. The contract that Innes prepared contained the following provision: "If title is found objectionable and is not cleared within a reasonable time, upon failure of Seller to comply herewith for any reason, Purchaser may demand back the earnest money, thereby releasing Seller from this contract, or Purchaster may enforce specific performance hereof or seek such other relief as may be provided by law."

Webb gave a check to Innes in the amount of $2,000.00 payable to Innes Realty, Trustee, to be held by Innes as stakeholder. According to Webb, Innes was to hold this check and return it to him in the event Huller failed to perform the contract.

Huller failed to perform the contract and left town. Webb demanded the $2,000.00. Innes answered that he had turned the money over to Huller in accordance with another clause in the contract which stated:

"$2,000.00 escrow to be turned over to Seller for initial deposit for materials and administrative costs. . . ."

Webb then sued Innes for the $2,000.00.

YOUNG, J.

At the trial the questioned contract was introduced into evidence. The contract is a printed form with blanks for appropriate interlineations. At the top of the instrument is the heading: "TEXAS STANDARD FORMS, EARNEST MONEY CONTRACT, CONVENTIONAL." Typewritten in were the names of the purchaser and the seller; the description of the property; the price to be paid. Then followed several printed clauses covering matters such as terms; closing; title insurance; default of the parties; broker's fee (with amount typewritten in); loss (fire and casualty); and special notices about several matters, one of which is that the instrument is the complete agreement of the parties. The default of the parties' clause contained the printed provision about return of earnest money. This is the clause that plaintiff alleged Innes breached. Then follows a typewritten clause about turning the $2,000.00 escrow over to the seller. This is the last clause and the one on which defendant relies to avoid liability.

All of which brings us to the construction of a contract that was prepared by a broker and that contains two apparently inconsistent clauses: one, a printed clause requiring the return by the broker to the purchaser of the earnest money on purchaser's demand if the seller fails to comply with the contract; and the other, a typewritten clause, requiring the broker to turn over the "$2,000.00 escrow" to the seller "for initial deposit for materials and administrative costs." Appellant urges that we should be guided here by the rule of construction which provides that the written or typewritten part of a contract controls in the event of any conflict thereof with the printed portion of the contract. This rule is set out in 13 Tex.Jur.2d Contracts § 144 (1960). The rationale for this rule is that the written or typed words are the immediate language of the parties themselves, whereas the language of the printed form is intended for general use only, without reference to the particular aims and objectives of the parties.

On the other hand, appellee contends that our case should be controlled by the rule which requires that an agreement be construed most strictly against the party who drafted it and thus was responsible for the language used. This rule can be found in 13 Tex.Jur.2d § 121 (1960).

When we attempt to apply these rules to our case, we find that we apparently have two conflicting rules urged by the parties. The question then arises which rule should prevail here. Our answer to that question is that the rule should be applied which says typed matter controls the printed instead of the rule which says that a contract will be construed against the author.

The rule of strict construction against the author has been dealt with in those authorities as follows: In *Daniel,* (citation omitted) our Supreme Court held that the rule applies only after ordinary rules of interpretation (such as the typed controls the printed) have been applied. In *KTRM, Inc.,* (citation omitted) that Court simply applied the typed controls the printed rule over the authorship rule. In *17A C.J.S.,* the statement is made that the authorship rule is the last one the courts will apply.

For all of those reasons, we hold that the typewritten clause in the contract determines the responsibility of the appellant for his disposition of the "$2,000.00 escrow"; that he delivered that money to the seller under the clause; that, therefore, he did not breach the contract in so delivering the money.

(Judgment for Innes.)

Questions for Discussion

1. *What is the reason for giving effect to the typewritten clause over a conflicting printed clause in a standard form contract?*
2. *Why did the Court not construe the terms of the contract against Innes since he prepared it?*

SUMMARY

Contracts are used to allocate risk between parties and to provide certainty in dealing with people. The Uniform Commercial Code has been adopted in all states and governs contracts for the sale of goods.

A contract is a type of promise that will be enforced in court. One definition of a contract is that it is a promise or set of promises for the breach of which the law gives a remedy, or the performance of which the law in some way recognizes as a duty.

In order to have a valid contract there must be an agreement made up of an offer and acceptance; the parties must have contractual capacity; the contract must have a legal purpose and object; and there must be consideration.

Contracts may be classified as written or oral; formal or simple; bilateral or unilateral; void, voidable, and unenforceable; and express or implied in fact. A contract implied in law is often called a quasi contract. It is used to avoid unjust enrichment of one party at the expense of another. Recovery under this theory is often called recovery in quantum meruit.

The question of which law governs when the laws of two or more jurisdictions are involved is covered by a field known as conflict of laws. The parties to a contract may specify which state law governs the terms of a contract. Otherwise the law of the state where the contract is made governs questions of interpretation and validity.

PROBLEMS

1. What is the difference between a contract implied in fact and one implied in law?
2. What is the definition of a contract?
3. Is it possible to have a unilateral executory contract? Why or why not?
4. Distinguish between a void and an unenforceable agreement.
5. What is meant by *bilateral contract*?

CHAPTER 5

Agreement: The Offer

The initial step in the formation of a contract is the making of an offer. It is important to be aware of the type of statement that will constitute a valid offer. Not all statements do. Some, as we shall see, fall short of the legal requirement and may best be described as negotiations or "dickering." The reason for this will become clear later, but for now realize that once an offer is made, all that remains is for the offer to be accepted in order for a binding contract to be formed (assuming, of course, that all the other elements of a contract are present). The result is that a great many statements may be made by the parties as part of the negotiation process before an actual offer is made.

An offer may be defined as ". . . a promise which is in its terms conditional upon an act, forbearance or return promise being given in exchange for the promise or its performance," (*Restatement Contracts,* § 24, 1932). There are a number of other definitions, but they all indicate that an offer is a promise made by one person (the *offeror*) to another (the *offeree*), stating what the offeror will do in return for an act, forbearance, or return promise.

5:1. Contractual Intent. In order to qualify as an offer, the statement made by the offeror must be made with the intent to enter into a contractual relationship with the offeree. If the contractual intent is lacking, there is no offer, regardless of what the specific words were. For example, if Harry Homeowner comes into a tavern after a frustrating day of engine repairs on his late-model car and loudly states, "I'd sell that car right now to anyone for ten bucks," the statement probably would not qualify as an offer because it was made in jest or in a state of excitement.

(1) Objective Versus Subjective Theories. How do we determine whether a statement has been made with the intent to enter into a contractual relationship? One possible solution is to attempt to determine what an individual's intent really was or, to put it another way, to find out what was actually going on in an individual's mind at the time the statement was made. This is the *subjective* theory. The difficulty is that there is really no way to find out what was in a person's mind at a particular time other than to ask him.

The *objective* theory does not depend upon a determination of what was actually in an individual's mind at the time an offer was made. In fact, what an individual actually thought at the time of the making of an offer is quite unimportant. An independent test is applied known as the *reasonable man*. Would the reasonable man have presumed that the individual made his statement with the intent to enter into a contractual relationship? This determination is made from the objective facts, actions, and circumstances surrounding the statement. If a statement was made under circumstances that would indicate to a reasonable man that the offeror intended to be contractually bound, there would be a valid offer even if the man had some secret mental reservation.

LUCY v. ZEHMER

196 Va. 493, 84 S.E.2d 516 (1954)

BUCHANAN, J.

This suit was instituted by W. O. Lucy and J. C. Lucy, complainants, against A. H. Zehmer and Ida S. Zehmer, his wife, defendants, to have specific performance of a contract by which it was alleged the Zehmers had sold to W. O. Lucy a tract of land owned by A. H. Zehmer in Dinwiddie county containing 471.6 acres, more or less, known as the Ferguson farm, for $50,000.

The instrument sought to be enforced was written by A. H. Zehmer on December 20, 1952, in these words: "We hereby agree to sell to W. O. Lucy the Ferguson Farm complete for $50,000.00 title satisfactory to buyer," and signed by the defendants, A. H. Zehmer and Ida S. Zehmer.

The answer of A. H. Zehmer admitted that at the time mentioned W. O. Lucy offered him $50,000 cash for the farm, but that he, Zehmer, considered that the offer was made in jest; that so thinking, and both he and Lucy having had several drinks, he wrote out "the memorandum" quoted above and induced his wife to sign it; that he did not deliver the memorandum to Lucy, but that Lucy picked it up, read it, put it in his pocket, attempted to offer Zehmer $5 to bind the bargain, which Zehmer refused to accept, and realizing for the first time that Lucy was serious, Zehmer assured him that he had no intention of selling the farm and that the whole matter was a joke. Lucy left the premises insisting that he had purchased the farm.

[Lucy testified that] on the night of December 20, 1952, around eight o'clock, he took an employee to McKenney, where Zehmer lived and operated a restaurant, filling station and motor court. While there he decided to see Zehmer and again try to buy the Ferguson farm. He entered the restaurant and talked to Mrs. Zehmer until Zehmer came in. He asked Zehmer if he had sold the Ferguson farm. Zehmer replied that he had not. Lucy said, "I bet you wouldn't take $50,000.00 for that place." Zehmer replied, "Yes, I would too; you wouldn't give fifty." Lucy said he would and told Zehmer to write up an agreement to that effect. Zehmer took a restaurant check and wrote on the back of it, "I do hereby agree to sell to W. O. Lucy the Ferguson Farm for $50,000 complete." Lucy told him he had better change it to "We" because Mrs. Zehmer would have to sign it too. Zehmer then tore up what he had written, wrote the agreement quoted above and asked Mrs. Zehmer, who was at the other end of the counter ten or twelve feet away, to sign it. Mrs. Zehmer said she would for $50,000 and signed it. Zehmer brought it back and gave it to Lucy who offered him $5 which Zehmer refused saying, "You don't need to give me any money, you got the agreement there signed by both of us."

Mr. and Mrs. Zehmer were called by the complainants as adverse witnesses. Zehmer testified in substance as follows:

He bought this farm more than ten years ago for $11,000. He had had twenty-five offers, more or less, to buy it, including several from Lucy, who had never offered any specific sum of money. He had given them all the same answer, that he was not interested in selling it. On this Saturday night before Christmas it looked like everybody and his brother came by there to have a drink. He took a good many drinks during the afternoon and had a pint of his own. When he entered the restaurant around eight-thirty Lucy was there and he could see that he was "pretty high." He said to Lucy, "Boy, you got some good liquor, drinking, ain't you?" Lucy then offered him a drink. "I was already high as a Georgia pine, and didn't have any more better sense than to pour another great big slug out and gulp it down, and he took one too."

At no time prior to the execution of the contract had Zehmer indicated to Lucy by word or act that he was not in earnest about selling the farm. They had argued about it and discussed its terms, as Zehmer admitted, for a long time. Lucy testified that if there was any jesting it was about paying $50,000 that night. The contract and the evidence show that he was not expected to pay the money that night. Zehmer said that after the writing was signed he laid it down on the counter in front of Lucy. Lucy said Zehmer handed it to him. In any event there had been what appeared to be a good faith offer and a good faith acceptance, followed by the execution and apparent delivery of a written contract. Both said that Lucy put the writing in his pocket and then offered Zehmer $5 to seal the bargain. Not until then, even under the defendants' evidence, was anything said or done to indicate that the matter was a joke. Both of the Zehmers testified that when Zehmer asked his wife to sign he whispered that it was a joke so Lucy wouldn't hear and that it was not intended that he should hear.

The mental assent of the parties is not requisite for the formation of a contract. If the words or other acts of one of the parties have but one reasonable meaning, his undisclosed intention is immaterial except when an unreasonable meaning which he attaches to his manifestations is known to the other party.

An agreement or mutual assent is of course essential to a valid contract but the law imputes to a person an intention corresponding to the reasonable meaning of his words and acts. If his words and acts, judged by a reasonable standard, manifest an intention to agree, it is immaterial what may be the real but unexpressed state of his mind.

So a person cannot set up that he was merely jesting when his conduct and words would warrant a reasonable person in believing that he intended a real agreement.

Whether the writing signed by the defendants and now sought to be enforced by the complainants was the result of a serious offer by Lucy and a serious acceptance by the defendants, or was a serious offer by Lucy and an acceptance in secret jest by the defendants, in either event it constituted a binding contract of sale between the parties.

QUESTIONS FOR DISCUSSION

1. *What objective factors in this case would indicate to a reasonable man that Zehmer intended to be contractually bound? What factors to the contrary?*
2. *Would it have made any difference if Lucy actually intended the offer as a joke?*
3. *What practical reasons can you give for the use of an objective rather than a subjective test?*

5:2. Negotiations Preceding the Offer. Many statements made by parties may not constitute an offer but are merely negotiations or invitations to deal. For example, if Tony Ticketseller says, "I will not sell my season ticket for less than $25," this would be an invitation to negotiate rather than an offer.

Advertisements are generally considered invitations to negotiate further rather than definite offers. Thus, if a merchant advertises a product in a newspaper at a certain price, that is not an offer but an invitation. Similarly catalogues, circulars, and handbills are generally not considered offers. There are exceptions. An offer of reward may be made by advertisement. Also, if one frames the advertisement in such a way as to convey an intent to be contractually bound, that may be construed as an offer. An advertisement soliciting bids is not considered to be an offer.

Auctions present a similar situation. When the auctioneer holds up Aunt Minnie's prized vase and asks, "What am I offered for this fine vase?" he is soliciting offers. The bid is the offer, which may be either accepted or rejected by the auctioneer. He may reject all offers. An exception is when the auction is "without reservation." In that case the object being auctioned off must be sold.

CRAFT v. ELDER & JOHNSTON CO.

38 N.E.2d 416 (Ohio 1941)

Martha Craft sued Elder & Johnston Company for breach of contract. The trial court dismissed the plaintiff's case and the plaintiff appealed.

BARNES, J.

On or about January 31, 1940, the defendant, the Elder & Johnston Company, carried an advertisement in the Dayton Shopping News, an offer for sale of a certain all electric sewing machine for the sum of $26 as a "Thursday Only Special." Plaintiff in her petition, after certain formal allegations, sets out the substance of the above advertisement carried by defendant in the Dayton Shopping News. She further alleges that the above publication is an advertising paper distributed in Montgomery County and throughout the city of Dayton; that on Thursday, February 1, 1940, she tendered to the defendant company $26 in payment for one of the machines offered in the advertisement, but that defendant refused to fulfill the offer and has continued to so refuse. The petition further alleges that the value of the machine offered was $175 and she asks damages in the sum of $149 plus interest from February 1, 1940. . . .

It seems to us that this case may easily be determined on well-recognized elementary principles. The first question to be determined is the proper characterization to be given to defendant's advertisement in the Shopping News. . . .

> It is clear that in the absence of special circumstances an ordinary newspaper advertisement is not an offer, but is an offer to negotiate—an offer to receive offers—or, as it is sometimes called, an offer to chaffer. Restatement of the Law of Contracts, Par. 25, Page 31.

Under the above paragraph the following illustration is given, " 'A,' a clothing merchant, advertises overcoats of a certain kind for sale at $50. This is not an offer but an invitation to the public to come and purchase."

"Thus, if goods are advertised for sale at a certain price, it is not an offer and no contract is formed by the statement of an intending purchaser that he will take a specified quantity of the goods at that price. The construction is rather favored that such an advertisement is a mere invitation to enter into a bargain rather than an offer. So a published price list is not an offer to sell the goods listed at the published price." Williston on Contracts, Revised Edition, Vol. 1, Par. 27, Page 54.

"The commonest example of offers meant to open negotiations and to call forth offers in the technical sense are advertisements, circulars and trade letters sent out by business houses. While it is possible that the offers made by such means may be in such form as to become contracts, they are often merely expressions of a willingness to negotiate." Page on the Law Contracts, 2d Ed., Vol. 1, Page 112, Par. 84.

"Business advertisements published in newspapers and circulars sent out by mail or distributed by hand stating that the advertiser has a certain quantity or quality of goods which he wants to dispose of at certain prices, are not offers which become contracts as soon as any person to whose notice they may come signifies his acceptance by notifying the other that he will take a certain quantity of them. They are merely invitations to all persons who may read them that the advertiser is ready to receive offers for the goods at the price stated." 13 Corpus Juris 289, Par. 97.

"But generally a newspaper advertisement or circular couched in general language and proper to be sent to all persons interested in a particular trade or business, or a prospectus of a general and descriptive nature, will be construed as an invitation to make an offer."

17 Corpus Juris Secundum, Contracts, Page 389, . . .

The judgment of the trial court will be affirmed and costs adjudged against the plaintiff appellant.

QUESTIONS FOR DISCUSSION

1. *What practical considerations are there in considering an advertisement as an invitation to negotiate rather than as an offer?*
2. *How would you frame an advertisement so that it constitutes an offer? How would you alter the advertisement in the above case?*

MARYLAND SUPREME CORPORATION v. BLAKE CO.

369 A.2d 1017 (Md. 1977)

Blake Company submitted a bid as general contractor to the Washington County Board of Education for the construction of a school. Maryland Supreme Corporation (Supreme) was a manufacturer of ready mixed concrete. It learned the identity of the general contractors who bid on the school construction job through a trade journal and submitted the following letter dated March 1, 1975, to Blake:

> The Blake Company
> P.O. Box 47
> Hagerstown, Maryland 21740
>
> ATTENTION: Mr. Vernon Tetlow
>
> RE: Western Heights Middle School
>
> Dear Sirs:
>
> We are pleased to submit a quotation on ready mix for the above mentioned project.
>
> Please take note that the price will be guaranteed to hold throughout the job. 3000 p.s.i. concrete $21.00 per yard, net.
>
> Hope that you are successful in your bid and that we may be favored with your valued order.
>
> Yours very truly,
>
> MARYLAND SUPREME CORPORATION
>
> /s/ Ben Wicklein
> Sales Representative

Blake was the successful bidder on the project and for a while Supreme delivered concrete to the job at the price of $21.00 per yard. Later Supreme raised the price to $27.00 per yard and Blake refused to pay contending that it had a contract for the whole project at the price mentioned in Supreme's earlier letter. There was no other written agreement. Supreme contended that its letter did not constitute an offer.

ORTH, J.

An offer must be definite and certain. To be capable of being converted into a contract or sale by an acceptance, it must be made under circumstances evidencing an express or implied intention that its acceptance shall constitute a binding contract. Accordingly, a mere expression of intention to do an act is not an offer to do it, and a general willingness to do something on the happening of a particular event or in return for something to be received does not amount to an offer. Thus, a mere quotation or a statement of a price or prices and an invitation to enter into negotiations, are not offers which may be turned into binding contracts upon acceptance. Such proposals may be merely suggestions to induce offers by others.

What this all boils down to is expressed in 17 Am.Jur.2d, *Contracts* § 33 (1964):

> From the nature of the subject, the question whether certain acts of conduct constitute a definite proposal upon which a binding contract may be predicated without any further action on the part of the person from whom it proceeds, or a mere preliminary step which is not susceptible, without further action by such party, of being converted into a binding contract, depends upon the nature of the particular acts or conduct in question and the circumstances attending the transaction. It is impossible to formulate a general principle or criterion for its determination.

Therefore, in its final determination, the question of whether an offer was made seems to be one dependent on the intention of the parties, and, being such, it depends on the facts and circumstances of the particular case. The UCC changes none of these principals of law. 67 Am.Jur.2d, *Sales* § 74 (1973).

Supreme's proposal was evidenced by its letter of 11 March 1975 to Blake. Supreme would now have it be merely a price quotation, and claims that did not contain many of the essential terms of an offer "such as the quality and quantity of the product to be supplied, the number and dates of the deliveries, the terms of payment, the costs of shipment and the time for performance." We do not agree. Considered in light of the facts and circumstances, the trial court could have found, as it obviously did, that the letter of 11 March 1975 constituted a definite and certain offer with the intent that, if accepted, it would result in a contract. By the letter, Supreme proposed to furnish Blake with ready mix 3000 p.s.i. concrete at $21 per yard, net, in such quantity as Blake required for the Western Heights Middle School project. . . . If Blake were awarded the general contract for the construction of the School and accepted Supreme's offer, there would be a binding contract. When viewed with reference to the method of operation of the construction industry and the prior course of dealings between Supreme and Blake, it is manifest that Supreme's letter was no mere price quotation or invitation to negotiate. It gave Blake the assurance that if Blake were the general contractor on the School project Blake could obtain from Supreme the concrete necessary for the job at $21 per yard. Thus, it was an offer, and the trial court did not err in so considering it.

5:3. COMMUNICATION. The offer must be communicated by the offeror to the offeree. The method of communication is up to the offeror, but the offeree cannot accept an offer until it is communicated. For example, A states to B that he intends to offer to purchase C's car. If B tells C of A's offer the offer has not been communicated because A did not communicate the offer to C.

A different situation arises in the case of implied contracts. One cannot accept an offer that he does not know about. If one purchases a ticket that he ordinarily would not examine carefully, he has not accepted an offer printed on the back merely by his use of the ticket, because the offer was not communicated to him. On the other hand, if for some reason he had a duty to read the ticket or the document, he would be bound.

GOLDSTEIN et al. v. RHODE ISLAND HOSPITAL TRUST NATIONAL BANK
296 A.2d 112 (R.I. 1972)

Burton Goldstein sued Rhode Island Hospital Trust National Bank, challenging interest rates and counsel fees charged delinquent debtors. The trial court held for Defendant on its motion to dismiss for failure to state a claim upon which relief could be granted. Reversed and remanded for new trial.

KELLEHER, J.

Sometime prior to February, 1970, plaintiff like thousands of the inhabitants of this state received in the mail from the defendant bank (hereafter called the bank) a plastic passport to, depending on one's point of view, instant credit or instant debt. This plasticized wonder is better known as a credit card. The credit cards were not solicited by the recipients but rather the defendant bank, together with other banks in this area, made a wholesale distribution of the cards as they attempted to make an auspicious entry into a new and presumably attractive field of consumer credit.

Goldstein used his credit card to make various purchases and in the process incurred a debt to the bank of $242. He did not or could not pay the debt. Consequently, on September 8, 1970, the bank obtained a judgment in the District Court against Goldstein in the amount of $323. This figure represented the $242 balance plus an attorney's fee of $81. Goldstein made a series of payments to the bank which reduced the amount of the judgment to $110. At this point, he consulted an attorney and in January, 1971, the District Court granted his motion to vacate the judgment and recall the execution.

Before proceeding, we shall detail the initial distribution plan referred to earlier. Goldstein received his credit card in the mail. It was attractively packaged. Two credit cards were inserted in a cardboard folder. The cardboard is folded in half. Three-quarters of the surface of the cardboard contains printed messages extolling the virtues of the card, such as, "It's free," "Extend your payments if you wish," "It's all you'll ever need!" These pleasant exhortations appear in boldface print. When the card is removed from its enclosure, there appears the word "Important" in large boldface print and a direction to read the "enclosed agreement." On the reverse side of this bottom-half of the card is set out the "Cardholder Agreement."

In essence, Goldstein contends in his first count that because of the packaging of the card, the placing of the terms on the Cardholder Agreement and the microscopic type which is found on the credit card and cardboard folder, he never had actual or constructive notice of the provisions relating to interest charges and other costs, particularly attorney's fees.

The trial justice, who dismissed the first count, relied on the general rule that the furnishing of a credit card which contains certain terms and conditions constituted an offer which ripened into an expressed contract by Goldstein's retention and use of the card. This was so, said the trial court, even though the terms relating to counsel fees and interest charges were set forth on a different document, to wit, the Cardholder Agreement.

In the Union Oil case, [*Union Oil* v. *Lull* 220 Or. 412, 349, P.2d 243 (1960)], the court observed that the credit cardholder had failed to raise either in his pleadings or by way of evidence the argument that he was not aware of the conditions printed on the back of the credit card and so he was bound by them. It noted that juries could pass on whether or not ". . . the terms of the contract were put in deceptive form which would mislead a reasonable person, and that defendant was so misled. . . ." 220 Or. at 420, 349 P.2d at 247. The plaintiff here has not made the same tactical error as Lull. Instead, he has alleged enough so that he may rely upon the cases which hold that a person is not bound by the terms of a written agreement if he has no knowledge of its terms because the manner in which they are embodied in the instrument would not lead a reasonable person to suspect that the terms are part of the contract.

Having in mind the liberality afforded the pleader, we believe that he has alleged enough facts which, if sustained by proof, would warrant relief. Whether he can prevail is a matter to be determined on the basis of his proof rather than on his pleadings. Any doubt as to his ultimate success is no cause at this stage of the proceeding for the dismissal of the first count. The initial dismissal of the first count was erroneous.

QUESTIONS FOR DISCUSSION

1. *Can you think of any situations in which one may not actually know of an offer but would be contractually bound by accepting the offer by implication?*
2. *In a credit-card situation, who typically makes the offer? What constitutes the acceptance? You should note that the bank in the above case could not today legally mail unsolicited credit cards.*

5:4. DEFINITENESS. An offer must be definite and certain. It must have sufficient terms from which a court can enforce the resulting contract after acceptance. If an offer lacks such essential terms as price, quantity, or identification of the subject matter, it is ineffective.

This does not mean that every conceivable bit of information must be contained in the offer. Furthermore certain terms may be implied and others may be stated in terms of market price. However, in that event some ascertainable standard must be available from which to calculate the price. Market price presents little problem, but suppose one is to receive a "fair share of the profits." What would that mean?

Requirements contracts are generally enforceable. They usually provide for one party to supply exclusively all of another party's needs for a particular product at a set price and for a specific period. For example, A offers to purchase all the steel he needs to manufacture cars for a particular year from B steel company. In return B agrees to supply all the steel needed by A at a set price. Most courts would hold that this is sufficiently definite because the quantity to be supplied is all the steel needed for the applicable period. Both parties are bound to a contract because steel is needed in the manufacture of cars. The Uniform Commercial Code (U.C.C.) recognizes requirements contracts in Section 2–306.

GREEN v. ZARING

222 Ga. 195, 149 S.E.2d 115 (1966)

GRICE, J.

The provisions of the agreement entered into by Green, Zaring, and a third party referred to herein as "Brown," insofar as material here, are those which follow.

Zaring is the owner of a described 43.3 acre tract of land, and Green is a general contractor engaged in the business of developing and improving real estate. They and Brown, who are instrumental in getting them together in this undertaking, desire to form a corporation through which this land will be developed and improved "with apparent buildings and related structures in accordance with the covenants and agreements" hereinafter set forth.

In consideration of the mutual covenants of the parties, they agree as follows.

A corporation shall be formed with specific powers to deal generally in all types of property, to acquire, develop, improve, mortgage, pledge, encumber and sell the same.

Green agrees to pay personally the legal expense involved in forming the corporation, and to advance for the corporation all monies required for architectural and engineering expense necessary to obtain building permits. He shall be repaid by the corporation from "loans obtained to develop and improve" the land. Green also agrees to pay personally "all costs of construction in excess of the construction loan or loans."

Zaring, upon the formation and authorization of the corporation to do business, agrees to convey fee simple title to the above mentioned land to the corporation, which shall hold title for the purposes set forth in this agreement.

Green further agrees to serve as "general contractor in the developing and improving of said land with apartment buildings and related structures without salary from the corporation," and also "to use his ability to obtain the necessary loans" for the corporation "at the best obtainable rates in order for the corporation to develop and improve said land."

The corporation shall assume obligation for all costs and expenses required in developing and improving the land "with apartment buildings and related structures." Where required by any lender of monies to the corporation, Green and Zaring agree to personally endorse "any note evidencing any such loan."

The corporation shall develop and improve the land in three stages, as suggested by the architect to be employed to prepare the plans and specifications, with only the title to each divided tract to be pledged as security for "a loan or loans" for development and improvement of that particular tract.

The agreement also contained various other provisions relating to such matters as continuance of an easement over a part of the property and issuance of corporate stock. None of these warrant further reference here.

Count 1 of the petition alleged that Zaring is capable of complying with the terms of this agreement but has failed and refused to convey the property in accordance therewith and has repudiated the agreement in its entirety.

It is axiomatic that the terms of a contract must be reasonably certain.

This requirement is especially applicable to agreements which are sought to be specifically performed. This court has held:

> "A court of equity will not decree the specific performance of a contract for the sale of land unless there is a definite and specific statement of the terms of the contract. The requirement of certainty extends not only to the subject matter and purpose of the contract, but also to the parties, consideration, and even the time and place of performance, where these are essential. Its terms must be such that neither party can reasonably misunderstand them. It would be inequitable to carry a contract into effect where the court is left to ascertain the intention of the parties by mere guess or conjecture, because it might be guilty of erroneously decreeing what the parties never intended or contemplated." *Williams* v. *Manchester Building Supply Co.*, 213 Ga. 99, 101, 97 S.E.2d 129.

Measuring the agreement now before us by the rule just recited, we find it lacking in certainty.

It appears that the parties envisioned that they would develop and improve designated real estate, through a corporation, by Green performing services, Zaring providing the land, and both obligating themselves along with the corporation on loans of money.

But nowhere in the document is there any statement with reasonable certainty as to what the parties were obligating themselves to do to effect what they envisioned, or as to what they envisioned. We refer to some of the deficiencies.

As to the improvements to be made, all that appears is "apartment buildings and related structures" and a reference to three stages of development. The number, size, design, or material content of the apartment buildings and related structures are not stated. Nor is their anticipated cost mentioned. The "related structures" are not identified in any manner.

As to financing for the improvement, the agreement says only "necessary loans," "at the best obtainable rates," with Green and Zaring agreeing to personally endorse "any note evidencing any such loans." Were the loans to amount to $5,000 or $5,000,000?

The foregoing deficiencies as to certainty are not alleged to have been cured by performance or conduct of the parties. Organization of the corporation by legal counsel procured by Green, to the apparent satisfaction of Zaring, is no remedy for those vital deficiencies.

(Judgment for Zaring)

QUESTIONS FOR DISCUSSION

1. *What were the subject matters in the contract that were too indefinite for enforcement by the Court?*
2. *How would you draft the provisions of the agreement so that they would be enforceable?*

5:5. TERMINATION. We have stated earlier that once an offer is made, one may be bound to a contract by the acceptance of the offeree. Does this mean that once one makes an offer, it hangs over his head for life like the sword of Damocles, waiting for the offeree to accept it in his own good time and bind the offeror to a contract? Obviously not. In addition to being accepted there are a number of ways that an offer may terminate: expiration of time, revocation, rejection, counteroffer, death or insanity, and illegality.

(1) Expiration of Time. The offeror may place a time limit on his offer, in which case the offer will expire at the end of that period of time. Often the offer does not expressly restrict acceptance to a stated period of time. In that case the offer will terminate at the expiration of a reasonable period of time. What constitutes a reasonable period of time depends on the circumstances and nature of the subject matter. A reasonable period of time for an offer to sell a parcel of land would very likely be quite different from an offer to sell a company's stock trading on the New York Stock Exchange.

(2) **Revocation.** As a general rule one may revoke his offer at any time prior to its acceptance. With the exception of options and certain firm offers under the U.C.C. discussed in Section (b), one may revoke an offer before it is accepted even when he promised to keep it open. The revocation may be made in any reasonable manner indicating that the offer is revoked. The revocation must be communicated to the offeree prior to acceptance in order to be effective. An offer made by advertisement, such as an offer of reward, may be revoked by the same method in which the offer was made. The revocation would then be effective even against an individual who read the original offer but not the revocation.

(a) *Option.* An exception to the general rule stated earlier is found in the case of an option. An option is an offer that the offeror promises to keep open for a stated time in return for some consideration being paid by the offeree. This distinguishes an option from other offers. An ordinary offer may be revoked prior to acceptance because no consideration was paid by the offeree in return for the promise to keep it open. Thus, if A promises to sell B Whiteacre for $5,000 and in return for $100 paid by B promises further to keep the offer open for one year, an option arises and cannot be revoked during that time. Without the consideration paid by B the offer could be revoked anytime prior to acceptance, despite A's promise to the contrary.

(b) *Firm Offer Under the U.C.C.* Another exception to the rule pertains to an offer by a merchant to buy or sell goods. When the merchant promises in a signed writing to keep an offer open for a stated period of time, it may not be revoked. In that case Section 2–205 of the U.C.C. provides that no consideration need be paid by B, the offeree, in return for this promise. If no stated time is mentioned, it is a reasonable time not to exceed three months. This, of course, does not apply to written offers made by a merchant in which he does not expressly promise to keep the offer open.

(3) **Rejection and Counteroffer.** A rejection by the offeree terminates the offer. It is effective when received by the offeror or his agent. The offer may not be accepted by the offeree after rejection. A counteroffer operates as a rejection. For example, A states, "I will sell you my motorcycle for $450." B replies, "I'll pay $400." B has made a counteroffer, which has the effect of terminating A's offer. B cannot come back and accept A's original offer of $450 unless A renews it.

Generally, a purported acceptance that varies the terms of the offer will operate as a rejection. Altering the method or time of payment or the place of performance are among those types of alterations. If one includes them in his acceptance, however, he runs the risk of its being ineffective. The rule has been considerably liberalized for those contracts subject to the U.C.C. These will be discussed more fully in Chapter 15, but in general the code, Section 2–207 provides that under certain conditions an acceptance that contains terms in addition to or different from those offered will not operate as a rejection.

(4) **Death or Insanity.** Death or insanity of either party will result in the termination of an offer. If the death of an offeror or an offeree occurs prior to acceptance, a contract cannot come into existence, even, for example, if the personal representative of the deceased offeree's estate attempts to accept the offer. One should be very careful to distinguish this situation from the case in which a contract is already in existence. Death of one of the parties to a contract does not necessarily terminate the contract.

(5) **Illegality.** If the legislature passes a law making the subject matter of a contract illegal, the offer is terminated. A court interpretation that would result in the subject matter of a contract's being illegal would have the same result.

FREY v. FRIENDLY MOTORS, INC.

129 Ga.App. 636, 200 S.E.2d 467 (1973)

Bess L. Frey executed a buyer's order for an automobile for $2,400 from Friendly Motors, Inc., and tendered it together with her check for $200. Friendly contended that there was no contract, and its motion for summary judgment was granted by the trial court. Plaintiff appealed. Reversed.

HALL, J.

The facts show that Mrs. Frey went to Friendly's premises accompanied by her husband, and there executed a buyer's order for an automobile, which itself stated that it should not be binding until signed by an agent of Friendly. She tendered it with her check for $200 earnest money to an agent of Friendly as an offer to buy the automobile. Subsequently, numerous oral communications were held by Mrs. Frey and her husband with a Friendly salesman, who in the conduct of negotiations went back and forth between the Freys and a Friendly sales manager empowered to accept the offer. Mr. Frey then told the salesman to bring him either an acceptance by Friendly or the return of the earnest money check. The salesman on his deposition testified that he then received the sales manager's consent to sell the car "tonight" for the amount written on the offer, and that he told the Freys they had the deal they wanted, and they had only to fill out and sign the bill of sale. Protesting that it was growing late in the evening and they would prefer to complete the paperwork the next morning, the Freys departed, apparently confident that the car could be picked up the next morning. In their affidavits, both Mr. and Mrs. Frey state that the salesman agreed to their plan to return the next day to pick up the car. Actually, when they returned they found that Friendly's position was that there was no contract. Friendly's evidence tends to show that the buyer's order was never signed by anyone on behalf of Friendly, and the earnest money check was torn up by Friendly after the Freys departed.

[Friendly] has staked its position on the failure of the parties to come to a meeting of the minds, there is no requirement that such meeting of the minds be manifested by a writing, except for such requirement to that effect as is imposed by the buyer's order. The buyer's order, however, as an offer to buy, is controlled by the offeror, Mrs. Frey. If an attempted acceptance is made in terms varying an offer, this may be considered a counteroffer and the counteroffer may be accepted by the offeror, making a contract. Thus, an attempted oral acceptance here by Friendly would have been a counteroffer because it would have varied the terms of the offer which originally required the acceptance to be in writing. However, if it had been agreed to by Mrs. Frey, it would have created the contract here sought to be sued on, absent the operation of the statute of frauds which is not claimed by Friendly to be relevant. Thus, Friendly may not prevail on its motion merely by showing that the buyer's order was never signed. These parties could have made an oral contract.

Whether the parties intended to make a contract is a question of fact. The record shows sworn allegations by the Freys that the parties reached a meeting of the minds and that the agreement was that they would return the next day to get the car. The position of Friendly as shown on the deposition of the salesman is that there was no such agreement and the departure of the Freys that evening manifested their breaking off negotiations. Thus, there is a sharp conflict in the evidence on the determinative question of whether there was a contract, and for this reason the grant of summary judgment for Friendly was erroneous.

5:6. Summary of Uniform Commercial Code Changes. The Uniform Commercial Code has made several changes in the general contract law discussed in this chapter. The purpose of this section is to highlight some of the more important changes at this time so that you may make a comparison between the U.C.C. provisions and the law of contracts not covered by the U.C.C. A full discussion of the U.C.C. regarding contracts for the sale of goods is found in Section Three, "Sales." Remember that the U.C.C. only covers contracts for the sale of goods and not other contracts.

The U.C.C. has a number of provisions that liberalizes the requirement of definiteness in an offer. The code specifically provides that a contract does not fail for indefiniteness if the parties intended to make a contract and there is a reasonably certain basis for giving an appropriate remedy. Section 2–204. In addition the code has a number of rules concerning particular aspects of a contract. For example, one provision states that if a price is not settled in a contract the price will be a reasonable price, if nothing is said as to price, so long as the parties intended that there be a contract. Section 2–305 (1) (a). The code has other provisions concerning delivery terms and the duration of the contract if the parties have not covered them in the agreement. Sections 2–308; 2–309. Not all of the provisions in the code change the general law of contracts but in general the effect of the code in this area is to allow offers to be somewhat less specific than under the general contract law.

This does not mean that any statement no matter how vague will qualify as an offer even under the U.C.C. For example, if there is no basis for determining a reasonable price a statement will not qualify as an offer even under the U.C.C.

The U.C.C. also has provisions regarding termination of the offer. Recall our earlier discussion concerning the written offer by a merchant who agrees to keep an offer open. That promise is enforceable even without consideration. Section 2–205. Another important provision regarding termination of the offer is found in Section 2–207. That section provides that an acceptance that contains terms in addition to or different from the offer does not automatically operate as a counteroffer terminating the offer as is the case under the general contract law. The terms may operate either as proposals for addition to the contract or, if the parties are merchants, may even be incorporated into the contract depending upon a number of circumstances more fully discussed in Chapter 15. In general, the thrust of the section is to give more flexibility to the parties in the process of arriving at a contract and to avoid some of the very strict rules of the common law.

SUMMARY

The initial step in contract formation is the offer. The offer may be defined as a promise made by one person (the offeror) to another (the offeree) stating what the offeror will do in return for an act, forebearance, or a return promise. An offer must be made with contractual intent as measured by the objective test of the reasonable man. Invitations to deal are not offers. Advertisements for the sale of goods are often invitations to deal as are the bids of an auction that is not "without reservation." An advertisement of a reward is an offer.

An offer must be communicated to the offeree to be effective. The method of communication is up to the offeror. The offer must be definite and certain so that the court can enforce the contract that results from acceptance. Requirements contracts are generally enforceable.

An offer may terminate when the time for acceptance, as stated in the offer, expires. If no time is stated, the offer terminates in a reasonable time. One may also revoke his offer at any time prior to acceptance unless the offeree has a valid option. An exception is a provision of the Uniform Commercial Code which requires a merchant to keep his offer open if he states in writing that he will do so.

If the offeree rejects an offer it is terminated. A counteroffer also operates as a rejection and terminates the offer. An acceptance varying the offer will operate as a counteroffer but the Uniform Commercial Code has liberalized the rules regarding the acceptance of an offer to sell goods.

Finally, the death or insanity of either party will result in the termination of an offer. A law making the subject matter of a contract illegal also terminates the offer.

PROBLEMS

1. In a case of the proverbial triangle the marriage of Mary Ogilvie and Byron Ogilvie came to an unhappy end because Mary believed that Byron's affections were alienated by Bertha May Welland. As part of the marriage settlement Byron signed a $3,000 note payable to Mary. Mary and Bertha signed a separate contract which stated that Bertha would "aid and assist Byron Ogilvie in the payment of (his $3,000 promissory note)." When Byron did not pay, Mary sued Bertha on her agreement to aid and assist Byron in paying the note. Was Mary entitled to recover from Bertha? Ogilvie v. Ogilvie, 487 S.W.2d 40 (Mo. App. 1972).
2. The Great Minneapolis Surplus Store published the following newspaper advertisement: "Saturday, 9 A.M., 2 Brand New Pastel Mink 3-skin scarfs selling for $89.50, Out they go, Saturday. Each . . . $1.00, 1 Black Lapin Stole, Beautiful, worth $139.50 . . . $1.00, First Come First Served." Lefkowitz demanded the lapin stole for one dollar. Was there a contract? Lefkowitz v. Great Minneapolis Surplus Store, 251 Minn. 188, 86 N.W.2d 689 (1957).

3. Owen wired Tunison, "Will you sell me your store property . . . for $6,000.00." Tunison replied: "Because of improvements . . . it would not be possible for me to sell it unless I was to receive $16,000.00 cash." Owen accepted. Was there a contract? Owen v. Tunison, 131 Me. 42, 158 A. 926 (1932).
4. The Wilhelm Lubrication Company agreed to purchase certain products from Brattrud according to the following schedule:

Quantity	*Description*	*Per Gal.*
5,000 gals.	Worthmore Motor Oil SAE 10–70	21–31
3,000 gals.	Beterlube Motor Oil SAE 10–70	26–36
2,000 gals.	Costal Motor Oil SAE 10–70	18–28

Brattrud repudiated and Wilhelm sued. Verdict for whom? Wilhelm Lubrication Co. v. Brattrud, 197 Minn. 626, 268 N.W. 634 (1936).
5. Frederick Loeser & Co., a department store, inserted an ad in a newspaper to the effect that it would sell, deliver, and install certain well-known standard makes of radio receivers. Among them were listed two De Forest receiving sets known as the "D-12 reflex radiophone receiving sets." Lovett offered to purchase two of the sets under the conditions stated in the advertisement, but Frederick Loeser refused to sell them. Lovett then tendered his certified check for the advertised price and when it was not accepted sued for breach of contract. Was there a contract? Lovett v. Frederick Loeser & Co., 207 N.Y.S. 753 (1924).
6. On July 7, 1903, the town council of Harrison, New Jersey, passed a resolution offering to purchase water from Jersey City. The council never directed anyone to communicate the offer to Jersey City. Jersey City officials came into possession of a certified copy of the resolution by a method never explained. Jersey City officials then purported to have accepted the terms of the resolution. Was there a contract? Mayor etc. of Jersey City v. Town of Harrison, 62 A. 765 (N.J. 1905).
7. Varney hired Ditmars as a draftsman at a salary of $35 per week. Varney then told Ditmars, "I'm going to give you $5.00 more per week if you boys will go on and continue the way you have been and get me out of this trouble and get these jobs started that have been in the office for three years, and on the first of next January I will close my books and give you a fair share of my profits." Ditmars then worked for Varney for a period of time. Was there an enforceable contract? Varney v. Ditmars, 111 N.E. 822, 217 N.Y. 223 (1916).

CHAPTER 6

Agreement: The Acceptance

6:1. Introduction. The next step in the formation of a contract is the acceptance of the offer. As you might expect, there are a number of legal considerations affecting the validity of the acceptance. These are extremely important because they affect such issues as the validity of the contract, the timing of its creation, and even the place of its creation. The acceptance is the assent of the offeree, either expressed or implied, to be bound to the terms spelled out by the offeror in his offer.

Once the acceptance is made, a contract arises. At the outset it is important to note several restrictions. In the first place, only the persons or their agents to whom the offer is addressed may accept the offer. For example, if Harry Hotrodder in speaking to Prospective Carowner and says, "I offer to sell you my car for $400 on May 1," a third party hearing the conversation cannot leap between the two, say "I accept," and bind the contract. On the other hand, an offer may be made to a group of persons or to the offeree "or his assigns." This would open the door for any number of people to accept the contract.

Another point to note is that although the offeror is at some disadvantage in making an offer and being the first to commit himself, he has the advantage of being able to set forth the requirements for acceptance. For instance, the offeror may specify the date, time, place, and method of acceptance. If he does so, his requirements must be met or the purported acceptance is not valid. Of course, it is not necessary for the offeror to specify any requirements at all, but he may wish to do so for his own protection.

DAVID J. TIERNEY, JR., INC. v. T. WELLINGTON CARPETS INC.

392 N.E.2d 1066 (Mass. App. 1979)

David J. Tierney, Jr., Inc. was the general contractor for the renovation of Pittsfield High School. T. Wellington Carpets, Inc. submitted a subbid on the project to Tierney for the installation of carpeting at a cost of $22,869.00. The bid provided that "the undersigned (T. Wellington's president) agrees to execute said contract within five working days thereafter on the standard form provided by Northeast Flooring Contractor's Association."

Tierney then purportedly accepted Wellington's bid but not on an NEFCA form as mentioned in the bid. Wellington refused to install the contract at the price of $22,869.00. It claimed that there was no contract since its bid was not accepted on the NEFCA form.

BROWN, J.

T. Wellington makes the . . . argument that . . . Tierney never accepted its offer because its subbid required a particular mode of acceptance which Tierney failed to comply with, and any other form of attempted acceptance was ineffective. T. Wellington thus asserts that Tierney could accept only by delivering to T. Wellington an executed contract on "the standard form provided by Northeast Flooring Contractor's Association (NEFCA)."

The offeror has full control over his offer and may prescribe a particular and exclusive mode of acceptance. To be effective the acceptance must be made in the manner required by the offer. . . . And if a purported acceptance substantially varies from the terms of the offer it is ineffective, creating no binding agreement. . . . In our view there was sufficient evidence from which the jury would conclude that Tierney accepted the offer of T. Wellington, thereby creating a binding agreement.

Although an offeror, as the creator of a power of acceptance, may demand an exclusive mode of acceptance, we think that other forms of acceptance are ineffective only "if the offeror clearly expresses, in the terms of the communciated offer itself, his intention to exclude all other modes of acceptance." Corbin, Contracts § 88, at 373 (1963). The language in the bid here is, at best, ambiguous as to whether it required an offeree to respond by tendering an executed contract on an NEFCA form. The bid provides that "the undersigned [T. Wellington's president] agrees to execute said contract within five working days thereafter on the standard form provided by Northeast Flooring Contractor's Association." The bid does not state that the offeree has an obligation to use the NEFCA form. Without a clear expression of intent by the offeror that acceptance of the bid could be made only by the use of the NEFCA form the jury could have found that Tierney had effectively accepted the offer in another manner.

(Judgment for Tierney)

QUESTIONS FOR DISCUSSION

1. *Who was the offeror in this case?*
2. *What should the offeror have done to make certain that the acceptance was only made on the desired form?*

6:2. METHOD OF ACCEPTANCE. An acceptance may be either expressed or implied. If it is expressed, it may be oral or written. An implied acceptance may result from an act of the offeree or by silence, if there is a duty to speak.

(1) Express Acceptance. As stated before, if the offeror elects to state the method of acceptance, that requirement must be followed. If no particular mode of acceptance is stated, any reasonable method is acceptable. For example, if the offeror makes his offer by mail, an acceptance by telegram, letter, or even an oral acceptance may be reasonable. On the other hand, an acceptance by mail might not be reasonable if the offer was telegraphed under circumstances in which time is important.

(2) Written Memorialization of Oral Agreements. A question often arises when the parties arrive at the terms of an oral agreement and then decide to reduce the agreement to writing. If the writing never comes about, is the oral agreement enforceable? The answer is that it depends upon the intent of the parties at the time the oral transaction was made. If the parties intended to be bound to the contract at the time of the oral agreement, then the contract is enforceable despite the absence of the written agreement. This is because the parties intended the written document only as a memorialization of their oral agreement. On the other hand, if the parties did not intend to be bound until the written document was executed, then no contract exists until that event occurs. The intent of the parties is, of course, determined by the objective test referred to in the previous chapter.

ABC TRADING CO. LTD. v. WESTINGHOUSE ELECTRIC SUPPLY CO.

382 F.Supp. 600 (1974)

Plaintiff, ABC Trading Co. Ltd, sued defendant, Westinghouse Electric Supply Co. (WESCO), for damages for breach of contract in the shipment of wire products to the plaintiff in Japan. Plaintiff filed a motion for judgment on an alleged settlement agreement entered into between itself and defendant. The defendant contests the existence of a binding settlement agreement.

NEAHER, D.J.

Plaintiff's position, simply stated, is that on June 12, 1972, WESCO's counsel, Mr. Dean, conveyed to plaintiff's attorney, Mr. Gottesman, an offer of settlement in the amount of $31,500, which offer Mr. Gottesman then accepted. Mr. Gottesman understood this offer to mean that in return for the $31,500, plaintiff would reship the wire and cable products to the United States and also enter into a stipulation of discontinuance.

Defendant, on the other hand, maintains that the negotiations were substantially more complicated than plaintiff's counsel describes. Defendant's attorneys point out that WESCO acted in effect as a middleman in both the sale of the disputed products and the settlement negotiations, the principal seller being a company named Eastern Wire and Cable Company of New Jersey (Eastern), which is not a party to this action. Defense counsel maintains that WESCO never had any intention of paying most of the cost of settlement, that Eastern was to contribute the major portion thereof, and that Mr. Gottesman was aware of these facts throughout the negotiations. Therefore, defendant's attorneys continue, when a settlement offer of $31,500 was communicated to Mr. Gottesman on June 12, 1972, it was understood that the offer was subject to reducing the terms of the agreement to a writing which would include a provision for payment of an amount by Eastern. The writing was necessary, according to WESCO's counsel, because without it WESCO would have no way of ensuring Eastern's participation in paying the cost of settlement.

Plaintiff's counsel does not dispute that a writing was contemplated but maintains that since all important points had already been agreed to, the writing would be merely a memorial of the agreement and without independent legal consequence.

Williston correctly states the general rule:

> It is . . . everywhere agreed that if the parties contemplate a reduction to writing of their agreement before it can be considered complete, there is no contract until the writing is signed. *Williston on Contracts* (3d ed.) § 28 at 66–67.

. . . This rule obtains even if the parties have orally agreed upon all the terms of the proposed contract. . . . Whether the parties intended to be bound without a formal writing is a question of fact. . . .

The court's inquiry must therefore focus on whether either of the parties manifested an intention not to be bound by an oral arrangement despite their apparent understanding of the terms upon which the action was to be settled. This intention (not to be bound) was manifested, however, in a letter, dated February 23, 1972, sent by WESCO's counsel to Mr. Gottesman:

> If your client finds this proposal agreeable in principle, we can proceed to reduce it to a written agreement incorporating, if you wish, a simple escrow arrangement. I am informed that this is the best offer that defendant and its supplier are prepared to make.

The mere fact that a written draft is referred to during negotiations has been held to be some evidence of an intention not to be bound until its execution. . . .

Furthermore, while it is true that many settlement agreements are in fact concluded orally, a situation made possible by their relatively uncomplicated nature, i.e., a sum of money is exchanged for a release, the involved nature of the instant settlement forecloses such an approach. . . .

The court must, therefore, conclude that WESCO had no intention of being bound until a written agreement, to which Eastern would be a party, was signed. . . .

(Plaintiff's motion denied.)

QUESTIONS FOR DISCUSSION

1. *Do you know from the decision in this case who finally won?*
2. *What objective facts were significant in the court's decision?*

(3) Implied Acceptance. (a) *Silence.* As a general rule, silence alone will never constitute acceptance. For example, Sonny Solicitor writes to Manny Mum, "I offer to sell you my stereo for $450. If I do not hear from you within two weeks I will assume that you accept my offer." Manny Mum need never reply. He would not be bound because the law will not allow another to foist a contract on him in this manner. Mum has no duty to reply and thus need not do so.

There are circumstances in which one does have a duty to reply and his failure to do so will constitute an acceptance. This duty arises most often in the commercial world because of the past history of dealing between the parties. It may also arise because of the relationship between the parties.

HOBBS v. MASSASOIT WHIP CO.
158 Mass. 194, 33 N.E. 495 (1893)

HOLMES, J.

This is an action for the price of eel skins sent by the plaintiff to the defendant, and kept by the defendant some months, until they were destroyed. It must be taken that the plaintiff received no notice that the defendants declined to accept the skins. The case comes before us on exceptions to an instruction to the jury that, whether there was any prior contract or not, if skins are sent to the defendant, and it sees fit, whether it has agreed to take them or not, to lie back, and to say nothing, having reason to suppose that the man who has sent them believes that it is taking them, since it says nothing about it, then, if it fails to notify, the jury would be warranted in finding for the plaintiff.

Standing alone, and unexplained, this proposition might seem to imply that one stranger may impose a duty upon another, and make him a purchaser, in spite of himself, by sending goods to him, unless he will take the trouble, and bear the expense, of notifying the sender that he will not buy. The case was argued for the defendant on that interpretation. But, in view of the evidence, we do not understand that to have been the meaning of the judge, and we do not think that the jury can have understood that to have been his meaning. The plaintiff was not a stranger to the defendant, even if there was no contract between them. He had sent eel skins in the same way four or five times before, and they had been accepted and paid for. On the defendant's testimony, it was fair to assume that if it had admitted the eel skins to be over 22 inches in length, and fit for its business, as the plaintiff testified and the jury found that they were, it would have accepted them; that this was understood by the plaintiff; and indeed, that there was a standing offer to him for such skins.

In such a condition of things, the plaintiff was warranted in sending the defendant skins conforming to the requirements, and even if the offer was not such that the contract was made as soon as skins corresponding to its terms were sent, sending them did impose on the defendant a duty to act about them; and silence on its part, coupled with a retention of the skins for an unreasonable time, might be found by the jury to warrant the plaintiff in assuming that they were accepted, and thus to amount to an acceptance. . . . The proposition stands on the general principle that conduct which imports acceptance or assent is acceptance or assent, in the view of the law, whatever may have been the actual state of mind of the party—a principle sometimes lost sight of in the case. . . . Exceptions overruled.

QUESTIONS FOR DISCUSSION

1. *What facts in the Hobbs case placed it outside the general rule that silence will not act as an acceptance? What should the Massasoit Whip Company have done if it did not want the eel skins?*
2. *Assuming that this was the first dealing between Hobbs and Massasoit, would there have been any liability on the part of the company if they had retained the skins without using them?*
3. *The judge in this case was Oliver Wendell Holmes (1841–1935). He was one of America's most famous jurists. He taught at Harvard Law School and then served for twenty years as justice and chief justice of the Supreme Judicial Court of Massachusetts. He was later appointed an Associate Justice of the U.S. Supreme Court, where he served for thirty years until he resigned in 1932.*

ROBERTS v. BUSKE

12 Ill.App.3d 630, 298 N.E.2d 795 (1973)

Plaintiff, an insurance agent, sued the defendant for breach of contract, alleging that defendant had accepted an automobile liability policy sent to him for renewal. The trial court rendered judgment in favor of plaintiff and the defendant appealed.

CREBS, J.

The facts are undisputed. In September, 1969, plaintiff, an insurance agent, sent defendant a policy which was a renewal of one defendant's father had previously held. Defendant had not ordered or requested issuance of this policy, but he accepted it and paid the premium. In September, 1970, just prior to the expiration date of the policy, a second unsolicited renewal was sent to defendant and attached to it was a printed notice stating that if defendant did not wish to accept it he must return it or be liable for the premium. Defendant made no response either to this notice or to two subsequent bills mailed to him. Finally, in December, the agent telephoned defendant personally to inquire about the premium. Defendant informed him that he had purchased a policy from another company in August and that since he had not ordered the renewal he felt no obligation to pay for it. The policy was then returned to the company and cancelled, resulting in a loss to the agent for the pro-rated portion of the premium which he had advanced.

In his brief plaintiff appears to accept the fact that the basic requisites of a valid contract are equally applicable to an insurance contract, and that a bare offer imposes no liability upon the person to whom it is made until it is accepted. However, he argues that because defendant had previously accepted a renewal policy and thereafter paid the premium his silence in replying to the second renewal constituted an implied acceptance thereof and obligated him to pay the premium.

We recognize that the practice of sending renewals oftentimes serves the best interests of an insured; but, likewise it serves the best business interests of the insurance agent. And it is the agent, as the offeror, who must assure himself that his offer has been accepted. Under ordinary circumstances silence cannot be relied upon to establish an acceptance of an offer to enter into a contract. . . . We do not preclude the possibility of an implied acceptance being established under certain circumstances. However, . . . we cannot find that acceptance of a single previous renewal in itself is sufficient to constitute an implied acceptance of a second renewal based solely on the silence of the offeree. Not only is the mailing of a first renewal policy insufficient to show a previous course of dealing between the parties, the plaintiff presents no evidence regarding the customary trade practice, if any, in situations similar to that presented here. It is obvious that plaintiff feels abused in that defendant did not inform him that he had bought another policy, but a simple telephone call could have revealed this fact and there would have been no reason for plaintiff to have incurred the loss caused by his unwarranted assumption.

(b) *Acts.* An acceptance may be made by an act of the offeree that indicates that he intends to accept the offer. One must be careful in this regard, for it is the act that indicates the acceptance. The thoughts inside the individual's head are not important because they are presumed from the act. The most common instance of an act's constituting an implied acceptance occurs when the offeree retains and uses goods that he has received. His use of these goods would constitute an implied acceptance of the offer.

One should note the changing law with regard to unsolicited goods. Several states—Maryland, Illinois, and New York, for example—have statutes declaring that unordered merchanidse may be treated as gifts. Their use by the recipient does not constitute an implied acceptance. Federal laws have also been passed in this area. The Federal Postal Reorganization Act of 1970 provides that one who receives unsolicited merchandise in the mail may treat it as a gift and ". . . shall have the right to retain, use, discard, or dispose of it in any manner he sees fit without any obligation whatsoever to the sender."[1] Note that the Federal Act applies only to goods received through the United States mails.

AUSTIN v. BURGE
156 Mo.A. 286, 137 S.W. 618 (1911)

O. D. Austin sued Charles Burge for breach of contract. Lower-court judgment was for the defendant and plaintiff appealed.

ELLISON, J.

It appears that plaintiff was publisher of a newspaper in Butler, Mo., and that defendant's father-in-law subscribed for the paper, to be sent to defendant for two years, and that the father-in-law paid for it for that time. It was then continued to be sent to defendant, through the mail, for several years more. On two occasions defendant paid a bill presented for the subscription price, but each time directed it to be stopped. Plaintiff denies the order to stop, but for the purpose of the case we shall assume that defendant is correct. He testified that, notwithstanding the order to stop it, it was continued to be sent to him, and he continued to receive and read it, until finally he removed to another state.

We have not been cited to a case in this state involving the liability of a person who, though not having subscribed for a newspaper, continues to accept it by receiving it through the mail. There are, however, certain well-understood principles in the law of contracts that ought to solve the question. It is certain that one cannot be forced into contractual relations with another and that therefore he cannot, against his will, be made the debtor of a newspaper publisher. But it is equally certain that he may cause contractual relations to arise by necessary implication from his conduct. The law in respect to contractual indebtedness for a newspaper is not different from that relating to other things which have not been made the subject of an express agreement. Thus one may not have ordered supplies for his table, or other household necessities, yet if he continues to receive and use them, under circumstances where he had no right to suppose they were a gratuity, he will be held to have agreed, by implication, to pay their value. In the case defendant admits that, notwithstanding he ordered the paper discontinued at the time when he paid a bill for it, yet plaintiff continued to send it, and he continued to take it from the post office to his home. This was an acceptance and use of the property, and, there being no

[1] Federal Postal Reorganization Act § 3009.

pretense that a gratuity was intended, an obligation arose to pay for it.

The preparation and publication of a newspaper involves much mental and physical labor, as well as an outlay of money. One who accepts the paper, by continuously taking it from the post office, receives a benefit and pleasure arising from such labor and expenditure as fully as if he had appropriated any other product of another's labor, and by such act he must be held liable for the subscripiton price. On the defendant's own evidence, plaintiff should have recovered.

The judgment will therefore be reversed, and the cause remanded. All concur.

QUESTIONS FOR DISCUSSION

1. *What should Burge have done if he did not want to be bound to the contract?*
2. *Would the result have been the same if the papers had been delivered to his home? How would the recent changes in the law have affected this case?*

6:3. Communication of Acceptance. (1) Communication in a Bilateral Contract. In order for there to be a binding contract in a bilateral contractual situation, the acceptance must be communicated to the offeror or his agent. When this occurs is most important, for it determines the time when a contract arises.

Obviously a mere mental determination by the offeree that he will accept the offer does not result in the formation of a contract. The acceptance must be communicated. When the parties deal face to face or even over the telephone, the time when communication occurs obviously presents no problem. When the parties communicate over a distance by mail or telegraph, a more complicated situation arises.

As a general rule, if the offeree accepts in a method authorized by the offeror, the contract is formed when the acceptance is placed in control of the transmitting agency. Most courts hold that when one accepts in the same manner as that in which an offer was made, the offeror has implied authorization of that method of acceptance. Consider this problem. Ollie Offeror mails an offer to Artie Acceptor. Artie places his acceptance in the mail, but prior to its receipt by Ollie, Ollie revokes the offer. The revocation is received by Artie after he has mailed the acceptance but before Ollie has received it. Is there a contract?

A majority of courts would hold that there is a valid contract because the acceptance was effective and communicated when mailed, whereas the revocation was not communicated until actually received. Note that this rule applies only to acceptances and only when the method used by the offeree is authorized by the offeror. The authorization may result either expressly or by implication, such as past practice, or by use of the same mode for communication of the acceptance as was used by the offeror. In addition, there is nothing to prevent the offerer from altering the effect of this rule by simply stating in his offer that the contract will not arise until the acceptance is received.

What is the reason for this rule? Why a different rule for acceptance than for revocations or rejections? There are a number of theories and much controversy over this question. Perhaps the reason most often given is that because the offeror determines the mode of communciation of acceptance, the agency selected is the agent of the offeror. Thus, when the acceptance is placed in the control of the postal authorities, the acceptance is in effect communicated to the agent of the offeror. We will learn in Section Six, "Agency," that the knowledge of an agent is imputed to his principal. In this instance the theory would be that once the offeror's agent learns of the acceptance, the offeror is deemed to know of it. Therefore, even if the offeror never receives the acceptance, a contract is formed.

Others argue against this reasoning on the ground that the postal regulations give one the right to retrieve a communication at any time before it is delivered to the addressee. That argument is answered with the explanation that the contract arises when the acceptance is mailed, and thus although one may have the *power* to retrieve it prior to delivery, he does not have the contractual *right* to do so.

Still others justify the rule on the ground that there must be some point when the contract arises. As somebody must bear the risk that the communication might be lost by the communicating agency, the risk should be placed on the party best able to protect himself. Because the offeror has the power to restrict the time, place, and method of acceptance, that is where the risk of loss is placed. In any event remember that the rule applies only when the method of acceptance is authorized by the offeror and that it does not apply to revocations or rejections. The offeror can alter the "rules of the game" by specifying that the acceptance will not be effective until received.

MORRISON v. THOELKE
155 So.2d 889 (1963)

On November 26, 1957, defendants executed a contract to purchase certain real property owned by plaintiffs. The contract was mailed to plaintiffs in Texas. On November 27, 1957, plaintiff executed the contract and mailed it back to defendant on that date. After mailing the contract but prior to its receipt by defendants, plaintiffs called defendants' attorney to repudiate the execution and contract. The defendants recorded the contract and plaintiffs filed suit to quiet title. The lower court found for plaintiff and defendant appealed.

ALLEN, J.

The question is whether a contract is complete and binding when a letter of acceptance is mailed, thus barring repudiation prior to delivery to the offeror, or when the letter of acceptance is received, thus permitting repudiation prior to receipt. Appellants, of course, argue that posting the acceptance creates the contract; appellees contend that only receipt of the acceptance bars repudiation.

The appellant, in arguing that the lower court erred in giving effect to the repudiation of the mailed acceptance, contends that this case is controlled by the general rule that insofar as the mail is an acceptable medium of communication, a contract is complete and binding upon posting of the letter of acceptance. . . . Appellees, on the other hand, argue that the right to recall mail makes the Post Office Department the agent of the sender, and that such right coupled with communication of a renunciation prior to receipt of the acceptance voids the acceptance. In short, appellees argue that acceptance is complete only upon receipt of the mailed acceptance. . . .

Turning first to the general rule relied upon by appellant some insight may be gained by reference to the statement of the rule in leading encyclopedias and treatises. . . .

A second leading treatise on the law of contracts, Corbin, Contracts 78 and 80 (1950 Supp.

1961), also devotes some discussion to the "rule" urged by appellants. Corbin writes:

> Where the parties are negotiating at a distance from each other, the most common method of making an offer is by sending it by mail; and more often than not the offeror has specified no particular mode of acceptance. In such a case, it is now the prevailing rule that the offeree has power to accept and close the contract by mailing a letter of acceptance, properly stamped and addressed, within a reasonable time. The contract is regarded as made at the time and place that the letter of acceptance is put into the possession of the post office department.

Like the editor of Williston, Corbin negates the effect of the offeree's power to recall his letter:

> The postal regulations have for a long period made it possible for the sender of a letter to intercept it and prevent its delivery to the addressee. This has caused some doubt to be expressed as to whether an acceptance can ever be operative upon the mere mailing of the letter, since the delivery to the post office has not put it entirely beyond the sender's control.

> It is believed that no such doubt should exist. . . . In view of common practices, in view of the difficulties involved in the process of interception of a letter, and in view of the decisions and printed discussions dealing with acceptance by post, it is believed that the fact that a letter can be lawfully intercepted by the sender should not prevent the acceptance from being operative on mailing. If the offer was made under such circumstances that the offeror should know that the offeree might reasonably regard this as a proper method of closing the deal, and the offeree does so regard it, and makes use of it, the contract is consummated even though the letter of acceptance is intercepted and not delivered.

As is abundantly clear from the quoted material excerpted from appellees' cases, the decision in each is predicated on an assumption, correct or incorrect, that the basis of the rule they reject was invalidated by changed postal regulations. The opinions cited by appellees each proceed on the theory that the "deposited acceptance" rule was based on a theory that the depositor lost control of his acceptance when it was deposited and that this fact rendered the acceptance complete upon deposit. To the extent that "loss of control" was the significant element in the "deposited acceptance" rule, the logic of appellees' cases is impeccable. On the other hand, if the rule is, in fact, not based on the "loss of control" element the fact that this element has been altered may in no way affect the validity of the rule. Determination of the question presented in this appeal cannot then be had merely by adoption or rejection of the logic of appellees' cases. Rather, the source and justification of the "deposited acceptance" rule must be found and appellees' argument considered in light of this finding. Should the proffered justification for the rule be other than the "loss of control" theory, adoption or rejection of the rule must be based on considerations other than those relied upon in appellees' cases. . . .

The rule that a contract is complete upon deposit of the acceptance in the mails, herein before referred to as "deposited acceptance rule" and also known as the "rule in *Adams* v. *Lindsell,*" had its origin, insofar as the common law is concerned, in *Adams* v. *Lindsell,* 1 Barn. & Ald. 681, 106 Eng. Rep. 250 (K.B. 1818).

Examination of the decision in *Adams* v. *Lindsell* reveals three distinct factors deserving consideration. The first and most significant is the court's obvious concern with the necessity of drawing a line, with establishing some point at which a contract is deemed complete and their equally obvious concern with the thought that if communication of each party's assent were necessary, the negotiations would be interminable. A second factor, again a practical one, was the court's apparent desire to limit but not overrule the decision in *Cooke* v. *Oxley,* 3 T.R. 653 [1790] that an offer was revocable at any time prior to acceptance. In application to contracts negotiated by mail, this latter rule would permit revocation even after unqualified assent unless the assent was deemed effective upon posting. Finally, having effectively circumvented the inequities of *Cooke* v. *Oxley,* the court, apparently constrained to offer some theoretical justification for its decision, designated a mailed offer as "continuing" and found a meeting of the minds upon the instant of posting assent. Significantly, the factor of the offeree's loss of control of his acceptance is not mentioned.

The unjustified significance placed on the "loss of control" in the cases relied upon by appellee follows from two errors. The first error is failure to distinguish between relinquishment of control as a factual element of manifest intent, which it is, and as the legal predicate for completion of contract, which it is not. The second error lies in confusing the "right" to recall mail with the "power" to recall mail. Under current postal regulations, the sender has the "power" to regain a letter, but this does not necessarily give him the "right" to repudiate acceptance.

> The additional reasons for holding that a different rule applies to an acceptance and that it is operative on mailing may be suggested as follows: When an offer is by mail and the acceptance also is by mail, the contract must date either from the mailing of the acceptance or from its receipt. In either case, one of the parties will be bound by the contract without being actually aware of that fact. If we hold the offeror bound on the mailing of the acceptance, he may change his position in ignorance of the acceptance; even though he waits a reasonable time because the letter of acceptance is delayed, or is actually lost or destroyed, in the mails. Therefore this rule is going to cause loss and inconvenience to the offeror in some cases. But if we adopt the alternative rule that the letter of acceptance is not operative until receipt, it is the offeree who is subjected to the danger of loss and inconvenience. He can not know that his letter has been received and that he is bound by contract until a new communication is received by him. His letter of acceptance may never have been received and so no letter of notification is sent to him; or it may have been received, and the letter of notification may be delayed or entirely lost in the mails. One of the parties must carry the risk of loss and inconvenience. We need a definite and uniform rule as to this. We can choose either rule; but we must choose one. We can put the risk on either party; but we must not leave it in doubt. The party not carrying the risk can then act promptly and with confidence in reliance on the contract; the party carrying the risk can insure against it if he so desires. The business community could no doubt adjust itself to either rule; but the rule throwing the risk on the offeror has the merit of closing the deal more quickly and enabling performance more promptly. It must be remembered that in the vast majority of cases the acceptance is neither lost nor delayed; and promptness of action is of importance in all of them. Also it is the offeror who has invited the acceptance. [Corbin, Contracts 578 (1950)]

In short, both advocates and critics muster persuasive arguments. As Corbin indicated, there must be a choice made, and such choice may, by the nature of things, seem unjust in some cases. Weighing the arguments with reference not to specific cases but toward a rule of general application and recognizing the general and traditional acceptance of the rule as well as the modern changes in effective long-distance communication, it would seem that the balance tips, whether heavily or near imperceptively, to continued adherence to the "Rule in *Adams* v. *Lindsell.*" This rule, although not entirely compatible with ordered, consistent and sometime artificial principles of contract advanced by some theorists, is, in our view, in accord with the practical considerations and essential concepts of contract law.

In the instant case, an unqualified offer was accepted and the acceptance made manifest. Later, the offerees sought to repudiate their initial assent. Had there been a delay in their determination to repudiate permitting the letter to be delivered to appellant, no question as to the invalidity of the repudiation would have been entertained. As it were, the repudiation antedated receipt of the letter. However, adopting the view that the acceptance was effective when the letter of acceptance was deposited in the mails, the repudiation was equally invalid and cannot alone support the summary decree for appellees.

The summary decree is reversed and the cause remanded for further proceedings.

Questions for Discussion

1. *What arguments can you think of against using the rule spelled out in this case? What reasons in support of the rule?*
2. *Does this case suggest any reasons that the offeror might want to insert a clause requiring receipt of the acceptance in order for it to be effective?*

(2) Communication is a Unilateral Contractual Situation. As you will recall, in a unilateral contractual situation there is an offer that calls for some sort of performance. Acceptance is made by the performance of the act requested. In general, no contract arises until the performance is completed. Thus the offeror could revoke his offer at any time prior to the completion of the performance. For example, Harry Homeowner states to Perry Painter, "I will pay $250 to you if you will paint my house." The offer can be revoked when the house is two thirds painted. This works an obvious hardship on Perry because there is no contract until the performance is completed. There is the possiblity that Perry might recover in quasi contract in order to avoid the unjust enrightment of Harry, but he would not be able to recover on the contract.

In general, it is not necessary for the offeree to notify the offeror prior to his performance if the offeror has knowledge of the performance. As a practical matter the offeror will have to be notified when he is asked to carry out his performance by the offeree. For example, offers of reward are unilateral offers. A performance is requested and not a return promise. It is not necessary to notify the offeror prior to performance of the requested act in order to bind him to the contract once the performance is completed. It is necessary that the offeree know about the offer at the time he performs the requested act, otherwise his acceptance would not be in response to the offer and would not be effective. It is not necessary that the single or primary motivation for performing the act, however, be to accept the offer.

Contracts under the Uniform Commercial Code are discussed more fully in Section Three, "Sales," but you should note at this time that Section 2–206(2) provides that when the beginning of performance is the proper mode of acceptance, an offeror may treat the offer as having lapsed unless he is notified by the offeree of acceptance within a reasonable period of time. This part of the code, of course, applies only to contracts for the sale of goods and not to other types of contracts. Obviously in these cases, however, the offeree should notify the offeror of his intent to perform in order to protect himself.

SIMMONS v. UNITED STATES
308 F.2d 160 (1962)

Diamond Jim III was a rockfish placed in the Chesapeake Bay on June 19, 1958, by employees of the American Brewery, Inc. His rise to stardom was occasioned by a publicity campaign announcing the Third Annual American Beer Fishing Derby.

Diamond Jim III was adorned with an identification tag. Anyone presenting the tag and an affidavit stating that the fish was caught on hook and line would be entitled to a cash prize of $25,000. The plaintiff, Simmons, caught Diamond Jim III, presented the tag, and was given $25,000. Plaintiff knew about the contest when he went fishing but catching Diamond Jim III was not uppermost in his mind.

The ever-vigilant Internal Revenue Service claimed that the $25,000 was includable in Simmons's gross income. He paid the tax and claimed a refund, which was denied. Simmons then sued the U.S. district court where he lost. Plaintiff appealed, claiming several grounds for relief.

SOBELOFF, C.J.

We do not understand the taxpayer to claim immunity from the tax on the ground that capture of the fish or the award of the prize had any religious, charitable, scientific, educational, artistic, or literary significance whatever. His argument is that the payment was made in recognition of a civic achievement. He attributes a civic purpose to the American Brewery, Inc., in offering a prize the effect of which, he says, is to popularize the recreation and resort facilities of the state of Maryland. Yet it requires a considerable flight of fancy to romanticize the Fishing Derby into a civic endeavor. A glance at the advertisement announcing first the Derby and later the capture of Diamond Jim III unmistakably reveals that the purpose of the contest and of the prize was to stimulate the sale of American beer.

Viewing the facts most favorably to the taxpayer, we hold that he was not rewarded for a civic achievement, properly interpreted. There was nothing meritorious in a civic sense in catching this rock fish. Simmons was not even rewarded for an extraordinary display of skill, if that could be considered a civic achievement, for catching Diamond Jim III was essentially a matter of luck. The case might be different if, for example, Simmons had at considerable risk to himself captured and destroyed a killer whale terrorizing the Maryland seashore. That could have been regarded as a genuine civic achievement. But catching this fish cannot reasonably be so denominated, for the only community interest in the event was one of idle curiosity. Innumerable are the rhapsodies uttered in praise of the delights and virtues of the piscatorial pastime, but never to our knowledge has it been seriously called a civic enterprise. The character of this fortuitous event is not raised to a civic level by being linked to an advertising campaign aimed at selling beer. Far from resembling a Nobel or Pulitzer prize-winner, Mr. Simmons fits naturally in the less-favored classification the legislators reserved for beneficiaries of "giveaway" programs.

The taxpayer's next point is that he was at least entitled to have a jury decide whether the $25,000 payment to him was a gift, excluded from gross income by section 102.

The established fact is that there was no personal relationship between Simmons and the brewery to prompt it to render him financial assistance. Nor was it impelled by charitable impulses toward the community at large, for the prize was to be paid to whoever caught Diamond Jim III, regardless of need or affluence. Rather, the taxpayer has apparently rendered the company a valuable service, for, by catching the fish and receiving the award amid fanfare, he brought to the company the publicity the Fishing Derby was designed to generate.

Moreover, under accepted principles of contract law on which we may rely in the absence of pertinent Maryland cases, the company was legally obligated to award the prize once Simmons had caught the fish and complied with the remaining conditions precedent. The offer of a prize or reward for doing a specified act, like catching a criminal, is an offer for a unilateral contract. For the offer to be accepted and the contract to become binding, the desired act must be performed with knowledge of the offer. The evidence is clear that Simmons knew about the Fishing Derby the morning he caught Diamond Jim III. It is not fatal to his claim for refund that he did not go fishing for the express purpose of catching one of the prize fish. So long as the outstanding offer was known to him, a person may accept an offer for a unilateral contract by rendering performance, even if he does so primarily for reasons unrelated to the offer.

(Affirmed.)

6:4. Variance from Offer. As a general rule the acceptance must be unequivocal and in the same terms as the offer. Any variation in terms, such as method of payment or place or time of performance, would operate as a rejection of the offer and constitute a counteroffer. Obviously merely stating terms implied by law that are not stated in the offer would not invalidate the acceptance. For example, Ollie Offeror states to Artie Acceptor, "I offer to pay you $250 upon completion of the painting of my house." Artie replies,

"I accept. Payment is to be made one half when the job is started and the balance upon completion." No contract would result because the time for payment has been altered by the offeree.

CITY OF ROSLYN v. PAUL E. HUGHES CONST. CO.

573 P.2d 385 (Wash. App. 1978)

Paul E. Hughes Construction Company submitted a bid to the City of Roslyn for construction of a sewage-collection and -treatment system. The bid submitted by Hughes was in accordance with requirements established by the city in its call for bids. The requirements stated that bidders would keep their offers open for 60 days to allow the City to complete financing arrangements with the U.S. Department of Agriculture. The document establishing the bid requirements also stated that any extensions beyond the sixty-day time limit were to be "mutually agreed upon."

On the sixtieth day the city sent a Notice of Award to Hughes containing the following additional terms:

> Because of need to obtain modification of building right-of-way permits, without which FHA's necessary approval and acceptance of the contract cannot be secured, there may be a delay. As you know, in the notice and instructions to bidders and in the contract documents the award and letting of the contract cannot be made without approval and acceptance of the FHA.
>
> Since there may be a delay, you need not submit the necessary contract performance and payment bond and insurance until required by FHA or until FHA approves and accepts this award of contract, whichever is earlier.
>
> If there is a delay, in fairness to you the City of Roslyn will not require you to work on the project between October 15, 1972 and May 15, 1973, and will extend the completion date accordingly.

Subsequently Hughes informed the city that it was withdrawing the bid. The City contended that it had accepted Hughes's bid and that there was a binding contract. Hughes contended that the Notice of Award was not a valid acceptance.

GREEN, J.

It is a fundamental rule that:

> An acceptance of an offer must always be identical with the terms of the offer or there is no meeting of the minds and no contract. *Owens-Corning Fiberglas Corp.* v. *Fox Smith Sheet Metal Co.*, 56 Wash.2d 167, 170, 351 P.2d 516, 518 (1960). . . .

An expression of assent and reply to an offer that purports to accept it, but changes the terms of the offer in any material respect, may be operative as a counteroffer; but it is not an acceptance and does not consummate the contract.

> Again, it is the rule that, where the contract is alleged to consist, as it is in this instance, of an offer on the one side and the acceptance of the offer on the other the acceptance to constitute a binding contract must be as broad as the offer, any conditions attached to the offer being in the nature of *new proposals* which themselves must be accepted to make a binding contract. [Citing authority.] (Italics ours.) *St. Paul & Tacoma Lumber Co.* v. *Fox*, 26 Wash.2d 109, 126, 173 P.2d 194, 204 (1946), quoting from *Coleman* v. *St. Paul & Tacoma Lumber Co.*, 110 Wash. 259, 188 P. 532 (1920).
>
> An acceptance, to be effectual, must be identical with the offer and *unconditional.* Where a person offers to do a definite thing, and another accepts conditionally or introduces a *new term* into the acceptance, his answer is either a mere expression of willingness to treat or it is a counter proposal, and in neither case is there an agreement. This is true, for example, where an acceptance varies from the offer *as to time of performance, place of performance, price, quantity, quality and in other like cases.* 13 C.J. 281 (Italics ours.) *Martinson* v. *Carter*, 190 Wash. 502, 505, 68 P.2d 1027, 1028 (1937).

However, we recognize that in order for an additional term expressed in the offer to change the purported acceptance into a counteroffer, the additional term must be a *material* modification. Further,

> if a "condition *added* by the intended acceptance can be implied in the original offer then it does not constitute a material variance so as to make the acceptance ineffective." *Johnson* v. *Star Iron and Steel Co.*, 9 Wash. App. 202, 205, 511 P.2d 1370, 1373 (1973).

Applying these rules to the present case, we find that (1) the Notice of Award purporting to accept Hughes' bid proposal changed the terms of the offer in a material respect, and (2) the new conditions were not *implied* in the original offer. While it is true that the documents, *i.e.*, the original offer by the City, put all bidders on notice that FHA approval of the contract, change orders, and the like would be necessary because the construction project was to be funded in whole or in part by FHA, the original offer did not allow a delay in *awarding* of the contract past the sixty-day time limit, *except in the event of a mutually agreed upon extension of time.*

It becomes clear, when reading the provisions together, that the overall effect of the added language in the Notice of Award was to delay the construction schedule. Postponed indefinitely was the commencement date of construction. Time of performance is a material term of a contract.

Consequently, an enforceable contract was not created, and Hughes' refusal to perform could not constitute a breach.

(Verdict for Hughes.)

QUESTIONS FOR DISCUSSION

1. *What statements in the purported acceptance by the City constituted a variance of the offer?*
2. *What was the practical result of the City's Notice of Award?*

6:5. Summary of Uniform Commercial Code Changes. The Uniform Commercial Code, Section 2–207, has made some important changes in this area. If the offeree states terms in addition to or different from those made in the offer, there still may be a valid acceptance. The additional terms are construed as proposals for additions to the contract. If the parties are merchants, the offeree's additional terms become part of the contract unless "(a) the offer expressly limits acceptance to the terms of the offer; (b) they materially alter it; or (c) notification of objection to them has already been given or is given within a reasonable time after notice of them is received." An offeror may still limit the acceptance to the exact terms of the offer. On the other hand, the offeree may make the acceptance conditional on the additional terms becoming part of the contract. In this case, it would operate as a counteroffer. Also, remember that in this section of the Uniform Commercial Code we are concerned only with contracts for the sale of goods.

The U.C.C. also clarifies the manner and method that should be employed in accepting an offer. Under the common law a few jurisdictions took the approach that acceptance had to be by the same mode as was used in extending the offer; for example, a mailed offer should be accepted by mail. Under the code, however, if the offer is silent as to the manner and method of acceptance, the acceptance may be made in any reasonable medium, such as by mail, by telegraph, or orally. Section 2–206 (1) (a). If the offer is silent as to time for acceptance, an acceptance made in a reasonable time is effective. Section 2–207 (1).

SUMMARY

The acceptance is the assent of the offeree, either expressed or implied, to be bound to the terms spelled out by the offeror in his offer. Once the acceptance is made, a contract arises.

Only the persons, or their agents, to whom an offer is made may accept an offer. The offeror may state the requirements for acceptance in his offer and these must be followed by the offeree. An express acceptance may be either oral or written. Any reasonable mode of acceptance may be used if the offeree does not require a specific mode. The intent of the parties determines whether a contract exists if the oral agreements between them is never reduced to writing.

An acceptance may be implied as a result of the actions of the offeree. In general, silence on the part of the offeree will not be construed as an implied acceptance unless circumstances place a duty upon him to reply. An acceptance may be made by an act of the offeree that indicates that he intends to accept the offer. A federal law provides that if one receives unsolicited goods in the mail he may treat them as a gift and his use of them will not be interpreted as an implied acceptance.

The acceptance must be communicated to the offeree to be effective. In a bilateral contractual situation, if the offeree accepts in a method authorized by the offeror, the contract is formed when the acceptance is placed in control of the transmitting agency. This way the mail is authorized and acceptance is considered communicated where mailed. This rule is limited to the acceptance. It is explained either by the agency theory or more simply as a method of risk allocation by the courts.

Acceptance of an offer in a unilateral contractual situation occurs when the performance of the requested act is completed. It is not necessary to notify the offeror prior to performance of the requested act in order to bind him to the contract once performance is completed. On the other hand, the offeree must know of the offer in order for his act to constitute an acceptance.

As a general rule, the acceptance must be unequivocal and in the same terms as the offer. A variance will operate as a counteroffer terminating the offer. The Uniform Commercial Code has made some changes to the strict rule against any variance in the acceptance.

PROBLEMS

1. D. G. Jurgensen submitted a written bid for the purchase of real property from the estate of Klauenberg on December 22, 1971. The bid was for $65,000 and was accompanied by a check of $6,500. On December 27, 1971, the executor orally notified the buyer of the acceptance of the bid and petitioned the probate court to confirm the sale. On January 13, 1972, the buyer filed objections alleging that he had never received a written acceptance of his offer and that his offer was withdrawn by a letter dated

January 11, 1972. The lower court found for the estate and Jurgensen appealed.

Was it necessary for the executor to accept the bid in writing in order for there to be a binding contract? In Re Estate of Klauenberg v. Jurgensen, 108 Cal. Rptr. 669 (Cal. App. 1973).

2. A veterans' organization offered a reward to the person furnishing information resulting in the arrest and conviction of the culprit who murdered one of its members. Mary Glover gave such information to the police concerning Reginald Wheeler, who was her daughter's boyfriend at the time. Wheeler and his partner Jesse James Patton were arrested and convicted based on Mary's information. She then learned of and applied for the reward. The offeror refused to pay and Mary sued. Verdict for whom? Glover v. Jewish War Veterans of the United States, Pool No. 58, 68 A.2d 233 (1949).
3. The Carbolic Smoke Ball Company ran the following advertisement in the local newspaper: "£100 reward will be paid by the Carbolic Smoke Ball Company to any person who contracts the increasing epidemic influenza, colds or any disease caused by taking cold, after having used the ball three times daily for two weeks according to the printed directions supplied with each ball. £1000 is deposited with the Alliance Bank, Regent Street showing our sincerity in the matter. . . ." Carhill, after reading the advertisement, bought the product, used it as directed, and promptly caught influenza. Carhill sued to recover the £100. Verdict for whom? Carhill v. Carbolic Smoke Ball Co., 1 Q.B256 (1863).
4. Lindsell wrote a letter to Adams on September 2, 1817, offering to sell wool. Adams received the letter on September 5, 1817, and replied by mail that evening. This letter was received by Lindsell on September 9. Meanwhile Lindsell, having expected an earlier reply, sold the wool in question to another party on September 8. Adams sued for breach of contract. Decide. Adams v. Lindsell, Court of Kings Bench, 1818, 1 Barnewall & Alderson 681.
5. Wilson went to an insurance agency that represented a number of companies to secure automobile insurance. He filled out an application, gave a premium of $30.80 to the agent, and in turn received a receipt. Below Wilson's signature was a statement that no coverage was in effect until a policy was issued by the home office. Wilson claimed that the agent told him that he was covered immediately. The insurance company contends that a policy was never issued and that there was no contract. Decide. See Turner v. Worth Insurance Company, 464 P.2d 990 (1970).
6. On August 23 and 24 Wilson, through Ammons's traveling salesman, Tweedy, ordered 43,916 pounds of shortening for prompt shipment. These orders were sent in by Tweedy. Wilson never received the shortening and was told when he inquired on September 4 that the orders were declined. During that time the price of shortening had risen from 7.5 cents/lb. to 9 cents/lb. Tweedy had represented Ammons for six to eight months and had previously taken orders from Wilson that had been accepted. Was there

a contract? Ammons v. Wilson & Co., 176 Miss. 645, 170 So. 227 (1936).

7. Roto-Lith, Ltd. manufactures cellophane bags for packaging vegetables. J. P. Bartlett & Co. makes emulsion for use as a cellophane adhesive. Roto ordered some emulsion from Bartlett. On the invoice an acknowledgment received from Bartlett prior to shipment was a disclaimer of any warranty. The emulsion failed to adhere and Roto sued Bartlett. Bartlett relied on the warranty disclaimer. Is Bartlett entitled to rely on the warranty disclaimer? Roto-Lith, Ltd. v. F. P. Bartlett & Co. 297 F.2d 497 (1962).
8. Weldon Hall and the Orange Boat Club sponsored a boat race, utilizing the race to advertise Hall's Marina. On the Thursday prior to the Saturday race a newspaper article stated that first prize was a fourteen-foot boat trailer and twenty-horsepower motor. Gerald Bean called Hall's Marina and verified the article. Bean then entered the race and won first prize. Hall then stated the article was incorrect and offered Bean a six-horsepower motor as first prize. Bean filed suit contending that he had an enforceable contract. Was he correct? Hall v. Bean 582 S.W.wd 263 (Tex. 1979).
9. Hendrickson ordered a broadcast seeder through a salesman for International Harvester who called on Hendrickson at his farm. Hendrickson signed a document indicating that the sale was subject to the approval of International Harvester at its office. The seeder was to be delivered in the spring but it never was delivered. International Harvester retained the order until the suit, never indicating acceptance or rejection. Hendrickson sued. Judgment for whom? Hendrickson v. International Harvester Company of America 135 A. 702 (1927).

CHAPTER 7

Agreement: Reality of Consent

The two previous chapters have discussed the agreement. One requirement is that the parties must have contractual intent at the time of the making of the agreement. We learned that an objective standard rather than a subjective standard is used in order to determine whether the requisite contractual intent is present. The reason for this measurement is so that one can rely on the outward manifestations of an individual's intent rather than worry about what might actually be secretly stored in the person's "mind."

This manifestation of assent must, however, be freely and voluntarily given. Furthermore, the parties must have the requisite contractual intent regarding the subject matter of the agreement. There are a number of circumstances that may interfere with these requirements. They are discussed traditionally under the subjects of misrepresentations, duress, undue influence, mistake, and unconscionable contracts. If the court finds that the contract was entered into under one of these circumstances, the usual remedy is rescission; that is, the court will allow the injured party to set aside the contract and recover anything he has paid.

7:1. MISREPRESENTATION. (1) Fraud. Fraud is a tort. One may sue another for damages if one is the victim of fraud. Fraud may also be a defense to a breach-of-contract action; that is, a person may validly refuse to perform a contract that he was induced to enter into as a result of fraud. The contract is voidable at the option of the injured party. Historically the courts have required that the following elements be proved in order for one to establish fraud: (1) There must be a false representation of a material fact. (2) There must be knowledge of the falsity on the part of the person making the statement, or a statement must be made with reckless disregard of the truth. This element is often technically known as *scienter.* (3) The person making the statement must intend that the other party act in reliance on the statement. (4) The innocent party must justifiably rely on the statement. (5) The innocent party must suffer damage as a result of his reliance. Let us examine each one of these elements.

The false representation must be of a fact. Ordinarily this means a present or past fact. Statements concerning the future are generally not considered statements of fact and opinions are not factual statements. For example, if one stated that he believed that a certain lot would be "zoned commercial" and it does not occur, there would be no fraud. The statement was simply an opinion about what might happen in the future. In addition, a misrepresentation of law is generally not considered to be fraud.

A question often arises as to whether one must disclose all the facts concerning a particular transaction. Where the parties are bargaining at arm's length and there is no relationship of trust and confidence between them, one is not required to disclose facts concerning the transaction. For example, Harry Homeowner is offering to sell his house to Barry Buyer. The basement has water in it frequently. Homeowner has no duty to disclose this fact to Buyer. On the other hand, if Buyer asks and Homeowner states that the basement is always dry, there would be fraud, assuming that the other elements are present.

One may not actively conceal facts and must not remain silent when there is a duty to speak. Such a duty may arise because of a confidential relationship existing between the parties. Examples of this would be the relationship of attorney–client, principal–agent, or guardian–ward. Furthermore fragmentary statements and half-truths that are designed to lead the victim to the wrong conclusion may result in fraud.

In order to constitute fraud the element of *scienter* must be present; that is, there must be knowledge of the falsity on the part of the party making the statment, or it must be made with reckless disregard of the truth. It is this element that distinguishes fraud from an innocent misrepresentation, discussed in Section (1), which may be grounds for setting aside a contract but would not be an actionable tort. If one makes a statement that he subsequently discovers to be untrue, he has a duty to correct it or he may commit fraud.

Of course, the misrepresentation must be material. A minor misrepresentation will not constitute fraud. For example, if one states that his car has been driven 75,000 miles when the distance is actually 75,250 miles, it is doubtful that a court would grant any relief to the innocent party. The reason is that the difference in mileage would not have affected the decision to enter into the contract. There would be no reliance on the falsity of the statement. Similarly, slight exaggerations will not constitute fraud. A statement that one's toothpaste is the "best-tasting toothpaste in the world" would not be considered fraud but rather that delightful concept of the advertising trade known as puffing. However, the more specific one becomes in his statement, the closer it may come to being fraud.

Closely related to the subject of materiality is the requirement that the innocent party must reasonably rely on the false statement in order for there to be fraud. If he does not rely on the statement, he is not deceived by it. If one is not induced to enter into the contract as a result of the false statement, he is not entitled to any relief. Furthermore, if the innocent party knows of the falsity of the statement, or reasonably should have known, a court will not grant relief because there is no reasonable reliance by the party and the statement has not induced the party into entering into the contract. Likewise, if a party is put on notice that facts are other than as represented, he may not rely on them but must determine them for himself. For example, if one is told that the basement of a house has always been dry but he sees moisture and water marks on the wall, he may have to inquire further. He may not rely on the statement.

Another problem area arises in cases in which a party signs a document without reading it and then claims that the contents of the document were misrepresented to him. Typically a party is charged with knowing and understanding the terms of the written contract that he has signed. If he does not understand it after reading it, he must find out what it means. One who signs a contract without first reading it does so at his peril and may not claim that he was induced to enter into it because of the fraudulent statements of the other party. An exception would be when there is an emergency or when the party is induced to sign by trick or artifice. The fact that one party tells another that the contract contains certain provisions when it does not will usually not result in the contract's being set aside. The reason is that all the individual has to do is to read the contract for himself. One who does not read a contract prior to signing it is not reasonably relying on the statements made by the other party.

Finally, in order to recover for fraud or use it as a defense to a contract, the innocent party must show that he suffered damages as a result of the fraud. If he has suffered no injury, he has no action.

GARDNER v. MEILING
572 P.2d 1012 (Or. 1977)

RICHARDSON, J.

Plaintiff, (Robert Gardner) a school teacher in The Dalles who had no experience in operating a tavern, journeyed to Portland in February 1976 and contacted Business Brokers, Inc. to inquire about the purchase of a tavern. He testified he was looking for a neighborhood tavern. James Stockard, a salesman for Business Brokers, Inc., showed plaintiff the Punjab Tavern which was owned by defendants and had been listed for sale with Business Brokers, Inc. The premises were owned by a third party and leased to defendants.

Plaintiff met on two occasions with James Stockard prior to signing an earnest money receipt. On one of these two occasions James Stockard, after determining plaintiff was new in the tavern business, explained some of the pitfalls of tavern operation to him. He explained specifically that the success of a tavern depended in large measure on the individual who operated the tavern and the persons he employed as bartenders and that the past performance of a tavern was no guarantee the volume of business would continue. Stockard did not make any representations as to the income of the tavern and testified Business Brokers, Inc. had, in fact, received no such information from defendants.

Plaintiff and Stockard obtained, from the wholesale beer distributor, a report of the number of kegs of beer delivered to the tavern during an eleven month period in 1975. The monthly average was approximately forty kegs. Stockard testified he told plaintiff as a rough rule of thumb a tavern could gross approximately $60 to $80 per keg. He explained this was a generalization about taverns and was contingent upon a number of factors.

Plaintiff testified, based on conversation with other tavern owners and with James Stockard, he calculated the tavern's gross income in 1975 was approximately $4,500 per month. This calculation was based solely on the average number of kegs of beer delivered in 1975. Plaintiff stated he was interested in the tavern's income but he did not depend on the keggage reports for income projections since they related to 1975 and that he was waiting for the income figures which would be given to the Oregon Liquor Control Commission (O.L.C.C.) to obtain transfer of the liquor license.

On February 20, 1976, the parties executed an earnest money agreement providing for a sale price of $40,000 to be paid $1,000 as earnest money, $7,000 on delivery of possessions of the business, $2,000 payable ninety days after possession, $2,000 on February 1, 1977 and the balance in monthly payments.

On May 20, approximately two months after the contract was executed, the plaintiff and defendant Jon Meiling met with an investigator from the O.L.C.C. to discuss transfer of the liquor license. The investigator asked for the gross income figures from the tavern business for the months of February, March and April, 1976. Although there is a conflict in the evidence regarding where the figures were obtained, Jon Meiling gave the investigator gross income figures for the three months. The figures given were $5,710.50, $4,918.57 and $5,009.51 respectively. The required O.L.C.C. license was subsequently granted on June 6, 1976, and plaintiff took possession of the tavern on June 10, 1976, after the transaction was closed.

The gross income figures given to the O.L.C.C. investigator in plaintiff's presence are the only representation received by plaintiff regarding the tavern's gross income and are the basis of his claim of fraud and misrepresentation.

After taking possession of the tavern and operating it for two days plaintiff discovered the income was very low and that there were few female patrons. He contacted the bookkeeping service utilized by the sellers and learned the gross income of the tavern had been approximately $1,400 to $2,000 during February, March and April of 1976. On June 16, 1976, a few days after taking possession, plaintiff, through his attorney, wrote to defendants demanding rescission on the basis of fraudulent misrepresentation. Rescission was refused by defendants. Plaintiff continued to operate the tavern, on advice of counsel, until January 9, 1977, as he stated, in order to protect the O.L.C.C. license.

Plaintiff seeks rescission based on three theories: fraud, misrepresentation and mistake. . . . Count I alleges that before he agreed to purchase the tavern and lease the premises defendants represented the gross income of the tavern to be $4,400 to $4,700 per month and that the tavern had valuable goodwill and a good reputation in the community. He alleged that he relied upon these representations and would not have purchased the tavern otherwise. He further alleges the representations were false and that defendants either knew they were false or made them recklessly without knowledge as to their truth or falsity.

Comprehensively stated, the elements of actionable fraud allowing avoidance of a contract consist of: (1) a representation, (2) its falsity, (3) its materiality, (4) the speaker's knowledge of its falsity or ignorance of its truth, (5) his intent that it should be acted upon, (6) the injured party's ignorance of its falsity, (7) his reliance upon the truth of the representation, and (8) his right to rely on the truth. The person alleging fraud has the burden of proving each of the elements and failure to prove any one or more of the elements is fatal to the cause of action.

Implicit in the element of reliance is a requirement the plaintiff prove a causal relationship between the representation and his entry into the bargain. In other words if he fails to prove that the fraudulent misrepresentations induced him to make the agreement he has failed to establish the element of reliance.

As we view the evidence an enforceable contract arose at the time the earnest money agreement was signed by both parties on February 20, 1976. The formal written contract on March 15, 1976, was merely a memorialization of this agreement. Prior to the execution of the earnest money agreement no representation regarding the gross income of the tavern had been made. Plaintiff conceded, during oral argument in this court, that the representations of May 20, 1976, made to the O.L.C.C. investigator in plaintiff's presence are the only representations respecting income that he received. Since these representations came after the binding agreement of February 20, 1976, and even after the execution of the written contract they could not have been relied upon by plaintiff in making the agreement.

(Decision for the defendant.)

QUESTION FOR DISCUSSION

1. *What element of fraud did the plaintiff fail to establish? What would the plaintiff have to prove in order to satisfy this requirement?*

WALSH v. EDWARDS

233 Md. 552, 197 A.2d 424 (1964)

This was action by plaintiffs Nathen and Doris Edwards against defendants Gerald and Grayce Walsh for fraud in the sale of a house. Prior to the purchase of the house the plaintiffs had inquired as to what a shallow creek in the rear of the property was likely to do "during a storm." *Gerald Walsh replied that the* "creek would come up over its banks in heavy rain but it never came near the house." *At settlement Doris Edwards expressed concern* "about living on a creek" *because of her small children. She was told by Grayce Walsh that* "during a storm the creek bed sometimes filled up but the water level went right down within an hour or so." *The plaintiffs then purchased the house.*

That summer the plaintiffs discovered to their dismay that during a heavy rainstorm part of the creek flowed into the garage, into the basement, under a door in the recreation room, and to the terrace in the rear of the house. They sued and received a favorable verdict by the jury. Defendants appealed.

HORNEY J.

Gerald Walsh admitted that the creek had flooded the property on four previous occasions in 1958 and 1959, and that he had protested to the county authorities about the matter. . . .

Grayce Walsh, who was well aware of the damage that had been caused by previous floods, admitted that she had signed her name and that of her husband to a neighborhood petition addressed to the county authorities, dated November 12, 1959, in which, among other things, it was stated that the creek "jumps its banks, flooding homes, inundating large areas of valuable properties, and leaves a trail of destruction, damage, debris and health-menacing mud and silt."

Ordinarily, of course, the seller of real property is not legally obliged to disclose to a prospective purchaser the objectionable or undesirable conditions or features of the property offered for sale, and mere silence or nondisclosure of material facts by the seller would not constitute actionable fraud. But where, as here, the seller, in addition to not disclosing the facts, made an active misstatement of fact, or only a partial or fragmentary statement of fact, which misled the purchaser to his injury, the legal situation of the seller was reversed and there was imposed on her a duty to disclose all that she knew as to the probability of the creek overflowing. Under the circumstances in this case, the failure to disclose the facts constituted actionable fraud.

In the instant case there was proof that the representation was false, that its falsity was known to the seller, that it was made for the purpose of deceiving the purchasers, that the purchasers relied on the misrepresentation and would not have purchased the property had the misrepresentation not been made, and that the purchasers actually suffered damage as a direct result of the fraudulent misrepresentation. That is all that was required in this case.

(Judgment affirmed.)

(2) Innocent Misrepresentation—Absence of *Scienter*. A contract may be set aside on grounds of innocent misrepresentation even in the absence of fraud. If there is a misrepresentation of a material fact, even though innocently made, the court will not enforce the contract as long as the other elements of fraud are present. In other words, the difference between fraud and innocent misrepresentation is the absence of the *scienter* element. Although the contract will not be enforced, because the false statement was made in the absence of knowledge of the falsity of the statement, there would be no tort action against the party making the representation. The remedy would be rescission of the contract at the option of the innocent party.

YORKE v. TAYLOR

124 N.E.2d 912 (Mass. 1954)

Plaintiff purchaser sued defendant for rescission of a contract to purchase real property. Plaintiff alleged that the assessment of the property was misrepresented to him as being $12,500 when in fact it was $26,000. The assessment had been increased during the previous year. There was no evidence that the defendant (seller) knew about the increased assessment. The lower court found in favor of defendant, and plaintiff appealed.

SPAULDING, J.

The [lower court] judge found that the defendants did not know that the assessed valuation had been increased when the information relating to the income and expenses of the property was submitted to the plaintiff. The judge further found that the defendants acted in good faith and had no intention of misleading or deceiving the plaintiff. These findings would not defeat the right to rescind. In this Commonwealth one who has been induced to enter into a contract in reliance upon a false though innocent representation of a material fact susceptible of knowledge which was made as of the party's own knowledge and was stated as a fact and not as a matter of opinion is entitled to rescission.

The judge, however, made other findings which must be considered. He found that the plaintiff was not misled by the representation of the defendants that the assessment was $12,500, and he also found that the plaintiff sustained no damage. These findings under the familiar rule must be upheld unless they are plainly wrong, but we are of opinion that evidential support for them is lacking. The plaintiff obviously could not have relied on any statement by his attorney as to the 1953 assessment because the attorney reported that he found only the 1952 assessments available in the records examined by him. The assessment on the property was undoubtedly a matter of materiality and it seems clear from the evidence that both the plaintiff and the defendants so considered it. The defendants included the information in the statement of income and expenses which they caused to be furnished to the plaintiff. And the plaintiff's inquiries to the defendants' brokers concerning the assessment show that he was seeking information on a matter he also considered important. Where, as here, the tax rate was $70.70 per thousand for the year 1953, concern as to what the property was assessed for is readily understandable. The judge in his findings seems to have been influenced by the fact that the property was worth what the plaintiff paid for it. But the plaintiff does not seek rescission on the basis of any misrepresentation touching the value of the property. Rescission is sought on the ground that the amount of the assessment was misrepresented.

But there is another finding which we must also consider. The judge found that the "assessed value . . . was a matter of public information equally within the reach of the plaintiff and the defendants." This suggests that even had the plaintiff relied on the defendants' representation such reliance would have been unreasonable and unjustified because the plaintiff could easily have ascertained the truth by recourse to the records in the assessor's office.

It is true that statements may be found in some of our decisions to the effect that a plaintiff ought not to obtain relief from the consequences of false representations where he has failed to use due care and diligence in protecting his rights. The reasoning of these cases appears to be that the court should exhibit no greater interest in protecting a plaintiff's rights than he himself has shown. But the trend of modern authority is opposed to this philosophy.

But whatever our rule has been formerly on the subject of diligence—and it is not easy to reconcile all that has been said—we prefer the rule of the Restatement that "The recipient of a business transaction of a fraudulent misrepresentation of fact is justified in relying on its truth, although he might have ascertained the falsity of the representation had he made an investigation." Restatement: Torts § 540.

The plaintiff here was not relying on a statement of opinion nor on a representation that was either preposterous or palpably false. He could reasonably rely on the representation as being a fact within the defendants' knowledge and he was not obliged to go further and ascertain its truth.

We are of opinion that, in the light of all the evidence the finding that the plaintiff was not misled by the representations of the defendants cannot stand. And that is likewise true of the finding that the plaintiff sustained no damage. By reason of the increased assessment in 1953 the plaintiff was faced with a tax bill which was larger by $954.45 than his bill would have been on the basis of an assessment of $12,500.

It follows that the decree dismissing the bill must be reversed and a new decree entered rescinding the sale, ordering the defendants upon reconveyance of the property to return the consideration paid by the plaintiff, and cancelling the $5,500 note and the mortgage securing it.

QUESTIONS FOR DISCUSSION

1. *Would the plaintiff's rights have been any different had the defendant been aware of the increased assessment when he made the misrepresentation?*
2. *What do you think of the lower court's statement that plaintiff could have checked the assessment records and that thus his reliance on defendant's statement was unreasonable?*

7:2. DURESS. When a party has entered into a contract under duress, it is voidable at his option. *Duress* involves wrongful coercion by one party against another so that the innocent party does not enter into the contract of his own volition. You must be careful to distinguish between those cases in which the contract is voidable because one has not been able to exercise his free will and those that are entered into as the result of persuasion or legal or economic pressures, which are nonetheless enforceable.

When it is to be determined whether a person has entered into a contract under duress, the question is whether he has entered into it under pressure that deprived him of contractual volition. There is no general standard of courage applied today to determine whether the contract was entered into under duress. The determination is made on a case-by-case basis, taking into account such factors as the age, sex, state of health, and mental capacity of the victim.

Duress obviously may be the result of physical force or the threat of it. Threats to one's relatives may also constitute duress, but what about threats to one's property or business? Generally, if the threat consists of an act that the party has no legal right to perform, it will constitute duress. In some cases, a threat of economic duress or business compulsion may result in a contract's being invalidated. The threat must generally be wrongful, unlawful, or unconscionable; otherwise the threatened action will not constitute duress, even though it could result in some harm to the other party.

Although there is not uniformity, many courts refuse to find duress from the threat of criminal prosecution if there may reasonably be criminal liability. Likewise the threat of a civil suit is generally not considered to constitute duress if the threat is made in good faith.

Increasingly the courts have been willing to rescind contracts on the grounds of what is termed "economic duress" or "business compulsion." This situation usually occurs where one party makes an agreement against its will because of its economic circumstances. Generally, in order to invoke the doctrine of economic duress the innocent party must show that the duress resulted from the wrongful act of the other party and that the innocent party had no other alternative.

KAPLAN v. KAPLAN

182 N.E.2d 706 (Ill. 1962)

Plaintiff, Leonard Kaplan, sued his former wife, Elaine Kaplan, for rescission of a property-settlement contract entered into between them prior to their divorce. Plaintiff contended that he had entered into the agreement under duress. The lower court found in favor of defendant and plaintiff appealed.

DALY, J.

Specifically, it was alleged that during the pendency of the separate maintenance action, private detectives employed by Elaine Kaplan or her attorneys forcibly broke into an apartment and took photographs of plaintiff and another woman, and that although there had never been immoral or improper behavior between plaintiff and the woman, the circumstances of the occurrence and the photographs were intended to cause great embarrassment to plaintiff and the other woman. Continuing, it was alleged that Elaine Kaplan thereafter expressly threatened to publicize the photographs by suing the other woman for alienation of affections, "by publicizing the said occurrence and photographs, and by other means calculated to cause great embarrassment to Leonard Kaplan, to the said woman in whose apartment he had been photographed, and to the family of said woman, all without cause." Immediately following, it was alleged that "in order to avoid the said embarrassment to himself and to the said woman, and under duress as aforesaid" plaintiff agreed to sign the two agreements previously described.

. . . The issue to be determined here is whether the alleged duress of which the plaintiff complains is such as would render the contract voidable at law and entitle him to equitable relief.

Duress has been defined as a condition where one is induced by a wrongful act or threat of another to make a contract under circumstances which deprive him of the exercise of his free will, and it may be conceded that a contract executed under duress is voidable. . . . Under modern views and developments, however, duress is no longer confined to situations involving threats of personal injury or imprisonment, and the standard of whether a man of ordinary courage would yield to the threat has been supplanted by a test which inquires whether the threat has left the individual bereft of the quality of mind essential to the making of a contract. . . . Any wrongful threat which actually puts the victim in such fear as to act against his will constitutes duress and, according to some authorities, a threat of personal or family disgrace may be of such gravity as to deprive the person threatened of the mental capacity necessary to execute a valid contract.

Considering the complaint in the light of these principles, it is our opinion that the allegations fall short of establishing grounds upon which to lay a claim of legal duress. The first allegation of the complaint is that Elaine Kaplan threatened to publicize the photographs by suing the woman in whose apartment they had been taken for alienation of affections. This allegation clearly presents no basis for a claim of duress inasmuch as it is well established that it is not duress to institute or threaten to institute civil suits, or for a person to declare that he intends to use the courts to insist upon what he believes to be his legal rights, at least where the threatened action is made in the honest belief that a good cause of action exists, and does not involve some actual or threatened abuse of process.

While the defense of duress has developed along liberal lines, we are not prepared to say as a matter of law that a threat of personal embarrassment, particularly under the circumstances reflected in the pleadings, was such as to control the will of the plaintiff or to render him bereft of the quality of mind essential to the making of a contract. We have said before that duress is not shown by subjecting one to annoyance or vexation. . . . and it is our belief that a threat of personal embarrassment does not rise above annoyance and vexation.

(Order affirmed.)

QUESTIONS FOR DISCUSSION

1. *Is the test applied by the court in this case an objective or a subjective test?*
2. *In some jurisdictions the action of alienation of affection has been abolished. If that were true in Illinois at the time of this case, would the outcome have been different?*

TOTEM MARINE T. AND B. v. ALYESKA PIPELINE SERVICES

584 P.2d 15 (Alaska 1978)

In June of 1975 Totem Marine Tug and Barge, Inc., entered into a contract with Alyeska Pipeline under which Totem was to transport pipeline construction materials from Houston, Texas, to a designated port in southern Alaska. In order to carry out this contract Totem chartered a barge and ocean-going tug. These charters and other initial operating costs of Totem were made possible by loans to Totem.

Difficulties were encountered during the trip. Totem charged that a number of these difficulties were caused by Alyeska. Because of these and other problems the trip from Houston to Alaska was proceeding more slowly than anticipated so Alyeska's agents unloaded the barge when it put into port in Long Beach, California, and terminated the contract. Totem then presented invoices to Alyeska totaling between $260,000 and $300,000.

Totem was in urgent need of cash to pay its creditors. Alyeska told Totem that payment might take anywhere from a day to six to eight months. Totem replied that it couldn't wait that long as if it didn't have immediate cash, it would go bankrupt. Totem then received a settlement offer from Alyeska in the amount of $97,500.00. Totem signed an agreement for that amount releasing Alyeska from all other claims.

Totem then filed this suit to rescind the agreement contending that it had signed the agreement under duress. Alyeska filed a motion for summary judgment based upon the signed settlement agreement. The trial court found in favor of Alyeska on the motion and Totem appealed.

BURKE, J.

This court has not yet decided a case involving a claim of economic duress or what is also called business compulsion. At early common law, a contract could be avoided on the ground of duress only if a party could show that the agreement was entered into for fear of loss of life or limb, mayhem or imprisonment. The threat had to be such as to overcome the will of a person of ordinary firmness and courage. Subsequently, however, the concept has been broadened to include myriad forms of economic coercion which force a person to involuntarily enter into a particular transaction. The test has come to be whether the will of the person induced by the threat was overcome rather than that of a reasonably firm person.

At the outset it is helpful to acknowledge the various policy considerations which are involved in cases involving economic duress. Typically, those claiming such coercion are attempting to avoid the consequences of a modification of an original contract or of a settlement and release agreement. On the one hand, courts are reluctant to set aside agreements because of the notion of freedom of contract and because of the desirability of having private dispute resolutions be final. On the other hand, there is an increasing recognition of the law's role in correcting inequitable or unequal exchanges between parties of disproportionate bargaining power and a greater willingness to not enforce agreements which were entered into under coercive circumstances.

Professor Williston states the basic elements of economic duress in the following manner:

1. The party alleging economic duress must show that he has been the victim of a wrongful or unlawful act or threat, and
2. Such act or threat must be one which de-

prives the victim of his unfettered will. 13 *Williston on Contracts,* § 1617 at 704 [footnotes omitted].

Many courts state the test somewhat differently, eliminating use of the vague term "free will," but retaining the same basic idea. Under this standard, duress exists where: (1) one party involuntarily accepted the terms of another, (2) circumstances permitted no other alternative, and (3) such circumstances were the result of coercive acts of the other party. The third element is further explained as follows:

> In order to substantiate the allegation of economic duress or business compulsion, the plaintiff must go beyond the mere showing of reluctance to accept and of financial embarrassment. There must be a showing of acts on the part of the defendant which produced these two factors. The assertion of duress must be proven by evidence that the duress resulted from defendant's wrongful and oppressive conduct and not by the plaintiff's necessities. (Citation omitted.)

As the above indicates, one essential element of economic duress is that the plaintiff show that the other party by wrongful acts or threats, intentionally caused him to involuntarily enter into a particular transaction. Courts have not attempted to define exactly what constitutes a wrongful or coercive act, as wrongfulness depends on the particular facts in each case. This requirement may be satisfied where the alleged wrongdoer's conduct is criminal or tortious but an act or threat may also be considered wrongful if it is wrongful in the moral sense.

In many cases, a threat to breach a contract or to withhold payment of an admitted debt has constituted a wrongful act. Implicit in such cases is the additional requirement that the threat to breach the contract or withhold payment be done in bad faith.

Economic duress does not exist, however, merely because a person has been the victim of a wrongful act; in addition, the victim must have no choice but to agree to the other party's terms or face serious financial hardship. Thus, in order to avoid a contract, a party must also show that he had no reasonable alternative to agreeing to the other party's terms, or, as it is often stated, that he had no adequate remedy if the threat were to be carried out. What constitutes a reasonable alternative is a question of fact, depending on the circumstances of each case. An available legal remedy, such as an action for breach of contract, may provide such an alternative. Where one party wrongfully threatens to withhold goods, services or money from another unless certain demands are met, the availability on the market of similar goods and services or of other sources of funds may also provide an alternative to succumbing to the coercing party's demands. Generally, it has been said that "[t]he adequacy of the remedy is to be tested by a practical standard which takes into consideration the exigencies of the situation in which the alleged victim finds himself." *Ross Systems,* 173 A.2d at 262.

An available alternative or remedy may not be adequate where the delay involved in pursuing that remedy would cause immediate and irreparable loss to one's economic or business interest.

Turning to the instant case, we believe that Totem's allegations, if proved, would support a finding that it executed a release of its contract claims against Alyeska under economic duress. Totem has alleged that Alyeska deliberately withheld payment of an acknowledged debt, knowing that Totem had no choice but to accept an inadequate sum in settlement of that debt; that Totem was faced with impending bankruptcy; that Totem was unable to meet its pressing debts other than by accepting the immediate cash payment offered by Alyeska; and that through necessity, Totem thus involuntarily accepted an inadequate settlement offer from Alyeska and executed a release of all claims under the contract. If the release was in fact executed under these circumstances, we think that under the legal principles discussed above that this would constitute the type of wrongful conduct and lack of alternatives that would render the release voidable by Totem on the ground of economic duress. We would add that although Totem need not necessarily prove its allegation that Alyeska's termination of the contract was wrongful in order to sustain a claim of economic duress, the events leading to the termination would be probative as to whether Alyeska exerted any wrongful pressure on Totem and whether Alyeska wrongfully withheld payment from Totem.

(Reversed in favor of Totem and remanded for trial.)

QUESTIONS FOR DISCUSSION

1. *What are the elements of economic duress or "business compulsion"?*
2. *Why was it necessary for Totem to accept Alyeska's offer of settlement? Would the outcome of the case have been different if Totem had been financially sound?*

7:3. UNDUE INFLUENCE. Undue influence is one of the more difficult concepts with which to deal because of the necessity of distinguishing it from mere persuasion. Essentially *undue influence* is the substitution of a dominant party's will for the dominated party's will. The result is that a contract entered into by the dominated party is not done so freely and of his own volition.

Typically cases involving undue influence arise from one of two situations: from actual domination of one party by another or from the confidential relationship of the parties. Actual domination occurs most often when a person is suffering from mental weakness, youth, illness, old age, or the like. Often the dominated person falls into the habit of doing exactly what he is told, without exercising any independent judgment of his own. For example, a contract may be placed before him. He is told to sign it and does so automatically, simply because of the instructions of the dominating person, rather than exercising his own judgment and free will.

One should be careful to distinguish between undue influence and mere persuasion or even high-pressure salesmanship. If one goes into a store and meets the proverbial supersalesman, he generally is not able to get the contract rescinded on the grounds of undue influence merely because of the sales pressure. This is because, although subject to persuasion, the party did exercise his own will in entering into the transaction.

Another typical problem area involves parties in a confidential relationship with one another. Contracts entered into between parties occupying a confidential relationship may be valid but are closely scrutinized by the courts. The most typical confidential relationships in which problems occur are attorney–client, guardian–ward, and trustee–beneficiary relationships. Among numerous other relationships are doctor–patient, clergy–parishioner, and husband–wife. Since the relationship involves the placing of trust and confidence in one party by another, the utmost good faith must be exercised by the party in whom such confidence is reposed. Any contract between the parties must be entered into fairly, openly, and honestly. Furthermore a confidential relationship may arise even without the more formal relationships, such as guardian and ward. When parties are bargaining at arm's length, each is expected to look out for his own interests. If he enters into a contract that is to his disadvantage, he has no one to blame but himself. When there is a relationship of trust and confidence, however, one is entitled to rely on the person in whom he places his trust. The nature of the relationship does not demand the wariness that must be exercised in the ordinary business transaction.

WENGER v. ROSINSKY

232 Md. 43, 192 A.2d 82 (1963)

Plaintiff, Wenger, was the sister of David Meyer Kurtz and the administratrix of his estate. Kurtz moved into a boardinghouse owned by the defendant, Rosinsky, on March 18, 1960. He paid $25 per week for room, board, and laundry. At the time he was sixty-one years old and had difficulty walking. He had $14,269.67 in a savings account and owned stocks valued at $9,136.79.

During the course of his stay at the boardinghouse, Kurtz turned over $7,700 in cash to Mrs. Rosinsky, which she and her husband used to pay off the mortgage on the boardinghouse. Meyer also purchased $5,500 worth of savings bonds that he attempted to turn over to Mrs. Rosinsky but was unable to do because they had not sufficiently matured beyond the purchase date. Mrs. Rosinsky was given access to Kurtz's safety-deposit box, in which he kept his securities. In order for Mrs. Rosinsky to have transportation while her husband was at work, Kurtz purchased a 1960 Chevrolet Impala, which was titled in her name.

On October 12, 1960, Kurtz became ill. He handed Mrs. Rosinsky the passbook to a savings account, which had been placed in their joint names, and told her that if he "checked out" *he wanted her to have it. Kurtz* "checked out" *within the hour. On October 17, 1960, Mrs. Rosinsky withdrew the balance of $5,300. Plaintiff sued to have the assets returned to the estate. The trial court found for the defendant and plaintiff appealed.*

MARBURY, J.

If this were a mystery story, the title would undoubtedly be "The Case of the Shorn Sergeant." Master Sergeant David Meyer Kurtz was retired from the United States Marine Corps in 1945, after approximately twenty years' service. He received a seventy-five percent disability allowance and his normal monthly retirement pay. The disability was caused by malaria, contracted by him while stationed in Haiti. Being somewhat of a "floater" since he was unmarried, he had lived in various places since his retirement.

Mrs. Wenger, as administratrix of Sergeant Kurtz's estate, brought this action by means of a bill in equity for an accounting and to impose a constructive trust in order to recover the amounts of money and property obtained by the appellees from Sergeant Kurtz, claiming the abuse of a confidential relationship. By way of defense, the appellees contended that the money used to pay the mortgage was in consideration for their agreement to provide him with board, lodging, and care for the rest of his life; that the car was a gift to Mrs. Rosinsky for the purpose of providing her with a means of transportation for him while her husband used their car in connection with his employment; and that the joint savings account was also a gift.

The question of confidential relationship has been before this Court on numerous occasions. In *Vocci* v. *Ambrosetti,* 201 Md. 475, 485, 94 A.2d 437, 441, we stated: "*'The relation requires the parties to abstain from all selfish projects.* The general principle is, if a confidence is reposed and that confidence is abused, courts of equity will grant relief. In such cases it is not necessary to prove the actual exercise of overweening influence, misrepresentation, importunity, or fraud *aliunde* the act complained of. . . . The general rule is that he who bargains in a matter of advantage with a person placing confidence in him is bound to show that a reasonable use has been made of that confidence, *a rule applying equally to all persons standing in confidential relations to each other.'* "

In Zimmerman v. *Hull,* 155 Md. 230, 240, 141 A. 531, we held that outside of recognized relationships from which undue influence may be presumed, it is a question of fact and not of law as to whether a confidential relationship does exist.

Applying these basic rules to the instant case we find, contrary to the conclusion of the chancellor, that such a confidential relationship did exist between Sergeant Kurtz and Mrs. Rosinsky. Almost from the outset of the relationship as boarder and landlady, he began to repose great trust and confidence in her, as she admitted in her testimony. She accompanied him on numerous trips to banking institutions and appeared to do the talking for him in his financial transactions. He relied on her to call a doctor of her choice during his illnesses and to contact two lawyers of her selection for the purpose of preparing his will. He gave her access to his safe deposit box and opened a joint savings account with her, using only his money. All these acts indicate a strong reliance on her judgment and integrity while he was concededly in poor health and unable to transact his own business without assistance. At no time did he have the advice of counsel or independent advice from any other source. This was far beyond a normal, business relationship between a boarder and his landlady.

If the arrangements and transactions between Kurtz and the Rosinskys were provident and in his best interests, and if appellees had not taken advantage of this confidential relationship for their own benefit, the existence of such a relationship would not, per se, be grounds for relief. However, where such a relationship does exist, and the party occupying the position of dominion or superiority, such as Mrs. Rosinsky, receives a benefit from the transaction, there is a presumption against its validity, placing upon the beneficiary the burden of showing by clear and convincing evidence that there has been no abuse of the confidence, that she acted in good faith, and that the act by which she was benefited was the free, voluntary, and independent act of the other party to the relationship. . . . We find that the appellees did not meet this burden. Furthermore, it is clear that all these transactions were not provident as to Sergeant Kurtz. Taken together, they had the effect of stripping him of most of his life savings. At his death he retained only the United States savings bonds which had been attempted to be cashed, and which would have passed to Mrs. Rosinsky under the draft of an unexecuted will prepared for Kurtz by an attorney of Mrs. Rosinsky's selection.

For the reasons above stated the findings of the chancellor were clearly erroneous, so that the decree must be reversed and the case remanded for further proceedings not inconsistent with this opinion.

QUESTIONS FOR DISCUSSION

1. *Of what significance is it that the court found a "confidential relationship" in this case?*
2. *Might the case have turned out the same if the court did not find that a "confidential relationship" existed between the parties?*

7:4. MISTAKE. A mistake may be of the type that will preclude the mutual assent necessary to the formation of the contract. When that type of mistake occurs, the court will set aside the contract. There are a multitude of mistakes that one may make when entering into a contract that will afford no ground for relief. In general, the mistake must be with respect to a past or present fact that is material to the contract. In addition, if the mistake is the result of the negligence of a party, the court will not grant relief.

(1) Unilateral Mistake. Often a mistake is made by only one party to the contract; that is, it is *unilateral.* Generally no relief is available when the mistake is made by one party, for it is often the result of carelessness or lack of diligence. For example, a contractor may neglect to allow enough money in his bid for excavation in a building contract because he did not bother to inspect the construction site. Another typical case involves the situation in which a contractor makes a minor arithmetical error in adding up his costs prior to submission of his bid. If a contract results, the court will generally not grant relief for this type of mistake.

There are exceptions to the rule. If the mistake is material and the offeree has knowledge of the mistake when he accepts, the offeror will not be bound. The difficult cases are those in which the offeror claims that the mistake is so large that it should have been apparent to the offeree whether he had actual knowledge of it or not. Generally, in these cases the court will not allow one party to take advantage of the other's error and will grant a rescission of the contract. The problem is in the determination of what constitutes a mistake so large as to place the offeree on notice of it.

(2) Bilateral or Mutual Mistakes. A *bilateral* or *mutual mistake* is one in which both parties to the contract have made false factual assumptions concerning the contract. Because the mutual mistake may preclude the required mutual assent necessary to the formation of a contract, it will often be the grounds for rescission.

There are a number of mutual mistakes for which the court will not set aside a contract. Generally the mistake must be with regard to the *subject matter* of the contract rather than the *quality* or *value* of the subject matter. For example, Penny Painter contracts to sell a painting to Connie Collector for a low price. The next day it is discovered that the painting is actually a rare masterpiece and is worth thousands. In the absence of fraud the contract would be enforceable even though each party was mistaken as to its true value at the time the contract was made.

If the mutual mistake concerns the subject matter, the court will grant relief. For example, Marvin Mechanic owns two blue cars of the same make and model, one at home and the other at his shop. Harry Hotrod contracts to buy Marvin's blue car. Harry has in mind the car located at Marvin's home and Marvin has in mind the car located at his shop. No contract results because there is no mutual assent as to the subject matter. A similar result would occur if the subject matter of the contract were destroyed prior to the contract without the knowledge of the parties.

SANTUCCI CONSTRUCTION CORP. v. COUNTY OF COOK

21 Ill.App.3d 527, 315 N.E.2d 565 (1974)

Plaintiff, Santucci Construction Corp., sued the defendant, Cook County, for rescission of a contract and the return of a $75,000 bid deposit submitted by plaintiff with its bid. The trial court found in favor of the plaintiff, and the defendant appealed.

SULLIVAN, J.

In November, 1966, defendant advertised for bids concerning the construction of the main drain of the Dan Ryan Expressway, and on December 14, 1966, plaintiff and three other construction companies submitted bid proposals. With its bid, plaintiff tendered a $75,000 deposit which was subject to forfeiture and retention by defendant in the event plaintiff was low bidder and failed to execute the contract.

Plaintiff's bid was $1,095,842, and the bids by the other three contractors were in the amounts of $1,693,353.85; $1,719,770; and $1,827,904.90. Each bid was required to specify a price for 7,132 lineal feet of drainpipe. In this regard, plaintiff's bid price, including labor, was $108.50 per lineal foot, with the other three contractors specifying $169.00; $180.00; and $200.00, respectively. The total price submitted by plaintiff for this item was $773,882 (its original estimate was $1,290,892); whereas, the prices specified by the other three contractors were $1,205,308; $1,283,760; and $1,426,400.

Carlo Santucci, General Manager of plaintiff, testified that on December 14, 1966, the day the bid was submitted, he was returning with Nicholas Santucci, President of plaintiff, from Indiana where they had been for two days and, upon radioing his office, he was informed that plaintiff was the low bidder on the project and that there had been a large discrepancy in bids. When he arrived at the company office, he examined the bid sheets and discovered that the original and intended material cost for the pipe of $142.75 per foot had been scratched out, and the figure of $74.00 had been substituted. He testified that he could not buy the pipe for the latter figure and that before he went to Indiana, the $142.75 cost figure had been used in their estimate, and he did not become aware of the substitution until his return.

It further appears that on December 15, 1966, plaintiff sent a letter to defendant requesting that the bid be withdrawn; that defendant accepted plaintiff's bid on January 3, 1967, and when plaintiff refused to enter into a contract, that defendant retained plaintiff's bid deposit.

Defendant first contends that plaintiff failed to sustain its burden of showing that a mistake justifying rescission had been made. It argues that if there was an error in the bid, it was the result of poor business judgment or negligence, neither of which would justify rescission.

Here we note that the bid eventually accepted was for $1,596,554, and there was a great disparity in the specified cost of drain in the various bids and estimates. Plaintiff's bid was (1) over $500,000 less than in its original estimate; (2) over $600,000 less than in defendant's estimate of drainpipe; (3) over $400,000 less than the next lowest bidder; and (4) over $650,000 less than the highest bidder. As a matter of fact, defendant's estimate for the entire project ($1,937,856.40) was $842,000 more than plaintiff's bid ($1,095,842), which figure was $300,000 less than defendant's estimate cost of the drainpipe alone ($1,408,213.40). This disparity resulted primarily from the change by Zoltani [employee of plaintiff] of the cost of drainpipe (base price and material cost) from $181.00 per lineal foot in plaintiff's original estimate to $108.50 in the submitted bid. In addition, Carlo Santucci testified that he could not purchase drainpipe for $74.00 per lineal foot, the base price used by Zoltani in reaching the $108.50 cost in the submitted bid, and furthermore, it appears that defendant had estimated a base price of $125.00 per lineal foot for drainpipe.

We turn now to the question as to whether the mistake justified rescission and it is noted that . . . the conditions generally required for rescission are (1) that the mistake relate to a material feature of the contract; (2) that it is of such grave consequence that enforcement of the contract would be unconscionable; (3) that it occurred notwithstanding the exercise of reasonable care; and (4) that the other party can be placed in *statu quo*.

Here, in view of the fact that the mistake involved the drainpipe cost, which was three-fourths of the cost of the entire project, it clearly involved a material element of the contract. It appears also that plaintiff stood to lose a substantial sum if it performed at the price quoted in its submitted bid of $1,095,842. This amount was almost $600,000 less than the next lowest bid and $500,000 less than the bid of $1,596,554, subsequently entered into by defendant for construction of the drain. Under those circumstances, we are of the opinion that it would have been unconscionable to enforce the contract.

Defendant contends that whether or not Zoltani was authorized to make the bid changes, he was negligent in doing so without further verification of the Vulcan price quote.

Here the evidence indicates that it was customary in the construction business to receive and act upon telephone price quotes. There is no testimony of any practice requiring verification or that there was some custom in that trade to prepare and submit bids in any other manner. In view thereof, it is our belief that the mistake was not the result of negligence.

We will next consider the propriety of the trial court's finding that defendant should have known of plaintiff's mistake. . . . We have previously noted the great disparity in the bids here, including the fact that plaintiff's submitted bid for the entire project was over $300,000 less than defendant's estimate for the cost of the pipe, and we cannot accept defendant's contention that they were not so disproportionate as to give notice that plaintiff's bid was the result of a mistake. Richard Cramer, head of the Estimating Division of defendant's Highway Department, testified that he considered plaintiff's bid to be "cheap, low." Moreover, the record discloses that plaintiff, after discovering the mistake and believing that it could not perform the contract as submitted, sent a letter to defendant on December 15, 1966, the day after its bid was submitted, requesting withdrawal of the bid. Approximately seventeen days later, defendant accepted the bid, even though it was clearly evident that there were large discrepancies between the amounts of the bids and particularly between plaintiff's bid and defendant's own estimate (over $842,000). We think the record amply justifies the trial court's finding that defendant should have been aware of plaintiff's mistake.

[The court affirmed that part of the case under consideration here.]

WOOD v. BOYNTON
64 Wisc. 265, 25 N.W. 42 (1885)

TAYLOR, J.

The defendants are partners in the jewelry business. On the trial it appeared that on and before the twenty-eighth of December, 1883, the plaintiff was the owner of and in the possession of a small stone of the nature and value of which she was ignorant; that on that day she sold it to one of the defendants for the sum of one dollar. Afterwards it was ascertained that the stone was a rough diamond, and of the value of about $700. After hearing this fact the plaintiff tendered the defendants the one dollar, and ten cents as interest, and demanded a return of the stone to her. The defendants refused to deliver it, and therefore she commenced this action.

The plaintiff testified to the circumstances attending the sale of the stone to Mr. Samuel B. Boynton, as follows: "The first time Boynton saw that stone he was talking about buying the topaz, or whatever it is, in September or October . . . I thought I would ask him what the stone was, and I took it out of the box and asked him to please tell me what it was. He took it in his hand and seemed some time looking at it. I told him I had been told it was a topaz, and he said it might be. He says, 'I would buy this; would you sell it?' I told him I did not know but that I would. What would it be worth? And he said he did not know; he would give me a dollar and keep it as a specimen, and I told him I would not sell it; and it was certainly pretty to look at. . . . Afterwards, and about the twenty-eighth of December, I needed money pretty badly, and thought every dollar would help, and I took it back to Mr. Boynton and told him I had brought back the topaz, and he says, 'Well, yes; what did I offer you for it?' and I says, 'One dollar'; and he stepped to the change drawer and gave me the dollar, and I went out." In another part of her testimony she says: "Before I sold the stone I had no knowledge whatever that it was a diamond. I told him that I had been advised that it was probably a topaz, and he said probably it was. The stone was about the size of a canary bird's egg, nearly the shape of an egg,—worn pointed at one end; it was nearly straw color—a little darker." She also testified that before this action was commenced she tendered the defendants $1.10, and demanded the return of the stone, which they refused.

The evidence on the part of the defendant is not very different from the version given by the plaintiff, and certainly is not more favorable to the plaintiff. Mr. Samuel B. Boynton, the defendant to whom the stone was sold, testified that at the time he bought this stone, he had never seen an uncut diamond; he had seen cut diamonds, but they are quite different from the uncut ones; "he had no idea this was a diamond, and it never entered his brain at the time." . . . The only question in the case is whether there was anything in the sale which entitled the vendor (the appellant) to rescind the sale and so revest the title in her. The only reasons we know of for rescinding a sale and revesting the title in the vendor so that he may maintain an action at law for the recovery of the possession against his vendee are (1) that the vendee was guilty of some fraud in procuring a sale to be made to him; (2) that there was a mistake made by the vendor in delivering an article which was not the article sold—a mistake in fact as to the identity of the thing sold with the thing delivered upon the sale. This last is not in reality a rescission of the sale made, as the thing delivered was not the thing sold, and no title ever passed to the vendee by such delivery.

In this case, upon the plaintiff's own evidence, there can be no just ground for alleging that she was induced to make the sale she did by any fraud or unfair dealings on the part of Mr. Boynton. Both were entirely ignorant at the time of the character of the stone and of its intrinsic value. Mr. Boynton was not an expert in uncut diamonds, and he made no examination of the stone, except to take it in his hand and look at it before he made the offer of one dollar, which was refused at the time, and afterwards accepted without any comment or further examination made by Mr. Boynton. The appellant had the stone in her possession for a long time, and it appears from her own statement that she had made some inquiry as to its nature and qualities. If she chose to sell it without further investigation as to its intinsic value to a person who was guilty of no fraud or unfairness which induced her to sell it for a small sum, she cannot repudiate the sale because it is afterwards ascertained that she made a bad bargain. There is no pretense of any mistake as to the identity of the thing sold. It was produced by the plaintiff and exhibited to the vendee before the sale was made, and the thing sold was delivered to the vendee when the purchase price was paid.

We can find nothing in the evidence from which it could be justly inferred that Mr. Boynton, at the time he offered the plaintiff one dollar for the stone, had any knowledge of the real value of the stone, or that he entertained even a belief that the stone was a diamond. It cannot, therefore, be said that there was a suppression of knowledge on the part of the defendant as to the value of the stone which a court of equity might seize upon to avoid the sale. . . . However unfortunate the plaintiff may have been in selling this valuable stone for a mere nominal sum, she has failed entirely to make out a case either of fraud or mistake in the sale such as will entitle her to a rescission of such sale so as to recover the property sold in an action at law.

The judgment of the circuit court is affirmed.

QUESTIONS FOR DISCUSSION

1. *The mistake in the Wood case was mutual, whereas the mistake in the Santucci Construction Co. case was unilateral. How do you account for the different results in these two cases?*
2. *Of what significance was the size of the mistake in the Santucci case? Was the size of the mistake in the Wood case of any consequence?*

7:5. UNCONSCIONABLE CONTRACTS. There are times when the bargaining power of the parties is so one-sided that the contract that results is really one that one party has imposed on the other. It often results when the party in the poor bargaining position must enter into the contract for one reason or another. He has no choice. Such contracts are said to be *unconscionable.* The Uniform Commercial Code, Section 2–302, incorporates this concept. If the contract is unconscionable, the courts may set it aside on the ground that the required mutual assent is not present because the oppressed party was in a sense forced to enter into the contract and the contract did not result from any bargaining.

7:6. SUMMARY OF UNIFORM COMMERCIAL CODE CHANGES. This is one area of contract law where the U.C.C. has not made many changes. There are no provisions that materially affect the already existing law of fraud, innocent misrepresentation, duress, and undue influence. Thus if a contract could be avoided because there was fraud under the law of a particular state, the fact that the contract was for the sale of goods and subject to the provisions of the U.C.C. would make no difference. The defense of fraud would apply as in any other contract.

The U.C.C. does make the statement that any remedies available to a party for breach of contract are also available for material representation or fraud. U.C.C. Section 2–721.

SUMMARY

The manifestation of assent to be contractually bound must be freely and voluntarily given. This requirement is not met if fraud is committed. Fraud consists of a false representation of a material fact; knowledge of the falsity on the part of the person making the statement, or a statement made with reckless disregard of the truth; intent; justifiable reliance by the innocent party; and damages suffered as a result of the reliance. One is presumed to have read and understood a document that he signs.

Innocent misrepresentation is similar to fraud but lacks the element of intent, sometimes known as scienter. It is not a tort but will operate as a defense to a breach of contract suit.

A contract entered into under duress is voidable at the option of the party suffering the duress. Duress is the wrongful coercion by one party against another so that the innocent party does not enter into the contract of his own volition. One type of duress is sometimes known as business compulsion or economic duress.

Undue influence is the substitution of a dominant party's will for the dominated party's will. Undue influence renders the contract voidable at the option of the innocent party. It is more than mere persuasion. If there is a confidential relationship and the dominant party receives benefit from a transaction, the dominant party must show that there has been no abuse of the confidence.

In general, the courts will not grant relief where a contract is entered into as the result of a unilateral mistake. An exception occurs when the mistake is material and the offeree has knowledge of the mistake when he accepts the offer. A court will not set aside a contract because of a mutual mistake of the quality or value of the subject matter. If the mutual mistake is as to the identity of the subject matter of the contract, a court will set aside the contract.

Under the Uniform Commercial Code, the court may set aside a contract when the bargaining power of the parties is so one-sided that the contract that results is really one imposed by one party on another. Such contracts are said to be unconscionable.

PROBLEMS

1. William Kalbaugh and his wife listed their house for sale with a real estate company. Joyce Barnes was an agent for the company and showed the house to Harry Wood and his wife. She told them that with the exception of the dishwasher "everything was in perfect shape." The Woods then contracted to purchase the house from the Kalbaughs.

   Before the time for settlement on the house arrived the Pacific Gas & Electric Company told Woods that the utilities to the house could not be turned on because there was a dangerous leak in the gas line. The Woods then refused to go through with the purchase of the house contending that they had a right to recission of the contract because of fraud or innocent misrepresentation on the part of the real estate agent. The leak was subsequently fixed by Kalbaugh. Are the Woods correct? Wood v. Kalbaugh, 114 Cal. Rptr. 673 (Cal. Apps.1974).

2. Loral Corporation was awarded a $6,000,000 contract by the U.S. Navy for production of radar sets. Loral then awarded a subcontract for construction of gear components to Austin Instrumental, Inc. Subsequently Loral asked for bids on a second group of gear components. Austin stated that it would cease delivery under the existing contract unless it received the contract for the second group of components at increased prices for both groups. Loral refused and Austin stopped delivery. Loral was then unable to find another manufacturer who could produce the gears in time so it agreed to the demands of Austin. After the gears were completed Loral refused to pay the increased price. May Austin collect? Austin Instrumental Inc. v. Loral Corp. 272 N.E.2d 533 (N.Y. 1971).
3. Raffles agreed to sell Wichelhaus wool that was to arrive on the ship *Peerless,* sailing from Bombay, India. There were two ships sailing from Bombay named *Peerless,* one in October and one in December. Raffles had in mind the ship sailing in December and Wichelhaus had in mind the ship sailing in October. Raffles sued Wichelhaus for breach of contract. Verdict for whom? Raffles v. Wichelhaus, Court of Exchequer, 1864, 2 Hurlstone and Coltman 906.
4. Schroeppel was a building contractor and Steinmeyer ran a lumber business. Steinmeyer gave Schroeppel a bid for lumber in the amount of $1,446. The list of items actually totaled $1,867, but the itemized list was not given to Schroeppel, just the total. Other bids were in the neighborhood of $1,890. Steinmeyer refused to supply the lumber. Schroeppel sued and Steinmeyer defended on grounds of mistake. Decide. Steinmeyer v. Schroeppel, 226 Ill. 9, 80 N.E. 564 (1907).
5. The sellers of a house knew that it was termite-infested. Prior to sale they whitewashed the basement and attached strips of wood to certain areas of the joists to cover up the more obvious damage. Other areas were obscured by shelves on which jars and clothing rested. The sellers had also treated the property for termites. Buyers inquired about the construction and were told that the joists were "as good as new." Buyer subsequently discovered the damage and sued for rescission of the contract. Decide. DeJoseph v. Zambelli, 392 Pa. 24, 139 A.2d 644 (1958).
6. Kronmeyer was employed by Buck, who was in the business of selling building materials. Buck's attorney accused Kronmeyer of stealing from Buck and said that he would have him jailed unless the matter was adjusted. Buck accused Kronmeyer of stealing $10,000 and threatened him. Although Kronmeyer contended that he did not steal from Buck, he deeded a home worth $5,000 to Buck and, together with his sister, Staehle, signed a $1,500 note payable to Buck while in an excited and frightened state. Kronmeyer and Staehle sued to set aside the deed and compel a refund of the proceeds of the note. Decide. Kronmeyer v. Buck, 101 N.E. 935 (Ill. 1913).
7. Hodde purchased land from Winn. Winn told Hodde that a particular stake

marked the corner of the property. Hodde then engaged Hall to sell the timber on the land and showed him the same stake. In fact, part of this land belonged to Hodde's neighbors. Chaney paid $6,500 for the timber rights and cut timber from Hodde's land as well as some from the neighbor's land, which had been described to Chaney as part of the Hodde property. Chaney paid the neighbor for the timber wrongfully cut and sued Hodde for the amount paid the neighbor. Hodde, in fact, believed that all the property was his. Decide. Hodde v. Chaney, 139 A.2d 510 (D.C. 1958).

8. Mrs. Trigg "nagged" Mr. Trigg continually and asked that he deed the ranch to her. She also threatened divorce, locked the bedroom door, and screamed for the servants if Mr. Trigg attempted to touch her. Finally the long-suffering Mr. Trigg transferred the property to her. Afterward he sought to have the conveyance set aside on the ground of undue influence. Decide. Trigg v. Trigg, 37 N.M. 296, 22 P.2d 119 (1933).
9. Joe Blanco purchased a car on credit and signed a conditional sales contract with General Motors Acceptance Corporation that reserved title to the car in them until the debt was paid. GMAC sued for possession of the car when Blanco did not make the payments. Blanco contended that he did not read this provision because it was on the reverse side of the contract and thus he was not bound. He also contended that the language was unintelligible, vague, ambiguous, and loaded in favor of GMAC and therefore should not be enforced. Decide. GMAC v. Blanco, 181 Neb. 562, 149 N.W.2d 516 (1967).

CHAPTER 8

Consideration

8:1. What Constitutes Consideration. One of the more troublesome concepts in the law of contracts is the requirement of consideration. It is a requirement unique to the common-law countries, and in fact some scholars contend that the concept is found in the common law only because of a judicial error that was perpetuated by the doctrine of stare decisis.

Without consideration there is no contract. For example, Danny Donor offers to give Ronny Recipient $2,500. Recipient accepts. There is no contract because Recipient has given no consideration in return for Donor's promise. All we have is a promise to make a gift. This promise is not enforceable in court, as without consideration it is not a contract.

There are a number of definitions given for consideration, but the one most frequently quoted is from the *Restatement of the Law of Contracts,* which states that consideration for promise is

(a) an act other than a promise, or
(b) a forbearance, or
(c) the creation, modification or destruction of a legal relation, or
(d) a return promise, bargained for and given in exchange for the promise.

From this definition it is obvious that consideration can be a number of things. It can constitute an act given in return for a promise, or it may be an exchange of promises. Consideration can also consist of giving up a legal right or the creation or destruction of a legal right. It need not be money or any other tangible thing. Nonetheless, in order to have a valid contract, some consideration must be given.

(1) Bargained for Exchange. Some define consideration as being a benefit to the promisor or a loss or legal detriment to the party to whom the promise is made. This definition is somewhat misleading, however, because the consideration given need not be to the detriment of anybody in the sense of harm. Consideration is also that which is bargained for in return for a promise; hence the definition also ignores the concept of a bargained-for exchange that is necessary in order for consideration to be present. For these reasons the *Restatement* definition would seem to be the more inclusive definition of the two. On the other hand, lawyers and courts frequently refer to the term *legal detriment* in discussing consideration. There is nothing wrong with using the term as long as you realize that the phrase is a term of art and is not synonymous with the usual or dictionary word *detriment.* As used in the law, it is much more inclusive. The act, forbearance, or promise given must be in response to a promise from the other party.

HAMER v. SIDWAY

124 N.Y. 538, 27 N.E. 256 (1891)

William E. Story, Sr., promised his nephew, William E. Story 2nd, that if he would refrain from drinking, using tobacco, swearing, and playing cards or billiards for money until he became twenty-one years of age, the uncle would pay him $5,000. The nephew agreed and did in fact refrain from engaging in the above activities. The uncle died without paying his nephew the money. The nephew's claim was eventually assigned to Hamer, and the executor of the uncle's estate, Sidway, refused payment. Hamer sued to recover the $5,000. The lower court found for defendant, and plaintiff appealed.

PARKER, J.

The question is whether by virtue of a contract defendant's testator, William E. Story, became indebted to his nephew, William E. Story, 2nd, on his twenty-first birthday in the sum of $5,000. The defendant contends that the contract was without consideration to support it, and therefore invalid. He asserts that the promisee, by refraining from the use of liquor and tobacco, was not harmed, but benefited; that that which he did was best for him to do, independently of his uncle's promise—and insists that it follows that unless the promisor was benefited, the contract was without consideration—a contention which, if well founded, would seem to leave open for controversy in many cases whether that which the promisee did or omitted to do was in fact of such benefit to him as to leave no consideration to support the enforcement of the promisor's agreement. Such a rule could not be tolerated, and is without foundation in the law. The exchequer chamber in 1875 defined "consideration" as follows: "A valuable consideration, in the sense of the law, may consist either in some right, interest, profit, or benefit accruing to the one party, or some forbearance, detriment, loss, or responsibility given suffered, or undertaken by the other." Courts "will not ask whether the thing which forms the consideration does in fact benefit the promisee or a third party, or is of any substantial value to any one. It is enough that something is promised, done, forborne, or suffered by the party to whom the promise is made as consideration for the promise made to him." Anson, Cont. 63. "In general a waiver of any legal right at the request of another party is a sufficient consideration for a promise." Pars, Cont. #444. "Any damage, or suspension, or forbearance of a right will be sufficient to sustain a promise," 2 Kent, Comm. (12th Ed.) 465. Pollock in his work on Contracts (page 166), after citing the definition given by the exchequer chamber, already quoted, says: "The second branch of this judicial description is really the most important one. 'Consideration' means not so much that one party is profiting as that the other abandons some legal right in the present, or limits his legal freedom of action in the future, as an inducement for the promise of the first." Now, applying this rule to the facts before us, the promisee used tobacco, occasionally drank liquor, and he had a legal right to do so. That right he abandoned for a period of years upon the strength of the promise of the testator that for such forbearance, he would give him $5,000. We need not speculate on the effort which may have been required to give up the use of those stimulants. It is sufficient that he restricted his lawful freedom of action within certain prescribed limits upon the faith of his uncle's agreement, and now, having fully performed the conditions imposed, it is of no moment whether such performance actually proved a benefit to the promisor, and the court will not inquire into it.

(Reversed.)

(2) **Exchange of Promises: Mutuality of Obligation.** In order for an exchange of promises to constitute a sufficient consideration, both promises must be legally binding. We are of course discussing a bilateral executory contractual situation. If one party or the other is not bound, then there is no contract. For example, Harry Hotrod promises to sell his car to Barney Buyer for $450, and Buyer promises to pay $450 for Harry's car if Buyer decides to spend that much money. There would be no contract because Buyer is not legally obligated to buy the car.

An area in which the question of mutuality of obligation often arises is with regard to requirements and output contracts. These are contracts in which one party agrees to supply all another party's requirements for a stated period of time or to purchase all of another party's output for a stated period. For instance, a steel company may agree to supply all the steel needed by an auto manufacturer for a specific period of time, or a tool manufacturer may agree to purchase all of the output of a particular type of steel manufactured by a steel company for a specific time period. Of course, each party must actually be bound to his promise. If the purchase of the output is strictly up to the buyer's whim or caprice, there would be no consideration. In our example the auto manufacturer needs steel in making its cars and would be bound to purchase that steel from the other party to the contract. Because it needs the steel in its business, the decision to purchase the steel is not up to its whim or caprice. Compare this situation to the case below.

OSCAR SCHLEGEL MFG. CO. vs. PETER COOPER'S GLUE FACTORY
231 N.Y. 459, 132 N.E. 148 (1921)

The defendant, Peter Cooper agreed to sell and deliver to the plaintiff, Oscar Schlegel, all its requirements of special BB glue for the year 1916 at nine cents per pound. During the year 1916, the price of glue increased from nine to twenty-four cents per pound. Peter Cooper refused to sell at the original price of nine cents, and Oscar Schlegel sued. The trial court found for plaintiff, and defendant appealed.

McLAUGHLIN, J.

The plaintiff at the time was engaged in no manufacturing business in which glue was used or required, nor was it then under contract to deliver glue to any third parties at a fixed price or otherwise. It was simply a jobber, selling among other things, glue to such customers as might be obtained by sending out salesmen to solicit orders therefore. The contract was invalid since a consideration was lacking. Mutual promises or obligations of parties to a contract, either express or necessarily implied, may furnish the requisite consideration. The defect in the alleged contract here under consideration is that it contains no express consideration, nor are there any mutual promises of the parties to it from which such consideration can be fairly inferred. The plaintiff, it will be observed, did not agree to do or refrain from doing anything. It was not obligated to sell a pound of defendant's glue or to make any effort in that direction. It did not agree not to sell other glue in competition with defendant's. The only obligation assumed by it was to pay nine cents a pound for such glue as it might order. Whether it should order any at all rested entirely with it. If it did not order any glue, then nothing was to be paid.

There are certain contracts in which mutual promises are implied; thus where the purchaser, to the knowledge of the seller, has entered into a contract for the resale of the article purchased, where the purchaser contracts for his requirements of an article necessary to be used in the business carried on by him, or for all the cans needed in a canning factory, all the lubricating oil for party's own use, all the coal needed for a foundry during a specified time, all the iron required during a certain period in a furnace, and all the ice required in a hotel during a certain season. In cases of this character, while the quantity of the article contracted to be sold is indefinite, nevertheless there is a certain standard mentioned in the agreement by which such quantity can be determined by an approximately accurate forecast. In the contract here under consideration there is no standard mentioned by which the quantity of glue to be furnished can be determined with any approximate degree of accuracy.

In the instant case, as we have already seen, there was no obligation on the part of the plaintiff to sell any of the defendant's glue, to make any effort towards bringing about such sale, or not to sell other glues in competition with it. There is not in the letter a single obligation from which it can fairly be inferred that the plaintiff was to do or refrain from doing anything whatever.

(Reversed.)

QUESTIONS FOR DISCUSSION

1. *How do you distinguish the Peter Cooper contract from the other requirements contracts mentioned in the opinion that the court said were enforceable?*
2. *In a valid requirements contract, how do you determine the quantity to be purchased by the buyer? How do you reconcile this with the requirement of definiteness in Chapter 5 concerning the offer?*

---

(3) **Forbearance.** A forbearance or promise to forbear from exercising a legal or equitable right is good consideration when given in exchange for the promise of another person. You should note that forbearance alone is not enough. It must be in response to a promise or request. Forbearance may take many forms. It may consist of a covenant not to compete, relinquishing a right to claim against an estate, giving up a right to go into bankruptcy, or refraining from bringing suit. There are obviously many other rights that one might give up that would constitute good consideration. The right given up may even be against a third person.

Frequently the forbearance requested is to refrain from suing the promisor or another person. This is good consideration as long as the person forbearing reasonably believes in good faith that he has a claim. It has generally been held that as long as the claim is doubtful, refraining from enforcing it constitutes consideration, but when it is obviously unenforceable, it cannot be consideration. For instance, giving up the enforcement of a gambling contract that is illegal would not be consideration. On the other hand, if the party believed that he had a cause of action at the time he agreed to give it up, the fact that it is subsequently determined that he would not prevail in the lawsuit is of no consequence.

FIEGE v. BOEHM

210 Md. 352, 123 A.2d 316 (1956)

Hilda Boehm testified that on January 21, 1951, she and Louis Fiege went to a drive-in movie and then to a restaurant and then had sexual intercourse in his automobile. When Boehm informed Fiege of her pregnancy and that he was the father, he agreed to pay all her medical expenses, compensate her for the loss of her salary caused by the child's birth, and pay her ten dollars per week for the child's support until it reached twenty-one years of age, provided that she would not institute bastardy proceedings against him as long as he made the payments. The child was born, and in May of 1953 Fiege inquired of Boehm's physician about blood tests to show the paternity of the child. The tests were made, and they indicated that it was not possible for Fiege to have been the father of the child. Fiege promptly ceased making payments to Boehm, who just as promptly filed bastardy proceedings against Fiege. Fiege was acquitted in the bastardy proceedings, whereupon Boehm brought this action for breach of contract. Judgment for plaintiff, and defendant appeals.

DELAPLAINE, J.

Defendant contends that, even if he did enter into the contract as alleged, it was not enforceable, because plaintiff's forbearance to prosecute was not based on a valid claim, and hence the contract was without consideration. He, therefore, asserts that the Court erred in overruling (1) his demurrer to the amended declaration, (2) his motion for a directed verdict, and (3) his motion for judgment n.o.v. or a new trial.

However, where statutes are in force to compel the father of a bastard to contribute to its support, the courts have invariably held that a contract by the putative father with the mother of his bastard child to provide for the support of the child upon the agreement of the mother to refrain from invoking the bastardy statute against the father, or to abandon proceedings already commenced, is supported by sufficient consideration.

We have adopted the rule that the surrender of, or forbearance to assert, an invalid claim by one who has not an honest and reasonable belief in its possible validity is not sufficient consideration for a contract. We combine the subjective requisite that the claim be bona fide with the objective requisite that it must have a reasonable basis of support. Accordingly a promise not to prosecute a claim which is not founded in good faith does not of itself give a right of action on an agreement to pay for refraining from so acting, because a release from mere annoyance and unfounded litigation does not furnish valuable consideration.

On the other hand, forbearance to sue for a lawful claim or demand is sufficient consideration for a promise to pay for the forbearance if the party forbearing had an honest intention to prosecute litigation which is not frivolous, vexatious, or unlawful, and which he believed to be well founded. Thus the promise of a woman who is expecting an illegitimate child that she will not institute bastardy proceedings against a certain man is sufficient consideration for his promise to pay for the child's support, even though it may not be certain whether the man is the father or whether the prosecution would be successful, if she makes the charge in good faith. The fact that a man accused of bastardy is forced to enter into a contract to pay for the support of his bastard child from fear of exposure and the shame that might be cast upon him as a result, as well as a sense of justice to render some compensation for the injury he inflicted upon the mother, does not lessen the merit of the contract, but greatly increases it.

Another analogous case is *Thompson* v. *Nelson*, 28 Ind. 431. There the plaintiff sought to recover back money which he had paid to compromise a prosecution for bastardy. He claimed that the prosecuting witness was not pregnant and therefore the prosecution was fraudulent. It was held by the Supreme Court of Indiana, however, that the settlement of the prosecution was a good consideration for the payment of the money and it could not be recovered back, inasmuch as it appeared from the evidence that the prosecution was instituted in good faith, and at that time there was reason to believe that the prosecuting witness was pregnant, although it was found out afterwards that she was not pregnant.

In the case at bar there was no proof of fraud or unfairness. Assuming that the hematologists were accurate in their laboratory tests and findings, nevertheless plaintiff gave testimony which indicated that she made the charge of bastardy against defendant in good faith. For these reasons the Court acted properly in overruling the demurrer to the amended declaration and the motion for a directed verdict.

QUESTIONS FOR DISCUSSION

1. *What standard did the court use in determining that the plaintiff believed in good faith that she had a claim?*
2. *Why did the court not set aside the contract on the grounds of "innocent misrepresentation"?*
3. *How could Fiege have framed the contract so that he did not suffer the result he incurred here?*

(4) Sufficiency of Consideration. It is often said that the courts generally will not inquire into the adequacy of consideration once they have determined that there is sufficient consideration to support the contract. In other words, once it is determined that there is some consideration to support the contract, the courts do not engage in a comparison of the relative values of the consideration. In the old and often quoted *Pinnel's Case,* 5 Coke, K.B. 111 (1602), the court stated that even the modest peppercorn would be sufficient consideration to support a contract. There are exceptions. If our friends' fraud, undue influence, or duress rear their ugly heads, the courts will inquire into the adequacy of consideration. Another situation occurs in which the consideration exchanged is of the same type, as in the case of money given in exchange for money or an exchange of fungible goods. Apart from these exceptions, however, the parties are free to make as good or as bad a bargain as they like. It is not the function of the courts to see that the bargain is fair in the economic sense.

One should note a word of warning. Although something of small value may be sufficient to support a contract, it should be paid. A mere recital of consideration in a contract without performance or payment of the consideration specified will often result in the contract's not being enforced. For example, one often reads a recital in a contract of the payment of one dollar as consideration. If this is not paid and there is no intent to pay it, the court will not enforce the contract.

BONNER v. WESTBOUND RECORDS, INC.
394 N.E.2d 1303 (Ill. App. 1979)

The defendants in this case, Westbound Records, Inc., and Bridgeport Music, Inc., were sued by Bonner and other individuals who were members of a rock music performing group known as The Ohio Players.

Westbound's business is making master recordings and selling them to others for sale and distribution. Bridgeport is a publisher and the capital stock of both companies is owned by the same person.

The Ohio Players and Westbound entered into a recording agreement that required The Ohio Players to record exclusively for Westbound for five years. The Ohio Players also entered into an agreement to render services exclusively for Bridgeport as long as the recording agreement was in effect. The Ohio Players received $4,000.00 for entering into the agreements.

During the next twenty-one months The Ohio Players recorded four single records and two albums for Westbound that were successfully distributed on a national basis. In fact one of the records, Funky Worm, received a gold record symbolizing sales in excess of $1,000,000.00. During this period Westbound advanced $59,390.00 for costs of recording sessions and other expenses to The Ohio Players and Bridgeport advanced $22,509.00 to enable them to pay income taxes and for other purposes. The companies were not obligated to make the payments and could not recoup the advances except out of royalties earned by The Ohio Players.

In January 1974 five of The Ohio Players repudiated the agreements and signed an agreement with Mercury Records, competitors of Westbound. On March 8, 1974, they filed this suit seeking a declaration that their agreements with Westbound and Bridgeport were invalid and unenforceable. The trial court found in favor of The Ohio Players on their motion for summary judgment and Westbound and Bridgeport appealed.

SIMON, J.

Contrary to the conclusion reached by the circuit court judge, it is our view that consideration passed to The Ohio Players when they accepted $4,000 to enter into the agreements. The fact that this payment was made by Westbound and Bridgeport by a check containing the notation that it was "an advance against royalties" does not disqualify the payment from being regarded as consideration. If sufficient royalties were not earned to repay Westbound the $4,000, The Ohio Players would not have been obligated to return it. By making the $4,000 advance, Westbound suffered a legal detriment and The Ohio Players received a legal advantage. Therefore, under both Michigan and Illinois law, the $4,000 payment constituted valid consideration. It is not the function of either the circuit court or this court to review the amount of the consideration which passed to decide whether either party made a bad bargain unless the amount is so grossly inadequate as to shock the conscience of the court. The advance The Ohio Players received, taken together with their expectation of what Westbound would accomplish in their behalf, does not shock our conscience. On the contrary, to a performing group which had never been successful in making records, Westbound offered an attractive proposal. The adequacy of consideration must be determined as of the time a contract is agreed upon, not from the hindsight of how the parties fare under it.

(Judgment reversed in favor of Westbound and Bridgeport and case remanded for trial.)

8:2. Special Problem Areas with Consideration. The law is well settled that the courts will not enforce a purely moral obligation without some other consideration. Thus if the son of Peter Paternal is rescued by Herman Heroic from a raging fire and Paternal promises that he will give Heroic $5,000 for his efforts, the promise is unenforceable. Although one may be of the opinion that the father has a moral obligation to honor his promise, he has no legal obligation. No consideration was given at the time Paternal made his promise, nor was any promised for the future.

There are some situations that appear to be exceptions to this general rule, although they really are not. They all involve situations in which an individual had a previously legally enforceable obligation that was rendered unenforceable by operation of law, such as bankruptcy or the running out of the statute of limitations. These situations will be discussed later in this chapter.

(1) Preexisting Duty. If one is legally bound to perform an act either because of law or contract, promising to perform that same act is not a valid consideration. Thus a promise to pay the sheriff $100 if he will serve a summons would be unenforceable because the sheriff has not given any consideration in return for the promise. He is already legally obligated to serve the summons.

The most frequent problems arise in situations in which there is a previously existing contractual obligation. For instance, a building contractor agrees to build a house for Harry Homeowner for $50,000. Because of rising costs he finds he is losing money. A promise to pay the contractor $55,000 to build the same house is not enforceable. It is not supported by any consideration from the contractor because he is already legally obligated to build the house for $50,000.

(a) *Rescission and Modification.* If both parties wish to make changes in an existing contract, there is no problem. There is obviously consideration given by both parties to the agreement since they are giving up a contractual right. If there is a change by just one party under the existing contract, it would not be binding for the reasons already stated. However, if both parties agree to a rescission of the agreement and then execute a new contract with the desired change, the new contract is enforceable. This process is called a *novation.* (A novation may also be defined as a substitution of one party for another in an existing contract.) Thus, when a building contract is mutually rescinded and a new contract agreed to that provides for the same work in return for more money, that contract is binding.

(b) *Unforeseeable Difficulties.* Particularly with regard to construction contracts a situation may arise in which one party finds performance of the contract substantially more difficult because of factors that were unknown or not anticipated by the parties. In some jurisdictions a promise by the other party to pay more money so that performance is continued will be enforced under these circumstances.

Great care should be taken in the application of this doctrine. It is not the law of all jurisdictions. Even those that follow the doctrine do so under limited circumstances. Generally the difficulty must have been something *unforeseeable* rather than something that the parties simply did not foresee. Thus, if a contractor is losing money because of an unexpected increase in labor costs, he would not be entitled to any relief under this doctrine. On the other hand, the discovery of bedrock in the excavation of a home located in a beach area where such a condition hardly, if ever, occurs would be unforeseeable and thus subject to this doctrine.

ANTHONY TILE & MARBLE CO., INC. v. H. L. COBLE CONSTRUCTION CO.

16 N.C.App. 740, 193 S.E.2d 338 (1972)

Under a contract of August 28, 1969, plaintiff Anthony was to execute and return to defendant Coble five copies of the contract, obtain a performance-and-materials bond, and deliver it to defendant. Plaintiff did not obtain the bond for over two months, although there was some discussion between the parties concerning the bond during that time. The bond was delivered on November 6, 1969. On November 4, 1969, plaintiff received a telegram from Coble stating that old negotiations were null and void. The plaintiff contended that the negotiations resulted in a modification of the contract extending the time for securing a performance-and-materials bond. Defendant contended that there was no consideration for the modification. Trial-court judgment for plaintiff, and defendant appeals.

MORRIS, J.

As stated above to be effective as a modification, a new agreement must possess all elements necessary to form a contract. Certainly, consideration is as much a requisite as it is in the initial creation of a contract.

Plaintiff contends that since the 28 August agreement was still executory with obligations remaining to be performed on both sides, no additional consideration was required for any modification. We do not agree.

As to executory contracts, it is generally held that "(A) modification can be nothing but a new contract and must be supported by a consideration like every other contract."

In support of its position plaintiff cites the following in its brief:

> Any executory contract which is bilateral in the advantage and obligations given and assumed may at any time after it has been made and before a breach thereof has occurred be changed or modified in one or more of its details by a new agreement also bilateral by the mutual consent of the parties without any other consideration.

While we agree that the basic soundness of the above principle as to the requisite sufficiency of consideration needed to support a modification, plaintiff has failed to show any modification that is indeed bilateral. Under the alleged modification, plaintiff incurred no new obligations or duties. No detriment was to be suffered by plaintiff nor new benefit to be received by defendant. Assuming mutual consent of the parties to the modification, only the time period in which to procure the bond was changed. Plaintiff simply promised to perform what it was obligated to do under the 28 August agreement. It is generally established that a promise to perform an act which the promisor is already bound to perform is insufficient consideration for a promise by the adverse party.

(Reversed.)

QUESTIONS FOR DISCUSSION

1. *If the parties wish to modify the obligations of only one of the parties to a contract, what must be done in order to make the modification enforceable?*
2. *Why wasn't the modification enforceable in this case?*

---

(2) Past Consideration. A party cannot use performance that he rendered in the past as consideration to support an agreement entered into in the present. Thus an agreement by a company to provide an employee with a pension because of his past forty years of loyal service would be unenforceable because no consideration is given by the employee at the time the contract is made or promised for the future. The reason for this rule is that the performance was not bargained for if it occurred prior to the promise. In order to be sufficient to support a contract, the consideration must be given at the time the promise is made or in the future, not in the past.

Past consideration should be distinguished from those situations in which a party requests another to perform work for him and leaves open the question of compensation for later determination. The later agreement to pay a specific sum is enforceable and supported by consideration because at the time the original agreement was made, the promisee was entitled to a reasonable compensation.

BROWN v. ADDINGTON
52 N.E.2d 640 (Ind. 1944)

The administrator of Brown's estate sued Addington to enforce an agreement. The plaintiff alleged that when Addington became homeless at age eight, Brown, his uncle, took him into his own home, where he fed, clothed, and educated him for six years without any remuneration. For a number of years thereafter, Addington stayed in Brown's home when he was unemployed and paid nothing. When Brown reached old age, he asked Addington to make certain payments to him. In response to the request, Addington executed the following agreement.

I, Claude L. Addington, remembering and appreciating the many favors and acts of kindness, rendered to me, during the years that have passed, by my beloved uncle William E. Brown, and desiring to express my gratitude to him in something more than empty words, hereby promise and pledge that I will pay to my said uncle William E. Brown, the sum of One Hundred Dollars ($100.00) during each year that the said William E. Brown shall live. Payment to be made on or about the first day of January of each said year, beginning with the year 1930.

(Signed) Claude L. Addington.

Addington failed to make any payments. When Brown died, his executor sued to enforce the agreement. Lower-court judgment for defendant, and plaintiff appealed.

CRUMPACKER, C.J.

If a person has been benefited in the past by some act or forbearance for which he incurred no legal liability and "afterwards, whether from good feeling or interested motives, he makes a promise to the person by whose act or forbearance he has benefited, and that promise is made on no other consideration than the past benefit, it is gratuitous and cannot be enforced; it is based on motive and not on consideration." 17 C.J.C., Contracts, P. 470 § 116, and cases cited.

By the great weight of authority a past consideration, if it imposed no legal obligation at the time it was furnished, will support no promise whatever. A past consideration is insufficient, even though of benefit to the promisor, where the services rendered or things or value furnished were intended and expected to be gratuitous.

Nowhere in the appellant's third paragraph of complaint is it alleged that the board and lodging furnished by said decedent to the appellee were furnished for any expected remuneration or that the appellee accepted them with any agreement or understanding that he was to pay for them. On the contrary, the reasonable construction of the pleading indicates that appellant's decedent had no thought of remuneration until he had reached old age and feared that he had insufficient means to meet his needs during the remainder of his life. Thus it would seem that the consideration, as pleaded, imposed no legal obligation on the appellee at the time it was furnished and was intended by the decedent and expected by the appellee to be gratuitous.

(Judgment affirmed.)

(3) **Payment of a Lesser Sum.** Where the debt is liquidated—that is, if for a specific amount and undisputed—the payment of a lesser amount will not discharge the debt. In the absence of any further acts or promise by the debtor he gives no consideration for the acceptance of the lesser amount by the creditor as he was already obligated to pay the full debt. On the other hand, should the debtor do something in addition to paying the money, he may be giving consideration. For example, he may pay the debt before it is due, pay in a medium other than that required in the contract, give some security, or pay at a place other than the one designated in the agreement. There are a few state statutes that provide that the acceptance of a lesser sum is satisfaction of an undisputed and liquidated debt and such acceptance will discharge the debt. Questions frequently arise as to whether the acceptance of a check marked "paid in full" will discharge the debt. If the debt is liquidated and undisputed, it will not. If the debt is unliquidated and disputed, it will be discharged if a notation is made with the payment that the payment is in full and final satisfaction of the debt and if the creditor knowingly accepts the payment or retains it for an unreasonable period of time.

(4) **Compromise, Settlement, and Composition Agreements.** Where a claim or debt is honestly disputed, the compromise of it will constitute good consideration. The dispute may concern the existence of a debt or the amount due. For example, Manny Mechanic does some work on Harry Hotrod's car. The bill rendered is $100. Hotrod contends that he didn't authorize the work and that in any event the cost of the work should not exceed $50. Mechanic claims that it was authorized and that the $100 is justified. One solution would be to let a court decide the issue, but rather than submit the issue to a court, the parties may compromise on a $75 charge. The consideration to support the agreement is found in the giving up of what each party honestly believes

his legal rights to be. However, if there is no foundation to the claim it will not constitute a valid consideration.

Sometimes a debtor may owe so much money to a large number of creditors that it would be extremely difficult ever to pay all the debts in full. Under these circumstances an attempt may be made to negotiate a composition agreement with the creditors. This usually involves the debtor's agreeing to pay a percentage of the debt to each creditor and refraining from filing for bankruptcy. The creditors, of course, agree to accept the partial payment in satisfaction of the debt and do not file an involuntary bankruptcy petition against the debtor. All the creditors must be given an opportunity to participate in the agreement. The consideration for the promises of each creditor to accept the lesser amount is found in the similar promises given by the other creditors, hence there always must be two or more creditors involved in a composition agreement. There is often an advantage in this approach as opposed to bankruptcy. The time-consuming bankruptcy process is avoided, often the creditors receive more money than they would as the result of a bankruptcy proceeding, and the debtor, if he is a businessman, is able to keep his business functioning and intact.

8:3. Equitable Enforcement and Substitutes for Consideration.

(1) Promissory Estoppel. Estoppel is an equitable doctrine that is applied in a number of contexts. It means literally that one is not permitted to deny. The doctrine of promissory estoppel is applied in those situations in which a promise is made and the promisee reasonably relies on that promise and suffers a detriment if the promise is not enforced. The doctrine is applied only in those cases in which there is no contract and thus no consideration flowing from the promisee. It is also most often applied in cases of promises made for subscriptions to charitable institutions. To illustrate, Gordon Gridiron promises to give $1 million to Doormat University to build a new football stadium. Doormat U. enters into construction of the facility expecting to use the million-dollar gift from Gridiron. Gridiron reneges on his promise. Although there is no consideration from Doormat U., the court would stop Gridiron from denying his promise and pleading lack of consideration because Doormat U. would suffer injury as a result of reasonable reliance on his promise. Of course, if Gridiron withdraws his promise before any change in position by Doormat U., he would not be liable. You should not make the mistake of assuming that every promise that does not achieve the dignity accorded to a contract will be enforced under this doctrine, as the courts do not grant relief except under unusual circumstances.

In more recent times the doctrine has come to be applied in the construction industry. If a general contractor relies on the promise of a subcontractor when the general contractor calculates its bid the courts will frequently enforce the promise of the subcontractor even if there is no contract between the general contractor and the subcontractor. In order to have the doctrine invoked by the courts, the general contractor must show that he reasonably relied on the subcontractor's bid in formulating his own bid. The general contractor must also prove that his reliance was foreseeable to the subcontractor and that the general contractor would suffer damages if the promise is not enforced.

JAMES KING & SON, INC. v. DESANTIS CONSTRUCTION, NO. 2 CORP.

413 N.Y.S.2d 78 (N.Y. App. 1977)

R. H. Macy & Co., Inc. (Macy's) announced its intention to build a new department store at the Sunrise Mall in Massapequa, Long Island. James King & Sons, Inc. decided to submit a bid to construct the store as the general contractor. In deciding on the amount of its bid King used a bid submitted to it by telephone from DeSantis Construction No. 2 Corp. for the concrete work. DeSantis's bid to King was for $275,000.00.

After receiving the bid for concrete work, the project manager for King called the president of DeSantis who assured King that DeSantis's bid to perform the work was in fact $275,000.00. King then used this figure in computing its bid as general contractor. After it was awarded the general contract, King immediately called DeSantis to inform it that its bid was successful.

DeSantis then refused to perform in accordance with its bid and King was forced to obtain the concrete work from another subcontractor and at additional cost of $44,000.00. King then sued DeSantis.

SCHWARTZ, J.

Defendant does not dispute that it placed an oral bid with plaintiff of $275,000 for the concrete work. Defendant claims that after plaintiff was awarded the contract by Macy's, plaintiff called defendant into their offices to sign a contract and defendant admits it failed and refused to sign an agreement.

Defendant contends that the bid admittedly made by the defendant was merely the beginning of negotiations and that the defendant could not reasonably anticipate that plaintiff would deem such bid as a firm bid embracing all the elements of a complicated construction job.

Defendant argues that plaintiff should have, in advance, secured a written contract conditioned upon the success of its bid to the owner. It contends that there was no acceptance by the plaintiff of defendant's bid and therefore no contract; that plaintiff's use of the defendant's figure for concrete in plaintiff's bid to the owner, was not the equivalent of an acceptance.

The doctrine of promissory estoppel is intended to avoid the harsh results of allowing the promisor to repudiate, when the promisee has acted in reliance upon the promise.

The elements of a promissory estoppel are a promise clear and unambiguous in its terms, reliance by the party to whom the promise is made, such reliance to be both reasonable and foreseeable; the party asserting the estoppel must be injured by his reliance.

The Appellate Division, Fourth Department in 1975, held that prices for valves given by a manufacturer with knowledge they would be used in bids submitted in bids for the plumbing subcontracts in construction of a sewage treatment plant, constituted offers. When the plumbing subcontractor, who used the quotation in his bid, was awarded the subcontract and thereafter, the valve manufacturer discontinued the valves, the plumbing subcontractor is entitled to recover its loss when obliged to purchase elsewhere. In the publication of the American Bar Association's National Institute on Construction Claims (1977) it is stated:

> As a general rule when a subcontractor refuses to perform at his submitted quotation, courts have usually held the subcontractor to his bid absent unusual considerations, or allowed the prime contractor to recover damages if he is forced to use another subcontractor at a higher price. The rationale for such decisions is the doctrine of promissory estoppel. . . .

The court finds that defendant made a clear and unambiguous bid on which plaintiff relied, to its damage and that such reliance was reasonable and foreseeable. The defendant, under the doctrine of promissory estoppel, is therefore liable for such damage.

(Judgment for King.)

QUESTIONS FOR DISCUSSION

1. *What are the elements of promissory estoppel as related by the Court?*
2. *By what action did King rely on the promise of DeSantis?*

(2) Promises Enforceable Under the U.C.C. in the Absence of Consideration. Section 2–209 of the U.C.C. provides that an agreement modifying a contract needs no consideration to be binding. You should note that this section is contained in the Article on Sales of the U.C.C. The only contracts covered by this article are contracts for the sale of goods. Other types of contracts are not affected. Goods are things moveable other than money, securities, and rights of action. Real estate, personal services, and debts are not goods. You should also consider how this provision affects our earlier discussion on modification.

Another point to remember is that when there is a signed written promise by a merchant to buy or sell goods that states that the offer will remain open for a stated period of time, that promise is not revocable for lack of consideration. This provision is found in Section 2–205 of the U.C.C.

(3) Promise to Pay Debt Discharged in Bankruptcy or Barred by Statute of Limitations. When one is discharged from his debts as the result of a bankruptcy proceeding, he is no longer required to pay them. This is true even if he subsequently becomes wealthy. However, if the debtor later promises to pay the discharged debt, this promise is enforceable in most jurisdictions, although some states require the promise to be in writing. The promise must be actual and not implied. For instance, a part payment unaccompanied by a specific promise to pay the remainder of the debt would not revive the original debt. However, the new Bankruptcy Code of 1978 has changed the law with regard to reaffirmation of the debt by the debtor. Prior to the new law a debtor frequently would become obligated for debts that were discharged as a result of the bankruptcy action. Under the new Bankruptcy Code a reaffirmation agreement entered into between the debtor and creditor is subject to a number of limitations. In the first place it is not enforceable unless it satisfies the requirements of nonbankruptcy law. Furthermore, in order to satisfy the requirements of the code, four conditions must be met: (1) the reaffirmation agreement must be made before the granting of discharge; (2) the debtor may rescind the agreement within 30 days after the agreement becomes enforceable; (3) a hearing must be held wherein the court advises the debtor that a reaffirmation agreement is not required, and advises the debtor of the consequences and legal effect of the agreement; and (4) if the debtor is an individual and the debt is a consumer debt not secured by real property of the debtor, the court must approve the agreement.

All jurisdictions have statutes of limitations that limit the amount of time within which a suit may be brought. Once the time expires, the right of action is lost. Many states have statutes that allow a revival of the right of action if there is a new promise to pay the debt. Most of these statutes require the promise to be in writing. Another point to consider is that the part payment of a debt will result in the limitations period's starting from the time of part payment. For example, if a debt is due January 1, 1978, and the limitations period is three years, any suit must be filed by January 1, 1981. However, if the debtor makes a partial payment December 30, 1980, the statute begins to run for another three years from that date.

(4) Seal. Under the old common law, prior to the development of the law of contracts, an agreement could be enforced if it was executed under *seal.* At that time the seal was imprinted in wax on the paper or some other suitable substance. The agreement was enforceable without consideration.

Today the use of the seal is greatly restricted. Some states have abolished its use altogether, and in most jurisdictions consideration must still be proved if a contract is to be enforceable. The chief function of the seal in many states is to extend the period of the statute of limitations. The Uniform Commercial Code is consistent with the notion that the seal is obsolete and provides that seals are inoperative in Section 2–203.

8:4. Summary of Uniform Commercial Code Changes. The code has not changed the general contract rule that in order to be binding an agreement must be supported by consideration. However, the Uniform Commercial Code has made some changes in the requirement of consideration to support a promise in a few special areas. Remember once again that the changes only apply to contracts for the sale of goods.

One change is with regard to the consideration needed to support a unilateral modification of a contract. Recall that if there is a change by just one party under an existing contract it would not be binding because of the absence of consideration. In contracts covered by the U.C.C., however, an agreement modifying a contract for the sale of goods needs no consideration to be binding. A limitation on this rule easing the requirement of consideration is that the change must be made in good faith. Section 2–209. The purpose of this change is to give effect to modifications entered into by the parties in good faith without regard to the technicalities of consideration.

The U.C.C. has also made a change in the consideration requirement with regard to offers made by merchants, as was discussed in Chapter 5. In short a promise made by a merchant, in writing, to keep an offer open for a specified period of time, needs no consideration in order to be enforceable. If no specific time is mentioned, the time is a reasonable time not to exceed three months. Section 2–205. Note however, that this provision only applies to written offers by merchants involving contracts for the sale of goods. The writing also must specifically promise to keep the offer open. The purpose of this section is to allow merchants, who presumably are knowledgeable in business affairs, to keep an offer open merely by his written promise to do so without regard to consideration.

SUMMARY

Consideration is that which is bargained for in return for a promise. It is defined as an act other than a promise; a forbearance; the creation, modification, or destruction of a legal relation; or a return promise, bargained for and given in exchange for the promise. Sometimes it is referred to as a legal detriment.

In order to have sufficient consideration both promises must be legally binding, that is, there must be mutuality of obligation. A forbearance or promise to forbear from exercising a legal or equitable right is good consideration when given in exchange for a promise.

In general, once the courts determine that there is sufficient consideration they will not inquire into the adequacy of consideration. Exceptions are fraud, duress, undue influence, or where the consideration exchanged is of the same type.

A court will not enforce a purely moral obligation without some other consideration. The consideration must be something in addition to what one is already legally bound to do. If one has a preexisting duty, a promise to perform that duty is not consideration.

A unilateral modification of a contract is not binding without further consideration. If both parties agree to a rescission of the contract and then a modification, there is consideration. This process is a novation. The Uniform Commercial Code provides for an exception for contracts for the sale of goods.

Performance rendered in the past will not suffice as consideration for an agreement entered into in the present. In general, the payment of a lesser amount due on a liquidated debt will not discharge the debt. If the debt is unliquidated and in dispute, a compromise of the amount due is supported by consideration. Composition agreements are also supported by consideration.

Sometimes courts will enforce agreements in the absence of consideration by applying the doctrine of promissory estoppel. The doctrine requires a promise, reasonable reliance by the promisee, and a detriment suffered by the promisee if the promise is not enforced.

PROBLEMS

1. George Slattery was a licensed polygraph operator. He was hired by the Florida States Attorney to question the suspect of a crime. A Wells Fargo agent was killed during the commission of a separate crime. Wells Fargo subsequently advertised a reward for information leading to the murder and arrest of the individual who committed the murder. In response to the advertisement, Slattery gave information to Wells Fargo which resulted in the conviction of the suspect for the murder. The information was

developed by Slattery in connection with his interrogation of the suspect for the crime not connected with the murder. When Wells Fargo refused to pay, Slattery sued for breach of contract. Was Slattery entitled to recover? Slattery v. Wells Fargo Armored Service Corp., 366 So.2d 157 (Fla. App. 1979).

2. Mary Dewitt made the following promise to her nephew, Ben Earle: "I want you to attend my funeral, Ben, if you outlive me, and I think you will, and I will pay all expenses and I will give you five hundred dollars." Ben agreed, attended her funeral, and demanded payment from the executor, who refused. Ben sued. Lower court for defendant and plaintiff appealed. Decide. Earle v. Angell, 157 Mass. 294, 32 N.E. 164 (1892).
3. Lucy, Lady Duff-Gordon was a creator of fashions. She employed Wood to place her endorsement on the designs of others. He was to have exclusive right to place the endorsement and to license others to market her designs. In return Lucy was to receive one half of all profits and revenues derived from contracts made by Wood. Wood sued, contending that Lucy placed her endorsement without his knowledge and without profits. Lucy contended that the contract was without consideration because Wood did not have to do anything; that is, it lacked mutuality. Decide. Wood v. Lucy, Lady Duff-Gordon, 222 N.Y. 88, 118 N.E. 214 (1917).
4. Kirshner operated an employment agency. Spinella was seeking employment and contracted with the agency. Under the contract Spinella was to inform Kirshner of any job that he applied for in person. Spinella applied to Fishers Island School on his own without informing Kirshner. Spinella got the job. In the meantime Kirshner expended time in trying to obtain an interview for Spinella at the same school. Of course, the job was not available because Spinella had already filled it. When Kirshner billed Spinella he refused to pay contending Kirshner gave no consideration since the job was already filled. Was there sufficient consideration to support the contract? Kirshner v. Spinella, 343 N.Y.S.2d 298 (1973).
5. Levi Wyman was twenty-five when he returned from sea and became very sick in Hartford. Mills took him in and cared for him from the 5th to the 20th of February, when Levi Wyman died. On February 24 his father, Seth Wyman, wrote to Mills promising to pay for the expenses of his son's illness. Seth subsequently had a change of heart, refused to pay, and was sued for breach of contract by Mills. Decide. Mills v. Wyman, 3 Pick. 207 (Mass. 1825).
6. Lee Taylor began to assault his wife when she knocked him down with an axe and was in the process of cutting his head open or decapitating him. Lena Harrington then intervened and caught the axe as it was descending, cutting and mutilating her own hand in the process. Lee Taylor promised to pay Harrington her damages but after one small payment made no more payments. Harrington sued to enforce the promise. Decide. Harrington v. Taylor, 255 N.C. 690, 36 S.E.2d 227 (1945).

7. Plaintiff was a widow of defendant's brother. Defendant wrote plaintiff: "If you will come down and see me, I will let you have a place to raise your family, as I have more open land than I can tend." Within a month or two plaintiff moved her family the sixty miles to the defendant's residence, where he put her in a house and gave her land to cultivate for two years. At the end of that time he notified plaintiff to leave. Lower court for plaintiff and defendant appeals. Decide. Kirksey v. Kirksey, 8 Ala. 131 (1845).
8. William Siegel stored his furniture with Spear & Company while he was away for the summer. He had originally purchased the furniture from that company and in fact still had not finished paying for it. Siegel did not agree to pay Spear & Company anything for the storage of the furniture, and Spear requested no payment. Spear's credit manager told Siegel that he would insure the furniture and bill Siegel. During the summer a fire destroyed the furniture, and it was not covered by insurance. Siegel sued Spear & Company. Decide. Siegel v. Spear & Company, 234 N.Y. 479, 138 N.E. 414 (1923).
9. Watkins agreed to excavate all material from a building site. No reservation appeared in the contract for unexpected conditions. Watkins encountered rock in excavating the property, and after many requests Carrig vocally agreed to pay an increased price for the entire excavation so that the work would not be delayed. When Carrig subsequently refused to pay the new amount, Watkins sued. Decide. Watkins v. Carrig, 91 N.H. 459, 21 A.2d 591 (1941).
10. Vischia agreed to sell Loewus all of certain specific wines that Loewus ordered for a sixteen-month period. Loewus was a wholesaler and agreed to purchase wines from Vischia from time to time "as it may require under labels bearing brand or trade names which are its own property." Loewus could also purchase wines from others under other brands. Vischia refused to fill orders for Loewus, and Loewus sued to enforce the contract. Decide. G. Loewus & Company v. Vischia, 2 N.J. 54, 65 A.2d 604 (1949).
11. Tucson Federal Savings & Loan is a federal savings-and-loan association and required by its charter to obtain fire and extended-coverage insurance as collateral security on all mortgage loans it makes. Aetna agreed to supply the insurance required by Tucson, and Tucson agreed to purchase exclusively from Aetna for a ten-year period. Subsequently Tucson sought to terminate the agreement and was sued by Aetna. Decide. Tucson Federal Savings and Loan Association v. Aetna Investment Corporation, 274 Ariz. 163, 245 P.2d 423 (1952).

CHAPTER 9

Capacity to Contract

In order for an individual to enter into a valid contract, he must have contractual capacity. One may lack legal capacity for a number of reasons: infancy, insanity, intoxication, or other legal disability. This chapter will examine a number of those reasons.

9:1. INFANCY. The courts have been protective of minors in contractual arrangements for centuries. Under the common law an infant or a minor was anyone under the age of twenty-one. Today the trend is toward a reduction of that age. In many states the age of majority is now eighteen, although there are some that retain the old age of twenty-one, and at least one jurisdiction has selected the age of twenty years as that point when one has full contractual capacity. Do not confuse the age of majority, when one has full contractual capacity, with age limits for other important matters. The federal law allowing those eighteen and over to vote and various state statutes that set an age limit for special protection given to juveniles who commit crimes are examples.

The purpose of setting an age limit for contractual capacity is to protect minors by preventing adults from taking advantage of them in contractual dealings. The age limit has been drawn somewhat arbitrarily and results occasionally in some problems of equity. For instance, some seventeen-year-olds are much more astute in their dealings than certain adults, and one may realistically question in a particular case who needs the protection. Over the years the courts have developed a number of legal approaches to protect the minor while at the same time preventing him from taking advantage of this protection. The fact that a minor may be emancipated does not affect the protection given him by the law. One is emancipated when he is self-supporting and released from the obligation of contributing his earnings to his parents.

(1) Liability for Necessaries and Nonnecessaries. The general rule is that infants are not liable on their contracts for either their necessaries or their nonnecessaries. The contract is voidable at the option of the infant. He may enforce it against the adult party or he may avoid it if he desires, subject to the restrictions discussed in Section (2). However, minors are liable in quasi contract for the *reasonable value* of their necessaries. The distinction between contractual liability for the contract price and quasicontractual liability for the reasonable value of the goods or services is often lost. Perhaps this is because there is no practical difference in many cases. On the other hand, there may be a significant difference. For example, Matilda Minor contracts with Clothing Shoppe to purchase a sweater for $75. The sweater sells elsewhere for $40, but Clothier is able to induce Minor to agree to the $75 price. Minor wears the sweater and then avoids the contract, having paid only $5 down on an installment plan. Assuming the sweater is a necessary, Matilda would be liable in quasi contract for its reasonable value, probably $40, rather than the $75 contract price. Most often, however, the reasonable value and the contract price are the same.

What are necessaries? They are those things that are required to maintain one's station in life and to provide for a person's continued life, health, and liberty. Clearly food, clothing, shelter, and medical expenses fall within this definition. Problems arise, however, in the determination of what quality and quantity of these items qualifies as a necessary. For an example, a cloth coat may qualify as a necessary, but what about a mink stole? Is a car a necessary, and if it is, does this include a new high-priced foreign sport model as well as a five-year-old used sedan? The answers to these questions depend upon the circumstances of the infant: the station in life that he occupies. A car may be a necessary for one individual and not for another. Whether an item is a necessary is a question of fact to be determined in each individual case. Of course as conditions change, what was once considered to be a luxury may become a necessary. Thus, while courts some years ago rarely held the automobile to be a necessary, today they are more likely to do so.

When a minor is living with his parents or guardian and already is supplied with a particular necessary, an additional purchase of the same item is not considered a necessary since the infant is already adequately supplied. The burden would be on the adult to show that the infant is not adequately supplied with the necessary and thus is required to contract for it on his own.

GASTONIA PERSONNEL CORPORATION v. ROGERS
172 S.E.2d 19 (N.C. 1970)

Bobby L. Rogers graduated from high school in 1966. When he was nineteen years old he was emancipated and married. In order to obtain employment, he went to an employment agency, Gastonia Personnel Corporation, and signed a contract. The contract provided that if Gastonia obtained employment for him, Rogers would pay them $295.00.

Gastonia did obtain employment for Rogers but after getting the job he refused to pay them. Gastonia sued Rogers who raised the defense of infancy. The trial court found in favor of Rogers on a motion for summary judgment and Gastonia appealed.

BOBBIT, J.

Under the common law, persons, whether male or female, are classified and referred to as ***infants*** until they attain the age of twenty-one years.

> By the fifteenth century it seems to have been well settled that an infant's bargain was in general void at his election (that is voidable), and also that he was liable for necessaries. 2 Williston, Contracts § 223 (3rd ed. 1959).

An early commentary on the common law, after the general statement that contracts made by persons (infants) before attaining the age of twenty-one "may be avoided," sets forth "some exceptions out of this generality," to wit: "*An infant may bind himselfe to pay for his necessary meat, drinke, apparell, necessary physicke, and such other necessaries,* and likewise for his good teaching or instruction, whereby he may profit himselfe afterwards." (Our italics.) *Coke on Littleton,* 13th ed. (1788), p. 172. The italicized portion of this excerpt from *Coke on Littleton* was quoted by Pearson, J. (later C. J.), in *Freeman* v. *Bridger,* 49 N.C. 1 (1856).

In accordance with this ancient rule of the common law, this Court has held an infant's contract, unless for "necessaries" or unless authorized by statute, is voidable by the infant, at his election, and may be disaffirmed during infancy or upon attaining the age of twenty-one.

In *Freeman* v. *Bridger, supra,* the opinion, referring to "such other necessaries," states: "These last words embrace boarding; for shelter is as necessary as food and clothing. They have also been extended so as to embrace schooling, and nursing (as well as physic) while sick. In regard to the quality of the clothes and the kind of food &c., a restriction is added, that it must appear that the articles were suitable to the infant's degree and estate."

In *Freeman,* the Court held that timber for the construction of a house on an infant's land was not a "necessary" and therefore the infant could disaffirm his contract for the purchase thereof.

When an infant purchased a motor vehicle, whether for pleasure or as necessary for use in his occupation or employment, the ancient rule of the common law was applied with full vigor.

In *Fisher,* [*Fisher* v. *Taylor Motor Co.,* 1075 F.2d 94 (1959)] the plaintiff, a minor, bought a 1953 Oldsmobile. The purchase price was $750.00, of which $600.00 was provided by the plaintiff. This car, while operated by the plaintiff, was involved in a wreck. Its value, after the wreck, was $50.00. The plaintiff elected to disaffirm his contract. In an action instituted in his behalf by a next friend, the plaintiff recovered $550.00 (the $600.00 he had paid less the value of the wrecked car).

Decisions in other jurisdictions which hold that a motor vehicle, under particular circumstances, may be a "necessary" for a minor, are reviewed in an article, Infant Contractual Responsibility: A Time for Reappraisal and Realistic Adjustment? Mehler, 11 University of Kansas Law Review 361, at 370 et seq. (1963).

In general, our prior decisions are to the effect that the "necessaries" of an infant, his wife, and child, include only such necessities of life as food, clothing, shelter, medical attention, etc. In our view, the concept of "necessaries" should be enlarged to include such articles of property and such services as are reasonably necessary to enable the infant to earn the money required to provide the necessities of life for himself and those who are legally dependent upon him.

The record before us contains only plaintiff's evidence and the stipulation. It may be that defendant can defeat plaintiff's claim on grounds other than the plea of infancy. His motion for nonsuit having been allowed, defendant has not offered evidence.

The evidence before us tends to show that defendant, when he contracted with plaintiff, was nineteen years of age, emancipated, married, a high school graduate, within "a quarter or twenty-two hours" of obtaining his degree in applied science, and capable of holding a job at a starting salary of $4,784.00. To hold, as a matter of law, that such a person cannot obligate himself to pay for services rendered him in obtaining employment suitable to his ability, education and specialized training, enabling him to provide the necessities of life for himself, his wife and his expected child, would place him and others similarly situated under a serious economic handicap.

In the effort to protect "older minors" from improvident or unfair contracts, the law should not deny to them the opportunity and right to obligate themselves for articles of property or services which are reasonably necessary to enable them to provide for the proper support of themselves and their dependents. The minor should be held liable for the reasonable value of articles of property or services received pursuant to such contract.

Applying the foregoing legal principles, which modify *pro tanto* the ancient rule of the common law, we hold that the evidence offered by plaintiff was sufficient for submission to the jury for its determination of issues substantially as indicated below.

To establish liability, plaintiff must satisfy the jury by the greater weight of the evidence that defendant's contract with plaintiff was an appropriate and reasonable means for defendant to obtain suitable employment. If this issue is answered in plaintiff's favor, plaintiff must then establish by the greater weight of the evidence the reasonable value of the services received by defendant pursuant to the contract. Thus, plaintiff's recovery, if any, cannot exceed the reasonable value of its services to defendant.

Accordingly, the judgment of the Court of Appeals is reversed and the cause is remanded to that Court with direction to award a new trial to be conducted in accordance with the legal principles stated herein.

QUESTIONS FOR DISCUSSION

1. *The court states that in deciding in favor of Gastonia it is expanding the concept of necessaries from the holdings of cases under the older common law. How has the court expanded the concept of necessaries?*
2. *If the old common-law standard had been applied to this case, do you think the court's decision would have been different?*
3. *On retrial, how do you feel a court might rule on the question of whether the contract was for a necessary?*
4. *If the court finds that the contract employing Gastonia was for a necessary, will Gastonia automatically be entitled to recover the contract price?*

(2) Time and Effect of Disaffirmance. Infants may disaffirm their contracts at any time prior to reaching the age of majority or within a reasonable time thereafter. Where the contract is wholly executory, this presents little problem, but what if there has been performance or partial performance? The courts generally hold that when the minor disaffirms the contract he must return any property or money that he has obtained. If the infant has used the property or no longer has it all in his possession, he must return what is left. Generally the fact that the property has been depreciated either in value or quantity or even the fact that the infant no longer has it in his possession does not prevent him from disaffirming the contract. Most jurisdictions hold that the minor is entitled to regain the consideration that he paid despite the fact that he has used or dissipated the consideration given to him. Some jurisdictions, such as New York, hold that the minor is entitled to his consideration less the depreciated value of the consideration he received from the adult. A few states will not allow disaffirmance when the infant cannot return the property he received under the contract. The following example illustrates these differences.

Monroe Minor, aged sixteen, contracts to purchase a sports car. He pays $3,000 for the car and signs an installment note for $2,000. Minor drives the car for six months and then decides to disaffirm the contract and seek return of his money. The car has depreciated $1,000 in that time. When he disaffirms the contract, he must return the car, and in most jurisdictions he will get his money back and the note canceled. In a few states the depreciation would be deducted from the amount of money returned to the minor. Unfortunately for the adult, in the majority of states the consideration would be returned to the minor even if he wrapped the car around a big tree and completely demolished it.

The prevailing view is that an infant can recover his property even if the other party has transferred it to an innocent third-party purchaser. The exception to this rule is the case of a contract for the sale of goods. Section 2–403 of the Uniform Commercial Code provides that one with a voidable title has the power to transfer a good title to a good-faith purchaser for value. Thus, if Danny Dealer accepts Monroe Minor's used car in trade for a new one and then sells Minor's old car to Inez Innocent, when Monroe avoids his contract with Dealer he cannot recover his old car from Innocent.

One exception to the rule on disaffirmance obtains in the case of real estate. At common law and in many states today an infant may neither disaffirm nor ratify a contract involving real estate until he reaches majority. The rule probably had its antecedents in the fact that real property was the primary means of wealth and power at the time the concept was developed, thus the minor could take no action whatsoever concerning such an important matter.

CENTRAL BUCKS AERO, INC. v. SMITH

310 A.2d 283, (Pa. Super. 173)

Smith, defendant, leased an airplane from Central Bucks Aero, Inc. He was twenty years of age at the time, a minor. In the process of landing, Smith damaged the airplane beyond repair, and to add insult to injury, he also damaged the landing field. Plaintiff sued in tort and Smith disaffirmed the contract. The trial court granted Smith's motion for summary judgment and plaintiff appealed.

SPAETH, J.

This is an appeal from the granting of defendant-appellee's motion for summary judgment. The issue is whether we should overturn the longstanding common law doctrine that a minor by disaffirming a contract can avoid liability under the contract.

When a minor disaffirms a contract, unless the contract is for necessaries the other party cannot recover the value of any item that the minor has obtained pursuant to the contract. The only remedy the other party has is an action in replevin to recover the item itself. If the minor no longer has the item, the other party is remediless.

An action in trespass, which is the form of action selected by appellant, will not lie. As stated in *Penrose* v. *Curren*, "The foundation of the action is contract, and disguise it as you may, it is an attempt to convert a suit, originally in contract, into a constructive tort so as to charge the infant."

A businessman may protect himself from loss incident to a minor's disaffirmance of a contract by finding out whether the person with whom he is dealing is a minor. Ordinarily this will present no difficulty. If the person is a minor, or if it is not clear that he is an adult, the businessman may decline to deal with him, or may require that someone he knows is an adult join in the contract. Inasmuch as appellant neglected such precautions, it has only itself to blame for its inability to recover for the damage to its airplane and landing field.

Apart from these considerations, to overrule the cases that permit disaffirmance would involve the court in a legislative function. Some age must be established as the age below which disaffirmance will be permitted; perhaps a twenty-year-old person should not be protected, but surely an eight-year-old should be. If the age is to be changed, the legislature is better equipped than the court to decide whether the change should be at age 19, 18, 16, or some other age. (Indeed, the legislature has recently made this determination, selecting age 18.)

(Affirmed.)

QUESTIONS FOR DISCUSSION

1. *Does the plaintiff have any remedy at all in this case? How could it have protected itself for the loss it sustained in dealing with the infant?*
2. *Why did Central Bucks try to sue the infant on the theory of tort in this case? Why does the tort action fail?*

(3) Time and Effect of Ratification. Although with the exception of real estate a minor may disaffirm his contract at any time before reaching majority, he cannot ratify a contract prior to majority. If the rule were otherwise, the policy of protecting minors could easily be circumvented and rendered worthless. Although a few states require the ratification to be in writing, in most jurisdictions the ratification may be oral or written, expressed or implied.

One can ratify a contract by his act. For instance, if a minor is making periodic payments on an installment contract and makes one after his majority, this may be construed by the courts as being an implied ratification. The infant could not successfully contend that he was ignorant of the right to disaffirm prior to his act. Furthermore, if the minor does not disaffirm the contract within a reasonable time of reaching his majority, he will be deemed to have ratified the contract if the contract has been executed. There is a division of opinion as to the effect of silence when the contract is executory, but most courts hold that if the infant received no benefit, mere silence does not constitute ratification.

JONES v. DRESSEL, et. al.
582 P.2d 1057 (Colo. App. 1978)

SILVERSTEIN, C.J.

Plaintiff sued to recover damages for personal injuries suffered in an airplane crash. He alleged that defendants were guilty of negligence and of malicious, oppressive, wilful and wanton misconduct. Defendants moved for a summary judgment of dismissal. The trial court granted the motion as to the negligence claim only, determined there was no just reason for delay, and directed entry of a judgment dismissing the negligence claim pursuant to C.R.C.P. 54(b). Plaintiff appeals from that judgment. We affirm.

The facts pertinent to this appeal are not in dispute. Plaintiff, then seventeen years old, entered into a contract with defendant, Free Flight Sport Aviation, Inc., under which Free Flight provided facilities for free-flight, or parachute jumping. Ten months after plaintiff's eighteenth birthday, a plane furnished by Free Flight crashed while carrying plaintiff and others preparatory to a parachute jump, causing the injuries to the plaintiff. The other defendants were owners or operators of the plane, acting as agents or employees of Free Flight Sport Aviation, Inc.

In granting the summary judgment the court relied on provisions of the contract, under which plaintiff exempted and released Free Flight, "its owners, officers, agents, servants, employees, and lessors from any and all liability [resulting from negligence and] . . . arising out of any damage, loss or injury to [plaintiff] . . . while upon the . . . aircraft of [Free Flight] or while participating in any of the activities contemplated by this agreement. . . ." The contract also contained a covenant not to sue for such injuries.

Plaintiff contends the trial court erred because the contract is void. . . .

It is well settled in Colorado that a contract entered into by a minor is not void but only voidable by the minor. *Fellows* v. *Cantrell*, 143 Colo. 126, 352 P.2d 289 (1960). Affirmance is not merely a matter of intent; it may be determined by the actions of a minor who accepts the benefits of a contract after reaching the age of majority, or by his silence or acquiescence in the contract for a considerable length of time. *Fellows* v. *Cantrell, supra.* Furthermore, § 13–22–101(1), C.R.S. 1973, states that "every person, otherwise competent, shall be deemed to be of full age at the age of eighteen years or older for the following specific purposes: (a) To enter into any legal contractual obligation and to be legally bound thereby to the full extent as any other adult person; . . ."

Thus, on reaching the age of eighteen plaintiff was required either to disaffirm the contract within a reasonable time, or be bound thereby. And, the undisputed facts establish that, after turning eighteen, plaintiff not only did not disaffirm the contract, but instead ratified it by accepting the benefits thereof. Hence, his being a minor when he entered the contract is without significance as to its present enforceability against him.

QUESTION FOR DISCUSSION

1. *Would the decision have been different had Jones not taken the flight after age eighteen, but merely attempted to get his money back?*

(4) Misrepresentation of Age. The situation frequently arises in which an adult has entered into a contract with a minor and the minor has misrepresented his age. Under this circumstance in most states the minor still has the protection afforded by law. The contract is voidable at the minor's option. A few states have statutes prohibiting avoidance of a contract when there has been a misrepresentation of age, and several jurisdictions have arrived at the same result by court decisions that apply the doctrine of estoppel.

Although the minor may avoid his contract despite the misrepresentation of age, in the majority of states the infant may still not get away with this misconduct because the misrepresentation usually results in the tort of fraud. Minors are liable for their torts. Thus, in many states where there is reasonable reliance by the adult on the misrepresentation of the minor, the injured party can sue the minor for damages on the theory of fraud. The adult can also raise the fraud as a defense if he is sued for breach of contract by a minor.

RICE v. BOYER

108 Ind. 472, 9 N.E. 420 (1886)

Rice sued Boyer to recover on a promissory note given for the sale of a buggy and harness. Rice claimed Boyer falsely and fraudulently represented that he was twenty-one years of age, and sued in tort for the value of the property delivered to Boyer. Boyer demurred to the complaint. The demurrer was sustained and Rice appealed.

ELLIOTT, C.J.

It is evident, from this brief reference to the authorities, that it is not easy to extract a principle that will supply satisfactory reasons for the solution of the difficulty here presented. It is to be expected that we should find, as we do, stubborn conflict in the authorities as to the question here directly presented, namely, whether an action will lie against an infant for falsely representing himself to be of full age.

Our judgement, however, is that, where the infant does fraudulently and falsely represent that he is of full age, he is liable where the consequence would be an indirect enforcement of his contract; for the recovery is not upon the contract, as that is treated as of no effect, nor is he made to pay the contract price of the article purchased by him as he is only held to answer for the actual loss caused by his fraud. In holding him responsible for the consequences of his wrong, an equitable conclusion is reached, and one which strictly harmonizes with the general doctrine that an infant is liable for his torts. Nor does our conclusion invalidate the doctrine that an infant has no power to deny his disability; for it concedes this, but affirms that he must answer for his positive fraud.

Our conclusion that an infant is liable in tort for the actual loss resulting from a false and fraudulent representation of his age is well sustained by authority, although, as we have said, there is a fierce conflict, and it is strongly entrenched in principle. It has been sanctioned by this court, although, perhaps, not in a strictly authoritative way; for it was said by Worden, J., speaking for the court, in *Carpenter* v. *Carpenter,* supra, that "the false representation by the plaintiff, as alleged, does not make the contract valid, nor does it estop the plaintiff to set up his infancy, although it may furnish ground of an action against him for tort."

It is worthy of observation that, in the cases which held that an infant's representation will not estop him to deny his disability, it is generally declared that he may, nevertheless, be held liable for his tort. It may often happen that the age and appearance of the infant will be such as to preclude a recovery for a fraud, because reasonable diligence, which is exacted in all cases, would warn the plaintiff of the nonage of the defendant. On the other hand, the infant may be in years almost of full age, and in appearance entirely so, and thus deceive the most diligent by his representations. Suppose a minor who is really twenty years and ten months old, but in appearance a man of full age, should obtain goods by falsely and fraudulently representing that he is twenty-one years of age, ought he not, on the plainest principles of natural justice to be held liable, not on his contract, but for the loss occasioned by his fraud? The rule which we adopt will enable courts to protect, in some measure, the honest and diligent, but none other, who are misled by a false and fraudulent representation; and it will not open the way to imposition upon infants, for in no event can anything more than the actual loss sustained be recovered, and no person who trusts where fair dealing and due diligence require him not to trust can reap any benefit. It will not apply to an executory contract which an infant refuses to perform, for, in such a case, the action would be on the promise, and the only recovery that could be had would be for the breach of contract, and the terms of our rule forbid such a result; but it will apply where an infant, on the faith of his false and fraudulent representation, obtains property from another, and then repudiates his contract.

(Judgment reversed, with instructions to overrule the demurrer to the complaint.)

MECHANICS FINANCE CO. v. PAOLINO

29 N.J. Super. 449, 102 A.2d 784 (1954)

Mechanics Finance Co. was the assignee of claims for merchandise sold to Florence Brenda on which she owed $960.62. When it threatened suit, Brenda suggested that she be allowed to make periodic payments. Mechanics Finance agreed on the condition that she get someone to guarantee the debt. Paolino agreed to guarantee the debt and signed a note along with Brenda. At the time Paolino was a minor but told the credit manager of Mechanics Finance that he was twenty-one years old. Brenda made only two $8 payments. Mechanics Finance sued Paolino. Lower court verdict for plaintiff, and defendant appealed.

GOLDMAN, J.

Defendant was twenty years, nine months and two days old at the time he executed the note. He was twenty-one years, three months and four days old when this action was instituted. The record supports the finding below that defendant deliberately misrepresented his age and that plaintiff relied on the misstatement. The trial judge was in a position to observe for himself whether defendant, only a few months older than when he signed the note, had the adult appearance to which McCaig testified. We give full weight to the court's findings that defendant did have that appearance.

Defendant contends that at common law and by the preponderance of American authority an infant is not estopped from setting up his infancy as a defense to an action on a contract, even though he induced the other party to enter into the contract by falsely representing himself to be of age. It is generally true that an infant may avoid his contract. However, in *LaRosa* v. *Nichols* . . . , the court of Errors and Appeals held there is an exception to that general rule; in a proper case an infant may be estopped from asserting infancy as a defense. Chancellor Walker, speaking for the court, said, "As applied to the facts in the case at bar, the law, as I view it, is, that if a youth under twenty-one years of age, by falsely representing himself to be an adult, which he appears to be, for the purpose of inducing another to enter into a contract with him, and, thereby, through such representation and appearance the other party is led to believe that such infant is an adult, and makes a contract with him, the benefit of which he obtains and retains, then, in a suit on that contract, the minor will not be permitted to set up the privilege of infancy, because by his fraudulent conduct he has estopped himself from so pleading; and this in a court of law as well as in a court of equity."

Defendant is on sounder ground when he urges that the doctrine of estoppel to plead infancy does not apply in this case because he obtained no personal benefit for his note. The LaRosa doctrine applies only where the infant received and retained a benefit under the contract he fraudulently induced.

The evidence is clear that defendant received no tangible personal benefit. Plaintiff's promise not to sue Miss Brenda, and its forbearance in bringing suit against her, was not a benefit running to defendant.

Plaintiff would avoid the absence of any benefit received or retained by defendant by arguing that an infant who misrepresents his age may never set up the defense of infancy. Such is clearly not the law of this State.

(The Judgment is reversed.)

QUESTION FOR DISCUSSION

1. *Note that in the Indiana case of Rice v. Boyer the defendant was allowed to raise the defense of infancy but was liable in tort for fraud. In the Paolino case New Jersey does not subscribe to this theory but uses an estoppel approach. Would the result in Paolino have been different under the Indiana approach?*

9:2. INSANITY. A contract entered into by a person lacking the mental capacity to form a contractual intent is either void or voidable. Bear in mind that we are talking about a legal definition and concept rather than a medical one. An individual may be legally capable of entering into a contract yet require medical treatment. Lack of mental capacity or insanity is a question of fact. Does the person understand the nature and effects of his act?

If there has been an adjudication of lack of mental capacity or insanity, any contract entered into by the mental incompetent is void because the adjudication would be a matter of public record. Typically this situation arises when the court has appointed a guardian or conservator to look after the affairs of the incompetent. If the incompetent regains his capacity to contract, a subsequent proceeding will establish this fact and he may then make his own contracts.

When there has been no prior court determination of mental incompetence a contract entered into by an individual suffering this disability is voidable in the same manner as the contract of a minor. The test is whether the individual lacks mental capacity at the time when the contract is formed. If the individual is competent at that point, the fact that he may have lacked mental capacity earlier would make no difference. As with a minor, a person who temporarily lacks mental capacity may subsequently ratify or disaffirm a contract entered into while he was suffering from that mental condition.

An individual lacking mental capacity is liable for the reasonable value of his necessaries. When the contract is for nonnecessaries, a majority of the states require the nonadjudicated incompetent to return the consideration he has received under the contract when he disaffirms it. Most states hold that this is a prerequisite to disaffirmance when the other party entered into the contract in good faith and in ignorance of the lack of capacity of the incompetent. If the incompetent was previously adjudicated an incompetent, the prevailing view is that the incompetent may disaffirm as long as he returns what is left of any consideration he received.

MILLS v. KOPF

216 Cal.App.2d 780, 31 Cal.Rptr. 80 (1963)

Shirley Mills's husband died in an automobile accident, leaving two insurance policies totaling $18,000. There was a dispute between Shirley and her husband's parents, Albert and Mary Kopf, as to who was entitled to the proceeds. An agreement was signed between the parties under which the parents were to receive $4,000 and Shirley the balance. Despite the agreement, the parents proceeded to collect additional money from the Veteran's Administration under the husband's government insurance policy. Shirley Mills sued to enforce the agreement, and the parents defended on grounds of the mental incompetency of Mary Kopf. The trial court awarded $5,100.21 to plaintiff and ordered the balance of the insurance proceeds paid to her. The Kopfs appealed.

PIERCE, J.

The primary point made by the parents on appeal is that the mother was incompetent when she signed the agreement.

A contract by a person of unsound mind, but who is not entirely without understanding, made before his incapacity has been adjudicated, cannot be rescinded unless the party purporting to rescind restores or offers to restore any consideration received.

Appellant, Mary Kopf, cannot restore and has not offered to. She has not been adjudicated an incompetent person. To avoid this contract, therefore, she would have to sustain the burden of proving that she was "entirely without understanding of any kind." In *Hellman Commercial T. & S. Bank* v. *Alden*, it is said:

> A contract which is challenged on the ground of incompetency is ordinarily not void, but merely voidable. The author of Norton on Bills and Notes (4th Ed.) 295, says 'A man of weak mind, if not a lunatic or fool, can contract. An epileptic or imbecile mind has been held competent to convey property, . . . and no mere want of business capacity or even monomania will, in the absence of fraud, prevent a party from being bound upon a bill, note, or endorsement. The mental incapacity to avoid such a contract must amount to an inability to understand the nature of the contract and to appreciate its probable consequences.'

Appellants here rely heavily upon the testimony of a psychiatrist who attended Mary Kopf during the period following her son's death. This expert first saw the patient February 18, 1956, and treated her until after May 7, 1956. He diagnosed her condition as "an agitated depression to a psychotic degree." At his recommendation the patient was hospitalized and was given shock treatments, was later released, returned to her home. In May 1956 she appears to have resumed many, if not all, of her household duties.

The psychiatrist testified that the patient, because of her condition, was unable to make decisions and was therefore incompetent in his opinion to transact her ordinary business affairs.

Plaintiff, on the other hand, produced Charles Miller, the parents' attorney, employed by them after Bernard's death to present their claims for insurance benefits and to prosecute the action in the federal court. During the period in question he had had frequent interviews with both the parents. These visits would last from 10 minutes to 2½ hours. During the visits both clients seemed to understand and discuss their problems intelligently.

Miller testified that in his opinion Mrs. Kopf was at all times when she consulted with him and when he advised her, competent to understand and transact these business affairs. He stated the only noteworthy thing he observed about Mrs. Kopf's condition was the grief exhibited whenever the death of the son was mentioned and Miller was concerned and expressed some doubts as to her ability to undergo the ordeal of a jury trial.

The trial court was not required, under the

circumstances of this case, to accept the testimony of the psychiatrist without question over the conflicting evidence and particularly over the proof, very convincing here, given by Mrs. Kopf's own attorney.

(The Judgment is affirmed.)

Questions for Discussion

1. *Is there a difference in the way the law handles infants' contracts as opposed to the treatment offered contracts entered into by mental incompetents?*
2. *What proof would have been necessary in this case in order for Mrs. Kopf to avoid the contract? Is the court saying it does not believe the psychiatrist?*
3. *Does the outcome in this case mean that the court does not feel that Mrs. Kopf was mentally ill from a medical standpoint when she entered into the contract?*

9:3. Intoxication and Drugs. A contract that is entered into by an individual who is suffering the effects of a large amount of alcohol and that he would not have made in a more sober condition is enforceable. The fact that his judgment is impaired by the alochol has no affect upon the validity of the contract. On the other hand, if a person is so intoxicated that he does not understand the nature of his act, he will be treated much the same as an insane person. He may disaffirm or ratify the contract when he sobers up, and the agreement is voidable. He would be liable for the reasonable value of his necessaries just as an infant would be.

If a court has declared a person to be a habitual drunkard, he is in the same situation as an individual judicially declared to lack mental capacity for other reasons. A guardian would be appointed to manage his affairs, and any contract entered into by the drunkard would be void. The law would be the same if the person is suffering the effects of drugs when entering into a contract or if a guardian is appointed in connection with drug addiction.

9:4. Aliens. An alien is a citizen of another country residing in this country. Aliens who are in this country lawfully have the same contractual rights as citizens. They may be sued and use the courts to enforce their contractual rights. An alien may not be denied access to the courts to enforce his rights under a contract. An enemy alien will not be able to enforce a contract, and at best the contract would be held in abeyance until the war is over.

9:5. Summary of Uniform Commercial Code Changes. The Uniform Commercial Code has not altered the general contract law regarding the capacity to contract. The general contract law of the state determines whether a contract is for the sale of goods, and, hence, covered by the Uniform Commercial Code. State law determines whether that contract is subject to the defense of lack of capacity.

The Uniform Commercial Code does limit the right of an infant to recover his property in one instance. Suppose as part of the contract that an infant places goods with the other party to the contract. If the goods have been transferred from the person dealing with the infant to an innocent third party purchaser, the U.C.C. limits the infant's right to recover the goods. The code states that one with a voidable title has the power to transfer good title to a good-faith purchaser for value. Therefore, in this situation the infant could not recover the goods even if he or she has the right to disaffirm the contract.

SUMMARY

Infants are not liable on their contracts for either their necessaries or their nonnecessaries. The contract is voidable at the option of the infant. Minors are liable in quasi contract for the reasonable value of their necessaries. Necessaries are those things that are required to maintain one's station in life and to provide for a person's continued life, health, and liberty.

Infants may disaffirm their contracts at any time prior to reaching the age of majority or within a reasonable time thereafter. Generally, when an infant disaffirms a contract he must return what is left of the money or property obtained under the contract. There are exceptions. An infant who disaffirms a contract may recover property transferred by him as part of the contract even if it has been transferred to an innocent third-party purchaser. This is not true of a contract for the sale of goods under the Uniform Commercial Code. If the subject matter of the contract is real estate, many states prohibit an infant from disaffirming the contract until reaching majority.

An infant may ratify a contract when he reaches the age of majority. In most states the ratification may be oral or written, expressed or implied.

In most states, an infant who misrepresents his age is not liable on the contract but may be liable in tort for fraud. Some states have statutes prohibiting avoidance of a contract when the infant misrepresents his age. Several jurisdictions apply the doctrine of estoppel.

If a person lacks mental capacity to form a contractual intent, the contract is either void or voidable. A prior adjudication of lack of capacity renders the contract void. An individual lacking mental capacity is liable for the reasonable value of necessaries.

A person so intoxicated that he does not understand the nature of his act will be treated much the same as an insane person. A person who is judicially declared to be a habitual drunkard is in the same position as one who has been judicially declared to lack mental capacity for other reasons.

PROBLEMS

1. Orval Robertson sued his son, James Robertson, for breach of an alleged contract to pay college expenses. Orval contended that he agreed to loan money to his son for his premedical education and dental school. In order to meet these expenses, Orval mortgaged his home and took out loans against his life insurance. When the son graduated from dental school, Orval requested that James repay the $30,000.00 expense in monthly installments of $400.00. After several conversations, the amount was reduced to $24,000.00. James stated in a letter that he was willing to make payments of $100.00 per month for 240 months. James paid $100.00 per month for

the next three years and then informed his father than he intended to make no more payments. Orval sued for the balance and received a verdict for $19,345. James raised the defense of infancy and appealed. Is Orval entitled to the money? (Robertson v. Robertson, 229 So.2d 642 (Fla. 1969).

2. Audrey H. Schmaltz was involved in an automobile accident in which she sustained minor injuries and had a number of antiques damaged. She subsequently signed a "release of claims" and accepted $2,000.00 as full payment from the insurance company. When she later incurred medical bills in connection with the accident, the insurance company refused to pay them because she signed the release. Schmaltz then sued the insurance company and others claiming she lacked mental capacity to sign the contract of release. Schmaltz contended she lacked mental capacity because of the "nervous tension and I was just upset over everything." She further stated that the tension was caused by not having the matter resolved. Is Schmaltz entitled to the defense of lack of mental capacity? Schmaltz v. Walder, et al., 566 S.W.2d 81 (Tex. App. 1978).
3. Kiefer was an emancipated minor when he paid the contract price of $412 for a car purchased from Fred Howe Motors. Kiefer then had difficulty with the car and asked the dealer to take it back. After the dealer refused, Kiefer's attorney wrote Howe, tendered return of the car, informed him Kiefer was a minor, declared the contract void, and demanded return of the $412. Howe refused, contending that an emancipated minor cannot disaffirm his contract. Kiefer sued. Decide. Kiefer v. Fred Howe Motors, Inc., 39 Wis.2d 20, 158 N.W.2d 288 (1968).
4. Nicolson opened a charge account at Goerke's department store. Although a minor, Nicolson stated that she was of legal age. Nicolson ran up a bill on the account that went unpaid. Goerke's sued, and Nicolson defended on the ground of infancy but did not return any of the goods purchased from plaintiff. Decide. R. J. Georke Co. v. Nicolson, 5 N.J.Supp. 412, 69 A.2d 326 (1949).
5. Joan Wheat sued James Montgomery, alleging that a release she had signed was unenforceable because she was taking the drug Librium at the time. The release was for money given her by an insurance company as a result of the death of her child. Wheat also alleged that she was suffering from grief and shock as well as taking the tranquilizer and thus had not read the release prior to signing. Defendant moved for summary judgment. Decide. Wheat v. Montgomery, 130 Ga.App. 202, 202 S.E.2d 464 (1973).
6. Pelham purchased an auto under a conditional-sales contract for $2,075.60 and paid $500 down. The bill of sale that he signed stated that he was over twenty-one. A few days later he returned the car, repudiated the contract, and demanded his money back. Pelham stated that he worked at a flower shop but did not make deliveries for the shop with his car. On the sales contract Pellham indicated that the car was to be for pleasure. The trial court decided that the car was a necessary as a matter of law. Defendant appealed. Decide. Pelham v. Howard Motors, Inc., 20 Ill.App.2d 528, 156 N.E.2d 597 (1959).

7. Stanley Barr, an infant, purchased a motor scooter from Allen Hurwitz, also an infant, for $240. A week later Barr attempted to rescind the contract on the ground that the scooter was defective. He tendered back the scooter, alleging infancy, but Hurwitz refused to return the money. Barr sued Hurwitz. Verdict for whom? Hurwitz v. Barr, 193 A.2d 360 (D.C.App. 1963).
8. William Little sued his grandnephew, Thomas Turner, Jr., for $670 alleged to have been loaned Turner in $30 monthly payments while he was in school. Turner signed promissory notes for these amounts. All the notes were signed and the money loaned while Turner was a minor except for the last three payments. Turner raised the defense of infancy. Decide. Turner v. Little, 70 Ga.App. 567, 28 S.E.2d 871 (1944).

CHAPTER 10

Illegality

10:1. EFFECT OF ILLEGALITY. In order for an agreement to result in an enforceable contract it must have a lawful purpose or objective. Agreements to commit torts or crimes are contrary to public policy and are generally void. If an illegal agreement is executory, one cannot invoke the power of the courts to enforce it, nor can he recover damages for its breach. If the agreement has been executed, the courts will usually not grant relief and will leave the parties as it finds them. This is true even when only one party has performed and the other has benefited from this performance. The courts will not give any force to the agreement because it is for an illegal purpose or objective. There are some exceptions to the general rule. For example, if the parties are not in pari delicto—that is, not equally guilty—the courts may grant relief to the less guilty party if it is in the interest of public policy. For instance, if one party is induced to enter into the agreement as the result of fraud or undue influence, a court may grant relief even though the agreement was for an illegal purpose.

Another exception obtains when there is an agreement for an illegal purpose but that purpose is not consummated. Thus, if one has paid money to another party to commit an illegal act, the agreement may be repudiated by one party and the money recovered if the act has not been performed. If there has been part performance, the majority rule is that it is too late to repudiate the agreement and invoke the court's aid in recovering the money.

A third exception applies in the case in which a statute is enacted for the protection of one of the parties to a contract, making that particular type of contract illegal. For example, some statutes prohibit attorneys from taking more than a limited sum for services rendered in the obtaining of a pension. If the applicant agrees to pay the attorney more than that amount, he may recover the excess paid even after the services have been rendered.

10:2. CONTRACTS TO COMMIT CRIMES OR CIVIL WRONGS. Contracts to commit crimes and civil wrongs are void. Thus agreements to commit assault, battery, arson, and criminal fraud are among those that will not be enforced. Contracts that involve the commission of a common-law crime as well as statutory crimes are treated similarly. If the contract involves the commission of a civil wrong or tort, it is also void. Agreements to slander another individual, to wrongfully appropriate a trademark or trade name or to knowingly induce another individual to breach a contract would be unenforceable. It should be obvious to anyone that an agreement to commit an act such as arson would be unenforceable. Most of the cases involve breaches of the more technical criminal statutes or civil wrongs.

(1) **Wager.** Most states have statutes making gambling illegal. Wager agreements and other forms of gambling are void and unenforceable. Even without such statutes the courts have generally refused to enforce wagering agreements as against public policy. As a general rule most states prohibit private lotteries in which, for example, one pays a sum of money or other consideration in return for the chance to recover a prize. There are many forms of lotteries, such as raffles and certain types of sales promotion schemes, but if the scheme does not require the payment of consideration and is a "giveway" program, it does not classify as a lottery and is not prohibited in most states. Many sales-promotion programs are of this type. Currently a number of states have instituted state-run lotteries, which are, of course, legal because they are passed by the state legislature.

(2) **Insurance.** In the broadest sense, insurance is a type of gamble. In the case of life insurance, for instance, if you die soon after purchasing the policy, you win; if you live for a long period of time, you lose! The analogy is not really accurate, however. An insurance contract is not the same as a gambling agreement because it involves transferring the risk of the occurrence of an event from one party to another. The agreement does not create a wholly new risk. Additionally, in order for the insurance agreement to be enforceable, a party must have an "insurable interest" in the property or the person insured. An insurable interest is an interest in a life on property so that one should suffer a pecuniary loss if there is death or destruction of the property.

(3) **Futures.** Another area in which the element of risk is present is the futures contract. Here one party agrees to sell a product or security in the future at a certain price. The seller usually does not have title to the product at the time the contract is made, so obviously the risk to the seller is that the market price may rise above the contract price prior to the delivery date. These contracts are legal provided that a sale is intended when the contract is formed. The fact that the seller does not possess or have title to the product at the time the agreement is made or that the parties do not believe delivery will ever actually be made does not render the contract invalid. On the other hand, if the intent of the agreement initially is to pay the difference between the contract and the market price on a specific date, then it is a gambling contract and void. If the parties later agree to pay the differential instead of making actual delivery, the agreement is enforceable.

10:3. Contracts to Interfere with Justice or Influence Governmental Action. An agreement to use corrupt influence to interfere with the performance of duties by a member of the executive, legislative, or judicial branch of government is void. Obviously an agreement to bribe an official so that he will make a favorable decision or engage in a specific course of conduct would be unenforceable. In addition, agreements to bribe a legislator to vote on certain legislation or to promise an official participation in the proceeds gained from his awarding a public contract would be unenforceable.

There are conflicting decisions regarding *lobbying* contracts. When one person hires another to influence legislative or other governmental action by threats or other improper means, the agreement is clearly contrary to public policy and unenforceable. However, courts will enforce agreements in which an agent or an attorney is hired to present the facts of his case while refraining from the use of threats or other illegal acts. Usually there are regulations for the registration of such lobbyists with the appropriate legislative body. In addition, some statutes prohibit the payment of a lobbyist on a contingent-fee basis, where his pay is dependent on the outcome of his efforts.

10:4. Public Policy. There are other contracts that are void because they violate public policy. Here we are entering an area where the illegality of the contract may not be as obvious, as, for instance, a contract to commit arson. Public policy is determined from acts of the legislature, court decisions, and more generally the attitudes of society. Thus a court may refuse to enforce an agreement as being violative of public policy even when it does not run afoul of any specific statute or the common law.

MARYLAND NATIONAL CAPITAL PARK AND PLANNING COMMISSION v. WASHINGTON NATIONAL ARENA

386 A.2d 1216 (Md. 1978)

The Maryland-National Capital Park and Planning Commission is a governmental agency that leased fifty acres to Potomac Sports, Ltd., which was Washington National Arena Limited Partnership's (the Arena) predecessor in interest. The purpose of the lease was for Arena to erect a major public athletic and recreational complex to be known as the "Capital Center."

The lease provided that Arena would relinquish its statutory right to challenge a determination by the local supervisor of assessments that the leased land was subject to real-property taxes to be assessed against Arena. When the land and improvements were assessed for tax purposes at a value of $13,630,000.00, Arena began to appeal that decision. The commission then filed this suit to enforce the lease provision prohibiting that action. The Arena contended that the noncontestability provision of the lease was void as against public policy. The trial judge agreed and the commission appealed.

LEVINE, J.

From the dawn of the common law tradition in England, courts have refused to implement those private contractual undertakings which, when measured against the prevailing mores and moods of society, contravene judicial perceptions of so-called "public policy." This Court stated early on that considerations of public policy are deemed paramount to private rights and where conflict between the two exists, private interests must yield to the public good. *Wildey* v. *Collier*, 7 Md. 273, 278–79, 61 Am.Dec. 346 (1854).

Nearly 150 years ago Lord Truro set forth what has become the classical formulation of the public policy doctrine—that to which we adhere in Maryland:

> Public policy is that principle of the law which holds that no subject can lawfully do that which has a tendency to be injurious

> to the public, or against the public good, which may be termed, as it sometimes has been, the policy of the law, or public policy in relation to the administration of the law. *Egerton* v. *Earl Brownlow,* 4 H.L.Cas. 1, 196 (1853).

But beyond this relatively indeterminate description of the doctrine, jurists to this day have been unable to fashion a truly workable definition of public policy. Not being restricted to the conventional sources of positive law (constitutions, statutes and judicial decisions), judges are frequently called upon to discern the dictates of sound social policy and human welfare based on nothing more than their own personal experience and intellectual capacity. Inevitably, conceptions of public policy tend to ebb and flow with the tides of public opinion, making it difficult for courts to apply the principle with any degree of certainty.

> [P]ublic policy . . . is but a shifting and variable notion appealed to only when no other argument is available, and which, if relied upon today, may be utterly repudiated tomorrow. *Kenneweg* v. *Allegany County,* 102 Md. 119, 125, 62 A. 249, 251 (1905).

Fearing the disruptive effect that invocation of the highly elusive public policy principle would likely exert on the stability of commercial and contractual relations, Maryland courts have been hesitant to strike down voluntary bargains on public policy grounds, doing so only in those cases where the challenged agreement is patently offensive to the public good, that is, where "the common sense of the entire community would . . . pronounce it" invalid. *Estate of Woods, Weeks & Co.,* 52 Md. 520, 536 (1879); . . . . This reluctance on the part of the judiciary to nullify contractual arrangements on public policy grounds also serves to protect the public interest in having individuals exercise broad powers to structure their own affairs by making legally enforceable promises, a concept which lies at the heart of the freedom of contract principle.

In the final analysis, it is the function of a court to balance the public and private interests in securing enforcement of the disputed promise against those policies which would be advanced were the contractual term held invalid. Enforcement will be denied only where the factors that argue against implementing the particular provision clearly and unequivocally outweigh "the law's traditional interest in protecting the expectations of the parties, its abhorrence of any unjust enrichment, and any public interest in the enforcement" of the contested term. Restatement (Second) of Contracts § 320, Comment b (Tent. Draft No. 12, 1977).

Armed with these fundamentals, we consider now the Arena's public policy contentions in the present appeal. . . .

At common law, agreements providing for the settlement of future legal disputes by means other than through conventional judicial proceedings were considered bargains obstructing the administration of justice and were consequently deemed unenforceable at law or in equity. This doctrine was later invoked to nullify provisions in contracts purporting to cut off, in advance of any actual controversy, the right of a party to prosecute an appeal from the decision of a trial court. Appellee (the Arena) would have us adopt and apply such a rule in the instant case. We decline, however, to do so.

Whatever may have been the basis for the doctrine at common law, the rule simply does not comport with contemporary thinking about the use of extra judicial modes of dispute resolution. The dramatic increase in the volume of litigation in the past 25 years and the resultant congestion of court dockets have generated a demand for alternative procedures to expedite the settlement of legal controversies. Accordingly, courts in recent years have grown more tolerant of the innovative contractual devices designed to terminate disputes quickly and equitably without the need for protracted formal litigation.

Of particular interest is *United States* v. *Moorman,* 338 U.S. 457, 70 S.Ct. 288, 94 L.Ed. 256 (1950), which involved a public policy challenge to a contract clause similar in effect to the noncontestability provision at issue in the present appeal. The contractor in *Moorman* sought to avoid compliance with a term in a standard form government agreement making the decision of the Secretary of War final and binding as to all disputes arising under the contract. Rejecting the contractor's contention that such a provision, if implemented, would wrongfully strip him of the right to have his controversy resolved by a court of competent jurisdiction, the Supreme Court, speaking through Mr. Justice Black, declared:

> It is true that the intention of parties to submit their contractual disputes to final determination outside the courts should be made manifest by plain language. . . . (citation omitted). But this does not mean that hostility to such provisions can justify blindness to a plain intent of parties to adopt this method for settlement of their disputes. . . . If parties competent to decide for themselves are to be deprived of the privilege of making such anticipatory provisions for settlement of disputes, this deprivation should come from the legislative branch of government. *Id.* at 462, 70 S.Ct. at 291.

The thrust of these and other modern decisions is that unless clearly prohibited by statute, contractual limitations on judicial remedies will be enforced, absent a positive showing of fraud, misrepresentation, overreaching, or other unconscionable conduct on the part of the party seeking enforcement.

(Reversed in favor of the commission.)

QUESTIONS FOR DISCUSSION

1. *In order to declare a contract void as against public policy is it necessary for the court to determine that the contract violates a statute or constitutional provision?*
2. *Had this case been decided under earlier common-law principals, would the outcome have been different? If so, why?*

MARVIN v. MARVIN
557 P.2d 106 (Cal. 1976)

TOBRINER, J.

Plaintiff avers that in October of 1964 she and defendant "entered into an oral agreement" that while "the parties lived together they would combine their efforts and earnings and would share equally any and all property accumulated as a result of their efforts whether individual or combined." Furthermore, they agreed to "hold themselves out to the general public as husband and wife" and that "plaintiff would further render her services as a companion, homemaker, housekeeper and cook to . . . defendant."

Shortly thereafter plaintiff agreed to "give up her lucrative career as an entertainer [and] singer" in order to "devote her full time to defendant . . . as a companion, homemaker, housekeeper and cook;" in return defendant agreed to "provide for all of plaintiff's financial support and needs for the rest of her life."

Plaintiff alleges that she lived with defendant from October of 1964 through May of 1970 and fulfilled her obligations under the agreement. During this period the parties as a result of their efforts and earnings acquired in defendant's name substantial real and personal property, including motion picture rights worth over $1 million. In May of 1970, however, defendant compelled plaintiff to leave his household. He continued to support plaintiff until November of 1971, but thereafter refused to provide further support.

On the basis of these allegations plaintiff asserts two causes of action. The first, for declaratory relief, asks the court to determine her contract and property rights; the second seeks to impose a constructive trust upon one half of the property acquired during the course of the relationship.

Defendant demurred unsuccessfully, and then answered the complaint. Following extensive discovery and pretrial proceedings, the case came to trial. Defendant renewed his attack on the complaint by a motion to dismiss. Since the parties had stipulated that defendant's marriage to Betty Marvin did not terminate until the filing of a final decree of divorce in January 1967, the trial court treated defendant's motion as one for judgment on the pleadings augmented by the stipulation.

After hearing argument the court granted defendant's motion and entered judgment for defendant. Plaintiff moved to set aside the judgment and asked leave to amend her complaint to allege that she and defendant reaffirmed their agreement after defendant's divorce was final. The trial court denied plaintiff's motion, and she appealed from the judgment.

Defendant first and principally relies on the contention that the alleged contract is so closely related to the supposed "immoral" character of the relationship between plaintiff and himself that the enforcement of the contract would violate public policy. He points to cases asserting that a contract between nonmarital partners is unenforceable if it is "involved in" an illicit relationship, or made in "contemplation" of such a relationship. A review of the numerous California decisions concerning contracts between nonmarital partners, however, reveals that the courts have not employed such broad and uncertain standards to strike down contracts. The decisions instead disclose a narrower and more precise standard: a contract between nonmarital partners is unenforceable only *to the extent* that it *explicitly* rests upon the immoral and illicit consideration of meretricious sexual services.

In the first case to address this issue, *Trutalli* v. *Meraviglia,* 215 Cal. 698, 12 P.2d 430, the parties lived together without marriage for 11 years and had raised two children. The man sued to quiet title to land he had purchased in his own name during this relationship; the woman defended by asserting an agreement to pool earnings and hold all property jointly. Rejecting the assertion of the illegality of the agreement, the court stated that "The fact that the parties to this action at the time they agreed to invest their earnings in property to be held jointly between them were living together in an unlawful relation did not disqualify them from entering into a lawful agreement with each other, so long as such immoral relation was not made a *consideration* of their agreement." (Emphasis added.) (215 Cal. at pp. 701–702, 12 P.2d 430, 431.)

Although the past decisions hover over the issue in the somewhat wispy form of the figures of a Chagall painting, we can abstract from those decisions a clear and simple rule. The fact that a man and woman live together without marriage, and engage in a sexual relationship, does not in itself invalidate agreements between them relating to their earnings, property, or expenses. Neither is such an agreement invalid merely because the parties may have contemplated the creation or continuation of a nonmarital relationship when they entered into it. Agreements between nonmarital partners fail only to the extent that they rest upon a consideration of meretricious sexual services. Thus the rule asserted by defendant, that a contract fails if it is "involved in" or made "in contemplation" of a nonmarital relationship, cannot be reconciled with the decisions.

The three cases cited by defendant which have *declined* to enforce contracts between nonmarital partners involved consideration that *was* expressly founded upon illicit sexual services. In *Hill* v. *Estate of Westbrook, supra,* 95 Cal.App.2d 599, 213 P.2d 727, the woman promised to keep house for the man, to live with him as man and wife, and to bear his children; the man promised to provide for her in his will, but died without doing so. Reversing a judgment for the woman based on the reasonable value of her services, the Court of Appeal stated that "the action is predicated upon a claim which seeks, among other things, the reasonable value of living with decedent in meretricious relationship and bearing him two children. . . . The law does not award compensation for living with a man as a concubine and bearing him children. . . . As the judgment is, at least in part, for the value of the claimed services for which recovery cannot be had, it must be reversed." (95 Cal.App.2d at p. 603, 213 P.2d at p. 730.) Upon retrial, the trial court found that it could not sever the contract and place an independent value upon the legitimate services performed by claimant. We therefore affirmed a judgment for the estate. (*Hill* v. *Estate of Westbrook* (1952) 39 Cal.2d 458, 247 P.2d 19.)

In the only other cited decision refusing to enforce a contract, *Updeck* v. *Samuel* (1964), 123 Cal.App.2d 264, 266 P.2d 822, the contract "was based on the consideration that the parties live together as husband and wife." (123 Cal.App.2d at p. 267, 266 P.2d at p. 824.) Viewing the contract as calling for adultery, the court held it illegal.

The decisions in the *Hill* and *Updeck* cases thus demonstrate that a contract between nonmarital partners, even if expressly made in contemplation of a common living arrangement, is invalid only if sexual acts form an inseparable part of the consideration for the agreement. In sum, a court will not enforce a contract for the pooling of property and earnings if it is explicitly and inseparably based upon services as a paramour. The Court of Appeal opinion in *Hill,* however, indicates that even if sexual services are part of the contractual consideration, any *severable* portion of the contract supported by independent consideration will still be enforced.

In summary, we base our opinion on the principle that adults who voluntarily live together and engage in sexual relations are nonetheless as competent as any other persons to contract respecting their earnings and property rights. Of course, they cannot lawfully contract to pay for the performance of sexual services, for such a contract is, in essence, an agreement for prostitution and unlawful for that reason. But they may agree to pool their earnings and to hold all property acquired during the relationship in accord with the law governing community property; conversely they may agree that each partner's earnings and the property acquired from those earnings remains the separate property of the earning partner. So long as the agreement does not rest upon illicit meretricious consideration, the parties may order their economic affairs as they choose, and no policy precludes the courts from enforcing such agreements.

In the present instance, plaintiff alleges that the parties agreed to pool their earnings, that they contracted to share equally in all property acquired, and that defendant agreed to support plaintiff. The terms of the contract as alleged do not rest upon any unlawful consideration. We therefore conclude that the complaint furnishes a suitable basis upon which the trial court can render declaratory relief. The trial court consequently erred in granting defendant's motion for judgment on the pleadings.

QUESTIONS FOR DISCUSSION

1. *What grounds did the defendant argue the court should use in order to refuse to enforce the contract because of public policy?*
2. *Why did the court reject the defendant's contention?*

10:5. CONTRACTS LIMITING LIABILITY FOR NEGLIGENCE. Often the question arises as to whether a contract limiting liability for negligence is valid. Generally such contracts are enforceable. Many states have an exception to this general rule for businesses that deal with the public and provide that they may not limit their liability for negligence by contract. Some state laws provide that one may enforce a contractual provision limiting liability for negligence as long as the agreement is not so one-sided as to be unconscionable.

10:6. CONTRACTS VIOLATING LICENSING STATUTES. Often a situation occurs in which one party to a contract has failed to comply with the terms of a licensing statute. For instance, a pawn broker, accountant, physician, or contractor may not have procured a license required by statute in order to do business. The question arises as to the status of a contract entered into between one of those individuals and another party. The answer depends upon the intent of the legislature when the statute is passed. If the legislative intent is primarily to raise revenue for the government, then the contract is enforceable despite the absence of the license. On the other hand, if the intent of the legislature is to regulate the activity in order to protect the health, welfare, and safety of the public, then the contract is void.

The intent of the legislature may be stated in the statute itself, but not always. It may be determined through an examination of the legislative history of the statute or by examination of the activity the statute is devised to control. Also an examination of the size of the license fee might be helpful. If a court decides that the licensing of electricians is regulatory and a particular electrician has not complied with the licensing statute, he may not recover for his services, but if the purpose of the license is merely to raise revenue, he may recover the fee. The answer to this inquiry might also involve a determination of whether there are any tests of competency administered in connection with the granting of the license or whether it is given to anyone who pays the license fee.

J. D. COLSTON et al. v. GULF STATES PAPER CORPORATION
282 So.2d 251 (Ala. 1973)

J. David and Willie Colston engaged in the business of cutting, hauling and selling pulpwood. They sold the pulpwood primarily to Gulf States Paper Corporation. The Colstons contracted to cut the pulpwood from land owned by Gulf States and paid a stumpage price for the wood cut. Most of the wood cut was in turn sold back to Gulf States for use in their mills. The contract also provided that the Colstons were to pay the Alabama Forest Products Severence Tax. The law required a tax to be paid on timber severed from the land by the "owner . . . of timber" or a "manufacturer of products derived therefrom."

Gulf States deducted the tax from the payments made to Colston for the timber sold them. The Colstons contended that the statute requires the owner of the timber, Gulf States, to pay the tax, and therefore the contractual provision requiring Colston to pay was void and unenforceable. The trial court found for Gulf States, and the Colstons appealed.

MERRILL, J.

It is apparent that the contractual arrangement consists of two separate contracts. The first is for the sale of the standing timber, which passes title, fixes a stumpage price per cord and provides that the "Buyer" (appellants) shall be liable for the severance tax on the timber severed.

The second contract governs the price which is paid the "Buyer" for the pulpwood when it is delivered in proper cuts to the mill, regardless of where the wood comes from or what price was paid for stumpage.

Appellants . . . argue that the contractual arrangement is void and unenforceable as against public policy.

The true test to determine whether a contract is unenforceable because of public policy is "whether the public interest is injuriously affected in such substantial manner that private rights and interests should yield to those of the public." Here, the public interest is not affected. The public interest demands that the severance tax be paid. Both appellants and appellee are liable to the State for the tax. They contracted that appellants would pay it and that appellee would collect it and pay it to the State.

Assuming, without conceding, that appellee would be liable for the tax under the statute, parties are still at liberty to contract where no principle of public policy is involved. This court, in *Ivey* v. *Dixon Investment Co.*, 283 Ala. 590, 219 So.2d 639, stated that "the general rule is that no contract or agreement can modify a law, the exception being where no principle of public policy is violated, parties are at liberty to forego the protection of the law." The exception is applicable here because there is no violation of public policy.

Finally, a cogent reason for upholding the decision of the trial court is that a contract which may not be made in compliance with statutory provisions is not void if the statute is merely a revenue measure and is not regulatory and for the protection of the public. The principle is stated in *Bowdoin* v. *Alabama Chemical Co.*, 201 Ala. 582,79 So. 4, as follows:

> The rule in this state is that, if a statute was not designed to prohibit the making of contracts without previous compliance with statutory provisions, but was intended merely to provide revenue, it is not void if no specific prohibition or penalty is provided or imposed. If the conditions of the statute were made for the benefit of the public, and not for the raising of revenue only, an agreement is void that does not comply with the statutory conditions.

The Forest Products Severance Tax Act was a revenue act and made no attempt to regulate the business of cutting or growing timber, nor to specify who might or might not engage in such business, or how it should be conducted, or what kind of contracts they should make. It provided a tax on any severance of trees and provided who should pay the tax, who would collect it and how often the collectors would report.

(Affirmed.)

Questions for Discussion

1. *If the court determined that the money raised from the taxes collected in this case was used to reforest the land, would it have affected the court's decision?*
2. *Suppose the tax was never paid by Gulf States after it was collected from the Colstons, would this have changed the outcome of the case?*

10:7. Usury. Every state has statutes setting the maximum amount of interest that may be charged for certain types of loans. These are called *usury statutes.* They apply only to certain types of loans, and in our credit-oriented society many exceptions to the usury laws have developed. In order to prove a violation of a usury statute, the following generally must be shown: (1) there must be a loan or forbearance of money (2) that is repayable absolutely, (3) in return for which a greater amount is charged than the interest rate allowed by statute (4) with an intent to violate the law. The last element may be implied from the commission of the first three. The usury statutes set the maximum *contract rate* of interest that may be charged for a loan and should not be confused with the so-called legal rate of interest, which is the rate charged on judgments or contracts when there may be no fixed rate of interest stated.

The penalties for violating the usury laws vary from state to state and may be grouped in three categories: (1) those statutes in which the full amount of interest is forfeited by the lender; (2) those statutes in which both the interest and the principal are forfeited; and (3) those statutes in which only that amount of interest that exceeds the usury limit is forfeited. The preferred approach and the one adopted by the largest number of states is the first. It could be argued that the third approach imposes no penalty at all upon one who violates the statute, as he forfeits only that to which he was not entitled anyway. The second approach may be too severe, as it is possible for one to innocently violate the statutes because of their complexity, particularly with regard to determining what charges are part of the interest calculation.

Because of the recent escalation of interest rates a number of states have raised significantly the interest limit in their usury statutes or, in some cases, have eliminated the limit entirely. Those usury statutes that retain an interest limit contain a number of exceptions. A typical exception would be where a higher limit is set for lenders loaning money in relatively small amounts, so-called small loan companies. The reason for this is that many of the debtors are in the high-risk category. If the amount of interest charged were restricted to a lower rate, people needing these loans would not be able to obtain them for the reason that no money would be available at the lower rate because of the risk factor. Many statutes also exempt corporate borrowers from their protection.

The rate of interest is obviously going to be affected by the manner in which charges made in connection with the loan are treated. For example, the effective rate of interest will be increased if the amount of interest is *discounted,* that is, paid before the proceeds of the loan are given to the borrower. Thus, if there is a 10-percent annual interest rate charged on a $1,000 loan, $100 interest would be deducted from the face amount of the loan and $900 given to the borrower. Today most usury statutes permit this practice and do not consider its effect upon the determination of the interest rate when a violation of the usury statute is being determined. Other charges also typically excluded from the calculation of the interest rate are recording fees, notary fees, credit-investigation fees, and fees for negotiating the loan.

These are fees that involve actual and necessary expenses to the lender. If charges are made that do not involve actual expense to the lender, they may simply be considered part of the interest charged and could result in a violation of the usury statute. Courts will closely scrutinize other transactions that are designed to circumvent the usury laws, such as loans that are made to look like sales of property, where an excessive amount is charged for the property and is actually interest.

Increasingly our society is dependent on the purchase of goods by consumers on installment sales contracts and by credit card. These loans are not subject to the typical usury limit but are regulated by separate statutory provisions. There is also a federal statute regulating this area known as the Federal Consumer Credit Protection Act of 1969, or the Truth in Lending Act. At this point you should note that the act is basically a *disclosure* statute so that comparison of various interest rates can be made by the consumer; it does not set a maximum interest rate. The act also authorizes the Federal Reserve Board to enforce the act and promulgate regulations. The board has passed Regulation Z, which among other things requires the interest rate to be conspicuously displayed and stated in specific terms. For example, if the interest rate is 1½ percent per month, the lender must also show it at an annual rate of 18 percent. All charges, credits, and the balance to be paid must be clearly stated. States that have adopted the Uniform Consumer Credit Code and adequate methods for enforcement are not also subject to the Federal Truth in Lending Act. Both of these statutes are more fully discussed in Chapter 47, "Consumer Protection."

KESSING et al. v. NATIONAL MORTGAGE CORP.

180 S.E.2d 823 (N.C. 1971)

Jonas Kessing, his wife, and Kessing Co. signed a note to National Mortgage Corp. in the amount of $250,000, payable in monthly installments of $500 beginning May 1, 1970, with interest at 8 percent per annum. The note was secured by a first deed of trust on the property purchased with the money. There were garden apartments on this land and as an additional requirement on the loan, National Mortgage Corp. required the establishment of a partnership between itself and Kessing, with National Mortgage Corp. as a limited partner. The partnership agreement provided that National Mortgage Corp.'s liability should be limited to its capital contribution of $25 and that it would have a 25 percent interest in the partnership and receive 25 percent of the profits. Kessing Co. was required to convey the apartments to the partnership.

This action was instituted by Kessing to recover alleged usurious interest paid to defendant and to have the deeds of conveyance to the partnership, as well as the partnership agreement, canceled and declared null and void. The lower court found for plaintiffs, and the defendant appealed.

MOORE, J.

In an action for usury plaintiff must show (1) that there was a loan, (2) that there was an understanding that the money lent would be returned, (3) that for the loan a greater rate of interest than allowed by law was paid, and (4) that there was corrupt intent to take more than the legal rate for the use of the money. The corrupt intent required to constitute usury is simply the intentional charging of more for money lent than the law allows. Where the lender intentionally charges the borrower a greater rate of interest than the law allows and his purpose is clearly revealed on the face of the instrument, a corrupt intent to violate the usury law on the part of the lender is shown. And where there is no dispute as to the facts, the court may declare a transaction usurious as a matter of law. *Doster* v. *English,* supra.

Under G.S. § 24–8 prior to the 1969 amendment, the legal interest allowed on the loan in question was 8%. The president of defendant corporation testified that the loan of $250,000 was secured by a deed of trust which had full warranty, and the loan was repayable to defendant under any circumstances. Defendant's president further testified that defendant would not have made this loan at the simple rate of 8% but that the added equity participation provided for by the creation of the Partnership and the conveyances to it were considerations for the making of the loan; that from the 25% of the profits to be realized by the Partnership the defendant had an expected or "hoped for" yield of between 16% and 20%—certainly over 8%. Our courts do not hesitate to look beneath the forms of the transactions alleged to be usurious in order to determine whether or not such transactions are in truth and reality usurious. Under G.S. § 24–8 before the 2 July 1969 amendments, this agreement would have been usurious, for as is said in *Ripple* v. *Mortgage and Acceptance Corp.,* 193 N.C. 422, 137 S.E. 156:

> . . . Where a transaction is in reality a loan of money, whatever may be its form, and the lender charges for the use of his money a sum in excess of interest at the legal rate, by whatever name the charge may be called, the transaction will be held to be usurious. The law considers the substance and not the mere form or outward appearance of the transaction in order to determine what it in reality is. If this were not so, the usury laws of the State would easily be evaded by lenders of money who would exact from borrowers with impunity compensation for money loaned in excess of interest at the legal rate.

G.S. § 24–8 as amended specifically prohibited the very type equity participation created by the Partnership formed in connection with this loan by providing "No lender shall . . . require . . . any borrower, directly or indirectly, to . . . transfer or convey . . . for the benefit of the lender . . . any sum of money, thing of value or other consideration other than that which is pledged as security . . ." A 25% interest in the Partnership (which owned the realty conveyed to it by Kessing Company) was a "thing of value." This made the partnership agreement unlawful. Under the statute, the loan was usurious.

The statutory penalty for charging usury is the forfeiture of all interest on the loan. The charging of usurious interest as provided for by the partnership agreement in this case is sufficient to cause a forfeiture of all the interest charged. The charging of such usurious interest strips the debt of all interest. It becomes simply a loan which in law bears no interest. Any payments of interest which have been made at a legal rate are by law applied to the only legal indebtedness—the principal sum. In the instant case Kessing Company has paid $25,000. Since all interest has been forfeited, the payments made should be credited on the principal amount of the loan.

(Judgment affirmed.)

Questions for Discussion

1. *Why do you suppose the court found the previous transaction to be a loan rather than a loan and a separate partnership agreement?*
2. *What facts can you cite to support the conclusion that the National Mortgage Corp. had a "corrupt intent to take more money than the legal rate for the use of the money"?*

10:8. Sunday Agreements. A number of state legislatures have enacted Sunday laws or "blue laws" prohibiting the making of certain types of contracts or sales on Sundays. Where these laws are in force, they vary widely, but generally they allow for the sale of certain types of items on Sundays, such as things that are related to the protection of life, health, or property. Generally "works of necessity" or "works of charity" are permitted. Furthermore, even if a contract is negotiated on a Sunday that would ordinarily be prohibited under a state's blue law, it will be enforceable if it is not consummated until a weekday. A contract made on a Sunday and ratified on a weekday is enforceable. These laws are rather liberally interpreted by the courts, which do not seem to be enthusiastic about them in many instances. This often leads to some confusion as to just what types of contracts and sales may be entered into on Sunday. Currently the trend seems to be in the direction of limiting or abolishing these laws in many states.

10:9. Contracts in Restraint of Trade. It is often desirable for the parties in certain business agreements to try to limit future competition. For example, in connection with the sale of a business the purchaser may want to make sure that the seller does not reopen a competitive business in his immediate trading area within a certain time after the purchase is consummated. This would diminish the goodwill that is presumably part of the purchase price. In the case of an employment contract the employer may wish to include a condition of employment in the agreement prohibiting the employee from taking subsequent employment with a competitor for a limited period of time in order to protect trade secrets, customer lists, and other similar assets.

On the other hand, the common law and more currently the antitrust laws have declared contracts in restraint of trade and those that limit competition to be generally against public policy. Under the old common law an agreement that placed a restraint on a man's right to pursue his trade or profession was void as against public policy. Early American cases permitted certain restrictive covenants as long as they were limited in their effect with regard to time and space. More recently the courts have permitted the enforcement of certain restrictrive covenants as long as they meet certain requirements. One is that the agreement must be *ancillary* to a larger transaction. For instance, Barney Buyer purchases a drugstore from Danny Druggist. As part of the contract Druggist agrees not to open a similar store on the same block for one year after the purchase. The agreement not to open a competing business is ancillary to the sale of the business. On the other hand, if Buyer paid Druggist $1,000 not to open a competing business without that agreement's being part of a larger transaction, it would be unenforceable.

Another requirement is that the restrictive covenant be reasonable with regard to the *time* and the *area* over which it is to operate. Thus as a condition of employment an employer may prohibit a salesman from working for a competitor in his sales territory for one year after ceasing employment. The courts may enforce such a covenant, particularly if the primary asset of the business is the customer contacts developed by the salesman. On the other hand, a covenant not to work for a competitor anywhere in the United States at any time could not be enforced. It would be unreasonable with regard to the time and the area of its effect. It places an unreasonable restraint upon the person restricted and goes beyond what is necessary to protect the goodwill of the business.

In sum, in order to be enforceable, the restrictive covenant should be (1) reasonable with regard to the time and the geographic area covered; (2) necessary for the protection of the goodwill of the business; and (3) not unduly burdensome on the person restricted. The covenant should also be ancillary to a larger transaction.

You should recognize the competing interests that the courts are trying to balance. On the one hand is the right of an individual to ply his trade or profession and on the other is the interest one may have in protecting his investment after purchasing a business or after training an employee. Restrictive covenants are most frequently used in connection with sales of businesses, employment contracts, and partnership agreements. Their value should now be obvious for the protection of goodwill, trade secrets, customer lists, and other similar assets.

Other types of restraints of trade involve the antitrust laws. These are discussed more fully in Chapter 46 on antitrust and trade regulation. Contracts in restraint of trade are among those prohibited by these statutes. Examples of these contracts would be price-fixing agreements and market-division schemes, either by product or by territory.

BARRETT-WALLS, INC. v. T. V. VENTURE, INC.

251 S.E.2d 558 (Ga. 1979)

T.V. Tempo, Inc., was a corporation that published and distributed a free weekly publication containing advertising and television schedules. T.V. Tempo, Inc., granted a distributorship to Barrett-Walls, Inc. with the right to sell associate publisher franchises throughout the United States except for Virginia and the District of Columbia. The distributorship agreement contained a noncompetitive provision which provided:

> 10(a) Neither party hereto nor any principal of either nor any assignee, shall enter into or associate with in any manner, or have any financial or other interest, in any other publication providing television scheduling in conjunction with advertising during the term and time of this Agreement anywhere in the United States, which is the area covered by the subject matter of this Agreement.

Through a series of transfers of the ownership of the business, Barrett-Walls, Inc. and a newly named company known as T.V. Venture, Inc., became parties to the distributorship agreement and the above noncompetitive clause. T. V. Venture then began producing a free publication called T. V. Venture which contained advertising and television schedules.

Barrett-Walls, Inc. filed a petition for injunctive relief against T.V. Venture, Inc., alleging that it was bound by the noncompetitive covenant contained in the distributorship agreement. The trial court held that the restrictive covenant was unenforceable and Barrett-Walls appealed.

BOWLES, J.

The contractual restraints sought to be enforced in this suit are those tending to lessen competition and are disfavored. Code Ann. §§ 2–2701 and 20–504. Noncompetitive provisions in contracts will be enforced only if the restraints are reasonable in time, reasonable and definite in territorial extent and reasonable and definite in the nature of the business activities proscribed.

Paragraph 10(a) of the distributorship agreement provides for the non-competitive restraint to apply to either party and their assigns for the duration of the agreement, the initial term of which is fifteen years, automatically renewable for a maximum of 95 years. The trial court found this provision to be unreasonable as to the time of restraint citing *Shirk* v. *Loftis Bros. & Co.*, 148 Ga. 500, 97 S.E. 66 (1895); *Aladdin, Inc.* v. *Krasnoff*, 214 Ga. 519, 105 S.E.2d 730 (1958). We find this determination unnecessary to our decision in this case.

The restriction in this case would prohibit appellees from competing in any state anywhere in the United States in which a franchise had been granted under the distributorship agreement. This provision of the covenant is overly broad and uncertain, rendering the entire covenant void on public policy grounds.

We find the provision in question is also unreasonable as to the nature of the business activities proscribed in that the publication sought to be restricted is not unique, nor is it a national publication such that it would require broad territorial protection. Therefore, the restrictive covenant is unreasonable as to the nature of the business activities proscribed.

Under the law of Georgia, covenants in restraint of trade may be enforced if they are reasonable as to time and place and are not overly broad in the activities proscribed, taking into consideration the interests of the individuals and commercial concerns as well as the public policy in promoting competition. A rule of reason will be applied by the courts in reviewing such contracts. If any one provision of a covenant not to compete is found to be unenforceable, the entire covenant will be struck down. We find the covenant in restraint of trade in this case, when read in its entirety, to be overly broad and unenforceable in Georgia. The trial court was correct in refusing to grant appellant's motion for injunctive relief.

(Judgment affirmed.)

Questions for Discussion

1. *How could the covenant have been drawn in this case so that the court would have enforced it?*
2. *Since this clause is obviously designed to restrain trade between competitors, what is necessary for it to be enforceable in addition to the fact that its time and territorial restrictions must be reasonable?*

SUMMARY

A contract must have a lawful purpose or objective or a court will not enforce it unless a party was induced to enter into the agreement by fraud or if the illegal purpose is not consummated. Another exception applies when a statute is enacted that protects one party to an illegal contract.

Contracts to commit crimes and civil wrongs are void. Wage agreements and other forms of gambling are void and unenforceable unless permitted by state legislation. Insurance policies are enforceable so long as the policy holder has an insurable interest in the property or person insured. Futures contracts are enforceable so long as a sale is intended when the contract is formed.

Lobbying contracts are generally enforceable. Some jurisdictions prohibit the payment of a contingent fee to a lobbyist. Contracts that violate public policy are void. A contract may violate public policy even if it does not violate any statute or the common law.

A contract limiting liability for negligence is generally enforceable. Generally courts will not enforce contracts limiting liability for negligence for businesses dealing with the public.

If a contract violates a licensing or similar statute whose primary purpose is to raise revenue, the courts will enforce the contract. Contracts that violate statutes whose primary purpose is to protect the health, safety, and welfare of the public will not be enforced.

Usury statutes limit the amount of interest that may be charged for a loss. The majority of statutes require the lender to forfeit the interest if the statute is violated. Usury statutes set the maximum contract rate of interest that may be charged. The legal rate of interest is the rate charged on judgments or contracts where no fixed rate of interest is stated. Typically a higher rate of interest may be charged by small loan companies than by commercial banks.

In general, contracts of restraint of trade are illegal. However, in some circumstances, some contracts restraining trade are enforceable if they are ancillary to a larger transaction and reasonable with regard to the time and area over which they operate. They usually must be necesary to protect the goodwill of the business and not unduly burdensome to the person restricted.

PROBLEMS

1. Food Machinery and Chemical Corporation's predecessor in interest entered into a contract which among many other things provided that it would pay a royalty of $25.00 for each cornhusking machine sold by it. The royalty was to be paid to Lothe Sells for her life and upon her death to her daughter Dorothy S. Landsheft, provided that she had not married. When Lothe Sells died Dorothy had married. The company refused to pay the royalty to Dorothy because of the clause prohibiting her marriage. Dorothy sued the company for the royalties. Decide. Shackleton v. Food Machiner Company, Inc. 166 F. Supp. 636 (C.D. Ill. 1958).
2. Jo Fay Kennedy went to a beauty school for a permanent and signed an agreement waiving her right to sue for any injury she received. The student operator's negligence caused her to have second degree chemical burns of the neck and shoulders and extending from the chin to the clavicle. The beauty school owner raised the agreement in defense. Decide. Smith v. Kennedy, 195 So.2d 820 (1966).
3. Avers was employed as a salesman by National Hearing Aid Centers, Inc. He signed an employment contract that provided that he would not sell hearing aids or accessories for two years after termination of the contract in Maine, New Hampshire, Vermont, and six counties in Massachusetts. National did business in this area. Avers left National and began selling hearing aids in this territory but had no contact with former customers. National sued to enjoin Avers from selling hearing aids in the territory in question. Decide. National Hearing Aid Centers, Inc. v. Avers, 311 N.E.2d 573 (Mass. 1974).
4. Rex Wulfhorst signed an employment contract with Hudgins which stated that upon its termination Wulfhorst would not compete with Hudgins in the general demolition business for two years. The noncompetition areas were the entire State of Georgia; Jacksonville, Florida; Chattanooga, Tennessee; Birmingham, Alabama; Wilmington and Charlotte, North Carolina; and Memphis, Tennessee. Hudgins conducted its business operations in part of the State of Georgia and some of the cities mentioned above. Wulfhorst quit his job and began competing with Hudgins. Hudgins sued to enjoin Wulfhorst from competing with him. May Hudgins enforce the noncompetition clause? Wulfhorst v. Hudgins and Company, Inc., 200 S.E.2d 743 (Ga. 1973).

5. Vitek, Inc., entered into a contract to build an ice-skating rink for Alvarado Ice Palace, Inc. A state statute required that a contractor should be licensed at all times during performance of the contract in order to have the right to sue for compensation for work performed under the contract. Vitek sued for $89,062.33 alleged to be due for work performed under the contract. Alvarado defended on the ground that Vitek did not have a license when the contract was made, although the license was obtained before construction began. Decide. Vitek, Inc. v. Alvarado Ice Palace, Inc. et al., 34 Cal.App.3d 586, 110 Cal. Rptr. 86 (1974).
6. Miller and Radikopf had an agreement to sell Irish Sweepstake tickets pursuant to which they received two free tickets for every twenty sold. They agreed to divide their winnings. One of the tickets won almost a half million dollars, which Radikopf kept. Miller sued for his share under the agreement, contending that there was nothing illegal about receiving the winnings of a lottery. Radikopf contended that lotteries were against the public policy of the state. Decide. Miller v. Radikopf, 214 N.W.2d 897 (Mich. 1974).
7. Paul Laos was hired by Joseph Sable as an expert witness in a condemnation proceeding. Sable agreed to pay Laos $1,500 if the recovery was $200,000 or less and $2,500 for anything over $200,000. Sable later refused to pay the fee and Laos sued. Was the agreement against public policy? Laos v. Sable, 503 P.2d 978 (Ariz. 1972).
8. Luther Wood, a pawnbroker, sued to foreclose a chattel mortgage given as security on a note executed by J. E. Krepps. Krepps defended on the ground that Wood did not have the license required of pawnbrokers by law. There were no particular requirements for license, but there were criminal and civil penalties for doing business without it. Was Krepps entitled to a judgment? Wood v. Krepps, 168 Cal. 382, 143 P. 691 (1914).
9. Beavers and his partner Dyett borrowed $5,000 from Taylor. The note provided for "no interest." Simultaneously the parties signed a separate contract whereby Beavers and Dyett, in consideration of the loan, agreed to pay Taylor 1 percent of the first $10,000 gross sales per month, 0.75 percent of the next $15,000, and 0.5 percent of everything over $25,000. The payments under the agreement amounted to 20 percent of the note annually which was higher than the usury limit. Beavers sued Taylor to recover damages for the alleged usurious interest collected by Taylor. Did this arrangement constitute usury? Beavers v. Taylor, 434 S.W.2d 230 (Tex. 1968).

CHAPTER 11

Writing Requirements

We have previously seen that both oral and written contracts are valid. As a practical matter it is wise to have any important agreement in writing. This reduces the problems that frequently occur with faulty memories concerning specific contractual provisions. Reducing an agreement to writing also requires the parties to think about the details of their transactions, thus avoiding later misunderstanding of the obligations of each party. In addition to the very practical advantages there are also statutes that require certain types of contracts to be in writing. Also a well-known rule of evidence, the parol evidence rule, places restrictions on the ability to contradict unambiguous terms in written contracts. These statutes and rules of evidence are the subject matter of this chapter.

11:1. Introduction and History of the Statute of Frauds. The procedure used for proving a contract in the seventeenth century was considerably different from that in use today. It was not uncommon to hire witnesses or to use friends in order to establish an agreement; in fact, one could not testify himself. It was possible to bind an individual to a contract by the use of perjured testimony or at least to put him to considerable burden to prove that there was no contract. In 1677 the English Parliament enacted a statute entitled, "An Act for Prevention of Frauds and Perjuries." This is commonly known today as the Statute of Frauds. It required that certain types of contracts be in writing or that there be a written memorandum concerning the contract, signed by the party being sued. The types of contracts in which the statute required a writing involved those situations in which there was often temptation to commit fraud, usually because they involved matters that were considered very important.

Today the statute has been adopted, in one form or another, in almost every state in this country. You should keep several factors in mind when studying this chapter. In the first place, the purpose of the Statute of Frauds, as should be obvious, is to prevent fraud. Second, over the years the courts have been loath to allow one to raise the Statute of Frauds as a successful defense when there obviously was a contract. As a result the statute has been narrowly construed, with a resulting large number of "exceptions." No doubt the statute has been successful in stopping a number of frauds, but it has also resulted in a number of contracts' being unenforceable when it was quite clear that there was a valid oral contract and no question of fraud. It is noteworthy that the British Parliament repealed all of the statute in 1954—after 277 years—except those sections dealing with promises to answer for the debt, default, or miscarriage of another and contracts for the sale of land. Among the reasons given was that the statute was a product of conditions that have long since passed away. As you read this chapter, see if you have an opinion as to whether the statute ought to be more restricted in its application in this country, and observe the rationale for the exceptions that have been developed by the courts and the legislatures.

When an agreement is subject to the Statute of Frauds and needs to have some writing in order to be enforceable, it is said to "fall within the statute." If the statute does not apply, the contract is said to be "outside or without the statute." You should also remember that the statute does not technically render oral agreements that fall within its ambit void but only unenforceable. The agreement may be a valid contract. Likewise remember in studying the chapter that once it is determined that the Statute of Frauds is satisfied, the plaintiff still has the burden of proving the existence of a valid contract, just as in any other case.

The parts of the old Statute of Frauds that have been adopted in this country are Section 4, dealing with a number of different situations, and Section 17, which deals with contracts for the sale of goods. They are similar in their effect except that Section 17 allows for satisfaction of its requirements by payment and acceptance. Section 17 has been largely incorporated in Section 2–201 of the Uniform Commercial Code, which you should study in connection with the sale-of-goods section of the Statute of Frauds.

Section 4 of the original Statute of Frauds provides as follows:

> No action shall be brought.
> (1) whereby to charge any executor or administrator upon any special promise to answer for damages out of his own estate;
> (2) or whereby to charge the defendant upon any special promise to answer for the debt, default, or miscarriage of another person;
> (3) or to charge any person upon any agreement made in consideration of marriage;
> (4) or upon any contract or sale of land, tenements, or hereditaments, or any interest in or concerning them;
> (5) or upon any agreement that is not to be performed within the space of one year from the making thereof;
>
> Unless the agreement upon which such action shall be brought, or some memorandum or note thereof, shall be in writing and signed by the party to be charged therewith, or some other person thereunto by him lawfully authorized.

Section 17 of the original statute is as follows:

Be it further enacted that no contract for the sale of any goods, wares, or merchandise for the price of ten pounds sterling, or upwards, shall be allowed to be good, except (a) the buyer shall accept part of the goods so sold, and actually receive the same; (b) or give something in earnest to bind the bargain, or in part payment; (c) or that some note or memorandum in writing of the said bargain be made and signed by the parties to be charged by such contract, or their agents thereunto lawfully authorized.

11:2. SECTION 4. (1) Promise by Executor or Administrator. When an individual dies, he leaves an estate, which consists of the assets he owns at death. If there is a will, it may name an executor of the estate, or in the absence of a will the court will appoint an administrator. In some jurisdictions this person is called a *personal representative.* Basically the function of these people is to collect the assets of the deceased, pay off creditors, and disburse the balance of the estate, if any, to the heirs and beneficiaries. Obviously the personal representative is not personally liable for the debts of the decedent merely because he fills this position. Should the estate prove to be insolvent, the personal representative normally is not liable for the deficiency.

In order for a creditor or other person to bind a personal representative to a contract requiring him to pay the decedent's debts out of his own assets, the contract must be written or there must be a written memorandum signed by the personal representative. On the other hand, if a personal representative makes a contract on behalf of the estate, there is no requirement of any writing under the statute because he is binding the estate and not himself. If there is a primary obligation incurred by the personal representative—that is, one on his own behalf—it does not fall within the statute. This is the case even when part of the consideration by the personal representative involves personally paying the debts of the estate.

(2) Promise to Answer for the Debt of Another. When one promises to pay the debt incurred by another, the contract must comply with the writing requirements of the statute in order to be enforceable. The courts have limited the applicability of this provision to those situations in which the promise is "collateral" to the primary obligation of the debtor. If the promise in question is "primary" rather than collateral, it does not fall within the statute.

The distinction between a collateral and a primary promise is sometimes difficult to ascertain. Typically the collateral promise in question is in the nature of a guaranty arrangement: "If Jones does not pay, I, Smith, will." The primary obligation is that of Jones, and if he does not pay, then the creditor may look to Smith. Because Smith's obligation arises only if Jones does not pay and thus is collateral to Jones's promise, it falls within the statute. Smith's is the collateral obligation.

Obviously, if an agreement is to fall within this section, there must be a principal debtor and a guarantor. Thus, if Jones tells a storekeeper, "Let Smith have the goods, and I will pay," the statement is without the statute. Although

Smith gets the goods, Jones's promise is primary and not collateral. Smith is not liable for the goods, Jones is liable.

When two parties become jointly liable on an obligation, the statute does not apply. Their obligations are both primary and the creditor may seek recovery from either one of them. Suppose Jones and Smith both orally agree jointly to charge a stereo receiver in a local department store for $150. There is no guaranty and the oral agreement is enforceable.

Another circumstance in which the courts hold that a promise is primary rather than collateral occurs when the main purpose of the person promising to pay another's obligation is for his own benefit. The courts apply what is sometimes called the *main-purpose doctrine* to this situation. If the main purpose of the promisor is to guarantee a debt in order to further or accomplish an objective he would not otherwise obtain, the oral promise falls without the statute and is enforceable. For instance, suppose Larry Launderer uses certain equipment in his business. The equipment belongs to Murry Machine, but Machine owes Carol Creditor for the equipment. Creditor has threatened to attach the equipment, which is essential to Launderer's business. Launderer promises Creditor that if she refrains from attaching the equipment, should Machine not pay, he, Launderer, will. This promise would not fall within the statute because the primary purpose of the promise is to benefit Launderer and only incidentally Machine. The courts view those promises as a result of which the promisor benefits from a pecuniary or business interest in the transaction as in the nature of a primary promise rather than a guaranty.

Indemnity contracts do not fall within this provision of the Statute of Frauds. An indemnity contract is one in which the promisor promises to protect the promisee from loss that may be incurred as the result of a transaction, an event, or an act. In other words, if the promise is made to the debtor, it does not constitute a contract of guaranty. Thus an oral promise of Inez Indemniter to pay any losses sustained by Barney Buyer, should Buyer agree to purchase certain goods from Sammy Seller, would fall without the statute and would be enforceable. Inez's promise is a promise of indemnity made to Buyer and is not a guaranty because it is not a promise made to a third person to pay the debt should the primary debtor default. The promise is made only to Buyer, the potential debtor.

Another situation that falls without the statute occurs when one party is substituted for another who has an obligation under the contract. Suppose that Danny Debtor owes Charles Contractor $800 for the construction of a patio that is yet to be completed. Harry Homebuyer purchases Debtor's home and makes an agreement with Contractor whereby Contractor agrees to discharge Debtor from the contractual obligation and substitute Homebuyer. This situation is encompassed under the term *novation;* that is, one party to a contract is substituted for another. The oral promise would be enforceable. The promise between Homebuyer and Contractor is considered primary and original.

GRILLO v. CANNISTRARO
155 A.2d 919 (Conn. 1959)

Plaintiff, Lucian Grillo, was a bail bondsman. Rosario was in jail charged with bigamy. The defendant, Joseph Cannistraro, was his father and orally promised Grillo that he would stand behind any bail bond and guarantee it. As a result of this promise Grillo agreed to act as surety on a recognizance entered into by Rosario on condition that he would appear in court to answer the charge. Grillo agreed to charge $150 rather than his normal fee of $200 for the bond. Rosario did not appear and the court ordered Grillo to pay $2,500 in satisfaction of his liability. Grillo then sued defendant Cannistraro to enforce the agreement set forth above. Cannistraro raised the Statute of Frauds as a defense. The trial court found in favor of plaintiff, and defendant appealed.

KING, A.J.

The basic claim of the defendant in his brief is that the Statute of Frauds is an effective defense as matter of law. It is an effective defense if the defendant's undertaking admittedly, not in writing, was, within the words of the statute as they have been interpreted, a "special promise to answer for the debt, default or miscarriage of another." Rev. 1958, § 52–550.

The defendant's claim that the complaint failed to state a valid cause of action in view of the Statute of Frauds is not a sound one. The complaint leaves something to be desired, but the fair import of its allegations, taken as a whole, is that the defendant agreed that if the plaintiff would enter into the recognizance the defendant would guarantee Rosario's appearance so that the plaintiff would sustain no loss if Rosario defaulted in his obligation under the recognizance, and that the plaintiff, in reliance on the defendant's promise, became surety on the recognizance. Under the agreement, the plaintiff could lose nothing by his undertaking. Either Rosario would appear, in which case neither the plaintiff nor the defendant would lose anything, or he would not appear, as proved to be the fact, in which event the plaintiff would, as he did, make good to the state on the recognizance and then would have a right of action against the defendant for the amount paid. Thus, the complaint alleged a promise to indemnify the plaintiff if he entered into the recognizance, and also alleged his action in reliance on the promise.

Smith v. Delaney, discusses the English cases and our own. Its holding is in harmony with what is now generally considered to be the majority rule. "(T)he prevailing rule . . . is that a promise to indemnify the promisee for becoming surety for a third person, at the request of the promisor, is not within the statute." In *Smith v. Delaney,* supra, we discussed the Connecticut cases and said: "(W)here the inducement (to the promisor) is a benefit to . . . (him) which he did not before or would not otherwise enjoy, and the act (of becoming surety) is done upon his request and credit, such promise is an original undertaking (as distinguished from a collateral undertaking) and not within the statute."

As in that case, there is nothing here to indicate that the plaintiff placed any confidence in the principle on the recognizance, Rosario. The only reasonable conclusion is that the plaintiff entered into the recognizance because of, and in reliance upon, the agreement and financial ability of the defendant to indemnify him against loss by reason of the recognizance. The fact that the plaintiff told the defendant that, if he made the guarantee, the charge would be reduced from $200 to $150 gives added support to the finding of the plaintiff's reliance on the promise.

The defendant makes much of a claim that there has to be some personal pecuniary benefit running to the promisor in order for the case to be taken out of the Statute of Frauds. In the first place, the statute has nothing to do with consideration, as such. Here, the defendant obviously received a legal benefit which he did not enjoy before, and otherwise would not have enjoyed, in that the plaintiff entered into the recognizance, thereby freeing Rosario from jail.

(Judgment affirmed.)

QUESTIONS FOR DISCUSSION

1. *How would you state the promise of Cannistraro so that it would have come within the Statute of Frauds?*
2. *Why does the court find that the promise made by Cannistraro is not a guaranty? Does the court find that the promise constitutes an indemnity, or does it apply the main-purpose doctrine?*

(3) Agreement Made upon Consideration of Marriage. The most notable aspect of this section of the Statute of Frauds is that it does not apply to the type of contract that is most common: mutual promises to marry or engagement contracts. Mutual promises to marry are outside the statute. You might note that most states have prohibited suits for breach of promise to marry under so-called antiheart-balm acts. The type of agreement that falls within the statute occurs when the marriage is part of the consideration given in return for some other performance. For instance, antenuptial agreements fall within the statute. A typical agreement of this type might provide for the husband to give up certain rights in the wife's estate, should she die first, in return for marriage. In such an agreement the consideration given is the marriage and the contract would have to be in writing.

(4) Contracts for the Sale of an Interest in Land. The statute provides that any contract for the sale of land or an interest in land should be in writing or there should be a written memorandum signed by the party to be charged. This provision applies not only to contracts involving the actual sale of land but also to any interest in the land. Thus easements, mortgages, rights-of-way, and in most cases leases extending for more than a year must all comply with the writing requirements of the statute. Although these transactions do not involve the sale of absolute title, they do involve the sale of an interest in land. A license, which is merely permission to use land, is not the sale of an interest in the land and thus would not be subject to the statute under this provision.

Questions often arise as to the sale of things that are attached to the land or are under the land, such as trees and minerals. Is a contract for their sale a contract for the sale of land or is it a contract for the sale of goods? The Uniform Commercial Code, Section 2–107, provides that contracts for the sale of items such as timber, minerals, or structures, where they are to be severed by the seller, are considered contracts for the sale of goods and outside this provision of the statute. If the buyer is to sever the items, the contract is one affecting land and is within the Statute of Frauds. Furthermore the code provides that a contract for the sale, apart from the land, of growing crops or other things that may be severed without material harm to the property is a sale of goods, regardless of whether the buyer or the seller severs.

Obviously, if one contracts to sell timber rights to a stand of oak trees under terms that provide for severence by the buyer, it is a contract for the sale of land. The trees cannot be removed without a depreciation in the value of the land; hence the contract would have to be written. On the other hand, it might be possible that the removal of corn could be accomplished without material harm to the land. Such an oral agreement would be enforceable even though the corn is to be removed by the buyer.

FREMMING CONSTRUCTION CO. v. SECURITY SAVINGS & LOAN

566 P.2d 315 (Ariz. App. 1977)

Security Savings and Loan orally agreed with Fremming Construction Co. to provide loans to buyers of homes constructed by Fremming at a certain rate of interest. Fremming was delayed in completing the homes and when the new homeowners obtained loans from Security they were at a higher rate of interest than had been agreed upon by Fremming and Security.

The homeowners sued Fremming contending that the delay in construction caused them to pay a higher rate of interest for their mortgage funds. Fremming filed a third-party complaint against Security alleging that Security breached its agreement to supply funds at a fixed rate of interest.

Security moved for summary judgment on the ground that its oral agreement with Fremming violated the Statute of Frauds. The trial court found in favor of Security and Fremming appealed.

HOWARD, J.

(The Arizona Statute of Frauds required any agreement for the sale of real property or an interest therein to be in writing.)

A.R.S. § 33–702(A) defines a mortgage as "every transfer of an interest in real property . . . made only as a security for the performance of another act. . . ." In Arizona a mortgage is not a conveyance and neither legal nor equitable title passes to the mortgagee. *Cooley* v. *Veiling*, 19 Ariz.App. 208, 505 P.2d 1381 (1973). Appellant contends that since a mortgage is not a "sale" of real property or an interest therein but is merely a lien under Arizona law, A.R.S. § 44–101(6) does not apply and we should follow the case of *Martyn* v. *First Federal Savings & Loan Association of West Palm Beach*, 257 So.2d 576 (Fla.App.1971). We decline to do so. Instead, we choose to follow the better reasoning by Judge Cardozo in *Sleeth* v. *Sampson*, 237 N.Y. 69, 142 N.E. 355, 356 (1923):

> . . . A contract to give a mortgage is a contract for the sale of an interest in real property within the meaning of [the statute]. No doubt the word 'sale,' when applied to such a transaction, is inexact and inappropriate. Our present statute comes to us by descent from the English statute (29 Car. II, c. 3, § 4), which speaks of 'any contract or sale of lands, tenements or hereditaments or any interest in or concerning them.' The change of phraseology has not worked a change of meaning. One who promises to make another the owner of a lien or charge upon land promises to make him the owner of an interest in land, and this is equivalent in effect to a promise to sell him such an interest. The meaning is fixed by an unbroken series of decisions. The agreement here was within the Statute of Frauds.

QUESTION FOR DISCUSSION

1. *Why does the court conclude that the agreement is subject to the Statute of Frauds when under Arizona law a mortgage is not a conveyance and neither legal or equitable title passes to the mortgagee?*

(5) Agreements Not to Be Performed Within One Year. The statute provides that contracts that are not to be performed within the space of one year from the time of their making must comply with the writing requirements of the statute. You should note that the applicable time is the period from the date following the day the contract is made until performance is completed; it is not the length of time required for the performance. Thus an agreement that will only take two weeks to perform but that requires the performance to begin one year from the date of making would have to comply with the writing requirements of the statute.

The courts have generally interpreted this clause in such a way that only those contracts on which performance within a year from the date of making the contract is impossible fall within the statute. Any contract on which it is possible for the performance to be completed within a year, even though it is not probable, does not come within this provision of the Statute of Frauds. This approach can lead to some strange results under certain circumstances. Suppose Ancient Al, aged eighty years with terminal cancer, agrees to mow Smith's lawn every week for two years. The contract must be in writing to be enforceable because by its terms the lawn must be mowed for more than one year from the date of contract. Contrast this situation with the case of the vigorous twenty-one-year-old individual who agrees to mow Smith's lawn "for the rest of my life." Although it is probable that the performance might take place over many years, it is possible that it could be completed within the space of one year from the date of making because one's life span is indeterminable and can end at any time. A few states do not take this view, but the majority of them do. The fact that an oral contract is not actually performed within one year has no effect upon its enforceability as long as it possibly could be performed. Obviously a contract to be performed upon the happening of some contingent event is enforceable as long as it is possible for that event to happen and the performance rendered within the year.

Another question often arises in the situation in which performance may be completed by one party to a contract but not by the other within a year from its making. Most jurisdictions will enforce an oral contract if performance is possible on one side within the year, particularly if the performance to be completed involves the repayment of money. Suppose Larry Lender loans Danny Debtor a sum of money to be repaid in installments lasting more than a year. Most courts would enforce such an oral agreement because there was performance on one side. An additional reason for enforcement would be applicable if the contract allowed for the option of payment within a shorter period of time, because performance within a year would be possible.

GENERAL FEDERAL CONSTRUCTION, INC. v. JAMES A. FEDERLINE, INC.

393A.2d 188 (Md. 1978)

General Federal Construction, Inc. is a general contractor. James A. Federline, Inc. submitted an oral bid to General for a subcontract for the mechanical work in the construction of a hospital. General used Federline's bid in calculating its bid to construct the hospital. When General was awarded the bid, it awarded the subcontract to someone other than Federline. Federline sued General for damages and General contended the agreement violated the Statute of Frauds.

SMITH, J.

There was testimony adduced on behalf of Federline flatly stating that the work could be performed within one year. Although the contract documents considered by prospective general contractors and subcontractors contain no specific statement of the time to be consumed in the project, there is no question but that it was *estimated* that 600 days would be required for completion. In support of its argument here on the matter of a directed verdict General points to this fact plus certain warranties which the subcontract bid documents required and work as to certain items which those documents specified should be done subsequent to completion of the project.

Under the contract documents a subcontractor who did the work Federline says it was to do was required in certain instances to guarantee that all material and apparatus used would "be new, of first-class quality . . ." There were a number of instances in which certain work was specified to be performed upon the completion of the project such as demonstrating "proper operation of the plumbing controls," and seeing that "all construction dirt [was] removed" from roof drains, water closets, urinals and the like. The mechanical contractor was specifically mandated to "provide the necessary skills and labor to assure the proper operation and to provide all required current and preventative maintenance for all equipment and controls provided under [one section of the contract] for a period of one year after substantial completion of the contract."

In *Ellicott v. Peterson,* 4 Md. 476, 487–91 (1853), Chief Judge Le Grand carefully reviewed for our predecessors the law relative to the Statute of Frauds and contracts to be performed within one year. Judge McWilliams again examined the law for this Court in *Sun Cab Co.* v. *Carmody,* 257 Md. 345, 349–51, 263 A.2d 1 (1970). He found that since the time of *Ellicott* the law has been well established in this State that where there is a *possibility* that a contract may be performed within a year the remedy for breach of the contract is not barred by the Statute of Frauds, notwithstanding the fact that the parties might have intended that operation should extend through a much longer period. The old Maryland authority, W. Brantly, *Law of Contracts* § 59 at 139 (2d ed. 1922), states, "The statute does not apply if the contract can by any possibility be completed within a year, although the parties may have intended that its operation should extend through a much longer period and in fact it does so extend." (Footnote omitted.)

Campbell v. *Burnett,* 120 Md. 214, 87 A. 894 (1913), is correctly cited in 2 A. Corbin, *Contracts* § 444 at 535 (1950), for the proposition that it makes no difference how long the parties expect the performance to take or how reasonable and accurate those expectations are, if the agreed performance can possibly be completed within a year. The full Corbin statement is:

> It makes no difference how long the agreed performance may be delayed or over how long a period it may in fact be continued. It makes not difference how long the parties expect performance to take or how reasonable and accurate those expectations are, if the agreed performance can possibly be completed within a year. Facts like these do not bring a contract within this provision of section 4. A provision in the contract fixing a maximum period within which performance is to be completed, even though that period is much in excess of one year, does not make the statute applicable. A building contract is frequently such that it can be fully performed within one year, even though the fixed time limit is in excess of one year. If so, it is not within the one-year clause, however long the parties may expect to take or actually do take. *Id.* at 535–37 (footnotes omitted).

H. Wood, *Statute of Frauds* § 269 at 463 (1884), states relative to contracts not to be completed within a year, "In order to bring a contract within this clause of the statute it must be one *which from its very terms shows that the parties intended that it was not to be completed within the year,* and therefore part performance within the year will not take the case out of the statute." (Emphasis in original). 2 Corbin, *supra.* § 444 states:

> In its actual application, however, the courts have been perhaps even less friendly to this provision than to the other provisions of the statute. They have observed the exact words of this provision and have interpreted them literally and very narrowly. The words are 'agreement that is not to be performed.' They are not 'agreement that is not in fact performed' or 'agreement that may not be performed' or 'agreement that is not at all likely to be performed.' To fall within the words of the provision, therefore, the agreement must be one of which it can truly be said at the very moment that it is made. 'This agreement is not to be performed within one year'; in general, the cases indicate that there must not be the slightest possibility that it can be fully performed within one year. *Id.* at 534–35.

It will be noted that by the terms of the contract document the mechanical subcontractor was required "for a period of one year *after* substantial completion of the contract" to provide "all required preventative maintenance for all equipment and controls" in this area. (Emphasis added.) This, presumably, means certain positive actions such as oiling certain equipment. It will further be noted that when the condensor water system was placed in operation for the period of "one calendar year *after* Owner's acceptance of installation" this subcontractor was to "furnish a complete water treatment service for control of corrosion, carbonate scale and slime or algae. . . ." (Emphasis added.) Again, this was a positive action to be taken by the mechanical subcontractor. Given these contract provisions, we conclude that there is a clear demonstration by the terms of the contract that it could not be performed within the period of one year. Hence, the "one year clause" of the Statute of Frauds is applicable and General was entitled to a directed verdict in its favor.

(Judgment for General Federal Construction, Inc.)

Questions for Discussion

1. *According to the court, the parties estimated that 600 days would be required for completion of construction. Given this factor alone, would the court have held this contract to be within the provisions of the Statute of Frauds?*
2. *Which provision in the contract caused the court to conclude that the contract could not possibly be completed within one year from the date of its making?*
3. *Do you think that the court's attitude that contracts are outside the Statute of Frauds if performance within a year is possible, even though improbable, is consistent with the reason for passing the statute in the first place?*

11:3. Contracts for the Sale of Goods Under the Uniform Commercial Code. The old Statute of Frauds, Section 17, provided that in order to be enforceable, contracts for the sale of goods in excess of a certain amount of money should be in writing or there should be a written memorandum signed by the party to be charged. The courts have engrafted a number of exceptions on this provision over the years. The Uniform Commercial Code, in Section 2–201, has incorporated old Section 17 of the Statute of Frauds as well as many of the exceptions. It has changed the old statute in some respects in order to reflect the realities of doing business in the modern commercial world. At this point you should review that section in the appendix.

The code provides that any contract for the sale of goods at a price of $500 or more should be in writing or there should be in writing an indication of the terms of the agreement, signed by the person against whom enforcement is sought. This provision is similar to the old Statute of Frauds provision. The code makes some significant alterations and additions, however. Under the code the writing is effective even when a term is misstated or left out, although the contract is not enforceable beyond the quantity stated in the writing or the quantity actually delivered and accepted in the case of part performance. *Goods* are defined by the code as including all things movable at the time of identification to the contract. The unborn young of animals, growing crops, and specially manufactured goods are included. Money and investment securities are excluded.

In the situation in which both parties are merchants and there is no written contract but only a written memorandum, such as a letter, Section 2–201(2) provides that the person writing the letter may use it to satisfy the terms of the statute if (1) both parties are merchants; (2) the letter is in confirmation of the contract; (3) it is received within a reasonable time; (4) the party receiving has reason to know its contents; and (5) the party receiving it does not object within ten days after receiving it. This provision changes the old Statute of Frauds provision by allowing the person writing and signing the letter to use it as a plaintiff under the circumstances stated previously. The old statute required that the defendant sign the writing. Suppose Caron Clothier orders over the phone from Sammy Supplier 100 pairs of pants at $10 per pair. Clothier then sends a letter to Supplier confirming their telephone conversation and setting forth the terms of the agreement. If Supplier does not object to the letter within ten days, Clothier may use her own letter to satisfy the statute if Supplier breaches the contract.

The code provides that an oral contract may be enforced (1) if goods are to be specially manufactured under circumstances that reasonably indicate that they are for the buyer, (2) if the goods are not suitable for sale to others in the ordinary course of the seller's business, and (3) if the seller has either made a substantial beginning of their manufacture or made commitments for their procurement (4) prior to receiving any notice of repudiation by the buyer. Observe that it is highly unlikely that one would try to defraud an individual by the use of goods requiring a special order. Also remember that once the Statute of Frauds is satisfied, the plaintiff is still required to prove his contract just as in any other contract action.

Under the old Statute of Frauds it was common practice when defending a contract action to admit the existence of a contract and then raise the statute in defense. This approach is certainly not consistent with the purpose of the statute, which is to protect persons from loss by having another party create a contract through fraud and perjury. No protection is needed if the contract is admitted. Thus the code provides that the requirements of the statute are satisfied if the existence of the agreement is admitted in court or in the pleadings. However, the contract is not enforceable beyond the quantity of goods admitted.

Finally Section 2–201(3)(c) of the code provides for the enforcement of an oral agreement for goods for which payment has been made and accepted or that have been received and accepted. Obviously the act of paying for and accepting goods or accepting payment for goods delivered indicates, apart from any writing, that some sort of agreement exists. The enforceability of the oral agreement is limited, however, to the goods actually accepted or paid for. Thus, if the oral contract involves a greater quantity than was actually delivered or paid for, the full amount could not be enforced.

BAGBY LAND AND CATTLE COMPANY v. CALIFORNIA LIVESTOCK COMMISSION COMPANY

439 F.2d 315 (Tex. 1971)

Bagby Land bought Mexican cattle and brought them across the Texas border for sale. California Livestock purchased approximately 1,600 cattle from Bagby in November and early December of 1968. Bagby failed to invoice 80 of these cattle.

An additional agreement between Bagby and California Livestock involved the purchase by the latter of 2,000 head, all to be delivered by January 15, 1969. In the meantime, California Livestock had obliged itself to deliver 2,000 head to a third party,

Fat City Cattle Company. Bagby delivered 222 of the 2,000 head and failed to deliver any more. California Livestock demanded delivery of the remaining 1,778 head and refused to pay for the 80 cattle previously received until full delivery was made.

Bagby Land sued for the price of the 80 cattle, and California Livestock counterclaimed for the undelivered 1,778. Bagby recovered $10,741.56 in damages and $4,000.00 attorney fees on its motion for summary judgment. California Livestock's claim was dismissed on the ground that it would be unenforceable under the applicable Texas Statute of Frauds, the same as U.C.C., Section 2–201, even if proved. California Livestock appealed.

GEWIN, J.

California Livestock seeks to show on several grounds that the admittedly oral contract does not come within the Statute of Frauds. First, it argues that delivery of 222 of the 2000 head of cattle contracted for was "part performance" of the contract and is sufficient to take the entire contract out of the Statute of Frauds. We cannot agree. Under Section 2.201 (c) (3) of the Texas Business and Commerce Code an otherwise invalid contract "is enforceable . . . (3) with respect to goods . . . which have been received and accepted." Although we have found and cited to no Texas cases interpreting this recently adopted legislation, we have no hesitancy in holding that the effect of this statute is to validate an oral agreement only as to those goods actually received and accepted by the buyer. While the U.C.C. Comments on this section are not final authority, we treat them as authoritative support for this holding, particularly in the absence of other precedents.

"Partial performance" as a substitute for the required memorandum can validate the contract only for the goods which have been accepted or for which payment has been made and accepted.

Receipt and acceptance either of goods or of the price constitutes an unambiguous overt admission by both parties that a contract actually exists. If the court can make a just apportionment, therefore, the agreed price of any goods actually delivered can be recovered without a writing or, if the price has been paid, the seller can be forced to deliver an apportionable part of the goods. The overt actions of the parties make admissible evidence of the other terms of the contract necessary to a just apportionment. This is true even though the actions of the parties are not in themselves inconsistent with a different transaction such as a consignment for resale or a mere loan of money.

California Livestock's acceptance of 222 cattle delivered pursuant to the alleged contract does not, therefore, give rise to the right to have delivery of any other cattle for which it orally contracted. Section 2.201 (c) (3) does not sustain the contention made.

(The judgment of the district court is affirmed.)

QUESTION FOR DISCUSSION

1. *Why do you suppose the Statute of Frauds prevented California Livestock, but not Bagby Land, from collecting on its claim?*

11:4. WRITING OR MEMORANDUM AND SIGNATURE. As has previously been noted, the Statute of Frauds does not require that the contract be in writing in order to be enforceable. If the contract is not written, then a writing "signed by the party to be charged" may suffice. With the exception of the Uniform Commercial Code provision with respect to goods already discussed, it is necessary that the defendant sign the writing.

The writing may be a letter containing the terms of the agreement, or it may consist of a series of letters or telegrams exchanged by the parties. Receipts, purchase orders, or invoices are among the many types of writings that satisfy the requirement. In short, there is no particular form that the writing must take in order to comply with the statute.

The writing should specify the terms of the contract as well as the parties.

If there is more than one writing to be used to satisfy the statute and only one is signed, they should be attached, or one should refer to the other or indicate in some other way that the second sheet is a continuation of the first. Remember that the Uniform Commercial Code provisions concerning contracts for the sale of goods are somewhat less stringent. Under the code a writing is not insufficient because it omits certain terms, but the agreement can be enforced only to the quantity of the goods stated in the writing.

The writing must be signed by the party being sued. Although some states require that the signature appear at the bottom of the agreement, most states do not. The "signature" may consist of initials, a stamp, or even a printed or typewritten name, as long as the party intended the form used to indicate that the writing was his own.

HARRIS et al. v. HINE
232 Ga. 183, 205 S.E.2d. 847 (1974)

H. E. Harris, Jack Harris, and Ray Harris, doing business as H. E. Harris & Sons, raised cotton. They contracted with Hine to sell him all the cotton produced on their land in 1973. Hine, in turn, had obligated himself to supply the cotton to several cotton mills. A shortage of cotton occurred and the market price rose from thirty cents a pound in March to seventy cents a pound in October. Harris sent Hines a letter repudiating the contract. Hines filed a complaint seeking an injunction preventing the sale of cotton to anyone else and for specific performance. Harris moved to dismiss on the ground that the complaint was defective, as the writing attached was insufficient to comply with the Statute of Frauds provision of the Uniform Commercial Code. The trial court overruled Harris's motion and granted an interlocutory injunction. Harris appealed.

GRICE, C.J.

The document in controversy is dated March 9, 1973, and recited as follows: "This agreement is entered into this date wherein Hine Cotton Company, 103 East Third Street, Roma, Georgia agrees to buy from H. E. Harris and Sons, Route 1, Taylorsville, Georgia all the cotton produced on their 825 acres. The rate of payment shall be as follows: 1) All cotton ginned prior to December 20, 1973 and meeting official U.S.D.A. Class will be paid for at 30¢ per pound. Below Grade Cotton at 24½¢ per pound. 2) All cotton ginned on or after December 20, 1973 will be paid for at the rate of 1000 over the CCC Loan Rate with Below Grades being paid for at 24½¢ per pound. Settlement will be made on net weights on Commercial Bonded Warehouse Receipts with U.S.D.A. Class cards attached with $1.00 per bale being deducted from the proceeds of each bale."

The writing was signed by E. W. Hine for Hine Cotton Company and H. E. Harris for H. E. Harris & Sons.

The Statute of Frauds provision of the Uniform Commercial Code as adopted in Georgia, which is pertinent here, states that "a contract for the sale of goods for the price of $500 or more is not enforceable by way of action or defense unless there is some writing sufficient to indicate that a contract for sale has been made between the parties and signed by the party against whom enforcement is sought or by his authorized agent or broker. A writing is not insufficient because it omits or incorrectly states a term agreed upon but the contract is not enforceable beyond the quantity of goods shown in such writing."

The Court of Appeals in applying this section quoted approvingly from the Comment to Section 2–201(1) of the 1962 Official Text of the Uniform Commercial Code from which the Georgia statute was taken, to the effect that "The required writing need not contain all the material terms of the contract and such material terms as are stated need not be precisely stated. All that is required is that the writing afford a basis for believing that the offered oral evidence rests on a real transaction

. . . Only three definite and invariable requirements as to the memorandum are made by this subsection. First, it must evidence a contract for the sale of goods; second, it must be 'signed,' a word which includes any authentication which identifies the party to be charged; and third, it must specify a quantity."

Growing crops, including cotton, are "goods" within the contemplation of the above section. It is undisputed that the defendant H. E. Harris signed the writing on behalf of H. E. Harris & Sons. A quantity was sufficiently shown for an "output" contract for the sale of all the defendants cotton produced on their 825 acres. The document indicated that an agreement to sell had been made.

Clearly this was more than the mere unaccepted offer the defendants contended it to be, and was sufficient to satisfy the requirements of the Uniform Commercial Code statute of frauds for an enforceable contract.

Therefore we unhesitatingly hold that the trial court correctly denied the defendants' motions to dismiss upon the ground that the complaint failed to state a claim upon which relief could be granted.

(Judgment affirmed.)

COHODAS v. RUSSELL

289 So.2d 55 (Fla. App. 1974)

Samuel Cohodas orally agreed to sell certain land to Harry Russell. Cohodas later repudiated the agreement. Russell sued to enforce the agreement and attached a letter signed by Cohodas to the complaint. A legal description of the property was attached to the letter. Cohodas raised the Statute of Frauds in defense, contending that the letter was insufficient to satisfy the requirements of the statute. The lower court denied a motion by defendant, Cohodas, to dismiss the complaint and he appealed.

GRIMES, J.

The letter, which was signed by Cohodas and mailed to the plaintiff, read as follows:

September 7, 1972

Mosby & Russell Engineering Associates, Inc.
Box 1779
Sarasota, Florida

ATTENTION: *Mr. Harry Russell*

Dear Mr. Russell:

In reference to our phone conversation, I am enclosing you a legal description of the property in question.

It consists of 120 acres net. The price is $2200 an acre.

We are willing to give you an option of six months upon receipt of a $5000 down payment—an extension of three months if necessary. The $5000 advance against the option is to be deducted from the purchase price, if no purchase price the $5000 is forfeited.

Let us hear from you immediately.

Sincerely,
/s/Sam Cohodas
S. M. Cohodas

SMC/ma
Enclosure

By its wording, the Statute of Frauds specifically encompasses a contract for the sale of land. The written memorandum relied upon to meet the requirements of the Statute must disclose all of the essential terms of the sale. The provisions of the contract cannot rest partly in writing and partly in parol.

Applying these principles to the instant case, we find a written memorandum signed by the party against whom the action is brought, which clearly sets forth the names of the parties, the purchase price and the property description. The defendants point out that among the items not covered by the memorandum are the manner of conveyance, the warranties to be given, the responsibility for the payment of documentary stamps, intangible tax and recording fees, the responsibility for procuring an abstract or title insurance and the responsibility for insuring improvements on the property while the alleged contract remained executory. Rather than being essential elements of the contract, these matters could properly be characterized as details determinable by reference to the customs incident to local real estate transactions. In the absence of a specified closing date, it could be assumed that closing would take place within a reasonable time.

In order for the Cohodas letter to meet the requirements of setting forth the terms upon which the purchase price was to be paid, we would have to presume that since credit was not mentioned, the parties must have intended the purchase price to be paid in cash.

Nevertheless, this court does not believe that it ought to be presumed that the parties intended a cash transaction when there is absolutely nothing in the written memorandum to suggest whether cash or credit was contemplated. There may have been a time when most real estate transactions were consummated by the payment of cash, but this was before the impact of income taxes became such a major consideration in the sale of land. We think it no more reasonable to presume that Cohodas intended to sell his land for $264,000 in cash than it would be to presume that he intended to sell on an installment basis secured by a purchase money mortgage. Since the letter does not include one of the essential provisions of the alleged agreement, it fails to satisfy the Statute of Frauds.

(Reversed and remanded.)

11:5. THE PAROL-EVIDENCE RULE. One reason people have their agreements put into writing is to provide for certainty in their dealings as well as to avoid subsequent disputes over the terms of the agreement. The advantage of the written over the oral agreement is that one does not have to rely so heavily on memory in order to determine the terms of a particular contract. In order to help assure that parties can continue to rely on written contracts with a high degree of certainty, the courts have developed a rule of evidence known as the *parol,* or oral, *evidence rule.* This rule states that one may not introduce oral evidence at a trial to contradict the clear, unambiguous terms of a written agreement.

The parol-evidence rule is often misunderstood. It does not prevent all oral testimony concerning a contract from being introduced at a trial. It means only that clear and unambiguous contract terms cannot be orally contradicted. Suppose Bill Buyer and Sam Seller enter into a written contract for the sale of a house for $50,000. Seller could not testify in court that they orally agreed to $55,000 absent other circumstances. The term, $50,000 is clear and unambiguous. On the other hand, one may introduce oral testimony to explain contractual terms that are not clear or ambiguous. Oral testimony concerning the oral cancellation or modification of a written contract may be introduced. Of course, one may always introduce oral testimony concerning the events surrounding the formation of the contract in order to show fraud, duress, or undue influence. Oral testimony may be introduced to show the existence of a condition precedent to a contract, that is, that an event occurred that was necessary to the legal effectiveness of the contract. Likewise oral testimony concerning contractual performance is permissible.

HATHAWAY v. RAY'S MOTOR SALES, INC.

247 A.2d 512 (Vt. 1968)

Edmund and Rose Hathaway purchased a mobile home from Ray's Motor Sales, Inc. doing business as Ray's Mobile Homes. They paid $5,700 for the home. They later began to have trouble with the home, and it developed that the unit was improperly insulated, which resulted in condensation and freezing *inside during the winter months. The Hathaways sued for damages, and Ray's raised the written-disclaimer-of-warranty provision set forth below in defense. The Hathaways testified to an additional oral agreement over objection by Ray's. The jury awarded plaintiffs $502, and defendant appealed.*

SHANGRAW, J.

We shall first consider the evidentiary phase of the case. The foregoing purchase agreement was a printed form on a pad. At the bottom of this agreement reference is made, in very small print, to other terms and conditions printed on the back of the agreement. Such printed terms in part provided, "It is mutually agreed there are no warranties, either expressed or implied, made by either the seller or the manufacturer of the trailer, mobile-home, or the parts furnished hereunder, except as follows:" Here, we are not concerned with the enumerated exceptions.

Contemporaneous with the execution of the purchase agreement, Ray's Mobile Homes, and Mr. Bessette, while acting on behalf of the plaintiffs, also signed a further instrument, Defendant's Ex. A which, so far as here material, reads:

"It is understood between the Seller and the Purchaser of said Mobile Home that the Seller, which is Ray's Mobile Homes of 1700 Williston Road, South Burlington, Vermont, shall do no free service of any kind on said Home. The only warranty is that of the Manufacturer."

The foregoing disclaimer of warranty is in bold print and is so connected with the entire transaction as to form a part of the sales agreement. . . . Mr. Hathaway testified that at the time the mobile home was purchased the defendant promised that if plaintiffs had any problems with the mobile home, the defendant "would take care of them." This testimony is the only evidence we have found in support of plaintiffs' claim of express warranty. This line of inquiry was objected to by the defendant. The case was tried and submitted on this theory.

It is the claim of the defendant that the parol evidence rule, on the facts present, precluded the introduction of evidence tending to establish an express oral warranty. The admission of this testimony over objection was crucial and failed to come within any of the exceptions to the parol evidence rule. . . .

It is a well-established general rule that when contracting parties embody their agreement of sale in writing, evidence of a prior or contemporaneous oral agreement is not admissible to vary or contradict the written agreement.

The facts do not present an exception to the parol evidence rule. This rule applies to a written contract of sale and oral testimony is inadmissible to add to or contradict the written provisions. The above quoted evidence upon which plaintiff relies directly contradicts the written disclaimer set forth in Defendant's Ex. A. and violates the parol evidence rule. Error appears.

(Judgment reversed, and case remanded.)

QUESTION FOR DISCUSSION

1. *What should Hathaway have done to avoid the parol-evidence rule and get the protection of the oral warranty?*

11:6. Summary of Uniform Commercial Code Change. With regard to the Statute of Frauds, the primary changes made by the U.C.C. are with regard to contracts for the sale of goods in Section 2–201. Those changes are discussed extensively earlier in this chapter in text Section 11:3.

Section 2–202 of the Uniform Commercial Code adopts the principle of the parol-evidence rule in contracts for the sale of goods. It is somewhat less restrictive, however, in that it allows for the explanation of contract terms or even a supplementation of the agreement by the introduction of evidence of a prior course of dealing between the parties. The agreement may also be explained by introducing evidence of the customs in the particular trade in question if that is appropriate. Oral evidence of consistent additional terms may also be introduced unless the court finds the writing to be the complete and exclusive statement of the terms of the agreement.

SUMMARY

The Statute of Frauds was first enacted in 1677 in England. It provides that certain types of agreements must be in writing to be enforceable, or there must be a written memorandum of the agreement, signed by the party who is being sued.

An agreement by an executor or administrator to personally pay the decedent's debts out of his own assets falls within the statute. A promise to pay the debt of another, that is a guaranty, falls within the statute. Such agreements are limited to situations where the agreement is collateral to the primary obligation of the debtor. Indemnity agreements and those coming under the main-purpose rule fall outside the statute. An agreement binding two persons jointly does not fall within the statute, novation falls outside the statute.

The statute applies to agreements made in consideration of marriage. Mutual promises to marry are outside the statute. Antenuptial agreements are within the statute.

The statute also applies to contracts for the sale of an interest in land. Easements, rights-of-way, mortgages, and leases extending for more than a year are all construed as interests in land. The Uniform Commercial Code provides that growing crops are goods and not part of the land. Sale of minerals and structures which are to be severed from the land by the buyer are considered sales of an interest in land.

Agreements that cannot possibly be performed within a year from the time the contract is created fall within the statute.

Contracts for the sale of goods over $500 must generally comply with the statute and Section 2–201 of the Uniform Commercial Code. One exception applies to contracts between merchants. In that case a writing signed by the party being sued may satisfy the statute so long as the other requirements of the code are met. Other exceptions apply to specially manufactured goods, the admission of the existence of a contract for them in a court proceeding, and if the goods have been received, their having been accepted or paid for.

The contract need not be in writing to satisfy the statute. A writing signed by the person being sued will suffice. It must contain the terms of the agreement and may be a letter or a series of letters.

The parol-evidence rule states that one may not introduce oral evidence at a trial to contradict the clear, unambiguous terms of a written agreement. The rule does not prohibit all oral testimony concerning a contract from being introduced at a trial.

PROBLEMS

1. John Shearon, a lumber wholesaler, sued Boise Cascade Corporation, a lumber producer, for breach of an alleged oral contract giving Shearon the exclusive right to sell Boise lumber products to Capp Homes of Iowa. In the fall of 1963, Gordon King, a sales manager for Boise, told Shearon that

if he could induce Capp to purchase more Boise products, Shearon would be protected against other wholesalers in his sale of Boise products to Capp and Boise would not sell direct to Capp.

Shearon contacted Capp and induced it to greatly increase its purchase of Boise products by gaining acceptance of a program for providing Capp's lumber requirements for a one-year period, 1964. Unfortunately much of this increase was supplied through a wholesaler other than Shearon. The next year Boise lumber was sold to Capp through Shearon, but from 1966 to 1969 Boise sold directly to Capp, eliminating Shearon and other wholesalers.

Shearon sued Boise for breach of contract. The jury awarded Shearon $100,000 in damages, but the district court granted Boise's motion for judgment n.o.v. Shearon appealed.

May Shearon enforce the contract against Boise? Shearon vs. Boise Cascade Corporation, 478 F.2d 1111 (Iowa, 1973).

2. Mezzanotte signed a contract to purchase a tract of land from Freeland known as the Daniel Boone Complex. The contract mentioned the property but did not describe it. It did refer to an "attachment," which consisted of photocopies of five deeds never physically attached to the written instrument. Freeland refused to go through with the sale, and Mezzanotte sued for specific performance. Freeland contended that the contract did not satisfy the Statute of Frauds because the property was not described adequately in the contract. Is he correct? Mezzanotte v. Freeland, 20 N.C.App. 11, 200 S.E.2d 410 (1973).
3. Rhodes took out an automobile liability policy with Southern Guaranty Insurance Company. The policy excluded coverage of the insured's son. When Rhodes purchased a new car, he went to Southern's agent and added the new car to the policy. Rhodes testified that he was told by the agent that his son would be covered. The agent denied making the statement. Rhodes's son had an accident and Southern denied coverage. At trial Rhodes contended that Southern was bound on an oral contract by the statement, set out above, of its general agent. The jury returned a verdict for Rhodes, and Southern appealed, contending that the policy was intended to embody all the terms of the contract and that Rhodes's testimony should not have been admitted. Decide. Southern Guaranty Insurance Company v. Rhodes, 243 So.2d 717 (Ala. 1971).
4. Andrew Brown operated a funeral parlor. Romano was a supplier of merchandise to whom Brown owed money when he died. Brown's wife, Adele, was appointed administratrix of his estate. Adele continued to operate the business. She orally promised Romano that she would become personally liable for Andrew Brown's debt to Romano if he would extend credit to her to operate the business. Subsequently Adele refused to pay the debt owed by her husband after Romano had extended further credit. Romano sued and Adele raised the Statute of Frauds as a defense. Decide. Romano v. Brown, 15 A.2d 818 (N.J. 1940).

5. Burlington Industries, Inc., extended credit to Colonial Fabrics, Inc., for the sale to Colonial of yarn. Fowler and Foil were major stockholders of Colonial. Burlington's credit manager contacted Fowler before allowing sales to Colonial on credit. Fowler stated verbally that he and Foil "would guarantee the account." Foil also stated verbally that he understood that Burlington was looking to him for payment. Colonial subsequently filed for bankruptcy and Burlington sued Foil for $55,577.58. Foil raised the Statute of Frauds in defense. Burlington contended that Foil had a pecuniary interest in Colonial. Decide. Burlington Industries, Inc. v. Foil, 284 N.C. 740, 202 S.E.2d 591 (1974).
6. Klymyshyn orally agreed to purchase an apartment from Szarek and gave him a $500 deposit. Szarek gave Klymyshyn a receipt which stated: "May 2, 1969. Received from Mr. Stephen Klymyshyn the sum of $500.00 as deposit to purchase apt. at 20001 Conant, for $94,000.00. /s/ Alex Szarek and Elsie Szarek." Szarek refused to sell and, when sued by Klymyshyn, contended the receipt was not sufficient to satisfy the Statute of Frauds. Was he correct? Klymyshyn v. Szarek, 29 Mich.App. 638, 185 N.W.2d 820 (1971).
7. Dean entered into an oral employment contract with Co-Op Dairy, Inc. Co-Op agreed to pay Dean's moving expenses and employ him "for a minimum period of one year." Dean began work two weeks after the agreement was made, although he could have started immediately. Nine days later all the delivery and supervisory personnel of Co-Op resigned and refused to work until Dean and the general manager were fired. Co-Op promptly complied and Dean sued to enforce the agreement. Co-Op raised the Statute of Frauds in defense and conceded that Dean's work had been satisfactory. Decide. Co-Op Dairy, Inc. v. Dean, 102 Ariz. 573, 435 P.2d 470 (1967).
8. Cohen ordered certain merchandise from Arthur Walker and Co., Inc. The order had been filled in by Walker's agent on an order form with Walker's printed name in the body of the order form. The goods were not shipped and Cohen sued for damages. Arthur Walker contended that the order form was not sufficient to satisfy the Statute of Frauds because it was not signed. Was Arthur Walker correct? Cohen v. Arthur Walker and Co., Inc., 192 N.Y.Supp. 228 (1922).
9. Hall & Taysey purchased a shipment of whiskey from John T. Barbee & Co. Hall & Taysey sold their business to Tom LaDuke, at which time there was still an unpaid balance on the whiskey shipment. When Barbee & Co. sued LaDuke for the debt, Taysey testified that all three parties orally agreed that Hall & Taysey would pay one half of the original amount due, LaDuke would assume the balance, and Barbee & Co. would release Hall & Taysey from the balance of the debt. LaDuke contended that this was a guaranty and thus unenforceable under the Statute of Frauds. Decide. LaDuke v. John T. Barbee & Co., 73 So. 472 (Ala. 1916).

CHAPTER 12

Third-Party Rights

Up to this point we have been considering contractual arrangements with just two sides. Situations do arise, however, in which a third party who is not an original party to the contract wishes to enforce rights under the agreement. These situations may be grouped into two categories: assignments and third-party beneficiaries. Do not confuse the two. You will see that an assignment involves transferring a contractual right to a third party. A third-party beneficiary does not involve the transfer of a contractual right; rather, the original parties contemplate that the agreement will benefit a third party.

12:1. Third-Party Beneficiaries. A third-party beneficiary is one whom the parties to a contract intend to benefit directly. He is not a party to the contract and need not be consulted concerning it. The contract need not ever use the name of a specific beneficiary but may refer to a class or group of people. It is not necessary that the only motivation for the contract be to benefit the third party, only that there be an intent to confer a benefit on the third party as part of the transaction. This does not mean, however, that just anyone who has an interest in the provisions of a contract can sue to enforce them. There must be some relationship of the third party to the contract other than simply an interest in its performance.

For many years courts would not enforce the rights of third-party beneficiaries. There were several reasons generally given for this refusal. In the first place, the third party was, by definition, not a party to the agreement and thus lacked what the courts refer to as *privity of contract.* Second, the courts reasoned that the third party normally gives no consideration in return for the benefits conferred by the contract. Finally, the courts were troubled by the fact that whereas the third-party beneficiary was suing for benefits under the agreement, the parties had no right of action against him. Today the courts do enforce certain types of contracts involving the rights of third-party beneficiaries, the most common example being life-insurance contracts. Third-party beneficiaries whose rights are enforceable are classified as third-party *creditor* beneficiaries and third-party *donee* beneficiaries. The courts will not enforce third-party rights that are only *incidental.*

(1) Third-Party Creditor Beneficiary. The situation of a third-party creditor beneficiary occurs when the promisor obligates himself to satisfy a duty owed by the promisee to his creditor, in other words, when there is a debt running from the promisee to the creditor. The promisor agrees to satisfy that duty in return for some performance by the promisee. In this case, the creditor would be the third-party creditor beneficiary. Note that there exists a debtor–creditor relationship between the creditor and promisee. The creditor is a direct beneficiary of the agreement made between the promisor and the promisee because the promisor agrees to pay the debt.

This situation frequently occurs in mortgage-assumption cases. Suppose Harry Homeowner borrows $50,000 from Citizens Bank to buy a home. In return he signs a note and gives the bank a mortgage on the home as security. Three years later Barney Buyer agrees to purchase Homeowner's house and further agrees to assume and pay Homeowner's mortgage. The agreement between Homeowner and Buyer does not discharge Homeowner from his obligation to pay the debt. However, the bank may sue Buyer as a third-party creditor beneficiary of the contract between Buyer and Homeowner because Homeowner obtained Buyer's promise in order to pay his debt to Bank and also to confer a benefit on Bank.

(2) **Third-Party Donee Beneficiary.** If there is no debtor–creditor relationship between one of the parties and the third-party beneficiary but the intent of the promisor is to confer a gift upon a third person, the third person is a donee beneficiary. Life-insurance contracts are in this category. Suppose Inez Insured takes out a policy of insurance on her life with Long Life Insurance Company and names Barry Beneficiary as the beneficiary. Because there is no creditor relationship between Beneficiary and either of the parties and the obvious intent of Inez is to confer a benefit on the third party, Barry would be a third-party donee beneficiary. Should Long Life Insurance refuse to pay over the proceeds of the policy on the death of Inez, it could be sued by Beneficiary to enforce his rights as a third-party beneficiary. It is not necessary that the donee beneficiary know of the existence of the policy in order to have rights under it. Note, however, that in the case of a donee beneficiary, the right of action is only against the party who has promised to perform for the benefit of the third person. In the case of a creditor beneficiary, recovery may be had against either party to the contract.

BASS v. JOHN HANCOCK MUTUAL LIFE INSURANCE CO.
10 Cal.3d 792, 518 P.2d 1147 (1974)

J. T. Bass was employed by the Ford Motor Company from 1947 to his death in 1968. Prior to 1964 Ford offered its employees a group disability-insurance plan written by John Hancock upon payment by the employees of an additional premium. Bass refused to participate in this plan. In 1964 the Hancock plan was amended to provide life insurance as well as disability benefits free of charge to Ford employees. Bass was contacted in 1964 but apparently still indicated that he did not want the free insurance. In 1967 the plan was amended once more to provide increased benefits, but this time Bass was not contacted. Bass did receive a booklet that stated, "This Certificate–Booklet becomes applicable to you on October 25, 1967 if you are then at work. . . ." *Bass died in 1968 in a car accident. John Hancock refused to pay and was sued by Belva Bass as beneficiary of the policy in order to recover the benefits. The lower court found for John Hancock on the ground that J. T. Bass had refused the policy and that Ford had paid no premiums to the insurer to cover Bass. Belva Bass appealed.*

BURK, J.

It seems evident to us that a waiver which occurred in 1964 is not necessarily controlling with respect to rights under a 1967 insurance program. Traditional legal principles convince us that the trial court erred in concluding otherwise. Bass was a third party beneficiary under the Ford–UAW union contract and the Ford–Hancock group insurance program, and as such he initially acquired valuable rights to insurance coverage.

In the absence of a renewed rejection or disclaimer from Bass, or other "clear and convincing" indicium of waiver, we must conclude that Bass retained his group insurance rights under the 1967 plan.

Defendant Hancock contends that even if Bass did not waive his rights to coverage, plaintiff's cause of action would be against Ford rather than Hancock, since Ford paid Hancock no premiums on Bass' account. To the contrary, nonpayment of premiums may have constituted a breach of Ford's obligations to Hancock, but would not affect the right of Bass (or plaintiff) to enforce coverage which Hancock agreed to extend to all Ford employees. In a series of recent cases we have held that the employer is the agent of the insurer in performing the duties of administering group insurance policies, including the payment of premiums, and that accordingly the insurer shares responsibility for the employer's mistakes. We see no reason to deviate from this principle in the instant case. Indeed, we note that Hancock's own representative admitted at trial that if a Ford employee were not included in the calculation of premiums by reason of clerical error or mistake, Hancock would nevertheless honor his claim under the group policy and backcharge Ford for the unpaid premiums.

In the instant case, Ford's "mistake" was in assuming that Bass had rejected coverage under the 1967 group plan, without verifying that fact with Bass himself. We think that this kind of mistake is one for which Hancock should share responsibility despite the nonpayment of premiums.

The judgment is reversed and the cause remanded to the trial court. . . .

QUESTIONS FOR DISCUSSION

1. *Was Bass a creditor beneficiary or a donee beneficiary?*
2. *Which Bass was actually the third-party beneficiary of the contract between Ford and the UAW, Belva or J. T. Bass? Who was the beneficiary under the contract with John Hancock? Under which contract was Belva seeking recovery?*

---

(3) Third-Party Incidental Beneficiary. Contracts are often made in which a third party may benefit incidentally but in which there is no intent to benefit anyone directly other than the parties to the agreement. Third parties who are only incidental beneficiaries of contracts may not sue to enforce any of the contractual provisions. For example, a contract to build a shopping center in a particular area may enhance the value of surrounding commercial properties, but the owners of those properties would have no right to enforce the contract because the original parties did not intend to benefit the surrounding owners directly.

√WILLIAMS v. FENIX & SCISSON, INC.

608 F.2d 1205 (1979)

J. D. Williams was an employee of Parco, Inc., and was injured while working on a drilling rig at Amchitka Island, Alaska. Parco was a prime contractor for the Atomic Energy Commission (AEC) and had the job of drilling test holes at Amchitka in connection with hydrological testing.

Fenix & Scisson, Inc., was another prime contractor employed by the AEC at the same location to render engineering services. While Williams and some other Parco employees were using an air-hoist line to lift pipe, the pipe slipped severely injuring Williams.

Williams claims that because of its contract with the AEC, Fenix & Scisson had a contractual obligation to him to see that the pipe was handled in a safe manner.

Williams claims that he is the beneficiary of two clauses of the contract between Fenix & Scisson and the AEC. One clause required Fenix & Scisson to inspect drilling operations at Amchitka and recommend any improvements to the AEC. The second paragraph required Fenix & Scisson to take reasonable safety precautions in the performance of its work.

TANG, C.J.

This contract was one entered into solely between the AEC and Fenix & Scisson, an independent contractor. The plaintiff presumes that if these contractual provisions were breached by Fenix & Scisson, then liability flows to him as a matter of course. However, neither Williams, nor his employer, Parco, Inc. was a party to this contract. Since Williams was not a party to the contract, the primary question is whether Williams was an intended third party beneficiary of the contract between the AEC and Fenix & Scisson. It is a general rule of law that before recovery can be had under a contract by a third party, he must show that the contract was made for his direct benefit. There are three types of third party beneficiaries: donee beneficiaries, creditor beneficiaries, and incidental beneficiaries. Only creditor and donee beneficiaries have potential rights under a contract:

> A third party . . . has an enforceable right by reason of a contract made by two others (1) if he is a creditor of the promisee or of some other person and the contract calls for a performance by the promisor in satisfaction of the obligation; or (2) if the promised performance will be of pecuniary benefit to him and the contract is so expressed as to give the promisor reason to know that such benefit is contemplated by the promisee as one of the motivating causes of his making the contract. A third party may be included within both of these provisions at once, but need not be. One who is included within neither of them has no right, even though performance will incidentally benefit him. 4 Corbin, Contracts § 776 at 18, 19 (1951) (footnotes omitted).

It is clear that the plaintiff was not a creditor of the promisee (the AEC). Furthermore, plaintiff was not a donee beneficiary since there was no indication in the contract to give the promisor, Fenix & Scisson, "reason to know" that such benefit was contemplated by the AEC as a motivating cause for making the contract.

The contractual provisions required Fenix & Scisson to perform certain inspection and safety duties and then report back to the AEC but there was no duty under any provision of the contract to supervise the work of Parco, Inc. It is clear, moreover, that Parco was an independent contractor who was to supervise completely its own employees and drilling operations. In neither the Fenix & Scisson contract nor the Parco contract was there any contractual provision suggesting that either Fenix & Scisson or Parco would provide the other with supervision, personnel, materials, tools or anything else. Any benefit derived from the Fenix & Scisson contract was to the AEC, not Parco or Williams.

The right to inspect to see that provisions of the contract are carried out and the right to require safety regulations to be followed is not such control as to impose liability to the employee of an independent contractor.

We agree with the district court determination that at best Williams was merely an incidental beneficiary of the Fenix & Scisson and AEC contract. As an incidental beneficiary, Williams has no rights under the contract against Fenix & Scisson. The district court properly withdrew this issue from the jury.

QUESTION FOR DISCUSSION

1. *Why did the court determine that Williams was not entitled to the rights of a third-party beneficiary when the contract between the AEC and Fenix & Scisson required Fenix & Scisson to take reasonable safety precautions in the performance of its work?*

(4) Rescission and Vesting. Does there ever come a time when the rights of a third party become legally vested so that the original parties cannot change the contract in order to extinguish these rights? The answer is yes. Although there is some variance among jurisdictions, generally when a third-party beneficiary has accepted, adopted, or acted upon the agreement, the parties may not change it without his consent. The rights of the third party have vested. This is true in the case of a third-party donee beneficiary of life-insurance contracts. Once the beneficiary has accepted the terms of the contract, the insured cannot change the beneficiary without the consent of the present third-party beneficiary unless the terms of the contract reserve the right to the insured to make the change. Today most policies contain such a provision as standard form. Of course, if the third-party beneficiary has not accepted or acted upon the contract, it may be changed so as to divest him of its benefits.

12:2. ASSIGNMENTS. An assignment under a contract occurs when one party transfers his rights under the agreement to a third person. The person who transfers his contractual right is called the *assignor.* The person to whom the right is transferred or assigned is called the *assignee.* The other party to the contract against whom the right may be exercised is called the *obligor.*

Under a bilateral contract the parties have corresponding rights and duties. Suppose Roy Rider contracts to sell his bicycle to Tiny Twowheel for $50, with the provision that Twowheel is to pay sixty days after receiving the bicycle. Suppose further that Rider gives Twowheel the bicycle on May 1. At this point Rider has fulfilled his duty under the contract to give Twowheel the bicycle. He has the right to receive $50 from Twowheel, or, to put it another way, Twowheel is Rider's *obligor.* Rider may assign his right to receive the $50 from Twowheel to a third party, Annie Assignee. Rider is the assignor, Annie the assignee, and Twowheel the obligor.

As a general rule one may assign contractual rights, but he may not assign contractual duties. Although one may never assign contractual obligations and duties, he may delegate them if they do not involve personal service or the personal attention of the obligor. Thus, in the above example, while Rider may assign his right to receive $50 under the contract, Twowheel may not delegate his duty to pay $50. One reason for this limitation is that the credit was extended to Twowheel because of his financial position and reputation.

No particular form is required to have an assignment, but in order for it to be enforceable as between the assignor and the assignee, it must constitute a contract. Although most assignments are contracts and, as such, must have all the elements required for a valid contract to be enforceable, all that is necessary for a valid assignment is a statement indicating an intent to make the assignee the owner of the claim. If the assignment is not a contract, then the assignor may revoke it just as one may revoke a promise to make a gift. The assignment may be made either orally or in writing.

(1) Assignment of Claims for Money. Perhaps the most frequently assigned rights involve claims for money. Because the right to receive money is not personal in nature, it is generally assignable. Examples of these rights are accounts receivable, debts due, and wages and salaries.

When accounts receivable or other debts are assigned, the question often arises as to what the responsibility of the assignor is to the assignee for those accounts that cannot be collected. Typically this matter is resolved in the assignment agreement between the assignor and the assignee. The assignor may guarantee the collectability of the accounts, but lacking such an agreement, the mere fact that the debt is not collected would not give the assignee any recourse against the assignor.

The question arises as to whether there may be a valid assignment of accounts that are not yet due. The Uniform Commercial Code permits the assignment of accounts to come due in the future under an existing contract and thus adopts the general rule (U.C.C., Section 9–318). This section of the code also provides that a contract term between an account debtor and an assignor that prohibits assignment of an account is ineffective. In gencral, however, a contract may contain an antiassignment clause, and the obligor may recover damages against the other party if it is breached.

Earned wages and salary as well as future wages under an existing contract may be assigned, although there is an increasing trend toward restricting this practice on public-policy grounds. State statutes that regulate this practice entirely prohibit wage-and-salary assignments in some cases. In others the amount or percentage of wages that may be assigned is limited. A few states have adopted the Uniform Consumer Credit Code, which provides that an assignment of earnings to a seller or a lender is ineffective to pay a debt arising out of a consumer-credit sale. Generally a public officer may not assign the unearned salary or fees of his office.

(2) Personal-Service Contracts and Delegation of Duties. Personal-service contracts that involve the exercise of personal skill, judgment, taste, or knowledge are not assignable and may not be delegated. Thus contracts with artists, physicians, dentists, lawyers, or authors may not be assigned by either party. The person rendering the personal service may not assign the contract, nor may the employer require him to work for someone else by assigning the agreement. The obvious reason is that it is the ability of the particular individual that is contracted for, not that of someone else. Thus, if Valerie Vain contracts to have Archie Artist paint her portrait, Archie cannot delegate this task to Color Blind because it is Archie's skill that Valerie has contracted for. For the same reason, contracts that involve personal trust and confidence, such as extending credit, may not be assigned.

On the other hand, if the service or labor does not involve personal skill or trust and confidence that are peculiar to the obligor, the duties may be delegated. This situation occurs frequently with regard to construction contracts. The contractor may hire various subcontractors, such as plumbers or electricians, while remaining personally liable for the performance of the work.

The performance is not of a personal nature in the sense that it would be substantially different if rendered by some other party, and thus it may be delegated.

MACKE COMPANY v. PIZZA OF GAITHERSBURG, INC.

259 Md. 479, 270 A.2d 645 (1970)

Pizza of Gaithersburg, Inc., arranged to have cold-drink vending machines owned by Virginia Coffee Service, Inc., installed in each of their locations. On December 30, 1967, Virginia's assets were purchased by Macke Company and the contracts assigned to Macke by Virginia. Pizza attempted to terminate the contracts and was sued by Macke. From a judgment for Pizza, Macke appealed.

SINGLEY, J.

We cannot regard the agreements as contracts for personal services. They were either a license or concession granted Virginia by the appellees, or a lease of a portion of the appellees' premises, with Virginia agreeing to pay a percentage of gross sales as a license or concession fee or as rent, and were assignable by Virginia unless they imposed on Virginia duties of a personal or unique character which could not be delegated. . . .

The appellees earnestly argue that they had dealt with Macke before and had chosen Virginia because they preferred the way it conducted its business. Specifically, they say that service was more personalized, since the president of Virginia kept the machines in working order, that commissions were paid in cash, and that Virginia permitted them to keep keys to the machines so that minor adjustments could be made when needed. Even if we assume all this to be true, the agreements with Virginia were silent as to the details of the working arrangements and contained only a provision requiring Virginia to "install . . . the above listed equipment and . . . maintain the equipment in good operating order and stocked with merchandise." We think the Supreme Court of California put the problem of personal service in proper focus a century ago when it upheld the assignment of a contract to grade a San Francisco street:

> All painters do not paint portraits like Sir Joshua Reynolds, nor landscapes like Claude Lorraine, nor do all writers write dramas like Shakespeare or fiction like Dickens. Rare genius and extraordinary skill are not transferable, and contracts for their employment are therefore personal, and cannot be assigned. But rare genius and extraordinary skill are not indispensable to the workmanlike digging down of a sand hill or the filling up of a depression to a given level, or the construction of brick sewers with manholes and covers, and contracts for such work are not personal, and may be assigned.

Restatement, Contracts § 160(3) (1932) reads, in part:

> Performance or offer of performance by a person delegated has the same legal effect as performance or offer of performance by the person named in the contract, unless,
>
> (a) performance by the person delegated varies or would vary materially from performance by the person named in the contract as the one to perform, and there has been no . . . assent to the delegation. . . .

In cases involving the sale of goods, the Restatement rule respecting delegation of duties has been amplified by Uniform Commercial Code § 2–210(5), . . . which permits a promisee to demand assurances from the party to whom duties have been delegated.

As we see it, the delegation of duty by Virginia to Macke was entirely permissible under the terms of the agreements.

(Judgment reversed and remanded.)

QUESTIONS FOR DISCUSSION

1. *Could Pizza have assigned its side of the contract to another organization?*
2. *Pizza apparently felt strongly about dealing with Macke. Are there any steps it might have taken to prevent the result that occurred in the previous case?*

(3) Contractual Rights That May Not Be Assigned. Although as a general rule contract rights may be assigned, if such an assignment increases the corresponding duty of the obligor, it is ineffective. Thus if Sammy Supplier agrees to supply all of Manny Manufacturer's steel requirements for a particualr period, Manny cannot assign the right to receive steel to another manufacturer who has substantially greater requirements. The reason is that this assignment would increase the burden on the obligor, Supplier in this case.

Future accounts or earnings that are not founded on an existing contract are generally not assignable because the account or debt is merely speculative and has no potential existence. Be sure you distinguish this case from an account or debt that may arise in the future under an *existing* contract. Such obligations are generally assignable.

CRANE ICE CREAM CO. v. TERMINAL FREEZING & HEATING CO.

147 Md.588, 128 A. 280 (1925)

Terminal Freezing and Heating Co. contracted to supply W. C. Frederick, an ice-cream manufacturer, ice under terms set out in the opinion below. Frederick subsequently assigned the contract to Crane Ice Cream Company as part of a transaction in which Crane acquired the plant equipment, goodwill, and other assets of Frederick's ice-cream business. Crane carried on business in Philadelphia and Maryland and was a much larger company than Frederick. Terminal did not consent to the assignment and refused to perform for Crane. Crane sued Terminal, and Crane appealed from an adverse judgment of the lower court.

PARKE, J.

The contract imposed upon the appellee, Terminal, the liability to sell and deliver to Frederick such quantities of ice as he might use in his business as an ice-cream manufacturer, to the extent of 250 tons per week, at and for the price of $3.25 a ton of 2,000 pounds on the loading platform of Frederick.

The basic facts upon which the question for solution depends must be sought in the effect of the attempted assignment of this executory bilateral contract on both the rights and the liabilities of the contracting parties, as every bilateral contract includes both rights and duties of each side, while both sides remain executory. If the assignment of rights and the assignment of duties by Frederick are separated, they fall into these two divisions: (1) The rights of the assignor were (a) to take no ice, if the assignor used none in his business, but, if he did, (b) to require the appellee to deliver, on the loading platform of the assignor, all the ice he might need in his business to the extent of 250 tons a week, and (c) to buy any ice he might need in excess of the weekly 250 tons from any other person; and (2) the liabilities of the assignor were (a) to pay to the appellee on every Tuesday during the continuance of the contract the stipulated price for all ice purchased and weighed by the assignor during the week ending at midnight upon the next preceding Saturday, and (b) not directly or indirectly, during the existence of this agreement, to buy or accept any ice from any other person, firm, or corporation than the said the Terminal Freezing & Heating Company, except such amounts as might be in excess of the weekly limit of 250 tons.

The assignor had his principal plant in Baltimore. The assignee, in its purchase, simply added another unit to its ice-cream business which it had been, and is now, carrying on "upon a large and extensive scale in the city of Philadelphia and state of Pennsylvania, as well as in the city of Baltimore and state of Maryland. . . ." The appellee knew that Frederick could not carry on his business without ice wherewith to manufacture ice-cream at his plant for his trade. It also was familiar with the quantities of ice he would require, from time to time, in his business at his plant in Baltimore, and it consequently could make its other commitments for ice with this knowledge as a basis.

There can be no denial that the uniform delivery of the maximum quantity of 250 tons a week would be a consequence not within the normal scope of the contract, and would impose a greater liability on the appellee than was anticipated.

Moreover, the contract here to supply ice was undefined except as indicated from time to time by the personal requirements of Frederick in his specified business. The quantities of ice to be supplied to Frederick to answer his weekly requirements must be very different from, and would not be the measure of, the quantities needed by his assignee, and, manifestly, to impose on the seller the obligation to obey the demands of the substituted assignee is to set up a new measure of ice to be supplied, and so a new term in the agreement that the appellee never bound itself to perform. It was argued that Frederick was entitled to the weekly maximum of 250 tons, and that he might have expanded his business so as to require this weekly limit of ice, and that therefore the burdens of the contract might have been as onerous to the appellee, if Frederick had continued in business, as they could become under the purported assignment by reason of the increased requirements of the larger business of the assignee. The unsoundness of this argument is that the law accords to every man freedom of choice in the party with whom he deals and the terms of his dealing. He cannot be forced to do a thing which he did not agree to do because it is like and no more burdensome than something which he did contract to do.

Under all the circumstances of the case, it is clear that the rights and duties of the contract under consideration were of so personal a character that the rights of Frederick cannot be assigned nor his duties be delegated without defeating the intention of the parties to the original contract.

While a party to a contract may, as a general rule, assign all his beneficial rights, except where a personal relation is involved, his liability under the contract is not assignable inter vivos, because anyone who is bound to any performance whatever or who owes money cannot by any act of his own, or by any act in agreement with any other person than his creditor or the one to whom his performance is due, cast off his own liability and substitute another's liability. . . . If this were not true, obligors could free themselves of their obligations by the simple expedient of assigning them.

However, the analysis of the facts on this appeal leaves no room for doubt that the case at bar falls into the category of those assignments where an attempt is made both to transfer the rights and to delegate the duties of the assignor under an executory bilateral contract, whose terms and the circumstances make plain that the personal qualification and action of the assignor, with respect to both his benefits and burdens under the contract, were essential inducements in the formation of the contract, and, further, that the assignment was a repudiation of any future liability of the assignor. The attempted assignment before us altered the conditions and obligations of the undertaking. The appellee would here be obliged not only to perform the subsequent stipulations of the contract for the benefit of a stranger, and in conformity with his will, but also to accept the performance of the stranger in place of that of the assignor with whom it contracted, and upon whose personal integrity, capacity, and management in the course of a particular business he must be assumed to have relied, by reason of the very nature of the provisions of the contract and of the circumstances of the contracting parties. The nature and stipulations of the contract prevent it being implied that the nonassigning party had assented to such an assignment of rights and delegation of liabilities. The authorities are clear, on the facts at bar, that the appellant could not enforce the contract against the appellee.

(Judgment affirmed.)

Questions for Discussion

1. *If Frederick had agreed to remain liable for the liabilities on the contract, would the assignment of the contractual rights alone have been effective?*
2. *Under the terms of the contract Terminal's maximum duty was to supply 250 tons of ice per week. How, if at all, did the assignment from Frederick to Crane increase Terminal's duties?*

---

(4) Rights and Duties of the Parties. (a) *The Assignor.* When the assignor transfers a contractual right to the assignee, he gives up the ability to exercise that right against the obligor. If the assignment is by contract, he is also subject to the terms of the agreement made with the assignee. Thus, if in an assignment of accounts the assignor agrees to guarantee their collectability, he is liable on the assignment contract for all the accounts that are uncollectable.

The assignor also gives certain implied warranties to the assignee, regardless of any express warranties contained in the assignment contract. Although the courts have not clearly defined the extent of the warranties implied in all cases, some area of general agreement can be determined. In general the assignor warrants by implication that the right he is assigning exists and is valid and that lacking a disclaimer, there are no defenses against it. He also warrants that he has the right to assign the claim. In addition the assignor warrants by implication that he will do nothing to interfere with the right of the assignee to collect from the obligor. However, the assignor gives no warranty that the obligor will perform, and he is not liable to the assignee for such nonperformance, absent a specific guarantee. If the assignor breaches any of these implied warranties or the terms of the assignment agreement, he may be sued for damages by the assignee. For instance, suppose that Alice Assignor makes an assignment of the right to collect a $50 debt to Norman Assignee. Alice then collects the debt herself from the unwitting debtor. Alice would be liable to the assignee for the breach of warranty. The same result would obtain if Alice made a second assignment of the same account or procured the obligation as the result of fraud.

(b) *The Assignee.* The assignee takes the assignment of the contractual right subject to all the defenses that the obligor could raise against the assignor. The assignee is in no better position vis-à-vis the obligor than the assignor, and the obligor may raise any defense against the assignee that he could raise against the assignor. The rule is sometimes quaintly stated that the "assignee stands in the shoes of the assignor." The defenses that may be raised are those that may be raised in any breach-of-contract action. Suppose, for example, that Harry Hotrod purchases an automobile from Carlos Cardealer. Cardealer agrees to wait thirty days for payment of the balance of the purchase price and then assigns the account to Friendly Finance Company. Hotrod gets the car home and then discovers that it will not run properly. Cardealer refuses to honor the "thirty-day guaranty" that he gave with the car. When sued by Friendly Finance for the balance of the purchase price, Hotrod can raise this matter as a defense against Friendly.

HUDSON SUPPLY & EQUIPMENT COMPANY v. HOME FACTORS CORP.

210 A.2d 837 (D.C. 1965)

HOOD, C.

Home Factors Corp., to which Eastern Brick & Tile Co., Inc., had assigned two accounts receivable for brick sold and delivered to Hudson Supply & Equipment Company, brought this action against Hudson for the amount due under the accounts, namely, $1,324.25.

Hudson's defense was that Eastern was indebted to it in an amount in excess of that sued for, and that it was entitled to a set-off for the full amount claimed.

At trial Hudson offered testimony that at the time of the assignment of the two accounts by Eastern to Home Factors, Hudson had claims of over $2,200 against Eastern growing out of other purchases.

The trial court ruled that there was a proper assignment from Eastern to Home Factors and that Home Factors was entitled to judgment for the full amount of its claim. However, the trial court stated "that the evidence indicated that the problem was between Hudson and Eastern and that defendant Hudson was entitled to credits of $229.22 and $172.31 from Eastern and in addition had other claims for credits against Eastern—all of which indicated that Eastern and Hudson should litigate separately the issues between them."

The general rule here and elsewhere is that the assignee of a chose in action takes it subject to all defenses, including set-offs, existing at the time of the assignment. Since it is undisputed in this case that the asserted claims of Hudson existed at the time of the assignment, it is apparent that the trial court misconceived the law relating to assignments. When it was found that Hudson was entitled to certain credits, those credits should have been set off against the claim of Home Factors; and Hudson's "other claims for credits against Eastern" should have been determined, and, if established, should also have been set off against Home Factors' claim.

(Reversed with instructions to grant a new trial.)

QUESTIONS FOR DISCUSSION

1. *After the assignment from Eastern, could Hudson still recover directly from Eastern, the assignor, in this case?*
2. *Did Home Factors Corporation have any claim against Eastern in this case? If so, on what grounds?*

X **(5) Notice of Assignment.** Once an assignment has been made, it is generally wise for the assignee to notify the obligor of the assignment, even though there may be no legal requirement to do so. By giving notice, the assignee protects himself against at least two potential problems: (1) that the obligor will pay the assignor after the assignment is made; (2) that the assignor will assign the claim to more than one assignee. The danger to the assignee is that an unscrupulous assignor will not only collect the consideration from the assignee under the assignment contract but also collect from the debtor and/or subsequent assignees. The fact that the assignor would be liable to the assignee either under the implied warranty provisions made with the assignment or for breach of the assignment contract is of little solace when the assignor has absconded or become insolvent.

If there are multiple assignments of the same claim, the question arises as to which assignee is entitled to recover the claim from the obligor. There are two rules that govern this situation. A majority of jurisdictions follow what is known as the *English rule.* In these states the first bona fide assignee for value to notify the debtor of the assignment prevails, regardless of whether he is the first person to receive the assignment in point of time. Many jurisdictions have adopted the so-called American rule, which states that the first assignee in point of time prevails, regardless of which assignee gives notice first. The logic behind this approach is that once an assignor assigns his claim, he has nothing left to transfer, and thus the subsequent attempts at assignment transfer no right at all.

Suppose Danny Debtor owes William Wrongdoer $500 and Wrongdoer assigns the right to this claim to First Assignee in return for a consideration.

Suppose further that Wrongdoer makes subsequent transfers of the same claim to Second and Third Assignees. Second Assignee notifies Debtor of the assignment one day prior to the time First Assignee gives notice. In those states where English "notice rule" applies, Second Assignee's right to recover the claim would prevail. First Assignee would prevail where the American rule is followed. In either case, should the debtor pay the assignor after receiving notice, he would still be liable to the assignee for the claim. Most jurisdictions hold that a debtor need only be aware of the assignment in order to fix his liability to the assignee. It is not necessary that he be notified by the assignee, and no special form of notice is required.

Some states have statutes that require that certain types of assignments must be recorded in order to be effective against subsequent assignees for value. For instance, a few jurisdictions require assignments of accounts receivable to be recorded. Recordation operates as constructive notice to subsequent assignees.

BOULEVARD NATIONAL BANK OF MIAMI v. AIR METALS INDUSTRIES, INC.

176 So.2d 94 (Fla. 1965)

WILLIS, C.J.

Tompkins-Beckwith was the contractor on a construction project which had entered into a subcontract with a division of Air Metal Industries, Inc. Air Metal procured American Fire and Casualty Company to be surety on certain bonds in connection with contracts it was performing for Tompkins-Beckwith and others. As security for such bonds, Air Metal executed, on January 3, 1962, a "Contractor's General Agreement of Indemnity" which contains an assignment to American Fire of "all monthly, final or other estimates and retained percentages; pertaining to or arising out of or in connection with any contracts performed or being performed or to be performed, such assignment to be in full force and effect as of the date hereof, in the event of default in the performance of—any contract as to which the surety has issued, or shall issue, any (surety bonds or undertakings)."

On November 26, 1962, the petitioner (Boulevard National Bank) lent money to Air Metal and to secure the loans Air Metal purported to assign to the bank certain accounts receivable it had with Tompkins-Beckwith which arose out of subcontracts being done for that contractor.

In June, 1963, Air Metal defaulted on various contracts bonded by American Fire. On July 1, 1963 American Fire served formal notice on Tompkins-Beckwith of Air Metal's assignment. Tompkins-Beckwith acknowledged the assignment and agreed to pay. On August 12, 1963, the petitioner bank notified Tompkins-Beckwith of its assignment and claim thereunder. The claim was not recognized and on September 26, 1963, this action was filed in the trial court. On October 9, 1963, Tompkins-Beckwith paid all remaining funds which had accrued to Air Metal to American Fire.

The "question" which was passed upon by the certifying court is whether the law of Florida requires recognition of the so-called "English" rule or "American" rule of priority between assignees of successive assignments of an account receivable or other similar chose in action. Stated in its simplest form, the American rule would give priority to the assignee first in point of time of assignment, while the English rule would give preference to the assignment of which the debtor was first given notice.

The American rule for which petitioner contends is based upon the reasoning that an account or other chose in action may be assigned at will by the owner; that notice to the debtor is not essential to complete the assignment; and that when such assignment is made the property rights become vested in the assignee so that the assignor no longer has any interest in the account or chose which he may subsequently assign to another.

It seems to be generally agreed that notice to a debtor of an assignment is necessary to impose on the debtor the duty of payment to the assignee, and that if before receiving such notice he pays the debt to the assignor, or to a subsequent assignee, he will be discharged from the debt. To regard the debtor as a total non-participant in the assignment by the creditor of his interests to another is to deny the obvious. An account receivable is only the right to receive payment of a debt which ultimately must be done by the act of the debtor. For the assignee to acquire the right to stand in the shoes of the assigning creditor he must acquire some "delivery" or "possession" of the debt constituting a means of clearly establishing his right to collect. The very nature of an account receivable renders "delivery" and "possession" matters very different and more difficult than in the case of tangible personalty and negotiable instruments which are readily capable of physical handling and holding. It would seem to follow that the mere private dealing between the creditor and his assignee unaccompanied by any manifestations discernible to others having or considering the acquiring of an interest in the account would not meet the requirement of delivery and acceptance of possession which is essential to the consummation of the assignment. Proper notice to the debtor of the assignment is a manifestation of such delivery. It fixes the accountability of the debtor to the assignee instead of the assignor and enables all involved to deal more safely.

We do not hold that notice to the debtor is the only method of effecting a delivery of possession of the account so as to put subsequent interests on notice of a prior assignment. The English rule itself does not apply to those who have notice of an earlier assignment. The American rule is not in harmony with the concepts expressed. It seems to be based largely upon the doctrine of caveat emptor which has a proper field of operation, but has many exceptions based on equitable considerations. It also seems to regard the commercial transfers of accounts as being the exclusive concern of the owner and assignee and that the assignee has no responsibility for the acts of the assignor with whom he leaves all of the indicia of ownership of the account. This view does not find support in the statute or decisional law of this State.

(Judgment for defendant.)

QUESTIONS FOR DISCUSSION

1. *Under the English rule is it necessary to give notice to the obligor in order for the assignee to have a right to collect the debt if there is only one assignee?*
2. *What reasoning did the court give for its reliance on the "English rule" rather than the "American rule"? What is the theory underlying the "American rule"?*

12:3. Summary of Uniform Commercial Code Changes. The Uniform Commercial Code does have a section that deals with assignments. Section 9–318 deals with the assignment of accounts and is found in the article dealing with the subject of secured transactions. The official comments to this section indicate that the section makes no substantial changes in the prior law.

This section of the U.C.C. reiterates the existing law that we have already discussed. For instance, it provides that the rights of an assignee of a contract are subject to all the terms of the contract between the account debtor and assignor and any defense or claim arising out of that contract. In other words, the assignee stands in the shoes of the assignor when suing on the contract. Furthermore, the rights of the assignee are subject to any other defense or claim of the account debtor against the assignor that arise prior to the time the account debtor is notified of the assignment.

Section 9–318(3) provides that the account debtor is authorized to pay the amount due to the assignor until the account debtor receives notification of the assignment that that payment is to be made to the assignee. The account debtor may require the assignee to furnish reasonable proof that the assignment has been made. Unless the assignee furnishes such proof upon request, the account debtor may continue to pay the assignor.

The code also states that a contract term prohibiting assignment of an account is ineffective. A contract term requiring an account debtor's consent to assign an account is also ineffective.

Section 9–206 of the code modifies some of the statements made previously concerning the right of an account debtor to raise against the assignee any defenses that he might have against the assignor. This section provides that an agreement by a buyer that he will not assert against an assignee of the debt any claim or defense that he may have against the seller is enforceable by an assignee who gives value for the assignment and takes it in good faith without any notice of the claim or defense. The purpose of the section is to make the purchase of accounts more acceptable to potential assignees by including such clauses in the contracts. Of course, if the assignee is aware of a claim or defense when taking the assignment, that particular claim or defense may still be raised by the account debtor even with such a clause in the contract.

Section 2–210 clarifies assignments and delegation in contracts for the sale of goods. This section of the code provides that a party may perform his duty through a delegate unless the other party has a substantial interest in having the original promisor perform. The section also incorporates the principle that rights may be assigned so long as this does not materially change the duty of the other party to the contract. In addition, the section clarifies the meaning of certain language typically found in contracts. For example, unless there is a contrary intention, a prohibition of assignment of "the contract" is to be construed as barring only the delegation to the assignee of the assignor's performance.

SUMMARY

A third-party beneficiary is one whom the parties to a contract intend to benefit directly. Third-party creditor beneficiaries and third-party donee beneficiaries may enforce contractual rights. One who is merely an incidental beneficiary may not enforce any contractual rights. When the rights of a third-party beneficiary to a contract have vested, the original parties to the contract cannot change the contract to terminate or alter those rights.

An assignor is one who transfers his contractual rights to a third person. The person to whom the right is transferred is the assignee. The party against whom the right may be exercised is the obligor. As a general rule, one may assign contractual rights, but he may not assign contractual duties. One may delegate contractual duties if they do not involve personal service or the personal attention of the obligor.

When accounts receivable are assigned, payment is not guaranteed unless the contract of assignment specifically says that it is. In general, except for a provision of the Uniform Commercial Code, a contract may contain an antiassignment clause. Personal-service contracts that involve the exercise of personal skill, judgment, taste, or knowledge are not assignable and may not be delegated. If the assignment increases the corresponding duty of the obligor, it is ineffective.

The assignor impliedly warrants to the assignee that the right he is assigning exists, is valid and, that lacking a disclaimer, there are no defenses against it. The assignor also warrants that he has the right to assign the claim and that there are no defenses against it. Finally the assignor impliedly warrants that he will not interfere with the right of the assignee to collect from the obligor.

The assignee takes the assignment subject to all defenses the obligor has against the assignor. If there are multiple assignments of the same right, the first assignee to give notice to the obligor prevails in those jurisdictions where the English rule is applied. In a minority of jurisdictions the American rule is applied, which gives preference to the first assignee who receives the assignment.

PROBLEMS

1. Mrs. Matternes and her son were driving on Interstate 40 when they came to a bridge where there was an accumulation of snow and ice as well as a sharp curve on the bridge itself. The automobile went out of control, killing Mrs. Matternes and injuring the minor child. James R. Matternes sued the City of Winston-Salem for the wrongful death of his wife, for injuries to his child, and for property damage. The plaintiff alleged that they were third-party beneficiaries of a contract between the State Highway Commission, now the Board of Transportation, and the City of Winston-Salem. Under the terms of the contract the city was to assume maintenance of the bridge, which it failed to do, thus allegedly breaching its contract. Can Mr. Matternes recover for the death of his wife and injuries of his son on the grounds that they were third-party beneficiaries of the contract between Winston-Salem and the Board of Transportation? Matternes v. City of Winston-Salem, 209 S.E.2d 481 (N.C. 1974).
2. Rensselaer Water Company had a contract with the city in which it agreed to supply water to the city. Moch Co. had a fire in its building. When the fire company responded, it found that there was inadequate water pressure at the fire hydrants, and as a result Moch's building burned. Moch contended that it should recover from Rensselaer as a third-party beneficiary of the contract between the city and Rensselaer. Was Moch correct? Moch Co. v. Rensselaer Water Co., 247 N.Y. 160, 159 N.E. 896 (1928).
3. Langel contracted to sell Hurwitz and Hollander certain real property. The purchasers assigned the contract to Benedict, who in turn assigned it to Betz. There was no delegation of duties to the assignee. Betz, the assignee,

subsequently refused to go through with the purchase of the property and was sued by Langel for specific performance. What result? Langel v. Betz, 250 N.Y. 159, 164 N.E. 890 (1928).

4. The marriage of John and Faith Vrendenburgh came to an unhappy end with their divorce and the execution of a separation agreement, pursuant to which John agreed to make periodic payments to Faith, provided that she complied with certain provisions of the agreement. Faith assigned her rights to the money to Alexis I. duPont de Bie. He sued Faith and John for money due under the agreement. The trial court refused to award Alexis the money due after the assignment was made, although it did give him judgment for the $375,827.50 due prior to the assignment. Was the trial court correct? DuPont de Bie v. Vrendenburgh, 490 F.2d 1057 (1974).
5. Holly loaned Fox $300, stating that he owed Lawrence that amount and had agreed to pay him the next day. Fox agreed to pay Lawrence the amount he borrowed from Holly the next day. Fox did not pay the next day and was sued by Lawrence. Fox raised the defense of lack of privity of contract and want of consideration. Was he correct? Lawrence v. Fox, 20 N.Y. 268 (1859).
6. Anna Gift contracted with Midland Chautauqua, a talent bureau, to supply performers, a tent, advertising, and other similar services for a Chautauqua assembly at Mankato, Kansas. Midland then assigned the contract to Standard Chautauqua System. Gift refused to assist Standard, as the contract with Midland had required, and was sued by Standard for $950 damages. Was Standard entitled to the damages? Standard Chautauqua System v. Gift, 120 Kan. 101, 242 P. 145 (1926).
7. Superior Brassiere Company agreed to manufacture brassieres and corsets for Francbust, Inc. Francbust, Inc., assigned to Superior all moneys due and to become due for goods shipped by Superior to customers of Francbust, Inc. Unfortunately for Superior, Francbust made a subsequent assignment of the same accounts to Finance Trust and others. Finance Trust collected from most of the customers of Francbust, Inc. The debtors received no notice from Superior prior to paying Finance Trust. Superior Brassiere sued Finance Trust and others, contending that it was entitled to the money. Was Superior correct? Superior Brassiere Co. v. Zimetbaum, 212 N.Y.S. 473 (1925).
8. Mr. Wing placed himself in the hands of Allstate Insurance. When he was sued after a collision with another vehicle, Allstate refused to settle the case. Judgment was entered against Wing for $20,000, twice the amount of his policy. Wing claimed against Allstate for wrongfully refusing to settle, went into bankruptcy, and sold the claim to Selfridge. Selfridge sued Allstate. Judgment was rendered in favor of Allstate on grounds that the claim could not be assigned. Selfridge appealed. Decide. Selfridge v. Allstate Insurance Co., 219 So.2d 127 (Fla.App. 1969).

CHAPTER 13

Contractual Performance and Discharge

This chapter examines a number of legal concepts that may result in the termination or discharge of the parties' obligations under the contract. Consideration will also be given to contractual clauses that may shift the risks of contractual performance from one party to another.

13:1. CONDITIONS. One useful way of shifting the risk that certain events may or may not occur is by the use of conditions. A condition is a clause in a contract that will shift, suspend, modify, or rescind an obligation based upon some operative fact or event. Conditions may also be express, implied in fact, or implied in law. Express conditions may be either conditions precedent, conditions subsequent, or conditions concurrent. No particular words or phrases are necessary to create conditional clauses, although they are frequently preceded by such words as *provided, as soon as, upon, while,* and many others.

(1) **Conditions Precedent.** A condition precedent is a specified condition that must occur before the agreement between the parties can become binding or before a party is required to perform a duty or obligation under the contract. In other words, a condition precedent is an event that must occur before or preceding that point in time when a party's duty or liability under the contract arises.

Conditions precedent are frequently used to shift the risk that a certain event will or will not occur from one party to another. For instance, Manny Manufacturer may enter into a contract with Carl Clothier that by its terms provides that Manufacturer is to ship 100 pairs of slacks at $10 per pair "ninety days after Manufacturer receives the necessary cloth from Sammy Supplier." Manufacturer's liability to Clothier does not arise until after and unless he receives the necessary cloth from Supplier. The risk that Supplier will not perform is shifted by the condition precedent from Manufacturer to Clothier. The necessity of Supplier's performance precedes Manufacturer's liability to Clothier.

(2) **Conditions Subsequent.** Whereas a condition precedent is often an event that must occur prior to or preceding a contractual obligation, a condition

subsequent is an event that occurs after or subsequent to that point in time when one is obligated under the contract. A condition subsequent operates to cut off contractual liability. The contract may specify that in the event that the condition is breached, one of the parties may treat the contract as discharged. For example, goods often may be purchased under a "sale or return" policy. The buyer of the goods is immediately obligated to pay, but if he returns them within a certain time period, this event operates to cut off his obligation. Note that the condition terminating his obligation occurs subsequent to the obligation arising under the contract, whereas a condition precedent is an operative factor occurring prior to the duty or the obligation's arising.

(3) **Conditions Concurrent.** When the parties under a contract are required to perform simultaneously, the performances are conditions concurrent. Thus, in order to fix the liability of one party, there must be a tender of performance by the other. Tender will be discussed more fully in Section (1) (e) of this chapter, but a typical contract of this type occurs when Frank Farmer agrees to deliver 100 bushels of corn to Warren Wholesaler for cash, with delivery and payment to take place concurrently. Each party must be ready, willing, and able to perform his part of the contract, or at least tender performance, before he can place the other party in default and maintain an action for breach of contract.

(4) **Conditions as Express, Implied in Fact, or Implied in Law.** Just as with contracts in general, conditions may be either express or implied in fact. They may also be created by the courts, that is, implied in law. These are sometimes called *constructive conditions.* As previously stated, no specific words are necessary to create an express condition as long as the words used manifest the intent of the parties that the clause operate as a condition.

Conditions implied in fact are also considered to be part of the contract and result from the express language of the agreement. Such conditions are implied from the contract terms as actually being intended by the parties to be part of the agreement.

A condition implied in law is one that the court creates, when necessary, in order to make the performances of one party depend on the performances of the other, where this dependence is not inconsistent with the terms of the contract. Thus, if Alphonse contracts to buy a motor scooter from Easy Rider for $300 and no mention is made in the agreement as to the time of delivery and payment, the court will imply that the performances are not independent of one another. The justification for implying mutual promises to be dependent is to be fair to the parties and to relieve any potential hardship.

MASCIONI v. I. B. MILLER, INC.

261 N.Y. 1, 184 N.E. 473 (1933)

Mascioni sued Miller, Inc., to recover money that it contended Miller, Inc., owed for work performed by Mascioni as a subcontractor for Miller, Inc. The trial court decided in favor of Miller, Inc., but this decision was reversed by the intermediate level appellate court. Miller then appealed to the highest appellate court.

LEHMAN, J.

The plaintiffs and the defendant entered into a written contract whereby the plaintiffs, described in the contract as the "Sub-Contractor," agreed to provide all the materials and all the work for the erection of concrete walls, and the defendant, described in the contract as the "Contractor," agreed to pay therefore the sum of 55 cents per cubic foot. The concrete walls were to be erected as "specified in a certain contract between the Contractor and Village Apartments, Inc., described therein as Owner," and the defendant's promise to pay contained the proviso, "Payments to be made as received from the Owner." In spite of the fact that the Owner has made no payments to the defendant for the work and materials, or any part thereof, performed and furnished by the plaintiffs, the plaintiffs have recovered a judgment against the defendant for the agreed price.

The problem presented on this appeal is whether the defendant assumed an absolute obligation to pay, though for convenience payment might be postponed till moneys were received from the owner, or whether the defendant's obligation to pay arose only if and when the owner made payment to the defendant.

A provision for the payment of an obligation upon the happening of an event does not become absolute until the happening of the event. Whether the defendant's express promise to pay is construed as a promise to pay "if" payment is made by the owner or "when" such payment is made, "the result must be the same; since, if the event does not befall, or a time coincident with the happening of the event does not arrive, in neither case may performance be exacted."

Here on its face the contract provides for a promise to perform in exchange for a promise to pay as payments are "received from the Owner." Performance by the plaintiff would inure directly to the benefit of the owner and indirectly to the benefit of the defendant, because the defendant had a contract with the owner to perform the work for a stipulated price. The defendant would not profit by the plaintiff's performance unless the owner paid the stipulated price. That was the defendant's risk, but the defendant's promise to pay the plaintiffs for stipulated work on condition that payment was received by the defendant shifted that risk to the plaintiffs, if the condition was a material part of the exchange of plaintiff's promise to perform for defendant's promise to pay.

Here we are not called upon to decide whether the language of the contract, read in the light of the situation of the parties and the subject-matter of the contract, shows clearly and unambiguously that the condition attached to the debt or obligation to pay, and did not merely fix the time of payment. Certainly on its face it is open to the construction that the plaintiffs accepted the condition as a material part of the exchange of their own promise or performance. The trial judge, after receiving parol evidence of the actual intention of the parties, gave it this construction, and that construction was not erroneous as matter of law.

(Judgment for defendant.)

QUESTIONS FOR DISCUSSION

1. *Does this case involve a condition precedent, concurrent, or subsequent?*
2. *What function does the condition in this case perform? Does the decision in the case mean that Miller, Inc., was discharged from its obligation to Mascioni?*

13:2. DISCHARGE. There are a number of ways in which a contract may be discharged or terminated: by performance, by impossibility or commercial frustration, by agreement, and by operation of law.

(1) Discharge by Performance of Condition. One of the purposes of a contract is to set forth the conditions for fulfilling the agreement. Obviously, when those conditions have been fulfilled, the rights and obligations of the parties cease. One of the problems facing the courts is to determine when the performance rendered by one of the parties satisfies the requirements or conditions of the contract and when it does not. Another problem is to determine when nonperformance of a condition will discharge the innocent party from liability under the contract, or alternatively whether a nonperformance, although not discharging the innocent party from liability, will give him a right to sue for damages.

(a) *The Doctrine of Substantial Performance.* In some cases it is not difficult to comply with the terms of certain types of agreements, such as obligations to pay money. It is easy to determine what performance is required under the contract and to pay the sum due. Other types of contractual performance are not so simple to perform. The best examples are with regard to building contracts. According to the old common-law rule, strict performance of the contract terms was required in order to allow recovery under the contract. Today most jurisdictions have developed the doctrine of substantial performance, particularly with regard to building contracts and other analagous agreements. Because it is almost impossible to complete a construction project absolutely free from defects of some kind, the question rises as to whether the existence of relatively small deficiencies in performance will allow the innocent party to rescind the contract and be discharged from liability.

Obviously to allow one to be discharged from a contract for a small defect in performance, when the deficiency does not result intentionally, would be to place an unfair burden on the other party to the agreement and would often result in an unjust outcome, particularly if the innocent party receives a benefit from the performance.

The doctrine of substantial performance is applicable where full or exact performance of every detail of the contract has not been rendered but where there has been a good-faith attempt to perform the agreement according to its terms, resulting in substantial compliance with the contract. Substantial performance is a matter of degree, and each case must be resolved under its own facts and circumstances. If the deficiencies are so great as to pervade the entire contract or so material that little benefit is conferred on the innocent party, or when the objects of the contract are not accomplished, then the doctrine will not be applied. In addition, the doctrine generally will not be applied where the nonperformance is the result of willful or intentional omissions on the part of the promissor.

Where the doctrine of substantial performance is applicable, the party substantially performing under the contract is entitled to recover but subject to the right of the innocent party to be compensated for the damages occasioned by the defects in performance. The doctrine of substantial performance does not discharge the innocent party from liability under the contract, and if he fails to pay, he may be sued by the party rendering the substantial performance. We will see in the next section that in order for the innocent party to be discharged from liability, there must be a material breach of the agreement.

KICHLER'S INC. v. PERSINGER

24 Ohio App.2d 124, 265 N.E.2d 319 (1970)

Kichler's Inc., was engaged in the business of selling custom slipcovers, bedspreads, draperies, shades, and valances for homes.

Ann Persinger selected two bedspreads, two valances, and two laminated shades from a catalog. She paid Kichler's $40.00, leaving a balance due of

$82.54. The bedspreads were to reach from the top edge of the mattress to the floor, a distance of twenty-four inches. All items were to match, and Persinger suggested that the bedspreads, valances, and shades were to be made from the same bolt of cloth.

Alas, when the bedspread arrived, it was twenty-two inches instead of twenty-four. Kichler's valiantly attempted to rectify this fault by adding a two-inch ruffle. Persinger refused to accept the spread and stated that the material in the bedspread did not match or harmonize with the valances and shades. She refused to pay the balance and demanded return of the deposit. Kichler's sued for the balance, and Persinger filed a cross-petition for her deposit. The lower court found Kichler's had substantially performed and awarded it judgment. Persinger appealed.

HESS, J.

Upon the essential facts presented the trial court found the following:

1. The plaintiff, by adding the ruffle to the second set of bedspreads in order to make them the required depth of 24 inches, did not, in the court's opinion, materially change to a substantial degree the type and design of bedspread for which the defendant—cross petitioner contracted; and,
2. The variation in brightness and vividness of color between the shades and valances on the one hand and the bedspreads (both the original set and the second set submitted) on the other, was not of such sufficient nature that it should have been objectionable to the defendant.

Substantial performance of a contract is interpreted to mean that mere nominal, trifling, or technical departures are not sufficient to break a contract, and that slight departures, omissions and inadvertences should be disregarded.

There is a great difference in performing an interior decorating contract wherein the purchaser of materials is concerned with the motif and decor of the premises and an ordinary building or service contract where slight variances in the use of materials do not change the appearance of the premises.

In the instant case, the trial court found the bedspreads, shades and valances did not harmonize in color, and that the bedspreads were not made in keeping with the "Mary Beth" spread the defendant purchased, and concluded that such disparity was "not of sufficient nature that it should have been objectionable to the defendant"; and further, that the addition of a ruffle did not ". . . materially change to a substantial degree the type and design of bedspread."

It is not within the power of the court to determine whether the defendant should have been satisfied with the change in color and style of the bedspreads. The defendant was entitled to receive the style and color of furnishings in decorating her bedroom that she desired and specified. The plaintiff was obliged by his contract to provide the materials desired and ordered. It is common knowledge that color and style of interior decorations vary according to the desires and aesthetic pleasure of individuals. That which may be acceptable to one individual could be wholly unacceptable to another.

It is evident from the record in this case that there was a breach of contract by the plaintiff and that the defendant should not be required to accept and pay for the materials described therein, and that the Defendant is entitled to judgment on her cross petition.

(Judgment reversed.)

QUESTIONS FOR DISCUSSION

1. *How was the court's decision affected by the fact that this was not a construction contract?*
2. *Does the court's decision mean that the bedspread would have had to measure exactly twenty-four inches in order to satisfy the terms of the contract, or would there have been any variance for substantial performance to have been rendered?*

HUNTER v. ANDREWS
570 S.W.2d 590 (Tex. App. 1978)

John Hunter contracted with F. D. Andrews to have Andrews construct a swimming pool. The total price of the pool was $8,000.00, of which Hunter paid $4,300.00 upon signing the contract. After the pool was constructed, Hunter furnished Andrews with a list of defects concerning the pool and told Andrews he would not pay the balance owed on the pool. Andrews then sued Hunter for the amount still due *on the contract. The trial court found in favor of Andrews for the balance due less $400.00 to repair minor defects. Hunter appealed.*

McDONALD, J.

The court heard and weighed the evidence; and observed and passed on the credibility of the witnesses. We think the evidence ample to support the court's findings that plaintiff substantially complied with the contract, completed the swimming pool, that defects were minor; and that such findings are not against the great weight and preponderance of the evidence.

The law concerning the right of a building contractor to recover for work done and performed under a building contract is stated by our Supreme Court in *Graves* v. *Allert & Fuess*, 104 Tex. 614, 142 S.W. 869, and is: In the case of building and construction contracts, strict and literal performance is not essential to enable the contractor to recover on the contract; if he has substantially performed, he may recover the contract price less the reasonable cost of remedying the trivial defects and omissions so as to make the structure comply with the contract.

(Judgment in favor of Andrews affirmed.)

QUESTION FOR DISCUSSION

1. *Does this decision mean that Andrews is entitled to the entire $3,700.00 due on the contract?*

(b) *Partial Performance.* Substantial performance should not be confused with partial performance of a contract. Partial performance results when a portion of the required performance is not completed, generally intentionally. When the innocent party has done nothing to prevent full performance by the promisor and has not waived his right to full performance, there can be no recovery under an entire contract by the promisor. If the innocent party accepts the benefits of a partially performed contract, he thereby waives his right to full performance. The party performing will be allowed to recover the reasonable or fair value of the performance. Suppose Bill Builder contracts with Harry Homeowner to build a house for Homeowner on a lot owned by Builder. Builder stops work on the house when it is half-completed. Homeowner would not have to accept or pay anything under the contract. Builder cannot even recover for the work performed. However, Homeowner could accept the partially completed house from Builder and pay him the reasonable value of the labor and the materials furnished. There has been no substantial performance by Builder because the house was not completed and he intentionally stopped work on it.

(c) *Performance to Satisfaction of Promise or Third Party.* Without a specific contractual provision, one may not refuse to pay for contractual performance on the sole ground that he is not satisfied with the results. If there is a provision inserted in a contract to the effect that the performance must be to the satisfaction of the promisee, the courts will, as a general rule, enforce it. Contracts containing this provision can be grouped into two types: those contracts in which one's taste, judgment, and fancy are involved and those contracts involving operative fitness or mechanical utility.

In contracts involving one's personal taste, the courts will generally enforce a contractual provision requiring performance to the promisee's satisfaction. Perhaps the best example would be a case in which one is to have his picture painted and the artist is to be paid conditioned on the subject's being satisfied with the results. If the subject is not satisfied, he is not required to pay. It is the personal decision and satisfaction of the promisee that is important. It does not matter that a "reasonable man" would be satisfied. The promisee's satisfaction is the determinative factor. The reasons for the dissatisfaction are not important, although some jurisdictions hold that the court may inquire into whether the expressed dissatisfaction has been made in good faith.

Court decisions are somewhat in conflict when the performance involves mechanical utility. Some courts treat this type of case similarly to cases involving personal taste and hold that the performance must be to the satisfaction of the promisee when there is a contractual provision to that effect. Many courts have, however, been reluctant to take this approach, no doubt because there is a greater element of possible objectivity when a contract involves mechanical utility, such as of a machine, as opposed to the ability of a painter. A number of courts have held that when there is a provision involving mechanical or operative fitness, the test is whether the reasonable man would be satisfied with the performance.

On occasion, the acceptance of contractual performance may be subject to the approval of a third person. Such cases most frequently occur in building and construction contracts, where the contract provides that the work must be approved by an architect prior to payment. These provisions are valid and enforceable even when the third party is the agent of one of the parties to the contract. Such a requirement must be stated in the agreement in order to be enforceable, and one may not refuse to pay on the ground that a third party has not approved the work unless that is a provision of the contract. But if the architect does not approve the work, it must be an honest disapproval and not one induced by the party to the agreement.

COLUMBIA CHRISTIAN COLLEGE v. COMMONWEALTH PROPERTIES, INC.
594 P.2d 401 (Or. 1979)

Columbia Christian College owned a 268-acre tract of land in Multnormal County, Oregon. Commonwealth Properties expressed an interest in purchasing the property for development. In order to do this a number of zoning changes were required and a plan had to be approved by the city planning staff of the City of Gresham.

Commonwealth signed an option agreement to purchase the property from Columbia Christian College. Part of the agreement stated:

> 3. Upon completion of paragraph 2 above, and still within the 180-day option period, Commonwealth shall (a) attempt to secure approval to annex the site to the City of Gresham; (b) apply for and obtain satisfactory zoning; (c) apply for permission to connect to city utility services, which may include the formation of an improvement district to accomplish this end. . . .
>
> . . .
>
> In the event a decision satisfactory to Commonwealth Properties cannot be obtained to conditions 3(a), 3(b) or 3(c), then the option shall be terminated. . . .

Commonwealth was successful in getting the land annexed to the city. It also applied for and received a number of zoning changes after a period of time. After some further negotiations, Commonwealth decided not to go through with the purchase stating that the zoning was not satisfactory. The college filed suit to enforce the contract.

HOWELL, J.

Plaintiff contends that inability to obtain satisfactory zoning is not the "real" reason defendant refused to proceed with the transaction. There was evidence that in late 1974 defendant no longer considered the planned development economical, and that due to the general decline in the housing market, defendant intended to limit itself to smaller projects. There was also evidence, however, that in December of 1974, defendant's vice-president, Ernest Platt, told Paul Zeger, plaintiff's real estate broker, that defendant would "pass up" its option to purchase the property due to the "unsatisfactory zoning."

It is true that where a contract is made subject to the occurrence of a condition to the "satisfaction" of one party, the party's dissatisfaction must relate to the specific subject matter of the condition. *Western Hills* v. *Pfau,* 265 Or. 137, 144–45, 508 P.2d 201 (1973). It does not follow, however, the dissatisfaction with other aspects of the bargain as well means a party is acting in bad faith. A party may be dissatisfied with a number of aspects of a bargain, some of which allow him to repudiate the contract and some of which do not. If one of the sources of dissatisfaction gives him a right under the contract to repudiate, the fact that there are other sources of dissatisfaction is immaterial.

The present case illustrates the artificiality of requiring that a party be dissatisfied only with the subject matter of a condition, and not with anything else. Plaintiff contends that the "real" reason defendant called off the contract was its dissatisfaction with the general economics of the transaction. It is clear, however, that one of the reasons the project had become uneconomical for defendant was precisely because of the conditions the city attached to the zoning ordinance. This is demonstrated by Platt's testimony at trial:

> Q. Now, were these conditions as they were ultimately drafted into the Ordinance, by the City Council satisfactory to Commonwealth?
>
> A. No.
>
> Q. Explain in some detail, if you will, why they were not?
>
> A. Well, they are ambiguous, at best, as to what it is we are to satisfy. But to even attempt to satisfy them could cost nothing but a considerable sum of money. And my latest, most recent computations in performance of the project prior to that Council meeting was that the project was borderline at that time.

Upon de novo review of the evidence present at trial, we conclude that defendant's dissatisfaction with the zoning was in good faith. The fact that this dissatisfaction also was tied to the economics of the project as a whole does not mean defendant acted in bad faith. As defendant observes in its brief, "The suitability of available zoning is not determined in a vacuum."

Plaintiff also contends that defendant's actions after the ordinance was passed demonstrate that defendant was not dissatisfied with the zoning. Plaintiff notes that after the ordinance was passed defendant "proceeded to act as if it were prepared to complete the transaction by having its lawyers draft a final contract." We do not think this evidence is sufficient to show that defendant was satisfied with the zoning it obtained. Considering the magnitude of the transaction involved in this case, defendant was entitled to consider its options (proceeding with the purchase or refusing to proceed) after the city acted on its zoning application. The fact that it took steps toward a possible purchase did not mean it was satisfied with the zoning. Nor does that fact bar it from repudiating the contract on that basis.

For all of these reasons, we conclude that defendant performed its duty under the contract to reasonably pursue satisfactory zoning, and that the City's refusal to grant defendant satisfactory zoning excused defendant from further performance. It follows that the plaintiff is not entitled to equitable compensation for having its property tied up during the period of the option. Plaintiff assumed the risk of the possibility that its property might be removed from the market for a certain period of time without a sale being closed when it entered into the contract with defendant. Since plaintiff bargained for that contingency, it cannot now claim compensation simply because that contingency occurred.

(Judgment in favor of Commonwealth Properties, Inc.)

QUESTIONS FOR DISCUSSION

1. *What test was used by the court in order to determine that Commonwealth could avoid the agreement because it was not satisfied with the zoning?*
2. *Does the repudiation of the agreement based upon dissatisfaction of the zoning have to be measured by a "reasonable man" standard?*

(d) *Time of Performance.* The question often arises as to the rights of the parties when performance is not rendered by the time specified in the contract. Does the failure to render performance within the time specified give the other party the right to avoid the agreement? The answer depends to a large extent upon the intent of the parties. If there is a time for performance stated in the contract and the parties consider "time to be of the essence," then the failure by a party to perform within that time will discharge the other party to the agreement. When time is of the essence, it operates as a condition precedent to the performance of the other party, the failure of which will discharge the contract.

Whether the parties consider time to be of the essence is a question of fact to be determined from the subject matter of and the circumstances surrounding the agreement. For instance, time may be considered to be of the essence for a contract involving the shipment of perishable vegetables. Time may not be of the essence in a contract for the construction of an office building. The parties may state their intent in the body of the contract by stating that they consider "time to be of the essence" or by using other words of similar import. The language is usually given great weight by the court in its determination of the intent of the parties, so when the parties feel that time is important to the contractual performance, this should be stated clearly in the agreement.

If no time for performance is stated in the contract, the courts imply that performance is to be rendered within a "reasonable time." Once again, the subject matter of the contract as well as the facts and circumstances surrounding the agreement are considered in this determination.

CARTER v. SHERBURNE CORP.
315 A.2d 870 (Vt. 1974)

Garth Carter entered into several contracts with Sherburne Corp., a developer. Sherburne contended that the work was not performed within the time specified in the contracts and that some of the work was defective. Carter sued for the balance due under the contracts and for other work done. The lower court found for Carter, and Sherburne appealed.

SHANGRAW, C.J.

There were four written contracts between the parties covering (a) the furnishing and placing of gravel on one road, (b) the drilling and blasting of rock on various residential roads, (c) road construction, and (d) the cutting and grubbing of a gondola lift-line. The contracts called for weekly progress payments based upon work completed with a provision for retaining 10% until ten days after final acceptance. The billings from the plaintiff to the defendant amounted to $52,571.25 of which $41,368.05 was paid by the defendant. The difference between the $52,571.25 billed, and $41,368.05 paid, comprised $4,596.45 retained by the defendant under its holdback provision, and adjustments claimed by the defendant of $6,606.75. The Court found that adjustments in the amount of $4,747.25 were improperly taken by the defendant and that amount was decreed to the plaintiff. In addition the Court found that the plaintiff was entitled to all the retainage held by the defendant.

The defendant's primary contention is that the Court's ruling that the plaintiff was in substantial compliance under his contracts is in error. The contention is that this ruling was based on the erroneous conclusion that time was not of the essence of the contracts, and that as time was of the essence and plaintiff failed to perform within the time specified, plaintiff was not in substantial compliance and defendant is entitled to the amounts withheld as retainage.

Where time is of the essence, peformance on time is a constructive condition of the other party's duty, usually the duty to pay for the performance rendered. This may be made of the essence of a contract by a stipulation to that effect, or by any language that expressly provides that the contract will be void if performance is not within a specified time. Where the parties have not expressly declared their intention, the determination as to whether time is of the essence depends on the intention of the parties, the circumstances surrounding the transaction, and the subject matter of the contract.

As a general rule, time is not of the essence in a building or construction contract in the absence of an express provision making it such. "Construction contracts are subject to many delays for innumerable reasons, the blame for which may be difficult to assess. The structure . . . becomes part of the land and adds to the wealth of its owner. Delays are generally foreseen as probable; and the risks thereof are discounted. . . . The complexities of the work, the difficulties commonly encountered, the custom of men in such cases, all these lead to the result that performance at the agreed time by the contractor is not of the essence." 3A A. Corbin, Contracts § 720, at 377 (1960).

We conclude, then, that time was not of the essence of any of the contracts considered here. None of the four contracts included express language making time of the essence, and we can find nothing in the circumstances surrounding these contracts that would lift them out of the operation of the general rule. Two of the contracts called for completion dates and forfeitures for non-completion on schedule, but the inclusion of dates in construction contracts does not make time of the essence. Moreover, the inclusion of penalty or forfeiture provisions for noncompletion on schedule is strong evidence that time is not of the essence and that performance on time is not a condition of the other party's duty to accept and pay for the performance rendered. 3A A. Corbin, Contracts § 720 (1960).

Ordinarily, in contracts where time is not of the essence, a failure to complete the work within the specified time will not terminate the contract, but it will subject the contractor to damages for the delay. However, in this case, most of the delays were due to the actions of the defendant corporation in constantly shifting the plaintiff's activities from one contract to another, and in improperly withholding the plaintiff's payments. Delay in the performance of a contract will, as a rule, be excused where it is caused by the act of default of the opposite party, or by the act or default of persons for whose conduct the opposite party is responsible. Where this is the case, the contractor will not be held liable, under a provision for liquidated damages or otherwise, for his non-compliance with the terms of the contract; and his non-compliance will not be considered a breach. An obligation of good faith and fair dealing is an implied term in every contract, and a party may not obstruct, hinder, or delay a contractor's work and then seek damages for the delay thus occasioned.

(Judgment affirmed.)

QUESTIONS FOR DISCUSSION

1. *What steps could Sherburne have taken to protect itself against the possibility that the court would determine that time was not of the essence?*
2. *What rights would Sherburne have had if time was of the essence?*

(e) *Tender of Performance.* Tender of performance is simply offering to render the performance due under a contract. Tender is necessary in order to fix the other party's liability under the agreement when the acts to be performed are concurrent or when one's contractual right is dependent upon the performance of duties on his part. Tender shows that one is ready, willing, and able to perform. It is not necessary actually to perform in order to place the other party in default, but an offer to perform must be made. When the tender would be a vain and idle act it will be excused. For instance, if the other party to an agreement is not present at the place of performance and his presence is necessary if he is to perform the required act, tender may be excused.

The fact that one has tendered performance may also be a defense. Assume that Bill Buyer agrees to purchase goods from Sam Seller on a certain date. Seller offers the goods to Buyer on that date and Buyer wrongfully refuses delivery. Seller disposes of the goods elsewhere and is later sued by Buyer for nonperformance. Seller's tender woud be a good defense to an action for breach of contract brought by Buyer.

If one's contractual obligation consists of paying money, an offer to pay the debt when due has several effects if acceptance of the tender is refused by the creditor. Although the debt will not be discharged, (1) the interest will stop running as of the date of tender; (2) any security interest in property securing the debt will be extinguished; and (3) the debtor is relieved from the obligation to pay any court costs should the creditor subsequently sue.

In order to be effective against the creditor, the tender cannot be made before the date due. In addition, the creditor may refuse to accept a check on the ground that it is not legal tender, and if a check is accepted, the tender is conditional upon its payment. Naturally the tender must be in the amount due the creditor and not a lesser sum.

VANDER REALTY CO. v. GABRIEL

134 N.E.2d 901 (Mass. 1956)

Plaintiff sought to recover a deposit made by it under an agreement to purchase real estate. The lower court found for defendant, and plaintiff appealed.

SPALDING, J.

The facts pertinent to this appeal are these. On January 26, 1953, the parties entered into a written agreement for the sale by the defendant and the purchase by the plaintiff of eight lots of land in the town of Sharon. The purchase price was $8,000, of which $1,000 was paid by the plaintiff to the defendant upon the execution of the agreement. The agreement called for performance on or before June 15, 1953, but it contained a provision that "Either party may have thirty days extension to cure any defect found in title." The agreement also contained the following provisions: "If seller is prevented from performing by defect (of title) not caused by him, this agreement shall terminate and the seller shall return the deposit." "If the buyer fails to tender the entire consideration and accept conveyance of all the lots prior to June 15, 1953, this agreement shall terminate and the seller may retain the sums paid as a deposit as liquidated damages. . . . The seller agrees to tar surface of roadway of town . . . (specifications)."

Neither the deed nor the consideration was tendered on or before June 15, 1953; nor was the agreement extended. "The defendant did not tar the surface of the roadway, neither did the plaintiff call upon him to do so before the expiration of the agreement or in any manner state that it was going to void the agreement because that had not been done. As far as observing the terms of the agreement went, both parties let the matter drop. Nothing further was done by either until after the plaintiff had the title examined in 1954. Why it caused the examination to be made at that late date did not appear. Then upon report of the examiner that the title was not marketable, which was so, the plaintiff demanded return of its deposit."

The judge concluded that "Since no defect (in the title) was called to the attention of the seller within the term of the agreement, he had no opportunity to exercise the right given him to cure it." He ruled that the plaintiff having failed "to tender the entire consideration and accept conveyance of all the lots prior to June 15, 1953," the agreement was terminated and the defendant was entitled under the agreement to retain the deposit as liquidated damages.

In support of the request the plaintiff argues in substance that it was not in default because on June 15, 1953, when the agreement was to be performed, the defendant was unable to perform for the reason that he had not tarred the roadway which under the agreement he was obligated to do; that the promises in the agreement were mutual and dependent and the tarring of the roadway was a condition precedent to performance by the plaintiff; and that the failure to tar the roadway prior to the time of performance was a renunciation of the agreement by inconsistent conduct and relieved the plaintiff from the necessity of offering performance. It is apparent that the judge in failing to act on the plaintiff's fourth request and in finding for the defendant took the view that the promise to surface the roadway was not a condition precedent to the plaintiff's obligation to perform. We think that this was error and that the plaintiff was entitled to the requested ruling and to a finding in its favor on the facts found.

We are of opinion that the covenant of the plaintiff to pay the balance of the purchase price and the covenant of the defendant to tar the roadway were mutually dependent covenants and were to be performed not later than June 15, 1953.

The general rule is that when performance under a contract is concurrent one party cannot put the other in default unless he is ready, able and willing to perform and has manifested this by some offer of performance. *Leigh* v. *Rule*, 331 Mass. 664. But as we said in that case "the law does not require a party to tender performance if the other party has shown that he cannot or will not perform." Here at the time for performance the defendant had demonstrated his inability to perform by not having tarred the roadway as he had agreed to. No contention is made that the plaintiff waived this provision; and it did not lose the right to have it performed by not calling upon the defendant to do so prior to the date of performance. The plaintiff in these circumstances was not obliged to tender performance, and was entitled to rescind the contract and to the return of its deposit.

(Reversed in favor of plaintiff.)

QUESTIONS FOR DISCUSSION

1. *What were the significant conditions in this case? Were the conditions mentioned in the case precedent, concurrent, or subsequent?*
2. *Did the plaintiff have to tender performance? Why or why not?*

---

(2) Discharge by Impossibility of Performance and Frustration of Purpose. If performance of the contract is physically or legally impossible at the time of its making, neither party is liable. The courts generally excuse performance under this circumstance on the ground of mutual mistake or, if the parties were aware of the impossibility of performance, lack of contractual intent.

(a) *Subsequent Foreseeable Events.* The situation becomes vastly more complicated when an event that makes performance impossible occurs subsequent to the creation of the contract. Remember that one of the functions of a contract is to allocate risk between the parties. One of these risks is that some event will occur that may render performance impossible. If it is reasonably foreseeable that certain events might occur that would render performance impossible, there would seem to be little question that the actual occurrence of those events will not discharge the contract. It is assumed that without a clause excusing performance under such circumstances, the party chose to run the risk that such an event beyond his control rendered performance impossible. Obviously, when possible, it is important to insert contractual language excusing performance, should these forseeable events occur.

(b) *Subsequent Unforseeable Events.* As a general rule, so-called acts of God or even unavoidable accidents rendering performance impossible will not allow one to escape liability for damages, for the reasons stated previously. There are some exceptions, however, when the event is not foreseeable.

(i) DESTRUCTION OF SUBJECT MATTER. When the subject matter of a contract is destroyed so that performance of the contract is impossible, the contract will be discharged. The subject matter destroyed must be the specific subject matter of the contract, however, and not just the subject matter one of the parties intended to use to perform the contract. For instance, suppose that Frank Farmer contracts to sell Super Market 1,000 bushels of corn. He intends to use the corn located in his fields, but the night before performance is due, fire destroys his crop. Farmer would still have to supply the corn or be liable to Market in damages because the specific field of corn was not the subject matter of the contract. On the other hand, should the agreement state that the corn is to be the corn in Farmer's "north forty acres," which is destroyed by fire, the contract would be discharged.

(ii) DEATH OR INCAPACITY OF THE PARTIES. The death of a party may discharge a contract if the contract is for personal services or involves personal skill or a special relationship with the other party. Thus a contract with an artist to paint a portrait or with a singer to perform would be discharged at the death of the party who is to render the service. On the other hand, if the services are such that they might be provided by a personal representative, then the contract is not discharged. Obviously a contract calling for the payment of money by one of the parties will not be discharged at his death, as it can be paid by this estate. Likewise, if the party has a claim to money, his estate has a right to collect it and his death does not discharge the obligation of the other party. The same rules apply when one of the parties becomes insane subsequent to the formation of the contract.

IN RE STORMER'S ESTATE

385 Pa. 382, 123 A.2d 627 (1956)

CHIDSEY, J.

The question presented by this appeal is whether a contract for the construction of a sewerage system is of such nature that the obligations of the contractor thereunder are discharged by his death.

On March 18, 1952, Frederick A. Stormer entered into a contract with Shippenburg Borough Authority for the construction of a sewerage system. Mr. Stormer died on August 12, 1953, when the work was approximately 60% complete. His executors continued the work until January 21, 1954. When they ceased work on that date, the contract was declared in default by the Authority. Until October 3, 1953 the continuance of the work by the executors was under the direction of Hugh B. Stormer (one of the executors and a son of the decedent) who had been superintendent of the work during his father's lifetime. On the same date the executors petitioned the Orphans' Court of Blair County for a declaratory judgment to determine whether said contract and the performance bond securing it had been discharged by Stormer's death. Following the default declared by the Authority when the executors ceased work in January, the Authority which had brought in as a party plaintiff Fidelity and Deposit Company of Maryland, the surety on the contractor's bond, moved to have the petition for declaratory judgment dismissed on the ground that the question had become moot. On June 18, 1955, the lower court decreed that the contract was a binding obligation on the executors and the surety. A rehearing was ordered, following which the court affirmed its June 18th decree. This appeal by the executors followed.

The general rule is that to the extent of the assets that come into his possession, the personal representative of a decedent is responsible on all contracts incurred by decedent in his lifetime. An exception to this rule is: ". . . Where the agreement is for services which involve the peculiar skill of an expert, by whom alone the particular work in contemplation of the parties can be performed, or more generally, where distinctly personal considerations are at the foundation of the contract, the relation of the parties is dissolved by the death of him whose personal qualities constituted the particular inducement to the contract. . . . But where a party agrees to do that which does not necessarily require him to perform in person, that which he may, by assignment of his contract or otherwise, employ others to do, we may fairly infer, unless otherwise expressed, that a mere personal relation was not contemplated. . . ." Building contracts generally do not involve a peculiar skill or ability on the part of the person who is to perform them, and hence do not terminate on the death of the contractor. "It is otherwise, of course, where it is made to appear . . . that the character, credit and substance of the party contracted with was an inducement to the contract. . . ."

Appellants contend that the contract in question had at its foundation personal considerations and thus would terminate on decedent's death. Appellants base their contention on the fact that the award of the contract was predicated on the personal responsibility and competency of the contractor selected, and that performance of the work was deliberately restricted to a qualified contractor so selected, whose duties were nondelegable, in order that this permanent municipal improvement be well constructed.

We are unable to agree with appellants' contention. The formation of this contract was not induced by any peculiar ability or skill possessed by the decedent. What did induce the contract was the fact that decedent made the lowest bid. Price, not personal considerations, was the inducement. True, decedent had to qualify as a responsible bidder. But once in the group of qualified bidders, it was his price, not his qualifications, that induced the Authority to issue the contract to him.

An inspection of the contract documents substantiates our conclusion that the contract was not induced by personal considerations. The documents specify in detail the type of quality of the material to be used in the work, what the work was to consist of, where it was to be done, and the general methods of construction to be used. The only things left to the decedent were the mechanical details of excavation and construction. The evidence discloses, and the court below found, that the decedent's personal attention to the details of construction, his personal skill or taste, or his exclusive ability were not required, as the work progressed in charge of others, and that he was not on the job except periodically.

(Judgment affirmed.)

QUESTIONS FOR DISCUSSION

1. *In this case the death of the contractor did not discharge a contract to construct a town sewerage system. Do you think the decision would have been different if the contract had been for the construction of a private home?*
2. *Does the same rule apply on the death of an offeror prior to acceptance?*

(iii) CHANGE OF LAW. When a change in the law renders performance under the contract illegal and impossible, the contract is discharged. In this area, and the others discussed, the change must render legal performance impossible and not simply more difficult. Furthermore, if the change was foreseeable at the time of the formation of the contract, the courts assume that the risk of change was assumed, and the party bearing the risk is liable to the other party for any damages he sustains as a result of nonperformance.

(c) *Frustration of Purpose.* In more recent years the courts have eased the requirement that performance must be rendered impossible in order for a contract to be discharged under the circumstances previously discussed. When circumstances arise subsequent to the creation of the contract that makes performance fruitless and when the purpose of both parties to the contract cannot be accomplished, the contract will be discharged. The difference between this doctrine, known as *frustration of purpose* or *commercial frustration,* and impossiblity of performance is not always clear. Under the doctrine of frustration of purpose, although a literal performance of the contract might be possible, it will not be required if the purpose of the contract has been destroyed or frustrated. The courts often hold that such a condition is implied in the contract.

There are many limitations on the application of this doctrine. It will not be applied merely to extricate one from what turns out to be a bad bargain. In the first place, the purpose of the contract must be destroyed for both parties, not merely for one of them. In addition, if the event that occurs was reasonably foreseeable, the doctrine will not be applied. Also the doctrine generally will not be invoked except when the contract is executory. In sum, the same restrictions apply as in the doctrine of impossibility of performance except that the concept of strict impossibility has been relaxed somewhat. In one case, for example, there were elaborate preparations made for the coronation of Edward VII of England, including a grand procession that was to pass along Pall Mall. Henry agreed to rent a flat from Krell along the route in order to watch the procession, but this purpose was not stated in the contract. Unfortunately the king got sick and the procession was canceled. Krell sued Henry for the rent but was unable to recover. The court held that both parties were discharged from the agreement because the purpose for the contract was eliminated (*Krell* v. *Henry,* 2 K.B. 740 [1903]).

LLOYD v. MURPHY

25 Cal.2d 48, 153 P.2d 47 (1944)

Caroline S. Lloyd leased a showroom to William J. Murphy on August 4, 1941, "for the sole purpose of conducting thereon the business of displaying and selling new automobiles (including the servicing and repairing thereof and of selling petroleum products of a major oil company) and for no other purpose" except "to make an occasional sale of a used automobile." On January 1, 1942, the federal government prohibited the sale of new automobiles and modified this order on January 8, 1942, to permit sale to those engaged in military activities as well as persons having a rating of A-1-j or higher.

On March 15, 1942, Murphy vacated the premises after giving oral notice or repudiation of the lease. Lloyd agreed to waive the restrictions in the lease, but Murphy refused to continue to occupy the premises. Lloyd brought this action for declaration of rights under the lease and for rent. Defendant contended that the purposes of the lease were frustrated. The lower court found for Lloyd. On hearing before a panel of the Supreme Court of California, the case was reversed. Plaintiff petitioned for a rehearing in banc, which petition was granted.

TRAYNOR, J.

Although the doctrine of frustration is akin to the doctrine of impossibility of performance since both have developed from the commercial necessity of excusing performance in cases of extreme hardship, frustration is not a form of impossibility even under the modern definition of that term, which includes not only cases of physical impossibility but also cases of extreme impracticability of performance. Performance remains possible but the expected value of performance to the party seeking to be excused has been destroyed by a fortuitous event, which supervenes to cause an actual but not literal failure of consideration.

The question in cases involving frustration is whether the equities of the case, considered in the light of sound public policy, require placing the risk of a disruption or complete destruction of the contract equilibrium on defendant or plaintiff under the circumstances of a given case, and the answer depends on whether an unanticipated circumstance, the risk of which should not be fairly thrown on the promisor, has made performance vitally different from what was reasonably to be expected. The purpose of a contract is to place the risks of performance upon the promisor, and the relation of the parties, terms of the contract, and circumstances surrounding its formation must be examined to determine whether it can be fairly inferred that the risk of the event that has supervened to cause the alleged frustration was not reasonably forseeable. If it was foreseeable there should have been provision for it in the contract, and the absence of such a provision gives rise to the inference that the risk was assumed.

The doctrine of frustration has been limited to cases of extreme hardship so that businessmen, who must make their arrangements in advance, can rely with certainty on their contracts. The courts have required a promisor seeking to excuse himself from performance of his obligations to prove that the risk of the frustrating event was not reasonably forseeable and that the value of counterperformance is totally or nearly totally destroyed, for frustration is no defense if it was forseeable or controllable by the promisor, or if counterperformance remains valuable.

Thus laws or other governmental acts that make performance unprofitable or more difficult or expensive do not excuse the duty to perform a contractual obligation. It is settled that if parties have contracted with reference to a state of war or have contemplated the risks arising from it, they may not invoke the doctrine of frustration to escape their obligations.

At the time the lease in the present case was executed the National Defense Act, Public Act No. 671 of the 76th Congress, authorizing the President to allocate materials and mobilize industry for national defense, had been law for more than a year. The automotive industry was in the process of conversion to supply the needs of our growing mechanized army and to meet lend-lease commitments. Iceland and Greenland had been occupied by the army. Automobile sales were soaring because the public anticipated that production would soon be restricted. These facts were commonly known and it cannot be said that the risk of war and its consequences necessitating restriction of the production and sale of automobiles was so remote a contingency that its risk could not be foreseen by defendant, an experienced automobile dealer. Indeed, the conditions prevailing at the time the lease was executed, and the absence of any provision in the lease contracting against the effect of war, gives rise to the inference that the risk was assumed. Defendant has therefore failed to prove that the possibility of war and its consequences on the production and sale of new automobiles was an unanticipated circumstance wholly outside the contemplation of the parties.

Nor has defendant sustained the burden of proving that the value of the lease has been destroyed. The sale of automobiles was not made impossible or illegal but merely restricted and if governmental regulation does not entirely prohibit the business to be carried on in the leased premises but only limits or restricts it, thereby making it less profitable and more difficult to continue, the lease is not terminated or the lessee excused from further performance. (Citations omitted.) Defendant may use the premises for the purpose for which they were leased. New automobiles and gasoline continue to be sold. Indeed, defendant testified that he continued to sell new automobiles exclusively at another location in the same county.

(The judgment of the trial court in favor of Lloyd is affirmed.)

QUESTIONS FOR DISCUSSION

1. *What is the difference between impossibility and frustration of purpose?*
2. *What facts are present in the previous case that would indicate that Murphy chose to bear the risk of the government's limiting the sales of new automobiles?*

(d) *Prevention by Act of Party.* There is an implied condition in every contract that a party will do nothing to interfere with performance by the other. It would be absurd to allow one party to prevent performance by the other party and then to take advantage of his act. When one is prevented from performing by the other party, he may treat this as a breach of the contract and sue for damages. Suppose a contractor has begun construction of a house but is prevented from finishing it by the owner of the property. The contractor may sue the owner immediately for any damages resulting from the breach.

(3) **Discharge by Agreement.** Two parties to a contract may discharge them selves from that contract by a subsequent agreement. There are a number of ways that this may be accomplished.

(a) *Mutual Rescission.* When both parties agree to discharge one another from the obligations of a contract, they have entered into a mutual rescission of the agreement. The mutual rescission may be either express or implied and either oral or in writing except where a statute requires a mutual rescission to be written. There is no need to have a term in the contract allowing for mutual rescission. All that is needed is the agreement of the parties.

You should note the difference between mutual rescission by agreement of both parties and one party's right to rescind. In the latter situation the right arises because of the breach of the contract by the other party, whereas a mutual rescission does not involve a breach of the contract but merely involves an agreement by both parties to terminate it.

(b) *Substituted Contract and Novation.* The parties may agree to substitute one contract for another and thus terminate their obligations under one agreement by substituting another one for it. The parties may not expressly state that the new contract is discharging the old one, but if the substituted contract is clearly inconsistent with the terms of the prior agreement, the earlier contract will be discharged.

A *novation* occurs when one party is substituted for another under the terms of the same contract. This has the effect of discharging the original party. Obviously all parties to the agreement must agree to the substitution if a novation is to occur. Compare this situation to an assignment of contractual rights, where the obligor does not have to agree to the assignment and the assignor is not discharged from his contractual obligations and liabilities. Suppose Harry Hotrod purchases a car from Delbert Dealer and agrees to pay Dealer $100 per month for twenty-four months. After the first month Hotrod loses his job and gives the car to Fat Cat, with Cat agreeing to make the payments. At this point there is no novation. Hotrod is still bound to Dealer for the payments. Suppose further that Hotrod and Fat Cat approach Dealer, who agrees to discharge Hotrod of his liability and to accept Fat Cat in his place. At this point there would be a novation, with Fat Cat substituted for Hotrod under the same contract.

(c) *Waiver.* When one does not exercise the rights that he has under a contract, he may be said to have waived them. A waiver may be either express or implied. If one fails to exercise a contractual right within a reasonable time or accepts some performance other than that required by the contract, he may have waived his rights under the contract by implication and may not later enforce those rights under the agreement. Thus, if one is entitled to payment on the first of the month but accepts payment on the fifteenth, he may have waived his right to the earlier payment by implication if he does not object.

RYLANDER v. SEARS ROEBUCK & CO.

302 So.2d 478 (Fla.App. 1974)

Stella Rylander entered into a contract with Sears Roebuck for the installation of a bathroom with all accessories, a kitchen sink, a refrigerator, a range, lights, a living-room couch, and living-room lighting. The contract also provided for the installation of new wiring and light switches. When Rylander did not complete payment of her bill, Sears sued to recover $3,581.90. Rylander counterclaimed on the grounds set forth below. Judgment was entered for Sears, and Rylander appealed.

PER CURIAM.

Appellant filed her answer wherein she alleged that (1) plaintiff had not credited her account with all her payments, (2) many of the items supplied by Sears were defective and had broken, (3) that appellee had charged her twice for some items, and (4) plaintiff had refused to repair the defective refrigerator and range. Defendant–appellant also counterclaimed to recover as a setoff damages for holes left by Sears when Sears' employees attempted to install air-conditioning in defendant's business premises pursuant to another contract entered into in 1969 and, as a result thereof, defendant had to expend extra money to have the installation completed. The cause proceeded to a non-jury trial at the conclusion of which the trial judge entered final judgment in favor of the plaintiff for the sum of $3,581.90 and awarded the defendant a setoff in the sum of $91.95. Defendant Ms. Rylander appeals therefrom.

Defendant–appellant first argues that the trial judge erred in construing the May 4, 1968 contract between the parties to provide that the term "install new wiring" meant only the new wiring required to install the appliances purchased under the contract rather than all new wiring in the upstairs apartment in addition to the wiring necessary to install the appliances. We cannot agree.

In construing a contract, the intention of the parties is ascertained from the language used in the instrument and the objects to be accomplished and unless clearly erroneous, the construction placed upon a contract by the trial judge should be affirmed. After a close examination of the subject contract, we conclude that the trial judge's construction thereof was correct. Further, the appellant having failed to complain of plaintiff's inadequate performance of the May 1968 contract until the filing of the instant action almost four years later, we conclude that defendant–appellant's conduct constituted a waiver of her right of action for damages as alleged in her counterclaim.

(Affirmed.)

(d) *Accord and Satisfaction.* An accord occurs when the parties agree to performance different from that specified in the contract. Once the accord is performed, the original agreement is discharged by the accord and satisfaction. The original agreement is not discharged until the satisfaction occurs and may be enforced at any time up to that point.

(4) Discharge by Operation of Law. (a) *Merger.* Sometimes contract rights are merged into a greater right or one of a higher order. When this occurs, the contract is discharged but the parties will still have the greater right. Examples of this are when contracts are merged into deeds, court decrees, or judgments. In some cases one can prevent the contract from being merged by placing a clause in the agreement stating that the contract shall not be merged in the document of a higher order. Such a clause is frequently inserted in contracts to purchase real property and in separation agreements.

(b) *Alteration.* If one party to an agreement willfully and intentionally alters an agreement in a material respect without the consent or approval of the other party, the contract will be discharged. If the alteration is by accident, is not material, or is one that the law would imply, the contract is not discharged.

(c) *Bankruptcy.* The process of filing for bankruptcy is discussed extensively in Chapter 48. At this point it is enough to know that if the bankruptcy court approves the debtor's petition, then his or her debts will be discharged. This discharge in bankruptcy operates as a bar to enforcement of the obligation by creditors. The debtor may later waive this bar by following certain procedures found in the Federal Bankruptcy Code. Certain debts, such as taxes and alimony payments, are not subject to discharge.

(d) *Statute of Limitations.* All jurisdictions have statutes that limit the time within which a suit may be filed for breach of contract, as well as other actions. When the period elapses, the action is barred. Generally the period begins to run when the right of action accrues. The length of the period varies from one jurisdiction to another, as well as with the type of action. For instance, the period of breach of a simple contract is usually between three and five years. The passage of the statutory period does not discharge the contract but merely bars its enforcement.

A number of events may stop the statute of limitations from running. When this happens, the running of the statute is said to be *tolled.* Among those events that may toll the statute in some jurisdictions are the incapacity of the party with the right of action, partial payment of a debt, and the absence of the potential defendant from the jurisdiction. The statutes require that suit be filed within the applicable period, not that the case be tried and reduced to judgment. Thus, if one files the suit the day before the statutory period runs, the statute is tolled.

13:3. SUMMARY OF UNIFORM COMMERCIAL CODE PROVISIONS. The Uniform Commercial Code has several sections that affect the performance of contractual obligations. These sections are similar to general contract principles regarding discharge by impossibility of performance or the more recent doctrine of frustration of purpose. The basic approach of the common law is that strikes, embargoes, governmental regulations, or the occurrence of other events that make performance impossible or impractical do not relieve the parties of their contractual obligations. The rule of the common law is that failure to provide for the occurrence of these events in the contract indicates that one is bearing the risk of loss. The U.C.C. modifies this approach. It states that if there is a delay in delivery of goods, a nondelivery by the seller because performance "has been made impracticable by the occurrence of a contingency the nonoccurrence of which was a basic assumption on which the contract was made or by compliance in good faith with any applicable foreign or domestic governmental regulation," then the seller has not breached the contract of sale. Section 2–615 (a).

Note that this does not radically change the concept of frustration of purpose that has been discussed earlier. The event still has to be one that the parties assumed would not occur. Also the fact that an event merely increases the cost of performance would not excuse performance of the contract.

The code also has a provision governing the rights of the parties provided that the goods are damaged before the risk of loss passes to the buyers. If goods that have been identified to the contract are totally destroyed, the seller is excused from performing. The loss must not be the fault of either party. This rule is similar to the general contract principles that have already been discussed. The code continues, however, and gives the buyer an option if the goods are only partially destroyed under these circumstances.

If the goods are partially destroyed the buyer may demand inspection and at his option either (1) treat the contract as avoided or (2) accept the goods with allowance for deteriorations or the deficiency in quantity but without any further right against the seller. Section 2–613.

SUMMARY

A condition is a clause in a contract that will shift, suspend, modify, or rescind an obligation upon some operative fact or event. Conditions may be express, implied in fact, or implied in law. They may be either precedent, subsequent, or concurrent.

A condition precedent is a specified condition that must occur before the agreement between the parties becomes binding or before a party is required to perform a duty or obligation under the contract. A condition subsequent is an event that occurs after that point in time when one is obligated under the contract and operates to cut off contractual liability. When parties under a contract are required to perform simultaneously, the conditions are concurrent. Conditions implied in law are constructive conditions.

When the conditions of a contract have been fulfilled, the rights and obligations of the parties cease. Under the doctrine of substantial performance, the party substantially performing the contract is entitled to recover but is subject to the right of the innocent party to be compensated for defects in performance. The doctrine will not be applied where there is willful nonperformance on the part of the promisor.

Partial performance results when a portion of the required performance is not completed, generally intentionally. A party accepting part performance may waive his right to full performance of the contract.

A contract requiring performance to the satisfaction of the promisee may be of two types: those involving one's taste, judgment, and fancy and those involving operative fitness or mechanical utility. In the first instance, the courts do not agree but some apply a rule-of-reason test.

If time is of the essence in a contract, performance must be completed within the time specified. If no time is stated in a contract performance is to be rendered in a reasonable time.

Tender of performance is an offer to render the performance due under the contract. Tender that is a vain and idle act will be excused. A tender of payment of money when the debt is due stops the running of interest, extinguishes security interests, and relieves the debtor from paying court costs.

A contract will not be discharged by the occurrence of forseeable events that occur after the contract is made. If an unforeseeable event destroys the subject matter of a contract, the contract will be discharged. The death of a party to the contract may discharge the contract if it is for personal services or involves personal skill or a special relationship. An unforeseeable change in the law will result in the discharge of a contract if performance is made impossible.

The doctrine of frustration of purpose is applied when the purpose of the contract is destroyed or frustrated for both parties. The doctrine will not be applied if the event is foreseeable. A party must not interfere with contractual performance by the other party. Contracts may be discharged by agreement. Both parties may agree to discharge a contract or to mutually rescind the agreement. If the parties agree to substitute a new party under an existing contract a novation occurs. Parties may also agree to substitute one contract for another. Parties to a contract may waive their rights either expressly or by implication. A final method of discharge by agreement is accord and satisfaction.

Contracts may be discharged by operation of law. This may occur either through merger, alteration, bankruptcy, or the running of the statute of limitations.

PROBLEMS

1. Radlo contracted with Little to deliver to him sows and boars and to provide food, medicine, and veterinary care, as well as to pay for and move the young when they reached forty pounds. Little agreed to prepare facilities and care for the animals according to standards set by Radlo. Radlo terminated the contract, contending that the provisions allowed it to terminate based solely on its personal dissatisfaction. Little sued for lost profits, contending that the decision under the contract should be judged by what a reasonable producer would decide under the circumstances. Verdict was for Little, and Radlo appealed. Is Little correct? Radlo of Georgia, Inc. v. Little 199 S.E.2d 835 (Ga. App. 1973).
2. Alan Baldwin agreed to construct a house for Roland D. Smith in a good and workmanlike manner according to certain plans and specifications.

The agreed price of the house was to be $31,000.00. When the house was completed Smith refused to pay Baldwin, contending that the house was not constructed in a workmanlike manner. For example, a four-foot damper was installed in the fireplace when the contract specified that a five-foot damper should be installed. The cost of remedying these defects amounted to $1,500.00. To what, if any, remedy is Baldwin entitled? See Baldwin v. Smith, 586 S.W.2d 624. (Tex. 1979).

3. Triple M. Roof Corp. contracted to roof two buildings being constructed by Greater Jericho Corporation. Roof Corp. began work on both buildings, completing 85 percent of the work on one and 78 percent on the other. Roof Corp. failed to return for a substantial period of time, and the job was completed by another subcontractor. Jericho refused to pay the bill submitted by Roof Corp., who in turn filed suit. The lower court found for Roof Corp., and Jericho appealed. What result on appeal? Triple M. Roof Corp. v. Greater Jericho Corp., 349 N.Y.S.2d 771 (1973).
4. William and Paula Kruck contracted to sell real property to Ralph and Rose Cantrell and John and Darleen Krol. The price was $67,500, and a deposit of $6,200 in the form of a note was given. The closing date specified in the contract was December 1, 1971. The printed-form contract stated, "Time is of the essence of this contract." The sellers orally agreed to extend the selling date beyond December 1, 1971. When the buyers attempted to complete the sale on December 11, 1971, the sellers refused, contending that the buyers had breached the contract and forfeited their deposit as provided therein. Were the sellers correct? Cantrell v. Kruck, 324 N.E.2d 260 (Ill. 1975).
5. Boone conveyed to Eyre an interest in a West Indies plantation. Included were a number of slaves. Boone was to receive a lifetime annuity as part of the consideration for the interest conveyed. The deed provided that Boone had lawful possession of the slaves and good title to the land. Eyre decided not to go through with the deal on the ground that Boone was not lawfully possessed of all the slaves. Boone sued for nonpayment of the annuity. Decide. Boone v. Eyre, 126 Rep. 160 (K.B. 1777).
6. Waldemor Lach signed a contract to purchase a house from James J. Cahill and paid a $1,000 deposit. The contract contained a provision that, "This agreement is contingent upon buyer being able to obtain mortgage in the amount of $12,000.00. . . ." Lach, a young attorney, was unable to obtain financing after applying at several banks. Cahill contended that he was entitled to retain the deposit. Was he correct? Lach v. Cahill, 138 Conn. 418, 85 A.2d 481 (1951).
7. Republic Creosoting Company contracted with the City of Minneapolis to supply a quantity of creosoted wood paving blocks. Before the blocks were shipped, a nationwide car shortage occurred. Republic failed to deliver the blocks, and the City of Minneapolis sued for damages. Republic contended that performance was impossible because of the car shortage. Should Republic have been excused from performing on this ground? City

of Minneapolis v. Republic Creosoting Company, 161 Minn. 178, 201 N.W. 414 (1924).

8. Johnson entered into a contract with School District No. 4 (later merged with No. 12) of Wallowa County, Oregon, for the operation of a school bus for two school years. The contract also provided, "The said second party (Johnson) is to have option the next three years if a bus is run and his service has been satisfactory." When Johnson tried to exercise his option, the school district refused and operated its own bus. At trial there was no evidence other than that Johnson had provided reasonable service. Was Johnson entitled to exercise his option? See Johnson v. School District No. 12, 210 Or. 585, 312 P.2d 591 (1957).
9. The defendants agreed to furnish an opera troupe for performances at the plaintiff's theater. The star of the show, a tenor of some fame named Wachtel, was unable to sing because of illness. The show did not go on, and the plaintiff sued the defendants for damages. Was the plaintiff entitled to damages? Spalding v. Rosa, 71 N.Y. 40 (1877).
10. Guisezzse Palladi was a real-estate developer who contracted to build a house for Mr. and Mrs. Ohlinger on a lot he owned. The Ohlingers made a down payment of $1,500. An additional $6,000 was to be paid upon completion of the house. The contract gave Palladi the option to cancel the contract and return the money if prices increased prior to the date of settlement. The Ohlingers then built a wall on the property with Palladi's consent. The prices of material rose significantly, and Palladi canceled the contract and returned the Ohlingers' $1,500. Palladi knew of the rise in prices when the Ohlingers built the wall. Ohlinger sued Palladi for specific performance of the contract. What result? Palladi Realty Company v. Ohlinger, 190 Md. 303, 58 A.2d 125 (1948).

CHAPTER 14

Contractual Breach and Remedies

It is important not to get so bogged down in the niceties of determining contractual liability that you forget one of the major purposes of a contract: to provide a remedy for the innocent party if he is injured as the result of a contractual breach. There are many types of contracts, and the remedies and computation of damages for breach of contract vary according to the type of contract and the nature of the breach. Thus the remedy would differ depending upon whether the contract was a contract for the sale of goods, a construction contract, or an employment contract. The purpose of this chapter is to acquaint you in a general way with the various remedies available to the innocent party in the event of a contractual breach. The remedies available to the buyer or the seller of goods under the Uniform Commerical Code will be dealt with in Section Three, "Sales."

14:1. Materiality. A breach of contract occurs when a party fails to perform one of the duties or conditions of the agreement. If the breach is *material or total,* the innocent party may rescind the agreement and sue for damages. If the breach is *partial,* the innocent party may not rescind the contract and must still perform his promise. His remedy is to bring an action for damages caused by the partial breach. The determination of whether a breach is material or partial is difficult but often critical. If the innocent party erroneously determines that a breach is material and unilaterally rescinds the contract, he may be guilty of a material breach himself. On the other hand, a failure to rescind promptly when the right exists may result in a *waiver* of this right and preclude the availability of the remedy at a later date.

A material breach is one that goes to the essence of the contract. It may be thought of as a failure to perform a condition that operates to prevent the accomplishment of the object the parties had in mind when they formed the contract. The intention of the parties plays a large part in the determination of whether the failure to perform a condition constitutes a material breach. If the intent of the parties is stated in the contract, then the matter is simplified, but if the contract is silent on this point, the court will determine the intent of the parties based on the language of the contract and the circumstances of the case.

One factor that the court considers in determining whether a breach is material or partial is whether the contract is *severable* (divisible) or *entire* (indivisible). A contract is considered entire when all of its parts are considered to be interdependent and each part is common to the other. A contract is severable when it is judged that its parts may be performed separately. If a contract is severable, a failure to perform one of its parts will not entitle the innocent party to rescind the agreement. He must perform his obligation under the contract and sue for damages. On the other hand, a breach of an entire contract may be material and allow rescission. For instance, suppose Frank Fueldealer contracts to deliver one truck of fuel oil to Harry Homeowner's house once a month for a year. Fueldealer delivers for three months and then stops. If the contract is severable, Homeowner must pay for the oil delivered. His remedy is to sue for damages caused by Fueldealer's failure to make further delivery. If the contract is considered to be entire, then the breach would be material, and Homeowner would have the right to rescind the agreement; that is, he would not have to pay for Fueldealer's partial performance.

The question of whether a contract is severable or entire is answered by a determination of the intent of the parties. Thus in the previous example the parties can state in the agreement that the contract is considered to be entire. If their intent is not stated in the agreement, then the court will determine the contractual intent of the parties by examining the nature of the contract and the circumstances surrounding its formation. Some contracts are obviously more susceptible to severability than others. Nevertheless even those contracts that by their nature would be severable may be rendered entire by the intention of the parties.

In sum, the innocent party may with rare exception sue for damages as the result of a breach of contract. Whether he may also rescind is a function of whether the breach is material or partial. A factor in whether the breach is material or partial is whether the contract is severable or entire, which is determined by the intent of the parties.

NEW ERA HOMES CORP. v. FORSTER et al.

299 N.Y.303, 86 N.E. 2d 757 (1949)

New ERA Homes contracted with Forster to make extensive alterations to his home. The contract read as follows:

All above material and labor to erect and install same to be supplied for $3,075.00 to be paid as follows:
$150.00 on signing of contract
$1,000.00 upon delivery of materials and starting work
$1,500.00 on completion of rough carpentry and rough plumbing
$425.00 upon job being completed.

The work was partly finished and the first two stipulated payments were made. When the "rough work" was done, New ERA asked for the third payment of $1,500, but Forster would not pay it, so New ERA stopped work and brought suit. The jury awarded New ERA the $1,500, and Forster appealed.

DESMOND, J.

Defendants conceded their default but argued at the trial, and argue here, that plaintiff was entitled not to the $1,500 third payment, but to such amount as it could establish by way of actual loss sustained from defendants' breach. In other words, defendants say the correct measure of damage was the value of the work actually done, less payments made, plus lost profits.

The whole question is as to the meaning of so much of the agreement as we have quoted above. Did that language make it an entire contract, with one consideration for the doing of the whole work, and payments on account at fixed points in the progress of the job, or was the bargain a severable or divisible one in the sense that, of the total consideration, $1,150 was to be the full and fixed payment for "delivery of materials and starting of work," $1,500 the full and fixed payment for work done up to and including "completion of rough carpentry and rough plumbing," and $425 for the rest. We hold that the total price of $3,075 was the single consideration for the whole of the work, and that the separately listed payments were not allocated absolutely to certain parts of the undertaking, but were scheduled part payments, mutually convenient to the builder and the owner. That conclusion, we think, is a necessary one from the very words of the writing, since the arrangement there stated was not that separate items of work be done for separate amounts of money, but that the whole alteration project, including material and labor, was "to be supplied for $3,075.00." There is nothing in the record to suggest that the parties had intended to group, in this contract, several separate engagements, each with its own separate consideration. They did not say, for instance, that the price for all the work up to the completion of rough carpentry and plumbing was to be $1,500. They did agree that at that point $1,500 would be due, but as a part payment on the whole price. To illustrate: it is hardly conceivable that the amount of $150, payable "on signing of contract" was a reward to plaintiff for the act of affixing its corporate name and seal.

We find no controlling New York case, but the trend of authority in this State, and elsewhere, is that such agreements express an intent that payment be conditioned and dependent upon completion of all the agreed work. We think that is the reasonable rule—after all, a householder who remodels his home is, usually, committing himself to one plan and one result, not a series of unrelated projects. The parties to a construction or alteration contract may, of course, make it divisible and stipulate the value of each divisible part. But there is no sign that these people so intended. It follows that plaintiff, on defendants' default, could collect either in quantum meruit for what had been finished, or in contract for the value of what plaintiff had lost—that is, the contract price, less payments made and less the cost of completion.

The judgments should be reversed, and a new trial granted, with costs to abide the event.

QUESTIONS FOR DISCUSSION

1. *Did the court decide the contract was divisible or entire?*
2. *What measure of damages did the court say should be applied? What measure did the jury apply?*
3. *Two judges dissented in this case and thought the contract was divisible. What arguments can you give to support that conclusion?*

14:2. Anticipatory Repudiation. An *anticipatory repudiation* of a contract occurs when one of the parties in clear and unequivocal terms indicates that he does not intend to render the contractual performance when it is due. A qualified statement that one "possibly may not perform" would not be certain enough to qualify as an anticipatory repudiation. Of course, the repudiation occurs prior to the time when performance is required and may be either express or implied. For instance, if Sam Seller contracts to sell Bill Buyer his car on May 1 and then sells it to someone else prior to that time, this act would constitute an anticipatory repudiation.

When an anticipatory repudiation occurs, the innocent party has three options: (1) he may sue immediately for breach of contract and treat the repudiation as an immediate breach; (2) he may treat the contract as still binding, do nothing until the time for performance arrives, and then seek damages for breach of contract; (3) he may rescind the contract and recover the value of any performance rendered by him. The course of action selected depends upon the circumstances and may be affected by the ability to prove damages prior to the time performance is due.

As a general rule the doctrine of anticipatory repudiation applies only to executory bilateral contracts. In this application a contract that involves the payment of a debt for money already loaned would not be subject to the doctrine. Thus, if Danny Debtor owes Larry Lender $100 on a promissory note, Danny's statement that he does not intend to pay would not be subject to action until the date that the note is due.

An anticipatory repudiation may be withdrawn at any time before performance is due under the contract, provided that the innocent party has not yet treated the repudiation as a breach. If the nondefaulting party has acted on the repudiation, the breach cannot then be cured by the defaulting party. The innocent party may indicate his intent to treat the repudiation as a breach by changing his position in some manner, as by filing suit or by some other act or statement. Suppose Sammy Supplier has indicated that he will not honor a contract to supply Manny Manufacturer with the steel needed by Manufacturer. Once Manufacturer contracts with another party to supply the steel, he has changed his position, treated the repudiation by Supplier as a breach, and closed the door on Supplier's ability to cure the breach, even though the time for performance under the contract has not yet arrived. If the law were otherwise, Manufacturer would not be able to risk contracting with another steel company, as he could possibly be obligated under two contracts in the event that Supplier elected to retract his repudiation.

BUILDER'S CONCRETE v. FRED FAUBEL & SONS
373 N.E.2d 863 (1978)

Builder's Concrete Co. entered into a contract with Fred Faubel and Sons, Inc., on August 1, 1974. Under the terms of the contract, Builder's was to lease a number of trucks and other pieces of equipment to Faubel for use in Faubel's concrete business. The lease was to extend for one year and Faubel was to make payments of $8,500.00 to Builder's on the first of each month. The contract also gave Faubel an option to purchase the equipment upon proper notice to Builder's and apply $5,000.00 of each rent payment toward the purchase price. The agreement also provided that Faubel should provide maintenance and insurance coverage on the vehicles.

Faubel was late in each of its first four monthly payments but they were accepted by Builder's. On November 26, Faubel received a letter from Builder's which stated:

Based on recent management decisions by your company, Builder's Concrete and Material, Inc., refuses to transfer any assets to your company, Fred Faubel & Sons, Inc.

The letter further demanded that Faubel return all the leased property to Builder's by December 31, 1974. Faubel complied with the request.

Builder's then sued Faubel for damages to the returned vehicles. Faubel counterclaimed contending that by its letter Builder's breached its agreement to sell the equipment to Faubel. Faubel requested damages at the rate of $5,000.00 per month which was that part of the rent to be applied to the purchase price.

STENGEL, J.

(The court first determined that Faubel did not breach the contract.)

Having determined that defendant did not breach the contract, the next question is whether plaintiff breached the contract by renouncing its duty to convey the leased property. Defendant (Faubel) contends plaintiff's November 26 renunciation was an anticipatory repudiation and that, when acted upon by defendant on December 31, the repudiation became a breach of contract by plaintiff (Builder's). We agree. Unless justified, a definite statement to the promisee that the promissor will not perform its contractual duties constitutes an anticipatory repudiation. Under the option to purchase clause of the agreement, plaintiff clearly had a duty to convey the leased property to defendant if defendant so requested. Plaintiff's letter of November 26 contained a definite and unequivocal statement that plaintiff would not comply with the terms of the contract granting defendant an option to purchase. It further demanded that defendant return the leased property to plaintiff by a specified date. We have already determined that plaintiff's action was not justified by defendant's nonperformance or breach of contract. Thus, plaintiff's November 26 letter constituted an anticipatory repudiation of its duties under the contract.

Upon receiving an anticipatory repudiation by the promissor, the promisee has a choice of pursuing three alternative remedies. The promisee may (1) rescind the contract and seek quasi-contractual relief; (2) attempt to keep the contract in force by awaiting time for the promissor's performance and then bringing suit; or (3) elect to treat the repudiation as a breach putting an end to the contract for all purposes of performance. The promisee may evince his election to treat the repudiation as a breach by either promptly filing suit or by detrimentally changing his position in reliance on the repudiation. (17 Am.Jur.2d *Contracts* § 456.) In the instant case, defendant chose to follow the latter course. Complying with plaintiff's demand, defendant relinquished its rightful possession of the leased property to plaintiff on December 31, 1974. By so doing, defendant manifested its decision to treat the contract as ended and thereby put plaintiff in breach of contract. Plaintiff contends that even if its repudiation was unjustified, failure of defendant to object to the repudiation and defendant's acquiescence in returning the property caused the contract to be rescinded by mutual assent. We do not agree.

> A continued willingness upon the part of the injured party to receive performance is an indication that, if the repudiator will withdraw his repudiation, but not otherwise, the contract may proceed. It is not an irrevocable election not to treat the renunciation as a breach. (*Bu-Vi-Bar Petroleum Corp.* v. *Krow, et al.* (10th Cir. 1930), 40 F.2d 488, 492; *Williston on Contracts,* § 1334.)

Defendant was not required to object or to attempt to keep the contract in force. Plaintiff stated unequivocally that it would not fulfill its part of the bargain and defendant was justified in taking plaintiff at its word. *(Restatement of Contracts,* § 306, *comment a.)* An anticipatory repudiation continues in effect until affirmatively retracted by the repudiator. *(Restatement of Contracts,* § 320, *comment a.)* Plaintiff's theory would allow a party to announce repudiation of its contractual duty and then be held blameless unless the other party objected and attempted to change repudiating party's mind. Such is not the law. As noted above, upon receiving plaintiff's November 26 repudiation, defendant had an option of objecting and attempting to keep the contract in force or treating the repudiation as a breach ending the contract. Defendant was not required to pursue both options nor was it required to raise objection before treating the contract as ended. Defendant properly treated the contract as ended and plaintiff was placed in breach of contract on December 31, 1974.

QUESTION FOR DISCUSSION

1. *Prior to December 31, 1974, could Builder's have "cured" its anticipatory repudiation by retracting its letter of November 26?*

14:3. MONEY DAMAGES. (1) Compensatory and Nominal Damages. As a general rule money damages are available as a remedy for breach of contract. In order to be eligible for damages, the nonbreaching party must have suffered some injury from the contractual breach. If he has been placed in a better position as a result of the other party's breach, he is not entitled to anything other than *nominal* damages—that is, a trivial amount—for the technical breach of the contract. Thus, if a supplier breaches his contract to deliver goods to a retailer and the retailer can purchase the same goods on the open market at a price cheaper than the contract price, there has been no damage to the retailer. On the other hand, if the retailer purchases the goods for a greater price on the open market, he would be entitled to the difference between the contract price and the market price of the goods. In other words, the injured party is entitled to be compensated for the actual loss he has sustained from the contractual breach. This is one approach for the computing of damages in contracts for the sale of goods and is covered in the code, Section 2–713. The computation of damages for other contracts, such as construction and service agreements, would, of course, be done in a different manner, but the same principles apply. These damages are called *actual* or *compensatory* damages.

As a general rule the parties to a contract are entitled only to those damages that are the natural and probable consequence of the breach or that were contemplated by the parties. This principle is perhaps best stated in the famous old case of *Hadley* v. *Baxendale,* an English case decided in 1854. The plaintiffs operated a flour mill that was required to cease operation because of a broken crankshaft from the steam engine attached to the mill. The plaintiff delivered the shaft to defendants, who were common carriers, for shipment to a distant city in order to be repaired. The defendants promised to deliver the shaft the following day and were informed by the plaintiffs that their mill was stopped. The shaft was not delivered the following day as promised, which caused the mill to remain inoperative, with resulting loss of profits to the plaintiffs. At trial the jury was permitted to take the lost profits into account in computing damages. In reversing the case, the Court of Exchequer said:

> Where two parties have made a contract which one of them has broken, the damages which the other party ought to recover, in respect of such breach of contract, should be such as may fairly and reasonably be considered as arising naturally, i.e., according to the usual course of things, from such breach of contract itself; or such as may reasonably be supposed to have been in the contemplation of both parties, at the time they made the contract, as the probable result of the breach of it.

The court observed that the stoppage of the profits of the mill during this period had never been communicated to the defendant and that this was something a common carrier could not reasonably be expected to foresee. Lost profits were thus not within the contemplation of the parties in this case.

A further restriction on the award of compensatory damages is that they may not be speculative or hypothetical. Thus, although lost profits may be recovered under certain circumstances, suppose Ruth Retailer is contemplating opening a new store. A breach of contract by Sammy Supplier causes her to delay the opening of the store for a month. No recovery for lost profits would be allowed because there is no way to determine the amount of loss for that period of time. The store had not been in operation and the court could only speculate on the amount of profits that would be earned during that period, if any.

ERICSON v. PLAYGIRL, INC.

140 Cal. Rptr. 921 (Cal. App. 1977)

FLEMING, J.

Were damages awarded here for breach of contract speculative and conjectural, or were they clearly ascertainable and reasonably certain, both in nature and in origin?

The breach of contract arose from the following circumstances: plaintiff John Ericson, in order to boost his career as an actor, agreed that defendant Playgirl, Inc., could publish without compensation as the centerfold of its January 1974 issue of *Playgirl* photographs of Ericson posing naked at Lion Country Safari. No immediate career boost to Ericson resulted from the publication. In April 1974 defendant wished to use the pictures again for its annual edition entitled *Best of Playgirl*, a publication with half the circulation of *Playgirl* and without advertising. Ericson agreed to a rerun of his pictures in *Best of Playgirl* on two conditions: that certain of them be cropped to more modest exposure, and that Ericson's photograph occupy a quarter of the front cover, which would contain photographs of five other persons on its remaining three quarters. Defendant honored the first of these conditions but not the second, in that as the result of an editorial mixup Ericson's photograph did not appear on the cover of *Best of Playgirl*. Ericson thereupon sued for damages, not for invasion of privacy from unauthorized publication of his pictures, but for loss of the publicity he would have received if defendant had put Ericson's picture on the cover as it had agreed to do.

On appeal the sole substantial issue is that of damages, for it is clear the parties entered a contract which defendant breached.

Plaintiff's claim of damages for breach of contract was based entirely on the loss of general publicity he would have received by having his photograph appear, alongside those of five others, on the cover of *Best of Playgirl*. Plaintiff proved that advertising is expensive to buy, that publicity has value for an actor. But what he did not prove was that loss of publicity as the result of his non-appearance on the cover of *Best of Playgirl* did in fact damage him in any substantial way or in any specific amount. Plaintiff's claim sharply contrasts with those few breach of contract cases that have found damages for loss of publicity reasonably certain and reasonably calculable, as in refusals to continue an advertising contract. In such cases the court has assessed damages at the market value of the advertising, less the agreed contract price.

A yawning gulf exists between the cases that involve loss of professional publicity and the instant case in which plaintiff complains of loss of mere general publicity that bears no relation to the practice of his art. His situation is comparable to that of an actor who hopes to obtain wide publicity by cutting the ribbon for the opening of a new resort-hotel complex, by sponsoring a golf or tennis tournament, by presenting the winning trophy at the national horse show, or by acting as master of ceremonies at a televised political dinner. Each of these activities may generate wide publicity that conceivably could bring the artist to the attention of patrons and producers of his art and thus lead to professional employment. Yet none of it bears any relation to the practice of his art. Plaintiff's argument, in essence, is that for an actor all publicity is valuable, and the loss of any publicity as a result of breach of contract is compensable. Carried to this point, we think his claim for damages becomes wholly speculative. It is possible, as plaintiff suggests, that a television programmer might have seen his photograph on the cover of *Best of Playgirl*, might have scheduled plaintiff for a talk show, and that a motion picture producer viewing the talk show might recall plaintiff's past performances, and decide to offer him a role in his next production. But it is equally plausible to speculate that plaintiff might have been hurt professionally rather than helped by having his picture appear on the cover of *Best of Playgirl,* that a motion picture producer whose attention had been drawn by the cover of the magazine to its contents depicting plaintiff posing naked in Lion Country Safari might dismiss plaintiff from serious consideration for a role in his next production. The speculative and conjectural nature of such possibilities speaks for itself.

Assessment of the value of general publicity unrelated to professional performance takes us on a random walk whose destination is as unpredictable as the lottery and the roulette wheel. When, as at bench, damages to earning capacity and loss of professional publicity in the practice of one's art are not involved, we think recovery of compensable damages for loss of publicity is barred by the Civil Code requirement that damages for breach of contract be clearly foreseeable and clearly ascertainable.

Plaintiff, however, is entitled to recover nominal damages for breach of contract.

QUESTIONS FOR DISCUSSION

1. *Did Ericson prove that his contract with Playgirl, Inc., had been breached?*
2. *What would Ericson have to prove in order to recover compensatory damages?*

---

(2) **Punitive or Exemplary Damages.** Punitive or exemplary damages are those that may be assessed against a defendant beyond that amount necessary to compensate the plaintiff for the injuries suffered by him. They are not an appropriate remedy for breach of contract when the damages are limited to those that will compensate the plaintiff for his loss.

The purpose of punitive damages is to make an example of the defendant in order to demonstrate to others that society frowns on the type of behavior engaged in by the defendant. Punitive damages may be awarded in civil cases when there is malice or vindictive behavior shown on the part of the defendant. Generally the award of punitive damages is limited to intentional torts, such as battery, slander, and fraud where malice is proved. The punitive damages awarded are in addition to any compensatory damages to which the plaintiff may be entitled.

(3) **Liquidated Damages.** Under certain circumstances it is possible for the parties to stipulate the specific amount of damages to be paid by a party should the contract be breached. This is called a *liquidated-damages* clause. Such a provision is a particularly useful device where the damages would be difficult to ascertain or calculate. The amount specified as liquidated damages must not be so large as to constitute a penalty, and there must be some reasonable probability that the actual damages could total the amount specified in the contract as liquidated damages. If the court determines that the damages specified are so large as to constitute a penalty, the provision will not be enforced. This is true regardless of whether the parties characterize the damages as "liquidated" or "compensatory" or in some other manner. The court will look to the contract, its subject matter, and the circumstances surrounding the agreement in order to determine whether the provision in fact constitutes a penalty. On the other hand, once the court is satisfied that the liquidated damages are not in the nature of a penalty, the contract will be enforced and a party will not be able to escape the provision by showing that the actual damages are less. Liquidated-damages clauses are often used in construc-

tion contracts and in partnership agreements among professionals, such as doctors, lawyers, and accountants.

The Uniform Commercial Code, Section 2–718(1), provides for the use of liquidated damages, but only in an amount that is reasonable considering the anticipated or actual harm caused by a contractual breach, the difficulties of proving loss, and the feasibility of obtaining another remedy. A provision that fixes the damages at an unreasonably large amount is void as a penalty.

√ GARRETT v. COAST AND SOUTHERN FEDERAL SAVINGS & LOAN ASS'N.

511 P.2d 1197, 9 Cal.3d 731 (1973) In Bank

Plaintiff Roberta Garrett and others brought a class-action suit against Coast and Southern Federal Savings and Loan Association. Plaintiffs were obligors under promissory notes secured by deeds of trust in favor of defendant. The notes provided for late charges to be assessed in the event of a late payment at the rate of 2 percent per annum on the entire unpaid balance. Each of the members of the class had been assessed late charges, which they contended amounted to a penalty and could not qualify as liquidated damages. The lower court dismissed plaintiffs' suit on the ground that it did not state a cause of action. Plaintiffs appealed.

WRIGHT, C.J.

The plaintiffs allege that of approximately 32,000 obligors some 5,000 have paid late charges totaling $1,900,000 during the four year period immediately preceding the filing of the complaint. The promissory note signed by each obligor allegedly includes the following or similar provisions: "The undersigned further agrees that in the event that payments of either principal or interest on this note becomes in default, the holder may, without notice, charge additional interest at the rate of two (2%) percent per annum on the unpaid principal balance of this note from the date unpaid interest started to accrue until the close of the business day upon which payment curing the default is received."

In order to evaluate the legality of a provision for late charges we must determine its true function and character. . . . Defendant seeks to avoid the question of damages by maintaining that the lending agreement, to the extent that it requires the payment of additional interest, merely gives a borrower an option of alternative performances of his obligation. If he makes timely payments, interest continues at the contract rate, if, however, the borrower elects not to make such payments, interest charges for the loan are to be increased during the period of optional delinquency.

The mere fact that an agreement may be construed, if in fact it can be, to vest in one party an option to perform in a manner which, if it were not so construed, would result in a penalty does not validate the agreement. To so hold would be to condone a result which, although directly prohibited by the Legislature, may nevertheless be indirectly accomplished through the imagination of inventive minds. Accordingly, a borrower on an installment note cannot legally agree to forfeit what is clearly a penalty in exchange for the right to exercise an option to default in making a timely payment of an installment.

In the instant case, the only reasonable interpretation of the clause providing for imposition of an increased interest is that the parties agreed upon the rate which should govern the contract and then, realizing that the borrowers might fail to make timely payment, they further agreed that such borrowers were to pay an additional sum as damages for their breach which sum was determined by applying the increased rate to the entire unpaid principal balance.

The validity of a clause for liquidated damages requires that the parties to the contract "agree therein upon an amount which shall be presumed to be the amount of damages sustained by a breach thereof. . . ." Civ. Code, § 1671. This amount must represent the result of a reasonable endeavor by the parties to estimate a fair average compensation for any loss that may be sustained.

It is abundantly apparent for the reasons which follow that the parties here have made no "reasonable endeavor . . . to estimate a fair average compensation for any loss that might be sustained" by the delinquency in the payment of an installment. They have, in fact, contracted for the imposition of an additional sum to be paid by the borrower under the guise of an interest charge but which, in the absence of a showing that the same bore a relationship to any loss which may be suffered, must be construed as a penalty.

The fundamental difference between interest and penalty charge is that interest is a measure of compensation to which an obligee is entitled while a penalty is punitive in character. A penalty provision operates to compel performance of an act and usually becomes effective only in the event of default upon which a forfeiture is compelled without regard to the actual damages sustained by the party aggrieved by the breach. The characteristic feature of a penalty is its lack of proportional relation to the damages which may actually flow from failure to perform under a contract.

Late charges in home loan contracts are presumably imposed because borrowers fail to make timely payments of their obligations. Such charges serve a dual purpose: (1) they compensate the lender for its administrative expenses and the cost of money wrongfully withheld; and (2) they encourage the borrower to make timely future payments. Whether late charges represent a reasonable endeavor to estimate fair compensation depends upon the motivation and purpose in imposing such charges and their effect. If the sum extracted from the borrower is designed to exceed substantially the damages suffered by the lender, the provision for the additional sum, whatever its label, is an invalid attempt to impose a penalty inasmuch as its primary purpose is to compel prompt payment through the threat of imposition of charges bearing little or no relationship to the amount of the actual loss incurred by the lender.

The contractual provision as alleged in the complaint in the instant case provides that in the event of a late payment a borrower is to be charged an additional amount equal to 2 percent per annum for the period of delinquency assessed against the unpaid principal balance of the loss obligation. We are compelled to conclude that a charge for the late payment of a loan installment which is measured against the unpaid balance of the loan must be deemed to be punitive in character. It is an attempt to coerce timely payment by a forfeiture which is not reasonably calculated to merely compensate the injured lender. We conclude, accordingly, that because the parties failed to make a reasonable endeavor to estimate a fair compensation for a loss which would be sustained on the default of an installment payment, the provision for late charges is void.

Questions for Discussion

1. *Did the court state that all "late charges" amount to a penalty? If not, what specific provisions of the clause in this case caused the court to view it as a penalty?*
2. *How could the provision regarding a late charge have been redrafted so that it would amount to a valid liquidated-damages provision?*

---

(4) **Duty of Mitigation.** One has a positive duty to keep the amount of loss he sustains as a result of the other party's contractual breach to a minimum. In other words, he has a duty to mitigate damages. This concept operates in several ways. First, a party may not continue his activities so as to increase the damages once he is aware of the other party's breach. For instance, an accountant cannot continue to accumulate billable hours once he knows that a client has no intention of paying him. The accountant would be entitled to recover for those hours worked prior to learning of the breach but not those worked after he learned of the breach. The same would be true in other fields. Suppose a contractor learned that the owner did not intend to pay for a house being built for him by the contractor. If the contractor continues to purchase supplies and materials after learning of the breach, he will not

be able to recover that amount as damages because of his duty to mitigate.

In other cases a party may have a duty to take positive steps in order to mitigate his damages. Suppose one has an employment contract as an accountant for one year and is wrongfully discharged after three months. There is a duty on such an individual to mitigate his damages by seeking alternative employment. If he does not do so, he cannot recover the entire nine months' salary. On the other hand, he would not be required to take employment radically different from that for which he was trained in order to satisfy this duty.

ROCKINGHAM COUNTY v. LUTEN BRIDGE CO.

35 F.2d 301 (1929)

The commissioners of Rockingham County, North Carolina, entered into a contract with the Luten Bridge Co. to construct a bridge as part of the construction of a road. The composition of the board of commissioners changed and a resolution was passed rescinding the action for the construction of the bridge and the road. This resolution was sent to the Luten Bridge Co. Two weeks later a resolution was passed that the company be notified that any work done on the bridge would be done at its own risk, that the contract was invalid, and that the board did not desire to construct the bridge. This resolution was also given to the company.

At the time of the passage of the first resolution, little work had been done on the bridge. The estimated cost of labor and material to that point was $1,900. Despite the action of the commissioners the bridge company continued to work on the bridge. The board refused to pay for the bridge, and the company sued for $18,301.07. The lower court found for the Luten Bridge Co., and defendant appealed.

PARKER, C.J.

As the county now admits the execution and validity of the contract, and the breach on its part, the ultimate question in the case is one as to the measure of plaintiff's recovery, and the exceptions must be considered with this in mind.

Coming, then, to the third question—i.e., as to the measure of plaintiff's recovery—we do not think that, after the county had given notice, while the contract was still executory, that it did not desire the bridge built and would not pay for it, plaintiff could proceed to build it and recover the contract price. It is true that the county had no right to rescind the contract, and the notice given plaintiff amounted to a breach on its part; but, after plaintiff had received notice of the breach, it was its duty to do nothing to increase the damages flowing therefrom. If A enters into a binding contract to build a house for B, B, of course, has no right to rescind the contract without A's consent. But if, before the house is built, he decides that he does not want it, and notifies A to that effect, A has no right to proceed with the building and thus pile up damages. His remedy is to treat the contract as broken when he receives the notice, and sue for the recovery of such damages as he may have sustained from the breach, including any profit which he would have realized upon performance, as well as any other losses which may have resulted to him. In the case at bar, the county decided not to build the road of which the bridge was to be a part, and did not build it. The bridge, built in the midst of the forest, is of no value to the county because of this change of circumstances. When, therefore, the county gave notice to the plaintiff that it would not proceed with the project, plaintiff should have desisted from further work. It had no right thus to pile up damages by proceeding with the erection of a useless bridge.

It follows that there was error in directing a verdict for plaintiff for the full amount of its claim. The measure of plaintiff's damage, upon its appearing that notice was duly given not to build the bridge, is an amount sufficient to compensate plaintiff for labor and materials expended and expense incurred in the part performance of the contract, prior to its repudiation, plus the profit which would have been realized if it had been carried out in accordance with its terms.

(Reversed.)

QUESTIONS FOR DISCUSSION

1. *What did the court state was the proper measure of damages in this case?*
2. *Assume that the Luten Bridge Co. had never expended any money on the bridge when the board canceled. What would be the measure of damages at that point?*
3. *Now assume that the bridge was halfway constructed when the board breached, but that at that point Luten was losing money on the contract. What would be the measure of damages?*

14:4. NONMONETARY REMEDIES. As a general rule the appropriate remedy for breach of contract is to be compensated by money damages. There are situations, however, in which it is not possible to effect a remedy by an award of money damages. It is to these remedies that we now turn. Where the distinction between law and equity has not been abolished, these are called *equitable remedies.*

(1) Specific Performance. The courts will require specific performance of a contract only when the subject matter is *unique.* If the subject matter is unique, it would of course be difficult if not impossible to determine money damages. For example, suppose Metro Museum agrees to purchase a famous old painting from Carol Collector. There is only one of its kind; it is unique, and because one could not go out on the open market and purchase a similar painting, money damages would be inadequate. The court would therefore award specific performance of the contract and require Collector to sell the painting to Museum.

The Uniform Commercial Code, Section 2–716, provides for the remedy of specific performance in the sale of goods when they are unique or in other proper circumstances.

The courts have traditionally considered each piece of real property to be unique, even though there may be other parcels of equal quality and characteristics. A contract for the sale of real property is therefore specifically enforceable.

Contracts for personal service are never specifically enforceable, as a court will not require one individual to work for another against his will. The court will in certain cases, however, enforce a contractual provision prohibiting one's employment by another in a similar capacity. The limitations on this remedy are discussed in the next section.

(2) Injunction. An injunction is an order issued by a court requiring a person to do something or prohibiting him from doing something. If the injunction requires some affirmative action, it is called a *mandatory* injunction. A *prohibitory* injunction is issued to stop some activity.

An injunction will be issued only if the remedy of money damages is inadequate or if the harm to the party seeking the injunction would be irreparable without the issuance of the injunction. One of the major uses of an injunction in contract law is to enforce clauses that are sometimes called *restrictive negative covenants.* These are contractual clauses that forbid a certain type of behavior. For instance, we said earlier that a court will not specifically enforce a contract for personal services. In a proper case, however, the court may enforce a negative covenant that prohibits an individual from working for a competitor. Suppose a famous singing group is booked into a nightclub for an evening. The contract provides that they will not work for a competitor on that evening. Suppose further that the group indicates that it will not honor the contract and intends to sing that evening in the club located next door. A court might issue an injunction prohibiting such a performance, although it would not require the group specifically to perform the contract. Before issuing the injunction, the court would have to be convinced that the group was unique, so that it could not be replaced by another group with the remedy of money damages available to the innocent party. Injunctions are used in many other fields, such as labor disputes and cases involving civil disobedience.

ERVING v. VIRGINIA SQUIRES BASKETBALL CLUB
468 F.2d 1064 (1972)

Julius W. Erving, also known as "Dr. J.," signed a contract with the Virginia Squires Basketball Club of the American Basketball Association. After playing for them for one year, he defected to another team, the Atlanta Hawks of the National Basketball Association, and signed a contract to play for them. Erving's contract with the Squires provided that the club would have the right to injunctive relief to prohibit him from playing for any other club pending arbitration of any disputes before the commissioner of the American Basketball Association. Erving sued to have his contract with the Squires set aside for fraud. The Squires counterclaimed, seeking an injunction pending enforcement of the arbitration clause. Erving contended that the arbitration clause lacked mutuality and that an injunction should not be granted because the damage to the Squires was not irreparable. The lower court granted the injunction and Erving appealed.

MEDINA, C.J.

This case presents another chapter in the history of contract jumping by famous American athletes. As usual the amounts paid by the competing teams are fantastic. Julius W. Erving, we are told, was playing a remarkable game of basketball as an undergraduate at University of Massachusetts when, after his junior year, he agreed to turn professional and he signed a contract with the Virginia Squires to play exclusively for the Squires for four years commencing October 1, 1971 for $500,000.00. He made an extraordinary record in his first year as a pro, but he seems, for one reason or another, to have defected and in April, 1972 he signed a contract to play for the Atlanta Hawks. This contract with the Hawks is not before us but we were informed on the oral argument that it called for payments to Erving, or "Dr. J." as he was generally called by the fans, aggregating $1,500,000.00 or more.

In view of the large sums of money involved, and the publicity generated by the reputation of "Dr. J." as a highly talented basketball player with a brilliant future, we need not be surprised at the amount of perhaps pardonable exaggeration and bombast in the claims of the respective parties. On the one hand we are assured that "Dr. J." was, as stated in the opinion below, "for all practical purposes" the Squires' "whole team," that he was featured in the Squires' advertisements as "fabulous" and that the fans were deserting in droves when told that "Dr. J." had switched to the Hawks. On the other hand we are told that there is no showing of irreparable harm to the Squires if "Dr. J." plays with the Hawks, and the charge of fraud in inducing this innocent collegian to leave college and play for the Squires for four years for the inadequate sum of $500,000.00 is repeated ad nauseam.

We think, however, that irreparable damage to the Squires is plainly proved even if we assume that "Dr. J." is not the Squires' "whole team" and even if we doubt, as we do, that in the absence of "Dr. J." the Squires will collapse and with them the whole American Basketball Association.

Just as counsel for "Dr. J." repeat in various colorful phrases the claim that "Dr. J." was defrauded, counsel for the Squires insist that this is just a plain, ordinary case of contract jumping to get more money, that the claim of fraud did not originate until two months or more after "Dr. J." had signed his contract with the Hawks, and that the whole sorry business is nothing more nor less than the usual maneuvering by a greedy young athlete to sell out to the highest bidder. We do not pass upon these conflicting claims as they are the very issues to be determined by the arbitration of the dispute.

We find no lack of mutuality in the arbitration clause of the contract. Both parties are required to arbitrate any disputes arising between them. The provision relative to "obtaining an injunction or other equitable relief" is merely declaratory of existing legal rights.

(Affirmed.)

QUESTIONS FOR DISCUSSION

1. *Did the injunction issued in this case require "Dr. J." to play for the Squires?*
2. *Would the Squires have had a right to an injunction without a clause providing for this relief in the contract?*

14:5. RESCISSION AND REFORMATION. When a written contract does not accurately reflect the actual agreement of the parties, a court may reform the contract so that the agreement correctly states the parties' actual intent. This does not mean that a court will rewrite a contract in order to fill in missing terms or other errors, for it will not. The burden of proving the actual agreement and contractual intent is, of course, on the party seeking the reformation.

We have already seen that in certain cases one may have the right to a rescission of the contract. Rescission of a contract may be granted because of the lack of reality of consent or because of illegality. If a party is entitled to rescission, he is generally also entitled to *restitution;* that is, he is entitled to recover anything that he has paid or transferred to the other party under the contract. Generally, when there has been a rescission, the innocent party does not have a right to sue for damages. Under the Uniform Commercial Code, Section 2–271, however, a rescinding party may also collect damages. Remedies provided under the code for contracts for the sale of goods will be discussed more fully in Section Three, "Sales."

14:6. SUMMARY OF UNIFORM COMMERCIAL CODE CHANGES. The Uniform Commercial Code has extensive provisions regarding the rights and remedies afforded buyers and sellers if a contract for the sale of goods is breached. These provisions are discussed extensively in Chapter 19 of this text. At this point a few of the most significant provisions might be mentioned.

As far as the buyer's right to collect damages is concerned the various provisions of the code are summarized in Section 2–711. Basically the calculation of damages for breach of contract under the code is the same as is found in general contract law. However, there are some liberalizations of the common-law rules. For instance, under the general law of contract the buyer may only avail himself of the remedy of specific performance if the subject matter of the contract is unique. Under the code this approach is liberalized and the buyer may obtain specific performance if the goods are unique or in "other proper circumstances." Section 2–716. The inability of the buyer to purchase a similar good on the market would be evidence of "other proper circumstances."

The code also provides several methods for calculating the buyer's damages. In the event the seller breaches the contract, the buyer may purchase substitute goods within a reasonable time. He may then recover the difference between the contract price and the cost of the substitute goods, plus any other damages. Section 2–712. The buyer need not follow this procedure and may simply seek damages incurred by him because of nondelivery of the goods. This would be measured by calculating the difference between the contract price and the market price at the time the buyer learns of the breach. Section 2–713. Of course, the buyer also has the right to recover any money he has already paid to the seller for the price of the goods. Section 2–711.

In addition, the code affords the buyer the right to replevy or take the goods under certain circumstances. If the goods are identified to the contract and the buyer is unable after a reasonable effort to purchase similar goods, he may replevy the goods from the seller that are identified to the contract. Section 2–716. Furthermore, if the buyer has paid part of the price of the goods and discovers the seller to be insolvent within ten days after receipt of the first installment, he may recover goods identified to the contract upon payment of the balance due for them. Section 2–502.

The code also has a summary section concerning the seller's remedies for breach of contract for the sale of goods. Section 2–703. These are analagous to the remedies available to the buyer. For instance the seller may sell the goods to another buyer and recover the difference between the resale price and the contract price. Section 2–706. The buyer must be given credit for any expenses saved by the seller as a result of the buyer's breach.

An alternative is that the seller may retain the goods and recover the difference between the market price for the goods at the time and place for tender and the contract price. Section 2–708(1). If this remedy doesn't place the seller in as good a position as contractual performance would have done he may recover his lost profit. Section 2–708(2).

The code also has provisions under which the seller may recover the price of the goods and stop delivery upon discovery of the buyer's insolvency. These are similar to the buyers' remedies discussed above.

In general then, the code has clarified and in some ways liberalized the common remedies available for breach of contract. The code also has a provision concerning liquidated damages and takes an approach similar to that of the common law discussed earlier. Section 2–718. A specific provision for anticipatory repudiation is also contained in the code. Section 2–610. It is similar to the common-law approach.

Finally, the code follows the rule of *Hadley* v. *Baxendale* and allows for the recovery of consequential damages by the buyer. Consequential damages are those resulting from general or particular requirements and needs of which the seller had reason to know at the time of contracting. Section 2–715.

This discussion of remedies under the code is not exhaustive. All of the remedies have not been mentioned here but are more fully discussed in Chapter 19. However, you should see that while the code generally follows the common-law approach it liberalizes some rules in order to bring the law into conformity with the manner in which business is actually practiced. Remember once again however that the Uniform Commercial Code only applies to contracts for the sale of goods.

SUMMARY

If the breach of a contract is material or total, the innocent party may generally rescind the agreement and sue for damages. A consideration in determining whether a breach is total or partial is to determine whether parties intended the parts of the contract to be severable or entire.

An anticipatory repudiation of a contract occurs when one of the parties in clear and unequivocable terms indicates that he does not intend to render the contractual performance when it is due. The doctrine applies to executory bilateral contracts. The anticipatory repudiation may be withdrawn at any time before performance is due provided the innocent party has not acted on the repudiation.

As a general rule, money damages are available for breach of contract or the commission of a tort. Money damages are either nominal, compensatory, or punitive. Compensatory damages for breach of contract are those that are the natural and probable consequence of the breach or those that were contemplated by the parties. Compensatory damages may not be speculative.

Punitive damages are those assessed against a defendant beyond that amount necessary to compensate the plaintiff. They are not given for breach of contract and are generally limited to intentional torts. Liquidated damages are those stipulated by the parties in a contract. They must not constitute a penalty. One also has a duty not to intentionally increase damages and to try to keep the other party's damages to a minimum. This is known as the duty to mitigate damages.

In some cases nonmonetary damages are appropriate. The courts will require specific performance of a contract only when the subject matter is unique. Land is presumed to be unique but personal-service contracts are never specifically enforceable. Under certain conditions a court may order either a mandatory or prohibitory injunction. An injunction will be issued only if the remedy of money damages is inadequate or if the injury to the party seeking the injunction is irreparable.

A contract that does not accurately reflect the actual agreement of the parties may be reformed to conform to that agreement. If a court grants the remedy of rescission, it may also require restitution to the innocent party.

PROBLEMS

1. Drs. G. Gordon McHardy and Donovan C. Browne entered into a contract with Dr. Frank H. Marek that provided that after three years service to

their medical group, Dr. Marek was to be made a partner and receive 10 percent of the income to the partnership. After thirty-four months' service Marek was informed by McHardy that the group had no intention of granting him a 10 percent interest and that he would have to purchase this interest from them. At that time Marek left the service of the group and filed suit for $124,585.35.

McHardy and Browne claimed that Marek was not entitled to anything since he did not complete 36 months service. Are they correct? Marek v. MacHardy, et. al. 101 So.2d 689 (La. 1958).

2. Joe D. Ivester contracted with Family Pools, Inc., to install a swimming pool in his yard. After installation, problems developed so that the pool could not be filled with water. Ivester sued for damages after Family Pools failed to correct the problem. Family Pools contended that Ivester was under a duty to minimize his damages by repairing the defects and suing for the cost of repairs. Was Family Pools correct? Ivester v. Family Pools, Inc. Bosco Industries, 202 S.E.2d 362 (S.C. 1974).
3. Lewis and Grace Strong contracted to have carpet installed in their home by Commercial Carpet Company. It became apparent that there would not be enough carpet to complete the hallway, and because it would take three days to receive additional carpeting, it was agreed that the workmen would return then to finish. When they departed, the workman left an exposed metal tacking strip nailed to the floor between the hallway and another room. Mrs. Strong later tripped over the tacking strip and sustained severe back injuries. The Strongs then sued Commercial for the injuries sustained by Mrs. Strong on the theory of breach of contract. Could they recover? Strong v. Commercial Carpet Co., Inc., 322 N.E.2d 387 (Ind.App. 1975).
4. On April 12, 1852, Hochster agreed to enter the employ of De La Tour as a courier beginning June 1, 1852, when De La Tour was to begin a three-month tour of Europe. The pay was to be ten pounds per month. Prior to June 1, 1852, De La Tour told Hochster that he had no intention of taking him along. Hochster obtained another job but was not to begin until July. Hochster sued De La Tour on May 22, 1852. De La Tour contended that Hochster could not recover because he was not ready, willing, and able to perform the contract, and that he filed suit before performance was due. Was De La Tour correct? Hochster v. De La Tour, 2 E. & B. 678, 118 Eng. Rep. 922 (1853).
5. Bethlehem Steel Company contracted with the city of Chicago to furnish and erect the steel work for the construction of a superhighway for $1,734,200. The contract provided for $1,000 "liquidated damages" for each day of delay. The work was completed fifty-two days late, and the city deducted $52,000 from the price. Bethlehem contended that the city incurred no damages and that it was entitled to the money. Was Bethlehem correct? Bethlehem Steel Company v. City of Chicago, 234 F.Supp. 726 (D.C.N.D.Ill. 1964).

6. Carl E. Berke negotiated an employment agreement with Edward Bettinger as a sales manager in the temporary placement division of Bettinger's business. The contract provided that upon Berke's termination of employment he would not enter into the employment-agency business for one year within fifty miles of City Hall, Philadelphia, Pennsylvania. Berke subsequently left his employment with Bettinger and opened a competing business within the proscribed time and territory. Bettinger obtained an injunction restricting Berke, and Berke appealed, contending that Bettinger failed to show irreparable harm. Decide. Bettinger v. Carl Berke Associates, Inc., 455 Pa. 100, 314 A.2d 296 (1974).
7. Blaze Lane agreed to lend Pinkie Smith $500. Smith agreed to provide a spot for the operation of Lane's jukeboxes in his two places of business. The contract provided that no other jukeboxes were to be permitted on the premises, and if this provision was violated, $500 was to be paid as "liquidated damages." The contract was breached, and Lane was awarded the damages as provided in the contract. Smith appealed, contending that the damages constituted a penalty. Was he correct? Smith v. Lane, 236 S.W.2d 214 (1951).

SECTION III

Sales

15. Creation of the Contract for Sale of Goods Under the Uniform Commercial Code
16. Risk of Loss, Delivery Terms, and Transfer of Title
17. Warranties Under the Uniform Commercial Code, the Magnuson–Moss Warranty Act, and Products Liability
18. Performance of the Contract
19. Rights and Remedies
20. Financing the Sale: Secured Transactions

CHAPTER 15

Creation of the Contract for Sale of Goods Under the Uniform Commercial Code

15:1. **INTRODUCTION AND DEFINITIONS.** The subject matter of this chapter concerns the formation of contracts for the sale of goods under the Uniform Commercial Code. At the outset you should have a number of things clear in your mind. First, this section of the Uniform Commercial Code applies only to contracts for the sale of goods. It therefore does not apply to and has no effect on the law relating to many other types of contracts, such as construction contracts, real-estate contracts, or employment contracts. Thus the contract principles discussed in Section Two of this text remain unchanged for contracts other than for sale of goods. Second, although the code has altered some of the contract principles learned in Section Two, the general law of sale of goods is only a part of general contract law. Therefore, unless altered by the code, the general principles of contract law apply. The changes made by the code tend to be for the purpose of reflecting the realities of the commercial marketplace and to clarify some ambiguities in the general law of contracts as applied to sale of goods. The code does not, for instance, eliminate the general law requirements of agreement, consideration, legality, and capacity to make a valid contract.

The threshold question in this discussion is, Just what are "goods"? The code definition, Section 2–105(1), states that *goods* means all things movable at the time of identification to the contract. This includes specially manufactured goods, the unborn young of animals, and growing crops, as well as certain things to be severed from the realty by the seller. It does not include investment securities, things in action, and money in which the price is to be paid.

Another important question is, What do we mean by a *sale*, a *present sale*, and a *contract for the sale of goods?* Once again the code sheds some light. Section 2–106. A contract for sale includes both a present sale and a contract to sell goods at a future time. A sale is simply the passing of title from the seller to the buyer for a price, and a present sale is a sale accomplished by the making of a contract.

In order for there to be a present sale of goods, the goods must be both existing and identified. If both these requirements are not met, the goods are designated as *future goods.* An interest—for example, title—in such goods cannot pass; rather a present sale of future goods operates as a contract to sell as opposed to a sale. Section 2–105(2).

Several of the sections of this part of the code apply to "merchants." The code defines a merchant as one who deals in goods of the kind or holds himself out as having knowledge or skill concerning the goods or practices involved in the transaction. Section 2–104(1). To put it another way, a merchant is one who has specialized knowledge of certain goods or of business practices. The code provides further that the phrase "between merchants" means any transaction between parties who are chargeable with the knowledge or skill of merchants.

HELVEY v. WABASH COUNTY REMC

278 N.E.2d 608 (Ind.App. 1972)

Helvey purchased electricity from REMC for his home. At one point REMC furnished electricity of more than 135 volts, which damaged certain of Helvey's 110-volt appliances. Helvey filed suit against REMC to recover for these damages. REMC filed a motion for summary judgment predicated upon the statute of limitations. REMC contended that a statute of limitations of four years under the provisions of the Uniform Commercial Code was applicable. Helvey contended that this contract did not amount to a sale of goods and thus the general statute of limitation of six years applied. The lower court found for REMC and Helvey appealed.

ROBERTSON, J.

In order for the Uniform Commercial Code statute of limitations to apply, electricity must possess the following qualities:

> (1) "Goods" means all things (including special manufactured goods) which are movable at the time of identification to the contract for sale other than the money in which the price is to be paid, investment securities and things in action. . . .
>
> (2) Goods must be both existing and identified before any interest in them can pass. . . ." IC 1971, 26–1–2–105, Ind. Ann. Stat. § 19–2–105 (Burns 1964).

Helvey is of the opinion that electrical energy is not a transaction in goods but rather a furnishing of a service, which would make the following statute of limitations applicable:

> The following actions shall be commenced within six (6) years after the cause of action has accrued, and not afterwards.
>
> First. On accounts and contracts not in writing.

It is necessary for goods to be (1) a thing; (2) existing; and (3) movable, with (2) and (3) existing simultaneously. We are of the opinion that electricity qualifies in each respect. Helvey says it is not movable and in this respect we do not agree, if for no other reason than the monthly reminder from the electric company of how much current has passed through the meter. Logic would indicate that whatever can be measured in order to establish the price to be paid would be indicative of fulfilling both the existing and movable requirements of goods.

We further take note that one of the principle underlying purposes in adoption of the Uniform Commercial Code is "to make uniform the law among the various jurisdictions." IC 1971, 26–1–1–1–2(2)(c), Ind.Ann.Stat. § 19–1–102(c) (Burns 1914). With this in mind, we rely upon the authority of *Gardiner* v. *Philadelphia Gas Works* (1964), 413 Pa. 415, 197 A.2d 612, wherein natural gas was determined to be goods within the scope of the Uniform Commercial Code; therefore, the four-year statute of limitations was applicable.

(Judgment affirmed.)

EPSTEIN v. GIANNATTASIO et al.

25 Conn.Sup. 109, 197 A.2d 342 (1963)

Betty Epstein visited a beauty parlor on October 5, 1962, operated by Giannattasio. During the beauty treatment Giannattasio used a product called Zotos 30-day Color, which was manufactured by Sales Affiliates, Inc. A prebleach manufactured by Clairol, Inc., was also used. Epstein alleged that she suffered acute dermatitis, disfigurement resulting from loss of hair, and other injuries and damages as a result of the "treatment." Epstein sued Giannattasio, Clairol, and Sales Affiliates, Inc., for negligence as well as breach of warranty under Section 2 of the Uniform Commercial Code. All three defendants demurred to the complaint on the ground that the transaction did not constitute a sale of goods.

LUGG, J.

The issue reduces itself to the simple one of whether or not the use of the products involved in the course of the beauty treatment amounts to a sale or a contract for sale of goods under the pertinent sections of the Code. Section 42a–2–102 provides: "(T)his article (Sales) applies to transactions in goods. . . ." The word "transaction" is not defined in the act. "Goods" is defined in § 42–a–2–105 as follows: " 'Goods' means all things, including specially manufactured goods, which are movable at the time of identification to the contract for sale. . . ." Section 42–a–2–106 limits the words "contract" and "agreement," as used in the article, to the present or future sale of goods. "Contract for sale" includes a present sale of goods. § 42–a–2–106. "A 'sale' consists in the passing of title from the seller to the buyer for a price as provided by section 42–a–2–401." § 42–a–2–106.

As the complaint alleges, the plaintiff asked Giannattasio for a beauty treatment, and not for the purchase of goods. From such language, it could not be inferred that it was the intention of either party that the transaction be a transaction in goods within the meaning of the code. This claim of the plaintiff is hence distinguished more by the ingenuity of its conception than by the strength of its persuasion.

There is another line of cases which involves blood transfusions received by patients in the course of medical care and treatment in hospitals. These concern the claim that injuries caused by such transfusions ground a recovery under the Sales Act. This claim has been universally rejected. "Such a contract is clearly one for services, and, just as clearly, it is not divisible. . . . It has long been recognized that, when service predominates, and transfer of personal property is but an incidental feature of the transaction, the transaction is not deemed a sale within the Sales Act."

There are other cases, involving differing facts, which have decided that "when service is the predominant; and transfer of title to personal property the incidental, feature of a transaction, the transaction is not a sale of goods within the application of statutes relating to sales."

Building and construction transactions which include materials to be incorporated into the structure are not agreements of sale.

The language in the second count against Giannattasio alleges that "(t)he Defendant #1, in recommending and applying said products to the plaintiff thereby warranted. . . ." This amounts to a claim of implied warranty. The fourth and sixth counts aver implied warranty in terms. When this plaintiff made her arrangement with the beauty parlor, she did so as the complaint sets forth: ". . . for the purpose of receiving a beauty treatment." Obviously, the subject of the contract was not a sale of goods but the rendition of services. The materials used in the performance of those services were patently incidental to that subject, which was a treatment and not the purchase of an article.

(All three demurrers sustained.)

Questions for Discussion

1. *What standard do you apply in order to determine whether a transaction is a contract for the sale of goods? How do you distinguish the outcome in the Helvey case from the Epstein decision?*
2. *Is the sale of food in a restaurant a contract for the sale of goods? How about the sale of blood by a blood bank or the changing of the oil in one's car by a gas station that also sells you the oil?*

Now that we have some basic definitions from the code, we can examine further its impact on the basic law of contracts. Let us begin by analyzing the impact on the agreement process.

15:2. THE AGREEMENT. Just like general contract law the code provides that a contract may be expressed either orally or in writing or it may be created by implication by acts of the parties that indicate that they recognize the existence of an agreement. Sections 2–204(1), 2–207(3), 1–201(3). The code also has a Statute of Frauds provision that does require certain types of contracts for the sale of goods to be in writing. Section 2–201. We shall examine this section later.

15:3. THE OFFER. (1) Indefinite Offer. One of the basic propositions of the common law regarding an offer is that it must be stated with sufficient clarity and definiteness so that the parties are informed of their obligations and so that a court, in the event of litigation, knows what are the legal obligations of the parties. Leaving out a material term, such as price, could be fatal to the validity of the offer. The code has made a number of changes concerning the validity of such offers in contracts for the sale of goods by greatly liberalizing the degree of definiteness required for a valid offer. Section 2–204(3) provides that a contract shall not fail for indefiniteness if the parties intended to make a contract and there is a reasonable basis for fashioning a remedy. Let us look at some specific provisions.

(a) *Open Price.* One of the major considerations in a contract for the sale of goods is the price. The general attitude of the code is to give effect to an agreement that has actually been made in good faith. For example, a valid contract may arise without a statement of price if the parties so intend and if one of the following conditions is met: (1) nothing has been said as to price; (2) the parties fail to agree to price although the price is to be agreed to by the terms of the contract; or (3) the contract provides that the price is to be determined by some agreed market price or other standard set by third persons and that price is not set by the standard or market. If any of these circumstances occur, the price is a reasonable price. For instance, suppose two parties agree that a prize hog is to be sold by one to the other on a certain date at the market price for hogs established that day by a certain local hog-auction market. If the market is closed that day or has ceased to do business, the contract would not fail for lack of a price, but the price would be a reasonable price.

The code also provides that if a price is left to be fixed by a method other than the agreement of the parties and fails to be fixed by fault of one of the parties, the innocent party may (1) treat the contract as canceled or (2) fix a reasonable price himself. Section 2–305(3). Thus, if one party is to fix the price and refuses, the other party may himself fix a reasonable price or treat the agreement as canceled.

(b) *Open Delivery Terms.* Another important facet of contracts for the sale of goods concerns such terms as the time, place, and manner of delivery. Once again the code takes a more liberal attitude than the common law.

(i) PLACE. Unless the parties agree otherwise, the place of delivery of goods is the seller's place of business, unless the goods are identified to the contract and the parties know at the time of contracting that they are located at some other place. In that event the place of delivery is that place rather than the seller's place. Section 2–308. Remember that the code frequently uses the phrase "unless otherwise agreed"; thus the parties are free to change the "rule of the game" as they desire by stating a contrary intention in the contract.

(ii) TIME. If the parties do not state a time for delivery in the contract, the code provides that the delivery shall be in a reasonable time. Section 2–309(1). What constitutes a reasonable time depends, of course, on the facts and circumstances of each case.

(iii) MANNER. The seller is obligated to put and hold the goods at the buyer's disposition as well as to give him reasonable notification in order to enable him to take delivery. Section 2–503(1). The seller is also obligated to deliver all goods that are the subject of the contract in a single delivery unless agreed otherwise. Section 2–307.

(c) *Duration of Contract.* Occasionally parties will enter into a contract for successive performances and the contract will be silent as to its duration. For instance, Sammy Supplier agrees to supply Ronny Retailer with fifty dozen grade A eggs weekly at a set price. Nothing is stated as to the duration of this agreement. Under the common law, one might raise the question as to whether there is any mutuality of obligation. The code provides that such a contract is valid for a reasonable time but may be terminated at any time by either party, unless agreed otherwise. Section 2–309(2). However, when either party may terminate at will, reasonable notification must be received by the other party, and if an agreement dispenses with such notification, that provision is invalid if it is unconscionable. Section 2–309(3). The position of the code is that principles of good faith and sound commercial practice require notice of termination when the parties have been dealing with each other over a period of time in order that the other party is afforded an opportunity to seek a substitute arrangement upon termination.

(d) *Quantity: Output or Requirements Contracts.* We have learned that the modern common-law approach is to enforce requirements and output contracts when, for example, a supplier agrees to sell all the steel of a certain type required by a manufacturer for a year at a set price. As you might expect by now, the fact that the quantity term in a contract for sale of goods is measured by the output of the seller or the requirements of the buyer presents no problem under the code; such a contract is enforceable when the measure is the good-faith requirements or output. The official comment to this section takes the position that a requirement or output contract does not lack mutuality because the good-faith operation of the plant or business results in requirements or output that approximates a reasonably foreseeable figure. Comment 2 to Section 2–306. A contract that fails to state any quantity or a standard for determining such is still too indefinite to enforce and constitutes an invalid contract.

It should be clear by now that the code has liberalized the common-law approach and applies standards of good faith and commercial reasonableness in many situations. Thus, where reasonable, the absence of one of the discussed terms will not cause the contract to fail if that term can be determined by some commercially reasonable standard.

JAMESTOWN TERMINAL ELEVATOR, INC. v. HIEB
246 N.W.2d 736 (1976)

Archie Hieb and representatives of Jamestown Terminal Elevator, Inc. (Terminal) had a number of telephone conversations on July 3, 1973, concerning the sale of wheat by Hieb to Terminal. Terminal contends that these conversations resulted in a contract for Hieb to sell Terminal 10,000 bushels of wheat at $2.65 per bushel and to deliver the wheat "within a couple of weeks."

Hieb contends that no agreement was reached on that date. Hieb asserts that all that transpired was an offer by Terminal to purchase 10,000 bushels of wheat at $2.65 per bushel, such offer to be accepted by Hieb upon his signing a written contract with Terminal specifying the amount of grain, the price, and delivery date.

On the same date that Terminal contends it had a contract with Hieb, it resold the wheat to another party. When Hieb refused to deliver the wheat, Terminal purchased wheat on the open market to honor its contract of sale with a resulting loss of $37,500.00. It sued Hieb to recover this loss.

PAULSON, J.

The second issue raised by Hieb is that the trial court erred in failing to grant his motion for a directed verdict on the ground that as a matter of law a contract could not be found to have existed because an essential element—the delivery date—was not agreed to by the parties. We find no merit in this contention.

Section 41–02–11(3), N.D.C.C. [2–204(3), U.C.C.] provides as follows:

> FORMATION IN GENERAL.
> 3. Even though one or more terms are left open, a contract for sale does not fail for indefiniteness if the parties have intended to make a contract and there is a reasonably certain basis for giving an appropriate remedy.

Further, § 41–02–26(1), N.D.C.C. [2–309(1), U.C.C.], provides:

> *Absence of specific time provisions—Notice of termination.*—1. The time for shipment or delivery or any other action under a contract if not provided in this chapter or agreed upon shall be a *reasonable time.* [Emphasis added.]

Section 41–01–14(2), N.D.C.C. [1–204(2), U.C.C.], helps to interpret the Code's use of "reasonable time," stating:

> *Time—Reasonable time—'Seasonably.'*
> 2. What is a reasonable time for taking any action depends on the nature, purpose and circumstances of such action.

Finally, § 41–01–15, N.D.C.C. [1–205, U.C.C.], in pertinent part, provides the following guides to establish the "nature, purpose and circumstances" of an agreement:

> COURSE OF DEALING AND USAGE OF TRADE.
> 1. A course of dealing is a sequence of previous conduct between the parties to a particular transaction which is fairly to be regarded as establishing a common basis of understanding for interpreting their expressions and other conduct.
> 2. A usage of trade is any practice or

method of dealing having such regularity of observance in a place, vocation or trade as to justify an expectation that it will be observed with respect to the transaction in question. The existence and scope of such a usage are to be proved as facts. If it is established that such a usage is embodied in a written trade code or similar writing the interpretation of the writing is for the court.

3. A course of dealing between parties and any usage of trade in the vocation or trade in which they are engaged or of which they are or should be aware give particular meaning to and supplement or qualify terms of an agreement.

. . .

The foregoing statutory provisions, when applied to the instant case, provide a basis for a jury to determine the delivery date intended. Although the testimony of Terminal representatives clearly indicated that Terminal desired delivery within two weeks of July 3, 1973, the testimony adduced throughout the trial provided the jury with an ample basis for determining that the previous "course of dealing" between the parties, in addition to the industry's "usage of trade," placed Hieb on notice that Hieb's delivery date would have to be on or before Terminal's delivery date, pursuant to Terminal's resale agreement of July 3, 1973 with its commission firm, unless another date was agreed upon by the parties. Thus, based upon the previous "course of dealing" of the parties and the industry's "usage of trade," the jury could determine that August 31, 1973, was a "reasonable time" for delivery. We cannot say as a matter of law that there was no basis for a determination of the delivery date intended by the parties; therefore, the motion for a directed verdict was properly denied.

QUESTIONS FOR DISCUSSION

1. *How would this case have been decided under the common law?*
2. *What do you think of the approach taken by the code by which it supplies the missing term in the absence of a contrary agreement? Does this mean that the code provision is supplied in all cases in which there is no agreement?*

(2) **Firm Offer by Merchant.** We learned in our discussion of offers that under the common law an offeror may withdraw his offer at any time before it is accepted, notwithstanding that a promise to keep the offer open is unsupported by any consideration. If such a promise is supported by consideration, it, of course, constitutes an *option* and can be enforced.

The code provides for an exception to this basic principle when an offer is made by a merchant. Under certain circumstances a promise by a merchant to keep an offer to buy or sell goods open is enforceable in the absence of consideration. The circumstances are (1) the promise must be made by a merchant; (2) it must be made in a writing sent by the merchant; (3) the writing must state that the offer will be held open; (4) if not time is stated, the length of time must be reasonable; (5) the terms of irrevocability cannot exceed three months; and (6) if the promise of irrevocability is made on a form supplied by the offeree, the promise must be separately signed by the offeror. Section 2–205. The purpose of this approach by the code is to give effect to a merchant's intent to make a current firm offer binding, but it is limited to merchants who are, presumably, experienced in business matters. The reason for the last provision concerning forms is to avoid the situation in which a merchant might inadvertently make the promise by signing someone else's standard form without any real intent to promise irrevocability of the offer.

E. A. CORONIS ASSOCIATES v. M. GORDON CONSTRUCTION COMPANY

90 N.J.Super. 69, 216 A.2d 246 (1966)

M. Gordon Construction Company contracted to construct two buildings for the Port of New York Authority and in connection with that contract received bids from subcontractors. One of those submitting bids was E. A. Coronis Associates, who sent the following letter on April 22, 1963, to David Ben Zvi of Gordon Construction:

Dear Mr. Ben Zvi:

We regret very much that this estimate was so delayed. Be assured that the time consumed was due to routing of the plans through our regular sources of fabrication.

We are pleased to offer:

All structural steel including steel girts and purlins

Both Buildings delivered and erected ---
------------------ $155,413.50

All structural steel equipped with clips for wood girts & purlins

Both Buildings delivered and erected ---
------------------ $98,937.50

NOTE:

This price is predicated on an erected price of .1175 per pound of steel and we would expect to adjust the price on this basis to conform to actual tonnage of steel used in the project.

Thank you very much for this opportunity to quote.

Very truly yours,

E. A. CORONIS ASSOCIATES

/s/ Arthur C. Pease

Arthur C. Pease

Gordon made no reply to this letter but contended that it informed Coronis the same day that it found that it had submitted the lowest bid to the Port of New York Authority, April 19, 1963. Gordon was officially awarded the contract May 27, 1963. On June 1, 1963, Coronis revoked its offer, and on June 3, 1963, Gordon informed Coronis that it intended to hold Coronis to its bid.

Coronis sued Gordon on three contracts not pertinent here, and Gordon counterclaimed, claiming breach by Coronis of this alleged agreement. The lower court found for Gordon, and Coronis appealed.

COLLESTER, J.

Gordon contends that the April 22 letter was an offer and that Coronis had no right to withdraw it. Two grounds are advanced in support. First, Gordon contends that the Uniform Commercial Code firm offer section, J.J.S. 12–A:2–205, precludes withdrawal and, second, it contends that withdrawal is prevented by the doctrine of promissory estoppel.

Prior to the enactment of the Uniform Commercial Code an offer not supported by consideration could be revoked at any time prior to acceptance. The drafters of the Code recognized that the common law rule was contrary to modern business practice and possessed the capability to produce unjust results. The response was section 2–205, which reverses the common-law rule and states:

> An offer by a merchant to buy or sell goods in a *signed writing which by its terms gives assurance that it will be held open* is not revocable, for lack of consideration, during the time stated or if no time is stated for a reasonable time. . . .
> (Emphasis added)

Coronis' letter contains no terms giving assurance it will be held open. We recognize that just as an offeree runs a risk in acting on an offer before accepting it, the offeror runs a risk if his offer is considered irrevocable. In their comments to section 2–205 of the Code the drafters anticipated these risks and stated:

> However, despite settled courses of dealing or usages of the trade whereby firm offers are made by oral communication and relied upon without more evidence, such offers remain revocable under this Article since authentication by a writing is the essence of this section.

We think it clear that plaintiff's writing does not come within the provision of section 2–205 of a "signed writing which by its terms gives assurance that it will be held open."

Having so concluded, we need not consider the question of whether the Coronis letter was an offer or whether the letter dealt with "goods." We note in this connection that Coronis quoted the price for structural steel delivered and erected.

(The court also rejected Gordon's contention that the doctrine of promissory estoppel was applicable.)

(Reversed and remanded.)

QUESTIONS FOR DISCUSSION

1. *In this case Gordon was forced to hire another subcontractor to do the work at an additional cost of $55,857. What should it have done to protect itself against this outcome?*
2. *The court raised but did not answer the question of whether or not this was truly a contract for the sale of goods. What do you think? Why would the answer to this question be important?*

15:4. THE ACCEPTANCE. We have learned previously that under the common law the offeree had to accept the exact terms of the offer; otherwise his reply would be construed as a counteroffer. Any change in the terms of the offer or additional terms would result in the attempted acceptance's being construed as a rejection. This approach presents certain obvious difficulties, particularly in the commercial world, where there are a number of exchanges between parties in an effort to establish a satisfactory contractual arrangement. The code has made a number of changes in the common-law approach, particularly with regard to the manner of acceptance and the terms contained in the acceptance.

(1) Manner and Method of Acceptance. Under the common-law approach, if the offeror stipulates the time, place, and method of acceptance in his offer, the offeree is bound by those constraints. Thus, if the offer requires that the acceptance be made in writing, by mail, and within five days and is effective only when received by the offeror, the offeree must comply in order to accept. The code does not change this rule. What happens, however, if the offer is silent as to the manner and method of acceptance? A few jurisdictions took the approach under the common law that the acceptance had to be by the same mode as was used in the extending of the offer; for example, if the offer was by mail, the acceptance should be by mail. The code has clarified this area. If the offer is silent as to manner and method of acceptance, the acceptance may be made in any reasonable medium, such as by mail, by telegraph, or orally. Section 2–206(1)(a). If the offer is silent as to the time for acceptance, an acceptance made in a reasonable time is effective. Section 2–207(1).

In the commercial world offers are frequently made that invite acceptance either by a promise or by prompt performance. This procedure results in another problem. For instance, Randy Retailer sends Manny Manufacturer a purchase order saying "ship at once the following goods. . . ." Does this offer envisage acceptance by performance or acceptance by a *promise* to perform? The code provides that acceptance may be either by performance or by prompt promise of performance, thus avoiding the common-law problem presented by an improper response to an offer. Section 2–206(1)(b). Manufacturer may accept by promptly shipping the goods or by promising to ship. When acceptance is by performance, the offeror may withdraw his offer if he is not notified by the offeree within a reasonable time that there has been acceptance by performance. Section 2–206(2).

Occasionally the seller may not have the goods requested by the buyer. A problem arises if he ships goods that are nonconforming in order to accommodate the buyer. Under the common law this performance would constitute an acceptance and result in a contract, yet there would be a breach because the goods do not conform to the terms set forth in the offer. The code solves this problem by providing that if the seller notifies the buyer that the shipment is offered only to accommodate the buyer, an acceptance does not result from this performance. Section 2–206(1)(b). Thus the seller is not liable for breach of contract, and at the same time he is able to attempt to accommodate the buyer without exposing himself to the risk of a lawsuit. If there is no notification of an attempt to accommodate the offeror, the common-law approach remains unchanged, and the shipment of nonconforming goods could result in acceptance and breach.

(2) **Qualified Acceptance and Acceptance Varying Offer.** The common law required that an acceptance be in the exact terms as the offer in order for the negotiations to ripen into a contract. A purported acceptance varying the terms of the offer acted as a counteroffer and a rejection of the offer. The code takes a different approach in contracts for the sale of goods. If the purported acceptance states terms in addition to or different from those contained in the offer, a contract may still arise unless the acceptance is conditional on acceptance of the additional or different terms. Section 2–207(1). The reasoning behind this approach is that a contract should result if the parties in commercial understanding believe a deal is closed. Typically the parties have engaged in oral discussions or writings that state terms agreed upon and add additional terms, or send a letter confirming an oral conversation that includes minor additions, such as terms of shipment.

The treatment of these additional or different terms depends on the status of the parties. If they are not merchants, the terms are construed as proposals for addition to the contract. Section 2–207(2). If the parties are merchants, the terms become part of the contract unless one of the following exceptions applies.

(1) If the offer expressly limits acceptance to the terms of the offer, such additional terms are not included. Section 2–207(2)(a). This provision allows the offeror to keep complete control of the contract terms should he so desire.

(2) Terms that materially alter the contract are not automatically included. Section 2–207(2)(b). The official comments to the code state that the test of materiality is whether undue hardship or surprise would result if the terms were incorporated without express awareness by the other party. For instance, suppose a term excludes any implied warranties when in the usage of the trade such warranties are not normally excluded. This would be such a material change as not to be automatically included in the contract.

(3) If the merchant offeror gives notification of objection to the original terms within a reasonable time after receiving notice of them, or if objection

has already been given, such terms do not become part of the contract. Section 2–207(2)(c).

When merchants are concerned, a contract is often the result of an exchange of forms that may contain different or varying terms. This situation has frequently been referred to as the *battle of forms.* Under the common law the party submitting the last form would be considered the offeror, as his form would be construed as a rejection of any previously submitted form with different terms. If the last form is sent by the seller, the acceptance of the shipped goods would result in a contract, and the buyer would be struck with the terms of the seller's form. Under the Uniform Commercial Code this is not the case. Suppose that Bill Buyer submits a form purchase order to Sam Seller, including a provision that by accepting the order, Seller agrees to give certain warranties. Seller responds with a form acknowledging the order, on the back of which is a printed statement excluding certain of the warranties contained in Buyer's form and also stating that Seller is not liable for any damages from the use of his product beyond the contract price of the goods. What are the terms of the agreement? Obviously under the old common law no agreement would have resulted from this exchange because of the conflicting terms. If the goods were shipped and accepted, a contract with Seller's terms would have been formed. Not so under the code. In our example the clause limiting seller's liability to the contract price is an *additional* term, and the warranty limitation is a different term. Assuming Seller's proposals do not reflect the standard usage of the trade, they would *materially* alter the contract and hence would not become part of it. Buyer's form would prevail, although under the old common-law analysis, Seller's terms would have prevailed. However, any terms in Seller's form that do not materially alter the contract—as, for example, stating a particular date for shipment—would be included.

You can probably think of a number of variations on this example. The important thing to remember is that (1) under the code an acceptance with additional or different terms does not automatically constitute a rejection and counteroffer, as it would under the common law, and (2) therefore, the last form submitted is not necessarily the one whose terms form the basis for the contract.

McAFEE v. BREWER

214 Va. 579, 203 S.E.2d 129 (1974)

Jack and Mary Brewer contracted to buy a house from Don D. McAfee. The parties entered into negotiations for purchase by the Brewers of a number of items of furniture used by McAfee in the house. A dispute arose as to whether a contract existed for the purchase of the furniture. McAfee filed suit and the verdict was for defendants. McAfee appealed.

PANSON, J.

On April 30, 1971, plaintiff sent the defendants a letter containing the following: a list of furnishings to be purchased by the defendants at specified prices; a payment schedule of $3,000 due upon acceptance, $3,000 due 60 days after the acceptance date, and $2,635 due 120 days after the acceptance date; a blank space for the defendants' signatures and the date the signatures were affixed; and a clause reading, "If the above is satisfactory please sign and return one copy with the first payment."

On June 3, 1971, the defendants sent the following letter to plaintiff:

> Exams were horrible but Florida was great! Enclosing a $3,000 Ck.—I've misplaced the contracts. Can the secretary send another set? We're moving into Dower House on June 12—please include the red secretary on the contract for entrance foyer. I'll have to stop by sometime during the month & order a coffee table.
>
> Hope all is well—
>
> Sincerely—
>
> /s/ Va. & Jack

Plaintiff, in turn, sent the defendants a letter dated June 8, 1971, in which he enumerated the various items of furniture purchased by them. Except for several additionally approved items, the list on the June 8 letter corresponded precisely with the list in the April 30th letter. Believing he had a contract with the Brewers to sell them the listed items of furniture, the plaintiff purchased new furniture to furnish his new home.

Defendants testified that no agreement ever existed on what furniture was to be purchased. Defendant Jack R. Brewer further testified that he sent the $3,000 check only to buy several of the listed items, which totaled approximately $2,600, not to accept plaintiff's offer comprising all of the items listed. Brewer said he was not concerned about the overpayment because plaintiff was a friend, and he desired to buy some additional items from him.

Plaintiff argues that the defendants accepted his April 30th offer by their letter of June 3rd, the accompanying $3,000 check, and the request that the red secretary be included on the contract. We agree.

It is elementary that an agreement based on mutual assent is essential to a valid contract.

The Uniform Commercial Code is applicable here since the alleged contract is for the sale of "goods." Code § 8.2–105.

Code § 8.2–207 of the U.C.C. provides, in pertinent part:

> (1) A definite and seasonable expression of acceptance or a written confirmation which is sent within a reasonable time operates as an acceptance even though it states terms additional to or different from those offered or agreed upon, unless acceptance is expressly made conditional on assent to the additional or different terms. (2) The additional terms are to be construed as proposals for addition to the contract. . . .

Here the defendants' letter of June 3rd constituted a definite and reasonable acceptance or written confirmation sent within a reasonable time after receipt of plaintiff's offer to sell. The enclosure of the $3,000 check, the amount due upon acceptance of the contract, and the request to "include the red secretary on the contract," manifested defendants' assent or confirmation of the specific items enumerated in the April 30th letter. The reference to the red secretary was not expressed in language making acceptance conditional upon inclusion of the secretary. This item was merely a proposal for an addition to the contract.

While it is true that defendants did not sign and return one copy of the contract in the manner requested by plaintiff, their acceptance of the offer by letter was reasonable under the circumstances because they had misplaced the contract and the copy thereof. Moreover, there was no indication by the plaintiff that if the offer was not accepted in the suggested manner it would not be acceptable to him. Section 8.2–206 of the U.C.C. rejects the technical rules of acceptance in providing that "an offer . . . shall be construed as inviting acceptance in any manner and by any medium reasonable in the circumstances." See Virginia Comment, in the footnotes to that Code section.

For the reasons stated, the judgment of the court below is reversed and final judgment in the amount of $5,635 is here entered for the plaintiff.

QUESTIONS FOR DISCUSSION

1. *In what respect did the terms of the acceptance differ from the offer in this case?*
2. *Was the additional item listed in the acceptance part of the contract?*

15:5. CONSIDERATION REQUIREMENTS. The code has not eliminated the common-law requirement of consideration, but it has made some alterations in certain areas.

(1) **Firm Offers.** We have already considered the question of firm offers by a merchant. Thus, when a merchant states in writing that he will keep an offer open for a stated period of time, or for a reasonable time if no period is specified, that promise does not fail for lack of consideration. Section 2–205. Under the common law the promise could be revoked because of lack of consideration, but the code changed this rule for merchants to give effect to the deliberate intention of a merchant to make a firm binding offer.

(2) **Modification, Rescission, and Waiver.** The common law required the modification or rescission of an existing contract to be accompanied by consideration in order to be enforceable. Suppose Manny Manufacturer contracted to make certain goods and to sell them to Randy Retailer for $1,000. Manny's costs escalate to a point where he cannot manufacture the goods for less than $1,500. Randy tells him to go ahead. Randy would not be bound under the common law because Manny gave no consideration for the modification; he is under a preexisting duty to perform for $1,000.

The code provides that an agreement modifying a contract for sale of goods needs no consideration to be binding. Section 2–209(1). The purpose is to give effect to modifications without regard to the technicalities of consideration. A limitation is that the change must be made in good faith. Section 1–203. Thus, if in our example Manny extracted Randy's promise knowing of Randy's need for the goods, and merely to increase his own profit, rather than because his costs had escalated, the modification would not be enforceable. In addition, one may insert a provision in a contract requiring that further modifications or waivers be in writing. When this provision is included in a form supplied by a merchant, it must be separately signed by the other party unless the other party is also a merchant. Section 2–209(2).

What about an attempted oral modification or rescission where the contract requires a writing? This may operate as a *waiver.* Section 2–209(4). Suppose a contract prohibiting oral modification requires delivery by Supplier of goods to Retailer within ten days. The parties orally agree to delay delivery to twenty days, although Supplier is ready, willing, and able to perform. As an oral modification this change would be ineffective under the terms of the contract. However, Supplier in fact delays delivery to twenty days. Retailer has *waived* the ten-day requirement, although the code also allows him to retract the waiver by reasonable notification that strict performance will be required of any term waived (here the ten-day rather than the twenty-day period) unless this retraction would be unjust. Section 2–209(5). The purpose of this provision is to prevent contractual provisions excluding modification and rescission except by writing from operating to limit in other respects the parties' subsequent later conduct; as in this case, where it results in a waiver.

(3) **Output and Requirement Contracts: Lack of Mutuality.** In order for there to be a valid consideration, both parties must be bound; in other words, there must be mutuality of obligation. This is the general law and the code does not change it. One question that arises frequently concerns output and requirements contracts. The code states that these contracts do not lack mutuality. Section 2–306(1). Thus a contract in which Manufacturer agrees to sell

Buyer all Manufacturer's output of certain goods for a stated period of time at a certain price would be enforceable. Buyer must purchase the output and Manufacturer must sell his output to Buyer. Likewise, a contract in which Supplier agrees to sell Buyer all Manufacturer's output of certain goods for a stated period of time at a certain price would be enforceable. Buyer must purchase the output and Manufacturer must sell his output to Buyer. A contract in which Supplier agrees to sell Manufacturer all Manufacturer's requirements of a certain good for a set period of time at a certain price would be enforceable. Supplier must supply all Manufacturer's requirements and Manufacturer must purchase from Supplier. The contract is not too indefinite or illusory because the quantity is the actual good-faith output or requirements of a particular party. The code comments indicate that the party who will determine quantity is required to operate his plant or conduct his business in good faith so that his output or requirements approximate a reasonably foreseeable figure when usual standards of the trade are applied. Section 2–306(1) and Comment 2.

FARMERS ELEVATOR COMPANY OF RESERVE v. ANDERSON

552 P.2d 63 (Mont. 1976)

Dale Anderson was a farmer who had been a patron of Farmers Elevator Co. for several years and during 1972 had made numerous contracts with it for the sale and purchase of his durum wheat. On October 28, 1972, Anderson contracted with the Farmers Elevator Co. for the sale of 18,000 bushels of durum wheat at a price of $1.80 per bushel. Farmers Elevator Co. then contracted to resell this wheat for $2.44 per bushel. The contract between Anderson and Farmers Elevator Co. was strictly oral. The approximate delivery date contemplated by the parties was February 1973.

Pursuant to his contract, Anderson delivered 8,802 bushels of durum wheat in 36 truckloads between March 27, 1973, and May 30, 1973. The reason for the variance between the proposed delivery date and the dates the delivery actually took place is that Farmers Elevator Co.'s ability to accept delivery from its patrons is wholly dependent upon the availability of elevator space. In turn, the availability of elevator space depends directly on the availability of transportation for outgoing shipments. When sufficient rail cars or trucks cannot be found to transport the grain, the elevator becomes backlogged for space and, consequently, the delivery dates of established contracts must often be extended until transportation is secured.

During the fall of 1972 and early 1973 a serious regional boxcar shortage existed. Because of the resultant shortage of elevator space, Anderson was allowed to deliver less than half of the durum wheat which he had sold until transportation was obtained in June 1973. Anderson was well aware of this as he was contacted by phone as well as by letters dated July 31, August 13, and September 18, 1973. On September 27, 1973, the Chairman of the Board of Directors of Farmers Elevator Co. learned by telephone that Anderson refused to deliver the rest of the wheat due on his contract. The next day Farmers Elevator Co. was forced to purchase 9,198 bushels of wheat on the open market at a price of $6.50 per bushel in order to honor its contracts. Meanwhile, Anderson had sold his wheat to another elevator at $5.35 per bushel. Farmers Elevator Co. then sued Anderson for its damages.

HARRISON, J.

Is enforcement of the oral agreement and the alleged modification as to the date of delivery barred by the Statute of Frauds? (The Court first decided that Anderson had made a judicial admission of the original contract which satisfied the Statute of Frauds.)

In that the original oral contract is not rendered unenforceable by the Statute of Frauds, we next examine the effect of the alleged modification as to the date of delivery. Section 87A–2–209(3), provides:

> The requirements of the statute of frauds section of this chapter. . . . must be satisfied if the contract as modified is within its provisions.

While the record is clear as to Anderson's "admissions" regarding the original contract, no such admission was shown concerning the alleged modification of the delivery date. Anderson argues that if the original contract is enforceable that the exception which brought the original oral contract out of the Statute of Frauds, cannot be applied to the alleged oral modification and enforcement of the contract as modified under section 87A–2–209(3) is barred. However, this does not complete the analysis under the facts here. Anderson's deliveries after the set date must be considered.

Section 87A–2–209(4) provides that an attempt at modification which is void under the Statute of Frauds may still operate as a waiver to assert the defense through the course of performance engaged in by the parties under section 87A–2–208, R.C.M. 1947. While the trial court did not make a specific finding holding there was a waiver, its findings when read as a whole, do find there was a waiver and consent by Anderson. . . . The facts presented in the instant case, require that we find that a waiver such as is contemplated by section 87A–2–209(4) occurred and we agree.

By delivering, pursuant to contract, approximately 36 truckloads of wheat to the elevator between March 27 and May 30, 1973, Anderson established a course of conduct sufficient to constitute a waiver of his right to assert a defense under the Statute of Frauds. Anderson began to deliver on a date well after the delivery date originally contemplated, and his actions certainly induced plaintiff's apparent belief the contract would be honored. The enforcement of the oral contract, as modified, is not barred under section 87A–2–201, R.C.M. 1947.

(Judgment for Farmers Elevator.)

QUESTION FOR DISCUSSION

1. *If Anderson had raised as a defense lack of consideration for the modification, would he have been successful? Why or why not?*

15:6. FORM REQUIRED (1) Statute of Frauds and Parol-Evidence Rule. In order for a contract for sale of goods of $500 or more to be enforceable, the code requires that there be some sufficient writing to indicate that a contract for sale of goods has been made between the parties. This writing must be signed by the party against whom enforcement is sought. Section 2–201(1). The code provides for a number of exceptions to this rule. This section is discussed in detail in Chapter 11, dealing with the Statute of Frauds. You should refer to this portion of the text at this time for this analysis as well as an examination of the code's treatment of the parol-evidence rule. Section 2–202.

(2) Seal. Under the old common law a sealed contract did not have to be supported by consideration. Most jurisdictions have eliminated the seal as a substitute for consideration for contracts generally. The code provides that seals are inoperative with regard to contracts for the sale of goods. Section 2–203.

S U M M A R Y

Goods, with some exceptions, are all things movable at the time of identification to the contract. A contract for sale includes both a present sale and a contract to sell goods at a future time. A merchant is one who has special knowledge of certain goods or business practices.

The code has liberalized the common law requirements for an offer. A valid offer may be made, under certain conditions, which does not contain a price. The code also provides for open delivery terms and states rules, which operate in the absence of a contrary agreement, as to place, time, and manner of delivery. The code also has rules concerning contract duration if that matter is not addressed in the contract. As to the quantity of goods, the code specifically approves output and requirement contracts.

A promise by a merchant to keep the offer open is enforceable in the absence of consideration if it is written. If a specific time is not mentioned, the offer is to be held open for a reasonable time.

The code also eases the restrictions for accepting an offer. If the offer doesn't state a method for acceptance, any reasonable method is satisfactory. A seller may ship nonconforming goods to the offeror and not accept the offer if he notifies the offeror that the goods are only being shipped to accommodate the offeror.

The code has a number of rules concerning acceptances which vary the offer provided the offeror has not limited the acceptance to the exact terms of the offer. If the parties are not merchants, additional or different terms are construed as proposals for addition to the contract. If the parties are merchants, the terms become part of the contract unless the offer strictly limits the acceptance to its terms or the changed terms are material or the merchant offeror gives a timely notification of objection to the change.

The code has eliminated some common law consideration requirements. An agreement modifying a contract for the sale of goods needs no consideration. Under certain circumstances offers by merchants to keep an offer open do not require consideration.

PROBLEMS

1. Skoog Construction Company had a contract with Community Unit School District No. 1, Coles County, Illinois, to build a school. J and R Electric contracted with Skoog Construction to do the electrical work. An addendum to the original specifications for the school construction required the installation of an exterior unit switch gear at an agreed price of $9702. J and R, the subcontractor, paid the price of the switch gear and installed it. J and R contends that its original contract with Skoog was modified to require Skoog to reimburse it for the purchase which was over and above the amount provided in the original contract between J and R and Skoog. Skoog contends that there was no consideration to support the modification of the contract and that the provisions of the Uniform Commercial Code do not apply to the case. Is Skoog correct? J and R Electric Division of J. O. Mory Stores, Inc. v. Skoog Construction Company, 348 N.E.2d 474 (Ill. App. 1976).
2. Southwest Engineering Company obtained a contract to construct runway lighting facilities at McConnell Air Force Base. Prior to submitting the bid, Martin Tractor Company was contacted by Southwest regarding the price of a standby generator and accessory equipment. Martin quoted a price of $18,500. After obtaining the contract, officials of Southwest and Martin met and Martin quoted a higher price than its original figure for the generator. The officials also discussed the possibility of using a cheaper

generator. The Martin official then listed the component parts of the two generators on a piece of paper and set out the price of each item and totaled them. The memorandum was handed to the Southwest official. Later, Southwest sent Martin a letter directing them to proceed with drawings for the job. Martin responded that it was withdrawing all verbal quotations. Southwest purchased the equipment elsewhere at a higher price and sued Martin for damages. Martin contended that there was no contract since the terms of payment had not been agreed to. Is Martin correct? Southwest Engineering Company v. Martin Tractor Company, 205 Kan. 684, 473 P.2d 18 (1970).

3. Doughboy Industries, Inc., contracted to purchase certain goods from Pantasote. Doughboy sent a purchase-order form to Pantasote. The form used by Pantasote in accepting Doughboy's offer contained a general arbitration provision, whereas Doughboy's form did not. Doughboy's form contained a provision that only a signed consent would bind it to any additional terms. Pantasote's form provided that silence or failure to object in writing would be an acceptance of the terms and conditions of its form. Doughboy never objected to Pantasote's form. Did the arbitration provision become part of the contract? Application of Doughboy Industries, Inc., 223 N.Y.S.2d 488 (1962).
4. Bateman orally agreed to supply L.T.V. Aerospace Corporation with packing cases to L.T.V.'s specifications for overseas shipment of certain vehicles manufactured by L.T.V. After it received specifications, Bateman made a substantial beginning of the production of the packing cases. Subsequently L.T.V. stopped producing its vehicles and refused to take delivery of the packing cases. Bateman sued L.T.V., who defended on the ground that there was no writing as required by the Statute of Frauds. Decide. L.T.V. Aerospace Corporation v. Bateman, 492 S.W.2d 703 (Tex.Civ.App., 1973).
5. Paul Fallis, a farmer, verbally agreed to sell and deliver 5,000 bushels of soybeans to Cook at $2.54 per bushel. Following the discussion, Cook sent Fallis a written agreement signed by Cook. Fallis did not sign the written agreement. Later Fallis refused to deliver the soybeans and Cook sued. Fallis raised the Statute of Frauds as a defense, Cook countered by introducing the contract signed by itself as satisfying the provisions of the statute under UCC 2–201(2). Was Cook correct? Cook Grains, Inc. v. Fallis, 395 S.W.2d 555 (1965).
6. HML Corporation and General Foods Corporation entered into a contract whereby HML agreed to sell and General Foods agreed to purchase for thirty-two months at least 85 percent of its requirements of salad dressing in a designated geographic area at a certain price. Four months after the initiation of the agreement, General Foods notified HML that it could not profitably market the salad dressing and therefore would not require further production. HML sued, contending that General Foods had assumed a duty to continue to promote the product or at least to maintain requirements by continuing to distribute it. Was HML correct under the code? HML Corporation v. General Foods Corporation, 365 F.2d 77 (1966).

CHAPTER 16

Risk of Loss, Delivery Terms, and Transfer of Title

16:1. Introduction. This chapter seeks to answer two important questions that are raised whenever there is a contract for the sale of goods: (1) When does the risk of loss or damages to the goods pass from the seller to the buyer? (2) When does title to the goods pass from the seller to the buyer? In order to avoid confusion you should remember that these are *separate* questions under the Uniform Commercial Code.

Prior to the U.C.C. its predecessor, the Uniform Sales Act, answered the question concerning risk of loss with the artificial concept of passage of title; that is, risk of loss passed with the passage of title. The U.C.C. generally ignores the concept of title in determining when the risk of loss passes; rather it answers the question in terms of legal consequences flowing from the contract and actions taken under it. This is not to say that the U.C.C. ignores the concept of title but that the concept has been significantly reduced in importance and is resorted to only when other provisions of the code do not provide a solution to a particular problem.

First, let us examine the concept of risk of loss more closely; then various types of shipment terms will be analyzed, particularly as they affect risk of loss. Finally, the concept of passage of title will be treated.

16:2. Risk of Loss. When damage has occurred to goods, the question to be answered is whether the loss is to be placed upon the buyer or the seller. There are a number of rules governing this issue, and as will be observed, the code solves the problem generally without regard to title. We will examine three situations: risk of loss when there has been agreement of the parties as to risk allocation, risk of loss in the absence of a breach of contract, and risk of loss when there has been a breach of contract.

(1) Agreement of the Parties. The general policy of the code is to allow the parties to fashion the terms of their own agreement, and the provisions governing risk of loss are no exception. The code provides that the parties may state in the contract, when the risk of loss passes. Section 2–509(4). If the parties so desire, they may divide the risk or burden of loss rather than shift it entirely. Section 2–303. Obviously, if there is a contractual provision governing risk of loss, then one need look no further, as the contract determines who bears the risk. In the absence of such a contractual provision it is necessary to move to the next step in our analysis, risk of loss in the absence of contractual breach.

(2) **Risk of Loss in Absence of Breach of Contract.** A contract may require a seller to take several different actions with regard to the goods that he has contracted to sell. He may be required to ship the goods to the buyer by carrier, he may be required to deliver them into the possession of a bailee without being moved, or the buyer may himself take possession of the goods at the seller's place of business.

(a) *Seller Required or Authorized to Ship Goods by Carrier.* There are two general types of contracts concerning the seller's obligation to ship the goods by carrier: *shipment contracts* and *destination contracts.*

(i) SHIPMENT CONTRACTS. A *shipment contract* is one that requires the seller to deliver the goods to a carrier and put them in his possession after contracting for their delivery to the buyer, notify the buyer of the shipment, and deliver to the buyer any form needed for the buyer to obtain the goods from the carrier. Section 2–504. Because the contract does not require the seller to deliver the goods at a particular destination, it is called a *shipment contract,* and the risk of loss passes from the seller to the buyer when the goods are duly delivered by the seller to the carrier. Section 2–509(1)(a). As a result, any loss sustained because of damage to the goods occurring after their delivery to the carrier will be borne by the buyer. Note that the risk passes regardless of the use of any concepts of title, nor is the risk affected by a provision concerning when payment is to be made for the goods. Also observe that we are discussing the ultimate allocation of risk between the buyer and the seller regardless of any liability that the carrier might sustain for damage to the goods.

(ii) DESTINATION CONTRACTS. A *destination contract* is one that requires the seller to tender delivery of the goods at a particular destination. *Tender of delivery* means that the seller must put and hold conforming goods at the buyer's disposition and give him any reasonable notice necessary to enable him to take delivery, as well as tender of any documents of title necessary to take possession. Section 2–503(3). Thus, if the agreement is a destination contract, the risk of loss passes from the seller to the buyer when the goods are tendered to the buyer so as to enable him to take delivery. Section 2–509(1)(b). If any damage or loss occurs to the goods while in transit between the buyer and the seller, the loss is the seller's.

(b) *Goods in Possession of Bailee Without Being Moved.* A bailee is one to whom goods are delivered under an agreement that the property will be returned or delivered according to the agreement. An example might be the delivery of goods to a warehouseman for storage until the buyer takes delivery of the goods. The warehouse would be the bailee. Sometimes goods are delivered from seller to buyer through a third party, such as a warehouse. For instance, suppose Seymour Seller has goods located in Warren's Warehouse.

He contracts to sell those goods to Barney Buyer. Buyer does not want to move them until he has reduced his current inventory in order to make room for them. At what point does the risk of loss pass from Seller to Buyer? The code, Section 2–509(2), provides that the risk of loss passes to Buyer when one of three events occurs: (1) when the bailee acknowledges the buyer's right to possession of the goods; (2) when the buyer receives a negotiable document of title covering the goods; or (3) when the buyer receives a nonnegotiable document of title and he has a reasonable time to present the document or direction to the bailee. If the bailee refuses to honor the document or direction, the risk does not pass. Section 2–503(4). Note that a negotiable document of title causes risk of loss to pass when it is received by the buyer, but if the document of title is nonnegotiable, the risk of loss does not pass until the buyer has a reasonable time to present it to the bailee.

A document of title is an instrument that is used as evidence that the person in possession of it is entitled to receive, hold, and dispose of the document and the goods it describes. A document of title is negotiable if its terms state that the goods are to be delivered to the bearer or to the order of a named person. Bear in mind also that we are still discussing the allocation of risk between the buyer and the seller and no one else.

(c) *Other Situations: Delivery at Seller's or Buyer's Premises.* The previous discussion relates to contracts that involve third parties, either shipment of the goods by carrier or their control by a bailee. Obviously many contracts involve other methods of delivery, as, for example, when the buyer picks up the goods at the seller's place of business or the seller delivers the goods to the buyer. The code differentiates between a merchant and a nonmerchant in determining when the risk of loss passes in situations other than those mentioned in Sections (2)(a) and (b) of this discussion. If the seller is a merchant, the risk of loss passes when the buyer is in receipt of the goods, that is, when he takes them into actual physical possession. If the seller is not a merchant, the risk passes upon the tender of delivery by the seller. Section 2–509(3). The code attempts to place the risk of loss in this situation on the party most likely to protect himself against such loss. Thus the risk remains on the merchant seller until the buyer is in receipt of the goods, at it is unlikely that one would insure goods before they come into his control, whereas the merchant who maintains control can best be expected to insure them. Section 2–509, Comment 3.

LUMBER SALES, INC. v. BROWN

469 S.W.2d 888 (Tenn.App. 1971)

Lumber Sales, Inc., agreed to sell and deliver five carloads of lumber to Julius Brown at a certain railroad siding near Radmor Yards in Nashville, Tennessee, designated as track location 609–A. Lumber Sales contended that it delivered all five carloads of lumber to Brown at the railroad siding. Brown admitted that four carloads were received but denied receiving the fifth carload. Lumber Sales sued Brown and was awarded a judgment of $5,163.20 and costs. Brown appealed.

PURYEAR, J.

The uncontroverted evidence shows that during the early morning hours of November 27, 1968, the Louisville and Nashville Railroad Company, to which we will hereinafter refer as the carrier, placed a boxcar loaded with lumber consigned to the defendant on this siding at track location 609–A.

This boxcar was designated as NW 54938 and it was inspected by an employee of the carrier between 8:00 A.M. and 8:30 A.M. on November 27, 1968, at which time it was found loaded with cargo and so designated upon the carrier's records.

At 11:07 A.M. on November 27, 1968, the carrier notified one of defendant's employees that the carload of lumber had been delivered at track location 609–A.

At approximately 4:00 P.M. on that same day an employee of the carrier again inspected this boxcar at track location 609–A, found one of the seals on it to be broken and resealed it at that time. The evidence does not show whether the car was still loaded with cargo at that time or not.

The following day, November 28th, was Thanksgiving Day and the record does not disclose that the carrier inspected the boxcar on that date. But on November 29, 1968, between 8:00 A.M. and 8:30 A.M. an employee of the carrier inspected the car and found it empty.

From evidence in the record before us, it is impossible to reach any logical conclusion as to what happened to this carload of lumber without indulging in speculation and conjecture, but the defendant earnestly insists that he did not unload it and there is no evidence to the contrary.

Counsel for defendant insists that the transaction involved here is governed by provisions of the Uniform Commercial Code and we agree that it is.

The particular code section applicable here is Subsection (1) of T.C.A. § 47–2–509, as follows:

> 47–2–509. Risk of loss in the absence of breach.—(1) Where the contract requires or authorizes the seller to ship the goods by carrier (a) (this portion not applicable) (b) if it does require him to deliver them at a particular destination and the goods are there duly tendered while in the possession of the carrier, the risk of loss passes to the buyer when the goods are there duly so tendered as to enable the buyer to take delivery.

The trial court held that the risk of loss in this case did, in fact, pass to the defendant buyer.

There is competent evidence in the record which shows that on November 27, 1968, at 11:07 A.M. the carrier notified the defendant's employee, Mr. Caldwell, at defendant's business office, that the carload of lumber had been delivered at track location 609–A. Mr. Caldwell did not testify, so this evidence is uncontroverted.

There is no evidence in the record to the effect that the defendant declined to accept delivery at that time or asked for a postponement of such delivery until a later time.

The defendant testified that it would normally require about four or five hours for him and his employees to unload a carload of lumber and that on November 27, 1968, he and his employees were so busily engaged in other necessary work that he could not unload the lumber on that day and since the following day was Thanksgiving, he could not unload it until November 29th, at which time, of course, the carrier found the car to be empty.

From evidence in the record, a trier of fact could logically form one of two inferences:

(1) That the lumber was either stolen or unloaded by mistake by someone other than the defendant at some time between 8:30 A.M. and 11:07 A.M. on November 27th; or (2) that it was stolen or unloaded at some time after 11:07 A.M. November 27th and 8:30 A.M. November 29th.

If the first inference should be formed then the issue should be found in favor of defendant, but if the second inference should be formed, then the issue should be found in favor of plaintiff if it could also be found that the loss occurred after defendant had sufficient time to protect himself against loss after notice of delivery.

We think the second inference is the more logical of the two, especially in view of the difference between the two intervals of time and also in view of Mr. Crye's testimony to the effect that on Thanksgiving Day, November 28th, he observed some activity at track location 609–A, which he believed to be unloading of a railroad car at that location.

Of course, we recognize and adhere to the

rule that the burden of proof is upon plaintiff to prove delivery of the lumber and we are not required to indulge either of the above mentioned inferences because the trial court apparently concluded that the plaintiff had successfully carried the burden of proof with which he was onerated and the evidence does not preponderate against that court's conclusion.

Counsel for defendant argues that the lumber in question was not duly "so tendered as to enable the buyer to take delivery" as required by T.C.A. § 47–2–509.

However, this argument seems to be based upon the premise that it was not convenient for the defendant to unload the lumber on November 27th, the day on which it was delivered at track location 609–A and defendant was duly notified of such delivery.

This was an ordinary business day and the time of 11:07 A.M. was a reasonable business hour. If it was not convenient with the defendant to unload the lumber within a few hours after being duly notified of delivery, then he should have protected himself against risk of loss by directing someone to guard the cargo against loss by theft and other hazards.

To hold that the seller or the carrier should, under the circumstances existing in a case of this kind, continue to protect the goods until such time as the buyer may find it convenient to unload them would impose an undue burden upon the seller or the carrier and unnecessarily obstruct the channels of commerce.

The language of Subsection (1)(b) of T.C.A. § 47–2–509 does not impose such a burden upon the seller, in the absence of some material breach of the contract for delivery, and we think a reasonable construction of such language only requires the seller to place the goods at the buyer's disposal so that he has access to them and may remove them from the carrier's conveyance without lawful obstruction, with the proviso, however, that due notice of such delivery be given to the buyer.

(Affirmed.)

QUESTIONS FOR DISCUSSION

1. *Why was it critical to determine whether the lumber was removed from the car after 11:07* A.M. *on November 27?*
2. *Was the contract in this case a shipment contract or a destination contract? Does the answer to this question have any affect upon the outcome of the case?*
3. *Assume that the lumber delivered in this case was nonconforming to the contract. After reviewing the next section of the chapter, determine whether this would have affected the outcome of this case.*

---

(3) **Risk of Loss When Contract Is Breached.** (a) *Effect of Seller's Breach.* A breach by the seller may be detected by the buyer either prior to acceptance of the goods or in some cases subsequent to it. If the seller has breached the contract by shipping nonconforming goods and the buyer discovers the nonconformity prior to acceptance, the risk of loss remains on the seller until he cures the nonconformity or until the buyer accepts the goods despite the nonconformity. Section 2–510(1). Suppose Sam Seller contracts to sell goods to Bill Buyer under a shipment contract. The risk of loss would normally pass when the goods are delivered to the carrier. The goods shipped in this case, however, are not of merchantable quality and thus are nonconforming. Seller has breached the contract, and the risk does not pass to Buyer until the nonconformity is cured or until Buyer accepts them despite their nonconformity. In other words, the breach by Seller has stopped the risk from being shifted to Buyer upon delivery of the goods to the carrier, as it normally would under a shipment contract.

In some instances the buyer may accept the goods under circumstances that allow him to revoke his acceptance. The code provides that the buyer may revoke his acceptance when the nonconformity substantially impairs the value of the goods to him (1) if the goods were accepted on the reasonable assumption that the nonconformity would be seasonably cured and it has not been, or (2) if the goods were accepted by seller's assurances or under circumstances that make discovery of the nonconformity difficult. Section 2–608. If the buyer rightfully revokes his acceptance because of the nonconformity of the goods, he may treat the risk of loss as having remained on the seller from the beginning. To the extent that the buyer's effective insurance coverage is deficient, he may recover the deficiency from the seller. Section 2–510. Once again, the seller's breach has prevented the risk of loss from being shifted to the buyer when the goods are delivered to a carrier under a shipment contract or arrive at their destination under a destination contract. The policy of the code in the event of breach is to allocate the risk of loss between the party who has breached and the one with effective insurance coverage. The standards for determining nonconformity are discussed in Chapter 18.

UNITED AIRLINES INC., v. CONDUCTRON CORP.

69 Ill. App. 3d 847, 387N.E.2d 1272 (1979)

United Air Lines Inc. contracted to purchase an aircraft flight simulator from Conductron Corp., a division of McDonnell Douglas Corporation. The contract required Conductron to deliver a Boeing 727 digital flight simulator to United. Because flight simulators are used for training pilots, they must meet the requirements necessary for approval by the Federal Aviation Administration and the contract so provided.

While the simulator was still in the possession of Conductron, United's personnel noted some 647 deficiencies in its operation. The simulator was then delivered to United's facility and reassembled by Conductron. Since the machine had not received FAA approval it could not be used as a flight simulator.

On April 18, 1969, the simulator was tested for ten hours by two of United's test pilots. About 10 P.M. a fire was discovered in the machine which substantially damaged it. United then sued Conductron seeking rescission of the contract and damages. The trial court granted United's motion for summary judgment on its contention that the risk of loss remained on Conductron. Conductron appealed.

GOLDBERG, J.

Defendants contend that the trial court erred in entering summary judgment for plaintiff because the risk of loss of the simulator was upon plaintiff at the time of its destruction.

Plaintiff contends that at the time the simulator was destroyed the risk of loss was upon defendants. In this regard, plaintiff urges that defendants defaulted under the terms of the contract by failing to deliver a conforming aircraft flight simulator; this default was never cured; the simulator was never accepted by plaintiff because of its deficiencies and plaintiff at all times retained the right of rejection.

To evaluate the risk of loss issue, attention must be given to the impact of both the Uniform Commercial Code and the contract terms. Section 2–510 of the Code (Ill. Rev.Stat.1977, ch. 26, par. 2–510(1)), provides:

> (1) Where a tender or delivery of goods so fails to conform to the contract as to give a right of rejection the risk of their loss remains on the seller until cure or acceptance.

Few cases involve this section of the Code and those that do merely cite the Code with little explanation. The official Uniform Commercial Code Comment provides some guidance by stating that the purpose of this section is to make clear that "the seller by his individual action cannot shift the risk of loss to the buyer unless his action conforms with all the conditions resting on him under the contract." (S.H.A. ch. 26, par. 2–510(1) at page 398.) Of primary importance, then, is the determination of whether or not the simulator so failed to conform to the contract provisions as to vest the right of rejection in plaintiff and whether or not there was acceptance of the simulator by plaintiff.

The purchase agreement provided in part that the simulator would "accurately and faithfully simulate the configuration and performance of . . ." a certain specified Boeing aircraft and that final acceptance of the simulator would be "subject to satisfactory completion of the reliability demonstration requirements . . ." and to Federal Aviation Administration certification. The affidavit of John Darley, an employee of plaintiff who tests and evaluates flight simulators to determine whether they meet plaintiff's and the FAA specifications, states that the simulator at no time met those requirements and that plaintiff was never able to begin acceptance testing. His affidavit states clearly that the machine was destroyed by fire "before that time when United [plaintiff] was to begin acceptance testing . . ."

This record shows that the simulator at no time conformed to the specifications agreed to in the purchase agreement and it remained nonconforming until its destruction. Although plaintiff had use and possession of the simulator for six weeks, that arrangement was expressly sanctioned by the contract to allow testing. Retention of the simulator for testing purposes did not constitute acceptance so as to shift the risk of loss to the plaintiff. The simulator was destroyed before completion of acceptance testing and before receipt of FAA certification. Both were conditions precedent to acceptance of the simulator. In this situation the risk of loss remained on the defendants as seller. On the issue of risk of loss, there is no genuine issue regarding any material fact. On the contrary, in our opinion, plaintiff's right to summary judgment in this regard is clear beyond question. We conclude that the plaintiff is entitled to summary judgment as a matter of law.

QUESTIONS FOR DISCUSSION

1. *Why did the risk of loss for damages to the simulator remain with Conductron even after it had been delivered to United's facility?*
2. *What role does location of title to the simulator play in determining who had the risk of loss?*
3. *Suppose that the simulator was not destroyed by fire but was destroyed while being shipped back to Conductron for corrections to the equipment. Between United and Conductron, who would bear the loss?*

(b) *Effect of Buyer's Breach.* When the buyer breaches the contract for sale of goods prior to the time risk of loss would normally pass to him, the seller may, under certain circumstances, look to him for a commercially reasonable time for any loss sustained by the seller due to any deficiency in the seller's effective insurance coverage. Section 2–510(3). These circumstances referred to are that (1) the goods must have been identified to the contract and (2) the buyer repudiates or is otherwise in breach before risk of loss passes to him. Identification means that specific goods are identified as those to be used as the subject matter of the contract. Identification can occur at any time and in any manner agreed to by the parties. Section 2–501(1). In the absence of any agreement, identification occurs when the contract is made if it is for goods already existing and identified or, if the contract is for the sale of future goods, when the seller ships, marks, or otherwise designates the goods to which the contract refers. Section 2–501(1)(a) and (b).

Suppose Sam Seller contracts to sell goods to Bill Buyer under a shipment contract. Seller moves the goods to be shipped to Buyer to a separate portion of Seller's warehouse. Buyer then notifies Seller not to ship the goods and wrongfully repudiates the agreement. That evening, fire destroys Seller's warehouse, including the goods. If Seller's effective insurance covers only one third of the value of the goods, the remaining loss would fall on Buyer. Note, however, that under these circumstances Buyer bears the risk only for a commercially reasonable period of time; thus, if prior to the fire the goods had remained separated in the warehouse for a month after Buyer's repudiation, a month would probably exceed a commercially reasonable time and the loss would not fall on Buyer. Observe also that the breach by Buyer affects the rule that the risk of loss would ordinarily not fall on him until the goods are delivered to the carrier.

Before moving on, you should realize certain points. In the first place, this discussion has concerned the allocation of risk between buyer and seller. No consideration has been given to the culpability of third parties, such as the carrier. Second, you should have noted that the risk is allocated regardless of any concepts of passage of title and despite any agreement or understanding as to when the price for the goods is to be paid. Because the risk of loss often depends upon the terms of shipment, it is appropriate to examine those terms at this time.

16:3. Delivery Terms. People are often confused by the terms *F.O.B.*, *F.A.S.*, *C.I.F.*, and *C. & F.* The important thing to remember is that they are delivery terms.

Perhaps the most frequently used term is *F.O.B.*, which means "free on board." When the term is used in connection with a named place, it is a delivery term unless otherwise agreed. Section 2–319. Suppose that Sam Seller is located in New York and Barney Buyer is located in Atlanta. Their contract for sale of goods may provide that the goods are to be "shipped F.O.B. New York," that is, to where Seller is located. Because the contract provides that the goods are F.O.B. place of shipment, this is a shipment contract and the seller must bear the expense and risk of putting them into possession of the carrier. Section 2–319(1)(a). We already know that because this is a shipment contract, the risk of loss passes to Buyer once this is accomplished.

If the goods are shipped "F.O.B. Atlanta" the contract is a destination contract. Because the contract provides that the goods are F.O.B. the place of destination, the seller must transport the goods to that place at his expense and risk and tender delivery to the buyer. Section 2–319(1)(b). The seller bears the risk of loss during shipment. The contract may provide that the goods are also F.O.B. vessel, car, or other vehicle. In that case the seller has the additional risk and expense of loading the goods on board. Section 2–319(1)(c).

F.A.S. vessel at a named port is a delivery term meaning "free alongside." Under this term the seller must deliver the goods alongside the vessel at his own expense and risk in a manner usual at that port or at a dock provided by the buyer. Section 2–319(2)(a). In addition, the seller must obtain and tender a receipt for the goods from the carrier. The carrier then has a duty to issue a bill of lading for the goods. Section 2–319(2)(b).

Other delivery terms frequently used are *C.I.F.* (cost, insurance, and freight) and *C. & F.* (cost and freight). These are shipment contracts that result in the risk of loss's passing from the seller to the buyer upon shipment unless otherwise agreed. Under a C.I.F. contract the seller is obligated to put the goods into the possession of a carrier and obtain bills of lading to a named destination. The seller is also obligated, among other things, to load the goods and obtain a receipt from the carrier showing that the freight has been paid for, as well as to obtain a policy of insurance providing for payment to the buyer. Section 2–320(2).

A C. & F. contract means that the price includes in a lump sum the cost of the goods and the freight to the named destination, but the term does not include insurance. Risk of loss passes to the buyer upon shipment, provided that the seller has performed all his obligations with respect to the goods.

Sometimes a delivery term will be for delivery of goods *ex-ship.* This means that the goods are delivered "from the carrying vessel" and requires delivery from a ship that has reached a place at the named port of destination. Section 2–322(1). Under this term, unless otherwise agreed, the risk of loss does not pass to the buyer until the goods leave the ship's tackle or are otherwise properly unloaded. Section 2–322(2)(b).

If the contract is a *no arrival, no sale* agreement, the risk of loss does not pass until after the seller makes a tender of the goods to the buyer at their destination. On the other hand, the seller assumes no liability to the buyer for failure of the goods to arrive unless he has caused their nonarrival. Section 2–324(a).

In summary then, it is obvious that if the contract for the sale of goods embodies the above delivery terms, the risk of loss passes to the buyer upon delivery to the carrier in the following shipment contracts if there is no contrary agreement: F.O.B. seller's city; F.O.B. vessel; C.I.F.; and C. & F. The risk of loss passes to the buyer when the goods are tendered to him so as to allow him to take delivery under the following destination contracts: F.O.B. buyer's city; F.A.S., ex-ship; and no arrival, no sale.

NINTH STREET EAST, LTD. v. HARRISON

5 Conn.Cir. 597, 259 A.2d 772 (1968)

Philimore T. Harrison operated a clothing store in Westport, Connecticut, known as The Rage. He ordered $2,216 worth of clothing items from Ninth Street East, Ltd., located in Los Angeles, California. Ninth Street shipped the clothes by common carrier from Los Angeles, and all invoices bore the notation "F.O.B. Los Angeles," *as well as the phrase* "Goods Shipped at Purchaser's Risk." *When the carrier attempted delivery at Harrison's store, his wife requested the truck driver to deliver the merchandise inside the door of the store. The driver refused, and the dispute not having been resolved, he left the store with the goods. The shipment was never located thereafter, and Harrison never received the goods. When Harrison refused to pay for the lost goods, Ninth Street sued.*

LEVINE, J.

The sole special defense pleaded was, "The Plaintiff refused to deliver the merchandise into the Defendant's place of business." Therefore defendant claimed that he is not liable for the subsequent loss or disappearance of the shipment, or the purchase price thereof, and that the risk of loss remained with plaintiff.

The basic problem is to determine the terms and conditions of the agreement of the parties as to transportation, and the risks and hazards incident thereto. The court finds that the parties had originally agreed that the merchandise would be shipped by common carrier F.O.B. Los Angeles, as the place of shipment, and that the defendant would pay the freight charges between the two points. The notations on the invoices, and the bill of lading previously described, make this clear. The use of the phrase "F.O.B.," meaning free on board, made this portion of the agreement not only a price term covering defendant's obligation to pay freight charges between Los Angeles and Westport but also a controlling factor as to risk of loss of the merchandise upon delivery to Denver and subsequently to Old Colony as the carriers. General Statutes § 42a 2–319(1)(a). Uniform Commercial Code § 2–319, comment 1. . . . Title to the goods, and the right to possession, passed to defendant at Los Angeles, the F.O.B. point. Upon delivery to the common carrier at the F.O.B. point, the goods thereafter were at defendant's sole risk. § 42a–2–509(1). . . . See also *Electric Regulator Corporation* v. *Sterling Extruder Corporation*, 280 F. Supp. 550, 557 (D.Conn.), where the court, in commenting on § 42a–2–319, stated: "Thus, an F.O.B. term must be read to indicate the point at which delivery is to be made unless there is specific agreement otherwise and therefore it will normally determine risk of loss."

It is highly significant that all the invoices sent to defendant contained the explicit notation "Goods Shipped at Purchaser's Risk." This was, initially, a unilateral statement by plaintiff. The validity of this phrase, as expressing the understanding of both parties, was however, never actually challenged by defendant, at the trial or in his brief. The contents of the invoices therefore confirm the statutory allocation of risk of loss on F.O.B. shipments.

The arrangements as to shipment were at the option of plaintiff as the seller. § 42a–2–311(1). Plaintiff duly placed the goods in possession of a carrier, to wit, Denver, and made a reasonable contract for their transportation, having in mind the nature of the merchandise and the remaining circumstances. Notice of the shipment, including the F.O.B. provisions, was properly given to defendant, as required by law, pursuant to the four invoices. § 42a–2–504; Uniform Commercial Code § 2–504, comment 5.

The law erects a presumption in favor of construing the agreement as a "shipment" contract, as opposed to a "destination" contract. § 42a–2–503; Uniform Commercial Code § 2–503, comment 5. Under the presumption of a "shipment" contract, plaintiff's liability for loss or damage terminated upon delivery to the carrier at the F.O.B. point, to wit, Los Angeles. The court finds that no persuasive evidence was offered to overcome the force of the statutory presumption in the instant case. Thus, as § 42a–2–509(1) indicates, "(w)here the contract requires or authorizes the seller to ship the goods by carrier (a) if it does not require him to deliver them at a particular destination, the risk of loss passes to the buyer when the goods are duly delivered to the carrier." Accordingly, at the F.O.B. point, when the risk of loss shifted, Denver and Old Colony, as carriers, became the agents or bailees of defendant. The risk of subsequent loss or delay rested on defendant, and not plaintiff. A disagreement arose between defendant's wife and the truck driver, resulting in non-delivery of the merchandise, retention thereof by the carrier, and, finally, disappearance of the shipment. The ensuing dispute was fundamentally a matter for resolution between defendant and the carriers, as his agents. Nothing in the outcome of that dispute could defeat or impair plaintiff's recovery against defendant.

(Judgment for plaintiff.)

QUESTION FOR DISCUSSION

1. *Did the defendant have any remedy for the loss of the shipment?*

Sometimes goods are sold under contract terms that provide for "sale on approval" or "sale or return." When goods are sold on approval, the buyer gets possession of the goods but not title. Risk of loss does not pass to the buyer until he actually accepts the goods. Section 2–326 and Comment 1. By implication the buyer may approve the goods by any use of them inconsistent with the seller's ownership, but use of them in a manner consistent with trial of them is not approval.

A sale or return is the delivery of the goods to the buyer with an option to return them to the seller. This procedure generally involves a sale to a merchant whose unwillingness to buy the goods is overcome by the seller's promise to take them back in lieu of payment in the event that they are not sold. Section 2–326 and Comment 1. The risk of loss is on the buyer until the goods are returned to the seller. The buyer bears the risk of loss during the return trip.

If the seller wishes to maintain control over the goods until payment has been received he may ship *C.O.D.*, meaning "cash on delivery." When this procedure is used, the carrier acts as an agent of the shipper for collecting the money and would be liable for failing to do so. The payment must be in cash rather than by check. This provision has no effect other than to maintain possession until payment is made by the buyer. It has no affect on the determination of when risk of loss or title passes.

HAROLD KLEIN & CO., INC. v. LOPARDO
308 A.2d 538 (N.H. 1973)

DUNCAN, J.

The plaintiff, a wholesale jeweler, seeks to recover from the defendant, a retail jeweler, the wholesale price of two diamonds which were sent to the defendant and never returned. The trial in the superior court resulted in a verdict for the defendant. The plaintiff excepted to the denial of its motions for summary judgment and to set aside the verdict. The questions of law raised by those exceptions were reserved and transferred to this court by Morris, J.

There was evidence that the parties had a long standing business relationship whereby plaintiff would deliver jewels to defendant who would in turn sell the jewels to retail customers and pay plaintiff the agreed price. If unable to sell the jewels, defendant would return them to plaintiff. This was a commercial transaction, governed by the Uniform Commercial Code (RSA ch. 283–A), rather than by common law principles of bailment, as suggested by the defendant. Under the code the transaction was a "sale or return" as therein defined. RSA 382–A: 2–326 (Supp. 1972). Approximately 10 days after defendant received the diamonds in November 1968, they were stolen from his jewelry store in Exeter. The primary issue before the trial court was whether plaintiff or defendant should bear the loss resulting from the theft.

RSA 382–A:2–327(2) provides that "(u)nder a sale or return unless otherwise agreed . . . (b) the return is at the buyer's risk and expense." Plaintiff contends that the effect of this subsection is to place the risk of loss upon the buyer once the goods are delivered to him. We think this result is consistent with the intent of the U.C.C. draftsmen (RSA 382–A:2–327, Uniform Law Comment 3), and with the general rule allocating risk of loss between parties, RSA 382–A:2–509(3). . . . When the buyer fails to pay the price as it becomes due, the seller may recover the price of conforming goods lost or damaged within a commercially reasonable time after risk of their loss has passed to the buyer. RSA 382–A:2–709(1)(a). See RSA 382–A:2–606. The plaintiff thus is entitled to the contract price of the two diamonds.

A memorandum from plaintiff which accompanied the shipment of these two diamonds was introduced at the trial. The memorandum provided that the jewels were delivered at the defendant's risk from all hazards regardless of negligence, that title to the jewels would remain in the plaintiff, and the defendant should have no power to sell the jewels without plaintiff's approval. Whether the terms of this memorandum were accepted by defendant, or otherwise became the final written expression of the parties' agreement we need not decide. If the memorandum was binding on the defendant, it specifically and unambiguously allocated the risk of loss to the defendant. If it was not binding upon him, he was liable under the Uniform Commercial Code.

(Plaintiff's exceptions sustained; remanded.)

QUESTIONS FOR DISCUSSION

1. *After the plaintiff delivered the goods to the defendant, who had title to them? Does the answer to this question determine who bore the risk of loss in this case?*
2. *Would the absence of an agreement concerning risk of loss have affected the outcome of this case?*

16:4. TRANSFER OF TITLE. Under the Uniform Commercial Code the concept of the passage of title has been diminished significantly in importance. As we have seen already and will see in Chapter 19, such important questions as risk of loss and the rights, obligations, and remedies of the buyer and seller, as well as third parties, are determined under the code irrespective of the location of the title to the goods. Section 2–401. Formerly these issues were answered by a determination of the location of title. The problem with this latter approach is that the issues are all answered at one point, that is, when title passes. However, it is often not commercially practical to lump them all together. For instance, the risk of loss does not necessarily have to pass from buyer to seller at the same time the buyer has the right to replevy goods. Under the U.C.C. these various rights and liabilities can be determined at different times. This could not be done when these questions were all answered by a determination of where the title to the goods was located. Thus the code provides that under a destination contract the buyer may have the right to replevy goods in the possession of the carrier if they are unique. On the other hand, the seller will have the risk of loss for damages to the goods while they are still in the hands of the carrier.

One may question whether the concept of title is still necessary under the code. It is still necessary because the concept is important in the application of statutes other than the U.C.C., and it also is necessary to cover any unforeseen situations that may arise under the code where title is important. For example, title is often relevant to a determination of guilt or innocence in the commission of certain crimes. It may also be important in certain civil questions such as determining what state is entitled to collect a sales tax when goods cross state lines.

The basic provision of the code covering title is Section 2–401. The section provides that title to goods cannot pass prior to their identification to the contract. This is a matter of common sense. Until the goods to which the contract refers are identified, how can title to them pass? Section 2–401(1). Once the goods are identified to the contract, however, the parties are free to determine by agreement the point and manner in which it passes. Section 2–401(1). If there is no agreement concerning the passage of title, then the code has a number of rules that are applied.

As a general rule, in the absence of an agreement, title to goods passes to the buyer at the time and place at which the seller has completed his performance with reference to physical delivery of the goods. Section 2–401(2). Thus, if the contract is a shipment contract, such as those referred to earlier, title passes to the buyer at the time and place of shipment. Section 2–401(2)(a). If the contract is a destination contract, title passes to the buyer upon tender of the goods at the destination. Section 2–401(2)(b). If the delivery of goods is to be made without the goods' being moved, unless otherwise agreed, title passes at the time and place of contracting if (1) the goods are identified at the time of contracting and (2) no documents are to be delivered. Section 2–401(3)(b). If documents of title are to be delivered by the seller, then title passes when he delivers such documents. Section 2–401(3)(a).

Sometimes a situation may occur in which the goods are rejected by the buyer or he refuses to accept delivery or he revokes acceptance of the goods. Regardless of whether the rejection or refusal or revocation is justified or not, title revests in the seller. Thus, if a seller ships goods under a shipment contract, title passes to the buyer when the goods are shipped, if there is no contrary agreement. However, if the buyer refuses to accept delivery of the goods, then the title revests in the seller at that point. Section 2–401(4).

In summary, the rights and liabilities of the parties for risk of loss are not determined by the concept of title. When there is no agreement, the code has a number of rules that allocate the risk of loss. There are a number of terms governing the parties' respective contractual obligations for delivery and shipment of goods. These terms state, among other things, whether a contract is a shipment contract or a destination contract. If title would become an issue, the required performance may be determined by the rules previously alluded to as well as by the delivery terms. As a general rule then, title passes to the buyer at the time and place of the seller's completing his performance, if there is no contrary agreement.

MECHANICS NAT. BANK OF WORCESTER v. GAUCHER

386 N.E.2d 1052 (Mass. App. 1979)

Wauwinet Development Corporation (Wauwinet) was a dealer in mobile homes. Its inventory was financed by the Mechanics National Bank of Worcester and the bank had a perfected security interest in that inventory.

Wauwinet contracted to sell a mobile home to Charlene M. Garneau for $16,500. Garneau made a $2,000 down payment and the contract was contingent on Garneau's being able to obtain financing for the balance of the purchase price. In due course Garneau obtained credit from Westover Credit Union and made her first monthly payment on February 4, 1975. At that time she had not taken possession of the mobile home and physical delivery of the home had not yet been made when Garneau died on February 9, 1975.

The Credit Union obtained payment of Garneau's debt to it from proceeds of life insurance on Garneau's life. The Bank claimed that it had a right to the funds held by Mr. Belluci, an attorney for the Credit Union. These funds resulted from the sale of the mobile home. The attorney had not yet turned these funds over to Wauwinet when Garneau died. Garneau's estate claimed it was entitled to the funds because the sale of the home was never made.

KASS, J.

(The court first stated that the bank's claim depended upon whether it had a security interest in the funds as proceeds from the "sale" of the mobile home).

A "sale" consists in the passing of title from

the seller to the buyer for a price. G.L. c. 106, § 2–106(1). In this same definition, the code refers to § 2–401, which instructs us in an introductory sentence that the rights and obligations of parties under the code should be sorted out without traditional dependence on the concept of title. A similar statement of policy appears in the comment to § 2–101:

> The arrangement of the present Article is in terms of contract for sale and the various steps of its performance. The legal consequences are stated as following directly from the contract and action taken under it without resorting to the idea of when property or title passed or was to pass as being the determining factor. 1 Uniform Laws Annot., U.C.C., Comment to § 2–101 (Master ed. 1976).

Since, however, the passing of title may offer clues to the unraveling of the puzzle the litigants pose, we examine their positions under § 2–401.

Unless parties explicitly agree otherwise (the record does not favor us with the terms of the agreement between Wauwinet and Garneau, and so we must proceed on the basis that there is no explicit agreement), ". . . title passes to the buyer at the time and place at which the seller *completes* his performance with reference to the *physical delivery of* the goods . . ." G.L. c. 106, § 2–401(2) (emphasis supplied).

Before determining, however, whether physical delivery of the goods took place, we must pause to consider the impact of § 2–401(3) of the code, which provides:

> Unless otherwise explicitly agreed where delivery is to be made without moving the goods
>
> *(a)* if the seller is to deliver a document of title, title passes at the time when and the place where he delivers such documents; or
>
> *(b)* if the goods are at the time of contracting already identified and no documents are to be delivered, title passes at the time and place of contracting.

As we have noted, the record is silent about what the parties may have agreed. A mobile or modular home of the kind we have here is an item of considerable bulk and is unlike a tractor or car, which a customer can drive away. With property of the later kind, courts have applied § 2–401(3). . . . It is not to be imagined that Garneau, a consumer, could simply have dropped by the dealer to pick up a house. We think that where property is of a nature which requires special handling to enable the buyer to take possession, § 2–401(3) is inapplicable and § 2–401(2), requiring physical delivery, does apply, except in cases where the purchase price has been paid and the seller is holding the goods for the convenience of the buyer.

Unless otherwise agreed, "the place for delivery of goods is the seller's place of business." G.L. c. 106, § 2–308. Although a usage of trade may be tantamount to "otherwise agreed," there is no finding in the master's report from which we could conclude whether mobile home dealers generally deliver their product at the customer's site or at the seller's place of business. Wauwinet, therefore, could have made physical delivery at its place of business to complete the sale. Particularly in the light of the commercial facts later recited in this opinion, we are of the opinion that it failed to do so. While delivery and transfer of possession are not synonymous, . . . nonetheless, the "physical location of contract goods is a significant indicator of statutory rights, particularly when third-party interests are at stake." Wauwinet at all times had the goods.

The Bank urges that delivery of the bill of sale was the talisman of title and, hence, entitlement to the proceeds of the sale. But to impart significance to the document of title takes us back to § 2–401(3), which we have already concluded does not apply.

Analysis of the title question, therefore, points to retention of title to the mobile home by Wauwinet. But, in view of the code's renunciation of reliance on title, which we noted above, we rely for resolution of the controversy on factual signposts.

A series of commercial facts etches the inchoate nature of this transaction:

1. Garneau never tendered, much less paid, the balance of the purchase price. In this respect the instant case differs from other specimens of the genre in which full consideration had been paid, either in cash or in paper, and the dispute was between rival secured parties or persons claiming the right to possess property which was the subject of a security interest.

2. Wauwinet never notified Garneau (or her

estate) that the mobile home was at her disposition and that Wauwinet proposed to tender it. G.L. c. 106, § 2–503(1).

3. The security agreement which Wauwinet entered into with the Bank provided that "the Borrower will not sell or offer to sell or otherwise transfer or encumber the collateral or any interest therein without the prior written consent of the Bank." The record does not suggest this ever happened. To the contrary, within days after Garneau died, the Bank took steps under § 9–504 of the code to realize on its security interest in the mobile home, demonstrating the conviction of the Bank that the collateral belonged to Wauwinet.

4. No termination of the Bank's security interest was delivered to Mr. Bellucci. Obviously, the Credit Union would have insisted on receiving a termination statement.

However close the parties may have been to closing their transaction, the facts are that the deal never closed. There was not a completed sale nor a right to payment which generated proceeds for Wauwinet or the Bank, as its secured lender, to follow.

(Judgment affirmed.)

QUESTIONS FOR DISCUSSION

1. *Why did the court conclude that the title to the mobile home was still with Wauwinet?*
2. *Of what significance was the transfer of the bill of sale between the parties?*

---

(1) Title of Purchasers. A question frequently arises when goods are sold by one who does not have title to them and the purchaser is innocent of any wrongdoing. Does the innocent purchaser have a good title, or may the true owner recover the goods? If the seller has no title to the goods, he cannot transfer title, even to a good-faith purchaser. The code provides that a purchaser of goods acquires all title that his transferor had or had power to transfer. Section 2–403(1). Thus, if one innocently buys a watch from a thief, the true owner may recover the watch. This rule on occasion evokes sympathy for the innocent purchaser. The code provides remedies for the innocent purchaser under certain conditions.

(2) Voidable Title. A person with a voidable title has the power to transfer a good title to a good-faith purchaser for value. Section 2–403(1). Suppose Inez Innocent sells her watch as a result of certain fraudulent representations made by the purchaser, William Wrongdoer. Wrongdoer has a voidable title, and before Inez takes any action to avoid the title, Wrongdoer sells the watch to Penny Purchaser. Provided Purchaser is a good-faith purchaser for value, Inez cannot recover her watch. On the other hand, had Wrongdoer stolen the watch from Inez, she could recover it from Purchaser.

In order to be a good-faith purchaser for value, one must meet certain qualifications. *Good faith* means honesty in fact in the conduct or transaction concerned. Section 1–201(19). A purchase includes any voluntary transaction creating an interest in property, such as a sale, negotiation, or mortgage. Section 1–201(32). And one gives value for rights if he acquires them generally in return for any consideration sufficient to support a simple contract or if he gives security in total or partial satisfaction of a preexisting claim. Section 1–201(44). There are other methods for giving value not relevant here. Thus, in our example, if Wrongdoer gave the watch to Purchaser or if Purchaser knew of Wrongdoer's fraud in the transaction, Purchaser would not be an innocent purchaser for value.

(3) Entrusting by Delivery to Merchant. Another problem occurs when one entrusts goods to a merchant who, in turn, sells them to a buyer in the ordinary course of business. Suppose Terry Timekeeper gives his watch to James Jeweler, a merchant, to repair. Jeweler sells new and used watches in his store and sells Timekeeper's watch in the ordinary course of business to Connie Customer. Timekeeper cannot recover his watch. The code provides that any entrusting of possession of goods to a merchant who deals in goods of the kind gives him the power to transfer all the rights of the entruster to a buyer in the ordinary course of business. Section 2–403(2). *Entrusting* is defined by Section 2–403(3) as any delivery and any acquiescence in retention of possession, regardless of any conditions expressed between the parties concerning delivery or acquiescence, and regardless of whether the entrusting was procured or the goods disposed of under conditions that would be larcenous under the criminal law.

A *buyer* in the ordinary course of business "means a person who in good faith and without knowledge that the sale to him is in violation of the ownership rights or security interest of a third party in the goods, buys in the ordinary course from a person in the business of selling goods of that kind but does not include a pawnbroker." Section 1–201(9).

Note that in order for the purchaser to maintain title in the goods, he must show that the original owner entrusted goods to a merchant and that the goods were purchased in the ordinary course of business. Thus, if the goods are not purchased from a merchant or are goods that the merchant normally does not sell, then the purchaser must give them up to the true owner.

GRIMM v. PRUDENCE MUTUAL CASUALTY COMPANY
243 So.2d 140 (Fla. 1971)

Charles B. Grimm's troubles began when he purchased a Cadillac convertible from Mr. Ash. Grimm gave Ash his check for $1,500 and agreed to obtain an additional $1,000 in financing from a bank. Ash advised Grimm that the Bank of Clearwater had a lien for $700 on the car. The parties agreed that the lien would be satisfied out of the $1,500 check and Grimm would pay the final $1,000 when clear title was delivered to him.

Grimm took possession of the car and applied to Prudence Mutual Casualty Company for an insurance policy. Subsequently Grimm was unable to get Ash to deliver a clear title and pay off the lien, which was actually $2,800 rather than $700. Grimm got a default judgment against Ash and was notified by the Bank of Clearwater that he either had to assume the debt or have the car repossessed. Before the bank could repossess the car, it was stolen. Grimm filed a claim against Prudence for the theft of the car. Prudence claimed that Grimm had no insurable interest in the car and refunded his premium. Grimm sued Prudence. From an adverse verdict Grimm appealed.

PER CURIAM.

In the present case the bank had not repossessed the car. Grimm was still responsible for it and would have been liable for damages caused by it under Florida's dangerous instrumentality doctrine. Grimm could have satisfied the lien, thereby acquiring absolute title, and he could have sold the car, subject to the bank's outstanding lien. The mere fact that Grimm had agreed to the bank's peaceful repossession of the car in no way reduces his ownership or responsibility pending that repossession.

The bank had the power to terminate Grimm's interest in the car, but until it exercised that power, Grimm remained the legal owner.

The significance placed by the court below on Grimm's lack of a Florida title certificate was equally misplaced. This Court has clearly stated that the absence of a Florida title certificate does not indicate the absence of a valid title, but rather the absence of a marketable title. *Motor Credit Corporation* v. *Woolverton*, 99 So.2d 286 (Fla. 1957). See Fla. Stat. § 672–401(2) (1969).

This Court recently decided the case of *Smith* v. *State Farm Mut. Automobile Insurance Co.*, 231 So.2d 193 (Fla. 1970). In that case we approved the holdings of the First and Third District Courts of Appeal that bona fide purchasers for value of stolen automobiles have an "insurable interest" therein. The district court in the present case distinguishes these cases saying that in each case the insured had an equitable interest in the car because of the money he had paid. This reasoning is not only inaccurate but is also inapplicable to the present case.

Fla. Stat. § 672,403 (1969) states that a purchaser of goods acquires all title which his transferor had or had power to transfer. In the case of the stolen cars, the insured parties could not have any equitable interest therein since the transferors had no power to transfer. Their insurable interest, then, was their right to mere possession against all but the rightful owner. *Barnett* v. *London Assurance Corporation* . . . *Norris* v. *Alliance Insurance Company of Philadelphia* . . . and *Skaff* v. *United States Fidelity & Guaranty Co.*, supra, hold that the purchaser of a stolen automobile's right to possession against all but the rightful owner is the right that gives the purchaser an insurable interest. This rule of law was approved by us in *Smith* v. *State Farm Mutual Automobile Insurance Co.*, supra. The district court misread this case when it held that the stolen automobile purchaser's equitable interest therein was insurable interest.

There is no doubt but that Grimm had a right to possession of the automobile until such time as the bank's prior right cut his off. Nor is there any doubt but that Grimm had more than just an equitable interest in the automobile. He was not purchasing a stolen automobile, but rather an automobile with a larger lien than he had been aware of. Grimm was the legal owner of an automobile which was about to be repossessed.

With respect to respondent's argument that the sale to Grimm was never completed so no title passed, it is sufficient to point out Fla. Stat. § 672,401(2)(1969), which provides that title passes ". . . at the time and place at which the seller completes his performance with reference to the physical delivery of the goods, despite any reservation of a security interest and even though a document of title is to be delivered at a different time or place"; Grimm's default judgment against Ash did not cause the title to revert to Ash, so Grimm remained the legal owner.

(Reversed and remanded.)

QUESTIONS FOR DISCUSSION

1. *On what basis did the court rule that Grimm had an insurable interest in the car?*
2. *The court referred to several other cases involving stolen automobiles purchased by third parties. Would it make any difference to the true owner's right to recover the autos if the stolen autos were purchased by the third party from car dealers?*

MEDICO LEASING COMPANY v. SMITH et al.
457 P.2d 548 (Okla. 1969)

Medico Leasing Company gave possession of an automobile to a used-car dealer, Smith, to show to a prospect but did not give him the certificate of title. Smith, representing himself as the owner of the car, sold it to Wessel Buick Company, who later resold it to Country Cousins Motors, Inc., who resold it to W. C. and Dorene Carter. Smith never gave Medico the money for the car, and Wessel, believing that it had lost the title after talking with Smith, obtained a new one and used it in selling the car to Country Cousins.

Medico traced the car to the Carters and filed suit to recover possession. Defendant, W. C. and Dorene Carter and Country Cousins Motors, each filed cross-

petition against Medico for wrongful taking. From a verdict in favor of the Carters and Country Cousins, Medico appealed.

HODGES, J.

The principal issue presented by this appeal is the trial court's ruling that the defendant, Wessel Buick Company, was a "buyer in the ordinary course of business" in good faith as defined by Sections 1–201(9) and 2–403(2) of the Oklahoma Uniform Commercial Code when they purchased the automobile from defendant Smith, a used car dealer, whom plaintiff had entrusted possession of the car. The trial court found that Wessel Buick acquired sufficient legal title even though the sale was made without the actual transfer of the automobile's certificate of title, and therefore the defendants, Carters and Country Cousins Motors, were entitled to damages for the wrongful taking.

Plaintiff does not contest the fact that Smith is their agent, but they maintain that he was a limited agent and did not have title to the automobile or authority to convey title. They assert that Smith could not convey any better title than he had, and as he had no title, none was conveyed. It is further asserted by plaintiff that defendant Wessel was not a buyer in good faith as required by our Uniform Commercial Code, because they purchased the car without a certificate of title under facts and circumstances which would have put an ordinary prudent businessman on inquiry.

The provisions of the code pertinent to the issues in this case are in part set out:

> 12A O.S. § 2–403(2) Any entrusting of possession of goods to a merchant who deals in goods of that kind gives him power to transfer all rights of the entruster to a buyer in ordinary course of business.
>
> 12A O.S. § 1–201(9) "Buyer in ordinary course of business" means a person who in good faith and without knowledge that the sale to him is in violation of the ownership rights or security interest of a third party in the goods buys in ordinary course from a person in the business of selling goods of that kind but does not include a pawnbroker. "Buying" may be for cash or by exchange of other property or on secured or unsecured credit and includes receiving goods or documents of title under a preexisting contract for sale but does not include a transfer in bulk or as security for or in total or partial satisfaction of a money debt.

The "entruster" in the instant case, the plaintiff, had good title to the Buick, which was a used car. The one to whom the automobile was entrusted, Smith, as a used-car dealer, is a "Merchant who deals in goods of that kind" within the meaning of the statute. Smith was known by the plaintiff to be a used-car dealer. The Uniform Commercial Code has not changed the law in this state regarding clothing an agent with apparent authority to convey title, especially if the agent is one who ordinarily deals in the goods which the principal has entrusted to him. A recognized principle of estoppel consistently followed by this court is that if a principal or owner of an automobile permits a dealer in automobiles to have an automobile under circumstances indicating authority to sell, he is estopped to assert title against bona fide purchaser for value without notice.

Since Smith had apparent authority to convey legal title, the question then arises as to whether Wessel was a buyer in good faith when they purchased the car without a certificate of title.

It has long been held by this court that a certificate of title to an automobile issued under the motor vehicle act is not a muniment of title which establishes ownership, but is merely intended to protect the public against theft and to facilitate recovery of stolen automobiles and otherwise aid the state in enforcement of its regulation of motor vehicles. This rule was not changed with the passage of the Uniform Commercial Code. Under Section 1–201(15) the certificate of title of an automobile is not listed as a "document of title." It was not necessary for the defendant Smith to deliver the certificate of title before he conveyed ownership of the Buick Automobile, and the absence of a certificate does not invalidate the sale or prevent title from passing. The sale of the automobile was complete upon delivery of the car with the intent to sell. Title 12A O.S. Section 2–401(2) states:

> Unless otherwise explicitly agreed title passes to the buyer at the time and place at which the seller completes his performance with reference to the physical deliv-

ery of the goods, despite any reservation of a security interest and even though a document of title is to be delivered at a different time or place. . . .

The plaintiff introduced no evidence to show a lack of good faith upon the part of defendant Wessel. . . .

(Judgment affirmed.)

SUMMARY

Risk of loss and location of title are separate questions under the code. The parties are free to state in the agreement when risk of loss passes from the seller to the buyer. If there is no agreement as to when the risk of loss passes from the seller to the buyer, the code sets forth several rules if there is no breach of the contract.

A contract that requires the seller to deliver the goods to a carrier is a shipment contract. Risk of loss passes from the seller to the buyer when the goods are delivered to the carrier. A contract that requires the seller to tender goods at a particular destination is called a destination contract. The risk of loss passes from the seller to the buyer when the goods are tendered to the buyer.

If the goods are in the possession of a bailee, the risk of loss passes to the buyer when the bailee acknowledges the buyer's right to the goods or when the buyer receives a negotiable document of title covering the goods or when the buyer receives a nonnegotiable document of title and has a reasonable time to present the document.

If the goods are to be delivered by the seller, either at his premises or the buyer's, and not by a third party, then the following rules apply. If the seller is a merchant, then the risk of loss passes when the buyer is in receipt of the goods. If the seller is not a merchant, then the risk of loss passes upon the tender of delivery by the seller.

If the seller ships nonconforming goods, the risk of loss remains with the seller until he cures the nonconformity or until the buyer accepts the goods despite the nonconformity. If the buyer has accepted the goods and then rightfully revokes that acceptance the buyer may treat the risk of loss as having remained on the seller from the beginning.

If the buyer breaches the contract prior to the time risk of loss would normally pass to him, the seller may look to the buyer for a commercially reasonable period of time for any loss because of a deficiency in the seller's insurance coverage.

There are numerous delivery terms. F.O.B. seller's city; F.O.B. vessel; C.I.F.; and C&F are shipment contracts in the absence of a contrary agreement. F.O.B. buyer's city; F.A.S.; exship; and no arrival, no sale are destination contracts absent a contrary agreement.

The parties are free to agree when title to goods passes. Title to goods and risk of loss may pass at different times. As a general rule, title to goods passes to the buyer at the time and place at which the seller has completed his performance with respect to physical delivery of the goods.

The code provides that a person with a voidable title has the power to transfer a good title to a good-faith purchaser for value. Any entrusting of goods to a merchant who deals in goods of the kind gives him the power to transfer all the rights of the entruster to a buyer in the ordinary course of business.

PROBLEMS

1. Plaintiff contracted to purchase a motorcycle from the defendant's motorcycle shop. The seller was to deliver the motorcycle by June 30, 1977 for the agreed price of $893.00. The plaintiff paid for the motorcycle in full, was given the papers necessary for registration and insurance and did in fact register and insure the vehicle. The motorcycle was subsequently stolen by looters on July 11, 1977, during a power blackout. The buyer never exercised dominion or control over the motorcycle. Who bears the loss? Ramos v. Wheel Sports Center, 409 N.Y.S.2d 505 (1978).
2. Shook ordered three reels of burial cable from Graybar Co. Graybar mistakenly sent one reel of burial cable and two reels of aerial cable, although each carton was marked "burial cable." Shook rejected the two nonconforming reels and left them on the ground at the construction site because of their size. Graybar was notified of the rejection but did not collect the cable. Shook was unable to ship it to Graybar because of a strike of truck drivers. After the passage of four months, the cable was stolen. Graybar sued Shook for the purchase price, claiming that Shook had agreed to return the nonconforming reels. Decide. Graybar Electric Co. v. Shook, 283 N.C. 213, 195 S.E.2d 514 (1973).
3. S-Creek Ranch, Inc., purchased a number of older pregnant ewes from Monier & Company in order to raise a lamb crop. The ewes were shipped on railroad cars from Worland, Nebraska, on March 26. One of the cars was unloaded enroute in Alliance, Nebraska, because of a hot box, and the sheep were fed there and reloaded. These sheep reached their destination on March 28, whereas the others arrived on the 27th. The sheep were billed to S-Creek Ranch, who paid the freight. Upon arrival, a number of ewes were dead, and there were several aborted lambs as well. It was later determined that the sheep were suffering from vibriosis, an infection of the uterus often resulting in the death of the fetus. A large number of the lambs were lost. There was no agreement as to when the risk of loss should pass, nor was there any statement as to whether the contract was a shipment or a destination contract. Assuming the sheep picked up the disease during shipment, who should have borne the loss? See S-Creek Ranch, Inc. v. Monier & Company, 509 P.2d 777, (Wyo. 1973).
4. James Mercanti owned a yacht that lost a mast in a squall. Seth Persson agreed to construct a ninety-two-foot mast for the yacht. Mercanti paid Persson $1,000 to purchase the lumber and was also to supply the hardware to be affixed to the mast. It was also agreed he would pay Persson at the

rate of $4 per hour as billed. The woodwork on the mast was substantially completed and was to be finished after Mercanti delivered the hardware. Mercanti in fact delivered $1,200 worth of hardware on June 15. On June 17, Persson's boat yard and its contents, including the mast and hardware, were totally destroyed by fire. Prior to the fire Mercanti had paid $4,588 for labor and materials. It was determined that there was no breach of the contract up to the time of the fire. Mercanti sued Persson for the loss. Should Mercanti recover? Mercanti v. Persson, 160 Conn. 468, 280 A.2d 137 (1971).

5. Donaho, a salesman for B & B Parts Sales, Inc., delivered stereo equipment and tapes to H. F. Collier for sale in his service station. The contract provided "Terms 30–60–90 this equipment will be picked up if not sold in 90 days." Collier's station was burglarized and the merchandise stolen. Who bore the loss? H. F. Collier v. B & B Parts Sales, Inc., 471 S.W.2d 151 (Tex.App. 1971).
6. Lair Distributing Company sold a TV antenna and tower to Eugene Crumps for $900 under a conditional-sales contract payable in consecutive monthly installments of $7.50 for ten years. The contract provided that title would remain with Lair Distributing until the purchase price was paid in full. It was orally agreed that Lair would maintain the system during the ten-year term of the contract. After the tower was erected, it was struck by lightning and the entire antenna had to be replaced. Who bore the loss? Lair Distributing Company v. Crump, 261 So.2d 904 (Ala.App. 1972).
7. Dana Debs, Inc., a dress and suit manufacturer, received a written order from Lady Rose Stores, Inc., for the purchase of 288 garments. The order was on Lady Rose's printed form and advised Dana Debs to "ship via Stuart, 453 W. 57th Street." Stuart Express Company, Inc., picked up the garments and issued its printed receipt to Debs, naming Lady Rose as consignee for the garments. Debs issued a bill of sale on which was printed "Terms, F.O.B., NYC." Debs and the carrier were located in New York City. Stuart later notified Debs that the entire shipment was lost. As between Dana Debs and Lady Rose, who bore the loss? Dana Debs, Inc. v. Lady Rose Stores, Inc., 319 N.Y.S.2d 111 (1970).
8. Plaintiff alleged in his petition that he purchased from defendant "One houseboat, Model No. 924" for a stated price, that he paid the price, and that defendant failed to deliver a houseboat or to return the purchase price upon demand by plaintiff. When the plaintiff went to take possession of the boat, he could not because it had sunk. Was plaintiff entitled to recover the purchase price? Chatham v. Clark's Food Fair, Inc., 106 Ga.App. 648, 127 S.E.2d 868 (1962).
9. Guigin and Scott Oldsmobile had agreed on a "trade" of cars on Friday. Each agreed that the trade was final. Later that night Guigin's son smashed up the car received from Scott Oldsmobile. All that remained at the time was for title papers to be processed and for Guigin to return the next

day with a check for cash due on the trade. The trial court determined that Scott "still owned" the car at the time of the accident "despite any oral negotiations." Was the trial court correct? Motors Insurance Corporation v. Safeco Insurance Company of America, 412 S.W.2d 584 (Ky. 1967).

10. Cummins Diesel Engines delivered a generator to a job site for William F. Wilke, Inc. The contract provided that the generator was to be complete in all respects including the performance of certain tests. When the generator was delivered it lacked two batteries so Wilke could not use it. The generator remained on the site through the winter. When it was hooked up in the spring, it was discovered that the water in the cooling system had frozen damaging the engine. Cummins took the engine away to repair the damage and then billed Wilke. Is Wilke liable? William F. Wilke, Inc. v. Cummins Diesel Engines, Inc. 252 Md. 611, 250 A.2d 886 (1969).

CHAPTER 17

Warranties Under the Uniform Commercial Code, the Magnuson-Moss Act, and Products Liability

The law concerning warranties has had a curious evolution over a period of hundreds of years. The guilds and the Church in the Middle Ages combined to impose high standards for quality upon sellers. Later, under English law, the doctrine of *caveat emptor,* "let the buyer beware," gave little protection to buyers. When goods were purchased by the buyer, he had little, if any, recourse against the seller for defective merchandise. Today we have come full circle, and the Uniform Commercial Code provides significant protection to buyers in the form of express and implied warranties. In addition, state laws, as well as federal legislation in the form of the Magnuson-Moss Warranty Act, give the buyer more protection against defective merchandise than ever before. In this chapter we will examine the U.C.C. provisions governing warranties as well as the federal legislation that affects the U.C.C. Finally, we will examine the seller's liability for injuries that result from defective goods under the developing area of the law known as *products liability.*

17:1. Warranty Protection Under the Uniform Commercial Code. The code provides for three broad types of warranty protection: express warranties, implied warranties, and warranty of title and infringement.

(1) Express Warranty. An express warranty comes into being when the seller in some manner creates a warranty. An express warranty may be created by affirmation, promise, description, or sample. Section 2–313(1). Furthermore, in order to create a warranty under the U.C.C., it is not necessary that the seller use the term *warrant* or *guarantee* or even that he have an intention to create a warranty. Section 2–313(2). Let us examine these concepts more closely.

The code provides that any affirmation of fact or promise made by the seller to the buyer that relates to the goods and becomes part of the "basis of the bargain" creates an express warranty. Section 2–313(1)(a). The affirmation may be either oral or written. For instance, if the seller states to the buyer that the "goods are 100 percent cotton," an express warranty by affirmation would be created. A statement that "the cloth supplied by seller will not shrink under any circumstances" is a promise that creates an express warranty. In either case the statements must be part of the "basis of the bargain." The comments to the code state that affirmation of fact made by a seller about goods during a bargain are regarded as part of the description of the goods and thus part of the "basis of the bargain." The consequence of this fine legal terminology is that the buyer need not show that he relied on the affirmation in entering into the contract in order for an express warranty to be created. On the other hand, if the cloth in our example was in fact not 100 percent cotton and the seller could prove that the buyer knew this and entered into the contract solely for other reasons, no warranty would be created.

One may also create an express warranty by merely describing the goods without making any promise or affirmation of fact. A seller may agree to supply "five eight-by-four sheets of finish-grade half-inch plywood, free of knots." This creates an express warranty that the goods will conform to the description. Section 2–313(1)(b). If the plywood delivered is full of knots, the seller has breached the warranty. The description need not be limited to words; thus technical specifications or blueprints may create an express warranty under this section.

The creation of a warranty may result from the display of a sample or model of the goods to be supplied. Such a display creates an express warranty that the goods will conform to the sample or model, provided that it is part of the "basis of the bargain." Section 2–313(1)(c).

From the preceding discussion, it is obvious that an express warranty can be created in a number of ways. It may result from advertising, from salesmen's statements, from the typical printed "warranty," by a display, or by samples or models. It is not necessary that the seller have the intention of creating a warranty. Thus the retail seller may be bound by the statements placed on the can or package by the manufacturer. The seller's act may also create an express warranty. For example, if a customer asks for "oil-based outside house paint" and is handed a can of paint, this creates an express warranty that the paint is oil-based outside house paint.

(a) *Time of Making.* Although a sale of goods is necessary to create a warranty under the code, it is not necessary that the statement creating the warranty be made prior to the sale. The time is not really material as long as the language or the samples or models are fairly to be regarded as part of the contract. Section 2–313, Comment 7. For instance, suppose a buyer seeks further assurance before taking delivery of the goods. A sales clerk, on handing hair coloring to a wary customer who has paid, may say, "This product is harmless to the hair." The warranty becomes a modification to the contract and, presuming that it is reasonable, need not be supported by consideration. Sections 2–209(1), 2–313, Comment 7.

(b) *Fact or Opinion.* Statements of opinion or of the value of the goods or a commendation of them do not create a warranty. The problem, of course, is to distinguish these statements from those that do create a warranty. The basic question is, What statements of the seller have in the circumstances and in objective judgment become part of the basis of the bargain? Section 2–313(2) and Comment 8. Thus a statement that "this is the best-tasting toothpaste on the market today" is considered "puffing" or "sellers' talk." Buyers do not rely on these statements and they do not create a warranty. An opinion by an expert art dealer that a particular painting is a genuine Picasso that results in its purchase would create a warranty.

Under certain circumstances a breach of an express warranty can also constitute fraud. Suppose the seller of a car states that the used car he is selling has been driven 40,000 miles when in fact he knows it to have been driven 140,000 miles. This affirmation would constitute a warranty that the car has been driven only 40,000 miles. The seller has breached the warranty and also committed the tort of fraud.

EDDINGTON v. DICK
386 N.Y.S.2d 180 (1976)

Joanne Dick offered a used refrigerator for sale in the Geneva Times *newspaper classified section. The advertisement read as follows:". . . also large Frigidare refrigerator. Good condition. $150.00."*

Amos Eddington responded to the advertisement and purchased the refrigerator. It was delivered to his home but never would operate. When he demanded his money back, Dick refused. Eddington then sued Dick.

BRIND, J.

The Uniform Commercial Code in section 2–313(1)(a) provides that any affirmation of fact or promise made by the seller to the buyer which relates to the goods and becomes part of the basis of the bargain creates an express warranty that the goods shall conform to the affirmation or promise.

The courts of this state have heretofore held that the buyer's right to recovery is not limited to express representations made in a contract and the buyer may reasonably rely upon representations made by the seller through mass media advertisement.

This section is interpreted by the Federal Circuit Court of Appeals in *Borowicz* v. *Chicago Mastic Co.*, 367 F.2d 751, 7 Cir. The court there held that before liability on the theory of express warranty can be imposed upon a seller the following elements must be established:

1. An affirmation of fact, or promise by the seller.
2. The natural tendency of the said affirmation or promise was to induce the buyer to purchase goods.
3. That the buyer purchased goods in reliance thereon.
4. A breach of the expressed warranty by the seller.
5. Such breach proximately caused injury to the Plaintiff.

These requirements raise a number of questions of fact to be determined. In applying these principles to the case at bar, the court finds that there was in fact an affirmation or promise by the seller in that the refrigerator was represented as in "good condition."

See 51 N.Y.Jur. (Sales) section 159 at page 158 wherein the text-writer states: "it is not necessary to the creation of an express warranty that the seller use formal words such as 'warrant' or 'guarantee' or that he have a specific intention to make a warranty," citing *Hawkins* v. *Pemberton*, 51 N.Y. 198 and *Fairbank Canning Co.* v. *Metzger*, 118 N.Y. 260, 23 N.E. 372.

This decisional law is embodied in U.C.C. 2–313(2), with the further provision that "an affirmation merely of the value of the goods or a statement purporting to be merely the seller's opinion or commendation of the goods does not create a warranty."

The court finds that the printed advertisement goes beyond mere affirmation of value, opinion or commendation but definitely relates inherently to the condition of the goods and thus creates an express warranty.

This court does not dispute the good faith on the part of the defendant with respect to the advertisement or the sale. However, good faith on the part of the seller is no defense to an action for breach of warranty, . . . Nor is the knowledge on the part of the seller that the representation is false an essential part of the warranty. . . .

The court finds that the natural tendency of this affirmation in the advertisement was to induce this plaintiff to purchase the refrigerator. In fact, the plaintiff, under oath, stated that he purchased the goods in reliance upon this affirmation that the refrigerator was in fact in good condition. The court is convinced, after hearing the testimony and examining the exhibits, that the statements that the refrigerator was in "good condition" became a part of the basis of the bargain.

A purchaser is not under a duty to amply investigate property before purchasing it where the contract of sale contains an express warranty as to the condition of the property, but he may rely upon the representations made, and if the goods are defective and do not conform to the warranty, he may recover for the breach.

The court finds that there was a breach of this express warranty. The term "good condition" under these circumstances legally requires the goods sold to be in reasonable working order. Where the item was sold for a fair price but required additional expenditure of more than one hundred per cent of the purchase price to put it in working order, the item did not comply with such warranty.

There can be no doubt that the breach of warranty was the proximate cause of the injury to the plaintiff as requested in his demand for judgment.

Therefore, the court finds in favor of the plaintiff and awards the plaintiff judgment against the defendant in the sum of $180.00. The defendant is entitled to the return of the refrigerator at her cost.

Questions for Discussion

1. *Is it necessary to use the word "warranty" or "guarantee" in order to create an express warranty? What words in the advertisement created the express warranty?*
2. *Is it necessary to the creation of an express warranty that the seller have an intention to create a warranty?*

---

(2) **Implied Warranties.** In addition to an express warranty that may be created by the seller in a sale of goods, there may also be certain implied warranties created as a matter of law unless specifically excluded as part of the sale. We shall see later that the power of a seller to exclude these implied warranties has been restricted. These developments are quite a change from the days when the doctrine of caveat emptor was the rule. The burden has now shifted to the seller, although many laymen do not seem to appreciate this fact. The scope of the implied warranties varies, depending upon whether or not the seller is a merchant.

(a) *Implied Warranty of Merchantability or Usage of Trade.* A Warranty that the goods will be of merchantable quality is implied if (1) there is a sale of goods; (2) the seller is a merchant with respect to goods of that kind; and (3) the warranty was not excluded or modified. Section 2–314(1). If these conditions are met, the warranty is created automatically by the sale regardless of any statements by the parties.

Historically the courts have required that there be a sale, as distinguished, for example, from the leasing of goods, and that the sale must have been of goods as opposed to a contract for services of which the goods are but an incidental part. Thus, at one time in many jurisdictions, the purchase of food or drink in a restaurant was not considered a sale of goods but rather the sale of a service, and hence no implied warranty was created. The code specifically provides that the serving of food or drink for consumption on or off the premises is a sale. Section 2–314(1).

The courts' decisions have been in a state of flux and are not in agreement in this area. A more current problem involving the case of the purchase of blood from blood banks is a good illustration. Is such a sale governed by the provisions of the U.C.C.? Many courts have held that the implied warranty of merchantability does apply, as this is a sale of goods. Others have held that it is the sale of a service and not subject to the rules of the Uniform Commercial Code. The problem is particularly troublesome because in many cases it is impossible to determine whether the blood contains certain diseases, such as hepatitis, until it is used. Many legislatures have passed statutes providing that the U.C.C. does not apply in this instance. On the other hand, a number of courts have pointed out that there need not be a sale, in the technical sense, to the person injured in order for that party to recover for breach of the implied warranty of merchantability.

NEWMARK v. GIMBEL'S, INC.

102 N.J.Super. 279, 246 A.2d 11 (1968)

On November 16, 1963, Mrs. Newmark went to Gimbel's beauty salon and was told by one Valente, a beauty technician, that her fine hair was not right for the special permanent and that she needed a "good" permanent wave. She agreed. About three to five minutes after the last of the waving solution had been applied, she experienced a burning sensation on the front part of her head. After she complained, Valente applied more salve along the hairline. While she was under the dryer, the burning sensation returned, and when she complained, Valente turned down the heat. The following day Mrs. Newmark's forehead became red and blisters appeared. In addition, much of her hair fell out.

Valente admitted that the permanent-wave procedure followed was at his suggestion and that the permanent-wave solution he used was "Candle Glow," a product of Helene Curtis. He testified that the product had been applied as directed and that it was common for a customer to feel a burning sensation during the process.

The trial court ruled as a matter of law that the plaintiff, Newmark, was not entitled to recover on the basis of breach of either express or implied warranty. The jury found for the defendant on the issue of negligence, and plaintiff appealed.

LABRECQUE, J.A.D.

The core question here presented is whether warranty principles permit a recovery against a beauty parlor operator for injuries sustained by a customer as a result of use on the customer of a product which was selected and furnished by the beauty parlor operator. In ruling that warranty did not apply here the trial judge reasoned that the transaction between the parties amounted to the rendition of services rather than the sale of a product, hence defendant could be held liable only for negligence in the performance of such services. Our consideration of the question convinces us that his ruling was a mistaken one.

It would appear clear that the instances in which implied warranties may be imposed are not limited to "sales" that come strictly within the meaning of c. 2 of the Uniform Commercial

Code. In the comment to the warranty provision of the latter, N.J.S. 12A:2–313, N.J.S.A., it is said:

> Although this section is limited in its scope and direct purpose to warranties made by the seller to the buyer as part of a contract for sale, the warranty sections of this Article are not designed in any way to disturb those lines of case law growth which have recognized that warranties need not be confined either to sales contracts or to the direct parties to such a contract. They may arise in other appropriate circumstances such as in the case of bailments for hire, whether such bailment is itself the main contract or is merely a supplying of containers under a contract for the sale of their contents. . . . (at p. 190).

The policy reasons applicable in the case of sales would likewise justify the extension of liability for breach of warranty to any commercial transaction where one person supplies a product to another, whether or not the transaction is technically considered as a sale. The rationale underlying the liability of a retailer for defects in a product obtained from a reputable supplier and sold to a customer, has been explained as follows:

> The argument for the prevailing view points to the strict nature of warranty at common law and under the broad terms of the Sales Act, which certainly do not suggest an exception for latent defects. If reliance upon the seller is needed, it may be found in the customer's reliance on the retailer's skill and judgment in selecting his sources of supply. Broader considerations are also urged. The retailer should bear this as one of the risks of his enterprise. He profits from the transaction and is in a fairly strategic position to promote safety through pressure on his supplier. Also, he is known to his customers and subject to their suits, while the maker is often unknown and may well be beyond the process of any court convenient to the customer. Moreover, the retailer is in a good position to pass the loss back to his supplier, either through negotiations or through legal proceedings. 2 Harper and James, Torts, § 28.30, p. 1600 (1956).

Weighing the foregoing policy considerations, we are satisfied and hold that, stripped of its nonessentials the transaction here in question, consisting of the supplying of a product for use in the administration of a permanent wave to plaintiff, carried with it an implied warranty that the product used was reasonably fit for the purpose for which it was to be used.

Defendant also cites *Perlmutter* v. *Beth David Hospital*, 308 N.Y. 100, 123 N.E.2d 792 (Ct. of App. 1954) in support of its position that the transaction here in question did not amount to a sale and hence could not be the subject of a warranty. The holding in the former case, that the furnishing of blood and the giving of it in a transfusion (which later caused hepatitis) did not constitute a sale by a hospital, has not been followed in this state. Assuming that it had been, we incline to the view that that case may be better explained on the basis of special policy considerations in favor of hospitals, and on the further ground that there was no way in which the blood could be tested to ascertain that all of the hepatitis strain had been removed.

It follows that the issue of defendant's liability for breach of implied warranties of fitness for purpose and merchantability should have been submitted to the jury.

(Reversed and remanded.)

QUESTIONS FOR DISCUSSION

1. *The defendants in this case stated that there was no sale and that therefore the UCC implied warranties should not apply. How did the court answer this contention?*
2. *Note that the court mentioned the Perlmutter case, in which a New York court held that the furnishing of blood from a blood bank was not a sale and that therefore the UCC implied warranties did not apply. Can you distinguish this case from the Perlmutter case? How did this court treat that case?*

The implied warranty of merchantability arises only when the sale is made by a merchant "with respect to goods of that kind." Thus, when you buy a used car from a "friend," no implied warranty of merchantability arises. Nor is any implied warranty created when one who is in the jewelry business sells his own car to a customer who comes to the store to purchase a watch. The seller is a merchant, but not with respect to "goods of that kind."

The code, in Section 2–314(2), states that in order for goods to be merchantable they must

(a) pass without objection in the trade under the contract description; and
(b) in the case of fungible goods, are of fair average quality within the description; and
(c) are fit for the ordinary purposes for which such goods are used; and
(d) run, within the variations permitted by the agreement, of even kind, quality and quantity within each unit and among all units involved; and
(e) are adequately contained, packaged and labeled as the agreement may require; and
(f) conform to the promises or affirmations of fact made on the container or label if any.

The chief qualification is that the goods be fit for the ordinary purposes for which such goods are used. Suppose Tony's Tirestore sells Connie Carowner two brand-new tires. There is no express warranty given either by the tire manufacturer or by the retailer. Carowner places the tires on her car, drives normally for 5,000 miles, and discovers that the tread has separated from the casing. She would probably be able to recover for breach of the implied warranty of merchantability because she purchased the tires from a merchant and used them in an ordinary manner. Comment 2 to Section 2–314 states that "goods delivered by a merchant in a given line of trade must be of a quality comparable to that generally acceptable in that line of trade. . . ." The tires in this case obviously fail to meet that standard. On the other hand, had Carowner placed the tires on a four-wheel drive vehicle and driven exclusively over the renowned and torturous Baja Peninsula, she would in all likelihood not be able to assert a claim under the implied warranty of merchantability, for her use of the tires would not be for the ordinary purposes for which normal tires are used.

When the goods are fungible, such as apples, corn, wheat, and the like, the implied warranty of merchantability provides that the goods should be of fair and average quality. Section 2–314(2) (b) is aimed at agricultural bulk products. "Fair average quality" means goods around the middle belt of quality. The buyer is not entitled to have all units meet the highest standard, nor is he obligated to take the dregs. There may be a mixture of quality, but if all units are of the worst quality, they are not of fair and average quality and breach the implied warranty of merchantability. Section 2–314, Comment 7.

(b) *Trade Usage.* An implied warranty may also arise because certain practices are normally a matter of custom in the trade. The buyer is entitled to assume that the practice will be adhered to. When both parties are aware of the trade usage, the law will imply that the parties intended the usage to be part of their contract, if there is no evidence to the contrary. Section 2–

314(3). Thus, where it is the custom and usage of new-car dealers to lubricate new cars prior to delivery, a failure to do so would result in a breach of the implied warranty by the car dealer. Likewise, where pedigree dealers normally provide pedigree papers for the animal, they may be obligated to do so under this provision of the code.

Of course, in order to recover for breach of an implied warranty, it is necessary to show not only the existence of the warranty but also the fact that the breach was the proximate cause of the damage sustained. Thus, if one purchases hair dye and subsequently loses his or her hair, it must be shown that there was a warranty, express or implied, and that the breach of that warranty resulted in the loss of hair.

(c) *Implied Warranty of Fitness for Particular Purpose.* Occasionally one may purchase goods for a specific purpose and seek the advice of the seller as to the appropriate goods to use for that purpose. Under certain circumstances there may be created an implied warranty that the goods are fit for that particular purpose. In order for this implied warranty to arise, the seller must have reason to know (1) any particular purpose for which the goods are being used and (2) that the buyer is relying on the seller's skill or judgment to select suitable goods. Section 2–315.

In order for the buyer to recover for breach of an implied warranty for fitness for a particular purpose, he must have been relying on the seller's skill or judgment. From the seller's viewpoint it is enough if the circumstances are such that the seller should realize the purpose for which the goods are intended and that the buyer is relying on his advice. It is not necessary that the buyer specifically inform the seller of the particular purpose. Suppose Harry Homeowner enters Seller's Hardware Store and informs the salesperson that he is going to paint his outside concrete patio and needs paint for that purpose. The salesperson sells him five gallons of paint, saying, "This will do the job." In fact the paint turns out to be interior wall paint and shortly after application peels off Homeowner's patio. Homeowner would be able to recover for breach of the implied warranty of fitness for a particular purpose.

You should be careful to distinguish the warranty of fitness for a *particular* purpose created under Section 2–315 of the code, which envisages a specific use of the product, from the warranty of merchantability, which warrants that the goods are fit for the *ordinary* purposes for which such goods are used. Thus, in our example, although the warranty of fitness for a particular purpose is breached, there is no breach of the warranty of merchantability, assuming the paint is satisfactory when applied to interior walls, its *ordinary* purpose. The two warranties are not mutually exclusive, however; they both may arise under the same contract.

You should also note another difference between the implied warranty of merchantability and the implied warranty of fitness for a particular purpose. In order for the warranty of merchantability to be implied, the sale must be by a merchant. Although the warranty of fitness for particular purpose is most often created under circumstances in which the seller is a merchant with the "appropriate skill or judgment," it can arise under the right circumstances when the seller is a non-merchant. Section 2–315, Comment 4.

VLASES v. MONTGOMERY WARD & CO.

377 F.2d 846 (1967)

McLAUGHLIN, C.J.

This case revolves around the charge that defendant—appellant, Montgomery Ward, was liable for the breach of implied warranties in the sale of one day old chickens to the plaintiff—appellee, Paul Vlases. The latter came to this country from Greece when he was sixteen and until 1954, his primary occupation was that of a coal miner. He had always raised chickens but because of his job as a miner his flocks were small, ranging from between twenty-five to one hundred chicks. In 1958, plaintiff began the construction of a two-story chicken coop large enough to house 4,000 chickens and a smaller side building where he could wash, grade and sell the eggs. Vlases worked alone on the coop, twelve hours a day, fifty-two weeks a year, until its completion in 1961. In November of 1961, plaintiff placed an order at defendant's outlet store in Brownsville, Pennsylvania for the purchase of 2,000 one-day-old chicks. The chickens selected by the plaintiff from Ward's catalogue were hybrid Leghorns and were noted for their excellent egg production. On December 21, 1961, plaintiff received the 2,000 chickens and placed them on the first floor of the coop which had been equipped with new brooders, feeders and within a short time, waterers. As a further hygienic precaution wire and sugar cane were placed on the ground so the chickens would not come in contact with the dirt floor. For the first six months Vlases slept in the coop in order to give the new chicks his undivided attention.

During the first few weeks after delivery the chickens appeared to be in good health but by the third week plaintiff noticed that their feathers were beginning to fall off. This condition was brought to the attention of Mr. Howard Hamilton who represented the Agway Corporation which was supplying the plaintiff with feed on credit. In February of 1962 Mr. Hamilton took five chickens to the Bureau of Animal Industry Diagnostic Laboratory where they were examined by Dr. Daniel P. Ehlers. The examination revealed signs of drug intoxication and hemorrhagic disease involving the weakening of blood vessels. Four chicks were brought to Dr. Ehlers in May of 1962 and were found to be suffering from fatigue. On the 14th of August 1962 Mr. Hamilton brought three chickens to the laboratory where Dr. Ehlers' report noted that two of the chicks were affected with visceral leukosis, one with ocular leukosis, one had bumble foot and one had been picked. Visceral and ocular leukosis are two types of avian leukosis complex or bird cancer which disease infected plaintiff's flock either killing the chicks or causing those remaining to be destroyed.

Appellant takes the position that an action for breach of implied warranties will not lie for the sale of one-day-old chicks where there is no human skill, knowledge or foresight which would enable the producer or supplier to prevent the occurrence of this disease, to detect its presence or to care for the sickness if it was present. The jury was instructed by the court that recovery on behalf of the plaintiff required a finding that the chickens were afflicted with leukosis at the time defendant made delivery. The expert testimony for both sides indicated that there was no way of determining whether newly hatched chicks have leukosis and that there is no medication available to prevent the disease from occurring. Assuming the chickens were diseased upon their arrival the thrust of appellant's argument questions the sufficiency of the law to support a finding that Ward is liable under Pennsylvania law for the breach of implied warranties.

The two implied warranties before us are the implied warranty of merchantability, and the implied warranty of fitness for a particular purpose. Both of these are designed to protect the buyer of goods from bearing the burden of loss where merchandise, though not violating a promise expressly guaranteed, does not conform to the normal commercial standards or meeting the buyer's particular purpose, a condition upon which he had the right to rely.

Were it to be assumed that the sale of 2,000 chickens infected with avian leukosis transgressed the norm of acceptable goods under both warranties, appellant's position is that the action will not lie in a situation where the seller is unable to discover the defect or cure the damage if it could be ascertained. That theory does not eliminate the consequences imposed by the code upon the seller of commercially inferior goods. It is without merit.

The fact that avian leukosis is nondetectable could be an important issue but only as bearing on the charge of negligence, which is no longer in this suit. . . . Plaintiff's task was to develop facts and circumstances from which the jury could reasonably conclude that the chickens were diseased at the time of delivery. Here substantial support of that decisive element in the claim was furnished by the expert testimony connecting the progressively lower probabilities of avian leukosis infecting flocks under proper care and testimony underlining the sanitary precautions taken by Vlases to insure the flock's health. Under all of the testimony the plaintiff clearly met his required burden of proof and that evidence was adequate to uphold the verdict reached by the jury.

The judgment of the district court will be affirmed.

QUESTIONS FOR DISCUSSIONS

1. *Did Montgomery Ward know of the avian leukosis prior to the delivery of the chickens? Could they have determined that the birds had the disease? Is this of any consequence in the determination of the liability of Montgomery Ward?*
2. *What implied warranties was Vlases relying upon for recovery? What are the differences in proof required for these two warranties?*

(3) **Warranty of Title and Infringement.** Section 2–312(1) of the code provides that, subject to exclusions or modifications that will be discussed in Section (6), there is in a contract for sale of goods, a warranty that

(a) the title conveyed shall be good, and its transfer rightful; and
(b) the goods shall be delivered free from any security interest or other lien or encumbrance of which the buyer at the time of contracting has no knowledge.

Suppose Ulysses Usedcardealer purchases a stolen car and then sells it to Inez Innocent. The true owner of the car takes it back, Inez can recover from the seller for breach of the warranty of title. Whether Usedcardealer thought he had good title does not matter, only the fact that he did not convey a good title.

The warranty of title also applies to situations in which the goods sold are encumbered by a lien or other security interest. Suppose Inez Innocent purchases a car from her neighbor, who neglects to tell her that the car was financed through a bank that has a security interest in the car and has duly recorded the required financing statement, which gives *constructive* notice to all. Inez does not know of the bank's security interest until the bank repossesses the car for nonpayment of the loan. Inez would be able to recover from her neighbor under the warranty of title because she did not have *actual* notice of the bank's security interest. On the other hand, if the buyer purchases the goods with full knowledge of the security interest, the warranty is not breached. This section of the code also provides that a merchant regularly dealing in goods of the kind warrants that the goods are delivered free of the rightful claim of any third person for infringement unless the product is produced to the buyer's specifications, in which case the buyer must hold the seller harmless against any claim of infringement made by a third party.

Section 2–312(3). This warranty protects against possible claims arising out of patent or copyright infringement.

Interestingly, although the warranty created under Section 2–312(1) is created as a matter of law, it is not considered an "implied warranty," such as the warranty of merchantability or fitness for particular purpose. This is so primarily because there is a separate method for excluding the warranty of title as opposed to the other implied warranties, and hence it is not lumped with the other "implied" warranties. Section 2–312(2) and Comment 6. Exclusion and modification of warranties is discussed in Section (6).

AMERICAN CONTAINER CORPORATION v. HANLEY TRUCKING CORPORATION

111 N.J.Super. 322, 268 A.2d 313 (N.J. 1970)

On March 16, 1968, Hanley Trucking Corporation arranged to purchase from Herschel Trucking Company a 1965 Fruehauf semi-trailer. A week later, Hanley sold the semi-trailer to American for $2,800. Hanley forwarded $2,500 to Herschel and retained the $300 balance. A bill of sale and a certificate of legal title were given to American by Hanley; however, legal title on the certificate was still in the name of Herschel, which had endorsed the document in blank. Hanley had never taken physical possession of the semi-trailer, but delivery was made directly from Herschel to American.

American used the trailer for sixteen months before it was impounded by the New Jersey state police as a stolen vehicle. American then notified Hanley that it desired to rescind the sale between them and tendered the receipt given them by the state police. Hanley refused to return the purchase price. American sued Hanley, who in turn filed a third-party complaint against Herschel. All parties moved for summary judgment.

HERBERT, J.S.C.

American contends that it bought the semi-trailer from Hanley, who impliedly but falsely warranted and represented good title. Hanley denies that it sold the semi-trailer to American and that it warranted good title. Rather, says Hanley, Herschel sold the semi-trailer directly to American and warranted good title to both American and Hanley. Herschel asserts not only that it never warranted good title, but also that in fact good title was conveyed and the seizure of the semi-trailer improper. Moreover, Herschel argues that rescission is an inappropriate remedy here, where American cannot tender the semi-trailer itself, but merely the police receipt.

Hanley's assertions of caveat emptor notwithstanding, there can be no doubt that Hanley impliedly warranted good title when it sold the semi-trailer to American. By the same token, Hanley had received from Herschel an implied warranty of good title with its acquisition of the vehicle. The Uniform Commercial Code provides that in every contract of sale there is implied a warranty of good title unless such warranty is specifically excluded by the language or circumstances of the agreement:

> (1) Subject to subsection (2) there is in a contract for sale a warranty by the seller that
> (a) the title conveyed shall be good, and its transfer rightful; and
> (b) the goods shall be delivered free from any security interest or other lien or encumbrance of which the buyer at the time of contracting has no knowledge.
> (2) A warranty under subsection (1) will be excluded or modified only by specific language or by circumstances which give the buyer reason to know that the person selling does not claim title in himself or that he is purporting to sell only such right or title as he or a third person may have.

The facts of this case demonstrate no specific exclusions, and the general rule implying warranties of good title applies.

Herschel argues that good title was in fact conveyed to Hanley, and then to American.

This appears doubtful in light of the certification filed by Thomas Cornelissen, a detective with the New Jersey State Police. Cornelissen states that when he examined the semi-trailer, its serial plate had been removed and the die-stamped frame serial number had been ground off. Moreover, the title certificate was invalid, since the serial number it listed was a number used for trailers manufactured in years prior to 1965 and could not have been the proper number for this 1965 vehicle.

Herschel's claim that good title was conveyed would, if sustainable, not only negate any breach of warranty, but also indicate that return of the vehicle could probably have been secured by American within the time (90 days after seizure) allowed by N.J.S.A. 39:5–47.

The purchaser of goods warranted as to title has a right to rely on the fact that he will not be required, at some later time, to enter into a contest over the validity of his ownership. The mere casting of a substantial shadow over his title, regardless of the ultimate outcome, is sufficient to violate a warranty of good title. The policy advanced here has found expression in the past in cases where courts of equity have refused to order specific performance of contracts for the sale of land.

The true rule is stated in *3 Pars, on Con.* (6th ed.) 380, that if the character of the title be doubtful, although the court were able to come to the conclusion that, on the whole, a title could be made that would not probably be overthrown, this would not be good title enough, for the court (sic) have no right to say that their conclusion, or their opinion, would bind the whole world, and prevent an assault on the title. The purchaser should have a title which shall enable him not only to hold his land, but to hold it in peace; and if he wishes to sell it, to be reasonably sure that no flaw or doubt will come up to disturb its marketable value.

I am satisfied, under the circumstances, that American's inability to return the semi-trailer should not operate as a bar to rescission. Hanley never had perfect title. Possession of the vehicle, whether by Herschel or by Hanley or by American, was always subject to the possibility of seizure by the police. Once the vehicle was seized in fact, and for 90 days thereafter, the police receipt represented title equivalent in quality to that originally conveyed from Hanley to American. With the timely notice given by American, Hanley could have sought to establish the validity of its ownership and to secure repossession of the vehicle. That Hanley did not do so cannot weigh against American in its suit for rescission.

(The court granted summary judgment in favor of American against Hanley and in favor of Hanley against Herschel and remanded for a determination of damages.)

QUESTIONS FOR DISCUSSION

1. *Does it make any difference whether Hanley actually believed it conveyed good title in determining its liability to American?*
2. *What steps could Hanley have taken to protect itself in the event that it suspected that it did not have good title?*

X **(4) Cumulation of Warranties.** There may be both express and implied warranties existing at the same time. The code provides that if there are both express and implied warranties, they should be construed as consistent with each other and cumulative. If it is impossible or unreasonable to construe them as consistent with each other, the following rules are applied: under Section 2–317 of the code:

(a) Exact or technical specifications displace an inconsistent sample or model or general language of description.
(b) A sample from an existing bulk displaces inconsistent general language of description.
(c) Express warranties displace inconsistent implied warranties other than an implied warranty of fitness for a particular purpose.

(5) Persons to Whom Warranty Protection Is Extended. Several questions should occur to you at this time. May the implied warranties be enforced only by the party who purchases the product? May the party protected assert his rights only against the seller, or may he claim under the warranty further back in the marketing chain, for example, against the manufacturer as well as the retailer? These questions raise the problem of privity of contract, that is, whether there is a direct contractual relationship between seller and buyer. The first will be analyzed here, and the second will be treated in Section 17:3, "Products Liability."

Historically the law required privity of contract between the seller and the party seeking to recover for breach of warranty. The code alters this approach and provides that the seller's express and implied warranty "extends to any *natural* person who is in the family or household of his buyer or who is a guest in his home if it is reasonable to expect that such person may use, consume or be affected by the goods and who is injured *in person* by breach of the warranty." (Emphasis added.) Section 2–318.

This is the approach adopted by a majority of the states. There appcars to be no national concensus as to the scope of warranty protection however, and some states have extended the warranty protection beyond family members or household guests. In fact, the drafters of the code have provided alternative approaches to the problem. While the majority of states have adopted the section that extends recovery to a natural person who is a family member or household guest, a number of states have adopted the most liberal approach which extends protection to "any person who may reasonably be expected to use, consume or be affected by the goods. . . ." This alternative reflects the trend of more recent decisions by the courts.

The purpose of the code provision is to extend protection by freeing the beneficiaries from the technical rules of "privity." Note, however, that this extension applies only to natural persons—as distinguished from corporations—who suffer personal injuries, not property damage. Without this provision this class of users would have no right to recover. Just as with other third-party beneficiaries those who assert rights under this section are limited to the provisions of the express or implied warranty under which they seek to recover, which may contain limitations and exclusions. This section does provide, however, that once a warranty is created, it may not exclude coverage of this class of individuals by its terms.

Suppose Harry Homeowner has just purchased a new toaster from his local store. His mother-in-law, who is a guest in the home, plugs it in and is injured from a severe electrical shock. Even though she has no privity with the seller because she did not purchase the toaster, she could recover for her injuries under any warranty that came with the toaster, either express or implied. The drafters of the code made clear in their comments to this section that it was intended to be neutral and not to limit or enlarge the expanding case law in this area. Thus in some states, one who is not a member of the household or a guest may recover for injuries received on the theory of breach of the implied warranty of merchantability.

WOLFE v. FORD MOTOR CO.

376 N.E.2d 143 (Mass. App. 1978)

Alexis McLaughlin, her daughter Mary Ann McLaughlin, and her niece, Jocelyn Wujcik Wolfe, were traveling in a Ford truck on which a camper was mounted. While they were travelling at about 45 miles per hour, there was a blowout in the left rear tire; the truck went out of control and the camper became completely detached from the body of the truck. Wolfe, who was riding in the camper at the time, was seriously injured.

James McLaughlin, the husband of Alexis McLaughlin, had purchased the truck and camper unit from Harold R. Donahue, doing business as Donahue Mobile Homes. Donahue assembled the truck and camper unit.

Wolfe sued Ford and Donahue on a number of theories, one of which was an action for breach of implied warranty against Donahue. A jury found in favor of Wolfe against Donahue for failure to give adequate warning and instructions as to the gross-vehicle-weight requirements. Donahue appealed contending, among other things, that Wolfe was not within the class of persons who may recover for breach of warranty of merchantability.

GOODMAN, J.

1. We hold that Wolfe, the buyer's niece, is a "person who is in the family . . . of his [Donahue's] buyer." G.L. c. 106, § 2–318. The interpretation of § 2–318 in this respect has not come before the Supreme Judicial Court or this court. We are persuaded by the analysis in *Miller* v. *Preitz,* 422 Pa. 383, 221 A.2d 320 (1966), overruled on other grounds, *Kassab* v. *Soya,* 432 Pa. 217, 226, 246 A.2d 848 (1968), in which a nephew of the buyer who, like Wolfe, was not a member of the buyer's household was held to be "in the family" of the buyer under the identical Pennsylvania version of § 2–318 of the Uniform Commercial Code. . . . The court held that "considering the remedial nature of the provision and the natural connotations of the word, its meaning was not intended to be unduly restrictive. . . . This interpretation of the word 'family' is not too burdensome on the seller who makes the warranty because not only must the beneficiary be in the buyer's family but also it must be 'reasonable' to expect that such person may use, consume or be affected by the goods.' " 422 Pa. at 390, 221 A.2d at 323. The Pennsylvania court pointed out that the statute makes a distinction between "person . . . in the family" and "person . . . in the buyer's household," indicating that the former phrase was intended to extend beyond the members of the household. 422 Pa. at 389, 221 A.2d 320.

This construction is consistent with and finds support in *Dodge* v. *Boston & Providence R. R.,* 154 Mass. 299, 301, 28 N.E. 243, 244, (1891), in which the court held that "[t]he word 'family' has several meanings. Its primary meaning is the collective body of persons who live in one house, and under one head or management. Its secondary meaning is those who are of the same lineage, or descend from one common progenitor. Unless the context manifests a different intention, the word 'family' is usually construed in its primary sense." In this case the context does indicate otherwise; the statutory opposition of the terms "family" and "household" supports the broad construction of the term "family" to indicate those not ordinarily thought of as living in the buyer's household.

Further, the application of § 2–318 to such persons as Wolfe seems to us within the legislative intent to liberalize the "technical rules as to 'privity' " (comment 2 to § 2–318, as appearing in 1A Uniform Laws Annot., U.C.C. [Master ed. 1976]) consistent with the "remedial nature" (422 Pa. at 390, 221 A.2d 320) of the statute.

(Verdict for Wolfe affirmed.)

QUESTION FOR DISCUSSION

1. *You should note that the U.C.C. provides several alternatives of 2–318. Would Wolfe have recovered under all alternatives? Which one has Massachusetts adopted?*

(6) **Disclaimer or Exclusion or Modification of Warranties.** An area of the law that has been changing concerns provisions limiting warranties, most particularly the implied warranties of merchantability and fitness for particular purposes. The code provides for the disclaimer of warranties, but lately a

number of states have altered the code provisions allowing for this, and more recently there has been federal legislation limiting this right under certain circumstances. Let us first examine the basic code provisions, then what alterations some states have made, and finally, in the next section, let us take a brief look at the federal legislation that regulates in this area.

(a) *Warranty of Title.* The warranty of title may be excluded only by specific language or by circumstances that give the buyer reason to know that the seller does not claim title himself or that he is selling only such right or title as he has. Section 2–312(2). Note that the provisions of Section 2–316, a discussion of which follows, do not apply to the warranty of title.

(b) *Express Warranty.* The seller may exclude an express warranty by not making any promise, description of goods, affirmation of fact, or sale by sample or model that would give rise to such an express warranty. The seller can also place a clause in the contract stating that no express warranties are given. However, suppose a sales contract provides that the "cloth supplied by the seller will not shrink under any circumstances" and elsewhere in the agreement there is a provision that states that the seller "excludes all warranties express or implied." Does the last phrase exclude the express warranty created by the statement that the cloth will not shrink? The code states that words or conduct creating a warranty will be construed where reasonable as consistent with words or conduct negating a warranty. If this is impossible—that is, if it is unreasonable to construe them together—the limitation is inoperative. Section 2–316(1) and Comment 1. Thus, despite the limiting language, the warranty against shrinkage would be enforceable.

(c) *Implied Warranty.* (i) BY EXAMINATION. The code provides that no implied warranty is given for those defects that an examination of the goods ought to have revealed to the buyer if he has in fact examined the goods as fully as he desired or if he refuses to examine them. Section 2–316(3)(b). This provision contemplates something more than the buyer's mere "inspection" of the goods to see if they conform to the contract. There must be a demand by the seller that the buyer examine the goods fully. This demand by the seller puts the buyer on notice that he is assuming the risk of defects that the examination ought to reveal. Section 2–316, Comment 8.

(ii) BY COURSE OF DEALING, COURSE OF PERFORMANCE, OR USAGE OF TRADE. An implied warranty can be excluded by course of dealing, course of performance, or usage of trade. One frequently quoted example is the usage of trade in the cattle industry by which once the buyer inspects the cattle and cuts out those that are unsuitable, his acceptance of the others is "irrevocable and without recourse." Thus implied warranties are excluded.

(iii) BY EXPRESS LANGUAGE. The code in the area of express language is not a model of clarity. In general it does allow specifically for the exclusion or modification of the implied warranties of merchantability and fitness for a particular purpose.

To exclude the implied warranty of merchantability, (1) the language must mention merchantability, and (2) if the disclaimer is written, it must be conspicuous. An example would be "SELLER MAKES NO WARRANTY OF MERCHANTABILITY FOR THESE GOODS." Note that the exclusion need not be written. Section 2–316(2). On the other hand, in order to exclude the warranty for fitness for particular purpose the exclusion must be in writing and conspicuous.

The requirements for exclusion seem simple enough. However, the code goes on to erode them. In Section 2–316(3), it provides that despite the above requirements, unless circumstances indicate otherwise, language that in common understanding calls the buyer's attention to the exclusion of warranties, such as "with all faults" or "as is," is effective to exclude the implied warranties. Subsection (4) also provides that remedies for breach of warranty can be contractually limited under Sections 2–718 and 2–719, which will be dealt with in Chapter 18.

The changes in attitudes toward consumer protection have focused on the problem of the limitation and exclusion of implied warranties. For instance, a manufacturer may give a very limited express warranty and, coupled with it, disclaim all implied warranties. The result, of course, is that the warranty that the seller is "giving" the buyer is in fact limiting the protection he might have had in the absence of any express warranty at all. Some states, such as Maryland, Maine, and California, have altered their version of the code so as to eliminate the possibility of disclaiming the implied warranties of merchantability and fitness for particular purpose. Some have also restricted the right of sellers to limit the remedies available to the buyer for breach of warranty. The code itself provides that any limitation on the right of buyers to recover for physical injuries is prima-facie unconscionable and under those circumstances will not be enforced by a court. Section 2–719(3).

MOBILE HOUSING, INC. v. M. F. STONE
490 S.W.2d 611 (Tex. App. 1973)

By a written contract dated June 6, 1970, Mobile Housing, Inc., sold a mobile home to Mr. and Mrs. M. F. Stone for $1,000.46. The written conract contained a provision stating that the unit was taken "as is" and specifically disclaiming the warranties of merchantability or fitness for particular purpose. It also stated that the parties agreed that there were no descriptions, samples or models used as part of the contract.

Before buying the trailer, the Stones visited Mobile's lot numerous times, and on each occasion the salesman used a sample mobile home as an example of the one he was trying to sell them. The salesman admitted that in writing the contract, he took the description of the home from the sample on the lot. The Stones never saw the home they were buying until it was delivered, at which time they immediately rejected it. The new home differed in a number of respects from the sample, including the carpeting, the furniture, and the location of windows. The Stones sued to recover their down payment. Mobile repossessed the house and filed a counterclaim. From an adverse verdict in the lower court, Mobile appealed.

BATEMAN, J.

The conclusion is inescapable, it seems to us, that appellees' agreement to buy the mobile home was induced by and based upon their numerous inspections of Unit No. 103 on appellant's sales lot and the representations of appellant's salesman that the home he was trying to sell them would be precisely like it. There is no evidence that they had ever seen the home they were buying, or even a picture of it, or that the salesman described it in any other manner than by referring to Unit 103. The salesman testified that in drawing the contract they took the description from this Unit 103.

Section 2.313 provides:

> (a) Express warranties by the seller are created as follows:
>
> (1) Any affirmation of fact or promise made by the seller to the buyer which relates to the goods and becomes part of the basis of the bargain creates an express warranty that the goods shall conform to the affirmation or promise.
>
> (2) Any description of the goods which is made part of the basis of the bargain creates an express warranty that the goods shall conform to the description.
>
> (3) Any sample or model which is made part of the basis of the bargain creates an express warranty that the whole of the goods shall conform to the sample or model.
>
> (b) It is not necessary to the creation of an express warranty that the seller use formal words such as "warrant" or "guarantee" or that he have a specific intention to make a warranty, . . .

We hold that appellant made express warranties, within the meaning of § 2.313(a) (2),(3), that the mobile home would conform to the description given by the salesman and the model called Unit No. 103. It is true, the written contract provides that the article is sold "as is" and disclaims all warranties, either express or implied, and the use of descriptions, samples or models as a part of the contract. However, we hold that these express warranties rest on "dickered" aspects of the individual bargain, and go so clearly to the essence of that bargain that words of disclaimer in the purchase agreement are repugnant to the basic dickered terms. See comment 1 under § 2.316.

The code according to comment 4 under § 2.313 seems to take the view that the whole purpose of the law of warranty is to determine what it is that the seller has in essence agreed to sell, and refuses except in unusual circumstances to recognize a material deletion of the seller's obligation. Thus, a contract is normally one for a sale of something describable and described. A clause generally disclaiming "all warranties, express or implied," cannot reduce the seller's obligation with respect to such description and therefore cannot be given literal effect under § 2.316.

Section 2.316 provides in subsection (a) that if words or conduct creating an express warranty cannot reasonably be construed as consistent with words or conduct tending to negate or limit the warranty, the warranty shall predominate and negation or limitation is inoperative to the extent that it is unreasonable.

(Judgment affirmed.)

QUESTIONS FOR DISCUSSION

1. *What steps could Mobile have taken so as to not have been bound to a warranty?*
2. *How was the warranty by Mobile created?*
3. *Why was the disclaimer provision on the contract not effective?*

Many have thought that the Uniform Commercial Code approach to warranties is confusing. In an attempt to simplify warranties, the U.S. Congress passed legislation regulating the use of warranties that affect interstate commerce. This legislation became effective in early 1975. Because the law has some impact upon the warranty provisions of the code, let us examine it briefly.

17:2. THE MAGNUSON–MOSS WARRANTY ACT. For the layman the concept of an implied warranty for fitness for a particular purpose or of merchantability is somewhat difficult to grasp. When the manufacturer gives a "warranty" or "guarantee" with his product, the layman often feels that he is getting something to which he would not otherwise be entitled. This may not be the case. Often the seller is using the "warranty" to limit his liability by implied warranty. Under these circumstances the consumer would be far better off if there were no warranty statement at all, for many of the warranties given are "in lieu of all other warranties, expressed or implied." The warranty then given is very limited in scope. Because of this confusion Congress passed the Magnuson–Moss Warranty Act.

The provisions of the Magnuson–Moss Warranty Act are somewhat complex, and its impact upon the Uniform Commercial Code not entirely clear at this point, but several things should be noted at the outset. First, it does not affect every contract, only those that come within the jurisdiction of the act. Basically, when the goods are in interstate commerce or when there is an affect upon interstate commerce, contracts are subject to the terms of the act. As the law is construed at the present time by the courts, this means that the statute governs the large majority of contracts for the sale of goods by most merchants. Second, the act does not eliminate the Uniform Commercial Code warranty provisions. The only limitation is that when those provisions conflict with the Magnuson–Moss Warranty Act, they must give way. The trick is to determine where the conflict exists. First, let us examine the basic provisions of the act.

In order to simplify the understanding of warranties the act provides that, where applicable, warranties shall be called either "full (statement of duration) warranty" or "limited warranty." This designation requirement is limited to goods costing the consumer more than $10. If the seller decides to give a "full warranty," he must comply with the requirements for giving a "full warranty" as mandated by the act. For example, in order to give a "full warranty," the seller must agree to

1. Repair the product or, after a reasonable number of attempts, replace the product or grant a refund at the option of the consumer.
2. Not limit the duration of any implied warranty on the product.
3. Not exclude or limit consequential damages unless this provision appears conspicuously on the warranty.
4. Not impose any duty upon the consumer other than notification unless the warrantor has demonstrated that the duty imposed is reasonable.

If the warranty does not meet these minimum standards, it must then be designated a "limited warranty." The act also has a section prohibiting the express written disclaimer of implied warranties to a purchaser. The purpose of this provision is to eliminate the practice of substituting a limited express warranty for the implied warranties of fitness for particular purpose and merchantability that are created under the Uniform Commercial Code. The act is limited to written warranties and would not, for example, apply to the oral statements of salespersons made at the point of sale.

In addition, the act provides for the Federal Trade Commission to issue regulations requiring inclusion in the written warranty of such information as the step-by-step procedure that the purchaser should follow in order to obtain performance, the days and times that the warrantor will perform his obligations, and the products or parts covered, as well as additional requirements if the product actually costs the consumer more than $5.

To aid in making the proposal easier for consumers to enforce, the act provides that as part of the judgment recovered against a seller, there may be included costs and expenses, including an attorney's fee based upon actual time expended on the case. However, recovery of attorney's fees is not mandatory but is left to the discretion of the court and can be denied if the court "determines that such an award would be inappropriate." An additional provision requires the consumer first to submit the dispute to any dispute-settlement procedure established by the warrantor, prior to initiating suit, provided such a procedure meets the requirements of the act.

(1) Conflicts with the Uniform Commercial Code. When a written warranty comes within jurisdiction of the act, the Uniform Commercial Code provisions that conflict are preempted. There is a provision in the act to the effect that if a state's law is more protective of consumers than the act, the state law may be enforced.

The most obvious conflict is with the disclaimer provision of the U.C.C., which allows one to disclaim the implied warranties. The Magnuson–Moss Act provides that one may not disclaim implied warranties, although he may give a "limited warranty" and therein limit the duration of implied warranties to the period of the written warranty, provided that it is of reasonable duration. Naturally the code provision permitting exclusion of the implied warranties of merchantability and fitness for particular purpose would be unenforceable.

There are other provisions of the code that will obviously be affected by this legislation. For example, the code provides that the words *warrant* or *guarantee* need not be used to create an express warranty. On the other hand, the act requires that applicable written warranties must be specifically designated "full warranties" or "limited warranties." However, an express warranty is not necessarily written, and thus nonwritten express warranties such as those created by the use of a sample or model would presumably not be covered under this provision of the act.

Other potential conflicts between the act and the Uniform Commercial Code exist, particularly with regard to the limitation of remedies, which will be discussed in Chapter 19. A number of other potential conflicts await resolution by the courts and are beyond the scope of this text. Suffice it to say that for those written warranties that fall within the jurisdiction of the Magnuson–Moss Warranty Act there will be some significant effect upon the provisions of the Uniform Commercial Code. Let us now examine a number of grounds for recovery from the seller of a defective product that causes injury to a person.

PRATT v. WINNEBAGO INDUSTRIES, INC.

463 F. Supp. 709 (W.D. Penna., 1979)

James and Mary Jane Pratt lived in Erie, Pennsylvania. They purchased a Winnebago motor home from Gene Norris Oldsmobile near Cleveland, Ohio, in July, 1979 for $16,745.00. Winnebago manufactured the motor home on a chassis manufactured by General Motors.

After using the motor home two or three times, the Pratts had a number of complaints about the vehicle including a problem with the transmission and a claim that the brakes were "spongy." On August 10, 1979, GM arranged to pick up the vehicle and tow it to Dave Hallman Chevrolet in Erie for replacement of the transmission and adjustment of the brakes. They gave the Pratts a "loaner" vehicle while the work was performed on the Pratt's motor home. The Pratts used the loaner vehicle for over four weeks and never attempted to register their vehicle in Pennsylvania.

The work on the motor home was finished on September 10, 1977. Sometime before the repairs were finished, the Pratts decided they no longer wanted their mobile home and asked Winnebago and Gene Norris Oldsmobile for a replacement or their money back. Gene Norris offered to replace the motor home with a slightly more expensive model if the Pratts would pay the difference in value between the motor home they purchased and the new one. The Pratts refused and filed suit.

WEBER, C.J.

General Motors, the Gene Norris Agency, and Winnebago stood ready to repair all of the defects with the motor home which the Pratts had itemized. At trial, representatives of General Motors estimated that the complaints related to the chassis could be remedied in a few hours at a cost of only $78 to GM. Representatives of Winnebago estimated that the rest of the complaints could be repaired in only two or three days of labor time and a cost of about $400 to Winnebago or Gene Norris. The plaintiffs did not attempt to rebut the evidence.

Aside from the replacement of the transmission, no Defendant had an opportunity to repair the motor home. From the time the motor home was purchased, Gene Norris and Winnebago offered to repair the defects which the Pratts had itemized on the premises of the Gene Norris Agency in Middleburg Heights, Ohio, near Cleveland, if the Pratts would transport the motor home back to the agency. At trial, Mr. Gene Norris, President of the Norris Agency, testified that he would still repair the motor home.

The Plaintiffs contend that they are entitled to a refund of their purchase money under 15 U.S.C.A. § 2304(a)(4) of the recently enacted Magnuson–Moss Act, which provides:

> In order for a warrantor warranting a consumer product by means of a written warranty to meet the Federal minimum standards for warranty—
>
> (4) if the product (or a component part thereof) contains a defect or malfunction after a reasonable number of attempts by the warrantor to remedy defects or malfunctions in such product, such warrantor must permit the consumer to elect either a refund for, or replacement without charge of, such product or part (as the case may be). . . .

The statute specifies with utter clarity that a consumer has the right to elect to receive a refund only if the product has defects which survive a reasonable number of attempts by the warrantor to repair them. Except for their successful repair of the transmission, the defendants in this case have not had a single opportunity to repair the plaintiffs' motor home. Before the vehicle was even returned from the Hallman agency and while the plaintiffs were using the "loaner" vehicle supplied by Winnebago at their request, the plaintiffs had decided that they no longer wanted the motor home. Plaintiffs, by not paying the Pennsylvania sales tax and securing the Pennsylvania registration plates, have created their own obstacle to the return of the vehicle to the dealer's garage.

Finally, the plaintiffs contend that under 15 U.S.C.A. § 2304(b)(1) the plaintiffs are required only to notify Gene Norris of the defects and cannot be required to return the motor home to Cleveland for repairs. We do not believe that, in this case, the responsibility of the plaintiffs to return the motor home to Ohio for repairs is an impermissible duty under 15 U.S.C.A. § 2304(b)(1). First, the plaintiffs bought the motor home in Ohio and have testified that they knew that they would have to return the motor home to Ohio for repairs and regular maintenance. Second, the drafters of the Magnuson–Moss Act clearly contemplated that written warranties could be conditioned upon the obligation of consumers to return the defective item to its place of purchase for repair, see 15 U.S.C.A. § 2302(a)(5). Both the Itasca and General Motors' warranties obligate the plaintiffs to transport the motor home to the selling dealer or other authorized Itasca or Chevrolet dealer for repairs covered by the warranties. (plaintiffs' exhibits 2 and 3). The plaintiffs' reasoning would require faraway dealers to incur the expense of transporting motor homes from the owner's residence for repair. To so interpret the Magnuson–Moss Act would impose an unfair and unreasonable burden on dealers.

In short, we find that the plaintiffs are not entitled to rescission under 15 U.S.C.A. § 2304(a)(4) because the defendants did not have the opportunity to make a reasonable number of attempts to repair the vehicle. We also find that the obligation of the plaintiffs to return their motor home for repairs is a reasonable and necessary incident of owning a motor vehicle which the plaintiffs in this case voluntarily accepted and does not constitute an impermissible duty under 15 U.S.C.A. § 2304(b)(1).

QUESTION FOR DISCUSSION

1. *At what point do you believe the Pratts would have been able to successfully invoke the provisions of the Magnuson–Moss Act?*

17:3. PRODUCTS LIABILITY. With our increased industrialization came a corresponding increase in injuries suffered by those who used manufactured products. Until relatively recently, in order to recover damages for those injuries, it was necessary to prove that the manufacturer was negligent or that there existed a warranty between the manufacturer and the person injured; in other words, there had to be "privity." These requirements are frequently difficult to meet. The fact that a product is defective does not necessarily mean that the manufacturer was negligent in its production. In order to recover for breach of warranty, the law rquired that there be privity of contract between the manufacturer and the person injured, which would more often than not be absent. Today one injured by a defective product may recover on a number of different theories in what is one of the most rapidly developing and changing areas of the law. These theories are (1) express or implied warranties, (2) innocent or fraudulent misrepresentation, (3) negligence, and (4) strict liability in tort and by statute. Some of these theories are based on contract and some in tort, and some are a hybrid of both.

(1) **Negligence.** One obvious theory, already mentioned, upon which one may base a recovery for injuries suffered from a defective product is negligence. Recall that in order to recover for the tort of negligence, one must prove that there is a duty of care that is breached by a failure to act in a reasonable manner and that the breach is the proximate cause of the injury.

The area of products-liability law grounded in negligence was given impetus in 1916 by a landmark decision handed down by Judge Cardozo in the case of *McPherson* v. *Buick Motor Company,* 217 N.Y. 382, 111 N.E. 1050(1916). Until that case, the general rule was that there was no duty on the part of the manufacturer to a third-party user of the product because of the absence of any privity between the manufacturer and injured party. There were even some exceptions to this rule at that time, particularly in a case in which the defective product was "inherently dangerous" to human safety. The McPherson case involved a manufacturer of an automobile who purchased and installed a defective wheel on the car. The car was bought from a dealer and not the manufacturer by the ultimate purchaser, who was injured as a result of the defective wheel. The court held that the ultimate purchaser could recover from the manufacturer despite the absence of privity. According to Professor Prosser, although the case at that time purported to extend the doctrine of dangerous instrumentality, by its reasoning it clearly indicated that the "manufacturer, by placing the car upon the market, assumed a responsibility to the consumer, resting not upon the contract but upon the relation arising from its purchase, together with the foreseeability of harm if proper care were not used."[1]

Although a third party may now sue the manufacturer for negligence if it is foreseeable to the manufacturer that a user could be injured as the result of negligence in the manufacture of the product, it is still necessary to prove the negligence, that is, that the duty of care was breached. In addition, the defenses of contributory negligence and assumption of risk are available, making recovery difficult. This theory is still used, but in light of the development of other theories of product liability in which the burden of proof on the plaintiff is somewhat lighter, its importance in this field as a theory of recovery has diminished.

(2) **Breach of Warranty.** As we have previously discussed, the requirement of privity of contract for warranty recovery has been greatly limited by the Uniform Commercial Code, which provides that the seller's warranty, whether express or implied, extends beyond the purchaser to anyone who is in the family or the household of the buyer, as well as to a guest, if it is reasonable to expect that such a person may use, consume, or be affected by the goods. Section 2–318. Beyond this statement the code is neutral and leaves it to the developing case law to determine whether privity is needed. The trend, of course, is opposed to the requirement of privity, and a number of states have altered this section of the code to reflect this fact.

The advantage of seeking recovery for breach of warranty as opposed to negligence, of course, is that it is not necessary to establish the manufacturer's negligence, nor is the injured party subject to the defenses of contributory negligence or assumption of risk. Now, however, that there are some limitations on the product user, such as those illustrated by the *Johnson* v. *Erdman* case, infra. It is necessary, of course, to establish that the breach of warranty was the proximate cause of the injury in question.

There is a divergence of opinion among the courts as to whether one may recover for injuries received when they occur from substances found in food. If the substance is a "foreign" one, such as a paper clip in hamburger, there is little question that the party injured may recover. The problem occurs when the substance is a "natural" one, such as pits in cherry pie and chicken bones in chicken pot pie. Some states, such as Massachusetts, Louisiana, and North Carolina, apply what is known as the "foreign-natural" test and conclude that if the substance is a "natural" one there can be no recovery. Other states, such as Wisconsin, New York, Pennsylvania, and Maryland, have rejected this test in favor of what has become known as the *reasonable-expectation* test. This test determines liability based upon what it is reasonable for the consumer to expect in the food as served as opposed to what is natural to the food

[1] William L. Prosser, *Law of Torts*, p. 661 (3rd ed., 1964).

prior to preparation. Thus one might recover for injuries received from a chicken bone found in a chicken sandwich if it is not reasonable to expect to find a bone under the circumstances.

ERDMAN v. JOHNSON BROTHERS RADIO & TELEVISION CO.
260 Md. 190, 271 A.2d 744 (1970)

Erdman and Pfaff sued Johnson Brothers Radio & Television Co., alleging breach of warranties, breach of a service contract, and negligence. From an adverse verdict, plaintiffs appealed.

FINAN, J.

It has been said that "the seller's warranty is a curious hybrid, born of the illicit intercourse of tort and contract, unique in the law." A further reading of this opinion will show why.

On June 24, 1965, the appellants (Erdman and Pfaff) purchased a color television–radio-stereo console from the appellees (Johnson Brothers) for approximately $1,000. As events unfolded this proved to be a most unfortunate investment. The set was put into operation by one of Johnson Brothers' repairmen, and Erdman and Pfaff looked forward with great expectations to many hours of pleasant viewing. Their joy soon turned to consternation, however, as they began experiencing difficulty with the set almost from the outset. In an act of great foresight, the appellants had purchased a service policy from Johnson Brothers; they had many occasions to avail themselves of its benefits.

Approximately one month after the purchase, the set was sent back to Johnson Brothers for repairs and was returned to the appellants' home about a week later. Sometime after that, Erdman noticed a "crackling sound" in the television; the noise was often accompanied by a "tear" in the picture. This, of course, precipitated complaints by the appellants to Johnson Brothers, and resulted in some two dozen service calls to the appellants' house. Sometime in September, 1966, Erdman and Pfaff for the first time noticed sparks and heavy smoke shooting out of the back of the set and the smell of burning rubber, wire, or some other substance. Another complaint was made. Johnson Brothers' serviceman examined the set on September 30, 1966, and stated that whatever had happened had "fused itself together again," and that if anything serious developed he would be able to fix it.

For the next few months the television operated in its usual (cantankerous) manner and there was no difficulty serious enough to warrant another complaint, at least not until December 7, 1966, a Wednesday. On that date Erdman called Johnson Brothers and for the second time complained about having seen actual sparks and smoke emanating from the rear of the television. The person taking this complaint ventured no opinion as to the cause of the trouble, and merely noted that there would be a serviceman out to the appellants' house on Saturday, December 10, 1966. (Inasmuch as Erdman and Pfaff both worked during the week, the usual practice of the parties was to have the set serviced on Saturdays, as a matter of convenience to the appellants. The very fact that it was necessary to establish a "policy" for making service calls to the plaintiffs' residence perhaps describes the condition of the set more eloquently than this court ever could.)

On the fateful evening of Thursday, December 8, 1966 (after the second complaint and prior to the day on which the repairs were to be performed), Erdman and Pfaff watched television from approximately 11:20 P.M. until 1:30 A.M. of Friday, December 9, 1966, at which time they observed for the third time that there were sparks and smoke coming from the set. They turned off the television, and retired for the night. About half an hour later they were awakened by the barking of one of their eleven dogs, and discovered that a fire was very much in progress in the vicinity of the television set. The fire spread rapidly, and by the dawn's early light Erdman and Pfaff saw, tragically, that their residence had been completely destroyed. The total loss in real and personal property was $67,825.91.

The Uniform Commercial Code (U.C.C.) governs the sale of the television in this case, and it provides that anyone who sells goods and who is a merchant with respect to that kind of goods, impliedly warrants in his contract for sale that the goods sold are "merchantable." In order for goods to be considered merchantable, they must be "fit for the ordinary purposes for which such goods are used." Code (1964 Repl. Vol.) Art 95B (U.C.C.), § 2–314(2)(c). It would appear that Johnson Brothers most assuredly is a merchant within the meaning of the U.C.C. (§ 2–104) and that they gave an implied warranty to the appellants that the television in question was fit for the ordinary purposes to which a television might be put. The Official Comments to § 2–314 state that protection under the "fitness for ordinary purposes" aspect of the implied warranty of merchantability extends not only to a person buying for resale to the ultimate consumer (e.g. a retailer), but also, as here, to the ultimate consumer for his own use.

The comments to the U.C.C. speak in terms of "causation" with respect to the implied warranty of merchantability. Comment 13 to § 2–314 (implied warranty of merchantability) indicates that the buyer must show not only a breach of warranty, but also that the breach was the "proximate cause of the loss sustained." Comment 5 to § 2–715 (consequential damages) treats "proximate causation" in more explicit terms, reiterating the fact that the section allows damages for injuries resulting "proximately" from the breach of warranty. In further delineation, the comment states that if the buyer did in fact discover a defect in the goods prior to his using them, then the injury suffered from the use of the goods would not proximately result from the breach of warranty.

Judge Turnbull in his opinion makes it abundantly clear that he was of the opinion after hearing all the facts that the appellants used the television set after discovering the defect (when they noticed the burning and sparks and made their second complaint on December 7, 1966), and that therefore the implied warranty of merchantability did not apply.

Our reading of the U.C.C. in light of the record supports this interpretation. It would appear that an individual using a product when he had actual knowledge of a defect or knowledge of facts which were so obvious that he must have known of a defect, is either no longer relying on the seller's express or implied warranty or has interjected an intervening cause of his own, and therefore a breach of such warranty cannot be regarded as the proximate cause of the ensuing injury. Such an interpretation gives effect to the true nature of the action involved and the intention of the U.C.C. without needlessly involving the courts in a discussion of whether the implied warranty is founded in contract, tort, or both.

(Judgment affirmed.)

QUESTIONS FOR DISCUSSION

1. *Were the plaintiffs, Pfaff and Erdman, covered by an implied warranty of fitness for a particular purpose or of merchantability?*
2. *Why were the plaintiffs not allowed to recover for breach of an implied warranty? What was the proximate cause of the loss of their property?*

WEBSTER v. BLUE SHIP TEA ROOM

198 N.E.2d 309 (Mass. 1964)

REARDON, J.

This is a case which by its nature evokes earnest study not only of the law but also of the culinary tradition of the Commonwealth which bear so heavily upon its outcome. It is an action to recover damages for personal injuries sustained by reason of a breach of implied warranty of food served by the defendant in its restaurant.

The jury could have found the following facts: On Saturday, April 25, 1959, about 1 P.M., the plaintiff, accompanied by her sister and her aunt, entered the Blue Ship Tea Room operated by the defendant. The group was seated at a table and supplied with menus.

This restaurant, which the plaintiff characterized as "quaint," was located in Boston "on the third floor of an old building on T Wharf which overlooks the ocean."

The plaintiff, who had been born and brought up in New England (a fact of some consequence), ordered clam chowder and crabmeat salad. Within a few minutes she received tidings to the effect that "there was no more clam chowder," whereupon she ordered a cup of fish chowder. Presently, there was set before her "a small bowl of fish chowder." She had previously enjoyed a breakfast about 9 A.M. which had given her no difficulty. "The fish chowder contained haddock, potatoes, milk, water and seasoning. The chowder was milky in color and not clear. The haddock and potatoes were in chunks" (also a fact of consequence). "She agitated it a little with the spoon and observed that it was a fairly full bowl. . . . She did not see anything unusual about it. After three or four spoonfuls she was aware that something had lodged in her throat because she couldn't swallow and couldn't clear her throat by gulping and she could feel it." This misadventure led to two esophagoscopies at the Massachusetts General Hospital, in the second of which, on April 27, 1959, a fish bone was found and removed. The sequence of events produced injury to the plaintiff which was not insubstantial.

We must decide whether a fish bone lurking in a fish chowder, about the ingredients of which there is no other complaint, constitutes a breach of implied warranty under applicable provisions of the Uniform Commercial Code, the annotations to which are not helpful on this point. As the judge put it in his charge, "Was the fish chowder fit to be eaten and wholesome? . . . (N)obody is claiming that the fish itself wasn't wholesome. . . . But the bone of contention here—I don't mean that for a pun—but was this fish bone a foreign substance that made the fish chowder unwholesome or not fit to be eaten?"

The defendant asserts that here was a native New Englander eating fish chowder in a "quaint" Boston dining place where she had been before; that "(f)ish chowder, as it is served and enjoyed by New Englanders, is a hearty dish, originally designed to satisfy the appetites of our seamen and fisherman"; that "(t)his court knows well that we are not talking of some insipid broth as is customarily served to convalescents." We are asked to rule in such fashion that no chef is forced "to reduce the pieces of fish in the chowder to miniscule size in an effort to ascertain if they contained any pieces of bone." "In so ruling," we are told (in the defendant's brief), "the court will not only uphold its reputation for legal knowledge and acumen, but will, as loyal sons of Massachusetts, save our world-renowned fish chowder from degenerating into an insipid broth containing the mere essence of its former stature as a culinary masterpiece." Notwithstanding these passionate entreaties we are bound to examine with detachment the nature of fish chowder and what might happen to it under varying interpretations of the Uniform Commercial Code.

Thus, we consider a dish which for many long years, if well made, has been made generally as outlined above. It is not too much to say that a person sitting down in New England to consume a good New England fish chowder embarks on a gustatory adventure which may entail the removal of some fish bones from his bowl as he proceeds. We are not inclined to tamper with age old recipes by any amendment reflecting the plaintiff's view of the effect of the Uniform Commercial Code upon them. We are aware of the heavy body of case law involving foreign substances in food, but we sense a strong distinction between them and those relative to unwholesomeness of the food itself, e.g., tainted mackerel, and a fish bone in a fish chowder. Certain Massachusetts cooks might cavil at the ingredients contained in the chowder in this case in that it lacked the heartening lift of salt pork. In any event, we consider that the joys of life in New England include the ready availability of fresh fish chowder. We should be prepared to cope with the hazards of fish bones, the occasional presence of which in chowders is, it seems to us, to be anticipated, and which, in the light of a hallowed tradition, do not impair their fitness or merchantability.

(Judgment for the defendant.)

QUESTION FOR DISCUSSION

1. *Was the court saying that there was no warranty of merchantability in this case? If there was, why did the plaintiff not recover?*

---

(3) **Strict Liability in Tort.** In recent years the concept of strict liability in tort has been developed by the courts as a ground for recovery in products-liability cases. This theory has gained, although it is not universal, acceptance by the courts because a section governing strict liability was adopted by the

American Law Institute, *Restatement of the Law of Torts, Second,* Section 402A provides:

> (1) one who sells any product in a defective condition unreasonably dangerous to the user or consumer or to his property is subject to liability for physical harm thereby caused to the ultimate user or consumer, or to his property, if
> (a) the seller is engaged in the business of selling such a product, and
> (b) it is expected to and does reach the user or consumer without substantial change in the condition in which it is sold.
> (2) The rule stated in Subsection (1) applies although
> (a) the seller has exercised all possible care in the preparation and sale of his product, and
> (b) the user or consumer has not bought the product from or entered into any contractual relation with the seller.

This is a ground for recovery separate from breach of warranty or negligence. Basically a manufacturer or seller is strictly liable in tort when the product that he places on the market, knowing that it is normally not examined for defects, causes injury to a human being as a result of that defect. Note that the seller is liable despite the fact that he exercised all possible care; that is, there is no necessity of proving negligence. Also there is no need to show privity of contract; rather any user or consumer who is injured as a result of the defect may recover. One constraint is that the seller must be engaged in the business of selling the product. If the sale is an isolated one and not part of the principal business of the seller, the doctrine would not apply.

As should be evident, the scope of this doctrine is sweeping. The liability may not be disclaimed or excluded and is not governed by the Uniform Commercial Code. The liability is imposed by law as a matter of public policy for several reasons: (1) because of the public interest in human life and safety, which requires legal protection against defective products; (2) because the manufacturer, by placing the goods in the stream of commerce, represents that they are safe to use; and (3) because it avoids a multiplicity of suits that would ultimately lead to the same result, the liability of the maker of the product. The doctrine is separate from breach of contract, warranty, or negligence and is not subject to the traditional defenses available for suits based upon those theories.

Liability under the theory of strict liability is applicable to manufacturers, retailers, and wholesalers. Thus, if the manufacturer is beyond the jurisdiction of the court, a viable suit could still be brought against the retailer from whom the product was purchased. A question sometimes arises when the injury results from a defective component part of the product. A number of courts have held that the supplier of a component part is liable, particularly if the component is incorporated into the finished product without change. On the other hand, if the product is part of a mixture, the courts are reluctant to find the supplier of the component liable.

The doctrine of strict liability has not relieved the plaintiff of sustaining the burden of proof. He must establish that the injury in question resulted from a defect in the product that existed when it left the possession of the defendant and that the product caused his injury.

GREENMAN v. YUBA POWER PRODUCTS, INC.

59 Cal.2d 57, 377 P.2d 897 (1962)

TRAYNOR, J.

Plaintiff brought this action for damages against the retailer and the manufacturer of a Shopsmith, a combination power tool that could be used as a saw, drill, and wood lathe. He saw a Shopsmith demonstrated by the retailer and studied a brochure prepared by the manufacturer. He decided he wanted a Shopsmith for his home workshop, and his wife bought and gave him one for Christmas in 1955. In 1957, he bought the necessary attachments to use the Shopsmith as a lathe for turning a large piece of wood he wished to make into a chalice. After he had worked on the piece of wood several times without difficulty, it suddenly flew out of the machine and struck him on the forehead, inflicting serious injuries. About ten and a half months later, he gave the retailer and the manufacturer written notice of claimed breaches of warranties and filed a complaint against them alleging such breaches and negligence.

After a trial before a jury, the court ruled that there was no evidence that the retailer was negligent or had breached any express warranty and that the manufacturer was not liable for the breach of any implied warranty. Accordingly, it submitted to the jury only the cause of action alleging breach of implied warranties against the retailer and the causes of action alleging negligence and breach of express warranties against the manufacturer. The jury returned a verdict for the retailer against plaintiff and for plaintiff against the manufacturer in the amount of $65,000. The trial court denied the manufacturer's motion for a new trial and entered judgment on the verdict. The manufacturer and plaintiff appeal. Plaintiff seeks a reversal of the part of the judgment in favor of the retailer, however, only in the event that the part of the judgment against the manufacturer is reversed.

The manufacturer contends, however, that plaintiff did not give it notice of breach of warranty within a reasonable time and that therefore his cause of action for breach of warranty is barred by Section 1769 of the Civil Code.

Like other provisions of the uniform sales act (Civ.Code, §§ 1721–1800), Section 1769 deals with the rights of the parties to a contract of sale or a sale. It does not provide that notice must be given of the breach of a warranty that arises independently of a contract of sale between the parties. Such warranties are not imposed by the sales act, but are the product of commonlaw decisions that have recognized them in a variety of situations.

The notice requirement of Section 1769, however, is not an appropriate one for the court to adopt in actions by injured consumers against the manufacturers with whom they have not dealt. We conclude, therefore, that even if plaintiff did not give timely notice of breach of warranty to the manufacturer, his cause of action based on the representations contained in the brochure was not barred.

Moreover, to impose strict liability on the manufacturer under the circumstances of this case, it was not necessary for plaintiff to establish an express warranty as defined in Section 1732 of the Civil Code. A manufacturer is strictly liable in tort when an article he places on the market, knowing that it is to be used without inspection for defects, proves to have a defect that causes injury to a human being. Recognized first in the case of unwholesome food products, such liability has now been extended to a variety of other products that create as great or greater hazards if defective.

Although in these cases strict liability has usually been based on the theory of an express or implied warranty running from the manufacturer to the plaintiff, the abandonment of the requirement of a contract between them, the recognition that the liability is not assumed by agreement but imposed by law . . . , and the refusal to permit the manufacturer to define the scope of its own responsibility for defective products . . . , make clear that the liability is not one governed by the law of contract warranties but by the law of strict liability in tort. Accordingly, rules defining and governing warranties that were developed to meet the needs of commercial transactions cannot properly be invoked to govern the manufacturer's liability to those injured by their defective products unless those rules also serve the purposes for which such liability is imposed.

We need not recanvass the reasons for imposing strict liability on the manufacturer. They have been fully articulated. The purpose of such liability is to insure that the costs of injuries resulting from defective products are borne by the manufacturers that put such products on the market rather than by the injured persons who are powerless to protect themselves. Sales warranties serve this purpose fitfully at best. In the present case, for example, plaintiff was able to plead and prove an express warranty only because he read and relied on the representations of the Shopsmith's ruggedness contained in the manufacturer's brochure. Implicit in the machine's presence on the market, however, was a representation that it would safely do the jobs for which it was built. Under these circumstances, it should not be controlling whether plaintiff selected the machine because of the machine's own appearance of excellence that belied the defect lurking beneath the surface, or because he merely assumed that it would safely do the jobs it was built to do. It should not be controlling whether the details of the sales from manufacturer to retailer and from retailer to plaintiff's wife were such that one or more of the implied warranties of the sales act arose. "The remedies of injured consumers ought not to be made to depend upon the intricacies of the law of sales." (*Ketterer* v. *Armour & Co.*, D.C., 200 F. 322, 323.) To establish the manufacturer's liability it was sufficient that plaintiff proved that he was injured while using the Shopsmith in a way it was intended to be used as a result of a defect in design and manufacture of which plaintiff was not aware that made Shopsmith unsafe for its intended use.

(Judgment affirmed.)

QUESTIONS FOR DISCUSSION

1. *Upon what legal theory did the appellate court hold that Greenman had a right to recover: negligence, breach of express warranty, breach of implied warranty, or strict liability?*
2. *Why did the court find that it was not necessary for Greenman to comply with the notice requirement of the California statute?*

---

(4) Misrepresentation. A number of jurisdictions have adopted as law a ground of liability stated in the *Restatement of Torts, Second,* Section 402B. That section states:

> One engaged in the business of selling chattels who, by advertising, labels, or otherwise, makes to the public a misrepresentation of a material fact concerning the character or quality of a chattel sold by him is subject to liability for physical harm to a consumer of the chattel caused by justifiable reliance upon the misrepresentation, even though
> (a) it is not made fraudulently or negligently, and
> (b) the consumer has not bought the chattel from or entered into any contractual relation with the seller.

Note once again the aspects of strict liability. It is not necessary to show that the statement was made with the intent to deceive or made negligently, nor is any privity required. Once again, the public-policy reason for this position is that the consumer relies to a large degree upon the advertising of the seller for information about the product.

In a given case recovery may be sought based upon any or all of the theories already discussed. Obviously, however, in those jurisdictions where recovery based upon strict liability is well developed, the importance of relying on a theory of negligence or even breach of warranty is diminished in personal-injury cases.

SUMMARY

An express warranty may be created by affirmation, promise, description, or sample. It is not always necessary that a statement creating the warranty be made prior to the sale. Statements of opinion or of the value of the goods or a commendation of them do not create a warranty. To create a warranty the statement must be part of the basis of the bargain.

Implied warranties are the implied warranty of merchantability and the implied warranty of fitness for a particular purpose. For the warranty of merchantability to be implied there must be a sale of goods by a merchant with respect to goods of that kind and the warranty must not be excluded or modified. An implied warranty may also arise because certain practices are the custom of a trade. There are a number of requirements stated in the code as to what is necessary for goods to be merchantable. One of these requirements is that the goods must be fit for the ordinary purposes for which such goods are used.

The implied warranty of fitness for a particular purpose will arise if the seller of goods has reason to know any particular purpose for which goods are being used. The seller must also know that the buyer is relying on the seller's skill or judgment to select suitable goods.

Subject to exclusion or modification the code also provides for a warranty of title and infringement. The warranty is that the title conveyed is good and that there are no liens, security interests or other encumbrances of which the buyer has no knowledge.

The code in many states provides that the seller's express and implied warranties extends to any natural person who is in the family or household of his buyer or guest in the house if it is reasonable to expect that person will use the goods and that person is injured by breach of the warranty. Some jurisdictions have expanded this rule.

The code provides that the various warranties may be excluded. Some states have limited a seller's right to exclude the warranty of merchantability or fitness for a particular purpose.

The Magnuson–Moss Warranty Act provides, that where applicable, warranties shall be called either full warranties or limited warranties. To give a full warranty the seller must comply with the requirements of the act. Other warranties are called limited warranties. Under the act a seller may not exclude the implied warranty of merchantability. If the act conflicts with the U.C.C. the provisions of the act control.

Products liability actions may be brought on a theory of negligence, warranty, misrepresentation, or strict liability in tort. For cases brought on a warranty theory involving foreign substances in food the test for determining a breach of warranty is either the reasonable expectation test or the foreign-natural test depending upon the jurisdiction.

Under a theory of strict liability, a manufacturer or seller is strictly liable, without a showing of negligence, when the product he places in the market, knowing that it is normally not examined for defects, causes injury to a human being as a result of that defect. The doctrine is applicable to manufacturers, retailers and wholesalers.

PROBLEMS

1. Gillespie went shopping at a store owned by the Great Atlantic & Pacific Tea Company. While there he picked up two bottles of Sprite soft drink, which exploded as they were being carried by him to the checkout counter. Gillespie handled the bottles of Sprite normally while they were in his possession until the explosion. A & P denied liability to Gillespie for his injuries. Was it correct? Gillespie v. Great A & P Tea Company, 187 S.E. 441 (N.C. 1972).
2. Hunt purchased a cherry pie from Ferguson-Paulus Enterprises through a vending machine. When Hunt bit into the pie, one of his teeth broke off as it encountered a cherry pit. Hunt sued for damages. Could he recover? Hunt v. Ferguson-Paulus Enterprises, 243 Or. 546, 415 P.2d 13 (1966).
3. Michael Catania asked Charles J. Brown, who was engaged in the retail paint business, to recommend a paint to cover the exterior stucco wall of his house. Brown advised Catania to "wire-brush" any loose particles from the wall and apply Pierce's shingle and shake paint after adding a thinner. Five months after its application the paint began to peel, flake, and blister from the walls, despite the fact that Catania had complied with Brown's instructions. Was Brown liable to Catania? Catania v. Brown, 231 A.2d 668 (Conn. 1967).
4. Rooney, a city engineer, was overcome by sewer gas and died when his gas mask failed to function. Guarino and others also died when their masks failed to function during their attempt to rescue Rooney. The masks were manufactured by Mine Safety Appliance Company. Mine Safety Appliance Company was sued for the death on the theory of breach of implied warranty. Was the company liable? Guarino v. Mine Safety Appliance Company, 25 N.Y.2d 460, 255 N.E.2d 173 (1969).
5. Greenco was operating a forklift truck for his employer when he was injured as a result of an alleged defect in the truck. Greenco sued the manufacturer of the truck, Clark Equipment Company, on the theory of strict liability. Clark moved to dismiss the complaint on ground of lack of privity. Was Clark entitled to have its motion granted? Greenco v. Clark Equipment Company, 237 F.Supp. 427 (1965).
6. Bernstein, while a patient in a hospital, was served a hot drink in a paper cup manufactured by the Lily-Tulip Cup Corp. The cup came apart, and the contents spilled on Bernstein, causing her to be scalded. She sued for

breach of implied warranty. The trial court struck the count of Bernstein's suit based on implied warranty because there was no privity between Bernstein and the Lily-Tulip Cup Corp. What result on appeal? Bernstein v. Lily-Tulip Cup Corp., 177 So.2d 362 (Fla. 1965).

7. Jackson and Roy sued Gifford after Gifford stopped payment on a check given for hogs purchased at plaintiff's auction. At the time of sale plaintiff's auctioneer said, "We have a veterinarian. If he finds a hog that is sick we do not sell him. We do not sell sick hogs." Gifford testified that most of the hogs he purchased died within a few days of sale and defended on the ground of breach of express warranty. Plaintiffs claimed that no express warranty was intended or given. Decide. Jackson v. Gifford, 264 P.2d 313 (1953).
8. Newsom was a carpenter–foreman employed by Oak Creek Development Company. A fellow employee was using a power-loaded gun to drive studs through a steel I-beam for the purpose of attaching a two-by-six piece of wood to it. The head and shank of the stud separated with the shank ricocheting out of the wood, sticking Newsom in the abdomen and finally lodging in a nerve center of the pelvic region. The stud that was defective was purchased by Newsom's employer from Speed Fastners for breach of the implied warranty of merchantability. Can he recover? Speed Fastners, Inc. v. Newsom, 382 F.2d 395 (1967).
9. Doyle Connelly made repairs on a trailer. After the repairs were made, no one claimed it, so Connelly became the owner of the trailer as a result of his mechanics lien for repairs and storage. He then sold the trailer to J. T. Marvin for $6,045.62. Unknown to both Marvin and Connelly, the trailer had been stolen from its rightful owner at some point prior to Connelly's possession of it. When the true owner reclaimed the trailer, Marvin sued Connelly for breach of warranty of title. Connelly claims that he should win since he acquired a voidable title that would have ripened into an indefeasible title after the sale to Marvin. Decide. J. T. Marvin v. Connelly, 252 S.E.2d 562 (S.C. 1979).

CHAPTER 18

Performance of the Contract

In this chapter we are concerned with the contractual obligations of the buyer and the seller. In Chapter 19 the rights and remedies of the buyer and the seller will be discussed. The Uniform Commercial Code states that the obligation of the seller is to transfer and deliver the goods and that that of the buyer is to accept and pay for them in accordance with the contract. Section 2–301. The following discussion concerns the particulars of those obligations.

18:1. SELLER'S OBLIGATION TO DELIVER GOODS: TENDER. The first problem is to determine the nature of the seller's obligation regarding delivery of goods. Obviously it is impossible to physically deliver goods to a buyer who will not accept them. Therefore the obligation of the seller is that he tender delivery of goods. *Tender,* in this sense, means an offer to deliver the goods, coupled with a present ability to fulfill all of the conditions resting on the party making tender. Tender must be followed by actual performance if the other party, the buyer in this case, shows himself ready to proceed. Section 2–503, Comment 1. If the seller makes an appropriate tender and the buyer refuses delivery, the buyer has breached the contract and the seller may then rely on the rights and remedies available to him. Now let us examine what the seller must do in order to make a valid tender or delivery.

(1) Time, Place, and Manner of Delivery. Time for delivery or tender shall be a reasonable time if none is specified in the contract; otherwise the time specified in the contract governs. Section 2–309(1). What constitutes a reasonable time is, of course, a question of fact and may vary with the facts and circumstances of each case.

Tender of delivery requires that the seller put and hold *conforming* goods at the buyer's disposition and give the buyer any notification reasonably necessary to enable him to take delivery. If the goods are to be delivered to the buyer, the buyer must furnish the facilities for accepting the goods unless it is otherwise agreed. The tender by the seller must be made at a reasonable hour, and the goods must then be kept available for a reasonable period of time in order to allow the buyer to take possession. Section 2–503(1). In other words, backing one's truck up to the buyer's store at 2 A.M., knocking once, and leaving would not constitute a valid tender under normal circumstances.

The contract will frequently state the place of delivery. In the absence of such a provision, however, the place for delivery or tender is the seller's place of business or, if he has none, his residence. If the contract is for sale of goods identified to the contract and the parties have knowledge that the goods are at some other place, then the parties are presumed to intend that place as the place for delivery if there is no contrary contractual provision. Section 2–308.

The seller's obligation is to deliver all ordered goods at one time, if there is no contrary agreement. He may not dribble the goods into the buyer's place of business piecemeal and expect part payment each time. His obligation is to deliver the goods in a single delivery. If there are circumstances that give either party the right to make or demand delivery in separate lots, the price, if it can be apportioned, may be demanded for each lot. Section 2–307. For example, it may not be commercially feasible to deliver or receive the goods in one lot, as when a contract calls for the sale of brick or other building materials for the construction of a building and the buyer does not have the storage space to receive the entire load at one time, or when ten carloads of coal are to be delivered and only three are available.

A review of the requirements for tender under shipment contracts, destination contracts, and contracts on which delivery is to be made without the goods being moved should be helpful at this point to an understanding of the obligations of tender. Recall that they were discussed in some detail in Chapter 16. Tender under a shipment contract requires that the seller comply with those requirements for tender previously stated in Section 2–503(1) and (2). Tender under a destination contract requires the seller, in addition to complying with the requirements of 2–503(1), to tender any necessary documents in an appropriate case. Section 2–503(3). These documents may be similar to those required for a valid tender when the goods are in the possession of a bailee and are to be delivered without being moved or when the contract specifically requires the seller to deliver documents.

When the goods are in the possession of a bailee and are to be delivered without being moved, tender requires that the seller either (1) tender a *negotiable* document of title for the goods or (2) procure acknowledgement by the bailee of the buyer's right to possession of the goods. Tender of a *nonnegotiable* document of title or of written directions to the bailee to deliver the goods to the buyer is sufficient tender unless the buyer reasonably objects or unless the bailee refuses to honor the documents. Section 2–503(4).

If the contract specifically requires the seller to deliver the documents, they must be delivered in proper form to constitute a valid tender, but the tender may be through customary banking channels. Section 2–503(5).

Once the seller has made a valid tender, the next move is up to the buyer. He has an obligation to accept the goods upon a valid tender by the seller. However, prior to his obligation to accept there are circumstances under which the buyer has a right to inspect the goods. There may also be conditions of sale that do not allow for the buyer's inspection as a prerequisite to acceptance.

18:2. Buyer's Right to Inspection as a Prerequisite to Payment or Acceptance. The reason for the buyer's right to inspect the goods prior to payment or acceptance is to assure himself that the goods conform to the contract before he pays for or accepts them. The code provides that the buyer has a right to inspect the goods tendered, delivered, or identified to the contract before payment or acceptance. The goods may be inspected at any reasonable place and time in any reasonable manner. Inspection may be after the arrival of the goods when the seller is authorized or required to send the goods to the buyer. The right of inspection is available to the buyer unless otherwise agreed or unless the goods are shipped C.O.D. or for payment against documents of title. Under these circumstances, there is no provision allowing for a right of inspection prior to payment. Section 2–513(1) and (3).

The seller cannot negate the buyer's right to inspect by shipping the goods C.O.D. unless the contract provides for shipment under these terms. Shipment by the seller C.O.D. without such a provision constitutes a breach of contract. The code also provides that any expenses of inspection shall be borne by the buyer unless the goods do not conform and are rejected, in which case they may be recovered from the seller as damages. Section 2–513(2).

18:3. Situation of No Right to Inspect Before Payment. There are situations in which the buyer does not have the right to inspect prior to payment and acceptance. The contract may require that payment is to be made before inspection. This procedure does not, however, constitute an acceptance of goods or impair the buyer's right to inspect or any of his remedies. Section 2–512. The effect of such a provision is to shift certain risks to the buyer from the seller. Even in this case, however, the buyer does not have to make payment when nonconformity of the goods appears without inspection. For instance, suppose Norma Newlywed orders crystal glasses from a mail-order house. The contract provides that payment is to be made before there may be any inspection. When the package arrives, Norma shakes it and hears the sound of broken glass. She is excused from making payment. Section 2–512(1)(b).

As previously discussed, shipment of goods C.O.D. denies the buyer the right of inspection prior to payment. Section 2–513(3)(a). The buyer is also foreclosed from inspection when the contract provides for payment upon tender of the documents of title, except when the contract provides for payment only after the goods are available for inspection. Section 2–513(3)(b).

18:4. Acceptance by Buyer. After the seller has performed his obligations under the contract, the buyer must accept the goods if they conform to the agreement. Acceptance means that the buyer, pursuant to the contract, takes particular goods as his own whether or not he is obligated to do so. He may accept by words, action, or silence when there is a duty to speak. Section 2–606, Comment 1.

It is important to know what constitutes an acceptance. Section 2–606(1) of the code provides that an acceptance occurs when the buyer

(a) After a reasonable opportunity to inspect the goods signifies to the seller that the goods are conforming or that he will take or retain them in spite of their non-conformity; or
(b) fails to make an effective rejection (Subsection (1) of Section 2–602), but such acceptance does not occur until the buyer has had a reasonable opportunity to inspect them; or
(c) does any act inconsistent with the seller's ownership; but if such act is wrongful as against the seller, it is an acceptance only if ratified by him.

The importance of determining whether there has been an acceptance is that if there is no contrary agreement, acceptance gives rise to the second obligation of the buyer, to pay for the goods. Thus the buyer may specifically indicate acceptance by stating, "I accept the goods," or he may use and pay for the goods. Alternatively, if he fails to reject the goods within a reasonable time after tender or delivery of the goods, he will be deemed to have accepted them.

If the goods are to be delivered in separate commercial units, the buyer may accept any unit or units and reject the rest. Section 2–601(c). On the other hand, acceptance of a part of a commercial unit is acceptance of the entire unit. Section 2–606(2). The buyer must exercise his right only as to whole commercial units. For example, suppose the seller ships six crates of oranges. The buyer inspects and determines that three cases are rotten. He may accept the good crates and reject the three rotten ones. He may not, however, keep certain oranges out of a single crate and reject the rest. He must either accept or reject the crate, as it is the commercial unit.

ZABRISKIE CHEVROLET, INC. v. SMITH

240 A.2d 195 (N.J. 1968)

Alfred Smith purchased a new 1966 Chevrolet from Zabriskie Chevrolet. On February 9, 1967, he tendered Zabriskie his check for $2,069.50, having previously made a deposit of $124. Delivery was made to Smith's wife on the evening of February 10. Mrs. Smith started out for home, but unfortunately the newly purchased auto went only a short distance when it developed transmission trouble, and the journey had to be completed at a rate of five miles per hour in "low-low" gear. The next day Smith called Zabriskie to announce that they had sold him a lemon, that he had stopped payment on the check, and that the sale was canceled. Zabriskie sent a wrecker to Smith's home and picked up the car. Zabriskie then replaced the transmission, but Smith refused to take delivery of the repaired car. Zabriskie then sued Smith on the check, for the purchase price plus incidental damages. Smith counterclaimed to recover his $124 deposit.

DOAN, J.

Plaintiff urges that defendant accepted the vehicle and therefore under the Code (N.J.S. 12A:2–607(1), N.J.S.A.) is bound to complete payment for it. Defendant asserts that he never accepted the vehicle and therefore under the code properly rejected it; further, that even if there had been acceptance he was justified under the code in revoking the same. Defendant supports this claim by urging that what was delivered to him was not what he bargained for, i.e., a new car with factory new parts, which would operate perfectly as represented and, therefore, the code remedies of rejection and revocation of acceptance were available to him. These remedies have their basis in breach of contract and failure of consideration although they are also viewed as arising out of breach of warranty. The essential ingredient which determines which of these two remedies is brought into play is a determination, *in limine,* whether there has been an "acceptance" of the goods by the buyer. Thus, the primary inquiry is whether the defendant had "accepted" the automobile prior to the return thereof to the plaintiff.

N.J.S. 12A:2–606, N.J.S.A. states in pertinent part:

> (1) Acceptance of goods occurs when the buyer
> (a) after a reasonable opportunity to inspect the goods signifies to the seller that the goods are conforming or that he will take or retain them in spite of their non-conformity; or
> (b) fails to make an effective rejection (subsection (1) of 12A: 2–602), but such acceptance does not occur until the buyer has had a reasonable opportunity to inspect them, or
> (c) does any act inconsistent with the seller's ownership; but if such act is wrongful as against the seller it is an acceptance only if ratified by him.

It is clear that a buyer does not accept goods until he has had a "reasonable opportunity to inspect." Defendant sought to purchase a new car. He assumed what every new car buyer has a right to assume—and, indeed, has been led to assume by the high-powered advertising techniques of the auto industry—that his new car, with the exception of very minor adjustments, would be mechanically new and factory-furnished, operate perfectly, and be free of substantial defects. The vehicle delivered to defendant did not measure up to these representations. Plaintiff contends that defendant had "reasonable opportunity to inspect" by the privilege to take the car for a typical "spin around the block" before signing the purchase order. If by this contention plaintiff equates a spin around the block with "reasonable opportunity to inspect," the contention is illusory and unrealistic. To the layman, the complicated mechanisms of today's automobiles are a complete mystery. To have the automobile inspected by someone with sufficient expertise to disassemble the vehicle in order to discover latent defects before the contract is signed, is assuredly impossible and highly impractical.

Consequently, the first few miles of driving become even more significant to the excited new car buyer. This is the buyer's first reasonable opportunity to enjoy his new vehicle to see if it conforms to what it was represented to be and whether he is getting what he bargained for. How long the buyer may drive the new car under the guise of inspection of new goods is not an issue in the present case. It is clear that defendant discovered the nonconformity within seven tenths of a mile and minutes after leaving plaintiff's showroom. Certainly this was well within the ambit of "reasonable opportunity to inspect." That the vehicle was grievously defective when it left plaintiff's possession is a compelling conclusion, as is the conclusion that in a legal sense defendant never accepted the vehicle.

Even if defendant had accepted the automobile tendered, he has a right to revoke under N.J.S. 12A:2–608, N.J.S.A.:

(1) The buyer may revoke his acceptance of a lot or commercial unit whose non-conformity substantially impairs its value to him if he has accepted it. . . .

Accordingly, and pursuant to N.J.S. 12A:2–711, N.J.S.A., judgment is rendered on the main case in favor of defendant. On the counterclaim judgment is rendered in favor of defendant and against plaintiff in the sum of $124, being the amount of the deposit, there being no further proof of damages.

Questions for Discussion

1. *Did the buyer ever accept the car from Zabriskie?*
2. *Did the buyer have a reasonable opportunity to inspect prior to taking delivery of the car? What period do you think the court would have held to be a reasonable opportunity to inspect in this case?*

CAN-KEY INDUSTRIES, INC. v. INDUSTRIAL LEASING CORP.

593 P.2d 1125 (Or., 1979)

Can-Key Industries, Inc. manufactured a turkey-hatching unit. This unit was purchased by Industrial Leasing Corporation (ILC) which in turn leased it to Rose-A-Linda Turkey Farms. The ILC purchase order agreement with Can-Key Industries conditioned its final acceptance of the equipment upon the willingness of Rose-A-Linda to accept the equipment. When Rose-A-Linda indicated that it was dissatisfied with the equipment, ILC refused to proceed with the contract of sale and was sued by the seller, Can-Key Industries. The trial court found in favor of Can-Key Industries and ILC appealed.

HOWELL, J.

The equipment was shipped to Rose-A-Linda in parts during January, February and March of 1976. On February 27, 1976, plaintiff sent an invoice to ILC stating that payment was to be made "immediately after acceptance" and that the equipment had been received by Rose-A-Linda.

Rose-A-Linda's president, Chester Gibson, testified that the first hatch came off on April 1. Gibson characterized the results as "a disaster" and said he immediately notified plaintiff that the equipment was unsatisfactory and asked that it be removed. Plaintiff replied by letter dated April 8,

> . . . Although our philosophy on 'acceptance' and 'warranty' hasn't changed, in light of your new problems, please disregard our request for acceptance until we have solved your present problem.

The letter was signed by Gil Martini, plaintiff's president, and a copy was sent to defendant.

Gibson testified that after the first hatch came off, plaintiff's employees made some changes and Rose-A-Linda made three more sets. These sets, according to Gibson,

> . . . were likewise a disaster, and we had poor results.
>
> Then we wrote Mr. Martini and told him absolutely they were of no value to us, we wanted them removed. The fact is, we made a claim for damages and wanted him to sustain and take care of damages he caused.

Rose-A-Linda then employed Robert Cannon, the original developer of the turkey-hatching equipment, to modify it. According to Gibson, the equipment was used four times in 1977. Eggs were set in the equipment on February 3, March 4, and April 3, each time following modifications or suggested modifications by Cannon. Gibson testified that each time the results proved unsatisfactory. The last set occurred on April 8, in order to make what Cannon termed "a complete comparative test." The resulting hatch occurred on May 6, 1977, and the equipment has not been used since.

The sole issue in this case is whether defendant "accepted" the equipment manufactured by plaintiff. The contract between plaintiff and defendant provided that defendant's obligation to pay would be conditioned upon acceptance of the equipment by its lessee. Consequently, the trial court could properly find that defendant accepted the equipment only if there is evidence that Rose-A-Linda, the lessee, accepted the equipment.

In these particular circumstances, we hold that the uncontradicted testimony of Gibson is conclusive of the facts involved in this issue. We further hold that Gibson's statements constituted an "effective rejection" as that term is used in ORS 72.6060(1)(b). Plaintiff's evidence that Rose-A-Linda accepted the equipment is therefore sufficient only if it shows that Rose-A-Linda performed acts inconsistent with the seller's ownership under the terms of ORS 72.6060(1)(c).

What constitutes "any act inconsistent with the seller's ownership" has proved to be one of the trouble areas under Article 2 of the Uniform Commercial Code. Courts that have applied the provision have reached inconsistent results and commentators have termed the provision an "obstreperous" one.

A reasoned application of the section requires that the court recognize the existence of two competing policies. A buyer who verbally rejects goods should not in all cases be allowed to use the goods as if he were the owner and effectively "have it both ways." On the other hand, there are many cases in which use of the goods after rejection is not only reasonable in that it minimizes economic waste, but may be required under the buyer's statutory duty to mitigate consequential damages. *See* ORS 72.7150(2)(a). The court must consider both policies when defining the scope of "any act inconsistent with the seller's ownership."

Nearly all the evidence plaintiff relies upon to demonstrate that Rose-A-Linda performed acts inconsistent with plaintiff's ownership was provided by Gibson, Rose-A-Linda's president. Plaintiff did introduce testimony that Rose-A-Linda used the equipment in March and April of 1976, but it is clear from the record that these uses related to Rose-A-Linda's initial inspection of the equipment and plaintiff's efforts to solve the "problems" with the equipment that it recognized in its April letter to Rose-A-Linda. Neither of these uses was inconsistent with plaintiff's ownership of the equipment. Rose-A-Linda's initial inspection cannot be considered inconsistent with plaintiff's ownership because ORS 72.6060(1)(b) assures a buyer of goods a "reasonable opportunity to inspect them." Nor can the use of the equipment during April be considered inconsistent, because that use was approved by plaintiff, which was attempting to solve the problems with the equipment.

Plaintiff's primary reliance is on the modifications and alterations performed by Rose-A-Linda "after the equipment was installed and functioning." These acts all occurred after Gibson notified plaintiff that the equipment was unacceptable and asked that it be removed. The only testimony concerning these acts is Gibson's. That testimony shows that Rose-A-Linda employed the original developer of the equipment in an attempt to remedy the defects. It used the equipment four times during 1977. Three of those uses followed modifications or suggestions for modifications by the developer, and the final use was for the purpose of conducting a comparative test. Although the equipment apparently remains in Rose-A-Linda's possession, it has not been used since May of 1977.

We hold that this evidence does not demonstrate a use inconsistent with the seller's ownership under the terms of ORS 72.6060(1)(c). To hold otherwise would have the effect of penalizing Rose-A-Linda for its apparent good faith efforts to cure the defects in the equipment. We do not believe such a holding is compelled by the language of ORS 72.6060(1)(c). On the contrary, we think such a holding might be inconsistent with other provisions of the code, specifically the statutory duty to mitigate consequential damages and the statutory obligation of good faith. ORS 72.7150(2)(a), 71.2030.

The Official Comments to Oregon's Commercial Code state that courts should avoid "the pinning of a technical 'acceptance' on a buyer who has taken steps toward realization on or preservation of the goods in good faith," and that a rejecting buyer should be afforded "all reasonable leeway" in dealing with goods in the absence of instructions from the seller. Viewing the evidence in the present case in a light most favorable to the plaintiff, we nevertheless conclude that, as a matter of law, Rose-A-Linda did not perform any act inconsistent with plaintiff's ownership within the meaning of ORS 72.6060(1)(c).

Because there is no evidence that Rose-A-Linda ever accepted the equipment in this case, the judgment against the defendant Industrial Leasing must be reversed.

QUESTIONS FOR DISCUSSION

1. *What acts of Rose-A-Linda did the plaintiff contend were inconsistent with its ownership of the goods?*
2. *Why did the court conclude that Rose-A-Linda's use of the machines was not inconsistent with the plaintiff's ownership of them?*

---

(1) Effect of Acceptance. As previously mentioned, the acceptance of the goods by the buyer gives rise to the obligation to pay for them. In addition, an acceptance by the buyer precludes rejection of the goods even if the acceptance was made with knowledge of their nonconformity, unless the acceptance was made on the reasonable assumption that the nonconformity would be cured and it has not been seasonably cured. Section 2–607(2) and 2–608(1)(a). The effect of this provision is to preclude rejection of the goods. After this point any return of the goods must be under circumstances allowing for revocation of acceptance.

Acceptance may be revoked when the nonconformity was not discovered and the acceptance was reasonably induced either by difficulty of discovery prior to acceptance or by the seller's assurances or on the reasonable assumption that the nonconformity would be cured and it has not been seasonably cured. Section 2–608(1)(b). Suppose Harry Homeowner purchases a storm door from Sam Seller. When the door is delivered, it does not have the type of lock specified. Seller assures Homeowner that if he accepts the door, the proper lock will be installed later. Homeowner waits a reasonable period and the lock is not changed. He may revoke his acceptance and return the door.

ED FINE OLDSMOBILE, INC. v. KNISLEY

319 A.2d. 33, (Del. 1974)

Thomas S. Knisley, III, purchased a car from Ed Fine Oldsmobile, Inc. Knisley then sued the dealer and was allowed to revoke acceptance. Ed Fine Oldsmobile appealed this decision.

O'HARA, J.

In February, 1970, the buyer, visiting dealer's place of business, expressed an interest in a 1968 Oldsmobile, 4–4–2, convertible, which was on display. He was seeking reliable transportation and, knowing that such models had high performance reputations, sought to ascertain the history of the used vehicle in question. He feared that a vehicle of this sort might have been used for racing or contain racing equipment, and if so, did not intend to buy the automobile. Buyer clearly stated this to the defendant's sales agents. They assured him that it had been well cared for by its previous owner, one of the dealer's mechanics, and that he had neither raced it nor installed any racing equipment in it.

Relying on those assurances, buyer purchased the automobile, taking delivery on February 26, 1970. He immediately began to have difficulties. It burned an excessive amount of oil, hesitated, and ran poorly. On April 3, 1970, dealer took the automobile into its shop for repairs. It was at this time that buyer discovered that several parts of his automobile had, in fact, been altered for racing. Among other things, it had a special racing transmission, a nonstock cam shaft, and racing weights in the carburetor. The dealer replaced these parts without charge and assured the buyer that all defects caused by racing or the use of racing equipment had been remedied.

Thereafter, buyer continued to have difficulties with the automobile. On May 5, 1970, it was placed in dealer's shop for additional repairs, including replacement of the starter. On June 9, 1970, shortly after those earlier repairs had been completed, it was towed to dealer's for the last time. The engine, which had locked after throwing a rod, was completely inoperative.

On condition that he need pay only for the labor costs, buyer agreed to allow the dealer to install a new engine. However, when that work was completed, in late July or early August, dealer told the buyer that he would also be required to pay an additional $364.00, to replace the transmission which had been stolen while the automobile was in dealer's possession. Confronted with what he considered a final act of bad faith, buyer refused to cooperate any further and left the automobile with the dealer. He then filed suit in the Court of Common Pleas for return of the $2,456.75 he had paid for the automobile. Essentially, dealer bases this appeal on two grounds: 1) that there was no breach of warranty, express or implied; and 2) that buyer failed to make an effective revocation of his acceptance of the automobile.

In the case at bar, the trial judge properly concluded that dealer should not be allowed to profit from his own wrongdoings by strictly holding buyer to the written terms of the sales and warranty agreements. The evidence amply supports the legal conclusion that buyer, at the time he discovered the deception, could have rejected the automobile and had his purchase price refunded.

This leads, however, to the other prong of dealer's attack on the trial court's decision. Buyer did not reject the automobile when he learned that it actually contained several pieces of racing equipment and that it did not otherwise fulfill his reasonable expectations or the dealer's representations. Instead, faced with a dealer who would not return the purchase price and who promised to remedy the nonconformities, buyer continued to use the vehicle while frequently bringing it to dealer for replacement of racing parts and other repairs. Given this history of buyer patience, dealer now claims that buyer failed to make a timely revocation of his acceptance as required by 5A Del.C. § 2–608.

The provision of the Uniform Commercial Code provides, in pertinent part:

> (1) The buyer may revoke his acceptance of a lot or commercial unit whose nonconformity substantially impairs its value to him if he has accepted it
>
> (a) on the reasonable assumption that its non-conformity would be cured and it has not been seasonably cured: . . .
>
> (2) Revocation of acceptance must occur within a reasonable time after the buyer discovers or should have discovered the ground for it and before any substantial change in condition of the goods which is not caused by their own defects. It is not effective until the buyer notifies the seller of it.

There is no doubt that buyer was persuaded to withhold his revocation from early April, when the non-conformities were discovered, until that summer by dealer's refusal to rescind the sale and its assurances that those non-conformities could and would be corrected. Nor is there any question that the final failure of the engine was due to the vehicle's own defects. Therefore, the question must turn on the timeliness of the revocation.

In this regard, it is of limited utility to look to other cases. What may have been too much delay before revocation in one case may be reasonable in another context. Hence, dealer's reliance on *Waltz* v. *Chevrolet Motor Division*, Del.Super., 307 A.2d 815 (1973), is misplaced. There, no allegation of fraud was made, much less proven. There, the vehicle in question was suitable for general purposes, failing only to properly haul a horse trailer. There, the buyer continued to use the vehicle after all attempts to cure had failed. There, the buyer waited more than a year before revocation.

Here, on the other hand, although the vehicle had been driven some 1,500 miles, the buyer had had little opportunity for inspection or continued use, since it was so frequently being repaired. Every attempt at revocation was forestalled by dealer's assurances and foot-dragging. Finally, dealer's attempt to foist the cost of the stolen transmission upon buyer is hardly support for now permitting dealer to profit thereby. Patience is admittedly a virtue. Dealer would here change it into a vice.

(Affirmed.)

Questions for Discussion

1. *What reason did the court give for allowing the buyer to revoke his acceptance?*
2. *Why was the buyer entitled to revoke acceptance after having driven the car over 1,500 miles?*

HAYS MERCHANDISE, INC. v. DEWEY
474 P.2d 270 (Wash. 1970)

Dale R. Dewey and his wife Jane Doe Dewey operated Dewey's Fuller Paint Store. In Autumn of 1967, it was decided to stock a "Toyland" in the store for Christmas trade. The Deweys visited Hays Merchandise, Inc., a wholesale toy company in Seattle. While there, they discussed the purchase of toys with a Mr. Woodring. It was agreed that a selection of toys, picked by Mr. Woodring, would be shipped to the Deweys at a cost of between $2,500 and $3,500. The Deweys were particularly interested in a stock of stuffed animals.

Several shipments were received, but they did not include the quantity of stuffed animals anticipated by the Deweys. Mr. Dewey contacted Hays on a number of occasions and was assured that more would be sent. After a final call without receiving the desired *toys, an exasperated Mr. Dewey called Hays and said that they wanted no more toys. This call was received too late to stop another shipment, which was received several days later. Mr. Woodring advised Dewey to send this latest shipment back unopened. Dewey kept this shipment at his store unopened. Several months later, Dewey informed another of Hays's salesmen that he had authority to ship a considerable quantity of unmarked toys, which he did. Hays refused the returned goods and sued Dewey for the amount due on the account. The lower court granted judgment to Hays for $3,436.36 less a credit of $299.98 for the shipment received by Dewey after the cancellation. Dewey appealed.*

FINLEY, A.J.

All of the transactions involved took place after the effective date of the Uniform Commercial Code in Washington. . . . The trial court held that the delivery of less than one half of the stuffed animals was not a "material breach" of the sales contract. Appellant Dewey contends that this finding is in error. His contention is based largely upon RCW 62A.2–608, which provides as follows:

> Revocation of acceptance in whole or in part. (1) The buyer may revoke his acceptance of a lot or commercial unit whose nonconformity substantially impairs its value to him if he has accepted it
>
> (a) on the reasonable assumption that its non-conformity would be cured and it has not been seasonably cured; or
>
> (b) without discovery of such non-conformity if his acceptance was reasonably induced either by the difficulty of discovery before acceptance or by the seller's assurances.
>
> (2) Revocation of acceptance must occur within a reasonable time after the buyer discovers or should have discovered the ground for it and before any substantial change in condition of the goods which is not caused by their own defects. It is not effective until the buyer notifies the seller of it.
>
> (3) A buyer who so revokes has the same rights and duties with regard to the goods involved as if he had rejected them.

There is no question but that Dewey accepted the toys in question. The issue before this court is whether there was an effective revocation of acceptance. This, in turn, is dependent upon (1) whether the nonconformity substantially impaired the value of the total order to Dewey, and (2) whether notice of the revocation took place within a reasonable time.

We are convinced that the question of "substantial impairment of value to (the buyer)" is best determined as a factual question by the trial court based upon all objective evidence properly before that court. In the instant case, the trial court's finding that there was no "material breach," while perhaps inartfully phrased, is in essence a finding that there was no substantial impairment. Our general rule is that findings of fact must be accepted as verities unless there is no substantial evidence in the record to support them. Appellant has not presented any considerations in the instant case which would lead this court to depart from that rule.

There remains the question of whether the notice of revocation was given within a reasonable time. Under the code, there is a distinction between notice of breach (RCW 62A.2–607(3) and notice of revocation of acceptance (RCW 62A.2–608(2). There is no question but that there was adequate and timely notice of breach. That, however, is not the question before this court. The notice of revocation of acceptance need not be in any particular form, but it must at least inform the seller that the buyer does not want the goods and does not desire to retain them. With the exception of the last small lot, there is no indication that the Deweys gave this notice prior to mid-February, when they attempted to return the unmarked and unsold toys. Indeed, the Deweys advertised the "Toyland" and attempted to sell the toys during December. It was not until a considerable time after Christmas that they attempted to return the toys. In view of the seasonal nature of the toy business and the somewhat faddish demand for certain toys, this delay in giving notice was unreasonable.

Even if the notice of revocation had been given in early December and if this were considered timely, the buyer's subsequent acts of dominion over the goods are inconsistent with such claimed revocation. The buyer's acts of pricing, displaying, advertising and selling were for his own account and were not in keeping with his duty to use reasonable care in holding the goods at the seller's disposition for a reasonable time.

(The judgment of the trial court is affirmed.)

18:5. Buyer's Obligation to Pay for Goods. The tender of delivery entitles the seller to acceptance of the goods and to payment according to the contract unless otherwise agreed. Section 2–507. Likewise the buyer's tender of payment is a condition to the seller's duty to tender and complete any delivery unless otherwise agreed. Section 2–511. These two sections of the code result in concurrent conditions of tender of payment and of delivery if there is no contrary agreement by the parties. The result is that unless there is agreement otherwise, delivery and payment are to be performed at a single place or time. Section 2–511, Comment 2. Recall that the code also provides that if the contract is silent, that place is the seller's place of business or residence. Section 2–308(a). Obviously the condition of payment is not concurrent with the condition of delivery under many contractual arrangements, such as when the goods are sold on credit. At this point an example illustrating some of these basic points may be helpful. Suppose Sam Seller agrees to sell Bill Buyer goods for $2,000. The contract contains no other provisions. Under the code, Seller must deliver the goods at Seller's place of business within a reasonable time. The delivery is to be tendered in a single delivery and is tendered by Seller's putting and holding the goods at Buyer's disposition and giving him any notification necessary to take delivery. Buyer must now accept and pay for the goods, after inspecting, or be in breach of contract, assuming that the goods are conforming. Similarly, a tender of payment by Buyer is a condition to Seller's duty to tender and complete delivery. Thus, if neither party tenders, the other has no obligation to perform. Now let us examine the code provisions concerning time and place of payment as well as the manner of payment.

(1) Time and Place of Payment. (a) *Cash Sales.* When the sale of goods is for cash—that is, when there is no agreement as to payment or the contract states that the sale is for cash, then payment is due at the time and place at which the buyer is to receive the goods. Section 2–310(a). In other words, when the buyer receives the goods, he pays. Don't forget that the buyer normally has the right to inspect the goods prior to acceptance or payment.

(b) *Documentary Sales.* The contract may provide for payment by the buyer when the documents of title to the goods are presented, regardless of the fact that the goods may still be in the process of shipment. Under these circumstances the buyer waives his right to inspection prior to his duty to accept and pay for the goods.

When the contract is silent, the seller may have a documentary sale, but the buyer still has the right to inspect the goods prior to payment. Section 2–310(b) provides that the seller may ship the goods "under reservation" and may tender documents of title, but the buyer may inspect the goods after their arrival, before payment is due, unless such inspection is inconsistent with the terms of the contract. The effect of this provision is that the buyer may inspect before he pays, but the seller does not have to give up possession of them prior to payment. This procedure is accomplished in the following manner. The seller obtains a bill of lading from the carrier and attaches it to a "sight draft" for the purchase price. The documents are sent to a bank for collection with the bill of lading marked "hold until arrival, inspection allowed." When the goods arrive, the buyer inspects them but is not given possession. He then goes to the bank to pay the draft and obtain the bill of lading for presentation in order to obtain the goods. Thus, when the contract is silent, this procedure of shipment "under reservation" allows the seller to retain control of the goods and gives the buyer the right of inspection prior to payment, but payment is due where and when the buyer is to receive the documents. Section 2–310, Comment 2.

If the seller wishes to have payment before inspection, he must insert a term providing for that procedure in the contract. If the terms are C.O.D. or C.I.F. or, as previously mentioned, cash against documents, then the buyer has no right of inspection prior to payment. The parties may provide in a C.I.F. contract for inspection by the buyer prior to payment, however.

(c) *Credit Sale.* If the sale of goods is on credit terms, then the time when the buyer must make payment is, of course, controlled by the contract. When the seller is required or authorized to ship the goods on credit, the credit period runs from the time of shipment. Section 2–310(d).

(2) **Manner of Payment.** The buyer may tender payment by any means or in any manner current in the ordinary course of business, unless the seller (1) demands payment in legal tender and (2) gives the buyer any extension of time reasonably necessary to procure it. Section 2–511(2). The purpose of this approach by the code is to avoid a commercial surprise at the time of performance. For instance, suppose Sam Seller is to deliver goods worth $3,000 to Bill Buyer's place of business on August 31. Seller arrives at 4:30 P.M. with the goods and Buyer tenders his check for $3,000, a manner of payment current in the ordinary course of business. Seller refuses the check and demands legal tender, but Buyer's bank is closed. Seller would not be able to force Buyer into a breach in this manner, as Buyer would have an extension of time reasonably necessary to procure the money, until Buyer's bank opens, in this case. Thus the seller can still demand cash but cannot surprise the buyer with demand and force him into breach if payment by check is the customary method of payment.

MODERN AERO SALES, INC. v. WINZEN RESEARCH, INC.

486 S.W.2d 135 (Tex. 1972, rehearing denied)

Winzen Research, Inc., had a contract to perform certain services for University Corporation for Atmospheric Research (UCAR). This contract gave UCAR an option to purchase an aircraft from Winzen that was registered to D. R. Williams, vice-president of Winzen. The contract and option ended May 15, 1970. On May 11, UCAR awarded a new contract for these services to Modern Aero Sales, Inc., and assigned the option to purchase to Modern Aero.

On May 12, Modern Aero attempted to exercise the option by mailing to Winzen a letter with an envelope draft containing a bill of sale. The draft directed the National Bank of Commerce of Dallas to pay to the order of D. R. Williams $8,161.03, the price of the aircraft.

The letter and draft were received by Winzen on May 14, 1970. D. R. Williams of Winzen rejected the draft by mail as not being legal tender but noted that a bill of sale would be delivered upon a timely tender. The letter from Williams reached Modern Aero on May 16. On the same day, Winzen sent a pilot to Modern Aero to pick up the aircraft, which had been previously delivered. Modern Aero refused to release it on the ground that the option had been exercised. Modern Aero then sued Winzen for breach of the option agreement.

GUITTARD, J.

Our questions are (1) whether a bank draft was a means of payment "current in the ordinary course of business" under the Uniform Commercial Code, § 2.511(b), . . . (2) whether the seller's rejection of such draft and demand for legal tender had the effect of extending the time of payment for a reasonable time after the purchaser received such demand and (3) whether further tender was excused by statements of the seller to purchaser's agent indicating unwillingness to accept payment on grounds other than the medium of payment.

The trial court found that tender of the envelope draft "was not a tender made by any means or in any way current in the ordinary course of business." Plaintiff contends that this finding is contrary to the undisputed evidence. We agree. Harold Weiser, an officer of Modern Aero, testified that Modern Aero was in the business of buying and selling aircraft, among other businesses, and that such drafts were frequently used in sales of aircraft. Louie Robinson, vice-president of National Bank of Commerce, testified that Modern Aero had maintained a continuing line of credit for purchase of aircraft at his bank since 1965, and that the draft in question, if presented for payment, would have been honored. He had frequently seen similar drafts drawn by Modern Aero. According to Robinson, the envelope draft was a method used in ordinary course of business by his bank and other banks for simultaneous transfer of title and disbursement of funds when documents needed to be checked. In fact, he said that the envelope draft was the only method in normal use for transfer of title to aircraft. Defendant presented no rebutting evidence on this point.

Defendant argues that the envelope draft was not a means or manner of payment "current in the ordinary course of business" because it had not previously been used in business transacted between defendant and plaintiff and was not familiar to defendant, who was not in the business of buying and selling aircraft. We do not construe this statutory language as restricted to the business of the parties in question or to previous dealings between them. Neither do we construe it as limited to a common-law "custom," which is binding only on parties who contract with knowledge of it. The transaction involved was sale of an aircraft, and in this context we interpret "ordinary course of business" to mean the ordinary course of business of selling aircraft. The seller's ignorance of practices in that business is not controlling. The purpose of § 2.511(b) is avoidance of commercial surprise at the time of performance. This purpose would be defeated, at least in part, if a purchaser who relies on a means of payment current in the ordinary course of the business involved does not discover until after time for tender has expired that the seller is unwilling to accept payment by that means because he is ignorant of the current business practices. Our construction imposes no hardship on the seller, since § 2.511(b) preserves his right to demand payment in legal tender, although it requires him in that event to give "any extension of time reasonably necessary to procure it."

However, we cannot agree with plaintiff's further contention that defendant failed to grant the extension required by the statute and that such failure excused any further tender. Although defendant's letter of May 14 did not expressly grant such an extension or specify any time, such an extension is necessarily implied from the statement: "If a timely legal tender is presented for exercise of the purchase option . . . we will execute and deliver a bill of sale." This language can be construed only as giving plaintiff a reasonable time after receipt of the letter to make a legal tender. The trial court found that a reasonable time for plaintiff to have procured and made a legal tender was the period from Saturday, May 16, to and including Tuesday, May 19, since the banks were closed on Saturday and Sunday. We accept this finding as establishing the period of extension allowed.

The next question is whether plaintiff's failure to make a legal tender by May 19 is excused by defendant's conduct. On this point there is some conflict in the testimony, but none we find to be material. Weiser testified that as soon as the bank opened on Monday, he purchased from Hampton State Bank a cashier's check payable to D. R. Williams and delivered it to a pilot, Richard Morrison, with instructions to fly to Minneapolis, present the cashier's check to Williams, and obtain his signature on the bill of sale.

The trial court found that in the telephone conversation with Morrison on May 19, Williams stated that Winzen "would not accept tender of the said check for the reasons that plaintiff still was not making legal tender, that the purchase option had expired on May 15, 1970, and that the $8,161.03 amount was insufficient to reimburse additional costs incurred by Winzen Research, Inc. after May 15, 1970." Plaintiff challenges this finding insofar as it states that one of the reasons given was "that plaintiff still was not making legal tender." We sustain this contention. Neither Williams nor Morrison testified that Williams told Morrison he would not accept the check because it was not legal tender. According to Williams' own testimony, he did not tell Morrison he would not take the check and the only reasons he gave Morrison as to why it would do no good to come to his office were (1) that the option had already expired on May 15 and (2) that he had incurred additional expenses since that date. This statement by Williams can be reasonably understood only as an unqualified refusal to accept a further tender. Defendant argues that plaintiff was nevertheless required to confront Williams or some other officer of Winzen in Winzen's office and physically lay before him the amount of currency required. The law is well settled that formal tender is excused when the tenderor is able to pay and the tenderee has signified unwillingness to accept the money.

Defendant also contends that it was justified in refusing to accept tender of the cashier's check because of additional expenses incurred after May 15. We do not agree. Defendant's letter of May 14 demanding legal tender mentioned no such expenses. If a seller demands legal tender and the purchaser attempts a further tender within the extension period allowed by § 2.511(b), the seller is not justified in refusing to accept the amount originally due on the ground of expenses previously unspecified. If the seller knows that the extension will involve additional expenses, he should specify the amount of such expenses and give the buyer a reasonable time to tender the additional amount. When the seller refuses to accept tender on grounds of its inadequacy to cover previously unspecified expenses in an undetermined amount, the buyer is justified in treating such nonacceptance as an unqualified refusal.

We hold that Williams' statement to Morrison was a refusal of the tender on grounds other than the medium of payment and excused further tender by plaintiff, and that his refusal to transfer title to the aircraft was a breach of the option contract, which entitled plaintiff to recover damages measured by the difference between the option price and the market value. In this suit for such damages, as distinguished from one for title to the aircraft or for specific performance, plaintiff was not required to show that it had made a further tender or had kept the tender good, since after defendant's breach plaintiff had no obligation to pay.

(Reversed and remanded.)

Questions for Discussion

1. *Was Winzen justified in originally rejecting the letter draft as legal tender? Why did Modern Aero receive additional time?*
2. *Was Modern Aero's tender on May 19 timely under the circumstances, considering the fact that the option expired May 15?*
3. *On May 19, Aero tendered a cashier's check. Might Winzen have validly rejected this tender? How? Why was Aero excused from making a further tender?*

18:6. Seller's Right to Cure on Improper Tender or Delivery. What happens if the time for contractual performance has not yet expired and the goods delivered by the seller are rejected as nonconforming? The code provides that the seller may seasonably notify the buyer of his intention to cure and may make a conforming delivery within the contract period. Section 2–508(1). This rule would apply even if the seller has taken back the goods and refunded the buyer's money, but note the requirement of "seasonable" notification. The closer it is to the date for contractual performance, the greater the necessity for promptness in notification. Section 2–508, Comment 1. For instance, suppose Buyer orders a white storm door for his home for delivery November 1. On October 15, a black one is delivered by Seller. Seller may notify and cure by delivery of a white door by November 1.

A more difficult provision of the code in terms of its application is Section 2–508(2), which provides that if the buyer rejects a nonconforming tender, the seller may have further reasonable time to substitute a conforming tender even if the time for performance has expired, if he had reasonable grounds to believe that it would be acceptable with or without money allowance. The reason for this provision is to avoid injustice to the seller by a surprise rejection by the buyer when, for example, the seller delivers goods of a higher quality than called for by the contract. Note that the seller must have reasonable grounds for believing that the nonconforming goods would be acceptable. This belief might result from a prior course of dealing between the parties by usage of the trade. This provision has *not* been construed to allow the seller to substitute merely because he wants to after he has delivered nonconforming goods.

WILSON v. SCAMPOLI

228 A.2d 848 (D.C. 1967)

Nick Scampoli purchased a new color television set from Willie Wilson TV. When the set was delivered and turned on, the picture had a red cast to it. Several days later a service representative came to examine the set. After he examined it for over an hour, he concluded that he would have to remove the chassis from the cabinet and take it to the shop for repair. Mrs. Kolley, who was Scampoli's daughter, refused to allow this and instead demanded a new television. Later she demanded return of the purchase price. Wilson refused but renewed his offer to adjust, repair, or, if the set could not be made to function properly, replace. Scampoli then filed suit against Wilson for a refund of the purchase price. The trial court found for Scampoli and Wilson appealed.

MYERS, J.

Appellant does not contest the jurisdiction of the trial court to order rescission in a proper case, but contends the trial judge erred in holding that rescission here was appropriate. He argues that he was always willing to comply with the terms of the sale either by correcting the malfunction by minor repairs or, in the event the set could not be made thereby properly operative, by replacement; that as he was denied the opportunity to try to correct the difficulty, he did not breach the contract of sale or any warranty thereunder, expressed or implied.

D.C. Code § 28:2–508 (Supp.V, 1966) provides:

> (1) Where any tender or delivery by the seller is rejected because non-conforming and the time for performance has not yet expired, the seller may seasonably notify the buyer of his intention to cure and may then within the contract time make a conforming delivery.
>
> (2) Where the buyer rejects a non-conforming tender which the seller had reasonable grounds to believe would be acceptable with or without money allowance the seller may if he seasonably notifies the buyer have a further reasonable time to substitute a conforming tender.

A retail dealer would certainly expect and have reasonable grounds to believe that merchandise like color television sets, new and delivered as crated at the factory, would be acceptable as delivered and that, if defective in some way, he would have the right to substitute a conforming tender. The question then resolves itself to whether the dealer may conform his tender by adjustment or minor repair or whether he must conform by substituting brand new merchandise. The problem seems to be one of first impression in other jurisdictions adopting the Uniform Commercial Code as well as in the District of Columbia.

Although the Official Code Comments do not reach this precise issue, there are cases and comments under other provisions of the Code which indicate that under certain circumstances repairs and adjustments are contemplated as remedies under implied warranties.

While these cases provide no mandate to require the buyer to accept patchwork goods or substantially repaired articles in lieu of flawless merchandise, they do indicate that minor repairs or reasonable adjustments are frequently the means by which an imperfect tender may be cured. In discussing the analogous question of defective title, it has been stated that:

> The seller, then, should be able to cure (the defect) under subsection 2–508(2) in those cases in which he can do so without subjecting the buyer to any great inconvenience, risk or loss.

Removal of a television chassis for a short period of time, in order to determine the cause of color malfunction and ascertain the extent of adjustment or correction needed to effect full operational efficiency, presents no great inconvenience to the buyer. In the instant case, appellant's expert witness testified that this was not infrequently necessary with new televisions. Should the set be defective in workmanship or parts, the loss would be upon the manufacturer who warranted it free from mechanical defect. Here the adamant refusal of Mrs. Kolley, acting on behalf of appellee, to allow inspection essential to the determination of the cause of the excessive red tinge to the picture defeated any effort by the seller to provide timely repair or even replacement of the set if the difficulty could not be corrected. The cause of the defect might have been minor and easily adjusted or it may have been substantial and required replacement by another new set—but the seller was never given an adequate opportunity to make a determination.

We do not hold that appellant has no liability to appellee, but as he was denied access and a reasonable opportunity to repair, appellee has not shown a breach of warranty entitling him either to a brand new set or to rescission. We therefore reverse the judgment of the trial court granting rescission and directing the return of the purchase price of the set.

(Reversed.)

18:7. Excuse of Performance. The code contains several provisions that excuse or alter allowable performance under certain circumstances. These situations are if there is a casualty to identified goods, if there is a failure of presupposed conditions, and if a substituted performance may be rendered.

(1) Casualty to Identified Goods: Impossibility. If the contract *requires* goods for performance that are identified in the contract at the time the contract is made and the goods suffer a casualty without fault of either party *before* the risk of loss passes to the buyer, then the contract is avoided if the loss is total. If the loss is partial or the goods have so deteriorated as to no longer conform to the contract, the buyer may still demand inspection and at his option either (1) treat the contract as avoided or (2) accept the goods with allowance for deterioration or the deficiency in quantity, but without further right against the seller. Section 2–613.

Note that the section applies only if the loss occurs *prior* to the risk's passing to the buyer. After that time, the loss is the buyer's. Also the loss must occur *without fault* of either party. If the loss occurs because of the intentional act or negligence of either party, the innocent party may still recover damages regardless of this section. Furthermore, the casualty must be to goods *required* for performance of the contract. If the loss occurs to goods that the seller intends to use for performance, then this section does not apply. For instance, suppose Sam Seller agrees to sell a car, Brand X, to Car Buyer. Seller has five of them, but the one he intends to use to satisfy the contract is destroyed by fire. This destruction would not excuse Seller because it was not the specific car required to perform the contract. On the other hand, if the car were damaged and it was the specific car identified to be delivered to Buyer, then Seller would be excused from performance, although he sustains the loss for damage to the car. Buyer would have the option to avoid the contract or to take the car, less an allowance from the contract price for the damages to the vehicle.

This section also applies under a "no arrival, no sale" contract if the arrival of the goods is delayed rather than suffering a casualty. The buyer has the options set forth under this circumstance, just as he or she would with a physical change in the goods.

(2) **Failure of Presupposed Condition: Impracticality.** Often events can occur, such as strikes, embargoes, or the passage of governmental regulations, that make contractual performance either impossible or impractical. Normally impossibility or impracticality of performance is not an excuse for discharge of the contract unless there is a provision in the contract covering this eventuality. The common-law presumption is that the failure to provide for such an event in the contract indicates that one is bearing the risk. However, the code takes the position that if there is nondelivery or delay in delivery by the seller because performance "has been made, impracticable by the occurrence of a contingency the nonoccurrence of which was a basic assumption on which the contract was made or by compliance in good faith with any applicable foreign or domestic governmental regulation," then the seller has not breached the contract of sale. Section 2–615(a).

Note that the event has to be one that the parties assumed would not occur. The code comment to this section states this in a more traditional manner as being an event that was unforeseen and not within the contemplation of the parties at the time of contract. Obviously an event that was *foreseen,* although not provided for in the contract, would not result in the invocation of this section. Comment 4 to this section is most helpful in clarifying it: Increased cost alone does not excuse performance unless the rise in cost is due to some unforeseen contingency which alters the essential nature of the performance. Neither is a rise or a collapse in the market in itself a justification, for that is exactly the type of business risk which business contracts, made at fixed prices, are intended to cover. But a severe shortage of raw materials or of supplies due to a contingency such as war, embargo, local crop failure, unforeseen shutdown of major sources of supply or the like, which either causes a marked increase in cost or altogether prevents the seller from securing supplies necessary to his performance, is within the contemplation of this section.

(3) **Failure of Presupposed Condition: Allocation.** Often the events under discussion may affect only a part of the seller's capacity to perform. In that case, he must allocate production and deliveries among his customers, but the code gives him an option to include regular customers not then under contract. This preserves the right of the seller to give attention to the needs of regular customers who are probably relying on him as a source of supply. Section 2–615(b). If an allocation is required, then the seller must seasonably notify the buyer of the estimated quota to be made available to the buyer. In the event that there is to be a delay or nondelivery, the buyer must also be seasonably notified. Section 2–615(c).

(4) **Failure of Presupposed Condition: Procedure on Notice Claiming Excuse.** Once the buyer receives notification of a material or indefinite delay, he may notify the seller, in writing. The notification may be as to any delivery concerned or, if the delivery impairs the value of the whole contract, to the whole contract. This notification informs the seller of the buyer's election of one of two options. The buyer may either (1) terminate and discharge any unexecuted portion of the contract or (2) modify the contract by agreeing to take his available quota in substitution. Section 2–616(1).

The contract lapses with respect to any affected delivery if the buyer fails to modify the contract within a reasonable time—not exceeding thirty days—after receiving notification from the seller. Section 2–616(2).

(5) **Substituted Performance: Delivery and Payment.** If without fault of either party the agreed manner of delivery becomes commercially impracticable, the seller must tender a commercially reasonable substitute if available, and it must be accepted by the buyer. Section 2–614(1). For instance, the agreed shipment might be by a certain truck company. If a strike shuts down that company, shipment by another company or by rail might be tendered and accepted.

Likewise, if the agreed means or manner of payment fails because of foreign or domestic governmental regulation, the seller may withhold or stop delivery unless the buyer provides a means or manner of payment that is a commercially substantial equivalent. If delivery has already been made, payment as prescribed by the regulation discharges the buyer unless it is discriminatory, oppressive, or predatory. Section 2–614(2).

Note that this provision concerning substituted delivery and payment goes to an incidental matter of the agreement. The provisions concerning failure of presupposed conditions and casualty to identified goods goes to the very heart of the agreement.

MANSFIELD PROPANE GAS COMPANY, INC. v. FOLGER GAS COMPANY

231 Ga.868, 204 S.E.2d 625 (1974)

GUNTER, J.

This is an "energy crisis" case involving the validity and interpretation if valid of a contract between the parties for the sale and delivery of propane gas.

The appellants (Mansfield) were the defendants in the trial court. The appellees (Folger) brought an action in the trial court against Mansfield seeking a declaration of rights pursuant to the alleged contract and injunctive relief requiring the parties to abide by the contract as interpreted.

The trial judge found that the contract between the parties was a valid and binding contract, and we agree. The contract provided that Mansfield would sell and deliver to Folger during a five-year term (or possible seven-year term) the latter's propane gas requirements up to a maximum of ten million gallons during the entire term at the price specified in the contract. While Mansfield had many other customers, Folger was its only customer having a written contract with Mansfield that required the sale and delivery of a maximum number of gallons of propane gas during a stipulated term.

The contract had been in effect for approximately one year and sales and deliveries pursuant to its terms had gone smoothly until the beginning of the "energy crisis" and a shortage of propane gas supply. On June 18, 1973, Mansfield notified Folger that it intended to "allocate" its supply of propane gas among all of its customers pursuant to Section 109A–2–615 of the Georgia Uniform Commercial Code. Folger contended that its requirements of propane gas were not subject to "allocation," and since Mansfield had an adequate supply to furnish all of Folger's requirements, then Mansfield had to do so first and could then allocate its remaining supply among its other customers.

With respect to "allocation" Code Ann. § 109A–2–615 provides as follows: "Except so far as a seller may have assumed a greater obligation and subject to the preceding section on substituted performance: (a) Delay in delivery or non-delivery in whole or in part by a seller who complies with paragraphs (b) and (c) is not a breach of his duty under a contract for sale if performance as agreed has been made impracticable by the occurrence of a contingency the nonoccurrence of which was a basic assumption on which the contract was made or by compliance in good faith with any applicable foreign or domestic governmental regulaton or order whether or not it later proves to be invalid. (b) Where the clauses mentioned in paragraph (a) affect only a part of the seller's capacity to perform, he must allocate production and deliveries among his customers but may at his option include regular customers not then under contract as well as his own requirements for further manufacture. He may so allocate in any manner which is fair and reasonable. (c) The seller must notify the buyer seasonably that there will be delay or non-delivery and, when allocation is required under paragraph (b), of the estimated quota thus made available for the buyer."

Folger argues that Code Ann. § 109A–2–615 is not applicable in this case because Mansfield, the seller, has assumed a greater obligation to Folger by virtue of the written contract between them than Mansfield has assumed with any of its other customers. None of Mansfield's other customers have written contracts providing for a maximum number of gallons of propane gas to be sold and delivered over a term of years.

Therefore, Folger argues that its contractual relationship with Mansfield makes it an exception to the "allocation" rule provided for in Code Ann. § 109A–2–615.

The other side of the argument is that since all sales are sales by contract, whether written or oral, and since subsection (b) of Code Ann. § 109A–2–615 provides that when only a part of the seller's capacity to perform is affected, the seller "must allocate production and deliveries among his customers but may at his option include regular customers not then under contract," then any exception or exemption from the "allocation" rule must be supported by an affirmative and explicit provision in the contract to the effect that the seller will not "allocate" in the event that his supply or capacity is affected so that his "performance as agreed has been made impracticable by the occurrence of a contingency the nonoccurrence of which was a basic assumption on which the contract was made."

As stated before, we have found no reported case adopting either of these arguments as against the other in interpreting this provision of the Uniform Commercial Code. Ronald Anderson, in his treatise on the Uniform Commercial Code, seems to adopt a construction favorable to Mansfield as follows: "The Code Section (2–615) expressly recognizes the right to impose, by the terms of the contract, a higher standard upon the seller, with the result that the parties may restrict the excusing contingencies to those specified in the contract or may eliminate the protection given by the code section by imposing upon the seller an absolute contractual duty to make delivery." Anderson, Uniform Commercial Code, Vol. II, Section 2–615: 5 (2d ed.). An obscure remark in 67 Am.Jur.2d p. 512 also seems to substantiate a construction favorable to Mansfield's position when in talking about this code section the authors state: ". . . (1) The seller must not have assumed 'a greater obligation' such as giving an unconditional guarantee of performance. . . ."

We therefore construe Code Ann. § 109A–2–615 to mean that in order for there to be an exception to and an exemption from the rule of allocation applicable to a contract of sale, such a contract must contain an affirmative provision that the seller will perform the contract even though the contingencies which permit allocation might occur.

This holding means that the parties to the contract in the instant case are bound by the rule of allocation contained in Code Ann. § 109A–2–615, and the judgment of the trial court must be reversed.

Since Code Ann. § 109A–2–615 also provides that allocation must take place in a manner that is "fair and reasonable," this issue remains for decision in the trial court. We therefore reverse the judgment and remand for further proceedings consistent with this opinion.

(Judgment reversed.)

(Motion for rehearing was denied.)

QUESTIONS FOR DISCUSSION

1. *What conditions must be fulfilled in order for a seller to justifiably "allocate" supplies among his or her customers?*
2. *How could Folger have protected itself against the situation in which its supplier was justified in allocating to it a lesser quantity than the amount stated in the contract?*

18:8. BULK TRANSFERS. When a merchant sells the major part of his materials, supplies, merchandise, or other inventory, the transfer is governed by the "bulk transfer" provisions of the Uniform Commercial Code. A "bulk transfer" is defined by the code as "any transfer in bulk and not in the ordinary course of the transferor's business of a *major portion* of the materials, supplies, merchandise or other inventory of an enterprise. . . ." Section 6–102(1) (emphasis added). "A transfer of a substantial part of the equipment of such an enterprise is a bulk transfer if it is made in connection with a bulk transfer or inventory, but not otherwise." Section 6–102(2). A typical example of a bulk transfer would be the sale by a retail merchant of his inventory to someone purchasing his business. Whether the sale is a "major part" of the inventory is determined by the value of the inventory rather than the quantity.

The purpose of the bulk-transfer provisions is to protect creditors of the transferor. Suppose, for instance, that Manny Merchant is heavily in debt. He sells his inventory to anyone for virtually any price and then promptly disappears, leaving his creditors unpaid. The bulk-transfer provisions are designed to protect those creditors by requiring notice to them prior to such a bulk transfer.

(1) Compliance Requirements. Basically the code protects creditors by holding any bulk transfer ineffective against creditors unless proper notice is given to those creditors prior to the transfer. In particular Section 6–104 of the code provides that a bulk transfer is ineffective unless (1) the transferee requires the transferor (the seller) to furnish a list of his existing creditors; (2) the parties prepare a schedule of the property to be transferred; and (3) the transferee (buyer) preserves the list and schedule for six months following the transfer and permits inspection and copying of it by any creditor or files the list in a public office.

The list must be signed and sworn to by the transferor. The responsibility for the completeness and accuracy of the list rests on the transferor, and the transfer is not rendered ineffective by errors unless the transferee has knowledge of the errors. Section 6–104(3). Because the transferor prepares the list under oath, the sanctions for lack of accuracy are the criminal laws relative to false swearing. On the other hand, a transferee takes a chance in not requiring the list as a condition of purchase, because the creditors of the seller can levy on the purchased goods even after they are transferred to the buyer and paid for.

In addition to the above requirements the code places a burden on the transferee to notify the creditors of the impending transfer at least ten days prior to the time he takes possession of the goods or pays for them, whichever happens first. Without this notification the transfer is ineffective against the creditor. Section 6–105.

The notice given to creditors must state (1) that a bulk transfer is about to be made; (2) the names and addresses of the transferor and the transferee, as well as any other business names used by the transferee within the last three years; and (3) whether the debts of the transferor are to be paid when they fall due and if so where the bills should be sent. If the bills are not to be paid when they fall due or if the transferee is in doubt, the notice must state the location and description of the property transferred and an estimate of the transferor's debts, together with a statement of where the property and list of creditors may be inspected. In addition, the notice must state whether the transfer is to pay the debts and whether the transfer is for new consideration. The notice is to be delivered personally or by registered or certified mail. Section 6–107.

The effect of the notice requirement is to give the creditors an opportunity to move against the property of the debtor—if they are entitled to do so—prior to transfer. If they are notified and do not act within the specified time, then the transfer will usually cut off their rights to the property. The code provides for exceptions to this procedure when the transfer is by auction. Section 6–108.

The code also provides that a bona fide purchaser for value from the transferee takes the goods free from any claim of the creditors if the requirements are not followed. If a purchaser from the transferee pays no value for the goods or takes them with notice of the noncompliance with the requirements, then he is subject to the rights of the transferor's creditors.

You should note that these requirements are for the protection of the creditors of the transferor. Failure to comply with them has nothing to do with the rights and liabilities between the transferor and transferee.

CORNELIUS v. J & R MOTOR SUPPLY CORPORATION

468 S.W.2d 731 (Kent. 1971)

Morris B. Costello was indebted to J & R Motor Supply Corporation for $1,370.83, and to Orgill Brothers and Company for $1,491.33, for merchandise he had purchased for his automobile accessory and appliance business. On November 22, 1966, Costello sold his business to Eldon and Jerry Cornelius. At that time, the debts were unpaid. The parties did not comply with the provisions of the bulk-sales law.

J & R Motor Supply and Orgill Brothers claimed that the Corneliuses should pay the claims. Meanwhile Costello filed a petition in bankruptcy. J & R Motor Sales and Orgill Brothers then sued Cornelius contending that the Corneliuses were indebted to them because of the failure to comply with the bulk-sales law.

STEINFELD, J.

KRS 355.6–104 declares that ". . . a bulk transfer subject to this article is ineffective against any creditor . . . unless: (a) The transferee requires the transferor to furnish a list of his existing creditors . . ." prepared as statutorily required.

KRS 355.6–105 provides:

> In addition to the requirements of KRS 355.6–104, any bulk transfer subject to this article except one made by auction sale (KRS 355.6–108) is ineffective against any creditor of the transferor unless at least ten days before he takes possession of the goods or pays for them, whichever happens first, the transferee gives notice of the transfer in the manner and to the persons hereafter provided. . . .

KRS 355.6–106 provides:

> In addition to the requirements of KRS 355.6–104 and KRS 355.6–105:
>
> (1) Upon every bulk transfer subject to this article for which new consideration becomes payable except those made by sale at auction it is the duty of the transferee to assure that such consideration is applied so far as necessary to pay those debts of the transferor which are either shown on the list furnished by the transferor (KRS 355.6–104) or filed in writing in the place stated in the notice (KRS 355.6–107) within thirty days after the mailing of such notice. This duty of the transferee runs to all the holders of such debts, and may be enforced by any of them for the benefit of all.
>
> (2) If any of said debts are in dispute the necessary sum may be withheld from distribution until the dispute is settled or adjudicated.
>
> (3) If the consideration payable is not enough to pay all of the said debts in full, distribution shall be made pro rata.

The intent of the Bulk Sales Law is to protect the rights of all creditors existing at the time of the transfer. . . .

We now approach the question of whether a personal judgment was authorized. We believe the act did not so contemplate except under special circumstances. As stated in 6 *Corbin, Contracts,* sections 1514 at 977–8 (1951), the ordinary bulk sales statute ". . . merely makes it (the sale) voidable against both parties for the benefit of creditors." With respect to section 6–104 the Official Code Comments state: "Any such creditor or creditors may therefore disregard the transfer and levy on the goods as still belonging to the transferor, or a receiver representing them can take them by whatever procedure the local law provides." The status of the buyer is as a trustee or receiver for the benefit of creditors of the seller existing at the time of the sale. The rule is well expressed in *Southwestern Drug Corp.* v. *McKesson & Robbins,* 141 Tex. 284, 172 S.W.2d 485, 155 A.L.R. 1056 (1943):

> Failure of the purchaser to comply with the Bulk Sales Law fixes his liability as that of a receiver, and he becomes bound to see that the property, or its value, is applied to the satisfaction of claims of the creditors of the seller. In other words, he becomes a trustee, charged with the duties and liabilities of a trustee. Under the law he is charged with liability only to the extent of the value of the property received by him, and his liability is to all of the creditors pro rata. *Gardner* v. *Goodner.* However, if a purchaser or receiver disposes of or converts to his own

> use property acquired in violation of the Bulk Sales Law, placing it beyond the reach of creditors, he will be held personally liable for the value thereof.

From the foregoing and the rationale of other cases and texts we will refer to we conclude that since the purchasers so comingled the merchandise that it could not be segregated personal liability ensued.

The measure of recovery under the circumstances was the fair market value of the merchandise. . . .

(Judgment for the plaintiffs.)

QUESTIONS FOR DISCUSSION

1. *Would the Corneliuses have been liable for the debts of Costello had they complied with the bulk-sales law?*
2. *Explain why the Corneliuses became personally liable for the debts in this case?*

SUMMARY

The obligation of the seller is that he tender conforming goods. If the contract fails to mention a time and place for the delivery, the delivery is to be at a reasonable time, at the seller's place of business. If there is no contrary agreement, the seller is obligated to deliver all the goods at one time.

The buyer generally has a right to inspect the goods prior to payment or acceptance. An acceptance occurs when the goods are shipped C.O.D. After the seller has performed, the buyer is obligated to accept the goods. The code has rules as to what constitutes acceptance by the buyer. Acceptance gives rise to the buyer's obligation to pay for the goods and generally precludes rejection. However, acceptance may be revoked when a nonconformity in the goods was not discovered and the acceptance was reasonably induced either by difficulty of discovery prior to acceptance or by the seller's assurances or by the reasonable assumption that the nonconformity would be cured.

Unless there is agreement otherwise, delivery and payment are to be performed at a single place and time. Payment may be by any means or manner current in the ordinary course of business unless the seller demands payment in legal tender and gives the buyer a reasonable time to procure it.

If the seller delivers nonconforming goods prior to the time performance is due, the seller may make a conforming delivery within the contract period. Under certain conditions a seller may cure after the contract period has run.

Performance may be excused if there is a casualty to identified goods, without fault of either party, prior to the time the risk of loss passes to the buyer. Performance may also be excused under the code if there is a failure of presupposed conditions. If such a failure occurs, the seller may allocate production and deliveries among his customers. If the agreed manner of delivery becomes impossible without the fault of either party, then a commercially reasonable substitute must be tendered and accepted.

A bulk transfer occurs when a merchant sells a major part of his materials, supplies, merchandise, or other inventory. The purpose of the bulk-transfer act is to protect the creditors of the transferor. To be effective against creditors, the transferee must require the transferor to furnish a list of his existing creditors; the parties must prepare a schedule of property to be transferred and the transferee must preserve the list and schedule for six months after the transfer. In addition, the transferee must notify creditors of the impending transfer at least ten days before he takes possession of the goods or pays for them.

PROBLEMS

1. Bertram Fleet and Sidney Danowitz purchased an ice cream freezer from F. W. Lang Co. on April 30, 1957. A written installment contract of purchase was signed by the parties. Fleet and Danowitz moved to a new location about a year later and disconnected the compressor from the freezer and connected it to an air conditioner. Fleet and Danowitz never paid the amount due on the installment contract and were sued by F. W. Lang Co. Fleet and Danowitz filed their own suit contending that the equipment was defective and unusable for the purpose intended. They contended that they had a right to rescission of the contract. Are they correct? F. W. Lang Co. v. Fleet, 165 A.2d 258 (Pa., 1960).
2. Marine Mart, Inc. sold a boat, which was on display, to L. D. Pearce on June 2, 1970. It was then agreed that a boat in the warehouse could be delivered because it was "identical." When the boat was delivered, it differed in several ways from the display model and was damaged in a number of respects, including scratches on the windshield, a tear in the upholstery, and damage to the hull. The damage was reported to Marine, which agreed to repair it on June 20. Pearce was not satisfied with the repairs and told the repairmen not to come back. Pearce used the boat only three times for five hours. After a number of other communications Pearce filed suit for rescission, alleging that the repairs had not been made and that the boat that was delivered was not identical to the one displayed. Marine defended on the ground that there was acceptance of the goods and that it had a right to cure defects. Decide. Marine Mart, Inc. v. L. D. Pearce, 480 S.W.2d 133 (Ark. 1972).
3. Plaintiff agreed to purchase four used airplanes from defendant. Payment was to be made by certified check upon delivery of the planes to the plaintiff at the airport on June 1. The evidence at trial showed that the defendant was ready, willing, and able to perform on that date, although it did not actually deliver the planes. When the planes were not delivered, the plaintiff filed suit. The evidence at trial showed that the plaintiff never tendered payment or made a demand for the planes. Decide. Vidal v. Transcontinental & Western Air, Inc., 120 F.2d 67 (1941).
4. Crawford purchased fuel equipment from Fram Corporation for use in a building that it was constructing for the U.S. Navy. Fram furnished the fuel equipment and Crawford installed all the units. Crawford refused to pay $6,298.50 of the purchase price of $55,564.20 because it contended that the equipment was defective. Fram contended that Crawford could not revoke its acceptance of the equipment after it was installed and used.

Was Fram correct? United States for the Use of Fram Corp. v. Crawford, 443 F.2d 611 (1971).

5. Square Deal Machine Co. ordered one 36-tooth and one 72-tooth gear from Garrett Corp. in order to modify a lathe. Collins went to pick up the gears from Garrett Corp. They were handed to him in a paper bag. The number of teeth was not counted, and the gears were installed in Square Deal's machine. When the lathe did not function properly, it was discovered that the gears had the wrong number of teeth. Square Deal refused to pay and rejected the gears. It sued Garrett for the financial loss it suffered. Garrett contended Square Deal could not prevail because it had a duty as a matter of law to examine the gear upon delivery to see that it complied with the contract. Was Garrett correct? Square Deal Machine Co. v. Garrett Corp., 275 P.2d 46 (Cal. 1954).
6. Frank Mark purchased all the stock and inventory of a saloon business conducted by Peter Escalle in Reno, Nevada. At the time of the purchase the requirements of the Bulk Sales Act were not complied with. Escalle sued Mark for the balance of the purchase price, and Mark defended on the ground that Escalle had not complied with the Bulk Sales Act and therefore the transfer of inventory was void. Was Mark correct? Escalle v. Mark, 43 Nev. 172, 183 P. 387 (1919).
7. State whether the seller of goods may be excused from performing under each of the following circumstances:
   a. A strike shutting down the seller's production lines.
   b. A strike shutting down the carrier on which the goods were to be shipped by the seller.
   c. Increased cost of production due to increased materials costs.
   d. A shutdown of the seller's plant due to lack of fuel caused by extreme weather conditions.

CHAPTER 19

Rights and Remedies

In Chapter 18, we examined the performance requirements of the parties. Of necessity some of the discussion alluded to the remedies to which the parties are entitled. In this chapter the remedies of the seller and the buyer will be examined further.

The buyer breaches his contract by wrongfully repudiating, rejecting, or revoking acceptance of the goods as well as by failing to make payment when due. The remedies for the seller, or for the buyer when the seller is in breach, are compensatory and intended to place the innocent party in as good a position as if there were no breach. Damages allowed under the code do not include penal or exemplary damages, but consequential and incidental damages are allowed as we shall see later. Furthermore, there is no requirement that the seller, or the buyer, make an election of remedies. He or she may recover any remedies appropriate to the particular case.

19:1. The Seller's Remedies. The seller's remedies include the right to (1) cancel; (2) withhold delivery of goods; (3) stop delivery of goods by a bailee; (4) identify goods to the contract and finish or salvage them; (5) resell and recover money damages; (6) recover damages for nonacceptance or repudiation; (7) recover the price; and (8) reclaim the goods on buyer's insolvency. Some of these remedies are mutually exclusive, and others may be used in combination. The appropriate remedies, of course, depend upon the circumstances of each case. In reading this section, be sure to distinguish between the seller's right to recover money in the form of the price, to recover damages for nonacceptance, to resell and recover money damages, and to recover lost profits. Observe when each remedy may be applied and when it is appropriate. Also note the seller's different rights as to how he may act with respect to the goods. For instance, in a proper case the seller may cancel the contract, without delivery of the goods, resell them, and obtain money damages.

(1) Cancellation. If there is a breach by the buyer, the seller may cancel the contract. The breach might be a wrongful rejection, a revocation of acceptance, a failure to make payment when due, or a repudiation of the contract by the buyer. Section 2–703(f).

Cancellation by the seller occurs when the seller puts an end to the contract as the result of breach by the buyer. Cancellation results in the discharge of all obligations that are still executory on both sides. It is important to note that despite cancellation the seller retains any remedy for breach of the whole contract or any unperformed balance. Section 2–106(4). The effect of cancellation is to put an end to the obligations of the seller for any future performance of the contract while retaining his right to other remedies. Thus, if the buyer breaches the contract, the seller may cancel and then recover damages suffered as a result of the contractual breach.

(2) **Seller's Rights as to Goods: Withholding Delivery.** The seller may withhold delivery of the goods if the buyer wrongfully repudiates, rejects, or revokes acceptance of the goods or fails to make a payment due on or before delivery. If the breach is of the whole contract, all undelivered goods may be withheld. Section 2–703(a). Suppose that the buyer fails to make a payment that is due prior to delivery of the goods. The seller has the right to withhold delivery under this circumstance. If the buyer's failure to pay is the result of financial difficulty, it would not be fair to require the seller to deliver goods to him in which the seller may not even have a security interest.

(3) **Seller's Right as to Goods: Stopping Delivery of Goods by a Bailee.** Occasionally a problem with the buyer's contractual performance may surface after the goods have left the seller's possession but while they are still in the possession of a bailee, such as a warehouseman, under the control of the seller. Under what circumstances may the seller stop the delivery of the goods in the possession of a bailee? First, he may stop delivery when he discovers the buyer to be insolvent. Second, if the buyer is not insolvent and repudiates or fails to make a payment due before delivery or if for any other reason the seller has a right to withhold or reclaim the goods, the seller may stop delivery of a carload, a truckload, a planeload, or a larger shipment of express or freight. Section 2–703(b) and 2–705(1). Note that in the second instance, because stoppage is a burden on carriers, the right to stop shipment for reasons other than insolvency is limited to larger shipments. Section 2–705(1), Comment 1. There is no such limitation when the reason is insolvency. Section 1–201(23) of the code states that a person is insolvent "who either has ceased to pay his debts in the ordinary course of business or cannot pay his debts as they become due or is insolvent within the meaning of the federal bankruptcy law."

(a) *When Right of Stoppage Ceases.* The right of the seller to stop goods continues until (1) receipt of the goods by the buyer or his designated representative; (2) acknowledgement to the buyer by any bailee of the goods, except a carrier, that the bailee (such as a warehouseman) holds the goods for the buyer; (3) such acknowledgement to the buyer by a carrier by reshipment or as warehouseman; or (4) negotiation to the buyer of any negotiable document of title covering the goods. Section 2–705(2). In essence the seller's right to stop the goods continues until the goods reach the place of final delivery by the seller.

(b) *Manner of Stoppage and Duty of Bailee.* When the seller desires to stop delivery he or she must notify the bailee by reasonable diligence in order to prevent delivery of the goods. After notification in a proper manner the bailee must hold and deliver the goods according to the directions of the seller. The danger to the carrier, of course, is the potential liability to the buyer for *wrongful* withholding of the goods. Therefore the bailee has a right of indemnity against the seller in case the withholding is wrongful. Section 2–705(3)(a) and (b).

If a *negotiable* document of title has been issued for goods, the bailee is under no obligation to obey a notification to stop until the document is surrendered. Section 2–705(3)(c). If the carrier has issued a nonnegotiable bill of lading, he or she is not obligated to obey a notification to stop received from a person other than the consignor. Section 2–705(3)(d). The purpose of these provisions is to make clear that the carrier is under no duty to obey a stop order of a person who is a stranger to the carrier's contract. On the other hand, the provisions have no effect upon the seller's right against the buyer. Therefore, if the carrier does obey the stop order despite the fact that it is not required to do so, the buyer has no right to complain.

(4) Seller's Right on Buyer's Insolvency: Withholding Delivery or Reclamation. If the seller discovers the buyer to be insolvent, he may refuse delivery of the goods except for cash, including payment for all goods previously delivered under the contract. As previously mentioned, the seller may withhold delivery on discovery of the buyer's insolvency. Section 2–702(1).

The purpose of allowing the seller to withhold delivery is so that he does not find himself in the unenviable position of delivering goods to the buyer and joining the ranks of unsecured creditors of an insolvent buyer. The code goes one step further. If the buyer purchases goods on credit, there is at least a tacit assumption that he is solvent. If the seller discovers that the buyer has received goods on credit while insolvent, he may reclaim them upon demand within ten days after receipt. The ten-day time limit does not apply if the buyer has misrepresented his solvency to the seller in writing within three months before delivery. Section 2–702(2). The seller's right to reclaim is subject to the rights of a buyer of the goods in the ordinary course or to the rights of other good-faith purchasers. Section 2–702(3). Protection of lien creditors was eliminated by a 1966 amendment to the code.

AMOCO PIPELINE COMPANY v. ADMIRAL CRUDE OIL CORPORATION
490 F.2d 114 (1974)

Certain oil producers agreed to sell oil to Admiral Crude Oil Corporation. This oil was to be transported by Amoco Pipeline Company and stored by it. From November 1971, through January 1972, Admiral did not pay Amoco its charges for gathering, transporting, and storing the crude oil. Amoco then asserted a lien against 49,953.98 barrels, which it had received from the producers and stored in its facilities.

On February 10, 1972, Amoco tendered the crude oil to Admiral, conditioned on the payment of the lien claim. Admiral refused to accept the oil. On February 16, 1972, Admiral filed for bankruptcy in the Federal District Court for the Northern District of Texas. Admiral's parent, Tulsa Crude Oil Company, had previously filed in the district court for Oklahoma. The trustee in bankruptcy claimed the right to the oil in Amoco's possession. Amoco then filed an action of interpleader in the New Mexico District Court for a decision as to who was entitled to the oil. The New Mexico court rendered a decision adverse to the trustee, who appealed, contending that the New Mexico court lacked jurisdiction because of a previous injunction by the Oklahoma court staying all proceedings against Tulsa and Admiral.

SETH, C.J.

Prior to February 10, 1972, the oil producers discovered that Amoco had received oil from them for Admiral when Admiral was insolvent; the checks which had been tendered to the oil producers by Admiral in payment for oil sold to it in December 1971, were dishonored by the drawee bank and returned marked "insufficient funds." The oil producers therefore notified Amoco and other interested parties to stop delivery of crude oil to or for the benefit or account of Admiral, thus exercising a right of stoppage in transitu and reclamation. Amoco notified the producers that it would enforce its lien against the oil in its possession as Admiral had refused to pay the lien claim and accept delivery of the oil.

Amoco thereafter sold 13,000 barrels of crude oil which it had in its possession on March 15, 1972, to satisfy its common carrier lien against Admiral of $39,564.52, and continued to assert a lien against the balance of the proceeds of sale and 36,953.98 barrels in storage for unpaid demurrage and storage charges against Admiral.

The New Mexico court found that at all material times the crude oil, or the proceeds from its sale, was in the sole and exclusive possession of Amoco, the carrier, or in the registry of the court. This finding is supported by the record.

In the case at bar, the New Mexico district court necessarily had to consider the merits of at least one portion of the case before it. It had to decide who had possession, and who had title to the oil in issue, in order to ascertain if it had jurisdiction of the case and whether it should stay its proceedings. Thus as often happens, a determination of jurisdiction necessarily also decides a substantive issue.

As decided by the district court, the sale of the crude oil by the producers was a sale of goods, and was thus governed by Article 2 of the Uniform Commercial Code. If the Uniform Commercial Code is applied, Admiral, when it refused on February 10, 1972, to accept the tender of the crude oil from Amoco conditioned upon payment by Admiral of Amoco's common carrier lien, caused thereby title to the oil to revest, if indeed it ever passed to Admiral, in the oil producing sellers. Section 2–401(4). Similarly, upon the notice given by the sellers to Amoco, prior to February 10, 1972, to stop delivery of the crude oil to Admiral based upon the previous dishonoring by the drawee bank of Admiral's "insufficient funds" checks to the sellers, the sellers thereby timely exercised their rights of stoppage in transitu under sections 2–702 and 2–705. Thus, regardless of whether title ever passed to Admiral or whether it had a special property interest in the oil, section 2–501, the sellers could reclaim the oil upon demand and notice to Amoco as given herein. Sections 2–702(1), (2) and 2–705(1). This was effective for several reasons. The trial court found, and it is borne out by the record that the tender of the "insufficient funds" checks constituted a written misrepresentation of solvency. From the record it is clear, as the trial court found, that the oil at all material times was in possession of a third party, the carrier. Thus Admiral never had possession, constructive or otherwise, of the oil and under the above-cited provisions of New Mexico's Uniform Commercial Code even if it could be said to have had constructive possession, it did not so have it after February 10, 1972, when Admiral refused tender of the oil and the sellers exercised their rights of stoppage of delivery in transitu and reclaim. The same considerations govern the question of Admiral's "property" or "title" in the oil. The bankruptcy petition was filed, at the earliest, on February 16, 1972, so that upon the date of billing, the title to the oil was not in Admiral. It was not the "property" of the debtor. The Oklahoma district court's stay order thus could not reach the oil, not in the debtor's possession nor its property. 11 U.S.C. § 511. Likewise the New Mexico district court's exercise of jurisdiction and refusal to stay its proceedings was correct.

(Accordingly, the judgment of the district court is in all respects affirmed.)

Questions for Discussion

1. *Why were the oil producers entitled to order the carrier, Amoco, to stop delivery of the oil to Admiral?*

2. *Did Admiral misrepresent its solvency to the sellers? If so, how was this insolvency misrepresented? Upon a misrepresentation, how long does the seller have to act?*
3. *Why did the appellate court conclude that the action in the New Mexico district court was not subject to the injunction against further proceedings issued by the bankruptcy court?*

(5) Seller's Rights as to Goods: Identifying Goods to the Contract and Finishing or Salvaging. If the buyer breaches a contract for goods already identified and finished, we shall see that the seller's primary remedy is to dispose of the goods elsewhere and seek damages from the buyer for the difference between the contract price and the resale price. However, if the goods are not identified or finished at the time of breach, the law prior to the code was that the goods could not be finished because of the seller's duty to mitigate damages. The code has altered this approach in order to eliminate certain wasteful results. Now, in the event of breach by the buyer, the seller may identify conforming goods not already identified that are in his or her possession or control. He may treat these goods as the subject of resale if they have demonstrably been intended for the particular contract, even if the goods are unfinished. Section 2–704(1). As to goods that are unfinished, the code allows the seller to exercise reasonable commercial judgment either to complete the manufacture of the goods or to resell the goods without completion for scrap or salvage value. Alternatively the seller may proceed in any other reasonable manner in order to avoid loss. Section 2–704(2). The seller is given this option unless it is clear that the completion of the goods would result in a material increase in damages. The buyer has the burden of establishing that the completion would result in such an increase.

(6) Seller's Right to Money Damages: Resale and Recovery of Damages. Under the conditions to which we have previously alluded in Section 2–703, the seller may resell the goods and recover the difference between the *resale price* and the *contract price* from the buyer. Section 2–706(1). In addition, any incidental damages may be recovered from the buyer, such as expenses incurred in stopping delivery or in the transportation, care, and custody of the goods after the buyer's breach. Section 2–710. The buyer must be given credit for any expenses saved by the seller as the result of the buyer's breach. Most importantly, in order for the seller to recover the difference between the contract price and the *resale* price, the resale must be made in a commercially reasonable manner. The seller cannot, for example, resell the goods at a grossly inadequate price, knowing that he has a right of recovery against the buyer. He must resell at a commercially reasonable price. Thus suppose Sam Seller contracts to sell goods to Bill Buyer for $500. Buyer wrongfully repudiates the contract while the goods are still in Seller's control but are in the possession of a carrier on route to Buyer's place of business for delivery. Seller orders the delivery stopped and finds another purchaser of the goods for $450, a commercially reasonable figure. The costs for stoppage and delivery to the new purchaser are $25. Seller would be entitled to the difference between the contract price, $500, and the resale price of $450, or $50. In addition,

Seller is entitled to the $25 incidental costs that were incurred as the result of Buyer's breach.

The resale may be either a private sale or a public sale. It may be the sale of a unit or a sale in parcels and may be made at any time, place, and terms as long as every aspect of the sale is commercially reasonable. Section 2–706(2). If the resale is private, the seller must give the buyer reasonable notification of his intention to resell. Section 2–706(3). If the sale is public, only identified goods can be sold except when there is a recognized market for public sale of futures in goods of the kind. The sale must be at a usual place or market for public sale if one is reasonably available, and notices must be given to the buyer of the time and place, except if the goods are perishable or threaten to speedily decline in value, so that the buyer may have an opportunity to bid or secure other bidders at the resale. Section 2–706(4).

Even if the seller fails to comply with all the requirements stated, the good-faith purchaser at resale takes the goods free from any rights of the original buyer. Section 2–706(5). The seller is not accountable to the buyer for any profit made on resale. Section 2–706(6).

ALCO STANDARD CORPORATION v. F & B MANUFACTURING CO.

51 Ill.2d 186, 281 N.E.2d 652 (1972)

SCHAEFER, J.

In 1968 the defendant, F & B Manufacturing Co., entered into a contract to buy an industrial heat treating furnace from Ipsen Industries, a division of the plaintiff, Alco Standard Corporation, for a price of $66,595. The contract described the dimensions and other specifications of the furnace. The furnace was delivered to F & B's plant in Phoenix, Arizona, on or about July 6 of the same year. F & B rejected it on the ground that it was not large enough to meet its needs. By agreement of the parties the defendant paid the cost of shipping the furnace to a warehouse in California, where the plaintiff resold it to another manufacturer for $57,500.

Thereafter the plaintiff brought this action to recover the difference between the contract price and the price for which the furnace was resold. A jury was waived, and after a trial judgment was entered for the plaintiff in the sum of $9,905. Interest at 5 percent in the sum of $505, was also allowed, because payment had been withheld "by unreasonable and vexatious delay." The Appellate Court Second District, affirmed the judgment for $9,905, and reversed the judgment insofar as it provided for the payment of interest. We allowed leave to appeal.

In this court the buyer contends that in reselling the rejected furnace, the seller was required to comply with section 2–706(3) of the Uniform Commercial Code, but failed to do so. That provision states: "Where the resale is at private sale the seller must give the buyer reasonable notification of his intention to resell." Ill.Rev.Stat.1967, Ch. 26, Par. 2–706(3).

The buyer asserts that before a seller is entitled to recover damages under that section of the Uniform Commercial Code he must prove his compliance with it, and that the appellate court erroneously held that section 2–706(3) establishes an affirmative defense, which must be pleaded and proved by the buyer.

The opinion of the appellate court states: "The defendant was not notified of the proposed resale of the furnace as required by section 2–706(3) of the Commercial Code (citation), and now complains that this failure is a complete defense to the action for the deficiency. Once again, the failure to notify defendant of the resale was not pleaded affirmatively." These statements strongly suggest a misinterpretation of the statute, which does not require that the defaulting buyer be notified of "the proposed resale," but instead requires only that the seller "give the buyer reasonable notification of his intention to resell." It is not necessary however to determine the proper allocation of the burden of pleading and proof under the code for the reason that the testimony is undisputed that the buyer's plant manager in Phoenix was told by the seller's west coast manager that there was a potential buyer for the machine in California and the buyer then paid for shipping the machine there. If we assume the buyer's contention to be correct this is sufficient proof of notice under the provisions of the code.

(Affirmed in part and reversed as to that part denying plaintiff interest.)

QUESTIONS FOR DISCUSSION

1. *What notice was given to the defendant of the resale?*
2. *What information was the seller required to give the buyer concerning the resale in this case?*

MOTT EQUITY ELEVATOR v. SVIHOVEC
236 N.W.2d 900 (N.D. 1975)

Rudy Svihovec, a wheat farmer, contracted to sell 4,000 bushels of spring wheat at $1.82 to $1.92 per bushel to Mott Equity Elevator. The exact price was to be based upon the protein content of the wheat. Mott Equity Elevator refused to take delivery of the wheat within the time allotted under the contract. Svihovec then took his wheat elsewhere and sold it for the price of $2.20 per bushel without notifying Mott of the resale. Later, the price of wheat increased to $4.00 per bushel. Mott demanded delivery from Svihovec who refused, stating that Mott had breached the contract. Mott sued Svihovec contending that Svihovec was required to notify it under the provisions of the Uniform Commercial Code controlling the right of a seller to resell the goods. The trial court decided in favor of Svihovec and Mott appealed.

VOGEL, J.

The elevator strenuously argues that Svihovec was not entitled to resell his grain under . . . (UCC § 2–706), without giving reasonable notice of his intent to resell. The argument also is made that . . . (UCC § 2–309), imposes a duty on Svihovec to give reasonable notice to the other party that he was terminating the contract.

We find these arguments to be without merit. Before discussing these questions, it may be helpful to reiterate the remedies available to the seller following breach by the buyer. Under section . . . (UCC § 2–703), . . . the seller is entitled to, among other remedies, withhold delivery, resell and recover damages, recover damages for nonacceptance, or cancel. Svihovec pursued the remedy of cancellation, as was his right. He thereafter resold his grain to another buyer, as was his right. The parties have confused Svihovec's right to dispose of his grain as he wished under a canceled contract with the code remedy allowing a seller to "resell and recovery damages" under section . . . (UCC § 2–703), and section . . . (UCC § 2–706).

The seller's right to resell and recover damages is, of course, available to a seller in addition to his right to cancel; subsection 1, b of Section . . . (UCC § 2–719), creates a presumption that clauses prescribing remedies are cumulative rather than exclusive. Official Code Comment, UCC § 2–719.

The only condition precedent to the seller's right to resell is a breach by the buyer within Section . . . (UCC § 2–703). The trial judge found that Svihovec had a right to pursue this remedy when he sold his grain directly to the Grain Terminal Association. We would agree that Svihovec did have such a right if it were necessary to apply this section to the seller's conduct in reselling his grain in this case. But the section does not apply. In a falling market Svihovec would probably have desired to resell and recover damages. To recover damages under this section he would be required to act in good faith, sell in a commercially reasonable manner, and give reasonable notice to the buyer of his intention to resell (if the sale was at private sale). Failure to act properly under this section merely deprives the seller of the measure of damages provided in subsection 1.

Official Code Comment, UCC § 2–706. In any event, the seller is not accountable to the buyer for any profit made on any resale under section 41–02–85, N.D.C.C. (UCC § 2–706), where, as here, the resale occurred in a rising market and the contract had been canceled. In this case, involving a rising market, Svihovec suffered no damages and thus did not need to resort to this code remedy.

QUESTIONS FOR DISCUSSION

1. *When Svihovec resold the grain after Mott Equity Elevator breached the contract, why did he not have to give the notice required by 2–706?*
2. *For what practical reason did Svihovec simply choose the remedy of cancellation? Do you think his actions would have been different in a wheat market where the price was falling?*

(7) Seller's Right to Money Damages: Damages for Nonacceptance or Repudiation. In Section (5) the damages were computed as the result of a *resale* of the goods. The seller is not required to resell the goods. If the buyer does not accept conforming goods or if he repudiates, the measure of damages may be the difference between the *market price* and the unpaid *contract price,* together with any incidental damages but less expenses saved as a consequence of the buyer's breach. The *market* price is the market price at the time and place for tender of the goods. Section 2–708(1).

If this remedy is inadequate to put the seller in as good a position as performance would have done, then the measure of damages is somewhat different. In this case, the measure is the lost *profit* that the seller would have made from full performance by the buyer, together with any incidental damages, allowance for costs reasonably incurred, and credit for payments or proceeds of resale. The seller is to include reasonable overhead in calculating the lost profit. Section 2–708(2).

Note that the seller in some cases may have the opportunity of selecting among three methods of calculating damages. He may resell the goods and take the difference between that amount and the contract price. If this is not a satisfactory remedy, he may recover the difference between the contract price and the market price as outlined above. If this is not an adequate remedy, the seller may recover lost profits. Sometimes none of these remedies is appropriate, in which case the seller may recover the price of the goods.

DETROIT POWER SCREWDRIVER COMPANY v. LADNEY
25 Mich.App. 478, 181 N.W.2d 828 (1970)

Michael Ladney, Jr., contracted to purchase a machine described as a double-spindle stud driver from Detroit Power Screwdriver Company. The original specifications were changed, and Ladney indicated that he needed the machine no later than March 31. There was no evidence indicating that Detroit made a binding acceptance of the last proposed date. On April 11, Ladney notified Detroit to stop work on the machine on the ground that it had not been delivered by March 31. On May 11, Detroit submitted a bill for $12,017 for work completed through April 11.

Detroit then sued Ladney for breach of contract. At trial it submitted evidence of its price quotation, Ladney's purchase order, and the invoice, as well as evidence that the scrap value of the machine was $1,500. Evidence was also adduced as to the nature of the machine and the fact that this machine differed from the standard stud driver usually built by Detroit. The lower court found that Ladney had breached *the contract but held that Detroit had failed to prove damages sufficiently certain to permit recovery. Detroit appealed.*

LESINSKI, C.J.

Of the sections providing remedies for breach by the buyer, plaintiff places principal reliance on M.C.L.A. § 440.2709(1) (U.C.C. 2–709(1)). The authority uncovered in our research, however, indicates that since the machine was not completed, § 2709(1) is inapplicable.

Since plaintiff is not entitled to the price under § 2709(1), its right to recovery is controlled by § 2708.

The measure of damages under § 2708(1) is:

> (T)he measure of damages for nonacceptance or repudiation by the buyer is the difference between the market price at the time and place for tender and the unpaid contract price together with any incidental damages provided in this article (section 2710), but less expenses saved in consequence of the buyer's breach.

However, § 2708(2) provides:

> If the measure of damages provided in subsection (1) is inadequate to put the seller in as good a position as performance would have done then the measure of damages is the profit (including reasonable overhead) which the seller would have made from full performance by the buyer, together with any incidental damages provided in this article (section 2710), due allowance for costs reasonably incurred and due credit for payments or proceeds of resale.

One of the few reported cases relevant to the question of which paragraph in § 2708 is applicable is *Anchorage Centennial Development Co.* v. *Van Wormer & Rodrigues, Inc.* (Alaska, 1968), 443 P.2d 596. In *Anchorage Centennial,* 50,000 gold-colored metal coins had been ordered by defendant in conjunction with a state-wide celebration of the 100th anniversary of the purchase of Alaska by the United States. The order was cancelled after the seller had already manufactured 29,000 of the coins. At p. 599 the Court stated:

> In the case at bar, we are in accord with Van Wormer's argument to the effect that since there is no market for these made-to-order coins the proper measure of damages is governed by A.S. 45.05.208(b).

The cited section is the equivalent to our § 2708(2).

We are in accord with the conclusion in *Anchorage Centennial.* A formula basing damages on the difference between market price and contract price is without meaning in the context of a contract for a specialty item which has no market.

Thus, the question of whether or not the machine in the instant case is a specialty item within the meaning of the code must be determined in order to know which measure of damages applies. The question, however, also has significance in the instant case beyond the correct measure of damages.

M.C.L.A. § 440.2610 (Stat.Ann. 1964 Rev. § 19.2610), provides in part:

> When either party repudiates the contract with respect to a performance not yet due the loss of which will substantially impair the value of the contract to the other, the aggrieved party may
>
> . . .
>
> (b) resort to any remedy for breach (section 2703 or section 2711), even though he has notified the repudiating party that he would await the latter's performance and has urged retraction; and
>
> (c) in either case suspend his own performance or proceed in accordance with the provisions of this article on the seller's right to identify goods to the contract notwithstanding breach or to salvage unfinished goods, section (2704).

The right to suspend performance in § 2610, however, must be read in light of M.C.L.A. § 440.2704(2) which states:

> Where the goods are unfinished an aggrieved seller may in the exercise of reasonable commercial judgment for the purposes of avoiding loss and of effective realization either complete the manufacture and wholly identify the goods to the contract or cease

manufacture and resell for scrap or salvage value or proceed in any other reasonable manner.

Whether plaintiff's decision not to complete the machine in the instant case was the result of the "exercise of reasonable commercial judgment" thus also depends, at least in part, on whether a market exists for the finished product.

Our review of § 2708(2) satisfies us that the proofs offered meet the requirements of the statutory formula. Plaintiff's invoice was clearly intended to cover expected profits and costs incurred during the work done. That amount would then be reduced by the salvage value of $1,500.

Yet, § 2708(2) applies only "if the measure of damages provided in subsection (1) is inadequate to put the seller in as good a position as performance would have done." As noted above, however, § 2708(1) does not apply where the contract is found to involve a specialty item without a reasonably accessible market. Such a finding is crucial since it determines whether § 2708(1) or § 2708(2) controls the measure of damages. The determination of which paragraph of § 2708 controls is, in turn, important since plaintiff has proven its loss under paragraph (2) but has not proven the amount of its loss under paragraph (1).

The ruling of the court below did not include a factual determination of whether the machine here involved is a specialty item without a reasonably accessible market. It is the function of the trial court sitting without a jury to make findings of fact. CCR 1963, 517.1. Although the record on appeal may contain sufficient details to make a finding possible, this Court is not the proper forum to make such initial determinations.

We therefore remand the case for further findings. If the trial court finds that the machine is a specialty item without a reasonably accessible market, then it shall award damages pursuant to § 2708(2). If it does not so hold, the trial court shall dismiss the action due to plaintiff's failure of proof under § 2708(1).

QUESTIONS FOR DISCUSSION

1. *The court concluded that Detroit was not entitled to recover the price under Section 2–708(1). Why?*
2. *In the event that the trial court failed subsequently to find that the machine was a specialty item, why would Detroit have failed to prove damages under Section 2–708(1)—the difference between the market price and the contract price?*

---

(8) Seller's Right to Money Damages: Price. There are certain circumstances in which the seller is to recover the price of the goods from the buyer regardless of any considerations of market price. When the buyer fails to pay the price as it comes due, the seller may recover it plus any incidental damages when the goods are (1) goods accepted by the buyer; (2) conforming goods lost or damaged after risk of loss has passed to the buyer; or (3) goods identified to the contract that the seller cannot, after a reasonable effort, resell at a reasonable price or that circumstances indicate could not be sold by a reasonable effort. Section 2–709(1). Essentially this remedy provides for a "specific performance" of the contract insofar as the buyer is required to pay for the goods at the agreed price. It applies to those cases in which resale of the goods is impractical, except when the buyer has accepted the goods or when they have been destroyed after risk of loss has passed to the buyer. Section 2–709, Comment 2. If the seller sues for the price and recovers, the rational decision is for the buyer to accept the goods. Therefore the seller is required to hold for the buyer any goods that have been identified to the contract and are still in his control. If resale becomes possible, the seller may resell the goods prior to judgment, but the net proceeds must be credited to the buyer. Payment of the judgment by the buyer entitles him to the goods. Section 2–709(2).

Suppose Sam Seller agrees to sell certain goods to Bill Buyer for $500. The goods are shipped to Buyer and are lost during shipment after the risk of loss passes to Buyer. Seller would be entitled to the price of $500. If the goods arrive and are accepted by Buyer, Seller is thus entitled to the price. Assume that the goods consist of silver objects with Buyer's name engraved on them. Prior to shipment Buyer repudiates the contract. Seller is entitled to the price from Buyer.

CHICAGO ROLLER SKATE MANUFACTURING CO. v. SOKOL MANUFACTURING CO.

185 Neb. 515, 177 N.W.2d 25 (1970)

NEWTON, J.

Defendant, Sokol, purchased of plaintiff truck and wheel assemblies with plates and hangers for use in the manufacture of skate boards. The skate board fad terminated and several weeks later, defendant returned, without plaintiff's consent, a quantity of the merchandise purchased. There was due plaintiff the sum of $12,860. The merchandise was not suitable for other uses and could not be resold. It was held by plaintiff for 7 months. Plaintiff offered a credit of 70 cents per unit which defendant neither accepted nor rejected. Plaintiff then disassembled, cleaned, and rebuilt the units to make them suitable for use on roller skates. The undisputed evidence shows the rebuilt units had a reasonable value of 67 cents and 69 cents. In the salvage operation plaintiff incurred an expense of $3,540.76. Profits lost amounted to an additional $2,572. Plaintiff, disregarding its expense, credited defendant with 70 cents per unit and brought suit for the balance due of $4,285 for which sum it recovered judgment in the trial court. . . .

Section 2–718(4), U.C.C., provides:

> Where a seller has received payment in goods their reasonable value or the proceeds of their resale shall be treated as payments. . . .

In accordance with section 2–709, U.C.C., plaintiff was entitled to hold the merchandise for defendant and recover the full contract price of $12,860. Plaintiff did not elect to enforce this right, but recognizing that there was no market for the goods or resale value and that they were consequently worthless for the purpose for which they were designed, it attempted to mitigate defendant's damages by converting the goods to other uses and credited defendant with the reasonable value of the goods as converted or rebuilt for use in roller skates. In so doing, plaintiff was evidencing good faith and conforming to the general rule requiring one damaged by another's breach of contract to reduce or mitigate damages.

The Uniform Commercial Code contemplates that it shall be supplemented by existing principles of law and equity. It further contemplates that the remedies provided shall be liberally administered to the end that an aggrieved party shall be put in as good a position as it would have been in if the contract had been performed. Here the buyer was demanding of the seller credit for the full contract price for goods that had become worthless. The seller was the aggrieved party and a return of worthless goods did not place it in as good a position as it would have been in had the contract been performed by the buyer paying the contract price. On the other hand, the crediting to defendant of the reasonable value of the rebuilt materials and recovery of the balance of the contract price did reasonably reimburse plaintiff. This procedure appears to be contemplated by section 2–718(4), U.C.C., which requires that a seller paid in goods credit the buyer with the reasonable value of the goods.

It is the defendant's theory that since the

goods were not resold or held for the buyer, the seller cannot maintain an action for the price. We agree with this proposition. We also agree with defendant in its contention that the controlling measure of damages is that set out in section 2–708(2), U.C.C. This section provides that the measure of damages is the profit which the seller would have made from full performance by the buyer, together with any incidental damages resulting from the breach and costs reasonably incurred.

Defendant overlooks the provision for allowance of incidental damages and cost incurred. The loss of profits, together with the additional costs or damage sustained by plaintiff amount to $6,112.76, a sum considerably in excess of that sought and recovered by plaintiff. Although the case was tried by plaintiff and determined on an erroneous theory of damages, the error is without prejudice to defendant. There being no cross-appeal, the judgment of the district court is affirmed.

(Affirmed.)

QUESTIONS FOR DISCUSSION

1. *What theory of damages did the plaintiff use to recover in the trial court? Why was this theory inappropriate in this case? On what grounds did the defendant contend that it owed nothing?*
2. *On what theory did the appellate court allow the plaintiff to recover?*

19:2. BUYER'S RIGHTS AND REMEDIES WHEN BUYER HAS NOT ACCEPTED GOODS OR WHEN BUYER HAS JUSTIFIABLY REVOKED ACCEPTANCE. Section 2–711 of the code summarizes the remedies of the buyer if he either has not accepted the goods or has justifiably revoked acceptance. Under these circumstances he may (1) cancel; (2) recover any price paid; (3) cover; (4) recover damages for nondelivery or repudiation; (5) obtain specific performance; (6) replevy the goods; (7) recover goods on seller's insolvency; (8) resell the goods; and (9) obtain consequential and incidental damages. You should note that the buyer is not entitled to all of these remedies in each case. However, just as with the seller, he does not have to elect one remedy to the exclusion of all others. For instance, in a particular case the buyer may cancel, cover, and recover any part of the price paid. More than one remedy is often appropriate in a particular case. Let us examine them individually.

(1) **Buyer's Right to Cancel.** The buyer has the right to cancel if the seller fails to make delivery or repudiates or if the buyer rightfully rejects or justifiably revokes acceptance. If the breach goes to the whole contract, the buyer may cancel as to the whole; otherwise the buyer may cancel with respect to the goods involved. section 2–711(1). The code specifically provides that this right is in addition to other remedies. The effect of cancellation is that all executory obligations are terminated and the canceling party retains his right to recover damages for nondelivery or may avail himself of other remedies as discussed in Section (4).

(2) **Buyer's Recovery of Price Paid.** If the buyer has paid the price or part of it to the seller, he may of course recover it. Section 2–711(1). Frequently the buyer may have paid for the goods prior to delivery or prior to inspection. If the seller then fails to deliver or if the buyer rightfully rejects the goods after inspection or justifiably revokes acceptance, the price paid may be recovered. This remedy is in addition to the right to recover damages for nondelivery or the remedy of "cover" discussed next.

(3) Buyer's Remedy of "Cover." If the seller fails to make delivery or repudiates or if the buyer rightfully rejects or justifiably revokes acceptance, then the buyer may "cover." Section 2–711(1)(a). A buyer accomplishes cover by making, in good faith, any reasonable purchase of or contract to purchase goods in substitution for those due from the seller. The cover must be accomplished without unreasonable delay. Section 2–712(1). The damages recoverable by the buyer consist of the difference between the cost of cover and the contract price, plus any incidental or consequential damages. The seller is to be given credit for expenses saved as a consequence of the breach. Section 2–712(2).

Suppose Bill Buyer contracts to purchase goods from Sam Seller for $500. Seller fails to deliver the goods. Buyer purchases similar substitute goods from another supplier for $550. Buyer is entitled to recover $50 from Seller plus incidental and consequential damages which are discussed in Section (9).

The remedy of cover essentially results in the buyer's getting the difference between the *contract price* and the *costs of cover.* It is analogous to the seller's remedy of the difference between the contract price and the resale price. It is not mandatory for the buyer to cover. There are times when cover is not desirable from the buyer's point of view, and of course if the goods are "unique," cover is not possible and failure to cover does not bar him from any other remedy. Section 2–712(3). The buyer may elect instead to seek damages for nondelivery of goods, as described in Section (4). When the buyer does elect to cover, he may purchase commercially reasonable substitute goods rather than identical goods. The test is whether he has acted in good faith and in a reasonable manner, not whether hindsight indicates that another method of cover may have been cheaper.

THORSTENSON v. MOBRIDGE IRON WORKS

208 N.W.2d 715 (S.D. 1973)

DOYLE, J.

Plaintiff Adolph Thorstenson brought an action for damages for breach of contract against defendant Mobridge Iron Works Company. The trial court directed a verdict in favor of defendant and plaintiff appeals.

On December 1, 1967, plaintiff and defendant entered into a written agreement whereby defendant agreed to sell to the plaintiff a Case 730 farm tractor and a mounted F-11 Farmhand loader with certain attachments. The contract provided for a trade-in of a used tractor and loader owned by the plaintiff with an agreed cash difference of $3900 to be paid when the seller delivered the equipment to the buyer's farm. No delivery date was specified in the contract. In the fall of 1968, defendant notified the plaintiff that there would be no delivery of the equipment as specified in the contract. It is undisputed that the tractor, loader and attachments were not delivered to the plaintiff in accordance with the contract. However, there is a considerable dispute between the plaintiff and defendant as to why the equipment was not delivered. The defendant contends that the F-11 Farmhand loader could not be mounted on the 730 Case tractor, that plaintiff refused to accept the tractor and loader unit if mounting required a remodeling or working over of the tractor, and that plaintiff would not accept any other replacement. On the other hand, the plaintiff contends that the loader could be mounted on the 730 Case tractor without remodeling the tractor or loader to the extent that defendant claimed was necessary. He further contends he offered to mount the loader himself and that he later purchased a Case 730 tractor and an F-11 Farmhand mounted as he desired. The plaintiff purchased this equipment in December 1968, from a Case dealer in Aberdeen, South Dakota, at a price increase of $1000 which the plaintiff claims is a "cover" purchase as provided in SDCL 57–8–31. Uniform Commercial Code (U.L.A. § 2–712). Plaintiff testified the cover purchase was similar equipment while defendant claims it was an "entirely different tractor" from the one specified in their contract. In our view, these disputed questions of fact should have been submitted to a jury.

The trial court limited its directed verdict in favor of the defendant to the issue of damages and found that the plaintiff failed to introduce evidence of any damages sustained.

When the seller fails to make delivery or repudiates, or the buyer rightfully rejects or justifiably revokes a contract, the buyer has certain remedies available by statute. Uniform Commercial Code (U.L.A.) § 2–711 through § 2–725. In SDCL 57–8–28, it is provided that the buyer may:

> (1) "Cover" and have damages under §§ 57–8–31 to 57–8–33, inclusive. . . .

SDCL 57–8–31 provides that:

> . . . the buyer may "cover" by making in good faith and without unreasonable delay any reasonable purchase of or contract to purchase goods in substitution for those due from the seller.

SDCL 57–8–32 provides:

> The buyer may recover from the seller as damages the difference between the cost of cover and the contract price together with any incidental or consequential damages as hereinafter defined . . . but less expenses saved in consequence of the seller's breach.

It is stated in Uniform Commercial Code (U.L.A.) § 2–712,460):

> This section provides the buyer with a remedy aimed at enabling him to obtain the goods he needs thus meeting his essential need. This remedy is the buyer's equivalent of the seller's right to resell.
>
> The definition of "cover" . . . envisages . . . a single contract or sale; goods not identical with those involved but commercially usable as reasonable substitutes under the circumstances of the particular case. . . . The test of proper cover is whether at the time and place the buyer acted in good faith and in a reasonable manner, and it is immaterial that hindsight may later prove that the method of cover used was not the cheapest or most effective. . . .
>
> This section does not limit cover to merchants, in the first instance. It is the vital and important remedy for the consumer buyer as well. Both are free to use cover; the . . . non-merchant consumer is required only to act in normal good faith. . . .

(Reversed and remanded for trial by jury.)

QUESTIONS FOR DISCUSSION

1. *In order to "cover" is it necessary that the buyer purchase equipment identical to that specified in the contract?*
2. *What other options might the buyer have pursued in his dispute with the seller in this case?*

---

(4) Buyer's Recovery of Damages for Nondelivery or Repudiation. As noted above, the buyer is under no obligation to cover, that is, purchase substitute goods. He may simply seek damages for the nondelivery or repudiation. The measure of damages is the difference between the *contract price* and the *market price* at the time when the buyer learns of the breach, plus any incidental and consequential damages, less expenses saved in consequence of the seller's breach. Section 2–713(1). The code states that the market price is determined as of the *place* for tender. In cases of rejection after arrival or revocation of acceptance the market price is determined as of the place of arrival. Section 2–713(2).

Suppose Bill Buyer contracts to purchase goods from Sam Seller for $500. Seller fails to deliver. Buyer elects not to cover. The market price at the time Buyer learns of Seller's nondelivery is $600. Buyer is entitled to $100, plus any incidental and consequential damages, less any expenses saved as a result of Seller's breach.

If the current market price is difficult to prove, a comparable one may be used. If the difficulty stems from scarcity of goods, a case could be made for the remedy of specific performance, which is discussed in Section 5. Note that the remedy discussed here is similar to the seller's remedy to recover the difference between the market price and the contract price.

CARGILL, INC. v. FICKBOHM
252 N.W.2d 739 (1977)

John Fickbohm agreed to sell Cargill's elevator 10,000 bushels of corn at $1.26 per bushel for delivery in June or July 1973. The exact date was to be determined by Fickbohm. Cargill hedged this purchase by selling a futures option of 10,000 bushels at $1.39 per bushel.

Fickbohm failed to deliver the corn as required by the contract. Cargill then filed suit against Fickbohm. At the trial, Cargill's manager testified that the market price of corn in that area on July 31, 1973 was $2.49 per bushel for corn for sale by the elevator and $2.45 for corn for purchase by the elevator.

The trial court ruled that Cargill had not introduced sufficient evidence to prove its damages and Cargill appealed.

UHLENHOPP, J.

A buyer's usual remedy for nondelivery is set forth in § 554.2711(1) of the Uniform Commercial Code, which provides in pertinent part:

> Where the seller fails to make delivery . . . the buyer may cancel and . . . may . . . recover damages for nondelivery as provided in this Article. . . .

Under § 554.2713(1), a buyer's usual measure of damages for nondelivery

> is the difference between the market price at the time when the buyer learned of the breach and the contract price together with any incidental and consequential damages provided in this Article (section 554.2715), but less expenses saved in consequence of the seller's breach.

The measure of damages which would normally apply here, then, is the difference between the per-bushel contract price and the market price of corn for purchase by the elevator at the time it learned defendant was not going to deliver in accordance with the agreement—which the fact finder could find was July 31, 1973. The Commercial Code draftsmen state in Comment 1 to § 554.2713, I.C.A.: "The general baseline adopted in this section uses as a yardstick the market in which the buyer would have obtained cover had he sought that relief. . . ."

Cargill proffered substantial admissible evidence that the per bushel market price of corn for purchase was $2.45 on July 31, 1973.

Section 554.2713(1) requires deduction of expenses saved, but none appears here.

Cargill could recover its normal damages without covering, since § 554.2711(1) gives that option (use of disjunctive "or"). See also § 554.2712(3).

Cargill's manager did not testify to mere offers, . . . He presented substantial testimony as to market price.

Cargill could recover damages although it hedged its purchase by selling 10,000 bushels on the futures exchange. That transaction is of no concern to defendant. Anyway, Cargill would have to buy itself out of its short position and do so at the advanced price of corn.

The trial court erred in refusing Cargill's proffered evidence on the market price of corn for purchase and in sustaining defendant's motion for directed verdict. Cargill is entitled to another trial.

(Reversed.)

QUESTIONS FOR DISCUSSION

1. *In this case Cargill "hedged" the purchase for Fickbohm. What does this term mean and what was the purpose of this action on the part of Cargill? Did it have any affect on the calculation of damages in this case?*
2. *What other remedies were available to Cargill?*

(5) Circumstances in Which Buyer Has Right to Specific Performance. You will recall that under the common law, when the goods are unique the court will decree specific performance of the contract because money damages would not compensate the buyer. When the seller fails to deliver or to repudiate, the code maintains this basic approach, although liberalizing it. One can still obtain specific performance only under certain conditions. Specifically the code provides that specific performance may be decreed if the goods are unique or in "other proper circumstances." Section 2–716(1). It is through this latter phrase, "other proper circumstances," that the liberalization is attempted. Historically, if the goods were heirlooms or priceless works of art, the courts decreed specific performance. Today, although specific performance would be decreed under these circumstances, the most typical commercial contracts involving specific performance are requirements or output contracts. An inability to cover is strong evidence of "other proper circumstances." Section 2–716, Comment 2. Suppose Sam Seller agrees to sell and Bill Buyer agrees to purchase all of Buyer's requirements of a component part of Buyer's product at a set price for a certain year. Seller is the only available source for this product. Upon a default by Seller, the court would decree specific performance of the contract. The code provides that the decree may also include such items and conditions as to payment of the price, damages, or other relief as the court may deem just. Section 2–716(2).

(6) Circumstances in Which Buyer Has Right to Replevin. The buyer has a right to replevy goods from the seller upon seller's failure to deliver or repudiation (1) when the goods are identified to the contract *and* (2) if after reasonable effort the buyer is unable to effect cover *or* circumstances reasonably indicate that such effort would be unavailing. Section 2–716(3). The distinction between this remedy and specific performance is the absence of the requirement of uniqueness of goods and the fact that before this remedy is available, the goods must have been identified to the contract. The remedy is comparable to the seller's right to the price if he is unable to sell the goods.

Suppose Sam Seller contracts to sell certain goods to Bill Buyer for $500 and sets aside the goods in his warehouse for shipment to Buyer. Seller then repudiates the contract and Buyer unsuccessfully attempts a purchase the goods elsewhere. Buyer has a right to replevy the goods from Seller.

(7) Buyer's Right to Recover Goods on Seller's Insolvency. Sometimes the buyer will pay for the goods prior to their delivery. He then discovers to his dismay that the seller is insolvent and that the goods have not been shipped.

The code takes the position that if the seller becomes insolvent within ten days after receipt of the first installment on their price, the buyer may recover the goods from the seller (1) if he has a special property interest in those goods; (2) if he has paid all or part of the price for the goods; and (3) provided that he tenders any unpaid portion of their price. Section 2–502(1). The buyer may obtain a special property interest in goods by their identification to the contract. Section 2–501. If the goods are identified to the contract by the buyer, he may recover the goods only if they conform to the contract. Section 2–502(2). The purpose of this latter requirement is to prevent the buyer from unjustly enriching himself by identifying goods that are greatly superior to those called for by the contract for sale.

Suppose Bill Buyer contracts to purchase a refrigerator from Sam Seller's Appliance Store. A pink one is selected and set aside in Seller's warehouse, and Buyer pays a deposit of 25 percent of the price, balance due on delivery. Six days later Seller, as a result of being unable to pay his bills, closes his doors. Buyer may recover the refrigerator by tendering the 75 percent of the price. Note that the buyer's remedy puts him in the position of recovering the identified goods rather than participating in a division of the seller's assets as an unsecured creditor. On the other hand, this advantage to the buyer is limited to ten days. If the buyer has doubts about the seller's solvency and desires greater protection for insolvency that may take place after the ten-day period, he must get a security interest in the goods under Article 9 of the code as discussed in Chapter 20.

(8) **Buyer's Right to Resell the Goods: Security Interest.** If the buyer rightfully rejects the goods or justifiably revokes acceptance of them, he has a security interest in the goods in his possession or control for (1) any payments made on their price and (2) any expenses reasonably incurred in their inspection, receipt, transportation, care, and custody. The buyer may hold such goods and resell them in order to recover the above costs in the same manner as an aggrieved seller. Section 2–711(3). The buyer must, however, account to the seller for any amount over his security interest. Section 2–706(6).

(9) **Consequential and Incidental Damages.** The code provides for the recovery by the buyer of incidental and consequential damages in certain cases. Incidental damages are intended to provide reimbursement for the buyer if he incurs reasonable expenses in connection with the rightful rejection of goods or if there is justifiable revocation of acceptance in addition to the situation in which the remedy of cover is involved. Incidental damages are those that are incurred in "inspection, receipt, transportation and care and custody of goods rightfully rejected, any commercially reasonable charges, expenses or commissions in connection with effecting cover and any other reasonable expenses incident to the delay or other breach." Section 2–715(1). Note that these damages are in addition to other remedies that may be available to the buyer, such as cover. You will also recall that incidental damages are available to the seller in many circumstances.

Consequential damages are those damages resulting from general or particular requirements and needs of which the seller at the time of contracting had reason to know and that the buyer could not reasonably prevent by cover or otherwise. Section 2–715(2)(a). They also include injury to the person or property proximately resulting from any breach of warranty. Suppose Buyer is a retailer and contracts to purchase goods from Sam Seller for resale. Loss of profits would be an element of consequential damages because Seller has reason to know of these potential damages. Note also that in order to recover damages for injury to person or property, the buyer must show that the breach proximately caused the injury.

TRAYNOR v. WALTERS

342 F.Supp. 455 (E.D.Pa. 1972)

George and Ruth Walters agreed to sell a number of different types of Christmas trees to David Traynor, a wholesaler who supplied New York City florists with trees. The Walters warranted that the trees would be of "top quality" and knew that the trees were to be resold to quality florists in New York. The trees were delivered in several lots and were baled in such a manner as to make inspection impossible prior to delivery. After delivery the trees were inspected by Traynor, who immediately notified the Walters that a number of the trees did not conform to the contract. The rejected trees were dry, poorly colored, and unsheared and suffered from needle drop and few needles on the lower branches. Traynor then rented a construction site in Manhattan, hired a night watchman, and sold the rejected nonconforming trees for the seller's account. He then sued the Walters for damages.

MUIR, D.J.

Damages. The measure of damages for non-delivery by the sellers is the difference between the market price at the time when the buyer learned of the breach and the contract price together with any incidental and consequential damages, less expenses saved in consequence of the seller's breach. 12A P.S. § 2–713(1). In this case, the buyer did not cover; therefore, his sole claims for damages fall into the categories of incidental and consequential damages. The buyer's failure to effect cover does not bar these other remedies. 12A P.S. § 2–712(3).

A. Incidental Damages:

Section 2–715(1) provides:

(1) Incidental damages resulting from the seller's breach include expenses reasonably incurred in inspection, receipt, transportation and care and custody of goods rightfully rejected, any commercially reasonable charges, expenses or commissions in connection with effecting cover and any other reasonable expense incident to the delay or other breach.

The Plaintiff established that by virtue of the sellers' delivery of non-conforming trees he was forced to incur the following additional expenses:

Rental of storage and sales lot	$100.00
Night Watchman	60.00
Additional labor in selling rejected trees, 8 hrs/day for 5 days at $3.00/hr.	120.00
	$280.00

B. Consequential Damages:

The bulk of the plaintiff's claim falls into the category of consequential damages. On this subject the Uniform Commercial Code provides as follows in § 2–715(2):

(2) Consequential damages resulting from the seller's breach include

(a) any loss resulting from general or particular requirements and needs of which the seller at the time of contracting had reason to know and which could not reasonably be prevented by cover or otherwise

. . .

Count III of the Complaint is a claim for damages for loss of future profits resulting from loss of good will occasioned by the buyer's inability to perform his contracts with florists in New York City whom he was to have supplied during the Christmas season of 1967. Damages of this nature are entirely too speculative for reasonable calculation and none are included in the award herein.

The balance of the plaintiff's claim is for loss of profits for the 1967 Christmas season occasioned by inability to fulfill his contracts with New York florists for that season. Prior to the adoption in Pennsylvania of the Uniform Commercial Code, the general rule in Pennsylvania governing recovery of damages for loss of profit in contract actions was that set forth in *Taylor* v. *Kaufhold*, 368 Pa. 538, 546, 84 A.2d 347,351 (1951), where the Pennsylvania Supreme Court stated:

> Where one party to a contract without any legal justification, breaches the contract, the other party is entitled to recover, unless the contract provides otherwise, whatever damages he suffered, provided (1) they were such as would naturally and ordinarily result from the breach, or (2) they were reasonably foreseeable and within the contemplation of the parties at the time they made the contract, and (3) they can be proved with reasonable certainty.

In the instant case, the plaintiff sustained his burden of proving each of the elements of this test of recoverability of lost profits.

(The court found that the buyer was able to recover incidental damages of $280.00 and consequential damages consisting of lost profits of $4,801.25.)

19:3. Damages When the Buyer Has Accepted the Goods. The damages discussed thus far refer to situations in which the buyer has not accepted the goods but has rejected or justifiably revoked acceptance. If the buyer has accepted the goods and notified the seller, he may recover as damages for any noncomformity of tender the loss resulting in the ordinary course of events from the seller's breach. Section 2–714(1). This remedy is applicable if the goods have been accepted by the buyer and it is too late to revoke. Note the condition of notification to the seller.

If there is a breach of warranty, the measure of damages is the difference at the time and place of acceptance between the value of the goods accepted and the value they would have had if they had been as warranted, unless special circumstances show proximate damages of a different amount. Section 2–714(2). Incidental and consequential damages may also be recovered in a proper case.

Suppose Sam Seller sells tires to Bill Buyer at $20 per tire. The tires are retreads rather than new tires and thus do not conform to the contract. The value of the retreads is $12 per tire. Buyer's damages would be $8 per tire, the difference between the goods accepted at the time and place of acceptance and the price if they had been as warranted. Buyer may notify Seller and deduct the difference from any price still owed to Seller. Remember that Buyer might have rejected the goods as nonconforming, but if he does not do so or does not revoke acceptance, this remedy is available.

19:4. Rights and Remedies Affecting Both Buyer and Seller: Liquidated Damages. The code takes the position that the parties may provide for liquidated damages in the contract as long as the amount is reasonable in light of the anticipated or actual harm, the difficulties of proof of loss, and the inconvenience or nonfeasibility of obtaining an adequate remedy. This approach is similar to the common-law approach. If the liquidated damages are unreasonably large, the term is construed as a penalty and void. Section 2–718(1).

Frequently the contract will provide for forfeiture of the buyer's down payment in the event that he breaches the contract. However the buyer is entitled to restitution by the amount his down payment exceeds 20 percent of the value of the total performance for which the buyer is obligated under the contract, or $500, whichever is smaller. Section 2–718(2)(b). Thus, if the contract price is $600 and the buyer has made a payment of $150, the buyer is entitled to restitution of $30 (20 percent of $600 is $120, with a difference of $30 from $150). If there is a liquidated-damages provision and the amount of payments by the buyer exceeds that provision, then the buyer is entitled to restitution by the amount of the difference. Section 2–718(2)(a). The buyer's right to restitution may be offset by the seller, if he can establish the right to recover damages under other provisions of this article, or by any benefits received by the buyer by reason of the contract. Section 2–718(3).

19:5. Limitation of Remedies. The parties may limit their remedies for breach of contract. This is consistent with the code philosophy allowing for the parties to create their own contracts with a maximum of freedom. The parties may limit the buyer's remedy to return of goods and repayment on the price or to repair and replacement of nonconforming goods. Section 2–719(1)(a). The code treats remedies stated in contracts as cumulative rather than exclusive unless the contract provides for the stated remedy to be exclusive. Section 2–719(1)(b). One must take care, therefore, to state that the remedy provided in the contract is exclusive if that is the desire of the parties.

The code states specifically that the parties may limit or exclude consequential damages unless the limitation is unconscionable. A limitation of consequential damages for personal injury in the case of consumer goods is prima-facie unconscionable. Section 2–719(3). A number of states, such as Maryland and Maine, have greatly restricted the right of the parties to limit remedies, particularly in the case of consumer goods.

SUMMARY

A seller has a number of remedies. The seller's remedies for money damages include the right to: resell the goods and recover money damages; recover damages for nonacceptance or repudiation; and recover the price. The seller may also, under the proper circumstances: cancel; withhold delivery of the goods; stop delivery of goods by a bailee; identify goods to the contract and finish or salvage them; and reclaim the goods on the buyer's insolvency.

The code states the circumstances when each of these remedies are appropriate. More than one remedy may be applicable at one time.

The buyer's remedies for money damages include the right to: cover; recover damages for nondelivery or repudiation; resell the goods; and obtain consequential and incidental damages. The buyer may also, under the proper circumstances: cancel; recover any price paid; obtain specific performance; replace the goods; and recover goods on the seller's insolvency. As with the seller, more than one remedy may be applicable at one time.

The buyer is only entitled to specific performance if the goods are unique or in "other proper circumstances." Consequential damages which may be recovered are those damages resulting from general or particular requirements of which the seller at the time of contracting had reason to know and which the buyer could not reasonably prevent. If the buyer has accepted the goods and notified the seller, he may recover as damages for any nonconformity of tender the loss resulting in the ordinary course of events from the seller's breach.

The code allows the parties to provide for liquidated damages in the contract. The code limits the amount of the buyer's down payment that may be forfeited by the seller. It also allows the parties to limit their remedies.

PROBLEMS

1. Corkle agreed to sell thirty Holstein milk cows to Putnam. Corkle placed thirty cows in a separate corral for delivery to Putnam. On the morning of September 9, 1967, Putnam paid the purchase price of $5,700. That afternoon, someone other than Putnam picked up the cows and Putnam never received them.

   Three years later, at a sale of livestock advertised to be held by Corkle, the sheriff, as a result of an action by Putnam, executed an order of replevin by seizing thirty cows. There was no evidence that these were the same thirty cows set aside by Corkle three years earlier. Was Putnam entitled the cows? Putnam Ranches Inc. v. Corkle, 189 Neb. 533, 203 N.W.2d 502 (1973).
2. On January 5, 1966, Ned Downing agreed to purchase a color television from Arnold Wood's store for the price of $450. Downing paid $100 down and agreed to pay the balance upon Wood's delivery at a later date. In early May of 1966, Downing went to the store to pay the balance due and pick up the television. Wood claimed that under the agreement Downing was to have picked up the television and thus he had sold the set to another person. Wood refused to return the $100 deposit and did not notify Downing that he intended to sell the television. Did Wood properly adhere to the procedures under the Uniform Commercial Code in pursuing his remedy? Wood v. Downing, 243 Ark. 120, 418 S.W.2d 801 (1967).
3. Neri contracted to purchase a boat from Retail Marine Corporation for the price of $12,587.40. A deposit of $4,250.00 was made in consideration for immediate delivery of the boat. Six days later Neri's lawyer sent Retail Marine a letter rescinding the contract because Neri was about to be hospitalized, which would make payments impossible. Retail Marine declined to refund the deposit and Neri filed suit for the deposit plus attorney's fees. Retail Marine counterclaimed for lost profits, costs of storage, and

insurance. Decide what damages, if any, each was entitled to. Neri v. Retail Marine Corp., 30 N.Y.2d 393, 334 N.Y.S. 165 (1972).

4. Jack Richards Aircraft Sales contracted to sell an aircraft to J. C. Vaughn for $30,950. Vaughn later decided not to go through with the sale. Jack Richards sued Vaughn for breach of contract and then, after repainting the aircraft, sold it to another buyer for $22,000. The sale was made after the plane had been advertised and had been shown to several prospective customers. Defendant contended that the $22,000 price was substantially below the market price of the aircraft and that therefore the resale price should not be considered in the computation of damages. Was Vaughn correct? Jack Richards Aircraft Sales v. J. C. Vaughn, 203 Kan. 967, 457 P.2d 691 (1969).
5. The Anchorange Centennial Commission contracted with Van Wormer & Rodrigues for the manufacture and sale to the commission of 50,000 metal anniversary coins. After partial performance the commission canceled the contract. Van Wormer sued for damages caused by the breach as well as for lost profits of 3 cents per coin on each of the 21,000 coins not delivered. The trial court awarded Van Wormer 15 cents for the cost of each of the 29,000 coins already manufactured but refused an award of lost profits. Van Wormer appealed. Decide. Anchorage Centennial Development Co. v. Van Wormer & Rodrigues, 443 P.2d 596 (Ark. 1968).
6. Gulf Chemical and Metallurgical Corp. sold Sylvan Chemical Corp. certain goods to be delivered in three installments. The first shipment of the goods was delivered to Sylvan, who in turn sold the shipment to its customers. Sylvan claimed the goods were nonconforming and Gulf adjusted the price. Sylvan refused to pay for the delivered goods and, when sued by Gulf, counterclaimed for consequential damages incurred as a result of Gulf's refusal to deliver the last two shipments until paid for the first. Gulf contended that its obligation to deliver the remaining two shipments was terminated because of Sylvan's failure to pay for the first shipment. Sylvan contended that Gulf never canceled the contract and that Sylvan's failure to pay for the first installment did not substantially impair the value of the whole contract. Decide. Gulf Chemical and Metallurgical Corp. v. Sylvan Chemical Corp., 122 N.J.Super. 499, 300 A.2d 878 (1973).
7. Plotnick contracted to sell and did make numerous shipments of lead to Pennsylvania Smelting & Refining Co. in Philadelphia. Plotnick complained of slow payment by the buyer, and Pennsylvania Smelting demanded more lead. Plotnick then refused to ship more lead unless a recently shipped carload was paid for. After threats by buyer's attorney, Plotnick replied that the contract was "canceled" as a result of buyer's failure to pay for the lead already delivered. Was Plotnick justified in this refusal? Plotnick v. Pennsylvania Smelting & Refining Co., 194 E.2d 859 (1952).
8. Keystone Diesel Engine Co. sold a diesel engine to Irwin. The engine was installed on one of Irwin's tractors and was then repaired by Keystone when it did not work properly. Irwin was not charged for these repairs,

but when subsequent repairs were made, Keystone charged Irwin for them. Irwin refused to pay and Keystone sued. Irwin counterclaimed for damages for loss of use of the tractor for the time it was out of commission as a result of the breakdowns. There had never been any discussion between the parties as to possible damages for loss of use of the vehicle. Could Irwin collect on this counterclaim? Keystone Diesel Engine Co. v. Irwin, 191 A.2d 376 (Pa. 1963).

CHAPTER 20

Financing the Sale: Secured Transactions

One of the more mystifying areas of commercial law to the layman is the field of secured transactions. Part of the problem arises from a misunderstanding of terminology brought about by the large number of methods of obtaining a security interest in goods prior to the Uniform Commercial Code. Some of these methods were called the *chattel mortgage, conditional sale, pledge, assignment of accounts,* and *trust receipt.* There were technical differences between these types of security interests. For example, in a conditional sale the title to the goods would remain in the seller until the goods were paid for, although the purchaser–debtor could use them for this period. In most states the chattel mortgage gave the seller-mortgagee a "lien" on the collateral of the debtor. On the other hand, the pledge provided the creditor with a security interest in the property by giving him possession of it. The U.C.C. eliminated the distinctions between these various forms of financing and calls them all a *security interest.* The parties involved are called *debtors* and *creditors.* Although the old terminology is still frequently used, the distinctions as to form have been abolished. The distinctions made by the code are based upon functional distinctions in the type of property used for collateral. These functional distinctions are in turn based upon the nature of the collateral and its use, a subject that will be treated more extensively in the following pages.

The purpose of creating a security interest is to allow the creditor who makes the loan to the debtor to obtain an interest in specific property of the debtor so that he may collect from that property in the event that the debtor does not repay the loan. One way to view this problem is to visualize the creditor as beset by numerous enemies in his attempts to collect from the debtor. That is, there are others who may be competing with the creditor for the right to obtain payment by using the debtor's assets. Among these enemies are the debtor, other creditors, secured creditors, the trustee in bankruptcy, common-law lienors, and others, such as federal and state governments, who may have an interest in the debtor's property as the result of nonpayment of taxes.

Consider what happens when one purchases an item at a local store on an open account. Bill Bydebt purchases a color television set from Selcred Department Store. Bydebt does not pay cash but charges the full amount of the purchase on an open account. Bydebt then does not pay for the television and has numerous other creditors whom he also does not pay. May Selcred take any action toward collecting the debt by taking the television from Bydebt? What if another creditor seeks to take the television set? May Selcred prevent this because he financed the sale of the television? The answer to all of these questions in this case is no. Selcred has no security interest in the television and therefore no particular rights in it at this time. He must first reduce his debt to judgment by a court action and then seek to collect by court process through the sheriff. He may then go after Bydebt's property, but he may be too late—other creditor's may have moved first, with the result that all Bydebt's property, including the television, is already gone. In the alternative, Bydebt may have filed for bankruptcy. In this event Selcred must join the ranks of other unsecured creditors and be paid out of any assets that may be left after Bydebt's secured creditors have been paid from the secured property. Selcred's prospects are rather dim in this situation also. What can Selcred do to avoid this problem? He may obtain a security interest in certain property of Bydebt. This chapter concerns how one obtains a security interest under the U.C.C. and how that security interest ranks against the competing interests of others in the debtor's property.

20:1. Creating the Security Interest. If one wishes to obtain a preferred position as to certain of the debtor's assets as opposed to those who may have competing interests, certain steps must be taken. In the first place it is necessary for the creditor to obtain *attachment* of the security interest on certain of the debtor's property. This gives the creditor rights as to specific property of the debtor vis-à-vis the debtor. (Remember that on a mere open account the creditor does not have this right). However, there are other competing interests in the debtor's property, such as the interests of other creditors. How does our creditor obtain rights against these competing interests? The answer is that the security interest must be *perfected* in order to obtain preferential rights against the competing interests. Another question arises. Once the creditor obtains a perfected security interest against other competing interests, just what property of the debtor does this perfected security interest cover—that is, what is the *range of perfection?* In this section, then, we will first examine the requirements for attachment of the security interest, then how one obtains perfection of this security interest, and finally the range of perfection. The following sections will deal with priorities among competing security interests, as well as the rights and duties of the debtor and creditor.

(1) Attachment. As a general rule three things are necessary in order for the creditor to obtain attachment of a security interest against property of the debtor: (1) a written agreement; (2) value given by the creditor to the debtor; and (3) the debtor's having rights in the property used as collateral. The agreement need not be in writing if the creditor has possession of the collateral. Section 9–203(1), 1962, 9–204(1). When these three requirements are met, the security interest attaches, giving the creditor rights in the debtor's property. Let us examine these requirements further.

(a) *The Security Agreement.* In order for a security interest to attach, there must be agreement that it attach. This agreement must be written unless the collateral is in possession of the secured party, in which case the evidentiary requirement of a writing is eliminated. The *security agreement* sets forth the rights that exist between the debtor and the creditor as to the collateral. Be careful not to confuse it with the *financing statement,* which is used to record the agreement and which will be discussed in Section 20:3(1)(a).

In order to be valid, (1) a security agreement must be signed by the debtor; (2) it must create a security interest; and (3) the collateral must be described. Section 9–203(1). Any sign will suffice as a signature if it is executed by a party with the intent that it be his or her signature. The security may be created by the use of any appropriate words, such as "The debtor grants to the creditor (herein after called the 'secured party') a security interest in the following goods. . . ." As for the description of the property, it is sufficient if the agreement reasonably identifies the property that is described. Section 9–110.

The purpose of the security agreement is to set forth the rights of the creditor in the debtor's property. The typical security agreement may contain many provisions other than the bare requirements necessary to satisfy the code. For example, the agreement may cover such matters as the amount of indebtedness and terms for payment, what constitutes default of the debtor, whether insurance must be maintained on the collateral, the debtor's promise to keep the collateral free from other liens, and so on. The agreement may also provide for after-acquired property coverage of proceeds and for future advances by the creditor. These subjects are covered in detail later, but remember that if the creditor wants these things covered, they must be mentioned in the security agreement.

(b) *Value Given.* In order for the security interest to attach, value must be given by the creditor to the debtor. This requirement is usually fulfilled when the creditor makes the loan to the debtor, as when he gives him $500 in cash or extends credit for the sale of a television by the creditor to the debtor. Value may also be given when the creditor agrees to extend credit in the future. Note that the term *value* as used by the code is *not* synonymous with *consideration.* Thus the creditor may obtain a security interest in property of the debtor to secure a loan made at a prior time. This is known as obtaining a security interest for an antecedent debt.

(c) *Debtor's Interest in Property.* The debtor must have an interest in the property that is the subject of the security interest. If the debtor owns the property in which the creditor seeks a security interest, this requirement is satisfied. On the other hand, the debtor would also have an interest in the property if a contract is signed for the sale of goods that are identified to the contract. Recall that when the goods are identified to the contract, the buyer has an insurable interest in them. Section 2–501(1). This interest is sufficient to satisfy this requirement for a security interest. On the other hand, if the goods are not yet identified to the contract of sale, the buyer–debtor has no interest in the goods until they are identified to the contract.

Once these requirements are met, the creditor has an interest in certain property of the debtor, but what about the creditor's positions as to this property against third parties, that is, others not a party to the security agreement? In order to obtain protection against these parties, the creditor must obtain *perfection* of the security interest.

(2) **Perfection.** The creditor has a perfected security interest in the property of the debtor when it is attached and when the steps required to obtain perfection have been taken under Section 9–302. In a determination of whether there is perfection of a security interest, three factors must be considered: the nature of the collateral, whether the interest is a purchase–money security interest, and the method of perfection.

(3) **The Nature of the Collateral.** The method of perfection is affected by the nature of the collateral used. It is therefore important to categorize the type of property used as collateral.

(a) *Tangible Personal Property.* Tangible personal property is frequently used as collateral. Items such as an automobile, a refrigerator, or jewels may all be used as collateral and are classified as *goods* by the code. Goods may be classified as *consumer goods, equipment, farm products,* or *inventory.* In determining the category into which the goods fall, one should examine the nature of the collateral and how it is *used.* Obviously any type of item might fit into more than one category, although not at the same time. For example, if Pearl Player owns a piano for use in her home for personal pleasure, it would be a consumer good. On the other hand, if Player is a piano teacher, the piano would be equipment. If Player has the piano for sale in her music store, it is classified as inventory.

(i) CONSUMER GOODS. The code defines goods as consumer goods if they are used or bought for use primarily for personal, family or household purposes. Section 9–109(1). Thus if Bydebt gives a security interest in his or her household furniture, that security interest is on consumer goods.

(ii) EQUIPMENT. Goods are classified as equipment if they are used or bought for use primarily in business or by a debtor who is a nonprofit organization or a governmental subdivision or agency or if the goods are not included in the definitions of inventory, farm products, or consumer goods. Section 9–109(2). Thus a hammer purchased by Paul Pounder for use on his job as a carpenter would be equipment. The fact that Pounder uses the hammer around his own house occasionally would not alter the classificaton of the collateral, as it is the primary use of the collateral that is important in the determination. Note the negative aspects of the definition. Goods are equipment if they are not inventory, farm products, or consumer goods.

(iii) INVENTORY. Goods are classified as inventory if they are held by a person who holds them for sale or lease or to be furnished under contracts of service or if he has so furnished them. Inventory also includes raw materials, work in process, or materials used or consumed in business. A person's inventory cannot be classified as his equipment. Section 9–109(4). The major test is whether the goods are held primarily for immediate or ultimate sale. Thus the refrigerators on the floor of Sam Seller's Appliance Store would be inventory if they are to be sold.

(iv) FARM PRODUCTS. Goods are farm products if they are crops or livestock or supplies used or produced in farming operations. Goods are also farm products if the goods are products of crops or livestock in their unmanufactured state (such as milk, eggs, and syrup) and if they are in the possession of a debtor engaged in farming operations. Section 9–109(3). Farm products cannot be equipment or inventory. Thus the corn growing in Frank Farmer's field, his chickens, and their eggs are all farm products.

(b) *Intangible Property.* Intangible property may be the subject of commercial transactions, but it has no physical existence. Intangible property represents a right or right to receive tangible property. The code specifically mentions documents, instruments, chattel paper, accounts, contract rights, and general intangibles as intangible property. The term *documents* refers to all documents of title, such as warehouse receipts and bills of lading. *Instruments* include negotiable instruments, investment securities, or any writing that evidences a right to the payment of money. *Chattel paper* is defined as a writing that evidences both a monetary obligation and a security interest in specific goods. An example might be when one sells an item under a conditional sales contract and then the seller transfers that contract to a lending institution to secure a loan. Section 9–105.

An *account* is defined as any right to payment for goods sold or leased or services rendered that is not evidenced by chattel paper or by an instrument. An example would be the ordinary account receivable. *Contract right* is any right to payment under a contract not yet earned by performance and not evidenced by an instrument or chattel paper. *General intangibles* means any personal property, including things in action, other than the goods and intangibles already mentioned. Section 9–106.

(c) *Proceeds.* Proceeds include whatever is received when collateral or proceeds are sold, exchanged, collected, or otherwise disposed of. Money, checks, and the like are cash proceeds. All other proceeds are noncash proceeds. Section 9–306(1). Suppose Delbert Cardealer sells a new car to Bill Bydebt. Cardealer has given a security interest in his cars to his bank. Bydebt pays cash for the new car in addition to trading in his old car. The cash would obviously be cash proceeds and the used car would be noncash proceeds of the inventory collateral.

20:2. PURCHASE-MONEY SECURITY INTEREST. Once the nature of the property to be used as a security interest is classified, it is useful to determine whether the security interest involved is a purchase-money security interest or a non-purchase-money security interest. A *purchase-money security interest* occurs when (1) the security interest is taken or retained by the seller of the collateral to secure all or part of its price or (2) if the security interest is taken by a person who by making advances or incurring an obligation gives value to enable the debtor to acquire rights in or the use of the collateral. Section 9–107. A common situation that results in a purchase-money security interest occurs when Connie Carbuyer purchases a car using the money received from a loan with Friendly Bank. Bank takes a security interest in the car that was purchased with the money loaned by Bank. Carbuyer is the debtor, Bank is the creditor, and the car is collateral. Bank's security interest is a purchase-money security interest. The importance of the distinction between a purchase-money security interest and a non-purchase-money security interest will become more apparent as you read this chapter. Basically the reason is that purchase-money obligations have priority over non-purchase-money obligations. A non-purchase-money security interest obviously occurs when the value given by the creditor is not used to purchase the collateral. Suppose Delbert Debtor borrows $200 from Friendly Bank. Bank wants security for the loan and Debtor gives the bank a security interest in some family diamonds. Because the $200 was not used to purchase the collateral, it is a non-purchase-money security interest.

20:3. Method of Perfection. Once the classification of the type of goods has been accomplished and a determination has been made as to whether the security interest would be a purchase-money security interest or a non-purchase-money security interest, we are in a position to determine the steps that should be taken in order to *perfect* our security interest. Remember that the purpose of perfection is to give the creditor rights against other claimants to specific goods of the debtor. Attachment has given the creditor rights to specific goods vis-à-vis the debtor. The code provides for four methods of perfection under Section 9–302 and the sections referred to by that section. These methods are perfection by filing, perfection by possession, and automatic and temporary perfection.

(1) Filing. The code provides that a financing statement must be filed in order to perfect all security interests, except for a number of situations discussed in Section (2). Section 9–302(1). Perfection of the security interest by filing may be used for most types of collateral, but not for money or instruments because they are negotiable.

(a) *The Financing Statement.* The document that is filed is called a *financing statement.* The purpose of filing the financing statement is to put people on notice that there is a security interest in certain property of the debtor held by the creditor. The information contained in the financing statement is for the purpose of allowing a person to determine the nature and terms of the security interest by further investigation. From the financing statement one can determine who the secured party is and how he or she may be contacted. It is sufficient if the collateral is described sufficiently so that it may be identified. It need not be described with any greater degree of particularity.

The financing statement must contain the names of the debtor and the secured party, give the address of the secured party from whom information concerning the security interest may be obtained, and give the mailing address of the debtor. The debtor must sign the financing statement. The 1962 code required that the secured party also sign, but the 1972 amendment has eliminated this requirement. Finally, the financing statement must contain a statement indicating the types or describing the items of collateral. When the financing statement covers items such as timber to be cut, growing crops, or minerals to be mined, the land must also be described. Section 9–402(1).

Do not confuse the security agreement with the financing statement. They serve two different purposes. The security agreement may be filed as a financing statement, provided that it meets the requirements for a financing statement already outlined. Section 9–402(1).

(b) *Where Filed: Mistake.* The code provides three alternative provisions for determining where the financing statement should be filed. It is necessary to determine which alternative has been adopted in a particular state before one knows when to file. Section 9–401.

Under the first alternative the financing statement is filed in most cases with a central office in the state, usually the office of the secretary of state. An exception applies to securied items that are to become fixtures. In those cases, the filing should be in the office where the mortgage on real estate would be filed.

The second alternative provides that the financing statement should be filed in the county where the debtor maintains his residence or, if the debtor is not a resident, then the county where the goods are kept. Those types of collateral covered by this rule are farm equipment or products, accounts, consumer goods, or general intangibles arising from or related to the sale of farm products by a farmer. If the goods are growing crops or crops to be grown, then the statement must also be filed in the county where the land is located. If the collateral is timber, minerals, or the like or a good that is to become a fixture, the financing statement should be filed where a mortgage on the real estate is to be recorded. In all other cases filing is to be in a central office in the state, such as the office of the secretary of state. The county office for filing is frequently the office of the clerk of the county court or the registrar of deeds. The third alternative is similar, except that when filing is to be with a central office, such as the office of the secretary of state, then if the debtor has a place of business in only one county of the state, the statement must also be filed in that county. If the debtor has no place of business in the state but has a residence in the state, the statement is filed in the county of residence.

The basic difference between the first alternative and the second and third is that the first alternative provides for one central filing location in the state, whereas the second and third alternatives provide for filing within the local jurisdiction. The important thing to know, of course, is where one should look, as a prospective creditor, in order to determine whether the property the prospective debtor intends to use as collateral is already subject to a recorded security interest.

The code also provides that the filing is effective if it is made in good faith, even if the financing statement is filed in the wrong place or not in all the required places. The key phrase, of course, is *good faith,* so one could not purposefully file in the wrong place and still have the protection of a perfected security interest. Even if there is no filing, the financing statement is effective against any person who has knowledge of its contents; that person would not be prejudiced by the lack of filing because he or she is already aware of the security interest. Section 9–401(2).

A change in the debtor's residence or a change in the location of the collateral subsequent to a valid filing does not render the filing ineffective. If the filing was originally made in the proper place, it remains effective even after such a change. Section 9–401(3). There is an alternative provision of the code that requires a new filing within four months of a change of residence of the debtor or a change in the location of the collateral if there is a change to another county.

Once the financing statement is filed, it is effective for five years from the date of filing. When the period expires, the security interest becomes unperfected. If one wishes extension beyond the five-year period, a *continuation statement* must be filed prior to the lapse. This statement may extend the time for another five years and may be repeated. Section 9–403(2).

AMERICAN RESTAURANT SUPPLY CO. v. WILSON

371 So.2d 489 (Fla. App. 1979)

American Restaurant Supply Company, (American), entered into a security agreement with Robert Morkland and Wilmark Inc. who were the debtors. The security agreement described the collateral as: "Food service equipment and supplies delivered to San Marco Inn at St. Marks, Florida". Another creditor of Morkland and Wilmark Inc. sought to attach these same assets and contended that American's security agreement was not enforceable against him because the description was inadequate.

MILLS, J.

A security interest cannot be enforced against the debtor or third parties unless the collateral is in the possession of the secured party or the security agreement contains a description of the collateral. Section 679.203(1), Florida Statutes; Section 9–203(1), U.C.C. The description of collateral ". . . is sufficient whether or not is is specific if it reasonably identifies what is described. . . ." Section 679.110, Florida Statutes; Section 9–110, U.C.C. The Comment to Section 9–110 of the Uniform Commercial Code states that the test of sufficiency of a description ". . . is that the description do the job assigned to it—that it make possible the identification of the thing described."

Although Section 679.110, Florida Statutes, (Section 9–110, U.C.C.), sets forth the test for sufficiency of the description of collateral in both the security agreement and the financing statement, a description of collateral sufficient for a financing statement might not be sufficient in a security agreement. This is because the financing statement and the security agreement serve different purposes.

The purpose of the financing statement is merely to provide notice of a possible security interest in the collateral in question. Official Comment, Section 9–402, Uniform Commercial Code. Section 679.402, Florida Statutes, (Section 9–402, U.C.C.), requires that the financing statement contain a description indicating the types of collateral in which the secured party may have a security interest. The description of collateral in a financing statement is sufficient if it reasonably informs third parties that an item in the possession of the debtor may be subject to a prior security interest, thus putting the parties on notice that further inquiry may be necessary.

The security agreement is the contract between the parties; it specifies what the security interest is. Because of its different functions, greater particularity in the description of collateral is required in the security agreement than in the financing statement. A description of collateral in a security agreement is sufficient if the description makes possible the identification of the items in which a security interest is claimed.

The Comment to the Uniform Commercial Code, Section 9–208 (Section 679.208, Florida Statutes), clearly indicates that the description of collateral in a security agreement is required to be more precise than in the financing statement. Section 9–208 permits the debtor to submit a list of collateral to the secured party who must approve or correct the list in writing within two weeks. The Comment to this Section states:

> 2. The financing statement required to be filed under this Article (see Section 9–402) may disclose only that a secured party may have a security interest in specified types of collateral owned by the debtor. *Unless a copy of the security agreement itself is filed as the financing statement third parties are told neither the amount of the obligation secured nor which particular assets are covered.* Since subsequent creditors and purchasers may legitimately need more detailed information, it is necessary to provide a procedure under which the secured party will be required to make disclosure. (Emphasis supplied)

Thus, a security agreement should describe the collateral with details sufficient for third parties to be able to reasonably identify the particular assets covered.

The security agreement under consideration describes the collateral as: "Food service equipment and supplies delivered to San Marco Inn at St. Marks, Florida." Many courts have held that a description of collateral is sufficient when the agreement covers all of a certain type or types of assets.

However, the agreement before us does not cover all of the food service equipment and supplies located at San Marco Inn or owned by the debtor. The agreement attempts to cover some food service equipment and supplies, but the description does not do its assigned job of making possible the identification of the equipment and supplies in which American claims a security interest.

We agree with the trial court that ". . . the description of the property pledged as security in the security instrument was not legally sufficient" to enable the security interest to be enforced. Therefore, the judgment of the trial court is

(Affirmed.)

QUESTIONS FOR DISCUSSION

1. *If a security agreement describes the collateral as "all accounts," "all inventory," or "all goods," is the description adequate? How do these descriptions differ from the description in this case?*
2. *What is the difference in purpose between a financing statement and security agreement?*
3. *Is it possible to use one document as both the financing statement and security agreement?*

---

(2) Possession. If a financing statement has not been filed, there are still a number of other situations that give rise to perfection of the security interest. One of these situations is when the creditor has *possession* of the collateral. This is the form of security interest known as *pledge* under the common law. Most types of collateral are susceptible to possession by the creditor. In fact, a security interest in a few types of collateral, such as money and some types of instruments, can be perfected only by possession. On the other hand, a security interest in general intangibles and accounts cannot be perfected by possession because this type of intangible collateral cannot really be possessed. Section 9–302,305.

The rationale for allowing the creditor's security interest to be perfected without filing if he or she has possession is that the creditor's possession is notice to others of his interest in the collateral. When the debtor does not have possession of property, the lack of possession should also indicate to third parties that the property may be subject to a security interest. The security interest is perfected from the time possession is taken and continues only as long as possession is maintained. Section 9–305.

The creditor may also have possession when the collateral is in the hands of a third party, such as a warehouseman. All the creditor need do in order to obtain possession in this situation is notify the bailee of his security interest in the collateral.

Possession may be an appropriate method of perfection, for example, if the debtor has a quantity of bulk goods in his warehouse that he wishes to put up as collateral for a loan and that are to remain in place for a period of time. In that case, the creditor may fence off the goods that are to be used as collateral and post a sign indicating the creditor's security interest. Another example would be the case of a pawnbroker, who holds the goods in exchange for money loaned to a debtor. There is no need to file a financing statement in order to get perfection.

(3) Automatic Perfection. There are circumstances in which automatic perfection of a security interest occurs, that is, in which nothing need be done beyond obtaining attachment of the security interest in order to get perfection.

(a) *Purchase-Money Security Interest in Consumer Goods.* The most common instance of automatic perfection is in the case of a purchase-money security interest in consumer goods. In that case, perfection is automatic, except that filing is required if perfection of a security interest is to be obtained in a motor vehicle required to be registered or in a fixture. Section 9–302(d). The reason for this exception is that filing offices would be overwhelmed with financing statements because of the large number of such transactions. Also, the value of the collateral is often not large when compared to the costs of filing.

For example, suppose Sam Selcred sells a stereo on an installment basis to Bill Bydebt for use in Bydebt's home. Selcred has Bydebt sign a security agreement giving Selcred a security interest in the stereo for the unpaid purchase price. Selcred's security interest is automatically perfected. The result would be the same if Bydebt borrowed the money from a lending institution in order to pay Selcred for the stereo and the institution took a security interest in the stereo.

The 1962 version of the code provided for automatic perfection of a purchase-money-security interest in farm equipment having a purchase price not in excess of $2,500. This provision has been eliminated from the 1972 code.

(b) *Assignment of Accounts.* Normally, in order to obtain a perfected security interest in the accounts of a debtor, such as accounts receivable, the creditor must file a financing statement. (Remember that this form of collateral is not even susceptible to perfection by possession.) However, if the assignment of the accounts (and contract rights under the 1962 code) does not constitute a transfer of a significant part of the outstanding accounts of the assignor, no filing is required and perfection is automatic. Section 9–302(1)(e). The purpose of this provision is to save casual or isolated assignments from ex post facto invalidation. On the other hand, any person who regularly takes assignments of any debtor's accounts should file.

(4) Temporary Automatic Perfection: Proceeds. When a debtor disposes of collateral, the creditor who has a perfected security interest in that collateral has an automatically perfected security interest in the *proceeds* for ten days after the receipt by the debtor. Section 9–302(b). If the creditor wants a perfected security interest in the proceeds beyond the ten-day period it is necessary that he (1) perfect a security interest in the proceeds before the expiration of the ten-day period or (2) have provided for coverage of the proceeds in the financing statement that was filed to cover the original collateral. Section 9–306(3).

For example, suppose Creditbank finances Cardebt's inventory of new cars. The financing statement filed by Creditbank does not provide for coverage of the proceeds. A customer purchases one of the new cars and trades in his old car. Creditbank would have a perfected security interest in the used cars automatically for ten days. In that time the bank would have to perfect a security in the used car or lose the security interest. On the other hand, if the original financing statement states that proceeds are covered, the perfection is not limited by the ten-day period.

There are other situations in which the creditor may have automatic temporary perfection of his security interest. These involve security interests in instruments or negotiable documents when there is automatic temporary perfection for twenty-one days after attachment when new value is given under a written security agreement. Section 9–304(4). Temporary automatic perfection may also occur for a period of twenty-one days in an instrument, a negotiable document, or goods in possession of a bailee other than one who has issued a negotiable document therefor when the secured party (1) makes the goods or documents available to the debtor for the purpose of ultimate sale or exchange or (2) to prepare for their sale or exchange, delivers the instrument to the debtor for the purpose of ultimate sale or exchange or of presentation, collection, renewal, or registration of transfer. Section 9–304(5). In these situations the secured party may find it appropriate to release the collateral to the debtor for a brief period, and no useful purpose would be served by a cluttering of the files with records of such short-term transactions.

(5) Fixture Filing. A special problem arises with regard to fixtures. The code does not specifically define fixtures but leaves the definition to the law of the individual states. In general a fixture is an item or good that becomes so firmly affixed to or is such an integral part of the real estate that it is considered part of it. Examples would be furnaces, air conditioners, and stoves. On the other hand, an air conditioner might be of the portable variety and hence would not qualify as a fixture. The code states specifically that goods incorporated into structures, such as bricks and lumber, are not subject to a security interest. The reason should be obvious.

The problem is illustrated when one purchases a furnace on credit and subject to a security interest. The furnace then becomes part of a house, which is itself subject to a mortgage. How does the seller of the furnace obtain a security interest in the furnace that is superior to the interest of the financer of the house after the furnace is installed?

Under the code a security interest that attaches to goods before they become fixtures takes priority over the claims of those who have security interests in the real estate. Section 9–313(2). The 1972 amendment requires filing before or within ten days after the goods have become fixtures. If the attachment is made after the goods become fixtures, there is priority against interests subsequently acquired in the real estate, except for a purchase for value of the real estate, a lien creditor on the real estate obtained by judicial proceedings, and a prior secured real-estate creditor to the extent that he makes further advances. Section 9–313(3) and (4).

The 1972 amendment to the code also requires that there be a "fixture filing" in the office where a mortgage on real estate would be recorded. If one has doubt as to whether he is dealing with a fixture or a nonfixture, he should file in the normal place and also in the appropriate place for filing for a fixture. The 1972 amendment makes some other technical changes designed to make a shift in the law in favor of construction mortgages. For example, it accords priority to a construction mortgage recorded before the goods became fixtures if the goods become fixtures before the completion of construction. Section 9–313(4), 1972 Amendment.

There are a few other situations involving perfection, but they are beyond the scope of this section. Note that Section 9–302(1) of the code is the summary section for the various methods of perfection; it refers to other sections of the code for more detail. Once the method of perfection has been determined, it is necessary to know just what goods are subject to the perfected security interest. In other words, what is the range of perfection?

KIMBRELL'S FURNITURE CO. v. FRIEDMAN
198 S.E.2d 803 (S.C. 1973)

Charlie O'Neal purchased a new television on July 11, 1972, and his wife purchased a tape player on July 15, 1972. Both items were purchased from Kimbrell's Furniture Co. on credit, and in each instance a conditional sales contract was executed. On the third day of each purchase, O'Neal took the item to Friedman, a pawnbroker doing business as Bonded Loan. Friedman retained possession of the items, and Kimbrell never filed a financing statement. Friedman contended that his lien was superior to Kimbrell's. The lower court agreed and Kimbrell appealed.

LEWIS, J.

The question to be decided is correctly stated by appellant as follows: Is a conditional seller of consumer goods required to file a financing statement in order to perfect his security interest against a pawnbroker who subsequently takes possession of such goods as security for a loan?

Goods are classified or defined for purposes of secured transactions under Section 10.9–109. Subsection 1 defines "consumer goods" as those "used or bought for use primarily for personal, family or household purposes." The property here involved was a television set and tape player. They are normally used for personal, family or household purposes and the purchasers warranted that such was the intended use. It is undisputed in this case that the collateral involved was consumer goods within the meaning of the foregoing statutory definition.

Kimbrell clearly held a purchase money security interest in the consumer goods sold to the O'Neals and, by them, subsequently pledged to Bonded Loan. Section 10.9–107(a).

When filing is required to perfect a security interest, the U.C.C. requires that a document designated as a financing statement must be filed. Section 10.9–302. The U.C.C. does not require filing in order to perfect a purchase money security interest in consumer goods. Pertinent here, Section 10.9–302(1)(d) provides:

> (1) A financing statement must be filed to perfect all security interests except the following:
>
> . . .
>
> (d) a purchase money security interest in consumer goods; . . .

Since filing was not necessary, the security interest of Kimbrell attached and was perfected when the debtors executed the purchase money security agreements and took possession of the property. Sections 10.9–204, 10.9–303(1). Therefore, Kimbrell's security interest has priority over the security interest of Bonded Loans by virtue of Section 10.9–312(4) which provides:

> (4) A purchase money security interest in collateral other than inventory has priority over a conflicting security interest in the same collateral if the purchase money security interest is perfected at the time the debtor receives possession of the collateral or within ten days thereafter.

This result is consistent with and confirmed by the residual priority rule of 10.9–312(5) (b) providing for priority between conflicting security interest in the same collateral ". . . in the order of perfection unless both are perfected by filing. . . ."

(Reversed and remanded.)

Questions for Discussion

1. *Did the pawnbroker have perfection of his security interest? Did Kimbrell have perfection? Who had a purchase-money security interest?*
2. *Would the outcome have been different if O'Neal had purchased the goods as a retailer, for resale?*

BRAMBLE TRANSPORTATION, INC. v. SAM SENTER SALES, INC.

294 A.2d 97 (Del. Super. 1971, affrm'd. 294 A.2d 104, 1972)

Sam Senter Sales, Inc., owed Bramble Transportation, Inc., $10,825.20 for freight charges. Draper-King Cole, Inc., of Delaware owed Sam Senter Sales, Inc. $7,053.25. Bramble attached this account on May 12, 1970, and Draper deposited that amount with the court pursuant to the attachment. On June 2, 1970, Walter E. Heller & Co. filed a motion to intervene, claiming that it had an assignment of the accounts owed by Draper to Sam Senter Sales, Inc., and that it therefore was the owner of the debt attached on May 12 by Bramble.

Heller filed copies of three sets of assignments with its motion, all dated prior to May 12, 1970, by which Sam Senter Sales assigned accounts receivable from Draper to Sam Senter Farms, Inc., which in turn assigned them to Heller. Through this claim Heller claimed a perfected security interest in the accounts and requested that Bramble's attachment of the account be quashed.

Because the offices of Sam Senter Farms and Sam Senter Sales were located in Florida, the court applied the law of that state, found in favor of Heller, and ordered that the funds be turned over to it. Bramble appealed.

QUILLEN, J.

Under the Uniform Commercial Code, a security interest in an account receivable is perfected when it has attached and when a financing statement has been filed. If filing occurs before the security interest attaches, it is perfected at the time when it attaches.

Heller, in support of his contention that he holds a perfected security interest in the Draper-King Cole accounts receivable, has submitted the following documents to the Court:

1. Financing statements (Form UCC–1) covering present and future accounts from Sales (debtor) to Heller (secured Party) and from Farms (debtor) to Heller (secured Party) dated November 1, 1966 and recorded in the office of the Clerk of the Circuit Court, Palm Beach County, Florida.
2. Financing statements (Form UCC–1) covering present and future accounts from Farms to Heller, Sales to Farms, and Sales to Heller recorded in the office of the Florida Secretary of State.
3. Specific written assignments of the three Draper-King Cole accounts totaling $7,053.25 from Sales to Farms (signed by both parties) and from Farms to Heller (signed by both parties) dated April 24, April 28 and May 5, 1970, respectively.

In order to determine the legal effect of these documents, it is necessary to distinguish between the concept of security "attachment" and the concept of security "perfection" under the Uniform Commerical Code. Attachment deals with those steps legally required to give the secured party an interest in the subject property effective against the debtor. Perfection deals with those steps legally required to give the secured party an interest in the subject property against the debtor's creditors. Furthermore, as noted above, attachment is a prerequisite to perfection.

The following statutory criteria govern the attachment of a security interest:

> A security interest cannot attach until there is agreement . . . that it attach and value is given and the debtor has rights in the collateral. It attaches as soon as all of the events in the preceding sentence have taken place unless explicit agreement postpones the time of attaching. F.S.A. § 679.9–204(1).

In the present case, the security agreements from Sales to Farms and the subsequent agreements from Farms to Heller were each in writing, signed by the debtor, value in the form of an advance was given, and the debtor had rights in the account at the time it was sold. Therefore, this court concludes that the security interest properly attached in each case. Keeping in mind the distinction between attachment and perfection, however, it does not necessarily follow that Heller is protected against the claims of Sales' and Farms' creditors. The proper place designated by Florida law in which to file a financing statement required to perfect a security interest in accounts receivable is in the office of the clerk of the circuit court in the county of the debtor's residence. F.S.A. § 679.9–401. Therefore, since the present case deals specifically with accounts, this court may consider only the effect of the financing statements from Sales to Heller and from Farms to Heller on file with the circuit court of Palm Beach County. The documents on file with the Secretary of State, which include an additional financing statement from Sales to Farms, cannot be properly considered. The fact that there is no financing statement properly filed between Sales and Farms with the clerk of the circuit court adds an element of difficulty to this case.

The U.C.C. deals specifically with the assignment of perfected security interests, stating that no further filing is required in order to continue the perfected status of the security interest against creditors of the original debtor. See F.S.A. § 679.9–302(2). That provision contains no requirement of privity between the original debtor and the subsequent assignee. In the present case, however, Farms' security interest in Sales' accounts receivable was not perfected due to the fact that the financing statement between those parties was not filed in the proper location. The U.C.C. does not directly address the problem present here of whether a financing statement signed by the original debtor and the assignee of the original secured party may function to perfect a series of properly attached security agreements which in fact places the parties in the position of debtor and secured party. Inasmuch as judicial research has disclosed no cases on point, the court must turn to an examination of the intent of the code's filing provisions to determine if the code provides a satisfactory indirect answer to this question.

The U.C.C. system of notice filing, in effect

in Florida, requires that only a financing statement need be recorded in order to perfect an otherwise valid security interest. It is not necessary to file the underlying security agreement itself.

The code defines a sufficient financing statement as follows:

> A financing statement is sufficient if it is signed by the debtor and the secured party, gives an address of the secured party from which information concerning the security interest may be obtained, gives a mailing address of the debtor and contains a statement indicating the types, or describing the items, of collateral. . . . F.S.A. § 679.9–402.

The key to perfection under Article 9 is notice. The purpose of a financing statement is to put a searcher on notice that an underlying security agreement may be outstanding. A properly filed financing statement would thus serve its intended purpose if a subsequent party would have been put on notice of an outstanding security agreement. In the present case, had plaintiff searched the Florida records he would have discovered the financing statement from Sales to Heller, containing a statement of the type of collateral involved and the signatures and addresses of both parties to the statement. Plaintiff then had a duty to inquire of Heller exactly what property, if any, was held under the statement. There is no indication in the record that such an inquiry was ever made. Were it made, it would have disclosed that Heller had an assignment of the accounts receivable from Draper-King Cole. It thus appears that the financing statement from Sales to Heller fulfilled the notice function envisaged by the drafters of the Code. . . . Therefore, having concluded that Sales' creditors were properly protected by the notice provisions of the code through the Sales–Heller financing statement and having taken notice of the fact that the chain of agreements constituted an agreement in fact between Sales, Farms and Heller that a security interest attach to the Draper-King Cole accounts, this Court finds that Heller holds a perfected security interest in the accounts in question.

(The writ of attachment should be quashed.)

Questions for Discussion

1. *Distinguish between attachment and perfection.*
2. *How had the filing in this case satisfied the requirements of the U.C.C.?*

20:4. Range of Perfection and the "Floating Lien." One may provide for various types of property to be covered by the security agreement. For example, the secured creditor may wish to have the security agreement cover the proceeds of the sale of collateral, or he may wish to cover property that may be acquired by the debtor after the security agreement has been executed. On the other hand, the creditor may be making loans to the debtor on a continuing basis. The security agreement may be made applicable to the future loans without the preparation and execution of a new agreement for each loan. Note, however, that a security agreement does not automatically cover these situations. The security agreement must mention them specifically in order for them to be subject to it. Let us examine each of these subjects briefly and then see how, when used together, they can be used to obtain a *floating lien,* over a merchant's inventory.

(1) Proceeds. There has already been extensive discussion of proceeds. *Proceeds* include whatever is received when collateral or proceeds are sold, exchanged, collected, or otherwise disposed of. Proceeds may be cash proceeds or noncash proceeds. Section 9–306(1). One way to obtain a perfected security interest in the proceeds of collateral, beyond the ten-day automatic period already mentioned, is to provide for the coverage of proceeds in the original security agreement. Then, when the financing statement is filed and there is perfection of the security interest in the original collateral, any proceeds of that collateral will also be covered. Coverage of proceeds is a particularly important provision if the collateral is of the type that is likely to be sold, exchanged, or otherwise disposed of. In this event the creditor may not have any rights to property of the debtor without coverage of proceeds.

SALZER v. VICTOR LYNN CORPORATION

315 A.2d 185 (N.H. 1974)

On November 4, 1970, George W. Salzer attached the funds of Victor Lynn Corporation, which were in a bank account in the name of "Victor Lynn Corporation, Oven Division." The attachment was as the result of a judgment obtained by Salzer against Lynn. Matrix intervened, claiming ownership of the account under an agreement with Lynn resulting from the sale to Lynn of Matrix's micro oven business. The lower court held that Salzer was entitled to satisfy his judgment out of the funds in the account. Matrix appealed this decision.

LAMPRON, J.

Under an agreement dated May 28, 1970, Matrix agreed to sell to Lynn all assets of its microwave oven business including inventory, tools, supplies, finished ovens, work in process and contracts. Matrix was to retain possession of the existing finished goods inventory which was to be released to Lynn as sales were made. The resulting accounts receivable were to be assigned to Matrix and collections thereon held in trust by Lynn and delivered in specie to Matrix to be deposited in a bank account under the name of Lynn on which Matrix was to be the sole signatory. As further security for Lynn's note to Matrix Lynn was to pledge all of the remaining inventory to be acquired under the agreement and any after acquired inventory connected with the microwave oven business. All resulting accounts and their proceeds were to be assigned to and held in trust for Matrix and handled in the manner above described.

A security agreement was executed and a financing statement covering: "All of the Debtors now owned or hereafter acquired, inventory, accounts receivable, contract rights, general intangibles relating to the production and sale of microwave ovens" was duly recorded on June 8, 1970, in accordance with the requirements of article 9 of the Uniform Commercial Code. RSA 382–A:9–103. The bank account in question, bearing the designation "Victor Lynn Corporation, Oven Division" was created exclusively to carry out the purposes of this agreement. All the proceeds of sales were segregated in this account and all disbursements were made for the benefit of Matrix, either directly to it or to others for purposes related to the oven business. The only variation from the agreement was that the funds were paid out on the signatures of an official of Lynn and an official of Matrix. It is agreed that Salzer's claim against Lynn is not connected with its oven business. Also agreed is that Victor Lynn has filed a Chapter XI proceeding under the Bankruptcy Act. . . .

The security agreement of the parties when recorded on June 8, 1970, was effective according to its terms between the parties and against creditors. RSA 382–A:9–201. At this point all persons were put on notice that Matrix had reserved a priority interest in all articles specified in the financing statement as well as "whatever (was) received when collateral . . . (was) sold, exchanged, (and) collected. . . ."

Matrix's perfected security interest in the proceeds of the collateral sold persisted in spite of Lynn's recourse to the Bankruptcy Act. RSA 382–A:9–306(4)(b)(c) provided that in the event of insolvency proceedings instituted by or against a debtor a secured party with a perfected security interest in proceeds has a perfected security interest in identifiable "cash proceeds" (money or checks) which are not commingled with other moneys or which are deposited in a segregated bank account. Contrary to plaintiff's contention the segregated account was to continue beyond six months from the date of the agreement (May 28, 1970) as it was to secure the amounts due on Lynn's promissory note payable in thirty equal monthly installments commencing August 4,

1970. The moneys in this account constituted identifiable noncommingled cash proceeds under the code.

Matrix complied with all code requirements for perfecting its security interest in Lynn's after acquired inventory accounts receivable, contract rights and general intangibles related to the production and sale of microwave ovens. In accordance with their agreement cash proceeds arising from the above were deposited in trust for Matrix in the segregated account in question under their dual control. The funds in this account were protected by Matrix's perfected security interest against the claims of Lynn's creditors including the plaintiff who cannot reach these funds to satisfy his judgment.

(Reversed.)

QUESTIONS FOR DISCUSSION

1. *Explain the steps that Matrix took to establish a security interest in the assets that it sold to Victor Lynn Corporation.*
2. *What was the purpose of establishing the bank account?*

---

(2) After-Acquired Property. *After-acquired property* of the debtor is property acquired by him subsequent to the execution of the security agreement. A creditor may provide for coverage by the security agreement of any collateral of the kind subject to the agreement that the debtor may acquire at any time during the continuation of the agreement. Section 9–204. As we shall see in Section (4), the recognition of the right to a security interest in after-acquired property greatly facilitates inventory financing if the collateral is constantly turning over.

There is a limitation on the right to obtain a security interest under an after-acquired property clause. Such a clause does not result in the attachment of a security interest in consumer goods—other than accessions to the collateral—when given as additional security unless the debtor acquires rights in the collateral within ten days after the secured party gives value. Section 9–204(2), 1972, 9–204(4)(b), 1962. The purpose is to protect the consumer who may be in a bad financial condition from encumbering all his present and future property.

One might note at this point that under the 1978 Federal Bankruptcy Code a transfer of property may be set aside under certain circumstances if it is a preferential transfer. The Bankruptcy Code has some provisions relating to the situation where a creditor obtains a security interest in after-acquired property. These provisions are discussed later in this chapter.

(3) Future Advances. The security agreement may also include *future advances* or other value given by the creditor whether or not the secured party is obligated to make the advance. Section 9–204(3), 1972. Thus the debtor may have a line of credit with the creditor under which the creditor is frequently making loans to the debtor. If the security agreement provides that future advances are covered, it is not necessary to execute and perfect a new security agreement every time an advance is made to the debtor.

(4) Floating Lien. By providing for coverage of after-acquired property, proceeds, and future advances in the security agreement, one may construct what is commonly referred to as a *floating lien.* The concept is typically referred to in connection with inventory financing. One of the problems in the financing of inventory is that although one may desire a lien on a merchant's "inventory," the individual units or items in the inventory are constantly changing. New items are being added to the inventory while others are being sold. In addition, the merchant may have a line of credit with the creditor under which new loans are constantly being made or payments against the debt being received. Under these constantly changing conditions it is often not feasible to execute a new security agreement and take the necessary steps for attachment every time a new unit of inventory is sold or acquired or a new cash advance is made. What is needed is a lien on the "inventory" in general that allows for a flow of individual items in and out of it. This need is fulfilled by the floating lien, which is created by a provision for the coverage of proceeds, after-acquired property, and future advances in the security agreement and the financing statement.

For example, suppose that Auto Bydebt is a car dealer with a line of credit to finance his inventory of cars with Creditbank. Bydebt and Creditbank enter into a security agreement that provides for coverage of proceeds, after-acquired property, and future advances.

One day Bydebt sells a new car for which he receives a used car in trade and purchases two new cars from the manufacturer with an additional sum of money obtained from Creditbank. Creditbank gets a perfected security interest in the used car under the proceeds clause, has a perfected security interest in the new cars under the after-acquired property clause, and has the new sum of money advanced to Bydebt secured by the future-advance clause—all this under the original perfected security agreement. In other words, even though the various items in the inventory have changed, Creditbank has a perfected security interest in that thing known as *inventory,* hence the term *floating lien.* See Figure 1.

Next it is necessary to examine who prevails when there are a number of creditors competing for the property of the debtor. We shall also examine situations that result in the loss by the creditor of the security interest.

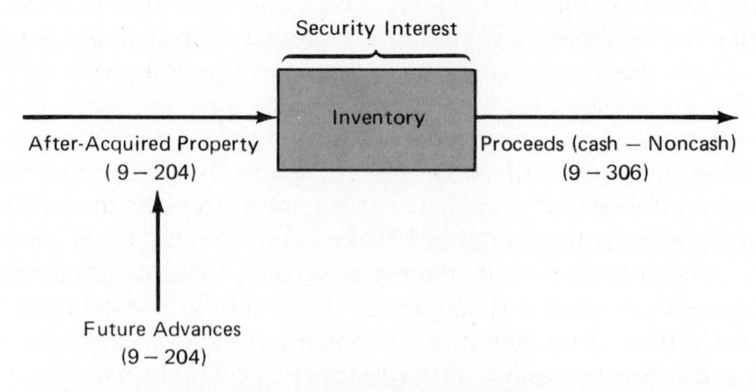

FIGURE 1. *The Floating Lien.*

20:5. PRIORITIES. Section 9–312 of the code is the most significant section of the code dealing with priority of security interests. It is divided into seven subsections, each governing a different situation. Subsection (1) refers to other parts of the code that cover specialized situations, such as the security interest of collecting banks in items being collected. Subsection (2) deals with conflicting security interests in crops, (3) with purchase-money security interests in inventory, (4) with purchase-money security interests in goods other than inventory, and (5), (6), and (7) with general rules of priority for those situations not covered by the prior sections.

In the determination of priorities it is important to determine (1) what type of collateral is involved, such as inventory or equipment; (2) whether the security interest is a purchase-money or a non-purchase-money security interest; and (3) who has the conflicting interest in the collateral, that is, another creditor or a buyer from the debtor. Now let us examine the priority rules of Section 9–312, particularly Subsections (3)–(7).

(1) Priority of Purchase-Money Security Interest. The code generally accords priority among competing interests to those creditors who have a purchase-money security interest. The purchase-money security interest may be either perfected or nonperfected. It may be an interest in inventory or other types of collateral. The following discussion examines this narrow situation.

(a) *Perfection of Purchase-Money Security Interest as Collateral Other than Inventory.* Section 9–312(4) states that a purchase-money security interest in collateral *other than inventory* has priority over a conflicting security interest in the same collateral *if* the purchase-money security interest is perfected (1) at the time the debtor receives possession of the collateral or (2) within ten days thereafter. The 1972 amendment to the code includes proceeds of the collateral. For example, suppose Bydebt purchases a car with money borrowed from Creditbank. Creditbank perfects by filing a financing statement within five days after Bydebt gets possession of the car. The same day as he gets possession of the car Bydebt borrows $500 from Loanshark and gives Loanshark a security interest in the same car, which Loanshark perfects by filing the next day. Creditbank has priority because it has a purchase-money security interest perfected within ten days of the time Bydebt got possession of the car. Loanshark, of course, does not have a purchase-money security interest.

(b) *Inventory Collateral.* Special problems arise if the collateral is inventory. If the purchase-money security interest is *not* inventory, it takes priority over other security interests, provided that it is perfected prior to the time when the debtor receives possession or within ten days thereafter. If the purchase-money security interest is in inventory collateral, the same rule applies, but without the ten-day grace period and with an additional requirement that the party seeking the purchase-money security interest in inventory collateral notify any other secured party who has previously filed against the inventory. Section 9–312(3).

The reason for the additional requirement is the nature of inventory and the commercial practice that is often followed in connection with a floating lien. Recall that a floating lien applies to after-acquired property as well as to future advances of money made by the creditor. Thus a merchant may have a line of credit with a financer under which he is constantly borrowing money for the purchase of the inventory. The danger is that an unscrupulous debtor might apply to a second financer for loans against the same after-acquired inventory. Because the second financer would have a purchase-money security interest in the *newly acquired* inventory, this transaction would normally cut off the security interest in that inventory for the future advances made by the first financer. The first financer might still make loans to the debtor ignorant of the existence of the second financer, thinking that he, the first financer, has a purchase-money security interest. Because of this problem the code requires the second financer to give notice in writing to the holder of a conflicting security interest, the first financer in our example. The notice must be given if the holder of the conflicting security interest has filed a financing statement covering the same types of inventory. The notice must be given before the filing by the purchase-money–secured party (our second financer) or before the beginning of the temporary period of perfection under Section 9–304(5). The notice must state that the person giving the notice expects to acquire a security interest in inventory of the debtor, describing it by item or type. The 1972 amendment to the code provides that such notice must be within five years before the debtor receives possession of the inventory.

BABSON CREDIT PLAN, INC. v. CORDELE PRODUCTION CREDIT ASSOC. et al

146 Ga. App. 266, 246 S.E.2d 354 (1978)

Cordele Production Credit Association brought a foreclosure action against its debtor the Vienna Dairy. The Vienna Dairy was also indebted to Babson Credit Plan who then intervened as a third party in the foreclosure suit. Babson was the assignee of Floyd A. Brewer Company who originally made the loan to the Vienna Dairy.

QUILLIAN, J.

On April 12, 1976, Cordele made a loan in the principal amount of $300,000 to six individuals d/b/a the Vienna Dairy, and took from them a promissory note secured by a deed to secure debt on the dairy's real estate. This loan was further secured by an additional security agreement on all of the dairy's cattle and other personal property associated with the operation of a farm and a dairy farm in particular. This contract contained an "after-acquired" clause on "personal property" and "equipment of every description" of the Vienna Dairy. The deed to secure debt and the financing statement were filed for record on the same date as the loan.

On June 10, 1976, The Floyd A. Brewer Company sold to the Vienna Dairy the milking equipment which is the property involved in this suit. The property consisted principally of two vacuum pumps, twenty pulsators, forty milk nipples, a "cow flow trainer," and other associated dairy milking equipment having a total value of $16,369.30. The seller retained a "Uniform Commercial Code security interest" in the equipment with a retail installment constract and security agreement. A "financing statement" was executed by the seller and the purchaser. On that same date Brewer sold the contract to the intervenor, Babson Credit Plan, Incorporated. During 1977, the Vienna Dairy defaulted on both promissory notes to Cordele and Babson. Cordele filed a petition for a writ of possession and the court permitted Babson to intervene.

After hearing evidence by both parties (Vienna Dairy did not answer or appear), the trial court reached a finding of fact that the installed milking equipment became fixtures upon affixation to the realty and became a part of the real estate. The court also concluded as a matter of law, that the intervenor's (Babson) financing statement was not perfected as to the milking equipment since no description of defendant's real estate upon which the fixtures were to be located was attached to and made a part of intervenor's recorded financing statement. The court also stated, as a conclusion of law, that plaintiff's security interest was entitled to priority under plaintiff's deed to secure debt from defendants, both by reason of "the property's attachment to the real estate and by virtue of the fact that the property would be after-acquired property as contemplated by the security instrument."

Code Ann. § 109A–9–301 (U.C.C., Ga.L.1962, pp. 156, 397) provides: "Except as otherwise provided in subsection (2) [Not applicable here.], an unperfected security interest is subordinate to the rights of (a) persons entitled to priority under 109A–9–312." Code Ann. § 109A–9–312 (U.C.C., Ga.L.1962, pp. 156, 405) sets forth criteria for priority of conflicting security interests. The sections enumerated in subsection (1) list *specific statutes* for specific problems in priority and *take precedence over* the *general rules* or priorities between conflicting security interests under subsections (2) through (6). Uniform Laws Annot., U.C.C. (Official Comment), § 9–312(1).

(1) We turn first to the priority to be accorded as between plaintiff's (Cordele) prior deed to secure a debt and intervenor's unperfected purchase money security interest. Subsection (1) of Code Ann. § 109A–9–312 refers us to Code Ann. § 109A–9–313 (U.C.C., Ga.L.1962, pp. 156, 407) for priority of security interests in fixtures. Subsection (2) thereof provides that "[a] security interest which *attaches* to goods before they become fixtures takes priority as to the goods over the claims of all persons who have an interest in the real estate except as stated in subsection (4)." (Emphasis supplied.) Thus, the criterion is *attachment* and not perfection of a security interest. Subsection (4) of Code Ann. § 109A–9–313 states that "[t]he security interests described in subsections (2) and (3) do not take priority ever . . . (c) a creditor with a prior incumbrance of record on the real estate to the extent that he makes subsequent advances . . ." Accordingly, Babson's security interest in the goods sold Vienna Dairy, although not perfected, attached to the goods before they became fixtures, and took priority over the prior recorded deed to secure debt to the extent that advances were made by the prior creditor to the common debtor under the deed to secure debt, before the goods were delivered—but not advances subsequent to attachment of the latter security interest. There was no evidence that any advances were made to Vienna Dairy by Cordele after the purchase money security interest attached. However, "[i]f the secured party perfects his interest by filing, which he may do in advance of affixation, he takes priority over subsequent realty claims as well. So long as he fails to perfect his interest he may, however, be subordinated to the claimants in subsections (4)(a), (b) and (c)." Uniform Laws Annot., U.C.C. (Official Comment) § 9–313, p. 220. Cordele does not come within the enumerated sections.

The purchase money creditor (Babson) prevails "even if the conditional seller did not file notice of his security interest." Thus, we find intervenor's security interest in the fixtures takes priority over Cordele's prior deed to secure debt.

Our last issue is the priority between Cordele's prior collateral security agreement on the dairy's cattle, farming and milking equipment, and any similar "after-acquired [personal] property," and Babson's assignor's unperfected purchase money security interest.

We return to Code Ann. § 109A–9–312, supra, to determine priority of conflicting interests in the same collateral. Subsection (a) does not have a specific section that applies to the instant factual base. Thus we must turn to the remaining general rules. Under subsection (4), the code provides that "[a] purchase money security interest in collateral other than inventory has priority over a conflicting security interest in the same collateral if the purchase money security interest is perfected at the time the debtor receives possession of the collateral or within 10 days thereafter." Babson's assignor had a purchase money security interest in the collateral—the milking equipment, effective upon delivery. It was not "inventory" collateral. Code Ann. § 109A–9–109(4). Cordele also had a security interest in the same collateral which attached as soon as the dairy received possession of the milking equipment, under the "floating lien" theory of the "after-acquired" property clause of the prior security agreement.

The "floating lien" theory that all subsequently acquired property comes under the earlier security instrument has been approved by Code Ann. § 109A–9–204(3). . . . U.C.C. (Official Comment), § 9–204(1). However, the statute provides the seller, under a purchase money contract, with a right to retain his priority *provided* he perfects his security interest before delivery or within ten days after delivery. The security interest of Babson's assignor was not perfected. Does this failure permit Cordele's security interest to prevail? We find it does not.

It would appear that we have the anomalous situation wherein the unperfected security interest of the seller of equipment, subsequently affixed to realty, takes priority over the holder of a prior security interest in the real estate, but the holder of that prior interest in the realty also has a collateral security interest in "after-acquired" personalty which would appear to prevail over an unperfected security interest of the seller. However, we must remember that this personality was affixed to the realty and became a fixture, thus passing with the realty and is subject to the same holding we stated above that an unperfected purchase money security interest prevails over a prior interest in the realty to the extent of advances made prior to attachment of the latter security interest but not those advances made subsequent to attachment.

The court erred in both conclusions of law that Cordele's security interest took priority and the judgment must be reversed.

(Judgment reversed.)

QUESTION FOR DISCUSSION

1. *Why did Babson's unperfected security interest take priority over Cordele?*

---

(2) Priority in Other Cases. Subsection (2) of Section 9–312 deals with a security interest in crops. This provision will not be covered in detail here because it has a narrow application. Subsection (5) of Section 9–312 deals with priority among conflicting security interests in those cases that are not otherwise covered by the section. Thus one must first determine that his problem is not covered by one of the other sections, such as a purchase-money security interest under either Subsection (3) or Subsection (4). If it is not covered, the rules stated in Subsection (5) apply. For ease of analysis the following discussion will treat the 1972 amendment to the code, as it has simplified the language and eliminated certain ambiguities that were present under the 1962 code without significantly changing its meaning for purposes of this discussion.

(a) *Perfection by Filing or Otherwise.* If the conflicting security interests are perfected, then the priority goes to the interest that was filed or otherwise perfected first. The priority dates from the time a filing is first made covering the collateral or the time the security interest is first perfected, whichever is earlier, provided that there is no period thereafter when there is neither filing nor perfection. Section 9–312(5)(a), 1972 Amendment.

For example, suppose Selcredit files against Bydebt on March 1. Latecredit files against Bydebt on April 1. Latecredit makes a non-purchase-money advance against certain collateral on May 1. Selcredit makes an advance against the same collateral on June 1. Selcredit would have priority even though Latecredit's advance was made earlier and was perfected when made, and it makes no difference whether or not Selcredit knew of Latecredit's interest when he made the advance. If we alter the example and assume that Selcredit and Latecredit both make non-purchase-money advances against the same collateral, the first to perfect by filing or by some other method would prevail, regardless of which party makes the loan first.

One should note at this point that the 1972 amendment to the code states that for purposes of Subsection (5) of Section 9–312 the date of filing or perfection as to collateral is also a date of filing or perfection as to proceeds. Section 9–312(6), 1972 Amendment.

(b) *No Perfection.* The code provides that as long as security interests are not perfected, then the first to attach has priority. Section 9–312(5)(b), 1972 Amendment. However, if two parties have conflicting security interests, once the conflict becomes apparent it is likely that at least one of the parties would then perfect by filing. When that event occurs, the party who perfects first would prevail, regardless of who achieved attachment first. It thus appears that the rule has little practical application because it is hard to imagine a conflict in which one party would not take the additional step of perfecting the security interest.

To recapitulate: in order to determine the priority among conflicting security interests, it is necessary to determine what is the nature of the collateral, whether the security interest is a purchase-money security interest or a non-purchase-money security interest, and finally whether there is perfection. When these steps have been followed, it becomes a rather simple matter to apply the rules stated in Section 9–312 in order to determine priority.

(3) **Cutting Off the Rights of the Secured Creditor.** There are a number of situations in which the rights of the secured creditor may be cut off even if the security interest is perfected. Two of the most common situations are the case in which the secured collateral is sold as a consumer good and the case in which there is created what is commonly known as a *common-law lien.* The rights of the parties against the trustee in bankruptcy should also be mentioned.

(a) *Debtor's Sale of Consumer Goods.* There are two situations in which the rights of a secured creditor may be cut off from collateral as a result of its sale: a sale of consumer goods in the ordinary course of business and a sale of consumer goods that have not been purchased in the ordinary course of business. Remember that the creditor may still protect himself in this event by covering the proceeds of such a sale in his financing statement.

(i) BUYER IN THE ORDINARY COURSE OF BUSINESS. If one purchases consumer goods in the ordinary course of business, the goods are purchased free of any security interest in them created by the seller, even though that security interest is perfected and even though the buyer knows of the security interest. Section 9–307(1). Suppose one buys a new car from an automobile dealer whose inventory is financed by a bank and the bank has a perfected security interest in the dealer's inventory. The perfected security interest is known to the consumer. The consumer takes the car free of the bank's lien even if the dealer subsequently defaults on the loan. The purpose of the rule is so that consumers can purchase goods from merchants without concern as to whether the merchant's inventory is financed. Have you ever considered whether the product you are purchasing from a local department store is the subject of a perfected security interest on the merchant's inventory? If the law were otherwise, one would have to determine the answer to that question before he could purchase goods with any degree of confidence.

(ii) BUYER NOT IN THE ORDINARY COURSE OF BUSINESS. If one purchases consumer goods not in the ordinary course of business, he takes the goods free of a perfected security interest if he has given value for the good and if it is for his own personal, family, or household purpose. In addition, he must buy without knowledge of the security interest. The goods purchased are subject to a security interest if a financing statement is filed prior to the purchase. Section 9–307(2). For example, suppose Bill Buyer purchases a car from his neighbor, Sam Seller. Seller has financed the car through Friendly Bank, who has filed a financing statement against the car. Seller never pays off the debt after selling the car to Buyer. The car Buyer has purchased is still subject to Friendly's security interest. The result would be the same if Friendly had not filed but Buyer was aware of the security interest when he made the purchase. The reason Subsection (1) of Section 9–307 does not apply is that Buyer has not purchased the car in the ordinary course of business.

BALON v. CADILLAC AUTOMOBILE CO. OF BOSTON
CADILLAC AUTOMOBILE CO. OF BOSTON v. GIBERT

303 A.2d 194 (N.H. 1973)

Cadillac Automobile Co. of Boston sold two Cadillacs to Russell Saia, who in turn sold them to John Balon and Stanley Gibert under the circumstances set forth below. Upon default in payment on the debt owed by Saia for the two cars, Cadillac Automobile repossessed one car and filed a replevin action to recover the other. The trial court ruled that Balon and Gibert had clear title to each of the cars. Cadillac Automobile appealed.

LAMPRON, J.

On September 28, 1965, Charles Pernokas was a salesman for Cadillac Automobile in Boston, Massachusetts. He had previously worked for another employer as a car salesman with Russell Saia. On that day Saia came to Cadillac Automobile looking for a convertible for a customer. Pernokas showed him two cars and Saia said he would hear from him shortly. He telephoned soon thereafter saying that his customer, Peter J. Russell, would take one of the Cadillac convertibles and gave the required credit information and references. On the next day, Saia telephoned Pernokas again and told him another customer, Joseph P. DeLuca, would take the other convertible and gave credit information on him.

Pernokas testified that in each instance he delivered the car to Saia's place and that Saia and another person, supposedly Russell in one instance and DeLuca in the other, identified himself and signed the security agreements. The selling price, $5300 for each car, was paid by a $1000 cash down payment on each and the balance financed on a conditional sale agreement.

At about that time, an individual named Arthur Freije told Balon in Manchester that "somebody" had a "friend" who could get a good deal on Cadillacs. "Somebody" was Fred

Sarno who had accompanied Saia on his visit to the Cadillac garage after which the two cars in question were bought. The "friend" was Russell Saia. Balon passed the information along to his stepfather, Gibert, and eventually both Balon and Gibert purchased a Cadillac convertible through Freije for $4300 each.

The October and November payments on these Cadillacs were not made to Cadillac Automobile. As a result of these defaults, Simons, its credit manager, made an investigation and concluded that Peter J. Russell and Joseph P. DeLuca, the apparent purchasers, did not exist and decided that "both of these were two straw deals. . . ." The trial court properly found on the evidence that the two Cadillac automobiles in question were purchased by Russell Saia and that he was the principal in their sale to Balon and Gibert.

RSA 382–A:9–307, insofar as material here, reads as follows: (1) "A buyer in the ordinary course of business . . . takes free of a security interest created by his seller even though the security interest is perfected and even though the buyer knows of its existence. (2) In the case of consumer goods . . . a buyer takes free of a security interest even though perfected if he buys without knowledge of the security interest, for value and for his own personal, family or household purposes . . . unless prior to the purchase the secured party has filed a financing statement covering such goods." Under Massachusetts law the secured interest of Cadillac Automobile was perfected when the agreement of the parties was executed. Mass. Gen.Laws. Ann.C. 106 § 9–302. However the security agreements covering these two automobiles were never filed. RSA 382–A:9–403.

The buyer protected by § 9–307(1) is one who purchases in the ordinary course of business from a person in the business of selling goods of the kind involved. § 1–201(9). Hence § 9–307(1) applies primarily to purchases from the inventory of a dealer in the type of goods sold. The buyer protected under § 9–307(2) is one who purchases goods for consumer use, that is, for personal, family or household purposes, from a consumer seller. In order to fall within the protection of this section the goods must be consumer goods in the hands of both the buyer and the seller.

We hold that the trial court properly found and ruled that Russell Saia was a dishonest consumer purchaser of these automobiles. We further hold that these cars remained consumer goods in his hands at the time of the sales to Balon and Gibert who are protected by RSA 382–A:9–307(2) if they were good faith consumer buyers for value without knowledge of Cadillac's security interest.

The company maintains, however, that the only conclusion which can be reached on the evidence is that Balon and Gibert "could not have conceivably held honest convictions that these transactions were legitimate." In support it cites § 1–201(19) which provides: "'Good faith' means honesty in fact in the conduct or transaction concerned." By its terms this is a subjective standard of good faith, that is, whether the particular purchaser believed he was in good faith, not whether anyone else would have held the same belief. The test is what the particular person did or thought in the given situation and whether or not he was honest in what he did.

There was evidence that Gibert had known Freije, who made the approaches which culminated in these sales, in a social way for about fifteen years. His wife had known him all her life. Balon knew him also and had purchased a 1963 Cadillac from him without any untoward incidents. Balon and Gibert learned from inquiries made to dealers known to them that the asking price of $4300 was consistent with prices at which such cars could be bought. The explanation advanced that these convertibles sold in September, when the new models were due, could be found by them plausible reasons for the price quoted. Simon testified that when Balon and Gibert came to Boston after their cars were taken they seemed genuinely concerned and trying to figure out what happened. There was no evidence that they had actual knowledge of the status of the title to these cars. The fact that others might have acted differently, made more inquiries, or been more suspicious does not require a conclusion that they lacked good faith when they purchased these cars. The evidence is clear that they paid value and bought for personal, family, or household purposes.

We hold that the trial court properly found and ruled that Balon and Gibert were good faith consumer buyers for value from a consumer seller without knowledge of Cadillac Automobile's security interest which had not been filed.

(Affirmed.)

QUESTIONS FOR DISCUSSION

1. *Distinguish between the protection afforded under Section 9–307(1) and Section 9–307(2).*
2. *In this case would the outcome have been different if Cadillac Automobile Co. had perfected its security interest by filing? Would filing affect the outcome of a case when Section 9–307(1) is applicable?*

(b) *Common-Law Lienors: Liens Arising by Operation of Law.* Some liens arise by operation of law either by statute or from the common law. The most common of these involve claims arising as the result of work that has enhanced the collateral. Examples are mechanics' and materialmen's liens and garagemen's liens. The code provides that if a lien arises as the result of the services or materials supplied in the ordinary course of business with respect to goods subject to a security interest, that lien prevails over the perfected security interest. The only exception is if a statute expressly provides for the lien and states a different rule. Section 9–310.

The most common example today arises in the case of automobiles. Suppose Harry Hotrod has purchased a car that is financed by Friendly Bank. Friendly Bank has a perfected security interest in the car. Hotrod takes the car to Manny Mechanic for repairs but does not pay. As long as Mechanic keeps possession of the car, he has a garageman's lien in most jurisdictions that would have priority over the perfected security interest of Bank, unless there is a statute expressly providing otherwise.

GABLES LINCOLN-MERCURY, INC. v. FIRST BANK AND TRUST COMPANY OF BOCA RATON

219 So.2d 90 (Fla. App., rehearing denied, 1969)

Delray Motors, Inc., sold an automobile for $2,070.62 on credit. It then assigned the credit balance to First Bank and Trust Company of Boca Raton. The bank perfected a security interest in the car. Subsequently the car was taken to Gables Lincoln-Mercury, Inc., for repairs. The bill for repairs was not paid, and Gables retained possession of the car, asserting a garageman's lien under the applicable Florida statute. The bank filed an action of replevin against Gables. The trial court found in favor of the bank, and Gables appealed.

CARROLL, C.J.

This is an appeal by the defendant below from an adverse judgment in replevin. The determinative question is whether a garageman's lien for repairs on an automobile has priority over an existing security interest represented by a retain title contract.

All transactions in this matter occurred after January 1, 1967, and therefore the provision of the Uniform Commercial Code relating to priorities is pertinent here.

Under that statutory provision it is apparent that a garageman's lien is entitled to prevail over such an existing security interest when the statute which confers the former does not expressly provide otherwise.

The lien held by the garageman in this instance is one provided for under Chapter 713, Part II, Fla.Stat., F.S.A., entitled "Miscellaneous Liens." Section 713.50 thereof reads: "Liens prior in dignity to all others accruing thereafter shall exist in favor of the following persons, upon the following described property, under the circumstances hereinafter mentioned in Part II of this chapter, to wit:" Among the liens provided for thereunder in subsequent sections of the statute is one "in favor of persons performing labor or services for any other person, upon the personal property of the latter upon which the labor or services is performed." § 713.58.

Following the listing of the liens conferred by the statute for labor, services and materials furnished under a variety of circumstances, there is a section entitled "Priority of foregoing liens" (§ 713.73), which reads as follows:

> Liens for labor and liens for material provided for by this law shall take priority among themselves according to the times that the notices required to create such liens respectively were given or were recorded in the cases where record is required; that is to say, each such lien which shall have attached to the property shall be paid before any such lien which shall have subsequently attached thereto, shall be entitled to be paid.

It is obvious that the last above quoted statutory provision, dealing with priorities of liens, has no bearing on this case, since it has reference to priorities among liens for labor and material which are provided for by that chapter, and the lien relied upon by the bank is not one created or provided for therein, but is a contract lien.

Appellant argues, and we agree, that under the express provision of § 679.9–310 Fla.Stat., F.S.A. its garageman's lien for repairs to the automobile, provided for by statute, has priority over the bank's (contract) security interest inasmuch as the statute which conferred the garageman's lien for repairs did not expressly provide otherwise.

The appellee argues that the language of the opening paragraph of Part II of Chapter 713 (§ 713.50) should be construed as "expressly providing" that the garageman's lien for repairs shall not have priority over the contract lien of the bank because it accrued after the latter. We must reject that argument. What the statute there provides is that a lien conferred thereunder will be prior in dignity to other liens accruing thereafter. That provision is not inconsistent with the position taken here by the garageman. There is nothing in that section, or elsewhere in the statute creating the garageman's lien, which expressly provides that such a lien may not be superior to an existing "security interest," as distinguished from a labor and material lien conferred by statute.

On the other hand, section 679.9–310 expressly provides that a lien, such as the garageman's lien for repairs to the automobile in this case, acquired by "a person in the ordinary course of his business" on "goods subject to a security interest" is superior to the latter, subject only to the proviso that where the lien is statutory and the statute expressly provides otherwise, such priority will not exist. Since the proviso did not apply here, the clear intent of the statute (§ 679.9–310) that the garageman's lien shall have priority over the bank's security interest, must be observed.

(Reversed and remanded.)

QUESTIONS FOR DISCUSSION

1. *What policy considerations are there for granting priority to a garageman's lien over a previously perfected security interest?*
2. *What must the garageman do in order to obtain and maintain his lien?*

(c) *Trustees in Bankruptcy.* Under the Bankruptcy Code of 1978, a trustee in bankruptcy may set aside certain transfers of assets by the debtor if they result in a preference of one creditor over another. The conditions that must be shown for a transfer to be set aside as a preference are as follows: (1) the transfer must be to or for the benefit of a creditor, (2) the transfer must be for or on account of an antecedent debt owed by the debtor prior to the transfer, (3) the transfer must be made while the debtor is insolvent, (4) the transfer must have been made during the ninety days immediately preceding the commencement of the case, (5) if the transfer is to an insider who had reason to know of the debtor's insolvency the period is one year, and (6) the transfer must enable the creditor to or for whose benefit the payment was made to receive a greater percentage of his claim than he otherwise would have received.

These provisions have some application to the secured transactions section of the Uniform Commercial Code. Remember that the U.C.C. security interests may automatically become perfected on after acquired property such as a new inventory. If such a security interest is perfected within ninety days of the time when the debtor files a petition in bankruptcy, then may this security interest be set aside as a preference? Under the Bankruptcy Act that preceded the 1978 Bankruptcy Code, the courts generally found that a voidable preference did not exist in this situation and, therefore, the security interest in the after-acquired property took precedence over the claims of a trustee in bankruptcy. The new Bankruptcy Code of 1978 changes this law.

Under the 1978 Bankruptcy Code, the creditor who receives a security interest in inventory may have that security interest set aside if by reason of that security interest the creditor has improved his position at the expense of the bankrupt's estate during the ninety-day period before the date of filing the petition. For example, assume that ninety days before filing a petition in bankruptcy the inventory which is collateral for a debt of $10,000 has a value of $7,000. On the date of filing the bankruptcy petition, the value of the inventory has increased to $9,000 and the debt has remained at $10,000. The $2,000 increase in value of the inventory represents an improvement in the position of the creditor with the security interest in the inventory. To the extent that this improvement results from a contribution by the bankrupt's estate, a preference exists. On the other hand, if the improvement in the estate was the result of a further loan by the creditor in the amount of $2,000, a preference would not exist. Likewise, if the increase in value of the estate was the result of market forces, such as seasonal price fluctuations, then a preference also would not exist.

The new Bankruptcy Code also has another provision affecting secured transactions. If a creditor loans the debtor funds for the purchase of new property and perfects his security interest in that new property within ten days after the security interest attaches, then the security interest will not be set aside as a preference. However, if the lender's security interest is not perfected within ten days after attachment, it may be set aside as a preference.

20:6. Rights and Remedies of Debtor and Creditor. (1) Rights Before Default. The security agreement itself will spell out the terms of the security interest, and it is, of course, binding on the parties. Such matters as the terms of payment, who bears the risk of loss for damage to the collateral, and the debtor's obligation to maintain or repair the collateral are frequently covered in the security agreement.

In addition, the code provides for certain rights and obligations of the parties, particularly when the collateral is in the secured party's possession. In general the code requires that the secured party use reasonable care in the custody and preservation of the collateral in his possession. Section 9–207(1). Furthermore, unless otherwise agreed, reasonable expenses incurred (such as cost of insurance) in the custody, preservation, and use of the collateral are chargeable to the debtor. The risk of loss is on the debtor to the extent of deficiency in insurance coverage. The secured party may hold any increase or profit from the collateral (except money) received from the collateral, and if money is received, it should be applied toward the reduction of the debt. The secured party must keep the collateral identified, although he may commingle fungible collateral. Finally, the secured party may repledge the collateral as long as the repledging does not interfere with the debtor's right to redeem it. Section 9–207(2).

If there is any loss occasioned by the secured party's failure to meet any of his obligations, he is responsible for the loss, but this does not cause him to lose his security interest. Section 9–207. The secured party may use or operate the collateral for the purpose of preserving its value. Section 9–207(4). Note that these obligations and rights are in effect only if there is no contrary agreement between the parties. Therefore any of the risks may be shifted to the other party by a placement of the appropriate clause in the security agreement.

The code also provides a procedure whereby a debtor may obtain from the secured party a statement as to his indebtedness and in some cases a statement of collateral. The secured party must comply with such a request within two weeks after receipt. Section 9–208. The need for such a statement may arise in a situation in which a debtor has much of his inventory financed and wishes to sell his business to a third party who demands a statement of indebtedness. The debtor may obtain such a statement under this procedure, but the third party has no right to make such a demand of the secured creditor.

(2) **Rights After Default.** The whole purpose of obtaining a security interest in specific collateral of the debtor is to allow the secured creditor to have the right to use that specific property for the satisfaction of the debt. In general these rights are governed by the security agreement and the applicable sections of Article 9. The secured party may also reduce his claim to judgment, foreclose, or otherwise enforce the security interest by any available judicial procedure. Section 9–501(1).

One of the most important rights of the secured creditor is to take possession of the collateral upon default by the debtor unless it is otherwise agreed. If it is possible to do so without breaching the peace, the secured creditor may take the collateral without resorting to judicial process. This right is the source of many apocryphal stories concerning repossessions, mostly of automobiles, in the wee hours of the morning. The secured creditor also has the right to render equipment unusable and to dispose of it on the debtor's premises. Furthermore, if the security agreement provides, the secured party may require the debtor to assemble the collateral and make it available to him at a place to be designated that is reasonably convenient to both parties. Section 9–503. All of this presupposes that the debtor will not act in such a way as to require a breach of the peace in order to enforce these rights. If he does, then the secured party must resort to judicial process.

After default by the debtor the secured party may sell, lease, or otherwise

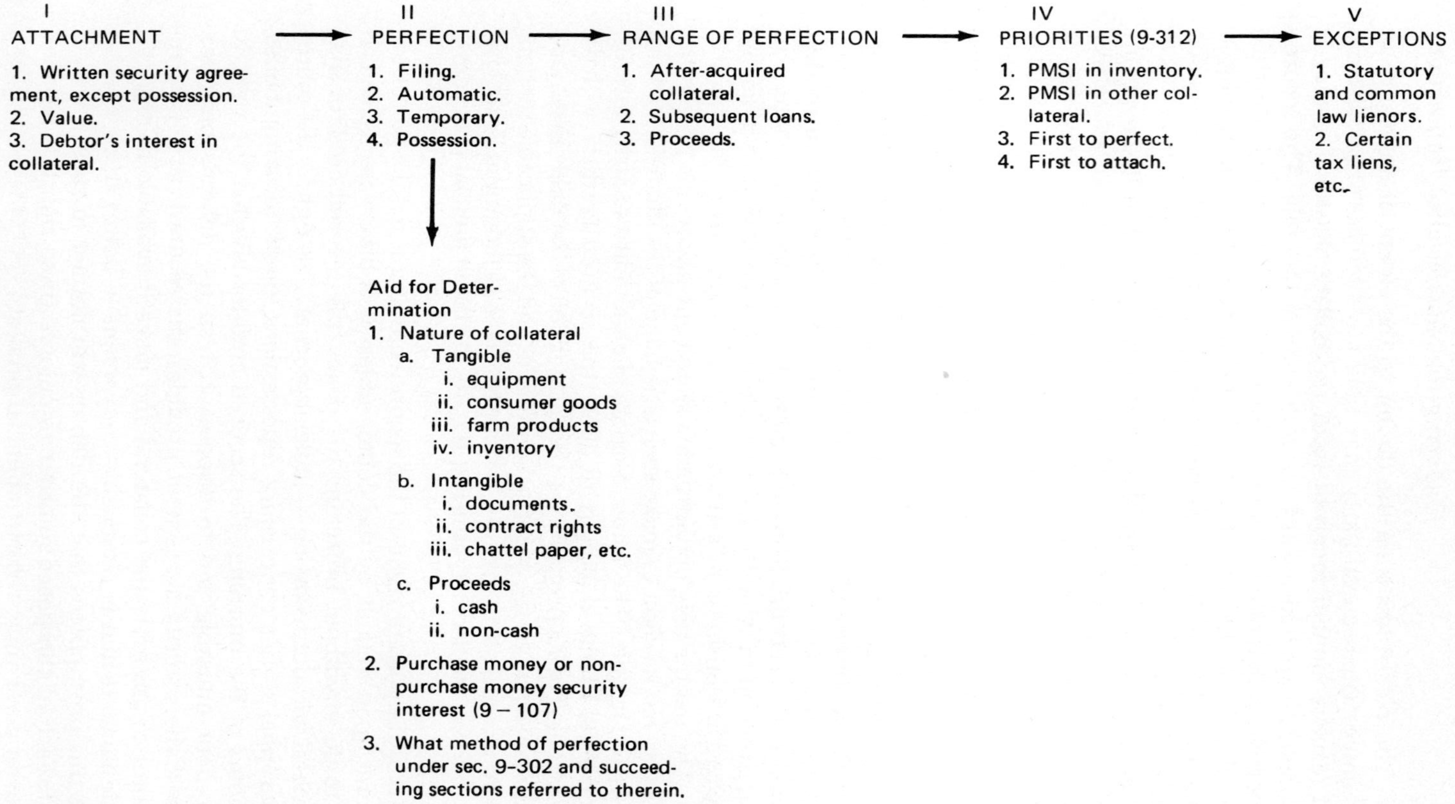

I
ATTACHMENT →

1. Written security agreement, except possession.
2. Value.
3. Debtor's interest in collateral.

II
PERFECTION →

1. Filing.
2. Automatic.
3. Temporary.
4. Possession.

↓

Aid for Determination

1. Nature of collateral
   a. Tangible
      i. equipment
      ii. consumer goods
      iii. farm products
      iv. inventory
   b. Intangible
      i. documents.
      ii. contract rights
      iii. chattel paper, etc.
   c. Proceeds
      i. cash
      ii. non-cash
2. Purchase money or non-purchase money security interest (9 – 107)
3. What method of perfection under sec. 9-302 and succeeding sections referred to therein.

III
RANGE OF PERFECTION →

1. After-acquired collateral.
2. Subsequent loans.
3. Proceeds.

IV
PRIORITIES (9-312) →

1. PMSI in inventory.
2. PMSI in other collateral.
3. First to perfect.
4. First to attach.

V
EXCEPTIONS

1. Statutory and common law lienors.
2. Certain tax liens, etc.

FIGURE 2. *Steps in Determination of the Quality of a Security Interest. (Note that this chart is not all-inclusive. It is intended to give an outline of the steps one might follow in determining the quality of a security interest.)*

dispose of any or all of the collateral. It may be disposed of in its present condition, or it may be prepared for resale. Once the goods are disposed of, the reasonable expenses of retaking, holding, preparation for sale, and so on, as provided for in the agreement and not prohibited by law, may be deducted from the proceeds. Reasonable attorney's fees and legal expenses may also be deducted. Then the amount of the indebtedness is deducted from the proceeds. Next, any subordinate interest must be paid if the holder of the subordinate interest notifies the secured party of this interest before disposition of the collateral. Section 9–504(1). The secured party must account to the debtor for any surplus as a result of the disposition. On the other hand, if the disposition of the collateral does not produce proceeds sufficient to satisfy the debt the debtor is liable for any deficiency.

The property may be disposed of in a public or private sale and in a single unit or in parcels, as long as the method, manner, place, and time are commercially reasonable. Except when the collateral is perishable or threatens to decline in value speedily or is to be sold in a recognized market, the debtor is entitled to reasonable notification of the time and place of any public sale or to reasonable notification of the time after which a private sale or disposition of the collateral is to be made. Except in the case of consumer goods, a similar notice must be sent to any person who also has a security interest in the collateral and who has filed a financing statement or who is known to the secured party. Section 9–504(3).

There are some instances in which the secured creditor is compelled to dispose of the collateral. If the debtor has paid 60 percent of the cash price of a purchase-money security in consumer goods or 60 percent of the loan of another security interest in consumer goods and has not signed a statement renouncing his rights in the goods, the secured party who has taken possession of the collateral must dispose of it within ninety days or he is liable to the debtor for conversion. Section 9–505(1). This provision, of course, prevents a creditor who has only a small financial interest in collateral of this type from seizing it and holding it indefinitely without returning to the debtor the debtor's large interest in the property.

In other cases involving consumer goods or other collateral, after default the secured party may propose to keep the collateral in satisfaction of the obligation. He may do so by notifying the debtor and, except in the case of consumer goods, any other secured creditor who has filed or who is known to the secured creditor. In the absence of a timely written objection, the secured party may retain the property in satisfaction of the obligation. Section 9–505(2). There may be situations in which this arrangement is more satisfactory to all parties concerned. Section 9–507 provides sanctions against the creditor for failure to comply with the requirements of this part of the code.

The debtor may redeem the property at any time before the secured party has disposed of the collateral or entered into a contract for its disposition unless otherwise agreed in writing. The debtor may redeem the property by tendering fulfillment of all obligations secured by the collateral, including any costs the secured party has incurred as a result of retaking the property or otherwise preparing for its disposition. Section 9–506.

FROST v. MOHAWK NATIONAL BANK et al.
347 N.Y.S.2d 246, 74 Misc.2d 912 (N.Y. 1973)

Judith Frost purchased a 1971 Fiat automobile from Hickey Ford Sales on July 7, 1971. Mohawk National Bank financed the sale under a retail installment contract that provided for Mohawk's right to repossess the car upon default by Frost. Frost failed to keep up the payments and the bank repossessed the car by removing it from its parking place.

Frost filed a motion for a preliminary injunction to keep the Mohawk National Bank from continuing to deprive her of the use of the car during the pendency of the full case. Mohawk contended that it had the right to remove the vehicle without judicial process as long as it was done peacefully. Frost contended that the provisions of the Uniform Commercial Code allowing for the repossession are unconstitutional.

CONWAY, J.

There is no longer any question but that where repossession is obtained through judicial process under the replevin statutes, a person must be afforded notice and an opportunity to be heard prior to any taking, except in extraordinary circumstances. *Sniadach* v. *Family Finance Corp.*, 395 U.S. 337, 89 S.Ct. 1820, 23 L.Ed.2d 349; *Fuentes* v. *Shevin*, 407 U.S. 67, 92 S.Ct. 1983, 32 L.Ed.2d 556.

In the opinion of this court, the rationale of the Sniadach and Fuentes cases is that in the exercise of "due process" by the State, notice and an opportunity to be heard must be given where the object of the action is the taking of property in which the possessor has a significant property interest except in extraordinary circumstances.

The court has examined the retail installment contract in the instant case, and finds that there is a clause which allows the seller, in the event of a default, the right without notice or demand, to enter upon any premises where the vehicle may be found and take possession of and remove the vehicle without process of law.

Section 9–503 of the Uniform Commercial Code provides in part:

> Unless otherwise agreed a secured party has on default the right to take possession of the collateral. In taking possession a secured party may proceed without judicial process if this can be done without breach of the peace or may proceed by action. . . .

Many text writers on the Uniform Commercial Code have expressed opinions that "self help" under section 9–503 of the code may very well be struck down under the rationale of *Sniadach* v. *Family Finance Corp.*, . . . and *Fuentes* v. *Shevin*. However, they all make a distinction between "self help" and replevin prior to a hearing even though both involve a "taking" of the debtor's property before any judicial determination of the validity of the taking. Most of them make a distinction when self-help repossession is authorized by a private contract and the consumer has contracted for the security interest.

It is the opinion of this court that an agreement entered into in compliance with section 9–503 of the Uniform Commercial Code must be held to be legal especially where the debtor has expressly authorized the "self help" provisions of the contract. Such an agreement cannot be held to be unconscionable when such terms are expressly authorized by statute. To stretch the rationale of the Sniadach case and later cases following it to include a prohibition of "self help" under a security agreement would require this court to go beyond the point permitted by good reason under the circumstances of this case.

Question for Discussion

1. *Note that the court discussed the Fuentes and Sniadach cases, which prohibited repossession by judicial process without a hearing. How did the Frost case differ from those two cases?*

SUMMARY

The code has eliminated the distinctions between various forms of financing and calls them all a security interest. The parties are debtors and creditors.

The first step in creating a security interest is for the creditor to obtain attachment of the interest on the debtor's property. This requires a security agreement, value given by the creditor to the debtor, and the debtor must have rights in the property to be used as collateral.

In order to get protection against parties other than the debtor, the creditor must obtain perfection of the security interest. One primary method of accomplishing this is to file a financing statement covering the property. Another method is for the creditor to retain possession of the collateral. There are times when perfection is accomplished in other ways. For instance, there may be times when there is temporary or automatic perfection of the security interest.

In deciding what type of perfection applies, consideration may be given to the nature of the collateral. Tangible collateral may be either consumer goods, equipment, farm products or inventory. Intangible collateral may be such things as documents, instruments, chattel paper, accounts or contract rights. Collateral may also consist of proceeds. The security interest may be either a purchase-money security interest or a non-purchase-money security interest.

One method of inventory financing involves the concept of a "floating lien" over the inventory. This may be accomplished by having the security agreement cover any proceeds from sale of the collateral, after-acquired property, and any future advances to be made by the lender.

A purchase-money security interest in collateral that is not inventory takes priority over the competing security interests, provided that it is perfected prior to the time the debtor receives possession or within ten days thereafter. For purchase-money security interests in inventory collateral the same rule applies but without the ten-day grace period and with an additional requirement that the party seeking the purchase-money security interest in inventory collateral notify any other secured party who has previously filed against the inventory. If these, or some other rule, do not apply then if the conflicting security interests are perfected, priority goes to the interest that was filed or otherwise perfected first. If there is no perfection, the first security interest to attach has priority.

If one purchases consumer goods in the ordinary course of business, the goods are purchased free of any security interest in them created by the seller, even though that sercurity interest is perfected and even though the buyer knows of the security interest. If one purchases goods not in the ordinary course of business, he takes the goods free of a perfected security interest if he has given value for the goods and if it is for his own personal, family, or household purpose. In many jurisdictions liens, such as mechanic's, materialman's and garageman's liens, prevail over a perfected security interest.

The secured party has certain obligations to care for the debtor's property that is in his possession. Any loss occasioned by the secured party's failure to meet these obligations is the responsibility of the secured party. After default, the secured party may take possession of the property, unless otherwise agreed, and may dispose of it at a public or private sale.

PROBLEMS

1. Mogul Enterprises Inc. loaned money to A-1 Industries Inc. Its security agreement stated that the agreement applied to "all assets of A-1 Industries regardless of type or description now owned by A-1 Industries. . . ." The agreement also listed certain specific items of collateral but Mogul claimed that this language was sufficient to bring A-1 Industries' accounts receivable and inventory within the agreement. Was Mogul correct? Mogul Enterprises, Inc. v. Commercial Credit Business Loans, Inc. 585 P.2d 1096 (1978).
2. Able files against Charlie (debtor) on February 1. Baker files against Charlie on March 1. Baker makes a non-purchase-money advance against certain collateral on April 1. Able makes an advance against the same collateral on May 1. As between Able and Baker, who has priority? What is the difference if Baker's advance created a purchase-money security interest?
3. On February 1, Able makes advances to Charlie under a security agreement that covers "all the machinery in Charlie's plant" and contains an after-acquired property clause. Able promptly files his financing statement. On March 1, Charlie acquires a new machine, and Baker makes an advance against it and files his financing statement. On April 1, Able, under the original security agreement, makes an advance against the machine acquired on March 1. Assume that Baker's advance created a purchase-money security interest. Who has priority? Assume that Baker's advance did not create a purchase-money security agreement, who has priority?
4. St. Petersburg Bank loaned money to Bob King, Inc., and Turk, its president, an automobile dealer. The loan was secured under a security agreement by the inventory of Bob King, Inc., and gave the bank the right to repossess the collateral in the event of default. When a default occurred, the bank repossessed the collateral and sold it without giving notice to Turk. The sale of the collateral resulted in an amount substantially less than the balance due on the debt. The bank sued Turk for the deficiency. Decide. Turk v. St. Petersburg Bank and Trust Co., 281 So.2d 534 (Fla.App., 1973).
5. Parlor Piano Company sells a grand piano to Harry Homeowner under a conditional-sales contract, which Harry signs. Harry has the piano placed in his living room to use for his own enjoyment. A judgment creditor of Harry's levies on the piano. Who has priority? Would it make any difference if Harry were a music teacher?
6. In June 15, 1960, National Cash Register Co. sold Edmund Carroll doing business as Kozy Kitchen, a cash register under a conditional-sales contract signed by Carroll. The cash register was delivered between November 19

and 25, and a new contract was entered into. A financing statement was filed on December 20. On November 18, 1960, Firestone and Co. made a loan to Carroll, who put up certain personal property as collateral under a security agreement that provided for coverage of "all contents of luncheonette including equipment such as: booths and tables. . . ." There also was an after-acquired property clause. This agreement was the subject of a financing statement filed on November 18. Carroll defaulted on both loans. In a contest between National and Firestone over the cash register, who prevailed? National Cash Register v. Firestone and Co., 191 N.E.2d 471 (1963).

7. National Cash Register Co. entered a conditional-sales contract for the sale of goods to Excel Stores, Inc. The filed financing statement indicated that the debtor was "Excel Department Stores" instead of "Excel Stores, Inc." When Excel Stores went bankrupt, the trustee contended that the goods covered under the financing statement were not subject to National's security interest because of the mistake. Decide. In Re Excel Stores, Inc., 341 F.2d 961 (1965).
8. First National Bank loaned Safety Plastics, Inc. $9,217 and secured its loan by filing a financing statement on June 6, 1973. The statement provided that future advances were covered. On March 8, 1974, Associated Business Investment Corporation loaned Plastics $54,000 and filed a financing statement on the same property that day. Burkhart–Randall loaned Plastics $101,000 and filed its security agreement against the same property on April 26, 1974. First National made two other loans to Plastics. One on March 21, 1975, and the other on March 1, 1976.

   The first loan to First National Bank was paid in full in December 1973, but the security interest was not released. Associated claims its interest should prevail over First National's subsequent loans. First National contends its 1975 and 1976 loans are protected by the 1973 security agreement. Is First National correct? Associated Business Investment Corporation v. First National Bank of Conway, 573 S.W. 328 (Ark. 1978).

SECTION IV

Commercial Paper

CHAPTER 21

Introduction, Nature, Types, Parties

21:1. Introduction. Perhaps one of the most important concepts developed over the years for use by people in business is the concept of commercial paper. In fact, today our whole system of commercial enterprise is virtually founded on various types of commercial paper. Among the variety of commercial paper are notes, drafts, checks, and certificates of deposit, as well as some other types. As you might expect, a rather complicated set of rules has developed to govern these types of instruments. The purpose of this section is to examine some of the more important concepts and rules that relate to commercial paper.

Some types of commercial paper can be traced as far back as 2000 B.C. The laws and concepts that govern our commercial paper have their antecedents in the *lex mercatoria* or law merchant developed in medieval Europe. The law merchant later became absorbed by the common law and was transferred to this country by the English colonists. Eventually the laws affecting negotiable instruments were codified when the National Conference of Commissioners on Uniform State Laws wrote the Uniform Negotiable Instruments Act in 1896. This act was referred to as the Negotiable Instruments Law or NIL and was adopted by all the states. As you already have learned, the Uniform Commercial Code was first adopted in 1952 by Pennsylvania, and later subsequent amended texts were adopted by all the states except Louisiana, which itself has adopted parts of the code. The code now covers and replaces the law formerly treated by the old NIL in Article 3. Our discussion of commercial paper will focus on this article of the code.

21:2. The Nature of Negotiable Instruments. The term *negotiable instrument* implies that there must be an attribute of transferability of the instrument between parties. The need for such an instrument soon became apparent to the medieval traders. After all, it was rather risky to transport gold bullion from England to Italy in those days. What was needed was a document that could be exchanged for goods without fear that it could be freely transferred to other parties who did not participate in the original transaction. The negotiable instrument therefore developed certain attributes that were not present in the ordinary contract.

(1) Negotiation Versus Assignment. Perhaps the best way to understand the nature of a negotiable instrument is to compare it with an ordinary contract. We have already seen that certain contract rights can be transferred from one person to another by *assignment.* It is not always so, but more often than not what is transferred is the right to receive money under the contract. For instance, suppose Alice Assignor agrees to deliver her motorbike to Ollie Obligor, immediately, in return for $100, which Ollie agrees to pay in six months. Alice then assigns her right to receive the $100 to Artie Assignee in return for a consideration. Recall that there are several rules that govern such an assignment. In the first place, one can assign only the rights that he has, and the assignee is subject to any of the defenses that the obligor could have raised against the assignor. In other words, the assignee "stands in the shoes of the assignor" in enforcing the claim against the obligor and has no greater rights than did the assignor. Thus, if Ollie Obligor has a defense against the $100 claim of Alice based on failure of consideration, fraud, duress, or some other theory, that defense is not cut off by the assignment and may be raised by Ollie against the assignee.

The assignee may be subject to any defenses that arise *after* the assignment is made but *before* the obligor is notified of the assignment. It is therefore important that the obligor be notified of the assignment by the assignee.

A further characteristic of an assignment is that it carries with it certain implied warranties. Among these are a warranty that the right assigned actually exists, that the right is not subject to a defense or counterclaim, that the assignee will not hinder the enforcement of the rights given by the assignment, and so on. This warranty, of course, extends only to the assignor's immediate assignee. If the assignee, in turn, assigns the right to a third party, the original assignor is not liable to that person for any breach of implied warranty. In addition, there is no implied warranty that the obligor will actually perform the contract.

The transfer by *negotiation* of a negotiable instrument is quite a different matter. In the first place, the law has set up a number of rules to enhance the transferability of negotiable instruments so that they will be traded in the marketplace between strangers without fear that the instrument cannot be collected because of some unknown defense.

The negotiation of a negotiable instrument has several consequences different from the assignment of a contract. For example, suppose that Manny Maker signs a negotiable instrument promising to pay $500 to the order of Paula Payee. Paula then negotiates the instrument to City Bank in exchange for $450. Maker has signed the note to Paula as consideration for Paula's motorbike. We shall see later that, provided that certain conditions are met—the note is taken for value, in good faith, without notice of a defense or that it is overdue or has been dishonored—City Bank would qualify as a *holder in due course* of the negotiable instrument. If Paula's bike is other than as represented to Maker, can Maker raise this as a defense and avoid paying City Bank? The answer is no. City Bank's position vis-à-vis Maker is superior to Paula's. Maker cannot raise the typical defenses, such as failure of consideration, against the holder in due course. There are some defenses, known as *real defenses,* that could be raised, but the typical and more common ones cannot be. These defenses will be discussed more fully in Chapter 24, "Holder in Due Course." The consequences obviously would have been different had this been an assignment of a contract instead of the negotiation of a negotiable instrument to a holder in due course.

What if Maker refuses to pay City Bank? May City Bank look to Paula for payment? The answer in most cases is yes, although once again, had this been the assignment of a contract, no warranty of performance would be given by the assignor. The holder of a negotiable instrument in most cases, if the negotiation is unqualified, can look to its transferor for payment in case the primary obligor does not pay.

Finally, as contrasted with an assignment of contract rights, the holder in due course is not required to give the obligor notice of the negotiation of the instrument in order to avoid defenses that may arise after the transfer. Clearly the person to whom a negotiable instrument is negotiated is in a far superior position to that of the assignee of an ordinary contract. Therefore the negotiable instrument can be transferred much more freely in the commercial world than can an ordinary contract by assignment.

Next let us examine some of the various types of negotiable instruments and the names and characteristics of the various parties to the types of instruments. In Chapter 22 the requirements for qualifying as a negotiable instrument will be analyzed.

21:3. Types of Negotiable Instruments and Parties. There are a number of different types of negotiable instruments that are designed for different purposes. These instruments involve either two parties or three parties. Those that involve two parties are referred to on occasion as *two-party paper* and are best exemplified by promissory notes and certificates of deposit. Instruments that involve three parties are sometimes referred to as *three-party paper* and are best exemplified by drafts and checks. In this section we will examine certain of the characteristics of the various types of negotiable instruments and identify the terms the law uses to describe the various parties to these instruments.

(1) Two-Party Paper. A *promissory note* is the least complicated of the various types of negotiable instruments. A promissory note is sometimes merely referred to as a *note.* It is a written promise, which is signed by the promisor, to pay money and is other than a certificate of deposit. The promisor, who obligates himself to pay money, is known as the *maker.* The person to whom the money is to be paid is known as the *payee.* These are the only two parties to the instrument, hence the term *two-party paper.* Note that in this form of instrument the promisor (maker) agrees to pay money directly to the payee.

You should also realize that a promissory note may be either negotiable or nonnegotiable. In order to be negotiable, it must meet a number of requirements, which will be discussed in Chapter 22.

Promissory notes may express the obligation of the *maker* in a number of ways. The figure gives an example of a promissory note.

The note is written on a printed form obtained from a bank. In fact many notes are prepared by private parties and need not be printed formally, just so they are written. The note illustrated specifies that the money should be paid to the payee on a specific date, which is known as the *maturity date.* This kind of note is known as a *time note.* If there is not time specified but payment is to be made when it is demanded by the payee, the note is known as a *demand note.* If periodic payments are to be made, it is known as an *installment note.*

$ 500 April 4, 19 78

One Year after date (I) (we) promise to pay to the order of Paula Payee

Five Hundred Dollars

at Franklin State Bank

No. 2 Due April 4, 1979 Manny Maker

In addition to the simple terms stated in the note shown in the figure, a number of other items may be included, such as provisions concerning security interests, interest, acceleration clauses, place of payment, and other details of the transaction. Many of these terms will be discussed in Chapter 22 along with whether or not they affect the negotiability of the note. At this point it is enough to know that a note may be very simple, as for example, "I promise to pay to the order of Paula Payee, $1,000.00," or it may contain several pages.

A *certificate of deposit,* sometimes called simply *C.D.,* is another form of two-party paper. The essential difference between a promissory note and a certificate of deposit is that a certificate of deposit is always issued by a bank, whereas a note is not. It is a promise by the bank, which may be called the maker, to pay a specific amount of money to the party who has previously deposited money with the bank. In addition to the promise to repay, the C.D. may also state a rate of interest and the date when payment is to be made.

(2) Three-Party Paper. Drafts differ from promissory notes and other types of two-party paper in that the person signing the draft is not promising to pay the payee directly but rather ordering some third party to pay the payee.

The concept of the draft has a long history. It was developed to a large degree by the traders of the Middle Ages. Consider the plight of the English merchant, traveling to Italy with a load of gold to exchange for goods purchased there. His chances of keeping the gold until his arrival in Italy were not high, to say the least. How much better it would be if the gold could be deposited with an individual in the business of dealing with gold, or as later developed, banks. Then when goods were purchased, our merchant could merely give an order to the seller requiring the holder of the merchant's gold to pay the amount specified on the order to the seller. For example, Terry Trader might deposit a quantity of gold with Sammy Silversmith, located in Venice. Trader would then travel to Florence and purchase a quantity of leather from Manny Merchant. In exchange for the leather Trader would give Merchant an order directing Silversmith to pay Merchant the designated quantity of gold upon presentation of the instrument. In this way Trader would not have to haul the gold all over the country, and Silversmith would pay because Trader had previously made arrangements with him. It is from these antecedents that our modern draft developed.

The *draft,* also called a *bill of exchange,* is known as three-party paper because there are, as already stated, three parties involved: the *drawer,* the *drawee,* and the *payee.* The draft is an order to pay a specific amount of money. The person making the order is the *drawer.* The person to whom the order to pay is directed is the *drawee,* and the person to whom the drawee is ordered to make the payment is the *payee.* Note that the person who issues a draft is a *drawer,* whereas a promissory note is issued by a *maker.*

There are differences between a promissory note and a draft other than the number and names of the parties involved. In the first place, when the drawer issues a draft he does not promise to pay the payee himself but rather is ordering the drawee to pay. Indeed the drawer probably has no expectation of paying, although we shall see later that the payee may have recourse against the drawer should the drawee fail to pay the order. The maker of a note, on the other hand, does promise to pay the payee.

There are some similarities between notes and drafts. In the first place, both may be due at a specific time or upon demand. They may also be made payable to a specific person or to his order or to the bearer of the document. Furthermore, although a draft may be very simple, it may also contain many different terms. An example of a draft is given in the figure.

(a) *Sight and Time Drafts.* There are a number of different types of drafts, and some of the classifications overlap. For instance, one type of draft is called a *sight draft,* for the very simple reason that it is payable at sight. If the draft is payable at a specific future time, it is known as a *time draft.* The time may be stated as a specific date or as a specific period of time—such as twenty days—after sight or presentment.

(b) *Trade Acceptance.* The *trade acceptance* is a form of draft that is frequently used in commercial transactions in order to extend credit. For example, assume that Sally Seller wants to sell goods to Bill Buyer. Buyer wishes to purchase the goods on credit, but Seller is not sure enough of Buyer's credit

$ 500 College Park, Md. January 19, 1978

Ninety days after date (I) (we) promise

to pay to the order of Paula Payee

Five Hundred Dollars

Value Received and Charge to Account of

To Delbert Drawer

No. 2 Baltimore, Md. Doris Drawer

worthiness to sell him the goods on open account and in addition needs the money herself to restock her inventory. Seller then draws a draft on Buyer for the price of the goods, obligating Buyer to pay to the order of Seller at some future date. In other words, Seller would be the drawer, Buyer the drawee, and Seller, or "to her order," the payee.

The draft is then presented to Buyer by a bank located in his city that represents Seller. In order to obtain the goods, Buyer "accepts" the draft, thereby obligating himself to pay Seller on the draft. The acceptance is usually written across the face of the draft. The draft is then returned to Seller, who then may discount the draft to a financing institution and obtain cash or, if nothing else, hold it as evidence of the debt. Furthermore the draft is a negotiable instrument and more readily accepted by the bank in return for cash or security for a loan because the bank may qualify for the protection afforded a holder in due course.

(c) *Bank Acceptance.* A *bank acceptance* is an even more secure and marketable instrument than a trade acceptance because it is drawn on the buyer's bank rather than on the buyer personally. Otherwise, as a practical matter, its use and functions are similar to those of a trade acceptance. In the case of a bank acceptance the draft is prepared by Seller in an amount drawn on Buyer's bank with Seller as payee. The draft is then sent to Buyer's bank, which accepts the draft. The bank may require Buyer to put up collateral or to keep deposits on hand in an amount sufficient to cover the draft.

(d) *Bank Draft.* Another form of the draft is the *bank draft.* Its primary use is to transfer funds from one bank to another. For instance, if a customer needs to have funds available in another city, his bank may prepare a bank draft drawn on a bank in that city. The customer will then have the funds available in that location for any commercial transaction for which they are required.

(e) *Checks. Checks* are by far the most frequently used type of commercial paper. A check is really a special type of draft by which the *drawer,* who has an account at the bank, draws the instrument on a *bank* as the drawee. A check is payable on *demand* rather than at some future date. The form for the check is almost always supplied by the bank because it should contain a combination of numbers and symbols required by the check-collection system, which uses computers to sort the checks and charge them to the proper account. Finally, a check differs in character from the typical draft in that it is usually intended to pay a particular individual immediately rather than to be discounted and negotiated through a number of parties, as a draft might be.

Our discussion has focused primarily on the typical personal check, but there are some other forms of checks. One such form is the *cashier's check*. The cashier's check is a check drawn by the bank on itself. The bank therefore is both the drawer and drawee.

The characteristics of various types of negotiable instruments have been discussed in this chapter in a very broad manner. As intimated previously, there are a number of important rules that govern whether an instrument qualifies as "negotiable." It is on that subject that we focus our attention in Chapter 22.

SUMMARY

Negotiable instruments may be classified as either two-party paper or three-party paper. Examples of two-party paper are promissory notes and certificates of deposit. The parties to a promissory note are the maker and payee. Examples of three-party paper are drafts, checks, and trade acceptances. The parties to a draft are the drawer, drawee, and payee.

Under certain circumstances the rights of the transferee of a negotiable instrument may be greater than the rights of his transferor. The position of the assignee of the rights of a contract are not greater than the rights of the assignor.

PROBLEMS

1. How did the use of commercial paper originate?
2. What is the difference between an instrument that is negotiable and one that is nonnegotiable?
3. What is the difference between the negotiation of an instrument and the assignment of a contract?
4. What advantages may the holder of a negotiable instrument, who is a holder in due course, have over the assignee of contract right?
5. Distinguish between:
   a. Two-party and three-party paper.
   b. A promissory note and a draft.
   c. Trade acceptance and bank acceptance.
   d. A personal check and cashier's checks.
6. Explain the use of a trade acceptance to facilitate the sale of goods between two merchants. What advantages accrue to the seller from the use of trade acceptances?

CHAPTER 22

Negotiability

We have seen that a negotiable instrument has characteristics that distinguish it from the ordinary contract. In this chapter we will examine specifically the requirements that must be met in order for an instrument to be negotiable. The difference between an ordinary contract and a negotiable instrument is one of *form.* Both contracts and negotiable instruments can be used to evidence a debt, for instance, but the difference in their characteristics arises because of the form that the negotiable instrument must take.

Article 3 of the Uniform Commercial Code governs negotiable instruments. Section 3–104 states the form that an instrument must take in order to be negotiable. In order to be negotiable, the instrument must (1) be in writing and signed by the maker or drawer; (2) contain an unconditional promise or order to pay; (3) be for a sum certain in money; (4) contain no other promise, order, obligation, or power except as authorized by Article 3; (5) be payable on demand or at a definite time; and (6) be payable to order or to bearer. If the instrument meets all these requirements, it is negotiable and has all the characteristics of a negotiable instrument. If any one of the requirements is not met, the instrument is nonnegotiable. In order for the instrument to be negotiable, all these requirements must be met on the face of the instrument, that is, without reference to any other document. On the other hand, the fact that an instrument meets the requirements for negotiability does not necessarily mean that the instrument is valid and enforceable. These are two different questions. In other words, an instrument may meet all the requirements for negotiability and still be unenforceable. A negotiable instrument must meet the requirements of a contract. As between the original parties, for example, the instrument would have to be given for a legal purpose in order to be enforceable. On the other hand, an instrument that is not negotiable may be enforced between the original parties if it is otherwise valid.

22:1. Instrument Must Be in Writing and Signed. The requirement of a writing usually presents no problem. Obviously there can be no such thing as an oral negotiable instrument, and the requirement is easily met. The instrument may be written on anything and in any form, such as being in handwriting or in print. Most negotiable instruments are executed on printed forms of an appropriate size, but such a form is not required. It would be absurd, of course, to print a negotiable instrument on anything not readily transferable, but there is no requirement in this regard.

The code requires that a note be signed by the maker and a draft must be signed by the drawer. Normally this requirement is met when the appropriate party signs his name in longhand in the lower-right-hand corner of the instrument. However, an instrument is "signed" when it includes any symbol executed or adopted by the maker or the drawer with the present intention to authenticate the instrument. Section 1–201(39). Hence any mark or symbol that one wishes to adopt as his or her signature is sufficient to satisfy this requirement. Thus one may adopt a stamp, trade name, or other symbol as his signature. However, before accepting a negotiable instrument with an unusual signature, you should note that the burden of proving the authenticity of the signature when it is denied is upon the person who claims under the signature. Section 3–307(1)(a). Because one of the major advantages of a negotiable instrument is its free transferability and acceptance, it obviously does not make sense to get too "cute" in signing the instrument, as this may materially reduce its acceptance in commercial circles. There are circumstances, of course, in which one has no choice other than to use a symbol as his signature.

The instrument may be signed anywhere, although it is customarily signed in the bottom-right-hand corner. Courts have held that even if the "signature" is contained in the middle of the instrument it is sufficient, as, for example, in the handwritten statement, "I Danny Debtor promise to pay. . . ." Even a letterhead on the paper on which the instrument is written has been held to be sufficient when the court determined that that letterhead was intended as the maker's or the drawer's signature.

22:2. Instrument Must Contain Promise or Order to Pay. The words contained in a note or draft must obligate someone to pay. If the instrument is a note, the maker of the note must promise to pay. For example, the note may state, "I promise to pay. . . ." Other words may be used, but they must clearly indicate an undertaking to pay. Merely acknowledging an obligation without promising to pay it is not sufficient. Section 3–102(c). Thus a paper stating "Received $50" would not be a promise, nor would a simple statement that "I.O.U. $50."

An *order* is a direction to pay and also must be more than an authorization or a request. Section 3–102(b). To qualify as a negotiable instrument the draft must contain an order by the drawer to the drawee to pay a sum of money. The drawee who is ordered to pay must be identified with reasonable certainty, and as the code states, the language used must be more than an authorization or a request to pay. The obligation to pay cannot be uncertain. Thus language such as "I wish you would pay" clearly would not qualify as an order, with the result that the instrument would be nonnegotiable. The requirement of identifying the drawee is usually satisfied if the name of the drawee is printed on the draft following the word *To.* On a check the name of the drawee bank is printed on the face of the check.

Although the code allows different words to be used, the most common words used are "I promise to pay . . ." for a note and use of the word *pay,* as in "pay bearer," for a draft. Because the purpose of a negotiable instrument is to have it readily accepted, it is unwise to use other words that may call the validity of the instrument into question.

22:3. The Promise or Order Must Be Unconditional. The two primary functions of negotiable instruments are to serve as a basis for credit and as a substitute for currency. Clearly, if these functions are to be served, the obligations contained on the face of the instruments must be unconditional. If conditional promises were allowed, no one could safely take the instrument without determining whether the condition had been fulfilled, which would be both time consuming and costly. At best, such an instrument would be taken only at a very high discount rate. If a high discount rate is necessary, the use of the negotiable instrument as a credit tool is destroyed. Therefore, in order to be negotiable, the promise or order to pay must be unconditional. Section 3–104(b).

The determination of whether the promise or order is conditional or unconditional must be made from the face of the instrument itself. No extrinsic information, such as other documents or words spoken when the instrument is executed, can be considered because a negotiable instrument may eventually pass between parties who have no knowledge of the original transaction and who have no way to ascertain what went on at the time the instrument was executed. If the requirement were otherwise, people would not freely accept negotiable instruments.

The code is fairly explicit in stating the factors that may or may not render an instrument conditional. Section 3–105. At this point an examination of some of the more common problem areas would be appropriate.

(1) Reference to Other Agreements or Documents. Frequently negotiable instruments are executed in connection with some other document, such as a contract. If the instrument merely refers to the other document or agreement without making payment of the instrument conditional upon performance of the other agreement, the instrument is negotiable, that is, unconditional. Thus an instrument is unconditional if it merely recites that it is drawn under a letter of credit or states that it has arisen out of a separate agreement or states that it is secured by a mortgage.

A question sometimes arises as to whether the language used in reference to an outside document makes the instrument conditional upon the performance of that document. The most troublesome language, which resulted in conflicting court decisions prior to the passage of the code, involved the use of the phrases "as per" and "subject to" the agreement. The code clarified this problem by stating that the phrase "as per" does not make the instrument conditional. Thus an instrument would still be unconditional and negotiable if it stated, "This note is given for payment as per contract for goods of even date. . . ." The phrase "as per," therefore, merely refers to another agreement. Section 3–105(b). On the other hand, the code states that the use of the phrase "subject to" or "governed by" another agreement makes the instrument conditional and therefore nonnegotiable. Thus, if an instrument states that it is "subject to terms of contract between maker and payee of this date," it is conditional and nonnegotiable—Section 3–105(2)(a)—because one cannot determine all the terms and conditions of payment merely by looking at the instrument itself. If one must examine documents other than the instrument in order to determine the conditions of payment, its negotiability is destroyed.

If the instrument contains an *acceleration* or *prepayment* clause its negotiability is not destroyed. Section 3–105(1)(c). Thus the instrument is unconditional and negotiable if it states that the obligation becomes due in its entirety if an installment payment is missed. The reason is that there is no dependence upon performance of conditions established by a document other than the instrument. The only effect of the prepayment or acceleration clause is to speed up the time of payment. The obligation to pay is not in doubt.

(2) **Reference to an Account or Fund.** The general rule is that if an instrument states that it is to be paid *only* out of a particular fund or source, it is conditional and therefore nonnegotiable. Section 3–105(2)(b). This concept is sometimes referred to as the *particular fund doctrine.* The condition, of course, is that the instrument will be paid only if there is money in the particular fund. On taking the instrument, one would have to inquire into whether there is money in the fund before knowing whether the instrument would be paid. An instrument is unconditional and therefore negotiable only if the entire credit of the maker or the drawer is behind it.

On the other hand, if the instrument merely indicates the fund or account that is to be charged without limiting the payment of the instrument to the funds in that account, it is unconditional and negotiable. Section 3–105(1)(f). Thus the drawer may instruct the drawee of a draft to "charge the payroll account" when it is presented by the payee. The general credit of the drawer is still behind the instrument, and the reference to a particular account is only for accounting purposes.

There are two exceptions to the general rule established by the code. One is when a government agency draws checks or issues other short-term instruments limiting payment to a particular fund or to proceeds from certain taxes. Such an instrument is still negotiable as long as the maker or the drawer is a governmental agency. Section 3–105(1)(g). The second exception applies when an instrument is limited to payment out of the entire assets of a partnership, an unincorporated association, a trust, or an estate by or on behalf of which the instrument is issued. Section 3–105(1)(h). The effect, of course, is to limit the liability on the instrument to the assets of a partnership, for example, rather than making the partner's personal assets liable. In other words, as far as the liability aspects are concerned, the partnership would be treated like a corporation.

BOOKER v. EVERHART

294 N.C. 146, 240 S.E.2d 360 (1978)

Koyt W. Everhart, Jr., and Jane C. Everhart were husband and wife. On May 1, 1972, they separated and entered into a property-settlement agreement. As part of that agreement Koyt signed a promissory note payable to Jane in the amount of $150,000. Jane then signed a document purporting to assign one-third of the note to James J. Booker and Oren W. McClaim, her attorneys, as payment for their legal services. That same document also named them as Jane's agents for collection of the note.

Koyt failed to keep up the payments on the note and was sued by Booker and McClaim. The note stated, in part, that Koyt agreed to pay $150,000 to Jane ". . . in lieu of a property settlement supplementing that certain Deed of Separation and Property Settlement, dated May 1, 1972, the terms of which are incorporated herein by reference. . . ."

Booker and McClaim contended on appeal that as holders of the note they were entitled to payment of the note in their own names without the necessity of joining the owner of the note, Jane, in the law suit. Koyt contended that the note was not a negotiable instrument, that as result Booker and McClaim were not "holders" under the Uniform Commercial Code and that it was therefore necessary to make Jane a party to the suit.

MOORE, J.

Under the law of this State prior to the adoption of the Uniform Commercial Code, it was clearly established that a conditional promise or contingent condition contained in the instrument itself had the effect of defeating the negotiability of the instrument. This prior law is carried forward in G.S. 25–3–104(1)(b).

G.S. 25–3–105. *When promise or order unconditional,* states in part:

> (2) A promise or order is not unconditional if the instrument
>
> (a) states that it is subject to or governed by any other agreement

The official comment to 25–3–105 says that, as far as negotiability is concerned, the conditional or unconditional character of the promise or order is to be determined by what is expressed in the instrument itself. When the instrument itself makes express reference to an outside agreement, transaction, or document, the effect on the negotiability of the instrument will depend on the nature of the reference.

In the present case, the instrument marked Exhibit A says, after the promise to pay Jane Crater Everhart $150,000: ". . . *in lieu of a property settlement supplementing that certain Deed of Separation and Property Settlement, dated May 1, 1972, the terms of which are incorporated herein by reference.* . . ." (Emphasis added.)

Incorporation by reference has been defined as:

> The method of making one document of any kind become a part of another separate document by referring to the former in the latter, and declaring that the former shall be taken and considered as a part of the latter the same as if it were fully set out therein. *Black's Law Dictionary* (Revised 4th Ed.), Incorporation.

To incorporate a separate document by reference is to declare that the former document shall be taken as part of the document in which the declaration is made, as much as if it were set out at length therein.

By incorporating into the note in question the Deed of Separation and Property Settlement, the parties made the note "subject to" any and all possible conditions contained in those prior documents. Under 25–3–105(2)(a), this renders the promise to pay the sum certain conditional. Whether or not the documents incorporated contained any such conditions or contingencies is a matter beside the point. The essential point is that all of the essential terms of the note in question cannot be ascertained from the face of the instrument itself. Because separate documents have been made a part of the note by its express terms, the promise contained therein is conditional, and the note nonnegotiable.

Under G.S. 25–3–105(1)(b) and (c), it is clear that mere reference in a note to the separate agreement or document out of which the note arises does not affect the negotiability of the note. But to go beyond a reference to the separate agreement, by incorporating the terms of that agreement into the note, makes the note "subject to or governed by" that agreement, and thus, under G.S. 25-3-105(2)(a), renders the promise conditional and the note nonnegotiable.

Since the instrument in the present case is nonnegotiable plaintiffs cannot be "holders" under article 3, and thus cannot argue that, under G.S. 25-3-301, they have the power to enforce this note as collection agents for the owner.

(Held that Jane must be made a necessary party in a new trial.)

QUESTIONS FOR DISCUSSION

1. *Would the decision have been different if the note had said "This note refers to an agreement dated May 1, 1972."*
2. *If a note incorporates by reference another document that itself is not conditional may the note qualify as a negotiable instrument?*

22:4. PROMISE OR ORDER MUST BE FOR A SUM CERTAIN IN MONEY. In order for an instrument to be negotiable, the holder must be able to ascertain from an examination of the instrument itself exactly what he will get in money. If the amount is uncertain or if reference must be made to an outside source, the instrument is nonnegotiable. The promise or order, in other words, must be for a *sum certain,* a specific amount. Section 3-104(1)(b).

The requirement of a sum certain does allow for some flexibility. If the amount of money that the obligor eventually will pay on the instrument can be computed from the instrument itself, without reference to an outside source, the instrument is negotiable. Thus an instrument drawn to pay a sum certain on demand "plus interest at 7 percent" meets the requirement of a sum certain because the holder can calculate the specific amount due at any time from the instrument itself. If the interest rate before and after maturity (the date the payment is due) differs, the sum due is still certain. Likewise, if the instrument provides for installment payments, the amount due can be calculated and the sum certain requirement is not breached. Section 3-106(1)(a). Thus a note would be negotiable that provides for a promise to "pay $1,000 with interest at 6 percent, but if not paid at maturity the interest shall be calculated at 8 percent for the total period." Section 3-06(1)(b). Also, if the instrument merely provides that the face amount is to be paid "with interest," the sum certain requirement is met because this phrase is construed to mean the "legal interest rate" (not the maximum rate), which all people are presumed to know. However, if the note provides for payment of interest at the then "current rate," the sum would not be certain because one would have to look to a source outside the fact of the note in order to determine the "current rate." Section 3-118(d). An instrument is negotiable if no reference is made to interest whatsoever. There is simply no interest payable on the instrument.

A provision frequently found in promissory notes is for the maker to pay attorney's fees and other costs of collection if the maker should default in payment of the instrument. Such a provision does not render the note nonnegotiable—Section 3-106(e)—because the holder still can ascertain with certainty the minimum amount to which he is entitled. The costs of collection and attorney's fees are an additional amount. They must be reasonable, however, or the courts will hold the provision to be void as against public policy because it constitutes a penalty.

Finally, the code provides that an instrument is payable for a sum certain even though it is to be paid with a stated discount if paid before the due date or with an addition if paid after the date fixed for payment. Section 3–106(1)(c). The same is true if the instrument is to be paid with exchange or less exchange, whether at a fixed rate or at the current rate. Section 3–106(1)(d). The *exchange* referred to is the difference in values between two currencies, such as the United States dollar and the German mark. The relative values may fluctuate from time to time. Thus the code relaxes the rule somewhat here because one must look beyond the instrument in order to determine the "current" exchange rate. The exchange rate is readily ascertainable, however.

WALLS v. MORRIS CHEVROLET, INC.
515 P.2d 1405 (Okla.App. 1973)

BAILEY, J.

Plaintiff sued defendants for allegedly taking a negotiable instrument in conjunction with a consumer credit sale in violation of 14A O.S.1971, § 2–403. He sought the recovery of three times the amount of the credit service charge, the maximum penalty provided for such violations, for himself and for others similarly situated. The note and security agreement signed by the plaintiff were on the same sheet of paper and defendants demurred on several grounds, including that the papers did not include a negotiable instrument.

The trial court sustained the demurrer and dismissed the petition when the plaintiff declined to amend. Plaintiff appeals.

First, both parties assume that the note, considered by itself, is not negotiable. So do we. The sum payable from the face of the note does not appear to be a sum certain because of the privilege stated in the note of refund of any unearned finance charge based on the Rule of 78 upon prepayment of the balance. The amount of the finance charge is not apparent from the face of the note and therefore the sum to be paid is uncertain in the event of prepayment. Under 12A O.S. 1971, § 3–106: "(1) The sum payable is a sum certain even though it is to be paid . . . (c) with a stated discount . . . if paid before . . . the date fixed for payment. . . ." In this instance the amount of the discount is not stated in the note and cannot be computed from its face. As is stated in the Uniform Commercial Code Comment to this section: "A stated discount or addition for early or late payment does not affect the certainty of the sum so long as the computation can be made. . . . The computation must be one which can be made from the instrument itself. . . ."

To overcome the absence of a sum certain on the face of the note, the plaintiff argues that the amount of the finance charge appears in the accompanying security agreement, that the security agreement and the note should be considered one instrument because on the same sheet of paper, that so construed the missing term is supplied and both note and security agreement are negotiable.

It is our opinion that a note cannot depend upon another agreement for elements of negotiability whether that agreement is attached to the note or separate from it except in those rare instances where such an incorporation is sanctioned by the Uniform Commercial Code expressly or by necessary implication. Negotiable notes are designed to be couriers without excess luggage under both the prior law and under the code and so negotiability must be determined from the face of the note without regard to outside sources (with rare exceptions) so that the taker may know that he takes a negotiable instrument with the insurance of collectability provided by the code and not an ordinary contract subject to the possibility of all defenses by the maker.

We are particularly concerned about the incongruous result which might be reached in this case if we held that the security agreement here imparted the necessary element of negotiability to the promissory note so that it became negotiable. Though on the same sheet of paper, the security agreement and the note are so situated on the paper that they could be detached from each other and each would appear to be an agreement complete in itself. There is nothing in the terms of either agreement to make it illegal or even inappropriate for a holder to so separate them. In that event any subsequent holder of the note could not determine its negotiability from the note itself but this would depend upon a now separate instrument, the security agreement. Presumably such a note, though negotiable when attached (if we should so hold) would be nonnegotiable when separated, since under the code and before with rare exceptions none of which is relevant here, a separate agreement cannot supply elements of negotiability to a note which from its very nature must be negotiable or not from its face. This is a bad result to be avoided if possible.

Questions for Discussion

1. *Why was the note, considered by itself, not negotiable?*
2. *What reasons did the court give for not allowing reference to the security agreement to supply the missing element of negotiability?*

UNIVERSAL C.I.T. CREDIT CORP. v. INGEL
196 N.E.2d 847 (Mass. 1964)

Albert J. Ingel and Dora Ingel hired Allied Aluminum Associates, Inc., to put aluminum siding on their house. In connection with this transaction the Ingels signed a promissory note payable to Allied in the amount of $1,890. The note provided for "interest after maturity at the highest lawful rate." *The note also contained a provision stating that at the maker's request Allied would obtain group-credit life insurance for the maker. The terms of such insurance were also noted. Allied also gave the Ingels a "completion certificate" at the conclusion of the job. Allied then sold the note to Universal C.I.T. Credit Corporation. Ingels refused to pay the note and C.I.T. sued, claiming that it was a holder in due course. The Ingels sought to raise certain personal defenses that they had against Allied, contending that they had this right because C.I.T. was not a holder in due course but only an assignee of a nonnegotiable instrument. The Ingels attempted to introduce, through Charles Fahey, a credit report on Allied. The court excluded this as well as certain other evidence and found in favor of C.I.T. The Ingels appealed.*

SPIEGEL, J.

The defendants contend that the note was nonnegotiable as a matter of law and, therefore, any defense which could be raised against Allied may also be raised against the plaintiff. They argue that the note contained a promise other than the promise to pay, failed to state a sum certain, and had been materially altered.

It appears that the note was a form note drafted by the plaintiff. The meaning of Fahey's general testimony that the note and the completion certificate were "together" when given by the plaintiff to Allied is unclear. However, we see nothing in this testimony to justify the inference urged upon us by the defendants that in this case the note and completion certificate were "part of the same instrument" and that an additional obligation in the completion certificate rendered the note nonnegotiable under G.L. c. 106, § 3–104(1)(b). Similarly, we are not concerned with any variance between the written contract (entered into by Allied and the defendants) and the note, since there is nothing in the note to indicate that it is subject to the terms of the contract. We are equally satisfied that the insurance clause in the note does not affect negotiability under § 3–104(1)(b) since it is clear that the "no other promise" provision refers only to promises by the maker.

The provision in the note for "interest after maturity at the highest lawful" rate does not render the note nonnegotiable for failure to state a sum certain as required by § 3–104(1)(b). We are of opinion that after maturity the interest rate is that indicated in G.L. c. 107, § 3, since in this case there is no agreement in writing for any other rate after default. This being the case, we do not treat this note differently from one payable "with interest." The latter note would clearly be negotiable under G.L. c. 106 § 3–118(d).

The note in question provides that payment shall be made "commencing the 25 day of July,

1959." It appears that there is an alteration on the face of the note in that "July" was substituted for "June," the "ly" in the former word being written over the "ne" in the latter. The alteration has no effect in this case, where the defendants admitted that they had paid a particular sum on the note and where the sum still owing (assuming the note to be enforceable on its face) is not in dispute.

We thus conclude that the note in question is a negotiable instrument.

QUESTION FOR DISCUSSION

1. *What was the importance to the defendants' case of their contention that the instrument was nonnegotiable?*

(1) Meaning of Payable in Money. The code defines *money* as a "medium of exchange authorized or adopted by a domestic or foreign government as a part of its currency." Section 1–201(24). An instrument is payable in money if the medium of exchange stated in the instrument is money at the time the instrument is made. Section 3–107(1). Furthermore the instrument is negotiable if it is stated in terms of the money of any recognized government. Thus a note may be stated as payable in pounds, lira, yen, francs, marks, guilders, or other official exchange of a government. Section 3–107(2). Unless the note specifies otherwise, payment may be made in United States dollars at the stated or current exchange rate. However, if the instrument requires that it be paid in a specific currency, then it is payable in that currency. Section 3–107(2).

Sometimes local custom accepts certain goods as money because of peculiar local conditions. For example, cigarettes were widely exchanged in occupied Germany after World War II. In other areas gold dust or fur pelts might be exchanged as money by local custom. These media of exchange, however, are not money because they have not been recognized as such by any domestic or foreign government. Therefore, although there might be a certain charm to it, a promissory note given to a payee for "fifty beaver pelts" would not qualify as a negotiable instrument. On the other hand, an instrument payable simply in "currency" or "current funds" is payable in money. Section 3–107(1).

22:5. THE INSTRUMENT MUST CONTAIN NO OTHER PROMISE OR ORDER. We already know that a negotiable instrument must contain a promise or order to pay money. As a general rule, however, the instrument may contain no other promise, order, or obligation except as authorized by Article 3 of the code. Section 3–104(1)(b). Thus a negotiable instrument could not promise to pay money and "work for two weeks in Smith's store." The reason is that the value of the negotiable instrument must be readily ascertainable when being negotiated and discounted. There is no way to place a value on the promised services.

(1) Exceptions and Omissions. Section 3–112(1) lists the exceptions referred to in the preceding paragraph. It also clarifies certain omissions that will not affect the negotiability of the instrument. Thus a statement that collateral has been given to secure the instrument and a statement that upon default there may be a disposal of the collateral will not affect the negotiability of the instrument. In addition, a statement frequently made in connection with the foregoing one, that the obligor promises to maintain or protect the collateral or give additional collateral, will not render the instrument nonnegotiable. Additionally, a term authorizing confession of judgment, if the instrument is not paid when due, or a term waiving the benefit of a law intended for the benefit of the obligor is consistent with the concept of negotiability. Another term frequently contained particularly in insurance company drafts is one that provides that an endorsement by the payee acknowledges full satisfaction of all obligations. Such a term does not affect the negotiability of an instrument. Finally, an instrument is effective even if a statement is contained in a draft drawn in a set of parts, to the effect that the order is effective only if no other part has been dishonored.

You should note that all of these exceptions are used for the purpose of strengthening the promise or order to pay and are not promises that are independent of that obligation. However, the mere fact that a term is listed in this section of the code does not mean that it is legal if there is a statute making such a term illegal. Section 3–112(2). Perhaps the best illustration of this is a clause authorizing a confession of judgment in the event that the obligor defaults on the instrument. Many states declare by statute that so-called confessed-judgment clauses are illegal. In most of those states, however, if an instrument is otherwise valid, the courts will still enforce the instrument without giving effect to the confession-of-judgment clause.

This Section 3–112(1) also makes provisions for the negotiability of instruments that omit certain statements. Thus the negotiability of an instrument is not affected by the omission of a statement of any consideration or a statement of the place where the instrument is drawn or payable.

22:6. The Instrument Must Be Payable on Demand or at a Definite Time. A further requirement of negotiability is that the instrument must be payable on demand or at a definite time. Section 3–104(1)(c). Thus an instrument that is payable when the "shipment of goods from Atlanta arrives" would not be negotiable. The reason for the requirement is that the person to whom the instrument is negotiated would have to assess when the shipment would arrive or indeed whether it would arrive at all. This, of course, would take time and expense but would be necessary before the prospective holder could determine what value to give for the instrument. Obviously the necessity of such an investigation would destroy the whole purpose of the negotiable instrument. Indeed that is the reason for most of the requirements of negotiability.

(1) Payable on Demand. An instrument is payable on demand if it is payable "at sight" or "on presentation" or if no time for payment is stated in the instrument. Section 3–108. Many instruments, such as the ordinary check, fall into the latter category and reflect no specific time when the instrument is to be paid.

An instrument payable on demand satisfies the requirement of negotiability because the holder of the instrument controls when the instrument will be paid because he makes the demand on presentation for payment. There is therefore no uncertainty as far as the holder is concerned as to when the instrument can be collected. It follows, therefore, that the holder can assess the value of the instrument, at least as regards the time of payment, when it is negotiated to him.

DAVIS v. DENNIS

448 S.W.2d 495 (Tex.App. 1969)

Charlie Davis signed a note payable to Archie A. Dennis, Jr., on March 25, 1961. The note was as follows:

> $3,000.00 Palestine, Texas, March 25 A.D. 1961
>
> For Value Received, I, we, or either of us, the undersigned, promise to pay to Archie A. Dennis or Archie A. Dennis, Jr. on order, the sum of Three Thousand and no/one-hundred Dollars, with interest from date at the rate of nine (9) per cent per annum, interest payable bi-weekly both principal and interest payable at South Texas Producers Association, Houston, Texas
>
> This note is payable in Seventy-eight (78) installments of Thirty-eight dollars and forty-six cents (38.46) each. This is the note mentioned in the mortgage covering 15 head of dairy cows bearing the same date. . . .

Dennis subsequently sued Davis on September 2, 1966, to collect the note. The statute of limitations was four years and began to run from the time that the right of action on the note accrued. Davis contended that the note was a demand instrument and that the statute of limitations for filing suit on the note had expired. Dennis contended that the instrument was not one payable on demand because of the provision concerning payment in installments and that therefore the statute of limitations had not passed. From an adverse decision of the trial court awarding Dennis a $3,168.79 judgment, Davis appealed.

McKAY, J.

The note does not provide for any fixed time of payment. It provides "interest payable bi-weekly" and "This note is payable in Seventy-eight (78) installments of Thirty-eight dollars and forty-six cents ($38.46) each." Section 3.108, Uniform Commercial Code, Vernon's Texas Code Annotated, provides: "Instruments payable on demand include those payable at sight or on presentation and *those in which no time for payment is stated*," (Emphasis added) . . . It has been held from early Texas cases that if no time for payment is stated in a note, it becomes a demand note. . . . If it was a demand note, it was payable on demand and actionable immediately without demand.

Since the note provided "interest payable bi-weekly," the question arises whether that language makes the note payable at a fixed time. We believe it does not.

In 10 C.J.S. § 247, page 743, we find this statement:

> . . . The fact that notes indicating no time of payment are expressed to be payable with interest annually does not prevent them from being payable on demand. . . .

In another paragraph on the same page, it is stated:

> Paper is none the less payable on demand because it contains a provision as to interest, as where it is payable "on demand" with interest after a specified time, of "after maturity," "within six months from date," "without interest," or "without interest during the life of the promisor."

We are also of the opinion that the language "this note is payable in Seventy-eight (78) installments of Thirty-eight dollars and forty-six cents (38.46) each . . ." does not affect the demand character of the note because there is no maturity date or dates or fixed time of payment of any installment.

The note, being a demand note, the four-year statute of limitation would begin to run from the date of its execution or delivery. The note in this case, executed March 25, 1961, was barred by the statute of limitation on September 2, 1966, the date suit was filed.

(The judgment of the trial court is reversed and judgment is here rendered for appellant.)

QUESTIONS FOR DISCUSSION

1. *According to the court, when does the right of action for a demand instrument begin to run?*
2. *Do you feel that the decision of the court might have been different if the note had provided for a specific date for the payment of the installments?*

(2) Payable at a Definite Time. An instrument is obviously payable at a definite time if it is payable on a specific date, such as "payable on January 2, 1978." Similarly an instrument payable "thirty days after date" is payable at a definite time, provided the instrument is dated. Frequently an instrument will be payable "on or before" a stated date. Such an instrument is also payable at a definite time because the holder can control the time of payment. Section 3–109(1)(a). Such a clause may be inserted in the instrument so that the maker has the privilege of paying the instrument early if he finds himself in a position to do so. There are other arrangements that may affect whether an instrument is payable at a definite time. These require a more detailed explanation.

(a) *Fixed Period After Sight.* An instrument payable at a fixed period after sight also satisfies the requirement of definiteness. Section 3–109(1)(b). This phrase is frequently used in the trade-acceptance type of draft as a method of extending credit. Thus, if goods are shipped by the seller to a distant city, he may also send a draft to his bank in that city along with the bill of lading, which is necessary to obtain the goods from the carrier. The goods are shipped to the buyer's city. The buyer cannot get the bill of lading until he "accepts" the draft. The bank will then give him the bill of lading so that he can obtain the goods from the carrier. Once the buyer accepts the draft, he is obligated to pay the instrument but has a period of time to do so. Thus, if a draft is payable "sixty days after sight," it meets the requirement of definiteness. The drawee (the buyer) has sixty days after accepting the instrument in which to pay it.

(b) *Acceleration Clauses.* Frequently instruments that are to be paid in installments contain an *acceleration clause* making the full amount of the instrument due and payable immediately upon default in the payment of an installment. The practical purpose of such a clause, beyond encouraging payment of the installments, is so that suit may be brought immediately for the full amount of the instrument upon a default in payment of an installment. Without such a clause, suit cannot be brought on the full amount until the last installment is due. There also may be other reasons given for accelerating a note, such as the failure of the maker of the note to comply with some other agreement regarding secured property. An example would be the failure to make the payment of taxes on real property or to maintain insurance on it.

The code specifically provides that an instrument payable at a definite time, subject to any acceleration, is negotiable. Section 3–109(1)(c). Comment 4 to this section of the code notes that a note with an acceleration clause is no less payable at a definite time than a demand note and in fact is more certain because there is a definite time stated beyond which the instrument cannot run. Even if the instrument provides that the holder may accelerate payment "at any time he shall deem himself insecure," the requirements of negotiability are satisfied. At first blush it would seem that the holder may have the opportunity to abuse this ground for acceleration. The code, however, takes the position that such possible abuse has nothing to do with the negotiability of the instrument and that the obligor is protected from potential abuse in that the code also requires such an acceleration to be made in good faith. Section 1–208. Thus an acceleration clause is good whether it is at the option of one of the parties or automatic upon the occurrence of some event.

(c) *Extension Clauses.* The code also provides that an instrument is payable at a definite time if it is payable "at a definite time subject to extension at the option of the holder, or to extension to a further definite time at the option of the maker or acceptor or automatically upon or after a specified act or event." Section 3–109(1)(d).

The reason the extension by the *maker* does not impair negotiability is because it is for a definite time. If the maker were to have the ability to extend the time for payment for an indefinite period of time, the instrument would not be negotiable. On the other hand, if the instrument giving the maker the power to extend the time for payment is *silent* as to the time for extension, the extension is to be for a period not in excess of the original period of the instrument. Section 3–118(f).

If the *holder* of the instrument is given the power to extend the time for payment of the instrument, even if it is for an indefinite period, the instrument is still negotiable. This is because the time for payment is within the control of the holder, and therefore there is no indefiniteness as far as he is concerned as to when the instrument will be paid. Consequently the prospective holder is still easily able to calculate the potential value of the instrument in deciding whether to take it or not. However, the holder cannot extend the time for payment and refuse a tender of payment by the obligor. This provision prevents the holder from keeping the interest on the instrument accumulating when the obligor wants to stop it. Section 3–118(f).

(d) *Event Uncertain to Happen.* An instrument that is payable upon an event uncertain to happen is not negotiable even though the act or event has occurred. Section 3–109(2). Thus an instrument payable "thirty days after my twenty-second birthday" is not negotiable. Likewise the courts have held that an instrument payable "thirty days after my death" would not be payable at a definite time and would therefore not be negotiable. The reason is that although death will eventually occur, the time of its occurrence is not definite or controllable by any party. Interestingly enough one could probably achieve the same result by making an instrument payable "100 years from date" with an acceleration clause upon an individual's death. Such a note would be negotiable.

(e) *Antedated, Postdated, and Undated Instruments.* The code makes it clear that an instrument is negotiable regardless of whether it is undated, antedated, or postdated. Section 3–114(1). In the case of an antedated or postdated instrument the time when it is payable is determined by the date on the instrument. If the instrument is undated, it is considered to be a demand instrument. However, if the instrument is payable at a stated time after date and there is no date, it is not payable at a stated time and is not negotiable. Thus an instrument payable "sixty days after date" is nonnegotiable if there is no date. The code still presents a way out, however. An instrument without the date stated is an "incomplete instrument" under the code and the date may be inserted. Section 3–115(1). It is then payable at the stated time after that date and negotiable.

ECKLEY v. STEINBRECHER
482 P.2d 392 (Colo.App. 1971)

The Fergusons contacted Dworak Realty Co. for the purpose of purchasing a farm from Dorothy Steinbrecher. The Fergusons did not have the amount of money necessary for the down payment on the farm, so they sought a loan from Vincent D. Eckley in the amount of $2,000. On July 25, 1962, Eckley executed and delivered his check, postdated July 29, 1962, and payable to the Fergusons. On the face of the check was written "Loan on farm." The Fergusons then endorsed this check and delivered it to Steinbrecher. The check was dishonored for insufficient funds and the sale of the property was never completed. Suit was filed in July of 1967 against the Fergusons and Eckley. The Fergusons defaulted and a judgment was entered against Eckley. Eckley appealed, contending that the instrument was not negotiable.

COYTE, J.

Defendant's first argument is that the check was defective on its face and non-negotiable since it was postdated and bore the notation "Loan on farm." In order for the check to be non-negotiable, it would have to have a patent defect, which would mean that the check does not meet the requirements of negotiability. . . .

Since there is no denial of the fact that the check was signed by the defendant, or that it was for a sum certain, or that it was payable at a definite future time to a definite drawee, the defect assigned by the defendant must go to the requirement that it be an unconditional promise to pay. C.R.S. 1963, 95–1–3 (1) (c), states that a promise is not qualified or made conditional merely because it recites: ". . . the transaction which gives rise to the instrument." This notation on the check does not affect the negotiability of the check as this section of the statute is applicable to the notation "Loan on farm."

The mere fact that the check was postdated does not render it void, but merely defers negotiability to a subsequent time. Nor does it qualify or make the promise conditional. C.R.S. 1963, 95–1–12, specifically provides that postdated instruments are not made invalid unless done for some illegal purpose. Since the defendant does not claim, nor prove, that the postdating was done for such an illegal purpose, the check remains valid.

(Judgment affirmed.)

FERRI v. SYLVIA
214 A.2d 470 (R.I. 1965)

Antonio Sylvia and his wife signed a promissory note payable to Maria P. Ferri. The amount of the note was for $3,000 and was payable "within ten (10) years after date." *The note was dated May 25, 1963. Ferri then tried to collect the amount of the note within the next two years and the Sylvias refused to pay. Ferri filed suit. At the trial the trial judge found the term of the note that provided for payment* "within ten (10) years after date" *to be ambiguous. He therefore admitted extrinsic evidence to explain the term and based upon that evidence concluded that Ferri could have the balance due on the note at any time she needed it. The Sylvias appealed the decision, contending that the extrinsic evidence* *should not have been admitted because the term as to the time of payment was not ambiguous.*

JOSLIN, J.

The question is whether the note is payable at a fixed or determinable future time. If the phrase "within ten (10) years after date" lacks explicitness or is ambiguous then clearly parol evidence was admissible for the purpose of ascertaining the intention of the parties.

At the law merchant it was generally settled that a promissory note or a bill of exchange payable "on or before" a specified date fixed with certainty the time of payment. . . . The same rule has been fixed by statute first under the negotiable instruments law, G.L. 1956, § 6–18–10, subd. 2, and now pursuant to the Uniform Commercial Code. The code in § 6A–3–109 (1) reads as follows: "An instrument is payable at a definite time if by its terms it is payable (a) on or before a stated date or at a fixed period after a stated date. . . .

The courts in the cases we cite were primarily concerned with whether a provision for payment "on or before" a specified date impaired the negotiability of an instrument. Collaterally, of course, they necessarily considered whether such an instrument was payable at a fixed or determinable future time for unless it was, an essential prerequisite to negotiability was lacking.

They said that the legal rights of the holder of an "on or before" instrument were clearly fixed and entitled him to payment upon an event that was certain to come, even though the maker might be privileged to pay sooner if he so elected. They held, therefore, that the due date of such an instrument was fixed with certainty and that its negotiability was unaffected by the privilege given the maker to accelerate payment. Professor Chafee referred to it as providing "the simplest form of acceleration provision," 32 *Harv.L.Rev.* 747,757, and Judge Cooley in *Mattison* v. *Marks*, supra, observing that notes of this kind were common in commercial transactions, said 31 *Mich.* at page 423:

> It seems to us that this note is payable at a time certain. It is payable certainly, and at all events, on a day particularly named; and at that time, and not before, payment might be enforced against the maker. . . . The legal rights of the holder are clear and certain; the note is due at a time fixed, and it is not due before. True, the maker may pay sooner if he shall choose, but this option, if exercised, would be a payment in advance of the legal liability to pay and nothing more.

On principle no valid distinction can be drawn between an instrument payable "on or before" a fixed date and one which calls for payment "within" a stipulated period. This was the holding in *Leader* v. *Plante,* 95 Me. 339, where the court said at page 341, 50 A.54:

> "Within" a certain period, "on or before" a day named and "at or before" a certain day, are equivalent terms and the rules of construction apply to each alike.

We follow the lead of the Maine court and equate the word "within" with the phrase "on on before." So construed it fixes both the beginning and the end of a period, and insofar as it means the latter it is referable to the date the instrument matures. We hold that the payment provision of a negotiable instrument payable "within" a stated period is certain as well as complete on its face and that such an instrument does not mature until the time fixed arrives.

For the foregoing reasons it is clear that the parties unequivocally agreed that the plaintiff could not demand payment of the note until the expiration of the ten-year period. It is likewise clear that any prior or contemporaneous oral agreements of the parties relevant to its due date were so merged and integrated with the writing as to prevent its being explained or supplemented by parol evidence.

(The decision of the trial court was reversed.)

QUESTIONS FOR DISCUSSION

1. *Why were the parties prohibited from introducing extrinsic parol evidence concerning the time when the note was due?*
2. *Explain how the court arrived at the conclusion that despite the phrase payment "within ten (10) years," the payee could not demand payment prior to the expiration of ten years from the date of the instrument.*
3. *Would the maker be allowed to pay the instrument prior to the expiration of the ten-year period?*

22:7. Instrument Must Be Payable to Order or Bearer. Another requirement that must be met in order for an instrument to qualify as negotiable is that it must be made payable to "order" or to "bearer." Section 3–104(1)(d). These are what are known as *words of negotiability,* and they or their legal equivalent must be used for the instrument to be negotiable. In the following discussion a number of phrases that are typically used for *order* or *bearer* instruments will be examined. There is no question as to the acceptance of these words as words of negotiability. To use other words, even if they might be acceptable, is unwise because an instrument that does not use the typical phrases may not be as readily acceptable when the holder attempts to negotiate it. Next a brief examination will be made of the concept of an *order instrument* as opposed to a *bearer instrument.* The basic difference between the two is that an order instrument is negotiated by indorsement and delivery, whereas a bearer instrument may be negotiable merely by delivery.

(1) Payable to Order. The code provides that an instrument is payable to order "when by its terms it is payable to the order or assignor of any person therein specified with reasonable certainty, or to him or his order, or when it is conspicuously designated on its face as 'exchange' or the like and names a payee." Section 3–110(1). Typically the instrument will state "Pay to the order of Jones" or "To Jones or his order." As mentioned before, other phrases may be used but their use is risky. Clearly an instrument drawn "Payable to Jones" would not be negotiable because it is payable only to Jones and not to him or to someone else at his order. The words of negotiability are missing.

The payee may be any one of a number of parties, including the maker, the drawer, the drawee, or anyone else who is not one of the named parties to the instrument. In addition, the instrument may name two or more payees together or in the alternative. Thus it may be made out "To the order of Jones *and* Smith," or it may be made out "To the order of Jones *or* Smith." In the former example both Jones and Smith must indorse the instrument to negotiate it, whereas in the latter case either one of them may negotiate the instrument.

The code also provides that an instrument may be made payable to the order of an estate or trust, to an office or officer of an organization, or to a partnership or association. Section 3–110(1). If it is made payable to an estate or a trust, the person who holds the position of personal representative of the estate or trustee may take payment on the instrument even though it is not made out to him. He is, of course, responsible to the estate or trust for the money. Thus an instrument made payable to "the estate of Danny Decedent" would be negotiable and could be cashed by Jones if he were the personal representative of the estate. Similarly a negotiable instrument made payable to the "comptroller of the Zaney Corporation" would be negotiable and could be indorsed or transferred by the person holding that office acting as principal for the corporation. Additionally the fact that an instrument is made payable to an organization that is not incorporated does not destroy its negotiability. Thus an instrument made payable to "Smith and Brown Transfer Company," a partnership, would be negotiable. The instrument could be indorsed or transferred by any authorized member of the partnership.

Despite the apparent liberality of the code in this area it is necessary that the payee be designated with reasonable certainty. However, the fact that an individual's name is misspelled or that a corporation is not named with exact accuracy will not destroy the instrument's negotiability.

(2) Payable to Bearer. An instrument is negotiable if it is made payable to "bearer." Section 3–111 of the code defines when an instrument is payable to bearer.

> An instrument is payable to bearer when by its terms it is payable to
> (a) bearer or the order of bearer; or
> (b) a specified person or bearer; or
> (c) "cash" or the order of "cash," or any other indication which does not purport to designate a specific payee.

Although the code is fairly clear in this area, a few areas of confusion remain. These problems center primarily around the situation in which a printed form is used that states, "Pay to the order of ________ or bearer." If the name of the payee is written in the blank left on the printed form, the instrument is construed as an order instrument on the ground that this was the party's intent and that the word *bearer* was probably overlooked. On the other hand, if the word *bearer* is handwritten or typed, this would indicate a contrary intention and the instrument would be a bearer instrument. If neither of these situations apply, an instrument made payable to both order and bearer is order paper. Section 3–110(3). If, however, the printed instrument is made out, "Pay to the order of ________" and the blank is not filled in, it is not a negotiable instrument and is not treated as bearer paper.

HOSS v. FABACHER
578 S.W.2d 454 (Tex. App. 1979)

Roger Hoss sued Leo Fabacher in an action based upon a "bearer instrument" which was a form on which some of the blanks were filled in:

Freeport, Texas, 15 April 1971 $6002.19

For value received, I, we, or either of us, the undersigned, promise to pay to the order of ________In ________monthly installments of $________ each and one installment of ________, the first installment to become due and payable on or before the 16 day of July, 1971, and one installment to be due and payable on the ________day of each succeeding month until the whole of said indebtedness is paid with interest from date at the rate of 10 percent per annum.

All past due principal and interest.

It is understood and agreed that failure to pay . . .

(There was also a provision for payment of an attorney's fee and a waiver of presentment, notice, and protest).

Hoss alleged payment was due and that a demand for payment was made. Fabacher contended the note was unenforceable.

PEDEN, J.

Under the Texas Business and Commerce Code, " 'instrument' means a negotiable instrument." § 3.102(a)(5) (1967). To be a negotiable instrument, the writing must (1) be signed by the maker or drawer, (2) contain an unconditional promise or order to pay a sum certain in money and no other promise, order, obligation or power given by the maker or drawer except as authorized by the code, (3) be payable on demand or at a definite time, and (4) be payable to order or to bearer. § 3.104(a) (1967). An instrument is payable to bearer when by its terms it is payable to (1) bearer or to the order of bearer; or (2) a specified person or bearer; or (3) "cash" or the order of "cash," or any other indication which does not purport to designate a specific payee. § 3.111 (1967) The official comment to this section clearly states

2. Paragraph (c) is reworded to remove any possible implication that "Pay to the order of __________" makes the instrument payable to bearer. It is an incomplete order instrument and falls under Section 3.115.

Section 3.115(a) of the code, titled "Incomplete Instruments," provides that when a paper whose contents at the time of signing show that it is intended to become an instrument is signed while incomplete in any necessary respect, it is unenforceable until completed. Comment 2 following that section defines "necessary" as "necessary to complete instrument. It will always include the promise or order, the designation of the payee, and the amount payable." In our case the paper in question, stating: "Pay to the order of __________" is not a bearer instrument and it does not contain a promise to pay any amount. The trial judge was correct: it is unenforceable on its face and is incomplete.

QUESTION FOR DISCUSSION

1. *What additions to the wording of the instrument are necessary to make it a "bearer" instrument?*

22:8. RULES OF CONSTRUCTION. Section 3–118 of the code sets forth certain rules for use in the construing of instruments and for the interpretation of ambiguous terms. They may be summarized as follows:

1. Where there is doubt whether an instrument is a draft or a note, the holder may treat it as either.
2. Handwritten terms control typewritten and printed terms, and typewritten terms control printed terms.
3. If the words are not ambiguous, they control figures.
4. If an instrument provides for interest without stating the amount, the interest is calculated at the judgment rate at the place of payment from the date of the instrument. If the instrument is undated, the interest is calculated from its date of issue.
5. Unless the instrument provides otherwise, two or more persons who sign as maker, acceptor, drawer, or indorser and as part of the same transaction are jointly and severally liable. This is so even though the instrument contains such words as "I promise to pay."
6. Unless otherwise specified, consent to an extension of the instrument authorizes a single extension for not longer than the original period of the instrument.
7. Section 3–115 provides that when a paper whose contents show at the time of signing that it is intended to become an instrument and is incomplete when signed, the instrument cannot be enforced while incomplete.

However, such an instrument may later be completed in accordance with authority given to complete it and it is then effective.

SUMMARY

To be negotiable an instrument must be in writing and signed by the maker or drawer. The instrument must contain an unconditional promise or order to pay a sum certain in money. It must, with a few exceptions, contain no other promise, order, obligation, or power and be payable on demand or at a definite time to order or to bearer.

A negotiable instrument may refer to another document by using the words "as per." Its payment may not be conditional on the performance of requirements contained in another document or "subject to" that document. An acceleration clause does not destroy the negotiability of an instrument. The payment of a negotiable instrument may not be limited to a particular fund. There is an exception for government agencies and when the instrument is limited to the entire assets of a partnership.

A negotiable instrument is payable on demand if it is payable "at sight" or "after presentation." An instrument is payable at a definite time if it is payable at a fixed period after sight. An extension clause is permissible if the extension is at the option of the holder or to a further definite time at the option of the maker or acceptor or automatically upon or after a specified act or event. If the instrument is payable on the happening of an event uncertain to happen, it is not negotiable. An instrument that is antedated, postdated or undated may be negotiable.

An instrument must contain words of negotiability in order to be negotiable. It must be either an order instrument or a bearer instrument.

PROBLEMS

1. Ronald W. Schriber signed a promissory note in the amount of $40,000.00 payable to the Seattle First National Bank. The note provided that it was payable "On demand but no later than 180 days after the date. . . ." The trial court ruled that the note was a demand note as a matter of law. The result of this decision was that the statute of limitations ran from the date the note was signed and Seattle's suit was therefore filed too late. Was the trial court's decision correct? Seattle First National Bank v. Schriber, 282 OR. 625, 580 P.2d 1012 (1978).
2. Barbara Hall signed the following document for John L. Westmoreland, Jr.: "I agree to pay to your firm as attorney's fees for representing me in obtaining property settlement agreement and tax advice, the sum of $2,760.00, payable at the rate of $230.00 per month for twelve (12) months beginning January 1, 1970." Did Barbara sign a negotiable instrument? Hall v. Westmoreland, Hall & Bryan, 123 Ga. App. 809, 182 S.E.2d 539 (1971).

3. Erskine & Sons, Inc., signed a promissory note payable to the Gibson-Stewart Co. in the amount of $16,529.58. The note provided for payments in installments and also provided that the full amount of the note would immediately come due "In the event any installment hereof is not paid when due and/or in the event that the chattel mortgage securing this note is breached in any respect. . . ." Gibson negotiated the note to National City Bank of Cleveland, who sued Erskine & Sons, Inc., to recover on the note. Erskine contended that this provision rendered the note nonnegotiable. Was this contention correct? National City Bank of Cleveland v. Erskine & Sons, Inc., 110 N.E.2d 598 (Ohio 1953).
4. Incitti sued Ferrante on a note dated at Hackensack, N.J., and payable at the Bank Italia Company. The note provided for the defendant to pay the sum of "15,400 Italian lires." Ferrante contended that this provision rendered the note nonnegotiable. Was this contention correct? Incitti v. Ferrante, 175 A. 908 (N.J. 1933).
5. Davis Aircraft Engineering, Inc., signed a promissory note payable to a bank. The note was subsequently transferred to the plaintiff. It provided that "This note evidences a borrowing made under and is subject to terms of a loan agreement dated Jan. 3, 1952 between the undersigned and the payee thereof. . . ." Farrington indorsed the note, and Davis subsequently became insolvent. The holder of the note sued Farrington because of his indorsement. Farrington contended that the note was not negotiable. Was he correct? United States v. Farrington, 172 F.Supp. 797 (1959).
6. Joe Wellbanke and Richard Martin signed three trade acceptances payable to Chemical Products, Inc., which were subsequently transferred to Federal Factors, Inc. The instruments stated that "The transaction which gives rise to this instrument is the purchase of goods by the acceptor from the drawer." Federal Factors sued Wellbanke on the instruments. Wellbanke contended that they were not negotiable. Was he correct? Federal Factors, Inc. v. Wellbanke, 406 S.W.2d 712 (Ark. 1966).
7. Thomas D. Clines signed a note payable to C. P. Moore in the amount of $1,100. The note was dated May 1, 1924, and was payable four months after date. It contained a provision for interest and also stated that "It is agreed that this note is to be paid in Elks Club #8 Second Mortgage real estate bonds." Was the note negotiable? Moore v. Clines, 57 S.W.2d 509 (Ky. 1932).
8. Schleider signed a note to secure payment for dental work performed by his dentist, the payee. The dentist transferred the note to Reserve Plan, Inc. Schleider was apparently dissatisfied with the dental work and refused to pay the note. Reserve Plan sued. The note contained a clause that stated that "In case of death of maker all payments not due at date of death are cancelled." Schleider contended that the instrument was nonnegotiable and that he could raise the defenses of breach of warranty and breach of agreement. Was Schleider correct? Reserve Plan, Inc. v. Schleider, 145 N.Y.S.2d 122 (1955).

9. The Paddocks executed a note payable to Harper Realty Company in the amount of $12,388 as payment for a real-estate commission. Harper then negotiated the note to Alexander and William McLean. The note was dated August 9, 1958, and provided for payments in installments of $75 each "payable monthly after date beginning ______________ 1, 1958 and on the first day of each month thereafter until the whole amount . . . shall have been paid in full." Harper refused to pay the note and raised several defenses to which the McLeans contended that they were not subject as holders in due course. Harper contended that they could not be holders in due course as the note was not negotiable because the month was not filled in the blank, leaving the time for payment uncertain. Was Harper correct? McLean v. Paddock, 430 P.2d 392 (1967).
10. Zander lost a certificate of deposit with the New York Security and Trust Co. that stated "The New York Security and Trust Company, New York, July 11, 1901, has received from Caroline Zander the sum of $500 of current funds upon which the said company agrees to allow interest . . . and on five days notice will repay, in current funds, the like amount with interest, to the said Caroline Zander, or her assignee, on return of this certificate. . . ." The company contended that the instrument was negotiable and therefore required Zander to put up security before replacing the certificate. Was the instrument negotiable? Zander v. New York Security & Trust Co., 39 Misc. 98 (N.Y. 1902).

CHAPTER 23

Transfer and Negotiation

23:1. Introduction. The chief attribute of a negotiable instrument is the ease with which it may be transferred. The transfer of a negotiable instrument may be accomplished by one of two methods: *assignment* or *negotiation.* If the transfer is accomplished by assignment, the assignee takes no greater rights than the person who transferred the instrument to him. If the transfer is by negotiation, however, the transferee may have, particularly if he qualifies as a holder in due course, certain rights and advantages superior to those of the original party to the instrument. These rights will be discussed in Chapter 24, "Holder in Due Course," and Chapter 26, "Liability of the Parties and Discharge." At this point you should be aware of two potential advantages that a transferee may have over an original holder if the transfer is by negotiation. First, the transferee of a negotiable instrument receives the protection of certain warranties from the transferor. Second, if one is a holder in due course—that is, if he gives value for the instrument in good faith and without notice that it is overdue or subject to defenses, or claims—the instrument is taken free of all claims and free of most of the common defenses that might have been asserted against his transferor. Because the negotiation of an instrument can create and modify many of these rights, it is important to understand how a negotiation is accomplished as well as to examine the various conditions under which the negotiation may be made.

23:2. Negotiation of Bearer Instruments: Delivery. The negotiation of an instrument is its transfer in such form that the transferee becomes a *holder.* The status of a holder entitles him to transfer the instrument further and gives him certain other rights discussed in Section 23:3 (2). If the instrument is a *bearer* instrument, negotiation is accomplished by delivery alone. There is no need for an indorsement or any other action. On the other hand, if the instrument is *payable to order,* its negotiation is accomplished by indorsement *plus* delivery. Section 3–202. Delivery means very simply the voluntary transfer of possession of the instrument. Section 1–201 (14). Once there has been delivery of bearer paper, it has been negotiated and the transferee achieves the status of a "holder." The requirements for achieving that even more "exalted" status of a "holder in due course," with all its attendant advantages, will be discussed in Chapter 24. In Section 23:3 we will examine the process of indorsement. Indorsement is a necessary step for the negotiation of order instruments.

BROADWAY MANAGEMENT CORPORATION v. BRIGGS

30 Ill. App.3d 403, 332, N.E.2d 131 (1975)

CRAVEN, J.

Conan Briggs appeals from a circuit court's refusal to vacate an allegedly void judgment by confession against him, or to quash a garnishment summons against the holder of certain stock certificates belonging to him.

The note on which the confession of judgment was based reads in part "Ninety Days after date, I, we, or either of us, promise to pay to the order of Three Thousand Four Hundred Ninety Eight and 45/100 ———— Dollars." (The underlined words and symbols have been typed in; the remainder is printed.) There are no blanks on the face of the instrument, any unused space having been filled in with hyphens. The note contains clauses permitting acceleration in the event the holder deems itself insecure and authorizes confession of judgment "if this note is not paid at any stated or accelerated maturity."

Thus, the critical question of whether this is order or bearer paper is to be determined by Section 3 of the Uniform Commercial Code, which governs negotiable instruments. If this is bearer paper, the plaintiff's possession was sufficient to make it a holder, and this note on its face authorizes the holder to confess judgment against the maker.

On the other hand, if the instrument is order paper, it becomes apparent that the payee cannot be determined upon the face of the instrument. The power to confess judgment must be clearly given and strictly pursued. The warrant of authority having been given in favor of a named person, that warrant may be exercised only by the person named. If the warrant in this case cannot be read to extend to "bearer," then it may not be exercised, since the strict construction mandated by Illinois decisions will not allow a court to guess in whose name such a power may be exercised.

Under the code, an instrument is payable to bearer only when by its terms it is payable to:

(a) bearer or the order of bearer; or
(b) a specified person or bearer; or
(c) "cash" or the order of "cash," or any other indication which does not purport to designate a payee.

The official comments to the section note that an instrument made payable "to the order of ________" is not bearer paper, but an incomplete order instrument unenforceable until completed in accordance with authority. U.C.C., § 3–115. . . .

The instrument here is not bearer paper. We cannot say that it "does not purport to designate a specific payee." Rather, we believe the wording of the instrument is clear in its implication that the payee's name is to be inserted between the promise and the amount, so that the literal absence of blanks is legally insignificant.

Since the holder could not be determined from the face of the instrument, the trial court was in error in allowing plaintiff Broadway Management Corporation to exercise the warrant of attorney granted by this instrument to its holder. The judgment by confession therefore must be vacated.

(Reversed and remanded with directions to allow the motion to vacate the confession of judgment and quash the garnishment summons.)

Questions for Discussion

1. *Was the instrument in this case an order instrument or a bearer instrument?*
2. *What acts are necessary in order to negotiate a bearer instrument to another individual so that he or she becomes a "holder?" An order instrument?*
3. *Was the individual who attempted to confess judgment against the defendant in this case a "holder?" Why or why not?*

23:3. Negotiation of Order Instruments: Indorsement Plus Delivery. An indorsement on an instrument is accomplished very simply by the process of writing one's name on it. It is most frequently done in order to negotiate the instrument; however, as we shall see in Chapter 26, "Liability of the Parties and Discharge," the act of indorsement obligates the indorser to pay the instrument under certain conditions. Hence, by indorsing the instrument, one is doing much more than merely identifying himself or transferring the paper to some other person.

An indorsement is generally placed on the instrument itself, usually on the back, although one may indorse on the front of the instrument. To do this is dangerous, however, as unless it is clearly indicated that one is signing only as an indorser, he may be held to be a co-maker or drawer of the instrument and may be held liable in that capacity, which is quite different from the liability incurred by an indorser.

If one indorses an instrument on a paper separate from the instrument itself, the signature operates as an indorsement only if the paper is so firmly affixed to the instrument as to become part of it. Section 3–202(2). Such a piece of paper is called an *allonge* and is typically used when there is no room left on the instrument itself for an indorsement. For instance, if a paper were merely attached to the instrument by a paper clip, it would not qualify as an allonge and there would not be a valid indorsement. Stapling the paper to the instrument would meet the requirement of the code.

LAMSON v. COMMERCIAL CREDIT CORPORATION

531 P.2d 966 (Colo. 1975)

DAY, J.

I

A chronology of the transactions and the subsequent court trial and appeal draws the issues into focus. Originally, the drawer Commercial Credit Corporation ("the Corporation") issued the two checks payable to Rauch Motor Company ("Rauch"). Rauch indorsed the checks in blank, deposited them to its account in University National Bank ("the Bank"), and received a corresponding amount of money. The Bank stamped the checks "pay any bank," and initiated collection. However, the checks were dishonored and returned to the Bank with the notation "payment stopped." Rauch was obliged to return the money advanced. Its account with the Bank was then overdrawn, but through subsequent deposits Rauch regained a credit balance, which the Bank used to repay itself.

Some months later, to compromise a lawsuit, the Bank executed a special two-page indorsement of the two checks to the plaintiff Lamson. Lamson sued the defendant drawer Corporation on the checks. The Corporation pled a twofold defense. It affirmatively alleged fraud in the inducement and prior payment by Rauch.

The trial court found that the Corporation failed to prove fraud or any other defense. It concluded the defense of payment was unavailable under the Uniform Commercial Code, Section 1–3–306 . . . 1973. Judgment was entered for Lamson for the face amount of the checks plus the legal interest.

In reversing the trial court the Court of Appeals held as a matter of law that the plaintiff Lamson was not a holder of the checks. It arrived at the decision by ruling that the Bank's indorsement to Lamson was not in conformance with the Uniform Commercial Code because it was stapled to the checks.

II

When Rauch deposited the checks, it indorsed them in blank, transforming them into bearer paper. Section 4–1–201(5) and 4–3–204(2). The

Bank in turn indorsed the checks "pay any bank." That is a restrictive indorsement. Section 4–3–205(c). After a check has been restrictively indorsed, "only a bank may acquire the rights of a holder . . . (u)ntil the item has been specially indorsed by a bank to a person who is not a bank." Section 4–4–201(2)(b).

There is no question that the checks were indorsed to Lamson by name, thus qualifying as a special indorsement. Section 4–3–204(1). The problem is whether the special indorsement was correctly and properly affixed to the checks under Section 4–3–202(2). It provides inter alia that "(a)n indorsement must be written . . . on behalf of the holder and on the instrument or on a paper so firmly affixed thereto as to become a part thereof."

The subject indorsement was typed on two legal size sheets of paper. It would have been physically impossible to place all of the language on the two small checks. Therefore, the indorsement had to be "affixed" to them in some way. Such a paper is called an allonge. In this case the allonge was affixed by stapling it to the checks.

We agree with the Court of Appeals' statement that a separate paper pinned or paper-clipped to an instrument is not sufficient for negotiation. However, we hold, contra to its decision, that the section does permit stapling as an adequate method of firmly affixing the indorsement. Stapling is the modern equivalent of gluing or pasting. Certainly as a physical matter it is just as easy to cut by scissors a document pasted or glued to another as it is to detach the two by unstapling. Therefore we hold that under the circumstances described, stapling an indorsement to a negotiable instrument is a permanent attachment to the checks so that it becomes "a part thereof."

Section 4–201(20) defines a holder as "a person who is in possession of . . . an instrument . . . indorsed to him. . . ." The Bank's special indorsement stapled to the two checks, effectively made Lamson a holder, although not a holder in due course.

(The judgment is reversed, and the case remanded with directions to reinstate the judgment of the trial court.)

ESTRADA v. RIVER OAKS BANK & TRUST CO.

550 S.W.2d 719 (Tex. App. 1977)

Dr. William J. Estrada executed four promissory notes payable to George J. Lewis. Lewis then borrowed money from River Oaks Bank and Trust Co. The loan from River Oaks was secured by a single collateral assignment of those notes signed by Lewis in favor of River Oaks. Lewis did not indorse any of the Estrada notes by placing his signature on them although there was adequate space to do so.

When Lewis failed to repay the amount due to River Oaks it filed suit against him. River Oaks also sued Dr. Estrada to recover on the four notes it held as collateral. River Oaks contended that it was a holder in due course of the four Estrada notes. Estrada contended River Oaks was not a "holder" because the four notes were not indorsed by Lewis.

CAULSON, J.

This appeal presents the question of whether the transferee (Lewis) of four unindorsed promissory notes stapled to a single collateral assignment is a holder in due course.

The code definition of a holder applies to payees and transferees of negotiable instruments. A transferee, however, must also meet the negotiation requirements of section 3.202:

> (a) Negotiation is the transfer of an instrument in such form that the transferee becomes a holder. If the instrument is payable to order it is negotiated by delivery with any necessary indorsement; if payable to bearer it is negotiated by delivery.
>
> (b) An indorsement must be written by or on behalf of the holder and on the instrument or on a paper so firmly affixed thereto as to become a part thereof.

The Estrada notes were payable to the order of Lewis and, as order instruments, could be negotiated to River Oaks only by delivery with Lewis' indorsement. River Oaks is not in possession of instruments indorsed to it or its order unless the indorsement on the collateral assignment stapled to the notes is so firmly affixed thereto as to become a part thereof. If the signature on the collateral assignment is not an indorsement of the notes, River Oaks is neither a holder nor a holder in due course.

Comment 3 to section 3.202 is the only code explanation of when an indorsement on a separate instrument is sufficient for negotiation:

> 3. Subsection (2) follows decisions holding that a purported indorsement on a mortgage or other separate paper pinned or clipped to an instrument is not sufficient for negotiation. The indorsement must be on the instrument itself or on a paper intended for the purpose which is so firmly affixed to the instrument as to become an extension or part of it. Such a paper is called an allonge.

An allonge has been defined as: "[a] piece of paper annexed to a bill of exchange or promissory note, on which to write indorsements for which there is no room on the instrument itself." *Black's Law Dictionary* (rev. 4th ed. 1968).

. . . (U)nder the facts of this case, the collateral assignment would not be an indorsement of the notes even if the notes were so covered with previous indorsements that use of an allonge would be an absolute necessity. All four of the Estrada notes were transferred by a single collateral assignment bearing a single signature by Lewis. River Oaks argues that this single signature is an indorsement of all four notes. We disagree. The collateral assignment cannot possibly be so firmly affixed to four notes as to become an extension or part of each one. Although the assignment could conceivably be an indorsement of one of the notes, a court could not determine which note the parties intended to indorse.

River Oaks urges this court to adopt the holding of the Colorado Supreme Court in *Lamson* v. *Commercial Credit Corp.*, 187 Colo. 382, 531 P.2d 966 (1975). In *Lamson*, a special indorsement typed on two legal size sheets of paper was stapled to two checks. This stapling was held to be an adequate method of firmly affixing the indorsement when it would have been physically impossible to place all of the language of the indorsement on the small checks. The facts in *Lamson* differ significantly from those before this court. In *Lamson*, the transferor of the checks signed a document which was clearly intended to be a special indorsement of the checks. In the case at bar, River Oaks wants this court to construe the signature on a collateral assignment to be an indorsement of four notes. We do not believe that the legal principles announced in *Lamson* are applicable to the facts before this court. The problems inherent in negotiating checks, small in dimension, by a lengthy special indorsement inject policy considerations not present in the case at bar. We are aware of no case, based upon facts substantially similar to those before this court, that has adopted the position taken by River Oaks.

We hold that the signature of Lewis on the single collateral assignment is not an indorsement of the four Estrada notes attached thereto, and, therefore, River Oaks is not a holder in due course of those four notes.

QUESTIONS FOR DISCUSSION

1. *Were the instruments in this case order or bearer instruments?*
2. *Why did the court find that there was no valid allonge in this case? How did that decision affect the bank's contention that it was a holder in due course?*
3. *In the Lamson case the court found a valid allonge. In what respect is this case different from the Lamson case?*

---

(1) Wrong or Misspelled Name. The question oftens arises as to how one should indorse an instrument on which his or her name is incorrect or misspelled. The code clarifies this problem by stating that if an instrument is made payable to a person under a misspelled or wrong name, he may indorse it either by using the misspelled or wrong name or by using his correctly spelled name or both. However, a person who is paying or giving value for the instrument may require a signature in both names. Thus, if one receives a check made out with his name misspelled, he may have to indorse it with the misspelled name as well as with his correct name before a bank will cash it.

WATERTOWN FEDERAL SAVINGS & LOAN ASSOCIATION v. SPANKS

193 N.E. 333 (Mass. 1963)

Robert Spanks and his wife signed a promissory note payable to "Greenlaw & Sons Roofing & Siding Co." *The note was indorsed to Colony Distributors, Inc., by an indorsement signed* "Greenlaw & Sons by George M. Greenlaw," *a different name than appeared on the face of the instrument. Colony then indorsed the instrument over to the Watertown Federal Savings & Loan Association. The Spankses did not pay the note and were sued by the bank on the note. From an adverse decision of the trial court the Spankses took exceptions.*

CUTTER, J.

The defendants requested the trial judge to rule that their "demand for proof . . . of their supposed signatures and of the supposed indorsements . . . is constructively broad enough to come within" G.L. c. 106, § 3–203.[1] The judge denied this request "not because as an abstract statement of law it may not be correct, but because upon the facts as found by me . . . it has no bearing." Apart from an offer of proof mentioned later in this opinion, the defendants offered no evidence.

The trial judge found for the bank both as plaintiff and as defendant in set-off. The case is here on the defendants' bill of exceptions.

The trial judge correctly denied the defendants' requested ruling as immaterial. It does not appear that Greenlaw & Sons and Greenlaw & Sons Roofing & Siding Co. are not the same company. The indorsement by Greenlaw was not shown to have been in a name other than his own, nor is it shown that the name of the payee, as stated in the note, was not a name under which Greenlaw individually did business, identifiably repeated in the indorsement. Section 3–203 purports to give only an indorsee for value, and not the maker of a note, the power to require indorsement in both names in the circumstances stated in the section. No evidence was introduced with respect to the indorsement. It comes within G.L. c. 106, § 3–307 (an see the official comments on that section), which reads in part, "(1) . . . When the effectiveness of a signature is put in issue (a) the burden of establishing it is on the party claiming under the signature; but (b) the signature is presumed to be genuine or authorized (with an exception not here pertinent). (2) When signatures are . . . established, production of the instrument entitles a holder to recover on it unless the defendant establishes a defense." There was no evidence whatsoever to counter the presumption of the indorsement's regularity existing under § 3–307 (1) (b). Thus the signature of Greenlaw was established under § 3–307(2), and the bank, as the holder of the note . . . is entitled to recover.

(Exceptions overruled.)

Question for Discussion

1. *Why were the makers of the note unable to avail themselves of the provisions of Section 3–203 and require indorsement of the instrument in both names?*

(2) Transferee's Right to Indorsement. An indorsement generally provides the holder of the instrument with rights against the transferor as well as against the maker, the drawer, or the drawee of the instrument. The question then arises as to whether the transferee, who is a transferee of paper by delivery

[1] As inserted by St.1957, c. 765, § 1 (the Uniform Commercial Code, § 3–203 reads, "Where an instrument is made payable to a person under a misspelled name or one other than his own he may indorse in that name or his own or both; but signature in both names may be required by a person paying or giving value for the instrument."

alone, can require the transferor's indorsement in order to have the added rights against the transferor that the indorsement gives. For example, suppose Peter Payee holds a note payable to his order. Ted Transferee gives Payee $100 in return for the instrument. Payee gives the instrument to Transferee without indorsing it. Several problems should be apparent. In the first place, because this is an order instrument, there is no negotiation of it until it is indorsed by Payee, and Transferee has no indication on the instrument that he is the owner. Therefore the code provides that the transferee has the right to have the transferor indorse the instrument if the transfer is (1) for value and (2) of an instrument that is *not* then payable to bearer. This right is specifically enforceable in court. Negotiation of the instrument does not take effect until the indorsement is made. Section 3–201(3).

At this point several principles should be noted. First, there is no right to the indorsement of the transferor of a bearer instrument, an instrument indorsed in blank, or an instrument given as a gift. If the right to obtain the indorsement applies, the right is to an unqualified indorsement unless there is an agreement to the contrary. If the instrument is transferred without the required indorsement, all that the transferee really has is the right to the indorsement of the transferor; he has no rights on the instrument itself because if it is an order instrument, it cannot be negotiated without the proper indorsement.

(3) Effect of Right to Rescind Indorsement. An instrument may be negotiated even though the right may exist for the negotiation to be rescinded by the transferor. The code in Section 3–207(1) states that the negotiation is effective to transfer the instrument even though the negotiation is

(a) made by an infant, a corporation exceeding its powers, or any other person without capacity; or
(b) obtained by fraud, duress or mistake of any kind; or
(c) part of an illegal transaction; or
(d) made in breach of duty.

The result of this provision is that if one takes an instrument negotiated by an infant, he or she becomes a holder and may therefore negotiate it to yet another party. However, if the transferor, such as an infant, has the power to rescind, he may still rescind after he negotiates the instrument or he may seek another appropriate remedy. For instance, the infant may recover the instrument or recover proceeds from the holder to whom it was negotiated by the infant. However, if the instrument is negotiated to a holder who qualifies as a holder in due course, there can be no rescission effective against him. Thus the protection normally afforded when there is some lack of capacity involved in the transaction is not cut off by negotiation unless it is to a holder in due course.

(4) Types of Indorsement. There are a number of different types of indorsement that may be placed on a negotiable instrument. The manner in which an indorsement is made can affect the liability of the indorser. It will also determine the character of the negotiable instrument—that is, whether it is order or bearer paper—and the indorsement may restrict the course of future negotiations. In studying this section, you should realize that language may be used in an indorsement and will not affect the negotiability of an instrument that if used in the body of the instrument would render it nonnegotiable. There are some rules that must be followed in order for there to be a valid indorsement, however. For instance, in order to be valid, the indorsement must be an indorsement of the entire instrument and not just part of it. If the indorsement purports to be of only part of the instrument, it is nothing more than a partial assignment of the instrument and there is no negotiation. Section 3–202(3). Other types of indorsement may be discussed in the context of (1) blank or special; (2) restrictive or unrestrictive; or (3) qualified or unqualified. These categories are not mutually exclusive, and indeed an instrument is characterized by being slotted in a category in each group.

(a) *Blank Indorsement.* A blank indorsement specifies no particular indorsee to whom the instrument is being transferred. It may consist of a mere signature. One may indorse an instrument in blank if it is an order instrument. Section 3–204(2). The effect of such an indorsement is to convert the order instrument to a bearer instrument. The result, of course, is that the instrument may then be negotiated by delivery alone without further indorsements. For example, suppose a promissory note is made payable "to the order of Paula Payee." The instrument is obviously an order instrument. Paula then indorses it "Paula Payee." This is a blank indorsement because no particular indorsee is specified by Paula. The instrument is now a bearer instrument.

One needs little imagination to realize the risks involved in bearer paper. Anyone can negotiate it by delivery alone, even a thief or a finder of the paper if it is lost. By negotiation the thief may deprive the true owner of his property. These risks are greatly diminished if the paper is order paper. Just as order paper may be converted to bearer paper by the character of the indorsement, so may bearer paper be converted to order paper by special indorsement.

(b) *Special Indorsement.* A special indorsement specifies the person to whom or to whose order it makes the instrument payable. When an instrument is specially indorsed, it becomes payable to the person named in the special indorsement and can be negotiated only by his or her indorsement. Section 3–204(1). Suppose a promissory note is made payable "to the order of Paula Payee." She indorses it "pay to the order of Terry Transferee, Paula Payee." This is a special indorsement and only Terry Transferee can further negotiate the instrument by indorsement. The instrument is an order instrument. It is not necessary that the words "pay to the order of" be used. The indorsement may simply state "pay Terry Transferee" and the effect is the same.

Clearly a bearer instrument can be converted to an order instrument by special indorsement. The code also provides that one may convert a blank indorsement into a special indorsement by writing over the signature of the indorser in blank any contract consistent with the character of the indorsement.

Section 3–204(3). Assume a note is made payable to Paula Payee, which she indorses in blank by placing "Paula Payee" on the back of the note. She negotiates it to Terry Transferee by delivering the instrument to him after indorsing it. Terry now has a bearer note, which he feels nervous about keeping in that state, so he inserts the words "pay to Terry Transferee" above Paula's indorsement. The instrument is now converted to an order instrument and can be negotiated only by Terry's indorsement and delivery of the note.

KLOMANN v. SOL K. GRAFF & SONS
22 Ill.App.3d 572, 317 N.E.2d 608 (1974)

Fred Klomann held three promissory notes as payee. The notes were given by Sol K. Graff & Sons to Fred Klomann as the result of various real-estate transactions. Graff & Sons were real-estate brokers.

Fred Klomann held the three notes for a period of time, and then by special indorsement he gave the notes to his daughter, Candace Klomann. Fred Klomann signed the three notes and handed them to Candace. After examining them, she returned them to Fred for collection. In April 1970 Fred scratched out the name of Candace in the special indorsement, inserted the name of his wife, Georgia Klomann, and delivered the three notes to Georgia.

On May 18, 1971, Georgia Klomann filed suit for collection of the amount due on the three notes. A motion for summary judgment was filed by Georgia. The defendants opposed the motion on the ground that Georgia had no standing to sue on the notes because she was not a holder and had no right, title, or interest in the promissory notes. From a verdict in favor of Georgia the defendants appealed.

DIERINGER, J.

We believe that the plaintiff has no right, title or interest in the promissory notes, Section 3–204 of the Uniform Commercial Code provides:

> (1) A special indorsement specifies the person to whom or to whose order it makes the instrument payable. Any instrument specially indorsed becomes payable to the order of the special indorsee and may be further negotiated only by his indorsement.

A review of the record in the instant case reveals Fred Klomann specially indorsed the promissory notes to his daughter, Candace, in August, 1967. The notes, therefore, could only be further negotiated by Candace. Examination of the record further reveals Candace, the special indorsee, has never negotiated the notes. Fred Klomann, in April, 1970, improperly scratched out Candace's name in the special indorsement and inserted the name of his wife, Georgia. Section 3–201 of the Uniform Commercial Code provides in pertinent part:

> (1) Transfer of an instrument vests in the transferee such rights as the transferor has therein. . . .

When Fred Klomann assigned the notes in question to his daughter he no longer had any interest in them. His attempted assignment to Georgia approximately three years later conveyed only that interest which he had in the notes, which was nothing. Plaintiff, therefore, has no interest in the notes sued on in the instant case.

(Reversed and remanded.)

Questions for Discussion

1. *Explain why Fred Klomann could not substitute his wife's name for his daughter's in this case.*
2. *Would the case have been decided differently had Fred merely signed the notes naming his daughter without ever actually handing the notes to his daughter?*

(c) *Restrictive Indorsement.* Once it is determined whether the paper is bearer or order paper, the next determination to be made is whether any indorsement is a restrictive or a nonrestrictive indorsement. This will reveal whether any limitation has been placed on the further transfer of the instrument by the indorser. Section 3–205 of the code provides that

> An indorsement is restrictive which either
> (a) is conditional; or
> (b) purports to prohibit further transfer of the instrument; or
> (c) includes the words "for collection," "for deposit," "pay any bank," or like terms signifying a purpose of deposit or collection; or
> (d) otherwise states that it is for the benefit or use of the indorser or of another person.

The primary use of restrictive indorsements is in banking channels. Most of the paper used in the business community outside these channels is unrestrictive. Next an analysis of each of these types of restrictions would be appropriate.

(i) CONDITIONED INDORSEMENT. A condition may be placed in an indorsement without its negotiability being destroyed. Compare this with the fact that a condition placed on the face of an instrument renders it nonnegotiable. For instance, suppose a note is made payable "to the order of Paula Payee." Paula then indorses the instrument "pay to the order of Terry Transferee when he delivers 100 bushels of apples to me, Paula Payee." Because this condition appears in the indorsement of an instrument rather than on its face, the instrument is still negotiable. Terry Transferee does not have a right to be paid on the instrument unless the condition is met. If the condition is not met when the instrument matures, then the maker should not pay Transferee. If the maker does pay Transferee without the condition's having been fulfilled, the maker must again pay Paula because the condition placed in the indorsement has not been fulfilled.

The same rule applies if Transferee negotiates the instrument to another party, Holder Ofbag. Ofbag is required to pay any value given by him consistent with the indorsement. If he does not do so, Paula would also have to be paid. Section 3–603(1)(b) of the code adheres to this rule and provides that the liability of a party is not discharged if the party pays or satisfies the holder of an instrument that has been restrictively indorsed in a manner not consistent with the terms of a restrictive indorsement. Hence, if Holder Ofbag pays Terry and Terry has not delivered 100 bushels of apples to Paula, then Ofbag would also have to pay Paula because his payment of Terry was not consistent with the indorsement.

(ii) INDORSEMENT PURPORTING TO PROHIBIT FURTHER TRANSFER. On occasion an indorsement will attempt to restrict further negotiation of an otherwise negotiable instrument. Thus Paula Payee may indorse an instrument "pay Terry Transferee, only, Paula Payee." The indorsement may even state in clear and certain language that the instrument is not to be negotiated beyond the immediate transferee. In a flash of refreshing clarity, the code states in Section 3–206(1) that "No restrictive indorsement prevents further transfer or negotiation of the instrument." Thus, although such an indorsement may be classified as restrictive, it is ineffective in prohibiting further transfer.

As a practical matter, there are few reasons to prohibit the transfer of a negotiable instrument, and thus the code generally does not allow it.

(iii) INDORSEMENT FOR COLLECTION OR DEPOSIT. The most common type of restrictive indorsement is the indorsement that includes the words "for deposit," "for collection," "pay any bank," or similar words that indicate that the purpose of the indorsement is for deposit or collection. Thus, if an instrument, typically a check, is made out to the order of Paula Payee, she might indorse it "for deposit only, Paula Payee." Such an indorsement places on notice and requires those outside the banking collection chain who give value for the instrument to give value consistent with the indorsement.

Those inside the bank collection system do not have to worry about such an indorsement. Section 3–206(3) of the code provides that "An *intermediary bank*, or a *payor* bank which is *not the depositary* bank, is neither given notice nor otherwise affected by a restrictive indorsement of any person except the bank's immediate transferor or the person presenting for payment" (emphasis added). An example might help clarify the impact of this section. Suppose Paula Payee, located in Baltimore, sells goods to Merry Maker, located in Dallas. Merry Maker prepares a check payable to Paula Payee drawn on Maker's bank in Dallas. Paula indorses the check "for deposit only" and deposits it in her bank in Baltimore. The Baltimore bank then indorses it "for collection" and sends it through banking circles to the Dallas bank for collection. The Baltimore bank in which Paula deposited her check is known as the *depositary* bank. The Dallas bank on which Maker drew the check is the *payor* bank. The other banks through which the check passes are known as *intermediary* banks.

The problem with restrictive indorsements for the intermediary and payor banks is that they handle large quantities of checks daily. It would be almost impossible for them to consider the effect of each indorsement on each check. Therefore the code provision quoted earlier gives them immunity from the effect of the restrictive indorsements. In other words, they can ignore the restrictive indorsement of anyone other than the bank's immediate transferor or the person presenting the instrument for payment.

The same immunity is not granted to the depositary bank and transferees outside the collection process. Section 3–206(3) provides that any transferee under an indorsement that is conditional or includes the words "for collection," "for deposit," "pay any bank," or similar terms must pay or apply any value given by him for or on the security of the instrument consistently with the indorsement. The restrictive indorsement, such as "for deposit only," gives a subsequent transferee notice that the holder is an agent for collection and has no authority to use the proceeds for any other purpose. Thus suppose a check made payable to Paula Payee is indorsed by her "for deposit only." She gives it to Albert Agent to place in her account. The depositary bank is charged with notice that the proceeds must be applied to Paula's account. If it gives Albert Agent the cash or applies it to his account, the bank is liable to Paula because this action is not consistent with the restrictive indorsement.

IN RE QUANTUM DEVELOPMENT CORPORATION
397 F.Supp. 329 (D.V.I., 1975)

Quantum Development Corporation went bankrupt, and Charles Joy was appointed receiver of its assets. American Fidelity Fire Insurance Company was the surety and in this capacity prepared, through its agent, a check in the amount of $84,858.00 to the order of "Charles R. Joy, Receiver." On the back of the check was an indorsement that read:

For deposit in Quantum Acct.
Quantum Bankruptcy Charles R. Joy

Joy took this check to the local branch of the Bank of Nova Scotia (BNS), of which William Chandler was the manager, in order to obtain certificates of deposit. Three C.D.s were issued in the name of "Charles R. Joy" in the amount of $75,000. The balance of the money was placed in a checking account. Chandler denied having seen the indorsement on the reverse side of the check. Joy subsequently cashed these three C.D.s and BNS issued a check to "Mr. Charles R. Joy" in the amount of $77,664.54, which represented the original amount plus interest. After using these funds to purchase another C.D. in his own name from Citibank, Charles Joy subsequently absconded with the money, and at the time of this case none of the money had been recovered.

American Fidelity instituted this action to recover these funds from the Bank of Nova Scotia.

YOUNG, J.

Plaintiff's final claim focuses on the alleged failure of BNS to issue the CDs in accordance with the instructions given it by Charles Joy. As heretofore noted, the check which Joy presented to Mr. Chandler was payable to "Charles R. Joy, Receiver," and bore on the back the indorsement "For Deposit In Quantum Acct. Quantum Bankruptcy Charles R. Joy." A perusal of Section 3–205 of the Uniform Commercial Code, intended to provide a functional definition of the term "restrictive indorsement," leaves little doubt that the foregoing fits within the prescription: An indorsement is restrictive which either . . . (c) includes the words "for collection," *"for deposit,"* "pay any bank," or like terms signifying a purpose for deposit or collection; or (d) otherwise states that it is for the benefit or use of the indorser or of another person. IIA V.I.C. § 3–205 (emphasis added). Under either subdivision (c) or (d), the $84,858.00 presented to BNS bore a restrictive indorsement. In addition to stating "For Deposit," the indorsement clearly indicates that it was for the benefit or use of the Quantum bankruptcy estate.

The recognized purpose of a restrictive indorsement is to restrict the use to which the indorsee may put the proceeds of the instrument when a party pays them to the indorsee. No section of the Uniform Commercial Code specifically requires an indorsee–bank to examine the restriction and to ensure that its payment is not inconsistent therewith; nor does any provision set forth any liability on the part of a bank for payment inconsistent with a restrictive indorsement. But, despite the absence of any explicit reference thereto, such duty and resulting liability for the failure to carry out such duty can be fairly inferred from a number of code sections.

The common-law rule, sifted from both case precedent and the respected treatises in the field, suggest that if a bank receives a deposit with instructions to place it to the credit of a fiduciary in his representative capacity and instead credits it to the individual account of the fiduciary, the bank is liable in conversion if the deposit is later disbursed by the trustee for nontrust purposes.

Ignoring the restrictive indorsement inscribed on the back of a check in the substantial amount of over $84,000.00, the bank proceeded to issue to Mr. Joy in his personal and individual capacity the CDs in the amount of $25,000.00 each. Although the bank officials in no way participated in the extreme breach of trust of which Mr. Joy was guilty, their placing the CDs in Joy's individual name provided the opportunity and encouragement which Joy may have needed to perpetrate his scheme.

Plaintiff BNS urges the court that liability cannot be predicated on the restrictive indorsement theory, because the bank was a holder in due course. BNS ignores, however, the fact that if a court finds, as I have, that a bank has failed to abide by a restrictive indorsement, that bank is precluded from attaining holder in due course status. Section 3–206 of the code recognizes that a trust indorsement does not ipso facto give a payor–bank such notice as to deny it the status of a holder in due course. Such payor is, however, immunized from liability by 3–206(2) only insofar as its payment to the trustee or to a purchaser from the trustee is "consistent with the terms" of the trust indorsement. Such is not the case here and BNS is not a holder in due course.

(Judgment for American Fidelity Fire Insurance Company.)

QUESTIONS FOR DISCUSSION

1. *Did it make any difference regarding the liability of BNS whether it was a depositary, a payor, or an intermediary bank? Which was it in this case?*
2. *Under the terms of the restrictive indorsement, how should BNS have handled the funds given it by Charles Joy?*

(iv) INDORSEMENTS IN TRUST. Another type of restrictive indorsement occurs when the indorser indorses for the benefit of himself or some other person and thereby establishes a trust or fiduciary relationship. For example, an instrument made payable to Paula Payee is indorsed "pay Ted Trustee for the benefit of Betty Beneficiary, Paula Payee." Although such a restrictive indorsement does not prohibit the further negotiation of the instrument, the question arises as to whether the person to whom it is going to be transferred can achieve the status of a holder in due course because the trustee is being paid rather than the beneficiary. The code recognizes the fact that trustees frequently sell assets that they control for the benefit of the beneficiaries of the trust. Under the code the trustee may negotiate the instrument and make his transferee a holder in due course. The trustee, of course, is under an independent duty to apply the proceeds in a manner consistent with the trust, but the transferee has no independent duty to see that this is done and hence is unaffected by the restriction of the indorsement. However, the code goes on to state that if the transferee has knowledge that the value furnished for the instrument is obviously for the benefit of the trustee, he cannot qualify as a holder in due course. Section 3–206(4).

(d) *Qualified Indorsements.* Another category into which an indorsement must fit is with regard to whether it is qualified or unqualified. Chapter 26, "Liability of the Parties and Discharge," will discuss the liability incurred when one places an indorsement on a negotiable instrument. It is sufficient for our purposes at this point to recognize that in general, when one indorses an instrument, he promises he will pay the instrument, according to its tenor at the time of indorsement, to the holder or to any subsequent indorser. Section 3–414. The transferor of an instrument by indorsement also gives certain warranties to his and subsequent transferees. In short, the indorsor normally enters into a little "indorsement contract" concerning the conditions under which he will pay if the instrument is subsequently dishonored. This contract can be avoided, however, if the indorsement is *qualified.* The indorsement can be qualified if the indorsement is made *without recourse.* The effect of this qualified indorsement, then, is to eliminate the indorsor's obligation under the indorsement "contract" and to alter the warranty given by the indorser, as will be discussed in Chapter 26. It has no effect on the negotiability of the instrument.

(4) Summary. In summary then, one must first examine an instrument to determine whether it is a *bearer* instrument or an *order* instrument. If it is a bearer instrument, it may be negotiated by delivery alone. If it is an order instrument, it can be negotiated only by indorsement plus delivery. If it has a *blank* indorsement, it is a *bearer* instrument. If it has a special indorsement, it is an *order* instrument. Once the method of negotiation has been established, the next thing is to determine whether any indorsement is a *restrictive* or an *unrestrictive* indorsement. A restrictive indorsement may limit the rights and powers transferred under the indorsement. Finally, the indorsement may be examined for whether the indorsor is subject to the indorsement "contract." The indorsor is not subject to the indorsement if the indorsement is *qualified.*

23:4. ILLEGAL INDORSEMENTS. Negotiable instruments are tempting tools that are occasionally used in various schemes to defraud and steal money from people and businesses. Three basic methods are used to accomplish this: (1) the instrument is drawn in the name of a fictitious payee by one authorized to draw the instrument; (2) an impostor negotiates the instrument; and (3) an indorsement is forged or unauthorized.

(1) The Fictitious Payee. Typically the problem of the fictitious payee is created by an employee inside the organization. For example, suppose Barney Businessman gives William Wrongdoer, his employee, the authority to issue checks in order to pay creditors or other employees. Wrongdoer issues a check for $500 payable to Fictitious Payee who is neither a creditor nor an employee. Wrongdoer indorses the check using the name Fictitious Payee and naming himself as the indorsee. Wrongdoer then cashes the check at Citizens Bank, which pays him the $500. Barney Businessman then learns of the crime and demands that the bank recredit his account.

Obviously the check was drawn on Barney's account, and both the bank and Barney are innocent of any wrongdoing. The policy of the code in this situation is to place the loss on Barney Businessman. Section 3–405(1)(b) states that "an indorsement by *any person* in the name of a named payee is effective if . . . (b) a person signing as or on behalf of a maker or drawer intends the payee to have no interest in the instrument" (emphasis added). In this case, Wrongdoer signed on behalf of Barney and obviously intended that Fictitious Payee have no interest in the instrument. Nonetheless the indorsement was effective in making Wrongdoer a holder with the power to cash the check. The bank, therefore, was authorized, as drawee, in good faith to pay the holder, Wrongdoer.

The code provision in this case is really allocating the risk of loss between two innocent parties, Barney and the bank. The code places the loss on Barney because it was his employee who was responsible for the wrongdoing, and Barney is in a better position to protect himself than the bank. Barney can select the employee and insure against the risk of loss.

The same result obtains if instead of authorizing Wrongdoer actually to issue the check, Barney authorizes him only to prepare the check. Section 3–405(1)(c). The drawee is not liable as long as it pays in good faith.

(2) **Indorsement by Impostor.** Typically an indorsement by an impostor occurs when someone induces an innocent person to draw an instrument to the order of some honorable citizen. Often the citizen is being impersonated by a crook, who then indorses the name of the citizen and makes off with the proceeds. For instance, suppose William Wrongdoer induces Dopey Drawer to make out a check to Respectable Citizen—whom Wrongdoer is impersonating—for $2,500. Wrongdoer then indorses the name of Respectable and cashes the check at Drawee Bank. As between Dopey and the bank, who should sustain the loss?

Once again the code provides the answer by stating that "An indorsement by *any person* in the name of a named payee is effective if (a) an imposter by use of the mails or otherwise has induced the maker or drawer to issue the instrument to him or his confederate in the name of the payee" (emphasis added). Section 3–405(1)(a). It makes no difference whether the imposture was face to face or through the mails, the loss is placed on the drawer. The drawee is entitled to charge the drawer for the instrument regardless of whether the holder is the impostor or someone else because the payment is in accord with the order of the drawer.

The code draws the line, however, where an instrument is made payable to a proper payee and someone else indorses the check, falsely claiming to be an authorized agent of the payee. The drawer or maker is not liable on that false indorsement as he is entitled to the indorsement of the principal.

(3) **Forged Indorsement.** Suppose a check has been made to a proper payee but it falls into the hands of a third party, William Wrongdoer (again), who forges the indorsement of the payee and presents it to the drawee bank, which pays it and charges the drawer's account. In this instance the loss is then placed upon the drawee bank. The code provides that an instrument is converted when it is paid on a forged instrument. This provision adopts the view that payment on a forged indorsement is not an acceptance, even if made in good faith, but rather an exercise of domination and control over the instrument inconsistent with the rights of the owner, and that such payment results in liability to the drawee for conversion. Section 3–419(1)(c). The loss, therefore, rests with the drawee bank when it pays on a forged indorsement. There is an exception, however, if the rightful owner was negligent in his handling of the instrument. The forger, of course, is himself personally liable on his forged indorsement. Section 3–404(1).

FAIR PARK NATIONAL BANK v. SOUTHWESTERN INVESTMENT CO.

541 S.W.2d 266 (Tex. App. 1976)

Southwestern Investment Company agreed to lend James Impson $12,000.00 to finance the purchase of a "front-loader" machine. The sellers of the machine were identified by Impson as J. L. Williams and James L. Wilson doing business as Universal Contractors.

Southwestern then prepared a draft payable to J. L. Williams and James L. Wilson in the amount of $12,000.00. A man who identified himself as J. L. Williams appeared at the offices of Southwestern and picked up the draft. The draft was then presented to the Fair Park National Bank by Impson with the *indorsements of J. L. Williams and James L. Wilson in blank. The draft was sent to Amarillo National Bank which charged Southwestern's account and paid the draft with a cashier's check to Fair Park Bank. Fair Park Bank then issued its cashier's check to Impson for the amount of the draft.*

Subsequently, Impson failed to make the payments due on the loan from Southwestern. Southwestern undertook an investigation which revealed that the machine had not been purchased from Williams and Wilson but stolen. No trace could be found of Williams, Wilson, or any firm doing business as Universal Contractors. Southwestern then sued Fair Park Bank and others contending the bank wrongfully paid the draft.

The trial court found that Fair Park Bank was liable and the bank appealed contending that under the impostor rule the indorsements on the draft were effective whether or not they were the genuine signatures of J. L. Williams and James L. Wilson.

GUITTARD, J.

Southwestern responds that even though the man representing himself to be Williams may have been an impostor, so that the indorsement was effective as respects his purported signature, nevertheless there was no effective endorsement by or on behalf of the other payee, James L. Wilson, because no person representing himself as Wilson had any connection with the issuance of the draft. Southwestern argues that under § 3.116(2) of the Code, when an instrument is payable to two or more persons and not in the alternative, it must be negotiated by both. Consequently, Southwestern asserts that Fair Park Bank is liable for breach of warranty of the Wilson indorsement.

We conclude that Fair Park Bank is not liable for breach of warranty for two reasons. In the first place, the impostor rule applies to the person who signed the bill of sale as James L. Wilson as well as to the person who signed as J. L. Williams. The jury found that Wilson as well as Williams was an impostor. Southwestern does not attack this finding for lack of evidence, but contends rather that this finding is immaterial because whether Wilson is an impostor within the meaning of § 3.405(a)(1) is not a question of fact but a question of law, which the trial court properly resolved in Southwestern's favor. Apparently, Southwestern's position is that there was no imposture with respect to James L. Wilson because no one pretending to bear that name appeared before Southwestern's representative and joined with the purported J. L. Williams in inducing Southwestern to deliver the draft.

This argument erroneously assumes that an "impostor" under § 3.405(a)(1) must meet his victim face to face. That section does not so provide. Rather, it states the rule as follows:

> (a) An indorsement by any person in the name of a named payee is effective if
>
> (1) An imposter [sic] by use of the mails or otherwise has induced the maker or drawer to issue the instrument to him or his confederate in the name of the payee
>
> . . .

One of the purposes of drafting the rule in this language was to eliminate the requirement of a face-to-face meeting, which had been imposed by some of the pre-code cases. Under the code, it is only necessary that "an impostor by use of the mails or otherwise has induced the maker or drawer to issue the instrument to him or his confederate in the name of the payee."

Here the evidence and the verdict established that the person who signed the bill of sale as "James L. Wilson," as well as the person who signed as "J. L. Williams," was an impostor. Southwestern's own evidence showed that there never was any partnership composed of J. L. Williams and James L. Wilson doing business as Universal Contractors at the address given by Impson. This evidence supports the jury's finding that "the James L. Wilson in question"—that is, the person who signed the bill of sale in that name—was an impostor.

Although an impostor cannot be a fictitious person, since there must be a real person who impersonates someone else, an impostor may impersonate a fictitious person. Both of the persons who signed the bill of sale induced Southwestern to issue the draft, since, presumably, Southwestern would not have issued it without a bill of sale bearing both signatures. Consequently, § 3.405(a)(1) makes both indorsements effective to relieve Fair Park Bank of liability for breach of warranty, regardless of who wrote them.

(Reversed in favor of Fair Park National Bank.)

Questions for Discussion

1. *In order for the impostor rule to apply, is it necessary that the issuer of the negotiable instrument actually meet the impostor face to face?*
2. *May the person being imposed be a fictitious person?*
3. *Why do you suppose the U.C.C. provides that the drawee bank is not liable when it pays to an impostor?*

SUMMARY

If an instrument is a bearer instrument, negotiation is accomplished by delivery alone. If the instrument is an order instrument, negotiation is accomplished by indorsement plus delivery.

Indorsement of an instrument is done by writing one's name on it, usually on the back. An indorsement on a separate piece of paper firmly affixed to the instrument is called an allonge. If one's name is misspelled, the instrument may be indorsed by using the misspelled name or by using the correctly spelled name or both.

The transferee of an instrument has the right to have the transferor indorse the instrument if the transfer is for value and the instrument is not then payable to bearer. In general, an instrument may be negotiated even though the right may exist for the negotiation to be rescinded by the transferor.

A blank indorsement specifies no particular indorsee and such an indorsement may create a bearer instrument. A special indorsement specifies the person to whom or to whose order it makes the instrument payable. A restrictive indorsement places limitations on the further transfer of the instrument. A condition may be placed on a restrictive indorsement without destroying the negotiability of the instrument. A restrictive indorsement is not effective in prohibiting further transfer of the instrument.

A restrictive indorsement may be "for deposit only," limiting transfer to banking channels. Intermediary banks and payor banks are not restricted by such an indorsement but a depositary bank is.

A qualified indorsement avoids the indorsement contract. Such an indorsement may be given by indorsing "without recourse."

An indorsement by any person in the name of a named payee is effective if a person signing as or on behalf of a maker or drawer intends the payee to have no interest in the instrument. Thus the drawer suffers any loss for payment to a fictitious payee. Likewise, an indorsement by any person in the name of a named payee is effective if an impostor by use of the mail, or otherwise, has induced the maker or drawer to issue the instrument to him in the name of the payee. If a drawee bank pays on a forged indorsement, it suffers the loss rather than the drawer so long as the drawer was not negligent.

PROBLEMS

1. Elizabeth A. Odgers (now Elizabeth A. Salsman) retained an attorney, Harold Breslow, to handle matters arising from the death of her husband,

Arthur J. Odgers. Mrs. Odgers received a check in the amount of $159, 770.02 made out to her order as beneficiary of a profit-sharing plan in which her husband had participated. Mrs. Odgers was told by Breslow that the proceeds of the check must go into the estate for payment of taxes and other purposes. She indorsed the check, "Pay to the order of Estate of Arthur J. Odgers." After Mrs. Odgers left his office, Breslow indorsed the check, "Estate of Arthur J. Odgers—for deposit Harold Breslow, Trustee." Breslow's secretary then wrote under that, "For deposit Harold Breslow, Trustee." The proceeds were then deposited in Breslow's general trust account in the National Community Bank of Rutherford. The bank did not inquire into Breslow's authority to indorse the check for the estate, and there was no estate account with the bank. Breslow then misappropriated the funds and by the time of the case in question here was serving a prison sentence for his actions. After obtaining a judgment against Breslow, Mrs. Odgers sued National Community Bank. Is the National Community Bank liable to Mrs. Odgers? Salsman v. National Community Bank of Rutherford, 245 A.2d 162 (N.J.App. 1968).

2. Refrigerated Transport Company (RTC) retained United Accounting Systems (UAS), a collection agency, to collect some overdue accounts. UAS received a number of checks made payable to RTC and deposited them in its account with National Bank of Georgia. UAS indorsed these checks in a variety of ways, such as by merely stamping its name on the back or stamping its name and adding words "agent for RTC." RTC never gave UAS authority to indorse the checks. Is the bank liable to RTC? National Bank of Georgia v. Refrigerated Transport Company, Inc., 248 S.E.2d 496, (Ga. App. 1978).
3. William Hamrick and William Steitz were partners in the operation of a restaurant called the Desert Inn. Loans were made to the Desert Inn by Mrs. Hey and were evidenced by two promissory notes. Later Mrs. Hey desired that the name of her son be added as payee, so a new note was executed. It was signed by William Hamrick. Underneath his signature appeared the name *Desert Inn.* Upon default, suit was filed against Hamrick and Steitz. Steitz contended that the use of the words *Desert Inn* did not subject him to liability as a partner and that his name appeared nowhere on the note. Was Steitz liable? McCollum v. Steitz, 67 Cal. Rptr. 703 (1968).
4. A note was signed by Mrs. Martens promising to pay $1,500 to her daughter Mabel Martens Bonk on December 1, 1930. The back of the note contained the following indorsement: "This money is coming to her for teaching $1,000 and $500 is what the rest got also. Mother." The note was placed in Mrs. Martens's safe and was never given to Mabel. Mrs. Martens died on January 2, 1936, and the note was found in her safe in an envelope. On the outside of the envelope was written: "Please give this to S. Fisher in case of death. Mabel Martens from Mother." Mabel Martens Bonk filed a claim with the estate, contending that she was entitled to $1,500 because

of the note. Was she correct? In Re Martens Estate, 283 N.W. 885 (Iowa 1939).

5. Two promissory notes were executed by William C. White naming Citizens' Bank of Lane payee. Citizens' Bank then indorsed the notes "without recourse" and sold them to North End State Bank. The notes were not paid by White, and North End charged the notes back to Citizens'. Citizens' then sued North End, contending that North End had no right to charge it for the notes because it had indorsed them "without recourse." North End contended and proved that there was an oral agreement between the banks by which they agreed to take back notes that were unpaid at maturity even though they were indorsed "without recourse." The trial court found in favor of North End. Was it correct? Citizen's Bank of Lane v. North End State Bank, 226 P. 998 (Kan. 1924).
6. Robert W. Lesco and Willa Mae Lesco executed a note payable to Anthony Joseph Caruso and Marie Doris Caruso on July 15, 1958, in the amount of $4,298.26. The Carusos then assigned the note to the Puzinas. The Puzinas then delivered the note to John and Emanuela Lopez as part payment for property sold to the Puzinas. No indorsement or assignment was placed on or physically affixed to the note. When the note became due, Lesco refused to pay. Lopez then sued the Lescos as maker and the Puzinas as indorsers of the note. Lopez claimed that the Puzinas were indorsers because on a separate instrument entitled "Assignment of Deed of Trust" they had signed a statement transferring the deed of trust and promissory note to Lopez. Were the Puzinas liable on their indorsement? Lopez v. Puzina, 49 Cal. Rptr. 122 (1966).
7. Wilma and Clarence Cole signed a contract to purchase a home from Wyoming Homes. They signed a check as a deposit, payable to Wyoming Homes, for the sum of $4,000.00. The check was drawn on their account in the First National Bank of Buffalo, Wyoming. The salesman for Wyoming Homes took the check, without any indorsement, to the First National Bank of Gillette. The Gillette bank then deposited $4,000.00 to the credit of Wyoming Homes without the check's having any indorsement other than a stamp that read, "First National Bank Gillette, Wyoming, For Deposit Only," in addition to the Gillette bank's own transferring indorsement to any other bank. The check then cleared banking circles and was charged to the Coles' account with the Buffalo, Wyoming, bank. The Coles claimed failure of consideration for the check, contending that it was fraudulently obtained. They sued Gillette Bank on the theory that it was a transferee with the rights of only Wyoming Homes and not even a holder of the check and therefore subject to the defenses of fraud and failure of consideration. Were the Coles correct? Cole v. First National Bank of Gillette, 433 P.2d 837 (Wyo. 1967).
8. A promissory note is indorsed, "Pay Terry Trustee in trust for Benny Beneficiary, Danny Debtor." Trustee attempts to sell the instrument to Inez Innocent by indorsing it and delivering it to her in return for a

sum of money. Can Inez take the instrument and qualify as a holder in due course?

9. A negotiable instrument is made payable to Peter Payee. Payee indorses it, "Pay Terry Transferee one half of the amount of this note, Peter Payee." Is Terry a holder of the note? If the indorsement stated, "Pay Terry Transferee and Tony Transferee, Peter Payee," could Terry and Tony be holders?
10. J. Y. Barnes was the payee of a check drawn by the Portland Cement Association dated October 7, 1963, for $5,088.70. Barnes indorsed the check, "J. Y. Barnes, Jack Y. Barnes, For Deposit Only." The check was mailed for deposit in a Denver bank. It never reached its destination because it was stolen by Deuzil Arthur Woodward, who represented himself as Barnes and opened a checking account with the stolen check at the Cherry Creek National Bank of Denver. The check was indorsed by Woodward using Barnes's name. Woodward then rapidly drew all but a small amount of the money out of the account. The real Jack Barnes then sued the Cherry Creek Bank to recover the funds. Was the Cherry Creek Bank liable to Barnes? Barnes v. Cherry Creek National Bank of Denver, 431 P.2d 471 (Colo. 1967).

CHAPTER 24

Holder in Due Course

In the previous chapters we have examined the requirements that must be met in order for an instrument to be negotiable, as well as the manner in which it may be negotiated. The importance of determining whether an instrument has been properly negotiated is to see if the holder may qualify as a holder in due course. If one qualifies as a holder in due course, he or she is given a status unique to negotiable instruments, because the person who created the instrument may not raise a number of defenses against the holder in due course despite the fact that those defenses are otherwise valid. The purpose of this preferred status is to make negotiable instruments more readily acceptable to parties other than those who were originally involved in the transaction. Remember, however, that if one does not have the status of a holder in due course, any valid defense may be raised against him. First, let us examine the requirements of a holder in due course and then the defenses that may or may not be raised against a holder having that status. Finally, some attention will be given to the increasing tendency to protect consumers against one claiming to be a holder in due course.

24:1. Qualifications of Holder in Due Course. In order for one to qualify as a holder in due course, he must be the holder of an instrument that is negotiable. If he is not a holder or if the instrument is not negotiable, there can be no holder in due course. Beyond this, Section 3–302(1) of the Uniform Commercial Code states the requirements as follows:

> A holder in due course is a holder who takes the instrument
> (a) for value; and
> (b) in good faith; and
> (c) without notice that it is overdue or has been dishonored or of any defense against or claim to it on the part of any person.

(1) **Requirement of Value.** In order for the holder of an instrument to qualify as a holder in due course, he must give some value in return for receiving the instrument. The most obvious application of this requirement is in the case of a gift. Thus, suppose Manny Maker makes out a note payable to Paula Payee. She negotiates it to her nephew Harry Holder as a gift on his birthday. Harry would not have the status of a holder in due course, and Manny Maker

could raise any valid defense against him. Because Holder did not give any value for the instrument, he suffers no loss beyond the instrument if Maker does have a valid defense.

Still unanswered is the question of what constitutes value. Contrary to what you might otherwise expect, the concept of *value* under the code is not the same as the contractual concept of consideration. Value given for an instrument may be the equivalent of consideration but this is not necessarily so. These concepts differ in a number of important respects.

(a) *Executory Promise.* It is basic law that an exchange of promises may constitute consideration even though the promises are not yet performed; that is, they are executory. Thus, if Sam Seller delivers his car to Bill Buyer and Buyer promises to pay for the car in six months, that promise constitutes a valid consideration. Such an executory promise would not constitute value for Article 3 purposes, however. The code states that "A holder takes the instrument for value to the extent that the agreed consideration has been performed. . . ." Section 3–303(a).

As a general rule an executory promise does not constitute value. Thus suppose Paula Payee negotiates a promissory note to Terry Transferee in return for Transferee's promise to pay Payee $200 in the future. Transferee's promise would be consideration because he is obligating himself to do something he was not required to do prior to making the promise. The promise is not value, however, because the agreed consideration has not been performed. Transferee would not qualify as a holder in due course. Note that should Transferee learn of a defense by the maker, he has the remedy of rescinding the transaction. There is not the need for holder-in-due-course status to protect Transferee in this case because actual value was not given for the note. Another example is the bank credit not drawn upon that can be revoked by the bank when a claim or defense arises. The bank simply charges the account for the previous credit not drawn upon. Section 3–303.

The code makes an exception to this general rule when it states that one takes an instrument for value "when he gives a negotiable instrument for it or makes an irrevocable commitment to a third person." Section 3–303(c). The reason the negotiable instrument is value even though it really constitutes a promise to pay in the future is because it can in turn be negotiated to a holder in due course whom the party who gives the instrument cannot refuse to pay. The same reasoning applies when the commitment to a third party is irrevocable. Thus suppose a bank credits a depositor's account with a check and then certifies a check drawn on the amount of the deposited check. The bank then learns of a defense on the deposited check. It would be a holder in due course because the promise made on the certified check is irrevocable.

(b) *Instrument taken as Payment or Security for Antecedent Claim.* Section 3–303(b) of the code states that a holder takes the instrument for value "when he takes the instrument in payment of or as security for an antecedent claim against any person whether or not the claim is due." Here we have a situation in which the thing given as value would not qualify as consideration. If the instrument is given as *security* for an antecedent debt, there is no consideration given by the holder because he gives up nothing in return for the instrument. On the other hand, if the instrument is *payment* of the debt, there is consideration because the right to collect for the original debt is foregone.

Assume the following typical situation. Paula Payee holds a promissory note executed by Money Maker. Paula owes money to Crabby Creditor, who is pressing for payment. Finally, Paula agrees to transfer the note to Creditor as security for the debt. Creditor does not extend time for payment or incur any other legal detriment. Clearly Creditor has not given consideration for the note, but under the code he has given value and therefore may assume the status of a holder in due course in seeking payment from Maker.

If one acquires a lien or security interest in a negotiable instrument, he gives value for the instrument *unless* the lien was acquired by legal process. Thus a creditor who attaches a note that is in possession of a debtor by legal process does not give value for the note. Section 3–303(a).

(c) *Bank Credit for Check.* At this time the position of a bank who gives credit for a check deposited in a customer's account warrants some examination. The code provides that a bank has a security interest in an item—in the case of an item deposited in an account—to the extent to which credit given for the item has been withdrawn or applied. Section 4–208(1)(a). In the usual case, a check is deposited in a customer's account, but he or she is not allowed to draw against the item until it is collected. When this is the case, the bank has not given value for the instrument and thus is not a holder in due course. If the item cannot be collected, all the bank has to do is to deduct that amount from the depositor's account.

Occasionally, however, the depositor is permitted to draw or does draw against the deposited instrument prior to the time when it is collected. When this occurs, the bank gives value and may qualify as a holder in due course of the item. Section 4–208(1)(a). Thus if the bank must sue on the instrument, it may do so with the advantages of a holder in due course because it has given value.

If there is little or no money in the depositor's account, it is clear when the depositor has drawn against the item before it has been collected. The problem gets a little more difficult, however, if there has been a series of deposits placed in the account. In determining whether credit has been allowed on the item, the code follows the rule of "first in, first out" sometimes known as *fifo*. That is, the first funds deposited are the first withdrawn. For example, suppose Danny Depositor has a balance of $200 in his account at the bank. Danny then deposits the check of William Wrongdoer in the amount of $150, which the bank credits to Danny's account. Before Wrongdoer's check is collected Danny withdraws $150 from the account. Wrongdoer's check is then returned for insufficient funds. The bank has not given value for that check because the $150 withdrawn by Danny would be charged against the earlier deposit of $200. The bank would then notify Danny of the return of Wrongdoer's check and deduct $150 from Danny's account.

If the facts are altered slightly, the effect of the fifo concept may be seen more clearly. Assuming the same $200 balance in Danny Depositor's account, suppose that he then deposits a $1,200 check from William Wrongdoer. Later that same day, Danny makes two separate withdrawals, one for $200 and one for $1,200. Wrongdoer's check is then returned as uncollectable. Danny's first withdrawal of $200 would be charged against the earlier deposit of $200. The bank then applies the last $1,200 withdrawal against the last deposit of $1,200. Because it gave Danny money against this check before it was collected, the bank is considered to have given value for the check and as a holder in due course may pursue Wrongdoer if it meets the other requirements.

LEININGER v. ANDERSON
255 N.W.2d 22 (1977)

Larry L. Leininger contracted to buy a business known as Minneapolis Dexstrand Corporation from Howard C. Anderson who also was president of Dexstrand. Leininger gave Anderson cashier's checks in the amount of $20,000.00. The checks were made payable to Dexstrand and Wayzata Bank as copayees. Dexstrand indorsed the checks and they were subsequently paid. The bank used the proceeds of the cashier's checks to pay notes signed by Anderson and Dexstrand and credited the balance to Dexstrand's account.

Subsequently Leininger claimed that Anderson fraudulently misrepresented the value of Dexstrand's assets and that the bank, as receiver of the sale proceeds, converted Leininger's funds. Among other things Leininger contended that the bank could not be a holder in due course because it did not give value for Leininger's cashier's checks. The trial court found in favor of the bank and Leininger appealed.

SCOTT, J.

Leininger argues that Wayzata Bank did not give value for the following reasons: Leininger bargained for a bill of sale; the bank did not sign the sales agreement; Leininger did not receive a bill of sale; therefore the bank did not give value. The Uniform Commercial Code rule is otherwise. Minn.St. 336.3–303, defining "taking for value," reads as follows:

> A holder takes the instrument for value
>
> (a) to the extent that the agreed consideration has been performed or that he acquires a security interest in or a lien on the instrument otherwise than by legal process; or
>
> (b) when he takes the instrument in payment of or as security for an antecedent claim against any person whether or not the claim is due; or
>
> (c) when he gives a negotiable instrument for it or makes an irrevocable commitment to a third person.

The facts of this case clearly show the bank to have met these criteria. The trial court's finding of fact on this issue, which is well supported by the testimony of Boswinkel, indicates that a large portion of the proceeds of Leininger's cashier's checks was applied to satisfy the two notes on which Dexstrand was either primarily or secondarily liable, plus interest. This application falls under § 336.3–303(b) as "payment of . . . an antecedent claim against any person whether or not the claim is due." The remaining $4,284.99 was placed in the form of a check payable to Dexstrand. Anderson endorsed this check on behalf of Dexstrand and the check was deposited to Dexstrand's account. This transaction comes under § 336.3–303(c)—the bank gave a negotiable instrument for this part of the proceeds. Taken together, the bank retained no portion of the proceeds of the checks for its own benefit, but rather the checks were used to clear the obligations of Dexstrand to the bank with the remainder being credited to the account of Dexstrand. The bank further assigned its security interest in Dexstrand's assets to Leininger, which would come under § 336.3–303(a) as part of the "agreed consideration" pursuant to Anderson's instructions. The facts therefore show that the bank did give value for the cashier's checks as required by Minn.St. 336.3–303.

(Affirmed.)

QUESTIONS FOR DISCUSSION

1. *What was the "value" given by the bank?*
2. *Does the value given by the bank qualify as consideration under the law of contract?*

24:2. Requirement of Good Faith. In order to qualify for the status of a holder in due course, one must have taken the instrument "in good faith." This has proved to be one of the more troublesome requirements. How does one go about determining whether an instrument was taken "in good faith?" One possible approach is to apply an objective standard. If the instrument is taken under circumstances that "the reasonable person" would regard as being free from defects or defenses, then the instrument would be taken in good faith. Indeed this is the approach used in the solving of many legal problems. The code rejects this test, however, and applies a subjective test. It states that " 'Good faith' means honesty in fact in the conduct or transaction concerned." Section 1–201(19).

Under the subjective test one may qualify as having taken an instrument in good faith even if he is naïve or gullible to the point where the suspicions of a reasonable person would have been aroused. Remember, the question is whether the individual in question in fact took the instrument in good faith. Even if reasonable commercial standards are ignored, the holder may still have taken an instrument in good faith. The reason for this approach is, once again, to promote the free transferability of commercial paper. Thus one may take an instrument from a total stranger without inquiring into his background and still take it in good faith.

This test of good faith has led to some abuses, particularly in transactions involving consumers. Consequently there have been some court decisions that have found a lack of good faith on the part of the holder when his suspicions should have been aroused. There has also been some reaction by legislatures and government agencies, discussed in Section 24:11, dealing with the special treatment of consumers.

BOWLING GREEN, INC. v. STATE STREET BANK AND TRUST COMPANY
425 F.2d 81 (1970)

Bowl-Mor, Inc., was a manufacturer of bowling-alley equipment. Bowling Green, Inc., operated a bowling alley and obtained a United States check of $15,306.00 from a Small Business Administration loan. It contracted to purchase bowling-alley equipment from Bowl-Mor, Inc., and negotiated the check on September 26, 1966, to Bowl-Mor as a first installment on the purchase of the equipment.

On September 27, 1966, Bowl-Mor deposited the check in its account at the State Street Bank and Trust Company. The bank immediately credited $5,024.85 of the check against an overdraft in Bowl-Mor's account. Later that day the bank learned that Bowl-Mor had petitioned for reorganization under the Bankruptcy Act. It then transferred $233.61 of Bowl-Mor's funds to another account and applied the remaining $10,047.54 against debts that Bowl-Mor owed the bank. Thereafter Bowl-Mor was adjudicated a bankrupt. Bowling Green, Inc., never received the pin-setting machines for which it had contracted.

Bowling Green sued the bank on the grounds that the bank was constructive trustee of the funds depos-

ited by Bowl-Mor. The bank contended that it was a holder in due course. Bowling Green contended that Bowl-Mor knew that it could not perform when it made the contract and accepted payment and that the bank was aware of this fraudulent conduct and that its close working relationship with Bowl-Mor prevented it from being a holder in due course because it did not take the instrument in good faith. From an adverse verdict Bowling Green appealed.

COFFIN, J.

Plaintiff's second objection arises from the intimate relationship between Bowl-Mor and the Bank, a relationship which plaintiff maintains precludes a finding of good faith. The record shows that the Bank was one of Bowl-Mor's three major creditors, and that it regularly provided short term financing for Bowl-Mor against the security of Bowl-Mor's inventory and unperformed contracts. The loan officer in charge of Bowl-Mor's account, Francis Haydock, was also a director of Bowl-Mor until August 1966. Haydock knew of Bowl-Mor's poor financial health and of its inability to satisfy all its creditors during 1966. In the five months before the transaction in question, the Bank charged $1,000,000 of Bowl-Mor's debt to the Bank's reserve for bad debts. However, the record also shows that the Bank continued to make loans to Bowl-Mor until September 12.

The Bank was also aware of the underlying transaction between Bowl-Mor and the plaintiff which led to the deposit on September 26. During the week prior to this transaction, Bowl-Mor had overdrawn its checking account with the Bank to meet a payroll. In order to persuade the Bank to honor the overdraft, officials of Bowl-Mor contacted Haydock and informed him that a check for $15,000 would be deposited as soon as plaintiff could obtain the funds from the Small Business Administration. The district court found, however, that the Bank was not aware that the directors of Bowl-Mor had authorized a Chapter X petition or that Bowl-Mor officials planned to file the petition on September 27.

On the basis of this record, the district court found that the Bank acted in good faith and without notice of any defense to the instrument. The Code defines "good faith" as "honesty in fact." Mass.Gen.Laws Ann. c. 106 § 1–201(19), an essentially subjective test which focuses on the state of mind of the person in question. The Code's definition of "notice," Mass.Gen.Laws Ann. c. 106 § 1–201(25), while considerably more prolix, also focuses on the actual knowledge of the individuals allegedly notified. Since the application of these definitions turns so heavily on the facts of an individual case, rulings of a district court under §§ 3–302(1)(b) and 3–302(1)(c) should never be reversed unless clearly erroneous. In this case, the evidence indicated that Bowl-Mor had persevered in spite of long-term financial ill health, and that the event which precipitated its demise was the withdrawal of financial support by another major creditor, Otis Elevator Co., on the morning of September 27, after the deposit of plaintiff's check. Thus at the time of deposit, the Bank might reasonably have expected Bowl-Mor to continue its shambling pace rather than cease business immediately. The findings of the district court are not, therefore, clearly erroneous.

(Affirmed.)

QUESTIONS FOR DISCUSSION

1. *Did the court apply a "subjective" or an "objective" test in determining good faith in this case? What is the difference?*
2. *Note how the court in its discussion intertwined its discussion of "good faith" with the separate element of "notice."*

24:3. REQUIREMENT OF LACK OF NOTICE. The requirement of taking an instrument in good faith is often confused with the requirement that the instrument also must be taken without notice. The code lists three "danger signals" regarding an instrument. If the potential holder has notice of any one of them, he cannot be a holder in due course. These danger signals are (1) that the instrument is overdue; (2) that the instrument has been dishonored; or (3) that there is any defense against it or claim to it on the part of any person. Section 3–302)(1)(c). The concepts of notice and good faith are closely related, but they are separate and distinct requirements. For instance, notice of a defense to an instrument is quite clearly evidence of a lack of good faith.

The test for determining whether a taker has notice of a defect in the instrument is also different from the test used for establishing good faith. A subjective test is combined with an objective standard. It is not necessary to have actual knowledge in order to be charged with notice. The code states that "A person has notice of a fact when (a) he has actual knowledge of it; or (b) he has received a notice or notification of it; or (c) from all facts and circumstances known to him at the time in question he has reason to know that it exists." Section 1–201.

Not only must one have notice, but he must have an opportunity to act on it. Notice must be received at a reasonable time and in a reasonable manner. Section 3–304(6). Thus notifying the bank president of a defect in a check one minute before it is cashed by a teller would not constitute notice.

The purpose of the notice requirement is to prevent one from gaining the status of a holder in due course when he is aware of the fact that there may be a problem with the instrument. If one knows of the fact that there are potential problems with the instrument but chooses to ignore them, then he should assume the risk. There is then no reason to clothe him with the protections afforded a holder in due course. At this point an analysis of each of the three "danger signals" mentioned by the code is appropriate.

(1) Notice That Instrument Is Overdue. If one takes an instrument with notice that it is overdue, he or she cannot qualify to become a holder in due course. The application of this principal is relatively simple in the case of time instruments. If a promissory note is payable on January 1, 1978, then one cannot become a holder in due course by taking the note on January 2, 1978. The code provides that the principal amount, or part of it, must be overdue in order to reconstitute notice. Section 3–304(3)(a). Default in the payment of interest on the instrument is not notice. Section 3–304(4)(b).

Demand instruments present a more difficult problem because there is no specific time when the instrument is due. If the potential holder is aware or has reason to know that he is taking the instrument after a demand for payment has been made, then he has notice that it is overdue and cannot qualify as a holder in due course. A potential holder frequently does not have actual notice that a demand has been made, but the code charges him with such notice of demand after a reasonable time passes from the date that the instrument was issued. Section 3–304(3)(c). The crunch comes, of course, in determining whether a reasonable time has passed since the issue of the demand instrument.

The code creates a presumption, which is rebuttable, as to what constitutes a reasonable time in the case of *checks* drawn and payable within the United States and the District of Columbia. Thirty days after one takes a check, the presumption arises of notice that it is overdue. No guidance is given by the code as to what constitutes a reasonable time for other types of instruments. One must look to the nature, purposes, and circumstances of each transaction in order to determine whether a reasonable time has passed since the date when the instrument was issued. Thus a reasonable time in some cases may be sixty or ninety days, whereas in others in may be one or two years. Such matters as the type of instrument, the purpose of its issue, whether it bears interest, and whether the instrument is typically held for an investment should be considered in a determination of what constitutes a reasonable time. For instance, the reasonable time period for a check that does not bear any interest would probably be shorter than that for an interest-bearing demand note or an interest-bearing certificate of deposit, which is usually kept for an investment.

Sometimes notes are issued in a series of separate instruments, or an instrument may be payable in installments. One has notice that an instrument is overdue if he or she has reason to know that an installment is overdue or that any part of the principal amount is overdue or that there has been a default in payment of another instrument of the same series.

We have already learned that a negotiable instrument may contain a clause accelerating the due date of the entire instrument upon default in the payment of an installment. Logically, therefore, if one has reason to know that an acceleration of the instrument has been made, he or she has reason to know that it is overdue. This constitutes notice and prevents one from becoming a holder in due course of the instrument. Section 3–304(3)(a). Remember, however, that the potential holder must have reason to know of the acceleration in order to be charged with notice. The mere fact that the instrument has been accelerated does not automatically charge him with this notice.

(2) Notice of Dishonor. If one has notice that an instrument has been dishonored, he cannot qualify as a holder in due course because the potential holder knows before he takes it that the instrument will not be paid. There is, therefore, no reason to clothe him with the protection of a holder in due course.

In the case of a check or other instrument presented through ordinary banking circles, the determination that an instrument has been dishonored is relatively simple because it is frequently stamped "insufficient funds" or "payment stopped" or is marked in some other manner. If the instrument itself does not bear evidence that acceptance or payment has been refused, then whether a potential holder is charged with notice of dishonor is a question of fact.

(3) Notice of a Claim or Defense. The code states that one cannot attain the status of a holder in due course if he or she has notice of a claim or defense to the instrument. Thus one may contend that he was induced to sign the instrument because of fraud. One taking the instrument with knowledge of this defense could not be a holder in due course, and the defense could be raised by the obligor when the holder attempts to collect the instrument. On the other hand, notice of a counterclaim or set-off will not prevent the transferee from achieving the status of a holder in due course because the defense has nothing to do with the issuing of the instrument in the first place. Thus suppose Paula Payee owes Manny Maker $100 from a previous transaction. In a subsequent transaction Maker issues a promissory note to Paula for $200. Paula negotiates the note to Terry Transferee for value. Transferree knows of the $100 debt owed by Paula to Maker, that is, Maker's set-off. Transferee would still be able to qualify as a holder in due course.

(a) *Notice on the Face of the Instrument.* The code delineates a number of situations that give rise to notice of a claim or defense. For instance, there are a number of indications that may appear on the instrument itself that should indicate to the potential holder that there may be a claim or defense to it. A person has an obligation to examine the instrument before purchasing it and therefore is charged with notice if "the instrument is so incomplete, bears such visible evidence of forgery or alteration or is otherwise so irregular as to call into question its validity, terms or ownership or to create an ambiguity as to the party to pay." Section 3–304(1)(a).

(i) INCOMPLETE INSTRUMENT. Under certain circumstances taking an instrument that is incomplete in some material respect can amount to notice of a claim or defense. Material blanks in an instrument could be the sum to be paid, the name of the payee, or the time when the instrument is due. Prior to the code an instrument had to be "complete and regular on its face." However, there are times when in the normal course of business the holder of an instrument may have the authority to fill in certain blanks. A person in possession of an instrument is presumed to have the authority to fill blanks. Therefore one may take an instrument as a holder in due course even though a blank is filled in his presence as long as he has no notice that the filling is improper. Section 3–304(4)(d). For instance, an agent may have a check of his principal that has been signed with the amount left blank. In dealing with a third party, the agent may fill in the amount in the presence of the third party, and the third party may acquire the status of a holder in due course as long as he is without notice that the filling in is improper.

COOK v. SOUTHERN CREDIT CORP.

448 S.W.2d 634 (Ark. 1970)

BYRD, J.

The record here shows that appellant Joe Cook, on Oct. 22, 1966, purchased a 1966 Ford Thunderbird automobile from Gerald Harris. He traded in a 1964 Thunderbird and agreed to pay an additional $3700.00 to consummate the sale. On that date Joe Cook signed a blank conditional sales contract and note which he gave to Harris with the understanding that Harris would hold the contract and note until he could furnish Cook the registration title and pink slip for the 1966 automobile. After approximately three weeks Cook returned the automobile to Harris because Harris could not furnish the title and pink slip as he had agreed to do. When the '66 Thunderbird was returned, Harris had already sold or given up possession of the '64 Thunderbird. At that time Harris agreed that he would get Cook a cheaper automobile.

Cook says that to his knowledge the conditional sales contract and note were never filled out. It is admitted that on Oct. 22, 1966, Harris took Cook's conditional sales contract and note signed in blank to the offices of Southern Credit Corporation where Harris procured the services of a secretary of Southern Credit Corporation to fill in the blanks. Henry McClure, the Vice President of appellee Southern Credit Corporation, testified that he had no knowledge of any defense or claim against the note at the time his company obtained it in good faith and for value.

The trial court entered judgment for appellee Southern Credit Corporation on the conditional sales contract and note. For reversal

Cook raises three points, all of which are premised on the fact that appellee was not a holder in due course because it was aware that Harris had used a typewriter and one of the appellee's secretaries to fill in the blanks in the signed contract and note. We find appellant's contentions to be without merit. Ark.Stat. Ann. § 85–3–304(4) (Add. 1961) provides:

> Knowledge of the following facts does not of itself give the purchaser notice of a defense or claim
>
> (a) . . .
>
> (d) That an incomplete instrument has been completed, unless the purchaser has notice of any improper completion;

The committee comment to the Uniform Commercial Code with reference to subsection (4)(d) points out that the policy of the statute is premised upon the fact that a person in possession of an instrument signed in blank has prima facie authority to fill in the blanks. Therefore as we interpret the statute a purchaser of a commercial instrument is entitled to the status of a holder in due course even though he knows that the possessor of the signed instrument filled in the blanks.

(Affirmed.)

QUESTIONS FOR DISCUSSION

1. *What are the dangers of signing a note in which the terms are left blank? What should Cook have done to protect himself in his dealings with Harris?*
2. *Why did the court allow Southern Credit to raise successfully its status as a holder in due course when it was aware that Harris had filled in an instrument signed by Cook?*

(ii) IRREGULARITY OR ALTERATION. If an instrument bears such visible evidence of alteration or forgery or is otherwise so irregular "as to call unto question its validity," the potential holder has notice. This language does not mean, however, that any change or alteration of an instrument denies the taker the status of a holder in due course. The alteration must be such as to "call into question" the validity of the instrument. For instance, suppose an instrument is made payable to William Wrongdoer in the amount of $1,000. Wrongdoer squeezes in another zero and erases the word *one* and changes it to a *ten* in a not-so-subtle manner, so that the instrument is payable to him in the amount of $10,000. He then negotiates the instrument to Terry Transferee, who could not be a holder in due course because of the obvious evidence of tampering with the instrument. On the other hand, suppose that an instrument is made out and dated "January 1, 1977," when in fact it is January 1, 1978. An alteration to the correct date in this circumstance would not be evidence of the type of alteration that would "call the validity of the instrument into question" because this is a normal and frequent mistake, having nothing to do with the validity of the instrument.

The fact that an instrument is antedated or postdated does not mean that an instrument is irregular. Section 3–304(4)(a). A person may take a postdated or antedated instrument and still qualify as a holder in due course.

BRIAND v. WILD
268 A.2d 896 (N.H. 1970)

Norman Wild and Lois Wild were negotiating with Romeo Briand for the purchase of a piece of land. On March 31, 1965, Mrs. Wild sent Briand a letter that stated: "(C)onfirming our telephone conversation of yesterday, I am enclosing herewith my check in the amount of $400.00 (dated May 1st as suggested *by you) to cover . . . Lot #33." The postdated check was also enclosed in the letter. All the details of the agreement had not been worked out and no writing contained the purchase price. Mrs. Wild also stated that her husband was away and would contact Briand when he returned.*

Prior to May 1, 1965, the Wilds decided not to purchase the lot and stopped payment on the check. Briand sued for breach of contract and also sued on the check for $400.00, contending that he was a holder in due course. The trial court found for Briand and the Wilds appealed.

KENISON, C.J.

[After finding that the contract action of Briand could not be sustained because of the Statute of Frauds the court dealt with the postdated check.]

A payee of a negotiable instrument may be a holder in due course, RSA 382–A:3–302(2), if he meets the requirements of RSA 382–A:3–302(1). The negotiability of a check is not necessarily affected by the fact that it is postdated. RSA 382–A:3–114(1), 3–304(4)(a). But a purchaser of a postdated instrument who knows at the time of his purchase of an asserted defense or of facts or circumstances which may create a defense, is precluded from being a holder in due course.

The plaintiff Romeo Briand, the payee of the check, had notice that the check was postdated and that the sale might not be completed when defendant Norman Wild returned if the terms of payment could not be worked out. The plaintiff as payee does not qualify as a holder in due course for he is charged with notice of the claim of the defendants to the $400 if the sale were not completed. RSA 382–A:3–302. Plaintiff took the check subject to "all defenses" the defendants have "available in an action on a simple contract" and "the defenses of want or failure of consideration, (and) non-performance of any condition precedent." RSA 382–A:3–306. The plaintiffs are not entitled to a judgment for $400 in the suit on the check.

(Judgment for plaintiff reversed.)

(4) **Notice That Obligation Is Voidable or That Parties Have Been Discharged.** If one takes an instrument with notice that the obligation of any party is voidable, he cannot qualify as a holder in due course. An instrument may be voidable because a party entered into it as the result of undue influence, duress, or fraud. Section 3–304(1)(b). These defenses are normally not available against a holder in due course.

Likewise notice that all parties on the instrument have been discharged prevents one from attaining the status of a holder in due course. Section 3–304(1)(b). Observe that the notice of discharge must be as to *all* parties. The fact that a transferee has knowledge that some parties to an instrument have been discharged will not prevent him from acquiring the status of a holder in due course as to those parties of which he has no notice.

(a) *Fiduciary.* One does not have notice of a defense or claim merely because the person who negotiates the note to him is a fiduciary. Section 3–304(4)(e). There are many valid reasons for a fiduciary to negotiate an instrument and even paying cash to the fiduciary in return for an instrument is valid because one is entitled to the assumption that the fiduciary is acting properly. However, one cannot be a holder in due course if he has knowledge that a fiduciary "has negotiated the instrument in payment of or as security for his own debt or in any transaction for his own benefit or otherwise in breach of duty." Section 3–304(2).

24:4. Taking Through a Holder in Due Course. In an effort to assure that the holder in due course has a free market for the instrument, the code has what has commonly been referred to as a *shelter provision.* Section 3–201(1) of the code provides that "Transfer of an instrument vests in the transferee such rights as the transferor has therein. . . ." Thus one who holds an instrument as a holder in due course may transfer this protection to his immediate transferee. For example, assume that Paula Payee induces Manny Maker to make an instrument payable to her. She then negotiates the instrument to Harry Holder, who has no notice of the fraud and qualifies as a holder in due course. After the instrument is due, Harry gives the instrument to Terry Transferee, who has notice of but did not participate in the fraud. Transferee could collect the instrument from Maker as a holder in due course, cutting off Maker's defense of fraud. Note the number of reasons that Transferee would not qualify as a holder in due course in his own right.

There are limits to the shelter provision. The primary limitation is that one who participates in the fraud or illegality affecting the instrument or who as a previous holder has notice of a defense or claim against it cannot improve his position by taking from a later holder in due course. Section 3–201(1). To put it in a more quaint way, such a person is not able "to wash the paper clean by passing it into the hands of a holder in due course and then repurchasing it." Section 3–201(1), Comment 2. Thus, if we assume the same facts as the previous example except that Harry Holder gives the instrument back to Paula Payee, she would not have the protection of a holder in due course, having previously been a party to the fraud.

CANYONVILLE BIBLE ACADEMY v. LOBEMASTER

108 Ill.App.2d 318, 247 N.E.2d 623 (1969)

Louis and Bernice Lobemaster executed a promissory note in the amount of $12,694.22 payable to the order of Lobemaster Trailer Sales, Inc., in eighty-four installments. Trailer Sales indorsed the note to Jefferson Bank and Trust Company, who qualified as a holder in due course. The Lobemasters defaulted on the note and the Bank then assigned the note to Canyonville Bible Academy.

Canyonville Bible Academy sued the Lobemasters, who contended that they could raise certain defenses against it as the academy was not a holder in due course of the note because it had not given value for it. From an adverse decision at the trial level the academy appealed.

TRAPP, J.

The second affirmative defense is that the note was made payable to the Lobemaster Trailer Sales, Inc., and there was never a valid assignment or purchase of the note for value.

Under Section 3–201 of the Uniform Commercial Code transfer is provided for as follows:

> (1) Transfer of an instrument vests in the transferee such rights as the transferor has therein except that a transferee who has himself been a party to any fraud or illegality affecting the instrument or who as a prior holder had notice of a defense or claim against it cannot improve his position by taking from a later holder in due course.

In the Illinois Annotated Statute, the Illinois Code Comment under this section is the following interpretation:

> The first clause of this subsection changes the rule under the first part of § 49 of the Illinois NIL (§ 49 of the NIL with variations), which provided, "where the holder of an instrument payable to his order transfers it *for value* without indorsing it, the transferer vests in the transferee such title as the trans-

feree (transferer) had therein" (emphasis supplied). This subsection eliminates the requirement of value for purposes of transferring the rights of the holder.

Additionally, under the Uniform Commercial Code Comment in the Illinois Annotated Statute is the following:

> 2. The transfer of rights is not limited to transfers for value. An instrument may be transferred as a gift, and the donee acquires whatever rights the donor had.

Under Section 3–201 of Ch. 26, Ill.Rev.Stat. (1963), it is quite clear that any transfer of an instrument transfers all rights of the transferor. . . . Since the Bank was a holder in due course and plaintiff was not a prior holder the plaintiff by transfer from the Bank acquired the rights of a holder in due course irrespective of the question of value.

(Reversed and remanded.)

QUESTION FOR DISCUSSION

1. *Would the decision have been the same if Canyonville Bible Academy had been a prior holder of the note without notice of any defense or claim against it? What if the academy had had notice of a claim or defense as a prior holder?*

24:5. PAYEE MAY BE A HOLDER IN DUE COURSE. A provision that on the surface may be somewhat confusing is that a payee may be a holder in due course. Section 3–302(2). Although a payee frequently will not qualify for holder-in-due-course status because he or she is often a party to the transaction that gives rise to the instrument, it is possible. A payee may become a holder in due course to the same extent and under the same circumstances as any other holder, regardless of whether the instrument is taken by purchase from a third party or directly from the obligor. For instance, suppose that Manny Maker and Inez Innocent as co-makers sign a promissory note payable to Paula Payee. Manny induces Inez to sign as the result of fraud and delivers the note to Paula Payee without authority from Inez. Paula takes the note for value, in good faith, and without notice of the fraud. Paula would have the status of a holder in due course and would be able to enforce the note against Inez without being subject to the defense of fraud.

24:6. TRANSACTIONS PREVENTING THE STATUS OF HOLDER IN DUE COURSE. The purpose of according a qualified transferee the status of a holder in due course is to facilitate the flow of commercial paper. The code therefore lists several transactions that are normally not of a commercial nature. If one acquires or purchases the instrument under these circumstances, he is denied holder-in-due-course status. In these situations the reasons for the creation of holder-in-due-course status do not exist. Section 3–302(3) provides:

> A holder does not become a holder in due course of an instrument:
> (a) by purchase of it at judicial sale or by taking it under legal process; or
> (b) by acquiring it in taking over an estate; or
> (c) by purchasing it as part of a bulk transaction not in regular course of business of the transferor.

In each of these situations the purchaser is normally only a successor in interest to the prior holder and receives no better rights than the prior holder.

Of course, if the prior holder had holder-in-due-course status, then so would the purchaser under the umbrella provision of the code. Section 3–201(1).

24:7. Defenses. There are a multitude of defenses that may be raised by one who is otherwise obligated to pay on a negotiable instrument. As between the immediate parties to the instrument—for example, the maker and payee of a note—or against one who is not a holder in due course, any defense that would be available in an action of simple contract may be raised. Section 3–306(b). Unless one is a holder in due course, any person who takes the instrument is subject to all valid claims to it on the part of any person. Section 3–306(a). There is a technical distinction between a *defense* to an instrument and a *claim* to it. Thus one may raise a defense in a negative sense to protect himself against liability, as in the case of the defense of duress. A claim, on the other hand, might be asserted by an individual positively to contend, for instance, that he has a lien against the instrument.

The advantage of qualifying as a holder in due course has been frequently mentioned. The primary advantage is that a holder in due course is not subject to many of the defenses that may be raised against one who is simply a holder. However, the protection afforded a holder in due course is not absolute, and there are some defenses that may be raised against a person who has this status. These defenses are known as *real defenses* and are enumerated by the code. All other defenses that may normally be raised, except against a holder in due course, are known as *personal defenses.* The code makes no effort to enumerate these specifically.

24:8. Defenses Against Holder in Due Course. Section 3–305 of the code lists the rights of a holder in due course and what are commonly referred to as *real defenses.* These defenses are (1) infancy, if it is a defense to a simple contract; (2) lack of capacity that renders the instrument void; (3) duress that renders the instrument void; (4) illegality that renders the instrument void; (5) fraud in causing one to sign an instrument without knowledge of or a reasonable opportunity to obtain knowledge of its essential terms; (6) discharge in bankruptcy; (7) discharge of which the holder has notice when he takes the instrument; (8) material alteration; and (9) forgery. These defenses are, of course, also good against one who is not a holder in due course.

(1) Infancy or Other Incapacity. If under the local state law infancy would be a defense to a simple contract action, then it may be raised against a holder in due course. The code does not attempt to state when the defense may otherwise be raised, and it may vary from one state to another. The defense may be raised against a holder in due course regardless of whether it would render the contract void or voidable under local state law. Thus the policy of protecting the infant against those who might take advantage of him is extended even to the holder in due course.

Other types of incapacity, such as mental incompetence or lack of corporate capacity to do business, may be raised as a defense against a holder in due course if under state law the obligation entered into by one suffering such a lack of capacity is null and void. If the obligation is merely *voidable*, the defense may not be raised against a holder in due course. For example, in many states a contract entered into by one who lacks mental capacity because of insanity or for some other reason is voidable. This defense could not be raised against a holder in due course. On the other hand, if there has been a prior adjudication by a court of lack of capacity, as, for example, when a guardian is appointed, then in many states the obligation is void. That defense could be raised against a holder in due course. Thus other types of incapacity do not receive as strong protection from the code as does infancy.

(2) **Duress.** If the duress that forces one to sign a contract merely renders the obligation *voidable* at the option of the person who is the victim of the duress, then that defense may not be raised against a holder in due course. If the duress renders the obligation *void*, then that defense may be raised against a holder in due course. Whether the duress involved renders the obligation void or voidable is largely a matter of degree in most jurisdictions. For instance, if a person signs an instrument because of direct physical force, such as at the point of a gun, the instrument is void. The defense may be raised against a holder in due course. If one signs an instrument because of the threat of an unjustified lawsuit, the instrument is voidable. This defense may not be raised against a holder in due course.

(3) **Illegality.** Most frequently instruments are illegal because they violate a state's usury or gambling statutes, although there are many other reasons that the obligation may violate the law. The defense of illegality may not be raised against a holder in due course unless the state statute that renders the obligation illegal also makes the obligation *void*. If such an obligation is merely voidable, it may not be raised as a defense against a holder in due course.

GLASSMAN v. FEDERAL DEPOSIT INSURANCE CORPORATION
210 Va. 650, 173 S.E.2d 843 (1970)

Herbert S. Glassman executed a number of notes, three of which totaled some $30,000. One note was payable in the amount of $10,000 to "George Vantraub." *The other two notes were each in the amount of $10,000 and were payable to order of bearer. These notes were negotiated and eventually were taken by Crown Savings Bank as collateral security for loans. Glassman eventually refused to pay the notes. Crown Savings went into receivership, and the Federal Deposit Insurance Corporation as receiver became successor to the interests of Crown Savings.*

The FDIC sued Glassman, contending that Crown held the notes as a holder in due course. Glassman raised the defense of illegality, contending that the notes were given for a gambling debt. From an adverse decision in the trial court Glassman appealed.

SNEAD, C.J.

The trial court held, first, that if the notes were given for gambling losses, they would be invalid under Code, § 11–14, even in the hands of a holder in due course. Code, § 11–14 declares, in part, that "All . . . contracts and securities whereof the whole or any part of the consideration be money or other valuable thing, won . . . at any game . . . shall be utterly void."

In *Lynchburg Nat'l Bank* v. *Scott*, 91 Va.

652, 22 S.E. 487 (1895), we held that a usurious note, which the applicable Virginia statute declared "illegal," could be enforced against the maker by a holder in due course. We recognized, however, that the holder could not have enforced the note if the statute had declared it "void," rather than illegal.

We therefore agree with the trial court's holding that even though Crown Savings was a holder in due course. . . . the note should be held invalid under Code, § 11–14, if Glassman proved that the notes were given for gambling losses.

Glassman testified that the gambling losses resulted from a gin rummy game that took place during August and September of 1963 at the Golden Triangle Motor Hotel in Norfolk, Virginia. He stated that his losses totaled about $100,000. The participants in the game, besides Glassman, were George Vantraub, and the two Halprin brothers, Jack and Burt.

After losing a substantial sum, and after the notes were given, Glassman discovered that the cards with which the game had been played were marked.

Richard B. Keeley testified for Glassman. Keeley at the time of the game was manager of the Golden Triangle Motor Hotel. He stated that of his own knowledge he knew of the gin game and had been present at some of the sessions. He corroborated Glassman's testimony as to the participants in the game, the general period during which it was played, and that the cards were marked. He stated that he saw scores kept and knew that Glassman was losing. He also testified that he did not actually see the notes or any money passed.

Glassman stated that after he had lost all his cash, "I had no money coming in, and I gave these notes, and it was a gambling debt." On advice of counsel Glassman notified the banks with which he did business not to honor the notes. Crown Savings was not among those banks.

Although the trial court sat as a jury and heard the witnesses (with the exception of Ridley) we cannot agree with the determination that Glassman failed to prove by a preponderance of the evidence the facts on which his defense rests. To establish his case by a preponderance it is not necessary that he eliminate every doubt or question in the mind of the trier of fact. In our view his testimony established the existence of the gin game, its participants, its location, the general period during which it was played and the fact that it was rigged against him. All this was corroborated by Keeley.

(Reversed.)

QUESTION FOR DISCUSSION

1. *Does the statute declare contracts concerning gambling debts void or voidable? How did this provision affect the outcome of this case, if at all?*

(4) Fraud. If one signs an instrument as a result of fraud that prevents him from having knowledge, or the reasonable opportunity to obtain the knowledge, of the essential terms or character of the instrument, this constitutes a real defense and may be raised against a holder in due course. Section 3–305(2)(c). This type of fraud is known variously as *fraud in the essence, fraud in factum,* or *fraud in the execution.* Such fraud renders the instrument void because the person who signs as instrument under such circumstances really is not aware of what he or she is signing and therefore has no intent to sign such an instrument. For this reason it is a real defense. For example, one might be unable to read yet be told to sign a "receipt" for goods that in reality is a promissory note. At the extreme, one might be asked for his "autograph," and then the holder might write a promissory note around it.

This defense is available only if the ignorance of the contents of an instrument is reasonable. There must have been no reasonable opportunity to determine what was contained in the instrument. Such factors as the age, the intelligence, the education, and the experience of the party are considered, as well as his ability to understand English and the nature of the representations. Other relevant factors may also be considered.

Other types of fraud do not constitute a real defense and therefore may not be raised against a holder in due course. Sometimes this fraud is called *fraud in the inducement* or *fraud in the procurement.* This type of fraud results in the victim's signing of the negotiable instrument but does not result in the signer's ignorance of the nature of the instrument. Suppose one falsely alleges that aluminum siding will be placed on the house when in fact it is not made of aluminum but some sort of composition board. The homeowner signs a promissory note to pay for it. He has been induced to sign the note by fraud but was aware of the nature of the instrument. This fraud is a personal defense and may not be raised against a holder in due course.

RICKS v. BANK OF DIXIE

352 So.2d 798 (Miss. 1977)

The Bank of Dixie sued J. V. Ricks and others for the collection of a promissory note. The note was signed by J. V. Ricks and payable to the order of Dixie Machine Works. Dixie Machine Works then indorsed to the Bank of Dixie.

Rick contended that his signature was obtained by fraud because Dixie Machine Works represented that the document was an order, sold on open account, for additional materials and equipment.

SUGG, J.

A decision of the issue requires determination of this threshold question, did J. V. Ricks, Jr. sign the note on behalf of the defendants because of misrepresentation which induced him to sign the note with neither knowledge nor reasonable opportunity to obtain knowledge of its character or its essential terms?

Plaintiff filed an answer specifically denying every allegation of the defendants in their affirmative defenses. Defendants' first affirmative defense would relieve them for liability on the note, if supported by a preponderance of the evidence, under section 75–3–305 Mississippi Code Annotated (1972) which follows:

> To the extent that a holder is holder in due course he takes the instrument free from . . .
>
> (2) all defenses of any party to the instrument with whom the holder has not dealt except . . .
>
> (c) such misrepresentation as has induced the party to sign the instrument with neither knowledge nor reasonable opportunity to obtain knowledge of its character or its essential terms; . . .

The first affirmative defense was based on subsection (2) part (c) which permits defenses to be asserted against a holder in due course which were not available under the Uniform Negotiable Instruments Act. It allows the maker of an instrument to assert as a defense, "(S)uch misrepresentation as has induced the party to sign the instrument with neither knowledge nor reasonable opportunity to obtain knowledge of its character or its essential terms."

The defense authorized by section 75–3–305(2)(c) is a limited defense and may be asserted against a holder in due course only if a party was induced to sign an instrument because of misrepresentation coupled with the fact that the party signing the instrument had neither, (1) knowledge of its character or its essential terms, nor (2) reasonable opportunity to obtain knowledge of its character or essential terms. The comment pertaining to this defense found in Anderson's Uniform Commercial Code, section 3–305:(1) p. 595 (1961) states:

7. Paragraph (c) of subsection (2) is new. It follows the great majority of the decisions under the original Act in recognizing the defense of 'real' or 'essential' fraud, sometimes called fraud in the essence or fraud in the factum, as effective against a holder in due course. The common illustration is that of the maker who is tricked into signing a note in the belief that it is merely a receipt or some other document. The theory of the defense is that his signature on the instrument is ineffective because he did not intend to sign such an instrument at all. Under this provision the defense extends to an instrument signed with knowledge that it is a negotiable instrument, but without knowledge of its essential terms.

The test of the defense here stated is that of excusable ignorance of the contents of the writing signed. The party must not only have been in ignorance, but must also have had no reasonable opportunity to obtain knowledge. In determining what is a reasonable opportunity all relevant factors are to be taken into account, including the age and sex of the party, his intelligence, education and business experience; his ability to read or to understand English; the representations made to him and his reason to rely on them or to have confidence in the person making them; the presence or absence of any third person who might read or explain the instrument to him, or any other possibility of obtaining independent information; and the apparent necessity or lack of it, for acting without delay.

Unless the misrepresentation meets this test, the defense is cut off by a holder in due course.

In our opinion the above comment correctly states the factors to be considered by a trial court when it is called on to determine if a defendant is to be released from liability for signing an instrument. In order to determine if the defense in this case meets the above test, we must consider the evidence which was before the trail court.

The evidence shows without conflict that J. V. Ricks, Jr. signed a promissory note on October 7, 1974 payable to the order of Dixie Machine Works and plaintiff on behalf of all the defendants. The original note was introduced in evidence and is a negotiable instrument under the requirements of section 75–3–104 Mississippi Code Annotated (1972). Plaintiff purchased the note from Dixie Machine Works for a valuable consideration on October 21, 1974 and the note was negotiated on that date in accord with section 75–3–202 Mississippi Code Annotated (1972). Plaintiff met the requirements of section 75–3–302 and became a holder in due course.

The only witness offered by defendants was J. V. Ricks, Jr., who testified that, on the date the note was executed, he signed numerous purchase orders and he thought the note was a verification of terms. He signed all documents presented to him on that day without reading any of them. The witness has a college education and is a businessman with many years of experience. J. V. Ricks, Jr. did not use ordinary care when he signed the note without reading it and putting it into circulation: he was not prevented from reading the note; therefore, defendants' claim of misrepresentation has no legal substance. His testimony fails to establish that he was induced to sign the note. "[W]ith neither knowledge nor reasonable opportunity to obtain knowledge of its character or its essential terms."

In this case the trial court had before it testimony which was not in conflict on the issues to be resolved. It properly considered the evidence, not in conflict, of both plaintiff and defendants in arriving at its decision to grant the peremptory instruction, and correctly decided the issues of law from nonconflicting facts.

In sum, the defendants executed a promissory note through their partner, the partner was an educated businessman who signed the note without reading it, the note was negotiated to plaintiff for a valuable considersation, plaintiff became a holder in due course, and defendants did not establish their affirmative defenses. The trial court correctly granted the peremptory instruction for the plaintiff.

(Affirmed.)

Questions for Discussion

1. *Does the decision in this case turn on the question of whether Ricks actually knew the document was a promissory note?*
2. *What test does the court apply in determining whether there was reasonable opportunity to obtain knowledge of the contents of the document?*

---

(5) Discharge in Insolvency Proceedings. If one is discharged in bankruptcy or other insolvency proceedings, this defense may be raised against a holder in due course. Section 3–305(2)(d). The purpose of such proceedings is to allow a debtor who gets into a financial position from which there is virtually no hope of recovery to wipe the slate clean and get a fresh start. This public policy is considered superior to the right granted to a holder in due course, which is to facilitate the flow of commercial paper, and hence may be raised against a holder in due course.

(6) Discharge of Which Holder in Due Course Has Notice. A number of persons may be liable on an instrument. Occasionally one of them may be discharged, as, for example, when one party's indorsement is canceled. We have already seen that knowledge of such a discharge does not prevent a holder from becoming a holder in due course as to the other parties. But as to the person discharged, the holder does not take in due course if he has knowledge of the discharge, and this may be raised as a defense against him. Conversely, if a party is discharged but the holder is without knowledge of such an event, then such a discharge may not be raised against him if he is a holder in due course.

(7) Material Alteration. If an instrument is altered in such a manner that the obligation of a party to the instrument is changed, then the alteration is material. Section 3–407(1). Thus, if a co-maker is added to a note, the alteration would not be material because this would not alter the contract of any of the parties to the instrument. However, if an alternative payee is added, this would change the contract of the parties (the maker would be liable to the alternative payee), and thus the alteration would be material. Perhaps the most common alteration is of the amount due on the instrument. A change in the date that the instrument is due would also be a material alteration.

The defense of material alteration is available against a holder in due course as to the alteration only. The code states that "A subsequent holder in due course may in all cases enforce the instrument according to its original tenor. . . ." Section 3–407(3). For example, suppose an instrument is originally payable for $10. The amount is then changed to $100. The holder in due course may only enforce the instrument in the amount of $10 rather than in the changed amount of $100.

There is one exception. If an incomplete instrument is completed otherwise than as authorized, this is a material alteration. However, when an incomplete instrument is completed, a holder in due course may enforce it as completed. To put it another way, the unauthorized completion may not be raised against a holder in due course.

There is one other limitation to the defense of material alteration. If the person who relies on the defense has substantially contributed to the material alteration of the instrument or to the making of an unauthorized signature, then he is precluded from raising the defense against the holder in due course. Section 3–406. Thus if one prepares an instrument and large spaces are left in the body of the instrument that enable additional words or figures to be inserted, this could be found to constitute negligence in the preparation of the instrument and would preclude the defense of material alteration from being raised. Whether an instrument has been prepared in a negligent manner depends on the facts and circumstances of each case.

UNADILLA NATIONAL BANK v. McQUEER
277 N.Y.S.2d 221 (1967)

AULISI, J.

On March 17, 1966 McQueer purchased a truck and milk cooling tank from the defendant Johnson's Garage, Inc., for the sum of $10,698. He signed a note for this amount on one of plaintiff's printed forms supplied by the garage. Rex Hinman, general manager of Johnson's Garage, wrote in the terms of payment in his own hand before McQueer signed. Johnson's Garage, Inc., was the payee and there was a reference to the sale between Johnson's Garage and McQueer typed on the note. Affidavits supplied by McQueer, Hinman and one Lyle Bright, who was present at the transaction, state that the note was drawn to become due in five days from date which would have been on March 22, 1966. The note, as yet unpaid, was on March 28, 1966 negotiated by the payee to plaintiff bank which paid the face amount over to the payee and the check was deposited in the Johnson's Garage account in an Oneonta bank. Plaintiff's employees assert in their affidavits that the note was received, accepted and posted in the regular course of business and that when it was received it was payable 45 days from date; that the bank had a policy of not accepting overdue notes; and that no one at the bank, at least not anyone who had handled the note, inserted the figure 4 before the figure 5.

McQueer on April 1, 1966, allegedly unaware that Johnson's Garage had negotiated the note to plaintiff, paid to the garage the $10,698 for the equipment which he had purchased by personal check. This was also taken and deposited by Johnson's Garage in its account at the Oneonta bank. In his affidavit McQueer states that he demanded the return of the note at the time of his payment but that Mr. Johnson said it was locked up in the garage vault and would be returned to him by mail. Despite McQueer's calls the note was never returned—it having been negotiated to the plaintiff three days previously. On May 1, 1966, the due date of the note based on 45 days, the bank demanded payment by the maker, McQueer, and upon his refusal this action was commenced.

Upon the facts presented by defendant, we must assume that there was an alteration on the note from 5 to 45 days somewhere along the line. If the alteration was effected before the transfer to plaintiff, the bank might still be deemed a holder in due course. A holder in due course is one who takes an instrument "without notice that it is overdue or has been dishonored or of any defense against or claim to it on the part of the person." (Uniform Commercial Code, § 3–302, subd. (1)(c).) Where, as here, there is an altered instrument, the statute provides as follows: "A subsequent holder in due course may in all cases enforce the instrument according to its original tenor, and when an incomplete instrument has been completed, he may enforce it as completed." (Uniform Commercial Code § 3–407, subd. (3); see also § 3–304, subd. (1)(a).) So in this case, where the note was overdue as originally drafted, the bank may claim status as a holder in due course and enforce the note if the note came to it in such condition that alteration was not noticeable.

Special Term (the lower court) assumed that the bank took the note after the alteration was made. We can not accept this assumption as a fact beyond dispute. All the persons who have executed affidavits here deny any knowledge of the alteration, but the crucial question still remains—who made the alteration? When there are defenses such as payment or that the paper is overdue, the burden is upon the party claiming the rights of a holder in due course to prove his status as such (Uniform Commercial Code § 3–307, subd. (3)). We do not think the burden has been met where there remains, among others, the question as to the identity of the one who made the alteration. If it was not accomplished prior to the bank's acceptance then, as to the bank's reduced status as a mere holder, questions remain as to knowledge and authority surrounding the alteration. We, therefore, are not now ready to accept the bank's contention that even as a mere holder it would have a right to payment from McQueer because the latter made payment negligently to a party who was not a holder of the note. At any rate, whatever factual questions may arise it is our view that the bank must prove to the satisfaction of a trier of the facts that it took the note in its altered form in good faith before judgment may be entered in its favor as a holder in due course. We also leave open for trial the question whether the alteration was noticeable or not.

(Judgment reversed.)

QUESTIONS FOR DISCUSSION

1. *Would the bank have qualified as a holder in due course if the alteration occurred prior to its acceptance of the note? Would it have made any difference if the alteration was noticeable?*
2. *How would this case be affected if it were established that the alteration took place after the bank had the note? Can you think of any motive for an employee of the bank to alter the date?*

---

(8) **Forgery.** In order to be bound on a negotiable instrument, one must have signed it. One does not sign an instrument if his signature is forged or unauthorized. This defense may be raised against a holder in due course unless, of course, one's own negligence contributed to the forgery. Furthermore, the defense of forgery or unauthorized signature can be raised by persons other than the immediate parties to the instrument, such as an indorser.

24:10. DEFENCES THAT MAY NOT BE RAISED AGAINST A HOLDER IN DUE COURSE. No attempt will be made to analyze all the defenses that cannot be raised against a holder in due course. They consist of all the defenses to a breach of contract other than those that have been discussed. A few general comments concerning these defenses might be helpful at this point, however. Note that the typical defenses of lack of capacity, illegality, duress, and undue influence are not available against a holder in due course unless local state law renders such obligations *void.* If such obligations are merely *voidable,* the defenses may not be raised against a holder in due course. Fraud that *induces* one to sign an instrument may not be raised against a holder in due course. Fraud *in the execution* may be raised against a holder in due course because the victim does not intend to sign a negotiable instrument.

The most common defenses to a contract action are failure or lack of consideration, breach of contract, setoff or counterclaim, discharge, nondelivery of an instrument, and lack of authority of a corporate officer or partner or of an agent to sign. All these defenses are also personal defenses and may not be successfully raised against a holder in due course.

24:11. PROTECTION OF CONSUMERS AGAINST A HOLDER IN DUE COURSE. The protection afforded one who achieves the status of a holder in due course has been subject to some abuse. In the typical case a seller may make a sale of goods to a consumer, who in turn finances it by giving back a promissory note payable to the order of the seller. The seller quickly negotiates the note to a financial institution or some other individual who qualifies for the status of a holder in due course. The consumer then discovers that the goods are defective or not as represented. When the holder in due course presents the note for payment, however, these matters cannot be raised as a defense by the consumer. He must seek to recover in a separate action against the seller, who may have left town or who may be financially irresponsible. Because of these types of problems and because of the abuse of the protection of holder-in-due-course status, the concept of a holder in due course as applied to consumers has been under attack. The attacks have come from various quarters: the judiciary, the legislature, and the Federal Trade Commission.

(1) Court Decisions. On occasion the courts deny holder-in-due-course status to financial institutions seeking to collect on a negotiable instrument against a consumer. Most often these cases result when the court finds that the holder has not taken the instrument in good faith or that the seller and the financial institution are so intertwined as to be one party. For instance, the consumer may establish lack of good faith if he or she can show that there is collusion between the seller and the holder; that the holder has reason to be suspect because of its familiarity with the seller's methods of operation; that the holder's separation from the seller is only part of a scheme to cut off the purchaser's defenses; and other similar facts. The decisions in this area have varied, and, not surprisingly, some courts are much more likely to find a lack of good faith than others.

Sometimes a court will refuse due-course status because of the close relationship of the seller and the holder. These decisions are based on the fact that even a holder in due course may be subject to defenses raised by the person with whom he deals. Section 3–305(2). The court may render such a decision if it is established that the institution that dealt with the consumer by investigating his credit, furnishing the forms, is an owner of the seller or is in the same corporate family and so on. Once again the decisions in this area vary widely and do not provide any reasonable degree of uniformity. The result is that the legislatures in some states have acted.

(2) State Statutes. Some states—such as Massachusetts, New York, New Jersey, and Wisconsin—have enacted measures to abolish or severely restrict holder-in-due-course status in consumer transactions. In most of such legislation a consumer is defined as one who purchases goods and services primarily for personal, family, or household use. The states that have enacted legislation in this area usually follow one of several approaches. Some statutes may deny holder-in-due-course status to one who purchases consumer paper. In other states paper used in consumer sales may be declared nonnegotiable, or negotiable notes may be prohibited in consumer transactions. In still other jurisdictions the state may provide that the holder is subject to any defense of which he receives notice within a stated period of time.

(3) Federal Trade Commission Regulations. One of the more significant developments concerning the application of holder-in-due-course status to consumer transactions came in the form of a Federal Trade Commission regulation. This regulation makes it an unfair or deceptive act or practice within the meaning of Section 5 of the Federal Trade Commission Act to take or receive a consumer-credit contract unless it has notice printed in boldface type that "Any holder of this consumer credit contract is subject to all claims and defenses which the debtor could assert against the seller of goods or services obtained pursuant hereto or with the proceeds hereof. Recovery here-under by the debtor shall not exceed amounts paid by the debtor here-under." This notice then must be made part of any consumer-credit contract between the merchant and the consumer. The effect of the notice is to prevent anyone who takes such an instrument from achieving holder-in-due-course status. It also makes inoperative the former practice of having a consumer waive certain defenses as part of an installment sales agreement.

The regulation also attacks another problem in consumer financing. Frequently the seller will not enter into the financing agreement with the consumer but will direct him to a lender, such a financial institution or even a corporate subsidiary, that will lend the money directly to the consumer. The lender may not, therefore, be considered part of the sales transaction, making it difficult to raise defenses connected with the sale against the lender. The lender may also negotiate the instrument to another party, who then may be a holder in due course. Because of this practice the regulation also makes it an unfair trade practice to

> Accept, as full or partial payment for such sale or lease, the *proceeds* of any purchase money loan . . . unless any consumer credit contract made in connection with such purchase money loan contains the following provision. . . . "Any holder of this consumer credit contract is subject to all claims and defenses which the debtor could assert against the seller of goods or service obtained with the proceeds hereof. Recovery hereunder by the doctor shall not exceed the amounts paid hereunder." (Emphasis added.)

This regulation became effective May 1, 1976. It restricts the claim of holder-in-due-course status when a consumer transaction is the basis for the issuance of the negotiable instrument. It is directed at the sellers of the consumer goods as well as at any subsequent transferee of the instrument or any financial institution that deals directly with consumer but makes a loan for the purpose of purchasing consumer goods. Remember, however, that the regulation has nothing to do with any transaction other than a sale of goods to a consumer. It would not affect the sale of goods or supplies between a manufacturer, retailer, or wholesaler, for instance. Nor does it affect any transaction, consumer or otherwise, that does not come within the jurisdiction of the Federal Trade Commission. As a practical matter, however, the large majority of transactions would come within the commission's jurisdiction, as its jurisdiction extends to matters that *affect* interstate commerce.

SUMMARY

In order to qualify as a holder in due course one must take a negotiable instrument for value; in good faith; without notice that it is overdue or has been dishonored; and without notice of any defense against or claim to it on the part of any person.

Value is not synonymous with consideration. An exchange of promises is not value. Value may be given when an instrument is given as payment or security for an antecedent claim. Value for an instrument is given by a bank that permits a depositor to draw against the deposited instrument prior to the time the instrument is collected.

Good faith means honesty in fact in the conduct or transaction concerned. An objective test is not applied. A potential holder cannot qualify as a holder in due course if he has notice that an instrument is overdue, that the instrument has been dishonored, or that there is any defense against it or claim to it.

One has notice that a demand instrument is overdue if the instrument is taken more than a reasonable time after its issue. For checks issued in the United States or its territories, a reasonable time is presumed to be thirty days.

Notice of a claim or defense is given if the instrument has such visible evidence of alteration or forgery or otherwise is so irregular as to call into question its validity. Although a person in possession of an instrument is presumed to have the authority to fill blanks, under certain circumstances taking an instrument that is incomplete in some material respect can amount to notice of a claim or defense. Also, taking an instrument with notice that the obligation of any party is voidable prevents one from qualifying as a holder in due course.

One who holds an instrument as a holder in due course may transfer the protection to his immediate transferee. However one who participates in fraud or illegality concerning the instrument or, as a prior holder, has notice of a claim or defense to the instrument cannot take advantage of this provision.

A payee may become a holder in due course. But one who acquires an instrument in a number of noncommercial circumstances, such as at a judicial sale, may not become a holder in due course.

A holder in due course is subject to real defenses but not to personal defenses. Real defenses include infancy, if it is a defense to a simple contract; lack of capacity; duress; illegality that renders an instrument void; fraud in the essence; discharge in bankruptcy; forgery; material alteration; and discharge of which the holder has notice when he takes the instrument.

There are some state and federal statutes or regulations giving some protection to consumers against the holder-in-due-course rule. One is a regulation passed by the Federal Trade Commission making it an unfair or deceptive act or practice to take or receive a consumer credit contract unless a notice is given warning that the contract is subject to defenses by the consumer.

PROBLEMS

1. Sam Kay signed a promissory note in the amount of $220,000.00 payable to Investments S. A., Inc. Arthur Cunningham and Phillip T. Weinstein

were attorneys who represented Investments S. A., Inc., in various legal matters for which fees were due. Investments S. A. negotiated the note to the attorneys who presented it to Kay for payment. Kay refused raising several defenses. Cunningham and Weinstein contended they were holders in due course. They contended the note was negotiated to them in return for legal services but submitted no evidence as to the extent or reasonable value of those services. Are Cunningham and Weinstein entitled to the status of holders in due course? Fernandez v. Cunningham, 268 So.2d 166 (Fla. App. 1972).

2. Florida City Express, Inc., signed a promissory note for $3292.12 payable to Latin American Tire Company in return for certain merchandise. It also signed another note payable to Latin American in the amount of $3500.00. Latin American negotiated these notes to the Bank of Miami which qualified as a holder in due course. When the notes came due the Bank of Miami attempted to collect from Latin American and was unsuccessful. The Bank of Miami then demanded payment from Florida City. Florida City informed the Bank of Miami that it had already paid back the money owed on the notes to Latin American. No notation of payment appeared on the notes. Florida City contended that it was never notified that the notes were negotiated. Can Bank of Miami recover the amount due on the notes from Latin American? The Bank of Miami v. Florida City Express, Inc. 367 So.2d 685 (Fla. App. 1979).
3. Green Lane, Inc., signed three notes payable to the order of Comfort Cooling Co. These notes were then discounted to Factors and Note Buyers, Inc. Before the notes were negotiated to Factors, George Gottesman told Factors that Comfort Cooling Co. had not yet performed the contract for which the notes were given and that Comfort Cooling had orally agreed not to negotiate the notes. Gottesman did not state that Comfort Cooling was in breach of contract. Comfort Cooling subsequently went bankrupt without performing its obligations under the contract. Could Factors collect from Green Lane as a holder in due course? Factors and Note Buyers, Inc. v. Green Lane, Inc., 245 A.2d 223 (N.J.App. 1968).
4. On August 7, 1967, Robert F. Gonderman signed a promissory note payable to Hamlet V. Boisson in the sum of $2,500.00 "one year after date." On October 1, 1968, Boisson endorsed the note over to the State Exchange Bank. Gonderman contended that he was induced to sign the note because of the fraudulent conduct of Boisson. Could Gonderman raise this defense against the bank? Gonderman v. State Exchange Bank, Roann, 334 N.E.2d 724 (Ind. 1975).
5. Robert and Katherine Matthews were induced to agree to put aluminum siding on their home by a representative of All-Style Builders. They were told that the total price was $3,250.00, that their modest dwelling was to be a "model home," and that they would receive $100 for every potential customer All-Style brought to view their home. The contract stated that the price was $3,250.00, but when the smoke had cleared, the Matthewses

found they had signed a note and mortgage in the grand total of $5,127.36. The note and contract were assigned to Aluminum Acceptance Corporation. The Matthewses contended that they were unable to read the papers that they signed and that they were blank at that time. Aluminum Siding filed suit to collect on the note and foreclose on the mortgage, contending that it was a holder in due course. Was it correct? Matthews v. Aluminum Acceptance Corp., 137 N.W.2d 280 (Mich.App. 1965).

6. George Hernreich, an Arkansas jeweler, signed three promissory notes payable to W. F. Sebel Co., Inc., a foreign corporation that was not qualified to do business in Arkansas. An Arkansas statute makes any contract entered into by a foreign corporation void. Sebel Co. then negotiated the notes to Pacific National Bank, who sought to collect the notes from Hernreich as a holder in due course. Could Pacific recover? Pacific National Bank v. Hernreich, 398 S.W.2d 221 (Ark. 1966).
7. A representative of Leo's Used Car Exchange, Inc., purchased three cars at a car auction from Frederick Villa, for which two checks were given naming Villa's Auto Sales, Inc., as payee. Villa was a customer of Industrial National Bank of Rhode Island and presented the checks to that bank, where he maintained an account, and was immediately given cost for them. Meanwhile Leo's had stopped payment on the checks. Industrial Bank therefore looked to Leo's Used Car to pay the checks. Leo's had a defense, to which Industrial Bank responded that it was a holder-in-due-course. At trial there was evidence that Industrial had a practice of requiring approval of the branch manager before any corporate checks drawn on another bank were cashed. This was not done in this case. Leo's contended that Industrial could not be a holder in due course because it did not take the instrument in good faith: it had not exercised ordinary care in the transaction because it had not adhered to its own internal policies. The trial court decided that Industrial was not a holder in due course. What result on appeal? Industrial Nat. Bank of R.I. v. Leo's Used Car Ex., Inc., 291 N.E.2d 603 (Mass. 1973).
8. Woodward accepted a trade acceptance from Moody Manufacturing Company when it purchased certain goods. At the time of acceptance the amount remained blank because a final determination had not been made of the goods to be purchased from Moody. After the amount was filled in, the trade acceptance was negotiated to Illinois Valley Acceptance Corporation. Woodward contended that it could raise several defenses against Illinois because it signed the trade acceptance while it was blank. Was this contention correct? Illinois Valley Acceptance Corp. v. Woodward, 304 N.E.2d 859 (Ind. App. 1973).
9. Supreme Radio, Inc., gave two notes in the form of trade acceptances to Southern New England Distributing Corporation. Southern was a client of Korzenik, an attorney. The notes were transferred to Korzenik by Southern "as a retainer for services to be performed." Korzenik then attempted to collect on the notes from Supreme, which raised the defense of fraud

in the inducement. Korzenik contended that he was not subject to these defenses as a holder in due course. There was no evidence of the value of legal services rendered by Korzenik. Decide. Korzenik v. Supreme Radio, Inc., 197 N.E.2d 702 (Mass. 1964).

10. Parsons signed a note payable to Burton & Company, which in turn negotiated it to Jason Manchester and L. M. Elliott, who qualified as holders in due course. Parsons contended that he paid Burton & Co. prior to the negotiation of the note to Manchester and Elliott and that he therefore was not liable to them. Was this contention correct? Manchester v. Parsons. 84 S.E. 885 (W. Va. 1915).

CHAPTER 25

Presentment, Dishonor, Notice of Dishonor, and Protest

In the preceding chapters the requirements for a negotiable instrument as well as the steps that are necessary for its proper negotiation have been analyzed. Also qualifications for the status of a holder in due course have been examined along with its immunity from many common defenses that may be raised by those who are obligated to pay on an instrument. In this chapter and Chapter 26 the other side of the coin will be discussed, that is, how and under what conditions the various parties to an instrument may become liable on it. Before we examine these matters in detail it may be helpful to look at the general picture in order to get a "feel" for what lies ahead and how some of the concepts presented in these two chapters relate to each other. The following section is only a broad statement containing many exceptions that will be brought to light in the material that follows it.

25:1. Overview. We have previously seen that there may be a number of parties to a negotiable instrument. Depending on the type of instrument, these parties may be called variously a *maker,* a *drawer,* a *drawee-acceptor,* an *indorser,* and so on. Certain of the parties, such as the maker of a promissory note or the drawee who has accepted a draft, expect to be called upon eventually to pay the instrument. In order to fix his liability, the holder must—generally, but not always—take certain steps. Usually the instrument must be presented for payment or acceptance. This is known as *presentment.* Most often the person to whom the instrument is presented, such as a maker of a promissory note, pays it and that is the end of the matter. The persons to whom the instrument must first be presented for payment are said to be *primarily* liable. The maker of a note and the drawee who has accepted a draft are parties who are primarily liable.

If the party who is primarily liable fails to pay the instrument when it is presented (or fails to accept it, if that is all that is required), he is said to *dishonor* it. There may be other parties who have also been involved in the instrument. An indorser or drawer of a draft are examples. They do not expect to be called upon to pay because there are others who are primarily obligated to pay. However, if those parties who are primarily liable dishonor the instrument when it is presented, the holder may look for payment to certain other parties to the instrument who were associated with it previously. These other parties may be liable because of their prior actions concerning the instrument, as, for example, when the instrument is transferred with an indorsement. These parties are generally not liable on the instrument until after it has been established that those primarily liable will not pay after a proper presentment. These parties are therefore said to be *secondarily* liable on the instrument. In certain cases parties may be liable not only because of a prior indorsement but also because certain warranties are given when the instrument is transferred or presented for payment or acceptance.

The object of this chapter, then, is to examine in some detail what constitutes proper presentment of an instrument, what amounts to a dishonor, and in what manner and under what circumstances the notice of such a dishonor must be given. In Chapter 26 the basis for liability of the various parties to the instrument will be analyzed as well as the manner in which a party to an instrument may be discharged.

25:2. Steps That Fix the Liability of Parties. In order to fix the liability of various parties to the instrument, certain steps must be followed. These steps are presentment, dishonor, notice of dishonor, and in some few cases protest.

(1) Presentment. In order for an instrument to be dishonored, it must be presented for payment, unless for some reason presentment is excused. Presentment to the primary party is generally a necessary first step for establishing the liability of the secondary parties. Very simply stated, presentment is a demand for acceptance or payment made upon the maker, acceptor, drawee, or other payor by or on behalf of the holder of the instrument. Section 3–504(1). Note that the presentment may be either a demand for *payment* or, in the case of a draft, a demand upon the drawee for *acceptance* of the draft.

(a) *How Presentment Is Made.* As a general rule any demand upon the party to pay the instrument, regardless of where or how it is made, is a presentment as long as it is made by or on behalf of the holder. For instance, the code provides that presentment may be made by mail, through a clearinghouse, or at the place of acceptance or payment specified in the instrument. If no such place is specified in the instrument, then presentment may be made at the place of business or residence of the party. Section 3–504(2). If the presentment is made by mail, the presentment is effective when the mail is received. When presentment is properly made at the party's place of business or residence and neither he nor anyone authorized to act for him is present or accessible at such place, then presentment is excused. Section 3–504(2)(c). The code does make a specific requirement in the case of banks. If the draft is to be accepted or if the note is payable at a bank in the United States, then presentment must be at that bank. Section 3–504(4).

(b) *To Whom Presentment Is Made.* The code provides that the presentment may be made to any one of two or more makers, acceptors, drawees, or other payors. In general it may also be made to any person who has authority to make or refuse the acceptance or payment. Section 3–504(3). If more than one person is obligated on an instrument, the holder has a right to expect any of them to pay. The code does not, therefore, require the holder to present the instrument to more than one of those primarily obligated to pay or accept. As a practical matter, once presentment has been made to one individual primarily liable on the instrument, he or she will normally notify the others who are also primarily liable.

(c) *Rights of Party to Whom Presentment Is Made.* On the other side of the coin, the code gives the party to whom presentment is made a number of specific rights that he may exercise without fear of dishonoring the instrument. For instance, when presentment of the instrument is made, a party may require exhibition of the instrument. The person making the presentment may also be required to identify himself, and if he represents another individual, he may also be required to show evidence of his authority to make the presentment. The person to whom the instrument is presented also has a right to have the instrument presented at a place specified in the instrument. If no place is specified, then the person to whom it is to be presented has a right to have the instrument presented at any place reasonable under the circumstances. If a partial payment is made upon presentment, the person to whom it is presented has a right to a receipt for the partial payment. If full payment is made upon presentment, then the person making the full payment has a right to a receipt and surrender of the instrument itself. Section 3–505(1).

If the person making the presentment fails to comply with any of these requirements, then the presentment is invalidated. However, the person presenting the instrument has a reasonable time in which to comply with the requirement. The time for acceptance or payment of the instrument runs from the time of compliance. Section 3–505(2).

BATCHELDER v. GRANITE TRUST CO.

339 Mass. 20, 157 N.E.2d 540 (1959)

C. H. Batchelder was president and sole stockholder of C. H. Batchelder Co. On April 1, 1946, he sold the assets of the company to a corporation organized by George E. Felton, called C. H. Batchelder Co., Inc., and referred to as the "Corporation." The sale was partially financed by the execution of three promissory notes by the Corporation and indorsed by Felton individually. These notes were then delivered to Charles Batchelder, who in turn indorsed them in blank. None of the notes indicated either the place of payment or the address of the maker.

On Arpil 29, 1946, Batchelder placed the notes with Granite Trust Co. for collection. The first two notes were paid, but the third was not. The third note was presented to the Corporation by Granite by a process described below. The note was not paid and the Corporation was adjudged a bankrupt. Batchelder attempted to sue Felton on his indorsement contract and was unsuccessful. Batchelder then sued Granite Trust, contending that the procedure used by the bank to present the note was improper and that this accounted for his inability to recover against Felton. From an adverse ruling in the trial court Batchelder appealed.

SPALDING, J.

The following was agreed to by the parties: "At all times material to this action, it was the practice and usage of . . . (the defendant), when holding a note for collection such as the notes . . . (here) to send a notice by mail to the maker about ten days before the due date, informing him that the bank held the note, that it would fall due on a date specified, and inquiring what disposition he wished to make of it. . . . The bank continued to hold the note at its office through banking hours on the date of maturity. If it was not paid by the close of banking hours, the bank turned it over to a notary public to make a demand at the bank and send out notices of protest to the maker, payee and all indorsers. It was not the custom of the bank to have someone take the note itself to the maker's place of business or residence." It was also agreed that this procedure was followed with respect to the three notes placed with the defendant by the plaintiff.

On March 31, 1949, the third note was turned over to one Cameron, a notary public, who was also employed by the defendant. Cameron stamped on the note "Protested for Non-payment, Mar. 31, 1949." Notices of protest properly addressed were sent by mail to the corporation, to the company, and to Felton. The third note was returned to the plaintiff on April 21, 1949.

Since this case arose prior to the adoption of the Uniform Commercial Code the rights of the parties must be governed by the negotiable instruments law, G.L. c. 107, §§ 94–97. (The provisions are substantially the same for purposes of this case as the applicable Code provisions, 3–503 to 3–505). Section 96 of the statute provides that "Where no place of payment is specified, and no address is given" (p)resentment for payment is made at the proper place" when "the instrument is presented at the usual place of business or residence of the person to make payment." And § 97 requires that the instrument must be exhibited at the time payment is demanded. The plaintiff argues that this case is governed by these provisions, and since the defendant did not strictly comply with them, it failed to make proper presentment.

Prior to the adoption of the negotiable instruments law in 1898, it was well settled in this Commonwealth in situations similar to this that a written demand mailed by a bank to the maker of a note to pay the note at the bank on the due date was sufficient to make the offices of the bank the place of payment. And it has been said that "such previous notice to the promisor, and neglect on his part to pay the note at the bank, are a conventional demand and refusal, amounting to a dishonor of the note." *Mechanics' Bank at Baltimore* v. *Merchants' Bank*, 6 Metc. 13, 24; a physical exhibition of the note was not required. We are not aware of any case decided since 1898 which holds that the language of G.L. c. 107, §§ 96, 97 either approves or disapproves the earlier practices sanctioned by our common law.

. . . Thus . . . this court held that although a strict construction of G.L. c. 107, § 97, would require the actual exhibition of the instrument where payment is demanded, the adoption of the statute did not change the previously established rule that, where payment is to be made at a bank, it is presumed that the note was at the bank on the date of maturity and that proof of its exhibition was not required.

It could be argued that a strict construction of § 96 would call for a different presentment procedure than that employed by the defendant. But we are not disposed to construe that section as abrogating a rule which has been so deeply embedded in our law. . . . If conformity to custom need by shown, it was not lacking. The bank followed its usual practice in dealing with this note. Its practice was similar to that of the other local banks. The payee, maker and indorsers of this note were acquainted with this practice, since it was followed in making demand for payment of the two earlier notes, and they acquiesced in its use with regard to them.

It is worthy of note that the Uniform Commercial Code, which became law in this Commonwealth on October 1, 1958, although not applicable here, would sanction the presentment procedure followed by the defendant. See G.L. c. 106 §§ 3–504, 4–210.[1]

. . . The basic and decisive question in the case at bar was whether the presentment procedure followed by the defendant was proper. In order to prevail the plaintiff had to show that it was not. Since we hold that the procedure followed was proper the plaintiff has no case. . . .

(Judgment for defendant.)

QUESTIONS FOR DISCUSSION

1. *What did Batchelder contend was wrong with the method of presentment used by Granite Trust Co.?*
2. *Under the code would the method of presentment used by the bank be proper?*
3. *Under the code must one exhibit the instrument in order to make a valid presentment if not requested by the party to whom presentment is made? See Sections 3–504 and 3–505.*

(d) *Time of Presentment.* The code specifies a number of rules for determining when an instrument may be presented for payment or acceptance. Remember that an instrument may state a particular date when payment is due or it may be a demand instrument with no fixed time for payment. As to the hour of payment the code provides that presentment must be made at a reasonable hour, and if presentment is made at a bank, it is to be made during the banking day.

The day when presentment is to be made depends upon the type of instrument, whether presentment is for payment or acceptance, whether it is payable at a stated date or is a demand instrument, and the situation of the secondary party whose liability the presentment is designed to fix. In the following discussion be sure you note the difference in the rules as they apply to presentment for *payment* of the instrument or just for *acceptance* of it. When we are discussing acceptance, we are, of course, referring to the acceptance of a draft prior to payment.

(i) PRESENTMENT FOR ACCEPTANCE. Frequently a draft will be presented to the drawee for acceptance. The drawee may not be expected to pay the instrument at that time. The reason that the instrument may be presented to the drawee for acceptance is that the drawee is not liable on the instrument until he accepts it. He is then primarily liable. Hence the motivation for presenting only for acceptance. If the instrument is a *time instrument*—that is, if it is to be paid on a stated date or within a fixed period after a stated date—then any presentment for *acceptance* of the instrument must be made on or before the date that it is payable. Thus, if a draft is due on December 31, 1978, the holder may present it to the drawee for acceptance on December 20 but not after December 31. Obviously a presentment for payment could not be made prior to December 31. Section 3–503(1)(a).

If the instrument is payable *after sight,* it must be presented for acceptance or negotiated within a reasonable time after date or issue, whichever is later. Section 3–503(1)(b). If the instrument is payable on demand and the liability

[1] General Laws c. 106, § 4–210(1), provides: "Unless otherwise instructed, a collecting bank may present an item not payable by, through or at a bank by sending to the party to accept or pay a written notice that the bank holds the item for acceptance or payment. The notice must be sent in time to be received on or before the day when presentment is due and the bank must meet any requirement of the party to accept or pay under section 3–505 by the close of the bank's next banking day after it knows of the requirement."

of any secondary party, such as an indorser or drawer, is under consideration, then a presentment for acceptance must be made within a reasonable time after such secondary party becomes liable on the instrument. Section 3–503(1)(e). Generally an indorser's liability stems from the time the instrument is indorsed and delivered and a drawer's liability stems from the time he draws the instrument.

(ii) PRESENTMENT FOR PAYMENT. The code also states a number of rules for presentment for *payment.* If an instrument *shows the date* on which it is payable, presentment for payment is due on that date. Section 3–503(1)(c). Presentment for payment on that date is necessary to fix the liability of those entitled to a presentment. Presentment for payment when an instrument is accelerated is due within a reasonable time after the acceleration. Section 3–503(d). If the person who is primarily liable refuses to pay when the instrument is presented, the instrument is dishonored. If the instrument is a *demand* instrument, presentment for payment is due within a reasonable time after any secondary party becomes liable on the instrument. Section 3–503(1)(e).

(iii) DEFINITION OF A REASONABLE TIME. Note that Section 3–503(1) states that the time for presentment of an instrument for payment or acceptance is measured by the test of a "reasonable time" after date or issue of an instrument payable after sight, after acceleration, and after a party becomes secondarily liable on the instrument. It is important, therefore, to have some concept of what constitutes a reasonable time, because the consequence of a failure to present the instrument within the proper time may be a discharge of some parties on the instrument who might otherwise be liable.

The code states broadly that "A reasonable time for presentment is determined by the nature of the instrument, any usage of banking trade and the facts of the particular case." Section 3–503(2). Thus the purpose for which an instrument is held and whether the instrument bears interest are among the factors that are considered.

The code gets more specific with regard to the common check, or as stated by the Code an "uncertified check drawn and payable within the United States and which is not a draft drawn by a bank." A presumption, which is rebuttable, is established by the code as to the reasonable time to present the instrument for payment or to initiate bank collection. With respect to the liability of the *drawer,* thirty days after date or issue of the check, whichever is later, is presumed to be a reasonable time to present the check for payment or to initiate bank collection. With regard to the liability of an *indorser,* seven days after his indorsement is presumed to be a reasonable time. The time for presentment is longer for a drawer than for an indorser because the drawer issues the check and normally expects to have it paid and charged to his account. It is therefore reasonable to require him to stand behind it for a longer period of time than an indorser, who normally may have just received the check and passed it on with no expectation of paying. He is entitled to more prompt notice of whether the check will be dishonored so that he may go after the person with whom he has dealt. Section 3–503, Comment 3.

There are certain situations in which delay in presentment may be excused and certain situations in which it is not excused. Because these situations also generally apply to the requirement of giving notice of dishonor and protest, they will be treated together in Section 25:3.

(2) **Dishonor.** Before secondary parties can be charged on the instrument, there must be a dishonor upon presentment by the party who is primarily liable.

The code states that "An instrument is dishonored when a necessary or optional presentment is duly made and due acceptance or payment is refused or cannot be obtained within the prescribed time. . . ." Section 3–507(1)(a). Presentment for payment is necessary to charge any indorser, drawer, acceptor of a draft payable at a bank, or maker of a note payable at a bank unless presentment is excused. Section 3–501(1)(b), (c). Presentment for an acceptance is optional for a draft payable at a stated date. In those situations in which the presentment is excused, the instrument is dishonored when it is not duly paid or accepted when due. Section 3–507(1)(b).

The dishonor may occur because the party on whom the demand for acceptance or payment has been made simply refuses to comply. It is also a dishonor, however, to require the party presenting the instrument to do more than he is legally required to do prior to making payment or accepting the instrument. Thus a drawee dishonors a draft if upon presentment he refuses to pay the draft because the drawer has insufficient funds on deposit with the drawee. A maker of a note dishonors it if he refuses to pay—when the note was duly presented—until he has had a chance to determine whether he is satisfied with the consideration for which he gave the note.

Once the instrument is dishonored, the holder of the instrument has an immediate right of recourse against those parties secondarily liable; the drawers and indorsers. This right is subject to any necessary notice of dishonor and protest. Section 3–507(2).

(a) *What Does Not Constitute Dishonor.* A number of actions by the party upon whom demand is made do not amount to dishonor of the instrument. Insisting on some legal right provided under the code before accepting or paying does not constitute dishonor. Thus upon presentment, if one demands reasonable identification of the person making presentment before paying the instrument, he would not dishonor the instrument. Section 3–505(b).

There are other actions that do not constitute dishonor. Return of an instrument for lack of a proper indorsement is not dishonor. Section 3–507(3). Similarly a presentment that is not necessary, as for acceptance of a demand draft, would not constitute dishonor. An example would be the refusal of a bank to certify a check. However, refusal to pay a demand draft obviously would constitute dishonor.

The code provides that upon presentment for *acceptance,* one may defer acceptance until the close of the next business day following presentment without dishonoring the instrument. Section 3–506(1). Also one may defer *payment* of an instrument without dishonor pending a reasonable examination in order to determine whether it is properly payable. However, payment must be made in any event before the close of business on the day of presentment. Section 3–506(2). The purpose of these rules is to give the party a fair opportunity to determine whether he is, in fact, liable on the instrument before dishonoring it. Note the different rule for presentment for acceptance versus presentment for payment of the instrument.

The foregoing discussion applies primarily to presentment of instruments that do not involve a bank. These rules are somewhat different when banks are involved because of their peculiar characteristics. The differences will be referred to in Chapter 27.

(3) **Notice of Dishonor.** Once there has been a presentment and a dishonor of the instrument, then the next step required to fix the liability of secondary parties is to give notice of dishonor.

(a) *Who May Give and Receive Notice.* As a general rule any party who may be compelled to pay an instrument may give notice to any party who may be liable on it. Typically the holder of the instrument or an indorser who has himself received notice of dishonor gives notice to the other parties. The code provides that notice may be given "by or on behalf of the holder or any party who has himself received notice or any other party who can be compelled to pay the instrument." Furthermore an agent or a bank in whose hands the instrument is dishonored may notify his principal or customer of the dishonor of the instrument. Notice may also be given to another agent or bank from which the instrument was received. Section 3–508(1). Suppose a promissory note was executed by Mary Maker, payable to Paula Payee. Payee indorses the note to A, who in turn indorses it over to Harry Holder. Holder presents the note to Maker for payment. Maker dishonors the note by refusing to pay it. Holder then may give notice to A, who is secondarily liable on the note, as will be more fully explained in Chapter 26. A may then give notice to Payee, who is also secondarily liable. Holder could also give notice to both A and Payee, because both could be liable to him and A might not pay Holder, who would then have to look to Payee.

The code has a number of rules to deal with specific situations. If a note has been indorsed by the members of a partnership, one need give notice to only one of the partners in order to fix their liability. The code states that "Notice to one partner is notice to each although the firm has been dissolved." Section 3–508(5). A problem sometimes occurs when a party has begun insolvency proceedings—such as bankruptcy—after the instrument has been dishonored. The code allows for notice to be given either to the party or to the representative of his estate, such as a receiver or trustee in bankruptcy. Section 3–508(6). If a party is dead, notice may be given to his personal representative or may be sent to the deceased's last known address. Section 3–508(7).

There is no specific order in which notice of dishonor must be given. Usually the holder gives notice to his immediate transferor, but this is not necessary and he may give notice to any one or all of the prior transferors. Once notice of dishonor has been received that notice operates for the benefit of all parties who have rights against the party notified. Section 3–508(8).

(b) *Time for Giving Notice.* Once the holder of an instrument has an instrument dishonored, he or she must give notice of that dishonor before midnight of the third business day after dishonor. If a prior transferee of the instrument receives notice that an instrument has been dishonored, he or she also must then give notice of dishonor before midnight of the third business day after he or she receives the notice. The rule is somewhat different for a bank. It must give notice before its midnight deadline. Section 3–508(2). A bank's midnight deadline is midnight on its next banking day following the banking day on which it receives the relevant item on notice. Section 4–104(1)(h). Finally, written notice is given when sent even if it is not received. Thus notice is given when it is properly mailed although the notice may be lost before the addressee receives it. Section 3–508(4). Failure to give timely notice of dishonor will result in the discharge of the parties to whom timely notice was not given, unless notice was excused or waived. These situations will be discussed in Section 25:3.

The code does not prescribe any particular method for giving notice. Notice may be given in any reasonable manner. Section 3–508(3). There is no requirement as to the substance of the notice as long as the notice identifies the instrument and states that it has been dishonored. Even if there is a misdescription contained in the notice, it is valid as long as the party notified has not been misled.

The code provides specifically that the notice requirement is satisfied if the instrument itself is sent bearing a stamp, ticket, or writing stating that payment or acceptance has been refused or if a notice is sent of debit with respect to the instrument. Typically this is the practice of banks. For example, if one deposits a check in his account that is returned unpaid because of insufficient funds, the bank may then return the check to the depositor stamped, "Returned for insufficient funds." This would constitute adequate notice of dishonor of the check.

Finally, notice of dishonor may be oral or written. However, because proof that notice was given is needed, it is a good practice to give such notice in writing or to follow up oral notification with a written confirmation of the notice.

NEVADA STATE BANK v. FISCHER
565 P.2d 332 (Nev. 1977)

THOMPSON, J.

The Nevada State Bank charged Lucile Fischer's account $2,000 when it received notice that a check in that amount, endorsed by her, had been dishonored. Therefore, she commenced this action against the Bank for wrongfully so debiting her account.

The district court ruled that her liability as endorser was discharged since notice of dishonor was not timely given her. Judgment was entered in her favor together with interest, costs, and attorney fees. The Bank appeals from that judgment. We affirm.

The facts are not disputed. On May 1, 1970,

Mrs. Fischer endorsed a $2,000 check payable to the drawer and drawn on the Clayton Bank of Clayton, Missouri. She did this as an accommodation to the payee-drawer. The Nevada State Bank cashed the check for the payee-drawer and initiated collection through Valley Bank of Nevada that same day.

On July 28, 1970, the Valley Bank of Nevada notified Nevada State Bank that the check had been dishonored stating "original lost in transit—account closed." On July 29, 1970, the Nevada State Bank debited Mrs. Fischer's account for $2,000 and notified her in writing of the payor bank's dishonor of the check.

The record does not disclose which of the several banks involved in the collection process either lost the check or delayed action with regard to it. It is clear, however, that Nevada State Bank acted promptly upon receiving notice of dishonor. Whether it was permissible in these circumstances for that bank to charge its innocent depositor rather than to look to one of the other banks involved is the issue for our decision.

Lucile Fischer endorsed the check as an accommodation party and is liable in the capacity in which she signed. NRS 104.3415. By endorsing the check she engaged that upon dishonor and any necessary notice of dishonor and protest, she would pay the instrument according to the tenor at the time of her endorsement. NRS 104.3414(1).

An endorser is a secondary party, NRS 104.3102(1)(d), whose liability is subject to the preconditions of presentment, NRS 104.3501(1)(b), and proper notice of dishonor NRS 104.3501(2)(a). Where, without excuse, any necessary presentment or notice of dishonor is delayed beyond the time it is due, an endorser is discharged. Such is the command of NRS 104.3502(1).

It is the contention of Nevada State Bank that since it initiated collection within one day of the endorsement and notified the endorser of dishonor within one day of receipt of such notice by it, "delay" does not exist and Mrs. Fischer, as endorser, is not discharged from liability.

An uncertified check must be presented for payment, or collection initiated thereon, within a reasonable time, which in this case is presumed to be seven days. NRS 104.8503(2)(b). Although the record does not advise us when presentment was made to the proper party, NRS 104.3504, we do know that Nevada State Bank initiated collection within one day after cashing the check. Consequently, bank collection was timely initiated.

In our view, however, the second precondition to liability of the endorser, that is, timely notice of dishonor, was not met. Although the Nevada State Bank notified Mrs. Fischer within its midnight deadline, NRS 104.4104, after receipt of notice of dishonor from Valley Bank of Nevada, this fact, alone, does not resolve the timeliness issue.

The record does not disclose at what point in time the check first was dishonored. We know only that almost ninety days elapsed between Mrs. Fischer's endorsement of the check and her receipt of notice of its dishonor. It is apparent that one of the several banks involved in the collection process violated its midnight deadline in giving notice of dishonor. Had such bank given timely notice, Mrs. Fischer would have learned within a reasonable time that the check had been dishonored.

Prompt action by all parties to the transaction is contemplated before an endorser may be held liable. As stated in the official comment to sec. 3–503 of Uniform Commercial Code (our NRS 104.3503):

> The endorser who has normally merely received the check and passed it on and does not expect to have to pay it, is entitled to know more promptly whether it is to be dishonored, in order that he may have recourse against the person with whom he has dealt.

As already expressed, at sometime in the chain of collection a midnight deadline was violated. Notwithstanding such violation, we are asked to conclude that notice of dishonor given ninety days after initiation of bank collection was timely. We decline to so conclude. Mrs. Fischer's liability as an endorser was discharged when the violation of the midnight deadline by a bank, identity unknown, resulted in unreasonable delay in notice of dishonor. The Nevada State Bank may look to the violator for its recovery. Its customer-endorser should not be held responsible for a violation of law committed by another bank involved in the chain of collection.

(Affirmed.)

Questions for Discussion

1. *Note that Fischer was secondarily liable in this case as an indorser; that is, she could be charged only after demand had been made and notice of any dishonor given to her. How much time did the bank have to give Fischer notice of dishonor? What was the result of the failure to comply with this requirement?*
2. *Was presentment made within the required time in this case? How much time did the bank have to present the check to Fischer if it was going to require her to pay?*

SCHENECTADY TRUST COMPANY v. ESTATE OF SCIOCCHETTI

82 Misc.2d 1075, 371 N.Y.S.2d 36 (1975)

CERRITO, J.

Basically the facts are that on December 9, 1971 the plaintiff loaned the Tiber Construction Company, Inc., thirty thousand dollars payable on demand. This note was signed on behalf of the corporation by its president and vice president, August Sciocchetti and Edward R. Downey and endorsed by Sciocchetti and Downey in their individual capacities. On June 29, 1973, the plaintiff mailed a certified letter, return receipt requested, to the corporation demanding full payment of the note by July 16, 1973. This letter was returned to the plaintiff unclaimed; however, carbon copies of this letter were received by Downey and the executrix of the estate of August Sciocchetti. After the corporation defaulted on this note, the plaintiff, in September 1973, commenced this action to recover the balance due from the defendant.

The complaint contains two causes of action; the distinguishing feature between them is that the first cause of action alleges that notice of dishonor was given to the endorser while the second cause of action alleges that plaintiff was excused from giving notice of dishonor under provisions of the Uniform Commercial Code.

It is the defendant's contention that the fact that she is the executrix of the estate of August Sciocchetti does not excuse the plaintifff from giving her notice of dishonor and its failure to do so discharges her on this note.

The Uniform Commercial Code states that notice of dishonor is necessary to charge any endorser unless excused under one of the circumstances set forth in U.C.C. 3–511 (U.C.C. 3–501(2)(a)). The purpose of giving an endorser notice of dishonor is to give him prompt notification of the maker's default so that he might take steps to protect his rights against prior endorsers and makers. It is the court's opinion that the reason for giving notice of dishonor is obviated when it is shown that the endorser of a corporate note is an officer of and actively engaged in the affairs of the corporation, is fully cognizant of the corporation's financial affairs and is the party to whom the presentment was in fact made. Under these circumstances, the endorser would or should know of the corporation's default, therefore giving such an endorser notice of dishonor would be a redundant act.

Since this formulation raises several factual issues which can only be adequately resolved by trial, both parties' motions for summary judgment are denied without cost.

Question for Discussion

1. *Do you think that the case would have been decided differently had the note been presented to some other corporate officer than Sciocchetti?*

(4) Protest. The concept of protest has a limited application for many types of commercial transactions. Protest of an instrument is not necessary unless the instrument is drawn or payable outside of the states and territories of the United States and the District of Columbia. Section 3–501(3). Failure to protest an instrument when required will result in a discharge of the drawer and all indorsers.

A protest is very simply a certificate of dishonor. It is typically made under the hand of a notary public, but it may also be made under the hand of and seal of a United States consul or vice-consul or other person authorized to certify dishonor by the law of the place where dishonor occurs. Section 3–509(1). The code states that the protest must (1) identify the instrument, (2) certify that due presentment was made or why it is excused, and (3) certify that the instrument has been dishonored by either nonacceptance or nonpayment. Section 3–509(2). If protest is necessary, it is due by the time notice of dishonor is due.

25:3. Waiver, Excuse, or Delay of Presentment, Notice of Dishonor, or Protest. There are certain circumstances under which a delay in presentment or a notice of dishonor or protest is *excused.* These are usually situations in which the intended beneficiaries of these requirements would not benefit. Delay is excused if the delay is caused by circumstances beyond the holder's control and he exercises reasonable diligence in presenting the instrument or giving notice after the cause of the delay ceases to operate. Section 3–511(1). A good illustration of this situation occurs when an instrument is accelerated without the knowledge of the holder. He may not be in a position to know that a circumstance accelerating the instrument has occurred. However, when he learns of the acceleration, he must exercise due diligence in presenting the instrument, and if it is dishonored, he must give notice and any necessary protest. Basically an excusable delay results when the delay is beyond the control of the holder.

(1) Presentment, Notice, or Protest Entirely Excused. The code sets forth a number of situations in which presentment or notice is *entirely excused.* Here the steps of presentment and notice do not have to be taken at all, whereas in the circumstance illustrated in the previous paragraph, presentment, notice, or protest had to be given after the reason for the delay was eliminated.

If the party to be charged has waived the requirement of presentment, notice, or protest either expressly or by implication, then that action is entirely excused and need not be performed. Such a waiver may occur either before or after the instrument is due. Section 3–511(2)(a). Frequently an instrument will contain a provision that waives any right to presentment, notice, or protest. Such a waiver is binding on all parties when made in this manner. Section 3–511(6). Because these requirements are primarily for the protection of the drawer or indorsers, they may waive them. A waiver of protest is also a waiver of the right to presentment and of notice of dishonor, even though there is no requirement for protest of the instrument. Section 3–511(5).

If a party has no reason or right to expect that an instrument will be honored, then he is not entitled to presentment, notice, or protest. Such a circumstance occurs when the party has himself dishonored the instrument or countermanded payment. Section 3–511(2)(b) and Comment 4. For example, if the drawer of a check has issued a stop-payment order to his bank, then presentment and notice of dishonor by the bank are not required under this section in order for him to be charged. The reason is that the drawer in this circumstance knows that the instrument will be dishonored without receiving any notice of that fact. Furthermore it is the drawer who is responsible for the fact that the instrument would not be paid if duly presented for payment.

Presentment, notice, or protest is entirely excused if by the exercise of reasonable diligence the presentment or protest cannot be made or notice given. Section 3–511(2)(c). Proof must be offered that reasonable diligence has not been or would not be successful. And if a draft has been presented for *acceptance* and dishonored, then a subsequent presentment for *payment* and any necessary notice of dishonor or protest are excused unless in the meantime an instrument has been accepted. Section 3–511(4).

A. J. ARMSTRONG CO. INC. v. JANBURT EMBROIDERY CORP.

96 N.J.Super. 246, 234 A.2d 737 (1967)

On March 1, 1960, Janburt Embroidery Corp. obtained a loan from Robert Reiner, Inc., in return for which Janburt executed seventy-two negotiable promissory notes payable to Reiner's order at monthly intervals. The notes were secured by a chattel mortgage on machinery owned by Janburt. The signatures of James Lo Curto, president of Janburt, and Harry Josephs, George Josephs, and Bernard Josephs, officers and stockholders of Janburt, appeared on the reverse side of each of the notes. The following statement appeared above the four signatures: "We hereby waive notice of protest, presentment and dishonor."

These notes were subsequently negotiated by Reiner to A. J. Armstrong Co., Inc., although Reiner was still to collect the money from the notes. In January 1963 Janburt asked Reiner to refinance their notes. Pursuant to this request a "refinance note" was drawn which provided for forty-eight monthly payments. The refinance note contained the following:

> 1. Upon failure to make any payment as herein agreed, or in the event of death, insolvency or bankruptcy or failure in business of the maker, this note shall, at the option of the holder immediately become due and payable without demand or notice. . . .
> 2. The undersigned hereby waives notice of nonpayment, protest, presentment and demand. . . .

All four parties mentioned above signed the refinance note on the reverse side. Subsequently there was a default on payment of the note, the entire balance was accelerated, and the security was sold at sheriff's sales. The proceeds from the sale did not cover the balance due on the note. Armstrong sued Janburt and the other signers of the note for the deficiency.

PINDAR, J.

The second defense raised by counsel in order to release the accommodation indorsers from personal liability is that plaintiff failed to give notice of nonpayment, protest and presentment. There is no legal merit to this contention for several reasons. First, there is an express waiver of notice contained in the body of the refinance note and reproduced supra. The waiver appears not once but twice. N.J.S. 12A:3–511(2)(a), N.J.S.A. expressly provides that presentment or notice or protest, as the case may be, is entirely excused when the party to be charged has waived it expressly or by implication, either before or after it is due. New Jersey Study Comment #1 to N.J.S. 12A:3–511(2)(a), N.J.S.A. and cases cited therein, states that where such a waiver is stated on the face of the instrument, it is binding on all parties.

Secondly, N.J.S. 12A:3–511(2)(b), N.J.S.A. states that notice is unnecessary and excused when the party has no reason to expect that the instrument will be paid. Lo Curto, as president of Janburt, and Josephs as treasurer, certainly knew that there was default on the refinance note and that the corporation could not pay it. As to them notice was unnecessary.

Questions for Discussion

1. *Is a waiver of presentment and notice of dishonor that appears on the face of a note binding on indorsers?*
2. *Would Lo Curto and the Josephs have been liable even if there had not been a waiver on the face of the note? Why or why not?*

(2) Presentment Alone Entirely Excused. There are some other situations in which presentment is entirely excused, although notice of dishonor and protest would not be excused. These situations occur "where immediate payment or acceptance is impossible or so unlikely that the holder could not reasonably be expected to make presentment." Section 3–511(3)(a), Comment 6. Thus, if the maker, acceptor, or drawee of any instrument (except a documentary draft) is dead or in insolvency proceedings that were instituted after the instrument was issued, presentment is excused. The holder may immediately have recourse upon the drawer or indorser. That party may then file a claim against the deceased's estate or in the insolvency proceedings. This procedure essentially eliminates an unnecessary step.

Presentment is also entirely excused when it is clear that the maker, acceptor, or drawee will not pay. The holder is not required to go through the useless step of presentment for payment or acceptance. He may immediately give notice of dishonor or any necessary protest. Section 3–511(3)(b).

SUMMARY

Presentment to the primary party is generally a necessary first step for establishing the liability of the secondary parties. Presentment is a demand for payment or acceptance of the instrument made upon the maker, acceptor, drawee, or other payor.

The party to whom presentment is made has a number of rights, such as to have the instrument exhibited and to require identification by the person making the presentment. Failure to comply with these or other rights makes the presentment invalid.

There are a number of rules for determining when an instrument should be presented for payment or acceptance. For instance, presentment of a time draft for acceptance must be made on or before the date it is payable. If the instrument is payable after sight it must be presented for acceptance within a reasonable time after date or issue. If an instrument shows the date on which it is payable, presentment for payment is due on that date. If the instrument has been accelerated, presentment for payment is due within a reasonable time after acceleration. If the instrument is a discount instrument, presentment for payment is due within a reasonable time after any secondary party becomes liable on the instrument.

For an uncertified check drawn in the United States which is not a bank draft, a reasonable time to present for payment with respect to the liability of the drawer is presumed to be thirty days after issue or check date, whichever is later. With respect to the liability of the indorser, seven days after his indorsement is presumed to be a reasonable time.

Before secondary parties can be charged on the instrument, there must be a dishonor upon presentment by the party who is primarily liable. A dishonor occurs when a necessary or optional presentment is made and acceptance or payment is refused or cannot be obtained within a reasonable time. Some actions by the party upon whom presentment is made do not constitute dishonor. For instance, insisting on a legal right or return of an instrument for a proper indorsement does not constitute dishonor.

As a general rule, any party who may be compelled to pay an instrument may give notice to any party who may be liable on it. Once the holder of an instrument has an instrument dishonored, he or she must give notice of that dishonor before midnight of the third business day after dishonor. A different rule applies to banks. They must give notice before their midnight deadline.

Protest of an instrument is not necessary unless it is drawn or payable outside the states and territories of the United States. The code lists a number of circumstances under which presentment, notice, or protest is excused or entirely excused.

PROBLEMS

1. Theta Electronic Laboratories, Inc., signed a promissory note in the total amount of $15,377.07. The note was payable in monthly installments of $320.47 and contained a clause accelerating the note upon default in any monthly payment. Gerald M. Exten and several others indorsed Theta's note. Hane held the note for eighteen months after a default which accelerated the note. He then made a demand for payment from Exten. Is Exten liable on the note? Hane v. Exten, 255 Md. 668, 259A.2d 290 (1969).
2. Dale Eastman sold four cows to Leo Pelletier, who gave him a note in return signed by him and his wife. Pelletier then sold some of the cows to Carlton Achilles, who gave a check to Pelletier for $250.00, payable to Eastman and Pelletier jointly. Pelletier indorsed the check over to Eastman on July 10 and paid the balance due on the cows in cash. Eastman held the check until July 30 before depositing it in the bank, although he lived only one mile from the bank. On August 5 Eastman received notice that Achilles had ordered payment stopped on the check on July 13. Eastman then sued Pelletier on his indorsement. Pelletier contended that the check was not presented in a reasonable time. Was he correct? Eastman v. Pelletier, 47 A.2d 298 (Vt. 1946).
3. A signed a promissory note as an indorser and negotiated it to H. H presented the note to M, the maker, who refused to pay. H then telephoned A to tell him that the maker had dishonored the note. When sued by H, A contended that he was not properly notified of the maker's dishonor. Was A correct?
4. I Corporation indorsed a promissory note and negotiated it to H. The note was dishonored by the maker. H attempted to notify I Corporation of the dishonor by leaving the notice at the corporation's office during regular

business hours after failing to locate any of the corporate officers. Was notice of dishonor properly given to I Corporation?

5. C. C. Merrill was an indorser of a promissory note. Written notice of dishonor was mailed to him by Bank of America in an envelope addressed "2277 West Twenty-Third Street, Los Angeles." The letter was sent by registered mail with instructions to deliver to the addressee only. The letter was returned by the post office department with the indication that it had been readdressed by the Los Angeles post office to "622 North Arden, Beverly Hills, California" and marked "Due 1¢." It had then been forwarded to the Beverly Hills post office and marked "Unclaimed, Return to Writer." At that time the Los Angeles city directory listed a business address of Charles C. Merrill at "727 West Seventh Street, Los Angeles" and a residence address of "Beverly Hills." There was no other C. C. Merrill in the directory. When sued on the note, Merrill contended that he was discharged because he was not properly served with notice of dishonor. The bank contended that it used the address for Merrill that was penciled in on its records and that it used "due and reasonable diligence" in attempting to notify Merrill. Was the bank correct? Bank of America v. Century L. & W. Co., 19 Cal.App.2d 197.
6. M Corporation executed a promissory note by its president, A. A then personally signed the note as an accommodation indorser and negotiated it to H. The corporation then dishonored the note through its president, A. No notice of dishonor was ever given to A by H. A contended that he was not liable because he had never received notice of dishonor. Was he correct?
7. M signed a promissory note payable to P. P indorsed the note to H. H later presented the instrument to M for payment. M refused to pay the instrument unless H produced some identification indicating that he was in fact H. Did M dishonor the instrument?
8. On January 1 Delbert Drawer executed a draft drawn on Danny Drawee and payable to Paula Payee in the amount of $100. Paula then negotiated the draft, for value, to Harry Holder. Holder waited until March 1 to present the instrument to Danny Drawee for payment. Drawee refused to honor the draft, although Delbert Drawer had given him $100 to cover the draft. When Holder notified him of the dishonor, Delbert refused to pay, contending that he was discharged by Holder's delay in presentment to Drawee. Is Delbert correct?
9. Assume the same facts as in Question 8 except that after Drawer gives Drawee $100 and prior to Holder's presentment, Drawee becomes insolvent. What rights, if any, does Drawer have if Holder attempts to collect the draft from him?

CHAPTER 26

Liability of the Parties and Discharge

A party who signs an instrument may be either a *primary* party or a *secondary* party. If one is a primary party, he is expected to pay the instrument when it is due; one need not demand payment from him in order to fix his liability. Parties that are secondarily liable are not normally expected to pay unless the party expected to pay does not. Normally, therefore, in order to fix the liability of the secondary party, one must first present the instrument to the primary party, who dishonors it. One must then give notice of dishonor or any necessary protest to the secondary party in order to establish his liability. These requirements and exceptions to them were discussed in Chapter 25.

The parties who are primarily liable on an instrument are the maker of a promissory note and the drawee who has accepted a draft, frequently referred to simply as an *acceptor.* A drawee has no liability on a draft unless he accepts it. An individual who is the drawer of a draft and all indorsers are secondarily liable on the instrument.

26:1. Parties Primarily Liable. (1) Liability of the Maker. The maker engages that he will pay the instrument according to its tenor at the time he signs it. If he signs an incomplete instrument, he agrees to pay it as completed in accordance with the principles discussed in Chapter 25. Section 3–413(1). For instance, if the instrument is completed in an amount that complies with the authority for completion, the maker must pay it. If the completion exceeds the authority given by the maker, he is liable for the full amount if his negligence contributed to the alteration.

Because the maker is primarily liable, there is no need for the holder of the instrument to make a demand upon the maker in order to fix his liability. Furthermore, if demand is made upon a maker who refuses to pay, the secondary parties then become liable on the instrument when the proper steps are followed. However, the primary party is always the one who ultimately bears the burden of finally paying the instrument. Thus, if a secondary party must pay, he still may come back against the individual who has primary liability in order to obtain payment.

When the maker signs an instrument, he makes two admissions. He admits as against all subsequent parties (1) the existence of the payee and (2) the payee's capacity to indorse. Section 3–413(3). The result of these admissions is that when the maker is sued on the instrument, he cannot defend on the ground that the payee if fictitious or that the payee lacks capacity, as in the case of infancy. Thus, although the payee may have the defense of lack of capacity, if called upon to pay the instrument on an indorsement, the maker could not raise it.

(2) **Liability of the Drawee–Acceptor.** As previously mentioned, the drawee of a draft is not liable on the instrument until he accepts it. Once the drawee accepts the draft, however, he or she becomes an *acceptor* and is primarily liable on the instrument. Most of the principles of liability discussed with regard to the maker also apply to an acceptor. He engages that he will pay the instrument according to its tenor at the time of his engagement, or as completed, according to the principles previously discussed. Section 3–413(1). The acceptor also admits to all subsequent parties the existence of the payee and his capacity to indorse.

Not only is a drawee not liable on an instrument until he accepts it, but the mere drawing of a draft does not operate as an assignment of any funds in the hands of the drawee. Section 3–409(1). However, if an instrument is presented to the drawee for acceptance or payment and he refuses to accept it or pay and also refuses to return the instrument, then he may be liable in tort for conversion of the instrument. Such an act does not operate as an implied acceptance, however. Section 3–419(1)(a),(b).

26:2. Parties Secondarily Liable. (1) **Liability of the Drawer.** The parties who are secondarily liable on an instrument do not usually expect to pay it because normally an effort must first be made to collect the instrument from any party who is primarily liable. The drawer of a draft is secondarily liable. By drawing the instrument, he agrees that upon dishonor of the draft and any necessary notice of dishonor or protest he will pay the amount of the draft to the holder or to any indorser who takes it up. Section 3–413(2). Thus, suppose Danny Drawer drafts an instrument on Dorothy Drawee, payable to Paula Payee in the amount of $500. Payee presents the instrument to Drawee for payment. Drawee refuses to pay and hence dishonors the draft. Payee then notifies Drawer that the instrument has been dishonored. Drawer is then liable to Payee on the instrument as a secondary party.

One may escape liability as a secondary party by drawing the instrument "without recourse." Whoever takes the instrument then knows that the drawer intends to incur no liability on it. Section 3–413(2).

The drawer admits to all subsequent parties the existence of the payee and his capacity to indorse. This is the same admission made by makers and acceptors. Section 3–413(3).

(2) **Liability of the Indorser.** The act of indorsing an instrument was discussed in some detail in Chapter 25. An indorsement is often required in order to negotiate an instrument. When one indorses the instrument, he agrees

to be bound by the indorsement contract stated in the code unless he indorses "without recourse." This would of course be a qualified indorsement. The code provides that "every indorser engages that upon dishonor and any necessary notice of dishonor and protest he will pay the instrument according to its tenor at the time of his indorsement to the holder." Section 3–414(1). The indorser also makes the same promise to any *subsequent* indorser who takes the instrument even though the subsequent indorser was not obligated to do so.

Note once again that the indorser's liability is secondary. He agrees to pay the instrument only if the party who is primarily liable does not. In order to fix the liability of the indorser, the holder must normally present the instrument to the primary party for payment, have it dishonored, and give the indorser notice or protest if necessary. These steps may be waived or excused as discussed earlier, but unless they are, the indorser is not liable on the instrument until the requirements are met.

There may be a number of indorsements on an instrument. The indorsers are liable to one another in the order in which they indorse unless they agree otherwise. The code creates a rebuttable presumption that the instrument was indorsed in the order in which the signatures appear on it. Section 3–414(2). One may introduce evidence to show that the indorsements were made in some other order.

The indorser's liability extends to the tenor of the instrument at the time he indorsed it. In other words, if an instrument was payable in the amount of $500 when indorsed by Inez Innocent, then she is liable for that amount. This is true even if the instrument originally was for $50 and was altered by someone prior to the indorsement of Inez. She is liable for the altered amount.

An example might help to clarify some of these points. Suppose Manny Maker executes a promissory note to Paula Payee. Payee indorses it to Interim Indorser, who in turn indorses it to Harry Holder. Holder presents it to Maker for payment. Maker dishonors the instrument. Holder then notifies Interim Indorser, who is secondarily liable to Holder. (Payee would also be liable to Holder upon proper notice.) Interim then pays Holder and notifies Payee, who is liable to Interim because she indorsed the instrument to him. Payee pays Interim and then sues Maker. Unless he has a defense against Payee, Maker is, of course, ultimately liable to her on the instrument. Interim and Holder could also have sued Maker because he is primarily liable on the instrument.

MECHANICS NATIONAL BANK OF WORCESTER v. SHEAR

386 N.E.2d 1299 (Mass. App. 1979)

Maurice Shear, Richard H. Gins, and Hyman H. Silver became involved in a business venture with the Heywood Nursing Home, Inc. Shear negotiated a loan from the Mechanics National Bank of Worcester. The loan was made to the Heywood Nursing Home and the proceeds deposited in its account by the bank.

A note for $90,000.00 was signed on behalf of

Heywood by Shear. Shear also obtained the signatures of Gins and Silver on the back of the note. The note was not paid and the bank sued the three individuals.

GOODMAN, J.

From the note itself it is clear that the three individual defendants signed as indorsers. At the bottom of the instrument are the words, "Please endorse (over)"; and the note does not "clearly indicate[]" that the signatures were made in some other capacity. See § 3–402. Further, the indorsements are in blank; and in such a case, if any necessary notice of discharge and protest has been waived, "every indorser engages that . . . he will pay the instrument according to its tenor at the time of his indorsement to the holder. . . ." § 3–414(1). The defendants are therefore obligated to "pay the instrument" (official comment 1 to § 3–414) as written–as "set forth in the words and figures" of the instrument "at the time of [their] indorsement. Thus if a person indorses an altered instrument he assumes liability as indorser on the instrument as altered." Official comment 3 to § 3–414.

In *Community Natl. Bank* v. *Dawes*, 369 Mass. 550, 561, 340 N.E.2d 877 (1976), the Supreme Judicial Court also concluded that a defendant who signed a note on the back was an indorser under § 3–402 and pointed out: "As an indorser, [the defendant] contracted to pay the instrument according to its tenor at the time he signed. G.L. c. 106, § 3–414. The note [as in our case] explicitly provides for a waiver of presentment, protest, or notice of dishonor or delay therein, and on its face makes this waiver applicable to indorsers, among others. . . . With this information alone before us, we consider it an inescapable conclusion that, as matter of law [the defendant], is liable to the bank for the *debt represented by the note*" (emphasis supplied).

The liability under § 3–414 attaches to "all indorsers" whether or not for accommodation—"whether or not the indorser . . . received consideration for his indorsement." Official comment 1 to § 3–414. Further, the liability is joint and several among the three since each signed "in the same capacity" and as "part of the same transaction," each signing, as the trial judge found, before the note was retured to the bank. § 3–118(e), and official comment 6.

The defendants' obligations on the note "according to its tenor" make irrelevant Silver's contention that the trial judge was clearly erroneous in finding that Shear was authorized to sign the note on behalf of Heywood. The defendants' liability does not depend on whether the signature is operative as Heywood's (see § 3–404[1]), but on the defendants' undertaking as indorsers under § 3–414.

(Judgment for plaintiff.)

QUESTION FOR DISCUSSION

1. *In this case did the liability of the individual indorsers depend upon whether Shear had authority to sign on behalf of the corporation?*

26:3. OTHER PARTIES LIABLE. (1) Liability of Accommodation Parties. The code defines an *accommodation party* as "one who signs the instrument in any capacity for the purpose of lending his name to another party to it." Section 3–415(1). For example, suppose Manny Maker desires a loan from Friendly Bank. Bank is not satisfied with Maker's credit rating and desires someone else to sign the note already signed by Maker. Maker then has Al Accommodation indorse the note, which is then given to Bank. Al is an accommodation maker and Maker is the party accommodated.

One may accommodate as a maker, an acceptor, a drawer, or an indorser. The liability of an accommodation party is determined by the capacity in which he indorses when the instrument is taken for value before it is due. Section 3–415(2). If he indorses as co-maker, he is primarily liable. If he accommodates as an indorser, he is secondarily liable in that capacity. He would normally then have a right to have the instrument presented to the party who is primarily liable and receive notice of any dishonor before incurring liability himself. It is not necessary, however, for the holder to proceed against the party accommodated first, but the accommodation party does have certain rights against the party accommodated.

The person who is accommodated has no claim against the accommodation party. On the other hand, if the accommodation party is required to pay the instrument, he has a right of recourse on the instrument against the party accommodated. Section 3–415(5). Because the accommodation party is a surety, he also has the right of *subrogation, exoneration,* and, if there are other parties, *contribution. Subrogation* means that once the accommodation party has paid the creditor, he then has the rights of the creditor against other persons indebted to the creditor on the instrument. That person would normally be the accommodated party. Exoneration gives the accommodation party the right to require the party accommodated to pay the instrument when it is due. Contribution entitles the accommodation party to require other co-signers to pay their share of what is due on the instrument.

W. S. KERR et al. v. DeKALB COUNTY BANK

135 Ga.App. 154, 217 S.E.2d 434 (1975)

DEEN, J.

The appellee bank filed suit against Kerr, Mott and Thurston on a promissory note signed by these three individuals in the lower right hand corner, dated December 29, 1972, and stating in the body: "Time 180 days, note due June 18, 1973." Demand for payment and notice of intention to recover attorney fees was sent to each of these persons by letters dated January 15, 1974. Summary judgment was eventually granted against Thurston, and this appeal involves only the subsequent grant of a summary judgment against Kerr and Mott.

These appellants urge that they signed the instruments only in the capacity of accommodation parties and received no benefit from the loan. All three signed the note as makers, and the amount of the loan in the form of a certified check shows all three as named payees therein. While Kerr and Mott contend they were no more than sureties, and while the bank readily admits that Thurston's credit rating was insufficient to support a loan in any amount, the affidavit of its president shows that "the loan was made by me to Thurston, Mott and Kerr as co-makers and joint principals and each of them signed the instrument in this capacity" and also that both Mott and Kerr "indicated that they had a percentage interest in this production" for which Thurston, their business associate, was specifically seeking the funds. "When the instrument has been taken for value before it is due the accommodation party is liable in the capacity in which he has signed even though the taker knows of the accommodation." Code Ann. § 109A-3–415(2). Here Mott and Kerr signed in the capacity of makers and the check was made out to them equally with Thurston. In *Smith* v. *Singleton,* 124 Ga.App. 394, 184 S.E.2d 26, where one party, although known to be lending his credit for accommodation purposes, signed as a maker, it was held under this section that his plea that he in fact signed as a guarantor was unavailing, in an action brought by the payee against the co-makers. *"An accommodation maker . . . is bound on the instrument without any resort to his principal, while an accommodation indorser may be liable only after presentment, notice of dishonor and protest."* (Emphasis added.) Uniform Laws Anno., Uniform Commercial Code, Vol. II, Sec. 3–415, Official Comment. The knowledge of the payee that one is signing a promissory note as an accommodation maker would not relieve such signatory from liability thereon. Mott and Kerr were, along with Thurston, primarily liable on the instrument.

In opposing the motion for summary judgment, Mott and Kerr contended that the note was a 90-day note rather than a six-month note. This is a mere conclusory statement, in view of the instrument itself, admitted by these defendants to have been signed by them, which shows that it is a 180-day instrument. This is related to the contention that the maker delayed unnecessarily in demanding payment after maturity of the instrument, as a result of which the defendant Thurston removed from the state various assets which might have been seized and subjected to the payment of the indebtedness.

The note was due June 18, 1973. Suit notice was received January 15, 1974. The note stipulated that the note was payable on maturity at its office. "A demand for payment is not necessary in order to charge the maker of a promissory note; and hence . . . is not a prerequisite to the institution and maintenance of a suit on the note against the maker."

(Judgment affirmed.)

QUESTIONS FOR DISCUSSION

1. *Why were Kerr and Mott arguing that they had signed the note only as accommodation parties in a capacity that rendered them no more than sureties? In what capacity did they accommodate?*
2. *Assuming that no demand was made for payment of the note, why were Kerr and Mott liable for its payment?*

---

(2) **Liability of Guarantor.** If one indorses an instrument and adds the statement "payment guaranteed" or similar words, then the signer "engages that if the instrument is not paid when due he will pay it according to its tenor without resort by the holder to any other party." Section 3–416(1). The effect of this statement is for an indorser to waive presentment, notice of dishonor, and protest as well as any demand by the holder against the party primarily liable in order to charge the indorser. In other words, when the instrument comes due, the indorser must pay it without any other requirement on the holder.

If the words "collection guaranteed" or similar words are added to an instrument, the signer promises to pay the instrument according to its tenor when due but only after certain conditions are met. First, the holder must reduce his claim against the maker or acceptor to judgment and have an attempted execution of the judgment returned unsatisfied. This process is not necessary if the maker or acceptor becomes insolvent or if it otherwise appears that it would be useless to proceed against him. Section 3–416(2). In other words, under this set of circumstances the collection of the instrument is guaranteed only after an attempt is made to collect from the party primarily liable.

26:4. Liability Based on Warranty. Up to this point our discussion has involved the liability of a party that arises on the instrument because he has signed it as an indorser, a maker, or a drawer or has accepted it for payment. There is another basis upon which one may be liable, quite apart from liability based on the instrument itself. This liability comes about from certain warranties that may be implied merely by his dealing with the instrument. These warranties impose liability in addition to the liability imposed by reason of a party's relationship to the instrument itself, such as the obligation one may incur as the result of the indorsement contract.

The warranties are basically of two types. One group of warranties is given by the *transferor* of an instrument and is given regardless of whether the instrument is negotiated by indorsement or simple delivery or is transferred by assignment. Section 3–417(2). These will be called *transferor's warranties.* The other group of warranties is given by the party who obtains *payment* or *acceptance* to the person who pays or accepts the instrument. Section 3–417(1). For ease of identification we will call these *presenter's warranties.* These presenter's warranties are also given by a prior transferor of the instrument. A transferor thus gives not only the transferor warranties but also the presenter's warranties. The major difference between these two types of warranties is in the persons who benefit from them. The presenter's warranties run in favor only of a party who accepts or pays the instrument. The transferor's warranties, on the other hand, run in favor of the transferor's immediate transferee and subsequent holders, with one minor exception to be discussed in Section (1). For example, suppose Doris Drawer draws a draft on Delbert Drawee made payable to Paula Payee. Payee negotiates the draft to Albert Able, who in turn negotiates it to Bertha Bearer. Bertha then presents the draft to Drawee, who accepts the draft. The *presenter's warranties* are given to Drawee by Bertha because she *obtained* payment. The transferor's warranties are given by Payee to Albert because he is Paula's transferee. Albert gives the same warranties to Bertha for the same reason. If the transfer is by indorsement, Paula gives the warranty not only to Albert but to any "subsequent holder who takes the instrument in good faith." Section 3–417(2).

These warranties are given even if the holder fails to present the instrument and give notice of dishonor as required. He can still pursue a prior transferee, for example, on the basis of the warranties even if he cannot do it *on* the instrument. Of course, if the instrument has been properly presented and notice given, the holder may proceed against the secondary parties who preceded him, on both theories. An examination of the specific transferor's and presenter's warranties follows.

(1) Transferor's Warranties. The code provides that any person who (1) transfers an instrument and (2) receives consideration makes certain warranties to (1) his immediate transferee and (2) any subsequent holder *if* the transfer is by indorsement and the instrument is taken in good faith. Section 3–417(2). Note that those who benefit from these warranties are the transferee and the subsequent holder under certain circumstances. Thus the effect of indorsing an instrument is not only to obligate oneself under the indorsement contract but to extend his warranty obligations beyond his immediate transferee. A transferor who receives consideration gives the warranties.

The code states that the warranties given by the transferor are that

(a) he has a good title to the instrument or is authorized to obtain payment or acceptance on behalf of one who has a good title and the transfer is otherwise rightful; and
(b) all signatures are genuine or authorized; and
(c) the instrument has not been materially altered; and
(d) no defense of any party is good against him; and

(e) he has no knowledge of any insolvency proceeding instituted with respect to the maker or acceptor or the drawer of an unaccepted instrument.

Note that the warranty stated in (d) is that there is no defense of any party good against the transferor. This warranty is absolute. If there turns out to be a defense against the transferor that may be raised by a party and of which the transferor was ignorant at the time of the transfer, the warranty is still given. This warranty may be limited if the transfer is made by an indorsement "without recourse." When the transfer is made in this manner, not only does an indorser escape liability under his indorsement contract but he also limits this warranty obligation to a warranty that he has no *knowledge* of a defense that might be raised against him. Section 3–417(3). An indorser "without recourse" gives all the other transferor's warranties, however, regardless of the fact that he or she might not be liable on the indorsement contract.

(2) Presenter's Warranties. One who obtains payment or acceptance of an instrument, or any prior transferor, gives three warranties to the person who in good faith pays or accepts the instrument. These warranties are (1) that he has good title to the instrument; (2) that he has no knowledge that the signature of the maker or drawer is unauthorized; and (3) that the instrument has not been materially altered. These warranties require some further analysis.

The warranty of title given by a presenter is that he has good title to the instrument or is authorized to obtain payment or acceptance on behalf of one who has good title. Section 3–417(1)(a). This warranty allows one who pays or accepts an instrument to recover from a person presenting an instrument if an *indorsement* turns out to be forged, which would result in the presenter's not having a valid title to the instrument. Note, however, that the warranty does not extend to the drawer's signature. Hence, once the drawee accepts the instrument, he can recover back against a presenter or prior transferor for a loss suffered as the result of a forged indorsement but not for a forged signature of a drawer. Theoretically, at least, the reason for the distinction is that a drawee can verify a drawer's signature but is not in a position to check the signature of an indorser.The second warranty given by the prior transferor or person obtaining payment is that he has no *knowledge* that the signature of the maker or drawer is unauthorized. Section 3–417(1)(b). Normally, if one takes an instrument with knowledge of the unauthorized signature, then he could not qualify as a holder in due course. The problem arises, however, when the holder discovers after having taken the instrument that the signature of the maker or drawer is unauthorized. Therefore this section also provides that the warranty is not given by a holder in due course to a drawer or maker with respect to the drawer's or maker's own signature. The maker or drawer should know his own signature; hence if he pays on his own forged signature without detecting the forgery, the loss should remain with him as against a holder in due course. Another exemption to the warranty is made for a holder in due course as against an acceptor of a draft if the holder in due course took the draft *after* it was accepted by the drawee, as when one becomes a holder in due course of a certified check. The warranty is also not given by a holder in due course who obtains acceptance of a draft without knowledge that the drawer's signature was unauthorized.

The third warranty given by one who obtains payment or acceptance of an instrument to the person who in good faith pays or accepts is that the instrument has not been materially altered. Section 3–417(1)(c). Once again, this warranty is not given by a holder in due course acting in good faith to the maker of a note or the drawer of a draft. Nor does the holder in due course give the warranty to the acceptor of a draft if the alteration is made after acceptance by the acceptor. A holder in due course also does not give the warranty to the acceptor if the alteration occurred *prior* to the acceptance and the holder in due course took the instrument *after* it was accepted.

One final note. As to both the transferor's and the presenter's warranties, if the transfer or presentation is made by a selling agent or a broker and that individual does not reveal that he is acting as a selling agent or a broker then he gives all the warranties we have discussed earlier in the chapter. If the selling agent or broker reveals his capacity, then he warrants only his good faith and authority. Section 3–417(4).

OAK PARK CURRENCY EXCHANGE, INC. v. MAROPOULOS
363 N.E.2d 54 (Ill. App. 1977)

John Bugay had a certified check drawn on American National Bank in the amount of $3,564.00. The check was payable to the order of "Henry Sherman, Inc." and indorsed "Henry Sherman" on the reverse side.

J. Maropoulos agreed to aid Bugay in cashing the check at the Oak Park Currency Exchange where Maropoulos had previously done business and was known to the tellers. When Maropoulos and Bugay arrived at the Exchange the teller, Mrs. Panveno, agreed to cash the check if Maropoulos indorsed it. Maropoulos did so and gave the money he received from the teller directly to Bugay. Later it developed that the signature of "Henry Sherman" was forged and the Exchange had to pay the claim of the Belmont National Bank where it had deposited the check.

Oak Park Currency Exchange then sued Maropoulos for the amount of the check. The trial court directed a verdict in favor of Maropoulos and Oak Point Currency Exchange appealed.

GOLDBERG, J.

In this court, plaintiff urges that defendant breached his warranty of good title when he obtained payment of a check on which the payee's endorsement was forged and that there was sufficient evidence to support a directed verdict in favor of plaintiff. Plaintiff's contentions are based exclusively on section 3–417(1) of the Code. Defendant contends that an accommodation endorser does not make warranties under Ill.Rev.Stat.1975, ch. 26, par. 3–417(1) and that the trial court properly directed a verdict for defendant.

The portion of the Code upon which plaintiff seeks to hold defendant liable is section 3–417 entitled "Warranties on Presentment and Transfer." (Ill.Rev.Stat.1975, ch. 26, par. 3–417.) As shown above, the parties both confine their arguments to subsection 3–417(1) of the Code and the judgment order refers specifically thereto. Section 3–417(1) sets out warranties which run only to a party who "pays or accepts" an instrument upon presentment. We note that presentment is defined as "a demand for acceptance or payment made upon the maker, acceptor, drawee or other payor . . . ." (Ill.Rev.Stat.1975, ch. 26, par. 3–504(1).) As applied to the instant case, the warranties contained in section 3–417(1) are limited to run only to the payor bank and not to any other transferee who acquired the check. In the case before us, plaintiff (Exchange) is not a payor or acceptor of the draft.

This interpretation is strongly supported by the official comment which details the reasons for distinguishing warranties made to a payor or acceptor of an instrument from those made to a transferee. The case before us involves a transferee, not a party who paid or accepted the instrument. Thus it appears that reliance by plaintiff upon subsection 3–417(1) was misplaced. The authorities cited by plaintiff do not support its contention as all of these cases were decided before the effective date of the Code.

Defendant has cited no case bearing directly upon the situation before us. Our research has not disclosed any case construing this aspect of the pertinent subsection of the Code. Aside from the enactment itself and the comments above cited, we found only dicta to the effect that by this subsection the warranties pertaining to authenticity of endorsements run only "to a person who in good faith pays or accepts . . ." the instrument.

An additional theory requires affirmance of the judgment appealed from. Subsection 3–417(2) of the Code provides that one "who transfers an instrument and receives consideration warrants to his transferee . . ." that he has good title. (Ill.Rev.Stat.1975, ch. 26, par. 3–417(2).) The Illinois comments to this portion of the Code confirm that this warranty is made only by any party who transfers an instrument for consideration. In *First Bank & T. Co. of Boca Raton* v. *County Nat. Bank* (Fla.App.1973), 281 So.2d 515, 517, by way of dicta, the court noted the presence of the phase "and receives consideration" in this subsection of the Code.

The evidence presented in the case at bar establishes that defendant received no consideration for his endorsement. Though Mrs. Panveno testified that she saw Bugay hand defendant some money as the two left the currency exchange, she also testified that defendant stated that he was doing a favor for his friend; that she was not paying close attention to the two men and that she did not watch them as they walked away from her. Thus her testimony was considerably weakened by her own qualifying statements and it was strongly and directly contradicted by the positive and unshaken testimony of defendant that he received nothing in return for his assistance. The simple fact standing alone that this witness saw Bugay hand some money to defendant, even if proved, would have no legal significance without additional proof of some type showing that the payment was consideration for defendant's endorsement.

In our opinion, the trial court correctly determined that no contrary verdict based on the testimony offered by plaintiff could ever stand and properly directed a verdict for defendant.

(Judgment affirmed.)

QUESTIONS FOR DISCUSSION

1. *Which of the presentment warranties or transferor's warranties were relied upon in this case?*
2. *Why did the court hold that the Exchange was not entitled to rely on the transferor's warranties?*
3. *Why was the Oak Park Currency Exchange not entitled to the benefit of the warranties given by UCC 3–417(1)?*

26:5. DISCHARGE. The code provides for a number of ways in which the obligations of the parties to an instrument terminate. Before examining these individual methods of discharge, let us note several general principles. In the first place, no discharge of any party is effective against a subsequent holder in due course unless he has notice of the discharge when he takes the instrument. Section 3–602. The code also provides that a holder in due course or a person who has changed his position in reliance on the payment has no further liability to one who pays or accepts once the instrument has been paid, except for the presentment warranties. Section 3–418. Thus, if the drawee pays on the instrument and later discovers that the drawer's signature was forged, he cannot recover his money. The payment is final if the party changed position because of the payment or was a holder in due course.

In general, any party is discharged from his liability on an instrument to another party by any act or agreement with such party that would discharge his simple contract for the payment of money. Section 3–601(2).

VALLEY BANK OF NEVADA v. BANK OF COMMERCE
343 N.Y.S.2d 191 (1973)

SANDLER, J.

In this action by Valley Bank of Nevada, successor to the Bank of Las Vegas (Las Vegas), to recover a certain sum of money paid to the Bank of Commerce (Commerce) allegedly under a mistake of fact or law, both parties have moved for summary judgment.

The case concerns a check for $4,000.00, drawn on Commerce, dated July 17, 1969, payable to the order of M & R Investment Company, Inc., a Las Vegas corporation, and purportedly signed by Commerce's customer, Matthew Carpelow. On or about July 19, the check was deposited with Las Vegas, and then forwarded by Las Vegas for collection through its correspondent bank in New York City, Morgan Guaranty Trust Company (Morgan).

Commerce paid the check on or about July 22, and charged Carpelow's account. Las Vegas received a credit from Morgan in the sum of $4,000.00, and in turn credited the account of M & R Investment Company, Inc. in that amount.

Thereafter, on or about Aug. 7, Carpelow informed Commerce that his signature on the check was forged, and signed an affidavit of forgery. Commerce credited Carpelow's account for $4,000.00, and by letter dated Aug. 7, made claim on Las Vegas for reimbursement through Morgan.

Following receipt of the original documents on or about Aug. 22, 1969, Las Vegas debited the account of M & R for $4,000.00, informed Morgan of that action and that Morgan's account had been credited in that amount in connection with reimbursement to Commerce. On Aug. 28, 1969, Commerce received Morgan's Treasurer's check in the sum of $4,000.00 together with a transmittal memorandum that the check was given as reimbursement for the amount of the original forged check.

At this point there appears the only area of possible factual disagreement in the papers before me. Commerce asserts that it received no further communication in the matter until it received Las Vegas' letter dated Dec. 28, 1970 and signed by Plaintiff's Senior Vice President and Cashier, in which a formal claim for reimbursement was made.

Las Vegas states that M & R had refused the charge back, apparently on Aug. 27, according to the Dec. 28, 1970 letter, and that Las Vegas promptly sought an offset credit from Morgan. A delay followed because of a fire at Las Vegas in which its reconciliation records were destroyed, and Las Vegas did not learn the item was still outstanding until mid-April, 1970, when the work of reconstructing the accounts was completed. Thereafter, on April 15, 1970 and again on July 22, 1970, reconciliation tracers were sent to Morgan relating to outstanding items one of which was the $4,000.00 credit, to which there was no response. Eventually, in a telephone conversation on Aug. 6, 1970, Morgan took the position that Las Vegas was bound by its action of Aug. 22, 1969.

The circumstance that Commerce had heard nothing further of the matter after it received the Morgan check until the letter of Dec. 28, 1970 setting forth the Las Vegas claim, suggests that Morgan had not seen fit to transmit to Commerce the various communications it had received from Las Vegas in the intervening period.

A natural starting point for analysis is the obvious fact that Las Vegas was under no legal obligation whatever to reimburse Commerce for the $4,000.00, or to charge its customer's account. The rule of law—first set forth in *Price* v. *Neal,* 3 Burr. 1354 (1762), widely followed thereafter, and firmly accepted in the U.C.C.—is explicit that a drawee bank that pays an instrument on which the signature of the drawer is forged cannot recover back its payment from a collecting bank that acted in good faith. U.C.C. 3–418; see also U.C.C. 3–417(1)(b), 4–207(1)(b). The defense of forgery interposed by Commerce to this action is thus patently without merit.

Since Las Vegas was under no legal obligation to reimburse Commerce, the critical question presented is why Las Vegas in fact did so.

Unquestionably, as contended by Las Vegas, its action of Aug. 22, 1969 returning the money to Commerce was a mistake. What is not clear from the papers is the precise nature of the mistake, but the following seems to me to cover the realistic possibilities.

No such change of position occurred in this case or could have occurred since Commerce's liability to its own depositor for paying on the forged check was clear and acknowledged.

Second, Las Vegas may have acted under the erroneous legal understanding that it was obligated to reimburse Commerce.

. . . I am satisfied that Las Vegas returned money to Commerce that it was not legally required to do under a mistake of fact and law, and that under a mistake of fact and law, and that under all the realistic possibilities as to the nature of the mistake, Las Vegas is entitled to recover.

Approaching the issue in a somewhat different way, the underlying reality in this case was most clearly suggested in the affidavit submitted on behalf of Commerce by an Assistant Vice-President who said:

> On the basis of my personal experience, I am able to inform the Court that requests for reimbursement of forged checks initially honored are not uncommon and are frequently granted by the collecting bank and its correspondent . . . where the collecting bank is satisfied that it will not need to assume the loss.

Undoubtedly, the uniform practice is for the collecting bank to reimburse the drawee bank when its depositor consents to being charged, and to refuse such reimbursement when its depositor does not consent. When a drawee bank is reimbursed the inescapable inference must be that the collecting bank's depositor has consented to being charged.

Las Vegas responded to Commerce's request in a spirit of good will and friendly cooperation but on the basis of a mistake of fact or law. As a result, Commerce received a windfall which in fairness it should not be permitted to retain, and which the law does not permit it to retain.

Accordingly, Plaintiff's motion for summary judgment is granted, and the Defendant's motion is denied.

Questions for Discussion

1. *Note that in this case the court cited the rule established in a landmark English case, Price v. Neal, that dealt with the drawee's liability when it pays on a drawer's forged indorsement. Was it required to pay when the check was presented for payment?*
2. *If the signature of an indorser rather than the drawer had been forged, would Bank of Commerce have been able to recover from Morgan Guaranty Trust or Bank of Las Vegas? See Society National Bank of Cleveland v. Capital National Bank, Section 26:4.*

---

(1) Payment or Satisfaction. By far the most common method of discharging a party who is liable on an instrument is for him to make payment or satisfaction to the *holder* of the instrument. Section 3–603(1). Note that the payment or satisfaction must be given to the proper person, that is, the holder. If the wrong person is paid, one is not entitled to be discharged.

The payment of the instrument will discharge the party making the payment even if the payment is made with knowledge of the claims of another person to the instrument. There is an exception to this rule, however. Another person having a claim to the instrument can stop the discharge either by supplying an indemnity deemed adequate by the person seeking the discharge or by obtaining a court order in which the holder and the other person are parties, enjoining payment or satisfaction of the instrument. The purpose of this provision is to avoid placing the party making payment in the middle of a dispute between two other parties without providing adequate indemnification for his inconvenience. For instance, suppose Manny Maker makes out a promissory note payable to Paula Payee. William Wrongdoer induces Payee to negotiate the note to him by fraud. Before the note matures, Payee notifies Maker of the fraud and requests that he not pay the note. Wrongdoer then presents the note to Maker for payment. When Maker pays the note, he is discharged because he paid the holder. In order to stop the payment, Payee would need a court order enjoining payment or she must supply Maker with an indemnity deemed adequate by Maker.

There are two other exceptions to the fact that payment or satisfaction given to the holder normally discharges the person making the payment. First, the liability of a party is not discharged if he "*in bad faith* pays or satisfies a holder who required the instrument by theft or who (unless having the rights of a holder in due course) holds through one who so acquired it." Section 3–603(1)(a). Normally, if one pays knowing that the instrument was stolen, his bad faith is established and he is not discharged. Second, a party who pays or satisfies the holder of an instrument in a manner inconsistent with the terms of a restrictive indorsement is not discharged from liability. Section 3–603(1)(b). An exception to this latter rule is the case of a payment by an intermediary bank or payor bank that is not a depositary bank.

(2) Tender of Payment. If one tenders full payment of an instrument to a holder when it is due or later, he is discharged to the extent of all subsequent liability for interest, costs, and attorney's fees. Section 3–604(1). If the law were otherwise, one could continue to collect interest, even at a high rate, as long as he desired merely by refusing to accept the tender. Note, however, that the obligation of the person making tender is not discharged.

Another result of the holder's refusal to accept a valid tender is to wholly discharge any party who has a right of recourse against the party making tender. Section 3–604(2). For example, suppose Manny Maker signs a note payable to Paula Payee, who indorses it over to Harry Holder. When the note is due, Maker tenders payment to Holder, who refuses payment. Maker is discharged as to future interest, costs, and attorney's fees. Payee is completely discharged because she is a party who had recourse against Maker. Had Harry accepted the payment, Payee would have been discharged from liability anyway. Why should Payee suffer any inconvenience because of Harry's refusal to accept the tender?

The code provides that unless the instrument is payable on demand, the fact that the maker or acceptor "is able and ready to pay at every place of payment specified in the instrument when it is due" is the equivalent to tender. Section 3–604(3).

(3) Cancellation and Renunciation. If the holder desires to allow a party who is liable on an instrument to be discharged from liability on the instrument, he may discharge that party even without consideration. This may be accomplished in any manner apparent on the face of the instrument. The instrument may be intentionally canceled, or the party's signature may be destroyed, mutilated, or crossed out. Section 3–605(1)(a). The same result may also be accomplished if the holder renounces his rights in a writing that he signs and delivers. He may also renounce his rights on the instrument by surrendering it to the party to be discharged. Section 3–605(1)(b).

(4) Reacquisition. There are a number of actions that may result in the discharge of the intervening parties in a chain of transferees. One such situation is the case in which an instrument is returned to or reacquired by a prior party. When that occurs, the reacquirer may cancel any indorsement that is not necessary to his title, and any intervening party is discharged as against the reacquired instrument except as to holders in due course. If an intervening party's indorsement has been canceled, he is discharged against holders in due course as well. Section 3–208. This rule makes sense because if the party who reacquires the instrument tries to enforce it against intervening parties, they normally would have an action against him in his status as a previous holder. For example, suppose Albert Able negotiates an instrument to Barry Baker, who in turn negotiates it to Connie Carter, who negotiates it back to Albert Able. Baker and Carter would then be discharged by Able's reacquisition. If Able were then to give the instrument to Harry Holder, Able and Baker would be discharged as to him unless he was a holder in due course. If Able canceled Baker's indorsement, Baker would be discharged even as to a subsequent holder in due course.

(5) Impairment of Recourse or of Collateral. There may be occasions on which the holder of an instrument does not desire to sue or otherwise pursue payment from one party to an instrument. However, if he were to release such a party, it would be unfair to other parties to the instrument. The code therefore states that the holder of an instrument discharges any party to an instrument to the extent that without that party's consent the holder "without express reservation of right releases or agrees not to sue any person against whom the party has to the knowledge of the holder a right of recourse or agrees to suspend the right to enforce against such person the instrument or collateral. . . ." Section 3–606(1)(a). The other parties may, of course, consent to the release of one party and thus remain liable on the instrument.

Note that the code does allow a holder to release a party and maintain his rights against others provided that in releasing one party, he reserves his rights against the other parties. The effect of reserving rights against a nonreleased party, however, is to allow that party still to proceed against the party who was released. The holder thus cannot totally release a party on an instrument without also discharging the other parties who also recourse against him. Those parties who do not have a right of recourse against the released party are not discharged. Section 3–606(2)(c).

An example will help clarify these principles. Suppose Manny Maker signs an instrument payable to Paula Payee, who in turn negotiates it to Albert Able. Able negotiates to Bill Baker, who negotiates it to Harry Holder. If Holder releases Baker, Able and Payee are not discharged because they had no recourse against him. However, if Holder releases Payee, Able and Baker are discharged because they did have recourse against her. Holder may release

Payee and reserve his rights against Able and Baker. If Holder then proceeds against Able and Baker, they may look to Payee, as she was not released from Able and Baker's right of recourse against her because of Holder's reservation of rights.

If the holder unjustifiably impairs any collateral for the instrument, then a party who has a right of recourse will be discharged. Section 3–606(1)(b). Suppose Manny Maker borrows $500 from Paula Payee and gives her his promissory note in that amount and his motorboat as security. Paula then negotiates the note to Harry Holder and also gives him the boat, which she holds as collateral. Harry then gives Maker his boat back. Maker then dishonors the note. Harry cannot recover from Paula on her secondary liability because he impaired the collateral to which Paula would have been entitled had she been required to pay the instrument.

PEOPLES BANK OF POINT PLEASANT v. PIED PIPER RETREAT, INC.

209 S.E.2d 573 (W.Va. 1974)

Frank R. Curatolo and P. A. Sayre were the sole stockholders of Pied Piper Retreat, Inc., a corporation that purchased a motel. In order to finance that purchase, they borrowed money from the Small Business Administration. This loan was secured by a security interest in the motel, coupled with a clause covering any after-acquired property.

Subsequently, in April 1968, Curatolo and Sayre negotiated a loan with the Peoples Bank of Point Pleasant in order to purchase television sets for the motel. A note was prepared stating ". . . we promise to pay . . ." and was signed "Pied Piper Retreat, Inc., by Frank R. Curatolo." *On the reverse side the note was also signed by Frank R. Curatolo and P. A. Sayre. Sometime later the Small Business Administration took possession of all the property, including the television sets, because of a default in payments. The bank's claim of priority for the television sets was disallowed because it had failed to perfect its security interest by filing a notice of it.*

The bank then sued Pied Piper, Curatolo, and Sayre for the balance due on the loan. Curatolo and Sayre contended that they were discharged from the note because the bank's failure to perfect its security interest impaired the collateral. The bank contended that the failure to perfect a security interest was not such an act as would entitle Curatolo and Sayre to this defense and that in any event the defense was not available to them because they had signed as primary parties to the instrument and not as accommodation parties or indorsers. From a judgment in favor of the bank, Curatolo and Sayre appealed.

SPROUSE, J.

The most serious issue presented on this appeal is whether the appellants should be excused from paying the balance due on the note because the bank "impaired the collateral" by not perfecting its lien on the television sets.

There is no doubt that the bank could have perfected its lien on the television sets so that it would have priority over the Small Business Administration lien, if the bank had filed the security instruments in accordance with the provisions of Code. . . .

There is likewise no doubt that the failure of the bank to file the financing statement at the office of the secretary of state resulted in its losing its lien's priority. That raises the question whether such loss, caused by the bank's neglect, releases Curatolo and Sayre from obligation on the note.

Code, 1931, 46–3–606, as amended, provides:

> (1) The holder discharges any party to the instrument to the extent that without such party's consent the holder
>
> . . .
>
> (b) unjustifiably impairs any collateral for the instrument given by or on behalf of the party or any person against whom he has a right of recourse.

In considering whether the defendants signed the note in such manner as to avail themselves of this defense, we must first determine whether the bank's failure to perfect the lien is an action which unjustifiably impairs any collateral. The few courts considering the question are not in agreement as to the applicability of the "impairment of collateral" defense to a transaction such as is involved here. The official comment to Code, 46–3–606, in discussing what action constitutes "unjustifiable" impairment, refers to Code, 46–9–207. That section relates to "Rights and duties when collateral is in secured party's possession," and governs direct impairment of the security rights in the collateral. A number of courts have interpreted this provision of the Uniform Commercial Code in this manner. Some courts have held, however, that such impairment of the security rights in collateral as was involved here is an unjustifiable impairment of collateral as contemplated by Code, 46–3–606. The better logic and the purposes of the Uniform Commercial Code would seem better served by holding that failure to so perfect a lien is an unjustifiable impairment as contemplated by Code, 46–3–606.

The appellant concedes that this defense of unjustifiable impairment of collateral is available only to secondary or accommodation parties. This is undoubtedly the rule—the defense is available to both secondary parties and accommodation parties, whether the latter are secondary accommodation parties such as accommodation endorsers, or primary accommodation parties such as accommodation makers. The defense is not available to principal debtors, i.e., makers. See the official comment to Code, 46–3–606, where it is indicated that this is a suretyship defense available to a party who has recourse against another party to the instrument.

The issue evolves, therefore, into the determination of the capacity of the defendants Curatolo and Sayre, in signing the notes. If they signed as accommodation parties, the defense is a valid one. If they signed the note as principal debtors, i.e., makers, and not as an accommodation maker or endorser, however, the defense of "impairment of collateral" is not available to them.

An accommodation party is defined in Code, 1931, 46–3–415(1), as amended, as: ". . . (O)ne who signs the instrument in any capacity for the purpose of lending his name to another party to it." Under Code, 1931, 46–3–415(2), as amended, an accommodation party is liable in the capacity in which he has signed. He may, therefore, be an accommodation maker, acceptor or endorser, etc.

It is clear, however, from the face of the negotiable instrument here that the appellants intended to sign the note other than as endorsers. The note, containing the language ". . . we promise to pay . . . ," while evidencing both a non-corporate and plural promise, bore only the signature of the corporation on the front of the instrument. Where it is clear from the face of a negotiable instrument that the parties signed in a capacity other than as endorsers, but there is a dispute as to which capacity, parol evidence is admissible to show the intention of the parties as to the capacity in which the instrument was signed. *First Bank and Trust Company* v. *Post,* supra.

Testimony concerning the capacity in which the appellants signed the note was so conflicting that a trier of fact could only have resolved it by believing some witnesses and disbelieving others. The trial court apparently chose to believe the evidence for the bank that the loan was made directly to the two individual defendants as principal debtors (makers). This evidence, together with the evidence concerning the structure of the corporation, with active solicitation of the loan from the bank by the individual defendants, and the position of the defendant, Sayre, certainly comprises evidence which the trial court could have properly considered in arriving at a finding of fact as to the intention of the parties concerning the capacity in which they signed the note.

We cannot say in reviewing the evidence in this case that the trial court was wrong in finding that Curatolo and Sayre were neither secondary nor accommodation parties.

(Affirmed.)

QUESTIONS FOR DISCUSSION

1. *Does the failure of a secured party to perfect his security interest constitute an unjustifiable impairment of collateral? Does this appear to be the law in all states?*
2. *Is the defense of discharge by unjustifiable impairment of collateral available to an indorser? Is it available to an accommodation maker?*

---

(6) Other Reasons for Discharge. Section 3–601(1) of the code summarizes the various circumstances under which a party may be discharged from an instrument. Most of them have been discussed in this section or elsewhere. For instance, we have seen that unexcused delay in presentment, notice, or necessary protest may result in discharge from liability on an instrument. If the holder of a draft assents to an acceptance of it that varies the terms of the draft, the drawer and the indorsers who do not assent to the draft are discharged. Section 3–412(3). Also, if a holder materially alters an instrument in a fraudulent manner and the alteration changes the contrast of a party to the instrument, then that party is discharged as against any person other than a holder in due course. Section 3–417(2)(a). Finally, if the holder of a check procures certification of a check by the drawee bank, this certification operates as an acceptance and discharges all prior indorsers and the drawer. Section 3–411(1).

SUMMARY

A party who signs an instrument is either a primary party or a secondary party. The maker of a note is primarily liable. No demand is necessary to fix his liability. The drawee of a draft is primarily liable only after he has accepted it.

The drawer of a draft is secondarily liable. He must pay the instrument after it has been dishonored and notice of dishonor has been given. An indorser is also secondarily liable. He is liable to the holder and any subsequent indorser who takes the instrument.

Accommodation parties are liable according to the capacity in which they indorse the instrument. One may accommodate an indorser, drawer, acceptor, or maker. Accommodation parties also have the right of subrogation and exoneration.

A guarantor of an instrument agrees to pay the instrument when it is due. Demand, presentment, and notice of dishonor are waived. On the other hand if one guarantees collection, a number of conditions must be met before payment is due from the person making the guarantee.

One who deals with an instrument may give certain warranties. One group of warranties, called transferor's warranties, is given by the transferor of the instrument regardless of whether the instrument is negotiated by indorsement or simple delivery or is transferred by assignment. The other group of warranties are known as presenter's warranties. These warranties are given the party who obtains payment or acceptance to the person who pays or accepts the instrument.

Any instrument may be discharged by payment or satisfaction. Tender of payment to a holder discharges the party making the tender of all subsequent liability for interest, costs, and attorney's fees. An instrument may be discharged by cancellation and renunciation. No consideration is necessary.

Reacquisition occurs when an instrument is returned to a prior party. If the reacquirer cancels an indorsement, any intervening party is discharged as against the reacquired instrument except as to holders in due course. Also, a holder of an instrument discharges any party to any instrument if the collateral or recourse is impaired unless there is an express reservation of rights.

PROBLEMS

1. Henry F. Billingsley, Paul J. Kelly, and Jack R. Huffner formed Urban Systems, Inc., for the purpose of procuring private and government contracts to perform studies of urban problems. Billingsley, Kelly, and Huffner each owned one third of the corporation. They each loaned the corporation money. In return for his loan the corporation signed a note payable to Billingsley and indorsed by Kelly and Huffner. The corporation also executed a note to Kelly indorsed by Huffner and Billingsley, as well as a note to Huffner indorsed by Kelly and Billingsley.

   When the note to Billingsley was not paid, he sued Huffner and Kelly on their indorsements. The trial court found in favor of Billingsley on his notes but also in favor of Huffner and Kelly on the notes payable to them, which Billingsley had indorsed. Unfortunately, for Billingsley, the amount owed by the corporation to Huffner and Kelly exceeded the amount the corporation owed Billingsley. Billingsley appealed the judgment, contending, among other things, that because he was the last indorser on each of the notes, he was not liable to either of them on his indorsement contract.

   Was Billingsley liable to Huffner and Kelly? Billingsley v. Kelly 174A.2d 113 (Md. 1971).
2. In October 1964, Fred Rzepka, a customer of Society National Bank of Cleveland (hereinafter Society) drew two checks totaling $4,500 on Society, payable to the ABS Company, and delivered them to William Mishler, a selling agent of the payee company. Mishler forged the payee's indorsements on the two checks, signed his own name, and in exchange obtained from Society two cashier's checks, payable as before to the ABS Company. He also forged the indorsements on the cashier's checks and deposited them in an account titled "Windows, Inc.," which he maintained at defendant Capital National Bank (hereinafter Capital). Capital transferred the checks to defendant Union Commerce Bank, which in turn presented them to Society for payment. Society paid the checks. After more than a year, Fred Rzepka informed Society that the indorsements had been forged. Whereupon Society promptly notified Union Commerce, the presenting bank, and Capital, the depositary bank, of the forgeries. Society reimbursed its customer's account and brought this action against defendant–appellant banks to recover the amount paid.

   Is Society entitled to benefit from any of the warranties and if so which one? Society National Bank of Cleveland v. Capital National Bank 281 N.E.2d 563 (Ohio App. 1972).
3. Jelks Taylor was a subcontractor for Deen and Yarborough, who were part-

ners in the contracting business. Taylor owed DeSoto National Bank of Arcadia $2,500. Thinking that they were indebted to Taylor, Deen and Yarborough allowed Taylor to give the bank a draft in that amount drawn on them with interest at the rate of 8 percent per annum. When the bank presented the note to them, Deen and Yarborough accepted it. However, when the maturity date arrived, they refused to pay, having discovered that Taylor was in fact indebted to them. Were Deen and Yarborough liable to the bank? Was Taylor liable to the bank? Deen et al. v. DeSoto National Bank of Arcadia, 97 Fla. 862, 122 So. 105 (1929).

4. Universal Lightning Rod, Inc., held a note in the amount of $590.00 signed as follows: "Rischall Electric Co., Inc., Harold M. Rischall." Rischall contended that he was not liable on the note, having signed in only a representative capacity. Was he correct? Universal Lightning Rod, Inc. v. Rischall Electric Co., 192 A.2d 50 (Conn. 1963).
5. Joseph Mobilla held a note that he said had been signed by Theresa Piotrowski as maker when she purchased a used Ford from him. There was nothing on the face of the note to indicate that Piotrowski had not signed it. Mobilla then indorsed the note to the Union Bank "without recourse" as part of a financing plan. When the bank presented the check to Piotrowski, she denied having signed the note and in fact had not purchased a car from Mobilla. The bank sued Mobilla. He admitted that Piotrowski's signature was a forgery but denied liability on the ground that he indorsed "without recourse." Was he correct? The Union Bank v. Mobilla, 43 Erie Co. Leg. J. 45 (Pa. 1959).
6. Two drafts were drawn on John Price, payable on November 22, 1790, and February 1, 1761, respectively. These drafts were indorsed to Edward Neal, who presented them to Price for payment. Price paid both drafts but then discovered that both drafts were forged by one Lee, who had been hanged for the forgery. Price sued Neal to recover for the forgery of the drawer's signature. Was Price entitled to recover? Price v. Neal, 3 Burr. 1354, 97 Eng. Rep. 871 (K.B. 1762).
7. Neuse Engineering & Dredging Co. drew a check payable to the Kirbys in the amount of $2,500.00. The check was drawn on the First and Merchants National Bank, where the Kirbys also had an account. The Kirbys indorsed and deposited the check, and their account was credited by the bank. The bank then discovered that the Neuse Engineering & Dredging Co. did not have sufficient funds to cover the check. Were the Kirbys liable to the bank? Kirby v. First Merchants Bank, 210 Va. 88, 168 S.E.2d 273 (1969).
8. Hubert U. Moore executed a note payable to Lawrence Held in the amount of $15,000.00 with interest at 6 percent. At that time Lois Moore signed her name on the back of the note under the following indorsement:

   For value received the undersigned and each of them hereby forever waives presentment, demand, protest, notice of protest, and notice of dishonor of the within note and the undersigned and each of them guarantees the payment of said note at maturity and consents without notice to any and all extension of time or terms of payment made by holder of said note.

Hubert Moore refused to pay the note. The note was then presented to Lois Moore, who also refused to pay. Held then sued Lois Moore, who admitted signing the note but denied knowing in what capacity she had signed. She also contended that she had received no notice of dishonor or demand for payment. Was Lois Moore liable to Held? Held v. Moore, 59 Lancaster Law Review III (Pa. 1964).

CHAPTER 27

Checks and the Relationship of the Bank to Its Customers

Most of what has been said in the previous chapters regarding commercial paper also applies to the checks drawn on the drawer's bank. There is no need to repeat those principles here. However, there are certain characteristics unique to the relationship between banks and their customers when checks are involved. The Uniform Commercial Code devotes a special section to this topic. The approach of this chapter will be first to examine the character of the relationship between the bank and its customer. Then the process of the collection of checks will be examined, first from the standpoint of the depositary bank and then from the opposite side of the transaction, the payor bank. Finally, some specific rights and duties of the bank and its customers will be discussed. Our discussion will be limited insofar as possible to concepts not covered in the previous chapters.

27:1. Character of the Relationship of the Bank to Its Customer. Normally, when one opens a checking account at a bank, he deposits money with the bank to which the bank then has title and uses as part of its general funds. The depositor and the bank enter into a contractual relationship as a result of this transaction. The contract between the bank and its customer is usually expressed on a signature card that the customer signs. The terms of the contract generally incorporate the customary banking practices that have been established over time. Other terms of the contract, in addition to stating any changes, usually state no more than the law that would govern the relationship if there is no express agreement. This relationship is governed by the provisions of the code; however, the parties are free to alter the impact of the code by agreeing to other provisions in the contract. One exception is that there can be no provision disclaiming a bank's responsibility for its own lack of good faith or failure to exercise ordinary care. In addition, the contract cannot limit the measure of damages should the bank act without ordinary care or in good faith. However, the parties may determine the standards by which the bank's responsibility is to be measured, provided that the standards are not unreasonable. Section 4–103(1).

When the customer deposits money in an account with the bank, there arises a relationship of creditor–debtor. The depositor is the creditor and the bank the debtor. There is also a principal–agent relationship. The bank is the agent of the customer for the purpose of paying checks, collecting items deposited in the customer's account, and so on. At this point it is appropriate to examine the collection process, first from the standpoint of the collecting bank and then from the payor bank's viewpoint.

27:2. The Process of Collection of Items. In the process of collection a bank may fall into one of several categories. It may be a *depositary bank,* which means the first bank to which an item is transferred for collection. Thus, when a depositor deposits his pay check in his checking account with Deposit Bank, that bank is a *depositary bank.* The term *payor bank* means a bank by which an item is payable as drawn or accepted. *Intermediary bank* means any bank to which an item is transferred in the course of collection except the depositary or payor bank. Finally a *collecting bank* is any bank handling the item for collection except the payor bank. Section 4–105. Thus the collecting bank and the depositary bank may be one and the same.

27:3. Depositary Bank and Collecting. (1) Right of Charge Back. In the normal course of events the customer deposits a check in his account with the depositary bank and receives *provisional* credit or settlement for the check. The check is then sent through banking circles to be collected from the payor bank. Each bank through which the check passes gives a similar provisional credit. When the check arrives at the payor bank, the payor bank then pays the item and all the provisional credits become final. If a check is not paid when presented to the payor bank, then the provisional credits are reversed until the check is finally deducted from the customer's account with the depositary bank. The customer must then seek redress from the drawer or indorser of the check. Section 4–212(1). The code makes clear that a collecting bank is an agent for the owner of the item, and therefore the settlement given for an item is only provisional. Section 4–210(1).

The customer of a depositary bank has no right to draw against the provisional credit until it becomes final. Sometimes funds are drawn against checks that have been only provisionally credited either because the bank has made a mistake or because the depositor is a valued customer of the bank. The bank still has the right of charge back, however, when this is done. Section 4–212(4).

DOUGLAS v. THE CITIZENS BANK OF JONESBORO

244 Ark. 168, 424 S.W.2d 532 (1968)

HARRIS, J.

This litigation involves two separate causes of action, which however, by agreement, were set forth in one set of pleadings, and disposed of at one hearing. Appellants, Weldon Douglas, and Janie Chandler, each maintained a checking account in the Citizens Bank of Jonesboro. Rees Plumbing Company, Inc. (which is not presently a party to this proceeding), was a customer of the bank, and maintained checking accounts. On August 19, 1966, the plumbing company delivered its check in the amount of $1,000.00 to Douglas. On that same day Douglas presented the check to the bank for deposit to his own checking account; an employee at the teller's window prepared a deposit slip, dated as of that day, reflecting that the check was being deposited to Douglas' account. He was given a duplicate of the deposit slip, and an employee of the bank thereafter affixed to the back of the check a stamp in red ink, denoting the August 19th date, and stating, "Pay to any bank—P.E.G., Citizens Bank of Jonesboro, Jonesboro, Arkansas." Under date of August 20, 1966, the bank dishonored the check because of insufficient funds, and charged the amount back to the account of Douglas. This same statement of facts applies to Mrs. Chandler, except that the check she presented was originally made payable to a Richard R. Washburn (in the amount of $1,600.00) by the same Rees Company, and this check had been properly endorsed by Washburn before coming into the hands of Mrs. Chandler.

. . . After first demurring, and moving to make the complaint more definite and certain, the bank filed an answer setting out that the accounts of Rees were insufficient on August 19 to honor the checks, and further, that both were charged back to the accounts of the respective appellants on August 20, and the appellants so notified. The bank further denied that the endorsement stamp, heretofore mentioned, constituted an acceptance stamp. The bank asserted that the stamp was no more than a method of identification. . . .

The principal question at issue is, "Did the bank, by stamping the endorsement upon the checks deposited by appellants, and by delivering to appellants the deposit slips, accept both of said checks for payment?" The answer is, "No.". . . This case is controlled by the following sections of the Code: Ark.Stat.Ann. § 85–4–212(3), § 85–4–213, and § 85–4–301(1) (Add.1961).

Subsection (3) of Section 85–4–212 reads as follows:

> A depositary bank which is also the payor may charge back the amount of an item to its customer's account or obtain refund in accordance with the section governing return of an item received by a payor bank for credit on its books (Section 4–301 (§ 85–301)).

Subsection (1) of Section 85–4–301 provides:

> Where an authorized settlement for a demand item (other than a documentary draft) received by a payor bank otherwise than for immediate payment over the counter has been made before midnight of the banking day of receipt the payor bank may revoke the settlement and recover any payment if before it has made final payment (subsection (1) of Section 4–213 (§ 85–213)) and before its midnight deadline it
> (a) returns the item; or
> (b) sends written notice of dishonor or nonpayment if the item is held for protest or is otherwise unavailable for return.

Section 85–4–213 simply sets out the time that a payment becomes final, not applicable in this instance.

When we consider the statutes above referred to, it is clear that appellants cannot prevail. Clark, Bailey and Young, in their American Law Institute pamphlet on bank deposits and collections under the Uniform Commercial Code (January, 1959), p. 2, comment as follows:

> If the buyer–drawer and the seller–payee have their accounts in the same bank, and if the seller–payee deposits the check to the credit of his account, his account will be credited provisionally with the amount of the check. In the absence of special arrangement with the bank, he may not draw against this credit until it becomes final, that is to say, until after the check has reached the bank's bookkeeper and, as a result of bookkeeping operations, has been charged to the account of the buyer–drawer. (The seller–payee could, of course, present the check at a teller's window and request immediate payment in cash, but that course is not usually followed.) If the buyer–drawer's account does not have a sufficient balance, or he has stopped payment on the check, or if for any other reason the bank does not pay the check, the provisional credit given in the account of the seller-payee is reversed. If the

seller–payee had been permitted to draw against that provisional credit, the bank would recoup the amount of the drawing by debit to his account or by other means.

The comment of the commissioners is also enlightening. Comment 4, under Section 85–4–213, states:

A primary example of a statutory right on the part of the payor bank to revoke a settlement is the right to revoke conferred by Section 4–301. The underlying theory and reason for deferred posting statutes (Section 4–301) is to require a settlement on the date of receipt of an item but to keep that settlement provisional with the right to revoke prior to the midnight deadline. In any case where Section 4–301 is applicable, any settlement by the payor bank is provisional solely by virtue of the statute subsection (1) (b) of Section 4–213 does not operate and such provisional settlement does not constitute final payment of the item.

(Affirmed.)

QUESTION FOR DISCUSSION

1. *At what time does the provisional credit granted on an item by the bank become final?*

---

(2) **Time of Receipt of Items.** The code provides that a bank may fix an afternoon hour of 2 P.M. or later as a cutoff time for the handling of money and items and making entries on its books. If the item is received after the time fixed for cutoff, such as 2 P.M., then that item is normally treated as being received at the opening of the next banking day. Section 4–107. Thus a check deposited with the bank at 4 P.M. on Wednesday would not be provisionally credited to the depositor's account until Thursday morning.

(3) **Customer's Missing Indorsement.** The code also allows a depositary bank to supply any indorsement of its customer that may be necessary for title when the depositary bank takes an item for collection. Section 4–205(1). Thus, if a depositor forgets to indorse his check, the depositary bank may supply this indorsement in order to collect the check. It would not have to return the check to its customer for indorsement.

(4) **Responsibility for Collection.** Because a collecting bank is an agent for the owner of the item, it must use ordinary care in the process of collection, such as presenting items or sending items for presentment, giving notice of dishonor, and so on. Section 4–202(1). The code states what actions the bank must take and when. For instance, a collecting bank acts seasonably if it takes proper action prior to its *midnight deadline.* If it takes action after that time, it may still act seasonably, but the bank then has the burden of establishing this. Section 4–202(2). As a general rule, a collecting bank must send items by a reasonably prompt method, taking into consideration the nature of the item, the costs of collection, any instructions, and the methods generally used to present such items. Section 4–204(1). The code specifies no specific methods that the bank must use because of the many types of methods available.

(5) **Warranties.** Section 4–207 of the code provides that a customer or a collecting bank must give substantially the same warranties as were discussed in Chapter 26. Sections 3–417 and 3–414. This section also provides that unless a claim for breach of warranty is made within a reasonable time after the person claiming learns of the breach, the person who is liable for the breach of warranty is discharged to the extent of any loss he suffers because of the

delay in the claim's being made against him. The most common illustration of the impact of this section would be the case in which a party who is liable might have recovered from another person who went bankrupt during the period of the delay. If, however, the bankruptcy occurred prior to the time when the claimant knew of the breach of warranty, the party liable would not be discharged to any extent because the delay did not result in the loss.

The code states a number of rules that govern the manner in which the payor bank may handle items presented to it for payment. It is not necessary for our purposes to go into these matters in detail, but several code provisions in this matter are worth highlighting.

(6) Posting. Section 4–301(1) of the code provides for the time when a payor bank may revoke any provisional settlement. Because of the great increase in collections, the code approves the practice of deferred posting and delayed returns. For instance, the payor bank will sort and prove items on the day they are received, for example, Monday. However, the payor bank does not post the items to its customer's account or return "not good" items until the next day, in this case Tuesday. Section 4–301, Comment 1. The purpose of this practice is to even the work load for the bank and avoid abnormal peak-load periods. Thus this section provides that if an authorized settlement for a demand item (such as a check) is received by a payor bank—otherwise than for immediate payment over the counter—before midnight of the banking day of receipt, the payor bank may revoke the settlement and recover any payment if several conditions are met. First, the payor bank must act before it has made final payment and before its midnight deadline. The action it must take is to return the item or give written notice of dishonor or nonpayment if the item is held for protest or otherwise unavailable for return. If the bank does not comply with the time limits outlined in Section 4–301(1), then it is accountable to its customer for the amount of the item. The bank is not liable, however, if it has a valid defense, such as breach of a presentment warranty. Section 4–302.

The payor bank may not honor an item for many reasons, all of which may freeze the funds in the account or eliminate them. For instance, the bank may have received notice of a stop order. It may have a right of setoff against the drawer's account, which it exercises, or it may be served with legal process, such as a garnishment or attachment of the drawer's account. Section 4–303(1) states various rules when knowledge of these actions or rights come too late to suspend or modify the bank's duty to pay the item. When these events occur, the item has priority and a charge may be made to the customer's account. Such knowledge or notice comes too late if a reasonable time for the bank to act has expired or the setoff is exercised after the bank has done any of the following:

(a) accepted or certified the item;
(b) paid the item in cash;
(c) settled for the item without reserving a right to revoke the settlement and without having such right under statute, clearing house rule or agreement;

(d) completed the process of posting the item to the indicated account of the drawer, maker or other person to be charged therewith or otherwise has evidenced by examination of such indicated account and by action its decision to pay the item; or

(e) become accountable for the amount of the item under subsection (1) (d) of Section 4–213 and Section 4–302 dealing with the payor bank's responsibility for late return of items.

Sometimes a number of items may arrive at the bank on the same day by a number of different ways. If the customer does not have enough funds on hand to pay all the items, the bank may charge them against his account in any order convenient to the bank. Section 4–303(2). If the customer has drawn all the checks, it is his responsibility to have the funds available to pay all of them, and he has no right to request that one item be paid before another.

BLAKE v. WOODFORD BANK & TRUST CO.

555 S.W.2d. 589 (Ky. App. 1977)

K & K Farm was the drawer of two checks on the Woodford Bank and Trust Co. totaling $27,649.84. The payee of these checks was Wayne Blake who deposited them in his account at the Morristown Bank for collection on December 19, 1973.

The Morristown Bank forwarded the two checks for collection through the Cincinnati Branch of the Federal Bank of Cleveland. From the Federal Reserve Bank, the two checks were delivered to the Woodford Bank by means of the Purolator Courier Corp. The check arrived at the Woodford Bank on December 24, 1973, shortly before the bank opened for business. The two checks were returned by the Woodford Bank to the Federal Reserve Bank by means of Purolator on Thursday, December 27, 1973, because of insufficient funds. Blake contends that Woodford Bank is accountable for the checks because it retained them beyond its midnight deadline. Woodford Bank contends that the delay was excused.

PARK, J.

The Model Deferred Posting Statute was the basis for the provisions of the Uniform Commercial Code. Under § 4 301(1) of the Uniform Commercial Code (UCC), a payor bank may revoke a provisional "settlement" if it does so before its "midnight deadline" which is midnight of the next banking day following the banking day on which it received the check. Under the Model Deferred Posting Statute, the payor bank's liability for failing to meet its midnight deadline was to be inferred rather than being spelled out in the statute. Under UCC § 4–302, the payor bank's liability for missing its midnight deadline is explicit. If the payor bank misses its midnight deadline, the bank is "accountable" for the face amount of the check.

Like the Model Deferred Posting Statute, the Uniform Commercial Code seeks to decrease, rather than increase, the risk of liability to payor banks. By permitting deferred posting, the Uniform Commercial Code extends the time within which a payor bank must determine whether it will pay a check drawn on the bank. Unlike the Bank Collection Code or the Uniform Negotiable Instruments Law as construed by most courts, the Uniform Commercial Code does not require the payor bank to act on the day of receipt or within 24 hours of receipt of a check. The payor bank is granted until midnight of the next business day following the business day on which it received the check.

EXCUSE FOR FAILING TO MEET MIDNIGHT DEADLINE

UCC § 4–108(2) provides:

> Delay by a . . . payor bank beyond time limits prescribed or permitted by this Act . . . is excused if caused by interruption of communications facilities, suspension of payments by another bank, war, emergency

conditions or other circumstances beyond the control of the bank provided it exercises such diligence as the circumstances require.

The basic facts found by the circuit court can be summarized as follows: a) the bank had no intention of holding the checks beyond the midnight deadline in order to accommodate its customer; b) there was an increased volume of checks to be handled by reason of the Christmas Holiday; c) two posting machines were broken down for a period of time on December 26; d) one regular bookkeeper was absent because of illness. Standing alone, the bank's intention not to favor its customer by retaining an item beyond the midnight deadline would not justify the application of § 4–108(2). The application of the exemption statute necessarily will turn upon the findings relating to heavy volume, machine breakdown, and absence of a bookkeeper.

Because of the cumulative effect of the heavy volume, machine breakdown and absence of a regular bookkeeper, the bank claims it was unable to process the two checks in time to deliver them to the courier from Purolator for return to the Federal Reserve Bank on December 26. As the bank's president testified:

> Because we couldn't get them ready for the Purolator carrier to pick them up by 4:00 and we tried to get all our work down there to him by 4:00, for him to pick up and these two checks were still being processed in our bookkeeping department and it was impossible for those to get into returns for that day.

The increased volume of items to be processed the day after Christmas was clearly foreseeable. The breakdown of the posting machines was not an unusual occurrence, although it was unusual to have two machines broken down at the same time. In any event, it should have been foreseeable to the responsible officers of the bank that the bookkeepers would be delayed in completing posting of the checks on December 26. Nevertheless, the undisputed evidence establishes that no arrangements of any kind were made for return of "bad" items which might be discovered by the bookkeepers after the departure of the Purolator courier. The two checks in question were in fact determined by Mrs. Stratton to be "bad" on December 26. The checks were not returned because the regular employee responsible for handling "bad" checks had left for the day, and Mrs. Stratton had no instructions to cover the situation.

Even though the bank missed returning the two checks by the Purolator courier, it was still possible for the bank to have returned the checks by its midnight deadline. Under UCC § 4–301(4)(b) an item is returned when it is "sent" to the bank's transferor, in this case the Federal Reserve Bank. Under UCC § 1–201(38) an item is "sent" when it is deposited in the mail. I. R. Anderson, *Uniform Commercial Code* § 1–201 pp. 118–119 (2d ed, 1970). Thus the bank could have returned the two checks before the midnight deadline by the simple procedure of depositing the two checks in the mail, properly addressed to the Cincinnati branch of the Federal Reserve Bank.

This court concludes that circumstances beyond the control of the bank did not prevent it from returning the two checks in question before its midnight deadline on December 26. The circumstances causing delay in the bookkeeping department were foreseeable. On December 26, the bank actually discovered that the checks were "bad," but the responsible employees and officers had left the bank without leaving any instructions to the bookkeepers. The circuit court erred in holding that the bank was excused under § 4–108 from meeting its midnight deadline. The facts found by the circuit court do not support its conclusion that the circumstances in the case were beyond the control of the bank.

(Judgment for Blake.)

Questions for Discussion

1. *By what date should Woodford Bank have dishonored the checks?*
2. *Why did the court fail to excuse the bank's delay based upon the volume of business and breakdown of the posting machines?*

27:4 Relationship Between Payor Bank and Its Customer. The payor bank has a duty to its customer to pay items that are properly presented to it. If it dishonors an item when it has sufficient funds on hand in the customer's account to pay it, the bank is said to dishonor the item wrongfully. Wrongful dishonor does not include a permitted or justified dishonor, such as when the item is lacking a proper indorsement. If the bank does wrongfully dishonor an item, it is liable to its customer for damages proximately caused by the wrongful dishonor. As long as the dishonor results from a mistake on the part of the bank, the damages for which the bank is liable are limited to actual damages proved. That is, the customer must prove some actual financial loss as a result of the dishonor. Damages that result from arrest or prosecution of the customer and other consequential damages may be recovered if proximately caused by the wrongful dishonor. Section 4–402. The reason for this damage provision is that before the code, if the wrongful dishonor was of the check of a merchant, this constituted a per se defamation of his business and he could recover damages for the defamation without regard to the actual damages suffered as a result of the defamation.

On the other hand, a bank may charge its customer's account and pay a check even if there are insufficient funds in the account and payment of the item creates an overdraft. Section 4–401(1). If the bank makes payment in good faith to a holder, it may charge its customer's account according to the original tenor if the item was altered. It may also charge the account of its customer according to the tenor of his completed item unless the bank has notice that the completion was improper. Mere knowledge than an item was completed would not prevent the bank from charging the customer's account. Section 4–401(2).

BANK OF LOUISVILLE ROYAL v. SIMS
435 S.W.2d 57 (Ky.App. 1968)

CLAY, J.

Appellee (Nancy Sims) recovered $631.50 for the wrongful dishonor of two small checks. This sum included the following items of damage: $1.50 for a telephone call; $130 for two weeks' lost wages; and $500 for "illness, harassment, embarrassment and inconvenience." The trial court, trying the case without a jury, found that the dishonor was due to a mistake and it was not malicious.

KRS 355.4–402 provides:

> A payor bank is liable to its customers for damages *proximately caused* by the wrongful dishonor of an item. *When the dishonor occurs through mistake liability is limited to actual damages proved.* If so proximately caused and proved damages may include damages for an arrest or prosecution of the customer or other consequential damages. Whether any consequential damages are proximately caused by the wrongful dishonor is a question of fact to be determined in each case. (Emphasis added.)

This statute does not define "consequential" damages but it is clear they must be proximately caused by the wrongful dishonor. It appears this statute codifies the common law measure of damages as it heretofore existed in Kentucky. As in other cases of breach of contract, "proximately caused" damages, whether direct or consequential, would be those which could be reasonably foreseeable by the parties as the natural and probable result of the breach.

The plaintiff deposited for her account with appellant a check for $756, drawn on an out-of-town bank. In order to permit such a check to clear, it was apparently customary for the bank to delay crediting the account for a period of three days. By mistake one of appellant's clerks posted a ten-day hold on this check, and during that period two of plaintiff's checks were dishonored and returned with the notation "Drawn Against Uncollected Funds." Apparently she had some difficulty getting the matter straightened out.

The plaintiff had respiratory trouble and, because of it and a case of "nerves," her doctor advised her to take a two-week leave of absence from her place of employment, which she did. She testified she was embarrassed, humiliated and mortified, but her principal complaint seems to be of the difficulty and delay in getting the bank to correct its mistake.

In *American Nat. Bank* v. *Morey,* 113 Ky. 857, 69 S.W. 759, 58 L.R.A. 956, it was held that if there was no basis of punitive damages for the dishonor of a check, recovery cannot be had for humiliation or mortification. It was also held that plaintiff's "nervous chill" was not the natural result of the dishonor or such a thing as could be reasonably anticipated. In *Berea Bank and Trust Co.* v. *Mokwa,* 194 Ky. 556, 239 S.W. 1044, it was recognized that loss of time could be a proper item for damages provided it was the direct and proximate result of the bank's refusal to honor a check.

On the authority of Morey, the plaintiff was not entitled to recover for her hurt feelings or for her "nerves." It follows, therefore, that she was likewise not entitled to recover for her two weeks' lost time from work even if her mental state actually contributed to this loss. From the proximate cause standpoint, these nebulous items of damage bore no reasonable relationship to the dishonor of her two checks and consequently they could not be classified as "actual damages proved." (Had the action of the bank been willful or malicious, justifying a punitive award, damages of this kind might have been recoverable as naturally flowing from this type of tortious misconduct, but we do not have that question here.)

The charge for the telephone call was a proper item of damages.

The judgment is reversed, with directions to enter judgment for the plaintiff in the sum of $1.50.

(1) Customer's Right to Stop Payment. Sometimes the customer of a bank may wish to stop payment of a check he has written. He may wish to stop payment because he has been notified that the check has been lost, because he discovers that he has been defrauded by the person to whom the check was issued, or for any one of a number of other reasons. Stopping payment on a check involves some degree of inconvenience and difficulty for the bank, but the drafters of the code felt that it is a service that is expected by customers and to which they are entitled. Therefore the code provides that a customer may order his bank to stop payment of any item payable for his account. For the order to be effective, the bank must receive it in a manner and at such a time as to afford the bank the opportunity to act on it prior to any action by the bank described in Section 4–303 with respect to the item. Section 4–403(1). Thus notifying a bank by telephone to stop payment on a check two minutes before it is presented to a teller for payment would not give the bank a reasonable opportunity to act on the order.

After the bank has credited a check, the customer has no right to stop payment on it. Furthermore, when the bank certifies a check, it accepts it and becomes primarily liable on the instrument. The bank, therefore, is not required to impair its credit by refusing payment for the convenience of the drawer. Section 4–403, Comment 5. If the holder of a check has obtained certification of it, the drawer is discharged. If the drawer has obtained certification, the bank is primarily liable but the drawer is not discharged. Section 3–411(1).

Most frequently a stop payment is first made orally over the telephone. An oral order to stop payment, whether made over the telephone or otherwise, is binding upon the bank for only fourteen calendar days unless confirmed in writing within that period. A written order is effective for only six months unless renewed in writing. Section 4–403(2).

If the bank pays an item after having received an effective stop-payment order, it is liable to its customer for the wrongful payment. Although it is inevitable that wrongful payment will occur because of the large number of items handled by banks, the code takes the position that this burden should be placed on the banks as a cost of doing business. However, the burden of establishing the fact of loss that results from the payment of an item contrary to a binding stop-payment order is on the customer. Section 4–403(3).

DINERMAN v. NATIONAL BANK OF NORTH AMERICA
390 N.Y.S.2d 1002 (1977)

HYMAN, J.

Man's inventive genius and explorations in the Twentieth Century have brought forth many innovations; and now, one of the most sensational of all such—one which if allowable must join the ranks of the great—is how to gamble at Las Vegas, or some other such place which permits gambling, losing money in and to their palatial casinos, without it costing the loser anything. The theory presented is an interesting and novel one, made more so in that it has not been conceived by any mathematical genius, but rather by a member of the Bar.

Plaintiff (Irving P. Dinerman) alleges that on or about April 2, 1973 and prior thereto, he maintained several checking accounts with defendant bank and that on or about the above date (actually March 30, 1973, three days earlier) he entered into an "agreement" or "condition of the account," a "contract" (affirmation of plaintiff dated Oct. 8, 1976) with defendant, wherein defendant agreed not to make payments drawn on plaintiff's checking accounts unless the instrument being presented for payment was on "printed checks of the bank," but that in September 1975 (two and one-half years later) the defendant violated such alleged "agreement-contract," without plaintiff's permission and paid out of his checking account the sum of $1,000 "on instruments which were not on the printed check forms of defendant bank"; thus causing the plaintiff $1,000 in damages.

The documentary proof indicates that on March 30, 1973 the plaintiff did put his instructions into "written" form and forwarded same to defendant; and such instructions were acknowledged to have been "received" by defendant in April 1973. The instructions read as follows:

> In accordance with your request for written authorization, I wish to advise you not to pay any instruments drawn on the above accounts unless they are printed checks from your bank.
>
> . . .
>
> Yours very truly,
> /s/ Irving P. Dinerman
> *Received* by National Bank of North America this ____ day of April, 1973
> By: [signature unreadable] 4/2/73
> (Emphasis supplied.)

Having left the tables of the Puerto Rican casinos, plaintiff apparently decided to try his luck at the casino tables of the hotels in Las Vegas, Nevada. This he did in August 1975, some two years and four months *after* his so-called prior written instructions. But, to his consternation, he was no better at the tables in his new found gambling haven than he had been at the former palaces of delight, nor, he contends, was his luck any better with the Las Vegas hotel as with the Puerto Rican hotel, insofar as their so-called agreements were concerned, because he did not obtain an "offset" for airfare and other adjustments (also unnamed) from the Las Vegas hotel at whose tables he gambled but lost $1,000, for which he executed two (2) so-called "markers" (actually printed checks) which read substantially as follows:

> To Natl. Bk. of N. America
>
> Customers Check 31502
>
> 911324983 (For Cash Only)
> Your Account Number
>
> N.Y.C., N.Y. Date 8–15–1975
> City and State
>
> PAY TO THE ORDER OF Sahara Hotel
> Five Hundred $500.00
>
> I represent that I have received cash for the above amount and that said amount is on deposit in said bank or trust company in my name, is free from claims and is subject to this check.
>
> Signed: Irving P. Dinerman
>
> (Emphasis supplied.)

Except for the fact that the second "check" (marker) bore number 29871, they both were dated the same date, drawn upon the same bank with the same numbered account, and were for the same amount.

The defendant opposes the plaintiff's cross motion and in support of its primary mention to dismiss the complaint, admits that a customer does have a limited right to make a "stop payment order" by statutory authority under section 4–403 of the Uniform Commercial Code, which reads as follows:

> (1) A customer may by order to his bank stop payment of any item payable for his account but the order must be received at such time and in such manner as to afford the bank a reasonable opportunity to act on it prior to any action by the bank with respect to the item described in Section 4–303.
>
> (2) An oral order is binding upon the bank *only for fourteen calendar days* unless confirmed in writing within that period. A *written order* is effective for only *six months unless renewed in writing.*
>
> (3) *The burden of establishing the fact and amount of loss* resulting from the payment of an item contrary to a pending stop payment order is on the customer. (Emphasis supplied.)

As stated in the Official Comment (Official Comment 7, McKinney's Cons.Laws of N.Y. Anno., Book 62½, Uniform Commercial Code, § 4–403, p. 612):

> The existing statutes all specify a time limit after which any direction to stop payment becomes ineffective *unless it is renewed in writing;* and the majority of them have specified six months. The purpose of the provision is, of course, to facilitate stopping payments by clearing the records of the drawee of accumulated unrevoked stop orders, as where the drawer has . . . settled his controversy with the payee, but has failed to notify the drawee. (Emphasis supplied.)

In the instant matter the bank merely acknowledged "receipt" of the written request (stop payment order) on April 2, 1973; as such it cannot be legally interpreted as a general, unlimited lifetime "agreement", but at best is a receipt of the notice to stop payment within the purview of subdivision (26)(b) of section 1–201 of the Uniform Commercial Code, which provides that, "A person 'receives' a notice or notification when . . . it is duly delivered at the place of business . . . [at] the place for receipt of such communications." At best it is an acknowledgement of the taking into possession of something, a mere admission of a fact in writing of the receipt of something, without containing any affirmative obligation of creating any contractual obligation, except that in the instant case it would create the statutory obligation under section 4–403 of the Uniform Commercial Code; a far cry from any contractual obligation.

In the instant matter, the plaintiff, an attorney learned in the law to an extent far greater than a mere layman, cannot now be permitted to claim a lack of knowledge of the statutory limitation placed upon "stop payment orders," particularly one "in writing," beyond a six-month period without written renewal thereof (Uniform Commercial Code, § 4–403, subd. [2]). To permit the plaintiff to now claim, as he does, that the stop payment order was to extend beyond the statutory limitation by virtue of an alleged "oral agreement" simultaneously made with the written stop payment order would do violence to the statute, the enacting legislative intent, and to the "depositor's contract" entered into when opening the account, which provides as follows:

> Written request for stop-payment shall be effective for six (6) months, but renewals may be made from time to time. No stop payment request, renewal or revocation shall be valid unless made in writing and served upon the bank.

(Decision for National Bank of North America.)

QUESTION FOR DISCUSSION

1. *What steps did Dinerman need to take in order to keep the stop payment order effective for 2½ years?*

---

(2) Bank's Right to Subrogation on Improper Payment. A bank is liable if it pays a check over a valid stop order. However, there are situations in which liability could result in unjust enrichment of the drawer or of other parties to the instrument. For example, suppose the holder whom the bank paid was a holder in due course. Had the bank properly adhered to the stop-payment order of its drawer, the holder in due course would then demand payment from the drawer, who, as we have learned, would not be able to raise any personal defenses against the holder in due course. Because this is the case, why should the bank suffer the loss? By way of another illustration, suppose that Paula Payee sells goods to Delbert Drawer, who gives her a check drawn on Drawee Bank. The goods are partially defective, and Drawer orders the bank not to pay the check. When Payee presents the check, the bank mistakenly pays it anyway. In the event that the bank had honored the stop-payment order, Paula would still have had an action for part of the contract price had Delbert retained the goods because they were only partially defective. To allow Delbert to retain all these goods after the bank has paid for them in damages because of its failure to honor the stop-payment order would be to enrich Delbert unjustly. In order to avoid this result the code provides that the payor bank, to the extent of its loss, shall be subrogated to the rights of a holder in due course on an item against the drawer or maker. The bank is also subrogated to the same extent to the rights of a payee or any other holder of an item against the drawer or maker either on the item or under the transaction out of which the item arose. Section 4–407(a),(b). In other words, if either of those parties had a right against the drawer or maker, had the stop payment been invoked, then the bank could exercise its right against the drawer or maker when it mistakenly failed to stop payment, but only to the extent necessary to prevent loss to itself and avoid unjust enrichment of the drawer or maker. Of course, if none of those parties has any rights that are enforceable against the drawer or maker, then the bank sustains the loss for failing to honor the stop-payment order.

A final provision of this section makes the payor bank subrogated to the rights of the party on the other side of the transaction. The bank is subrogated to the rights of the drawer or maker against the payee or any other holder of an item with respect to the transaction out of which the item arose. Section 4–407(c). For instance, suppose a fraudulent salesman induces the buyer to draw a check payable to the salesman's order. The goods are never delivered or are totally defective, and the drawer orders the bank to stop payment on the check. If the bank mistakenly pays the check, it would be able to proceed against the salesman as subrogee of the rights that the buyer would have had against the salesman.

SOUTH SHORE NATIONAL BANK v. DONNER et al.

104 N.J.Super. 169, 249 A.2d 25 (1969)

Mark Donner and Bette Donner lost jewelry as the result of a burglary. They were insured by the Quincy Mutual Fire Insurance Company. In making a claim for the loss, they obtained an appraisal of the value of the stolen jewelry from Jules Bruskin, president of Ruth Satsky Jewelry, a corporation. Quincy issued its check to the Donners in the amount of $7,000.00, dated March 2, 1965, and drawn on the South Shore National Bank. Quincy then issued a stop-payment order on the check in the belief that the Donners, with the cooperation of Bruskin and Satsky Jewelry, had overvalued the stolen jewelry. South Shore National Bank paid the check by mistake when it was presented after the stop-payment order was given to the bank by Quincy. The bank then sued the Donners, Bruskin, and Ruth Satsky Jewelry as subrogee of the rights of Quincy. Bruskin and Ruth Satsky Jewelry filed a motion for summary judgment, contending they were not liable to the bank because they had received nothing from Quincy and thus were not unjustly enriched, as they contended Section 4–407 of the code requires in order for the bank to recover.

MILMED, J.

In memoranda submitted in support of the motion for summary judgment it is contended that section 4–407 of the Uniform Commercial Code, "Bank Deposits and Collections" (N.J.S.A. 12A:4–407), is the controlling statute; that this section limits plaintiff's cause of action as against parties to the check who would otherwise be unjustly enriched; that the only such parties in the case who can be unjustly enriched are defendant payees Mark M. Donner and Bette S. Donner; that neither Ruth Satsky Jewelry nor Jules Bruskin was a "holder of the item" within the meaning of this term as used in subsection (c) of N.J.S.A. 12A:4–407, and that accordingly there is no statutory or common law basis for the maintenance of the action as against defendants Ruth Satsky Jewelry and Jules Bruskin.

Section 4–407 of the Uniform Commercial Code (N.J.S. 12A:4–407, N.J.S.A.) provides that:

> If the payor bank has paid an item over the stop payment order of the drawer or maker or otherwise under circumstances giving a basis for objection by the drawer or maker, to prevent unjust enrichment and only to the extent necessary to prevent loss to the bank by reason of its payment of the item, the payor bank shall be subrogated to the rights (a) of any holder in due course on the item against the drawer or maker; and (b) of the payee or any other holder of the item against the drawer or maker either on the item or under the transaction out of which the item arose; and (c) of the drawer or maker against the payee or any other holder of the item with respect to the transaction out of which the item arose.

This section is broader in scope than the prior subrogation statute, N.J.S.A. 17:9A–225(D) (4), which apparently served as its model and was repealed by the enactment of the Uniform Commercial Code, e.g., including within its coverage not only payments made over stop payment orders but also payments made "under circumstances giving a basis for objection by the drawer or maker."

Section 1–102 mandates a liberal construction of the Uniform Commercial Code and its application to promote its underlying purposes and policies. Section 1–103, N.J.S. 12A:1–103, N.J.S.A., states that:

> Unless displaced by the particular provisions of this Act, the principles of law and equity, including the law merchant and the

> law relative to capacity to contract, principal and agent, estoppel, fraud, misrepresentations, duress, coercion, mistake, bankruptcy, or other validating or invalidating cause shall supplement its provisions.

In regard to this section, Uniform Commercial Code Comment 5 under N.J.S.A. 12A:4–407 points out that "The spelling out of the affirmative rights of the bank in this section does not destroy other existing rights (section 1–103,)" including "rights to recover money paid under a mistake." Similarly, Uniform Commercial Code Comment 8 under N.J.S.A. 12A:4–403 point out that:

> A payment in violation of an effective direction to stop payment is an improper payment, even though it is made by mistake or inadvertence. Any agreement to the contrary is invalid under Section 4–103(1) if in paying the item over the stop payment order the bank has failed to exercise ordinary care. The drawee is, however, entitled to subrogation to prevent unjust enrichment (Section 4–407): retains common-law defenses, . . . and retains common-law rights, e.g., to recover money paid under a mistake (Section 1–103) in cases where the payment is not made final by Section 3–418.

The gravamen of the sixth and seventh counts of the complaint in this action is an alleged conspiracy among defendants to defraud the Quincy Mutual Fire Insurance Company by means of "false estimates of the value of property for which claim was made upon the said insurance company." Here it is accordingly alleged that defendants' concerted tortious actions combined to produce a single indivisible result, i.e., issuance of the check of $7,000 by Quincy to the Donners. In such circumstances, joinder of the defendants would be sanctioned even under the strict common law rules limiting joinder of defendants in actions sounding in tort "to cases of concerted action, where a mutual agency might be found." See, Prosser, Torts (3d ed., 1964), § 44, p. 260. The joint alleged responsibility of the defendants to Quincy would be apparent and "the identity of the cause of action against each defendant" would be entirely clear. Ibid., p. 260.

It is evident from an analysis of the related section of the Uniform Commercial Code referred to above that by section 4–407, as supplemented by the broad provisions of section 1–103, the Legislature intended to grant to a payor bank which has mistakenly paid an item over the stop payment order of the drawer or maker, a full and effective remedy by way of subrogation to the rights and remedies of the drawer or maker against the payee as well as those who have, by their tortious acts in concert with him, combined to produce the single indivisible result, e.g., the issuance of the item to the payee, the remedy being furnished not only to prevent unjust enrichment but also to the extent necessary to prevent loss to the bank by reason of its payment of the item. See New Jersey Study Comment 1 and 2 under N.J.S.A. 12A:4–407. Comment 1 states in part that "Subrogation, in the case of the improperly paid check, permits the bank to step into the position of the person receiving payment, the payee, some other holder, or the drawer," and Comment 2 points out in part that section 4–407 would not permit a bank to subrogate "unless some party has been 'unjustly enriched.' "

While the remedy afforded by section 4–407 is in part expressly directed to the prevention of "unjust enrichment," nowhere in the statute is there any requirement that all of the defendants who may be joined in the action be "unjustly enriched." The reasonable implication is that the Legislature intended that the remedy granted be comprehensive and effective.

The editorial staff of the Banking Law Journal, in commenting on the objective of section 4–407, concisely pointed out that, "The aim of the Code provision is clear enough; the payor bank by paying over a stop order would have paid out its own money and would be given remedies to make itself whole." "Stopping Payment of Checks," 79 Banking L.J. 204 (1962).

It is not contended that the movants, Ruth Satsky Jewelry and Jules Bruskin, have in any way changed their position in reliance on the bank's payment of the check. Nor is there any suggestion that any injustice is done to them by the subrogation. Their position is the same as it would have been had they been impleaded along with the Donners as third-party defendants in an action by Quincy against the bank to have its account recredited. In such litigation the burden would be upon Quincy to establish the fact and amount of loss resulting from the payment of the check over the stop-payment order. . . . Plaintiff bank having requisite standing to maintain this suit, the factual issues between the parties, including the strongly contested charge of conspiracy to defraud, are for determination at a plenary trial. The motion for summary judgment is accordingly denied.

QUESTIONS FOR DISCUSSION

1. *Did the court rule in this case that the bank was actually entitled to recover from Ruth Satsky Jewelry and Bruskin?*
2. *What is meant by the phrase "subrogated to the rights of . . ."? Was the bank proceeding against Ruth Satsky Jewelry and Bruskin on its own rights against them?*

(3) **Stale Checks.** A bank is under no obligation to pay a check drawn by its customer if it is presented for payment more than six months after it was drawn. Thus failure by the bank to pay a check that is six months old does not constitute wrongful dishonor. However, if the check is certified, the bank must pay it because it is the bank's primary obligation. Remember that the time for presentment of a check for payment is presumed to be thirty days with respect to the liability of the drawer. Section 3–503(2)(a). Delay beyond this time will not discharge the drawer on the check unless the bank became insolvent in the meantime. Then the drawer is discharged to the extent that the delay caused any loss to him. Remember also, however, that the time for presentment with regard to the liability of an indorser of the check is seven days. Section 3–503(2)(b). After that time the indorser is discharged. Section 3–502(1)(a).

Even though a bank is not required to pay a stale check, it may be in a position to know that the customer would want the check paid. Therefore the bank has the option to pay the check, and it may do so as long as it acts in good faith. For instance, the dividend check of a bank's corporate customer may be presented more than six months after date. The bank has the option of paying it and may very well do so in this instance. Alternatively the bank may check with its customer to see whether the stale check should be paid. Remember, however, that the fact that the check is stale does not—in the absence of the bank's insolvency—discharge the drawer from his obligation to pay it, even though the bank is not required to do so. An indorser will be discharged, however.

ADVANCED ALLOYS, INC. v. SERGEANT STEEL CORP. et al.

72 Misc.2d 614, 340 N.Y.S.2d 266 (1973)

COHEN, J.

The question presented is whether a bank has a duty of inquiry before paying a check which is stale in that it was presented for payment 14 months after issuance. Prior to the enactment of the Uniform Commercial Code, such a duty existed. However, UCC 4–404 states that:

A bank is under no obligation to a customer having a checking account to pay a check, other than a certified check, which is presented more than six months after its date, but it may charge its customer's account for a payment made thereafter in good faith. This statute appears to change the common law rule.

Under this statute, it must be determined whether this payment was made "in good faith." Since no evidence was presented on this point and since both plaintiff and defendant The Chase Manhattan Bank, N.A. agree that there are no issues of fact—the case having been presented to the Court on affidavits prepared for a summary judgment motion—the Court must simply decide whether a payment of a check by a drawee bank 14 months after issuance is a payment "in good faith" when made without inquiry of the depositor. UCC 1–201(19) defines "good faith" as "honesty in fact in the conduct or transaction concerned." Under this definition, to which the Official Comment to UCC 4–404 makes reference, it appears that the payment of the stale check, without making such inquiry, constitutes a payment "in good faith."

Presumably, if it were intended to place a duty of inquiry upon a bank before it could safely pay a stale check, a broader definition of "good faith" would have been made applicable to this situation. It may very well be that in enacting the Code consideration was given, as defendant argues, to the vast number of checks being issued and the requirement that a bank accept or refuse to honor a check within a short, prescribed time limit (UCC 4–301, 302), leading to the conclusion that a bank should not be liable for paying stale checks as long as the bank was honest in fact.

The court realizes that a determination that there is no duty of inquiry puts a substantial burden upon one who issues a check, and then, even for a good reason—as in this case—does not want it to be paid. Since a stop payment order is good for only six months (UCC 4–403(2), it means that the issuer must, in order to protect himself, either continue to renew the stop payment order every six months or close the account. Apparently, in balancing the problems of the issuer and the bank in this situation, the Code resolved the matter in favor of the bank.

The court notes the statement in the Official Comment (. . . Uniform Commercial Code, § 4–404) that normally a bank will not pay a stale check without consulting the depositor and, further the bank ". . . is given the option to pay because it may be in a position to know, as in the case of dividend checks, that the drawer wants payment made." Plaintiff argues that this option to pay is given only when the drawee bank is in a position to know that the drawer wants the check to be paid; and in this case the bank could only know this if it made inquiry—something it did not do. However, the language of the Code itself, as indicated above, does not support this argument and does not impose a duty of inquiry upon the bank in the situation presented herein.

(Judgment for defendant.)

QUESTIONS FOR DISCUSSION

1. *Did the bank know that the depositor in fact wanted the check paid in this case?*
2. *Would the case have been decided differently if the bank had had knowledge that the plaintiff did not want the stale check paid? How does your answer square with the court's statement that its decision placed a burden on the depositor to renew a stop payment every six months or close the account? Did the bank have no knowledge that the depositor did not want a check paid after it received a stop-payment order?*

---

(4) Death or Incompetence of Customer. If the customer of a collecting or payor bank is incompetent when the item is drawn or dies before an item is collected or paid, the bank still has the authority to accept, pay, collect, or account from proceeds of its collection. A bank cannot be expected to verify the continued life and competence of all its customers. Section 4–405(1). Once the bank does know of the fact of death or incompetence and has a reasonable opportunity to act on it, then its authority to act on an item is revoked. The bank must have actual knowledge even of an adjudication of incompetency. Such an adjudication is not constructive knowledge to the bank.

An exception is provided for the payment of checks. A bank may pay or certify checks drawn by its customer for ten days after the date of his or her death unless it is ordered to stop payment by a person claiming an interest in the account. Section 4–405(2). Because of this provision those holding checks drawn shortly prior to the death of the "poor beloved departed" are not required to file a claim against his estate.

(5) Customer's Duty to Discover and Report Unauthorized Signature of Alteration. A drawee bank may charge its customer's account only for the original amount of a check if it has been altered. Additionally, if the drawer's signature is forged or if an indorsement is forged and the drawee bank pays the check, the drawer's account may not be charged if the drawer–customer discharges certain duties.

Periodically most banks send statements of account or make them available to their checking-account customers. The canceled checks that have been paid by the bank are also included with this statement. The code imposes a duty on the customer to inspect the statement and checks within a reasonable time after receiving them in order to detect any alterations or forgeries. Section 4–406(1). Once an alteration or forgery is detected, the customer must notify the bank promptly. Failure by the customer to adhere to these duties may result in the bank's passing to the customer any loss incurred as a result of the alteration or forgery.

If the bank proves that the customer failed to meet the duties of inspection and notification, then the customer cannot have restored to his account the funds that have been charged as the result of an unauthorized signature or alteration, provided that the bank also establishes that it suffered a loss as the result of the customer's failure. Section 4–406(2)(a).

One of the dangers of a customer's failure to inspect his or her bank statement and canceled checks is that a wrongdoer may repeat his act of forgery, or alteration. Therefore the code states that the customer is precluded from asserting the defense of forgery or alteration against the bank when the customer does not meet his duties of inspection and notification and when an unauthorized signature or alteration by the same wrongdoer on any other item is paid in good faith by the bank after the first item and statement were available to the customer for a reasonable period—not exceeding fourteen calendar days—and before the bank is notified by the customer of the first unauthorized signature or alteration. Section 4–406(2)(b). For example, suppose William Wrongdoer forges the signature of his employer, Delbert Drawer, on a payroll check. Wrongdoer cashes the check at Delbert's bank, Friendly Bank, on August 1, 1978. The bank could not charge Delbert with the check because his signature was forged. Now assume that Delbert receives his bank statement and canceled checks from the bank on August 15, 1978, including the forged check. On September 1, 1978, Wrongdoer again forges Delbert's signature and repeats the process. On September 3, 1978, Delbert inspects his statement, finds the first forgery and notifies Friendly Bank. Delbert's account would be charged with the second forgery because he did not inspect his statement and notify the bank within a reasonable period of time.

There is one other limitation on the bank's ability to charge a customer's account if the customer fails to meet the duties that have been discussed. If the bank fails to exercise ordinary care in paying the instrument, it is still liable despite the customer's negligence. So if the bank pays a check that has been marked in a manner and under circumstances that obviously indicate that it has been altered without the drawer's authority, it would be liable for the amount paid above its original tenor despite the drawer's negligence. Section 4–406(3).

The code places an absolute time limit upon the rights of a customer to assert an unauthorized signature or alteration against the bank. Regardless of the bank's or customer's lack of care, the customer may not assert his own unauthorized signature or alteration of an item unless he does it within one year from the time that the statement and items are made available to him. For an unauthorized indorsement the period is three years. Section 4–406(4).

Finally, if the payor bank fails to raise a valid defense against a claim of a customer resulting from the payment of an item, the payor bank may not assert a claim against a collecting bank or prior party based upon the unauthorized signature or alteration. Section 4–406(5).

ARROW BUILDERS SUPPLY CORP. v. ROYAL NATIONAL BANK OF NEW YORK

21 N.Y.2d 428, 235 N.E.2d 756 (1968)

BURKE, J.

Faye Zappacosta was employed as a bookkeeper by plaintiff Arrow Builders Supply Corp. Each month, she drew checks for plaintiff's president LaSala on its account in defendant Royal National Bank, payable to the order of that bank, representing income and social security taxes withheld from the wages of Arrow's employees. Between August, 1962 and June, 1964 Zappacosta drew and LaSala purportedly signed 54 such checks, of which only 23 were properly applied. The 31 remaining checks, totaling $132,000, were presented individually by Zappacosta to the bank whereupon the bank would issue its own check for the same amount payable to a person designated by Zappacosta. In most cases the payee she selected was either a friend or a relative who would either indorse over the check to her or cash it and deliver the proceeds.

In furtherance of her scheme, she concealed these unlawful withdrawals, as best she could. As bookkeeper, she was the one who first checked the cancelled vouchers and examined these bank statements. Before submitting a statement and the accompanying checks to plaintiff's accountant, she would withdraw the checks involved, write void on their corresponding checkbook stubs and then prepare a tape which, while excluding the amounts of any checks appropriated by her, nevertheless reflected the proper total for all the checks drawn for that month. Thus the total shown always exceeded the sum of the enumerated checks. The statement, with this tape, was then presented to the accountant who neither compared the checks he received against the debits on the statement, nor totalled these checks himself. Consequently, Zappacosta was able to perpetrate this scheme for 23 months whereas a proper review of the first statement would have exposed the first defalcation. Plaintiff brought this suit to recover the entire sum taken by its bookkeeper Zappacosta, alleging that the defendant bank paid these funds without making any appropriate inquiry as to her authority to direct the application of these checks. Defendant, conceding its carelessness, asserts as an affirmative defense plaintiff's negligence in failing to discover the embezzlement. After defendant had been given an opportunity of availing itself of the disclosure procedures, plaintiff successfully moved for summary judgment. That order, subsequently affirmed, can only be sustained where it clearly appears that no material and triable issue of fact is presented. It is our opinion that, upon this record, a triable issue is presented at least as to plaintiff's negligence in failing to discover these defalcations.

At the outset, it should be noted that the defendant was negligent in disbursing plaintiff's money without receiving proper instructions and without having made any inquiry. In receiving checks payable to itself, and signed by the plaintiff's president, the bank properly presumed that these checks had commercial significance. In such a situation, the bank cannot properly ascertain this significance by merely relying on the directions of one who is without either actual or apparent authority to represent the drawer. Accordingly, if this encompassed all the transactions between the parties, summary judgment for the drawer would indeed be proper.

A second pattern of activity existed between these parties, however, which also requires evaluation. After the bank began to supply these checks under the advisement of Zappacosta, it continued to supply the plaintiff with monthly statements reflecting all the transactions for this account. Thus, during the 23-month period involved, at least two checks were returned each month, each purportedly issued by the plaintiff to cover its monthly payment for income and social security taxes withheld from the wages of its employees. Indeed, in some months, plaintiff received as many as four cancelled checks, each allegedly drawn to cover this single monthly payment. The issuance of such a statement is not without purpose. Rather, it is intended to inform the plaintiff of the status of his account and to provide him with an opportunity to notify the bank of any discrepancies existing between the statement as presented and the balance as maintained by the drawer. In this regard, the duties of the depositor have been often enumerated in unequivocal terms. Justice Shientag's opinion in *Screenland Mag.* v. *National City Bank of New York,* 181 Misc.454 N.Y.2d 286 is illustrative. There, it was noted that, inter alia:

> A depositor is under a duty to his bank to examine canceled checks and statements received from the bank and to notify the bank promptly of any irregularities in the account. If a depositor disregards this duty, any further losses occurring as a result of such omission must be borne by the depositor unless the bank itself is guilty of contributory negligence. . . .
>
> The duty of verification and of reconcilement may be delegated to any employee who has proved himself competent and trustworthy. . . . A depositor cannot be charged with the knowledge which the dishonest employee had gained while he was stealing from him (. . .), but a "depositor must be held chargeable with knowledge of all the facts that a reasonable and prudent examination of the returned bank statements, vouchers and certificates would have disclosed had it been made by a person on the depositor's behalf who had not participated in the forgeries."

These rules, which have evolved basically with reference to forged signatures, are also applicable where the plaintiff's negligence may have made the embezzlement possible. Applying these standards to the instant case, it is apparent that there are questions of fact as to whether the depositor used due care in the examination of these statements and vouchers. Thus the depositor will be precluded from recovering from the bank, if the jury should find that the bank was damaged by the drawer's negligence in failing to properly examine the monthly statements.

The bank, of course, is liable in the amount of $5,237.55, which was improperly charged to plaintiff's account upon presentment of the first check. Summary judgment was, therefore, properly granted as to that check.

Accordingly, the order appealed from should be modified, with costs to abide the event, to the extent of reversing the grant of summary judgment, except as to the first check in the amount of $5,237.55, and as so modified, affirmed.

Questions for Discussion

1. *Why was the bank negligent in this case? What procedure should it have followed?*
2. *Do you think Arrow Builders can be held to be negligent when Zappacosta was first assigned to examine the statements? If so, what procedure should Arrow Builders have followed?*

SUMMARY

The creation of a checking account with a bank creates a creditor–debtor relationship and a principal–agent relationship. The bank is the debtor and agent. In the collection process a bank may be either a depositary bank, payor bank, intermediary bank, or collecting bank.

When a customer deposits a check, he receives provisional credit for the check. The customer has no right against the check until the provisional credit becomes final. If the bank pays the item before that time, it has the right of charge back.

A bank may fix an afternoon hour of 2 P.M. or later as a cutoff time. The bank also may supply a missing indorsement. A collecting bank must use ordinary care in the collection process and is presumed to act reasonably if it takes proper action before its midnight deadline.

The code approves the practice of deferred posting and delayed returns for a payor bank. If a payor bank receives an authorized settlement on a demand item (other than for immediate payment over the counter) before midnight of the banking day of receipt, the payor bank may revoke settlement and recover any payments made provided it meets several conditions mentioned earlier in this chapter. The code also states a number of rules that prevent the payor bank from revoking settlement.

The payor bank has a duty to its customer to pay items properly presented to it. Damages for wrongful dishonor are limited to actual damages proved if the wrongful dishonor was a result of the bank's mistake.

The customer of a bank may stop payment on a check if the bank is notified in a manner and time so that the bank can act before it is required to pay the item. An oral stop-payment order is effective for fourteen days. A written stop-payment order is effective for six months unless renewed in writing. If the bank pays a check over a valid stop-payment order, it is liable to its customer but may have a right of subrogation to the rights of its customer or to a payee, other holder of the item, or the drawer.

A check becomes stale six months after it was drawn. The bank is not required to pay an uncertified stale check but it may do so in good faith. A bank may also accept, pay, collect, or account for proceeds from its collection if the customer dies before the item is collected or paid.

A customer has a duty to discover and report unauthorized signatures and alterations. If the customer is negligent in doing this, the bank is not liable for any loss caused by the customer's negligence.

PROBLEMS

1. Billie Fittring wrote a check drawn on the Continental Bank in the amount of $800.00. After writing the check, she had second thoughts and contacted the bank about the possibility of stopping payment. An employee told her she could not stop payment on the check until the bank opened for business

the next day. Fittring then discussed with the bank's employee the possibility of deliberately creating an overdraft situation by withdrawing enough money so that there would be insufficient funds to pay the check. The employee indicated that in those circumstances the bank would not pay the check. The next morning Fittring removed the funds but as fate would have it the bank paid the check when it was presented even though there were insufficient funds in Fittring's account to cover it. Was the bank entitled to pay the check and seek recourse from Fittring for the amount that the account was deficient? Continental Bank v. Fittring, 559 P.2d 218 (Ariz. App. 1977).

2. At 9:00 A.M. Tusso stopped at the Security National Bank and placed a stop-payment order on a check he had written the previous day. At 10:40 A.M. on the same day the check was presented to the bank, the check was certified, and Tusso's account was charged with the check in the amount of $600.00. Tusso sued the bank which defended on the ground that he had failed to submit evidence of the bank's negligence. Was Tusso required to prove that the bank was negligent in order to recover? Tusso v. Security National Bank, 349 N.Y.S.2d 914 (1973).
3. A check drawn on the Trust Company of Georgia was made payable to Ed Samples Transmission Service in the amount of $170.75. The check was indorsed by Samples and deposited in his bank, Citizens and Southern National Bank, for collection. The check was then presented to Trust Company of Georgia. Trust Company paid the check on February 2, 1967, and mistakenly debited the check to the account of one of its employees, Carolyn Jordan. When Jordan received her bank statement, she notified Trust Company of the error. They then credited the account and returned the check for the reason that there was no such account. Citizens and Southern returned the check to Trust Company for the reason that Trust Company had not dishonored the check within the time allowed by law. Trust Company then sued Samples as an indorser of the check. The trial court found for Trust Company, and Samples appealed. What decision? Samples v. Trust Company of Georgia, 118 Ga.App. 307, 163 S.E. 325 (1968).
4. Charles Wolfe, Per Olof Holtze, and Wolfgang H. Altman were three general partners in the Watkins Glen Limited Partnership. The partnership had a checking account with University National Bank. The agreement with the bank required that checks be signed by two of the general partners, Wolfe and Holtze or Wolfe and Altman. Later, checks signed by Paul S. Waymouth and Wolfe instead of Altman were authorized. From October 27, 1969, to September 18, 1970, thirty-seven checks totaling $235,012.02, each bearing only one signature, were drawn on the Watkins Glen account. During this time monthly statements were sent to Watkins Glen's office and were accessible to Wolfe. No notice was given to the bank until March 1972, when Wolfe and Watkins Glen sued University National. University contended that it was entitled to judgment as a matter of law because its customer failed to exercise its duty to inspect its statement and notify it

of the improper payments within a reasonable period of time. Wolfe contended that the bank paid on the instruments without the presence of any proper signature and that therefore the case was not affected by Wolfe's failure to inspect the statements within a reasonable time. Was Wolfe correct? Wolfe v. University National Bank, 310 A.2d 558 (Md. 1973).

5. Huber Glass Co., Inc., maintained a checking account with the First National Bank of Kenosha. R. C. Huber and his wife, Bertha Huber, were the only ones authorized to write checks on the account. Kenneth J. Miller began working in Huber's bookkeeping department in 1959. In 1963 Miller was fired and killed himself the following day. It was later discovered that Miller had forged R. C. Huber's name and made out a number of checks payable to himself since August 1960. Huber sued First National for $23,875.42, alleging that the checks had been wrongfully paid. Huber further contended that there was no negligence on his part because Miller, as bookkeeper, examined the statements monthly as provided by the bank. During this inspection Miller removed the forged checks and then gave the statement to Huber. Huber never actually reconciled the statement and checkbook balances. Was Huber entitled to recover? Huber Glass Co. v. First National Bank of Kenosha, 138, N.W.2d 157 (Wisc. 1965).
6. Motor Contract Co. was the holder and payee of two checks drawn on National City Bank of Rome, which Motor Contract deposited in its account with First National Bank of Rome for collection. The checks were dated January 15, 1966, and issued to Motor Contract on January 18, 1966. Motor Contract deposited the checks and they were presented to the payor bank, National City, on January 19, 1966. They were returned to the depositing bank, First National, on January 21, 1966, without being paid. No notice of dishonor was given prior to that time. Motor Contract was not able to collect because the drawer was adjudicated bankrupt on February 7, 1966. Motor Contract then sued National City, contending that the bank had not dishonored the check in time. National City contended that the drawer's account did not contain sufficient funds. Was National City liable? National City Bank of Rome v. Motor Contract Co., 119 Ga. App. 208, 166 S.E.2d 742 (1969).
7. Harry Tenn drew a check in the amount of $6,000.00 on City Bank of Honolulu, payable to Sonic Educational Products, Inc. When he drew the check, Tenn knew that it was approximately $4,700 in excess of the amount he had on deposit. Tenn told this to Hawley and gave the check to Hawley on the oral understanding that Hawley would not cash it immediately. The check was to evidence Tenn's good faith in entering into a business deal with Hawley. Hawley deposited the check in his account despite the agreement, and City Bank of Honolulu cashed it after unsuccessfully trying to reach Tenn. Tenn then refused to reimburse City Bank for the excess paid on the check by the bank, contending that it had no authority to pay the check because there were insufficient funds in Tenn's account. City Bank of Honolulu v. Tenn, 469 P.2d 816 (Hawaii 1970).

8. Linsky gave a check for merchandise to the Board of Survey of the United States on February 8, 1921. On February 10, 1921, the board presented the check to Tremont Trust Company, the bank on which the check was drawn, for certification. The check was certified and the goods delivered to Linsky. On February 17, 1921, the bank was closed by the state bank commissioner. That same date the check was deposited for collection and returned the following day because the bank was closed. Linsky had sufficient funds at all times to cover the check. Was Linsky liable to the United States for the amount of the check? Linsky v. United States, 6 F.2d 869 (1925).

SECTION V

Property

CHAPTER 28

General Considerations and Personal Property

28:1. Introduction. The concept of *property* is extremely important in the law. It is also a term that is used in various ways. For instance, if one states that he "owns property in Florida," he is referring to some physical *object.* He is using the term as a noun to identify some tangible thing. On the other hand, the term *property* may be used to connote the *interests* or *rights* that an individual has in an object. Thus one may say, "That land is my property." By this the individual usually implies that he or she has certain rights or interests in the object that the law will protect.

The term *property* is really even broader than this. It may, for instance, apply to the rights and interests that an individual has in anything of value, whether that thing is tangible or intangible. These rights and interests, which are implied in the word *property* are the subject of protection in our federal as well as most state constitutions. For instance, the Fifth Amendment to our federal Constitution states that "no person shall be . . . deprived of life, liberty or property without due process of law; nor shall private property be taken for public use without just compensation." The Fourteenth Amendment to the federal Constitution makes this protection applicable to the states. It is these rights and interests that may be called *property* and that are the subject of the constitutional protection.

We shall see later in this section of the text that the various types of property interests may vary. For instance, one may lease land or an apartment. He may lease a typewriter or have a right-of-way over certain land. One may even have a cause of action against another party for breach of contract. All these are examples of property. Thus the word *property* has more than one meaning. At this point it is appropriate to examine a number of concepts that are applicable to property in general.

28:2. Classification of Property. (1) Real and Personal Property. All property can be broadly classified as either *real* or *personal* property. Broadly speaking, real property is land and whatever is growing or affixed to it, such as a house. All other property is personal. Generally property that is classified as personal remains personal, but there are circumstances under which personal property may become part of the real property because it becomes so firmly affixed to the real property. This change in character from personal property to real property is discussed more fully in Section 28:5, "Fixtures."

One might legitimately ask at this point whether the distinction between real property and personal property is of any practical significance. The differences between real and personal property are important (1) in determining the method to be used in transferring property; (2) in determining how the property passes at death; (3) in determining the applicable law; (4) in determining the necessity for written contracts concerning the sale of such property; and (5) in determining how property is taxed.

There are specific formalities that must be followed in the transferral of the title to real property. If real property is sold, the seller must give the purchaser a *deed* in order to transfer title. No such formality need be followed with regard to personal property. Sometimes the seller of personal property will give the buyer a bill of sale, but this is not always done and is not necessary. If one purchases a magazine in a drugstore, the title to the personal property is transferred from the store to the purchaser by the simple expedient of delivering the magazine to the customer upon receipt of his or her money. One reason for the need for a deed in the transfer of real property is that real property is not movable, whereas personal property can be moved.

There is also a difference in how the property passes at death, depending upon whether it is real or personal property. If one dies without a will naming who is to receive his real property, then in most states title to it still passes directly to the heirs of the deceased. On the other hand, with regard to the personal property, if a person dies without a will naming who is to receive his personal property, the title to the personal property passes to the personal representative of the deceased, who then distributes the property according to law, after paying the debts of the deceased. Some states have abolished this distinction by statute, and all property, both real and personal, passes to the personal representative.

The laws of the various states vary somewhat regarding the rights to property. For instance, the laws of Maryland may differ from the laws of Texas or New York as to what is to happen to property should the owner die without a will. If the property is real property, the law of the place where the real property is located determines how it is distributed. If the property is personal property, the law of the place where the owner is domiciled governs the property, regardless of where the personal property is physically located. For example, suppose Time Isup lives and is domiciled in New Jersey and owns real property in Pennsylvania. He also has a certain number of stocks and bonds in a safety deposit box in New York. When he dies, the law of Pennsylvania will govern the disposition of the real property located in that state. The law of New Jersey, where Time Isup was domiciled, would govern the personal property rather than the law of New York, where the property is located.

Recall that the Statute of Frauds requires contracts for the sale of land or an *interest* in land to be in writing, or it requires a written memorandum signed by the party to be charged. There is no such general requirement for the sale of an interest in personal property. The Uniform Commercial Codes does have a Statute of Frauds provision for contracts for the sale of goods, a type of personal property, under certain circumstances.

In most cases, the various states tax real property and personal property. The method of determining the tax on the property is usually quite different for real as opposed to personal property. It is important, therefore, to determine whether certain property is real or personal.

There are other reasons for distinguishing between real and personal property. The examples given should be sufficient, however, to illustrate the importance of the distinction.

(2) **Tangible and Intangible Property.** *Tangible* property consists of objects or things that have a physical existence. You can touch them, A car, a textbook, a record player, and a desk are all examples of tangible property in which one may have a property *interest.*

Intangible property does not have physical existence. It consists of rights that the law will protect, and therefore it has *legal* existence. Examples of intangible property would be the ownership of stock in a corporation or an easement or right-of-way over real property. The ownership of stock in a corporation is the right to participate in the profits, losses, growth, and other matters concerning the corporation. The rights of a stockholder are protected by law and are enforceable in court. Normally corporate ownership is evidenced by a stock certificate. But it is not the certificate that is the property. The certificate only represents the rights and interests that have legal but no physical existence and that the law protects. The case of an easement or right-of-way is similar. There may be a deed evidencing the legal right to use another's property in a particular manner, but it is the right to use the property that is the property and not the deed. Other examples of intangible property would be contract rights, patents, and copyrights.

(3) **Chattels Real and Chattels Personal.** Personal property may also be categorized as *chattels real* and *chattels personal.* The word *chattel* is another and perhaps more exotic word for personal property. *Chattels real* is the term used to describe an interest in land that is less than a freehold estate. The concept of a freehold estate will be discussed in Chapter 30, "Real Property." It is enough at this point to know that a freehold estate is a status of ownership and applies to ownership interests in land that last for the life of the owner or longer. Chattels real are interests in real property of lesser status than a freehold estate. Chattels real are treated as personal property and are best exemplified by the common lease of real property.

All other chattels are considered chattels personal. Chattels personal are divided into *choses personal in possession* and *choses personal in action.* Choses personal in possession are chattels that are tangible and that therefore may be physically moved from place to place. Choses personal in action, or *chose in action,* is intangible property to which one has a right, but that right may have to be enforced. Examples are such intangible property rights as negotiable instruments, contract rights, patents, copyrights, and corporate stock.

28:3. Forms of Multiple Ownership. At this point it is appropriate to discuss the implications of multiple ownership of property. What follows concerning multiple ownership applies equally to the ownership of real and personal property. The types of co-ownership are as tenants in common, joint tenants, tenants by the entirety, and, in a few states, community property.

(1) Tenants in Common. When two or more persons own property, they may own it as tenants in common. In some states there is a presumption that when two or more people own property, it is as tenants in common. In order to own property as tenants in common, the owners must have what is known as *unity of possession.* That is, all the owners have an equal right to the possession of the property. In legal parlance they are each said to have an *undivided interest* in the property. Thus, if two people own property as tenants in common, in equal shares, they would each own a one-half undivided interest in the whole of the property. If the property is a 200-acre farm, for example, they each have a possessory one-half interest in the whole 200 acres. It is *not* accurate to say that one party owns 100 acres and the other party owns 100 acres. They *both* own the entire 200 acres. It is possible for tenants in common to own property in unequal shares or interests.

Ownership as tenants in common does not result in the right of survivorship when one of the parties dies. When one of the tenants dies, his or her interest passes to his or her heirs or whoever is named in his will. Suppose Albert Able, Bill Baker, and Connie Cornball are tenants in common of Whiteacre farm. Cornball dies. Her interest would pass to her heirs and not to Able and Baker. It is also possible for a tenant in common to sell his or her interest to another person, who would then be a tenant in common with the original owners.

(2) Joint Tenancy. Whereas the only unity required of a tenant in common is unity of possession, for a joint tenancy four unities are required: (1) unity of time; (2) unity of possession: (3) unity of title: and (4) unity of interest. *Unity of time* means that the interest of the parties must be created at the same time. *Unity of title* means that the parties must have the same ownership interest in the property. For instance, one who is leasing property cannot be a joint tenant with the owner of that property. *Unity of interest* results when both parties have equal shares in the property. Whereas tenants in common may have unequal shares, the shares of joint tenants must be equal. *Unity of possession* has already been discussed.

The most significant characteristic of a joint tenancy is the right to *survivorship.* This concept means that upon the death of one of the joint tenants, his or her ownership interest passes to the other surviving joint tenants. The interest does not pass to the deceased's heirs and does not go into his or her estate. For example, suppose Able, Baker, and Cornball are joint tenants in Whiteacre. They each have a one-third undivided interest. Cornball dies. Cornball's interest goes to Able and Baker by survivorship, and they then each have an undivided one-half interest in Whiteacre. If Baker then dies, Able owns the property alone. He is then said to have an estate in *severalty;* that is, he is the sole owner of the property.

A joint tenant may sell his or her interest in the property, but he may not devise it by will because of the attribute of survivorship. If a joint interest is sold, the new owner is not a joint tenant but a tenant in common with the remaining joint tenants because there is only the unity of possession between the new owner and the remaining joint tenants. Suppose Able, Baker, and Cornball are joint tenants. Able then sells his interest to Zora Zell. Zell would then have an undivided one-third interest in the property as a tenant in common with Baker and Cornball. If Zora dies, her interest would pass to her heirs. Baker and Cornball would still remain as joint tenants to each other, and if, for instance, Baker dies, his interest would pass by survivorship to Cornball. Cornball would then have an undivided two-thirds interest in the property as a tenant in common with Zora.

In many states there is a presumption against joint tenancy. The document creating the interest must state that it creates the interest as joint tenants and not as tenants in common. It is also possible to destroy a joint tenancy by a suit in equity asking that the property be partitioned, that is, divided among the owners.

(3) **Tenants by the Entirety.** Broadly speaking, tenancy by the entirety, or by the entireties, is a joint tenancy between a husband and wife. This type of tenancy is most frequently referred to with regard to real property, but in most states personal property may be held by a husband and wife as tenants by the entirety.

A tenancy by the entireties has the same unity requirements as a joint tenancy, with the additional requirement that the owners be husband and wife. The right of survivorship applies to tenants by the entireties just as it does to joint tenants. The distinguishing characteristic between a joint tenancy and a tenancy by the entireties is the right of one of the tenants to destroy the tenancy. A joint tenant may destroy the joint tenancy by selling his interest in the property to a third party. One who owns property with his or her spouse as tenants by the entirety cannot destroy the tenancy by acting alone. Both parties must join in the sale in order to sell the property. Suppose Harry and Wanda Spouse own a house as tenants by the entirety. Harry wants to sell the house and Wanda does not. The house cannot be sold. Harry cannot sell his interest in it without Wanda's participation.

A tenancy by the entireties is destroyed upon the termination of the marriage. Thus, if Harry and Wanda get a divorce, they then own the property as tenants in common. The result of this change in the manner of ownership is that either party can sell his or her interest in the property, and there is then no right of survivorship. In the case of a house, the practical result is to force a sale of the house and a division of the proceeds of the sale between the tenants in common.

(4) Community Property. The concept of community property originated in the civil-law countries, such as Spain and France, and thus ownership of property in this manner is found only in those states where the Spanish and French had a strong influence. Theses states, of course, are located mainly in the West and the Southwest. Arizona, New Mexico, California, Nevada, Louisiana, and Texas are examples of community-property states. There are others.

The ownership of community property is governed by statute. Under this type of ownership the theory is that all property acquired during the course of a marriage is the result of the joint efforts of the husband and wife and therefore is their community property and shared equally. Any property owned by the husband and wife prior to the marriage remains their individual property. Also, any property acquired during the course of the marriage by gift, devise, bequest, or descent, as well as any income from such property, remains the individual property of the husband or wife who acquired it. Most of the statutes provide for the right of survivorship for property that qualifies as community property. In a few states the deceased spouse's interest goes to his or her heirs.

D'ERCOLE v. D'ERCOLE
407 F.Supp. 1377 (1976)

Mary E. D'Ercole filed suit in United States District Court for Massachusetts contending that the common-law tenancy by the entirety violates the United States Constitution because it discriminates against women.

TAURO, D.J.

I

The plaintiff and the defendant have been married for some thirty-five years. In November of 1962 they bought a residence at 61 Stone Road, Waltham, Massachusetts, for $20,000.

In 1971, the plaintiff and defendant determined they could no longer live together. When the defendant refused to leave the marital home, plaintiff departed, moving to a relative's home where she still resides.

Proceedings for legal separation and for divorce are now pending in the Middlesex County Probate Court. The defendant husband is seeking a divorce. The plaintiff wife is seeking a separation and is vehemently opposing the divorce on factual issues and because of religious beliefs.

Defendant has refused to share the marital home with plaintiff by allowing her sole occupancy for part of the year, by selling the house and dividing the proceeds, by paying plaintiff her share in the equity of the house, or by renting the premises and dividing the proceeds. In support of his position, defendant points out that thc property in question is held under a tenancy by the entirety, which gives both him and the plaintiff an indefeasible right of survivorship, but gives him exclusive right to possession and control in his lifetime. He has stated that he will grant plaintiff one-half the equity in the house is she will grant him an uncontested divorce.

II

When two or more persons wish to hold property together in Massachusetts they may select one of three common law forms of ownership: the tenancy in common, joint tenancy or tenancy by the entirety.

The tenancy in common is the holding of land "by several and distinct titles."

Each tenant owns an undivided fraction, being entitled to an interest in every inch of the property. With respect to third persons the entire tenancy constitutes a single entity. There is no right of survivorship as between tenants in common. Upon the death of a tenant in common his undivided interest in the property is transferred to his heirs or devisees, subject to liens, claims and dower. 28 Mass.Practice (Park) § 125, at 119–120. (Footnotes omitted.) There is a presumption in favor of the tenancy in common over the joint tenancy as a matter of construction in Massachusetts. Each tenant in common has a right to free usage of the whole parcel and may freely convey out his share of the property to a third party, who then becomes a tenant in common in relation to the remaining cotenants.

The joint tenancy is "a single estate in property owned by two or more persons under one instrument or act."

A joint tenancy is similar to a tenancy in common in that all tenants have an equal right to possession, but the joint tenants hold the property by one joint title and in one right, whereas the tenants in common hold by several titles or by one title and several rights.

The joint tenancy differs also in that there is a right of survivorship in a joint tenancy but not in a tenancy in common. On the death of one of the joint tenants his interest does not descend to his heirs or pass under his will as in a tenancy in common. . . . The widow of the deceased tenant has no dower rights and his creditors have no claim against the enlarged interest of the surviving tenants. 28 Mass.Practice (Park) § 126, at 121–122. (Footnotes omitted.) A joint tenant may convey out his share in the property via a legal partition. See Mass.Gen.L. ch. 241, § 1.

The tenancy by the entirety is designed particularly for married couples and may be employed only by them. Until 1973 unless there was clear language to the contrary a conveyance to a married couple was presumed to create a tenancy by the entirety. This form of property ownership differs from the joint tenancy in two respects. First, each tenant has an indefeasible right of survivorship in the entire tenancy, which cannot be defeated by any act taken individually by either spouse during his or her lifetime. There can be no partition. Second, the spouses do not have an equal right to control and possession of the property. The husband during his lifetime has paramount rights in the property. In the event of divorce the tenancy by the entirety becomes a tenancy in common unless the divorce decree reflects that a joing tenancy is intended.

III

The stage was set for the instant case by this court's decision in *Klein* v. *Mayo,* 367 F.Supp. 583 (1973) aff'd, 416 U.S. 953, 94 S.Ct. 1964, 40 L.Ed.2d 303 (1974). In *Klein* a three-judge panel heard a challenge to the constitutionality of Mass.Gen.L. ch. 241, § 1. This statute states that:

> Any person, except a tenant by the entirety, owning a present undivided legal estate in land, not subject to redemption, shall be entitled to have partition in the manner hereinafter provided.

The plaintiff in that case was in precisely the same position as the present plaintiff, having separated from her husband without dissolving her marriage and desiring an equal share in their property, which was held under a tenancy by the entirety. She contended that, given the bias of the tenancy by the entirety, the quoted statute barring partition descriminated against her on the basis of sex. In ordering judgment of the defendant, *Klein* separated the issue of partition from the issue of any underlying bias in the common-law tenancy by the entirety. Finding that partition was equally unavailable to men and women holding property by the entirety, the court concluded that the challenged statute was nondiscriminatory, while leaving open the possibility of future direct challenges to the tenancy by the entirety itself.

This action presents the challenge referred to in *Klein,* a direct attack on the common law tenancy by the entirety. No challenge to a statute is involved here. Plaintiff's original claim for an order of partition and sale by this court has been waived. Plaintiff asks this court to "restrain and enjoin the defendant from collecting any rents or profits from the above described premises," and to "restrain and enjoin the defendant from continuing to exercise exclusive possession and control over the above described premises."

IV

Defendant has not contested the court's jurisdiction in this case, but defends on the merits, contending that the Massachusetts tenancy by the entirety does not discriminate against women.

As was conceded in *Klein,* the common-law concept of tenancy by the entirety is male oriented.

It is true that the only Massachusetts tenancy tailored exclusively for married persons appears to be balanced in favor of males. There is no equivalent female-biased tenancy, nor is there a "neutral" married-persons' tenancy providing for indefeasible survivorship but not vesting paramount lifetime rights in the male. Married couples may, it is true, elect a joint tenancy, a tenancy in common, or a sole tenancy. However, the survivorship feature of a joint tenancy may be destroyed by partition. A wife who wants the security of indefeasible survivorship can achieve it only by means of a male-dominated tenancy. 367 F.Supp. at 585.

But, the dispositive issue is not merely whether the tenancy by the entirety favors males. Rather, the issue is whether it does so in a manner that creates a constitutionally impermissible classification. On the specific facts of this case, this court holds that tenancy by the entirety, being but one option open to married persons seeking to take title in real estate, is constitutionally permissible.

It may be possible in some future case for a plaintiff wife to demonstrate factually that a selection of tenancy by the entirety was made through coercion, ignorance or misrepresentation. But no such facts were presented here, nor does plaintiff advance such a theory. Rather, this record makes almost inescapable the conclusion that plaintiff freely entered into a contract along with her husband in 1961, selecting one among several options open to her. Events have not transpired as she expected and now she seeks to revise the terms of her contract, because she now feels that her husband got the better end of the bargain.

The Commonwealth permits tenancy by the entirety as one available option of property ownership. The plaintiff now complains of its consequences. But, as indicated in *Klein,* this is not to say that other married women would not prefer a vehicle giving husbands the right to possession, while at the same time foreclosing the possibility of partition, thereby preserving the family homestead even during bitter but temporary periods of separation.

This court is sympathetic with plaintiff's concern that the tenancy by the entirety is to some degree a legal artifact, formerly justified by the presumed incompetence of women to manage property. But this decision is not based on such an archaic and patently invalid stereotype. Rather, the fact is that, regardless of its roots, the tenancy by the entirety exists today as one of several options open to married persons seeking to purchase real estate. Its existence constitutes a matter of choice not discrimination. If there is a classification, it is one selected by the plaintiff, not one imposed by the Commonwealth. She is entitled to the benefit of her bargain, no more and no less. There being no evidence that the plaintiff in this case made her choice among then existing options other than freely, this court will not step in to re-write the agreement between her and her husband.

The court orders entry of judgment for the defendant.

QUESTIONS FOR DISCUSSION

1. *What is the difference between a joint tenancy and tenancy by the entirety?*
2. *Why did the court say that the common-law tenancy by the entirety is not unconstitutional although admitting that it is "male dominated"? Are you persuaded by the court's argument? Why or why not?*

28:4. Acquiring Title to Personal Property. It is appropriate at this point to discuss some concepts that have specific application to personal property. First, it is important to distinguish between *possession* of property and *title* to property. One may possess property without having title to it. Thus, if one leaves his or her car with a parking garage, the garage may have possession of the car but not the title to the property. There are a number of methods of acquiring title to property. Among these are the acquisition of title by purchase, which is covered in Section Three, "Sales." The acquisition of title by inheritance and by will are discussed in Chapter 32, "Wills, Trusts, and Decedent's Estates." In addition, one may acquire title to property by gift, accession, or confusion; by taking possession of wild, abandoned, lost, or mislaid property; and by creation.

(1) Title by Gift. One method by which an individual acquires the title to personal property is to be the recipient of a gift. The act of making a gift of property when one is feeling generous would seem to be rather uncomplicated, but as you may have realized by now, it is fraught with legal implications and potential pitfalls. If Danny Donor has a ring and he wishes to give it to Doris Donee now, so that it will be her personal property now and forever more, the procedure is not very difficult. Obviously he wants to make the gift *inter vivos,* while he and Doris are alive. However, if the gift is of an intangible, an additional complicating factor is introduced into the problem. More complicated still is the case in which a donor wishes to make a conditional gift of his property on the assumption that he is going to die from some disease in the relatively near future. Such a gift is called a *gift causa mortis.*

You are already armed with the knowledge that there are two types of gifts: (1) *gifts inter vivos* and (2) *gifts causa mortis.* A gift *inter vivos* is the most common gift situation: *Inter vivos* means "between the living," hence a gift *inter vivos* is a gift between the living. For a valid gift *inter vivos* several requirements must be met: (1) the parties must have contractual capacity: (2) the gift must be made voluntarily; (3) there must be an intent to make a gift, that is, to give title to the property to the donee; (4) the donee must accept the gift; and (5) there must be delivery of the property.

The elements of a valid gift *inter vivos* do not require much explanation except perhaps for the element of contractual capacity and delivery. If one does not have the capacity, he or she may not make a valid gift *inter vivos.* By definition a gift is not a contract; there is no consideration. Nonetheless the donor must have capacity to make a gift. Thus, if an infant attempts to make a gift, he or she may later recover the property from the donee.

Most of the legal problems concerning gifts *inter vivos* arise from the requirement of delivery. If there is no delivery of the property, there has not been a completed gift vesting title in the donee. Suppose Danny Donor is lying on his deathbed and states to Doris Donee, "I give you $1,000, which is in an envelope with your name on it located in my desk." Danny then expires without ever delivering the money to Doris. There is no valid gift, title to the money never passes from Danny to Doris, and it remains part of his estate. The reason is that there was no valid delivery. Had Danny handed the money to Doris prior to his death, the gift would have been valid, title to the money would have passed to Doris, and, incidentally, Danny could not have revoked the gift because title passed from him to Doris.

The next question to be considered is what constitutes delivery. Obviously physically handing the property to the donee constitutes delivery. However, what if the $1,000 Danny wanted to give Doris in the example was contained in an envelope located in Danny's safe-deposit box at his bank? If Danny gave Doris the keys to this box, would this constitute sufficient delivery? The answer given by most courts would be yes, if this was all that was necessary to gain admission to the box. Similarly delivery of the keys to a car, the passbook to a savings account, or the delivery of some other document of title, such as a warehouse receipt, would often constitute delivery. Thus there may be symbolic or constructive delivery of the gift.

Another problem that sometimes arises with regard to the making of a valid gift concerns donative intent. By a transfer of the property did the donor intend to make a gift? The problem is most troublesome when third persons are involved. As a general rule, if the donor transfers property to his own agent with instructions to turn the property over to the donee at some later time, then there is no valid gift because the donor can recover the property from his own agent. On the other hand, if the property is given to the agent of the donee, the gift is completed because the donor has parted with control of the property.

A gift *causa mortis* is a gift in contemplation of death. A gift *causa mortis* is made by the donor on the assumption that he or she will die from some peril that the donor considers imminent. It may be a serious illness or the fact that the donor is undergoing a serious operation or a dangerous trip. The gift is made on the condition that the donor dies from this peril or, if he dies from some other cause, at least expires within the time period during which he expects the "peril" to "get him." If the donor survives the operation or the trip or is miraculously cured, then the gift if revoked and may be recovered from the donee. This is different from a gift *inter vivos,* which is irrevocable. Just as with a gift *inter vivos,* the property must be delivered to the donee during the donor's lifetime, with the intent to make a gift conditioned on the donor's death.

IN RE ESTATE OF EVANS
356 A.2d 778 (Pa. 1976)

Vivian Kellow was the niece of Arthur Evans's deceased wife. She cared for Mrs. Evans for a number of years before her death and then cared for Mr. Evans. She cooked a hot meal for him every day, did his laundry, and made sure the house was tidy. She continued to do this after her marriage. In May of 1971 Mr. Evans moved into Mrs. Kellow's home. He was confined to bed on occasions but took walks and made trips to the bank.

On October 22, 1971, Arthur Evans went to the bank and spent some time there. Before leaving, he obtained both keys to his safety-deposit box. He told a number of people that he gave the contents of the safety-deposit box at the bank to Vivian Kellow. He gave her both keys to the box but did not change the name of the owner of the box on the records at the bank.

On November 5, 1971, Mr. Evans entered the hospital for the last time and died on November 23, 1971. Kellow reluctantly reliquished the keys to the safety-deposit box to a bank officer. When the box was opened approximately $800,000 in bonds and corporate stock was found, as well as a will. The will was dated September 16, 1965, and by its terms left Kellow $1,000.

The executor included the contents of the box as part of the estate. Kellow contended that the contents of the box were not part of the estate and were the subject of an inter vivos *gift to her from Evans. From a lower court verdict in favor of the estate. Kellow appealed.*

NIX, J.

The lower court correctly noted that the requirements for a valid inter vivos gift were donative intent and delivery, actual or constructive. With respect to donative intent, the court found:

> Turning to the facts of this case, certainly no one can reasonably argue that Arthur Evans lacked sufficient motive to make a gift to Vivian. The record clearly manifests, both by his conduct and his statements, donative intent, the first prerequisite.

Nevertheless, the court ruled that no delivery had been made. This result was predicated upon a finding that the deceased had not divested himself of complete dominion and control over the safe deposit box. After properly noting that constructive delivery is sufficient when manual delivery is impractical or inconvenient, the court reasoned:

> The record contains no evidence of circumstances which were such that it was impractical or inconvenient to deliver the contents of this box into the actual possession or control of Vivian.
>
> Arthur Evans, although suffering physical infirmities and apprehensive of death, was nonetheless ambulatory. On October 22, 1971, he appeared at the Nanticoke National Bank in the company of Harold Turley and Leroy Kellow and spent approximately one hour going over the contents of his safe deposit box in a cubicle provided in the bank for that purpose. He left the bank after redepositing the contents and took with him only the keys which independent testimony indicates he delivered to Vivian the next day. There was no manual delivery of the contents. The contents of the box remained undisturbed. The box, and its contents, were registered in the name of the decedent at the date of his death. The objects of the gift were not placed in the hands of Vivian, nor was their placed within her power the means of obtaining the contents.

In the instant case, the controversy focuses on whether there was an adequate delivery. In *Allshouse's Estate,* 304 Pa. 481, 487–488, 156 A. 69, 72 (1931), we elaborated on the requirement of delivery:

> As said in *Walsh's App.,* 122 Pa. 177, 187, 15 A. 470, 471, 1 L.R.A. 535, 9 Am. St.Rep. 83: "If there remains something for the donor to do before the title of the donee is complete, the donor may decline the further performance and resume his own," and again, at page 190 of 122 Pa., 15 A. 470, 472; "(i)t is not possible that a chancellor would compel an executor or administrator to complete a gift by the doing of any act which the alleged donor if living might have refused to do, and thereby revoked his purpose to give. . . ." Again, as we stated in *Clapper* v. *Frederick,* 199 Pa. 609, 613, 49 A. 218, 219: "Without a complete delivery during the lifetime of the donor there can be no valid gift inter vivos. 'Though every other step be taken that is essential to the validity of the gift, if there is no delivery, the gift must fail. Intention cannot supply it; words cannot supply it; actions cannot supply it. It is an indispensable requisite, without which the gift fails, regardless of consequence': Thornt. Gift. p. 105."

We have recognized that in some cases due to the form of the subject matter of the gift or due to the immobility of the donor actual, manual delivery may be dispensed with and constructive or symbolic delivery will suffice. In *Ream Estate,* 413 Pa. 489, 198 A.2d 556 (1964), for example, the Court found there had been a valid constructive delivery of an automobile where the donor gave the keys to the alleged donee and also gave him the title to the car after executing an assignment of it leaving the designation of the assignee blank. The assignment was executed in the presence of a justice of the peace and the evidence was overwhelming that the name of the donee was to be inserted upon the death of the decedent. . . . In the instant case, appellant did not have dominion and control over the box even though she was given the keys to it. The box remained registered in Mr. Evans' name and she could not have gained access to it even with the keys. Mr. Evans never terminated his control over the box, consequently he never made a delivery, constructive or otherwise.

Although appellant suggests that it was impractical and inconvenient for Mr. Evans to manually deliver the contents of his box to her because of his physical condition and the hazards of taking such a large sum of money out of the bank to her home, we need only note that the deceased was obviously a shrewd investor, familiar with banking practices, and could have made delivery in a number of simple, convenient ways. First, he was not on his deathbed. He was ambulatory and not only went to the bank on October 22, 1971, but took walks thereafter and did not enter the hospital until November 5, 1971. On the day he went to the bank he could have rented a second safe deposit box in appellant's name, delivered the contents of his box to it and then given the keys to appellant. He could have assigned the contents of his box to appellant. For that matter, he could have written a codicil to his will.

The lower court noted that the deceased was an enigmatic figure. It is not for us to guess why people perform as they do. On the record before us it is clear that regardless of Mr. Evans' intention to make a gift to appellant, he never executed that intention and we will not do it for him. On these facts, we are constrained to hold that there was not an inter vivos gift to appellant and that the contents of the safe deposit box were properly included in the inventory of Mr. Evans' estate.

(Affirmed.)

ROBERTS, J. (dissenting).

I dissent. The central issue in this case is whether donor made an adequate delivery of the gift to donee. The majority finds that adequate delivery was not made because the safe deposit box was leased solely in donor's name and supports this conclusion by pointing out that there were several alternative means of delivering the gift which would have been adequate. I believe that the inquiry should not be what form of delivery would have been clearly sufficient, but rather whether the delivery made by donor was adequate. I believe that it was.

QUESTIONS FOR DISCUSSION

1. *Why did the majority opinion determine that there was no delivery of the contents, of the safety-deposit box?*
2. *Under what circumstances, according to this case, would the delivery of the keys have been sufficient to transfer title of the contents of the box?*

CLARK v. O'NEAL

355 S.W.2d 68 (Mo. App. 1977)

Harold J. Hawkins was admitted to the hospital in Chillicothe on June 16, 1973, with a heart attack. Teri O'Neal was his niece and, with her husband, Merle, went to visit Hawkins in the hospital on several occasions. During those visits, he told them he had terrific pain and did not believe he would ever make it out of the hospital.

When Teri and Merle visited Hawkins on June 17, he asked if they had some certificates of deposit, a deed of trust, and a note. When they replied that they did not have them, Hawkins asked if they could get a key to his house and get them. Teri said she could. She then went to Hawkins's home and picked up an envelope containing thirteen certificates of deposit, a note, and a deed of trust.

When Merle and Teri returned to the hospital, Hawkins asked whether thay had the certificates and the note and Teri replied she did. Hawkins then said "Good, now you keep it, its yours." That night Hawkins died.

Leeta Clark as administratrix of Hawkins's estate filed this action, claiming that the certificates of deposit totaling $22,000.00 and a note secured by a deed of trust in the amount of $34,000.00 payable to Hawkins was being wrongfully detained from her by Teri O'Neal.

WASSERSTROM, J.

Plaintiff narrows the scope of inquiry on this appeal by limiting her attack to only two of the constituent elements required for a gift causa mortis. She states in her brief. "Thus, the issues before this court concerning the validity of the gift causa mortis are reduced to whether or not a sufficient delivery under the existing law together with donative intent has been established by the defendant's evidence."

Viewing the question of Hawkins' donative intent in light of the foregoing rule, there can be no doubt that he intended to make the gift, if the testimony of Merle O'Neal is believed. Plaintiff strives strenuously to discredit that testimony by pointing out minor alleged discrepancies, some slight variance from a prior extra-judicial statement, and the financial interest of this witness inasmuch as he is the donee's husband. . . .

It cannot be said that the evidence here on donative intent fails to measure up to the legal standard.

As to the issue of delivery, "[n]o absolute rule can be laid down as to what will constitute a sufficient delivery to support a gift in all cases, for in each case the character of the delivery requisite to sustain the transaction as a gift will depend very largely on the nature of the subject matter of the gift and the situation and circumstances of the parties . . . it is not necessary that there be a manual delivery, or an actual tradition from hand to hand, where . . . constructive or symbolic delivery would be sufficient." 38 C.J.S. Gifts § 82, p. 902–3. Under the facts here, the instructions from the donor to go to his house, find the house key, enter the house and take possession of certain documents is not too far different from handing over the keys to a safe deposit box, which has been held a sufficient delivery numerous times. Even more directly to the point, actual physical delivery to the donee by the donor has been held unnecessary when the donee already had the property in possession at the time of the donative expression.

(Judgment in favor of O'Neal affirmed.)

QUESTIONS FOR DISCUSSION

1. *How does the gift* causa mortis *differ from a gift* inter vivos?
2. *If Hawkins had survived and left the hospital would he have been able to recover the property from O'Neal?*

---

(2) **Accession.** The concept of accession dates back to a civil-law concept that the owner of property is entitled to all that is produced by or added to it whether naturally or throught the efforts of individuals. The word *accession* means literally an increase or addition. Problems arise in two basic factual situations. Suppose Harry is the owner of lumber and George makes the lumber into a cabinet. Who is the owner of the cabinet? In this case, the lumber changed its form, and value was added to it by the creation of the cabinet, but Harry is still the owner. Another type of accession occurs when a chattel owned by one individual is joined with that owned by another. Who is the owner of the new chattel?

In the first example the legal problems are usually not difficult because the issue of ownership is determined by a prior agreement and in fact rarely comes up. Harry is the owner of lumber and hires George to convert it into a cabinet. Harry is the owner of the cabinet and owes George for the value added to the cabinet. Complications arise, however, when there is no prior agreement and an individual innocently converts the property to his or her own use and changes its form. The owner of the property is always allowed to recover damages for the conversion of his or her property, regardless of whether the conversion is innocent. But the issue here is who has title to the property and therefore the ultimate right to possession of it. For example, let us assume that George uses the lumber not knowing that it belongs to Harry and that its value is $15. After George has finished, the new cabinet is worth $75. The older cases used an *identity* test in order to establish title; that is, if the original chattel was changed to such an extent that it lost its identity, then the innocent converter had title to the chattel. The modern cases and the majority rule today depend on a *relative-value* test. If the labor and the material added by the innocent converter greatly add to the value of the property, even if its identity is not lost, then the innocent converter gets title to the property, even if its identity is not lost, then the innocent converter gets title to the property, although he is liable for damages for the conversion. Thus George would probably get to keep the cabinet, although he would be liable to Harry for damages in converting the $15 worth of lumber.

If the conversion of the property was not innocent, the legal problems are simple. The willful converter cannot benefit from his intentional conversion and therefore does not get title to the property, regardless of how much value was added to it. The owner of the original property may recover either the property itself in its improved condition to recover damages for the conversion of the original property. As a general rule, when the property itself is recovered, the owner need not compensate the intentional converter for the value of the improvements.

A slightly different problem is presented when accession occurs as the result of the joining of two chattels owned by two different individuals. If the united chattels cannot be separated without damage to the original chattels, then an accession occurs and title to the joint product goes to the owner of the principal chattel. Thus, if Harry owns an automobile and George owns a muffler and exhaust system, when the muffler and exhaust system is joined to the automobile, Harry has title to the muffler and exhaust system because the automobile is the principal chattel. This rule applies regardless of whether there was any conversion of the property, although damages for the conversion may be recovered.

MOSSLER ACCEPTANCE CO. v. NORTON TIRE CO., INC.

70 So.2d 360 (Fla. 1954)

Bruton entered into a conditional sales contract for the purchase of an automobile from a Florida car dealer. Bruton then purchased a new set of tires under a retain-title contract from Norton Tire Center. Subsequently Bruton defaulted on the car payments and it was repossessed by Mossler Acceptance Co. Norton then sued Mossler to recover the unpaid balance of the purchase price for the tires. A verdict adverse to Mossler was rendered in the lower court. Mossler then petitioned the Supreme Court of Florida for a writ of certiorari to review that judgment.

SEBRING, J.

The question for decision is whether under the facts stated Mossler Acceptance Company obtained title to the tires and tubes that were on the car at the time of repossession.

The principle involved in this case is that of accession, which as a general doctrine, is not disputed: "Accessories added to a chattel, if so incorporated as to be incapable of serverance without injury to the whole, merge in the principal thing and become subject to the rights of ownership of the conditional vendor thereof. . . ." 1 Am.Jur. 201. As respects this principle, in the field of automobile law the rule is stated in varying language: "Under the rules of accession when attachments on a car can be easily distinguished and separated, no change of property takes place, providing the separation can be made without injury to the automobile." Blashfield, Cyclopedia of Automobile Law and Practice, permanent edition, Vol. 7A. § 4687.

There can be no doubt that "the ordinary repairs upon a personal chattel . . . become accretions to, and merge in, the principal thing, and become the property of the general owner." Blashfield, § 4687, supra.

In recognition of this principle, it has been held in Florida that materials and labor involved in repairs to an automobile become a part of the vehicle, and that claims by persons supplying such materials or labor are subject to prior liens on the car.

In the present case, however, there is no real question as to the priority of liens. The conditional vendor of the automobile undoubtedly has a prior lien on that vehicle, and the only issue is whether certain items also sold by conditional sale have become a part of the vehicle to which they were affixed, or whether they remain independent and separable chattels subject to their vendor's claim. In respect to this issue it is held that *"where the seller of an automobile under a contract for conditional sale retakes the automobile upon default of the buyer to keep the terms of the contract, he is entitled to any tires or other replacements which a purchaser placed on the machine while it was in his possession, provided the title to such parts passed to the purchaser when he acquired them."* (Emphasis supplied.) Berry on Automobiles (6th Ed.) Vol. 2, § 1806. *"Where, however, such equipment is furnished by a third person, in whom the title remains, they do not belong to the conditional seller upon retaking the car under a contract recognizing that such equipment is separable and not accessions."* (Emphasis supplied.) Berry, § 1806, supra.

Where, as in the case at bar, the controversy is between two sellers of chattels, both retaining title to secure payment of the purchase price, we think the more numerous and better reasoned cases hold that tires sold under a conditional sales agreement and placed on an automobile similarly pledged do not become part of the vehicle by the principle of accession.

As is stated in *Goodrich Silvertown Stores v. Caesar*, . . . "*'the doctrine of accession is inapplicable in cases where personal property is placed upon other personal property if the property so placed had not become an integral part of the property to which it was attached and could be conveniently detached . . . we think it plain that one who attaches tires which he does not own to a motor truck which he does not own does not thereby pass title in the former to the owner of the latter.'*" (Emphasis supplied.)

The rules herein set forth control the present case, in our opinion, and require that the writ of certiorari be denied.

(3) **Confusion.** Confusion of personal property occurs when the property of different owners is mingled so that the original units cannot be identified and returned to the original owner. This situation occurs most frequently with regard to fungible goods, that is, goods of which each unit is like every other unit. Coal and grain are the classic examples. There may also be confusion of nonfungible property, such as cattle and horses. If confusion of property occurs innocently, the original owners are entitled to recover their proportionate interest in the entire mass. This is particularly important if a partial destruction of the mass occurs. Suppose Albert, Barney, and Carl store, respectively, 5,000; 5,000; and 10,000 bushels of grain in a granary. The grain becomes mingled when half of the grain is destroyed. They would each be entitled to their proportionate shares of the remaining mass, or 2,500; 2,500; and 5,000 bushels.

If the confusion is the result of misconduct of one of the parties, it presents a different problem. In this case, the innocent party has a claim to the entire mass. However, if the entire mass has a value equal to or greater than the total per-unit value prior to the confusion, then the wrongdoer may recover his share of the mass, provided that he can establish this fact. On the other hand, if the per-unit value of the mass is less after the confusion than prior to the mingling of the property such that the innocent party cannot recover his prior per-unit value, then the wrongdoer can recover nothing.

(4) **Lost, Mislaid, and Abandoned Property.** Lost property results when one inadvertently or through negligence drops or misplaces property where he has no intent to place it and has no idea where the property is located. The basic law is that a finder of lost property has good title to the property against the whole world except for the true owner. The true owner is entitled to recover his property even if the finder of the property sells it to a bona fide

purchaser for value. The finder of lost property may prevail even against the person on whose property the lost property is found. Thus, if Hawkeye Finder discovers a diamond ring on the front lawn of a home belonging to Harry Homeowner, Finder has the right to the ring against everyone but the true owner. Many states have statutes allowing the finder of lost property to gain the absolute title to the property by advertising in the paper for a certain number of times. After the finder complies with the statute, he has title to the property, and the individual who lost it cannot recover the property from the finder.

Abandoned property is property to which the owner has given up all rights, claim, title, and possession with the intention of not reclaiming it. The first person to come along and take possession of the property then has title to it. Suppose Albert Abandoner deposits an old chest along the side of the road and then drives off, never intending to return for the chest. Pearl Possessor then sees the chest, stops, and places it in her car. Pearl now has title to the chest.

The case of mislaid property presents a more difficult problem. Mislaid property is distinguished from lost property because in the case of mislaid property the owner intentionally places it somewhere but forgets the location. In the case of lost property, the owner has no specific intent to place the property where it is found. When it is determined that property has been mislaid rather than lost, the person on whose premises the property is found may hold the mislaid property, even when he is not the finder, as bailee for the true owner. Thus suppose one lays his wallet on the counter of a jewelry store and then walks away without it. If another customer picks it up, the store owner could require that the wallet be turned over to him to hold until the owner comes to claim it. In this case, the owner of the premises has a superior right over the finder, whereas if the property were lost, the finder would have the superior right.

FAVORITE v. MILLER

407 A.2d 974 (Conn. 1978)

BOGDANSKI, J.

On July 9, 1776, a band of patriots, hearing news of the Declaration of Independence, toppled the equestrian statue of King George III, which was located in Bowling Green Park in lower Manhattan, New York. The statue, of gilded lead, was then hacked apart and the pieces ferried over Long Island Sound and loaded onto wagons at Norwalk, Connecticut, to be hauled some fifty miles northward to Oliver Wolcott's bullet-molding foundry in Litchfield, there to be cast into bullets. On the journey to Litchfield, the wagoners halted at Wilton, Connecticut, and while the patriots were imbibing, the loyalists managed to steal back pieces of the statue. The wagonload of the pieces lifted by the Tories was scattered about in the area of the Davis Swamp in Wilton and fragments of the statue have continued to turn up in that area since that time.

Although the above events have been dramatized in the intervening years, the unquestioned historical facts are: (1) the destruction of the statue; (2) cartage of the pieces to the Wolcott Foundry; (3) the pause at Wilton where part of the load was scattered over the Wilton area by loyalists; and (4) repeated discoveries of fragments over the last century.

In 1972, the defendant, Louis Miller, determined that a part of the statue might be located within property owned by the plaintiffs. On October 16 he entered the area of the Davis Swamp owned by the plaintiffs although he knew it to be private property. With the aid of a metal detector, he discovered a statuary fragment fifteen inches square and weighing twenty pounds which was embedded ten inches below the soil. He dug up this fragment and removed it from the plaintiffs' property. The plaintiffs did not learn that a piece of the statue of King George III had been found on their property until they read about it in the newspaper, long after it had been removed.

In due course, the piece of the statue made its way back to New York City, where the defendant agreed to sell it to the Museum of the City of New York for $5500. The museum continues to hold it pending resolution of this controversy.

In March of 1973, the plaintiffs instituted this action to have the fragment returned to them and the case was submitted to the court on a stipulation of facts. The trial court found the issues for the plaintiffs, from which judgment the defendant appealed to this court. The sole issue presented on appeal is whether the claim of the defendant, as finder, is superior to that of the plaintiffs, as owners of the land upon which the historic fragment was discovered.

Traditionally, when questions have arisen concerning the rights of the finder as against the person upon whose land the property was found, the resolution has turned upon the characterization given the property. Typically, if the property was found to be "lost" or "abandoned," the finder would prevail, whereas if the property was characterized as "mislaid," the owner or occupier of the land would prevail.

Lost property has traditionally been defined as involving an involuntary parting, i.e., where there is no intent on the part of the loser to part with the ownership of the property. Abandonment, in turn, has been defined as the voluntary relinquishment of ownership of property without reference to any particular person or purpose; i.e., a "throwing away" of the property concerned; while mislaid property is defined as that which is intentionally placed by the owner where he can obtain custody of it, but afterwards forgotten.

It should be noted that the classification of property as "lost," "abandoned," or "mislaid" requires that a court determine the intent or mental state of the unknown party who at some time in the past parted with the ownership or control of the property.

The trial court in this case applied the traditional approach and ruled in favor of the landowners on the ground that the piece of the statue found by Miller was "mislaid." The factual basis for that conclusion is set out in the finding, where the court found that "the loyalists did not wish to have the pieces [in their possession] during the turmoil surrounding the Revolutionary War and hid them in a place where they could resort to them [after the war], but forgot where they put them."

The defendant contends that the finding was made without evidence and that the court's conclusion "is legally impossible now after 200 years with no living claimants to the fragment and the secret of its burial having died with them." While we cannot agree that the court's conclusion was legally impossible, we do agree that any conclusion as to the mental state of persons engaged in events which occurred over two hundred years ago would be of a conjectural nature and as such does not furnish an adequate basis for determining rights of twentieth century claimants.

The defendant argues further that his rights in the statue are superior to those of anyone except the true owner (i.e., the British government). He presses this claim on the ground that the law has traditionally favored the finder as against all but the true owner, and that because his efforts brought the statue to light, he should be allowed to reap the benefits of his discovery.

Although few cases are to be found in this area of the law, one line of cases which have dealt with this issue has held that except where the trespass is trivial or merely technical, the fact that the finder is trespassing is sufficient to deprive him of his normal preference over the owner of the place where the property was found.

The basis for the rule is that a wrongdoer should not be allowed to profit by his wrongdoing. Another line of cases holds that property, other than treasure trove, which is found embedded in the earth is the property of the owner of the locus in quo. The presumption in such cases is that possession of the article found is in the owner of the land and that the finder acquires no rights to the article found.

The defendant, by his own admission, knew that he was trespassing when he entered upon the property of the plaintiffs. He admitted that he was told by Gertrude Merwyn, the librarian of the Wilton Historical Society, *before* he went into the Davis Swamp area, that the land was privately owned and that Mrs. Merwyn recommended that he call the owners, whom she named, and obtain permission before he began his explorations. He also admitted that when he later told Mrs. Merwyn about his discovery, she again suggested that he contact the owners of the property, but that he failed to do so.

In the stipulation of facts submitted to the court, the defendant admitted entering the Davis Swamp property "with the belief that part of the 'King George Statue' . . . might be located within said property and with the intention of removing [the] same if located." The defendant has also admitted that the piece of the statue which he found was embedded in the ground ten inches below the surface and that it was necessary for him to excavate in order to take possession of his find.

In light of those undisputed facts the defendant's trespass was neither technical or trivial. We conclude that the fact that the property found was embedded in the earth and the fact that the defendant was a trespasser are sufficient to defeat any claim to the property which the defendant might otherwise have had as finder.

Where the trial court reaches a correct decision but on mistaken grounds, this court has repeatedly sustained the trial court's action if proper grounds exist to support it. The present case falls within the ambit of that principle of law and we affirm the decision of the court below.

(Affirmed in favor of the plaintiff.)

QUESTIONS FOR DISCUSSION

1. *Did the trial court determine that the property had been lost, abandoned, or mislaid? Who is entitled to the property under each of these three circumstances?*
2. *Would the outcome have been altered had Miller obtained permission to come onto the property of Favorite?*

---

(5) Title by Creation. Title to personal property may be gained by the creation of the property. The classic example of this, of course, is the artist who creates a painting. The same is true of the writer or inventor. Our patent, copyright, and trademark laws recognize the property interest in these creations and protect the interest of the creator when the property requirements are met.

28:5. FIXTURES. A fixture is a classification of property that changes its nature, chameleonlike from personal to real property. A fixture may be defined as personal property that becomes so firmly affixed or attached to real property that it becomes part of it. Many disputes center around the problem of whether a certain article is a fixture, because if the property is a fixture, as a general rule, it goes with the realty. For example, suppose Barney Buyer contracts to purchase a house from Sam Seller. Do the wall-to-wall carpeting, television antenna, drapes, storm windows, kitchen cabinets, and mailbox go with the property? The answer depends upon whether the item is considered as fixture. The parties could, of course, resolve the issue by stating their intent in the contract, but frequently they do not. The intent of the annexor is determinative as to whether the property qualifies as a fixture.

The courts frequently state that the test for determining whether property qualifies as a fixture depends on three elements: (1) annexation to the realty; (2) adoption to the use to which the realty is devoted; and (3) the intent of the annexor that the property become a permanent accession to the realty. These elements really go to determining the intent of the parties.

In its most simple form annexation results from some method of actually attaching the object to the real property. Thus, if one bolts or nails kitchen cabinets to the wall, they are annexed to the realty. The purpose of the annexation requirement is to show that the annexor's intent is to make the object part of the realty. It is not necessary, however, that the object be permanently attached to the realty. The object may be "constructively annexed." For instance, specially made storm doors to storm windows may obviously be intended to be part of the realty because they are specially adapted to the realty. On the other hand, if the object is so firmly affixed that it cannot be removed from the realty without damage to the realty, then there would be little doubt that it is fixture. A kitchen sink or plumbing are obvious examples.

The intent of the party who annexes the object may be determined from his relation to the real estate. For instance, if a person owns the home in which he lives, it is likely that personal property that he annexes to the real estate is intended to be a fixture. On the other hand, if an individual merely rents an apartment or a house, it is less likely that he intends that the objects that he attaches to the realty should become fixtures and thus part of the realty.

The nature of the fixture is also a factor in determining whether the parties may remove it. If the fixture is a *trade fixture,* the courts are more likely to allow the annexor to remove the object. A trade fixture is an object that one uses in connection with his trade. Thus the shopkeeper who rents a store would be able to remove his display cases from the premises after the termination of the lease, so long as the realty is not materially injured.

SEARS, ROEBUCK & CO. v. SEVEN PALMS MOTOR INN
530 S.W.2d 695 (Mo. 1975)

When the Seven Palms Motor Inn was constructed, Sears, Roebuck supplied drapes and bedspreads to the motel as well as labor to install the drapes and traverse rods on which to hang the drapes. Seven Palms Motor Inn did not pay Sears, which then tried to establish a mechanic's lien on these items. Whether Sears had a right to the mechanic's lien depended upon the determination of whether any of these objects qualified as fixtures. The bedspreads and drapes throughout the motel were matching in color and design and were specially ordered and custom-made for the motel.

HENLEY, J.

Characterization of an item as a fixture, something otherwise personal but attached to realty under such circumstances as to become part of it, depends upon the finding of three elements: annexation to the realty, adaption to the use to which the realty is devoted, and intent of the annexor that the object become a permanent accession to the freehold.

Appellants (Seven Palms) contend that neither the drapes nor the bedspread are fixtures . . . and therefore not lienable, because they are not annexed or attached to the building.

In *Crane Co.* v. *Epworth Hotel Construction and Real Estate Co.*, 121 Mo.App. 209, 98 S.W. 795 (1906) the plaintiff sought a mechanic's lien for fire hose (and hose racks) sold to defendant and attached to standpipes installed to convey water for fire protection purposes and run up through the floors of the Epworth Hotel, the hose being attached by screwing it to a valve projecting from the pipe. The court noted that the hose could be either attached or detached by a man with his bare hand, and this could be done without marring the pipe.

Stating that it was intended that this attachment be permanent, the court held that the hose was a fixture and lienable. In reaching that conclusion the court said (98 S.W. at 797): "Of (the elements of annexation, adaptation, and intent) in modern times the latter two are more important than the one relating to the method by which the chattel is attached to the freehold. Such annexation though slight and easily displaced, will not prevent an article becoming a fixture which is adapted to the proper use of a building, and which was placed therein by the owner with the intent of forming a part of the special object and design for which the building was constructed."

The purpose of attaching the traverse rods to the realty was to hang drapes therefrom which could be opened or drawn across a window by the motel's guest to control the light in his room or secure his privacy. Of itself, the traverse rod attached to the wall above the window in the room did not accomplish this purpose. To serve this purpose it was essential that the drapes be provided and attached to the rod. They were provided and attached, and became an integral part of the instrument designed for use in connection with the window in the guest's room. As such, the drapes were as much a fixture as the traverse rod itself. It is obvious that the rod and drapes, as a unit, were adapted to the proper use of rooms in a motel and were placed therein with the intent they would form a part of the special purpose for which the building was designed to be used.

Not so the bedspreads. Respondent admits that those items are not physically attached to the realty in any way but insists that they have been "constructively annexed." In support of this proposition, respondent argues: the rods are physically fastened to the building; the drapes are affixed to the rods by hooks, the bedspreads match the drapes; a fortiori, the bedspreads "are at least 'constructively annexed' to the rooms . . . by their relationship with the drapes."

The doctrine of constructive annexation recognizes that a particular article, not physically attached to the land, "may be so adapted to the use to which the land is put that it may be considered an integral part of the land" and "constructively annexed" thereto. 36A C.J.S. Fixtures § 6, pp. 613–614. Since its development, the doctrine has ordinarily been applied to only three types of objects: (a) machinery placed in an industrial establishment for permanent use and necessary to the operation of the plant (sometimes referred to as "the integrated industrial plant rule"); (b) items that are essential to the use of what is clearly a fixture and cannot readily be used independently elsewhere; and, (c) items normally physically attached to the realty that are severed for a temporary purpose such as cleaning or repair.

The bedspreads are not essential to the use of what is clearly a fixture, nor has it been shown that they cannot readily be used independently elsewhere. Respondent asserts that because the bedspreads "were designed to match and to coexist with the drapes" they must be considered part of a matched set which is essential to the use of rods which are clearly fixtures. Respondent seeks support for this contention in cases that have held easily removable parts of machines and other fixtures may not be considered as separate items. . . . However, in each of these cases, the fixture would have been rendered absolutely useless by removal of the items in question, and such items could not readily be used independently elsewhere. There is no indication that the unit of rod and drapes could not serve its function, which respondent says is to "regulate the flow of light and serve the need for privacy," if the bedspreads were removed. That the decor of a guest room in a motel may be more aesthetically pleasing when bedspreads are made of the same material as drapes, falls far short of the functional relationship needed to justify "constructive annexation." Since the bedspreads were not annexed, physically or constructively, they cannot be characterized as fixtures and are, therefore, nonlienable items.

(Remanded to the trial court to impose a lien for the amount of the balance due respondent after deducting the amount charged for the bedspreads.)

Questions for Discussion

1. *What elements did the court state must be considered in order to determine whether an object qualifies as a fixture?*
2. *Which did the court consider the least important of these elements?*
3. *Because the bedspreads were specially designed for the motel, why did the court conclude that they were not fixtures?*

SUMMARY

Property may be either real or personal. Real property is land and that which is grown on or affixed to it. Other property is personal. Tangible property consists of objects or things that have a physical existence. Intangible property consists of legal rights. *Chattel* is another word for personal property.

The forms of multiple ownership are tenants in common, joint tenants, and tenants by the entirety. Tenants in common must have unity of possession. There is no right of survivorship. Joint tenants must have units of time, possession, title, and interest. There is right of survivorship. Tenants by the entirety are husband and wife. The same characteristics as joint tenants apply, with the additional characteristic that one tenant cannot sell his or her interest without the other tenant's permission. In community property states all property acquired during the marriage is shared equally.

Title to personal property may be acquired by gift. A gift may be either *inter vivos* or *causa mortis.* For a gift *inter vivos* there must be contractual capacity, voluntariness, an intent to make the gift, acceptance, and delivery. A gift *causa mortis* has the additional requirement that donor make the gift under the threat of some imminent peril. Personal property may be acquired by accession, which means literally an increase or addition. Title may also be obtained by confusion. This occurs when the property of different owners is mingled so that original units cannot be identified and returned to the original owner.

Lost property results when one drops property where he has no intent to place it and has no idea where the property is located. The finder has good title except against the true owner. On the other hand, when the owner gives up all rights or claim to the property, it is abandoned and the first person to take possession of the property has good title to it.

Mislaid property occurs when one intentionally places property somewhere but forgets the location. The person on whose premises the property is found may keep it for the true owner. Title to property may also be gained by creation, such as by an artist of his painting.

A fixture is personal property that becomes so firmly affixed or attached to real property that it becomes part of it. Special rules apply to trade fixtures.

PROBLEMS

1. William James Owen Barfield died on July 1, 1954. After the funeral on July 2nd, fourteen of his heirs assembled at his home, which stood on eighty acres of land. At that time, Archie Barfield, one of the heirs said, "Well, here is the place, what are we going to do with it. It ain't worth nothing. Well, R. C. has looked after him (the deceased) the most. . . . I am willing for him to have the place, it won't bring nothing." Two or three of the heirs said in the presence of the rest, "Well, it suits me." The other heirs said nothing but voiced no objection. Later the estate of

the deceased applied to have the property sold in order to distribute the property among all the heirs. R. C. Barfield, opposed the procedure, contending that the heirs had given the property to him. Was he correct? Barfield v. Hilton, 231 S.E.2d 755 (1977).

2. In 1973 Alfred C. Knudson was suffering of cirrhosis of the liver. Knowing this he told Bernyce E. Samuel that he had set aside money for her in his safe-deposit box. Knudson died shortly thereafter. An envelope was found in the safe-deposit box with Bernyce Samuel's name on it. Inside the envelope was $10,000.00 in cash. May Bernyce keep the money? Samuel v. Northern Trust Co., 340 N.E.2d 162 (Ill. App. 1975).
3. L. M. Beewar leased a building for his business. The lease provided that Beewar would supply his own heat. Beewar installed a boiler, an oil burner, and other equipment necessary to heat a portion of the building. This was done with the permission of the lessor and placed in the basement of the building. The heating system was attached with bolts and couplings. Its removal would not have damaged the building. Could Beewar remove the system when he left the building? Beewar et al. v. Bear et al., 246 P.2d 1110, 41 Wash. 37 (1952).
4. Betty Dolitsky went to the Dollar Savings Bank of the City of New York to gain access to her safe-deposit box. While in a booth inspecting the contents of her safe-deposit box, Dolitsky discovered a $100 bill in a folder advertising life insurance that was located on a rack attached to the wall of the booth. The $100 bill was turned over to the bank. After one year Dolitsky contended that she was entitled to the $100 because the rightful owner had not claimed the bill. The bank contended that it had a duty to retain the $100 for the rightful owner. Who was entitled to the $100? Dolitsky v. Dollar Savings Bank, 118 N.Y.S.2d 65 (1952).
5. About March 1, 1918, Emory S. Reynolds expected to die from pneumonia. He sent for his good friend Newell and gave him a diamond ring. Reynolds told a nurse that his affairs were in order except for handing over the ring to Newell. Reynolds then recovered from the disease and lived for more than four years. Upon his death the executor of the estate claimed the ring from Newell on the ground that it was a gift *causa mortis* and thus conditional upon Reynolds's death from pneumonia Newell contended that the ring was the subject of a gift *inter vivos* and his to keep. Who was entitled to the ring? Newell v. National Bank of Norwich, 212 N.Y.S. 158 (1925).
6. On May 18, 1945, Mae Stevenson gave Everett O. Newman a key to her safe-deposit box and instructed him to go "and get what was his out of the box." Stevenson died before Newman went to the box. When the safe-deposit box was opened, there was an envelope with Newman's name on the outside that contained a number of gold debenture bonds payable to bearer. There was no evidence of any other key to the box except the one given to Newman. Was there a valid delivery of the property? In Re Stevenson's Estate, 79 Ohio App. 315, 69 N.E.2d 426 (1946).

7. Richard A. Rothchild and Carol S. Cohn, both over twenty-one years of age, were engaged. Richard gave Carol a diamond ring worth $1,000. Unfortunately Richard was killed in an automobile accident shortly before the wedding. His estate claimed the ring contending that it was consideration for an implied contract that was not fulfilled. Could Carol keep the ring? Cohen v. Bayside Federal Savings and Loan Ass'n., 362 Misc.2d 738, 309 N.Y.S.2d 980 (1970).
8. Mark Ellsworth, Jess Ellsworth, and David Gibson, three small boys, were playing in a salvage yard owned by Dwayne Bishop. They were playing there without his permission. They discovered a bottle imbedded in some loose earth on top of a landfill. In the bottle they found $12,590. Who was entitled to the property—the boys or Bishop? Bishop v. Ellsworth, 91 Ill. App.2d 386, 234 N.E.2d 49 (1968).
9. Alejo Lopez was married to Soledad. In 1946 he married Helen Lopez while still married to Soledad. In 1947 real property was purchased and conveyed to "Alejo Lopez and Helen Lopez, his wife, as tenants by the entirety. . . ." In 1952 another parcel of land was deeded to them "as tenants by the entirety." In 1954 Soledad divorced Alejo and he remarried Helen a week later. When Alejo died, Helen claimed the land by right of survivorship. The beneficiaries of a trust under Alejo's will claim an interest in the property because Helen and Alejo had not been validly married when the property was purchased. Were the beneficiaries correct? Lopez v. Lopez, 250 Md. 491, 243 A.2d 588 (1968).
10. Lane purchased a dump truck on credit. Texas Hydraulic & Equipment Co. then placed a new dump body and hoist on the truck. Associates Discount Corp. filed an action to foreclose on the truck when the debt was not paid. Hydraulic contended that Associates was not entitled to the hoist or the dump body as they could be removed by a torch without damage to the truck body. Associates contended that they were entitled to the hoist and the dump body because they were accessions. Was Associates correct? Texas Hydraulic & Equipment Co. v. Associates Discount Corp., 414 S.W.2d 199 (Tex.App. 1967).

CHAPTER 29

Bailments

The word *bailment* is not a familiar one to the layman. However, people very frequently enter into bailment arrangements in their everyday transactions. Leaving a car with a parking garage, a topcoat in a cloakroom in a restaurant, clothes at the cleaners, or baggage with an airline are all common situations in which a bailment might be involved. For instance, if you drive your car into a parking garage and give your keys and possession of your car to the attendant, a bailment situation arises. You would be the *bailor* and the parking garage would be the *bailee.*

In essence a bailment consists of the bailor's giving possession of his property to the bailee without giving the title to the property to the bailee. A bailment may easily be distinguished from a sale because there is no passage of title as in a sale but only a transfer to the bailee of a right to possession of the property. On the other hand, a bailee is more than a mere custodian of goods because he not only has a right to possession of the goods but also to control them as long as the right to control is exercised in a manner consistent with the bailment relationship.

When a bailment is created, the bailor and the bailee have certain rights and duties. Section 29:1 will discuss how a bailment is created. Attention will then be directed to the rights and duties of the bailor and the bailee. Finally, some special types of bailments will be discussed.

29:1. How a Bailment Is Created: Elements. In order for there to be a valid bailment, (1) there must be a delivery of the property to the bailee and an acceptance of the property by the bailee so that he has possession; (2) the bailment must be for a limited time; and (3) the subject matter must be personal property.

(1) Possession. In order for there to be a valid bailment, the bailee must take possession of the property. This is accomplished by the bailor's delivering custody of the property to the bailee together with an acceptance of the property by the bailee. In order for there to be sufficient delivery of the property, the bailee must have such custody and control over the property as will allow him to have exclusive right to possession of the property as against all other persons. The bailee must also accept that possession. For example, suppose one walks into a restaurant and hangs his coat on a hook in an area set aside for that purpose. A bailment has not been created because the restaurant owner has not accepted control over the coat. Also the situation does not exclude all others from control. Now suppose the facts are altered so that the customer delivers his coat to a person who is employed by the restaurant to accept coats and that person then hangs the coat up in a cloakroom maintained for that purpose. A bailment would be created because the owner, through his employee, has indicated his acceptance of control over the coat that the customer has delivered to him. Another very common example is the parking-lot situation. Suppose one drives into a parking lot, parks his own car, locks the car, and keeps the keys. The courts typically hold that this is not a bailment but a lease of space or a license. The reason is that the parking lot does not take possession and control of the automobile. Contrast this with the situation in which one gives his keys to the parking-lot attendant, who parks the car and retains the keys. Here a bailment would arise because there has been a transfer of possession of the car to the parking lot. They have custody and control of the property.

From our discussion it should be obvious that there cannot be a bailment without an intent to control the property on the part of the bailee. This intent can be mainfested expressly or by implication through the actions of the parties. Taking our parking-lot example, there would be no bailment if the car is parked on the lot without the knowledge of those running the lot or the owner.

(2) **Limited Time.** In order for there to be a valid bailment, the bailee should have a right to possession of the property for a limited time. If the property is to be returned to the bailor in the same or some altered form, then there is a bailment of the property. However, if the same property is not to be returned to the purported bailor and money is given in exchange for it, then this would be a sale of the goods rather than a bailment. The same rule would hold true if some equivalent of money or goods other than those delivered is exchanged.

(3) **Personal Property.** In order for there to be a bailment, the subject matter must be personal property. It cannot be land. If one is given the possession and control of real property without a transfer of title, then a landlord–tenant or some other relationship may arise, but not a bailment.

JOHNSON v. B. & N. INC.

190 Pa.Super. 586, 155 A.2d 232 (1959)

HIRT, J.

In the late afternoon or early evening of January 28, 1956, the plaintiff, with a Miss Williams, went to the defendant's restaurant for dinner, and to see a style show. There were about 200 people present, which number was about the capacity of the club. Some time after they were seated plaintiff and her companion decided to check their coats and Miss Williams took the two coats to the check room for that purpose. There, according to her testimony, the check room was filled with coats; that the attendant, referred to as Patty, told her that there were no more checks, but that she knew both Miss Williams and the plaintiff, and for that reason would accept the coats without giving checks for their later identification when claimed. When the two women were ready to leave, there was no attendant at the check room. This was a small room, open to the restaurant, except for a "half gate" across the doorway. Patty appeared later and delivered Miss Williams' coat to her, but was unable to find the plaintiff's Persian Lamb coat. Plaintiff reported the loss of the coat to the manager of the club and to the police, but it was never recovered. In this action to recover the value of the coat, tried before a judge without a jury, the finding was for the plaintiff in the sum of $500. The lower court on entering judgment on the finding, discharged the defendant's rules for judgment n.o.v. and for a new trial; hence this appeal.

Check room service was an incident of the defendant's business and was necessary for the accommodation of its patrons. And the authority of Patty, the check room attendant, to accept plaintiff's coat, without giving her a check, must be implied from the circumstances. When a case is tried before a judge without a jury every finding of the trial judge—even a general finding—has the force and effect of a verdict of the jury, if supported by the evidence. And in the light of the finding of the trial judge in the present case, supported as it is by the evidence, we must take it that a bailment for hire resulted—in reality a mutual benefit bailment—notwithstanding the plaintiff paid nothing for the service. The defense was that the coat was never delivered to the check room attendant. But the question of credibility raised by that issue—a question for the trial judge—was resolved in plaintiff's favor. When the bailment was established, coupled with a demand for the return of the coat and the defendant's failure to produce it, the burden of proving the loss through the defendant's negligence remained with the plaintiff in this case. That burden was met.

Under the findings, the plaintiff is entitled to the most favorable inferences from the testimony and the defendant is chargeable with negligence under the circumstances. In accordance with the duties of her employment Patty served the defendant as a hostess in the restaurant as well as the only check room attendant at the time. Plaintiff's evidence was to the effect that during dinner, Patty was seen moving about the restaurant and that when plaintiff and her friend came to the check room to claim their coats, no attendant was there. On this evidence the trial judge properly concluded: "The evidence was clear and precise that the cloak room attendant was absent therefrom at times, leaving it and its contents unguarded and an easy prey to the covetous eyes of a purloiner, an early departing patron perhaps or one entering, taking the coat and leaving unseen." The defendant in this instance failed to exercise ordinary care as bailee of the plaintiff's coat and accordingly was liable for the loss.

(Affirmed.)

BROADVIEW APARTMENTS CO. v. BAUGHMAN

30 Md.App. 149, 350 A.2d 707 (1976)

Glenn H. Baughman rented an apartment from Broadview Apartments Co. Under a separate agreement he paid Broadview $15 per month to park his car in the Broadview Garage. The garage is beneath the apartment building and has an entrance and exit on the lobby level as well as a single entrance and exit to the basement. There is an attendant in the basement level but not at the lobby level. Each tenant was provided with the keys to the garage door. There was also a security guard on duty.

On November 23, 1966, Baughman parked his car in his assigned spot on the lobby level, locked the car, and took the keys with him. When he returned the next day, his car was gone and has never been recovered. Baughman then sued Broadview for the loss of the car, contending that it was a bailee. Broadview did not put on any evidence and contended that no bailment existed. The trial court held that there was a bailment and found in favor of Baughman. Broadview appealed.

MELVIN, J.

A bailment is "the relation created through the transfer of the possession of goods or chattels, by a person called the bailor to a person called the bailee, without a transfer of ownership, for the accomplishment of a certain purpose, whereupon the goods or chattels are to be dealt with according to the instructions of the bailor." Dobie on Bailment & Carriers as quoted in *Refining Co.* v. *Harvester Co.*, 173 Md. 404, 414 (196 A. 131) (1938). . .

"To constitute a bailment there must be an existing subject-matter, a contract with reference to it which involved possession of it by the bailee, delivery, actual or constructive, and acceptance, actual or constructive." *Refining Co.* at 415 (196 A. 131).

Once the bailment relationship is proven certain responsibilities flow from the relationship. The bailee in accepting possession of the bailed property assumes the duty of exercising reasonable care in protecting it. He, however, is not an insurer and owes only such duty of care as persons of common prudence in their situation and business usually exercise towards similar property belonging to themselves.

From the above it is readily apparent that the answer to the threshold question of whether a bailment exists will have not only a procedural but oftentimes a substantive impact on the outcome of the case. If no bailment is shown, and the owner of the property is a mere licensee or lessee of the storage space, then in order to recover against the defendant garage owner the plaintiff would have to prove specific acts of negligence on the part of the defendant proximately causing loss or injury of plaintiff's property. Absent a bailor–bailee relationship, the plaintiff would not have the benefit of any inference or presumption of negligence that arises from a mere showing of non-return of his property if that relationship existed.

The question of whether the owner or operator of a garage or parking lot is a bailee of the cars left there has never been decided by a Maryland Appellate Court. The issue, however, has arisen on numerous occasions in other jurisdictions.

The courts have uniformly found a delivery of possession to the parking lot operators, and therefore a contract for bailment, where the keys are surrendered with the car or where the car is parked by an attendant. Some courts have also found a bailment where the patron parks his car himself and retains the keys, but in most such cases there have been other factors present bearing on the parking lot operator's control over the car. Some other factors which have been considered to be important are: (1) whether there are attendants at the entrances and exits of the lots, (2) whether the car owner receives a claim check that must be surrendered before he can take his car, (3) whether the parking lot is enclosed, and (4) whether the parking lot operator expressly assumes responsibility for the car. No single factor has been viewed as determinative of the issue. The law has probably been best stated in *Osborne* v. *Cline*, 263 N.Y. 434. 189 N.E. 483, where the New York Court of Appeals stated that

> Whether a person simply hires a place to put his car (licensor–licensee relationship) or whether he has turned its possession over to the care and custody of another (bailee–bailor relationship) depends on the place, the conditions and the nature of the transaction, at 437, 189 N.E. at 484.

In the instant case we think the evidence is legally insufficient to establish that a bailor–bailee relationship existed. Appellee merely rented a parking space monthly; he parked his own car, locked the car and took the keys with him. There was no testimony that Broadview had a set of keys for the car or had any right or authority to move or exercise any control over the car. The parking garage was laid out in such a manner that it was possible for appellee to park his car without any attendant even being aware of his presence in the garage. Appellee entered into the monthly lease arrangement with full knowledge of how the garage operated. He was not required to check his car in or out, and there was no evidence whatsoever that control of the car was ever turned over to the operators of the garage, or that they ever accepted delivery or control. Nor was there any evidence that Broadview expressly contracted or asserted that the car would be safe from theft while in the garage.

On the record before us, as we have indicated, there is insufficient evidence to warrant a finding of a bailor-bailee relationship between the parties. The trial court was wrong in finding there was such a relationship. There being no bailment, the burden was upon the plaintiff below (appellee here) to produce sufficient evidence of specific acts of negligence on the part of Broadview or its employees (unaided by an inference of negligence from the fact that the car was missing) from which the trier of fact could predicate a judgment in his favor. The record is devoid of any such evidence. The mere fact that Broadview may have employed a security guard and that the appellee's car was missing is not enough. As the trial judge correctly stated, Broadview (even if it were a bailee) was not an insurer.

(Judgment reversed.)

QUESTIONS FOR DISCUSSION

1. *Why did the court find that there was no bailment in this case? How could the facts have been altered to establish a bailment?*
2. *Why did Broadview win this case when it presented no evidence? Why would a finding by the court that a bailment existed have resulted in a verdict for Baughman?*

29:2. CLASSIFICATION OF BAILMENTS AND STANDARDS OF CARE. Bailments may be classified according to the party who is to receive the benefit from the bailment relationship. To some extent this classification affects the rights and duties of the parties. A bailment may be (1) for the benefit of the bailor; (2) for the benefit of the bailee; or (3) for the mutual benefit of the bailor and the bailee. Bailments may also be classified as *ordinary* or *extraordinary.* Those bailment situations classified as extraordinary include common carriers and innkeepers. Extraordinary bailment situations place certain special rights and duties on the parties and will therefore be discussed separately.

(1) Solely for the Benefit of the Bailor. A bailment solely for the benefit of the bailor arises when the bailee agrees to keep certain property for the bailor without any compensation. For example, suppose Barney Boatowner stores his boat in Harry Homeowner's garage over the winter. Harry receives no compensation or any other reward for keeping Barney's boat. The bailment would be solely for the benefit of the bailor, Barney. Such a bailment often classified as a *gratuitous* bailment. The traditional attitude of the courts in this situation has been to a place a rather low standard of care on the bailee because he receives no benefit. The bailee is held to have the duty merely to exercise slight care and is liable only for gross negligence, bad faith, willful injury to the property, or fraud.

(2) Solely for the Benefit of the Bailee. A bailment solely for the benefit of the bailee arises when the bailor agrees to allow the bailee the use of personal property for a period of time without any compensation. The most common example of this type of bailment would be when the owner of a lawn mower agrees to allow his neighbor to use it for a period of time without any compensation. In this instance, the duty of care placed upon the bailee is much greater than when the bailment is for the sole benefit of the bailor. When the benefit is for the sole benefit of the bailee, he is bound to exercise extraordinary care with the property. The bailee is not liable, however, for normal wear and tear.

(3) For the Mutual Benefit of the Parties. Most commercial bailments are classified as bailments for the mutual benefit of the parties. Each party receives something. For example, when one parks his car in a parking lot, he receives the benefit of a place to park the car, and the owner of the lot is paid the fee that is charged for parking. It is not necessary that both parties benefit equally in order for there to be a bailment for the mutual benefit of the parties. The liability imposed on the bailee in this situation is one of ordinary care. Ordinary care is the care that a man of ordinary prudence and discretion would use in caring for the object if it were his own property. In order for this standard of care to be applicable, the bailee must be compensated.

(4) Another Standard of Care: The Modern Standard. The distinctions with regard to the standard of care owed by the bailee under the three circumstances discussed above are difficult to apply in practice. What is the practical difference, for example, between ordinary and extraordinary care in the handling of a lawn mower? Although there may be a difference, it is very difficult to draw the line in a real-life situation. The result of this difficulty is that many courts have eliminated the distinction and state that the bailee is held to a standard of reasonable or ordinary care under the circumstances. The circumstances then determine whether the actions of the bailee are reasonable. For instance, if one party gives another a very inexpensive piece of costume jewelry for safekeeping, it may be reasonable to place the jewelry in an unlocked drawer. If the jewelry is an expensive diamond necklace, ordinary care might require the bailee to place the jewelry in a safe. The size and nature of the object of the bailment as well as the business of the bailee would also be factors for consideration.

KOENNECKE v. WAXWING CEDAR PRODUCTS LTD.
543 P.2d 669 (Or. 1975)

Glenn Koennecke leased a sawmill to Waxwing Cedar Products Ltd. and others. With the permission of Waxwing, Koennecke left certain equipment at the mill and was not required to pay to store it. Sometime later a fire destroyed the sawmill and Koennecke's stored equipment. Koennecke sued Waxwing, contending that the fire was caused by Waxwing's negligence.

As to Koennecke's cause of action for the loss of the stored equipment, the trial court found that in order to prevail, Koennecke would have to prove that Waxwing was guilty of gross negligence. The parties had stipulated that this was a gratuitous bailment. Because Koennecke had also stipulated that he could not prove Waxwing grossly negligent, the trial court granted Waxwing's motion for an involuntary nonsuit against Koennecke. Koennecke appealed, contending that gross negligence was not the proper standard of care to apply.

BRYSON, J.

On appeal plaintiff argues "that it is not necessary to prove gross negligence in order to recover for destruction of property held by a gratuitous bailee." Plaintiff asserts that gross negligence means "lack of reasonable care under the circumstances." This particular subject has generated substantial confusion in the law of bailments.

Although a majority of the states adhere to the common law rule that a gratuitous bailee is liable only for gross negligence, . . . courts frequently circumvent this rule by altering the definition or modifying the application of the common law rule. A good example of this is found in *Wilson* v. *Etheredge,* 214 S.C. 396, 52 S.E. 2d 812 (1949), wherein the court stated " 'gross negligence' . . . (is) the failure to exercise reasonable care."

"A substantial number of jurisdictions have completely abandoned the concept of divisibility of diligence and negligence into degrees and, consequently, apply only one standard of care, that of the ordinary prudent men (sic) under the particular circumstances. (Citations omitted.) . . ." 8 Ark.L.Rev. at 504.

Although we have recognized the existence of gratuitous bailments, . . . we have not passed on the duty of care required of a gratuitous bailee. No cases have been cited to us and we can find none.

As stated in Brown's treatise at page 10, it is more advantageous to adopt a standard of reasonable care. Such a course of action avoids the inevitable confusion which would be otherwise created by any attempt to define "gross negligence" or "slight care."

We conclude that the standard of care required of a gratuitous bailee is such reasonable care as the particular circumstances of the bailment demand. "(T)he so-called distinction between slight, ordinary, and gross negligence over which courts have quibbled for a hundred years (in gratuitous bailee cases) can furnish no assistance." *Maddock* v. *Riggs,* 106 Kan. 808, 190 P. 12, 16 (1920). Thus the trial court erred in granting defendant's motions for involuntary nonsuit as to plaintiff's second cause of action.

Questions for Discussion

1. *What is the standard of care for a bailee in a bailment for the sole benefit of the bailor (called* a gratuitous bailment) *under the common law?*
2. *How does the modern view of the standard of care to be applied to a gratuitous bailee, as stated in this case, differ from the common-law standard?*

29:3. Rights and Duties of the Bailor and Bailee. (1) Bailee's Duty of Care and Contractual Limitations. In the previous discussion we have attempted to analyze the standard of care against which the actions of the bailee are measured. The degree of care that must be exercised by the bailee varies with the circumstances of the bailment. The care required of the bailee is also governed by the provisions of any contract entered into between the bailor and the bailee. One provision of such contracts that requires attention concerns the ability of the bailee to limit his liability for his own negligence.

Certain types of bailees, such as warehousemen, common carriers, and innkeepers, may not limit their liability for negligence by contract other than as provided by statute. These are special bailment situations and will be discussed in Section 29:4. In other bailment situations the general rule is that the bailee may limit his liability by contract. The bailee, for instance, may wish to limit his liability to a certain dollar amount in the event that the bailed property is lost or damaged. In order for the contractual limitation of liability to be effective, it must be called to the attention of the bailor so that he may expressly or impliedly assent to it. In a number of jurisdictions, however, a bailee who deals with the public, such as a public parking garage, may not limit its liability for damage to the bailed property that arises from its own negligence because such a limitation is against public policy. Even in these jurisdictions bailees who are not dealing with the public may limit their liability. Also the bailee who deals with the public is not liable for property of which he is not aware and that is lost or damaged while in his possession. Thus suppose one parks his car in a public garage and leaves on the back seat valuable photographic equipment that is stolen. The parking garage would not be liable for the loss.

(2) Bailee's Duty to Return Property. The bailee is under an obligation to return the bailed property to the bailor, or other person, upon the termination of the bailment contract. The property returned should be the property that was given to the bailee, in either its original or an altered form. This duty is imposed upon the bailee by operation of law even if there is no provision in the contract of bailment that requires redelivery of the property.

The bailee must return the property to the proper party or he is liable to the bailor for any loss that results from his wrongful delivery, regardless of the fact that he may be free from negligence. Sometimes a third person may claim that he is entitled to the property from the bailee as opposed to the bailor. When this happens, the bailee may delay redelivery of the property for the purpose of conducting an investigation in order to determine who is entitled to the property. It is important that the bailee be sure that the property is delivered to the right party because he may be liable to either of the competing parties if he makes the wrong decision. Sometimes the bailee may file a court action known as *interpleader*. This action requires the parties competing for the right to the goods to prove that they are entitled to them. Once the court decides who has the right to the property, then the bailee may safely deliver the property to him. An action of interpleader is not always technically possible, however.

There are a number of excuses for nondelivery of the property that will allow the bailee to avoid imposition of the general rule that the property must be delivered to the bailor or other party. First, if the property is lost or destroyed before time for redelivery without any fault on the part of the bailee, he is excused. If the property is destroyed in a fire or stolen, the bailee is excused from the requirement of redelivery as long as he is without fault. This latter point is important. The mere fact that property is stolen does not excuse the bailee. He must also be free from negligence. Thus, if the property is a piece of extremely valuable jewelry and is left on the bailee's desk while he goes to lunch, the fact that it is stolen would not excuse him from the obligation of redelivery. Another example would be if the bailee has an obligation under a bailment contract to insure the goods and fails to do so. If they are destroyed, then the bailee would be liable even though he is not otherwise negligent. Still another example would be the loss or destruction of the property when it is being used in an unauthorized manner by the bailee. Because such use is unauthorized, the bailee would be liable for the loss of the property regardless of any negligence on his part. An example would be a television repairman's taking home for his personal use a television that is left with him for repair. If it is stolen from his home, he is liable regardless of any negligence on his part.

If the property is seized from the bailee by legal process, he is excused from returning the property to the bailee. The bailee should notify the bailor of the seizure. This may happen when a judgment creditor of the bailor learns that his property is in the possession of a bailee and then obtains an attachment or writ of execution against the property.

(3) Bailee's Right to Compensation and Reimbursement. When a bailment is for the mutual benefit of the parties, the bailee is entitled to be compensated. Most often the compensation to be paid to the bailee is determined by the terms of the contract of bailment. However, if there is no express contract or provision, the law implies an obligation on the bailor to pay a reasonable amount of compensation for the work performed by the bailee.

If the bailee's negligence results in damage to the bailed property, then he may not be entitled to compensation. However, if the property is damaged or stolen without fault on the part of the bailee, then he is entitled to be compensated for work that he performed on the property up to that point unless the contract of bailment provides otherwise.

If the bailee incurs expenses in connection with the preservation of the bailed property, he is entitled to be reimbursed or indemnified for the amounts actually expended. If, on the other hand, the bailment contract places the duty of maintaining the property on the bailee, then he is not entitled to be reimbursed. Thus, if one rents an automobile, he is normally required to supply the gas and oil unless the contract states to the contrary. The bailor would be responsible for major repairs to the property. The lessor is under no obligation to indemnify the lessee of an automobile against liability that may result from an automobile accident.

If a bailee, in a fit of enthusiasm, makes repairs to bailed property without the permission of the bailor, then the bailee is not entitled to be compensated for any labor or materials provided in the repair. This is true even if the repairs enhance the value of the bailed property.

(4) Bailee's Right to Lien. In Chapter 20 the subject of the right to a lien on certain property by those who have performed services and added materials to that property was discussed. An example would be the right of a garageman to hold a car until the charges for the repairs are paid. If the repair bill is not paid within a certain period of time, the car could be sold to satisfy the lien. This is a type of bailee's lien.

Under the common law a bailee has a possessory lien in the bailed property for the value of his labor and the materials expended on the property that enhance its value. In order to exert such a lien, the bailee must maintain possession of the property. Once he parts with possession, the lien is lost and is not reinstated if the bailee regains possession of the property. Many states have statutes today that extend the right to a lien for the amount of the storage charged to those who merely provide storage for the goods. Many of these statutes, like the common law, also require the bailee to maintain possession of the property in order to preserve the lien. A number of statutes, however, provide that the bailee will not lose his lien by returning the property to the bailor as long as a notice indicating the existence of the lien is filed at the proper place, usually the county courthouse. If the charges are not paid, the lien may be foreclosed and the goods sold to satisfy the obligation owed to the bailee.

(5) Duty of Bailor to Inform of Defects. If a bailment is for the mutual benefit of the parties and is one for hire, such as the rental of a motor vehicle, then the bailor impliedly warrants that the property is reasonably fit for the purpose for which it is hired. For example, if one rents a car, the brakes are presumed to be in good condition. If they are not and the bailee is injured as a result of that defect, then the bailor is liable. If the bailor knows of the defect, he must inform the bailee. The warranty is implied by law regardless of whether any warranty is specifically mentioned in the bailment contract. If the bailment is a property that would pose a danger to the bailee if defective, the bailor has a duty to use reasonable care to see that the property is safe for use, such as by inspecting it for defects. This does not mean that the bailor in a bailment for hire is an insurer against all possible defects that result in damage or injury to the bailee. He is obligated only to see that the property is reasonably free from defects.

If the bailment is one soley for the benefit to the bailee, the bailor is obligated only to disclose those defects to the bailee of which he actually knows. Thus if one lends his car to a neighbor and knows that the brakes are defective, he has an obligation to inform his neighbor of this condition. He has no obligation, however, to inspect for defects.

(6) Right of Bailor and Bailee Against Third Parties. Both the bailor and the bailee may have a right of action against a third party who causes damage to bailed property. For example, suppose a third person damages and converts certain property to his own use that was bailed to Barry Bailee by Bob Bailor. Bailee may sue the third person for the tort of conversion. Remember that the tort of conversion applies where one interferes with another's *right to possession* of property. Because a bailee has a right to possession of the bailed property, Barry Bailee could sue for any damages that he sustains as a result of the conversion. Bailee must account to Bailor for any damages recovered from the third person in excess of his interest. Similarly, if Bailor sues, he must account to Bailee for any recovery in excess of his interest. It is also possible, of course, that the bailor alone may have a right of action against a third person.

A question arises when the property that is the subject of the bailment is damaged through the contributory negligence of the bailee. Does such contributory negligence preclude recovery by the bailor against the third person? To put it another way, is the negligence of the bailee imputed to the bailor? Most courts deal with this problem by allowing the bailor to recover for the value of the bailee's interest in the bailed property.

29:4. PARTICULAR SITUATIONS. (1) Innkeepers, Hotels, or Motels. The position of an innkeeper or the owner of a hotel or motel under the common law is not an enviable one. Although it is difficult to find a bailment when property is kept in the room of the guest, the common law makes the innkeeper almost an insurer of the property of the guest. Under the common law an innkeeper (and a common carrier) is an extraordinary bailee. If the property is destroyed or stolen while in the room of the guest, the innkeeper is liable. The few exceptions involve the loss of property as the result of the negligence of the guest, an "act of God," or the act of a public enemy. This rule creates a hardship when the guest keeps a particularly valuable piece of property, such as a diamond necklace, in his room.

Today most states have passed statutes to protect innkeepers and hotel and motel owners and to limit their liability. Although the statutes differ in detail from state to state they follow a pattern. If the guest desires his valuables to be protected by the innkeeper or hotel or motel owner he must deliver them to the innkeeper or hotel or motel operator for safekeeping. These valuables are then usually placed in a safe. If the guest fails to do this, then the innkeeper or hotel or motel owner is liable only if the property is lost or destroyed through his or her negligence. In most cases a notice of the statute must be posted in each room of the lodging.

de SARIC v. MIAMI CARIBE INVESTMENTS, INC.

512 F.2d 1013 (5th Cir. 1975)

Dalila Pardo de Saric and her daughter sued the corporate owner of a Miami hotel for losses sustained by them in a robbery in their hotel room. The trial court granted a defense motion for summary judgment, and de Saric appealed.

COLEMAN, J.

Dalila Pardo de Saric and her daughter, Luz-Maria Saric, registered at the McAllister. They were assigned Room 1049, were escorted to that room immediately after checking in, remained in the room for some forty-five minutes to an hour, and upon opening the door to leave for the first time since registering were met by two masked robbers, one of whom was armed with a handgun.

By the menace of the gun the robbers forced their way into the room. There is no evidence that either of the victims were [sic] ever struck or jostled with the weapon. One of the intruders forced Luz-Maria Saric into the bathroom, where she fainted and fell to the floor. Upon regaining consciousness, she was examined by one of the robbers for hidden jewelry. Dalila Pardo de Saric was required to lie on the floor, with the gun to her head, while jewelry, money, and travelers checks were taken from her purse.

Appellants have framed the appellate issues in the following language:

> . . . Whether Florida statute § 509.111 applies to the facts of this case. . .

The above mentioned Florida statute limits a hotel's liability for the loss of money or valuables by its guests to those items deposited with the hotel management in exchange for a written receipt stating the value of the property deposited. The hotel is mandatorily required to accept for safekeeping valuables aggregating no more than $1,000 in value. Liability for loss is limited to $1,000 unless the hotel voluntarily accepts a greater amount. Florida statutes, § 509.111 (1972).

Florida law further requires a hotel to post notice of this limitation of its liability. Florida statutes, § 509.101.

At common law, innkeepers were insurers of the property of their guests, and their liability was like that of a bailee. Since the innkeeper was being compensated for his services, he was a bailee for hire and thus subject to the standard of ordinary care. . . . Many jurisdictions, including Florida, have limited an innkeeper's liability by statute. Enacted, as they were, in derrogation of the common law, these statutes have been strictly construed. . . .

In its decision, the District Court apparently relied on *Ely* v. *Charellen Corporation,* 5 Cir., 1941, 120 F.2d 984. As it develops, this case is the last word in this Court or from the Florida Courts dealing with the issue which, thirty-four years later, now makes its second appearance.

In *Ely,* the plaintiff registered at a Florida hotel. Jewelry was stolen from her room. At that time, as today, Florida law provided that a hotel was not liable for the loss of valuables unless deposited with the hotel management. . . . Also in effect at that time, as today, was a requirement that the hotel post notice of this and other rules. . . . In *Ely,* as in the case sub judice, neither party complied with the statutes, in that the plaintiff left valuables in her room and the hotel had failed to post notice of its liability limitations.

The plaintiff argued that the hotel's obligation to post notice of the Act had to be met before the hotel could claim its benefits. This Court, in a one-page opinion, held that because § 40 listed no exceptions, no liability would attach to the hotel proprietor for the loss of jewelry unless it was deposited with the hotel. We noted that § 40 provided that *in no event* would a hotel be liable for loss of goods not deposited with the hotel (emphasis ours). This same "in no event" language appears in the present statute. The statutes have been amended by the Florida legislature three times since *Ely*, always retaining the "in no event" phraseology which *Ely* held to be decisive regardless of whether the statutory notice had, or had not been, posted.

The outstanding, unreversed decision in *Ely* is binding on us.

In their brief, appellants state their position on this point as follows:

> As the plaintiffs had not yet deposited their valuables with the management at the time of the robbery, we have stipulated that if the statute applies to the facts of this case, and is constitutional, then the plaintiffs are barred from recovering for their loss of personal property. If, on the other hand, the statute does not apply, or is unconstitutional, then the defendants are strictly liable for plaintiffs' loss of personal property.

We are of the opinion that the statute does apply because it refers to any loss, which could apply to losses by robbery as well as from any other source. Accordingly, we cannot accept the argument that the statute applies only to items which guests have carelessly left in their rooms, subject to the possibility of theft.

(The case was remanded for further factual findings by the District Court regarding whether the statutory notices were posted and if they were not posted whether the statute could be constitutionally applied to the plaintiffs.)

QUESTIONS FOR DISCUSSION

1. *Would the hotel have been liable under the common law for the loss suffered by de Saric?*
2. *What should de Saric have done with her valuables in order to make the hotel responsible for her loss? Under Florida statute was the hotel required to accept her valuables for safekeeping?*

---

(2) Common Carriers. One may enter into a bailment arrangement with carriers of goods for their transportation. In this case, the carrier would be the bailee. A carrier may be either a *private carrier* or a *common carrier.* A private carrier is a carrier who agrees to transport goods by special arrangement with the bailor and does not agree to carry the goods of the public in general. A common carrier is one that agrees to carry goods as a public service and is required to transport for all who apply. Carriers may transport goods or people, but our discussion here is limited for the most part to the transportation of goods.

Common carriers today are regulated for the most part by agencies of either the federal or the state government. For example, trucking firms operating in interstate commerce are regulated by the Interstate Commerce Commission. The rates charged by the carrier are regulated and set by the commission as are some other activities of the carriers.

The common carrier has a duty to carry the goods of the public. It is required to accept for transportation those goods that it normally carries. It may, however, specialize in the types of goods that it will carry and may reject for transportation types of goods with which it does not generally deal. For example, some common carriers may specialize in the transportation of fruit, vegetables, and other produce, whereas others limit themselves to the transportation of furniture. The common carrier may also reject goods for carriage that are not properly packaged and otherwise prepared for transportation, dangerous goods, and goods beyond its capacity.

(a) *Liability of the Common Carrier.* The liability of a common carrier is somewhat different from that of the bailee in other mutual-benefit bailment arrangements. Like the innkeeper, the common carrier under the common law is an extraordinary bailee. It is a virtual insurer of the safety of the goods while they are in its custody. There are five exceptions to this general rule. The common carrier is not liable for damage to the goods if the damage is the result of (1) an "act of God"; (2) an act of a public enemy; (3) the misconduct of the shipper; (4) the inherent nature of the goods; and (5) an act of a public authority. Clearly then, a common carrier is liable to the bailor for damage to the goods, even if a third person causes the damage, unless the damage results from one of these exceptions.

An *act of God* is damage as the result of some natural disaster, such as an earthquake, lightning, tornadoes, and the like. These are clearly beyond the control of the carrier. However, a fire, unless started by lightning, is not necessarily beyond the control of the carrier and is not considered an act of God.

An act of a *public enemy* that results in loss to the goods will excuse the carrier from liability. An example would be if the goods are lost because of the action of the military forces of another nation.

Negligence of the shipper that causes loss to the goods will excuse the carrier. Defective packaging of the goods by the shipper or damage to the goods in loading, when the loading is done by the shipper, are examples of such acts.

If the goods are damaged as a result of their inherent qualities, then the carrier is not liable. A good example would be the spoilage of produce that is not caused by delay or defective facilities of the carrier.

Finally, if the loss is caused by the act of a public authority, the carrier is not liable. For example, if goods spoil because of a delay in a required inspection by a public official, the carrier would escape liability.

If the carrier is a carrier of persons, it owes them the highest duty of care. The carrier is liable to passengers for injuries that result from any negligence on its part, but it is not an insurer. In other words, there must be more negligence before the carrier is liable. This standard of liability is different from the liability of a carrier for the transportation of goods. Persons boarding or alighting from a carrier are entitled to this degree of care from the carrier of passengers.

SOUTHERN PACIFIC COMPANY v. LODEN

508 P.2d 347 (Ariz. 1973)

Lou Loden shipped two refrigerated vans, each containing 725 crates of cucumbers, from Nogales, Arizona, to Los Angeles, California, on a train of the Southern Pacific Company. The shipment was delayed because of damage to Pacific's tracks caused by a week-long rain. As a result, when the cucumbers *reached Los Angeles, they were spoiled. Loden then sued Southern Pacific for $10,047.68 in damages. Southern contended that the loss was caused by an act of God. From an adverse decision of the trial court, Southern appealed.*

HOWARD, J.

Common carriers impliedly agree to carry safely, and at common law they are held to a very strict accountability for the loss or damage of goods received by them. In this state a common carrier's liability for damage to goods in transit is based on the substantive rule of law that the carrier is an insurer for the safe transportation of goods entrusted to its care, unless the loss is caused by an act of God, the public enemy, negligence of the shipper, or the inherent nature of the goods themselves.

Furthermore, common carriers undertaking to carry perishable goods are held to a higher degree of care than when engaged in the shipment of other articles not inherently perishable and a failure to comply with this duty which results in a loss or injury to the shipper renders the carrier liable for the loss sustained, unless a proper defense is alleged and proved.

In addition, common carriers undertaking to transport property must, in the absence of an express contract providing for the time of delivery, carry and deliver within a reasonable time. The carrier is required to exercise due diligence to transport and deliver the property and guard against delay. Mere delay in transportation does not create a liability to respond in damages, and the rule is that the carrier is bound to use reasonable diligence and care, and only negligence will render it liable.

The law recognizes various fact situations as an excuse for delay which constitutes a good defense. Examples of such facts are accidents or misfortunes without fault or negligence on the carrier's part, or, an act of God; an act or fault of the shipper or consignee; the press of business; strikes; extreme weather conditions, and accidents not amounting to acts of God. The delay, however, must have been due to an occurrence as could not have been anticipated in the exercise by the carrier of reasonable prudence, diligence and care.

In the instant case the only defense asserted by Southern Pacific was an act of God. A case may not be tried on one theory of law in the trial court and upon another theory on appeal.

The question on appeal therefore resolves itself to whether or not appellant sufficiently met its burden of proving that an act of God occurred excusing its delay in transportation and delivery of appellee's goods.

The various definitions of an act of God practically all require the entire exclusion of human agency from the cause of the loss or injury. A casualty cannot be considered an act of God if it results from or is contributed to by human agency, and that which may be prevented by the exercise of reasonable diligence is not an act of God.

The only acts of God that excuse common carriers from liability for loss or injury to goods in transit are those operations of the forces of nature that could not have been anticipated and provided against and that by their superhuman force unexpectedly injure or destroy goods in the custody or control of the carrier.

Extreme weather conditions which operate to foil human obligations of duty are regarded as acts of God. However, every strong wind, snowstorm, or rainstorm cannot be termed an act of God merely because it is of unusual or more than average intensity. Ordinary, expectable, and gradual weather conditions are not regarded as acts of God even though they may have produced a disaster, because man had the opportunity to control their effects.

Application of the above principles to the evidence presented at trial leads to the conclusion that Southern Pacific failed to prove its defense of an act of God. Two of appellant's employees testified that it had rained in the Los Angeles area for a week prior to January 25, 1969, and it was not until a week after the rain began that Southern Pacific dispatched its employee Smith to patrol the track between Los Angeles and Yuma. Appellant offered knowledge and evidence of a week-long rain but offered no evidence of precautions to avoid possible consequences of this rainfall. Southern Pacific, through the testimony of its own witnesses, demonstrated notice of a gradual weather condition which could foreseeably cause damage to its facilities and consequent delay in transportation.

In order to defend on the ground that a loss or damage complained of was caused by an act of God, the act must have been the proximate cause of the damage, whether the complaint is for a breach of contract or for a tortious injury.

A rainstorm of unusual duration or intensity is not necessarily a superhuman cause or an act of God. *Southern Pac. Co.* v. *City of Los Angeles,* 5 Cal.2d 545,55 P.2d 847 (1936). The rainfall in the instant case was not shown to be totally unforeseeable or of greater intensity than other rainfalls in the region so as to justify being called an "act of God" and, therefore, the judgment of the trial court is affirmed.

Questions for Discussion

1. *For what reasons may a common carrier be excused from liability for goods damaged in transit? Is it necessary for the shipper to prove negligence on the part of the common carrier in order to recover for damaged goods?*
2. *Why did the court hold that the rain in this case was not an act of God?*

(b) *Limitation of Liability.* A common carrier may limit its liability to a stated amount of damages. Section 7–309 of the Uniform Commercial Code states that a common carrier generally may not totally contract away its liability for damage to goods as a result of its own negligence. The carrier may limit its liability to the value stated by the shipper in the bill of lading. The rate for shipping may increase with the value stated in the bill of lading.

If the carrier operates in interstate commerce, it is governed by provisions of the Interstate Commerce Act. A provision of the act states that a carrier cannot be exempted from liability for the actual loss, damage, or injury to property caused by it. However, the act also states that the restriction on the right to limit liability shall not apply in certain cases. The carrier may limit its liability both as to amount and losses resulting from certain hazards for baggage on passenger trains or boats. It may also limit its liability for the value of property, except livestock, to the value declared by the shipper if the Interstate Commerce Commission has required or authorized the carrier to establish and maintain rates dependent upon the value declared by the shipper. There are similar limitations of liability for baggage in the case of airlines in connection with the filing of tariffs by air carriers with the Civil Aeronautics Board.

(c) *Duration of Carriage.* The carrier is liable for the goods when the goods are delivered to it for carriage and accepted. As long as something remains to be done by the shipper, the liability of the bailee as a carrier does not begin. Examples of things that may remain to be done by the shipper after the goods have been delivered to the carrier are the loading of the goods or the giving of shipping instructions. The carrier's liability does not attach until the goods are within its exclusive control. If the goods are delivered to the carrier's property but remain to be loaded by the shipper, the carrier would be responsible for the goods only as a warehouseman and not as a carrier; he is therefore liable only for his own negligence and is not an insurer of the safety of the goods.

The liability of the carrier continues until the carriage comes to an end. Once the carrier has completed the contract of carriage, its liability as an insurer of the goods ceases. While the carrier holds the goods for the party to whom they are to be delivered, *the consignee,* its liability is that of a warehouseman. As a general rule, the contract of carriage is completed when the carrier unloads the goods and notifies the consignee of their arrival. A majority of jurisdictions hold that the liability of the carrier ends a reasonable time after notice is given to the consignee. In other jurisdictions the liability of the carrier ends when the goods are unloaded. A few others hold the carrier liable until the consignee has had an opportunity to inspect and unload the goods. However, if under the contract of carriage the carrier is not expected to unload the goods, then its liability as a carrier ends when the goods are delivered to the place designated in the contract or to the place where such goods would normally be placed in the normal course of business. Thus a carload of wheat may be delivered to a given elevator to be unloaded by the operator of that facility. If the shipper has the right to stop the goods in transit and exercises that right, then the carrier's liability reverts to that of a warehouseman.

(d) *Duty to Deliver Goods.* Just as with other types of bailments, the carrier has an absolute duty to deliver the goods to the proper party. Normally the contract of carriage or bill of lading determines who is to receive the goods. Bills of lading have been discussed in Chapter 16. Recall that a bill of lading may be negotiable or nonnegotiable. A nonnegotiable bill of lading is known as a *straight bill.*

If the goods are shipped under a nonnegotiable bill of lading, the carrier must deliver the goods to the person named on the bill. Even if the consignee has sold the goods prior to receiving them, the carrier must deliver the goods to the original consignee unless notified by him to deliver them to some other party. If the goods are shipped under a negotiable bill of lading, the carrier is aware that such a bill may be transferred. In this case, the carrier is required to deliver the goods to the party unless the bill of lading is surrendered to the carrier with the proper indorsement, showing that that person is entitled to the goods.

Although the carrier is normally absolutely liable for misdelivery of the goods, there are some circumstances under which misdelivery is excused. These situations most often result from instances in which the consignee misleads the carrier for one reason or another. The Uniform Commercial Code, Section 7–403, states a number of excuses for misdelivery by a carrier or warehouseman:

(a) delivery of the goods to a person whose receipt was rightful as against the claimant;
(b) damage to or delay, loss or destruction of the goods for which the bailee is not liable (but the burden of establishing negligence in such cases is on the person entitled under the document);
(c) previous sale or other disposition of the goods in lawful enforcement of a lien or on warehouseman's lawful termination of storage;
(d) the exercise by a seller of his right to stop delivery pursuant to the provisions of the Article on Sales (Section 2–705);

(e) a diversion, reconsignment or other disposition pursuant to the provisions of this Article (Section 7–303) or tariff regulating such right;
(f) release, satisfaction of any other fact affording a personal defense against the claimant;
(g) any other lawful excuse.

(3) **Warehouseman.** A warehouseman is one who, for compensation, stores bailed goods and merchandise as a bailee. A bailment involving a warehouseman is a bailment for the mutual benefit of the parties. The rights and liabilities of a warehouseman are similar to those that result from other bailments of this character. However, today, because of the extensive use of warehousemen for storing goods and merchandise, there are a large number of federal and state regulations controlling their activities. These regulations govern the types of goods that must be accepted for storage as well as matters concerning health and safety.

Another distinguishing feature of a warehouseman is that the receipts issued by him in return for goods accepted have become the subject of consideration in the Uniform Commercial Code. These receipts are classified as *documents of title* and are governed in Article 7 of the code. Among the many provisions of the code in this area are a provision allowing the warehouseman to limit his liability to the value of the goods stated in the warehouse receipt, provided that for an increased storage rate he must increase the limit of his liability at the bailor's request. Also the limitation does not apply in the event of a conversion of goods to the warehouseman's own use or other intentional wrong. Other provisions of the code specifically allow for a warehouseman's lien, the separation of nonfungible goods, and other matters.

SUMMARY

A bailment consists of the bailor's giving possession of his property to the bailee without giving the title to the property to the bailee. A bailment is created by a delivery of the property to the bailee and acceptance of it so that the bailee has possession; the bailment must be for a limited time; and the subject matter must be personal property.

Under an older approach to bailments, they may be classified as solely for the benefit of the bailor; solely for the benefit of the bailee; or for the mutual benefit of the bailor and bailee. The duty of care owed by the bailee for the property of the bailor is different under each of these classifications. Many courts have eliminated these distinctions and hold the bailee to a standard of reasonable care under the circumstances.

Bailees not dealing with the public may limit their liability by contract. In general, the bailee is under a duty to return the bailed property to the bailor at the termination of the contract. Performance is excused if the property is lost or destroyed without fault of the bailee.

The bailee has a right to be compensated and reimbursed for expenses incurred in connection with the bailment. A bailee has a lien on the bailed property for the value of his labor and any materials expended on the property to increase its value. If the bailor knows of a defect in the goods bailed, he has a duty to inform the bailee. Both the bailor and bailee have a right of action against a third party who causes damage to the bailed property.

The liability of innkeepers and hotel owners is limited by statute in most jurisdictions. The liability of common carriers is regulated by statute and may limit its liability to a stated amount of damages, although under the Uniform Commercial Code the carrier may not totally contract away its liability for negligence. It may limit liability to the value stated by the shipper in the bill of lading.

The carrier is liable for the goods for the duration of carriage and has a duty to deliver the goods to the proper party. The Uniform Commercial Code states a number of circumstances when misdelivery is excused.

PROBLEMS

1. In February of 1977 Elwood Nelson purchased an aircraft for $5,000.00 and arranged to have it flown to an airport owned by Schroeder Aerosports, Inc. The parties entered into an informal agreement whereby Nelson rented a parking space for his aircraft from Schroeder for $10 per month. Nelson parked his plane at the facility and kept the keys. He also gave permission to another person to fly the plane. During a violent windstorm the plane was damaged. Nelson contended Schroeder was liable as bailee of the plane. Was Schroeder a bailee? Nelson v. Schroeder Aerosports, Inc., 280 N.W.2d 107 (S.D. 1979).
2. Stephen Miller drove up to a self-service parking garage owned by Central Parking System. He took a ticket from a machine when he entered the garage. Miller then parked his car and took his keys with him. The only employee present at the garage was the attendant who collected the money from the drivers when they exited. While Miller was away the wire wheels were stolen from his car. He sued Central Parking System. The trial court found that bailment existed and awarded Miller $838.20. Was the trial court correct? Central Parking System v. Miller, 586 S.W.2d 262, (Ky. 1979).
3. Allen parked his car in a parking lot owned by Houserman. The keys to the car, including the trunk key, were left in the car. In the trunk Allen left a set of golf clubs valued at $373.53. The car was stolen from the lot and later recovered, but the golf clubs were missing. Allen had never informed Houserman that the clubs were in the trunk of his car. Was Allen entitled to recover from Houserman? Allen v. Houserman, 250 A.2d 389 (Del.Super. 1969).
4. Frigid contracted with Brinke to ship fans to one of Frigid's customers in Miami, Florida, from Brooklyn, New York. The truck that Brinke supplied was an open or "rag" van. Frigid called Brinke's office and expressed concern about the open van but was assured that it was safe and that a plastic

liner would be installed to ensure that it was watertight. When the fans arrived at their destination, they were irreparably damaged from water. Frigid contended that the carrier was absolutely liable. Brinke contended that it was not liable because the fans were loaded by the shipper, who was aware of the physical condition of the van, and hence the damage resulted from the acts of the shipper. Decide. Federated Department Stores v. Brinke, 450 F.2d 1223 (Ca.Fla., 1971).

5. W. S. McCurdy took his goods to Wallblom Furniture & Carpet Company for storage. McCurdy took his goods to the warehouse owned by Wallblom, saw them stored, and was given a warehouse receipt that did not specify where the goods were to be kept. Wallblom then transferred the goods to another warehouse, which burned down. McCurdy's goods were destroyed by the fire. He had not been informed of the transfer of the goods. Wallblom contended that it was not negligent. Could McCurdy recover? McCurdy v. Wallblom Furniture & Carpet Co., 102 N.W. 873 (Minn. 1905).
6. New York law provides that a hotel is relieved from liability for loss to a guest of his money, jewels, ornaments, and precious stones if the hotel provides a place of safekeeping and the guests do not avail themselves of it. A guest of the Waldorf-Astoria Hotel had his gold cufflinks stolen. The guest's insurance company contended that the hotel was liable for the loss because the cufflinks were articles of ordinary wear and not jewels, ornaments, or precious stones. Was the hotel liable? Federal Insurance Co. v. Waldorf-Astoria Hotel, 60 Misc.2d 996, 303 N.Y.S.2d 297 (1969).
7. Rhodes parked his car in the Pioneer Parking Lot. There was no attendant to service or look over the lot. Rhodes drove his car onto the lot, parked it himself, locked it, and took the keys with him. He obtained a ticket from a meter, into which he put the fifty-cent parking fee. There was only one entrance to the lot. When Rhodes returned, he found that his car was stolen. It was later found "stripped." Rhodes sued Pioneer. The trial court found a bailor–bailee relationship and awarded a judgment to Rhodes. Was the court correct? Rhodes v. Pioneer Parking Lot, Inc., 501 S.W.2d 569 (Tenn. 1973).
8. Kingsley Sportswear, Inc., shipped two containers of merchandise with the Standard Hauling Co. Standard picked up the goods. While in the care of Standard, the goods disappeared without a trace. Standard contended that it was not liable because it exercised due care and that such losses are inevitable and equivalent to an act of God. Was Standard liable? Kingsley Sportswear, Inc. v. Standard Hauling Co., Inc. 374 N.Y.S.2d 19 (1975).
9. David Gibson was scratched by a raccoon. The raccoon died. The head of the animal was packaged for shipment to Jacksonville, Florida, to the Department of Health and Rehabilitative Services Laboratory to be analyzed for rabies. The package was delivered to an agent of Greyhound Bus Lines, Inc., for carriage. The agent was informed of the contents of the package but not of its importance. No higher value for shipment than that provided for by the tariff was declared. The head was lost and never arrived in

Jacksonville. Gibson then had to undergo a series of rabies vaccinations, from which he suffered an adverse reaction. He then sued Greyhound for damages. Greyhound contended that its liability was limited to $50 under the provision of the tariff, which states that $50 is the limitation "unless a greater value is declared at the time of shipment." Gibson contended that the state statute imposes a duty on carriers to exercise proper care in shipping goods and that failure to do so renders the limitation inapplicable. Decide. Gibson v. Greyhound Bus Lines, Inc., 409 F.Supp. 321 (U.S.D.C.Fla. 1976).

CHAPTER 30

Real Property

30:1. Introduction. Perhaps no area of the law has changed more slowly or has been more influenced by the historical developments of the past than the law of real property. The concepts and even the language that govern the law of real property in the United States can be traced directly to the Norman Conquest of England in 1066. Since that time there has been a slow but steady evolution of the law concerning land. Many of the legal concepts as well as the language concerning real property are unique to that area of the law. For instance, in the technical sense one does not buy and sell real property, he "conveys" it. One is not a seller or buyer of real property but a "grantor" or "grantee." The result is that for most people the simple act of purchasing a house or land is an adventure into the world of the unknown. Perhaps the reason the law has been so slow to change with regard to real property is that for centuries land has been the primary means of wealth and power. As a result, the acquisition of land and other dealings in it have become very complicated and are treated differently from dealings with personal property.

When William the Conqueror and his successors came to the English throne, they were the ultimate owners of all the land in England. Beginning with William, the lands of the kingdom were granted to various people in return for services to be returned to the king. Under these grants, title to the land did not pass from the king; rather the recipient of the grant, known as a *vassal,* held the land as a tenant. The king was the *lord.* Mutual rights and duties between the lord and his vassal were established with the grant. For instance, the tenant might be required to supply a certain number of knights for the king's army. Tenants who held their land directly from the king were known as *tenants-in-chief.* Over a period of time the tenants-in-chief in turn made grants of their interest in the land to some of their subjects in return for service to them as lords. This process was repeated a number of times through a number of subtenants and was known as *subinfeudation.* Thus each tenant owed services to his lord, who in turn owed services as tenant to the lord above him and ultimately to the king, who was at the top of the structure, which can best be visualized as a vast pyramid. The king, of course, was not a tenant. Under this system the same piece of land could be the subject of property rights of many persons. The services required were of the land rather than of the tenant personally.

Over the years this system has evolved to the point where today an individual may own his own property. Nevertheless many traces of the system remain. For example, under certain conditions the state may take one's land in order to build a road, or other public project, by a process known as *eminent domain.* The owner is, of course, entitled to be compensated.

30:2. Types of Estates in Land. The word *estate* is used in real-property law to define the *interest* that one has in the land rather than the land itself. The different types of interests or estates that one may have in land are measured in terms of time. Thus the type of estate that one has depends upon the length of time that he has an interest in the land. In this section our discussion will involve *freehold* estates, which in medieval times were the estates that were given protection in the king's courts. *Nonfreehold estates* usually involve a duty on the party of the tenant to make a periodic payment to the lord and are discussed in Chapter 31, "Landlord and Tenant." There are a number of different freehold estates that are recognized but our discussion here will be limited to the two most common estates today, the fee simple and the life estate.

(1) **Fee Simple.** A *fee simple* is the greatest estate in land. If all the possible rights and interests connected with the ownership of land can be visualized as a pie, the fee simple denotes the entire pie, that is, all the rights, interests, and privileges connected with the ownership of land. Out of the fee-simple estate, smaller pieces of the pie—that is, smaller and lesser estates or interests—may be carved. Today, if one purchases a home from another, he typically buys it in fee simple.

The major characteristic of the fee-simple estate is that (1) it is of potentially infinite duration and (2) the estate is inheritable. Remember that estates are classified by the length of time of the interest. Because the fee simple may be inherited by one's heirs, it may continue to be inherited by the family line infinitely. Suppose one owns a home in fee simple and dies without leaving a will, or if there is a will, he does not name who is to get the house. The title to the property descends to his heirs at law, who then own the property in fee simple.

At common law, in order to create a fee simple estate when conveying the property, one had to use very precise words to indicate that the property was being transferred not only to the grantee (buyer) but also to his heirs. Thus if A was selling property to B in fee simple, the deed would recite the conveyance as being from A "to B and his heirs." If the conveyance was just "to B," no fee simple would be created because there was no indication that the estate was inheritable. Today by court decision or statute almost all states have eliminated the need for *words of inheritance* in order to create a fee simple. Simply stating that the land is conveyed "to B in fee simple" or "to B forever" is sufficient to create an estate in fee simple in most jurisdictions.

An estate may be subject to termination upon the happening of an event.

Such an estate is a fee simple if the estate still may be of infinite duration. For example, suppose A conveys his land "to B and his heirs in fee simple so long as the land is used for residential purposes," with a provision that the land shall revert to A if it is no longer so used. This kind of arrangement is known as a *fee simple determinable.* It is a fee because it is inheritable and can last forever. On the other hand, if the grant specifies that the grantor may reenter and take back the property upon the happening of the event, as in the previous example, then it is called a *fee simple on condition subsequent,* rather than *determinable,* because the estate does not terminate unless the grantor exercises the right of reentry, whereas the fee simple determinable terminates automatically.

CUMMINGS v. UNITED STATES
409 F.Supp. 1064 (1976)

WARD, J.

In 1961, the Burlington Administrative School Unit (School Unit) began a search for real estate which would serve as a multiple school site. The School Unit soon located a 77.8 acre tract just outside the city limits of Burlington and proceeded to negotiate for its purchase. Various impediments, not material here, arose with the result that the School Unit lacked the funds to purchase the property from its owners, F. E. McPherson and Edna Lucille T. McPherson (McPhersons), and was prohibited by law from incurring a long term obligation in order to make the purchase. The plaintiff learned of the difficulty and offered to purchase the property in order to give it to the School Unit. However, since the plaintiff was also financially incapable of paying the purchase price of $180,000 in one lump sum, the following procedure was arranged by the plaintiff, the School Unit, and the McPhersons.

The McPhersons, by warranty deed dated November 30, 1961, conveyed the entire 77.8 acre tract to the plaintiff. On the same day, the plaintiff executed a deed of trust in order to pledge the purchased property as security for the unpaid purchase price. Under the provisions of the deed of trust, it was agreed that the $180,000 purchase price would be paid over a period of twelve years at the rate of $15,000 per year. The debt was evidenced by twelve promissory notes each in the amount of $15,000, one of which was due on December 31 of each year from 1961 to 1972. The property was divided into twelve separate but unequal tracts so that one tract could be released from the deed of trust as each note was paid off. Accordingly, twelve separate deeds corresponding to the twelve tracts and conveying the property to the School Unit were executed by the plaintiff, the trustee in the deed of trust, and the McPhersons. All twelve deeds were dated November 30, 1961. Thus, under this arrangement, the plaintiff could pay off one note to the McPhersons each year, release one tract of the property, and then donate the property to the School Unit by means of one of the twelve previously executed deeds.

Plaintiff's tax returns for 1964, 1965, and 1966 were audited by the Internal Revenue Service. It was determined that he was entitled to deduct only the actual purchase price of the property which he had donated to the School Unit and not its fair market value because of certain conditions in the warranty deed from the McPhersons to the plaintiff. Plaintiff paid the assessed deficiencies and timely filed claims for refund of taxes which he alleges were erroneously collected in 1965 and 1966.

It is undisputed that the plaintiff was entitled to a charitable deduction for the land which he donated to the School Unit. At issue is whether he was entitled to deduct the fair market value of the land, which he attempted to do, or the price that he actually paid for the land, which the government insists is correct.

The government contends that, although generally the deduction for the charitable contribution of property is the fair market value of the property, the general rule does not apply in this case because of certain provisions in the warranty deed from the McPhersons to the plaintiff. The pertinent provisions of the deed in question read as follows:

> FIRST . . . This deed is executed by the grantors and accepted by the grantees *upon the express condition that said real property will be conveyed by the grantees for school purposes only to the Burlington Administrative School Unit Alamance County, or its successors,* upon such terms and conditions as the parties of the second part may desire to incorporate in said deed or deeds of conveyance to said School Unit, and *that said real property, and no part of same will be rented or leased to any person, firm, or corporation or sold by the grantees herein except to said Burlington Administrative School Unit for school purposes only and upon condition that if said real property shall, at anytime, cease to be used for school purposes that the title to all said real property hereby conveyed shall revert and be vested in the parties of the second part or their heirs in fee simple.* (Emphasis added.)

The government's basic argument is that the language in the deed to the effect that the property may only be sold to the School Unit and may only be used for school purposes restricts the marketability of the property. Thus, the government contends that, since the property may only be sold to the School Unit for school purposes, the plaintiff could not sell it on the open market and receive fair market value. Therefore, the government maintains, plaintiff is only entitled to deduct what he actually paid for the property instead of the fair market value.

The government has maintained that the previously quoted language in the deed is either (1) a fee simple determinable; (2) a fee simple subject to a condition subsequent; or (3) a restrictive covenant. Any one of these elements, argues the government, causes the land to be so encumbered that its fair market value is negatively affected. The plaintiff, on the other hand, denies that any of those elements are present and contends that, if anything, the contested language constitutes an unlawful restraint upon alienation which could not have been enforced by anyone. Thus, plaintiff argues, the fair market value of the property was unaffected.

The Court is of the opinion that the intention of the parties, as interpreted from the face of the instrument, was to create neither a restrictive covenant nor fee simple determinable. The parties to the deed, the Court believes, attempted to include a fee simple subject to a condition subsequent but, in doing so, created an unlawful and void restraint upon alienation.

The language in the deed is simply not indicative of a fee simple determinable:

> The estate known as the fee simple determinable is created when apt and appropriate language is used by a grantor or devisor indicative of an intent on the part of the grantor or devisor that a fee simple estate conveyed or devised will expire automatically upon the happening of a certain event or upon the discontinuance of certain existing facts. Typical language creating such estates may specify that a grantee or devisee shall have land "until" some event occurs, or "while," "during," or "for so long as" some state of facts continues to exist. Upon the happening of the specified event, the fee simple determinable automatically terminates and reverts to the grantor or to his heirs. Until the occurrence of the limiting event the grantor (or devisor), or his heirs, have a future interest known as a "possibility or reverter." When the specified event occurs, the possessory estate of the grantee or devisee ends by operation of law automatically and without the necessity of any act of re-entry, without the institution of any lawsuit or the intervention of any court.

J. Webster, Real Property Law in North Carolina, § 35 (1971). There is no language in the deed in question which would indicate that automatic reversion upon the happening of an event was intended. Therefore, the deed does not contain a fee simple determinable.

The deed in question conveys land "on express condition that" and "on condition that" certain events occur or do not occur and contains what was intended to have been a reverter clause. This language is more indicative of a fee simple subject to a condition subsequent than it is of a fee simple determinable.

A fee simple subject to a condition subsequent, unlike a fee simple determinable, does not terminate automatically upon the happening of the stated event. The grantor, his heirs, or assigns must take affirmative action in order to terminate the estate. In order to create a condition subsequent, the instrument must contain language "showing an intent that the property shall revert to the grantor, his heirs or assigns, or that the grantor, his heirs or assigns, shall have the right to re-entry. . . ." *First Presbyterian Church* v. *Sinclair Refining Co.*, 200 N.C. 469, 157 S.E. 438, 440 (1931).

The reason for this distinction (between a fee simple determinable and a fee simple subject to a condition subsequent) is that in the case of a fee simple determinable the words that provide for termination of the estate are regarded as a part of the original limitation of the estate; i.e., the estate is to last "so long as," "while," "during," to terminate "when" something exists or occurs. Words of "condition," however, as distinguished from "limitation" words, are considered as words providing for the termination of an estate before its natural termination. Estates in fee simple determinable thus "expire"; estates of fee simple subject to a condition subsequent are "divested" at the option of the person having a "right of entry" or "power of termination." Typical words introducing the estate of fee simple subject to a condition subsequent are: "on condition that," "provided that," "to be null and void if," or "to be forfeited if" a certain event occurs or fails to occur. . . . J. Webster, Real Property Law in North Carolina, § 37 (1971).

The primary import of these contested provisions is that the plaintiff is precluded from disposing of the property to any party other than the School Unit. Under North Carolina law, it is an unlawful restraint on the alienation of land to specify in a deed that the property may only be reconveyed to certain parties.

Since the restrictive provisions in the deed in question are void and thus unenforceable, the fair market value of the property could not be affected by them. Therefore, the Court will deny the defendant's motion for partial summary judgment. The Court specifically finds that the plaintiff's charitable contribution deduction will be calculated according to the fair market value of the gifted property rather than the actual cost of the property to the plaintiff.

Therefore, it is ORDERED that the defendant's motion for partial summary judgment be, and the same hereby is, DENIED.

QUESTIONS FOR DISCUSSION

1. *What is the difference between a fee simple determinable and a fee simple subject to a condition subsequent?*
2. *Why was the condition unenforceable in this case?*

---

(2) Life Estate. The life estate is another freehold estate. It is a lesser estate than a fee simple in terms of the measure of time and may be carved out of the whole "pie" that the fee simple represents. A life estate is not inheritable and has its duration measured by the life of some individual or individuals. A life estate may be created by deed, by will, or by operation of law.

Life estates may be created for a variety of reasons. For example, suppose Albert desires to allow his sister, Harriet, to have the use of the family home for her life but ultimately intends that his son, Sam, should have the property. He would convey the property to Harriet for life and the fee simple to Sam and his heirs. Harriet has a life estate and Sam a fee simple. On the other hand, Albert may decide to convey to Sam and his heirs in fee simple and reserve a life estate for himself. One can think of many other examples.

It is also possible to convey a life estate to one person for the life of another person. This is known as a life estate *pur autre vie.* Thus A may convey "to B to have and to hold for the life of C." When C dies, B's estate ends.

Absent a restraint in the creating instrument, a life estate may be sold or mortgaged. Another way of saying this is to state that it is *alienable.* Thus if B owns a life estate in real property, he may sell this interest to C. However, upon the death of B the life estate ends.

When one occupies property by virtue of holding a life estate, the question arises of the rights and duties of the holder of the life estate versus the party to whom the property ultimately reverts. The life tenant has the right to the possession and enjoyment of the property during his lifetime. On the other hand, he must preserve the property and keep it in a reasonable state of repair. If structures are destroyed without his fault, however, he is under no obligation to repair or rebuild them. The life tenant must pay the current taxes on the property and must generally pay the interest on any mortgage to which the life estate is subject, but not the principal.

The life tenant is also subject to the law of waste. For example, if a life tenant takes possession of property and oil is subsequently discovered, he would have no right to deplete the supply. Likewise, if the land is forested, he would have no right to remove the timber unless this activity was carried on prior to the creation of his life estate.

Another form of life estate is the common-law concepts of *dower* and *curtesy.* At common law the wife was entitled, upon her husband's death, to a life estate, as measured by her own life, in one third of the lands that he owned during the marriage. This was known as a *dower interest,* and even a conveyance to a purchaser during the husband's life could not defeat the wife's dower interest. This concept was transferred to the United States when the colonies were settled by the English.

Curtesy was a common-law concept under which the husband was entitled to a life estate in all the lands owned by his wife at any time during the marriage, provided that a child was born alive and capable of inheriting. In the United States curtesy is obsolescent. Instead the concept of dower has been applied to both men and women. In many states the concepts of both dower and curtesy have been abolished. As a substitute the surviving spouse is often given a fractional share in fee simple in the property owned by the deceased spouse at the time of death. The spouse is generally entitled to this share despite any contrary provision in a will. However, in those states where dower has not been abolished, a purchaser can obtain clear title to property free of dower rights only if both the husband and wife sign the deed, even if the property is owned solely by either the husband or the wife.

DURRENCE v. DURRENCE

239 Ga. 705, 238 S.E.2d 377 (1977)

UNDERCOFLER, J.

The issue presented in this case is whether a life tenant may cut timber from her estate or whether such an act constitutes waste.

D. L. Durrence died in 1960 and left a life estate in two tracts to his wife, Pearl—the 250 acre "homeplace" and the 434½ acre "flatwoods"—with the remainder to their children. Segal and Bobby Durrence are two of the sons, and as remaindermen tried to prevent their mother from cutting timber off these tracts. She then filed this suit to enjoin their interference with her life estate. The jury found in her favor and they appeal. We affirm.

In enumerations of error 1, 2, 3 and 4, appellants complain that the trial court improperly charged the jury on the law concerning the rights of the grantee of a life estate. We, however, find no error. The trial court correctly charged that a life tenant could use the property as a judicious, prudent owner of the estate would use it as long as she commits no acts tending to injure the remaindermen permanently, Code Ann. § 85–604; that the property could be used for the same purposes it was used when the life estate was created; that the cutting and thinning of pine timber in accordance with good forestry practices is not waste, unless wilful injury to the remainder was shown by acts not essential to the legitimate use of the life estate.

The evidence included the facts that D. L. Durrence had sold timber off the land and that that was his business, along with farming. Pearl Durrence had also sold timber before and Segal had tried to stop her by court action, which had resulted in a settlement. A state forester had marked trees to be cut according to good forestry practices. Although some of the evidence was in conflict, the jury was authorized to find in favor of Pearl Durrence. Enumerations 1 through 4 thus present no cause for reversal.

The trial court, therefore, did not err in failing to direct a verdict for the appellants as alleged in Enumeration 5. Furthermore, the evidence presented and the verdict authorize Mrs. Durrence to cut timber off her life estate generally as long as it is done in a manner consistent with good forest husbandry practices. Therefore, enumeration of error 6, regarding a directed verdict as to the flatwoods tract, has no merit.

(Judgment affirmed.)

QUESTIONS FOR DISCUSSION

1. *Would the outcome of this case have been changed if the land had not previously been forested?*
2. *May Mrs. Durrence cut down all the trees on the land in order to use it for some other purpose?*

30:3. TYPES OF FUTURE INTERESTS. We have seen that various estates may be carved out of the pie that constitutes the entire ownership interest of real property. These estates may be successive periods of time. Thus Albert may have a life estate in real property for his life, and Bertha may have a fee simple. Although both Albert and Bertha have an estate in the land, only one of them, Albert, has the present right to possession. Bertha has a right to possession at some time in the future. In other words, she has a future interest. The field of future interests can be very complicated and confusing. For the purposes of this text, however, our discussion is limited to a brief discussion of two future interests, the *reversion* and the *remainder.*

One has a *reversion* whenever he transfers to another a legally lesser estate while retaining some ownership interest. That interest is called a *reversion* because eventually the property will "revert" to the transferor. For instance, suppose Albert owns property in fee simple. He gives a life estate in that property to Bertha for the life of Bertha. At Bertha's death the property is to revert to Albert or his heirs. Albert has a reversionary interest in the property. Although Bertha has the present right to possession, Albert has a future interest in the nature of a reversion.

The future interest is known as a *remainder* if the residue of the estate is to go to a third party other than the original transferor. Suppose Albert owns property in fee simple and transfers it to Bertha for life and then to Carl and his heirs. Carl has a future interest in the property after the death of

Bertha known as a *remainder.* A remainder may be either *vested* or *contingent.* If the remainder is vested, the right to it is definite and indestructible. In the example given, Carl's remainder is vested because he or his heirs are sure to get the property in fee simple upon the death of Bertha. A contingent remainder is indefinite and destructible upon the happening of a contingency. Thus, if we alter the example so that Bertha has a life estate and the property is to go to Carl and his heirs provided that Carl survives Bertha, we have a contingent remainder. Carl must outlive Bertha in order for him or his heirs to get the property. If Carl predeceases Bertha, the property would *revert* to Albert or his heirs. One may sell or mortgage a reversion or a remainder just as he may any other property interest.

30:4. Rights in the Land of Another. There are a number of rights that one may have in the land of another that fall short of the absolute right to possession. The rights that will be considered here are easement, license, *profit à prendre,* and covenants in deeds.

(1) Easement. An easement is an interest that one has in the land of another that does not involve possession. The right to possession is in one other than the holder of the easement. Typically an easement involves the right to use the land of another for some specific purpose.

Easements may be classified as *easements appurtenant* and *easements in gross.* An easement appurtenant involves the right to use land adjacent to the land owned by the individual that is for the benefit of his property. For instance, suppose Albert owns Whiteacre and grants a right-of-way over Whiteacre to Bertha, who owns Blackacre, for passage to and from Blackacre. Bertha then has an easement appurtenant in Whiteacre. In this case the land that is benefited, Blackacre, is said to be the *dominant tenement* and the land subject to the easement, Whiteacre, is the *servient tenement.* Such easements may, of course, be very important to whoever owns the dominant tenement. The sale of the property does not destroy the easement, for the easement is said to "run with the land." By this is meant that the easement is not personal to the owner of the land but rather is for the benefit of the land and therefore passes to a new owner. An easement appurtenant may be created in a number of ways. It may be created by will, deed, contract, or prescription. Prescription is a concept analogous to adverse possession, which is discussed in Section 30:5, "Acquisition and Transfer of Title."

An *easement in gross* exists when the easement is not created for the purpose of benefiting land owned by the holder of the easement. In other words, there is no dominant tenement. For example, if Albert owns land with a pond on it and grants to Bertha the right by deed to come on the land and sail on the pond or fish in it at any time, Bertha has an easement in gross for this purpose. Another common example is the case of a utility company that has an easement to put poles on one's property. One important distinguishing feature of the easement in gross is that it cannot be transferred from the person to whom it is granted.

The fact that one has granted an easement in his land does not mean that he is prohibited from using the land for his own purposes. The owner may use the land for any purpose that is not inconsistent with the use granted by the easement. Thus if one has granted a right-of-way over his land, he may still use that land for any purpose that does not interfere with the right-of-way.

EDGELL v. DIVVER

402 A.2d 395 (Del. Ch. 1979)

The owners of several lots in Dewey Beach, Delaware, created easements in order to provide access to certain of the lots. Robert Edgell owned the property that was subject to the easements. The Divvers owned the lots that benefited from the easements. Edgell sued Divvers to have the easements extinguished.

HARTNETT, J.

The easements were created by deeds, recorded in 1953 and 1961, which describe the easements with sufficient particularity to enable them to be located with certainty. A plot of the lands showing the location of the easements was recorded in 1967. The easements were all created by predecessors in title to the parties in this action and the plaintiffs acquired title to their lands in 1976 and 1977. At the time these easements were created the lands were primarily used for residential purposes and the Divvers' parcels still maintain this use and character. The lands of the plaintiffs, however, are presently zoned for general commercial use, although they are not presently so used. The plaintiffs first contend that the easements have been extinguished as a matter of law by their taking of title to all the parcels upon which the easements are located, thereby effecting a merger of title of the dominant and servient estates. It is elementary, however, that an easement consists of two separate estates: the dominant estate which has the benefit of the easement and to which it is attached, and the servient estate on which the easement is imposed or rests. The plaintiffs have title to the servient estate, while the Divvers still retain title to the dominant estate. The mere acquisition of title to the entire servient estate by the plaintiffs does not effect a merger of title sufficient to extinguish the easement. Ownership of the dominant and servient estate has remained separate and no unity of ownership of these interests which is essential to the termination of an easement by merger has occurred.

The plaintiffs next contend that the easements should be extinguished, modified or relocated since the general area in which they are located has changed in character from residential to commercial. They argue that since the purpose for which the easements were originally created no longer exists, there is no reason for continued existence of the easements.

While it is true that the land owned by the plaintiffs and subject to the easements (the servient estate) is now zoned for commercial use, the Divvers' land (the dominant estate) is still used for residential purposes and therefore retains its original character. Since easements exist for the benefit of the dominant estate it is difficult to see the relevance of plaintiffs' arguments. The mere fact that the servient estate has changed in character is irrelevant to the question of termination of the easements. The primary restriction placed upon the owner of the dominant estate is that the burden created by the easement upon the servient estate cannot be materially increased, nor may new or additional burdens be imposed. Furthermore, plaintiffs as the owners of the servient estate must exercise their rights with regard to the land in a manner consistent with the existing easement. They may use the premises as they choose, except they may not interfere with the proper and reasonable use by the Divvers of their dominant right.

The easements were expressly granted for use as driveways. Plaintiffs contend that this purpose no longer exists due to the general change in the character of the area and that therefore the easements should be extinguished. It appears from the record, however, that the original purpose for the granting of the easements still exists and its use is apparently consistent with that purpose. Therefore, in the absence of a clear intent to abandon the easements or the termination of their purpose, the easements cannot be extinguished.

Lastly, the plaintiffs contend that the easements should be relocated as an alternative to extinguishing them in order to prevent irreparable damage to their use of their land. The Divvers have consistently refused to agree to any relocation of the easements. The question, therefore, is whether the holder of the servient estate can have easements relocated, over the objection of the dominant estate, simply because the location and use have become inconvenient to the use and enjoyment of the servient estate.

The general rule is well established that the consent of the owners of both the dominant and servient estates. In *Hibbel* v. *Fitch*, Md.App., 182 Md. 323, 34 A.2d 773 (1943) the Court stated at 34 A.2d p. 774:

> A way once located cannot be changed by either party without the consent of the other. When the right of way has once been exercised in a fixed and definite course, with full acquiescence and consent of both parties, it cannot be changed at the pleasure of either of them.

A brief examination of the deeds and the plot shows that the easements have been described with certainty for many years. The Divvers have consistently refused to agree to a relocation of the easements and, without such agreement, there can be no relocation regardless of the inconvenience to the plaintiffs.

(Decision for Divver.)

QUESTIONS FOR DISCUSSION

1. *Was this an easement in gross? Why or why not?*
2. *Identify the dominant and servient estates in this case.*
3. *Had Edgell purchased the lots that were dominant in addition to the lots he already owned would the easement have been extinguished?*

---

(2) License. Sometimes an easement is confused with a license. A license is a right to use the land of another, but it does not constitute an interest in the land of another as does an easement. A license is permission granted by the owner of land to use the property for a limited purpose. This prevents the person to whom the license is granted from being a trespasser when he comes on the licensor's property. In most cases, a license can be revoked by the licensor even if contractual liability arises. An easement cannot be revoked unilaterally. There are some situations, however, in which a license cannot be revoked. Another distinguishing feature of the license is that it cannot be transferred. Thus, if Albert orally tells Bertha that she may fish in his pond, Bertha has a license that may be revoked by Albert at any time. If Albert grants the same right by deed, he may create an easement, which cannot be unilaterally revoked.

BUNN et al. v. OFFUTT et al.
222 S.E.2d 522 (Va. 1976)

Harvey W. Wynn and Rosebelle Wynn signed a contract to purchase a house from Temco, Inc., on July 9, 1962. The house was adjacent to an apartment house being developed at the time by Temco. In the contract was the following provision: "Use of apartment swimming pool to be available to purchaser and his family." *The Wynns were told by Temco's sales agent that subsequent purchasers would be able to use the pool. There was no reference to the right to use the pool in the deed from Temco to the Wynns.*

On May 31, 1969, Edward and Sandra Bunn contracted to buy the house from the Wynns. There was no mention of the right to use the apartment pool in either the contract or the deed, but the Wynns told them that the right to use the pool went with the property. After the purchase was made, the Bunns were denied passes to the pool. They sued Temco, T. J. Offutt, owner of Temco, and Dittmor Co., Inc., another company owned by Offutt and having an interest in the apartment. From an adverse decision by the trial court, the Bunns appealed.

HARRISON, J.

The testimony of Offutt, owner of appellee corporations, is unequivocal that at no time did he ever intend to extend the privilege of using the swimming pool beyond the original purchasers of certain houses (including 900 South Wakefield Street) which were located adjacent to his apartment development. He said that at one time he thought he would experience difficulty in selling the houses without an added inducement, and therefore included in his sales contract a clause to the effect that the use of the apartment swimming pool would be available to the purchaser and his family. Offutt further testified he never intended the right to run with the land and inure to successors in interest and in fact had never extended pool privileges to any one beyond the first purchasers.

The dispositive issue in this case is whether the language in the contract, "Use of apartment swimming pool to be available to purchaser and his family," amounted to a grant of a mere license to the Wynns and their family; or whether the Wynns acquired thereby a private easement across the land of appellees to the swimming pool and to the use of the pool, which easement was thereafter transferred to the Bunns.

A license has been described as "a right, given by some competent authority to do an act which without such authority would be illegal, a tort, or a trespass." 12 M. J., License to Real Property, § 2, p. 148. A license is personal between the licensor and the licensee and cannot be assigned. And a grant which creates any interest or estate in land is not a license. Such a grant creates an easement.

An easement has been described as " 'a privilege without profit, which the owner of one tenement has a right to enjoy in respect of that tenement in or over the tenement of another person; by reason whereof the latter is obliged to suffer, or refrain from doing something on his own tenement to the advantage of the former.' " *Stevenson* v. *Wallace,* 27 Gratt. (68 Va.) 77, 87 (1876).

Easements may be created by express grant or reservation, by implication, by estoppel or by prescription. The only rights acquired by the Wynns in the property of appellees were acquired by deed from Temco. The provisions of the contract were merged in this deed. However, the deed is silent as to the pool, and the contract made the use of the pool available only to "purchaser and his family." The trial court found this language consistent with appellees' theory that a mere license only was granted to the purchasers and their families, and not an interest in land or an estate of inheritance; that the absence of any provision regarding the swimming pool in the deed to the Wynns was sufficient to preclude any easement by grant or reservation; and that the evidence and exhibits failed to show that an easement was created by estoppel, necessity or prescription. The trial court further found that no easement had been created by implication for there was neither a showing of a preexisting use of the easement prior to the conveyance by Temco to the Wynns, nor any showing that the use of the swimming pool was essential to the beneficial enjoyment of the land conveyed.

. . . The deed from Temco to the Wynns did not purport to convey an easement to the swimming pool, and the language in the sales contract between the parties is not sufficient to create an express easement. The Wynns and their family were given a mere license to use the swimming pool. It was not an interest running with the land that could subsequently be transferred by them.

(The decree of the lower court under review is affirmed.)

QUESTIONS FOR DISCUSSION

1. *How did the court distinguish an easement from a license?*
2. *What would have been necessary in order for the Wynns to obtain an easement rather than a license?*

---

(3) **Profit à Prendre.** A *profit à prendre* is very similar to an easement in that it cannot be unilaterally revoked by the person who grants it. A *profit à prendre* is frequently simply called a *profit.* A profit is a nonpossessory interest in the land of another that gives one the right to go on the land and sever some part of it. Typical examples are the right to cut timber or mine coal on another's land. *Profits à prendre* are created in the same manner that easements are created, and the rules that are generally applicable to easements may be applied to profits.

(4) **Covenants.** One method of controlling the use of land is by the inclusion of restrictions on the use of land by a covenant in a deed. Such covenants are most frequently used when a new subdivision is created. The restrictions may be included in the deeds or in a plat of the subdivision. Covenants that involve the use to which the land may be put are said to "run with the land" and are binding on all subsequent purchasers of the land, even if the covenants are not contained in their own deeds. Examples of such covenants are restrictions that require the land to be used for residential purposes only, that require all buildings to be set back a certain distance from the street, that require fences to be of a certain type, and so on. These covenants are, of course, entirely separate from the zoning laws passed by the local government, which may also regulate the use of the land. Generally a court will enforce covenants in deeds or plats. It will not enforce a covenant, however, that violates the federal or state constitution. For instance, covenants that discriminate among people on the basis of race, creed, or national origin are unconstitutional and unenforceable. Also, if the conditions surrounding the land have changed dramatically since the creation of the covenant, then a court may not enforce it. The primary use of covenants is to benefit the surrounding property; hence the covenants may generally be enforced by those who own that property.

GUILFORD ASSOCIATION, INC. v. BEASLEY
29 Md.App. 694, 350 A.2d 169 (1976)

GILBERT, J.

The factual genesis of this case is simple. Guilford is the assignee of The Roland Park Company and, as such, may enforce the restrictive covenants and conditions embodied in an agreement dated June 26, 1913 and recorded among the Land Records of Baltimore City. Basically the restrictions are designed to preserve the residential character of Guilford.

Beasley acquired the property known as 3809 Greenway on January 16, 1964. The Beasley residence is within the geographical confines of Guilford, and is subject to the restrictions in the extended agreement. The evidence shows that Beasley parks or stores five or more motor vehicles upon his driveway. Guilford alleged that the vehicles were stored on the premises and pointed out, in support of that assertion, that the cars were without license plates for a stated period of time. Beasley testified that prior to trial, apparently in response to a housing violation notice from the appropriate Baltimore City agency, he had placed "tags" on all of the cars and that they were operable. Beasley denied "storing" vehicles, and testified that they were for the personal use of himself and his family. He characterized the vehicles as "special interest" cars.

Guilford alleged in the circuit court, and argues here, that the motor cars were stored and that "storing" of the vehicles constitutes a violation of Sub-Division III of the 1913 agreement, as extended. That section provides:

> *The land included in said tract, except as hereinafter provided, shall be used for private residence purposes only* and no building of any kind whatsoever shall be erected or maintained thereon except private dwelling-houses, each dwelling being designed for occupation by a single family, and private garages for the sole use of the respective owners or occupants of the plots upon which such garages are erected. (Emphasis supplied.)

The Chancellor, in his "Memorandum Opinion and Order," construing the above quoted restrictive covenant as applied to the facts of this case, said:

> . . . The provision in question is directed at the kind and use of buildings which can be erected and maintained on the land. It prohibits all buildings except private dwelling-houses designated for occupancy by a single family and private garages for the sole use of the owners or occupants. There is neither allegation nor evidence that the Beasleys have violated any of these building prohibitions. There is no contention that the automobiles are not the property of the Defendants, and kept for their private purposes. Rather, the Plaintiff seems to be attempting to apply this provision restricting the erection and maintenance of buildings to a situation involving parked cars on the property in question. It argues that the language the ". . . land . . . shall be used for private residence purposes only . . ." should be construed to mean that the Defendants' vehicles must be used with some frequency or be deemed stored or otherwise in violation of Sub-Division III.
>
> The Court finds no foundation in Sub-Division III for this conclusion. . . .

Guilford avers that the covenant prescribes the parking of an excessive number of motor vehicles upon the Beasley property. Guilford reads the quoted clause to mean that unless Beasley's vehicles are "used with some frequency" they should "be deemed stored or otherwise in violation of Sub-Division III."

The general rule with respect to the interpretation of restrictive covenants in deeds is that they will be construed more strictly against the grantors or those seeking to enforce them. All doubts are to be resolved in favor of those resisting enforcement of the covenants. When the words of the restrictive covenant sought to be enforced are logically susceptible of a construction which would not violate the covenant, as against a construction which would violate it, courts will place a construction upon the words that will result in no violation.

If the clarity of the restrictive covenant is dubious, the courts will hold the restriction to its narrowest limits. Put another way, if there is doubt as to the meaning of the restriction, the courts will generally rule in favor of the freedom of the property from the strictures of the restriction.

Guilford takes the tack that "(t)he original parties to the Guilford Deed and Agreement could not have intended by Sub-Division III to restrict the buildings to private dwelling houses and private garages, and meanwhile to leave the surrounding land unencumbered and available for any form of commercial exploitation short of building." We agree. As we read the covenant it is clear that all the land in the Guilford tract was placed under the restriction, that it is to be used ". . . for private residence purposes only. . . ." The construction placed upon the quoted clause by the Chancellor leaves room, disregarding local zoning regulations, for the opening of a neighborhood vegetable and fruit stand, most certainly intended by the restrictive covenant to be proscribed within the Guilford geographical confines.

The courts, it would seem, are under a duty to effectuate rather than defeat an intention which is clear from the context, the objective sought to be accomplished by the restriction and from the result that would arise from a different construction. Furthermore, courts in construing restrictive covenants must consider the circumstances surrounding the parties at the time the covenant was made and the fact that the restrictions were not imposed solely for the benefit of the grantor, ". . . but mainly for that of the grantee and those similarly situated with him. . . ." *Wehr* v. *Roland Park Co.*, 143 Md. 384, 392, 122 A. 363, 366 (1923). The restrictions in the case now before us were not imposed for the benefit of The Roland Park Company nor its assignee, the appellant. We believe it beyond serious question but that the restrictions on the Guilford tract are for the benefit of the residents of Guilford in that the restrictions protect their property value, maintain the *status quo* with respect to the neighborhood esthetics and generally aid in making the area a better place in which to reside.

We hold the meaning of the words, "The land included in this tract . . . shall be used for private residence purposes only . . ." to mean that the *land* is to be used for private residence purposes and no other purpose. It logically follows that all buildings erected upon the land are restricted to the same residential use.

(Order dismissing Guilford's bill of complaint reversed.)

QUESTIONS FOR DISCUSSION

1. *May restrictive covenants be more restrictive than local zoning laws?*
2. *Are restrictive covenants in deeds binding on the subsequent purchasers of property whose deeds do not contain the covenant?*
3. *Can you think of some typical restrictive covenants that might validly be contained in a deed?*

30:5. ACQUISITION AND TRANSFER OF TITLE. There are a number of methods by which one may acquire title to real property: (1) deed; (2) adverse possession; (3) accretion; (4) public sale; (5) will; and (6) descent. The transfer of property by will and descent are discussed in Chapter 32.

By far the most common method of transferring and acquiring title to real property is by the sale and purchase of the land or by gift. Such a transfer of the title to real property is accomplished by the signing of a deed by the *grantor,* that is, the transferor of the property. The other party to the deed is known as the *grantee.* This is the person to whom the property is being transferred. Naturally a deed must be in writing. In addition to naming the parties, the deed must fully describe the property that is being transferred. It will also set forth how the property is to be held by the grantee. For instance, if there are co-owners, the deed should state whether the title is to be held jointly or in common. The deed will also recite the type of estate that is being transferred, typically a fee simple.

(1) Warranty Deed. There are two types of deeds, *warranty* deeds and *quitclaim* deeds. A warranty deed conveys the grantor's interest in the property and also gives the grantee a number of warranties. Among these warranties is a warranty that the grantor has title to the property and that he has the power to convey it to the grantee. Under the common law this was known as the *covenant of seisin.*

The grantor also warrants that the property is transferred free from encumbrances. Encumbrances might consist of such things as mortgages or deeds of trust, tax liens, mechanic's liens, or judgment liens. If the property is not to be transferred free from certain encumbrances, this fact must be stated in the deed.

The third warranty made by the grantor under a warranty deed is known as a *warranty of quiet enjoyment.* This warranty has nothing to do with the noise level surrounding the property; it means, rather, that the grantee shall have good title to the property. The grantor agrees to defend the grantee legally against all who claim to have a superior title.

There are two types of warranty deed, a *general warranty deed* and a *special warranty deed.* An easy way to remember the distinction between the two is that there is nothing very special about a special warranty deed. It seems that the grantor warrants to the grantee only that he himself has done nothing to breach the warranties discussed previously. On the other hand, a general warranty deed gives the grantor's warranty that neither he nor anyone prior to him in the chain of title has violated the warranties. Whether a special or a general warranty deed is given depends to a great extent upon the custom in certain sections of the country. For instance, in parts of Maryland a special warranty deed is almost always given by the grantor, whereas in Virginia it is customary for the grantor to give a general warranty deed.

(2) **Quitclaim Deed.** The grantor makes no warranties when he or she transfers real property by a quitclaim deed. A quitclaim deed conveys all the interest that the grantor has, if any, in the property. He does not warrant that he has any interest. Quitclaim deeds are used in special situations. For instance, if it is possible that an individual other than the grantor has a claim to certain property, the execution by that person of a quitclaim deed will remove any doubt as to whether the grantee is actually receiving a clear title to the land. Obviously a grantee should not normally accept a quitclaim deed from a grantor.

(3) **Delivery.** In order for there to be a transfer of the property, in addition to being properly executed by the grantor, the deed must also be delivered to the grantee. Without a valid delivery of the deed there is no transfer of the property. Many of the problems concerning the delivery of deeds were discussed in connection with the delivery requirement for gifts in Chapter 28. Sometimes problems with the delivery of a deed arise in connection with the delivery of it to a third party. If a deed is delivered by the grantor to a third party who is to hold the deed until some condition or event occurs, then title does not pass until the condition is fulfilled. Thus delivery of a signed deed to an escrow agent to hold until the grantor fixes his heating system would not transfer title until the system is fixed and then only if the escrow agent is independent and there is no control exercised over him by the grantor. If the grantor delivers a deed to his own agent without an intent to surrender control over the deed, then there is no delivery. However, delivery to a third person that does divest the grantor of control is a valid delivery even if the deed is not to be delivered to the grantee until some later time, as long as the grantor does not have control over the occurrence of the condition upon which the deed is to be transferred.

30:6. Steps in the Purchase of Property Prior to Execution of the Deed. (1) **The Contract of Sale.** In the typical sale and purchase of property certain general steps are followed of which you should be aware. The first step is generally the signing of a contract of sale by the grantor and grantee. Typically the buyer first signs the contract of sale. This of course constitutes an offer that the grantor may accept by also signing the contract. One should not take the signing of a contract to purchase real property lightly. This is probably the single most important step involved in the purchase of real property. Remember that a contract for the sale of land is specifically enforceable. It also sets forth the rights and duties of the parties. This is important because as a general rule there are no implied warranties concerning the property given in a contract for the sale of land. Although some inroads have been made, the doctrine of *caveat emptor* ("let the buyer beware") still has vitality in the field of real property. Thus, if one wants a guarantee that the house he is purchasing is not riddled with termites or that the appliances work, this guarantee must be stated in the contract. The contract may also state the financing arrangements to be made by the parties, the remedies available to each party for breach of the contract, the date when the transfer of the property is to occur, and so on.

(2) Equitable Conversion. One concept with which many laymen are not familiar is the doctrine of *equitable conversion.* Typically there is a period of time between the point when the parties enter into a contract and the date when the property is transferred from the grantor to the grantee by the execution and delivery of a deed. During this period a number of things may happen to the property or the parties to the contract. For instance, the house located on the real estate may be destroyed, or one of the parties may die. When such an event occurs, it is important to determine whether the interest of a party is realty or personalty. At this point the doctrine of equitable conversion comes into play. Because the contract is specifically enforceable, the purchaser's contract right is considered realty, whereas the interest of the seller is considered personalty. The justification the courts typically give for this magical transformation is the implementation of the old equity maxim that "equity considers as done that which ought to be done." Therefore for many purposes the equity courts treat the purchaser as the equitable owner of the land as soon as the contract for sale is made.

The equitable conversion of the parties' interests in the property can have important consequences. For instance, suppose that a house is uninsured and burns to the ground after a contract is signed but before title is transferred to the grantee. Who bears the loss? A number of jurisdictions would place the loss on the grantee, even though there has not been a transfer of title in the property. The loss would be placed on the grantee by a granting of the remedy of specific performance to the seller. For this reason a contract for the sale of land should always state clearly who has the risk of loss and is responsible for insuring the property prior to the date of transfer.

The doctrine has been subject to much criticism, and some jurisdictions refuse to follow it in cases in which the risk of loss would be placed on a purchaser who is not yet in possession of the property. The reasoning is that the seller should not be given specific performance of the contract when he cannot perform himself, that is, deliver the house.

BRIZ-LER v. WEINER
171 A.2d 65 (Del. 1961)

WOLCOTT, J.

This is an appeal from a judgment of the Court of Chancery dismissing a complaint seeking recission of a contract, an accounting, and an equitable lien upon real estate or upon the proceeds of insurance. The basic question of the appeal is whether a loss occasioned by a fire should fall upon the seller or the purchaser of real estate under an installment contract.

In October, 1954, the plaintiff and defendants entered into a contract for the purchase by the plaintiff of the Hotel Grande property in Wilmington together with certain fixtures and equipment for a total price of $114,000, of which $80,000 represented the consideration for the real property and $34,000 the consideration for the fixtures and equipment. The plaintiff paid down the sum of $11,500 and agreed to pay the balance of $102,500 plus interest at 6% in monthly installments of $865. In addition, the contract required the plaintiff to pay the defendants monthly a further sum to be held in escrow out of which all taxes and fire insurance carried on the property were to be paid by the defendants as they became due. Upon payment of the full amount the defendants were to convey title to the plaintiff free and clear of all liens and encumbrances, and to deliver a bill of sale covering the fixtures and equipment.

Upon execution of the contract, the plaintiff entered into possession of the property. . . .

In December, 1957, a fire occurred on the premises causing substantial damage. At this time a substantial balance was still owing on the total purchase price, and plaintiff had substantially complied with the terms of the contract.

Plaintiff claims that it should be repaid all the money paid by it pursuant to the installment contract because defendants cannot now deliver what they contracted to deliver, viz., a four-story hotel structure. In the alternative, plaintiff claims that it is entitled to an equitable lien on the premises, or on the proceeds of insurance, in the full amount paid by it under the contract.

The basic question involved in this appeal is whether or not a loss occasioned by fire to premises under an installment contract of sale shall fall upon the seller or the purchaser. Presumably, if the loss as a matter of law falls upon the seller, then plaintiff should be entitled to relief of some nature. If, on the contrary, the loss falls upon the purchaser, the complaint was properly dismissed.

The rule followed in a majority of American jurisdictions is that an executory contract for the sale of lands requiring the seller to execute a deed conveying the legal title upon payment of the full purchase price works an equitable conversion so as to make the purchaser the equitable owner of the land and the seller the equitable owner of the purchase money. The result is that the purchaser, the equitable owner, takes the benefit of all subsequent increase in value and, at the same time, becomes subject to all losses not occasioned by the fault of the seller.

The rule followed by the majority of American states finds its origin in *Paine* v. *Meller,* 31 Eng.Rep. 1088. The basic reason for the rule is that if a party by a contract has become in equity the owner of land and premises, they are his to all intents and purposes and, as such, any loss caused to them must be borne by him.

The rule is criticized severely in 4 *Williston on Contracts* (Rev.Ed.), § 935 et seq., on the ground that the destruction of the subject matter of the contract renders it impossible of performance. And other textwriters have advocated differing views as to the extent and effect of the rule. 4 Williston, § 940, advocates the rule followed by a minority of American jurisdictions that any loss from destruction falls on the purchaser only if at the time of entry into the contract he is put into possession of the land and thereafter exercises full rights of control.

In the case at bar the plaintiff entered into possession of the premises sold upon execution of the contract. Thereafter, it exercised all the rights ordinarily incident to ownership. We think that under any view of the rule of equitable conversion of title to real estate the fact that the purchaser has possession of the land sold and exercises sole control over it requires that any loss occasioned accidentally to the premises must fall upon him. Since this is the fact in the case at bar, it follows that, if this contract is unconditional, the loss must fall upon the purchaser. We are required to go no further in this case in applying the rule of equitable conversion. We leave undecided whether or not the majority rule based upon *Paine* v. *Meller*, which, under all circumstances, places the loss upon the purchaser following an equitable conversion, is the law of this State.

Finally, plaintiff argues that the destruction of the building made it impossible for the defendants to convey what they had contracted to convey, viz., the premises as they existed before the fire. There is, however, no provision in the contract providing for such event. It follows, therefore, that plaintiff wants us to hold, with a small minority of the States, that destruction of the subject matter makes inoperative the doctrine of equitable conversion. Such a view, however, is a rejection of the doctrine which we have found to be the law of this State, at least under the circumstance of admitting the purchaser into possession. It follows, therefore, that plaintiff upon the execution of the contract for the purchase of the Hotel Grande became the equitable owner of it and, as such, subject to losses occasioned other than by the fault of the defendants.

(The judgment below is affirmed.)

QUESTIONS FOR DISCUSSION

1. *The court noted that there is much criticism of the doctrine of equitable conversion. What is the basis for the criticism? What justifications are there for the doctrine?*
2. *In light of the doctrine of equitable conversion, what precautions should the purchaser take when entering into a contract to buy real property?*

---

(3) **Examination of Title.** Before the date when the grantee actually pays for the property and receives title, an examination will usually be made of the history of the title among the local land records. The reason for this examination, or title search, is to make sure that the grantor can deliver a clear or *marketable* title. For instance, suppose Albert conveys the land to Gary Grantor in 1967. At that time Albert is married, but his wife does not join in the execution of the deed and the concept of dower is in force in that jurisdiction. Gary Grantor then contracts to deliver a marketable title to the land to George Grantee. George's title examination may reveal that Grantor cannot deliver a clear title and needs to take some action to remove the "cloud" on the title. George should discover this information when he has his attorney "search the title" or review an abstract of title prepared by someone else.

If the buyer finances the purchase of the property, the lending institution will require a title search. It may also require that the buyer purchase a policy of *title insurance* in its favor as a condition of the loan. There are title-insurance companies that will write such a policy after an examination of the property. It is also possible for the buyer to purchase a policy of title insurance to protect his own interest. Such a policy will protect the policyholder in the event there are defects in a title that a title examination may not reveal, such as a forgery of a party's name somewhere back in the chain of title.

Some states have enacted "marketable title acts" that simplify the determination of whether the owner of land possesses a marketable title. These statutes were first passed in the Midwest and generally provide that one has a marketable title if he or she can show an unbroken chain of title of record for a period of time, such as forty years. The effect of this statute is to cut off other interests or defects in the title that may have existed prior to the start of the period. Such a statute, of course, does not eliminate the effect of liens, easement, adverse possession, and other matters that affect the marketability of a title.

(4) Settlement and Financing. The date when the actual transfer of the property is to occur is known in many jurisdictions as the *settlement date.* It is at this time that the deed is usually delivered to the purchaser and the purchaser pays for the property. If a loan is involved, the note to the bank must be signed. The bank will probably also require that the borrower give it a security interest in the property, usually in the form of a *mortgage.* The borrower gives the bank a mortgage on the property and he is known as the *mortgagor.* The mortgage usually gives the bank, as the *mortgagee,* the right to sell the property in case of default by the borrower. The provisions of the typical mortgage also allow the lender to charge the borrower fees for foreclosing on the mortgage and require the borrower to keep the property insured and to pay the taxes on the property.

When the borrower defaults on the mortgage, the mortgagee (the bank) may file a suit to *foreclose,* asking that the mortgagor's (the borrower's) equity of redemption be foreclosed, which means that the borrower's right to come in and redeem the property by paying off the debt is cut off or foreclosed. There are many theories of the nature of the interests of the various parties in a mortgage and foreclosure situation. The practical result of a foreclosure from the mortgagor's standpoint is that the property is sold at either a private or a public sale to satisfy the debt. If after the sale the debt is still not paid, a *deficiency* is said to exist, and the borrower is liable for that sum. If there is an excess over the amount owed, the borrower is entitled to that sum.

Because the implications of the real-estate transaction may not be apparent to many people, a number of states require that certain matters such as the total dollar amount of interest to be paid over the life of the loan, the nature of title insurance, and other relevant matters be fully explained to the buyer at settlement. If a federal guarantee of the loan is involved, the federal government also requires disclosure of many facts to the borrower at settlement. The borrower is required to sign documents that indicate that the required disclosures have been made to him.

After settlement it is important that the grantee immediately record his deed among the land records of the jurisdiction. Once the deed is recorded, it is constructive notice to all that the grantee is the owner of the property. This notice prevents the grantor from subsequently selling the same property to an innocent purchaser who might claim a superior title to the first grantee.

30:7. ACQUISITION OF TITLE BY ADVERSE POSSESSION. One may acquire title to land under the doctrine known as *adverse possession.* Adverse possession consists of actual possession of another's property, hostilely, openly and notoriously, exclusively, and for a continuous statutory period of time. When these requirements are met, then the person holding by adverse possession gets title to the land.

In order to meet the requirement of actual possession, one must actually occupy the land. It is not sufficient to go on the land daily to fish or even to remove trees. One must exercise control over the land as an owner, such as by fencing it in and keeping others out. The acts of adverse possession should be hostile to the interest of the true owner. Thus, if one uses land with the permission of the owner, his possession is not hostile but rather that of a tenant or licensee.

The acts of adverse possession must be so obvious that the true owner should know of the adverse claim. It is not necessary that he actually know about it but only that he should know. In addition, the acts must be continuous for the statutory period, usually from ten to twenty-one years, depending upon the state. In the calculation of the time the concept of *tacking* is allowed. That is, two or more adjacent periods of adverse possession by different individuals may be used to calculate the period of adverse possession. Thus, if Albert occupies property adversely for fifteen years and Jones occupies it for the next five years, Jones would get title to the land if the statutory period were twenty years. Some states also require that the adverse possessor pay the taxes in order to obtain title to the property.

One important use of the doctrine of adverse possession is to clear defective titles. If one has occupied a home and paid taxes for twenty years, the fact that the title of record is defective for one reason or another would not be significant because he also has title by adverse possession.

RUSSELL v. GULLETT
589 P.2d 731 (Or. 1979)

BRYSON, J

The parties to this suit are neighbors and own adjoining property in Hines, Oregon. Plaintiffs, the Russells, brought the suit to establish their title by adverse possession to a portion of the property of defendants, the Gulletts. The following diagram, not to scale, shows the disputed area involved on appeal.

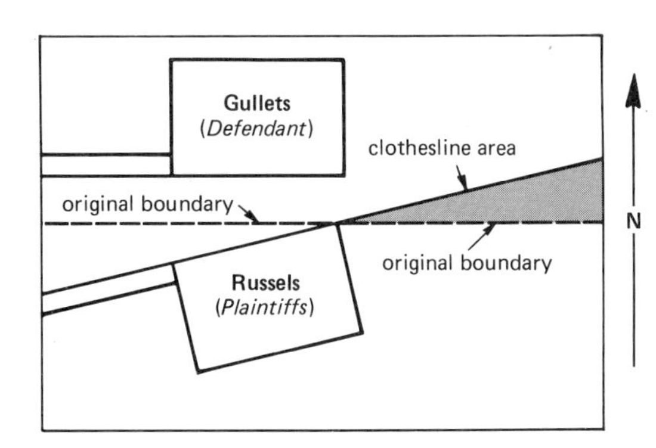

Plaintiffs claimed portions of defendants' front, side, and back yards. The trial court found that plaintiffs had proved their case only with respect to a strip of land in the back yard, where plaintiffs' predecessors in interest had built a clothesline, and a piece of land under the northeast corner of plaintiffs' garage. The defendants appeal. The only disputed land on appeal is the strip in the back yard. Defendants do not dispute the decree as to land under the garage. We review de novo.

Parties seeking to establish title to land by adverse possession must prove that their possession was "actual, open, notorious, hostile, continuous, and exclusive, under a claim of right or color of title, for a period of ten years." *Grimstad* v. *Dordan*, 256 Or. 135, 139, 471 P.2d 778, 780 (1970).

Defendants contend that plaintiffs did not prove that their use of the strip of land in the back yard was exclusive for the necessary period. They argue that the evidence shows that "plaintiffs and defendants were making joint or common use of the clothesline and clothesline area," which would defeat the "exclusive" element of plaintiffs' case.

The plaintiffs' predecessors in interest were the Nyleens, who bought the southerly parcel in 1949. Defendant Jane Gullett, formerly Jane Brewington, has lived on the northerly parcel since 1951. Mr. and Mrs. Nyleen both testified that by 1951 or 1952—it is not clear exactly when—they built a clothesline in the area shown on the diagram. The clothesline was a substantial affair, consisting of metal pipes embedded in concrete. The Nyleens believed that the clothesline was within their property. The Nyleens always maintained the strip of land under the clothesline. The Brewingtons never stored any of their belongings on the strip and never used the clothesline except with permission from the Nyleens. The Nyleens' use of the clothesline area continued from 1952 to 1973, when they sold to the plaintiffs.

The Nyleens learned about the boundary question after they sold their house to plaintiffs.

The Gulletts built a fence on the original boundary line and this suit followed. The former Mrs. Brewington, now defendant Mrs. Gullett, testified that the clothesline was not built until 1962 or 1963; that she and her deceased husband used the clothesline area to store wood and old junk; that she and her deceased husband watered the lawn in that area; and that she used the clothesline without permission from the Nyleens.

Given this conflict in the testimony, the trial court chose to believe that the Nyleens had a more accurate memory of the events. We have said repeatedly that we defer to the trial court's assessment of the credibility of witnesses. We have also reviewed the testimony and accept the Nyleens' version of the facts, as did the trial court. That version shows that from 1952 until 1973 the Nyleens used the property in a way that satisfied the test stated above to establish title by adverse possession. In particular, their use satisfied the exclusivity requirement, which does not require absolutely exclusive use, but only such use as would be " 'expected of an owner under the circumstances.' " *Nelson* v. *Vandemarr,* 281 Or. 65, 74, 573 P.2d 1232, 1237 (1978), quoting *Grimstad* v. *Dordan,* 256 Or. 135, 141, 471 P.2d 778 (1970). Allowing the neighbors to use the clothesline is such an expected use.

(Affirmed.)

QUESTIONS FOR DISCUSSION

1. *When did the Russell's period of adverse possession begin?*
2. *If the court accepted the testimony of Mrs. Gullett, would the case have a different outcome?*

30:8. ACQUISITION OF TITLE BY ACCRETION. One may obtain title to land by *accretion.* When one owns land on the bank of a river or other body of water and there is a gradual buildup of additional dry land by deposits made by the body of water, there is said to be an accretion of the land and the adjacent owner has title to the new land. On the other hand, if the addition to the land comes suddenly, as for example as the result of a flood, then there is no addition to the adjacent owner's property. Likewise, if the additional land is the result of the efforts of the adjacent owner, such as by dumping or dredging, he does not obtain title to the new land.

30:9. ACQUISITION OF TITLE BY JUDICIAL SALE. Real property may be sold by a public officer at a judicial sale. Such a sale may occur, for example, if a judgment creditor is not paid. The property of the defendant may be sold to satisfy the judgment. Other similar sales may result from the foreclosure of a mortgage, sale of the property for nonpayment of taxes, and foreclosure of a mechanic's or other lien. One who purchases real property at such a sale obtains good title to the property, provided that the applicable statute is followed.

SUMMARY

In the early English feudal system interest in land was created by the process of subinfeudation. The process has evolved to the point where today a person may own his own property.

The word *estate* defines the interest that one has in the land. The fee simple and life estate are types of freehold estates. A fee simple is of potentially infinite duration, and the estate is inheritable. A fee simple may be subject to termination upon the happening of an event. These arrangements are known as either a fee simple determinable or a fee simple on condition subsequent.

A life estate is not inheritable and its duration is measured by the life of an individual. If the measuring life is someone other than the person who has the life estate, it is called a life estate pur autre vie. The life tenant is subject to the law of waste and must pay taxes on the property. Another form of life interest is the common-law concept of dower and curtesy.

A reversion occurs when one transfers to another a legally lesser estate while retaining an ownership interest. If the residue of the estate is to go to a third party other than the original transferor, the future interest is known as a remainder which may be either vested or contingent.

Easements may either be appurtenant or in gross. An easement appurtenant has a dominant and servient tenement while an easement in gross does not. A license is a right to use the land of another but does not constitute an interest in the land of another. A profit a prendre is a nonpossessory interest in the land of another that gives one the right to go on the land and sever some part of it.

Covenants are sometimes used to control the use of land. Some covenants are said to "run with the land."

Title to land may be acquired by deed, adverse possession, accretion, public sale, will, and descent. Deeds may be either warranty deeds or quitclaim deeds. Warranty deeds may be either special or general warranty deeds. To be effective a deed must be delivered.

Adverse possession consists of actual possession of another's property, hostilely, openly and notoriously, exclusively, and for a continuous statutory period of time. In the calculation of time the concept of tacking is allowed. Accretion occurs when land is built up by deposits from a body of water.

When land is transferred by deed, there are several typical steps in the purchase. Once the contract of sale is signed, the doctrine of equitable conversion may apply. There may be a title examination to assure that there is a marketable title.

PROBLEMS

1. Annie L. Sauls conveyed certain lands to Don S. Crosby and Bertha Mae Crosby with the following reservation set forth in the conveyance: "The Grantor herein, reserves a life estate in said property." Annie then decided to cut timber for the first time and sell it. The Crosby's contend that she

could not do this. Are they correct? Sauls v. Crosby, 258 So. 2d 326 (Fla. App. 1972).

2. The Heads owned certain land adjacent to Yeomans. They had an easement across Yeomans' land for the purpose of egress and ingress to their property. When Yeomans damaged the road on the easement, the Heads filed suit. The court granted an injunction prohibiting Yeomans from using the road for any purpose. Was the court correct? Yeomans v. Head, 253 S.E.2d 746 (Ga. 1979).
3. James R. Costello signed a quitclaim deed conveying certain real property to his son and his brother. The deed was prepared by O'Brien, an attorney. The deed was signed by Costello in O'Brien's office and retained by O'Brien. After Costello's death the administratrix of his estate contended that the property was never transferred and should be part of the estate. Was the administratrix correct? Costello v. Costello et al., 136 Conn. 611, 73 A.2d 333 (1950).
4. In 1959 West acquired a tract of land. Part of that land was adjacent to Tilley's land. Part of West's land was enclosed by concrete walls erected by Tilley, dating originally from 1931 and altered in form several times since then. Within these walls Tilley planted a lawn and shrubs and used the land, which included a ten-foot strip of West's land. At trial Mrs. Tilley testified that when the land was enclosed she thought that only her own property was enclosed and that she did not intend to take West's property. She also testified that "I claim title to everything that is within the walls." West contended that Tilley's possession was not hostile. Was he correct? West v. Tilley, 306 N.Y.S.2d 591 (1970).
5. Hercules Life Insurance Company owned Lot 22 in Block Number 13/2028 in Dallas, Texas. In July 1936 Hercules sold one half of the lot to J. F. McNamara. In September of 1936 the other half of the lot was sold to George A. Ripley. In the conveyances Hercules created an "easement for drive-way purposes" over McNamara's half of the lot to the Ripley half. A subsequent owner of Ripley's lot contended that this gave her the right to park her car on the driveway crossing the McNamara lot. The owners of McNamara lot contended that she was permitted only to drive her car over the driveway and was not permitted to park it. Decide. Colborn v. Bailey, 408 S.W.2d 327 (Tex. 1966).
6. Skelly Oil Company signed a contract to purchase real property from Tom A. Ashmore and Madelyn Ashmore. A building on the property was destroyed by fire before the property was transferred to Skelly and while it was still occupied by the Ashmores. The Ashmores were not at fault. There was insurance on the building. Skelly claimed that it was responsible for the loss and thus was entitled to the proceeds of the insurance and specific performance of the contract. Was it correct? Skelly Oil Co. v. Ashmore, 365 S.W.2d 582 (Mo. 1963).
7. Jones purchased a residential lot. The deed stated that the lot was sold "subject to deed restrictions and easements of record." At the time of the

conveyance an open irrigation ditch ran across the lot. When Jones tried to fill it in, he was stopped because the ditch was established as a prescriptive easement. Jones then filed suit, contending that the seller was liable to him for breach of covenant against encumbrances. Could Jones collect? Jones v. Grow Investment and Mortgage Co., 358 P.2d 909 (Utah 1961).

8. Defendants attempted to erect two apartment buildings on a portion of a lot that was part of a subdivision. They were sued by residents of the subdivision to enjoin construction of the buildings because the construction would violate a restrictive covenant of the subdivision. That covenant restricted construction in the subdivision to single-family dwellings. Defendants argued that because the area had been rezoned to allow the construction of apartments, the injunction should not be issued. Decide. Lidke v. Martin, 500 P.2d 1184 (Colo. 1972).

CHAPTER 31

Landlord and Tenant

31:1. Introduction. The law concerning landlord and tenant relationships is another area of real-property law that in many respects has not kept up with the times and the many rapid changes in the field of commerce. The law governing landlord and tenant relationships in general is heavily loaded in favor of the landlord, although this appears to be changing. Many of the rules governing this relationship can be altered by the contract, known as a *lease*, that is entered into between the landlord and the tenant. However, the typical lease for the rental of an apartment also has many clauses that favor the landlord. If you are currently renting an apartment, now is a good time to examine the lease that you signed. If you do this, you will probably be horrified at the many provisions for the breach of which the landlord may toss you out on your ear. For instance, many leases prohibit putting nails or other holes in the wall, something that almost everybody does in order to hang pictures. The lease may also prohibit or limit the playing of record players, radios, or television after 11 P.M. and before 7 A.M. You also may not be pleased to learn that if the heat or air conditioning fails to function, you may not be relieved of your obligation to pay rent as long as you remain in the apartment. It is true that many jurisdictions have passed statutes that deal in various ways with the obligations of the landlord and the tenant. Nonetheless the ties to the past remain strong in this area of the law, and one should be very careful to understand the rights and obligations of the landlord and the tenant—both under the law and according to the terms of any lease—before entering into such a relationship.

31:2. Creation of the Landlord–Tenant Relationship. As has been mentioned, the contract that establishes a landlord–tenant relationship is called a *lease*. The landlord is the *lessor*, and the tenant is known as the *lessee*. The relationship between a landlord and a tenant is more than contractual. The lease creates an estate in land. The landlord has a fee and creates a lesser estate on behalf of the tenant known as a *leasehold estate*. By this relationship the landlord grants to the tenant the exclusive right to the possession of the property for a definite and limited period of time or term. In return, the tenant has the obligation to make periodic payments to the landlord.

If we apply some of the concepts that were learned in Chapter 30, the requirements for a leasehold become apparent. The estate of the tenant must be subordinate to that of the landlord. Also the landlord must have a reversionary interest; that is, the possession of the property must revert to the landlord at the expiration of the term. Finally, the tenant must have the exclusive right to possession and use of the premises. If this latter requirement is not met, there may be a license or an easement, but there is not a leasehold.

The lease may be either express or implied, just like other types of contracts. Most states have a Statute of Frauds provision in their statutes that requires a lease that is to run for more than a certain length of time to be in writing. The most common periods fixed for the requirement of a writing are one or three years.

31:3. Classification of the Relationship. The estate that a tenant holds in a landlord–tenant relationship may be classified as (1) an estate for years or other definite period; (2) an estate from year to year or other periodic tenancy; (3) tenancies at will; and (4) tenancies at sufferance.

(1) Estate for Years. An *estate for years* is a term used to describe all tenancies that are limited to a specific or definite period of time. Thus, if one rents an apartment for one year, one month, or one week, he would have an estate for years or for some other definite period. The estate terminates automatically upon the expiration of the period of time that is specified. There is normally no necessity that either party give notice of the termination of the relationship when the time expires.

When the landlord grants the tenant an estate for years in his property, a lesser interest is being given in the land that is owned by the landlord in fee simple. Although the interest given to the tenant is considered an estate in the land, under the common law that interest is considered personal property. Some statutes have modifed the common law to treat estates for years as real property for limited purposes.

(2) Estate from Year to Year. An *estate from year to year* or other periodic tenancy is one that is created to last for a specific period of time and then continue for successive subsequent similar periods of time unless one of the parties gives proper notice of its termination. For example, assume that Terry Tenant rents an apartment from Larry Landlord. The lease provides that the rental period is to be for one year and that at the expiration of that period the tenancy is to be renewed for subsequent one-year periods. This would be a tenancy from year to year. It is automatically renewed at the end of a year for another year of tenancy unless Larry or Terry gives proper notice that the tenancy is to terminate at the expiration of the period. The rental of many properties as residences is done under a lease that creates an estate from year to year. Perhaps the most common variation in many parts of the country is to rent the property initially for a year, with a month-to-month tenancy at the expiration of the year period. This tenancy may be created expressly or by implication. Thus, if one remains in the premises after the expiration of a year lease with the agreement of the landlord, an estate from year to year may be implied.

Many states have statutes requiring a minimum notice of intent to terminate such a tenancy, the most common period being thirty days. Such a notice then must usually be given thirty days prior to the last rent-payment date in order to terminate the tenancy. As a general rule, even in the absence of a statutory or contractual provision notice to quit the tenancy is necessary for its termination.

(3) Tenancy at Will. If a tenant leases premises from the landlord for an indefinite period of time, it is known as a *tenancy at will.* Such a tenancy may be terminated at anytime by either the landlord or the tenant. The permission of the landlord is required for a tenancy at will. The tenancy may be created either expressly or by implication. Thus a tenant who occupies the land with the permission of the owner without an agreement between the parties as to the period has a tenancy at will. It is not necessary that there be any payment of rent in order for a tenancy at will to be created. Some statutes require an express agreement for the creation of a tenancy at will, and many statutes require some notice for its termination.

(4) Tenancy at Sufferance. A *tenancy at sufferance* occurs when one comes into possession of property rightfully but holds over after the expiration of his term. It is necessary that the initial possession be rightful so that the tenant's possession is not adverse to the owner. This type of estate most frequently occurs when a tenant for a definite period of time does not quit the premises as he or she is obligated to do at the end of the term and the landlord fails to exercise his or her right to oust the tenant from possession. The tenancy continues only so long as the landlord does not exercise his right. Once the landlord exercises his right to oust the tenant, the tenancy at sufferance ends. Thereafter any occupancy of the property by the former tenant would be as a trespasser. You should also note that in the absence of a statute to the contrary the tenant at sufferance is not entitled to any notice to quit the premises. Some statutes, however, do require some notice for termination of the tenancy.

31:4. Rights and Duties of the Parties. When one is entering into the relationship of landlord and tenant, the provisions of a written lease are of primary importance. In the absence of a written lease there are a number of rules that govern the relationship of landlord and tenant under the common law. Some of these rules may have harsh consequences, particularly for the tenant. If a lease is properly drafted, these rules can be altered to reflect the intent of the parties and to accommodate their respective rights and duties to the needs of their own particular situation. It is important, therefore, to have at least a passing familiarity with the rights and duties of the parties in the absence of any contrary agreement. Many states have also passed legislation to partially regulate the landlord and tenant relationship.

(1) Tenant's Duty to Pay Rent. If the landlord and the tenant have entered into a formal written lease, it almost always will provide for the rent to be paid by the tenant to the landlord. Most often the rent will be paid in money, but it may take other forms. If there is no express agreement to pay rent, the obligation to pay a reasonable amount of rent is implied. There are obvious disadvantages to leaving a specific amount of rent unstated in any lease of real estate. One reason that may not be obvious is that in the absence of any express agreement to pay rent, the common-law rule is that the rent is not due until the end of the term. There are statutes in many states that alter this rule, however.

The typical lease will also provide that in the event the leased premises are destroyed, the tenant is discharged from his obligation to pay rent. The need for such a clause can once again be traced back to a common-law rule. Under the common law the various covenants in the lease are considered to be independent. That is, the tenant's right to possession of the property and his obligation to pay rent are not dependent upon one another. Thus under the common-law rule, if the building rented by the tenant burns to the ground, the tenant is not relieved of his obligation to pay rent for the rest of the term. The reason for this rule was that the tenant was considered to have rented the land on which the building happened to be erected. The loss of the building did not theoretically interfere with the tenant's right to occupy the land. As a practical matter, however, the use of the building is really the reason for the tenant's rental of the property. Hence most jurisdictions have changed the common-law rule by statute, discharging the tenant from his obligation to pay rent if the building is destroyed. It is wise, however, to place a clause in the lease that clarifies the rights and duties of the parties in the event that the premises are damaged or destroyed.

In most jurisdictions today failure of the tenant to pay rent entitles the landlord to evict the tenant and also to obtain a judgment for back rent. This was not true under the common law. In fact, because of the independence of covenants the failure of the tenant to pay rent did not necessarily result in the ability of the landlord to oust the tenant from possession. The landlord's remedy was an action at common law for "distress for rent." This action gave the landlord the right to seize any personal property located on the premises and sell it for the unpaid rent. Today this remedy has either been abolished or greatly modified by statute. In some jurisdictions the landlord is given, for the amount of unpaid rent, a lien on the tenant's personal property that is located on the premises.

Because of the independence of covenants the failure of the landlord to perform his duties under the lease does not allow the tenant to withhold payment of the rent unless the lease specifically provides for this remedy. In some jurisdictions there are statutes that give tenants the authority to withhold rental payments if the landlord fails to perform certain of his obligations. Generally these statutes allow for the withholding of rent if the landlord fails to perform obligations of a nature that may impair the health, safety, and welfare of the occupants of the premises. The obligation of the tenant to pay rent terminates upon his eviction by the landlord.

(2) Right to Possession. The landlord impliedly agrees to give possession of the premises to the tenant at the commencement of the term of the tenancy.

This right of the tenant to possession of the premises extends to the whole world, including the owner of the property, the landlord. The landlord has no right to enter the premises that he has rented at his own whim and fancy. The exclusive right to possession is in the tenant. Thus the landlord cannot enter the premises to show it to another prospective tenant or to install improvements to the property. If the landlord wants the right to do this, there should be a clause inserted in the lease to that effect. There are exceptions to this rule. In most jurisdictions the landlord may enter the premises to collect the rent, to inspect for or prevent waste, to determine whether damage has been committed, to comply with the orders of the authorities, and for other reasons. Many leases, of course, specifically reserve to the landlord the right to enter the premises for a number of different reasons.

If the landlord interferes with the tenant's right to possession of the premises, the tenant has a right of action against the landlord. The tenant may sue the landlord for any damages sustained as a result of the wrongful interference with his right to possession. This is not his sole cause of action, however. Because the landlord has no right on the premises, he may be a trespasser. In addition, in a proper case the tenant may be able to enjoin the landlord from continuing interference with his right to possession. Furthermore the tenant's right of action is not limited to the landlord. If a third person interferes with the tenant's right to possession, he may be liable to the tenant.

(3) Right to Use and the Duty to Repair. The tenant has the right to use the premises for any lawful purpose unless such use is restricted by a term of the lease. There is an implied obligation on the tenant not to use the premises for a purpose not contemplated by the parties. This is, of course, a matter that is typically stated in the lease. The lease should clearly state the purpose for which the premises are to be used.

The tenant is liable for any damages that he causes to the premises over and above normal wear and tear. This is frequently referred to as *waste.* Waste may be caused by the tenant's misuse of the property in such a manner as to cause damage to it. It may also be caused by an alteration to the property by the tenant that changes its character. The tenant is liable for damage caused by such alterations even if they enhance the value of the property, unless the landlord assents to them. Common examples of waste would be the tenant's removing a building or cutting trees without the permission of the landlord. The tenant would also be liable for damage to the premises that result from a fire caused by his negligence. If a tenant is committing waste, the landlord may seek an injunction to stop it and recover damages for the actual injury to the property.

Under the common law the tenant has the duty to make repairs to the premises that he controls. He is generally obligated to make those types of repairs that are necessary to prevent further damage to the property. Examples would be a leak in the roof, a broken window or lock, and so on. The tenant is not liable to make those repairs that result from ordinary wear and tear nor those that are substantial and that permanently damage the premises in a material way. Thus, if a roof blows off or a building collapses in a storm, the tenant is not obligated to repair. On the other hand the landlord has no duty to make repairs to that part of the premises not under his control. The general rule is that the landlord is obligated to maintain and repair those areas that are within his control and that may be used by the tenant. Examples would be halls, steps, stairways, and so on. If the premises are destroyed by fire or some other similar cause, neither the landlord nor the tenant is under an obligation to rebuild them. Nor is the landlord required to make repairs for defects that result from the normal deterioration of the premises. Obviously the question of which party is responsible for what repairs should be taken care of in the lease.

Increasingly this area of the law is becoming subject to statutory control. In many jurisdictions the public authorities have the right to order the landlord to make certain repairs to the premises, particularly when the defect may affect the health or the safety of the tenants. The provisions of such statutes vary widely, however, from one jurisdiction to another. Many of them require the landlord to put the premises in such condition as to be fit for human habitation if the premises are to be used as a dwelling. Some statutes allow the tenants to withhold rent for violations. As stated before, this is not a right at common law and does not exist in the absence of statute.

CHERBERG v. PEOPLES NATIONAL BANK OF WASHINGTON et al.

549 P.2d 46 (Wash. 1976)

James J. Cherberg and Arlene Cherberg rented a portion of the Lewis building in Seattle, Washington, in which they operated a restaurant. Joshua Green Corporation acquired that building subject to Cherberg's lease. The building abutting to the south was owned by Peoples National Bank of Washington. The bank began demolition of the Blue Mouse Theater located in that building in April 1972. This work resulted in exposing the south wall of the Lewis Building, revealing it to be structurally unsound and in need of substantial repairs in order to satisfy City of Seattle safety requirements. A dispute arose between the parties as to the landlord's duty to perform the repairs. The Cherbergs sued the bank and Joshua Green Corporation. From an adverse decision, the defendants appealed.

REED, J.

The pertinent lease provisions are as follows:

(5) Unless otherwise provided in this lease, Lessee, having ascertained the physical condition of said premises from a careful and complete inspection thereof, accepts said premises in present condition, no exceptions. Lessee shall place, maintain and keep the leased premises, including the store front, if any, in good, neat and sanitary physical condition, and at Lessee's sole expense, shall promptly make all repairs and do all acts and things necessary or incidental thereto; provided, however, that Lessee's said obligation shall not extend to the foundations, structural bearing parts, roof and outside walls of the premises unless repairs thereto or work thereon be necessitated by Lessee's act or negligence. . . .

. . .

(16) Lessee shall allow Lessor and Agent free access to said premises at all reasonable times for purpose of inspecting of the same or of making repairs, additions or alterations to said premises or to the building in which said premises are located *but this right shall not constitute or be construed as an agreement on the part of Lessor to make any repairs, which Lessee is required to make,* or to make any additions or alterations to said premises. . . .

(17) Unless otherwise provided in a rider to this lease, the use of the outside area of the walls and the roof of said premises or the building in which said Lessor who shall have the right to utilize the same for any purposes desired including sign purposes. . . . (Emphasis added.)

Washington adheres to the common-law rule that, in the absence of an express covenant, the landlord is under no duty to make repairs to the demised premises, even if they become defective through decay or deterioration. Nor may such a covenant be raised by inference.

In the lease before us it can readily be seen there is nothing rising to the status of an express covenant by the lessor to make repairs to either the "demised premises" or the Lewis Building proper. On the contrary, by express provision the lessee is obliged to make all repairs to the "demised premises," except for foundation, structural bearing parts, roof and outside walls. It is true the lessor has reserved the right to inspect and repair the demised premises. Without such a reservation or agreement it is questionable whether the lessor could enter to effect such repairs, even if it elected to do so, without being guilty of a trespass. Additionally, lessor has expressly reserved the right to repair the portions of the building not forming a part of the demised premises. Such a reserved right is far removed from an express undertaking to perform repair. We agree the language of paragraph (16) is somewhat curious, but it was obviously intended to stress the lack of any duty to make repairs, improvements and additions and it would be a strained construction to forge what was clearly intended as a shield into a sword. In *Refrigeration for Science, Inc.,* v. *Deacon Realty Corp.*, 70 Misc. 2d 500, 334 N.Y.S.2d 418, (1972), the court interpreted similar language as imposing no duty on either lessor or lessee to make major structural repairs to foundation and walls, saying at page 425:

The difficulty is that the lease contains no express covenant by the landlord to make structural repairs. "A landlord's obligation to repair in any case rests solely on express covenant. Without an express undertaking to repair the demised premises, the lessor is neither bound to do so himself nor to pay for repairs made by the tenant. . . ." Nor is plaintiff aided by the clauses of the lease granting the landlord permission to enter upon the premises "for the purpose of examining the same, or making such repairs or alterations therein as may be necessary for the preservation and safety thereof" and to remove signs when necessary to do so "for the preservation and safety thereof" and to remove signs when necessary to do so for painting and related purposes. *Such clauses do not impose on the landlord the duty to make repairs; they "simply give(s) permission to the landlord to inspect his premises and make such repairs as are necessary. It creates no duty upon him so to do and the tenant could not require it to be done. . . .* (Citations omitted. Emphasis added.)

The trial court erred, therefore, . . . in finding a lease-imposed duty by lessor to repair.

(4) Landlord's Liability For Injuries. The determination of the landlord's liability for injuries received from defects in the premises depends largely upon who has control of the area in which the injury occurs. In the first place, the landlord is not an insurer of the safety of those entering the premises, and if he has any liability at all, it must be the result of his negligence.

If the landlord has rented the entire building to a tenant, he is generally not liable for injuries that occur as the result of defects because he has no control over the premises once they are rented. Even if the landlord agrees to repair defects, he is not liable for injuries that result prior to his repairing them. However, once the landlord makes the repair, he may be liable for injuries resulting from his negligence in making the repairs. The landlord is not liable for injuries sustained as the result of defects that could be observed by the tenant. There is authority to the effect, however, that the landlord is liable to the party suffering injury from latent defects of which the landlord had knowledge but that were not disclosed to the tenant.

The landlord is liable for injuries that result from defects in those areas that remain under his control. For example, if the building has a number of apartments with a common hallway or stairway, those areas generally remain under the control of the landlord. If an injury results from a defect in that area, the landlord would be liable, providing that his negligence could be established. The problem of liability frequently occurs in shopping-center cases. In these cases, the lease between the landlord and the tenant may be important in a determination of who has control of a particular area. Thus if an injury results from a defect located in a tenant's store, the tenant would normally be liable for any negligence because that area is under his control. On the other hand, most leases provide that the landlord has control over the passageways between stores, and therefore he would be liable for injuries resulting from defects in that area, providing once again that the landlord's negligence can be established.

MANGAN v. F. C. PILGRIM COMPANY

32 Ill.App.3d 563, 336 N.E.2d. 374 (1975)

Catherine Mangan, age eighty-three, rented an apartment in an apartment building owned by Herbert J. Johnson and managed by the F. C. Pilgrim Company. One day, while she was opening her oven to remove biscuits, a mouse jumped out and frightened Catherine Mangan, causing her to fall and fracture her left hip. Suit was filed against the landlord and the management company, alleging negligence.

There was much testimony from other residents to the presence of mice in the complex and to the numerous complaints to the management concerning them. The defendants contended that no duty was owed Catherine Mangan for the extermination of mice. The trial court found for Mangan, and the defendants appealed. During the pendency of this action both Catherine and Johnson died, and their personal representatives were substituted in the suit on their behalf.

BURMAN, J.

The defendant argues first that no cause of action exists for the alleged injuries to Catherine Mangan, because the facts do not establish that defendant's deceased owed her any duty. After summarizing counts I and II of plaintiff's complaint (count III based on nuisance was never submitted to the jury), defendant argues that a landlord ordinarily owed no duty to a tenant to make any repairs for any defects unless he agrees to do so by the terms of some express covenant or agreement. An exception exists where there are latent defects of which the landlord either has knowledge or is chargeable with knowledge, in which case he has a duty to disclose them . . . but defendant notes that no hidden defect is alleged by plaintiff here.

We agree with these general propositions of law, but concur with the plaintiff that the allegation of an unsafe condition in the case at bar relates initially at least to areas of the building under common control (e.g., the incinerator system and walls). It is alleged in count I that defendant "owed a duty to the deceased and others lawfully upon the premises to so operate, manage, maintain and control those parts of the premises over which they retained control so as to avoid injury to the persons lawfully on said premises," including the deceased, Catherine Mangan, and that defendants "carelessly and negligently caused and permitted certain areas of the said building which remained under the control and responsibility of the Defendants, to become and remain infested with or inhabited by, mice and other vermin. . . ." It is established that where a portion of the premises is reserved for common use and is under the landlord's control, a duty is imposed upon him to use ordinary care to keep such portion in a safe condition. The fact that the actual injury occurred on the demised premises should make no difference if the cause of the injury was the landlord's negligent maintenance of a common area. Furthermore count II of the complaint alleges negligence on the part of the defendants in failing to fulfill the duty imposed by certain provisions of an ordinance of the Village of Oak Park prohibiting rodent infestation. (Village of Oak Park, Illinois, Housing Code §§ 578–16, 78–18, June 2, 1958.) The violation of a statute or ordinance prescribing a duty for the protection and safety of persons or property may constitute negligence such as gives rise to a cause of action on behalf of a person who suffers injury or damage by reason thereof. . . .

The gist of the plaintiff's allegation of negligence in count I is that the defendants permitted the apartment building to become infested with mice and/or failed to adequately rid the building of them. The question is whether as a result of this alleged negligent conduct defendant could be held liable for plaintiff's fall and consequent injury under the applicable tests of foreseeability. We feel that he could. The presence of the rodents gave rise to a variety of reasonably foreseeable risks of harm to occupants or visitors in the building. The spread of disease or bites might be a couple of commonly foreseeable risks. But it is not uncommon for any woman, let alone an 83 year old woman living alone, to be startled or frightened by the unexpected appearance of a rodent, especially as revealed by the circumstances of this case. As indicated in *Ney* v. *Yellow Cab Co.,* 2 Ill.2d 74, 117 N.E.2d 74, whether a defendant's conduct is to be considered a proximate cause of the injury is generally a question to be determined by the trier of the facts. The circumstances of the occurrence revealed by the record create a sufficient enough issue regarding proximate cause for the court to have submitted the question to the jury for a factual determination. The jury was instructed regarding proximate cause. We are of the opinion that the jury could properly find, as they did, that there was sufficient nexus between the negligent act and the injury to impose liability on the defendant.

QUESTION FOR DISCUSSION

1. *How did the court deal with the fact that Catherine Mangan's injury occurred in her own apartment rather than in the common areas of the building?*

(5) Right to Assign and Sublease. Under the common law, in the absence of a clause forbidding it, the tenant has a right to assign or sublease the premises. Today most leases contain a clause that either prohibits an assignment or sublease by the tenant or requires the landlord's consent for such an act. Therefore if one is renting property with the idea of eventually assigning his lease or subleasing the premises, this matter should be clarified in the original lease.

The distinction between an assignment and a sublease is important in a determination of the liabilities of the parties to each other. If the tenant transfers his entire interest in the premises, then we have an assignment. On the other hand, if the tenant transfers less than his entire interest in the premises—that is, he has a reversion—then we have a sublease. For example, suppose Terry Tenant has leased a building from Laura Landlord for two years. If Terry transfers the lease to Sammy Subtenant for one year, this action would constitute a sublease because the premises revert to Terry after one year. If Terry transfers the lease to Amy Assignee for the entire two-year period, then there would be an assignment of the lease from Terry to Amy.

In the case of an assignment, a privity of estate is created between the assignee and the original lessor—Landlord, in the previous example. The practical result is that the assignee has a right of action against the landlord for the breach of covenants that "run with the land." Just what covenants run with the land is not easy to define. Many of the old cases state that these covenants "touch and concern" the land. This definition itself is not particularly enlightening when it comes down to its application to a particular covenant. Covenants that run with the land may be distinguished from personal covenants between the landlord and the tenant. Thus examples of covenants that run with the land would be covenants to pay taxes, to repair the premises, and to pay rent. A promise by the landlord to loan his lawn mower to the tenant would be a personal covenant and not enforceable against the landlord by the assignee.

The assignee can enforce covenants that run with the land against the landlord. Likewise the assignee is liable to the landlord for breach of these covenants. The fact that the assignee is liable does not relieve the assignor, the original tenant, from his obligation, and he is also liable to the landlord for any breach of such a covenant, as, for example, the obligation to pay rent.

If there is a sublease of the premises, the sublease has no direct liability to the landlord. He is only liable to the original tenant for breach of any covenants contained in the original lease. The original tenant, of course, remains liable to the landlord for the breach of any covenants contained in the lease. For this reason it is always wise for the tenant who either subleases or assigns the premises to a third person to insist on a clause indemnifying the tenant for any damages caused by the third person for which the tenant might become liable to the landlord.

KROGER CO. v. CHEMICAL SECURITIES COMPANY
526 S.W.2d 468 (Tenn. 1975)

BROCK, J.

This case involves the construction of a shopping center lease to ascertain whether it contains an implied covenant by the tenant to occupy and operate a supermarket on the premises and an implied prohibition against subleasing to a nongrocery enterprise and without permission.

The Madison Square Shopping Center in Madison, Davidson County, Tennessee, was opened to the public in 1956. Located on thirty acres of land, the shopping center is composed of seven buildings and a parking lot for 2500 vehicles. Forty-six independent businesses occupy the 343,000 square feet enclosed in the buildings.

The Kroger Company was among the original tenants of the shopping center; it entered a lease on January 1, 1956, with Madison Square Shopping Center, Inc., for an area at the end of the largest building and opened a supermarket. The leases of several other original tenants required the landlord to acquire a minimum ten year lease for a Kroger supermarket.

. . . There is no clause in the lease concerning the right of the tenant to assign or sublet, but clause 26h provides that if the tenant voluntarily vacates the premises and they remain vacant for one year, the landlord has the right to cancel the lease and re-enter the premises.

On July 1, 1965, Chemical Securities Company, the plaintiff in this case, purchased all the real estate, store leases, and other assets of Madison Square Shopping Center, Inc., and the latter corporation was dissolved.

On July 16, 1973, shortly before the conclusion of the ten year term, Kroger entered into a sublease with Genesco, Inc., for the store, and on September 29 vacated the premises. According to Kroger the primary reason for the move was the declining sales at the supermarket. The subtenant assumed all of Kroger's obligations under the prime lease; the term of the sublease is equal to the three remaining extension periods of fifteen years and the subtenant's rental of $2,812.50 per month, to be paid to Kroger, is the same as the base rental under the prime lease. Genesco agreed not to assign or sublease the property without the consent of Kroger. The subtenant's use of the premises is limited to the sale of shoes, clothing and general merchandise, and he agreed not to use the premises in any manner which would compete with Kroger's business. The record reveals that Kroger operates another supermarket in Rivergate Shopping Center several miles away. The record also discloses that Winn-Dixie Louisville, Inc., has indicated to Chemical Securities Company that it is interested in leasing the Kroger store and opening a supermarket.

On September 19, 1973, the landlord filed a complaint in Chancery Court against the tenant and the subtenant claiming that Kroger's sublease of the property to Genesco for use as a commissary was inconsistent with the terms of the prime lease and a breach of the tenant's implied warranty to occupy the premises. The complaint requested a temporary injunction enjoining the transfer of the premises to Genesco during the pendency of the action, and that the sublease be annulled. The complainant further prayed that the prime lease be terminated or that the tenant be permanently enjoined from subleasing its premises to a nonretail grocery store and without prior approval of the landlord.

The Chancellor granted the temporary injunction and after a full hearing of the cause he found the prime lease vague and uncertain and admitted extrinsic evidence as to its meaning. . . . The Chancellor concluded that the intention of the parties to the prime lease was for the tenant to occupy and operate a grocery store on the demised premises and not to sublet. He ordered that the sublease be declared void and the prime lease terminated.

The Court of Appeals affirmed, finding the lease to be vague with respect to subletting. It concluded from evidence of the surrounding circumstances that if the subject of subletting by Kroger to a nongrocery operation had been mentioned the lessor would have insisted upon a covenant against subleasing without its permission and Kroger would have agreed. It also concluded that Kroger was guilty of bad faith in prohibiting its subtenant from going into the retail grocery business and that such bad faith warranted forfeiture of the prime lease as well as annulment of the sublease.

The primary issue to be resolved in this case is whether the Court of Appeals erred in construing the lease to contain an implied covenant of continuous occupancy and operation of a grocery business and an implied prohibition against subleasing without permission.

The courts have no rightful authority to make contracts for litigants. With regard to subletting and assignment in particular, this Court has long held that a lessee may freely transfer the demised premises without the lessor's consent, absent a covenant in the lease restricting the right of assignment. Also, the general rule is that covenants against subletting are strictly construed against the lessor.

It is also suggested that the implied covenants emanate from the landlord's promise not to lease any other premises in the shopping center, except the former A & P store, to any business for the sale of groceries. The Court of Appeals relied on this express covenant and the shopping center context and circumstances surrounding the execution of the lease as the source of the implied covenants.

However, in our opinion this restriction on competition written into the lease is not broad enough to give birth to implied covenants of continuous occupancy and operation of a grocery business and a prohibition against subleasing to a nongrocery and without the lessor's permission.

Unlike the leases in the cases above (omitted), we find no terms in this lease which justify the finding of the implied covenants. In fact the lease reveals that the parties considered departure from the premises by Kroger before the term expired to be a clear possibility. Clause 26h gives the landlord the right to cancel the lease and re-enter the premises if the tenant vacates them and they remain vacant for a year. In our opinion this clause recognizes Kroger's right to vacate the premises and the right to sublease them provided Kroger does so within one year of vacating.

(Reversed.)

QUESTIONS FOR DISCUSSION

1. *As a general rule, does a tenant have a right to assign or sublease the premises?*
2. *How could the landlord have drafted the lease to prevent Kroger from subleasing the premises in this case?*

(6) Right of Landlord to Sell Reversionary Interest. There is often confusion among laymen as to whether a landlord has the right to sell the leased property to a third person during the course of a tenancy and what affect such a sale has on the rights of the tenant. The landlord may sell his reversionary interest in the premises to a third person during the course of a tenancy. When he does so, the new owner has all the rights, interests, and liabilities of the original landlord. The rights, interests, and liabilities of the tenant are unchanged, and the tenancy continues.

31:5. TERMINATION. There are a number of ways that the relationship of landlord and tenant may be terminated: (1) by destruction of the premises; (2) by expiration of the term stated in the lease; (3) by notice, if there is no specific term; (4) by surrender of the premises; (5) by forfeiture; and (6) by eviction.

(1) Destruction of Premises. Under the common law, if the tenant leases an entire building, the destruction of the building would not relieve the tenant of his obligation to pay rent. In many jurisdictions the tenancy is terminated by statute if the building is destroyed. The lease, of course, may provide for the termination of the tenancy in the event of destruction or severe damage to the building. In view of the common-law rule, the inclusion of such a clause in the lease is wise, regardless of the existence of any statute. On the other hand, if the lease is of only part of a building, such as a room or an apartment, then if the building is destroyed, the lease is terminated even under the common-law rule.

(2) Expiration of Term. A tenancy for a fixed term, such as a tenancy for years, terminates upon the expiration of that term. No notice is necessary by either the tenant or the landlord for the termination of such a tenancy. In the event that either the landlord or the tenant dies during the term of the lease, the tenancy is not terminated. If the tenant dies, his interest passes to his personal representative because it is considered personal property. If the landlord dies, his reversionary interest is considered real property and passes to his heirs. Of course, the parties could stipulate in the lease that the tenancy terminates at the death of either one.

(3) Termination by Notice. In the case of a periodic tenancy, such as from year to year or month to month, notice must be given to terminate the relationship of landlord and tenant. Today most jurisdictions have statutes that specify the time when a notice must be given. Care should be taken to comply with the statute; otherwise the notice is defective and will not terminate the tenancy. In most jurisdictions, if the notice is not given properly and within the correct time, the expiration of an additional period of time will not cure the defect. Another notice must be given. For example, suppose that Terry Tenant has a lease from month to month and that the statute requires thirty days' notice to terminate the tenancy. Laura Landlord notifies Terry on February 14 that he is to vacate the premises on March 1. Such a notice would not comply with the statute. It is not effective and would not be effective even to require Terry to move on April 1. A new notice must be given.

Notice is frequently required by statute for termination of a tenancy at will but not for termination of a tenancy by sufferance. The parties, of course, can and should stipulate in the terms of the lease the length of notice that must be given and in what manner. In the absence of a statute or a lease provision, the common-law rule would prevail. The common law requires a six-month notice to terminate a tenancy from year to year.

SKLAR v. HIGHTOWER
342 A.2d 57 (D.C. 1975)

HARRIS, A.

Appellee rented an apartment from appellants on a month-to-month basis. A dispute arose between the parties, and the tenant vacated the apartment. In the Small Claims and Conciliation Branch of the Superior Court, judgments were entered directing that the landlords repay a $175 security deposit, less $30 in damages, to the tenant. The narrow question presented being one of first impression, we granted the application for allowance of an appeal. We affirm.

In early or mid-August of 1974, the tenant was informed that she would have to vacate her apartment by September 1 unless she would agree to a rent increase. On about August 22, the tenant orally notified the landlords that she would not agree to the increase and would vacate by September. She moved out on August 30. For reasons which are unclear, the tenant's attempts to return the keys to the apartment were ineffective, and the landlords spent $30 to change the locks.

The tenant sued for the return of her security deposit. The landlords counterclaimed for September's rent, contending that the rent was owed because the tenant had not given 30 days' written notice of her intention to vacate, as required by D.C. Code 1973, § 45–902.[1] The landlords also sought compensation for various items of alleged damages, including the cost for changing locks.

The trial court ruled that the tenant was entitled to the return of her deposit. As to the counterclaim, the court concluded that the tenant did not owe rent for September because the landlords had waived their right to insist upon strict compliance with § 45–902 and were estopped to assert the requirement of 30 days' written notice. The court found, however, that the tenant's efforts to return her keys had been "insufficient" and that she therefore was liable for the $30 expended to change locks.

We conclude that the trial court's finding that the tenant does not owe rent for September was neither plainly wrong nor without evidence to support it. A month-to-month tenant who fails to give 30 days' written notice may be liable for an additional month's rent if the lack of proper notice causes loss to the landlord. However, notwithstanding § 45–908 of the Code, it is not always necessary for a waiver of § 45–902 to be in writing; certain types of conduct also may constitute a waiver.

Under the circumstances, it was proper for the trial court to conclude that conduct constituting a waiver of the 30 days' written notice requirement had occurred, and that the landlords thereby were estopped to rely on the

[1] That statute provides: "A tenancy from month to month, or from quarter to quarter, may be terminated by a thirty days' notice in writing from the landlord to the tenant to quit or by such a notice from the tenant to the landlord of his intention to quit, said notice to expire, in either case, on the day of the month from which such tenancy commenced to run."

statute in their effort to collect an additional month's rent. The landlords orally informed the tenant that she must vacate by September 1, a date less than 30 days in the future, unless she agreed to a rent increase. That in effect was an attempted notice to quit by the landlords (with the tenant being given the option of staying and paying a higher rent), which itself did not comply with § 45–902 because it was not written and it did not furnish the full 30 days' notice. Nonetheless, the tenant did not assert her rights under § 45–902, but instead complied with the landlords' request. She notified them that she would not pay the increased rent, but would vacate, as they had asked her to do.

It would be grossly inequitable now to permit the landlords to collect an additional month's rent on the basis of the tenant's noncompliance with § 45–902. Considering the landlord's conduct, including their own disregard of § 45–902, and the fact that the tenant's failure to give 30 days' notice was due to her acquiescence in a specific request from the landlords, the trial court's decision was correct. The landlords' apparent inability to rent the apartment to another tenant for September is attributable to their own conduct, rather than to any dereliction on the part of the tenant.

(Affirmed.)

QUESTIONS FOR DISCUSSION

1. *What notice was required by the tenant in this case to terminate the tenancy? What were the consequences of failing to give such notice?*
2. *Why was the tenant excused from giving notice in this case?*

---

(4) **Termination by Surrender.** A term for years may be terminated at any time by agreement of the parties. The giving up of the leasehold by the tenant and acceptance by the landlord is called a *surrender.* It may be express or implied. In this context we are talking about the surrender and consequent termination of the leasehold as opposed to a mere giving up of possession of the premises, which does not necessarily terminate the relation of landlord and tenant.

BOVE v. TRANSCOM ELECTRONICS, INC.
353 A.2d 613 (R.I. 1976)

Transcom Electronics, Inc., leased certain premises from Michael J. Bove, Jr., and his wife. The lease began on September 15, 1965, and ran until September 14, 1970. On or about September 1, 1970, Transcom vacated the premises and delivered the keys to Joseph M. Hattub, an agent of the Boves, who informed Mr. Bove that he had received them. Hattub gave Transcom a receipt for the keys. Hattub and Bove inspected the premises in early September and found the only damage to be a few broken windows. When he next inspected the property on September 16 or 17, Mr. Bove discovered considerable damage inflicted by vandals. Bove contacted Transcom and demanded that the building be restored to its condition before it was leased to Transcom including the removal of certain fixtures installed by Transcom. Transcom refused, contending that it had surrendered the premises and was not liable under the lease. Bove contended that there was no valid surrender and sued Transcom. The trial court found in favor of Transcom, and the Boves appealed.

PAOLINO, J.

The narrow question raised by this appeal is whether the trial justice erred in finding that the acceptance of the keys by Mr. Hattub constituted an acceptance of the surrender by defendant of the demised premises on September 2, 1970, approximately 12 days prior to the formal lease termination date. For the reasons that follow, we affirm the judgment below.

The law governing the question raised by this appeal is well settled in this state. As this court pointed out many years ago, the relation of landlord and tenant, even where created by written instrument, can be terminated by the surrender of the premises by the tenant and the acceptance of such surrender by the landlord. Whether or not there has been such surrender and acceptance is determined by the intention of the parties. This intention is to be gathered from their acts and deeds.

But it is not essential to prove the contents of a written or oral agreement in order to establish a lessee's surrender and the lessor's acceptance thereof. Such consent need not be express in order to be effectual, but it may be implied from circumstances and from the acts of the parties. Finally, citation of authority is hardly needed for the proposition that a surrender may be made to, and accepted by, a duly authorized agent of the lessor.

With these guidelines in mind, we address ourselves to plaintiffs' arguments in support of their claim of error. The plaintiffs contend that there is no evidence that plaintiffs at any time manifested an intention to accept defendant's surrender and to thereby prematurely terminate the lease. They argue that the record is totally devoid of evidence of any act or statement on their part, or on the part of any authorized agent, evincing any intent to accept defendant's surrender. Since defendant has shown only that it delivered the keys to plaintiffs' business office, plaintiffs allege that it has failed to sustain its burden of proving a surrender which terminated the lease.

We address ourselves initially to the question of surrender. In our judgment there is competent evidence in the record to support the inference drawn by the trial justice that Mr. Hattub was plaintiffs' agent and that his acceptance of the keys on September 2, 1970, when viewed in the context of his work relationship with Mr. Bove, constituted an acceptance by plaintiffs of defendant's surrender.

Our reading of the cases cited by plaintiffs in support of their position reveals that those cases are factually distinguishable from the instant case and of no help to plaintiffs. In *Smith* v. *Hunt*, supra, this court found that mere surrender of a key did not constitute a surrender of premises when all the circumstances, including a written refusal of the proposed surrender by the plaintiff–lessor, indicated that such a surrender was not intended by the parties. Similarly, in *Newton* v. *Speare Laundering Co.*, 19 R.I. 546, 37 A. 11 (1896), this court held that the solitary act of sending a key to the lessor's office, without accompanying incidents of the parties' intention, is not such a surrender and acceptance as will discharge a tenant's liability for rent. These situations are to be contrasted to the facts of the instant case wherein the receipted delivery of the key, when properly viewed in the context of all relevant circumstances, displayed a clear intent on the part of plaintiffs through their agent to accept defendant's surrender of the premises.

(Affirmed.)

QUESTIONS FOR DISCUSSION

1. *What act of Transcom constituted surrender of the premises?*
2. *Why was Transcom not liable for the damage that occurred between September 2, 1970, and September 14, 1970, as the written lease ran until that time?*

---

(5) Termination by Forfeiture. A tenancy for years may be terminated by forfeiture. Frequently the lease will provide that in the event that the tenant breaches any of the covenants of a lease, the landlord has the right to terminate the lease. A common provision is that the landlord may terminate the lease if the tenant fails to make the required rental payments. The courts give effect to such provisions when they are clearly stated in the lease. However, forfeitures are looked upon with disfavor by the courts and are narrowly construed against the landlord and liberally construed in favor of the tenant. In some jurisdictions there are statutes that give the courts the power to grant relief against a forfeiture if it finds that the forfeiture would work an undue hardship on the tenant.

(6) Termination by Eviction. If the landlord evicts the tenant, the relationship of landlord and tenant is terminated. An eviction may consist of an actual ouster of the tenant by the landlord, as by legal process or by some other method. Basically anything done by the landlord to deprive the tenant of his use, occupation, and enjoyment of the premises may constitute an eviction. An eviction may be *actual,* which is the actual expulsion of the tenant from the premises, or it may be *constructive.* A constructive eviction occurs when the landlord does some act that materially interferes with the tenant's beneficial enjoyment of the premises and that causes the tenant to abandon possession of the premises within a reasonable period of time. When this occurs, the landlord has breached the implied covenant of the tenant's right to quiet enjoyment of the leased premises. You should note, however, that in order for there to be a constructive eviction, the tenant must actually abandon the premises. If he does not do so within a reasonable period of time, there is no constructive eviction and he may lose his right of action against the landlord. For example, suppose a tenant occupies an apartment to which the landlord has agreed to supply heat. If the landlord fails to supply the heat, this action would constitute a constructive eviction and justify the tenant in abandoning the premises. If the tenant fails to leave, however, the eviction is not complete, and in the absence of a statute giving him other rights, the tenant would be required to continue paying rent.

PETROLEUM COLLECTIONS INCORPORATED v. SWORDS

48 Cal.App.3d 841, 122 Cal.Rptr. 114 (1975)

Edward W. Swords leased a service station from Texaco, Inc., at the intersection of Adams Road and Freeway 99. The station also had a large "modular" *type of sign on top of the building that could be clearly seen from the freeway. Subsequently the Fresno County building inspector discovered that the sign had been installed by Texaco without a permit and that it constituted a fire hazard because it was too close to the vents to the gasoline storage tanks. Texaco was required to remove the sign to correct the hazardous condition. Texaco removed the sign and replaced it with an old sign that could not be seen from the freeway. Swords insisted that Texaco furnish him with a sign similar to the one that was on the property when he leased it. Texaco refused and Swords stopped paying rent. Texaco and Swords canceled the lease on March 13, 1970, and Swords subleased the premises. For the next eleven months neither Swords nor the sublessee paid any rent to Texaco. Texaco assigned its claim to Petroleum Collections Incorporated, who sued Swords for unpaid back rent. Swords defended on the ground of breach of the covenant of quiet enjoyment and constructive eviction. From an adverse decision, the plaintiff appealed.*

GARGANO, A.J.

It has long been the rule that in the absence of language to the contrary, every lease contains an implied covenant of quiet enjoyment. Initially, the covenant related solely to the right of possession and only protected the lessee against any act of molestation committed by the landlord or anyone claiming under him, or by someone with paramount title, which directly affected the tenant's use and possession of the leased premises; the covenant was construed to protect the lessee against physical interference only. . . . In recent years, the covenant of quiet enjoyment has been expanded, and in this state, for example, it insulates the tenant against any act or omission on the part of the landlord, or anyone claiming under him, which interferes with a tenant's right to use and enjoy the premises for the purposes contemplated by the tenancy. Under this view, the landlord's failure to fulfill an obligation to repair or to replace an essential structure or to provide a necessary service can result in a breach of the covenant if the failure substantially affects the tenant's beneficial enjoyment of the premises.

We consider next the question as to whether an implied covenant of quiet enjoyment and an express covenant to pay rent are mutually dependant; at common law, it was the traditional concept that covenants in leases were always independent.

The foundation for the tenant's obligation to pay rent is his right to use and possess the leased property for the purposes contemplated by the tenancy; rent is the compensation paid by the tenant in consideration for the use, possession and enjoyment of the premises. Consequently, when a tenant's possession of leased property has been interfered with, physically, by the landlord, or someone claiming under him, or by a person with paramount title, the tenant's covenant to pay rent is no longer supported by valid consideration and he is relieved of that obligation. The loss of possession, in short, goes to the very root of the consideration for the tenant's promise to pay rent, and after the eviction the obligation to do so ceases. It is in this sense that the implied covenant of quiet enjoyment and the covenant to pay rent are mutually dependent.

However, when the act of molestation merely affects the tenant's beneficial use of the premises, the tenant is not physically evicted and he has a choice in the matter. He can remain in possession and seek injunctive or other appropriate relief or he can surrender possession of the premises within a reasonable time thereafter. If the tenant elects to remain in possession, his obligation to pay rent continues unless the landlord has breached some other express or implied covenant which the covenant to pay rent is dependent upon. If, on the other hand, the tenant elects to surrender possession of the premises, a constructive eviction occurs at that time and, as in the case of an actual eviction, the tenant is relieved of his obligation to pay any rent which accrues thereafter. It is this doctrine, known as the doctrine of constructive eviction, ". . . which expanded the traditional 'covenant of quiet enjoyment' from simply a guarantee of the tenant's possession of the premises (citations) to a protection of his 'beneficial enjoyment' of the premises. . . ." *Green* v. *Superior Court,* supra, 10 Cal. 3d 616, 625, fn. 10. . .

Stated in another manner, the covenant of quiet enjoyment is not broken until there has been an actual or constructive eviction. . . . An actual eviction takes place when the tenant is physically dispossessed of the property; a constructive eviction occurs when the act of molestation merely affects the beneficial use of the property, causing the tenant to vacate the premises. If the tenant is evicted or if he surrenders possession of the premises within a reasonable time after the act of molestation has occurred, he is relieved of his obligation to pay rent accruing as of the date he surrendered; he also may sue for his damages or plead damages by way of offset in an action brought against him by the landlord to recover any unpaid rent that accrued prior to the surrender of the premises. If, in the case of an interference with the tenant's beneficial enjoyment of the premises, the tenant does not surrender the premises within a reasonable time after the date of the interference, he is deemed to have waived his right to abandon; what constitutes a reasonable period of time is a question of fact to be determined by the trier of fact after considering all of the circumstances.

In this case, there was sufficient evidence for the court to find that Texaco's failure to repair or replace the large modular sign, which was on top of the service station when the property was leased to defendant, substantially affected defendant's beneficial enjoyment of the property as contemplated by the parties; there was evidence to show that the parties knew that defendant would have to take advantage of the trade generated by the freeway in order to operate a service station successfully on the leased property at the monthly rental called for by the lease; there was also evidence to show that the sign was an integral part of the premises and that it was needed to attract drivers of vehicles traveling on the freeway. Nevertheless, we reverse the judgment. Defendant and his sublessee remained in possession of the service station for almost 11 months after the sign was removed by Texaco; albeit the trial judge found that during most of the period the parties continued to negotiate for a replacement sign, the judge made no finding as to whether defendant's conduct in staying in possession and in entering into a sublease with a third party was reasonable under the circumstances. Furthermore, even if we were to assume that defendant's conduct in remaining in possession during the 11-month-period in question was reasonable under the circumstances, the covenant of quiet enjoyment did not relieve him of his obligation to pay rent during the period; the covenant was not breached until defendant vacated the premises on March 13, 1970. Defendant should have pled his damages by way of offset in the action brought against him for the stipulated rental by Texaco's assignee.

(Reversed.)

QUESTIONS FOR DISCUSSION

1. *How did Texaco breach the covenant of quiet enjoyment? What were the tenant's remedies for such a breach?*
2. *Why did Swords have to pay the rent?*

SUMMARY

A contract that establishes a landlord–tenant relationship is called a lease. The lease creates an estate in land called a leasehold estate. The requirements of a leasehold are: the tenants' estate must be subordinate to that of the landlord; the landlord must have a reversion; possession must revert to the landlord at the end of the term, and the tenant must have the exclusive right to possession and use of the premises.

The type of estate that a tenant holds may be classified as an estate for years; an estate from year to year; a tenancy at will, or a tenancy at sufferance. An estate for years is a tenancy limited to a specific or definite period of time. The estate from year to year lasts for a specific period of time and then continues for successive subsequent periods unless one of the parties gives proper notice of termination. A tenancy at will extends for an indefinite period of time. A tenancy at sufferance occurs when one comes into possession of property rightfully but holds over after the expiration of his term.

The lease establishes the rights and duties of the parties. If there are no contrary provisions in the lease, the rights and duties are established by law. The tenant has a duty to pay rent and gets exclusive possession of the premises. The tenant has a right to use the premises for any lawful purpose not restricted by a term of the lease. The tenant is liable for any damages that he causes to the premises over and above normal wear and tear. Such damages are called waste. The landlord is obligated to maintain and repair those areas under his control. The landlord is liable for injuries caused by his negligence that result from defects in those areas that remain under his control. He is not liable for injuries resulting from defects within the area under the tenant's control.

In the absence of a contrary provision in the lease the tenant has the right to assign or sublease the premises. An assignment occurs if the tenant transfers his entire interest in the premises. A sublease occurs when the tenant transfers less than his entire interest in the premises. The landlord also has the right to sell his reversionary interest in the premises.

A landlord–tenant relationship may be terminated by destruction of the premises; by expiration of the stated term in the lease; by notice, if there is no specific term; by surrender of the premises; by forfeiture; and by eviction.

PROBLEMS

1. Sound City, U.S.A., Inc., a retailer, leased space in a shopping center from USIF Wynnewood Corporation. In the lease agreement Wynnewood agreed to repair ceiling tile, cover the light bulb fixtures, panel the south wall, and erect a partition. The lease was for a one-year period. Sound City took possession during October, although the agreed corrections had not been completed. By January, some of the repairs still had not been completed so Sound City vacated the premises with nine months remaining on the lease. Is Sound City liable for breaching the lease? American National Bank v. Sound City, U.S.A., 385 N.E. 2 144 (Ill. App. 1979).
2. Daniel Bruce Weber, Jr., aged three, wandered into the laundry room of an apartment building that was maintained for the use of the tenants of the building. He was injured when he placed his left hand in the drive mechanism of the clothes dryer. The guard that normally shielded the mechanism had fallen off and had never been replaced. The landlord had been notified of the condition. Was the landlord liable for Daniel's injury? Macke Laundry Service Co. v. Weber, 267 Md. 426, 298A.2d 27 (1972).
3. Mercedes Peters, a psychotherapist, rented space from Rockrose Associates for her office, in which she treated patients. The lease required the landlord to soundproof the premises. When he did not do so, Mercedes complained that the noise from neighboring tenants materially interfered with her patients' sessions. After two or three months without any action from the landlord, Mercedes moved out, some four years before the expiration of the lease. The landlord sued to recover lost rent. Decide. Rockrose Associates v. Peters, 366 N.Y.S.2d 567 (1975).
4. Mr. and Mrs. Court moved into an apartment leased to them by the Lamont Building Company. The lease provided that the apartment was to be occupied exclusively by adults. Shortly thereafter a child was born to them, and they were requested to move. They refused, contending that the provision in the lease was unenforceable. Were the Courts correct? Lamont Building Co. v. Court, 147 Ohio St. 183, 70 N.E.2d 447 (1946).
5. Nathan Seglin leased certain premises from Helen Bernstein. The lease prohibited assignment and subletting without the consent of the landlord. It also provided that the landlord would not withhold consent arbitrarily if the proposed assignee was financially responsible and an experienced grocer. Subsequently Seglin sold his business to Sharon and assigned the lease to him with the consent of Bernstein. Five months later Sharon moved without notice or surrender of the premises. There were four years remaining on the lease. The premises were not rerented for a year and one half after Sharon left. Could Bernstein recover lost rent from Seglin for this period? Bernstein v. Seglin, 184 Neb. 673, 171 N.W.2d 247 (1969).
6. Homsey rented an apartment and received a document that read as follows: "This is to certify that the rent at 294 Lawrence Street is $25.00 a month, and that immediate occupancy may be had by Mr. Homsey, and rent to begin December 1, 1945, and after a period of two years rent $35.00."

Did Homsey have a leasehold for two years? Farris v. Hershfield, 89 N.E.2d 636 (1950).

7. I. H. Heikens owned a mill that he leased to Hubert Cherry. Sparks from the engine of a passing train operated by the Nashville, C. & St. L. Railway caused a fire that destroyed the mill. The landlord sued the railroad company for the loss of the mill. The railroad contended that it was not liable to Heikens for the full value of the mill. Was it correct? What was the landlord's duty to Cherry? Nashville, C. & St. L. Ry. v. Heikens, 112 Tenn. 378, 79 S.W. 1038 (1904).

CHAPTER 32

Wills, Trusts, and Decedent's Estates

Some of the questions most frequently asked of the lawyer involve wills and the disposition of property at one's death. A thorough analysis of the field of estate planning is beyond the scope of a text in business law. However, because it is a subject of interest to almost everybody, a brief discussion is included here so that you may become aware of some of the general concepts and pitfalls. The field of estate planning can be extremely complex, particularly for the large estate, as consideration must be given to federal estate taxes and state inheritance taxes. Although the average individual may believe that he is without sufficient assets to cause the tax aspects of estate planning to be a concern, his presumption may be incorrect. Frequently a person underestimates the size of his estate because he fails to take into consideration the appreciated value of his home, which may be substantial, and the value of insurance and retirement plans that he may have individually or through his employer. It is appropriate to examine the two documents most frequently used in estate planning, the will and the trust, as well as the administration of the estate.

32:1. Wills. A formal will is a written instrument by which a person makes a disposition of his or her property to take effect after his or her death. The will may perform other functions, such as designating the party who is to be the executor or the personal representative of the estate. It may also state the person or persons who are to be the guardians of any minor children of the person writing the will. Without a will the disposition of a person's property as well as other related matters is governed by statute. These statutes will be examined in Section 32:2(2), and it will become clear that they may not always provide for the disposition of property in a manner consistent with the wishes of the deceased or the needs of those he or she leaves behind.

At this point it is appropriate to introduce a few common terms that are frequently used in any discussion of wills. One who writes a will is said to be a *testator,* or a *testatrix* if the person is female. If one dies without a will, he or she dies *intestate.* If there is a will, one is said to die *testate.* If the will designates a person to handle the affairs of the estate, that person is an *executor* or *executrix.* If there is no will and the person is appointed by the court to handle the affairs of the estate, he or she is an *administrator* or *administratrix.* In many jurisdictions the distinction between an executor and an administrator has been abolished, and the person is simply called a *personal representative.* If one makes a gift of real estate by will, that gift is called a *devise.* A gift of personal property is a *bequest.* Other terms will be explained in the following discussion.

(1) **General Requirements of Formal Wills.** Because a will controls the disposition of property after the person having an interest in that property is dead, all jurisdictions have statutes that have strict requirements that must be met to make a valid will. If these requirements are not strictly adhered to, the will is invalid and the property passes according to the statutes of descent rather than according to the will.

The requirements for a valid will vary considerably from state to state. Some statutes provide that any will valid in the state where it is executed is valid in the state that has the statute. What follows is a discussion of the most general requirements for executing a valid will. There may be exceptions to these rules on a state-by-state basis.

In order for there to be a valid formal will, it must be *written* by a person who has the *capacity* to execute his or her will. In order to have testamentary capacity, the person must be of some minimum age, most commonly eighteen or twenty-one. In addition to the age requirement, the testator must have the mental capacity to execute a will. The test of mental capacity in most jurisdictions is that the person writing the will must be capable of understanding the nature and extent of his or her property and also of appreciating the nature of the claims against him and the natural objects of his bounty. The natural objects of his bounty are those who would naturally receive his property. Clearly under this test one can lack mental capacity for many purposes and still have the capacity to write a valid will. Furthermore, the fact that the testator makes some strange disposition of his or her property, such as leaving all his property to a favorite cat, does not lead to the conclusion that the person lacks mental capacity to write a valid will.

In general, most statutes require that the will be signed by the testator or by some person acting for him in his presence and by his express direction. Note that most statutes allow for someone else to sign the will for the testator, provided that the requirements for doing so are met. This is to allow for the situation in which an individual may not be physically capable of executing the will.

Most statutes require that the will also be signed by credible witnesses in the presence of the testator. The number of witnesses required varies from state to state, but the most common number is two. Some states require that there be three witnesses. It is most important to observe the requirement that all the witnesses sign in the presence of the testator and preferably in the presence of each other. They should be of legal age and mental capacity and able to testify to the mental capacity of the testator. It is also important that the witnesses have no interest in the disposition of the property under the will in order to be credible. If a person does have such an interest, he may not qualify as a witness or may be required to give up any interest that he has in the will in order to testify.

It is not necessary that the witnesses read the will or know its contents. It is sufficient that they be aware that it is the testator's instrument that they are signing. Although the best practice is to have the testator sign the will in the presence of the witnesses, this is not required in many states as long as the testator acknowledges to the witnesses that the signature on the document is his.

Another common ground for challenging the validity of a will is lack of mental capacity and undue influence. The test for mental capacity has already been discussed. The concept of undue influence raises other problems. Undue influence has been discussed with regard to contracts in Chapter 7, and much of what was said there is applicable to wills. Remember that undue influence is more than mere persuasion or influence. It consists of another person's substituting his will in effect for that of the testator, so that the testator does not exercise his or her own judgment.

IN RE ESTATE OF KUZMA

408 A.2d 1368 (Pa. 1979)

Andrew Kuzma died on January 30, 1977. On February 23, 1970 he signed a will while a patient at the Corry Community Hospital. The will directed that Kuzma's property be given to Paul Kuzma, a son, and Martha Komenda, a daughter.

Kuzma's wife died before he did but four of Kuzma's surviving children received nothing under the will. They filed suit challenging Kuzma's capacity to execute the will. The trial court decided that Kuzma had the capacity to execute the will and they appealed.

EAGEN, C.J.

The instrument is a printed form captioned "Last Will and Testament" which contained blank spaces permitting inclusion of the designation of the names of the maker of the will and those who are to receive and administer his or her estate. The instrument in pertinent part directs that the residue of the estate (in effect the net estate, since no specific legacies are provided for) be given to Paul Kuzma, a son, and Martha Komenda, a daughter. These same two children are designated as executor and executrix, respectively. According to uncontradicted testimony in the trial court, the blanks in the instrument, providing for the disposition of the estate and appointments of the testator's representatives, were all filled in by Martha Komenda, one of the two beneficiaries, in the presence of the decedent and at his direction. Additional uncontradicted testimony established that an attorney was also present and that he read the completed instrument to the decedent following which the decedent expressed his satisfaction and signed his name thereto. It was witnessed by three disinterested witnesses.

Further uncontradicted testimony established the decedent was admitted to the hospital on February 16, 1970 after complaining of intense pain in the abdomen and while suffering nausea and fever. Following admission, an examination revealed, inter alia, jaundice, hernia, and an enlarged liver. The decedent was described as critically ill. On February 23, the day of the signing of the "Last Will and Testament," the decedent was given demerol to relieve pain. On the following day, February 24, he underwent surgery to relieve peritonitis, a hernia, and a stone in the common duct of the gall bladder.

First, the appellants complain the trial court erred in finding the decedent possessed testamentary capacity at the time of the execution of the instrument. They point with emphasis to the testimony of two physicians, Dr. Cruz and Dr. Tate, who attended the decedent in the hospital and who stated that, on the day in question, decedent was very ill physically, probably suffering severe pain, and lacked testamentary capacity. Dr. Tate said decedent was mentally confused on February 23. Appellants urge [that] the court capriciously disregarded this testimony.

The law concerning testamentary capacity is well-settled. A testator possesses testamentary capacity if he knows those who are the natural objects of his bounty, of what his estate consists, and what he desires done with it, even though his memory may have been impaired by age or disease. The burden of proving testamentary capacity is initially with the proponent; however, a presumption of testamentary capacity arises upon proof of execution by two subscribing witnesses. Thereafter, the burden of proof as to incapacity shifts to the contestants to overcome the presumption by clear, strong and compelling evidence. Furthermore, proof of execution does not require any representation of testamentary capacity by attesting witnesses. The condition of the testator at the very time of execution is crucial; however, evidence of capacity or incapacity for a reasonable time before and after execution is admissible as indicative of capacity. Finally, old age, sickness, distress or debility of body neither proves nor raises a presumption of incapacity.

While the testimony of the attending physicians supports appellants' position that the decedent lacked testamentary capacity, it is not the only testimony in the record pertaining to this issue. In fact, the physicians' testimony was flatly contradicted by several lay witnesses.

For example, Martha Komenda testified that, at the time involved, her father was mentally alert, cognizant of his children, and conversant with his property.

Paul Kuzma testified that decedent inquired about the care of the farm and cattle; that he knew who his children were; and, that decedent wanted the farm to remain with Paul and Martha. Paul testified that decedent did not want the farm sold, but that he wanted it to remain as it was, open to decedent's family.

Two of decedent's children, who were not named as beneficiaries under the will, testified in support of decedent's testamentary capacity. Alice Deming, decedent's daughter, visited decedent in Corry Community Hospital on February 22, 1970, the day before decedent executed his purported will. She testified that decedent recognized her, cautioned her to drive safely when she left the hospital, and remarked Martha and Paul were working on the farm. She testified decedent's mind was clear. Finally, she testified decedent always maintained that he did not want the farm sold, rather he wanted the farm to remain with the family.

Agnes Kiehart, decedent's daughter, testified decedent wanted the farm to remain in the care of Paul and Martha because they worked on the farm for many years. Agnes also testified that, in her presence, decedent directed Martha to fill out the form will and to leave the farm to Paul and Martha. Finally, Agnes testified decedent's illness did not in any way affect his mind. She stated decedent recognized his children.

Moreover, a careful study of the physician-witnesses' testimony indicates some material weaknesses.

Although Dr. Tate testified, on direct examination, that the decedent was mentally confused on February 23, he could point to no hospital or medical record to substantiate this. He also conceded that, due to the passage of time, it was very difficult at the time he testified to specifically remember the decedent's mental condition. He further conceded the testator executed a written consent to surgery on the same day the instrument under challenge was executed. While Dr. Cruz opined on direct examination that decedent lacked testamentary capacity, he admitted on cross-examination that he had no personal recollection of the purported testamentary incapacity and that his direct testimony was based solely on the severity of decedent's illness, as indicated by hospital records. He also admitted the decedent recognized him on February 23 and realized the necessity for surgery. Finally, he said the decedent was cognizant of his children and the farm.

It is clear that the findings of the trial court are supported by adequate and competent evidence and, hence, are binding in this appeal. It is also clear that the conclusion of mental capacity is amply supported by these findings. Finally, we find no error of law or abuse of discretion in the court's ruling.

(Affirmed.)

QUESTIONS FOR DISCUSSION

1. *What is the test for determining testamentary capacity?*
2. *Could Andrew Kuzma have effectively disinherited his wife if she were still living?*
3. *Why do you suppose the court was able to decide that Kuzma had capacity to write a will despite the testimony of two physicians that he lacked capacity.*

IN RE JEAN FISCHER WILSON'S ESTATE

402 A.2d. 197 (N.H. 1979)

MEMORANDUM OPINION

This is an appeal from a decree of the Cheshire County Probate Court *(Lichman, J.)* denying admission to probate of a document purporting to be the will of Jean Fischer Wilson.

Jean Fischer Wilson, at the time of her death on December 11, 1977, was a resident of Fitzwilliam, New Hampshire. The document purporting to be her will was filed for probate on December 13, 1977. It was attested by two witnesses and executed in the State of New Hampshire. The sole legatees under this purported will are the Smith College Alumnae Fund and two churches.

Roger H. Cole and David R. Cole, the first cousins of the decedent, are the only heirs at law and are entitled to the estate if the purported will is invalid. They seasonably objected to the allowance of the purported will on the grounds that it does not comply with the statute, RSA 551:2, which requires that a will be subscribed by three or more credible witnesses.

On June 27, 1978, a hearing was held in the Cheshire County Probate Court to determine if the purported will should be admitted to probate. Following the hearing, the court ruled that the purported will could not be admitted to probate and that the estate was to be administered on an intestate basis. The exceptions of the protesting charities were reserved and transferred to this court.

The legatees under the will ask us to depart from the holding of *In re Amor's Estate,* 99 N.H. 417, 112 A.2d 665 (1955), affirming the validity of our statutory three-witness requirement, RSA 551:2. They assert that New Hampshire is one of only four States that require three instead of two subscribing witnesses. That argument is for the legislature, which is the appropriate governmental body to amend or modify the statute as it deems appropriate.

(Exceptions overruled.)

QUESTIONS FOR DISCUSSION

1. *How many witnesses to a will are required by the laws of your state? Are there any restrictions as to who may serve as a witness?*
2. *What reasons can you give for the requirement that a will must contain the signatures of witnesses?*

(2) Modification of the Will. If one desires to modify a will, he or she must comply with the statutes that govern modification of wills. These requirements are generally similar to those that must be met in the execution of the original will. For example, if one has a valid will, he or she cannot change it by merely adding notes or making interlineations. The changes must be made in writing, be signed by the testator, and have the signatures of the required number of witnesses just as is required for the execution of the original will. A separate writing, properly executed, that modifies the original will is called a *codicil.* If one desires to modify an original will by interlineation he can accomplish this only by reexecuting it in the manner in which the original will is executed.

(3) **Revocation of the Will.** Because a will does not become effective until the death of the testator, it may be revoked by the testator at any time. Just as there are requirements for executing a will there are definite requirements for revoking it. If these requirements are not met, the will is still given full force and effect.

The execution of a subsequent will revokes all prior wills if there is a clause in the new will revoking prior wills or clauses that are inconsistent with clauses in a prior will. A will may also be revoked by its destruction, either by the testator or at his express direction and with his consent. Most statutes allow for the destruction of a will by burning, canceling, tearing, or obliterating. These acts must be done with the intent of the testator to revoke the will. If someone else tears the will without the consent of the testator, then it is not revoked.

If the testator marries after the will is executed, the marriage revokes the will in most states on the theory that the testator would want to dispose of his property differently after the marriage. In other states marriage and the birth or adoption of children subsequent to the execution of a will will result in its revocation. Also, if a will is executed after the testator is married and has children, the birth of subsequent children will not result in a revocation of the will. In most states the new children will participate equally with the previous children of the testator unless the will indicates clearly that the children are to be disinherited.

Divorce or annulment of a marriage that occurs subsequent to the execution of the testator's will is also a frequent ground for revocation of those parts of the will relating to the spouse. In most states the other parts of the will remain in force. The theory, of course, is that the testator would not want a former spouse to benefit from his or her estate.

Attempts to revoke a will by methods that are not condoned by statute have no effect. For example, one may tell any number of people that he is revoking his will and may in fact write on the will that it is revoked. These acts will have no effect on the will because they are not methods of revocation stated in the statute.

A will that has been revoked may be revised by reexecution of the will in the manner prescribed in the statute. In most states another writing providing for the revival of a previous will will be effective, providing that the writing is executed in the manner required for the execution of wills.

IN RE ESTATE OF SHIFFLET

170 So.2d 96 (Fla.App. 1964)

Florence Shifflet executed a will on July 21, 1961. Sometime later she also executed a codicil. After her husband died, Florence visited the office of her attorney to make changes in her will. At that time she crossed out some words in the body of the will and also made some interlineations. She did the same with the codicil. At each place where she made a change Florence placed her initials in the margin. Two wit-

nesses to the change also placed their signatures in the margin. Florence did not sign at the end of the will. The probate of the will was challenged on the ground that Florence's attempted revocation of part of the will was not effective because she failed to comply with the statute. From an adverse decision the administratrix appealed.

HENDRY, J.

It is our opinion that the attempted partial revocation of the will by the testratrix was inoperable because she failed to comply with the statutory requirements.

It is uncontested that the will in its original form was validly executed and would be entitled to probate, but for the attempted partial revocation. Before going further, it might be helpful to restate a valid generalization; the statute providing for revocation of will must be strictly complied with; partial or substantial compliance is insufficient.

The statute dealing with this proposition, § 731.13, Fla.Stat., F.S.A. provides:

> A will or any part thereof may be revoked or altered by a subsequent written will, codicil or other writing, declaring such revocation or alteration; provided, that the same formalities required for the execution of wills under this law are observed in the execution of such will, codicil or other writing.

It is obvious that this was the statute that the testatrix attempted to follow by her actions because partial revocation of a will may only be accomplished by strict compliance with the above statute and no other.

The partial, as opposed to complete revocation of a will, may not be accomplished by burning, tearing, cancelling, defacing, obliterating or destroying as provided in § 731.14 Fla.Stat., F.S.A. because that section does not have the necessary language "or any part thereof" as is found in § 731.13 Fla.Stat., F.S.A. Accordingly, unless the testatrix observed "the same formalities required for the execution of wills" then her attempted partial revocation must be held a nullity, and the original will unaltered be admitted to probate.

The remaining question necessary for resolution of the instant case is whether the attempted partial revocation of the will was accomplished with the same formality as the execution of the will. The statute providing for the execution of wills is § 731.07 Fla.Stat., F.S.A., which provides in relevant part:

> Every will, other than a nuncupative will, must be in writing and must be executed as follows:
>
> (1) The testator must sign his will at the end thereof, or some other person in his presence and by his direction must subscribe the name of the testator thereto.
>
> (2) The testator, in the presence of at least two attesting witnesses present at the same time, must sign his will or cause his name to be signed as aforesaid or acknowledge his signature thereto.
>
> . . .
>
> (6) No particular form of words is necessary to the validity of a will if it is executed according to the formalities required by law.

It is readily apparent that the testatrix did not strictly adhere to the statutory requirements above stated, in her attempt to partially revoke her will. Our statute requires that the will be signed at the end, and accordingly any partial revocation must similarly be so signed. Assuming for the purpose of this discussion but specifically not deciding, that the initialling by the testatrix was sufficient to constitute a signature, the revocation was signed at three separate places, none of which constituted the end of the will. Accordingly, the attempted partial revocation was ineffective and the will as originally written was entitled to probate. The order appealed is affirmed.

(Affirmed.)

QUESTION FOR DISCUSSION

1. *Under the Florida statute may one revoke a will in its entirety by burning, tearing, canceling, defacing, obliterating, or destroying?*

X **(4) Special Types of Wills.** The statutes of some states provide for exceptions to the requirements for the validity of wills. Be careful to note, however, that many state statutes do not allow for any of these exceptions, whereas some others allow for only one or the other of them.

In some states a will written entirely in the handwriting of the testator and signed by him is valid even in the absence of witnesses. Such a will is called a *holographic will.* The statutes of most of the states that recognize holographic wills require that all the words of the will, including the date, be in the handwriting of the testator. The writer of the document must also have the intention of creating a will by the handwritten document.

In a few states an oral will that meets certain statutory requirements is valid. This is called a *nuncupative will.* It is generally effective only to pass personal property and not real property. The will must be stated by the testator during his last illness. It must also be uttered in the presence of the required number of persons, who act as witnesses. The courts do not favor such wills even where they are allowed by the statute. Therefore the statute must be strictly adhered to. Such wills are difficult to establish.

Most states allow for soldiers' and sailors' wills under certain circumstances, even if nuncupative or holographic wills are otherwise disallowed. These wills are generally restricted to members of the armed forces who are outside of the states or territories of the United States. Some states limit these wills to written or holographic wills that do not meet other requirements of a valid will. Other states allow for oral wills uttered to a prescribed number of witnesses. Many of the statutes also limit the duration of such wills. For example, the Maryland statute declares a holographic will written by a member of the United States armed services void one year after the individual's discharge from the armed services unless he or she dies prior to that date or does not then possess testamentary capacity.

32:2. Administration of Estates. When an individual dies, the property that he leaves behind constitutes his or her estate. This property must be used to pay the creditors of the deceased and any taxes that are due. The balance is then distributed according to the instructions in the will or, if there is no will, to the persons named by the statutes that are applicable. It is the task of the personal representative to carry out these duties.

(1) **Probate of Will.** If the decedent has left a will, the personal representative must file it with the proper court. He does this typically by filing a petition requesting that the will be admitted to probate. The petition normally sets forth that the document is the last will and testament of the decedent, the size of the estate, the amount of debts, and the names and ages of the heirs and those having an interest in the estate. If there are any significant assets, the personal representative will normally also have to post a bond in an amount set by the court. The court will then issue *letters of administration* or *letters testamentary* appointing the proper individual as personal representative of the estate. It is then his duty to collect the assets of the decedent, pay off the lawful claims against the estate, and distribute the balance of the property to those who are entitled to it. Normally the personal representative is also required to publish a notice to creditors in a local newspaper announcing his appointment and informing them to submit their claim for payment within a certain period of time, usually ranging from six months to a year. The personal

representative may also have to get nonliquid property, such as real estate, appraised for taxation purposes. The final step after the distribution of property is to render a final accounting and to petition the court for an order discharging the personal representative from his position.

The personal representative is entitled to be compensated for his or her services. The amount of compensation to which the personal representative is entitled is usually limited by statute and must be approved by the court. If the personal representative fails to perform his or her duties faithfully and reasonably, he or she may be liable for any damages suffered by the estate that result.

(2) Administration Without Will: Statutes of Descent. The administration of an estate without a will and one with a will are similar. Two obvious differences are that the court must appoint a personal representative and that the persons who are to receive the property of the decedent after valid claims are paid are determined by statute.

One major disadvantage of dying without a will is that those who are to share in the decedent's property are determined by statute rather than by the wishes of the decedent. Often the provisions of the statute may differ from the decedent's wishes or from what would be the most reasonable disposition of the property under the circumstances. For example, a typical statute might provide as follows: if the decedent leaves a spouse and children the spouse takes one third of the property and the children two thirds; if there is a surviving spouse but no children and surviving parents of the decedent, the spouse takes one half and the parents one half; if there are no surviving children or parents but there are a spouse and surviving brothers or sisters of the deceased, one half goes to the spouse and one half to the brothers and sisters. The statute will normally continue to list how the estate is to be distributed as long as there are surviving relatives of the decedent to a certain degree. If there are no relatives, the statutes usually provide that the property is to go to the state.

These statutes also typically prohibit the testator from disinheriting his or her spouse by will. If this is attempted, the spouse is usually entitled to the share he or she would receive under the statute of descent. The balance of the estate would then be distributed according to the will. The order in which people are entitled to serve as personal representatives is also governed by statute. For instance, the typical statute may provide that the spouse is to have first priority, then adult children, and so on.

(3) Small Estates and Property Not Part of an Estate. The statutes of most states provide for a greatly simplified and inexpensive method of handling estates that are small in size, less than $2,500, for example. Many of the formalities are eliminated. A person is usually appointed to collect the assets, pay the claims, and distribute the balance of the estate to those who are entitled to share in the assets.

You should also note that certain property may not be included as part of the estate property and hence would not be controlled by the statutes of the decedent or the terms of a will. One common example is jointly owned property, which automatically passes to the survivor at death and does not pass to the estate. Another example is the proceeds of life insurance that are to go to a named beneficiary by the terms of the policy. If there is no named beneficiary, then the proceeds would go to the estate. You should also note that although these types of property may not go to the estate under the conditions outlined, they are still subject to the Federal Estate Tax Statute and may be included in the decedent's estate for federal estate-tax purposes.

32:3. Trusts. The trust is a legal concept that has a multitude of uses. It may be used to minimize the impact of taxes, provide for minor children after death, and limit the spending of money to a specific purpose after death. It may be used to accomplish a limitless number of other objectives.

The trust is designed to take advantage of the fact that there may be two types of interests in property: legal interests and equitable interests. Thus one may have the legal title to property but not be entitled to the equitable benefits of the property. Normally, of course, if one owns property, he has both the legal and the equitable interest in that property. In the trust these interests are divided between more than one individual. Basically a trust is a fiduciary relationship in which one person has the legal interest subject to the equitable interest that another individual has in the property. Thus, in a *private express trust,* one person, called the *trustee,* holds the legal interest in property, subject to the equitable interest of another person for whose benefit the property must be used, known as the *beneficiary.* The trustee is obligated to use the property solely for the benefit of the beneficiary.

The typical express trust is created by a person who transfers his property to a trustee to use for the benefit of the beneficiary. Such a person is called a *settlor.* The trust may be created by a transfer to a trustee during the lifetime of the settlor, in which case it is called an *inter vivos* trust. Trusts are frequently created by the will of the settlor. When the trust is created in this manner, it is called a *testamentary* trust. In the latter case, the trust does not come into effect until the death of the settlor.

In order to create an express trust, it is necessary that the settlor transfer the property with the *intent* to create a trust. A trust is not a contract, and so all the elements of a contract are not necessary for its creation. However, the intent to create a trust must be present. For example, in the case of *In Re Humphrey's Estate,*[1] John Longfield, by the terms of his will, left his house to his wife and expressed the "wish" that she in turn leave it to his son. The court held that Longfield's mere wish did not express an intent that the house was to be held by his wife in trust for the benefit of the son, and therefore no trust was created. Language that merely expresses a hope or wish rather than a definite intent is called *precatory language* and is not sufficient in the absence of other facts to establish the intent to create a trust.

[1] Ir. Rep. Chanc. Div. 21.

Any natural person or legal entity, such as a corporation, can be a trustee as long as it is capable of taking legal title to the property. For example, banks frequently are appointed trustees, particularly when a sizable amount of property is involved. One may be the settlor and also transfer the property to a trust with himself as trustee of that property for the benefit of another. It is also possible for there to be more than one trustee. For instance, a bank and a relative of the beneficiary may act as trustee. Furthermore, if there is more than one trustee or more than one beneficiary, a trustee may also be a beneficiary. However, the same individual cannot be the sole beneficiary and trustee because there would then be no separation of the legal and equitable interest in the property.

In creating a private express trust, the settlor must name a beneficiary or at least a method for ascertaining a beneficiary who can enforce the obligations owed by the trustee. Thus, if one leaves money in trust for the erection of a monument, a court might have difficulty in determining who is to be the beneficiary.

Any type of property that is capable of being transferred can be the subject of a trust. Such property can be tangible or intangible and may even be a future interest in real property.

One advantage of a trust is that the trustee can specify the manner and purposes for which the property in the trust can be used. For example, the settlor might specify that the income from a trust is to be used to pay for the beneficiary's education, or for medical expenses, or just for lollipops. The trustee is obligated to adhere to the instructions given in the document creating the trust. There may be more than one beneficiary, and the trust may provide for the transfer of the property to someone other than the beneficiary when the beneficiary dies or when he or she reaches a certain age. In short, the uses and terms that may be contained in a trust instrument are almost limitless, as long as public policy or the law is not violated. Indeed one of the major characteristics of a trust is its flexibility and the manner of uses to which the concept may be put.

ESTATE OF DeROCHE v. MITCHELL

330 So.2d 860 (Fla.App. 1976)

McNULTY, C.

Appellants Marie Pfaff DeRoche and Leonard M. DeRoche, as joint executors of the estate of LeRoy DeRoche, deceased, appeal from an order construing the testator's will as creating a precatory trust giving appellee, Violet Mitchell, a life estate in a mobile home and lot. We think the trial court erred in such construction and reverse.

The following circumstances existed both at the time of the will's execution and at the testator's death. Testator was the sole owner of the mobile home which was located on a lot owned by Venice Bay Trailer Park, Inc. Testator also owned fifty-three and one-half shares of the eighty outstanding shares of that corporation.

In his will, testator, among other things, bequeathed twenty of his shares in the corporation to appellant, Marie DeRoche, and the remainder equally to his brothers, Kenneth and the other appellant, Leonard DeRoche. Paragraph THIRD of the will contains the following language which the court construed to create a precatory trust in appellee, to wit: a life estate in the aforementioned mobile home and lot:

> THIRD: It is my desire that MRS. VIOLET MITCHELL be allowed to continue to live, rent free, on the lot and in the mobile home currently occupied by her, for and during the period of her natural life. I request that each of the three beneficiaries of my interest in Venice Bay Trailer Park, Inc., do their utmost to carry out my desire. In the event said beneficiaries sell the said corporation, I hope that provisions are made to insure that Mrs. Violet Mitchell shall have a life interest in the lot and mobile home currently occupied by her.

While a testator's use of precatory language may establish a trust, the language and surrounding circumstances must demonstrate an intention to impose a legal duty upon a legatee to carry out the desired purpose and not leave the legatee free to carry it out or not as he should choose even though clearly the testator had hoped that he would carry it out.

Nothing on the face of the will nor in the surrounding circumstances herein mandates the imposition of a legal duty on the legatees. Additionally, indeed, the courts have said that if the precatory words refer in whole or in part to property other than that bequeathed or devised by the testator, the words do not create a trust. Here, the language of Paragraph THIRD refers to a life estate in both the mobile home owned by the testator and the lot owned by the corporation. Clearly, the testator could have created a trust in the mobile home, but he totally lacked the power of disposition over the lot not owned by him. In neither case, in any event, was the present rule satisfied. No trust was created.

Questions for Discussion

1. *What was the "precatory language" in DeRoche's will?*
2. *How would you alter the language in the will in order to create a trust?*

---

(1) Trust Administration. The relationship between the trustee and the beneficiary is that of a fiduciary. The courts hold a trustee to the highest standard of behavior and loyalty. The trustee can do nothing to benefit personally from the trust either directly or indirectly and cannot occupy any position whose interests may be adverse to that of the trust. He is required to give the trust his undivided loyalty. If the trustee violates this duty, any transaction by which he does this will be set aside, and the trustee will be held liable for damages. The good faith of the trustee is not a defense to an act that is a breach of the trust relationship.

One function of the trustee, of course, is to invest the property of the trust. In making these investments, the trustee should make such investments "as a reasonable cautious man would use, having regard, not only to the interests of those who are entitled to the income, but to the interests of those who will take in the future."[2] This standard requires the trustee not only to make trust investments as a reasonable and prudent person would make of his own money but with the added consideration that the money invested is not his

[2] *In Re Whiteley,* Sup. Court of Judicature, Court of Appeals, 33 Ch.Div. 347 (1886).

but for the benefit of someone else. As a general rule, this standard requires the trustee to diversify investments and to avoid investments that are risky while giving consideration to the income that will be produced by the investment. These rules do not make the trustee an insurer of the trust property, however. If the trust suffers a loss, the trustee is not liable so long as he or she has adhered to the standards of behavior outlined previously.

The trust instrument may give the trustee certain powers in the administration of the trust in addition to those conferred upon him by statute. It may also limit the powers of the trustee that are granted by statute. For example, the trust instrument may give the trustee wide latitude in the types of investments that he may make, or it could restrict the trustee to very specific types of investments. The instrument may also spell out in detail the purposes for which the trust funds may be spent and may restrict spending to the income of the trust or may grant the trustee the right to invade the principal.

The settlor usually names a trustee because he has confidence in his or her skill and integrity. As a general rule, therefore, the trustee cannot delegate the performance of his trust duties to a third person. This does not mean that the trustee must personally perform every act of the trust. However, if he delegates discretionary acts, such as the making of investments, he is liable for any loss that results.

The trustee has a duty not to commingle any trust funds with his own. They must be kept entirely separate. Modern statutes do allow for *common trust funds,* however. For example, if a bank is acting as trustee for several small trusts, it may invest the money from these funds in the same investments. Each trust participates in the income from these investments in the proportion to which it contributed to the investments.

As a general rule, trustees are entitled to a reasonable compensation for their efforts in administering the trust. The amount of compensation may be fixed in the trust instrument. Many states also have statutes that fix the maximum fee that may be charged by a trustee. These fees are usually based on a percentage of trust income and the size of the trust principal. Other state statutes authorize a court in its discretion to allow the trustee a reasonable fee. A court may also deny compensation to a trustee if his neglect of duty, bad faith, or misconduct has caused loss to the trust.

The trust instrument may also provide for the termination of the trust. For example, a typical provision in a testamentary trust for the benefit of the settlor's children might provide for its termination when the youngest child reaches the age of twenty-one. A trust generally terminates when its purposes have been fulfilled. Sometimes there may also be a change in the law that renders illegal the purpose of a trust that was originally legal. In this case, the trust terminates.

An *inter vivos* trust may be revocable or irrevocable. If it is revocable, the settlor may terminate the trust and take control of the trust property. However, if the trust is irrevocable the settlor may not terminate the trust even if he later changes his mind concerning the desirability of the trust.

WICKMAN v. McGRAW
289 So.2d 788 (Fla.App. 1974)

In an action by minor beneficiaries of a testamentary trust to bring the trustees to account, the lower court found in favor of the trustees. The beneficiaries appealed.

MANN, C.

Carl Robert Wickman died in 1960. His widow and attorney were named executors of his will and trustees of a testamentary trust. The will left one-third of the estate to the widow and two-thirds to the trust, the widow being entitled to the trust income for life, and the sons to the remainder.

In 1963 the trustees qualified. In 1964 the probate estate was closed and the assets distributed. In 1969 the court, on its own motion, ordered the trustees to file an accounting, their first. When it was filed, the sons filed objections, alleging failure to file accountings, improper valuation of assets and failure to account for all of the assets of the corpus of the trust. They sought disapproval of the trustees' account, denial of compensation and removal as trustees.

Proffered testimony—not at this stage to be considered proven fact—will explain the sons' complaints. They assert that the executors distributed assets on the basis of the original probate inventory valuation. They distributed to Elizabeth McGraw, formerly Elizabeth Wickman, as her share, those assets which were, at the time of distribution, grossly undervalued and which were worth substantially more than their value stated in the inventory. Included are: (1) a 20% interest in an orange grove, inventoried at $49,000, which was distributed in kind to Mrs. McGraw in 1961 and shortly thereafter sold for $115,261, although the final probate accounting in 1964 showed her as receiving an asset worth $49,000; (2) the sloop Calusa, inventoried at $8,000 but said to be worth at least $22,500; (3) half interest in Bo-Wick Corporation, inventoried at $2,045 and said to be worth at least $12,500; (4) a personal note of C. E. Morgan, Jr., valued as of the time of the elder Wickman's death at $1,000, said ultimately to be worth its face amount of $4,000; (5) other property not accounted for.

On the other hand, the sons claim that the assets transferred to the trust at their original estate inventory value included: (1) Receivable from Southern Dunes Co. valued at $12,021, but worthless when distributed; (2) Southern Dunes lots, valued at $10,000, but worth no more than $2,500; (3) Bald Point Properties inventoried at $20,000, worth $16,000; (4) Gulf Terraces, Inc., receivable, inventoried at $4,862 but worthless; (5) Cyril Bayly receivable, valued at $2,537, but worthless; (6) Clearwater Yacht Club 50% Improvement Notes, valued at $700, but worthless; (7) the Bauknight tract, valued at $9,500, worth no more than $5,500; (8) other assets valued at $101,734, but of dubious value now, and in any event much less than the ascribed value.

These appellants are not seeking to recover specific property. They are seeking to bring trustees to account, and they have every right to. Loyalty is the fundamental duty of a trustee. Restatement 2d, Trusts, § 170 sums up the law:

> Duty of Loyalty
> (1) The trustee is under a duty to the beneficiary to administer the trust solely in the interest of the beneficiary.
> (2) The trustee in dealing with the beneficiary on the trustee's own account is under a duty to the beneficiary to deal fairly with him and to communicate to him all material facts in connection with the transaction which the trustee knows or should know.

What these beneficiaries are entitled to is a trust corpus worth two thirds of the residual estate at the time of distribution. The ascription of values as of the date of testator's death some three years earlier is part of the problem, assuming that the values as of that date were correct, an assumption by no means clear on this record.

The testator's provision that "The values which my trustees or executors may place upon any property so allotted, provided, and distributed shall be final and conclusive upon all persons having any interest therein" is not conclusive either. Such a clause has the effect of establishing a presumptive validity to those values, but does not absolve the trustees of their obligation to act in good faith and with fidelity to the trust. If Wickman had intended to give his widow a power of appointment he could have done so. Instead he created a trust.

The beneficiaries are entitled to adduce evidence which would prove their right to the relief sought. Should they desire to go behind the order closing the probate estate, the amendment of Article V of our Constitution, F.S.A., makes the Circuit Court an appropriate forum, but the record before us shows their contentment with an order holding these trustees to account for breach of fiduciary responsibility, making restitution from their own funds if need be, and leaving title to assets distributed as ordered by the probate court. If they can prove what they have proffered, they are entitled to relief.

(Reversed and remanded.)

QUESTION FOR DISCUSSION

1. *Assuming that the testimony that was offered by the beneficiaries was actually proved, how did the trustees violate their fiduciary duty in this case?*

---

(2) Special Trusts. The trust concept has been adopted by the courts in order to fashion remedies in certain cases in which equitable considerations require it. In these situations, the courts will create a trust where none previously existed. This is called a *constructive trust.* Generally there is no actual trust because of the lack of a transfer of property by a settlor with the intent to create a trust, so the courts create one at law in order to do justice and avoid unjust enrichment. The most common example of the application of a constructive trust is when one procures property by fraud. The courts may declare that the person who procures the property by fraud holds the property as the constructive trustee for the benefit of the rightful owner.

A trust may be implied from the acts of the parties. In this case, the trust is called a *resulting trust.* For example, if a minor purchases property in the name of his parent, the courts presume that the parent holds title to the property or is a trustee for the child.

A *charitable trust* is set up to benefit society as a whole rather than to benefit an individual or a specific group. Typical purposes of charitable trusts are to benefit education, to promote a religion, to benefit the arts, and so on. There are special rules applicable to charitable trusts that do not apply to private trusts. For example, if the purpose for which the trust was established cannot be accomplished, the courts may direct the application of the trust money to another similar charitable purpose. This is known as the *cy pres* doctrine. Furthermore a charitable trust may have an infinite duration, whereas a private trust must be of limited duration. Finally, charitable trusts escape many of the taxes that are applicable to private trusts.

SUMMARY

The person who writes a will is a testator or testatrix. The affairs of the estate of the deceased are handled by an executor or administrator who is sometimes called a personal representative. A devise is a gift of real estate by will; a gift of personal property is called a bequest.

In general, a will must be written by a person having testamentary capacity, signed by the writer (known as the testator) and signed by credible witnesses. Most jurisdictions require either two or three witnesses to the will. It is not necessary for the witnesses to know the contents of the will.

A modification of a will must be made by codicil. A revocation of a will must generally be made by either writing a new will or burning, tearing, canceling, or obliterating the will. The marriage of the testator after writing the will causes its revocation. In some states the divorce or annulment of a marriage that occurs subsequent to the execution of the testator's will revokes those portions relating to the spouse.

A few states recognize a holographic will, which is one written entirely in the handwriting of the testator. A nuncupative will is an oral will recognized by only a few jurisdictions under limited circumstances.

In the administration of an estate, the court will issue letters of administration or letters testamentary appointing the proper individual to act as personal representative of the estate. That person has the duty to take the necessary steps to administer and then wind up the estate. If there is no will, the distribution of the property of the deceased is governed by statutes of descent.

A trust is a fiduciary relationship in which one person has the legal interest in the property (trustee) subject to the equitable interest of another person (beneficiary). A trust created by will is called a testamentary trust. A trust created between living people is an *inter vivos* trust.

The relationship between the trustee and beneficiary is a fiduciary one. Violation of this duty by the trustee subjects him to a claim for damages by the beneficiary. For instance, the trustee must not commingle trust funds with his own funds.

The trust concept may be used by the courts who may create a constructive trust. A trust implied from the acts of the parties is a resulting trust. A charitable trust is set up to benefit society as a whole rather than to benefit an individual or a specific group.

P R O B L E M S

1. John Urbanovsky was eighty-seven years of age when he made his will. He suffered from high blood pressure, had sustained a stroke, and had been subject to some mental confusion but regained his mental clarity and was able to get around. John spent five and one-half years in the home of one daughter before he died and he left most of his property to her in his will. Two other children contested the validity of the will on the ground that Urbanovsky was unduly influenced by the daughter and also lacked mental capacity. Are the two children correct? In the matter of the Estate of Urbanovsky, 215 N.W.2d 74 (Neb. 1974).
2. W. R. Fletcher was convicted of rape and sentenced to be hung on February 15, 1907. On February 12 he wrote a letter to his daughter in his own handwriting. In the letter, after denying his guilt, he stated how his property was to be distributed after his death. He was duly hanged on February 15, and afterwards the letter was offered for probate as his will. Could the letter serve as a valid will in any state? Milam v. Stanley, 33 Ky. 783, 111 S.W. 296 (1908).

3. A was sick and signed his will in his bed with all the witnesses present. The doctor then suggested that the witnesses retire to the next room to sign the will. They left the door open between the bedroom and the room in which the witnesses signed the will, but the witnesses signed at a table that was not visible from A's bed. Was the will valid?
4. Lavinius B. Willden signed his will on August 24, 1894. The will gave all his property to Ella Gilliland and Lewis Lateer equally, provided that Lateer agreed to act as executor. Later, on the back of the will, Willden wrote "All I have I want to go to my sister Miss Ella Willden." Also written on the back of the will was a notation that stated "This is no good; will try to make another December 10, 1906." On the front of the instrument appeared the statement "Not any good—December 11, 1907." Was there a revocation of the will? Dowling et al. v. Gilliland et al., 286 Ill. 530, 122 N.E. 70 (1919).
5. Augustus M. Cosgrove brought a document, which was in the form of a will and signed in pencil, into an office and asked Miss Young to copy it for him. Before he left, Miss Young said to Cosgrove, "Would you like to sign this will in case anything should happen?" Cosgrove indicated that he would and signed it in front of Young and another woman, who then signed their names as witnesses. Cosgrove died before he could return to sign the copy. His heir contended that the penciled document was not a will but only a memorandum for the guidance of Miss Young in typing the will in final form. Was the signed document a valid will? In Re Cosgrove's Estate, 290 Mich. 258, 287 N.W. 456 (1939).
6. James G. Cantrell owned valuable property and wrote a will in which in the third paragraph he devised and bequeathed all his property to his wife. In the fourth paragraph Cantrell stated that it was his "request" that upon her death his wife should dispose of the property among certain relatives. These relatives contended that Mrs. Cantrell took the property as a trustee. Were they correct? Comford v. Cantrell, 117 Tenn. 553, 151 S.W.2d 1076 (1941).

SECTION VI

Agency

CHAPTER 33

Nature, Creation, and Termination of the Agency Relationship

It is important to be familiar with the legal concept of *agency*. The primary reason for this is that an agent may bind his principal by his acts. The law of agency is based on an old Latin maxim, "Qui facit per alium, facit per se," which roughly translates as, "One acting by another is acting for himself." It is therefore important to know when and under what circumstances one may be liable for the acts of another.

An agent may be broadly defined as one who acts for and under the contract of a principal. It has been defined more precisely as a fiduciary relation that results from the manifestation of consent by one person to another that the other shall act on his behalf and subject to his control and from manifestation of consent by the other so to act. 2A *Corpus Juris Secundum*, Agency 4. The person who consents to act for another is the *agent*, and the one for whom he acts is the *principal*. The acts of the agent may result in the principal's being bound to a third party contractually or his being liable to the third party in tort.

In this chapter and Chapters 34–36 we will examine the question of how the agency relationship is created, what are the consequences of its creation, and how the relationship may be terminated. First, let us distinguish the principal–agent relationship from some others.

33:1. Agency Distinguished from Other Relationships. The relationship between an employer and an employee is similar in many respects to that of a principal and an agent. The employee works for and is under the control of his employer and renders services to his employer. The rights and duties between the parties are also similar. In fact, an employee may also be an agent, but not necessarily. The critical difference is that the agent is authorized to act on behalf of his principal in dealing with third persons, whereas the employee is not.

For example, assume Pearl Purchaser and Arnie Assembler work for Manufacturing Company, Inc. Pearl is a purchasing agent authorized to make contracts with suppliers up to $5,000. Arnie works on the assembly-line. Both are employees of Manufacturing Company, but Pearl is authorized to deal with third parties and thus is also an agent. Arnie is obviously not an agent, and any contract that he enters into on behalf of Manufacturing Company would not be enforceable against Manufacturing Company.

The distinction between an agent and an independent contractor is largely a matter of the degree of *control* exercised over the party in the performance of his work. This distinction can be critical because, as we shall see later, a principal is liable for the torts of his agent committed within the scope of his employment, but the employer is not liable for the torts of the independent contractor.

There is no single set of criteria for determining what degree of control is necessary in order for one to qualify as an agent as opposed to an independent contractor. It is a factual determination that must be made on a case-by-case basis. In general, the courts will examine a number of factors. One of the most critical is who has a right to control the details of the work. If the employer only has a right to approve the result of the work, the party is most likely an independent contractor. Other factors to consider are who sets the hours of employment and the place of employment, supplies the tools used, determines the procedures to be used, and has the right to hire and fire. Analysis of these factors goes a long way toward determining whether the relationship is one of employer–independent contractor or principal–agent. Another factor to be considered is the method of payment. For instance, although not determinative, it would be important to know whether payment is by salary or by the job. The history of the relationship in the particular trade or business would also be significant.

Suppose, for example, that Ken Klean owns a drycleaning establishment. He hires Erwin Employee to drive a delivery truck owned by Klean. Erwin is paid by the hour and works on a schedule determined by Klean. Klean supplies Erwin with a uniform. Erwin is most likely not an independent contractor. On the other hand, assume that Klean hires Erwin to solicit for and pick up laundry. Erwin has his own truck, works any hours he desires, is paid by the pound for laundry collected, and also collects and delivers laundry for several other establishments in neighboring towns. Under these circumstances, Klean exercises very little control over Erwin, and Erwin would be an independent contractor.

DUMAS v. LLOYD et al.

6 Ill.App.3d 1026, 286 N.E.2d 566 (1972)

William Dumas sued to recover for personal injuries suffered from the negligence of the defendants, William G. Lloyd, Clarence Shaefer, Norman Oil Company, Inc., and William H. Frazier. A summary judgment was entered in favor of Lloyd and directed verdicts rendered in favor of Norman Oil and Shaefer. Dumas appealed the granting of the directed verdicts by the trial court.

ENGLISH, J.

Defendant Norman Oil Company owns and supplies gasoline stations in the Chicago area, one of which is located at 143 S. California Avenue, Chicago. In September, 1962, defendant Shaefer was hired by Dale Norman, an officer of Norman Oil Company, to operate that service station on behalf of Norman Oil. He was not given a written contract by the company, nor did he hold a license to operate the station or give Norman Oil money or security for the operation. Shaefer did not own anything at the station and was not permitted to sell any products other than those supplied by Norman Oil. He received a commission of four cents for each gallon of gasoline sold at the station, and each day he banked the receipts after deducting his commission, and sent copies of the deposit slips to Norman Oil.

Shaefer was empowered to hire other people to help him with work at the station but, in practice, did not hire anyone without telling Dale Norman. Defendant Frazier was one of those hired, and his employment continued at the times pertinent to this case. Each person hired by Shaefer was paid out of his own commissions.

The operating license for the service station was in the name of John Norman, president of Norman Oil Company. Each month Norman Oil paid the rent on the property, the electric bills, and sales tax on all products sold at the station. All of the equipment, such as the gas pumps, air compressor, underground tanks, and signs, including one which said "Norman Oil Products," were owned by Norman Oil, and all products sold there were furnished and delivered to the station by the company, usually through Dale Norman. Either he or John Norman would visit the station once or twice a week and would instruct Shaefer as to keeping the station clean.

The company supplied all sales books and had its name on all books and records used at the station. Shaefer had no control over the price set for gasoline, that being determined by Norman Oil. Nor could Shaefer draw on the bank account which was in the name of the company.

On various occasions Shaefer, in the presence of Dale Norman, accepted and held various items as security for products sold when the customer could not pay the full amount. On December 18 or 19, 1963, Shaefer took a revolver from a customer as security for a payment of $3.00 for gasoline. The gun had a belt wrapped around it which Shaefer did not remove, but he placed the gun with the belt in a desk drawer at the station. The desk had only one drawer and it was unlocked. He never unwrapped the belt and never looked to see if the gun was loaded.

The next day, plaintiff came to the gas station about 3:30 P.M. to visit Frazier and to have a grease job and oil change on his car, but it turned out to be too cold to do the grease job. He and Frazier had been pretty good friends for about five years and Frazier had been driving plaintiff's car all summer. About 5:30, while both men were in the station, along with several others, a man entered and asked Frazier to help him charge the battery in his car. Frazier said he couldn't do it right away and when the man asked how long he would have to wait, Frazier refused to do it and talked to the man in rough language. Plaintiff said he would help, and left the station and got the car started.

Plaintiff, in a joking manner, "told Frazier if he talked to me like he did to that old man, I would cut his throat off, and I did like this with my keys (indicating), and I walked out the door." As he went out, Frazier, also laughing, turned to a friend and said, "Watch me scare Red" (Plaintiff). He opened the drawer, took out the gun, and said, "Red, I'm going to shoot you," and shot him. Whereupon, Frazier immediately said, "Damm, look what I done did," and ran to help plaintiff, saying, "Man, I am sorry." They brought plaintiff back into the station, where Frazier called the police and told them he had accidentally shot a man.

Plaintiff declared that he having proved a prima facie case of agency between Norman Oil and Shaefer, the trial court erred in directing a verdict in favor of either Norman Oil or Shaefer because the latter's negligence is a question of fact for the jury.

Norman Oil makes three points: (1) that Shaefer was not its agent but an independent contractor. . . .

Of primary consideration in the determination of whether a person is acting as an independent contractor or as an agent or employee is the degree and character of control exercised over the work being done. When one undertakes to produce a given result without being in any way controlled as to the method used, he is considered an independent contractor and not an employee. But the relationship of principal and agent exists if the principal has the right or the duty to supervise and control, and also the right to terminate the relationship at any time. The test is in the right to control and is not dependent upon its exercise. The general rule of liability is that a principal is liable for the negligent acts of his agent, but not for those of an independent contractor.

We believe that the evidence as introduced by plaintiff did establish prima facie that Shaefer was acting as Norman's agent in the operation of the service station and was not an independent contractor. Although the day-to-day operating procedures were managed by Shaefer, his authority was limited by the interest of the owner whose representative frequently visited the premises and laid down for Shaefer certain rules as to buying and distribution methods. Norman owned all of the equipment used by Shaefer and set the prices for all the products sold. Signs, records and accounts were in the name of Norman Oil and Shaefer was powerless to change them. Furthermore, on January 5, 1965, without prior notice to Shaefer, the owners closed the station and terminated their relationship with Shaefer effective at that time. We realize, of course, that at this point in the trial Norman Oil had had no opportunity or need to introduce any countervailing evidence on this point in view of the court's directed verdict in its favor.

However, even though a principal–agent relationship between Shaefer and Norman Oil could have been found to exist, a principal cannot be made liable through the doctrine of respondeat superior when the actions of his agent in no way constitute negligence. Plaintiff contends that Frazier could not have discharged the gun were it not for the careless and negligent manner in which Shaefer permitted the loaded gun to remain at ready access in the desk drawer. Yet, the uncontroverted testimony of both Shaefer and Frazier discloses that Shaefer did not know if the gun was loaded, and Frazier definitely thought it was not. We believe Shaefer acted reasonably when he allowed the belt to remain wrapped around the pistol and its case and placed it in a drawer which, although unlocked, was out of sight from those persons who might enter the station. We also believe that the negligent or reckless act of Frazier was clearly not of a character which could be attributed to his employer. For both these reasons, therefore, we conclude that the trial judge acted properly in directing verdicts in favor of defendants Norman Oil and Shaefer. The judgments entered thereon are affirmed.

QUESTIONS FOR DISCUSSION

1. *Identify the principal and agent. Was Frazier an agent of Norman Oil?*
2. *What factors were significant in leading to the conclusion that there was a principal–agent relationship.*

33:2. CAPACITY OF PRINCIPAL AND AGENT. An individual generally has the capacity to be a principal if he has the capacity to act for himself. For instance, if one has the capacity to enter into a contract, he also has the capacity to do it through an agent. Thus, if a principal is an infant, a contract entered into with a third party by an agent on the principal's behalf is voidable at the option of the principal. The same rule applies if the principal is insane. To put it another way, if one lacks the capacity to act for himself, such as by entering into a contract, that disability is not removed simply because the act is committed on behalf of the principal by an agent. A corporation, of course, has the capacity to appoint agents.

As a general rule, a principal cannot complain of the lack of capacity of one whom he has chosen to act as his agent. Thus, if an agent is an infant or is otherwise incapable of acting for himself, this fact cannot be raised as a defense by the principal. After all, if the principal wants to select an infant to represent him, it is the principal's choice. We are, of course, talking about the ability of the agent to bind his principal in dealings with third parties.

As between the principal and the agent, however, if the agent is a minor, that defense could be raised by the minor in the event of a suit by the principal for breach of the agency contract.

There are some limitations. In order to be an agent, one must have some mental capacity. The courts have held that infants of tender years, as well as those obviously insane, lack the capacity to be an agent. One must at least have the capacity to follow instructions.

33:3. Classes and Kinds of Agents. A *general agent* is one who is empowered with all the authority that his principal has with regard to a particular business. He may be employed, for instance, to transact all the business of the principal with regard to that undertaking. The fact that he may be empowered to act with regard to only one line of business does not affect his status as a general agent.

A *special agent* is one who is employed for a special purpose and has limited powers. A special agent does not have the authority to bind the principal with regard to the business in general but only in matters concerning the special purpose for which he is appointed. For example, a general agent may be appointed by the principal to rent a house, collect the rent, take care of maintenance, and in general do all things required in the renting of the house. A special agent, on the other hand, might be hired for the specific purpose of collecting the rent. His authority is quite limited and circumscribed.

A *subagent* is one appointed by an agent to assist in transacting the affairs of the principal. He must be appointed with the knowledge and consent of the principal. A subagent has a fiduciary duty to the principal and thus should be distinguished from those who are merely servants of the agent.

Factors and *brokers* are generally considered to be special agents. A factor is one who is employed to sell merchandise consigned to him. The factor has possession of the merchandise and usually sells it in his own name for his principal. The function of a broker is to bring two parties, buyer and seller, together. He does not have possession of the property and does not have the authority to bind his principal contractually. A familiar example is the real-estate broker.

One may be a *gratuitous agent;* that is, he receives no compensation for his effort on behalf of his principal. If a gratuitous agent promises to act on behalf of his principal and then fails to do so, that promise cannot be enforced. However, once the gratuitous agent does begin to act, he must act properly and has a fiduciary duty to his principal.

CONTINENTAL CASUALTY CO. v. HOLMES et al.
266 F.2d 269 (1959)

RIVES, C.J.

Appellee, Mrs. Gladys Leith Holmes, as beneficiary, sued the appellant, Continental Casualty Company, for the proceeds of a purported accident and health insurance policy insuring her deceased husband, Oliver Wendell

Holmes, while flying as a pilot. She claimed that coverage was in force as a result of an oral binder or oral contract of insurance effected by the Company's agent, W. T. Musgrove, on January 3, 1957, and again on February 20, 1957. Her suit was primarily based on the proposition that W. T. Musgrove, a partner with W. R. Cadenhead in the firm of Musgrove Insurance Agency, had authority to orally bind the Company. Continental Casualty Company impleaded as third-party defendants by the Musgrove Insurance Agency and the two partners individually (appellees here), alleging that none of the third-party defendants had actual authority to make the oral agreement, and averred that, if Continental Casualty Company was liable to the plaintiff in this action (such liability being denied) on account of any apparent or ostensible authority, then the third-party defendants are in turn liable in the same amount to Continental Casualty Company.

The verdicts of the jury were in favor of the plaintiff, assessing her damages at $52,750, and in favor of the third-party defendants. Necessarily, therefore, the jury found that the agents had actual authority to make the alleged oral binder or contract.

Plaintiff's theory of the case rests upon the authority of the agent to effect a policy of insurance by an oral binder or by an oral contract of insurance, concluding that if sufficient authority existed in the agent Musgrove to bind the Company, it is immaterial for plaintiff's recovery whether the authority be actual (express or implied) or apparent (ostensible). Plaintiff also proceeds on the collateral theory that Musgrove was a general agent for the Company which position automatically gave him the power to orally bind the Company in various ways. In presenting this double-barrelled approach, the plaintiff claims that Musgrove had authority whether he is found to be a general agent or something less, such as a soliciting agent. . . .

We state at the outset that the test to determine whether a general agency exists, in both Alabama case law and in general insurance law, is whether the agent "has the power to bind the insurer by his contract of insurance, or to issue policies on his own initiative, or to accept risks, and if the agent has actual authority to do these things, he is a general agent"; he may then bind the company in other ways, such as by oral contract, by waiving certain policy or application provisions and the like. The general agent whose authority is limited in some way or an agent who is by certain limitations, less than a general agent may also bind his principal in the above ways as though he were a general agent by apparent or ostensible authority through the doctrines of waiver, estoppel, or ratification, and, of course, the injured party must have relied upon this apparent authority in some way to his detriment. He may also be less than a general agent (such as a soliciting agent) and have actual authority to bind the company by oral binders or contracts in certain types of insurance.

I. ACTUAL AUTHORITY.

(a) General Agent.

Therefore, if Musgrove had the power to bind the Company "by his contracts of insurance, or to issue policies on his own initiative, or to accept risks," or if the Company designated Musgrove to be a general agent by some action or manifestation, then he would be a general agent with the power to bind the Company in the manner claimed by Mrs. Holmes. It is elementary that actual authority can only be created "by written or spoken words or other conduct of the principal which, reasonably interpreted, causes the agent to believe that the principal desires him to act on the principal's account."

There was . . . undisputed evidence to show that Musgrove knew that he did not have actual authority to accept risks and make contracts for Continental and therefore was not a general agent. This evidence is: First, the two applications for coverage of Holmes and Culver while flying as passengers (which were sent to Musgrove for the signature of Holmes and Culver after an oral binder had been effected by Holland on the basis of the applications to Queen Insurance Company) contained the same question as the later application for pilot coverage, which was "Do you understand and agree that no insurance will be effected until the policy is issued?" Second, the two original policies issued on the above applications which insured Holmes and Culver while flying as passengers had as a normal policy provision the following:

> This policy, including the endorsements and the attached papers, if any, constitutes the entire contract of insurance. No change in this policy shall be valid until approved

by an executive officer of the Company and unless such approval be endorsed hereon or attached hereto. No agent has authority to change this policy or to waive any of its provisions.

Third, the application for pilot coverage signed by Holmes on February 20, 1957, stated: "12. Do you understand and agree that no insurance will be effected until the policy is issued?" This was answered. "yes.". . .

We are of the opinion that all of the evidence in this case can only show that Musgrove was merely a "local" or "resident" agent, having the powers normally conferred on a "soliciting" agent with no actual authority to bind the Company by oral contract or binder, and that his actions in assuring Holmes of immediate coverage were clearly without the scope of his actual authority. There were no "written or spoken words or other conduct of Continental Casualty Company which, reasonably interpreted, caused Musgrove to believe that Continental desired him to act on Continental's account as a general agent. . . ."

(The judgment is therefore reversed and the cause remanded with directions to enter judgment for the defendant.)

Questions for Discussion

1. *The court stated that because the jury found in favor of Mrs. Holmes and also in favor of the third-party defendant, Musgrove Insurance Agency, the agent, it must have necessarily determined that Musgrove had actual authority to make the alleged oral binder. Why was this conclusion necessary?*
2. *What facts were important in establishing that Musgrove was not a general agent?*

33:4. Creation of the Agency Relationship. No special requirements must be met in order for the relationship of principal and agent to arise. It may come into being in any number of ways: by contract, ratification, estoppel, necessity, or operation of law. The following material examines each of these methods.

(1) By Contract or Appointment. Although there is no particular form necessary to the creation of the principal–agency relationship, there must generally be a manifestation by one individual that another shall act as his agent and a consent by the other to do so. This manifestation most frequently takes the form of a contract between the parties. Just as with any other contract, a contract of agency may be expressed or implied. As noted before, however, one may be appointed to act as an agent and consent to that appointment without the creation of a valid contract. If, for example, one of the elements of a contract is missing, such as consideration, the person is still an agent. If the agency is created by the execution of a valid contract, both parties must obviously live up to its terms or be liable for its breach. A common example of an agency relationship created by contract is that of a real-estate agent hired to sell one's house.

(2) By Ratification. Sometimes an individual may act on behalf of another without the other's authority. When this occurs, the act may be ratified by the person for whose benefit it is performed. Such a ratification binds both the original party and the ratifying party to the act. The *Restatement, Agency Second,* Section 82, states:

> Ratification is the affirmance by a person of a prior act which did not bind him but which was done or professedly done on his account, whereby the act, as to some or all persons, is given effect as if originally authorized by him.

Ratification usually occurs under one of two circumstances. In one situation an individual in an existing agency relationship with a principal may perform an act beyond the scope of his authority. In this case, the principal may subsequently ratify the act. Sometimes the act is performed by someone who is not in an agency relationship with the one for whose benefit it is performed, and the act is subsequently ratified. In either case, the ratification results in a principal–agent relationship.

In order for the principal to ratify the act, it is necessary that the principal be in existence at the time that the act is performed. Thus if one performs an act on behalf of a corporation before it is created by the state, it cannot ratify that act when it does come into existence. (We shall see in Section Eight, "Corporations," that the act may be "adopted" but not ratified.) However, it is not necessary that the principal be present when the act to be ratified is performed in order to ratify it.

In order for the ratification to be valid, the principal must be in possession of full knowledge of the facts at the time that he ratifies an act. If material facts are not known to the principal, then the purported ratification is invalid. For example, if a principal ratifies a contract entered into by the agent, he must have full knowledge of the terms and conditions that are part of the transaction. It is possible, of course, to ratify a tortious act. If the principal decides to ratify a transaction, he must ratify it in its entirety. He cannot ratify a part of the transaction and reject the rest unless the third party consents. If the law were otherwise, the principal would be able to ratify that part of the transaction that is advantageous to him and reject the rest. This would obviously be unfair to the third party.

The principal must ratify the act within a reasonable period of time. What is reasonable, of course, depends upon the circumstances of each individual case. When the principal does ratify the act, he may do it expressly or by implication, which results from any acts or conduct that show his intent to ratify the transaction. This intent need not be communicated to the third party with whom the agent has dealt. For example, if the principal receives and accepts benefits that result from an unauthorized transaction by his agent, the principal will be deemed to have impliedly ratified the transaction.

When the principal ratifies the act of his agent, the ratification relates back to the time when the original act was performed. Suppose Artie Agent enters into a contract with Terry Thirdparty on January 1 on behalf of Perry Principal, but without Principal's authority. On January 10 Perry is made fully aware of the situation and ratifies the agreement. Although Perry was not bound to the contract during the period from January 1 until January 10, his act of ratification would relate back to January 1, and the rights and liabilities would be determined from that date. Note, however, that because Perry is not bound until he ratifies the contract, Thirdparty could withdraw from the contract any time prior to January 10.

(3) By Estoppel. Under certain circumstances a court will prohibit two parties from denying that a principal–agent relationship exists by invoking the doctrine of estoppel. There then results an *agency by estoppel,* which should be distinguished from an *actual* principal–agent relationship. When the court applies the estoppel doctrine, it is not holding that an actual principal–agent relationship exists, merely that the parties will not be permitted to deny its existence; that is, they are "estopped to deny" the relationship.

The typical situation in which this problem arises occurs when one individual represents himself to be another's agent. If the second individual becomes aware of this representation and allows it to continue without taking steps to stop it or to protect third persons who might be misled by such a misrepresentation, the court may invoke the doctrine of estoppel. Generally, for a court to invoke the doctrine, there must be conduct calculated to mislead under circumstances in which the alleged principal might have expected others to rely on it. In addition, the third person must reasonably act on the conduct, in good faith, thinking that the person with whom he is dealing is an agent of the alleged principal. Thus, if one is aware that another is dealing with third persons and is holding himself out as an agent, a failure to take steps to stop the practice may result in a court's applying the doctrine of estoppel even though no actual agency relationship exists.

(4) **By Necessity and Operation of Law.** An agency may arise by operation of law regardless of any intent or contractual relationship of the parties. For example, most states have a statute designating the secretary of state, or a person of similar capacity, as an agent for the purpose of accepting service of process for any nonresident motorist who is using the state's highways. The act of driving on the state's highway constitutes the appointment of the official as an agent by operation of law.

One may also be empowered to act as an agent on behalf of a principal because of necessity occasioned by an emergency. For instance, if one is an employee, an agency may be implied by necessity if the employee must act to protect the principal or his property because of the result of some emergency. The situation arises, of course, when the principal cannot be contacted in time to act for himself, and the emergency must be real. Thus the conductor of a train in a train wreck may act as an agent of the company and obtain a doctor. Under normal circumstances the conductor would not have this authority.

OSBORN v. GREGO
596 P.2d 1233 (Kan. 1979)

Raymond L. Osborn, Jr., is a contractor who built a house for Kenneth A. and Sharon L. Becker. Capitol Federal Savings and Loan Association loaned the money to the Beckers to purchase the home from Raymond Osborn, Jr., the builder.

When the settlement date for transferring the property arrived on May 21, 1976, Osborn still had some work to finish on the house. Accordingly, Capitol Federal agreed to act as an escrow agent and keep $7,000.00 of the money owed to Osborn until he finished the work which was primarily landscaping.

The items specified in the agreement were: two decks, two patios, sod, sidewalk, and 400 railroad ties. Later, on the same day as the closing, Osborn called Capitol Federal and informed an employee that the parties had agreed to use only 200 railroad ties and that the 400 figure was incorrect. In response to the call and without consulting the Beckers, the 400 figure was changed to 200 and the amount of money retained was $5,600.00. The difference of

$1,400.00 was paid to Osborn. The change in figures and additional release of funds were unauthorized by the Beckers.

By September 13, 1976, all the landscaping had been completed except for the use of some railroad ties. Some of the landscaping work was performed by George Grego, Jr., a subcontractor. Osborn had not paid him and he filed a mechanic's lien against the property. The Beckers then signed a document releasing any claims they might have against Osborn. They also instructed Capitol Federal to pay Osborn the $5,600.00 held in escrow.

Grego then filed suit to foreclose its lien. By the time of trial, all matters had been settled except a claim by the Beckers against Capitol Federal for $1,400.00. The Beckers contended that Capitol Federal's action in reducing the escrow by $1,400.00 was unauthorized. Capitol Federal contended its action was ratified.

FROMME, J.

Capitol Federal Savings and Loan Association became an agent of the Beckers when it accepted the duty of holding the $7,000.00 to assure completion of the contract. In fact, an escrow holder is the agent of both parties.

A depository may not perform acts or release funds while handling an escrow agreement if such acts are not authorized either by the agreement for escrow or by both parties. When Capitol Federal released the $1,400.00 at the request of Osborn without obtaining the approval of the Beckers the release of funds was unauthorized. A depository must be guided in carrying out its duty by what the contract says, for the agreement for escrow with its instructions constitutes the full measure of the duties and obligations assumed by the depository.

However, an unauthorized act of an agent may be ratified by a principal, and when ratified it is the equivalent of an original grant of authority. On acquiring knowledge of the unauthorized act of an agent, the principal should promptly repudiate the act; otherwise it will be presumed he has ratified and affirmed the act.

When the Beckers authorized the release of the balance in the hands of their agent, Capitol Federal, on September 13, 1976, they knew the railroad ties had not been installed on the back of their property. They knew that the Gregos as a subcontractor had not been paid by Osborn and were claiming a mechanic's lien. The written release submitted to Capitol Federal so stated. In releasing the $5,600.00 to Osborn and accepting Osborn's personal agreement to settle with the Gregos, which Osborn was required to do anyway, the Beckers ratified the act of their agent in releasing the first $1,400.00. The written release was given for the acknowledged consideration of $2,500.00 which the Beckers received from Osborn. Despite knowledge of the previous unauthorized release of the $1,400.00, the Beckers did not repudiate that payment, did not make demand for its return, and thereafter directed the release of the entire balance of the funds knowing that the contractor Osborn had not paid the Gregos for landscaping. Under these uncontradicted facts ratification occurred.

QUESTIONS FOR DISCUSSION

1. *What action of the Becker's constituted ratification of Capitol Federal's actions?*
2. *If the Beckers did not have knowledge of the previous release of $1,400.00 to Osborn, would their act of approving the payment of the escrow funds to Osborn have constituted a ratification?*

NOVICK et al. v. GOULDSBERRY

173 F.2d 496 (1949)

William H. Novick and Annette Novick owned and operated a bar in Seward, Alaska, known as Novicks' Cocktail Lounge. William Carroll was the bartender. A customer, Gouldsberry, came into the bar, and after some words with Carroll a fight ensued. The Novicks appealed from an adverse verdict of the trial court.

YANKWICH, J.

Gouldsberry walked into the bar and ordered beer. Carroll was the bartender in charge. Carroll's wife, who had been but recently divorced from Gouldsberry, was sitting at the end of the bar. Gouldsberry was not expecting any difficulties that evening. For he had taken along his small dog and had left his supper cooking in his home. When he entered the bar, he congratulated Carroll on his marriage, and offered to buy him and Mrs. Carroll drinks. A conversation ensued between Gouldsberry and Mrs. Carroll, during the course of which Mrs. Carroll said that people were "telling lies about her." After a little more talk, Gouldsberry told Mrs. Carroll: "Well, I don't suppose you bought another man a bathrobe and I had to pay for it." Carroll then struck Gouldsberry in the face with a bottle, knocking him to the floor. Gouldsberry was knocked partially unconscious. A scuffle ensued, with Gouldsberry on the floor. Mrs. Novick had been sitting at the bar. She and Mrs. Carroll and another person, Gouldsberry testified, were "hammering" on him when he was on the floor. He was not certain, however, whether Mrs. Novick struck and beat him. Nor did he know whether Mrs. Carroll took part in the scuffle on the floor. He suffered injuries, cuts on his lip, face and body and a broken ankle. He became quite "bloody." . . . The next morning, the Carrolls signed a criminal complaint against him. . . . The trial on the criminal complaint took place the following Monday, before the Municipal Magistrate at Seward. He was found guilty and sentenced to 75 days in jail or $150 fine.

After the trial, Novick said: "What's the matter with you, Gouldsberry? Are you crazy? *If I had been there, I would have broke your God damn neck.*"

[After further discussion the Court then took up the appropriateness of the lower court's instructions to the jury.]

In truth, only two specific objections were made. One related to Instruction No. 5. After objection, the Court modified the instruction by giving Instruction 5C, which defined fully the meaning of "scope of employment" in the following manner:

> An act done by an employee is within the scope of his employment and in the course of his employment where such act is or reasonably appears to be necessary, or proper, or suitable, to accomplish the purpose, or the work, or the duties of his employment, although in excess of the powers actually conferred on the employee by the employer. But an act is not necessarily done in course of employment or within scope of employment because done on the employer's premises and by use of the employer's property. An act cannot be said to be within the scope of the employment, where the employer himself, if present, would have no authority to do the act.

The other objection which, in reality, is the only one which, under the principles alluded to, is properly reviewable, relates to the modification which the Court inserted into Instruction 6. The instruction originally read:

> An employer is liable for the acts of his employee, even if such acts are willful or malicious, where they are done in the course of his employment and within its scope. But where an employee does a willful and malicious act resulting in injury to another while engaged in working for his employer, but outside of his authority, as when he steps aside from his employment to gratify some animosity, or private grudge, or to accomplish some unlawful purpose of his own, not in any manner connected with his employment or the duties thereof, and completely outside of the scope of his employment, the employer is not liable.

The modification which was inserted at the end of the instruction when it was re-read was in these words: "unless the jury finds by a preponderance of the evidence that the employer has ratified the acts of his employee as hereinbefore explained."

It is argued that this modifying language does not state the law correctly. Concede that if it stood alone, it might be inadequate. Nevertheless, when considered in conjunction with the other instructions, such as Instruction 5, we believe that it advised the jury properly of the circumstances and conditions from which ratification might be inferred. And, while the retention of Carroll in employment and Novick's admission after the trial are mentioned as acts which might be considered, the instructions, as a whole, allowed the jury to determine ratification "from all the evidence."

Ratification being a matter of intent, the jury was thus left free to deduce if from the entire conduct of the appellants with relation to the assault, after it occurred, including the retention of the bartender, and the approval of his act by their concern in the criminal complaint against Gouldsberry, and in using, after the trial, language indicating sanction of what Carroll did.

We conclude, therefore, that there was no error in the instructions given or refused.

(The judgment is affirmed.)

QUESTIONS FOR DISCUSSION

1. *What did Novick do, or fail to do, that led to the conclusion that the act of the bartender was ratified?*
2. *Would Novick have been liable if he had not ratified the act of the bartender? Why or why not?*

33:5. Types of Authority. The authority of an agent may be *actual* or *apparent.* Actual authority may be either *express* or *implied.*

(1) Actual Authority. Actual authority is the power that the principal intends the agent to have in order to carry out his assigned tasks. It results from the principal's manifestations to the agent, which may be either express or implied.

(a) *Express Authority.* Express authority is the power that the principal gives to the agent to use on his behalf. It may be given either orally or in writing. Suppose Perry Principal tells Artie Agent to sell his car for not less than $4,000. Perry has given Artie the express authority to sell the car for not less than a specific price. Artie may have to do certain things in order to accomplish this task. The fact that Perry has not given Artie the express authority to do these additional things does not mean that Artie lacks the actual authority to do them. Artie may also have implied actual authority.

(b) *Implied Authority.* Implied authority is actual authority that may arise by implication from the express powers conferred by the principal. Thus in the previous example Artie may have the power to advertise the car for sale in order to accomplish the objective set out in the express powers granted by the principal. The more general the express authority, the more authority the agent may have granted by implication. The agent may use those powers that are customary and usual under the circumstances. Thus, if the circumstances surrounding the sale of the automobile in the foregoing example are such that the sale is customarily on credit, then Artie may agree to credit terms despite the absence of express authority to do so. On the other hand, if the sale under similar circumstances never involves the use of credit, then Artie would not have this authority.

Implied authority may also arise from the acts and the conduct of the parties quite apart from those powers that arise from any grant of express authority. Thus the authority of an agent may be implied if the principal regularly permits an agent to engage in a particular course of conduct.

(2) Apparent Authority. Apparent authority is not *actual* authority, although if it is found to exist, the principal is no less bound by the act of his agent than if the agent acted with actual authority. Apparent authority consists of some act or conduct of the principal that results in a third person's reasonably believing that the alleged agent is acting with the authority of the principal. A further requirement is that the third person must in fact act reasonably on the belief that the agent is acting with the authority of the principal.

Note the distinction between apparent authority and implied authority. The latter is actual, whereas the former, although no less binding on the principal, is not actual authority.

The application of the doctrine usually results from something the principal does that would lead a reasonable person to believe that the agent has the authority to act as he does even though he in fact lacks that authority. Put more simply, the principal gives the alleged agent the trappings of authority although never giving him the actual authority. If the third person enters into a contract through this agent while reasonably believing that he has the authority to act, the principal would be bound.

For example, suppose Perry Principal is a car dealer and hires Alleged Agent as "vice-president of sales." He gives Alleged a paneled office complete with drapes, carpet, secretary, and brass nameplate on the door. Perry tells Alleged that he has no authority to agree to any sales contracts but that his position is purely for "public relations." Nonetheless, one day Alleged approves a contract for the sale of a car in a fit of exuberance. Assuming that agents in similar positions normally have the authority to approve such contracts, Alleged's act would bind Perry. Note that Perry expressly limited Alleged's authority; that is, he had no actual authority to approve contracts. Perry would still be bound to the contract because by his act he created apparent authority in Alleged. Alleged's liability to Perry for entering into the agreement in violation of his instructions will be covered in Chapter 34. You should note that there was no way that the third party could know about the restraint upon Alleged in this example, and hence he is not bound by it. On the other hand, if the third party was aware of the limitations on Alleged's authority, then Perry would not be bound to the contract because the third party did not rely on the appearance of authority that Perry created for Alleged.

PAILET v. GUILLORY
315 So.2d 893 (La.App. 1975)

Rae Abramson and Ruth E. Pailet leased certain premises to Cenla Equipment Company, Inc., on November 25, 1970, for a five-year term. The compliance of Cenla was personally guaranteed in the lease by Dr. Richard Michel and Twyman Guillory. The lease was later assigned, with permission of the lessors, to Guillory. No rent was paid under the lease after December 1, 1973. Ruth Pailet then sued Michel and Guillory for her half of the rent due, $1,980.00. The trial court found for the plaintiff, and defendants appealed.

FRUGE, J.

Defendants admit that no rent was paid after November of 1973. They contend, however, that the lease was cancelled and they are under no obligation to pay any rentals.

All matters concerning the leased property was [sic] handled for the lessors by Dr. Albert Abramson, who was the husband of Rae Abramson at the time the lease was entered into. The record shows that the defendants contacted Dr. Abramson about leasing the property, that they paid rent checks to Dr. Abramson (made out to the Estate of Anne Elster), and that Dr. Abramson handled minor repairs on behalf of the lessors on his own authority. The defendants have never had any direct contact with the lessors concerning the property. Their only contact was through Dr. Abramson.

Dr. Abramson made it clear to the defendants that he did not own the leased property and that on certain matters he could act only with the permission of the lessors. For instance,

Dr. Abramson made it clear that he did not have authority to lease the premises. The lease was made only upon approval of the terms by the lessors and they signed the lease themselves. Also, the assignment of the lease from Cenla to Twyman Guillory was approved in writing by the lessors.

In March of 1973 Guillory approached Dr. Abramson and asked that the lease be cancelled. Dr. Abramson informed him that he would have to check with the lessors, Mrs. Abramson and Mrs. Pailet, as he did not have the authority to cancel. A few days later Dr. Abramson notified Guillory and Michel that the lessors had agreed to cancel the lease.

Mrs. Rae Abramson testified that she had been contacted by Dr. Abramson and had consented to the cancellation. The plaintiff, Mrs. Pailet, however, vigorously denied at trial that she had been contacted by Dr. Abramson and denied that she ever agreed to the cancellation. At trial Dr. Abramson testified that he did contact Mrs. Pailet but admitted that there may have been some "misunderstanding" regarding cancellation.

Defendants have appealed, contending that the trial court erred in holding that the plaintiff had not agreed to a cancellation of the lease and, in the alternative, in holding that plaintiff had not vested Dr. Abramson with implied or apparent authority to cancel the lease.

Appellants contend that even if Mrs. Pailet did not agree to cancellation of the lease, she is bound by the act of Dr. Abramson who was vested with the implied and the apparent authority to cancel the lease.

An agency relationship may be created through either express or implied authority. Apparent authority, on the other hand, creates no actual agency relationship. However, where the principal clothes an agent with apparent authority to perform certain acts and a third party who has no knowledge of or reason to believe that there are limitations on that authority, deals with the agent, then the principal is bound by the acts of the agent, which although beyond the actual power delegated to him, are within his apparent authority.

In the case before us, although there was never any express agency relationship between Mrs. Pailet and Dr. Abramson, he clearly had the authority to act as her agent in some matters concerning the leased property. Dr. Abramson was given the authority to collect rents, to make minor repairs, and to represent the lessors in preliminary negotiations regarding the lease.

However, the question before us is whether he had the implied authority to cancel the lease. Implied authority is actual authority which is inferred from the circumstances and nature of the agency itself. "An agent is vested with the implied authority to do all of those things necessary or incidental to the agency assignment." In this case Dr. Abramson's agency authority included collecting the rent and making minor repairs. Certainly the authority to cancel the lease is not incidental or necessary to his authority to collect rent and make minor repairs. No such authority can be inferred in these circumstances, particularly where the authority to lease or permit assignment was not given.

We turn now to the issue of apparent authority. The concern here is whether the principal did anything to clothe the agent with apparent authority to perform the act though no actual authority was given. Because third persons are not privy to the actual terms of the agency agreement, they may rely upon indicia of authority with which the agent is vested.

We do not find that the lessors in this case clothed Dr. Abramson with the apparent authority to cancel the lease. Dr. Abramson himself denied that he had any authority to do so. The lease agreement and permission for the assignment to Twyman Guillory were signed by the lessors and not by Dr. Abramson. Thus all the indications were that Dr. Abramson did not have authority to cancel.

Appellants rely on the fact that all of their communications with the lessors concerning the leased property were through Dr. Abramson. This alone is not enough. The fact that Dr. Abramson failed to secure cancellation of the lease from Mrs. Pailet is not imputable to her, but rather to the appellants since they relied on Dr. Abramson to secure cancellation.

(Affirmed.)

Questions for Discussion

1. *Was Dr. Abramson an actual agent of Mrs. Pailet? Were any of his actions taken as the result of implied authority? If they were, what actions?*
2. *What do you think was the most important factor in the court's conclusion that Dr. Abramson had no apparent authority?*

33:6. Termination. The agency relationship may be terminated in a number of ways. It may be terminated by some act of the parties, such as by mutual agreement or by accomplishment of the agency purpose. It may also be terminated by operation of law caused by the death of one of the parties or by the bankruptcy of a party. Furthermore it is important in this connection to note the difference between the *power* to terminate the agency relationship and the *right* to terminate the relationship. The parties generally have the *power,* except in the case of an agency coupled with an interest, which will be discussed in Section (2). In the termination of the relationship a contract may be breached, however, in which case it cannot be said that there is the right to terminate it. Finally, an additional problem presents itself in an agency situation. This is the question of whether notice to third parties is required when the agency relation is terminated.

(1) Termination by Act of the Parties. The agency relationship may be terminated by revocation or renunciation, by mutual agreement, by fulfillment of the purpose of the agency, or by expiration of a stated or reasonable time.

(a) *Mutual Agreement.* Frequently the agency is created by an agreement. The law is quite clear that two parties to an agreement can terminate it at will. Thus the principal and the agent may mutually agree to terminate the agency relationship. As a general rule, with the exception of an agency by estoppel or in the case of apparent authority, one cannot be forced to maintain the relationship against his will. Because it is voluntary, the parties may terminate it.

(b) *Fulfillment of Purpose or Expiration of Time.* If the agency is established to accomplish a particular purpose, the agency terminates when that purpose is accomplished. A common example is the instance in which a real-estate agent is hired to sell a house for his principal. Once the house is sold, the agency is terminated. The agent would not have the right subsequently to sell the house to another.

Many times an agency contract will contain a clause stating that the agency terminates with the expiration of a specific period of time. When this is the case, the agency terminates even though the purposes of the agency have not been accomplished. It is often wise to have such a clause in certain types of agency contracts. For example, in the case of the real-estate agent, if he is not successful in selling the house after a period of time, the owner–principal may wish to be able to have another agent try to sell the house without fear of breaching an existing agency contract. If no time for expiration is stated in the contract, then it terminates after the passage of a reasonable time. What constitutes a reasonable time is, of course, determined by the facts and the circumstances of each case.

(c) *Revocation and Renunciation.* As a general rule, the principal has the power to revoke the agency relationship at any time, with or without cause, and thereby terminate the authority of the agent to represent him. There are some exceptions, such as in the case of an agency coupled with an interest discussed in Section (2), but as a general rule the law cannot force an agency relationship upon unwilling participants. Thus, even if the revocation breaches a contract that characterizes the agency as irrevocable, the principal still has the *power* to revoke it. He may not have the *right* and, in revoking the agency, may breach a contract. The result would be liability to the agent for damages sustained as a consequence of the breach, but nonetheless the principal has the power to revoke. On the other hand, a principal may have the right to revoke the agency if the agent has substantially failed to perform his obligations as an agent.

The law is similar from the agent's point of view. Because personal-service contracts are not specifically enforceable, the agent has the power to renounce the agency at any time except in the case of an agency coupled with an interest. As with the principal, if the renunciation breaches a contract, the agent may be liable to the principal for damages. On the other hand, if the agent has good cause, he may renounce the agency at any time. The agent might have this right, for instance, if the principal fails to compensate him for his previous services.

(2) Agency Coupled with an Interest. There are circumstances in which the agent has the authority to perform a certain act that is to be exercised in conjunction with his interest in the subject matter of the agency. This is known as an *agency coupled with an interest.* The effect of an agency coupled with an interest is to take away both the right and the power of the principal to revoke the agency without the consent of the agent.

An example of an agency coupled with an interest may be seen in the case of a real-estate mortgage. Assume that Perry Principal borrows money from Agent Bank. The loan is secured by a mortgage on Perry's home that gives Agent Bank the right to sell Perry's home in the event of default. In this case, Agent Bank has an interest in the subject matter as it made the secured loan to Perry. Under the terms of the mortgage Agent Bank has been made Perry Principal's agent for the purpose of selling Perry's home in the event of default. Because the bank is an agent coupled with an interest in the subject matter, Perry Principal has neither the power nor the right to revoke the agency. Were the law otherwise, Agent Bank could lose its security interest by the simple act of revocation by Perry. If there is an agency coupled with an interest, the agency is not revocable by act of the principal or by the death, insanity, or bankruptcy of a party.

Several further points should be made. First, the agent must have an interest in the thing to be sold or managed under the powers. Generally the interest must be such as will allow the agent to exercise the authority in his own name and not in the principal's. If the agent cannot exercise the power in his own name, it is not technically an agency coupled with an interest, although some courts hold that it is only necessary that the beneficial interest be exercised in the agent's name. For instance, if in the previous example Agent Bank had the right to sell the house only in Perry Principal's name and not in its own name, then this would not be a true agency coupled with an interest. Most courts would hold that the agency would terminate at the death of one of the parties, although it would not be revocable during their life. This is called an *agency given as security.* Some courts, however, do not make this distinction and treat these two situations as being the same.

(3) Termination by Operation of Law. There are situations in which the agency relationship may terminate automatically by operation of law. Among these are the death of either party, insanity, and bankruptcy, as well as other factors.

(a) *Death of Principal or Agent.* The death of either the principal or the agent terminates the agency relationship. Although the obligations of a deceased person may be a charge against his estate because the act of the agent is legally the act of the principal, the death of the principal ordinarily terminates the relationship, with the aforementioned exception of an agency coupled with an interest. Even if the agency is established for a specific period of time or to accomplish a particular purpose, the premature death of the principal will terminate the relationship. Likewise the death of the agent terminates the agency relationship.

Sometimes a problem arises if third parties act in ignorance of the death of the principal. For example, when the third person makes payments to the agent after the principal's death, is the obligation discharged if the former agent absconds with the payment? The majority of the courts seem to hold that the third person suffers the loss even if he had no notice. Other jurisdictions hold that the third party is credited with the payment if he had no notice or reason to know of the principal's death.

(b) *Insanity of Principal and Agent.* The insanity of the principal terminates the agency relationship. However, if a third person deals with the agent in ignorance of the principal's insanity, then the relationship is not terminated as to him. However, if there has been an adjudication of insanity of the principal prior to the time when the agent deals with the third party, the third party is usually charged with the knowledge of the insanity. The insanity of the agent terminates the agency relationship. If there has been a judicial determination of insanity, the decree will operate as notice of insanity against all who deal with him.

(c) *Bankruptcy of Principal or Agent.* The bankruptcy of a principal usually operates to terminate the power of an agent to act with respect to the property of the principal. An exception, of course, is an agency coupled with an interest because the assets of the bankrupt are normally placed under the control of a bankruptcy judge, who may appoint a trustee or a receiver who will have control over the property. Where the agent goes bankrupt in a business related to the agency, his authority to act for the principal is terminated.

(4) Other Factors Terminating Agency. Several other factors may operate to terminate the agency relationship as a matter of law. The loss or destruction of the subject matter of the agency will terminate it. Thus the agreement with a real-estate agent to sell his principal's house would be terminated if the house were totally destroyed by fire.

A change in the law that makes the agent's required performance illegal would, of course, result in a termination of the agency relationship. Similarly, if an agent requires a certain license in order to perform his duty, the revocation of that license would result in the termination of the agency. Thus, if a real-estate agent loses his license, the relationship is terminated as a matter of law.

BREMER ASSOCIATES, INC. v. M. D. INDUSTRIES, INC.

466 F. Supp. 111 (E. D. Mo., 1979)

On March 1, 1973, Bremer Associates, Inc., entered into an agency contract with the predecessor of M. D. Industries, Inc. for the purpose of becoming the selling agent for M. D. Industries' products. On December 13, 1976, M. D. Industries canceled its contract with Bremer. Bremer filed suit for $175,000 alleging that M. D. Industries breached the agency agreement by cancelling it. M. D. Industries then filed a motion for summary judgment.

NANGLE, J.

The contract in question herein appointed plaintiff's predecessor as a selling agent of defendant in a specified territory representing defendant for specified product lines. The contract set out the territory and product lines, the commission, credit and collection, a commission quota, and operation agreements. The contract is silent as to its duration or termination date.

The law is clear that

> Agreements between principal and agent for an indefinite time generally may be terminated at the will of either party. *Want* v. *Century Supply Company,* 508 S.W.2d 515, 516 (Mo.App.1974).

Since the agreement is silent on the question of its duration, or the procedure for termination, the Court concludes that the agreement was terminable at will.

A limitation on the power to terminate exists, however, by which the courts require compensation to the agent

> . . . if it appears that the agent, induced by his appointment has in good faith incurred expense and devoted time and labor in the matter of the agency without having had a sufficient opportunity to recoup such from the undertaking. . . . *Want* v. *Century Supply Company, supra,* at 516.

Plaintiff has not alleged the same in its complaint. Moreover, in answers to interrogatories requesting that plaintiff detail the basis of its claims for damages, plaintiff states that its claim for damages in the amount of $175,000.00 is the result of commissions lost. Thus, plaintiff does not base any of its claims for damages upon expenses, time and labor incurred in connection with the agency agreement.

Under these circumstances, the Court concludes that defendant's motion for summary judgment should be granted.

Question for Discussion

1. *If the contract had provided that the agency was to run for a three-year period, would the outcome of this case have been different?*

SUMMARY

An agent is one who acts for and under the contract of a principal. An agent is distinguished from an independent contractor by the degree of control exercised over his work. An employer is not liable for the torts of an independent contractor. All employees are not agents.

One has the capacity to act as a principal if he has the capacity to act for himself. Generally, a principal cannot complain of the lack of capacity of his agent. One must have some degree of mental capacity to act as an agent however.

A general agent is empowered with all the authority that his principal has with regard to a particular business. A special agent is employed for a special purpose and has limited powers. A subagent is appointed by an agent. Factors and brokers are special agents. One may be a gratuitous agent.

The agency relationship may be created by contract, ratification, estoppel, by necessity, or operation of law. In order to ratify the act of his agent, the principal must have full knowledge of the facts at the time he ratifies the act of an agent. The ratification relates back to the time when the original act was performed. Agency by estoppel occurs when the purported principal does not prevent a person from representing himself as the purported principal's agent.

The authority may be actual or apparent. Actual authority may be express or implied. Apparent authority is not actual authority and consists of an act of the principal that results in a third person's reasonable belief that another person is acting with the authority of the principal.

A principal or agent generally has the power to terminate the agency relationship even though the termination may violate a right. An exception is if there is an agency coupled with an interest.

An agency may terminate by an act of the parties. This may occur by mutual agreement, the fulfillment of the agency purpose, expiration of time, or by revocation and renunciation.

The agency relationship may also terminate by operation of law. This may be caused by the death of the principal or agent, the insanity of the principal or agent, or the bankruptcy of the principal or agent.

PROBLEMS

1. Atlanta Syndicates, Inc. was the maker of a promissory note payable to Renbaum. Attached to the note was a surety agreement signed by a number of people guaranteeing payment of the note. One of the signatures to the surety agreement was Robert T. Klingfeil, M. D. by Clifford T. Klingfeil, attorney-in-fact. When Renbaum sued Robert T. Klingfeil on the agreement, Robert T. Klingfeil contended that Clifford had no authority to sign his name at the time the agreement was executed. The evidence supported this allegation. The evidence also showed that Robert took no steps to repudiate Clifford's act and that he made several payments on the note. Is Robert Klingfeil liable on the agreement? Klingfeil v. Renbaum, 246 S.E. 2d 698 (Ga. App. 1978).
2. Davis Chipman was an employee of Robert Barrickman, a service-station owner, and drove his own truck on an authorized service run for Barrickman. Chipman gave Kim Lee Estell permission to ride with him in order to keep him company. The truck was involved in an accident for which Chipman was at fault and Estell was injured. Barrickman contends that

he is not liable to Estell since Chipman had no authority to invite Estell to ride with him. There was evidence that Barrickman had seen Estell and others ride with Chipman on occasion. Assuming Chipman did not have actual authority to take riders, is Barrickman liable to Estell? Estell v. Barrickman, 571 S.W.2d 650 (Kent. App. 1978).

3. Bonnie Eugene Coffin called the home office of Farm Bureau Mutual Insurance Company and was connected with Dick Pierson. When he told Pierson that he wanted to transfer liability and property-damage insurance on his 1953 Chevrolet to a new 1956 Buick and get additional collision and comprehensive insurance, Pierson replied, "O.K. you are covered." Pierson was not an insurance agent but a typing supervisor. The next day Coffin had an accident and filed a claim with Farm Bureau, which was denied. Was Coffin entitled to recover? Farm Bureau Mutual Insurance Co. v. Coffin, 186 N.E.2d 180 (Ind. 1962).
4. Eleanor Beard engaged women to make quilts. The women did the work in their homes within a twenty-five-mile radius of Beard's place of business. Beard supplied materials stamped with designs and specifications. Beard delivered the materials, thread, and specifications when a contract was signed between the individual workers and Beard. Under the terms of the contract the individual workers sewed the quilts according to the specifications. The work was performed at a time and place satisfactory to the worker. When the work was completed, it was to be delivered to Beard and a price paid to the worker, who was responsible for any damage to the materials. There was no supervision of the work, and Beard never called at the workers' homes. Were the workers Beard's agents? Glenn, Collector of Internal Revenue v. Beard, 141 F.2d 376 (1944).
5. Zummach was a manufacturer and dealer in storefronts, paints, and oils. James Biersach was his employee and called upon contractors and architects to promote sales. Biersach did some supervision of installation of storefronts but had no authority to make collections on behalf of Zummach except in the case of overdue accounts. Polasek was a contractor and began doing business with Zummach because of the solicitations of Biersach. When Polasek first went to Zummach's place of business, he was told that Biersach would take care of him, and all his subsequent dealings were with Biersach. Biersach also had authority to sign waivers of mechanic's liens, and he did sign some waivers for Polasek. After a period of time Biersach began to defraud Zummach. Zummach sued Polasek to recover the value of goods delivered to Polasek, for which Polasek had paid Biersach. Zummach claimed that Biersach had no authority to receive such checks. Decide. Zummach v. Polasek, 199 Wis. 529, 227 N.W.33 (1929).
6. Hugh B. Howland called at a theater owned by Tri-State Theaters Corporation, where his brother was employed as a stagehand. As part of his duties in closing the theater for the night the brother had to lower and then raise an asbestos fire curtain twelve feet above the stage door. The mechanism failed to work, and Hugh Howland was asked by his brother and

another employee to help. In the process Hugh was injured and sought to recover from Tri-State, contending that he was hired to raise the curtain by Tri-State's employees under conditions of an emergency. Was Hugh entitled to collect? Howland v. Tri-State Theaters Corp., 139 F.2d 560 (1944).

7. William Mubi sued Walter and Jane Tribble for personal injuries and property damage resulting from an automobile accident. Their attorney filed an offer to settle through the court. The rules of the court required acceptance within ten days or the offer was deemed withdrawn. William Mubi advised his wife to accept the offer on his behalf. She informed William's attorney, but before the acceptance was communicated to the court, William died. His death occurred prior to the expiration of the ten-day period, and his attorney then did accept the offer within that period. Was William Mubi's estate entitled to collect based on the offer of settlement? Mubi v. Broomfield, 108 Ariz. 39, 492 P.2d 700 (1972).
8. Koos Bros. was a large furniture store located in Rahway and other parts of New Jersey. Mrs. Hoddeson went to Koos to purchase some furniture. When she arrived in the store, she was greeted by a man who took her order and cash payment. A large number of salesmen were employed by Koos and were paid on commission. Unfortunately the person to whom Mrs. Hoddeson paid the money was an impostor. She never received her furniture, nor did she get a receipt for her cash payment. Was she entitled to recover from Koos? Hoddeson v. Koos Bros., 47 N.J.Super. 224, 135 A.2d 702 (1957).

CHAPTER 34

Duties and Liabilities of Agent and Principal

In the preceding chapter we examined the formation of the agency relationship. In this chapter and Chapters 35 and 36 the liabilities and duties of the parties to each other in this triangular relationship will be examined.

34:1. Duties and Liabilities of Agent to Principal. (1) Duty of Loyalty: The Agent as a Fiduciary. The relationship between the principal and the agent is primarily by contract, which, of course, may by its terms impose certain duties on each party. Beyond those terms, however, there is a fiduciary relationship between an agent and a principal. This relationship requires the agent to act with the highest degree of loyalty toward his principal in handling the principal's affairs. In general the principal is entitled to place the highest degree of trust and confidence in his agent, and the agent may not take any action that would adversely affect the best interests of his principal. This general duty of loyalty that results from the fiduciary relationship between the agent and principal can be further examined with regard to a number of specific types of behavior.

(a) *Dual Agency.* As a general rule, an agent cannot represent two principals in the same transaction. To do so would breach the agent's duty of loyalty, which requires the best efforts of the agent in the furtherance of the interests and welfare of the principal. Thus, if the interests of the two principals are adverse, to the extent that the agent advances the best interests of one principal, he cannot be acting in the best interests of the other. If the agent acts in a dual role without the knowledge of the principal, the transaction entered into is voidable at the option of that principal.

There is an exception to the prohibition of a dual agency. The exception obtains where the principals have full knowledge of the dual role and consent to it. When the agent acts in such a dual capacity, the duty of loyalty extends to both principals. If the positions of the principals are truly adverse, the agent is placed in a difficult position and he must walk a narrow line. For instance, on occasion a married couple desirous of a separation or a divorce will seek the services of a single attorney. Many attorneys are reluctant to represent both parties because of their adverse relationship in our legal system. This reluctance is caused by the fact that, as an agent, the attorney must represent each party as a fiduciary and with the highest degree of loyalty.

If the interests of the principals are not conflicting so that the loyalty of the agent to one of them does not constitute a breach of duty to the other, then an agent may represent the two principals. The courts have held that the relationship of buyer and seller, however, is adverse.

TABORSKY v. MATHEWS

121 So.2d 61 (Fla.App. 1960)

SHANNON, J.

This is an appeal taken by the defendants in a foreclosure suit from a summary final decree based upon the order of the chancellor below striking the defendants' affirmative defense and dismissing their counterclaim.

The defense and the counterclaim attempted to raise the fact that the real-estate agent who acted for the defendants was also the agent for the plaintiff in this transaction; that the agent had received a commission from both parties; and that the defendants had no knowledge of the dual agency. By their counterclaim the defendants sought to recover the portion of the purchase price which had been paid.

In this appeal the defendants have raised three points. Inasmuch as the first point comprehends all that is necessary to support our opinion, the only question which we will consider is:

In an action to foreclose a purchase-money mortgage upon certain real property where a real-estate broker without disclosing the dual nature of his agency to the purchasers, acted as agent for both parties in negotiating the sale of the property, have the purchasers the right, upon discovering the dual agency, to avoid the sale and purchase-money mortgage?

In our jurisprudence it is well established that an agent for one party to a transaction cannot act for the other party without the consent of both principals. Where an agent assumes to act in such a dual capacity without such assent, the transaction is voidable as a matter of public policy. Florida has unequivocally aligned itself with this principle by the case of *Quest* v. *Barge*, Fla. 1949, 41 So.2d 158, saying at page 164:

Perhaps the best statement of the law applicable to the inquiry at bar is that found in *Evans* v. *Brown*. . . .

"No principle is better settled than that a man cannot be the agent of both the seller and the buyer in the same transaction, without the intelligent consent of both. Loyalty to his trust is the most important duty which the agent owes to his principal. Reliance upon his integrity, fidelity, and ability is the main consideration in the selection of agents; and so careful is the law in guarding this fiduciary relation that it will not allow an agent to act for himself and his principal, nor to act for two principals on opposite sides in the same transaction. . . ."

It is evident from all these authorities that in cases of double agency the relief granted to the principal against the agent, or against the third party who has compromised the agent, is not made to depend upon the intention to defraud, the presence of actual misrepresentation or non-disclosure, or the presence of injury.

The law, in the words of Judge Cardozo, ". . . stops the inquiry when the relation is disclosed, and sets aside the transaction or refuses to enforce it . . . without undertaking to deal with the question of abstract justice in the particular case."

The law being well established, this court need only determine whether the allegations of the defendants' affirmative defense and counterclaim set forth facts sufficient to raise the issue of dual agency. Essentially, the defendants allege the dual agency, and the fact they were not aware of the agent's status with the plaintiff, and we find that this is sufficient to bring the rule in *Quest* v. *Barge,* supra, into play. Hence, the order of the chancellor below striking the defendants' affirmative defense and dismissing their counterclaim was in error, therefore, the final summary decree, which was based upon this order, is also in error.

(b) *Adverse Interests.* Because of his duty of loyalty to the principal, an agent cannot possess or represent any interests that are adverse to his principal's without full disclosure to and consent of the principal. If the agent stands to benefit in any way from the transaction in which he represents the principal, the principal must be fully informed and consent. This is true even if the agent receives gratuities from the third party with whom the agent is dealing. The classic case, of course, is the purchasing agent of an employer who receives gifts from a seller. Because the acceptance of gifts could affect his judgment, the agent must make full disclosure to his principal and obtain the principal's consent to the agent's acceptance of gifts.

It should be obvious by now that an agent may not enter into a business that is competitive with that of his principal's without the full knowledge and full consent of the principal. This rule extends to the situation in which the agent's business is not directly competitive with the principal's but nevertheless is of a type that will encroach on the time that the agent is to spend on his principal's business. For instance, suppose one works for an accounting firm and then begins to prepare tax returns on his own. This would violate the duty of loyalty unless the firm consents to this activity with full knowledge of the facts.

(c) *Appropriating Principal's "Secrets" and Customers.* The duty of loyalty requires that the agent not appropriate the principal's property. The property of the principal includes information of a confidential character, such as customer lists, or "trade secrets," such as special processes or formulas. The problem frequently arises when the agent leaves the employment of the principal in order to work for a competitor or to establish a competing business himself, using ideas learned while he was in the service of the principal. If the agent seeks to use the trade secrets in such a manner, his former employer may obtain an injunction prohibiting such use. In order to gain such relief, however, the employer must show that the trade secrets are his particular secrets and not merely general secrets of his trade or profession. The trade secrets need not amount to a patentable invention in order to be protected, but they must amount to discovery. On the other hand, the agent should not be prohibited from using knowledge gained from his experience in a new job or for himself as long as that knowledge is not of a peculiarly secret or confidential character.

The dividing line between a trade secret and important knowledge that does not amount to a trade secret is difficult to draw. The fact that an agent's use of knowledge acquired from his former employer causes the employer injury is not of itself determinative. For instance, there is nothing to prevent an agent who establishes a new business from contacting the customers of his former employer whom he came to know during the course of that employment. The employer can protect himself against this possibility by placing a clause in the employment contract reasonably restricting the employee's right to compete for a period of time after the relationship is terminated. This sort of restriction is particularly advisable if the primary asset of the employer is an intangible, such as a customer list.

(d) *Principal's Remedy.* If the agent breaches the duty of loyalty, he forfeits any profits that he has gained as a result of the actions that amount to the breach. He also forfeits the right to receive compensation from the principal and may be liable to the principal for any damages sustained by the principal as a result of the agent's wrongful act. Breach of the duty of loyalty by the agent also gives the principal the right to rescind the relationship. Suppose, for example, that a real-estate agent represents Sam Seller in the sale of Seller's house. The agent brings to see Seller, Bill Buyer, who is a friend of the agent, unknown to Seller. The home is sold to Buyer for $40,000, but at the time, the agent knew he could sell it to someone else for $50,000. Buyer purchases the home for the agent to resell. The agent then resells the home for $50,000. The agent would be liable to Seller for the profits gained above $40,000 and would forfeit his commission on the original sale. Seller could also terminate the agency relationship.

(2) **Duty to Obey Instructions.** The agent must obey the instructions given to him by the principal. If he fails to obey those instructions, he is liable to the principal for any damages that may result from his failure to obey. For example, suppose the principal instructs the agent to sell certain goods for cash. The agent finds a likely looking customer and sells the goods to him on credit at a somewhat higher price than might otherwise be obtained for the goods. The agent's judgment turns out to be faulty and the debtor never pays the principal. The agent would bc liable to the principal for failure to follow instructions. He may not use his discretion in the matter, and the fact that a somewhat better price was obtained is no excuse. On the other hand, if the principal has vested the agent with the power of discretion, then the agent may exercise that discretion. The agent would not be liable for any loss to the principal unless he was negligent. Furthermore, if the agent does violate the instructions of the principal, he is not liable if the principal, with knowledge of the facts, subsequently ratifies the act.

Not all of the instructions of the principal may be express. They may also be implied from the task assigned to the agent, from the circumstances of the case, from the previous course of dealing between the parties, or from the usages and customs of the trade. However, if the principal's express instructions run counter to the customary practice of the trade, then the express instructions must be followed.

If the instructions of the principal are vague or ambiguous, the agent is not liable to the principal if he makes an honest mistake in his interpretation of those instructions. On the other hand, the agent may not disobey his principal's instructions and substitute his own judgment even if the instructions are ambiguous. He must apply an interpretation that is reasonably derived from the instructions.

The duty to obey instructions extends even to a gratuitous agent who begins to carry out the business of the principal. On the other hand, if the principal gives instructions that require the performance of an illegal or immoral act, the agent will not be held liable for his failure to perform those acts. Furthermore, because the agent has a duty to act in the principal's best interest, he may deviate from the principal's instructions in the case of a sudden emergency that requires such a deviation to protect the principal. In fact, the agent would have a duty to deviate from the previous instructions under these circumstances. For example, the agent would be justified in ignoring previous instructions of a principal in order to save the principal's property from perishing.

(3) **Duty to Give Notice of Material Facts.** The agent has a duty to notify the principal of all the material facts concerning the transactions and the subject matter of the agency. The reason for this rule is that as a general rule the principal is chargeable with the knowledge of his agent. The rule most commonly stated is that the knowledge of the agent is "imputed to the principal." Because the principal is chargeable with this knowledge, the agent has a duty to inform the principal of the facts so that he may act upon them if necessary. The rule may seem unduly harsh in instances in which a principal may suffer some harm because of the agent's failure to notify him, but the rule may be explained on the basis of risk allocation. The third party has no control over the relationship, so the law places the risk on the principal and imputes the agent's knowledge to him. Another theory for explaining the rule is that there is legal identity of principal and agent; hence the knowledge of the agent is considered possessed by the principal. The risk-allocation explanation, however, seems more reasonable.

One question that frequently arises concerns the treatment of knowledge obtained by the agent prior to the time of the creation of the principal–agent relationship. The courts have taken the position that knowledge acquired prior to the agency relationship is imputed to the principal if the agent remembered it or "had it in his mind" after the creation of the relationship, when he should have communicated it. It is exceedingly difficult to prove what one "has in his mind" other than to ask him, so this test is somewhat difficult in application. Although there is some conflict, the courts seem to imply that they will presume that the knowledge was "in the agent's mind" if the circumstances are such that any other conclusion would be unbelievable.

The problem of the imputation of the agent's knowledge to the principal frequently arises in connection with the issue of whether the principal was notified of certain facts. For instance, in order to qualify as a bona fide purchaser of property, one must be ignorant of another's interest in that property. Thus, if the principal purchases certain property upon which a third person has an unrecorded mortgage, the principal takes the property free of the mortgage unless he is aware of its existence. If the agent has been notified of the existence of the mortgage, that knowledge is imputed to the principal whether he actually is aware of it or not.

If the agent does not inform the principal of certain facts in order to further his own personal interest or for his own gain, then the knowledge is not imputed to the principal. The same rule applies if the agent and the third person conspire to keep certain information from the principal.

A further question arises if the third party gives the agent notification or information at times other than in normal business hours. Suppose, for example, that the agent is given certain information by the third party at a cocktail party, unrelated to the business of the principal, that is, a purely social event. Is this knowledge imputed to the principal? The courts seem to take the position that if the notice is given in a reasonable manner, the agent has a duty to transmit it to the principal, and hence such knowledge is imputed to him. The fact that notification or information is transmitted at times other than in normal business hours does not of itself absolve the agent of his duty to notify his principal. On the other hand, if information is transmitted to the agent under circumstances in which it is not expected that the agent will inform the principal, then this knowledge is not imputed to the principal.

STATE FARM FIRE AND CASUALTY CO. v. SEVIER et al.

537 P.2d 88 (Or.Sup.Ct. In Banc, 1975)

Kenneth Sevier purchased an automobile and went to see Jack Henderson, an agent for State Farm Fire and Casualty Co., on May 19, 1970, to apply for insurance. The agent was in Arkansas. The application for insurance included a question asking whether the applicant had been convicted for traffic violations in the last five years or if the applicant's license had ever been suspended. Sevier testified that he was not asked those questions, that he had known Henderson for a long time, and that he told Henderson he had been picked up and convicted of drunken driving in Oregon, which resulted in his license's being suspended. The written application, filled out by Henderson, showed "no" as an answer to both questions. Henderson contended that he was never informed of the drunken-driving conviction or the suspension of the license.

The application for insurance was subsequently approved. Sevier then moved to Oregon and reapplied for a policy in Oregon as required by State Farm. The answer to the two questions concerning license suspension and drunken driving were reflected on the application as "no." Sevier testified that he did not recall being asked those questions. The trial court determined that he was asked those questions and answered them "no." A new policy was issued on January 7, 1971, for six months. On March 15, 1971, Sevier was involved in an accident in which two people were killed. On July 7, 1971, State Farm attempted to "rescind" the policy and mailed a check to Sevier for all premiums paid. Sevier refused to accept the check. State Farm contended that the policy was issued because of Sevier's misrepresentations. State Farm then filed suit for a declaration of rights under the policy, contending that it had "validly rescinded" the policy. Sevier and the personal representatives of the estates of the deceased persons opposed State Farm. The trial court found against State Farm.

TONGUE, J.

The general rules of contract and agency law are well established to the effect that notice to an agent is notice of grounds for the recission of a contract and who elects to rescind it must do so promptly or lose his right to rescind; that when a contract has been induced by misrepresentation by one party, so that the other party has an election whether to affirm the contract or to rescind it and return any payments received, one who retains the payments received and acts in a manner inconsistent with an intent to rescind the contract cannot later seek to rescind the contract; and that one who rescinds a contract must ordinarily rescind the entire contract and cannot, at the same time, recognize the continued existence and enforceability of a portion of the contract.

Consistent with these established rules, it has been held by other courts that an insurance company which has knowledge through one of its agents of the falsity of facts stated in an application for insurance and which nevertheless issues an insurance policy is either "estopped" from rescinding the policy based upon the alleged misrepresentation of such facts or cannot then establish that it acted in reliance upon such misrepresentations, as necessary for the rescission of a contract for misrepresentation.

In this case plaintiff, through its agent Henderson, acquired knowledge of the fact that Sevier had at least been stopped by the police for drinking and driving and had been given a "balloon" or "breathalyzer" test. Indeed, the trial court found that plaintiff's agent had knowledge of the fact that Sevier had been convicted for DUIL. As a corporation, plaintiff could act only through its agents and employees and once it acquired knowledge of that fact through one of its agents or employees plaintiff was chargeable with continued knowledge of that fact, regardless of whether or not its agent Henderson failed to make a written record of that fact or to convey that information to other agents or employees of the plaintiff.

It follows from this testimony, in our opinion, that at the time that both the Arkansas policy and the Oregon "renewal" policy were issued, plaintiff had knowledge of facts which constituted sufficient grounds for the rescission of both policies. It also follows, in our opinion, that if plaintiff desired to elect to rescind the Oregon policy it was required to do so promptly, and that in this case plaintiff did not undertake to rescind promptly, but did not do so until July 7, 1971, the date on which the Oregon policy expired by its terms. We hold that it was then too late to do so.

Plaintiff vigorously contends that the knowledge of its agent of Sevier's DUIL conviction was not imputed to it because of alleged collusion between its agent and Sevier to withhold knowledge of that fact from it.

Plaintiff's claim of collusion is based upon testimony that Sevier was given an opportunity to read the application before signing it and knew that the answers in the application, as filled in by plaintiff's agent, were false and upon the further testimony that, according to Sevier, when he told plaintiff's agent of his DUIL conviction the agent said "to Hell with it, maybe they'll never find it out."

The evidence was conflicting on both of these points. Plaintiff's agent denied that he said "to Hell with it, maybe they'll never find it out." He also testified that he was not sure whether or not Sevier read the application before signing it. At the time of his deposition, Sevier said that he knew that the agent wrote down false answers in the application before he signed it, but on trial he testified that he did not know what answers the agent wrote down. No findings were made by the trial court on these issues.

There may be good reasons why, in litigation between an insurance company and an insured who has acted in collusion with an agent of the company in concealing facts from it, there should be an exception to the general rule that knowledge to the agent of the company is to be imputed to it. The reason for that result in such cases is not so much that the knowledge of the agent was not imputed to the principal, but that a participant in a fraud should not be permitted to profit from his own fraud. Such reasons, however, have no proper application in litigation between an automobile insurance company and an innocent person who was injured by the negligence of the insured. The cases cited by plaintiff, all from other jurisdictions, either involve litigation between the insurer and insured or do not discuss the question whether a different rule should apply in cases involving innocent third parties.

The rule that knowledge of an agent is to be imputed to his principal, regardless of actual knowledge by the principal, is a rule based upon considerations of public policy to the effect that one who selects an agent and delegates authority to him should incur the risks of the agent's infidelity or want of diligence rather than innocent third persons.

Those reasons of public policy are not present in cases involving litigation between a principal and a third party who has acted in collusion with the agent. When, however, as in this case, the injured party is one who did not act in collusion with the agent, those same reasons of public policy are present.

(Judgment affirmed.)

Questions for Discussion

1. *What reason did the court give for refusing to allow State Farm to rescind the policy?*
2. *Would the decision have been the same if a third party other than State Farm and Sevier had not been involved?*

(4) Duty to Use Care and Skill. An agent has a duty to carry out his obligations with ordinary and reasonable care, skill, and diligence. If the agent performs these duties negligently, he is liable to his principal for any loss sustained as a result of such negligence. This is not a novel concept, for as we have seen, persons who commit negligent acts are liable for the damages that are proximately caused by those acts. The agent is not liable, therefore, by virtue of the position he holds but because of the negligence of his acts.

The agent does not guarantee the success of his actions on behalf of his principal. He is not an insurer. The agent is liable only when the loss is the result of some negligence on his part. If he is not negligent, there is no liability. Furthermore, if the loss is the result of the failure of the principal to protect himself from a loss when he is able to do so, the agent is not liable despite his negligence.

Normally the agent is liable only for the failure to use ordinary skill in carrying out his duties. However, if he professes to have special skills, he will be liable if he causes damage to his principal by failure to perform up to the standards required of one who possesses those skills. He is judged according to the standards applicable for those who possess these skills. Thus an orthopedic surgeon would be held to a higher standard in treating a compound fracture than would a general practitioner.

Chapter 35 will treat the subject of the principal's liability for injuries suffered by third parties as the result of the negligence of the agent. For now it is enough to know that a principal is liable for the torts of his agent when they are committed within the scope of his employment. When the principal suffers a loss as the result of the negligence of his agent, he may recover from the agent.

(5) Duty to Account. One of the more important duties of an agent is to account for any property or money that he holds for his principal or that in any way comes into his possession. In connection with this duty the agent is obligated to maintain true and accurate accounts of transactions. He should be able to render an accounting to his principal upon demand. Failure to maintain proper accounts raises a presumption that the agent has been unfaithful to the principal. Thus, if one is hired to manage apartments for his principal, he would have to keep accurate records of all sums collected and expended and turn over all monies due the principal upon demand or when he is otherwise obligated to do so.

A well-established rule is that an agent may not *commingle* the principal's funds or property with his own. If money is involved, for example, the agent may not deposit the principal's funds in his own account even though he has the best of intentions and notes the amount due the principal in his records. Failure to heed this rule may have grave consequences for the agent. Any losses experienced by the principal when the funds are commingled must be borne by the agent. If the agent deposits the principal's funds with a bank in a separate trust account and the bank then becomes insolvent, the agent will not be responsible for the loss. However, if the agent deposits the money in his own name, the agent is liable for any loss suffered by the principal.

The agent must account to the principal for any profits over and above his compensation. In addition, the agent may not use the subject matter of the agency in order to earn profits on his own behalf. If he does so, those profits must also be turned over to the principal. For instance, suppose the principal owns a stable. An agent is in charge of renting horses to the public from 9:00 A.M. to 5:00 P.M. On occasion the agent keeps the stable open beyond 5:00 to 6:00 P.M. and pockets the profits made during the extra hour. Those profits belong to the principal. A court would hold that the agent held these extra profits in trust for the principal and the agent would have to turn them over to him. Likewise, if the agent disposes of his principal's property in violation of his duty, the principal may trace the proceeds from such a misappropriation and impress a trust upon them.

If an agent fails to account to the principal after a demand is made or fails to turn over funds for which he should have accounted, the agent is liable, in addition to such funds, for interest at the lawful rate, calculated from the time when the accounting should have been made until the funds are turned over to the principal. If the agent has invested the funds at a rate of interest higher than the legal rate, the principal is entitled to all the interest received by the agent.

WESTINGHOUSE ELECTRIC CORP. v. LYONS et al.

125 N.Y.S.2d 420 (1953)

IRVING, J.

The plaintiff is seeking a decree requiring defendants to account for moneys collected from the sale of merchandise consigned to them for sale.

Beginning July 1, 1944, and annually thereafter to July 1, 1951, by virture of duly executed contracts, the defendants were designated by the plaintiff as their agents for the distribution of lamps. Monthly reports were to be submitted to the plaintiff setting forth direct sales to consumers and those also to "served agents," meaning smaller agents of the plaintiff, and in the latter instances defendants were given additional compensation ranging from five to six percent. The plaintiff now claims that the defendants by falsely reporting many of their "direct sales" as sales to "served agents" obtained for themselves very substantial allowances to which they were not entitled.

The plaintiff urges that the factors essential for the relief sought appear to be present: first, a fiduciary relationship between principal and agent; second, complicated and long accounts, covering an extended period of time; third, admitted collection of moneys by defendants, during the months of November and December, 1951, for which they have not accounted; and fourth, conceded falsifications in the reports submitted to the plaintiff from the very inception of the relationship, all without the knowledge or consent of the plaintiff.

The defendants maintain, however, that the plaintiff is not entitled to an accounting, contending that . . . (2) the acceptance of the defendants' monthly statements with knowledge of their falsity constituted a series of accounts stated.

As to the second defense, the plaintiff could not be bound by an account stated unless it had full knowledge or be [sic] offered an opportunity to gain such knowledge of all the true facts and circumstances.

It is well settled that a defrauded party with no knowledge of the facts is not bound simply by the acceptance of an account stated. Especially so is this true where a fiduciary relationship exists, as in the instant case, and this court will not lend its aid to enforce the alleged defense under such circumstances.

The defendants boldly admitted not only that a special, and apparently secret, consignment ledger was kept but also that their monthly reports were false and to such an extent that even they themselves were unfamiliar with all of the details. Those admissions cast a very severe and obvious pall over the defendants' position.

The evidence adduced reveals that it was not until late in 1950, or the early part of 1951, that the plaintiff was first put on notice that one or more of its agents were operating improperly, following with it forthwith secured undeniable knowledge to this effect by secretly marking lamps, which were traced subsequently to the doorstep of the defendants.

It may very well be that a full and thorough audit at an earlier date might have disclosed the fraudulent practices sooner, but the audits, prior to 1951, appear to have been for accountability and inventory reconciliation only and the failure, therefore, to uncover the defendants' course of conduct is understandable.

In view of all of the foregoing I have concluded that an accounting, as prayed for, is warranted and justified and accordingly I so decree.

QUESTIONS FOR DISCUSSION

1. *What was the purpose of the defendant, Lynn, in admitting the existence of a special and secret consignment ledger?*
2. *What was the purpose of the principal in seeking an accounting?*

H–B LTD. PARTNERSHIP v. WIMMER
257 S.E.2d 770 (Va. 1979)

In November 1976, Cheryl Switzer and George Vincent, as general partners of H–B Ltd. Partnership, entered into an oral agreement with Edgar Wimmer authorizing Wimmer to act as their agent in obtaining a sales contract for the purchase of commercial land in Stafford, Virginia.

A suitable parcel of land was discovered by Vincent, and Wimmer agreed to look up the owner's name and address and ascertain whether the property could be purchased. In January 1977, Wimmer advised Switzer and Vincent that he had met with the owner, who lived in Washington, D.C., and that she was asking $60,000.00 for the property. He also told them his commission would be $5,000.00, making the total price $65,000.00. Switzer and Vincent agreed to purchase the property and instructed Wimmer to draw up a contract.

In February 1977, Wimmer told Switzer that he feared the owner would change her mind about the sale of the property if he had to take the contract back to Stafford, Virginia, to get Switzer and Vincent to sign it; consequently he signed it himself as purchaser for $60,000.00. He told Vincent and Switzer he would assign it to them for $60,000.00 plus the $5,000.00 commission for acting as their agent.

Switzer and Vincent then discovered that a deed conveying the property in question had been recorded in the Clerk's Office of Stafford County. The deed indicated that the property had been sold to Wimmer for $36,000.00. Switzer and Vincent sued Wimmer to obtain the property from Wimmer on the theory that, because of his fraud, he held the property for them as constructive trustee. Wimmer contended that the suit should be dismissed because there was no written agency agreement between the parties and the agency contract was unenforceable under the statute of frauds. A Virginia statute provided that contracts for services by a real-estate agent had to be in writing in order to be enforceable.

I'ANSON, J.

An agent is a fiduciary with respect to the matters within the scope of his agency. A fiduciary relationship exists in all cases when special confidence has been reposed in one who in equity and good conscience is bound to act in good faith and with due regard for the interests of the one reposing the confidence.

When an agent is employed on an oral agreement to purchase real property for his principal, and buys the property with his own funds and takes a conveyance to himself, thereby violating his fiduciary relationship with his principal, he will be deemed in equity to hold the title thereto as a constructive trustee for the principal.

A constructive trust may be invoked even though the agreement out of which it arose is not legally enforceable. The invoking of a trust is not enforcing the contract but is invoking equitable relief from a fraud or breach of confidence. A trust arises by virtue of the relationship, and the cases do not require that the agency relationship be a legally enforceable one. Hence, a constructive trust may be proved by oral testimony and the statute of frauds is not a bar to the prosecution of such cases.

In the present case, the uncontradicted evidence shows that Wimmer was employed by Switzer and Vincent as their agent to ascertain the owner of the property in question and to negotiate on their behalf for its purchase. Wimmer fraudulently and in breach of faith and confidence reposed in him by his principals had the property conveyed to himself. By misleading Switzer and Vincent into believing that the sales price was $60,000, Wimmer would have made a secret profit of $24,000 for himself in addition to a real estate agency commission of $5,000. The unconscionable conduct of Wimmer was a breach of the confidence and faith reposed in him by his principals. Hence, under the uncontradicted evidence, a constructive trust was created by operation of law and Wimmer held the property as trustee for Switzer and Vincent. The object of this suit was to establish a constructive trust, not to enforce a contract.

(Lower-court decision reversed in favor of Switzer and Vincent.)

Questions for Discussion

1. *What price would Switzer and Vincent have to pay for the property under this decision? Would Wimmer be entitled to his commission?*
2. *Does this case mean that a real-estate agent cannot purchase property and then sell it to his principal under any circumstances?*

34:2. Duties and Liabilities of Principal to Agent. The primary duties of the principal to the agent are (1) to adhere to the terms of the agency contract; (2) to compensate the agent for his services; (3) to reimburse the agent for his expenditures on behalf of the principal; and (4) to indemnify the agent for losses sustained in the performance of his duty.

(1) Duty to Comply with Contract Terms. If a contract between the principal and agent spells out the terms of employment, the principal must adhere to those terms or he will be liable to the agent for damages, just as with any other contract. For instance, if the contract spells out a specific time for the existence of the relationship, a premature termination of the contract would result in the liability of the principal for any damage suffered by the agent. Contracts between real-estate agents and home owners typically spell out the length of time the relationship will continue. Premature termination of the contract by the home owner would allow the agent to recover any damages suffered, such as a lost commission if the home were sold during the contractual period.

As with any other contractual relationship the principal may be justified in refusing to perform if the agent has breached the contract. He may not, however, do anything to prevent the agent from effectively performing the objectives of the contract.

MONTGOMERY WARD, INC. v. TACKETT

323 N.E.2d 242 (Ind.App. 1975)

Thomas Tackett and Cassandra Tackett entered into a franchise agreement with Montgomery Ward that provided that the Tacketts would purchase merchandise from Ward, resell it to customers, and receive commissions from Ward. The agreement required that the Tacketts at all times operate the agency in accordance with the current policies and procedures of Montgomery Ward.

Ward unilaterally terminated the agreement in June of 1971, contending that the Tacketts submitted improper inventory clearance adjustments. These were claims for credit for merchandise previously ordered and paid for but not received. The Tacketts contended that Ward terminated the agreement without notice and without any attempt to consult, advise, or negotiate with the Tacketts and thus acted in bad faith. Ward sued the Tacketts for money due. They counterclaimed, seeking actual and exemplary damages. The lower court found for Ward in the amount of $8,000 and for the Tacketts in the amount of $11,000. Ward appealed.

LYBROOK, J.

The evidence reveals that the relationship between the parties was fraught with difficulty and misunderstanding from its inception. However, the core of the conflict revolved around the system of payments and credits for merchandise ordered by Tackett for sale to customers. Testimony revealed that the policy in effect during 1971 was that agents were to prepare a weekly report and remit payment for shipments of merchandise received during that week. An error in shipment such as damage to goods or a failure to receive certain items did not relieve the agent of liability for payment. The agent's recourse in the event of such an error was to file an ICA claiming credit. Tackett testified that many of the shipment errors were substantial, and that while the policy of Ward as explained to him was to act upon and return ICA's within ten days, his claims were in many instances delayed for months. He further testified that Ward refused to honor many of his valid claims for credit.

In the event that an agent eventually received an item for which he had previously claimed credit on an ICA, the proper procedure was to file an RNC through which the credit would be charged back to the agent. Tackett testified that he at times withheld RNC's due to problems he was experiencing with Ward in their handling of his ICA's claiming credit. A second reason given by Tackett was the failure of Ward to render certain aid and assistance which had been promised when he entered into the franchise agreement. Ward representatives had provided Tackett with a list of telephone numbers and assured him that he might make collect calls to solicit assistance with problems encountered in operating the store. However, Tackett experienced difficulty contacting proper persons and Ward required that he pay for the calls.

Tackett requested auditor assistance from Ward on several occasions, including one request as early as 1970. One reason for the requests was that the store, which had been opened several years prior to its acquisition by the Tacketts, had never undergone an audit, and Tackett desired to rid himself of extensive records which had accumulated. Secondly, Tackett desired the audit as a means through which to settle his financial difficulties with Ward.

The relationship of principal and agent is confidential and fiduciary, binding the agent to the exercise of utmost good faith. Likewise, the principal owes to the agent the obligation of exercising good faith in the incidents of their relationship and must use care to prevent the agent from suffering harm during the prosecution of the agency enterprise. Further, a contract of agency carries an implied obligation of the principal to do nothing to thwart the effectiveness of the agency.

In our opinion, sufficient evidence was adduced from which the jury could have found that Montgomery Ward failed to exercise good faith in its course of dealing with the Tacketts. Throughout the course of the agency, Tackett sought auditor assistance from Ward as a means through which to settle disputed financial problems. However, Ward appeared to have no corresponding desire to seek a resolution of the parties' disagreements. It was not necessary that the jury infer fraudulent intent from the withholding by Tackett of the RNC's or charges back to himself. Rather, it could have been inferred that such action was Tackett's only available response to Ward's failure to follow expressed policies and procedures by delaying and improperly denying Tackett's ICA's or claims for credit. Tackett explained to Bruzan his reasons for withholding the RNC's at the commencement of the audit. However, even when fully apprised of Tackett's position and motives, Ward was apparently unwilling to engage in any type of dialogue to determine whether the parties' difficulties could be resolved. Instead, Ward chose the expedience of termination.

The wrongful exercise of the principal's power to revoke an agency will render the principal liable in damages if substantial injury is sustained by the agent.

Montgomery Ward emphasizes that the franchise agreement did not accord to either party the absolute right to continue the agency relationship. In fact, we note that either party was accorded the right of termination in any year as of the anniversary date of the agreement. Further, the agreement limited Ward's liability to the agent upon termination to compensation for orders received and filled by Ward up to the date of termination.

In no respect do we question the validity of the terms governing termination of the franchise agreement. However, we do not believe it consistent with sound public policy to permit Ward to employ those provisions as a shield against liability for termination accomplished in breach of its duty to exercise good faith.

(Affirmed.)

QUESTIONS FOR DISCUSSION

1. *What reasons did the court give for sustaining an award of damages to Tackett? What duty did Ward breach?*
2. *Did the court say that Ward did not have the right to terminate the agency? Why or why not?*
3. *How did the court treat the provision of the agreement limiting Ward's liability to compensation for orders received and filled by Ward up to the date of termination?*

---

(2) **Duty to Compensate: In General.** The duty of the principal to compensate the agent is usually determined by the contract creating the relationship. As a general rule, the contract contemplates that the services rendered by the agent are in return for compensation. If the contract does not expressly provide for compensation for the agent, the right to receive compensation may be implied from the circumstances surrounding the transaction as well as the past relationship of the parties. In that event the agent is entitled to compensation for the reasonable value of his services. On the other hand, if there is no intent that the agent be paid and his services are gratuitous, then the principal has no duty to compensate him. In determinations of whether the agent should be compensated, in the absence of an express contractual provision, the customs of the trade, profession, or occupation in which the agent is engaged are relevant. These factors are also important, of course, in determining the reasonable value of services rendered. Also, if the agent is to be entitled to compensation, it must be for acts performed within the scope of his authority. If the agent has no authority to act, then he is not entitled to compensation unless the principal subsequently ratifies the act.

The time when the agent is entitled to receive compensation is determined from the terms of the contract. As a general rule, unless the contract states otherwise, the agent must perform his required duties before he is entitled to compensation. Problems frequently arise in connection with relationships in which the agent's compensation is contingent upon the performance of a particular act or duty. Common examples are provided by real-estate brokers and sales representatives.

(a) *Real-Estate Brokers.* The right of real-estate agents to receive compensation from their principals has been the subject of a great deal of litigation. In the absence of contractual provisions to the contrary, the agent earns his commission if he procures a buyer who is ready, willing, and able to purchase the principal's property and who enters into the contract. Once the buyer is located, the principal may not terminate the relationship with the agent and enter into a contract for the sale of the property with the buyer at a later time, thereby excluding the agent from his right to a commission. The agent is entitled to be paid if he is the "procuring and efficient cause of the sale." On the other hand, the prospective buyer must actually enter into the contract with the principal in order for the agent to be entitled to the commission. The fact that the prospective buyer indicates his willingness and ability to become contractually bound is not sufficient. However, the agent is still entitled to compensation even though the third party who has entered into a contract with the principal fails to perform. This arrangement is equitable because the agent at that point has done all that is expected from him under the agency contract. He is not a guarantor of the third party's performance. There is, of course, nothing to preclude an agency contract from providing that the agent's compensation is dependent upon the third party's actually performing the contract with the principal.

There are three different types of contracts that are normally entered into between real-estate brokers and their principals. The type used in an individual case is, of course, a matter of negotiation between the parties.

The first type is the simple property listing, according to which the real-estate broker agrees to attempt to find a buyer of the property. The property may be listed with more than one broker. The broker who finds a buyer first is entitled to the commission, or in the event that the owner–principal finds his own buyer, no commission is paid.

Another type of real-estate agency contract is often called an *exclusive agency* or *exclusive listing.* In this case, the principal agrees that the property will not be listed with any other broker during the period of the agency. If the agent is the "procuring and efficient cause" of the sale of the property, he is entitled to his commission. On the other hand, if the owner sells the property himself, he pays no commission. With any type of exclusive agency it is wise to have a time limit for the duration of the agency in the event that the principal is not satisfied with the performance of the broker–agent. Some state laws limit the length of an exclusive real-estate agent–principal relationship to a specific period of time, such as six months.

A third type of real-estate agency contract is called an *exclusive right to sell.* The terms of this type of contract provide that the agent is to receive a commission from the sale of the property for the period of the agency relationship regardless of who sells the property. In other words, if the principal sells his own house during the period of the agency, the agent is entitled to his commission even though he is not the procuring and efficient cause of the sale.

In some areas of the country real-estate brokers have established a *multiple-listing agreement.* Under this arrangement, if a home owner signs a listing agreement with a particular broker who is also part of the multiple-listing agreement, then any broker who is a member of that agreement may show and sell the house. The obvious advantage for the home owner is that more than one broker may sell the house, and he has greater exposure. The advantage for the broker is the greater number of houses that he may show. The commissions earned are split among the brokers according to the multiple-listing agreement.

The real-estate broker must be licensed by the state in order to offer his services. In most states one must pass an examination administered by the state before the license is issued. These state statutes also regulate the terms of the agreement between the property owner and the broker, including such matters as the maximum commission that may be charged and the maximum length of time the agreement is operable.

HECHT v. MELLER

23 N.Y.2d 301, 244 N.E.2d 77 (1968)

KEATING, J.

This appeal presents a question of first impression: Is a real estate broker entitled to commissions on the sale of real property if the purchaser asserts a statutory privilege to rescind the contract of sale because the property has been substantially destroyed by fire after the contract was executed, but before the buyer took title or possession?

Briefly, the facts submitted were the following: Helen Hecht, plaintiff–appellant, entered into a written contract with Herbert and Joyce Meller, defendants–respondents, by which she became the exclusive selling agent for the sale of their personal residence and an adjacent lot which they wished to sell for $75,000. Through the plaintiff's efforts suitable buyers were introduced to the Mellers, and on May 30, 1963 a contract for the sale of the property was signed which acknowledged that Mrs. Hecht had brought the parties together, established a sale price of $60,000, and set August 1 as the closing date.

On July 20, without fault of either the vendors or vendees, the dwelling house on the property was substantially destroyed by fire. The buyers elected to rescind the contract, as provided for by statute, and the Mellers, therefore, returned the buyers' down payment. The present action was commenced by the real estate broker when the seller refused to pay the $3,600 brokerage commission allegedly earned by the broker in bringing the contracting parties together.

This court has consistently stated that a real estate broker's right to commission attaches when he procures a buyer who meets the requirements established by the seller.

At the juncture that the broker produces an acceptable buyer he has fully performed his part of the agreement with the vendor and his right to commission becomes enforcible. The broker's ultimate right to compensation has never been held to be dependent upon the performance of the realty contract or the receipt by the seller of the selling price unless the brokerage agreement with the vendor specifically so conditioned payment. As we stated in *Gibber* v. *Davis* (. . . 137 N.Y. p. 506, 33 N.E. p. 600): "If from a defect in the title of the vendor, or a refusal to consummate the contract on the part of the purchaser for any reason in no way attributable to the broker, the sale falls through, nevertheless the broker is entitled to his commissions, for the simple reason that he has performed his contract."

In determining the extent to which the duties of sellers and brokers should be realigned, we must consider the fact that, simply because the contingency provided for by the statute arises, that is in no way determinative of the actual desire of the vendee to fully perform. Also, simply because a vendee elects to rescind the contract in no way reflects on the inadequacy of the broker's performance in finding an acceptable buyer.

The fact that our decisions decided before and after the enactment of the section have consistently held that a broker is entitled to commissions when he brings the parties together compels a finding in this case that the expense of the brokerage commission must be paid by the vendor who has contracted for the broker's services, even though the Legislature has given the vendee the privilege of rescinding the contract which the broker helped bring about. The sellers in this case, having failed to shift the possible loss, must be deemed to have assumed the risk themselves.

QUESTIONS FOR DISCUSSION

1. *Note that the broker was entitled to his commission in this case even though the sale was not completed. At what point was the broker entitled to the commission?*
2. *What could the seller–principal have done to protect himself against the circumstances that occurred in this case?*

(b) *Other Sales Representatives.* Questions frequently arise concerning the right of sales representatives to recover commissions from their principal. Sales representatives are normally entitled to commissions for the sale when an order from the buyer is obtained unless there is a contractual provision to the contrary. If the principal fails to fill the order, the agent is still entitled to his commission unless the principal rejects the order in the exercise of reasonable judgment, such as because of the lack of financial responsibility of the purchaser. If the principal rejects the order arbitrarily, the agent is still entitled to his commission.

Sometimes a sales representative will operate under a contract granting him an exclusive territory. Two questions frequently arise concerning the agent's right to recover a commission for goods sold in the territory. One question involves sales by the principal and the other involves sales by another agent of the principal. Unless the intention of the parties is contrary, a principal may himself sell in the agent's exclusive territory without paying a commission on the sale. On the other hand, if the intention of the parties is that the agent receive a commission on all sales, then he gets the commission regardless of who makes the sale. If the sale is by another sales representative in the territory, then the agent still is normally entitled to the commission for the sale. The place of sale is normally determined to be the place where the order is taken rather than the location of the residence or place of business of the purchaser, unless there is a trade custom to the contrary.

(3) **Duty to Reimburse and Indemnify.** The agent is entitled to be compensated for expenditures made on behalf of his principal within his scope of employment and for his principal's benefit. This is known as the *agent's right to reimbursement.* Such expenditures as the payment of rent, interest, legal fees, down payments on goods, and taxes are among the many expenditures for which the agent may be reimbursed when making them on behalf of the principal. The payments must, of course, be reasonably necessary and for the benefit of the principal.

An agent is entitled to be indemnified by his principal for damages sustained by the agent while acting within the scope of his authority. An agent is entitled to be indemnified if he is required to pay damages to a third party as the result of actions taken on behalf of his principal that constitute a tort or breach of contract if it results from the authorized performance of his duties. On the other hand, the principal does not have to indemnify the agent for acts outside the scope of his authority or those that are illegal. The agent's right to indemnification is limited to situations in which the agent has followed the instructions of the principal. The agent may not recover damages from the principal if the agent is at fault.

WOOD v. HOLIDAY INNS, INC.

508 F.2d 167 (1975)

Glen Wood checked into the Holiday Inn at Phenix City, Alabama, late in the afternoon on February 1, 1972. When he checked in he tendered his Gulf Oil Company credit card. It was returned to him after an imprint of it was made. Under a system established by Gulf, Holiday Inns are authorized to contact National Data Corporation, which disburses credit information concerning Gulf credit cards. Wood used his card for business as well as for personal expenditures, but he did not inform Gulf of this. When Gulf received his file, it became alarmed at the amount of credit extended to him in relation to his salary. As a result, Gulf, without telling Wood, notified National Data to give the following reply to anyone seeking credit approval on Wood:

> Pick up travel card. Do not extend further credit. Send card to billing office for reward.

During the early morning hours of February 2, 1972, Jessie Goynes, the "night auditor" of the Holiday Inn, checked with National Data to receive authority to extend Wood credit. The authorization was denied by the message from National Data. At 5:00 A.M.*, Goynes awakened Wood and told him he needed the card to make another imprint. Goynes came to Woods's room and picked up the card. At 5:30 Wood learned from Goynes that his card had been "seized" by authority of National Data. Wood paid cash for the room and left. When he returned home, Wood explained that he used his card for business and that his account was current. His credit was immediately reinstated. Wood's anger and frustration continued to build, and three days later, while explaining the incident to a friend, Wood had a heart attack. Wood sued Gulf Oil, Holiday Inns, Inc., Interstate Inns, Inc. (the owner of the Phenix City Inn), and Jessie Goynes. Interstate and Goynes filed a cross-claim against Gulf, alleging right to indemnification as its agents. The jury returned a verdict of $25,000 compensatory damages against Gulf, $25,000 punitive damages against Interstate and Goynes, and $10,000 punitive damages against Holiday Inns. The court then granted motions for judgments notwithstanding the verdict to Gulf and Holiday Inns and granted a motion made by Interstate and Goynes for a new trial. Wood appealed.*

Interstate and Goynes also appealed the district-court action in overturning a jury verdict on the cross-claim in their favor through the granting of Gulf's motion for judgment notwithstanding the verdict.

The court first found that there was sufficient evidence from which a jury could conclude that Interstate was Gulf's agent for the revocation of Gulf credit cards and that the jury could reasonably find Goynes was its subagent. It then dealt with Interstate's and Goynes's right to indemnification by Gulf.

MORGAN, J.

Cross-appellants Interstate and Goynes appeal the lower court's decision denying indemnification. We believe the district court was correct.

Under Alabama law, an agent is entitled to indemnification for any amounts he has been required to pay for his principal in the performance of his agency. However, the principal is not required to indemnify the agent for harm resulting solely from the agent's negligence. Hence, in the absence of agreements to the contrary, an agent has no right to indemnity for damages suffered by reason of his own fraud, misconduct or other tort, even if the wrong was committed within the scope of the agent's employment.

Goynes and Interstate have introduced no substantial evidence upon which a reasonable jury could conclude that Gulf was negligent. The only relevant action undertaken by Gulf was its directive to pick up Wood's credit card. However, by virtue of a statement contained on the card, and in the credit card application signed by Wood, Gulf retained the contractual right to revoke the credit card without notice.

Hence, Gulf's decision to terminate Wood's credit was not tortious, and Goynes and Interstate have shown no duty which Gulf is alleged to have breached.

The theory of Wood's case is based upon the harm he is alleged to have suffered because of the manner in which his card was revoked. Any harm to Wood must have resulted solely from the actions of Goynes, since Gulf has not been shown to have been negligent. We therefore believe no substantial evidence has been presented which would justify the indemnification of Goynes and Interstate.

For the reasons set forth above, we affirm in part, reverse in part, and remand for proceedings not inconsistent with this opinion.

34:3. Workmen's Compensation. At this point it is appropriate to mention the topic of workmen's compensation as a major means of recovery by employees for injuries suffered during the course of employment. It is applicable to *employees* in the sense that that word has a broader connotation than *agent.*

The traditional method of recovery by employees for injuries suffered while on the job was to sue the employer for negligence. This procedure, however, is subject to a number of problems from the standpoint of the injured employee. In the first place, the employer must be shown to have been negligent. Many accidents result not from the employer's negligence but from someone else's negligence, frequently that of a fellow employee. In the case of the negligence of a fellow employee, the employer would not be liable under what was known as the *fellow-servant doctrine.* Furthermore, even if the negligence of the employer could be established, there was always the defense of contributory negligence and assumption of risk.

With the advent of the increased industrialization of our society and the concomitant increase in industrial accidents, it became necessary to create some method to absorb this social cost. The answer was workmen's compensation, which placed the burden for accidents on the employer on the theory that such losses should be regarded as a cost of doing business because accidents are inevitable.

All the states have adopted workmen's compensation statutes in varying forms. Basically workmen's compensation acts impose liability without fault on the employer if the injury is job-related. In order for the employee to recover, he does not have to prove any negligence, and he can recover despite any contributory negligence or assumption of risk on his part. The fellow-servant doctrine is also inapplicable. Generally, if the employee is entitled to recover under workmen's compensation, he is paid according to a set schedule of benefits. He may also be able to collect for his medical costs. Benefits are usually computed as a percentage of weekly wages when the worker's disability is not permanent, but there are maximum amounts that may be paid under the schedule. If the injury is permanently and totally disabling, the weekly percentage is paid for a limited period of time, such as for four-hundred weeks. The schedule also provides so much for loss of life and lost hands, legs, and so on. In short, the liability under the workmen's compensation statutes is assessed without regard to fault, but there is a limitation on the amount that can be recovered by the employee. Often these amounts are low and have not kept up with the recent rate of inflation.

Not all employees are covered under workmen's compensation statutes. Agricultural and domestic workers are frequently excluded. If the employee is engaged in a job that is covered under a workmen's compensation statute, he must generally show that his problem (1) resulted from an "accidental injury" and (2) arose out of the course of employment. The courts have been somewhat liberal in interpreting these requirements, but still the employee must show that his injury was the result of more than a mere risk of life. Thus, if one has a gallbladder attack while on the job, he clearly would not be able to recover under a workmen's compensation statute. At the other extreme, if one loses his hand while operating a machine on an automobile assembly line, he would be able to recover. There are many cases in between that are not so clear-cut. A heart attack suffered while one is at work would not normally be compensable, but there have been cases in which the courts have found the employee entitled to recover, particularly if it could be shown that the job was particularly stressful.

In some states an employee may also sue his employer on the theory of negligence in addition to recovering under the workmen's compensation statute. In others, the employee must make an election between recovering under the workmen's compensation statute or suing on the basis of common-law negligence.

CLARK v. U. S. PLYWOOD

605 P.2d 265 (Or. 1980)

PETERSON, J.

This case involves a widow's claim for Workers' Compensation benefits. Her husband, George Clark, was killed while retrieving his lunch, which he had left to be warmed atop a hot glue press. The referee denied compensation. The Workers' Compensation Board reversed and ordered acceptance of the claim. The Court of Appeals reversed and denied recovery, and we granted review to consider the extent to which personal comfort activities of a worker will be deemed to arise out of and within the course of employment.

THE FACTS

Clark was employed at a Gold Beach plywood manufacturing plant. He worked a shift which began at 11 P.M. and ended at 7 A.M. During this shift Clark was paid for two 10–minute breaks and a 20–minute lunch period. The lunchrooms provided by the employer contained a table and vending machines, but no facilities for heating food brought by the employees.

On the night of Clark's death, he had brought a lunch which needed to be warmed. About two hours before his lunch break, he approached the assistant operator of a hot glue press and asked him to place Clark's food container on the top of the press to be warmed. The assistant press operator had done this before for Clark, and testified that two or three times a week he placed food on the press for other employees. The hot glue press was about 100 feet from Clark's work station.

When Clark returned to retrieve his lunch, the charger had just been loaded and the press operator and his assistant were getting ready to move the load into the press. The assistant press operator noticed that Clark was standing at the foot of a ladder which led to the top of the charger and heard him mention something about retrieving his lunch. The assistant press operator testified that he "didn't pay that much attention" to Clark because he had to go around to the back of the press to straighten panels. Nor could the press operator see Clark, because his control panel was on the opposite side of the charger. Clark possibly climbed the ladder, intending to ride the carriage over to the hot press whereupon he would reach over and retrieve his lunch. The press operator activated the charger and Clark was killed when the charger moved across the top of the carriage, crushing Clark between the charger and a stationary cross beam on the front of the carriage.

APPLICABLE WORKERS' COMPENSATION STATUTES

A "compensable injury," under ORS 656.005(8)(a), is "an accidental injury . . . arising out of and in the course of employment . . . resulting in disability or death . . . whether or not due to accidental means." Contributory fault or contributory negligence is no defense to a claim for compensation benefits, unless due to "the deliberate intention of the worker." ORS 656.156(1). All that a claimant must prove is that the injury arose "out of and in the course of employment." The worker has the burden of proving that the injury arose out of and in the course of employment.

The compensation act provides broad coverage, the boundaries of which are determined by the meaning of "arising out of and in the course of employment." As with most difficult questions, the delineation of the limits of the coverage is anything but knife-edge clear. But as in all difficult cases (this being one such case) the delineation must be made.

The Court of Appeals correctly characterized the issue as being ". . . whether the employee's death arose out of and in the course of his employment, as required by ORS 656.005(8)(a)."

MEANING OF "ARISING OUT OF AND WITHIN THE COURSE OF EMPLOYMENT"

The words "in the course of employment" have been repeatedly defined as relating "to the time, place, and circumstances under which the accident takes place." The words "arising out of" normally refer to the requirement of a "causal connection between the employment and the accident."

The following example will illustrate the difference: A machinist working at a lathe has an attack of appendicitis. The attack occurred in the course of his employment since it occurred while he was on the job performing his normal activities. On the other hand, it did not arise out of his employment. There was no causal connection between the work and the attack.

COMPENSABILITY OF ON-PREMISES INJURY CLAIMS

Most claims for on-premises injuries fall within one of two general categories:

Category 1. Injuries sustained while performing one's appointed task;

Category 2. Injuries sustained while engaged in other incidental activities not directly involved with the performance of the appointed task, such as preparing for work, going to or from the area of work, eating, rest periods, going to the bathroom, or getting fresh air or a drink of water.

Injuries sustained by a worker in doing the appointed task are normally compensable, absent self-inflicted injury. Contributory fault of the employee is no defense. Even when a worker is performing an appointed task in a prohibited manner, injuries are normally compensable. If a worker operates a machine with the guard removed, or fails to stop a machine before reaching into it, or oils machinery while it is running, injuries so sustained are normally compensable even though the specific act causing the injury was prohibited.

Many premises-related injuries are also normally compensable even though the worker is not engaged in the appointed task. A worker who trips over a step while walking to the bathroom on the employer's premises, or who falls on the way to the company locker room to change clothes, or who trips while going to get a breath of fresh air to escape the heat of working quarters—all normally are entitled to compensation.

Most courts allow recovery for injuries sustained while engaged in recreational activities during lunch hours or rest periods, if the activity is a normal or accepted one.

Lunchtime injuries are normally compensable, if they occur on the premises and arise from premises hazards such as building collapse, tripping on a hole in the floor, or falling on slippery steps.

In *Lamm* v. *Silver Falls Tbr. Co.,* 133 Or. 468, 277 P. 91, 286 P. 527, 291 P. 375 (1930), we held that a lumber camp worker returning from a holiday in Silverton was covered by the Workmen's Compensation Law when he sustained injury while riding on the company-owned train back to the logging camp. We quoted from *Cudahy Co.* v. *Parramore,* 263 U.S. 418, 423–424, 44 S.Ct. 153, 68 L.Ed. 366, 30 A.L.R. 532 (1923) as follows:

'Workmen's compensation legislation rests upon the idea of status, not upon that of implied contract; that is, upon the conception that the injured workman is entitled to compensation for an injury sustained in the service of an industry to whose operations he contributes his work as the owner contributes his capital—the one for the sake of the wages and the other for the sake of the profits. The liability is based, not upon any act or omission of the employer, but upon the existence of the relationship which the employee bears to the employment because of and in the course of which he has been injured. . . . No exact formula can be laid down which will automatically solve every case.' 133 Or. at 495–496, 277 P. at 94–95.

Respecting the quotation from *Cudahy*, we stated (133 Or. at 496–498, 277 P. 91):

The above being the test by which the right to compensation is determined, the propriety of the awards in the noon-hour lunch, sleeping upon the premises, shower bath, etc., cases becomes apparent. . . .

In all of the foregoing instances the injury, like that in our case, was sustained while the employee was doing something which was ancillary to his employment. We quote once more from *Larke* v. *John Hancock Mutual Life Ins. Co.*, supra [90 Conn. 303, 97 A. 320 (1916)]:

'The duty ancillary or incident to the employment has in some instances been held to include the doing of something primarily for the benefit of the employee, but ultimately it is assumed for the master, as the preparation of a noon-hour lunch, or the doing of something by the employee which he reasonably believes for the master's interest'. . . .

The above will suffice to establish the fact that in order to be entitled to compensation it is not necessary that at the time of the accident (1) the employee was doing something for the direct benefit of the master, (2) that he was at his place of duty, and (3) that he was injured during working hours. . . .

Benefits are payable for some on-premises injuries during the lunch hour, even though the ingestion of food may be no less valuable if consumed at home, because (1) the injuries normally result from some kind of on-premises hazard, and (2) the employee is within the time and space limits of the employment as set by the employer, i.e., the employer has expressly or impliedly allowed the conduct in question.

We believe that the compensability of on-premises injuries sustained while engaged in activities for the personal comfort of the employee can best be determined by a test which asks: Was the conduct expressly or impliedly allowed by the employer?

Clearly, conduct which an employer expressly authorizes and which leads to the injury of an employee should be compensated whether it occurs in a directly related work activity or in conduct incidental to the employment. Similarly, where an employer impliedly allows conduct, compensation should be provided for injuries sustained in that activity. For example, where an employer acquiesces in a course of on-premises conduct, compensation is payable for injuries which might be sustained from that activity. Acquiescence could be shown by showing common practice or custom in the work place.

(The case was remanded to the lower court for a determination of whether the employer allowed Clark's actions.)

Questions for Discussion

1. *May the employer defend this case by showing that Clark was negligent in placing his lunch on the glue press to get warm?*
2. *What rationale does the court give for passage of workmen's compensation legislation?*
3. *What issues would you have to consider in determining whether a worker who had a heart attack while on the job could recover under most workmen's compensation statutes?*

SUMMARY

The agent has a fiduciary relationship to the principal. As a general rule an agent cannot represent two principals in the same transaction unless the principals have full knowledge and consent to it. Nor can the agent possess or represent any interest adverse to the principal without the principal's consent. An agent cannot appropriate his principal's property including trade secrets and customers. If an agent breaches his duty of loyalty to the principal, he forfeits any profits gained from the breach and any compensation due him as a result of his employment as an agent.

The agent has certain other duties to the principal. Among these are the duty to obey instructions, the duty to give notice of material facts, and the duty to carry out his obligations with care and skill. An agent also has a duty to account for any money or property that he holds for his principal and may not commingle that property with his own.

By the same token the principal also has certain duties to the agent. Besides adhering to the terms of the agency contract, the principal must compensate the agent if he has agreed to do so. A special problem comes up with real-estate brokers and other sales people. If the contract with the real-estate broker or agent is a simple property-listing agreement or exclusive agency contract, the agent earns his commission if he is the procuring and efficient cause of the sale. If the contract is an exclusive right to sell, the agent receives a commission when the property is sold. The principal also has a duty to reimburse the agent and to indemnify him for expenses and damages incurred by the agent while acting within the scope of his authority.

Workmen's compensation statutes have altered the traditional basis of recovery for employees injured in job-related accidents. The basic concepts of negligence have been abandoned and the employee is compensated if the injury arose out of the course of employment and was an "accidental injury."

PROBLEMS

1. Robert Desfosses employed Steve A. Notis, a licensed real-estate broker, to assist generally in the work of Desfosses's Company with the special assignment of acquiring land for development as mobile-home parks. Notis received a weekly salary.

   Notis informed Desfosses that a particular tract of land was available to develop as a mobile-home park. Desfosses requested Notis to purchase it as a straw party, in the course of his employment, and then convey it to Desfosses. Notis told Desfosses that the land would cost $32,400.00, although he knew that the land could be purchased for $15,473.62. Prior to commencing to serve Desfosses, Notis had agreed to purchase the land in question from the owner and had participated in a deposit of $1,000.00. Desfosses knew nothing about this and, relying upon Notis's representation, delivered $32,400.00 to him, out of which Notis purchased the land and retained the balance. Is Desfosses entitled to a judgment and, if so, in what amount? Desfosses v. Notis, 333 A.2d 83 (Me. 1975).
2. John Kurkjian was chairman of the parish council of St. James Armenian Church, which was contemplating selling the church property and acquiring another site. He was also chairman of the building committee. Baxter Hall-

aian was a real-estate broker and a friend of Kurkjian. Foster Price, a real-estate broker representing the Mount Sinai Baptist Church, offered to purchase the St. James property for $750,000. The sale was made, including a commission of $37,500 paid to Price. Unknown to the council and concealed from it by Kurkjian was a payment of $22,000 of this commission by Price to Hallaian. In negotiating the purchase of a site by St. James, Kurkjian stated to the council that Hallaian had been helpful in selling their property and had received no commission. Hallaian then received a regular commission of $27,739.25 when St. James purchased the new site. The church then learned of the first commission of $22,000 and sued Kurkjian to recover. Was Kurkjian the church's agent and was he liable? St. James Armenian Church of Los Angeles v. Kurkjian, 47 Cal.App.3d 547 (1975).

3. Alexander S. Bacon had deposited with the Fourth National Bank of New York, in escrow, a mortgage and assignment to be forwarded to its correspondent bank in Boston, Maverick National Bank, for payment by Maverick to an individual in Boston. The payment was, of course, to be on behalf of Bacon. While the mortgage and assignment were in the hands of the Maverick Bank, an attachment was levied upon them. Maverick hired an attorney to defend against the attachment and paid his fee. Fourth National paid Maverick on behalf of Bacon and claimed that it was entitled to retain that fee out of Bacon's deposit. Was Fourth National correct? Bacon v. Fourth National Bank, 9 N.Y. 435 (1889).
4. Ralph and Karin Monty retained Lewis W. Shurtleff to sell certain property for them. He was then retained by Ward and Viola Peterson to sell a small parcel of land for them and to invest in some more extensive property. Shurtleff informed the Petersons of the Monty property. This resulted in a contract by the Petersons to purchase the Monty property for $55,000. The Petersons paid part of the price by transferring their smaller property to the Montys. This lot was zoned for duplex construction but was subject to a number of private covenants that restricted its use to single-family dwellings. Neither the Petersons or the Montys were aware of these covenants until a title report was received. Shurtleff, although acting in good faith, did not adequately call their attention to the effect that the covenant might have on the value of the property, which was determined to be $850. The Montys sued Shurtleff for the amount of his commission of $3,300. Were the Montys entitled to any recovery from Shurtleff, and, if so, in what amount? Monty v. Peterson et al., 85 Wash.2d 956, 540 P.2d 1377 (1975).
5. Charles DiMicco insured his boat through an insurance agent of the Fidelity Phenix Insurance Corporation. The boat was insured for $5,000 and was thirteen years old. The agent had been instructed by the company not to insure a boat over three years old or worth more than $5,000 without a condition survey. No condition survey was made of the DiMicco boat prior to the issuing of the insurance policy. The boat was lost in a storm. The

insurance company claimed that it should be indemnified by the agent in the event that it was liable under the policy. Was the insurance company correct? Crawford v. DiMicco, 216 So.2d 769 (Fla. 1968).

6. Hugh Neighbors, a real-estate agent, was hired by Bevins and Weaver, who operated the Harlem Bar in Pensacola, to purchase a building owned by Silar Barge. The building owned by Barge was in the same block as the Harlem Bar. When Neighbors contacted Barge, he told Barge that he represented some "out-of-town" people. Upon Barge's inquiry Neighbors also represented to Barge that he thought the property would be used as a secondhand store even though he knew in fact that this was untrue. Barge signed the contract to sell the property, and Neighbors signed on behalf of the purchaser and gave Barge a deposit. Barge subsequently learned the identity of the real purchasers and refused to sell. Neighbors assigned the contract to Quest, an employee of Bevins and Weaver, who sued for specific performance. Barge's property was currently being used as a café, selling beer and wine and in competition with the Harlem Bar. Barge contended that Neighbors was a dual agent because he also was asked to sell Barge's property and did not reveal the fact to him and that he was not bound to the contract. Was Barge correct? Quest v. Barge, 41 So.2d 158 (Fla. 1949).
7. Joe Edmondon was a local agent of Mechanics and Traders Insurance Company. Edmondon wrote a policy on the property of Sallie Young. Mechanics then instructed Edmondon to cancel the policy, but he neglected to do so for five weeks. As fate would have it, Sallie's property was destroyed by fire and Mechanics was compelled to pay. Mechanics sought to recover from Edmondon, who contended, among other things, that Mechanics had the power to cancel the policy itself. Was Edmondon liable to Mechanics? Washington et al. v. Mechanics and Traders Insurance Company, 50 P.2d 621 (Okla. 1935).
8. Rausch owned Black Acres Farm, which was subject to a first mortgage. Desirous of refinancing that obligation, he sent an agent to Citizens State Bank with the authority to make a new loan and to pledge certain stock as security for the new loan on condition that a second mortgage was obtained on the farm. Citizens knew of the condition of a second mortgage. When a second mortgage was not obtained, Rausch's agent and Citizens nevertheless made a new loan agreement, which required Rausch to deliver certain securities to the bank as security for the loan. Citizens sued Rausch to compel delivery of the securities. Was Rausch bound by the actions of his agent? If so, what, if any, liability did the agent have to Rausch? Citizens State Bank v. Rausch, 9 Ill.App.3d 1004, 293 N.E.2d 678 (1973).
9. Holt was in the process of liquidating the Lowell Bank and hired Joseph F. Dickmann, a real-estate agent, to sell a certain parcel of property. Dickmann presented an offer of one Cecelia Ross to purchase the property for $19,000. Holt accepted the offer. Holt later discovered that Cecelia Ross was merely a "straw party" and that the real purchaser was Dickmann,

who then resold the property for a $4,500 profit. Dickmann was paid a commission by Holt of $950 for the sale of the property. Holt sued Dickmann. Was Holt entitled to recover anything from Dickmann? If so, how much? Holt, State Commissioner of Finance v. Joseph F. Dickmann Real Estate Company, 140 S.W.2d 59 (Mo. 1940).

CHAPTER 35

Rights and Liabilities of Principal and Third Party

35:1. INTRODUCTION. One of the chief characteristics of the principal–agent relationship is the fact that the principal may become liable to and have certain rights against a third party as the result of the actions of his agent. We have already examined a number of ways that the agent may bind the principal. The agent may have express or implied authority, and he may bind his principal by apparent authority or under the doctrine of estoppel. The principal may also become liable to the third party as well as obtain certain rights against him by ratifying the acts of the agent.

In this chapter the rights and liabilities of the principal vis-à-vis the third party will be examined. The principal may, of course, be contractually bound to the third party as the result of the acts of his agent. In that event, the third party is also liable to the principal for any breach of contract. The tort liability of the principal that may result from certain acts of the agent will also be examined. In Chapter 36 the agent's rights and liabilities in each of these situations will be analyzed.

35:2. LIABILITY BASED ON CONTRACT. There are several different situations that may affect the rights and liabilities of the parties: where the principal is disclosed to the third party, where his existence is undisclosed, and where he is partially disclosed.

(1) Where the Principal Is Disclosed. In the most common situation the existence of the principal as well as his identity is known to the third party. The third party is fully aware that he is contracting with the principal through the principal's agent. Let us examine the position of each of the parties in the trio under this circumstance.

(a) *The Principal.* When the agent enters into a contract on behalf of his principal within the actual or apparent scope of his authority, the principal is bound to the third party. The rights and liabilities of the principal are just as they would be if he had entered into the contract directly with the third party. Furthermore the principal is liable to third persons on the contract even if he did not know specifically that the agent entered into it on his behalf. Even if the authority of the agent is only *apparent,* as discussed in Chapter 33, the principal is still liable to the third party.

If the agent acts without authority, either actual or apparent, the principal may ratify the acts of the agent and bind the contract. In this event the principal is then continually liable to the third party in the same manner as if the agent had acted within the scope of his authority. Remember, however, that in order to ratify the contract entered into by the agent beyond the scope of his authority, the principal must have knowledge of all the facts. On the other hand, the principal is liable to the third party for contracts made by his agent within the scope of authority even if the principal is ignorant of the facts.

If the agent had the authority to make a contract on behalf of his principal, the principal is liable to the third party for false statements knowingly made by the agent in order to induce the third party to enter into the contract. The third party must have reasonably relied on the statements. The principal is liable to the third party for such statements even if he instructed the agent not to make them. The law places this burden on the principal, even though he is innocent, because it is the principal who selects the agent, benefits from his acts, and is in the best position to control the agent.

If the agent clearly exceeds his authority in dealing with the third party, the principal is not liable to the third party. Under this circumstance the third party would be deemed to have notice of the agent's lack of authority. Thus, if the third party has actual knowledge or reasonably should know that the agent is exceeding his authority, the agent's actions do not bind the principal. Furthermore, if the act of the agent is clearly adverse to the best interests of the principal, the third party will be charged with this knowledge and hence cannot hold the principal to any contract entered into with the principal's agent under this circumstance.

(b) *The Third Party.* When the principal is disclosed to the third party, it is the principal with whom the third party is contracting. Consequently the third party may sue the principal for any breach of the contract entered into on his behalf by the agent. Likewise the third party is liable to the principal for any contractual breach and may be sued directly by the principal. Once the contract is made, the agent is not a party to it, and he has no part in any subsequent actions based upon the contract.

(c) *The Agent.* Although this chapter deals primarily with the rights and liabilities of the principal and the third party based upon contract, it is appropriate to mention the role of the agent at this juncture. When the principal is disclosed, the role of the agent may be analogous to that of a conduit of his principal's contractual intent. The agent is acting on behalf of his principal but is in no way party to the contract. Whereas the principal and third party each have rights and liabilities under the contract, the agent has no rights and liabilities. He may neither sue on the contract or be sued. In fact, as we shall see in Chapter 36, the agent normally does not even sign a written contract in his own name but, if he acts properly, only signs the contract on behalf of his principal. For example, he may sign, "Jones Corporation, by Richard Roe its agent."

Suppose Artie Agent enters into a contract with Terry Thirdparty on behalf of Peter Principal and Thirdparty knows that Agent is acting as an agent for Principal. Thirdparty may sue Principal for breach of the contract, and Principal has the same right against Thirdparty. But Thirdparty may not sue Agent, nor may Agent sue Thirdparty if the contract is breached, assuming, of course, that the contract entered into by the agent is authorized. Furthermore, this discussion has focused on the rights of the various parties to sue *on the contract.* However, the agent does expose himself to liability if he enters into a contract beyond the authority granted by his principal. The agent may be liable to either or both his principal or the third party under this circumstance, as will be more fully discussed in Chapter 36.

VERNON D. COX & CO., INC. v. GILES
406 A.2d 1107 (Pa. Super. 1979)

George E. Giles contracted with Vernon D. Cox in order to have Cox's firm do an appraisal of two tracts of land totalling 320 acres. The land was owned by LeChateau Inn and Country Club, a corporation. Cox wrote to Giles and quoted a price of $2,500.00 for the appraisal. Cox had been informed that the appraisal was necessitated by LeChateau's need to establish the fair-market value of the land in anticipation of a sale and lease-back arrangement.

Cox had been aware that Giles had some connection with LeChateau although he didn't realize Giles was chairman of the board of directors. At no time was Cox specifically told of Giles' position with LeChateau. Cox was orally informed by Giles to proceed with the appraisal in accordance with Cox's letter.

After the appraisal was completed, Cox sent his bill to Giles personally and not to LeChateau. No payment was forthcoming and Giles informed Cox that his services had been rendered to LeChateau and said any further billing should be forwarded directly to that corporation. Cox then sued Giles for the fee charged for the appraisal. Giles filed a demurrer with the trial court contending that since he was an agent he was not liable on the contract. From an adverse decision of the trial court, Giles appealed.

PRICE, J.

Appellant now maintains that the lower court erred in failing to sustain his demurrer, because the jury could not conclude from the evidence presented that he had personally entered into a contract with appellee. We agree. It is a basic tenet of agency law that an individual acting as an agent for a disclosed principle is not personally liable on a contract between the principle and a third party unless the agent specifically agrees to assume liability. Conversely, an agent who consummates a contract without disclosing either the fact of agency or the identity of the principle will be considered to have assumed personal liability. Instantly, the pivotal question is whether LeChateau was a disclosed principle in the dealings between appellant and Mr. Cox.

The Restatement (Second) of Agency § 4 defines a disclosed principal in the following terms: "If, at the time of a transaction conducted by an agent, the other party thereto has notice that the agent is acting for a principal and of the principal's identity, the principal is a disclosed principal." The Restatement subsequently explicates that a person has notice of a fact when he has actual knowledge of it, has reason to know it, should know it, or has been given notification of it. Restatement (Second) of Agency § 9. While it would ordinarily be for the trier of fact to determine whether the requisite notice of disclosure existed, to premise individual liability in the present case on the quantum of proof adduced by appellee would substitute conjecture and surmise for proof. Mr. Cox knew of the existence of LeChateau; he knew that it owned the land to be appraised and that it required the appraisal so as to acquire refinancing. He also knew that appellant was associated with the corporation on the basis of prior dealings. It is clear that on these facts, Mr. Cox knew or should have known that appellant was acting in a representative capacity.

(Reversed in favor of Giles.)

QUESTION FOR DISCUSSION

1. *What mistake did the plaintiff, Cox, make in this case?*

(2) Where the Principal Is Undisclosed. An undisclosed-principal situation occurs when the third party has no knowledge that the person with whom he is dealing is an agent. He not only is ignorant of the name of the principal but does not even know that one exists. This situation occurs most frequently in the case of land sales and other real-estate transactions. Frequently wealthy individuals or corporations will seek to acquire a block of land for one purpose or another. If they were to bid on the property in their own name, the sellers, knowing the means of the potential buyer, might not be willing to sell for a reasonable market price but would ask an inflated price for the property. In order to avoid this circumstance, such buyers may employ a real-estate agent or attorney to purchase the land without revealing the existence of the principal to the seller. Because the third party believes that he is actually contracting with the agent, the rights and liabilities of the parties are somewhat different than in the situation in which the principal is disclosed.

(a) *The Principal and the Third Party.* If the third party enters into a contract with the agent of a principal, not knowing of the existence of the principal, the principal is bound to the contract. The agent is also bound because he is the one with whom the third party makes the contract. In this event, when the third party discovers the existence of the principal, he may hold the principal to the agreement just as though he had been disclosed in the first place. The principal may, of course, reveal his existence after the contract is made and hold the third party liable.

How may one protect himself against the possibility that the individual with whom he is dealing is an agent for an undisclosed principal? The answer is to ask questions. If the agent falsely denies that he represents a principal, then fraud has been committed and the contract can later be set aside on this basis. The best protection, of course, is to insert a clause into the agreement that stipulates that the third party is contracting exclusively with the agent as the other party to the contract.

Because both the principal and the agent are liable to the third party in an undisclosed principal situation, the third party has a right of action against both. At some time, however, the third party must make an election. In most states he may sue both the principal and the agent for breach of contract. Also in most of those states the third party must make an election as to which party he wants to take judgment against. In a few states the third party may take a judgment against either the principal or the agent without making an election until recovery of the judgment. The latter approach is more advantageous to the third party because there are procedures available to a judgment creditor for obtaining information concerning the judgment debtor's assets that are not available prior to judgment. It is therefore easier under the latter approach to seek to satisfy the judgment from the party with the "deeper pocket." In the majority of states, however, if one takes a judgment against either the principal or the agent, with knowledge of the relationship, he has made his election. However, a judgment obtained against either party without full knowledge by the third party of the relationship does not generally constitute an election or bar a subsequent action against the other party.

It should be obvious by now that an undisclosed principal is not bound by a contract entered into by an agent if the agent acts beyond the scope of his authority or apparent authority. Nor is the doctrine of estoppel applicable in this situation in most jurisdictions. The reason is that because the principal's existence is unknown to the third party, there is no way that he could have relied on the agency relationship in entering into the contract. It is possible, however, that by ratifying it, the undisclosed principal may bind himself to a contract entered into by the agent beyond the scope of his authority.

There are some limitations on the right of an undisclosed principal under a contract. For example, an undisclosed principal cannot enforce a contract against a third person that involves personal skill, trust and confidence, or personal service. The reason is, of course, that the third person believes he is actually contracting with the agent. Thus, if one hires an individual to paint his portrait or to work in a beauty shop, the undisclosed principal could not come forward and claim the job. In addition, the third party may raise any defenses against the principal that he might have raised against the agent.

(b) *The Agent and the Third Party.* The agent's position in a contract involving an undisclosed principal is quite different than when the principal is disclosed. The major distinction is that if the principal is disclosed, the agent is not a party to the contract and normally has no liability to the third party. When the principal is undisclosed, the agent may be held liable on the contract by the third party. The reason is that when the principal is undisclosed, the third party believes that he is actually contracting with the agent, and in fact, if the contract is written, it is the agent's name that will appear on it. The liability of the agent is just the same as if no principal exists. However, the agent is liable only if the principal would have been liable under the circumstances of the case.

If the agent does not want to incur any liability on the contract, the burden is upon him to reveal and identify the principal to the third party. The third party has no duty to inquire as to the existence of the principal. On the other hand, if the third party is aware of the existence of a principal and knows his identity even if this information is not revealed to the third party by the agent, then the agent is relieved from liability. The reason is that the third party is not operating under the assumption that he is actually contracting with the agent. The agent is, of course, entitled to be indemnified by his principal for any loss suffered as a result of his liability to the third party.

It is only reasonable to assume that because the agent in an undisclosed-principal situation is liable to the other contracting party, the agent would also be able to sue the other party to the contract in his own name. This is in fact the case. Because the third party actually makes the contract with the agent, this right works no hardship on him. Furthermore the third party can raise any defense that he might have against either the agent or the principal.

The right of the agent to bring suit against the third party is subject to the principal's superior right to do so. However, the agent may bring suit against the third party without first obtaining the permission of the principal. But the principal and the agent cannot subject the third party to two suits on the contract. One or the other must sue. If the principal sues, then the agent may not.

BROWN v. OWEN LITHO SERVICE, INC.
384 N.E.2d 1132 (Ind. 1979)

SULLIVAN, J.

James J. Brown appeals from an adverse judgment entered after a bench trial in which he was held personally liable to Owen Litho Service, Inc. (Owen Litho) for what Brown claims to be the debts of J. J. Brown Publishing, Inc. (the Corporation). Brown contends that he incurred the debts while acting as officer and agent for the Corporation and therefore should be absolved from personal liability. The central issue on appeal is whether Brown revealed the capacity in which he contracted, i.e., whether Brown was acting on behalf of an undisclosed principal.

The evidence most favorable to the judgment discloses a series of transactions between Brown and Owen Litho for the printing of at least four issues of Brown's "Fishing Fun" magazine. Wayne Hicks, one of Owen Litho's salesmen, testified that he had heard Brown was a "prospect for a sale" and therefore called on Brown at his Speedway home. Brown expressed interest in Owen Litho's services and asked for an estimate of the cost involved in printing the magazine. As a result of these negotiations, the parties reached an oral agreement for Owen Litho to perform these services. Hicks testified that his dealings with Brown were conducted in Brown's home, that there were no signs or other indications that Brown's home served as a corporate office, and that he was not informed at any time that Brown was agent for the Corporation or even that the Corporation existed. In fact, Hicks testified, Brown represented that he owned the magazine.

The record indicates that Owen Litho printed four issues of "Fishing Fun": one each for August and September, 1973, and two combined issues for the months of October/November and December/January (1974). Full payment was remitted for the September issue (invoiced August 30, 1973) with a check dated September 21, 1973, and half payment for the August issue (invoiced July 30) with a check dated November 19, 1973.

It is well established that an agent, in order to avoid personal liability, must, at the time of contracting, disclose both the capacity in which he acts and the existence and identity of his principal. It is not sufficient that the third person has knowledge of facts and circumstances which would, if reasonably followed by inquiry, disclose the existence and identity of the principal. It is not the duty of third persons to seek out the identity of the principal. Rather, the weight of authority holds that the duty to disclose the identity of the principal is upon the agent. Thus, unless the third person knows or unless the facts are such that a reasonable person would know of the principal's existence and identity, the agent must be held to be acting for an undisclosed principal and is held liable in the same manner as if he were the principal. Actual knowledge brought by the agent or, what is the same thing—that which to a reasonable man is equivalent to actual knowledge—is the criterion of the law.

Most jurisdictions hold that disclosure which occurs subsequent to the execution of a contract has no bearing upon the relations created at the time of the transaction and will not relieve the agent from personal liability.

Whether a principal is disclosed, partially disclosed or undisclosed depends upon the representations of the agent and the knowledge of the third party at the time of the transaction. Thus, disclosure is essentially a question of fact to be determined by the facts and circumstances surrounding the transaction.

Under these facts, a trier of fact could reasonably conclude that Owen Litho did not know or that a reasonable person would not have known of Brown's agency and the existence and identity of his principal. Certainly we cannot say, as a matter of law, that Owen Litho knew or that a reasonable person would have known that Brown was at all times acting as agent for the Corporation.

QUESTIONS FOR DISCUSSION

1. *What should Brown have done in order to avoid liability on the contract?*
2. *At what point in the transaction is it necessary for an agent to reveal that he represents a principal in order to avoid liability on the contract?*

PANAMA REALTY, INC. v. ROBISON

305 So.2d 34 (Fla.App. 1974, rehearing denied, 1975)

In this suit by a real-estate broker, Panama Realty, Inc., to recover a commission, the trial court found in favor of the defendants, and plaintiff appealed. Affirmed.

Robison, Brown, and Miers owned the common stock of Bay Gulf Development Corporation. The corporation held title to certain lands. Robison—without revealing the existence of Brown, Miers, or the corporation—listed the property for sale with Panama Realty, Inc. To Panama's knowledge Robison was sole owner of the property. Miers had also listed the same property with another broker.

Panama Realty presented a buyer to Robison. In the meantime, Bay Gulf Development had conveyed this same property to Robison, Miers, and another. When the buyer was presented, Robison informed Panama that because Miers was a co-owner, his signature would be needed. When contacted, Miers advised Robison and Panama that the other realtor had presented a buyer first. The property was sold to that buyer, and Panama sued for its commission.

PER CURIAM.

The trial court, on the evidence and stipulations in the record, held:

5. Plaintiff (appellant) contends Robison is obligated to also pay it a fee because Robison represented an undisclosed principal. Representation of an undisclosed principal with authority of the principal does not create a personal liability on the part of the agent unless there would be liability against the principal. Here, both Robison and Miers had authority to act for the principal. As stated above, a purchaser was produced by Miers' broker before Robison's broker produced one. The principal was, therefore, obligated to sell to the first purchaser produced and was obligated to only that broker for a commission. An agent contracting with another and disclosing his principal is not liable for his authorized acts. Only the principal is liable. But if the agent does not disclose his principal, then he himself is liable under the contract as the principal. The party contracted with can hold either the agent or the principal liable. Here Plaintiff (appellant) has elected to go against the agent who is liable to him to the same extent, but only to the same extent as the principal. The principal not being liable to Plaintiff for the reason that another broker produced a purchaser ahead of Plaintiff, there is no liability on the part of the agent.

The findings of fact supported by substantial, competent evidence are binding on us. The conclusions of law are correct.

(The final judgment for the appellees, here reviewed, is affirmed.)

Questions for Discussion

1. *Why was there no liability of the principal to the third party in this case? Who were the principal and the agent the court was talking about?*
2. *Was this a disclosed- or an undisclosed-principal situation?*
3. *If there had been liability of the principal in this case, would the agent, Robison, have been liable to the third party?*

(c) *The Third Party.* The undisclosed principal may reveal himself and sue the third party on the contract unless the third party contracted exclusively with the agent. For this reason, if one suspects that he is dealing with an agent rather than with a principal, he may place a clause in the contract stating that he is contracting exclusively with the agent. As previously mentioned, if the principal sues the third party in an undisclosed-principal situation, he is subject to all the defenses that the third party might raise against the agent. Perhaps an exception should be noted at this point in the case of negotiable instruments. If an agent signs a negotiable instrument—such as a promissory note—on behalf of an undisclosed principal, the principal has no rights or liabilities to the third party on the instrument because his name does not appear on it. The agent would, of course, be liable on the instrument to the third party. There is, however, no reason that the principal cannot be sued by the third party on the agreement underlying the negotiable instrument. Suppose, for example, that Artie Agent contracts to purchase certain property from Terry Thirdparty and also signs a promissory note payable to Thirdparty for $10,000. Agent is representing an undisclosed principal. Should there be a breach of contract and failure to pay the note, Thirdparty, upon discovering the identity of the principal, could sue him on the contract but not for payment on the note.

COOPER v. EPSTEIN
308 A.2d 781 (D.C. 1973)

Warren R. Cooper arranged to have his employee Jimmie Buford negotiate the purchase of certain property from Benjamin and Molly Epstein. Buford negotiated a contract with the Epsteins and never revealed the interest of Cooper to them. The contract was essentially for the purchase of the real estate but also provided that Buford was to lease the property for up to thirty-six months prior to sale for a fixed monthly rental. Buford was also to pay all utilities, real-estate taxes, and insurance in addition to providing the Epsteins with a $1,000 deposit. The reason for the lease was because there was an encumbrance on the property. The sale provisions of the contract were to come into play when the encumbrance was terminated.

The contract provided for a sale price of $18,000 with a down payment of $3,000 in cash. The $1,000 deposit and rental payments were to be credited toward the cash down payment. The contract also provided:

> 4. That he (Buford) will not transfer nor assign this agreement, nor let nor sublet the whole or any part of said premises without written consent of Landlord first had and obtained such consent shall not be unreasonably withheld.

When Buford fell behind in the monthly rental payments, the Epsteins notified him that they intended to cancel the contract for nonpayment. Subse-

quently the encumbrance was removed, and the Epsteins sold the property to someone else for $27,000. Cooper then sued the Epsteins, alleging that Buford was his agent and was acting in that capacity when he signed the contract. The trial court ruled that as a matter of law the nonassignment clause precluded suit by an undisclosed principal. Cooper appealed.

FICKLING, A.J.

It is well settled, that, absent some special exception, an undisclosed principal may sue and be sued on a contract made by his agent. Even the fact that the agent denies that there is a principal, or represents himself to be the principal, is not sufficient to preclude suit by the undisclosed principal. However, where the express terms of the contract provide that it (the contract) is to be effective only between the agent and the third party, the undisclosed principal may not enforce the contract. The undisclosed principal simply cannot sue where to do so would violate a term of the written contract. Where the contract terms do not expressly exclude enforcement by or against an undisclosed principal, such liability may nevertheless be precluded if that was the intention of the parties. The resolution of the issue of whether the parties intended to exclude such liability is usually the duty of the fact finder.

We are of the opinion that when the motions judge ruled that Clause 4, supra at 2, precluded suit by an undisclosed principal as a matter of law, he erred. Comment c to Section 303 of the Restatement of Agency (Second) states:

> *Non-assignment clause.* A clause in the contract against assignment does not of itself prevent the principal from bringing suit upon the contract. The existence of such a clause, however, *may be considered as evidence* that the parties intended to exclude an undisclosed principal. . . . (Emphasis added.)

We adopt this statement of the law and, therefore, remand this cause for trial on the merits.

(Reversed and remanded.)

QUESTION FOR DISCUSSION

1. *Under what circumstances may an undisclosed principal be prevented from suing on a contract entered into between the agent and third party? Were those circumstances met in the Cooper case?*

(3) Where the Principal Is Partially Disclosed. There are occasions in which the third party is aware of the existence of a principal but does not know his identity. The third party knows he is dealing with an agent but does not know whom he represents. There is no duty or obligation on the part of the agent to reveal the identity of the principal. His only obligation is not to mislead the third party. If asked, the agent can refuse to identify the principal, but if he lies as to his principal's identity, the agent commits fraud and any subsequent contract may be set aside. The situation of a partially disclosed principal occurs frequently in the case of land acquisition. For example, many people may be aware that a real-estate broker is in the process of buying a number of tracts of land for somebody who is unidentified. In that case, the real-estate broker would be an agent for a partially disclosed principal.

The rules that have been discussed with regard to an undisclosed principal are also applicable to a partially disclosed principal. The third party has rights against both the agent and the principal and may in turn be sued by either the agent or the principal under the circumstances already discussed. However, the agent is not liable to the third party on the contract if the principal is partially disclosed and if the parties intend at the time of contracting that the principal and not the agent should be bound. The intent of the parties is critical in this situation because otherwise the agent is liable to the third party on the contract. It is, of course, always wise to state the intent of the parties in the contract itself, particularly if the agent for the partially disclosed principal does not wish to expose himself to potential liability.

35:3. Tort Liability for Agent's Torts. The question of the principal's liability for the acts of his agent is probably most frequently raised in connection with tort cases. Often this principal is a business organization. The liability of the principal, particularly if the principal is a substantial corporation, is important to the third party for several reasons. First, the principal may be in a better position financially to pay off a substantial verdict or settlement of the case if warranted. Second, whereas it may be difficult to locate the agent in order to recover compensation for one's injuries, a substantial principal is most likely to be well known and available. Finally, like it or not, there is the thought that in the event of a trial a jury is more likely to return a large verdict against a substantial principle, such as a large corporation. The fact that the principal may be liable for the torts of an agent under certain circumstances does not absolve the agent himself from liability, as one is always liable for his own torts.

(1) The Doctrine of Respondeat Superior. The principal is liable for the torts of his agent committed *within the scope of his employment* regardless of any fault of the principal. This is known as the doctrine of *respondeat superior,* "let the superior respond." This rule is derived from public policy and is justified by the fact that the principal is the one who generally benefits from the acts of his agent. Because the principal is best in a position to control the agent or protect himself in the event of loss, such as by the purchase of insurance, the law places this burden upon him.

For example, suppose Artie Agent is employed as a delivery driver by Paul Prin Corporation. While delivering packages for the corporation (the principal), Artie negligently runs through a red light and injures Terry Thirdparty. The corporate principal would be liable to Thirdparty under the doctrine of *respondeat superior* because the tort was committed within the scope of Agent's employment. Note that this liability attaches to the principal despite the fact that there is no fault on its part. Nor was the act of the agent authorized by the principal. If the agent is liable for a tort committed within the scope of his employment, the principal is liable without regard to the question of fault. Of course, if the agent is not liable, then there can be no liability for the principal.

The agent is liable to indemnify his principal for damages incurred by the principal as a result of the torts of the agent. Frequently the principal does not require indemnity from the agent—particularly if the principal is financially substantial—for several reasons. In the first place, the principal is most likely insured against the liability. Second, if the principal is a significant employer, it would be detrimental to morale and employee relations to collect from the employees.

The real question in this discussion is what activities are considered to be "within the scope of employment." It is to that subject that we now turn.

(2) The Scope of Employment and Independent Frolics and Detours. The general rule is that the principal is liable to the third party for the torts of his agent that the agent commits within the scope of employment. The phrase *scope of employment* is not susceptible to precise definition. For instance, the principal is liable for certain torts of an agent even if they are committed without authority. On the other hand if the agent is engaged in furthering his own business and not his principal's, the principal is not liable for the agent's torts. However, the concept of authority seems to have some relation to "scope of employment." The *Agency Restatement, Second,* states:

> § 229. Kind of Conduct within Scope of Employment
> (1) To be within the scope of the employment, conduct must be of the same general nature as that authorized, or incidental to the conduct authorized.
> (2) In determining whether or not the conduct, although not authorized, is nevertheless so similar to or incidental to the conduct authorized as to be within the scope of employment, the following matters of fact are to be considered:
> (a) whether or not the act is one commonly done by such servants;
> (b) the time, place and purpose of the act;
> (c) the previous relations between the master and the servant;
> (d) the extent to which the business of the master is apportioned between different servants;
> (e) whether the act is outside the enterprise of the master or, if within the enterprise, has not been entrusted to any servant;
> (f) whether or not the master has reason to expect that such an act will be done;
> (g) the similarity in quality of the act done to the act authorized;
> (h) whether or not the instrumentality by which the harm is done had been furnished by the master to the servant;
> (i) the extent of departure from the normal method of accomplishing an authorized result; and
> (j) whether or not the act is seriously criminal.

One authority, *Corpus Juris Secundum,* Agency § 424, states that the proper inquiry is "Was the act done in the course of the agency and by virtue of the authority as agent with a view to the principal's business?" In any event, it appears that the agent must be in the process of furthering the principal's business and that the activity should be of the nature of the job for which the agent was hired. Thus, although the term *scope of employment* is broader than the concept of authority and indeed may include liability for acts that are not authorized, the concept of authority does have some input in defining the limits of the activities that are within the scope of employment.

Acts that are clearly not for the purpose of furthering the principal's business are outside the scope of employment. Problems frequently arise when the agent is on the principal's business, then engages in a detour or "independent frolic" of his own, and then returns to the principal's business. What is the liability of the principal for a tort that occurs during the independent frolic?

At what point on the return of the agent from the detour does the principal's liability for torts resume?

If the agent combines the principal's business with his own while on the detour, then the principal remains liable for the torts committed by the agent during this time. If the detour of the agent is in order to take care of his own business, then any tort committed during this time will be committed outside the scope of employment. If the detour is slight, then any tort that occurs during this period is within the scope of employment. For instance, suppose an agent is to make a delivery to a customer some ten miles from the employer's plant. On the most direct route he deviates one block to his sister's house to get a pack of cigarettes and has an accident. Most courts would hold that the detour is not substantial and that the accident occurred within the scope of employment.

At what point the departure becomes so significant that the agent is operating outside the scope of employment is, of course, a question of fact. When this does occur the question is when the agent reenters the employment of his principal. The approach taken by the *Agency Restatement, Second,* is

> § 237. Re-entry into Employment
> A servant who has temporarily departed in space or time from the scope of employment does not re-enter it until he is again reasonably near the authorized space and time limits and is acting with the intention of serving his master's business.

BRINKLEY v. FARMERS ELEVATOR MUTUAL INSURANCE CO.
485 F.2d 1283 (1973)

William and Helen Brinkley sustained severe injuries when the vehicle in which they were riding collided with a vehicle owned and operated by Clarence N. Holeman. Holeman was killed in the accident. Holeman was intoxicated and the collision was solely the result of his negligence. Holeman was a practicing attorney and had been employed by Farmer's Elevator Mutual Insurance Co. to represent one of its insureds in a suit brought against the insured in Dodge City, Kansas.

The Brinkleys sued Farmers, alleging that it was liable for their injuries because Holeman was Farmers' agent, servant, and employee and therefore liable under the doctrine of respondeat superior. *The trial court dismissed the action, and the Brinkleys appealed.*

McWILLIAMS, C.J.

As indicated, Holeman was an attorney-at-law licensed to practice in Kansas, and he maintained his own law office in Wichita. In connection therewith Holeman employed a legal secretary and he had recently employed a young attorney to assist in the operation of his law office. Holeman had maintained his own law office since about 1955, when he had left a firm of lawyers with whom he had previously been associated for several years. Holeman held himself out to the public as engaging in the general practice of law, and he was so considered by other members of the Bar in the Wichita area.

Farmers is an insurance company incorporated in Iowa and maintaining its principal offices in Des Moines, Iowa. It has an office in Hutchinson, Kansas, which is headed by a claims adjuster who is a full-time employee of Farmers.

In the conduct of its business, Farmers always attempted to negotiate and settle any claim against one of its insureds through its claims adjuster. If a claim could not be settled, and a lawsuit was instituted against one of its insureds, then Farmers at that point would employ counsel to defend the suit. Prior to 1955,

Farmers had frequently employed the firm with which Holeman was then associated to represent its insureds. After Holeman terminated his association with the firm in 1955, Farmers began employing Holeman to represent its insureds against whom suit had been filed. This relationship grew to the end that in the ensuing years Holeman was traveling all over Kansas in his representation of Farmers' insureds and at one time the monies received by Holeman from Farmers represented nearly 90% of his total income from his practice of the law.

In traveling around the state of Kansas, Holeman generally used his own vehicle, although he sometimes traveled with others or used commercial airlines. There was no contract between Farmers and Holeman, and the latter was hired on a case-by-case basis, with his fees for services being rendered on an hourly basis at a standard charge. He was also reimbursed by Farmers for out of pocket expenses, including travel expense at 8¢ per mile.

The manner in which a particular case would be tried was in Holeman's hands, with Farmers reserving only the right to decide whether a given case should be settled or tried. Accordingly, though an adjuster sometimes sat in on a trial, Holeman was in command of the trial itself, as well as the pretrial investigation and preparation.

It was in this general factual setting that Farmers employed Holeman to represent one of its insureds who was being sued in Dodge City, Kansas, which is located some 150 miles west of Wichita. Pursuant to his employment, Holeman drove in his own automobile to Dodge City where, during the morning of October 11, 1967, he tried a case for Farmers' insured. This trial concluded before noon on the 11th. What Holeman did on the afternoon of the 11th is not known, although it is agreed that he performed no other services for Farmers during that time. En route back to Wichita, Holeman in the early evening of the 11th was involved in the accident out of which the present controversy arose.

Under Kansas law, the relation between an attorney and his client has been held to be one of agency to which the general rules of agency apply. . . .

However, even though the relationship between Holeman and Farmers be one of agency, it does not necessarily follow that under the doctrine of respondeat superior Farmers is vicariously liable for Holeman's tortious conduct toward the Brinkleys. Under Kansas law, the liability of a principal for the negligent acts of his agent is controlled by a determination as to whether, at the time in question, the agent was engaged in the furtherance of the principal's business to such a degree that the principal had the "right to direct and control" the agent's activities. If the principal had no such right to direct and control, then he is not vicariously liable to third parties for the consequences of the agent's tortious conduct.

Under the authorities above cited and based on the stipulated facts, we agree with the trial court that at the time he drove his automobile into the Brinkley vehicle, Holeman was not engaged in the furtherance of Farmers' business to such a degree that it could be said that Farmers, as the principal, had the right to direct and control Holeman's physical conduct. Accordingly, Farmers is not vicariously liable for the consequences of Holeman's misconduct. If Farmers had no right to direct or control Holeman's courtroom activities in Dodge City, Kansas, on the morning of October 11, 1967, which right it is agreed Farmers did not have even though Holeman in one sense was its agent, even less did it have a right to direct or control his homeward jaunt during the evening of that same day. Even as the messenger boy in *Kyle*, Holeman was free to go where he desired and could use his own mode and route of travel and he might well have chosen not to go home at all. In sum, Farmers had no right to direct or control Holeman's physical activities at the time he collided with the Brinkley vehicle.

(Judgment affirmed.)

Questions for Discussion

1. *What facts do you feel were important in the court's determination that Farmers did not have the "right to direct and control" Holeman at the time of the accident? Were there any facts that might support the Brinkley's position?*
2. *Would the outcome have been different if the accident had occurred when Holeman was on his way to the trial or at the lunch break?*
3. *Would the outcome have been different if Holeman had been a staff attorney for Farmers and had been paid a salary?*

RILEY v. STANDARD OIL CO. OF NEW YORK

231 N.Y.301, 132 N.E. 97 (1921)

Million was employed by the Standard Oil Company as a truck driver. He was ordered to go from Standard's mill to the freight yards of the Long Island Railroad, about two and one-half miles away, to pick up some barrels of paint. He was to return to the mill at once. After he picked up the paint and some scrap wood, Million left the freight yard and went four blocks in the opposite direction from Standard's mill to the home of his sister in order to drop off the scrap wood. After doing this, Million started to return to his employer's mill. At a point before he reached the entrance to the freight yards, which he was required to pass in order to return to the mill, Million was involved in an accident in which Arthur Riley was injured. Million was negligent and Riley sued his employer, Standard Oil Co. The trial court rendered judgment against Standard. The appellate division reversed the trial court, and Riley appealed.

ANDREWS, J.

A master is liable for the result of a servant's negligence when the servant is acting in his business; when he still is engaged in the course of his employment. It is not the rule itself but its application that ever causes a doubt. The servant may be acting for himself. He may be engaged in an independent errand of his own. He may abandon his master's service permanently or temporarily. While still doing his master's work he may be also serving a purpose of his own. He may be performing his master's work but in a forbidden manner. Many other conditions may arise.

A servant may be "going on a frolic of his own, without being at all on his master's business." He may be so distant from the proper scene of his labor, or he may have left his work for such a length of time, as to evidence a relinquishment of his employment. Or the circumstances may have a more doubtful meaning. That the servant is where he would not be had he obeyed his master's orders in itself is immaterial, except as it may tend to show a permanent or a temporary abandonment of his master's service. Should there be such a temporary abandonment the master again becomes liable for the servant's acts when the latter once more begins to act in his business. Such a re-entry is not effected merely by the mental attitude of the servant. There must be that attitude coupled with a reasonable connection in time and space with the work in which he should be engaged. No hard and fast rule on the subject either of space or time can be applied. It cannot be said of a servant in charge of his master's vehicle who temporarily abandons his line of travel for a purpose of his own that he again becomes a servant only when he reaches a point on his route which he necessarily would have passed had he obeyed his orders. He may choose a different way back. Doubtless this circumstance may be considered in connection with the other facts involved. It is not controlling.

We are not called upon to decide whether the defendant might not have been responsible had this accident occurred while Million was on his way to his sister's house. That would depend on whether this trip is to be regarded as a new and independent journey on his own business, distinct from that of his master, or as a mere deviation from the general route from the mill and back. Considering the short distance and the little time involved, considering that the truck when it left the yards was loaded with the defendant's goods for delivery to its mill and that it was the general purpose of Million to return there, it is quite possible a question of fact would be presented to be decided by a jury. At least, however, with the wood delivered, with the journey back to the mill begun, at some point in the route Million again engaged in the defendant's business. That point, in view of all the circumstances, we think he had reached.

(Judgment of the Appellate Division reversed in so far as it directed a dismissal.)

Questions for Discussion

1. *Would the outcome have been different if the accident had occurred after Million had left the freight yard but before he arrived at his sister's?*
2. *Why did the court find Standard liable when the accident occurred before Million returned to the point where he started his "detour" from his "master's" business?*

SUMMARY

When the principal is disclosed, the principal is liable to the third party for contracts entered into on his behalf by his agent acting within the scope of the agent's authority. Likewise, the third party is liable to the principal for any breach of such a contract. The agent is not a party to the contract and may not successfully sue or be sued on the contract.

If the principal is undisclosed, he is still bound to the contract as is the third party. The agent also may be liable to the third party for any breach of the contract and may sue the third party in his own name instead of his principal's. If the third party sues both the principal and agent, in most jurisdictions he must elect to take a judgment against one or the other, not both. In addition, an undisclosed principal cannot enforce a contract against a third person that involves skill, trust and confidence, or personal service. Where the principal is partially disclosed, the sales are similar to the situation where there is an undisclosed principal.

The principal is liable for the torts of his agent committed within the scope of his employment. The agent is liable to indemnify his principal for damages incurred by the principal as a result of torts of the agent. Generally, in order to be acting within the scope of employment the agent must be in the process of furthering the principal's business and the activity should be of the nature of the job for which the agent was hired. If the agent is on a significant detour when the tort occurs, then it is not within the scope of employment.

PROBLEMS

1. Julian Davis solicited the Fletcher Emerson Management Company for a contract to landscape "Emerson Center" in Atlanta. Fletcher Emerson replied that goods should be shipped to "Fletcher Emerson Management Co. . . . To be Used for Emerson Center." Business was conducted on an open-account basis. While $11,795.43 was still owing, an officer of Fletcher Emerson sent a letter to Davis stating that Fletcher would not be responsible for any further purchases for Emerson Center and that future invoices should be sent to Atlanta Venture No. 1. Davis sued Fletcher Emerson, contending that it was an agent for an undisclosed principal. Fletcher Emerson contended that it was merely a property-management agency and that its name indicated as much. Was this contention sufficient to constitute Fletcher an agent for an undisclosed principal? Fletcher Emerson Management Co. v. Davis, 134 Ga.App. 699, 215 S.E.2d 725 (1975).
2. Taylor stated to Mr. and Mrs. Taglino that he represented two individuals who wished to purchase their property for the erection of a dwelling house. The Taglinos entered into a contract with Taylor for the sale of the property. Taylor in fact was an agent for the White Tower Management Corporation, which was engaged in the restaurant business and intended to build a restaurant on the property. When the Taglinos discovered this, they refused to

sell the property. White Tower sued for specific performance of the contract. Was White Tower entitled to a judgment? White Tower Management Corp. v. Taglino et al., 302 Mass. 453, 19 N.E.2d 700 (1939).

3. An employee of Albright Transfer & Storage Co. was directed to haul a load of used telephone poles from Wichita Falls to Springtown, ninety-three miles to the southeast. The employee was then to return to Wichita Falls and was authorized to eat en route. The driver and his helper arrived at Springtown, unloaded the poles, and then proceeded ten miles further to a town named Agle to eat and have a beer. After leaving Agle, the driver began to return to Wichita Falls by a different route, thinking that it was a shortcut. He made a wrong turn and was attempting to find the proper route when an accident occurred involving Mosqueda, who sued the employer, Albright Transfer. Could Mosqueda recover from Albright Transfer? Mosqueda v. Albright Transfer & Storage Co., 320 S.W.2d 867 (Tex.Civ.App. 1958).
4. Kelly Asphalt Block Co. hired Booth as its agent to purchase asphalt paving blocks from Barber Asphalt Paving Co., a competitor. Kelly hired Booth because it feared that if Barber was aware that it was the purchaser, it would refuse to sell the blocks. Booth contracted with Barber Asphalt without revealing that he represented Kelly. The paving blocks were unmerchantable, and Kelly sued for damages. Barber contended that Kelly should not recover because it would never have delivered the blocks had it known that Kelly was the principal. Was Barber Asphalt correct? Kelly Asphalt Block Co. v. Barber Asphalt Paving Co., 211 N.Y. 68, 105 N.E. 88 (1914).
5. Hartford Ice Cream Co. had instructed its driver to collect payment on a delivery of ice cream to Louis Son, a shopkeeper. When the driver made a delivery on the day in question, Son claimed that the ice cream was not properly iced and refused to receive it. Undaunted, the driver insisted on leaving the ice cream and attempted to take the money from Son's cash register. Son locked the cash register, and the driver then attempted to carry away the cash register. A fight ensued in which Son was kicked and beaten. Son sued Hartford for damages. Hartford contended that the driver's assault was not within the scope of employment. Was Son entitled to judgment against Hartford? Son v. Hartford Ice Cream Co., 129 A. 778 (Conn. 1925).
6. Young purchased wood from Poretta. Young was acting as an agent for Superior Dowel Co. at the time but did not disclose this fact to Poretta. Poretta was not paid the purchase price for the wood and sued Superior Dowel when he learned of its existence. Superior Dowel defended on the ground that it had previously paid the purchase price to Young. Was this a good defense? Poretta v. Superior Dowel Co., 153 Me. 308, 137 A.2d 361 (1957).
7. Phillips was president of Metropolitan Designed for Living, Inc., a home-construction company. Henderson, a plumber, was contacted by Phillips to do some work on a particular house. Henderson sent two written contracts

addressed to "Design for Modern Living." The contracts were signed "James O. Phillips"and returned to Henderson. After Henderson did some work, he was given a partial payment on a check that was marked "Metropolitan Designed for Living, Inc." at the top. Henderson sued Phillips for the balance due him, alleging that Phillips was the agent for an undisclosed principal. Was Henderson entitled to judgment? Henderson v. Phillips, 195 A.2d 400 (D.C. 1963).

8. Beeman contracted to purchase certain real property. He negotiated a contract with May who represented the seller. May's principal was disclosed to Beeman. Beeman gave a $200.00 down payment on the property to May. The seller was unable to give a good title to the property. When the $200.00 down payment was not returned, as provided for in the contract, Beeman sued May. Can Beeman recover? Beeman v. May, 85 N.Y. Supp. 2d. 123 (1948).

CHAPTER 36

Rights and Liabilities of Agent and Third Party

We have already discussed some of the aspects of the agent's rights and liabilities to the third party. There are, of course, two areas that involve these rights and liabilities, namely, the field of contracts and the field of torts. Let us first examine the agent's liability and rights against the third party based on contract and then examine the relationship of the agent and the third party based on tort liability.

36:1. Liability of Agent to Third Party Based on Contract. (1) Liability Based on Writing. You should already be aware that as a general rule an agent is not liable to the third person on authorized contracts if the principal is disclosed. Frequently, however, the question arises as to whether a party to a written contract intended to sign the contract on his own behalf or as an agent for a principal. The problem is often complicated by the manner in which the party has signed the agreement. The proper way to sign a contract as an agent is to sign it in such a manner that there is no doubt that the parties know that it is being signed by an agent of a named principal. For example, one might properly sign a contract as an agent in the following manner:

Prin Corporation
by
Albert Able, its Agent

When the contract is signed in this way, there is no doubt that Albert Able is signing as an agent for Prin Corporation. Unfortunately agents often sign a contract "Albert Able, Agent" or "Albert Able, agent of Prin Corporation" or worse yet, "Albert Able, president." The words *agent, agent of Prin Corporation,* or *president* may be held by a court to be merely descriptive of the status of the person who signed the agreement. Thus Albert Able would be held to have signed the contract on his own behalf, and he happens to be a "president" of some kind. Although there is some conflict, most courts would hold that such a signature is ambiguous. Therefore it is possible for the agent, if sued, to establish through parol evidence that the parties to the contract knew and understood that he signed as an agent. This problem can be avoided if one executes the contract properly in the first place.

Special problems arise if the document signed by the agent is a negotiable instrument. If the name of the principal does not appear on the instrument and the name of the agent alone appears, the principal is not liable, but the agent is liable. In this case, parol evidence may not be used to show that the parties intended the signature to be that of an agent. The Uniform Commercial Code provides that an authorized agent who signs his own name to an instrument is liable if the instrument neither names the person represented nor shows that the representative signed in a representatives capacity. Section 3–403(2)(a).

An additional problem arises not only with negotiable instruments but also with contracts when the agent signs the principal's name, "Prin Corporation," and then signs his own name, "Albert Able," immediately following without indicating that he is signing in a representative capacity. The problem with regard to contracts is that it appears that Albert intends to be bound along with the principal, Prin Corporation. With regard to negotiable instruments this mistake can have disastrous results for the agent, as he will be held liable on the instrument along with the principal. The Uniform Commercial Code states that, "An authorized representative who signs his own name to an instrument . . . except as otherwise established between the immediate parties, is personally obligated if the instrument names the person represented but does not show that the representative signed in a representative capacity, or if the instrument does not name the person represented, but does show that the representative signed in a representative capacity." Section 3–403(2)(b). The code also states, however, that with regard to a negotiable instrument, the name of an organization preceded or followed by the name and office of an authorized individual is a signature made in a representative capacity. Section 3–403(3). Thus there would be no liability for the agent who signs in that manner because of this code provision. Remember that it applies only to negotiable instruments however.

DUNLOP v. McATEE

31 Ill.App.3d 56, 333 N.E.2d 76 (1975)

Crane M. Construction Co. was an Illinois Corporation of which Crane M. McAtee was president. Crane Construction undertook to do certain construction work on certain premises. Roy S. Dunlop, as subcontractor, proposed to furnish heating and plumbing materials for the property. The proposal was accepted in the following manner: "Crane M. Construction Co., by C. M. McAtee."

Subsequently Dunlop filed to foreclose a mechanic's lien against the property and named Crane M. McAtee as a defendant. The trial court rendered judgment in favor of McAtee, and Dunlop appealed.

HALLET, J.

This brings us to the plaintiff's *second* contention—that the contract between him as subcontractor and the contractor was so executed as to make both the corporation and its president individually liable thereon as a matter of law.

In 3 Am.Jur.2d—Agency, in section 190, it is said that:

> A signing by which the name of the principal appears "by" or "per" the agent is uniformly regarded as a proper method of executing the agency so as to impose liability upon the principal and, conversely, no personal liability upon the agent.

In Restatement of the Law of Agency, 2d, in section 156, it is said in comment:

> In the absence of a contrary manifestation in a document, the following signatures and descriptions, among others, created an inference that the principal and not the agent is a party: The principal's name followed by the agent's name preceded by a preposition such as "by" or "per"; the principal's name followed by the agent's name with the word "agent" added; the agent's name followed by the principal's name, the two names being separated by a phrase such as "agent of," "agent for," "on behalf of," "for" or "as agent of."

Illinois follows this view and holds that where an agent discloses the name of his principal or where the party dealing with the agent knows that the agent is acting as an agent, the agent is not personally liable on the contract unless he so agrees.

(Affirmed.)

(2) Liability Where Agency Is Not Disclosed. The discussion in Chapter 35 revealed that if the agency was not disclosed—that is, in an undisclosed-principal situation—the agent is liable to the third party on the contract. The reason is that it is with the agent that the third party thinks he is dealing and it is, of course, the agent who signs the contract in his own name if the contract is in writing. On the other hand, if the agency is undisclosed, the third party may be sued by the agent for breach of contract, although the agent's right is subordinated to that of the principal, should he choose to exercise it. Furthermore any defenses that the third party could raise against the principal, he could raise against the agent.

(3) Liability Where Agency Is Disclosed. You should also remember that if the principal is disclosed, the agent has no liability to the third party under normal circumstances. There are situations, however, in which the agent may be liable to the third party even if the principal is disclosed, for example, if the agent assumes liability under the contract or if the agent enters into unauthorized contracts. Let us examine these two situations further.

(a) *Agent's Assumption of Liability.* It is possible for the agent to obligate himself under the terms of a contract even if the principal is disclosed. This usually happens when the agent guarantees the performance of the principal or otherwise pledges his own credit and responsibility. Typically the third party may be unwilling to rely exclusively on the obligation of the principal, and in order to make the contract, the agent gives the guarantee or otherwise obligates himself. If the third party accepts, both the principal and the agent are liable for any breach of the contract. Note, however, that the liability of the agent is founded upon his contractual obligation and is not a result of his position as agent.

For example, suppose a contract of a corporation is signed on its behalf by its president as agent. The contract provides that "it shall be the duty of the undersigned officers to make the payments . . ." if the corporation does not perform.[1] If the corporation fails to perform, the president would be personally liable on the contract.

In order for the agent to sustain any liability in this case, it is necessary to clearly establish the intention of the agent to be bound on the contract because the law presumes that an agent who enters into a contract for a disclosed principal intends that his principal alone should be bound and that no liability is to be incurred by the agent. The manner in which the agent signs the contract is, of course, one factor to consider in a determination of the agent's intent.

NAGLE v. DUNCAN

570 S.W.2d 116 (Tex.App. 1978)

COLEMAN, J.

This is an appeal from a judgment for the plaintiff in an action for debt. The plaintiff, an official court reporter, sued the defendant, an attorney, for the cost of a transcript of the testimony adduced in a criminal trial which the attorney had ordered for use in an appeal.

As a general rule an agent is not personally liable on contracts made for a disclosed principal, in the absence of his express agreement to be bound. However, the agent of a known principal is not precluded from binding himself personally upon a contract if the agent has substituted his own responsibility for that of his principal, or has pledged his own responsibility in addition to that of his principal.

Although there are some cases for a contrary rule the majority rule has been stated by a commentator as follows:

> Applying the ordinary rule that an agent is not personally liable on a contract, made for his principal where the party with whom he is contracting knows that he is acting for another, and taking the view that an attorney incurring expenses in connection with litigation is ordinarily acting as an agent for the litigant, and that this is usually known to those with whom he deals, many courts have held that in such a situation the attorney is not personally liable for the charge in question, unless he expressly or impliedly assumes special liability therefor. Annotation, Attorney's Personal Liability for Expenses Incurred in Relation to Services for a Client, 15 A.L.R.3rd 531, 536.

In this case the plaintiff was the official court reporter at the trial of Cause No. 373826 entitled *State* v. *Excell Marks.* On February 5th defendant was found guilty of a misdemeanor and by letter dated February 20th, 1974, addressed to the official court reporter, County Criminal Court No. 1, Harris County Court House, and referring to the style of the case above mentioned, a transcript of the proceedings was ordered. The letter read: "Please accept this letter as our request to you to prepare the transcript of the proceedings in the above matter as quickly as possible, in order that we may continue with the appeal of the conviction rendered February 5, 1974." The letter was signed "Law Offices of David J. Nagle" followed by a signature line on which appears the signature of David J. Nagle.

Another letter dated March 13, 1974, addressed to Mr. Bill Duncan, Court Reporter, following the form used in the previous letter, reads:

> Enclosed is the firm's check in the amount of $150.00, as a deposit against the costs of the record to be prepared in the above matter.
>
> Please advise me when the record has been completed of the remaining charges.

There was also introduced into the record a statement dated April 24, 1974, billing Mr. Nagle for the costs of the statement of facts in Cause No. 373826 in the amount of $369.20, less $150.00 deposit, showing a balance due of $219.20.

1. *Steele* v. *Hollandale,* 125 So.2d 87 (Fla.App. 1960).

A letter from Mr. Duncan addressed to Mr. Nagle dated October 7, 1974, requesting payment of the sum of $219.20 was also introduced into evidence. Mr. Duncan testified that he received the $150.00 deposit and that no other money had been paid him for his services in preparing the statement of facts. He testified that it was his custom and generally the custom among court reporters in criminal courts that no transcript would be delivered to the attorney unless the charges were paid in full. He further testified that he knew the record was being ordered for the client and knew that Mr. Nagle was an attorney for the client. He did not deliver the transcript of testimony to Mr. Nagle, but he notified him that the record was ready and would be delivered on receipt of his check in the amount of $218.00.

Appellant contends by two points of error that there is no evidence that he agreed to be personally responsible for the costs of preparing the transcript of evidence. These points will not be sustained. The letter of March 13 constitutes sufficient evidence of the attorney's intent to be personally responsible for the cost of the record to raise an issue of fact.

In *Seale* v. *Nichols,* 505 S.W.2d 251 (Tex. 1974) the court said:

> Texas law provides that in order for an agent to avoid liability for his signature on a contract, he must disclose his intent to sign as a representative to the other contracting party. Uncommunicated intent will not suffice.

(The judgment is affirmed.)

QUESTION FOR DISCUSSION

1. *What should the attorney have done in order to make sure that he was not personally obligated to pay the bill?*

(b) *Agent's Liability on Unauthorized Contracts.* The agent may find himself liable if he enters into a contract with a third person that exceeds his authority. He may also incur liability if he makes a contract on behalf of his principal. If he makes a contract without the authority to do so or exceeds his authority, he is liable to the third party—not on the contract but for breach of the implied warranty. Thus, suppose that Albert Agent makes a contract on behalf of his principal with Terry Thirdparty. Albert has been told by his principal not to enter into any contracts exceeding $500. This one is for $2,000. Terry could sue Albert for breach of warranty of authority. The same would be true if Albert had no authority at all. Terry's right of recovery for breach of implied warranty of authority would be particularly critical if the situation did not warrant an action by Terry against the principal on the theory of apparent authority. The agent is liable on the contract regardless of whether he acted with the knowledge that he was exceeding his authority or honestly believed that he had the authority.

There is some divergence of opinion on this subject. A minority of jurisdictions would hold the agent liable on the contract itself when the contract is executed without authority. The theory for this view is that because the contract is intended to bind someone, the agent should be liable as a principal. This is not the preferred view, however, and is rejected by the majority of states on the ground that the courts would be creating a new contract for the parties.

Another theory of liability against the agent would be an action for deceit or fraud. The basis for this action would have to be the knowing misrepresentation by the agent of his authority or the misrepresentation of the existence of a principal.

There are some instances in which the agent is not liable to the third party even when he exceeds his authority. If the principal ratifies the action of the agent, he relieves the agent of any liability to third parties. Furthermore, if the third party is aware that the agent is acting without authority or has equal means of ascertaining the agent's authority, the agent is not liable to the third party. Also, if the agent manifests to the third party that he does not warrant his authority, the third party may not recover from the agent for breach of warranty of authority. If the agent indicates this fact to the third party either by stating that he is acting without authority or in some other manner, then the third party has not relied on the agent's warranty and has no reason to look to the agent for damages.

Finally, if the principal would not be legally bound to the contract entered into by the agent, the agent is not liable even if he did not have the authority to enter into the contract. Suppose that the agent of a corporation signs a contract obligating the corporation to loan money to certain individuals. The agent makes this contract on behalf of the corporation, although he does not have the authority to do so. If the contract can not be enforced against the corporation because it does not have the power to loan money or because it is illegal for it to do so, the agent would not be bound.

MOORE v. LEWIS

366 N.E.2d 594 (Ill.App. 1977)

Carole J. Lewis was vice-president of Calumet Federal Savings and Loan Association. On October 24, 1974, she telephoned William D. Moore and offered to sell him the mortgage on Moore's property for $10,000.00 which was less than half its amount. The property was in a state of disrepair. She informed Moore that he had until October 29, 1974 to accept the offer. During the conversation she held herself out as the authorized agent of Calumet Savings and Loan and did not indicate any limitations on her authority. Moore accepted the offer on October 29.

Between October 24 and 29, Moore expended large sums of money repairing the building. On October 31, 1979, Lewis informed Moore that the Savings and Loan would not sell him the mortgage for $10,000.00. Moore later learned that Lewis had no authority to enter into the agreement without authorization and approval of the Board of Directors of the Savings and Loan. Moore then sued Lewis for damages for breach of implied warranty of authority.

ROMITI, J.

The plaintiff (Moore) in his complaint and brief complained that he had been damaged by the defendant's concealment of her lack of authority to agree to sell the mortgage. The defendant's contention, in response, that she did not sign the contract is irrelevant. The plaintiff under this theory is not attempting to sue the defendant on the contract. Rather, the plaintiff is claiming that he was damaged because the defendant purported to make a contract in the name of the Savings and Loan which she was not authorized to make. It is well settled that one who purports as agent to enter into a contract upon which the principal is not bound because the agent has contracted without or in excess of the authority given is personally liable for the damage this occasions to the other contracting party because, in effect, the agent warranted his or her authority. Accordingly, accepting the plaintiff's allegations as true for purposes of review, the defendant could be found liable for a breach of the implied warranty of authority unless (1) the contract for some reason other than lack of authority was not binding on the Savings and Loan; (2) any damages which occurred were not caused by the breach of warranty or (3) the plaintiff was not damaged by the defendant's actions. We will consider each of these in turn.

The defendant has rested her defense mainly on the contention that the alleged contract with the Savings and Loan would not have been enforceable since it was oral, thus violating the Statute of Frauds. Of course, if this were true, the plaintiff could not complain that the defendant's lack of authority also rendered the contract unenforceable. (2 Restatement of Agency (Second) § 329, Comment J.) However, the Statute of Frauds has no application to this case. In effect, what the plaintiff was seeking was a discharge of the debt upon payment of $10,000, and a parol release or accord and satisfaction of the notes is sufficient to release the debt.

Since the Statute of Frauds was not applicable, the plaintiff's acceptance of the defendant's offer would have given rise to a contract if the defendant had been authorized to make the offer. Accordingly, the plaintiff would be entitled to recover for any damages sustained because of the defendant's lack of authority. However, the plaintiff cannot recover for the repairs because the loss was not caused by the defendant's unauthorized act.

According to plaintiff's complaint, the bill of particulars and his affidavit, the offer was made on October 24. The repairs were made between October 24 and October 29. But an offer does not ripen into a contract until it is accepted and the offer was not accepted until October 29 or October 31. In other words, at the time the repairs were made, there was no apparent contract. Thus, the plaintiff could not have made the repairs in reliance on the existence of a contract to sell him the mortgage. Since he did not act in reliance on the existence of the contract, it is immaterial whether the defendant had authority to make the offer or not, for he could not have been misled to his detriment by her concealment of any lack of authority until after he had accepted the offer.

(Judgment in favor of Lewis, affirmed.)

Questions for Discussion

1. *Did Lewis breach the implied warranty of authority in this case? If so, why was she not held liable by the court?*
2. *How should Lewis have handled this situation in order to avoid exposing herself to potential liability?*

(4) Liability Where Principal Is Nonexistent or Incompetent. An agent is liable to the third party if he contracts on behalf of a nonexistent principal, because the agent impliedly warrants the existence of the principal. This liability of the agent exists regardless of whether he acts in good faith or not, and it occurs most frequently in one of two situations. First, in order to create a corporation, the state must grant a corporate charter. This charter is usually granted at the behest of the organizers of the corporation, known as *promoters.* If, for example, a promoter enters into a contract for the leasing of office space prior to the time that the corporate charter is granted, the promoter would be liable on the lease. He would remain liable on the lease even after the corporation came into existence, when the corporate charter was granted by the state.

The other instance in which an agent enters into a contract for a nonexistent principal occurs when the agent represents an organization that is not incorporated and therefore has no legal existence. Social and service clubs, religious groups, and other organizations may have a large number of members and may be very active, but if the group is not incorporated or otherwise organized under state law, it has no legal existence. Thus, suppose that Albert Agent is a member of the building committee for a local church. He signs a contract with the builder as agent for the church, and the church defaults in payments. Albert is personally liable on the contract if the local church is not incorporated.

He may also find that the organization suddenly has fewer members than he previously thought once the contractor pressures for collection under the contract.

On the other hand, what if the principal is nonexistent because he has died and his death is unknown to the agent when he makes the contract? The courts generally hold that the agent is not liable in this case.

If the principal is incompetent, the agent is generally liable to the third party if he contracts on behalf of that principal. Thus, if the principal is insane or is a minor, the third party may look to the agent for any damage he sustains as a result of that lack of capacity.

If the third party has knowledge of the principal's nonexistence or lack of capacity, the agent is not liable. The same rule applies if the contract states or implies that the agent is not liable should it be determined that the principal lacks capacity. The reason obviously is that the third party is not relying on any representation of the agent in entering into a contract under these circumstances.

HARRIS v. STEPHENS WHOLESALE BLDG. SUPPLY CO., INC.
309 So.2d 115 (Ala.App. 1975)

BRADLEY, J.

Stephens Wholesale Building Supply Company, Inc., a corporation, filed an action in the Circuit Court of Jefferson County, Bessemer Division, against Norman R. Harris, individually and doing business as Bessemer Building and Improvement Company and also against Bessemer Building and Improvement, Inc., a Corporation, on an account. The amount claimed was $3,248.89 with interest. Attorneys fee of $812.22 was also sought. Defendant Harris denied owing plaintiff the amount of money claimed in the complaint. Bessemer Building and Improvement, Inc., a Corporation, admitted owing plaintiff an undetermined amount of money. Trial was held before the court sitting without a jury and a judgment was rendered in favor of plaintiff and against Norman R. Harris in the amount of $4,061.11 and costs. The motion for a new trial was overruled. Appeal was thereupon perfected to this court.

The evidence submitted to the court shows that on February 14, 1969 a credit application was made to Stephens Wholesale Building Supply Co., Inc. (hereinafter referred to as Stephens) by Norman R. Harris. The name of the account was to be "Bessemer Building & Improvement." The address given was "1920-8th Ave. No., Bessemer, Ala." The terms of the credit application specified that applicant was to pay all costs of collecting delinquent debts including a reasonable attorney's fee, and such agreement was to remain in force so long as there was an outstanding indebtedness. Applicant also waived any exemption rights that he might have to personal property. It was further provided in said application that Stephens was to deliver ordered goods and charge them to the applied-for account, and such arrangement would continue until there was written notice to the contrary. Acceptance of a proposed change was to be evidenced by Stephens in writing.

The certificate of incorporation for Bessemer Building and Improvement Co., Inc. was filed in the Jefferson County Probate Office, Bessemer Division, on April 2, 1969.

The first bill from Stephens was paid by a check drawn on the account of Bessemer Building & Improvement, Inc. on May 9, 1969. This business relationship lasted until February 23, 1973. The account was written off as a bad debt by Stephens on May 30, 1973. The evidence fails to reflect whether or not the first purchase was made before or after the incorporation. Appellant suggests in brief that it was probably made after the incorporation.

Appellant contends in brief that there is no evidence that he, in an individual capacity, contracted with Stephens for the purchase of building materials, but there is evidence that Stephens knew or should have known that it was dealing with a corporate entity.

The deciding question before this court, as it was before the trial court, is whether or not Stephens dealt with Harris in an individual capacity or as a representative of a corporation.

Prior to April 2, 1969, which was the day on which the articles of incorporation were filed in the Bessemer Probate Office, Bessemer Building and Improvement was not a de jure corporation.

Furthermore, Bessemer Building and Improvement was not a de facto corporation prior to April 2, 1969, for there is no evidence in the record that an attempt was made to incorporate it, either bona fide or colorable. . . .

The evidence does not support the theory that Bessemer Building and Improvement was considered by Stephens to be a corporation. To the contrary, the evidence preponderates to the opposite theory, i.e., Stephens dealt with Harris and Bessemer Building and Improvement as an individually owned proprietorship. We conclude therefore that there was no corporation by estoppel in this instance.

We would further observe that Harris did not sign the credit application in a representative capacity and even had he done so, he still would have been liable personally, for there was no evidence of an agency relationship. If a person professing to be an agent of another makes a contract, he thereby imposes personal liability upon himself when there is no principal to bind.

(No reversible error having been cited to us, the judgment of the trial court is affirmed.)

QUESTIONS FOR DISCUSSION

1. *What was the more important factor in the court's decision: the date when the certificate of incorporation was filed or the date when the first bill was paid?*
2. *If one contracts in the name of a corporation, prior to its formation, what is that individual's liability, if any, after the corporation is legally in existence?*
3. *The court discussed* de jure *and* de facto *corporations, as well as the doctrine of estoppel. From your reading of the case, do you have some idea of what these terms mean? They are covered in detail in Section Eight, "Corporations."*

(5) Agent's Liability for Money or Things Paid to Him. If an agent receives money from a third party and in turn pays it over to a disclosed principal, he has no liability to the third party, even if the third party is entitled to recover the money, because the agent has a duty to turn money over to his principal. Of course, the agent must have no knowledge of the third party's claim when the money is turned over to the principal. On the other hand, if the agent represents an undisclosed principal, he is liable for the return of the money to which the third party is entitled even though he pays it to the principal. The reason for this rule is that if the third party is ignorant of the existence of the principal, there can be no presumption of consent or instruction that the money be turned over to the principal. Finally, if the third party pays money to the agent as a result of the fraud of the agent, the agent is liable for its return even though the agent has paid it over to the principal.

The same rules generally apply if the agent has received things other than money. There is some difference in the duty of the agent to return things received from a third party if the principal is disclosed or partially disclosed, depending upon whether the right of the third party to the return of things or proceeds existed when the agent received the things or whether the third party's right arose subsequently. If the third party's right to the return of things existed when the agent received them, then the agent has a duty to return them to the third party upon notice by the third party to the agent, provided that the agent has not already paid the things over to the principal.

On the other hand, if the third party rescinds the transaction for reasons that arise subsequent to the receipt of the goods by the agent, the agent has no duty to return the things.

If the agent has received money from his principal for payment to a third party, the agent is liable to the third party for it. However, if he returns the money to the principal before the third party files suit, the agent has no such liability.

PELLETIER v. DWYER

334 A.2d 867 (Me. 1975)

Lloyd G. Dwyer contracted to sell certain real estate to Joseph Pelletier. Joseph A. Roy was a real-estate broker who represented Dwyer in making the contract. A $5,000 deposit was made by Pelletier as a "down payment." Roy took the deposit when the contract was signed. The purchase and sale of the land was never consummated, and Pelletier sued Dwyer and Roy for return of the deposit. The plaintiff lost in the trial court and appealed that decision.

DELAHANTY, J.

It remains to determine the nature of plaintiff's recovery and against whom it must run. We are of the opinion that return of the $5,000 together with interest and costs must be a joint and several liability of defendants Dwyer and Roy. The purchase and sale agreement between plaintiff and Dwyer recites that

> . . . in the event of the forfeiture of said sum of $5,000.00 as hereinabove provided, said Lloyd G. Dwyer agrees that fifty percent of said sum so forfeited shall be paid over to, and retained by said Joseph A. Roy as compensation for his services and disbursements in connection with the transaction and for bringing about the execution of this agreement.

By signing the purchase and sale agreement, Dwyer appears to have expressed an intent to remit to his broker Roy as a commission one-half of any deposit forfeited by the purchaser. Such a supplemental agreement, even as part of a purchase and sale agreement to which the broker is not a party, may be given due legal effect. From the record before us, we know only that the $5,000 "down payment" was received from plaintiff by Mr. Roy on November 16, 1971. There is no statement in the agreed facts that Roy ever paid over any portion of the $5,000 to Dwyer, and we cannot infer that he did. Neither is there any indication who holds the $5,000 at the present time, or whether it is in escrow or has been divided by the defendants as they anticipated and provided in the purchase and sale agreement.

There is no question that defendant Roy may be properly charged with liability for the retained deposit. So far as we known, he received it. We have held that where money has been paid to an agent for his principal, under such circumstances that it may be recovered back from the latter, the agent is liable as a principal so long as he stands in the original position and until there has been a change of circumstances by his having paid over the money to his principal, or done something equivalent to it.

Under the above principles, defendant Roy might have exonerated himself from liability by establishing at trial, as a matter of fact, that he had paid over all of the deposit to his principal. But the burden was on defendant Roy to go forward with evidence that would relieve him of liability by showing that prior to notice of plaintiff's claim he had paid the money to his principal. And when, as here, there is no evidence with regard to payment of the deposit to the principal, the broker is liable for the deposit.

However, Roy's liability affords no legal relief to Dwyer. The purchase and sale agreement stated that on buyer's default, the deposit was to be forfeited to Lloyd G. Dwyer. It was Dwyer who failed to tender performance and wrongly claimed a default against plaintiff. In such a case we have not hesitated to rule that a buyer may be entitled to a return of his deposit on the purchase price, and that either the real estate agency or the seller or both may be liable. Whether one defendant shall indemnify the other, or complain against him in a subsequent action, is of no concern to us here. Both defendants are properly subject to liability, and the plaintiff should be made whole.

(Reversed and remanded for entry of judgment for the Plaintiff.)

QUESTIONS FOR DISCUSSION

1. *If Roy had proved that the $5,000 deposit had been paid over to the principal, would it have affected the decision? How?*
2. *If it was established that Roy had not paid the $5,000 deposit to Dwyer, what would have been Dwyer's liability to Pelletier, if any?*

(6) Liability of Third Party to Agent Based on Contract. Our previous discussion in this chapter and Chapter 35 has covered most of the liability concepts under which the third party may be liable to the agent. There are a few additional points to be made, however, in addition to reiterating the previous points made in the context of our discussion here.

If the principal is undisclosed, the third party is liable to the agent for breach of the contract. Because the principal's right is superior to that of the agent, however, should the principal decide to sue the third party, the agent must withdraw. The third party cannot be subjected to suits by both the principal and the agent. Furthermore, the third party may raise those defenses against the agent that he has against the principal.

The agent who enters into a contract with a third party when the principal is disclosed generally has no liability to the third party. The contract is solely between the principal and the third person. The agent is not a party to the agreement. Remember, though, that the agent may bind himself to the contract in addition to the principal, and when he does so, he, of course, is liable to the third party.

Finally, if the third party enters into a contract with the agent that is meant to defraud the principal, the third person may not enforce the contract against the agent. Suppose that, in return for the payment of a sum of money by the third party to the agent, the third party succeeds in getting the agent to agree to a stipulation in a contract that would be adverse to the interests of the principal. If after receiving the compensation, the agent does not comply with the agreement, the third party can not recover damages from the agent. Nor can the agent receive the promised compensation from the third party in the event that the third party fails to pay the agent.

36:2. LIABILITY BASED ON TORT. (1) Liability of Agent to Third Party. An agent is liable to a third party for his torts just as is any person. The fact that the agent happens to commit the tort while he is acting as an agent generally has nothing to do with his liability. Thus, if Albert Agent injures a third person while driving a truck within the scope of his employment for Prin Corporation, Albert is liable if the injuries of the third person result from Albert's negligence. He is liable because he committed the tort, regardless of his relationship to Prin Corporation.

The agent is liable to the third party even if the tort is committed innocently or if he acted on the instruction of his principal. In that event, he may be entitled to be indemnified by his principal, but the agent is still liable to the third party. On the other hand, if the agent is not a party to a tort committed against the third party, then he has no liability.

Before the agent can be liable to the third party for his negligent acts, he must have a duty to the third party. The fact that the agent acts negligently regarding his obligations to his principal does not necessarily result in third-party recovery for damages. There must be a duty to the third party.

Historically the courts make a distinction between malfeasance and nonfeasance. If the agent acts negligently (malfeasance) and that act results in injury to the third party, the third party can recover. On the other hand, the courts hold that if the negligence of the agent is a failure to act, then the third party has no right of recovery against the agent. Thus under one old case of a railroad yardmaster whose duty was to inspect engines prior to their service, he was held not to be liable to a third person who was injured as the result of the explosion of an engine that the yardmaster had not inspected. *Kelly v. Chicago and A. Ry Co.* 122 Fed. 286 (W.D.Mo. 1903).

The modern decisions seem to place less weight on the distinction between malfeasance and nonfeasance—a distinction that is difficult to make in practice. Thus the law has gradually moved in the direction of holding an accountant, for example, liable not only to his principal for his negligence but also to third parties who see documents to which the accountant has attested and who in reliance upon those documents suffered monetary loss.

DAVI v. CABANA POOLS, INC.

135 N.Y.Super. 372, 343 A.2d 478 (1975)

Joseph Davi and Rose Davi entered into a contract with Cabana Pools, Inc., for the construction of a swimming pool at their home. Thomas Freda, an officer of Cabana, informed them that Cabana "would take care of everything." *Cabana applied to the borough for the necessary building and zoning permits on behalf of the Davis. Robert Stearns was employed part time to obtain these permits. He completed and signed the applications for the permits. The application stated that Stearns was an agent and would supervise the work and gave the lot and pool specifications. The application falsely stated the lot dimensions as well as other facts. As a result, the pool was installed in violation of the sidling requirements of the local zoning ordinance.*

The Davis were required to remove the pool at their own expense. They then sued Freda, Cabana, and Stearns for damages. Freda settled prior to trial. At trial the lower court directed a verdict against Cabana, which had defaulted, and gave judgment to Stearns on the ground that he did not owe a duty to the Davis. The Davis appealed.

MICHELS, J.

We are satisfied that in the light of the circumstances Stearns had a duty to exercise reasonable care in the preparation and filing of the applications for the building and zoning permits.

Stearns was paid at the rate of $20 an application and had processed approximately 60 to 70 such applications for Cabana alone. He also handled similar applications for other companies, as well as his own pool company. He knew, or at the very least should have known, that the issuance of the necessary building and zoning permits was dependent upon the truth and accuracy of the facts stated in the applications, and that if a pool or any other structure were built in violation of the zoning laws, the property owner would be compelled to remove the structure at his expense. In spite of this Stearns did nothing to verify the truth or accuracy of the facts he certified in the application and admittedly never gave any consideration to the consequences of his conduct. . . .

He merely assumed that the measurements furnished to him by Mr. Tholen, a vice-president of Cabana Pools, were truthful. Moreover, there was no proof that Stearns was to superintend the construction of the pool, and his false statement to this effect apparently was made without direction by or command of Cabana.

We are satisfied from our study of the record that reasonable minds could differ as to whether Stearns exercised that degree of care in the preparation and filing of the applications that the reasonably prudent person under similar circumstances would have exercised. Thus, a factual issue was clearly presented for resolution by the jury, and the trial judge therefore erred in granting Stearns' motion for judgment at the close of the evidence.

Furthermore, we point out that Stearns is not relieved from liability for his tortious conduct by the mere fact that he prepared and filed applications for the permits at the direction of Cabana, or was furnished measurements of the property and proposed location of the pool by one of Cabana's vice presidents. The general rule is set forth in 3 Am.Jur.2d, Agency, § 300 at 660–661:

> . . . In other words, if the agent is under a duty to third persons as well as to his principal, a breach of his duty to such third persons will render the agent liable to them. Thus, whether he is acting on his own behalf or for another, an agent who violates a duty which he owes to a third person is answerable to the injured party for the consequences. It is no excuse to an agent that his principal is also liable for a tort, inasmuch as the rights of a principal and agent inter se do not measure the rights of third persons against either of them for their torts, and the fact that an agent might have a right of exoneration or indemnity against his principal for a tort would not affect the rights of a third person against the agent. Nor is an agent who is guilty of tortious conduct relieved from liability merely because he acted at the request, or even at the command or direction, of the principal, unless he is exercising a privilege of the principal to commit the act. . . .

In accordance with these principles, an agent or other employee, merely because of his relationship as an agent or employee, or because of the additional fact that he has acted at the direction or command of his employer, cannot escape or exempt himself from liability to a third person for his own negligence or his own positive wrongs, such as a trespass, an assault, the conversion of property, fraud or misrepresentation, defamation, or other form of tortious conduct.

(Judgment reversed and remanded for trial.)

QUESTIONS FOR DISCUSSION

1. *Did Cabana have a right to be indemnified by Stearns for the damages awarded against it in this case?*
2. *In the event that a judgment should be recovered against Stearns, would he have a right to be indemnified by Cabana?*

---

(2) Liability of Third Party to Agent. One who is injured as a result of the tort of another may recover damages. If the person injured happens to be an agent, he or she can recover regardless of the factor of agency. The recovery is because of the injury and has nothing to do with whether or not one is an agent.

If an agent, such as a bailee, is holding property for his principal, he may sue a third party to recover for damages to the property. This is true even though the agent does not have title to the property but only the right to possession. The agent may recover for the full damage to the property and not just his interest. The agent's suit precludes suit by the principal, but the agent must account to the principal for any recovery.

SUMMARY

An agent should always sign a contract for his disclosed principal in such a manner that it is clear that the signature is by one acting as an agent. Where the principal is undisclosed, the agent is liable to the third party on the contract.

There are some situations where the agent may be liable to the third party where the principal is disclosed. This occurs when the agent assumes liability under the contract and if the agent enters into unauthorized contracts. In the latter situation the agent is said to have breached his implied warranty of authority. There are some jurisdictions that would hold the agent liable on the contract however, while still others would hold him liable for fraud. On the other hand, if the principal ratifies an unauthorized contact, the agent is not liable.

If the agent enters into a contract on behalf of a nonexistent principal, the agent is liable on the contract. The two most common situations where this occurs are where a corporation has not legally been created at the time of contracting and where the agent represents an organization that is not incorporated.

The agent has no liability for money paid to him by a third party that he pays over to a disclosed principal. He is liable if the principal is undisclosed or if the money was obtained by fraud.

If the principal is undisclosed, the third party may be sued by the agent for breach of contract, but the third party may raise any defenses that he had against the principal. In addition, an agent is liable for his torts just as any other person.

PROBLEMS

1. Frank C. Pierson was president of the Great River Steamboat Company. He signed a contract with Wired Music Inc. to supply a wired-in service aboard a steamboat and restaurant owned by the Great River Steamboat Company. The contract was signed by Pierson as follows:

   By /s/ Frank C. Pierson, Pres.
   Title
   The Great River Steamboat Co.
   ~~Port of St. Louis Investments, Inc.~~
   For the Corporation

   In signing, Pierson crossed out the incorrect name of the corporation and inserted the correct name. There were no discussions concerning Pierson's personal liability on the contract but it contained the following clause: "The individual signing this agreement for the subscriber guarantees that all of the above provisions shall be complied with." Wired Music Inc., sued Pierson, who testified that he had no intent to be personally liable on the contract and had not seen the guaranty clause. Is Pierson liable? Wired Music Inc. v. Great River Steamboat Co., 554 S.W.2d 466 (Mo. App. 1977).
2. Eula Pattee signed a contract for the sale of land to William Robinson on behalf of her husband, Walter J. Pattee, who was sick prior to his death. Eula Pattee signed the contract by authority of Walter's power of attorney.

She also signed the contract on her own behalf. The power of attorney did not authorize her to sign contracts to sell land. When she heard this, she stated that she would not go through with the sale. Was Eula liable to Robinson based upon her signature as Walter's agent? Robinson v. Pattee, 222 S.W.2d 786 (Mo. 1949).

3. Cathleen Russell, aged eight years, was injured while riding a horse on premises known as Anderson's Riding Stable. Suit was brought against Mary Downing, who helped her mother, the owner of the stable. On the day of the accident Mary was present, helped Cathleen mount the horse, saw her ride off, and witnessed the accident. At the trial Mary asked for dismissal on the ground that the plaintiffs could not prove that she was the owner and operator and manager of the stable. The motion was granted. Was the court correct? Russell v. Downing, 330 A.2d 454 (N.H. 1974).
4. The trustees of an unincorporated Catholic society in Eweb, Michigan, ordered building materials from Clark, Farnham & Company. The account was entered on the books as the "Catholic Church." When the debt was not paid, the trustees were sued by Clark. The trustees contended that they were not liable, as they ordered the materials as agents. Were the trustees correct? Clark et al. v. O'Rourke et al., 111 Mich. 108, 69 N.W. 147 (1896).
5. Samuel Goldfinger purchased stock as an agent of a minor from Doherty. The minor disaffirmed the purchase, and Doherty served a complaint upon Goldfinger, alleging that he had purchased the stock from Doherty "without disclosing the infancy of his principal." There was no allegation of Goldfinger's knowledge of the infancy of his principal. Goldfinger requested dismissal of the complaint. Should the request have been granted? Goldfinger v. Doherty, 276 N.Y.S. 289 (1934).
6. Aaron Bernstein was employed by Ruth Shops, Inc. On September 14, 1946, he sent a letter, on Ruth Shops letterhead, to Shoenthal, offering him a job as manager of one of the stores. The letter was signed, "Sincerely yours, A. Bernstein." Later, when he was fired, Shoenthal sued Bernstein, contending that he had been fired in breach of an employment contract. Was Bernstein liable to Shoenthal, assuming that the contract was breached? Shoenthal v. Bernstein, 93 N.Y.S.2d 187 (1949).
7. Berth Ness signed a promissory note payable to the order of Greater Arizona Realty, Inc. At the time, Ness represented three corporations as well as Louise Ness, but their names appeared nowhere on the note. Greater Arizona sued Berth Ness, Louise Ness, and the three corporations on the note on the theory that at the time Berth Ness was acting as the agent of Louise Ness and the corporations. The trial court granted judgment on the note against all the defendants. Was the court's decision correct? Ness v. Greater Arizona Realty, Inc., 21 Ariz.App. 231, 517 P.2d 1278 (1974).

SECTION VII

Partnership

CHAPTER 37

Definition, Creation, and Property

37:1. Introduction. The partnership is one of the oldest forms of business organization, predating the common law. Its origins have been traced back at least to ancient Babylonia, and references to it are found in the laws of that ancient civilization. Despite its ancient origins, the form of organization still creates a number of rather practical problems, particularly with regard to the relationship of the personal assets of the individual partners and the assets of the partnership. The law concerning this relationship as well as a number of other problems involving partnerships was confusing and contradictory until early in the twentieth century.

In 1902 the National Conference of Commissioners on Uniform State Laws began to consider a Uniform Partnership Act. The act was completed and first adopted by Pennsylvania in 1914. Today approximately forty-three jurisdictions in this country have adopted the act. The law in a number of jurisdictions that have not formally adopted the act parallels its significant features. Because of its overwhelming acceptance, the discussion in this chapter and in Chapters 38 and 39 will rely on the act.

37:2. Definition of Partnership. To define a partnership is not an easy thing. It is basically a relationship between two or more individuals arising out of a partnership contract. The Uniform Partnership Act defines a partnership as "an association of two or more persons to carry on as co-owners a business for profit." Section 6. Although this definition appears rather simple at first blush, it is more difficult in its application, as a further examination will reveal.

(1) "An Association of Two or More Persons." The term *association* implies that two or more persons have entered into a partnership on a mutually voluntary basis. Indeed, because a partnership arises out of a contract, it must of necessity be entered into on a voluntary basis in order to be valid. One cannot be forced into a partnership against his or her will. There must be an intent to enter into a partnership.

The phrase "two or more persons" requires further analysis. Questions frequently arise concerning what capacity is required in a partner and whether a corporation or a partnership can be a member of another partnership.

(a) *Infants.* Because a partnership arises out of contract, the law with regard to the capacity of an infant to enter into a partnership contract is similar to his capacity to enter into other types of contracts. In general, an infant may be a member of a partnership. However, the partnership agreement as to him is voidable, and he may disaffirm it just as he may any other contract. The practical consequence of this right is that the infant may withdraw from the partnership without incurring any liability to his partners for breach of contract. A question remains as to the rights of creditors of the partnership when the infant withdraws. If the firm is solvent, the infant may withdraw his capital contribution to the firm. If the firm is insolvent, a majority of courts hold that the creditors have a right to the capital contribution of the infant but that he has no personal liability to the creditors beyond that contribution. Note also that although the infant is protected regarding his contracts, in a partnership the partners are principals and agents of each other. Therefore any contract entered into by the infant on behalf of the partnership binds the adult partners as long as the infant is acting within the scope of the partnership business.

(b) *Insanity.* The mental capacity required for one to enter into a partnership relationship is the same as for any contract. If one has the mental capacity to contract, he may be a partner. If he lacks mental capacity when the partnership agreement comes into existence, he may avoid it. If at the time of the formation of the partnership there is a prior adjudication of insanity, the agreement is void. However, what happens if a partner becomes insane during the course of the partnership? The partnership does not automatically terminate in that event, but insanity is a ground for dissolution of the partnership upon petition by one of the other partners.

(c) *Corporations and Partnership.* There is nothing in the Uniform Partnership Act to prevent a corporation or a partnership from becoming a partner. The act defines *persons* as including individuals, partnerships, and other associations. However, a corporation is, as we shall see in Section Eight, a creature of the state and has limited powers. Therefore, unless the corporate charter and the state statutes allow a corporation to become a partner, the rule is that it may not. Recently the trend seems to have been in the direction of allowing for corporations to become partners. A 1969 addendum to the Model Business Corporation Act provides that each corporation shall have the power to be "a promotor, partner, member, associate or manager of any partnership, joint venture, trust, or other enterprise." Section 4(p), Model Business Corporation Act. Because the act is only a suggested law prepared by the Committee on Corporate Laws of the Sections of Corporation, Banking and Business Law of the American Bar Association, this act does not have the weight of law unless enacted by state legislatures. A number of states have adopted the act.

(2) "To Carry on as Co-owners." This phrase, taken from the definition of a partnership, raises several problems. Perhaps the most elusive is the notion that in order for a partnership to exist, there must be a "carrying on" of business. This seems to imply that an alliance of two individuals for a brief period in order to engage in a single transaction is not normally a partnership. There must be some ongoing transactions or series of transactions in a business. Usually an association between two or more people for a single transaction does not constitute a partnership; it may be a joint venture but not a partnership. For instance, suppose two people purchase a house as joint tenants, rent it, and share the profits. They would be joint owners but not partners because there is no sense of an ongoing business. On the other hand, suppose that the same two people buy and sell real estate on a continuous basis and also lease some properties. This would be a partnership as there is a carrying on of a "business."

The co-ownership referred to must relate to a business. Section 7(2) of the Uniform Partnership Act states that:

> Joint tenancy, tenancy in common, tenancy by the entireties, joint property, common property, or part ownership does not of itself establish a partnership, whether such co-owners do or do not share any profits made by the use of the property.

The co-ownership referred to in connection with a partnership is a co-ownership of the control of the acts that may result in a profit to the business, not merely co-ownership of property. Thus there is a sharing in the incidents of a partnership: the sharing of profits, losses, management, and control of the business. These elements are necessary in order for a partnership to exist. To be sure, one may have co-ownership of property and be a partner with the other co-owner. But that alone is not enough, even if profits are derived from the co-ownership, because the profits result from the co-ownership of property rather than the carrying on of a business, which must exist in order for there to be a partnership.

Consideration is also given to whether there is a sharing of the management and control of the business. Although the presence or absence of the sharing of management and control is not by itself decisive in determining the existence of a partnership, it is an element for consideration.

(3) "A Business for Profit." A requisite for a valid partnership is an agreement to share the profits (or losses) from a business. An association of individuals in a venture of a nonprofit nature is therefore not a partnership. For example, if a group of individuals associate themselves in order to raise money for a charitable institution, they are not partners.

The Uniform Partnership Act, Section 7(3), provides that the sharing of gross returns does not of itself establish a partnership. On the other hand, the next section, Section 7(4), states that a sharing of the profits of a business is prima facie evidence that one is a partner in a business. As you might expect, there are some exceptions listed in the statute that generally involve profits that are disbursed as payment for some obligation. Thus the inference of partnership should not be drawn if the profits are received in payment for the following purposes: (1) as a debt by installments or otherwise; (2) as wages of an employee or rent to a landlord; (3) as an annuity to a widow or a representative of a deceased partner; (4) as interest on a loan, though the amount of payment varies with the profits of the business; or (5) as consideration for the sale of the goodwill of a business or other property by installments or otherwise. Section 7(4).

The fact that one participates in the profits of a business is not conclusive but merely raises a rebuttable presumption of partnership. For example, if one were receiving a share of the profits merely as a method of computing salary, there might not be a partnership. However, when coupled with the other incidents of partnership, the sharing of profits may be conclusive. Although the act and our discussion have rather optimistically focused on the sharing of profits of a business, by implication at least the same concepts apply if the business experiences losses.

FULLER v. FULLER

518 S.W.2d 250 (Tex.App. 1974)

Betty R. Fuller sought an order against James S. Fuller's interest in the law firm of Fuller, Fuller, and McPherson. The court granted the order, and James Fuller appealed, contending that the firm was not a partnership under the Texas Uniform Partnership Act.

DIES, C.J.

This Act (art. 6132b, § 6) defines a partnership as "an association of two or more persons to carry on as co-owners a business for profit." The Source and Comments section tells us courts and commentators have had excessive difficulty in defining partnership.

The firm of which appellant is a member is a successful Port Arthur law firm. They refer to themselves as "partners," and file a partnership tax return; however, their manner of compensation changes from time to time. Essentially, appellant has a fixed sum per month as his "draw." From time to time, the members divide profits remaining after the "draws" and overhead expenses by a percentage previously agreed to by each member. This percentage is subject to change by the members. The law building is not owned by the firm. They have no written agreement. While the furniture and other personalty is [sic] owned by the four members, there seems to be no clear understanding what the percentage each owns. Any one of the four partners may sign a check on the bank accounts. Appellant's interest in the firm for purposes of distribution is 27½ percent. Appellant admits he has often referred to the firm as a "partnership." The firm had filed a lawsuit referring to itself as a "partnership." The "draw" each member gets is examined usually semi-annually. The earnings or fees of each attorney goes [sic] into a common fund from which the expenses, "draws," and percentages are distributed.

(1) Many Texas cases have spoken on the elements necessary to create a partnership. The Supreme Court of Texas quoted the early Texas case of *Goode* v. *McCartney*, 10 Tex. 193, 194 (1853):

> (T)he criterion by which to determine, in general, whether persons are partners or not, is to ascertain whether there is a communion of profit and loss, between them.

Intention of the parties is said to be important. But this intention may be implied, and is only one of the many factors to be considered. The use of a firm name has been given as one of these factors.

In *Conrad* v. *Judson*, 465 S.W.2d 819, 826 . . . this definition was approved:

> . . . a relationship between or among two or more persons where there is a common enterprise and a community of interest therein, the prosecution of the common enterprise for the joint benefit of the parties, and a right of each of the parties to participate to some extent in the profits as such and an obligation of each of the parties to bear some portion of the losses, if any, sustained by the business.

(The court concluded the firm was a partnership. Affirmed.)

CUTLER v. BOWEN
543 P.2 1349 (Utah 1975)

Dale Bowen leased premises on which the Havana Club had been operated for some years. He did not himself work in operating the club but owned the equipment, furnishings, and inventory. In June 1968 he discussed with Frances Cutler, who had worked for him as a bartender, the proposal that she take over management of the club. Their oral agreement included the following conditions: Cutler was to have the authority and responsibility for the entire active management and operation; to purchase supplies, pay bills, keep the books, and hire and fire employees; and to do whatever else was necessary to run the business. As for compensation, each was to receive $100 per week plus one half of the net profits.

This happy arrangement lasted for four years, until the building was taken over by the Redevelopment Agency in 1972. The agency was authorized to pay up to $10,000 to businesses for moving expenses and loss of patronage and goodwill occasioned by the move. The Havana Club qualified for the full $10,000. Cutler sued Bowen, contending that she was a partner and entitled to half of the money. The trial court found in favor of Cutler, and Bowen appealed.

CROCKETT, J.

The dispute giving rise to this lawsuit arose because the defendant contended that he was the sole owner of the entire business; and that the plaintiff's status was merely that of an employee, so defendant was entitled to the whole $10,000. Whereas, plaintiff took the position that, conceding the defendant was the owner of the physical assets of the business as above stated, insofar as the going concern and goodwill value, as a partner in the business, she was entitled to one half of the relocation fund.

One of the primary matters to consider in determining whether a partnership exists is the nature of the contribution each party makes to the enterprise. It need not be in the form of tangible assets or capital, but, as is frequently done, one partner may make such a contribution, and this may be balanced by the other's performance of services and the shouldering of responsibility.

When parties join in an enterprise, it is usually in contemplation of success and making profits, and is often without much concern about who will bear losses. However, when they so engage in a venture for their mutual benefit or profit, that is generally held to be a partnership, in which the law imposes upon them both liability for debts or losses that may occur. This basic principle of partnership law is set forth in our Uniform Partnership Act, Title 48 of U.C.A. 1953:

> Sec. 48–1–4. Rules for determining the existence of a partnership.—In determining whether a partnership exists these rules shall apply:
>
> . . .
>
> The receipt by a person of a share of the profits of a business is prima facie evidence that he is a partner in the business, but no such inference shall be drawn if such profits were received in payment:
>
> . . .
>
> (b) As wages of an employee or rent to a landlord.

On the question whether profits shared should be regarded simply as wages, it is important to consider the degree to which a party participates in the management of the enterprise and whether the relationship is such that the party shares generally in the potential profits or advantages and thus should be held responsible for losses or liability incurred therein. Section 48–1–12 of the Act provides that partners are [liable]:

> (1) Jointly and severally for everything chargeable to the partnership under sections 48–1–10 and 48–1–11. (Relates to torts and breach of trust.)
>
> (2) Jointly for all other debts and obligations of the partnership. . . .

It is not shown here that any occasion arose where the plaintiff's responsibility for debts or other liabilities of the business was tested. However, throughout the four years in which she operated and managed the Club, apparently with competence and efficiency, it was her responsibility to see that all bills were paid, including the rental on the lease, employees' salaries, the costs of the purchases, licenses and other expenses of the business. During that time she saw the defendant Bowen only infrequently for the purpose of rendering an accounting and dividing the profits. It is further pertinent that the parties reported their income tax as a partnership.

Under the arrangement as shown and as found by the trial court, a good case can be made out that it was largely through the capability, experience, and efforts of the plaintiff that, in addition to the physical plant, there existed a separate asset in the value of the "going concern and goodwill" of the business, which was being lost by its displacement. On the basis of what has been said above, we see nothing to persuade us to disagree with the view taken by the trial court; that the plaintiff's involvement in this business was such that she would have been liable for any losses that might have occurred in its operation; and that, concomitantly, she was entitled to participate in any profits or advantages that inured to it.

From the circumstances shown in evidence as discussed herein, there appears to be a reasonable basis for the trial court's view that, except for the physical assets, which belonged to the defendant and to which the plaintiff makes no claim, the further asset of the business: that is, the value of what is called going concern and goodwill, belonged to the two of them as partners in the enterprise; and that when the business could not be relocated, the $10,000 should properly be regarded as compensation for the loss by the forced relocation (which turned out to be a termination) of the business; and that the partners having lost their respective equal shares in the going business operation, they should also share equally in the compensation for its loss.

(Affirmed.)

QUESTIONS FOR DISCUSSION

1. *What factors did the court consider important in determining that a partnership existed? What factors can you cite that might lead to a contrary conclusion?*
2. *Assuming that Bowen did not intend to make Cutler a partner, what should he have done to protect himself?*

---

(4) Partnership as an Entity. The question is frequently raised as to whether a partnership is an *entity,* that is, whether it may be regarded as a single unit and whether it has any legal "existence." For example, we shall see in Chapter 40 that a corporation is a legal entity with the power to own property, buy and sell goods, make contracts, sue and be sued, and in general do other things legally that a person may do. The common law and the Uniform Partnership Act do not view the partnership as an entity distinct from its owners. Contrast this status with that of the corporation, which is a legal entity separate from its owners. On the other hand, there are some statutory provisions that treat the partnership as an entity for limited purposes. For example, a provision of the Uniform Partnership Act allows for real property to be titled in the name of the partnership.

HARTFORD ACCIDENT AND INDEMNITY CO. v. HUDDLESTON
514 S.W.2d 676 (Ky. 1974)

Clifford Huddleston and Orville Prewitt formed a partnership named City Motor Sales that engaged in the garage business. Hartford Accident and Indemnity Co. issued a garage liability policy to "City Motor Sales." *The policy provided protection against injury caused by an uninsured motorist. Carl Huddleston, the nineteen-year-old son of Clifford Huddleston, was killed as a result of the negligence of an uninsured motorist.*

Hartford contended that Carl was not covered because the policy covered only the partnership entity and also because the vehicle that Carl occupied at the time of the accident was one unrelated to the business. The trial court found for the estate of Carl Huddleston, and Hartford appealed.

REED, J.

It seems to us that . . . the uninsured motorist provisions require only that an injured party who is not a person insured be occupying an insured vehicle at the time he is injured by an uninsured motorist for the coverage to apply. No such requirement is exacted in the case of a "named insured" or a "designated insured" or, while residents of the same household, the spouse or relatives of either. It is apparent that the insurance contract expressly binds Hartford under the undisputed facts if Carl Huddleston was "an insured" under the uninsured-motorist coverage clauses of the policy.

Hartford's policy declarations page lists "City Motor Sales," a partnership, as the "named insured," while the uninsured-motorist coverage page calls "City Motor Sales" the "designated insured." Although not expressed directly, the basic premise of Hartford's contention concerning the identity of the insured raises a problem as old as the law of partnerships. There has always been considerable dispute as to whether a partnership is a legal entity or merely an aggregate of persons acting together. Vague expressions in our earlier cases are not helpful. Kentucky adopted the Uniform Partnership Act in 1954 as KRS Chapter 362. The adoption of the Act did not resolve the question as to the true nature of a partnership.

We are persuaded the better view is that although the Uniform Partnership Act regards the partnership as a legal entity for many purposes these purposes are, nevertheless, limited and the "entity" concept does not possess such attributes of public policy that it must be invoked to achieve an unjust result. The Uniform Partnership Act applies the "aggregate" concept when it makes partners jointly and severally liable; therefore, what public policy could be violated by knowledgeable parties contracting in a context of partnership liability insurance that they contemplate the partnership as an aggregate of persons rather than as a legal entity? The insurance contract with which we are here concerned plainly contracts for the "aggregate" concept to be applied.

A legal entity has no "spouse" nor "relatives" nor "household." A legal entity could not sustain "bodily injury." The uninsured motorist insurance contract plainly embraced the partners and their spouses and relatives living in the same household. The insurer framed the language of the contractual undertaking. The trial judge correctly imposed liability upon Hartford under its contract and the undisputed facts.

(The judgment is affirmed.)

QUESTION FOR DISCUSSION

1. *Does Kentucky apply the* entity *or* aggregate *theory of partnerships? If the court decided that the entity theory was Kentucky law, would that have affected the outcome of the case?*

37:3. TYPES OF PARTNERS, PARTNERSHIP, AND OTHER ORGANIZATIONS. Partnerships may be classified in a number of ways. There are trading and nontrading partnerships and general and limited partnerships. There are also a number of other similar types of business organizations, such as joint ventures, joint-stock companies, and mining partnerships.

(1) Trading Versus Nontrading Partnerships. The distinction between a *trading* and a *nontrading* partnership is predicated upon whether the enterprise is engaged in commercial transactions such as buying and selling. Thus a store or a manufacturer is a trading partnership, and a professional partnership is exemplified by an accounting or law firm. The importance of the distinction will become more apparent in Chapter 38, "Relationships Among Partners and Third Parties." For now it is enough to say that the apparent authority of a partner of a trading partnership is generally much greater than that of the partner of a nontrading partnership. For example, the partner of a trading partnership has the apparent authority to purchase stock for the firm on credit, whereas the partner of a nontrading partnership does not have that authority.

(2) General Versus Limited Partnerships. A *general partnership* is one that is organized to carry on a business over a period of years. It is the ordinary type of partnership and is the subject of most of our discussions. In a general partnership all the general partners have unlimited personal liability for obligations incurred by the firm.

A *limited partnership* is a form of business organization that may be established under a statute permitting this form of organization. Under most statutes a certificate must be filed with a public official and published in order for a limited partnership to be created. The purpose of the limited partnership is to allow one or more individuals to make capital contributions to the firm without incurring liability beyond the amount contributed. Every limited partnership must have at least one general partner, whose liability is, of course, not limited to his capital contribution.

Although the liability of the limited partners is limited, they give up something in return for the advantage and that is the right to manage and control the business. The general partners are vested with the control of the business. If a limited partner enters into the control or management of the firm, he becomes liable as a general partner. He cannot have it both ways. In exchange for limited liability the limited partner sacrifices a right to participate in the firm's management. This does not mean that the limited partner is without rights, however. He shares with general partners the right to inspect the books of the firm, to receive information concerning the partnership, and to dissolve and wind up the firm under certain circumstances. He also has a right to participate in the profits of the firm as provided for in the partnership certificate. A majority of states have adopted the Uniform Limited Partnership Act to govern this form of business organization.

FRIGIDAIRE SALES CORP. v. UNION PROPERTIES, INC.

562 P.2d 244 (Wash. 1977)

On January 15, 1969, Frigidaire Sales Corporation entered into a contract with Commercial Investors, a limited partnership, for the sale of appliances to Commercial. The contract was signed on behalf of Commercial by Leonard Mannon and Raleigh Baxter in their respective capacities as president and secretary-treasurer of Union Properties, Inc., the corporate general partner of Commercial Investors. Mannon and Baxter were also directors and officers of Union Properties, Inc., and each owned 50 percent of the outstanding shares of Union. Mannon and Baxter were also limited partners in the Commercial Investors limited partnership.

Frigidaire Sales, as creditor, sued Union Properties, Inc., as the general partner of Commercial Investors. Frigidaire also sued Mannon and Baxter individually. Baxter and Mannon were limited partners but also directors and shareholders of Union Properties, Inc., the corporate general partner.

HAMILTON, J.

We first note that petitioner (Frigidaire) does not contend that respondents acted improperly by setting up the limited partnership with a corporation as the sole general partner. Limited partnerships are a statutory form of business organization, and parties creating a limited partnership must follow the statutory requirements. In Washington, parties may form a limited partnership with a corporation as the sole general partner.

Petitioner's sole contention is that respondents should incur general liability for the limited partnership's obligations under RCW 25.08.070, because they exercised the day-to-day control and management of Commercial. Respondents (Mannon and Baxter), on the other hand, argue that Commercial was controlled by Union Properties, a separate legal entity, and not by respondents in their individual capacities.

Petitioner cites *Delaney* v. *Fidelity Lease Ltd.*, 526 S.W.2d 543 (Tex.1975), as support for its contention that respondents should incur general liability under RCW 25.08.070 for the limited partnership's obligations. That case also involved the issue of liability for limited partners who controlled the limited partnership as officers, directors, and shareholders of the corporate general partner. The Texas Supreme Court reversed the decision of the Texas Court of Civil Appeals and found the limited partners had incurred general liability because of their control of the limited partnership.

We find the Texas Supreme Court's decision distinguishable from the present case. In *Delaney,* the corporation and the limited partnership were set up contemporaneously, and the sole purpose of the corporation was to operate the limited partnership. The Texas Supreme Court found that the limited partners who controlled the corporation were obligated to their other limited partners to operate the corporation for the benefit of the partnership. " 'Each act was done then, not for the corporation, but for the partnership.' " This is not the case here. The pattern of operation of Union Properties was to investigate and conceive of real estate investment opportunities and, when it found such opportunities, to cause the creation of limited partnerships with Union Properties acting as the general partner. Commercial was only one of several limited partnerships so conceived and created. Respondents did not form Union Properties for the sole purpose of operating Commercial. Hence, their acts on behalf of Union Properties were not performed merely for the benefit of Commercial.

Further, it is apparently still undecided in Texas whether parties may form a limited partnership with a corporation as the sole general partner. See *Delaney* v. *Fidelity Lease Ltd.*, 526 S.W.2d 543, 546 (Tex.1975). The Texas Supreme Court was concerned with the possibility that limited partners might form the corporate general partner with minimum capitalization:

> In no event should they be permitted to escape the statutory liability which would have devolved upon them if there had been no attempted interposition of the corporate shield against personal liability. Otherwise, the statutory requirement of at least one general partner with general liability in a limited partnership can be circumvented or vitiated by limited partners operating the partnership through a corporation with minimum capitalization and therefore minimum liability. *Delaney* v. *Fidelity Lease Ltd., supra* at 546.

However, we agree with our Court of Appeals analysis that this concern with minimum capitalization is not peculiar to limited partnerships with corporate general partners, but may arise anytime a creditor deals with a corporation. See *Frigidaire Sales Corp.* v. *Union Properties, Inc., supra* 14 Wash.App. at 638, 544 P.2d 781. Because our limited partnership statutes permit parties to form a limited partnership with a corporation as the sole general partner, this concern about minimal capitalization, standing by itself, does not justify a finding that the limited partners incur general liability for their control of the corporate general partner.

If a corporate general partner is inadequately capitalized, the rights of a creditor are adequately protected under the "piercing-the-corporate-veil" doctrine of corporation law.

Furthermore, petitioner was never led to believe that respondents were acting in any capacity other than in their corporate capacities. The parties stipulated at the trial that respondents never acted in any direct, personal capacity. When the shareholders of a corporation, who are also the corporation's officers and directors, conscientiously keep the affairs of the corporation separate from their personal affairs, and no fraud or manifest injustice is perpetrated upon third persons who deal with the corporation, the corporation's separate entity should be respected.

Further, because respondents scrupulously separated their actions on behalf of the corporation from their personal actions, petitioner never mistakenly assumed that respondents were general partners with general liability. Petitioner knew Union Properties was the sole general partner and did not rely on respondents' control by assuming that they were also general partners. If petitioner had not wished to rely on the solvency of Union Properties as the only general partner, it could have insisted that respondents personally guarantee contractual performance. Because petitioner entered into the contract knowing that Union Properties was the only party with general liability, and because in the eyes of the law it was Union Properties, a separate entity, which controlled the limited partnership, there is no reason for us to find that respondents incurred general liability for their acts done as officers of the corporate general partner.

(Decision in favor of Union Properties.)

QUESTIONS FOR DISCUSSION

1. *What do you think of the public-policy consideration in allowing a corporation to be the sole general partner of a limited partnership? What is the legal objective in requiring a limited partnership to have at least one general partner?*
2. *According to the court, what should Frigidaire have done if it did not wish to rely solely on the solvency of Union Properties but as the only general partner?*

---

(3) **Joint Ventures.** Sometimes a number of individuals will associate themselves in order to accomplish a specific purpose or objective or a single transaction. For example, they may wish to purchase a piece of land and resell it for a profit. This would be called a *joint venture* or a *joint adventure.* The term is used to describe a business organization that is something less than the "ordinary" partnership, which carries on a continuing business. There is some confusion among the courts as to whether a joint venture is a partnership. In many respects, it is at least very similar. For instance, one of the joint adventurees acting within the scope of his authority may bind the other members of the venture. In addition, the liabilities of the parties to a joint venture are similar in other respects to the liabilities in a partnership. On the other hand, the definition of a partnership contemplates an ongoing business rather than a single transaction. Furthermore in a number of states it is held that a corporation may not participate in a partnership but that it may be a member of a joint venture. Also the apparent authority of a member of a joint venture would certainly be less than that of a general partner of the typical partnership. Finally, whereas a partner generally cannot sue a co-partner at law but must seek equitable relief from the courts, a member of a joint venture may sue the other ventures at law. For most practical problems, however, the joint venture may be viewed as a form of temporary partnership.

TRAVIS v. ST. JOHN
404 A.2d 885 (Conn. 1978)

In 1964 Edward Travis, Donald St. John, and others purchased 6.444 acres of land for investment purposes. The amount paid was $33,000.00. They also agreed to share equally the cost of the property, the expenses of maintaining it, and the profits or losses upon its resale. The property was titled in the name of St. John. It later became necessary to refinance the property, and St. John was the only one of the parties able to obtain refinancing. Through his personal credit St. John secured a mortgage in the amount of $25,000.00 which was used to pay off the former mortgages on the property.

The other parties in the investment paid their share of expenses in 1965 but thereafter defaulted in their payments. On October 1, 1969, St. John wrote the other parties that the venture was terminated and *offered to return their money if he sold the property at a profit. St. John then sold the property for $100,000.00 at a significant profit. The other parties filed suit against him alleging that the arrangement was a partnership and that they were therefore entitled to an accounting and a share of the profits. After the trial, St. John obtained a confession of judgment of $14,500.00 in order to settle the claims of the plaintiffs. The trial court adopted that amount as an equitable basis for the resolution of the dispute. The court also found that a partnership did not exist and that the plaintiffs were not entitled to an accounting and share of the profits. The plaintiffs appealed.*

BOGDANSKI, J.

The plaintiffs first claim that their arrangement constituted a partnership subject to the provisions of the Uniform Partnership Act. The distinction between a partnership and a joint venture is often slight, the former commonly entered into to carry on a general business, while the latter is generally limited to a single transaction.

Section 34–44 of the General Statutes defines partnership as an association of two or more persons to carry on a business for profit, and § 34–40 defines "business" to include every trade, occupation or profession. Section 34–45(2) further provides that coownership of property does not of itself establish a partnership whether or not the coowners share profits from the property. To find a true partnership, a mutual agency relationship is essential. Indeed, § 34–47 provides that to be a partnership within the statutes requires that every partner be an agent of the partnership for the purpose of its business.

To determine the nature of an association the court looks to the intent of the parties. In the present case, the association was established for the sole purpose of investing in a single parcel of real estate. The record does not indicate any intention by the parties to carry on a trade, occupation or business, or to create an agency relationship among themselves. The only indicium of a partnership in this case was the agreement to share profits or losses. We conclude that the agreement entered into by the parties did not constitute a partnership within the ambit of the Uniform Partnership Act.

The plaintiffs further contend that the court erred in refusing to find that an accounting of the profits was required. The relations and obligations in a joint venture are generally governed by the principles of common-law partnership; and an adventurer is accountable to his coadventurers, under the same equitable principles as require an accounting between partners at common law. At common law, "a partner who has not fully and fairly performed the partnership agreement on his part has no standing in a court of equity to enforce any rights under the agreement. *Marble Co.* v. *Ripley*, 10 Wall. [77 U.S.] 339, 358, 19 L.Ed. 955." *Karrick* v. *Hannaman*, 168 U.S. 328, 335, 18 S.Ct. 135, 138, 42 L.Ed. 484.

An action for an accounting calls for the application of equitable principles. The plaintiffs had the burden of showing that they were entitled to share in the profits. The court found that they failed in that burden. The confession of judgment was an offer by the defendant to compromise the controversy between the parties. The court adopted the amount offered as an equitable basis for the resolution of the dispute. Indeed, on the record before us, the defendant's confessed judgment constituted an amount well over the total contributions made by the plaintiffs.

(Judgment affirmed.)

QUESTIONS FOR DISCUSSION

1. *According to the court, what is the difference between a partnership and a joint venture?*
2. *Why was it important for the plaintiffs to establish a partnership?*

(4) **The Joint Stock Company.** A form of business organization, once popular but now virtually extinct, is the *joint-stock company.* It warrants some discussion because it does have historical significance as a transition form of organization between the partnership and the corporation. The joint-stock company was devised to meet one of the chief objections to the partnership, that is, the inability to transfer ownership of the firm without dissolving it. The phrase *joint stock* probably originated from the situation in which two or more mer-

chants would pool their stock in trade in order to do business. They then considered themselves a "company" and their individual interests were represented by a share of the stock. These shares could be transferred between individuals, and thus it became possible to transfer an interest in the business without affecting the existence of the company. The sale of these shares later became a method of attracting capital.

When a large number of investors were involved in the joint-stock company, the management of the firm became unwieldy, so the power to manage was delegated to a smaller group, who managed the company. If the membership retained a right to supervise this smaller group, the organization remained a joint-stock company and the members had unlimited liability. This form of organization has been almost entirely replaced by the corporation. On the other hand, if the membership turned the management over to a smaller group and did not retain a right to supervise that group, the organization might evolve into what is commonly known as a *business trust,* or a *Massachusetts trust,* as it became known because of its heavy use in that state.

37:4. The Business Trust. The business trust is designed to limit the liability of the owners of the business to the amount of their investment. For this to be accomplished, the control of the business and the assets are turned over to a trustee or trustees, who operate the business for the benefit of the investors. The ownership interest of the investors is evidenced by a trust certificate. Under this arrangement the trustees have *legal title* to the firm's assets, but because the business must be operated for the benefit of the investors, the investors have what is known as the *equitable title.* The investors can sell their ownership in the business by transfer of the stock certificate. Thus ownership can be transferred without the transfer's affecting the existence of the business. Because the primary benefit of this type of organization is limited liability, it is important that it be established properly so as not to be construed as a general partnership. It is critical that the investors have no control over the business. This form of organization is not used very frequently today because of the advantages of incorporation under the modern statutes, as well as because of the danger that in a particular case a court would hold the investors liable. For example, a number of courts have held that if the investors retain the right to elect the trustees, this is sufficient control over the business to impose partnership liability on the investors.

37:5. Other Partnership Terms. Mention has already been made in Section 37:3(2) of the distinction between a "general partner" and a "limited partner." The general partner is one who has the same rights, duties, and liabilities as other partners and conducts the firm's business. One also frequently hears the terms *dormant partner, silent partner, incoming* or *retiring partner, ostensible partner, surviving partner,* and *secret partner.*

An ostensible partner is not an actual partner but is an individual who appears to the public to be a partner and usually has consented to be held out as a partner. Frequently the ostensible partner is held to be liable just as is a general partner as a result of the application of the doctrine of estoppel, when people have in good faith relied on the holding out.

A silent partner is one who takes no role in the partnership business, whereas a secret partner is one whose membership in the firm is withheld from the public. A dormant partner is one who takes no role in the partnership and in addition keeps his membership in the firm secret. The terms *incoming, returning,* and *surviving* are all self-explanatory. All the partners discussed in this paragraph are real or actual partners, and all have unlimited liability for the period of time that they are members of the partnership.

(1) **Creation of the Partnership.** There is no particular formality required for the formation of a partnership. The partnership agreement may be express or implied, and except for some possible Statute of Frauds problems, it may be oral or written. If the partnership agreement provides that the partnership is to continue for more than a year or if it involves the sale of an interest in land, some courts hold that the Statute of Frauds is applicable, whereas others hold, for various reasons, that it is not.

However, as we shall see in Chapter 38, the Uniform Partnership Act provides for a number of rules concerning management, control, and the division of profits or losses of a firm in the absence of an agreement between the partners to the contrary. Therefore, unless one is satisfied to have these matters governed by statute, it is wise to have a partnership agreement, frequently called *articles of partnership.* It is also wise to have a written agreement in the event that disputes arise between the partners. Let us first examine some of the more important subject matter that might be covered in a written agreement.

(2) **The Written Agreement.** There is no necessity that a partnership do business other than in the names of the partners. However, for convenience it is often desirable to adopt a firm name. Frequently this name will be merely an amalgamation of the names of the partners. Sometimes the partnership will use the names of the partners and the phrase "and Company." Some states allow the use of the phrase as long as it represents an actual partner. That is, an individual in business alone could not use the phrase "and Company," because it implies a partnership and would therefore be deceiving.

There is generally no prohibition against using a fictitious name for the partnership. Most states have statutes, however, that require the name to be registered together with the names of the individual partners or owners of the business. One purpose of such statutes is to allow those who deal with a partnership or business using a fictitious name to identify the owners of the business. These statutes usually provide for criminal penalties in the event that one does not comply with them. Nevertheless contracts entered into by a party in violation of the statute may be enforced and damages recovered for their breach.

There are a number of advantages to having a written partnership agreement, not the least of which is that it forces the individuals involved to face a number of crucial questions concerning various aspects of the contemplated partnership in order that a proper agreement can be written. It also allows the parties to tailor the terms of the partnership to their particular situation and reduces the possibility of future disputes over the precise terms of the agreement.

As previously stated, the terms of each particular agreement will vary, but the typical agreement will cover the following matters: (1) the firm name and identity of the firm partners; (2) the place of the business and its purpose; (3) the duration of the partnership; (4) the capital contributions of the partners; (5) how profits and losses are to be shared; (6) matters concerning the management of the firm and the duties of the partners; (7) a provision for the payment of salaries to the partners or the maintenance of a drawing account; (8) restrictions upon the authority of partners to bind the firm with outside parties; (9) a provision concerning the dissolution of the firm and the right to continue the business in the event of the death or withdrawal of a partner; and (10) the right and conditions for a voluntary or involuntary withdrawal of a partner from the firm. Other provisions may, of course, be appropriate under the circumstances.

You should remember, however, that although the agreement may be binding on the partners, it is not binding upon innocent outside parties. For example, a provision limiting the authority of a partner would not be binding on an innocent outside party if the exercise of the authority appears to be reasonable. The partner would be liable to his other partners, of course, for damages that result from breaching the agreement.

(3) **Partnership Property.** A matter of critical importance is whether certain property is firm property or the property of an individual member of the firm. The issue frequently arises when a dissolution and winding up of the firm occurs. It is an issue that is most important when the firm is insolvent and there are creditors of the firm looking to the firm's assets to satisfy their claims. In addition, there may be creditors of individual members of the firm who are also looking for assets to attach in order to satisfy the obligations of individuals who are also partners. The concept of the *marshaling of assets* determines which creditors get what property, and it will be discussed in Chapter 39. As part of the implementation of that doctrine it is necessary to determine what property belongs to the firm and what property belongs to the individuals who happen to be partners. After an examination of partnership property, partners rights in specific property will be discussed, and finally the concept of partnership *goodwill* will be treated.

(a) *Partnership Property Versus Individual Property.* The partners frequently contribute property for permanent use by the firm after the partnership is created. Property contributed may be in money, in land, in equipment, or in other forms. This property is known as *partnership capital* and may not be withdrawn without the consent of the other partners unless there is a dissolution of the firm. Although the total partnership property may vary from day to day, the part of the property that is the partnership capital does not vary. On the other hand, it is not necessary that a partner contribute any property to the firm, and, in fact, there may be a partnership without any property.

The subject of partnership property is treated by the Uniform Partnership Act. Section 8 states in part:

> (1) All property originally brought into the partnership stock or subsequently acquired by purchase or otherwise, on account of the partnership, is partnership property.
> (2) Unless the contrary intention appears, property acquired with partnership funds is partnership property.

Note that the act defines partnership property in the absence of a "contrary intent." Frequently a partnership has the use of individual property, such as a building, whereas the property is to remain the individual property of a partner. The fact that legal title remains in an individual, however, does not alone determine whether the property is partnership property or individual property. It is a question of intent, and other factors may be considered in a determination of that intent. For example, in the case of a building, such factors as who pays the taxes and insurance, who uses the building and for what, who pays on the mortgage, and who has legal title are all considered. The question of what constitutes partnership property can often be solved easily by a statement in the partnership agreement of what property constitutes partnership property and what property, although used by the partnership, remains individual property.

The same reasoning applies when a loan is made to the partnership. In order to avoid future controversies, the agreement should reflect that the advance is a loan rather than a contribution to the partnership if that is the intent of the parties. If the advance is a loan, the partner is a creditor of the firm to that extent, and upon dissolution the debt is paid prior to the distribution of capital to the partners.

(b) *Title to Partnership Property.* Personal property may be held as well as purchased and sold in the partnership name. Real property has historically presented more of a problem. Because a partnership is not a legal entity at common law, real estate could not be purchased in the fictitious name of the partnership. If, however, the partnership name is composed of the surnames of the partners, such as Smith, Brown, and Jones, then real estate purchased in the partnership name is titled in the name of the individual partners. If the real estate is intended to be partnership property, the partners whose names appear on the deed hold legal title to the real estate. However, an equity court will hold that the property is held in trust by the individuals for the benefit of the partnership. Remember that title alone is not determinative of the issue of whether property is individual or partnership property.

The process described is somewhat cumbersome. Section 8(3) of the Uniform Partnership Act changes the common law and provides that real estate may be held and transferred in the name of the partnership. Any property held in the partnership name can be conveyed only in the partnership name. Section 10(1) enables any partner to convey real estate held in the name of the partnership by conveying the property in the partnership name. If the partner conveys the partnership real estate in his own name, the purchaser has an equitable interest in it. The right of the purchaser in these situations is predicated on the assumption that the selling partner has the authority to sell the partnership real estate or that the partner apparently has such authority. If the purchaser knows that the partner is acting beyond his authority, the property may be recovered by the partnership.

Obviously in those states that have adopted the Uniform Partnership Act or other similar enabling legislation, the common-law prohibition against ownership of real estate in the partnership name is no longer operable. This change has eliminated many problems in the conveyance of partnership real estate.

X (c) *Partner's Rights to Specific Property.* We have established the fact that a partnership is not a legal entity. The discussion has already highlighted some conceptual difficulties that this fact creates regarding the ownership of property. There are others under the common law. For example, the partners obviously are co-owners of the partnership property, but are they tenants in common or joint tenants with the right of survivorship? If the property is considered to be held in common, as it most often is, what are the partnership creditors' rights to that property as opposed to the rights of a partner's heirs upon the death of that partner? What about dower and curtesy rights?

The common-law treatment of this problem is somewhat cumbersome. The common law holds that title to personal property held by the partnership passes on the death of a partner to the surviving partners in order to pay off the partnership debts and for the winding up of the partnership. Real estate, as usual, presents a more complicated problem. The title to real estate normally goes to the deceased partner's heirs, subject to the rights of dower and curtesy. However, in order to protect the rights of partnership creditors, the courts have created a delightful legal fiction known as the *doctrine of equitable conversion.* Under this doctrine, upon the death of a partner the firm real estate is treated or "converted" to personalty for the purpose of paying the firm's debts and may be sold for this purpose free of any claims of dower, curtesy, or inheritance. Any real estate remaining after the debts are paid once again takes on the characteristics of real estate and descends to the heirs subject to the right of dower and curtesy.

The Uniform Partnership Act has eliminated the need for these legal gymnastics by abolishing the distinction between real and personal partnership property. Section 25(e). The act also states other rules with regard to partners' rights in specific property as against each other as well as persons outside the partnership. This type of ownership is often called *tenancy in partnership* to distinguish it from other types of co-ownership. Section 25 states as follows:

(1) A partner is co-owner with his partners of specific partnership property holding as a tenant in partnership.

(2) The incidents of this tenancy are such that:

(a) A partner, subject to the provisions of this act and to any agreement between the partners, has an equal right with his partners to possess specific partnership property for partnership purposes; but he has no right to possess such property for any other purpose without the consent of his partners.

(b) A partner's right in specific partnership property is not assignable except in connection with the assignment of rights of all the partners in the same property.

(c) A partner's right in specific partnership property is not subject to attachment or execution, except on a claim against the partnership. When partnership property is attached for a partnership debt the partners, or any of them, or the representatives of a deceased partner, cannot claim any right under the homestead or exemption laws.

(d) On the death of a partner his right in specific partnership property vests in the surviving partner or partners, except where the deceased was the last surviving partner, when his right in such property vests in his legal representative. Such surviving partner or partners, or the legal representative of the last surviving partner, has no right to possess the partnership property for any but a partnership purpose.

(e) A partner's right in specific partnership property is not subject to dower, curtesy, or allowances to widows, heirs, or next of kin.

(4) Goodwill and Other Property Rights. There are other property rights that one may have in a partnership other than the tangible real and personal property. One of the more elusive of these is the concept known as *goodwill.* There is no clear definition of the term. It seems to consist of the predictions of the future earning power of the business based to some extent on the firm's reputation. It is a somewhat nebulous term, and the calculation of its value, as a practical matter, involves a frightening amount of speculation and assumption. Nevertheless it is an intangible asset for which hard cash is paid when businesses are purchased. Obviously it must also be sold in connection with a business, not separately as one might sell the company truck. It cannot be sold without the consent of all partners, and it generally involves the right by the purchaser to use the name of the business or at least to advertise as a successor to the business. These are matters that should be made clear in any contract to purchase a business.

Section 24 of the Uniform Partnership Act states that in addition to a partner's rights in specific property, which have already been discussed, a partner has a right to his interest in the partnership and a right to participate in the management of the firm. The right to participate in the firm's management has previously been mentioned and will be discussed in more detail in Chapters 38 and 39. The partner's interest in the partnership is defined in Section 26 as his share of the profits and surplus that are personal property.

You should note that although Section 25(2)(b) prohibits the assignment of specific partnership property by an individual partner, he may assign his "interest in the partnership," that is, his right to profits and surplus. Such an assignment does not give the assignee the right to participate in the partnership in any way, such as in the management of the firm. It only entitles the assignee to the profits that would normally go to the assigning partner.

REINERS v. SHERARD
233 N.W.2d 579 (S.D. 1975)

DOYLE, J.

The facts of this case demonstrate the unfortunate results which accrue when good friends or relatives become partners and fail to keep adequate records. At the dissolution of the partnership, it too often happens that a clear view of the initial agreement is clouded by the years and there is suspicion on both sides that one partner has taken advantage of the other. So it appears to be here.

Reiners first came into contact with Sherard when Reiners was twelve or thirteen years old. Sherard was already a successful farmer and Reiners was a boy without a good home. Sherard took Reiners into his care and treated him as he would a son. In 1949, Sherard and Reiners entered into a livestock feeding partnership in which they were to share equally in profits and losses.

To finance the partnership operation, they obtained a line of credit from the Yankton Production Credit Association (PCA). During the next nineteen years the two men treated their account with PCA as a common fund. Whenever either sold cattle or sheep the proceeds were deposited with PCA to pay off current loans. Whenever either needed cash with which to pay for ongoing personal expenses, as for food or clothing, or to pay for expenses of the farming operation, he merely wrote a draft on the PCA account and deposited it in his own account. Each purchased real property in his individual name with funds from the PCA account. No annual accounting of profits and losses was ever made. No attempt was apparently made to assure that each partner withdrew an equal amount for personal or living expenses.

The first issue is whether the trial court erred in finding that the real property purchased with partnership funds became property of the individual partners and not of the partnership. We find that the trial court's determination was correct.

The applicable statutory law is stated in SDCL 48–4–2:

> Unless the contrary intention appears, property acquired with partnership funds is partnership property.

This statute thus creates a rebuttable presumption that property purchased with partnership money becomes partnership property. However, the "presumption . . . is rebuttable and the same is true with respect to the presumption that the partnership owns the property that it uses in the conduct of the firm business."

The trial court, then, was left to balance the facts presented to it. Several factors were urged on the court as favorable to the finding that the land was property of the partnership. First, of course, the land had been purchased with partnership funds and taxes on it *had* been paid with partnership funds. However, we believe that the probative value of these facts is severely limited in this case because of the manner in which the two men conducted their affairs. Each drew funds for every personal or business purpose conceivable from the same PCA account. To say that these funds could be used only to purchase property for the partnership would be to say that the partners had intended to have little or no property in their individual names. Thus, we agree with the trial court in its finding that the mere payment of taxes and of the purchase price of property is not persuasive when such payments come from a common fund used by both partners for all expenses. Furthermore, the cases are explicit in their view that the property purchased with partnership funds does not automatically become the property of the partnership.

The plaintiff further argues that the fact that the crops from the land in question were used in the partnership feeding business should lead to a finding that the land was intended to be partnership property. Again, however, the mere fact that the partnership uses property is not dispositive. In *Norcross* v. *Gingery,* 1967, 181 Neb. 783, 150 N.W.2d 919, the court held,

"Use of the property alone is not sufficient because an owner may intend to contribute only the use, as distinguished from the ownership, to the partnership."

Finally, the plaintiff would argue that even if these particular factors alone do not persuade the court that the property is partnership property, the facts together should. *Gertz* v. *Fontecchio,* 1951, 331 Mich. 165, 49 N.W.2d 121, adequately answers this contention. "Property acquired with partnership funds or by the partners individually for the use of the partnership does not necessarily constitute a partnership asset. In the absence of supervening rights of creditors, such property may, as between the partners, at least, be owned by them individually as tenants in common or otherwise, as distinguished from the partnership, *if such was their intention* in the acquisition and holding thereof." (Emphasis supplied.)

The trial court, in its review of all the facts and circumstances which evidenced the intent of the parties, found that the partners did not intend the land in question to be the property of the partnership. Several factors support this conclusion. First, none of the land was ever listed in applications to PCA as a partnership asset. Instead, the land was listed in the names of the individuals, i.e., "Sherard owns 560 acres and Reiners owns 320." Second, the partners once traded certain tracts of the land in question to each other. Such conduct would, of course, be illogical and unnecessary if the partnership owned the land. Cf., *Gertz* v. *Fontecchio,* supra. Third, there was no evidence that either man ever made a representation to anyone before the commencement of this litigation that the land farmed by the partnership was the property of the partnership. We find the factors cited above in support of the trial court's determination to be persuasive.

(Affirmed on this issue.)

QUESTION FOR DISCUSSION

1. *There is a presumption that property purchased with partnership funds is partnership property. What factors were presented in this case to rebut that presumption?*

SUMMARY

A partnership is an association of two or more persons to carry on a business for profit as co-owners. An infant may be a partner; the mental capacity required for one to enter into a partnership is the same as that required to enter into a contract. Corporations may be partners provided that a state's corporation laws do not prohibit it.

The business must be ongoing. A joint tenancy by itself does not establish a partnership. There must be a sharing of business, not merely property. The partnership must also be for profit and not a nonprofit or charitable venture.

The common law and Uniform Partnership Act do not view the partnership as a separate legal entity. Some statutes do treat the partnership as an entity for limited purposes, such as ownership of real property.

Partnerships may be categorized as trading or nontrading and general or limited partnerships. The apparent authority of a trading-partner is generally greater than that of a nontrading partner.

A limited partner's liability is limited to his investment. He may not participate in the management of the firm and remain a limited partner. There must be at least one general partner in a limited partnership whose liability is unlimited. All partners in a general partnership have unlimited liability. A joint venture is an association organized in order to accomplish a specific purpose, objective, or single transaction.

The joint-stock company consists of two or more individuals who pool their assets and then issue shares of stock to indicate their interest in the business. The owners have unlimited liability, but the business does not terminate when a share is sold. In a business, or Massachusetts trust, the control of the business and the assets are turned over to a trustee who operates the business for the investors who then have limited liability.

Partners may sometimes be characterized as dormant partners, silent partners, incoming or retiring partners, ostensible partners, surviving partners, and secret partners.

No particular formality is required for the formation of a partnership. There may or may not be a written agreement setting forth the rights of the partners.

Partnership property is contributed for the permanent use of the partnership and may not be withdrawn without the consent of the other partners unless there is a dissolution of the firm. Whether property is individual property or partnership property is a question of intent. In most states, partnership property may be held in either the name of the partnership or the name of the individual partner.

The Uniform Partnership Act has created a type of ownership known as a tenancy in partnership. A partner also has an interest in the goodwill of the partnership. Finally, while a partner may not assign specific partnership property he may assign his interest in the partnership.

PROBLEMS

1. Fred Koesling operated a tailoring and drycleaning store and James Bosamakis operated a shoeshine and hat-cleaning business. They then moved their businesses to the same location. The parties shared the cost of renting and maintaining the business premises. Koesling paid some of his profits to Bosamakis in reimbursement for certain expenses incurred by Bosamakis in connection with the business premises. Koesling claimed there was a partnership. Was he correct? Koesling v. Bosamakis, 539 P.2d 1043 (Utah 1975).
2. Fidelity Lease Limited was a limited partnership that leased property from Delaney and others to erect and operate a fast-food service restaurant. Fidelity Lease was made up of twenty-two limited partners. The only general partner was Interlease Corporation. The Interlease Corporation was organized by three of the twenty-two limited partners: Crombie, Sanders, and Kahn. These three men served as president, vice-president, and treasurer of the corporation and also served on its board of directors.

   Fidelity erected a building on the leased property but then failed to take possession of the premises or pay the rent required by the terms of the lease. The landlord sued Fidelity, the limited partnership, as well as the general partner, Interlease, and all of the limited partners, including Crombie, Sanders, and Kahn. The plaintiff contended that at least Crombie, Sanders, and Kahn of the limited partnership were personally liable.

Was the plaintiff correct? Delaney et al. v. Fidelity Lease Limited et al. 521 S.W.2d 543 (Tex. 1975).

3. Cornwell and Wilkinson were partners and owners of a one-half interest each in real estate. The real estate was the only partnership asset. Wilkinson, without Cornwell's knowledge, sold the real estate in the name of the partnership to Corey Carpenter, who purchased the property without knowledge of the partnership and the partnership agreement. That agreement gave one partner the first right to purchase the other partner's interest in the realty. Did Carpenter get title to the entire property? If not, what if any interest did he get? Carpenter v. Cornwell, 133 Ga.App. 797, 213 S.E.2d 56 (1975).
4. Walker, Mosby & Calvert, Inc., agreed to furnish three lots and the capital to erect three houses on the lots. Senseney agreed to furnish the labor and the material for the construction of the houses. It was agreed that the properties would be sold and that after expenses and commissions the profits would be divided equally between Senseney and Walker, Mosby & Calvert, Inc. Senseney contracted with C. L. Burgess for the plumbing and heating installations in the houses. Burgess knew of the contract between Senseney and Walker, Mosby & Calvert, Inc. Senseney did not pay Burgess, left town, and then was adjudicated as bankrupt. Burgess sued Walker, Mosby & Calvert, Inc., alleging that it was a partner of Senseney and therefore liable for the debt. Could Burgess collect? Walker, Mosby & Calvert, Inc. v. Burgess, 151 S.E. 165 (Va. 1930).
5. Rocco Rizzo, Sr., operated a wastepaper business in Chicago. Over a period of years his sons entered the business: Michael in 1910, Joseph in 1913, Rocco, Jr., in 1916, and John in 1920. Michael acted as general manager. The name of the business was originally Rocco Rizzo & Co., then Rocco Rizzo Son & Co., then Rocco Rizzo Sons & Co. In 1929 Rocco Rizzo, Sr., deeded the business property to Michael. All the brothers worked in the business, shared the profits equally, and went without pay if there were no profits. When Rocco Rizzo, Jr., died in 1931, his estate claimed one quarter of the business, contending that it was a partnership. The other brothers contended that the business was a sole proprietorship owned by Michael and that the division of profits was merely a method of determining compensation. Decide. Rizzo v. Rizzo, 120 N.E.2d 546 (Ill. 1954).
6. Mrs. Morris sued Mr. Morgan contending that by an oral agreement a partnership existed in which she agreed to contribute labor and services to a dairy business and to assume one half of his indebtedness. She contended that he agreed to contribute to the partnership his interest in real estate and personal property owned by him. This relationship lasted for fifteen years. Morgan inherited the business from his wife and placed the earnings from the dairy in his personal account and later in a joint account with Morris. No partnership account was ever opened. Morgan supported Morris and her children during the course of this relationship. Morris contended

that she was entitled to one half of the undivided interest in the real and personal property of Morgan and sought an accounting. Was she entitled to relief? Morris v. Morgan, 179 Cal.App.2d 463 (1960).

7. Curtis Cyrus wrote letters in 1934 and 1935 to Cecil Cyrus asking him to join him as a partner in a resort. At this time Curtis bought a sixty-acre tract of land with his own money and titled it in his name. In 1936 Cecil moved to the area and helped operate a resort on the land. He built a cabin on that same land. Curtis paid taxes on the land from his own funds, but Cecil contended that Curtis was compensated for the expense from resort profits. Later another forty acres was purchased and titled in Curtis's name, but it was paid for out of resort earnings. Cecil received living expenses from the resort earnings, and he and his wife did all the work at the resort. Curtis did not contribute any personal efforts or work to the resort business and lived elsewhere. Cecil's estate sued for one-half interest in the property, contending that there was a partnership. Could the estate recover? Cyrus v. Cyrus, 64 N.W.2d 538 (Minn. 1954).
8. Richard K. Chaiken was assessed for an unemployment-compensation tax by the State Employment Security Commission for two barbers working in his shop. Chaiken appealed the commission's ruling, contending that they were partners. Chaiken and the two barbers executed separate "partnership agreements." Chaiken was to supply all equipment and licenses, and the other two barbers were to supply the tools of the trade. Income was divided 30 percent–70 percent between Chaiken and one barber and 20 percent–80 percent between Chaiken and the other barber. The agreements provided that all partnership policy would be decided by Chaiken, whose decision was final. Another paragraph stated the hours of work and the holidays for each of the two barbers. The partnership name was registered, and federal information returns were filed. Was Chaiken required to pay the unemployment-compensation tax? Chaiken v. Employment Security Commission, 274 A.2d 707 (Del. Super. 1971).
9. Smith left an accounting firm and went to work for a firm of which Kelley and Galloway were partners. For three and one-half years Smith drew $1,000 per month plus $100 for travel. At the end of each year Smith received a bonus paid out of the profits of the firm. There was no partnership agreement, and Smith made no contribution to the partnership assets. Smith was held out to third parties as a partner, contracts with third parties indicated that Smith was a partner, and partnership tax returns listed him as a partner. Smith left the firm and contended that as a partner he was entitled to 20 percent of the profits. Kelley and Galloway contended that he was only an employee. Was Smith entitled to a share of the profits? Smith v. Kelley, 465 S.W.2d 39 (Ky.App. 1971).

CHAPTER 38

Relationships Among Partners and Third Parties

In this chapter attention will be concentrated upon the rights and liabilities of the partners to each other and, in turn, their rights and liabilities to persons outside the partnership, third parties. In studying this chapter, you should keep in mind the fact that much of what was discussed in the section on agency is equally applicable here. There are, however, as you may have come to expect by now, some complicating factors with regard to partnerships. Particular attention will be paid to such factors as the right to profits, rights to participate in firm management, the right to accounting, the duty to devote full time and energy to the firm, and so on.

38:1. RELATIONSHIPS AMONG PARTNERS. (1) Partner's Rights. (a) *The Partner as a Fiduciary.* The relationship of partners as principals and agents of one another has already been mentioned. Therefore the overriding right that a partner has, as well as a corresponding duty, is that of a fiduciary. As a prelude to the entire discussion in this chapter it can be said that a partner has a right to expect, and must in turn give, the highest degree of loyalty in his dealings in the partnership business. A partner may do nothing, either directly or indirectly, to interfere with or hinder the best interests of the business. For example, suppose that one is a partner in a small accounting firm. At night, on his own time, he sees his own clients for the preparation of their tax returns. Without the consent of all the partners, this would constitute a breach of the partner's fiduciary duty because he is taking business away from the firm and making a profit at its expense. Section 21 of the Uniform Partnership Act provides that a partner must account for any benefit and hold any profits that are derived without consent of the partners as trustee for their benefit. This applies to any profit or benefit that results from any transaction connected with the formation, conduct, or liquidation of the partnership or from any use of the partnership property. The rule also applies to personal representatives of a deceased partner involved in the liquidation of the partnership. In short, a partner may not benefit directly or indirectly at the expense of the partnership.

STARR v. INT'L. REALTY LTD.
533 P.2d 165 (Or. 1975)

Stanley G. Harris formed a partnership with a group of prominent Portland physicians for the purpose of purchasing an apartment house under construction. Harris told them that this would be a good "tax shelter." *The purchase price as stated by Harris to his partners was $1,010,000. Harris did not reveal that the property could have been purchased for $907,500, including $207,500 to the seller and the assumption of a $700,000 mortgage. Harris also did not reveal to his partners that a $100,000 commission together with an escrow fee of $2,500 was to be paid to International Realty, Ltd., of which he was the president and major stockholder. When the doctors learned of the commission, they sued to require Harris to account to the partnership for the commission received by him without their consent. Harris contended and proved that the doctors knew that someone was going to make a commission on the transaction; that Harris, although not informing them of the commission, did not conceal it from them; and that therefore, because this was not a secret commission, no accounting was required. The lower court found against Harris, and he appealed.*

TONGUE, J.
ORS 68,340 (1) provides:

> Every partner must account to the partnership for *any benefit,* and hold as trustee for it any profits derived by him *without the consent* of the other partners from any transaction connected with the formation, conduct, or liquidation of the partnership or from any use by him of its property. (Emphasis added.)

In *Fouchek et al.* v. *Janicek,* 190 Or. 251, 262, (1950), we said that this section from the Uniform Partnership Law states "the essence of the fiduciary (duty) of a partner," as stated by Justice Cardozo in *Meinhard* v. *Salmon,* 249 N.Y. 458, 463, . . . (1929), as follows:

> Joint adventurers, (and) copartners, owe to one another, while the enterprise continues, the duty of the finest loyalty. Many forms of conduct permissible in a workaday world for those acting at arm's length, are forbidden to those bound by fiduciary ties. A trustee is held to something stricter than the morals of the market place. Not honesty alone, but the punctilio of an honor the most sensitive, is then the standard of behavior. As to this there has developed a tradition that is unbending and inveterate. Uncompromising rigidity has been the attitude of courts of equity when petitioned to undermine the rule of undivided loyalty by the "disintegrating erosion" of particular exceptions.
>
> . . . Only thus has the level of conduct for fiduciaries been kept at a level higher than that trodden by the crowd. It will not consciously be lowered by any judgment of this court.

Real estate brokers are subject to potential conflicting interest in many transactions. Even when a real estate broker does not become a partner in a venture involving the purchase of property this court has held that he owes a fiduciary duty to protect his client's interests and also "to make a full, fair and understandable explanation to his client before having him sign any contract."

When, as in this case, a real-estate broker undertakes to join as a member of a partnership or joint venture in the purchase of real property on which he holds a listing, he is also subject to the fiduciary duties of undivided loyalty and complete disclosure owed by one partner to another. Indeed, one of the fundamental duties of any partner who deals on his own account in matters within the scope of his fiduciary relationship is the affirmative duty to make a full disclosure to his partners not only of the fact that he is dealing on his own account, but all of the facts which are material to the transaction.

It follows that the "consent of the other partners" required by ORS 68,340(1) before any partner may retain "any benefit" from "any transaction connected with the formation (or) conduct" of a partnership must necessarily be an "informed consent" with knowledge of the facts necessary to the giving of an intelligent consent.

In this case, Harris did not inform plaintiffs or disclose to them the fact that this property could have been purchased for $907,500 "net" to the seller or that upon its purchase for $1,010,000 Harris or International (of which Harris was the president) would be paid a commission in the amount of $100,000. In the absence of such a disclosure there could be no effective "consent" by plaintiffs to the payment or retention by Harris of any such "benefit" from that transaction, for the purposes of ORS 68,340(1).

For these reasons we must reject defendants' contention that the broker's commission paid to International was "neither secret nor concealed." For the same reasons, the trial court did not err in requiring defendants to account to the partnership for that commission.

(Affirmed.)

QUESTIONS FOR DISCUSSION

1. *If Harris wanted to obtain a commission for his services to the partnership, what was he required to do?*
2. *Was the court saying that a partner cannot make a personal profit at the expense of his partners under any circumstances?*

(b) *The Partner's Right to Share Profits.* The function of a partnership is to run a business with the objective of making a profit. In the absence of any agreement to the contrary, each partner has a right to share equally in the profits of the business. Uniform Partnership Act, Section 18(a). The other partners in the business may not vote to exclude a partner from his share of the profits. Furthermore there is a correlative right to have the profits distributed at reasonable intervals. The partners may not retain the profits in the firm indefinitely in order to force another partner out of the business.

As mentioned previously, the profits are to be distributed evenly in the absence of a contrary agreement. If the partners have made unequal contributions to the firm, the profits are still to be divided equally. If the partners desire that the profits should be divided on something other than an equal basis, this fact must be reflected in the partnership agreement. Such an agreement will, of course, then be given effect by the courts.

What if the hopes of the partners are not realized and the partnership shows a loss instead of a profit? How are the losses and expenses shared? The answer is the same as for profits. The rule is that losses and expenses are shared on the same basis as profits in the absence of a contrary agreement. Thus, if there is no agreement as to how profits are to be shared, the losses and expenses will be shared equally. If there is an agreement as to profits, it will be given the same effect as the sharing of losses. It is possible that the partners may agree to share profits and losses in different proportions, and such an agreement will be given effect.

It should be noted at this time that an agreement as to the division of profits does not have to be embodied in a written contract and may be shown by oral evidence. An agreement as to the division of profits may also be implied by the conduct and transactions of the parties. The best practice, however, particularly if the profits and losses are not to be divided equally, is to have the agreement reduced to writing.

(c) *The Partner's Right to Compensation.* The right of a partner to payment is to receive a share of the profits. Beyond that, he is generally not entitled to any compensation for his services. His reward for his efforts is his share of the profits. This is true even if a partner does a disproportionate amount of work or renders some special service to the partnership. He is not entitled to be compensated for these services because he is deemed to be performing them for his own benefit. The Uniform Partnership Act follows this view in Section 18(f), where it states that "No partner is entitled to remuneration for acting in the partnership business, except that a surviving partner is entitled to reasonable compensation for his services in winding up the partnership affairs."

You should note two things concerning the rule involved here. First, the partners may agree that a partner is to be specially compensated for some special or extraordinary service to the firm. They may agree, for example, to pay one partner a salary over and above his share of the profits for managing or handling the administrative details of a partnership such as a law firm. Such an agreement will, of course, be given effect. Second, a surviving partner may receive special compensation for services performed in connection with winding up the partnership affairs. This situation occurs when the other partners are deceased and the partnership is terminated.

KOENIG v. HUBER

210 N.W.2d 825 (S.D. 1973)

Edward Koenig and Ruben Huber entered into a partnership agreement on January 1, 1965, for the purpose of engaging in a plumbing and heating business. The agreement provided that the parties were to employ themselves diligently in the business for its greatest advantage. The agreement also provided for each partner to contribute equally to the capital of the firm in the form of equipment and machinery. One other provision of the agreement stated that all debts and obligations of the partnership would be shared equally and all profits would be distributed equally.

After a period of years Edward Koenig filed suit against Ruben Huber for the dissolution of the partnership and an accounting. Koenig then appealed the lower court's adoption of the accounting.

WINANS, J.

During the early years of the partnership business, few records were kept by the partners. Thus, after the plaintiff commenced this suit, the court appointed Thomas Vogel, a certified public accountant, to make an accounting of the partnership affairs. Vogel rendered a written accounting of the partnership which the court adopted in toto in its decision. In accordance with the partnership agreement, both parties were considered in the report to have a one half interest in the assets and profits of the business.

In this appeal it is the plaintiff's principal contention that the trial court erred in adopting the accountant's report in toto for the reason that it failed to consider "extra" time spent by the plaintiff in the partnership business. At the trial, the plaintiff introduced evidence that the defendant ran an oil station from the inception of the partnership until June of 1966; that the defendant served as deputy sheriff and sheriff from November 1970 to the cutoff date of the accounting, August 3, 1971; and that in the partnership income tax returns for 1965 and 1966, the plaintiff was reported as receiving more than half of the partnership income.

Based upon this evidence, the plaintiff argues that he should have been credited in the accounting with more than half of the profits for the years 1965 and 1966 as compensation for "extra" time that he spent in the partnership business while the defendant was running the oil station. We find this argument untenable. In their partnership agreement, the parties expressly stated that profits were to be divided equally. Such a provision is binding on the parties and must be given effect. We cannot rewrite the partnership agreement merely because the plaintiff claims that he has spent more time in the partnership business than the defendant.

We also reject the plaintiff's claim that he should have been allowed a salary of $600 per month for extra time that he spent in the partnership business while the defendant served as deputy sheriff and sheriff. In the absence of an agreement to the contrary, a partner is not entitled to compensation beyond his share of the profits. See SDCL 48–3–1 and 48–3–7. There was no such agreement to the contrary in the present case.

The plaintiff next contends that the trial court erred in failing to find that the defendant breached the fifth provision of the partnership agreement, as quoted above, by serving as sheriff and deputy sheriff. We find no merit in this contention. There was evidence presented at the trial that the defendant continued to perform his duties for the partnership after he began serving as sheriff and deputy sheriff. In view of this evidence, the trial court's findings will not be disturbed on appeal.

(Affirmed.)

QUESTIONS FOR DISCUSSION

1. *If Koenig desired compensation for his "extra work," what should he have done to assure it?*
2. *How do you rationalize the court's finding that Huber lived up to the agreement when during part of the life of the partnership he acted as deputy sheriff and sheriff and ran an oil station?*

(d) *The Partner's Right to Reimbursement and Indemnification.* If a partner personally incurs expenses on behalf of the partnership, he is entitled to be reimbursed or indemnified for those expenses. Section 18(b) of the Uniform Partnership Act states, "The partnership must indemnify every partner in respect of payments made and personal liabilities reasonably incurred by him in the ordinary and proper conduct of its business, or for the preservation of its business or property." Note that the expenses or expenditures must be incurred in the ordinary course of the firm affairs in order for the partner to be entitled to reimbursement. For example, a partner is entitled to be reimbursed for the personal payment of a firm debt or office expense or for taxes paid on firm property. There are, of course, many other expenses for which a partner may be entitled to reimbursement.

On the other hand, a partner is not entitled to reimbursement for expenses incurred on his own behalf or for his own personal business. Furthermore, if a partner incurs expenses and liability for damages to those outside the firm as a result of his own negligence, he is not entitled to indemnification. For example, if Percy Solicitor negligently prepares a will and trust for a client that results in loss to the client, Percy is liable to the client for the loss. He is not entitled to be indemnified by his partners for the money he must pay the client. In fact, because the law firm would be liable as a result of Percy's negligence, they are entitled to be indemnified by him.

(e) *The Partner's Right to Participate in Management.* One of the reasons for entering into a partnership is to take advantage of the personal talents of the individual partners. For example, in a body shop two of the partners may engage in the actual reconstruction and painting of damaged cars while a third partner keeps the firm's books, prepares tax returns, and other similar matters. However, simply because the individual partners work in a particular area of the firm's business does not mean that they are prevented from dealing with other aspects of the firm's operation. Both at common law and under the Uniform Partnership Act, Section 18(e), all partners have an equal right in the management and conduct of the partnership business unless there is an agreement to the contrary. As with the right to share in the profits, this equal right to participate in the management obtains regardless of the amount of the capital contributions of the partner. If one partner contributes the bulk of the assets to the firm and wants a greater than equal voice in the firm's management, he must make sure that there is a provision for this in the partnership agreement. Without such a provision the partners have a right to participate equally in the management.

On occasion a dispute may arise between the partners as to some aspect of the firm's management, such as where the business should be located or whether the location of the business should be moved. In the event that such a dispute exists, the majority of the partners rule with regard to ordinary matters connected with the partnership business. However, in order for any partner to engage in any act in contravention of the partnership agreement, the unanimous consent of all the other partners must be obtained. Uniform Partnership Act, Section 18(h). Section 9(3) enumerates certain acts that require unanimous consent of all partners unless they have abandoned the business. The act provides that:

> Unless authorized by the other partners or unless they have abandoned the business, one or more but less than all the partners have no authority to:
> (a) Assign the partnership property in trust for creditors or on the assignee's promise to pay the debts of the partnership,
> (b) Dispose of the good-will of the business,
> (c) Do any other act which would make it impossible to carry on the ordinary business of a partnership,
> (d) Confess a judgment,
> (e) Submit a partnership claim or liability to arbitration or reference.

What happens if the partners divide equally in voting on some question of the management of the firm? The Uniform Partnership Act does not provide for resolving such a stalemate, and unless the partners can resolve their differences, a dissolution of the partnership may result. One way to prevent such an occurrence is to have a clause in the partnership agreement stating how tie votes should be resolved. The agreement may provide, for instance, for the resolution of the dispute by some outside third party or arbitrator.

There are some limitations on the right of the majority to rule other than in the situations already enumerated. The courts generally require the majority to deal in good faith. If the majority fails to act in good faith and exploits the minority, the equity courts will act to protect the rights of the minority.

(f) *The Partner's Right to Information and Inspection of Books.* Partners are under a duty to keep books and accounts relating to the partnership business. Frequently the pivotal area in a dispute between partners is the right to have access to these records. Typically the partners have a falling out, and the partner in possession of the books and accounts refuses the other partner access to them. He may not act in this manner. The Uniform Partnership Act, Section 19, provides that, subject to any agreement, the partnership books shall be kept at the principal place of business of the partnership. It further provides that every partner shall have access to the books and may inspect and copy them. This right continues for a reasonable time after the dissolution of the firm.

The Uniform Partnership Act, Section 20, also makes clear the partner's right to information concerning the partnership. A partner is entitled, upon demand, to "true and full information of all things affecting the partnership." The legal representative of any deceased partner or any partner under a legal disability is also entitled to this information.

(g) *The Partner's Right to an Accounting.* The right of a partner to a formal accounting usually occurs in connection with the dissolution of the partnership. If the parties are unable to wind up the partnership on a basis agreeable to all of them, an action may be filed in equity seeking a formal accounting for the period of the partnership so that the assets may be properly divided.

There are other circumstances in which a court will grant an accounting. Section 22 of the Uniform Partnership Act provides that a partner has a right to a formal accounting if he is wrongfully excluded from the partnership business or possession of its property by his partners; if such an accounting is provided for in the partnership agreement; or if a partner derives profits from the business without the consent of the other partners during the formation, liquidation, or conduct of the business or by use of the partnership property. The act also provides for an accounting "whenever other circumstances render it just and reasonable."

Suppose that Percy Outpartner is excluded from the business by Inez Inpartner. He is in fact locked out of the premises and denied access to the books and to any profits. Upon a showing that he is a partner, Percy could petition a court of equity for an accounting, dissolution, and winding up of the business, thereby obtaining his interest in the partnership.

As a general rule, a partner cannot file a suit at law against the partnership because conceptually he would be suing himself. His remedy is to seek an accounting from his co-partners. A partner may not generally sue a partner in a law court over partnership matters. There are exceptions, however. A partner may bring a lawsuit against a partner who fails to make an agreed capital contribution to the firm. After a partnership is terminated, a partner may sue another at law to recover a final balance due from one partner to another.

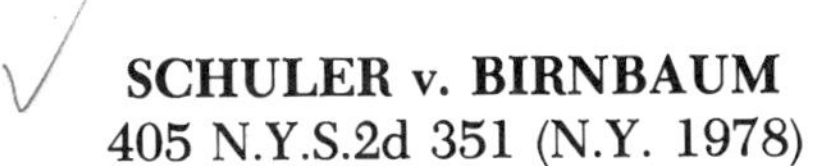

SCHULER v. BIRNBAUM
405 N.Y.S.2d 351 (N.Y. 1978)

Robert Hurlbut, Joseph Schuler, and Bernard P. Birnbaum formed a partnership in 1964 with the intention of constructing and operating a nursing home in Corning, New York. Preliminarily, the architectural firm of Corgan and Balistiere was retained by the partnership to prepare nursing home plans for submission to the appropriate state agency together with the partnership's license application. The state did not grant the license and the partnership was abandoned.

The partnership did not pay the architects, Corgan and Balistiere, and they sued the partnership for services rendered in preparing the plans. In 1975 the claim was settled. Hurlbut approved the settlement and paid his pro-rata share of the debt and legal expenses. Birnbaum did not approve the settlement or reimburse the plaintiffs. They then sued Birnbaum's estate (he had died during this period) to recover $2,475.12 which represented one third of the settlement and legal fees.

The defendants, Birnbaum's estate, contend that the action at law, prior to an accounting between the partners, is improper and should be dismissed.

SIMONS, J.

Generally, courts will not interfere in internal disputes between members of a partnership, preferring instead that the partners settle their differences among themselves or else dissolve and go out of business settling their affairs at that time by a final and full accounting with all partners joined. In this way, premature piece-meal judgments between partners which may later require adjustment when all the business of the partnership is reviewed are avoided. Thus, it is the general rule that partners cannot sue each other at law for acts relating to the partnership unless there is an accounting, prior settlement, or adjustment of the partnership affairs. One exception to this rule permits a partner to maintain an action at law against his co-partner when no complex accounting is required or when only one transaction is involved which is fully closed but unadjusted.

The evidence before the court upon the motion for summary judgment established that the parties had formed a partnership to operate a nursing home, that the partnership was abandoned after it failed to obtain a state license, and that the retention of the architects and filing of the license application was the sole business transacted by the partnership. This was not rebutted by defendants by factual allegations establishing the need for an accounting, nor did defendants counterclaim for an accounting as they could have. Accordingly, this action at law may be maintained.

Furthermore, we find no question of fact requiring a trial. The evidence established that decedent knew of the architects' claim for services in February, 1965; that he acknowledged that the services had been performed but disputed the value of them; that in 1969 he urged plaintiff to resolve the debt since "we are all in this deal"; that he was notified of the proposed settlement in 1975 but denied any legal responsibility for it or for plaintiff's legal expenses. Plaintiff's counsel offered to cooperate with decedent and his separate counsel if he chose to intervene in the architects' action, and when this was not done, counsel advised decedent and Hurlbut of the proposed settlement and requested approval. Mr. Hurlbut approved the settlement and the legal fees but decedent did not. On this appeal defendants do not question the amounts involved.

The proof established that payments by plaintiff were incurred in the reasonable operation of the partnership and became a partnership charge for which decedent became pro-rata responsible when the partnership assets were insufficient to satisfy plaintiff's claim. Plaintiff is entitled to indemnity for the sums expended to satisfy the architects' claim.

(Judgment for Schuler.)

QUESTIONS FOR DISCUSSION

1. *What is the rationale for requiring an accounting between partners prior to an action at law?*
2. *What reason did the court give for not requiring an accounting in this case prior to an action at law?*

(h) *The Partner's Right to Return Capital and Contributions.* Upon dissolution and winding up of a partnership, a partner is entitled to the return of the capital that he paid into the firm. This right is subordinate, of course, to the rights of any creditors. This process will be more fully discussed in Chapter 39. Suffice it to say at this point that a partner is not entitled to interest on his capital contribution to the firm unless there is a delay in its return to him upon the firm's dissolution. The rule makes sense when one remembers that the capital is used to provide profits that constitute the partner's return on investment.

If during the course of the partnership a partner makes loans to the partnership, he or she is entitled to have them paid. The question of interest is treated by the Uniform Partnership Act if there is no agreement to the contrary. It states in Section 18(c) that "A partner, who in aid of the partnership makes any payment or advance beyond the amount of capital which he agreed to contribute, shall be paid interest from the date of the payment or advance."

(2) **Partner's Duties.** Most of the rights of partners discussed previously involve correlative duties. Thus, although one partner has a right to expect another to act as a fiduciary, he also has a duty on his own part to act in good faith because of the fiduciary relationship. There are some duties, however, that have not been discussed. Among those duties are the duty to use care and skill, to devote full time and energy to the partnership, to account, and to comply with the partnership agreement.

(a) *The Partner's Duty to Use Care and Skill.* As a general rule a partner is not liable to the firm for losses occasioned by his bad judgment or from an honest mistake. However, if the loss to the firm is a result of the partner's negligence or wrongful conduct, then he has a duty to indemnify the partnership. If the partner acts in violation of the partnership agreement and such act results in a loss to the partnership, it is entitled to be indemnified for such loss unless, of course, the act is in some manner ratified by the partners.

The right to recover from a negligent partner applies either if the partner's negligence is directed toward the firm itself in the operation of the business or if the negligence is directed toward someone outside the partnership and the other partners become liable. Thus, if a partner negligently operates a car within the scope of the firm's business and a third person is injured, his co-partners will be liable under the agency theory previously discussed. In this case, the co-partners would be entitled to be indemnified by the guilty partner. Likewise, if the negligence of the partner is in the operation of the business and a loss results to the partnership from the negligence, then the partnership is entitled to be indemnified by the negligent partner.

(b) *The Partner's Duty to Devote Full Time and Energy.* It is sometimes said that a partner has a duty to devote his full time and energy to the business of the partnership unless there is an agreement to the contrary. If one is going to devote only part of his time to the partnership or no time at all, as a dormant partner, this should be spelled out in the agreement. In practice this duty is somewhat difficult to enforce. It certainly does not mean that all partners must devote equal time to the business. On the other hand, a partner clearly cannot devote half his time to a competing business—or any time for that matter—unless his co-partners agree. It has also been held that there is nothing inherently wrong with a partner's engaging in a separate business on his own behalf if it is in a different line of trade or commerce. Profits so derived are not subject to the partner's duty to account, which is discussed next.

(c) *The Partner's Duty to Account.* A partner's duty to account is somewhat different from his right to obtain an accounting in equity. The latter is a legal action brought in court in order to enforce certain rights. The former relates to the partner's fiduciary duty to account for any profits that may be made from a competing business or from any transaction connected with the formation, conduct, or liquidation of the partnership or from the use of partnership property unless the co-partners consent to his keeping such profits. Uniform Partnership Act, Section 21.

(d) *The Partner's Duty to Comply with the Partnership Agreement.* The partnership agreement is a contract. Just as with any contract, if there is a breach of the partnership agreement, the guilty partner is liable to the other partners for any damages sustained as a result of the breach. We shall see in Section 38:2 that the provisions of the agreement are not binding upon parties outside the partnership who are unaware of the agreement.

38:2. RELATION OF PARTNERS TO THIRD PERSONS. One thing to keep in mind while you are studying this section is that partners are principals and agents of one another; therefore much of the discussion regarding the principal's liability to third parties because of the acts of the agent has relevance to the discussion here. However, partners differ in their relationship in some respects from the simple relationship of principal and agent. Perhaps the most obvious difference is the common interest of the partners in the partnership property and business, whereas the agent does not have such an interest.

Section 9(1) of the Uniform Partnership Act provides:

> Every partner is an agent of the partnership for the purpose of its business and the act of every partner, including the execution in the partnership name of any instrument, for apparently carrying on in the usual way the business of the partnership of which he is a member binds the partnership, unless the partner so acting has in fact no authority to act for the partnership in the particular matter, and the person with whom he is dealing has knowledge of the fact that he has no such authority.

Attention is now directed to the subjects of partnership authority and liability.

(1) Authority of Partners. As in an agency relationship the authority of partners may be actual or apparent. Liability may also arise under the doctrine of estoppel. Actual authority may be either express of implied.

(a) *Actual Authority.* The *express authority* of partners may be either written or oral. Frequently the partnership agreement will provide for one partner to act in a particular area, or it may designate one partner as the "managing partner" and give him authority to handle such firm matters as the hiring and firing of personnel and the purchasing of supplies. If such an agreement expressly limits the authority of a partner, third persons are not bound by its terms unless they have notice of it. On the other hand, a partner may be liable to his co-partners for breaching the agreement if they suffer damage as a consequence.

A partner may bind the partnership because of his *implied actual authority.* A partner's implied authority extends to those acts that are reasonably necessary to the carrying on of the business of the partnership. Factors one should consider in ascertaining the extent of a partner's implied authority are the type of business, whether it is a *trading* or *nontrading* partnership, the usages and customs of the trade, and the customs or course of dealing of the particular partnership.

Section 9(3) of the Uniform Partnership Act specifically mentions a number of activities that an individual partner has no authority to perform unless authorized by the other partners. Among these activities are the assignment of partnership property; the disposal of the goodwill of the business; and the power to confess a judgment, to submit a partnership claim to arbitration, or to do any other act that would make it impossible to carry on the ordinary partnership business. Obviously then, a third party may not rely on the doctrine of apparent authority in order to bind a partnership in these areas if the partner has no actual authority to bind the firm. On the other hand, remember that the authority to bind the partnership on these matters may be implied.

(b) *Apparent Authority and Estoppel.* The doctrine of apparent authority has been fully discussed in Section Six, "Agency." It is not actual authority but stems from the action of a principal in clothing the agent with the indicia of authority so that a third person is reasonably entitled to believe that the agent is acting with authority even though in fact he is not. In the case of a partnership, it is the partnership that is bound by the acts of the individual partner where the doctrine of apparent authority is applicable. If the partnership suffers damages as a result of a partner's exceeding his authority, it may seek damages from the individual partner.

There are limitations to the application of the doctrine of apparent authority where a partnership is involved. In the first place, one may not rely on a partner's apparent authority if he is aware that the partner does not actually have that authority or believes that the partner may not have it. For example, if one knows that a partnership agreement limits a partner's authority to act in a certain area, then the doctrine of apparent authority can not be invoked for acts prohibited by the agreement if the third party has knowledge of the prohibition. Of course, if the third party has no such knowledge, the agreement between partners is not binding upon him.

The applicability of the doctrine of apparent authority is also conditioned by whether a partnership is a *trading* or a *nontrading partnership.* A trading partnership is generally considered to be one that is involved in the buying and selling of goods, merchandise, or other kinds of property in its ordinary course of business. The most common example would be a partnership engaged in a retail store, but there are many types of trading partnerships. Sometimes the line of demarcation between a trading and a nontrading partnership is not easy to draw. A clear example of a nontrading partnership would be a law firm or an accounting firm.

As a general rule, the apparent authority of partners in a trading partnership is much broader than the authority of a partner in a nontrading partnership.

For example, a partner of a trading partnership has the apparent authority and usually the implied authority to borrow money, extend the credit of the firm, and execute negotiable instruments. The reason is that a partner in a trading partnership usually engages in these kinds of activities as a normal course of business. Thus, unless put on notice to the contrary, a third person may reasonably believe that the partner in fact has this authority. On the other hand, these types of activities are not usually engaged in by nontrading partnerships, and thus partners of nontrading partnerships do not have the implied or apparent authority to act in these areas.

NATIONAL BISCUIT CO. v. STROUD
249 N.C. 467, 106 S.E.2d 692 (1959)

PARKER, J.

C. N. Stroud and Earl Freeman entered into a general partnership to sell groceries under the firm name of Stroud's Food Center. There is nothing in the agreed statement of facts to indicate or suggest that Freeman's power and authority as a general partner were in any way restricted or limited by the articles of partnership in respect to the ordinary and legitimate business of the partnership. Certainly, the purchase and sale of bread were ordinary and legitimate business of Stroud's Food Center during its continuance as a going concern.

Several months prior to February 1956 Stroud advised plaintiff that he personally would not be responsible for any additional bread sold by plaintiff to Stroud's Food Center. After such notice to plaintiff, it from 6 February 1956 to 25 February 1956, at the request of Freeman, sold and delivered bread in the amount of $171.04 to Stroud's Food Center.

The General Assembly of North Carolina in 1941 enacted a Uniform Partnership Act, which became effective 15 March 1941. G.S. Ch. 59, Partnership, Art. 2.

G.S. § 59–39 is entitled "Partner Agent of Partnership as to Partnership Business," and subsection (1) reads: "Every partner is an agent of the partnership for the purpose of its business, and the act of every partner, including the execution in the partnership name of any instrument, for apparently carrying on in the usual way the business of the partnership of which he is a member binds the partnership, unless the partner so acting has in fact no authority to act for the partnership in the particular matter, and the person with whom he is dealing has knowledge of the fact that he has no such authority." G.S. § 59–39(4) states: "No act of a partner in contravention of a restriction on authority shall bind the partnership to persons having knowledge of the restriction."

G.S. § 59–45 provides that "all partners are jointly and severally liable for the acts and obligations of the partnership."

G.S. § 59–48 is captioned "Rules Determining Rights and Duties of Partners." Subsection (e) thereof reads: "All partners have equal rights in the management and conduct of the partnership business." Subsection (h) hereof is as follows: "Any difference arising as to ordinary matters connected with the partnership business may be decided by a majority of the partners; but no act in contravention of any agreement between the partners may be done rightfully without the consent of all the partners."

Freeman as a general partner with Stroud, with no restrictions on his authority to act within the scope of the partnership business so far as the agreed statement of facts shows, had under the Uniform Partnership Act "equal rights in the management and conduct of the partnership business." Under G.S. § 59–48(h) Stroud, his co-partner, could not restrict the power and authority of Freeman to buy bread for the partnership as a going concern, for such a purchase was an "ordinary matter connected with the partnership business," for the purpose of its business and within its scope, because in the very nature of things Stroud was not, and could not be, a majority of the partners. Therefore, Freeman's purchases of bread from plaintiff for Stroud's Food Center as a going concern bound the partnership and his co-partner Stroud.

In Crane on Partnership, 2d Ed., p. 277, it is said: "In cases of an even division of the partners as to whether or not an act within the scope of the business should be done, of which disagreement a third person has knowledge, it seems that logically no restriction can be placed upon the power to act. The partnership being a going concern, activities within the scope of the business should not be limited, save by the expressed will of the majority deciding a disputed question; half of the members are not a majority."

(Affirmed.)

QUESTIONS FOR DISCUSSION

1. *Could Stroud have done anything to avoid becoming liable to National Biscuit?*
2. *Did Stroud have any rights against Freeman?*

(i) ESTOPPEL. Another basis for holding the partnership or one not actually a partner liable to a third person is the doctrine of estoppel. The doctrine is invoked if an individual who is in fact not a partner is held out to third persons by the partnership as a partner or if an individual represents himself to be a partner and the firm partners, knowing of the representation, do nothing to stop him. In both cases, the partnership may be held liable on the transaction even though the individual is not in fact a partner. The rule cuts the other way also. If an individual is represented by a firm as being a partner and is aware of this representation, he may be held liable as a partner even if in fact he is not.

If the third person is aware that the representation of partnership is untrue, then, of course, the doctrine of estoppel is not invoked. Likewise there generally must be an awareness of the holding out of partnership by the person not a partner but being held out as such, so that he has an opportunity to act to stop the misrepresentations. The invocation of the doctrine of estoppel is based upon the failure to put an end to the misrepresentation when one has an opportunity to do so. In some jurisdictions, in order for the doctrine to be invoked, there must be consent to the holding out. In others, however, consent is not necessary, merely failing to take action to deny the partnership is sufficient.

Section 16 of the Uniform Partnership Act provides that when the doctrine of estoppel is applicable, an individual is liable as though he were an actual member of the partnership if the partnership itself is liable. If there is no partnership liability, the individual is liable jointly with the other persons consenting to the contract or representation of partnership. The act requires consent to the false holding out of partnership in order for liability to be imposed, but many jurisdictions do not follow this rule.

An example may help clarify the doctrine. Suppose Albert and Burns are partners. Carl is not a partner. Nonetheless Albert and Burns represent to people in the course of business that Carl is a partner. They then call the firm "Albert, Burns, and Carl," and have signs, business cards, and invoices printed in this manner. In all states, if Carl knows of and consents to this representation, he would be liable to third persons reasonably relying on it. In some states, even if Carl did not consent to the holding out but was aware of it and did not make reasonable attempts to stop it, he would be bound.

Likewise, if Carl holds himself out as a partner when he actually is not and the members of the partnership consent to it, they would be bound by the representation.

McBRIETY v. PHILLIPS et al.
26 A.2d 400 (Md. 1942)

Ernest McBriety and others trading as R. J. Rhodes Distributing Company delivered beer to Hubert Phillips and Norman Wiley trading as Cozy Spot. Wiley was then adjudicated as bankrupt, and McBriety sought to collect from Phillips, alleging that he and Wiley were partners. Phillips denied that any partnership existed.

During the course of the trial McBriety attempted to introduce evidence to show that Phillips represented himself as Wiley's partner, as well as certain records showing that Wiley had also made this representation. The trial court excluded this evidence and directed a verdict in favor of Phillips. Plaintiffs appealed.

DELAPLAINE, J.

The principle of law is firmly established that a person, even though not a partner in fact, is liable as a partner to those persons who deal in good faith with the firm or with him as a member of it with the reasonable belief that he is a member of the firm, provided that he so holds himself out to such persons or the public, or is so held out by his authority or with his knowledge and assent, and credit is to some extent induced by this belief. The doctrine of partnership by estoppel is founded upon principles of justice for the purpose of preventing fraud. When a person has induced others to believe that he is a partner and to extend credit to the partnership by reason of such belief, he should not be permitted to deny that participation which he has asserted or permitted to appear to exist, even though not existing in fact. The statute provides that when a person consents to be represented in a public manner as a partner, he is liable to any person who gives credit on the faith of such representation, even though the representation is communicated to the creditor without the apparent partner's knowledge. In order to invoke the doctrine of estoppel, it must be shown that the creditor acted in reliance upon the representations at the time he entered into the transaction. If a person knows that he is held out as a partner by another, he is just as liable as though he had called himself a partner, unless he does all that a reasonable and honest man would do under similar circumstances to assert his denial in order to remove the impression and prevent innocent parties from being misled.

The question whether a partnership existed in fact or by estoppel is a question of fact for the consideration of the jury. Any conduct on the part of a person reasonably calculated to lead others to suppose that he is a partner in a particular business amounts to a holding out on his part. Evidence is competent to show that the plaintiffs trusted Phillips in good faith and sold the beverages with the understanding that he was interested in the business. The testimony of the sales agent, substantiated by the testimony of the other witnesses at the trial below, was sufficient to justify submission of the case to the jury. Even though it is shown that Phillips and Wiley were not partners in fact, the jury should be permitted to determine whether or not Phillips held himself out as a partner and thereby induced the extension of credit.

Since the withdrawal of the case from the jury was improper, the judgment in favor of Phillips must be reversed.

(2) Liability of Partners. (a) *Contract Liability.* It should be clear by now that a partner may bind the partnership or other partners to a contract in a number of ways. Now the question to be answered is just what is the nature of the obligations when this happens. Under the common law the liability of the partners on a contract is *joint* as opposed to *joint and several* liability. What this means is that because the obligation under the contract is an obligation of all the partners jointly, it then must be imposed against all the partners rather than against some and not others. The practical impact of the rule is to require suit against all members of a partnership for breach of contract and to get service of process against them all. In some states failure to obtain service against some of the partners will bar further action against them. Also, if the obligation is joint, release of one partner from the obligation operates to release the others.

The Uniform Partnership Act has adopted the common-law approach to contractual liability in Section 15. The common-law rule does make it somewhat more difficult to enforce contracts against a partnership because of the requirements set out before. Some states have eliminated the distinction between joint obligations and joint and several obligations by statute, sometimes called *joint-debtor statutes.* Where this is the case, one may obtain a judgment against an individual partner rather than having to proceed against all of them. The judgment may then be satisfied against the partnership's property and also the individual property of the partner against whom judgment was taken. The individual property of the partners not involved in the judgment would not be subject to the judgment, of course.

A few states have held that their joint-debtor statutes do not apply to the Uniform Partnership Act. Where this is the case, the obligations of partners under a contract are still joint.

(b) *Tort Liability.* The members of a partnership are liable for torts committed by their partners or other agents within the scope of the partnership business. The tort liability of the individual members of the partnership is *joint and several.* The basic distinction between joint liability and joint and several liability is that in the latter case the individual partner alone may be held separately liable for damages. A partner may be sued individually for damages, and release of one partner from liability does not release the others. If a single partner has a judgment imposed against him, the partnership assets and his individual assets are subject to attachment in order to satisfy the judgment, but the individual assets of the partners against whom no judgment is obtained are not subject to attachment. The partner who has a judgment against him may, of course, seek contribution from the other members and of the partnership. Sections 13 and 15 of the Uniform Partnership Act adhere to the common-law rule of joint and several liability for partner's tort liability.

MARTIN v. BARBOUR

558 S.W.2d 200 (Mo.App. 1977)

Marvin Martin sued his physician, Dr. Gene Barbour, and Dr. Barbour's partner, Dr. C. W. Egle, for malpractice.

HOUSER, J.

Plaintiff, having suffered for several weeks from stomach pains, first consulted Dr. Barbour on January 20, 1970. Suspecting stomach ulcer, Dr. Barbour placed plaintiff in Normandy Osteopathic Hospital for tests. After taking X rays and making other tests Dr. Barbour, an expert in the field of proctology, examined plaintiff's rectum. None of plaintiff's complaints was associated with his rectum. Plaintiff had never previously made any complaint with respect to or sustained any injury to or experienced any disease or treatment of any kind to his rectum. He did not complain of painful bowel movements in his interview with doctors at the hospital. Dr. Barbour tried to insert an examining pipe into plaintiff's rectum. He reported to plaintiff that he was unable to do so; that he found some sort of blockage in his rectum which would have to be opened up; that otherwise it would close; that if it closed all the way it would be "like locked bowels" and this would kill him; that surgery "had to be done." Dr. Barbour did not explain to plaintiff what surgery he proposed to perform; did not tell him he had hemorrhoids and a fissure; did not discuss the risks involved, the possibility of having the sphincter muscle cut or damaged, that if it were cut plaintiff might be incontinent, or that the operation might not be 100% successful. Having confidence in the doctor, plaintiff submitted to surgery on February 3, 1970. Following surgery his rectum was tight (a condition termed "stenosis"). After the operation he was examined weekly, either by Dr. Barbour or Dr. Egle, at their offices, and treated by digital dilation of the rectum and attempts to stretch the opening with an instrument. Dr. Barbour finally advised a second operation. No mention was made of any dangers or risks involved in the second surgery. Trusting the doctor, plaintiff was operated on a second time on April 16, 1970. Following the second surgery plaintiff could no longer control his bowels; he was incontinent.

Plaintiff's case against Dr. Egle, who did not assist or participate in either of the operations performed by Dr. Barbour and did not consult or treat plaintiff until after the operations, is based upon the contention that the two doctors were partners acting concurrently in treating plaintiff. There was evidence that on January 20, 1970 they were partners, but that at the time of the trial Lackland Clinic, owned by defendants, was a corporation. There is no evidence to show if and when the partnership was dissolved or when Lackland Clinic was incorporated. Plaintiff claims it is immaterial whether defendants treated him as partners or as coemployees of a corporation, in view of the evidence patients are seen back and forth between the two doctors; that patients know that either one or both of the doctors will see them, and that both doctors make notations on the file cards of patients and all their records show entries either by Dr. Barbour or Dr. Egle, or both. Conceding that under *Telaneus* v. *Simpson,* 321 Mo. 724, 12 S.W.2d 920 (1928), the negligence of Dr. Barbour in surgery could be imputed to Dr. Egle, if defendants were partners during the entire course of plaintiff's treatment, defendants say there was no evidence that defendants were partners beyond January, 1970. Further conceding that under *Baird* v. *National Health Foundation,* 235 Mo.App. 594, 144 S.W.2d 850 (1940), defendants would be jointly liable as employees of a common employer if they were acting concurrently in treatment of plaintiff which caused him injury, defendants say they were not acting concurrently on February 3 and April 16, the only dates or occurrences upon which negligence was charged; that since there was no charge of negligence with respect to postoperative care, plaintiff failed to make a case against Dr. Egle on the issue of concurrent treatment. Defendants further contend that any fraudulent concealment by Dr. Barbour constituted a wrongful act outside the scope of the partnership business, which could not be imputed to Dr. Egle, and therefore the statute of limitations as to the vicarious liability of Dr. Egle expired before suit was filed.

Having proved that a partnership between the two doctors existed in January, 1970 an inference arises that the partnership continued to exist until the corporate form of practicing medicine was adopted. The only evidence with respect to the corporate existence shows it in existence at trial time. There is no evidence that the corporation was in existence at any time prior to the filing of this lawsuit. For defendants to take the position that the law of partnership does not apply the burden was theirs to demonstrate when the partnership terminated. This they failed to do. "Pursuant to general rules, partners in the practice of medicine are all liable for an injury to a patient resulting from the lack of skill or the negligence, either in omission or commission, of any one of the partners within the scope of their partnership business; . . ." 70 C.J.S. Physicians and Surgeons § 54b., p. 977. "And where several physicians are in partnership, they may be held liable in damages for the professional negligence of one of the firm." 60 Am.Jur.2d Partnership § 166, p. 86. "It is plain, too, that when physicians and surgeons are in partnership, all are liable in damages for the professional negligence of one of the firm, for the act of one, within the scope of the partnership business, is the act of each and all, as fully as if each is present, participating in all that is done." 61 Am.Jur.2d Physicians, Surgeons, Etc. § 166, p. 295. On this record Dr. Egle is liable for Dr. Barbour's professional negligence on the basis of their partnership. There is no merit in the contention that fraudulent concealment by one partner may not be imputed to another partner. "The individual partners . . . are liable in a civil action for the fraudulent misconduct of a partner within the course or scope of the transactions and business of the partnership, whether such misconduct be by fraudulent representations or otherwise, even though the copartners had no knowledge of the fraud and did not participate therein; . . ." 68 C.J.S. Partnership § 170, p. 620; *Kearns* v. *Sparks*, 260 S.W.2d 353, 360[15] (Mo.App.1953).

(Judgment against Dr. Egle affirmed.)

PHILLIPS v. COOK

239 Md. 215, 210 A.2d 743 (1965)

Harris and Phillips entered into a partnership under the name of Dan's Used Cars for the purpose of buying and selling used automobiles. Each had an equal voice in the business. Neither partner owned a personal automobile, and it was agreed that Harris would use a partnership automobile for transportation to and from his home. Harris was authorized to demonstrate and sell automobiles, call on dealers, and go to the Department of Motor Vehicles on partnership business after leaving the lot in the evening and before returning the next day.

On January 7, 1960, at about 6:50 P.M. Harris was on his way home from the used-car lot when an accident occurred in which Delores Cook was injured as a result of Harris's negligence. The jury in the trial court found against Harris and Phillips on the theory that Harris was acting within the scope of employment when driving the car. Phillips appealed, contending that there was no legally sufficient evidence from which a jury could conclude that Harris was acting within the scope of employment.

MARBURY, J.

In a case involving a partnership, the contract of partnership constitutes all of its members as agents of each other and each partner acts both as a principal and as the agent of the others in regard to acts done within the apparent scope of the business, purpose and agreement of the partnership or for its benefit. It is clear that the partnership is bound by the partner's wrongful act if done within the scope of the partnership's business. Code (1957), Article 73A, Section 13 provides:

> Where, by any wrongful act or omission of any partner acting in the ordinary course of the business of the partnership, or with the authority of his copartners, loss or injury is caused to any person, not being a partner in the partnership, or any penalty is incurred, the partnership is liable therefor to the same extent as the partner so acting or omitting to act.

The text of the liability of the partnership and of its members for the torts of any one partner is whether the wrongful act was done within what may reasonably be found to be the scope of the business of the partnership and for its benefit. The extent of the authority of a partner is determined essentially by the same principles as those which measure the scope of an agent's authority. Partnership cases may differ from principal and agent and master and servant relationships because in the non-partnership cases, the element of control or authorization is important. This is not so in the case of a partnership for a partner is also a principal, and control and authorization are generally within his power to exercise.

Here, the fact that the defendant partners were in the used car business; that the very vehicle involved in the accident was one of the partnership assets for sale at all times, day or night, at any location; that Harris was on call by Phillips or customers at his home—he went back to the lot two or three times after going home; that he had no set time and worked irregular hours, coupled with the fact that he frequently stopped to conduct partnership business on the way to and from the lot; drove partnership vehicles to the Department of Motor Vehicles, and to dealers in Baltimore to view and buy used cars while on his way to or from his home; that one of the elements of the partnership arrangement was that each partner could have full use of the vehicles; that the use of the automobile by Harris for transportation to and from his home was admittedly "essential" to the partnership arrangement and the most practical and convenient way to operate; and that Harris conducted partnership business both at the used car lot and from his home requires that the question of whether the use of the automobile at the time of the accident was in the partnership interest and for its benefit be submitted to the jury. We find that the lower court did not err in refusing to grant appellant's motions for a directed verdict as to him in the capacity of a co-partner trading as Dan's Used Cars.

QUESTIONS FOR DISCUSSION

1. *What factors were significant in the court's determination that a jury could have found that Harris acted within the scope of employment?*
2. *Would the result have been the same had Harris first stopped at a bar for a "few drinks"? If the accident had occurred after he had been home and while he was on the way to the store to buy a pack of cigarettes?*

(c) *Criminal Liability.* As a general rule, in order for a partner to be liable for the criminal acts of his co-partner, he or she must also have participated in or authorized the crime. There must be something more than just the partnership relation in order to make a partner guilty of a crime. This is true even though the criminal act has benefited the partnership and was performed in the carrying out of the firm's business.

There are some statutes that do place criminal liability on partners or sometimes on the partnership itself. These are typically licensing statutes. These statutes may impose liability upon firm members if the business has failed to comply with the licensing statute even though there is no proof of a partner's personal participation in the failure to obtain such a license. The punishment for failing to comply with these types of statutes is usually—although not always—limited to a fine. The requirement of a license to sell liquor is a common example of such a statute.

(d) *Liability of Incoming Partners.* A question frequently arises as to the liability of a new partner who joins a partnership. Technically this creates a new partnership. Section 17 of the Uniform Partnership Act provides that the new partner is liable for all the obligations incurred by the firm prior to his arrival, just as though he or she had been a partner when the obligations arose. This liability, however, is limited to the new partner's contribution to the firm. His personal property is not subject to liability for the obligations of the partnership that were entered into before he became a partner.

Sometimes, as a condition of joining a partnership, an individual is required to assume all the firm obligations as of the time he or she joins the firm. In that event, of course, the new partner would have the same liability as the original partners.

MAGRINI v. JACKSON

17 Ill.App.2d 346, 150 N.E.2d 387 (1958)

In 1951 Arthur Magrini started a tavern business in premises leased from the Wilson Park Club. He sold the equipment in September 1952 to Earnest Englund, Jr., under a written conditional-sales contract. Under the contract Englund paid $7,000 down and agreed to pay the balance of $18,000 at the rate of $250 per month. Magrini's lease was also assigned to Englund. Subsequently Jackson entered into Englund's business as a partner under an agreement dated January 7, 1953. The agreement provided for the operation of a nightclub called the 8 O'Clock Club. The agreement also provided that:

> It is mutually understood that Arthur Magrini is the owner of a chattel mortgage covering the chattels enumerated in said mortgage, and that there is now due approximately Seventeen Thousand Two Hundred Fifty Dollars ($17,250.00) on said mortgage; that when said mortgage has been fully paid that said chattels shall belong to the parties hereto in equal shares, and that the payments on said mortgage shall be made from the funds in the partnership as part of the operating expense.

A number of payments were made to Magrini out of the partnership funds, and then the payments ceased. Magrini sued Jackson and Englund but failed to serve Englund as he was out of the country. Judgment was entered against Jackson in the amount of $14,318.39, and he appealed, contending that he was not liable because the debt was personal to Englund.

REYNOLDS, J.

The rule applicable to a situation as thus presented is succintly stated in 68 C.J.S. Partnership § 80, at page 520 as follows:

> The individual debts of partners, whether contracted by them before the establishment of the partnership, or in establishing it, or during its existence, may be converted into firm debts, by the mutual consent of the partners, if the firm is solvent. An agreement by partners taking over the business of one of them to assume the debts of the former owner may, in the absence of an express agreement, be inferred from facts and circumstances; but an express or implied agreement is necessary to create such liability, and a person becoming a member of an existing firm, or forming a partnership with another in the latter's existing business, does not thereby automatically become liable for the debts already incurred.

Thus, if the parties in forming a partnership agree to convert an individual debt into a debt of the partnership, they all become liable individually for the debt so converted, just as though the partnership had contracted the debt in the first instance, since it is provided by Section 15 of the Uniform Partnership Act that:

> All partners are liable
>
> (a) Jointly and severally for everything chargeable to the partnership under Sections 13 and 14.
>
> (b) *Jointly for all other debts and obligations of the partnership;* but any partner may enter into a separate obligation to perform a partnership contract. (Emphasis supplied.)

This rule finds support in *Hoffman* v. *Stewart,* 184 Ill.App.66. In that case an individual named Thompson had a lease on a theater. He also had a contract with a prima donna to sing at his theater on a stipulated weekly salary. With these two assets in hand, he entered into an agreement with two other individuals, construed by the court to be a partnership agreement. By the agreement Thompson was to be manager of the enterprise known as Thompson Opera Co. at a weekly salary; that funds received in the enterprise should be deposited in a designated bank and drawn out by Thompson on checks signed by him and countersigned by one of the other partners. The prima donna sang, and, not having been paid her weekly stipend as contracted for with Thompson, she sued all three partners. In affirming a recovery the court said:

> It is true that a person who enters into partnership with another does not thereby become liable to the creditors of his partner

for anything done before he became a partner, yet there are exceptions to the rule. One is that where a contract made by a person or firm remains executory until a partnership is created or a new partner is admitted into the firm, and such contract, while it remains executory, is adopted by the incoming partner, who acquires all the benefit as if he had been a partner in the original transaction, a promise may be implied to assume the liability of the partner or firm on such contract.

In the case at bar, as in the foregoing cases, the contract forming the basis of liability was executory. In the case at bar, as in these cases, the newly formed partnership received the benefit of the original contract, i.e. the use of the equipment. To these factors must be added the additional factors in the case at bar, that by the partnership agreement the debt to plaintiff was specifically recognized; that it was expressly provided that payments on the conditional sales contract should be made out of partnership funds; that payments were in fact made to plaintiff out of partnership funds; and that the partners agreed, in effect, that if the income to the partnership was not sufficient in any given month to pay the expenses of operation, including the payment to plaintiff, they would each contribute equally to make up the difference.

(The judgment will be affirmed.)

QUESTION FOR DISCUSSION

1. *In the absence of the written partnership agreement what would Jackson's liability have been to Magrini?*

(e) *Admissions of a Partner.* The Uniform Partnership Act, Section 11, provides that "an Admission or representation made by any partner concerning partnership affairs within the scope of his authority as conferred by this act is evidence against the partnership." This rule does not mean that a statement made by a single individual can be used to establish a partnership. However, once a partnership is established, then the admissions made by a partner within the scope of his authority are evidence against the partnership. Thus a statement made by William Wrongdoer that "Inez Innocent is my partner" could not be used to impose liability upon Inez on the basis of partnership. Statements made by Wrongdoer subsequent to the establishment of partnership could be used, however.

SUMMARY

Partners are principals and agents of one another and therefore have the duties of a fiduciary. In the absence of an agreement to the contrary, partners share profits and losses equally. Beyond receiving a share of the profits, a partner is not entitled to receive compensation for his services unless there is an agreement to compensate a partner for a special service.

Partners are entitled to be reimbursed for expenses incurred on behalf of the partnership. A partner also has an equal right to participate in the management of the business. In the event of a dispute among partners the majority rules so long as the partnership agreement is not violated. If there is no majority and a stalemate results, the partnership may be dissolved.

A partner has a right to inspect the partnership books and records and to obtain information concerning the partnership. Under the proper circumstances a partner is entitled to the remedy of an accounting. Frequently this is in connection with a dissolution of the partnership. Upon dissolution and winding up of a partnership, a partner is entitled to the return of the capital that he paid to the firm.

A partner also has certain duties. Among these are the duty to use care and skill; to devote full time and energy to the partnership; to account to partners; and, to comply with the terms of the partnership agreement.

Since partners are principals and agents of each other, the same rules that apply in agency law regarding dealings with third parties apply to partners' dealings with third parties. Thus the concepts of actual express authority, actual implied authority, apparent authority, and estoppel also apply to partners.

Under the common law the contract liability of partners is joint and not joint and several. The Uniform Partnership Act has adopted this approach, and, hence, a suit for breach of contract must be against all partners. Some states have eliminated the distinction between joint obligations and joint and several obligations by joint-debtor statutes. Tort liability of partners is joint and several.

Generally a partner is not liable for criminal acts of a partner unless he has participated in them. Some licensing statutes are an exception.

A new partner is liable for obligations incurred by the firm prior to his arrival. This liability is limited to the new partner's contribution to the firm.

PROBLEMS

1. Dr. Roberts was a specialist in obstetrics and gynecology and a member of the Astoria Medical Group, a partnership. Dr. Roberts was named as a defendant in a malpractice suit. When he checked his professional liability insurance policy, he discovered that the partnership had failed to procure the policy for which he had applied and failed to advise him of the difference between the policy he requested and the policy ultimately obtained.

   Dr. Roberts then sued the partnership for damages. The partnership contended that Dr. Roberts's proper remedy was first to get an accounting or dissolution of the partnership. Is the partnership correct? Roberts v. Astoria Medical Group. 350 N.Y.S.2d 159 (1973).
2. Sarret and Stevens was a partnership engaged in the business of raising and selling sheep. Boise Payette Lumber Company delivered lumber, fence posts, nails, barbed wire, and other merchandise to the partnership. Jules Sarret made the purchases in the name of the partnership without informing Stevens. When presented with a bill for the goods, Stevens refused to pay, contending that Sarret had had no authority to make the purchases. Was Stevens liable? Boise Payette Lumber Company v. Sarret et al., 221 P. 130 (1923).
3. The Jacqueline Shop was owned by Mrs. Boone, who retailed ladies' wear. In the same building one J. E. Jones sold shoes from the same entrance marked "The Jacqueline Shop." Boone and Jones were neither partners nor employees of one another. A salesman for General Shoe Corporation

sold shoes to Jones, relying on the credit rating of The Jacqueline Shop. Advertisments described the shop as selling shoes and ladies' wear. Invoices were sent to The Jacqueline Shop, and Boone had access to the mail. Was Boone liable for the bills incurred by Jones? Boone v. General Shoe Corporation, 242 S.W.2d 138 (Ark. 1951).

4. Waagen and Gerde owned and operated a fishing boat as partners. Gerde and his two sons designed and built shark nets used by the boat. Waagen did not help in any way with this job, and Gerde informed him that it would "be lots of work to fix them." When Waagen sued for an accounting, Gerde claimed $2,500 credit for the work with the nets. Was Gerde entitled to the credit? Waagen v. Gerde, 36 Wash.2d 563 (1950).
5. Hensley and seven other partners operated the Mary Gail Coal Company. A truck owned personally by Hensley was destroyed as the result of the negligence of an employee of the coal company. Hensley sued the partnership for the loss of the truck. The company contended that Hensley could not sue the partnership in an action at law and that his sole remedy was an accounting. Was the company correct? Smith v. Hensley, 354 S.W.2d 744 (Ky. 1962).
6. M. Neely and Jack McDonald were partners operating under the name McDonald Bros. In 1910 a note in the amount of $10,000.00 was executed by McDonald Bros. payable to Helena Hanks. The note was renewed in 1914. Subsequent to the execution of the note M. Neely joined the partnership. The holders of the note sought to hold Neely liable for this obligation. Was Neely liable? Stephens v. Neely, 255 S.W. 562 (Ark. 1923).
7. Savage and Webb had a partnership in certain horses trained at Lincoln Downs. Webb was a horse trainer and received $10 per day for training a horse plus 10 percent of the horse's purses, if any. J. Patrick Dowd stabled his horse at Lincoln Downs in a stall assigned to Webb, and in return Dowd's brother was to help Webb. One morning an altercation took place between Webb and Dowd over the quality of Dowd's work. Webb stabbed Dowd in the face with a pitchfork, blinding him in one eye, fracturing his nose, destroying his dental bridgework, and cutting his face. Dowd sued Savage for damages, contending that he was liable as a partner of Webb. Was Savage liable? Dowd v. Webb, 337 F.2d 93 (C.A.N.J., 1964).
8. Williams and Moser were partners for the purpose of selling six acres of land. Williams entered into a contract to sell the land in the name of the partnership at a price that included a substantial profit. He did not inform Moser of the contract. Moser then offered to sell his interest in the partnership to Williams. Williams did not inform Moser of the contract on the land and accepted Moser's offer. When Moser learned of the sale, he filed suit against Williams for one half of the profit made on the sale of the land. Was Williams liable to Moser? Moser v. Williams, 443 S.W.2d 212 (Mo.App.1969).
9. A partnership was formed by Summers and Dooley for the purpose of collecting trash. Summers became ill and unable to work, so he hired a man

to do his work, paying him with his own money. Dooley refused to pay the man from partnership funds. Over a period of time Summers paid the man more than $11,000. Summers sued Dooley for reimbursement for the costs and expenses incurred in hiring the third man. Could Summers recover? Summers v. Dooley, 481 P.2d 318 (Idaho 1971).

CHAPTER 39

Dissolution and Winding Up

Unlike the life of a corporation the life of a partnership is limited and must end sometime. The legal terms used to describe the events surrounding the demise of a partnership are often used in a loose and confusing manner. For instance, the terms *dissolution, liquidation and winding up,* and *termination* are often used interchangeably. They should not be, for they have very different legal meanings and consequences. The *dissolution* of a partnership results from a change in the partnership relationship caused by any partner ceasing to be associated in the carrying on of the business of the partnership. Uniform Partnership Act, Section 29. The legal event of a dissolution is followed by a process known as *winding up* or *liquidation and winding up.* This involves the liquidation of firm assets, the paying off of claims, and the distribution of remaining assets among the partners. When this process is completed, then the partnership is *terminated.* It is also possible that there may be a provision for the continuance of the business upon the dissolution of the partnership, in which event there may not be a process of liquidation and winding up at all. Rather, the partners who are to continue the business arrange to pay, according to the agreement, a sum for the former partner's interest in the business. The purpose of this chapter is to examine these various activities.

39:1. Causes and Grounds for Dissolution. The dissolution of the partnership may be caused by some act of the partners, by operation of law, or by judicial decree. Among the acts of the parties that will result in dissolution are dissolution by mutual consent or a dissolution by a partner of a partnership at will without consent of the other partners. Dissolution by operation of law may result from any of the following: the expiration of the contractual term of the partnership, death, bankruptcy, impossibility of performance, the withdrawal of a partner, or the admission of a new partner. A court may decree the dissolution of the partnership for a number of reasons, among which are the misconduct of a partner, the incapacity of a partner, or dissension among the partners.

(1) Dissolution by Act of the Partners. As with any other sort of contract even if the partnership is to continue for a specific period of time, the parties may consent to an earlier termination. This consent may be manifested either expressly or by words and acts from which such a consent may be inferred. On the other hand, if the partnership is to continue for a fixed period and one of the partners wishes to terminate the relationship prior to the expiration of the fixed period, he has the *power* to do so. Because partners are fiduciaries, the law does not require the maintenance of the relationship against their will. If the withdrawal of a partner violates the partnership agreement, the partner may be liable to his co-partners for breach of the partnership contract if, under the terms of the contract, he or she does not have the *right* to dissolve the partnership even though he or she has the power to do so. Thus, suppose that Perry Partner and Connie Copartner agree to form a partnership for two years for the purpose of carrying on a house-painting business. At the end of the year both parties agree to cease operating the business. There would be no liability to either party because the partnership was dissolved by mutual consent. On the other hand, suppose that Connie unilaterally decides to withdraw from the business in violation of the agreement. There is nothing Perry can do to prevent Connie's withdrawal. However, Perry would have an action against Connie for any damages suffered as a result of the breach of the partnership agreement.

Sometimes the partnership agreement provides for the dissolution of the firm after the passage of a fixed period of time or when a specific purpose or objective is obtained. If this is the case, the partnership is automatically dissolved upon the lapse of the fixed period or upon the accomplishment of the specific purpose. Obviously such a provision is appropriate for certain types of partnerships and generally not for a firm such as a retail partnership.

Frequently when a partnership is established, there is either no formal agreement or the terms of the agreement do not provide for the continuance of the partnership for any fixed period. Such a partnership is a *partnership at will*. Under this circumstance a partner may dissolve the partnership at any time without incurring any liability to the other partners. The right of a partner to dissolve a partnership does not require the consent of any of the co-partners, only a notice to them of the withdrawing partner's intent to dissolve the partnership. He or she need not have any particular reason for doing so. Under certain circumstances the election of a partner to withdraw from the partnership would have disastrous consequences for the business. It is wise, therefore, to have a partnership agreement that provides for this contingency.

Sometimes a partnership agreement may provide for the expulsion of a partner from the partnership under certain circumstances. If those circumstances occur, the expulsion of a partner in accordance with the agreement dissolves the partnership. Of course, no liability is incurred by the expelling partners as long as they adhere to the terms of the partnership agreement.

HUNTER v. STRAUBE
543 P.2d 278 (Or. 1975)

McALLISTER, J.

This suit was filed by the plaintiffs, Dr. Arthur F. Hunter and Dr. O. D. Haugen, to dissolve a three-man medical partnership in which the defendant, Dr. Kurt R. Straube, was the third member. The three doctors were radiologists practicing in Portland under the firm name of Lloyd Center X-Ray. The partnership was created by a partnership agreement dated July 26, 1969 and a written addendum dated November 24, 1971 which added the plaintiff Haugen to the partnership.

On September 11, 1974 the plaintiffs filed this suit in Multomah County to dissolve the partnership and prayed for the appointment of a receiver and the winding up of the partnership. The defendant counterclaimed, alleging that he was entitled to continue the partnership business, to recover damages from plaintiffs for the breach of the partnership agreement, and to settle with the plaintiffs as withdrawing partners as provided by the partnership agreement.

The trial court found that by the filing of this suit the plaintiffs "did not cause by express will a dissolution of their partnership with defendant." The court further found that since "the partnership continues as an entity," the court had no jurisdiction to wind up the affairs of the partnership. The court also dismissed the counterclaims of defendant because "no dissolution had occurred."

Plaintiffs contend that they expressed their will to dissolve the partnership by the filing of this suit. . . . We agree with this contention. However, we disagree with the contention of the plaintiffs that by the filing of this suit they are entitled to a dissolution in accordance with the Uniform Partnership Law. The plaintiffs ignore the provision of the partnership agreement that limits the dissolution to the withdrawing partners and expressly provides that "the retirement of any Partner shall not dissolve the partnership as to the other Partners."

The power to dissolve a partnership is governed by ORS 68,530 and provides for dissolution both without violation of the partnership agreement and in contravention of the partnership agreement. In either case, it is clear that if the partnership agreement provides for the distribution of the partnership property the rights of the partners are governed by the partnership agreement rather than by the Uniform Partnership Law.

In the case at bar the plaintiffs had the power to dissolve the partnership by electing to withdraw as partners, a choice which they made by filing this suit. Plaintiffs, however, did not have the right to dissolve the partnership without complying with the terms of the partnership agreement. As was succinctly stated in *Straus* v. *Straus,* 254 Minn.234, 94 N.W.2d 679, 686 (1959):

> A distinction must be recognized between the power to dissolve a partnership and the right to dissolve a partnership. Any partner may have the power to dissolve a partnership at any time . . . and this is true even though such dissolution is in contravention of the partnership agreement. . . . If a partner exercises his power to dissolve a partnership, but does not have the right to do so, he must suffer the penalties. . . ."

The pertinent provisions of this partnership agreement have been set out earlier verbatim in the opinion. In brief, the agreement required the plaintiffs to give defendant, the remaining partner, six months' notice of their desire to withdraw and provided that the "retirement shall take effect six months from the date of delivery, of such notice to the other Partners." The agreement also provided a specific plan of distribution to a withdrawing partner and contained a restriction on competition. It should be noted that Oregon law does uphold non-competition clauses if freely entered into by the partners and reasonable in effect. Such clauses are typical of professional partnerships.

The plaintiffs attempted by filing this suit to divest the defendant from his right to continue the partnership business, which he clearly had the right to do. They attempted to avoid the requirement of giving six months' notice. They also attempted to render nugatory the provisions of the partnership agreement governing the rights to the partnership property upon the withdrawal of a partner. We think the plaintiffs cannot, by merely calling their withdrawal a dissolution, escape from the liabilities which they assumed when they executed the partnership agreement. The plaintiffs have not cited a single authority in support of their contention.

We hold that the filing of this suit by the plaintiffs was an election by each of them to withdraw from the partnership in contravention of the partnership agreement. Under those circumstances the withdrawal entitles the defendant to continue the partnership in accordance with the terms of the partnership agreement. The defendant also has the right to any damages he may have suffered on account of the plaintiffs' breach of the provision for six months' notice of withdrawal.

(Reversed and remanded.)

QUESTIONS FOR DISCUSSION

1. *Why do you suppose Hunter and Haugen wanted a dissolution and winding up under the Uniform Partnership Act rather than under the partnership agreement?*
2. *What act of the partners resulted in the dissolution of the partnership?*
3. *What distinction did the court make between the power to dissolve a partnership and the right to dissolve it?*

(2) Dissolution by Operation of Law. There are a number of events that result in the dissolution of a partnership regardless of whether the partnership agreement provides for the termination of the firm after a fixed period. One of these events is the death of a partner. The Uniform Partnership Act specifically provides that the death of a partner causes a dissolution of the partnership. Section 31. This fact should be faced squarely by the partners at the time the partnership is established. If there is a desire on the part of the partners that the firm should continue after the death of one of them, it is necessary that they provide for that event in the agreement. The estate of a deceased partner can force dissolution and winding up of the firm in the absence of such an agreement, which, of course, could cause a serious interruption of the business. In order to preclude this, the partnership agreement may provide for the continuance of the business by the surviving partners, with compensation for the deceased partner's interest in the business to be paid by them to the deceased partner's estate. Such a provision may be funded by life insurance.

A partnership is dissolved by operation of law if one of the partners is adjudicated a bankrupt. It should be noted that the insolvency of the firm or of a partner does not of itself operate to dissolve the firm. Rather, there must be a formal adjudication of bankruptcy by the court. When this occurs, the firm is dissolved and a liquidation and winding up occurs. If there are assets, the bankruptcy court will appoint a trustee to preside over the liquidation of the assets of the bankrupt and the distribution of the remaining assets of the bankrupt to his or her creditors and other claimants.

If it becomes illegal to continue in the partnership business or if the association of the partners becomes illegal, then the partnership is dissolved. For instance, if the business requires a license to operate and that license is withdrawn, the firm would be dissolved. Dissolution would also occur in a firm of attorneys if one of the partners licensed or privileged to practice law was disbarred.

There are several other events that result in the dissolution of the partnership as a matter of law. Technically the existing partnership is dissolved upon the withdrawal of a partner or the admission of a new partner. It is wise for the partnership agreement to provide for such contingencies. However, only a complete withdrawal from the business can effect dissolution. The Uniform Partnership Act, Section 27, provides that the sale or assignment by a partner of his interest in the business is evidence only of an intent to withdraw and does not automatically result in the dissolution of the firm. A partnership may also be dissolved as a matter of law if events make it impossible to continue the business or if the entire firm property is disposed of, such as by the sale of the firm assets to a corporation.

GIRARD BANK v. HALEY
332 A.2d 443 (Pa. 1975)

Anna Reid formed a partnership with Haley and two other men. They entered into a written partnership agreement for the purpose of leasing for profit certain real property. The agreement provided that if a partner died during the period of the partnership, the surviving partners would have the right to purchase the interest of the deceased partner.

Subsequently Mrs. Reid notified the other partners that she was dissolving the partnership and requested that the partnership assets be liquidated as soon as possible. When the other partners failed to agree to a plan of liquidation, Reid filed suit praying for a winding up the partnership and a liquidation of the assets. After the suit was filed, Mrs. Reid died and her executors were substituted in her place.

The chancellor found that the partnership was not dissolved by Mrs. Reid's letter but rather by her death. He therefore ruled that the surviving partners were entitled to purchase her interest under the partnership agreement. The executors contended that the partnership was dissolved by Mrs. Reid's letter and that therefore the surviving partners had no right to purchase under the partnership agreement but rather were governed by the provisions of the Uniform Partnership Act.

POMEROY, J.

None of the parties disputes the chancellor's conclusion that the partnership has been dissolved; the dispute, as indicated at the outset, is when that event occurred. Dissolution of a partnership is statutorily defined as "the change in the relation of the partners caused by any partner ceasing to be associated in the carrying on, as distinguished from the winding up, of the business." Uniform Partnership Act § 29 ("the Act"). There is no doubt that dissolution of a partnership will be caused by the death of any partner, § 31(4) of the Act, . . . and if Mrs. Reid's death was the cause of the dissolution here involved, the chancellor was quite correct in looking to the provisions of paragraphs 11 and 12 of the agreement . . . as defining the rights and obligations of the surviving partners on the one hand and the estate of the deceased partner on the other. If, however, dissolution occurred during the lifetime of Mrs. Reid, those portions of the agreement, which are concerned solely with the effect of the death of a partner, are not germane. The agreement being otherwise silent as to winding up and liquidation, the provisions of the Act will control.

The chancellor was impressed with the fact that the decedent "was a strong willed person" who dominated the partnership enterprise, that the defendant partners had each contributed many thousands of hours of hard work and planning to the "joint venture," and that neither Mrs. Reid (who testified at the first hearing, but who then, according to the adjudication, "appeared confused and feeble") nor her personal representatives had offered "evidence to justify a termination." In supposing that justification was necessary the learned court below fell into error. Dissolution of a partnership is caused, under § 31 of the Act . . . "by the express will of any partner." The expression of that will need not be supported by any justification. If no "definite term or particular undertaking (is) specified in the partnership agreement," such an at-will dissolution does not violate the agreement between the partners; indeed, an expression of a will to dissolve is effective as a dissolution even if in contravention of the agreement. We have recognized the generality of a dissolution at will. If the dissolution results in breach of contract, the aggrieved partners may recover damages for the breach and, if they meet certain conditions, may continue the firm business for the duration of the agreed term or until the particular undertaking is completed.

There is no doubt in our minds that Mrs. Reid's letter effectively dissolved the partnership between her and her three partners. It was definite and unequivocal: "I am terminating the partnership which the four of us entered into on the 28th day of September, 1958." The effective termination date is therefore February 10, 1971, and Mrs. Reid's subsequent death after this litigation was in progress is an irrelevant factor in determining the rights of the parties.

The remaining question is whether or not the unilateral dissolution made by Mrs. Reid violated the partnership agreement. The agreement contains no provision fixing a definite term, and the sole "undertaking" to which it refers is that of maintaining and leasing real property. We thus conclude, on the record before us, that the dissolution of the partnership was not in contravention of the agreement.

In light of our conclusion that an inter-vivos dissolution took place, the provisions of the Act rather than the post-mortem provisions of the agreement, will govern the winding-up of the partnership affairs and the distribution of its assets.

(Reversed and remanded.)

QUESTIONS FOR DISCUSSION

1. *Why was it important to determine whether the date of dissolution occurred when Mrs. Reid wrote the letter or when she died?*
2. *What difference do you think the date of dissolution made to the surviving partners, as a practical matter?*

(3) **Dissolution by Court Decree.** There are a number of circumstances under which a court may decree a dissolution of the partnership. Among these are an adjudication of insanity of a partner, incapacity of a partner, misconduct of a partner, or the willful or persistent commission of a breach of the partnership agreement or if the business can be carried on only at a loss. Uniform Partnership Act, Section 32. In decreeing a dissolution of a partnership, the courts apply general equity principles in determining the rights of the parties and will not dissolve the partnership in all cases.

The insanity of a partner will generally warrant the court to dissolve the partnership. It is not necessary that there be an adjudication of insanity, merely establishment of the fact that the partner lacks the mental capacity to carry on as a partner. Furthermore an adjudication of insanity does not automatically result in dissolution but is usually a compelling reason for the court to decree dissolution. Other forms of incapacity of a partner may form the basis for dissolution if that incapacity results in the inability of a partner to carry on his or her duties as a partner. On the other hand, a mere temporary incapacity or illness will not result in a decree of dissolution.

Under certain circumstances the misconduct of a partner may form the basis of a court decree dissolving the partnership. Such a decree will usually result if the misconduct renders it impractical to carry on the firm's business or if the misconduct is prejudicial to the carrying on of the business. An example of such misconduct occurs when one partner wrongfully excludes another from the business, such as by locking him or her out of the firm's premises. Other examples of such misconduct might be the waste and willful mismanagement of the firm assets by a partner or consistent violations of the partnership agreement. The innocent partner's remedy for such behavior is an action for accounting and dissolution. In some cases it may be appropriate to couple that action with a petition for an injunction requiring the offending partner to cease his or her offensive behavior. The court may also allow damages to the innocent partner. The misbehavior and misconduct of a partner must be serious, however, before the court will decree dissolution.

If it can be shown that the business of the partnership cannot be carried on at a profit and is doomed to failure, the court may decree dissolution of the firm. However, merely showing that the partnership has sustained a temporary loss will not cause a court to dissolve the firm.

CREEL v. CREEL
73 F.2d 107 (D.C. 1934)

Robert T. Creel and Edwin J. Creel, brothers, were equal partners in an automotive jobbing and service business. Edwin J. Creel was also an inventor and over the course of the years spent quite a bit of time away from the business because of this interest and threatened litigation. Eventually these pressures caused problems between the partners. Robert charged that Edwin became nervous and arbitrary in his conduct relating to the partnership business. Arguments frequently developed between the partners, and, among other things, Edwin withdrew money from the business in excess of Robert, told the bookkeeper not to pay bills, and refused to join in the execution of a promissory note in order for the partnership to borrow funds from the bank.

Robert Creel filed a bill seeking the appointment of a receiver and the dissolution, winding up, and liquidation of the firm. Edwin answered the bill and suggested as an alternative to the appointment of a receiver, the appointment of one E. Quincy Smith as an arbitrator with absolute power of dictation over the actions of both brothers. From an order appointing Smith as receiver to operate the business for thirty days and at that time to report back to the court, Edwin appealed.

MARTIN, C.J.

We think that the action of the trial justice in appointing a receiver for the partnership was not erroneous. The record, fairly considered, discloses that there are such irreconcilable disagreements and dissensions between the partners in regard to the conduct of their affairs as to endanger the partnership good will and property. In such case the court is justified in appointing a receiver for the preservation of the assets and the liquidation of the partnership's affairs.

In *Jones* v. *Jones,* 229 Ky. 71, 16 S.W. (2d) 503,504, a receiver was appointed for a farm partnership conducted by two brothers. In upholding the validity of an order appointing the receiver prior to the hearing on dissolution, the court said:

> The relationship existing between the partners is so strained and bitter that it is impossible to even hope for the slightest cooperation between them. There is absolutely nothing to do but to appoint a receiver to take charge of the partnership property pending a final settlement of the partnership affairs.

In *Reed* v. *Beals,* 77 Fla. 801, 82 S. 234, the court said:

> From the examination which we have made of the authorities on this subject, we think the law may be considered as settled, that whenever the intervention of a court of equity becomes necessary, in consequence of dissensions or disagreements between the partners, to effect a settlement and closing of the partnership concerns, upon bill filed by any of the partners, showing either a breach of duty on the part of the other partners, or a violation of the agreement of partnership, a receiver will be appointed as a matter of course.

It may be observed that the receiver appointed by the court in this case is the person whom the appellant recommended for appointment as a dictator of the affairs of the firm. Nevertheless the present appeal was taken immediately upon his appointment, and before any action was taken by him in the conduct of his duties under his appointment.

(Affirmed.)

QUESTION FOR DISCUSSION

1. *Will the court appoint a receiver for the partnership when there is any disagreement between the parties? What do you think must be shown to the court before a receiver will be appointed?*

39:2. RESULTS AND EFFECT OF DISSOLUTION. The dissolution of the partnership changes the relationship of the partners to each other. They are no longer principals and agents, with the result that a partner no longer has the actual authority to bind his co-partner on new obligations. The partners retain the power to act for each other only in matters relating to the winding up and liquidation of the business. The dissolution of the firm, therefore, does not deprive a partner of the authority necessary to complete existing partnership contracts. In fact, the partnership contracts must be performed because they are not discharged by dissolution of the partnership.

Upon dissolution of the partnership, the partnership agreement may provide for the continuation of the business by some of the former partners. If this is the case the problem is somewhat different from the situation in which the winding up is to lead to a termination of the business. The problem is primarily one of valuation of the business and distribution to the withdrawing partner according to the terms of the partnership agreement.

If the partnership is dissolved, the termination of the agency relationship does not eliminate all the fiduciary obligations of the partners to each other. The partners are generally held to the standards of a fiduciary in their dealings with each other in matters involving the dissolution and winding up process. They must deal fairly with one another in this regard. For example, if a partner learns of a business opportunity during the course of the partnership and delays entering into it until the partnership is dissolved, the other partners have a cause of action against the offending partner for violation of his duty as a partner. On the other hand, there is no duty between the former partners concerning entirely new matters entered into subsequent to the dissolution of the partnership.

The powers of the partners to participate in the process of winding up after dissolution are shared equally, if there is no agreement to the contrary. Among these powers are the collection or payment of firm debts, the compromise of certain obligations, and other matters connected with the winding-up process. If the dissolution is caused by the death of a partner, these powers are shared by the survivors. If the dissolution results from the insolvency or bankruptcy of a partner, the solvent partners have the powers, subject to the control of the court. If the partnership itself is insolvent, the power to wind up the business may be taken away from the partners and given to a receiver appointed by the court.

A final word should be said about the liability of partners to each other for obligations entered into by a partner in ignorance of the dissolution of the partnership. Section 34 of the Uniform Partnership Act provides that if a partner enters into an obligation after dissolution, he may look to his partners for contribution to the payment of this obligation unless they can show (1) that the partner acting for the partnership had knowledge of the dissolution by act of another partner or (2) that the partner acting for the partnership had notice or knowledge of the death or bankruptcy of a partner that caused the dissolution. The obligation of the estate of a deceased partner to notify the other partners of his death should be obvious.

LAVIN v. EHRLICH
363 N.Y.S.2d 50 (1974)

BERTRAM, J.

A neighborhood tax preparing business depends on many of the same clients returning each year. To this end, the location of the business is important and forms part of its goodwill. Were a partner in the business to buy in his own name the building housing it and then eject the partnership at the end of the lease, that would be an intolerable breach of fiduciary relationship. But, what if that partner first serves notice to dissolve the partnership before buying the building; can he still be held to the same duties to his partners?

Plaintiffs Lavin and Dillworth and defendant Ehrlich were partners in a storefront tax preparing business. Ehrlich managed the business, Lavin and Dillworth were essentially investors. By letter dated October 9, 1973, Ehrlich announced his immediate withdrawal from and dissolution of the partnership. Later that month, he contracted to buy the storefront property from the landlord, and, in January 1974, he took title. The lease on the storefront ran until April 30, 1974 and Ehrlich would not negotiate a new lease with the partnership which he considered dissolved.

His partners claim that Ehrlich has breached his fiduciary duties in buying this property and they ask the Court to rule that he holds it in constructive trust for the partnership.

Ehrlich argues that the agreement provides no specific term of life for the partnership, and so it is a partnership at will, to be dissolved at the insistence of any partner. This is true, and Ehrlich's letter of October 9, 1973 effectively triggered the dissolution of the partnership, and should have set the winding up in motion towards final termination.

However, this dissolution did not free Ehrlich of all fiduciary relationship to the partnership. "On dissolution the partnership is not terminated, but continues until the winding up of the partnership affairs is completed." Partnership Laws 61. There may be a relaxation of a partner's duties to his co-partners in relationships that look to the future of the newly dissolved partnership. But, in dealings effecting the winding up of the partnership and the proper preservation of partnership assets during that time, "the good faith and full disclosure exacted of partners continues."

At the bottom, Ehrlich is doing something hopefully achievable from a technical standpoint, but sadly lacking in equity. He was the actual conductor of the business. He went out and grabbed the building, and now hopes, with this maneuver and his possession of the list of names and past tax return files (which he has undisputedly) to capitalize on the partnership location and goodwill to the exclusion of his partners. We cannot accept his strategy.

Fairness to all now requires the dissolution of the partnership, with the sale of its assets. This includes not only the building, but also the list of customer names and their permissible records. After the conference with the Court, the parties agreed on the division of cash on hand and payment of bills. The plaintiffs, accordingly, withdrew as resolved that portion of the complaint requiring an accounting by Ehrlich.

Accordingly, the Court finds Ehrlich holds the property in constructive trust for the partnership and must surrender his right, title, and interest to the partnership if offered two thirds of the purchase price he paid by the remaining partners Lavin and Dillworth on or before January 31, 1975.

On motion of either party, the Court shall appoint a receiver for the purpose of winding up the partnership, selling its assets, and disbursing the net proceeds to the partners.

Questions for Discussion

1. *Under the court's ruling would it have been feasible for any of the partners to carry on the business at the location in question against the wishes of the other partners?*
2. *Did it really make any difference whether Ehrlich contacted the landlord in order to purchase the building before or after dissolution of the partnership? At what point did Ehrlich become free to purchase the building for his own purposes?*

39:3. Liability of Partners and Effect on Third Parties. The dissolution of the partnership does not discharge any of the partners from existing firm obligations, and all the partners remain personally liable for their payment. In some circumstances some of the partners may agree to assume the obligation to pay the debts of the firm. Such an agreement is not binding on the creditors for the very simple reason that the partners cannot by agreement among themselves limit the rights of a creditor who is not a party to such an agreement. The same rule applies if one partner withdraws from the partnership and the remaining partners agree to assume the obligations to the firm creditors. Such an agreement would not bind the creditors, and the withdrawing partner would have a right of action against his former partners should he have to pay a creditor because of the failure of the former partners to honor the contract with the withdrawing partner. On the other hand, should the creditors participate in the agreement and release the withdrawing partner from his prior obligations, then they would be bound by the release and would have no further claim against the former partner.

Upon the dissolution of the partnership, the implied authority of partners to enter into new obligations on behalf of the firm is terminated. However, it is possible that a third person may bind the partners to new obligations after dissolution under the doctrine of apparent authority. It is important, therefore, that notification of the dissolution of the partnership be given. As a general rule, *actual notice* must be given to those persons who had dealings with the firm prior to the dissolution. If the notice is not given, such a person is entitled to the presumption that the partnership is still in business. The danger, of course, is that if actual notice is not given to a creditor, for instance, who dealt with the firm prior to its dissolution, then one of the partners could by his act bind his co-partners to an obligation that he incurs after the time of dissolution. Uniform Partnership Act, Section 35(b).

Notice of the dissolution of a partnership should also be given to the public. If there have been no previous dealings, the notice is sufficient if it is *constructive.* Constructive notice may be accomplished by the publication of the notice of dissolution in a newspaper with a circulation in the area. Some statutes require the filing of the notice of dissolution in a public office. On the other hand, if the dissolution of the firm occurs because of operation of law, for instance, as the result of bankruptcy or death, no notice is required. The reason for this exception is that these events are a matter of public record of which everyone is bound to take notice. Additionally, if one is a dormant partner—that is, his membership in the firm has been unknown—there is no obligation on his part to give notice of dissolution because third persons have not relied on the dormant partner's presence with the firm in extending credit to it. However, if the identity of a dormant partner becomes identified with the firm, then notice must be given.

It should be obvious by now that it is important for a withdrawing partner to give the actual and constructive notice referred to, particularly when the remaining partners continue the business. If the withdrawing partner fails to give the proper notice, he runs the risk of being liable for obligations entered into by the firm after he has withdrawn.

WOLFE v. EAST TEXAS SEED COMPANY

583 S.W.2d 481 (Tex.App. 1979)

Charles R. Wolfe and Nick Wolfe were brothers who formed a partnership and did business as Wolfe Construction Company. East Texas Seed Company had done business with Wolfe Construction Company on a number of occasions. During the period from February 9, 1974, through September 30, 1974, East Texas Seed Company sold certain goods and merchandise to Wolfe Construction on open account for the total amount of $6,323.05.

The construction company did not pay the account and East Texas filed suit against both brothers as partners of the construction company. Unknown to East Texas the partnership had been dissolved in 1973. Nick Wolfe ordered the materials from East Texas Seed Company.

The trial court entered a judgment against both Wolfe brothers. Charles R. Wolfe appealed the trial court's decision contending that he was under no duty to notify creditors of the dissolution of the partnership.

DOYLE, J.

A fair summary of the testimony concerning the partnership and its dissolution is as follows: Appellee's manager, Mr. Kirby, admits he did not know whether appellant's company was a partnership when it first began doing business with them. He stated that he first became aware of the partnership in 1974 after it had been dissolved. He further explained that if he had known of the dissolution he would have investigated the matter and that it would have made a "whole lot" of difference to his company.

Appellant strongly contends that since the partnership had been terminated before the inception of the debt, he should not be personally liable. As a general rule this would be true where notice of the dissolution has been given to third parties who had dealings with the partnership. On the other hand, absent any notice of dissolution, appellant can not escape personal liability for the subject debt unless he can show that prior to dissolution, he had been "So far unknown and inactive in partnership affairs that the business reputation of the partnership could not be said to have been in any degree due to his connection with it." Article 6132b, Sec. 35(2)(b). The facts before us show that appellee did business with the company prior to the partnership dissolution; that appellant was known as Wolfe Construction Company; that the company was composed of two brothers, Nick and Charles R. Wolfe; that either man had authority to purchase seed; and that both men bought and paid for the seed. Appellant denies receiving any benefit from these transactions personally, but admits the partnership could have purchased the seed and benefited from it. Business was continued by appellant under the same name of Wolfe Construction Company, Nick Wolfe still worked for the company and no notice of partnership dissolution was sent to appellee. Appellant's reputation was relied on in transactions with appellee, and even in the absence of any partnership agreement parties may be bound as partners as to third parties.

(The judgment of the trial court is affirmed.)

QUESTION FOR DISCUSSION

1. *What should Charles Wolfe have done in order to have avoided liability to East Texas Seed Company?*

39:4. WINDING UP THE BUSINESS. After the partnership dissolution occurs, the process of winding up leads to termination of the partnership. In this section we shall examine the partner's right to wind up and the manner of the disposition of the firm assets.

(1) Right to Wind Up. As previously mentioned, if the partners want to provide for the continuation of the business after a dissolution, they should provide for it in the partnership agreement. Without such an agreement, upon dissolution the partners who have not wrongfully dissolved the partnership or the legal representative of the last surviving partner who is not bankrupt has the right to wind up the affairs of the partnership. Uniform Partnership Act, Section 37. If a partner demonstrates to the court that there is a good reason, the court may appoint a receiver to wind up the partnership. Note that a partner who wrongfully causes a dissolution of the partnership or goes bankrupt does not have the right to wind up the partnership.

Once the partnership is dissolved, the partners may not continue to operate the business unless there is a contrary agreement. They must proceed with the winding-up process. If they fail to do so, a court of equity may be petitioned to force them to wind up. There are other consequences of failing to wind up the partnership. Section 42 of the Uniform Partnership Act provides that if the business is continued when it ought to be wound up, the retiring partner or the estate of a deceased partner has an option of one of two alternative remedies. One alternative is that the retired partner or the estate may have the value of his or her partnership interest, plus interest, at the date of dissolution. In this case, the retiring partner or the estate is an ordinary creditor against the partnership or persons who wrongfully continue to operate the business. The other alternative is to receive, in lieu of interest, the profits attributable to the use of the deceased's or the retired partner's right in the property of the dissolved partnership. Whether the retired partner or the representative of the deceased partner claims the interest on his share or the profit attributable to it obviously depends upon the success of the operation after dissolution. He has the best of two worlds. If the partnership is run at a loss, he escapes such a loss by obtaining the value of his share plus interest calculated from the time of dissolution. If the operation is successful, he may get the value of his share plus the profits attributable to it. This is a just remedy because there is no right, without an agreement, to continue the business after dissolution.

Some state statutes provide that the business may be continued until it can be sold as a going concern on a reasonable basis. In other states it may be possible to get court permission to do this. Otherwise the business may have to be liquidated at a price less than its normal market value. These matters should be given thorough consideration when one is entering into the partnership agreement.

KING v. STODDARD

29 Cal.App.3d 708, 104 Cal.Rptr. 903 (1972)

Lyman E. Stoddard, Sr., and Alda S. Stoddard owned 51 percent of the Walnut Kernel, a newspaper. Lyman E. Stoddard, Jr., a son, owned 49 percent. Within about a year of each other Lyman E. Stoddard, Sr., and Alda S. Stoddard died. John Stoddard, another son of Lyman, Sr., was appointed executor of both estates after February 1964.

The operation of the partnership continued for a number of years after the death of the two partners. The executor, John Stoddard, wanted to wind up the partnership as soon as possible, but Lyman Stoddard, Jr., continued to operate the business. In 1966 Lyman agreed to be responsible for all debts arising out of the business after February 1964. The business was unsuccessful and was eventually discontinued.

Prior to the death of Alda and Lyman, Sr., Harley King and Stanford White, accountants, had rendered accounting services to the partnership. They continued to render such services after Lyman and Alda's deaths and filed suit against their estates to recover fees for those services rendered after the deaths of Lyman, Sr., and Alda. The lower court found the partnership liable for the services in question and entered judgment against the estates in the amount of $12,370. The executors appealed. (The Corporations Code provisions referred to are identical to those of the Uniform Partnership Act.)

BROWN, A.J.

The question presented on this appeal is whether the continuation of the newspaper after the death of the partners was an act of winding up the partnership under Corporations Code section 15035 so as to render the estate of the partners liable for an accountant's bill incurred subsequent to the death of the partners.

The partnership was dissolved by operation of law upon the deaths of Alda and Lyman F. Stoddard, Sr. Corporations Code section 15029 provides that dissolution of a partnership is ". . . caused by any partner ceasing to be associated in the carrying on as distinguished from the winding up of the business." Death is one of the causes of dissolution. Dissolution, however, does not terminate the partnership which ". . . continues until the winding up of partnership affairs is completed. . . . In general a dissolution operates only with respect to future transactions; as to everything past the partnership continues until all preexisting matters are terminated." Although the general rule is that a partner has no authority to bind his copartners to new obligations after dissolution, section 15035 of the Corporations Code provides that "(a)fter dissolution a partner can bind the partnership . . . (a) By any act appropriate for winding up partnership affairs. . . ."

It is this latter provision upon which the court based its decision that the estates of the deceased partners were liable for the accounting services performed after dissolution. The court found that "LYMAN STODDARD, JR.'S continuation of the WALNUT KERNEL business was an appropriate act for winding up the partnership, since the assets of the business would have substantial value only if it was a going business. It was to the advantage of the partnership that the business be maintained as a going business."

We disagree with this finding. It is probably true that there might have been advantages to the partnership to sell the business as a going business, but the indefinite continuation of the partnership business is contrary to the requirement for winding up of the affairs upon dissolution. In *Harvey* v. *Harvey*, . . . 203 P.2d 112, 115–116, the court disapproved a finding that the business and assets of a partnership were of such character as to render its liquidation impracticable and inadvisable until a purchaser could be found. The court stated: "In effect it (the finding) authorizes the indefinite continuation of the partnership after the death of a partner, a procedure not in accordance with section 571 of the Probate Code. Respondents counter with the argument that the business is such that it cannot be wound up profitably, and the estate given its share. But this argument overlooks the distinction between winding up a business and winding up the partnership interest in that business."

Even if we assume that a situation might exist where continuation of the business for a period would be appropriate to winding up the partnership interest, such a situation did not exist here. The record reflects the fact that the surviving partner was not taking action to wind up the partnership as was his duty under Probate Code section 571, nor did the estates consent in any way to a delay. Rather, their insistence on winding up took the form of an effort to sell the business and a suit to require an accounting. There is nothing in the record upon which to base the argument made by respondents that appellants consented to their continued employment. The fact that they did not object is of no relevance. They had no right to direct and did not participate in the operation of the business. Therefore, the determination that the acts of the accountants were rendered during a winding up process is not based upon substantial evidence.

We conclude that the services of respondents were rendered after the dissolution resulting from the deaths of the partners, Lyman

Sr. and Alda Stoddard, and do not constitute services during the "winding up" processes of the partnership within the meaning of section 15031 of the Corporation Code. The claim for those services, therefore, are not chargeable to the partnership.

(The judgment is reversed.)

QUESTIONS FOR DISCUSSION

1. *What is the distinction between "winding up a business" and "winding up the partnership business" referred to by the court?*
2. *Does this decision mean that the accountants could not collect their fee from anyone?*

39:5. DISTRIBUTION OF FIRM ASSETS. The Uniform Partnership Act provides rules for the distribution of the firm assets. There are three separate situations that must be considered: (1) where the partnership and all partners are solvent; (2) where the partnership is insolvent and all partners are solvent; and (3) where the partnership and at least one partner are insolvent.

(1) Where Partnership and All Partners Are Solvent. If all parties are solvent, the distribution of firm assets is relatively uncomplicated. The Uniform Partnership Act, Section 40(c) provides that the assets shall be applied first to the payment of creditors other than partners. Next in line for payment are loans made by partners to the partnership, hence the obvious need to have clearly established whether a partner's payment is a loss or a contribution to partnership capital. Third are amounts owing to partners in respect of capital. In other words, the partner's capital contribution is returned. Fourth, anything left over is returned to the partners as profits equally or according to the partnership agreement if it provides for a different arrangement.

(2) Where the Partnership Is Insolvent and All Partners Solvent. On occasion the assets of the firm will be insufficient to pay off all the creditors. If the partners are solvent, they are liable to pay off the creditors out of their personal assets. Unless the partnership agreement provides otherwise, these losses are shared equally by the partners. Obviously, if the partnership is insolvent, there would be no distribution of firm assets beyond the firm creditors.

This arrangement does not mean that the partners need to come up with equal amounts of cash, because their capital contributions to the firm may have been different. For example, suppose Able, Baker, and Cable contributed \$3,000, \$2,000 and \$0, respectively, to the capital of the partnership. At the time of dissolution, partnership liabilities exceed assets by \$4,000, resulting in a total loss of \$9,000. Able would have to pay nothing additional because his previous \$3,000 contribution would be applied to his share of the loss. Baker would have to pay \$1,000, and Cable \$3,000.

(3) Where the Partnership and at Least One Partner Are Insolvent. Frequently, if the partnership is insolvent, at least one of the partners is also insolvent. Other partners may not be personally insolvent. There may be creditors of the partnership as well as separate creditors of the individuals. The question arises as to the rights of these creditors, as against each other; to the partnership assets; and to the individual assets of the partners. The answer is supplied by the *doctrine of marshaling of assets,* which will be applied by an equity court when it has assumed jurisdiction over the case.

The doctrine of marshaling of assets has been adopted by both the Federal

Bankruptcy Code and the Uniform Partnership Act. Basically this doctrine provides that partnership creditors must first look to partnership assets to satisfy their claim, and individual creditors of the partners must first look to the personal assets of the partner to satisfy their claims. If the assets of the partnership are insufficient to satisfy the partnership creditors, those creditors may not collect from the personal assets of a partner until the personal creditors have all been paid out of the partner's personal property. Conversely, if the personal assets of a partner are insufficient to pay the personal creditors of a partner, these creditors may not collect from the partnership assets until the partnership creditors have been paid. Once the partnership creditors have all been paid, any personal creditors who have not been fully paid from the personal assets of an insolvent partner may collect from partnership assets *after* the claims of creditors of the solvent partners have been paid. Finally, personal creditors may not claim against the personal assets of other partners.

There are three exceptions to the doctrine of marshaling of assets. The first involves dormant partners. If an individual holds himself out as operating a sole proprietorship when he has a dormant partner, the doctrine will not apply. The creditors who believe that they are dealing with the sole proprietor also believe that they have a claim not only against his personal assets but also against those of the business. Therefore, because the conduct of the dormant partner has led these creditors to rely on the business assets of the active partner in extending credit to him, the dormant partner is estopped from raising the doctrine of marshaling of assets. The result is that firm assets are shared equally by firm creditors and the personal creditors of the active partner. The dormant partners are still liable to any unpaid firm creditors.

A second exception applies if neither the partnership nor any living partners have assets. In that case, the firm creditors and any individual creditors share equally in the estate of a deceased partner. The third exception occurs when a partner has by his fraud converted the partnership assets to his own use. In this circumstance, the individual creditors and partnership creditors share equally in the individual assets of the partner who committed the fraud.

A final word about tort claims. The doctrine of marshaling of assets does not apply when a claim arises because of the tort of a partner. He is obviously individually liable for the tort because of the agency principles discussed earlier. Because the liability of partners is joint and several, the doctrine of marshaling of assets is inapplicable. Recovery may be had from firm assets or the individual assets of any partner against whom a judgment is obtained.

SUMMARY

A distinction should be made between the dissolution of a partnership, winding up the partnership, and termination of the partnership. Dissolution may be caused by an act of the partners such as mutual consent of the partners. A partner may dissolve a partnership without the consent of the other partners.

A partnership may be dissolved by operation of law. Death of a partner will dissolve the partnership, as will the bankruptcy of a partner. Dissolution by operation of law will also occur if the business of the partnership becomes illegal. Withdrawal of a partner or admission of a new one also technically dissolves a partnership.

The court may dissolve a partnership by decree. This may be caused by adjudication of insanity of a partner, incapacity of a partner, misconduct of a partner, willful or persistent commission of a breach of the partnership agreement, or if the business can only be carried on at a loss.

Upon the dissolution of a partnership, the partners are no longer principals and agents of one another. They may act for each other only in matters relating to the winding up and liquidation of the business. Although the implied authority of partners to enter into new obligations on behalf of the partnership is terminated upon the dissolution of the partnership, a third person may still bind the partnership by apparent authority.

Upon dissolution of the partnership, the partners who have not either wrongfully dissolved the partnership or gone bankrupt (or a deceased partner's legal representative) have the right to wind up the affairs of the partnership. If the partnership is wrongfully continued after dissolution, the returning partner may have the value of his or her partnership interest, plus interest, at the date of dissolution. Alternatively, the returning partner may receive, in lieu of interest, the profits that are attributable to the use of his or her right in the property of the dissolved partnership.

There are rules for the distribution of firm assets where the partnership and all partners are solvent; where the partnership is insolvent and all partners are solvent; and where the partnership and at least one partner are insolvent. Where the partnership and at least one partner are insolvent, the rights of personal creditors and partnership creditors are determined by the doctrine of marshaling of assets.

PROBLEMS

1. Claude Dreifuerst formed a partnership at will with several of his brothers. The brothers served notice of dissolution upon Claude and filed a court action for a dissolution and a winding up of the partnership. Claude contended that the assets of the partnership had to be liquidated and his share paid to him in cash. The brothers contended that the court was empowered to divide the physical assets of the partnership among the partners and that Claude did not have to be paid cash for his share of the partnership. Were the brothers correct? Dreifuerst v. Dreifuerst, 280 N.W.2d 339 (Visc. App. 1979).
2. Phillip J. Froess and Jacob Froess were partners engaged in the sale of pianos and other musical instruments. Phillip died on January 19, 1920, and Jacob continued the business for some time thereafter. After negotiations no satisfactory arrangement could be made with Phillip's widow, who was the administratrix of his estate, for payment of Phillip's interest in

the partnership. No permission to continue the business was granted. Phillip's estate then filed an action for an accounting and requested that the estate be paid the value of Phillip's share at the time of dissolution plus interest. Jacob contended that the Uniform Partnership Act, Section 42, allowed the estate this interest only as an ordinary creditor with a personal judgment against the surviving partner and that therefore the remaining firm assets were released from liability to satisfy all debts in whole or in part. Was he correct? Froess v. Froess, 131 A.276 (Pa. 1925).

3. Mr. Jones and Mrs. Jones were partners in a business. There came a time when Mrs. Jones sued Mr. Jones for divorce based on the grounds of adultery. She also filed a complaint alleging that the action for divorce had been filed based on the husband's adultery, that this caused a dissolution of the partnership, and that the husband, her business partner, refused to make a distribution of the firm assets. She did not allege in the complaint that any notice was given to the husband of an election to terminate the partnership. Was the partnership dissolved? Jones v. Jones, 179 N.Y.S.2d 480 (1958).
4. Under a 1960 agreement Allen, George, Donald, and Carlos Fisher were partners in the operation of an insurance business. The agreement required each partner to contribute to the partnership 15 percent of all fees and commissions he received for acting in any fiduciary capacity. In 1961 the partners discovered that Carlos had been receiving money for which he had not accounted. On June 23, 1961, the partners notified Carlos that he was "permanently suspended" from the partnership for failing to account to the partnership. Carlos then filed suit for an accounting in order to obtain his interest in the partnership assets. The other partners argued that by his actions Carlos withdrew from the partnership and that the provisions of the partnership agreement governing withdrawal were applicable. Was Carlos entitled to his interest upon dissolution? Fisher v. Fisher, 212 N.E.2d 222 (Mass. 1965).
5. Lunn and Kaiser were partners in the operation of a farming-and-livestock business. The partners had a number of arguments, and there were several other incidents, including arguments about walking across the lawn, the amount of cream furnished to Lunn, and the pounding on the house, which was being remodeled, while Kaiser's children were asleep. There were other similar incidents but the business was run successfully. Lunn filed an action to dissolve the partnership. The partnership agreement provided for the termination of the partnership on March 1, 1956. This action was brought more than a year prior to that time. Was Lunn entitled to a dissolution of the partnership? Lunn v. Kaiser, 72 N.W.2d 312 (S.D. 1955).
6. In 1958 the partnership of Timmerman and Company was organized by Ernest and Sylvia Timmerman and their two sons, Stanley and Lynne. The purpose of the partnership was for farming grain peas. Sylvia Timmerman died in 1968, leaving her interest to the remaining partners. In 1969 Lynne gave notice of intent to withdraw from the partnership and actually withdrew in August of 1970. The partnership was losing money and was near

insolvency after Lynne withdrew. Thereafter it began making money again. One and a half years after withdrawing from the partnership, Lynn filed suit for liquidation of the partnership and claimed that he was entitled to an equal share of the profits from the use of his share of the assets since withdrawal. The trial court applied this measure but deducted therefrom an allowance for Stanley and Ernest's labor and management services after the dissolution. Was the trial court correct? Timmerman v. Timmerman, 538 P.2d 1254 (Or. 1975).

7. Johnson and Signor were partners operating under the name of R. Johnson & Company and had purchased coal from Lyon & Burr. Subsequently the partnership was dissolved, and Signor carried on the business alone. Notice of the dissolution was published on three successive weeks in the local newspaper, but no notice was given to Lyon & Burr, and they never saw the advertisement. Signor then purchased additional coal under the name of R. Johnson & Company. This bill was never paid, and Lyon & Burr sued Johnson and Signor. Was Johnson liable? Lyon et al. v. Johnson et al., 28 Conn. 1 (1859).

SECTION VIII

Corporations

CHAPTER 40

Nature and Formation

The corporation is a more complicated form of business organization than the sole proprietorship or the partnership. Although its origins are somewhat uncertain, the concept in England can be traced back to Crown charters granted by the king to ecclesiastical and public bodies such as monasteries and boroughs. The merchant and craft guilds also achieved a degree of corporateness as a result of Crown charters, although the guild members sought these primarily to gain a monopoly rather than to do business in the corporate form. The concept of a corporation has evolved over a long period of time to the type of organization that we have today. Some of the steps in this evolutionary process, such as the joint-stock company, we have examined previously. The primary impetus for the most recent developments has been the need for a form of business organization that has the ability to raise large amounts of capital with characteristics that would entice investors to risk engaging in the business.

By the late nineteenth and early twentieth centuries most of the states in this country had passed statutes that permitted relatively easy formation of corporations. Today a large majority of the states have adopted, in whole or in part, the Model Business Corporations Act. Those that have not adopted the act have statutes substantially similar. Much of our discussion therefore will be based on this statute.

40:1. Nature of the Corporation. The corporation is a creature of the state and comes into being only as the result of an act of the state. It has a separate legal existence from its creators, that is, those who have contributed to its creation and control. Once created, the corporation may have a perpetual or a limited life; it may change its ownership without affecting its existence; it may enter into contracts, sue, be sued, and in general enter into the other types of activity required to do business.

The most well-known and often-quoted definition of a corporation is the one framed by Chief Justice Marshall of the U.S. Supreme Court:

> A corporation is an artificial being, invisible, intangible, and existing only in contemplation of the law. Being the mere creature of law, it possesses only those properties which the charter of its creation confers upon it, either expressly or as incidental to its very existence. These are such as are supposed best calculated to effect the

> object for which it was created. Among the most important are immortality, and, if the expression may be allowed, individuality; properties by which a perpetual succession of many persons are considered as the same, so that they may act as a single individual. A corporation manages its own affairs, and holds property without the hazardous and endless necessity of perpetual conveyances for the purpose of transmitting it from hand to hand. It is chiefly for the purpose of clothing bodies of men, in succession, with these qualities and capacities, that corporations were invented, and are in use. By these means, a perpetual succession of individuals are capable of acting for the promotion of the particular object, like one immortal being.[1]

Perhaps the most important characteristic of the corporation is the fact that it is a legal entity, separate and distinct from the individuals who own or manage it. Because of this characteristic the liability of those individuals who own the corporation is limited to their investment in it. Thus, if one purchases an ownership interest in a corporation for $100, that is the limit of one's financial exposure, under normal circumstances, should the corporation become insolvent. One generally obtains an ownership interest by owning corporate stock. The owners are therefore generally known as *stockholders* or *shareholders.* This concept of limited liability makes it possible for the aggregation of the large sums of capital that many types of businesses require today.

Although the corporation has often been described as a "person," a number of court decisions have held that it is not a "citizen," entitled to the protection of certain of the provisions of the federal Constitution. For example, Article IV, Section 2, provides that the citizens of each state shall be entitled to the privileges and immunities of the other states. Thus, although a citizen can go into any state and do business on the same basis as the state's own citizens, the corporation has no such right.

(1) Comparison with Partnership. When one goes into business, one of the considerations that he or she must deal with is the form of business organization to use; sole proprietorship or partnership or corporation. Because previous chapters have treated the partnership in detail, it might be feasible at this point to compare the partnership and the corporate form of organization in several respects: method of creation, liability of owners, duration, transferability of ownership interest, management and control, and ability to raise capital.

The partnership form of organization is created by the agreement of the partners. One creates a corporation by filing the proper documents with the state in order to obtain a corporate charter. Although it is more difficult to form a corporation than a partnership, in many cases it is still a relatively simple procedure.

Perhaps the most important distinction between a corporation and a partnership is the fact that partners have unlimited liability for the torts, contracts, and debts incurred by themselves and the partnership. On the other hand, the owners of a corporation have limited liability in the sense that their exposure is generally limited to the amount of their investment in the corporation.

[1] *Dartmouth College* v. *Woodward,* 4 Wheat (U.S.) 518, 636; 4 L.Ed. 629 (1819).

For example, once the assets of the corporation are exhausted, a creditor cannot under normal circumstances satisfy a corporate debt from the personal property of the owners of the corporation. This advantage of the corporate form of organization may be somewhat illusory for the small business person, however. It is extremely unlikely that a bank, for instance, will loan any significant amount of money to a new corporation without requiring its owners also to become personally obligated for the amount of the loan.

The existence of a corporation may be perpetual or limited to a specific period of time by its charter. Its life is not affected by the life or death of its owners. A partnership, on the other hand, ceases upon the death or the withdrawal of a partner, and the business may have to be terminated. Furthermore there is no way that one may transfer ownership in a partnership without dissolving the original firm. Ownership of a corporation is easily transferred, however, and has no effect upon the life of the corporation. In addition, the corporation is a separate legal entity, as we have seen. It may sue and be sued, make contracts, and do many other things in its own name. A partnership is not a separate legal entity and is very restricted in what it may do in its own name.

A further distinction between the two forms of organization has to do with the management of the firm. In a partnership, unless agreed otherwise, the partners all have equal voices in the management of the firm. Furthermore the act of a partner within the scope of his or her authority will bind the other partners. In the case of a corporation, the business is managed by a board of directors elected by the owners of the corporation. An individual owner of the corporation has no power to act for the firm or to bind it by reason of his or her ownership of an interest in the corporation. We shall see in Chapter 44 that even an individual member of the board of directors does not have the power to act for the corporation.

A further comparison between the corporation and other forms of business organizations, such as the partnership or the sole proprietorship, is frequently made in the firm's ability to raise capital. The corporation may raise capital by attracting additional investors in the firm who may be willing to buy an ownership interest in a corporation, and for other reasons, such as limited liability, investors may be willing to pay to purchase an interest in a corporation whereas they would not be willing to join a partnership. The partnership and the sole proprietorship are rather limited in their ability to attract capital. On the other hand, from a practical standpoint the small business person who forms a corporation is also frequently quite limited in the amount of capital that can be attracted.

One final note in the comparison of the forms of business organization. There are frequently differences in the way the organizations are treated by the state and federal governments. There are advantages and disadvantages to both the partnership form of organization and the corporation from a tax standpoint. Further discussion in this area is best left to a tax course. However, by way of example, many expenses such as health insurance and life insurance may be deducted as an expense of doing business by the corporation but not by the individual doing business as a partnership. On the other hand, the income of the corporation is taxed and then any dividends paid by the corporation to its owners may also be taxed to them, so that there may be an element of double taxation in some cases. If the corporation is a small business and otherwise qualifies, it may elect to be treated for tax purposes as a partnership or a sole proprietorship. A corporation that elects to do this is known as a *Subchapter S corporation* under the tax laws because of the applicable provisions of the Internal Revenue Code. Among the limitations on electing to be treated in this manner are that there may not be more than twenty-five shareholders of the corporation and no more than 20 percent of the corporate income can be derived from rents, interest, dividends, or royalties.

40:2. The Corporation Chartered by the State. As implied by the definition quoted from the Dartmouth College case, the corporation is a creature of the state and exists solely as a result of the grant of a charter by the state. The authority of the state to grant a corporate charter is provided by statute. Under these statutes the states may issue a *charter,* a *certificate,* or *articles of incorporation.* These terms are used interchangeably because the various states do not use them uniformly. For instance, the Delaware and New York statutes speak of a certificate of incorporation, whereas California and the Model Corporations Acts refer to articles of incorporation.

Because the corporation is created as the result of state action, most statutes provide for the withdrawal of the corporate charter under certain circumstances. For example, some state statutes provide for the forfeiture of the charter of a corporation if a corporate office, with the knowledge of the president and the board of directors, is engaged in organized crime. Also the statutes of almost every state provide that the legislature may amend, repeal, or modify its statute at its pleasure. The reason for this latter provision stems from the decision in the Dartmouth College case. The federal Constitution provides that no state may pass a law impairing the obligation of contract. Justice Marshall stated in the Dartmouth case that the grant of a corporate charter by the state creates a contract between the state and the corporation. Thus, without the statutory language that becomes part of the charter allowing for amendment of the statute by the legislature, the state would not be able to affect the rights of a corporation by amending the statute subsequent to the granting of a charter. However, with this statutory provision the state may, of course, amend, repeal, or modify its corporation law and subject any corporation created under its general corporate statute to such a change.

40:3. Consequences of Defective Incorporation. In order to create a valid corporation one must be careful to comply strictly with the requirements of the relevant state statute. When a corporation is created in this manner, as most are, it is known as a *de jure corporation,* a corporation by law. Such a corporation has all the rights provided under the general corporation statute. It has the right to exist for the period of time provided for in its charter, or it may have perpetual existence under the statute. Sometimes the organizers of the corporation do not comply with the statute in attempting to organize a corporation but nevertheless make a good-faith attempt to adhere to the statutory requirements. The following discussion examines the legal status of such efforts.

Unless the mandatory requirements of the statute are followed, a *de jure* corporation will not come into existence. Under certain circumstances the law will still treat the business organization as a corporation, with the result that those doing business have the protections afforded to the corporate form of organization. When the courts treat an organization in this manner, it is known as a *de facto corporation,* a corporation as a matter of fact.

The requirements of a *de facto* corporation are that (1) there must be a law under which a *de jure* corporation may be created; (2) there must be a bona fide attempt to incorporate under the statute; and (3) there must be actual use or exercise of the corporate powers in the belief that a corporation has been formed. All these requirements must be met or the courts will treat the organization as a partnership and hold the organizers liable as partners. Sometimes the concept is misunderstood to mean that the courts will treat a business as a corporation if it acts like one. Do not make this mistake. In the formation of the corporation there must be a good-faith *attempt* to comply with the statute that fails for some reason. The organizers then operate the business as a corporation unaware of the failure to form a *de jure* corporation.

A *de facto* corporation has the same status legally as a *de jure* corporation, including its relation to third parties and the limited liability of its owners. As a general rule, in most states third parties may not *collaterally attack* the existence of a *de jure* corporation. This rule also applies to a *de facto* corporation. By the term *collaterally attack* we mean that in a private suit, such as for the collection of a debt incurred by the corporation, the defendant cannot set up the defense that the corporation was never legally created. The Model Business Corporation Act, Section 56, provides that "upon the issuance of the certificate of incorporation . . . such certificate of incorporation shall be conclusive evidence that all conditions precedent required to be performed by the incorporators have been complied with. . . ." Even the state cannot collaterally attack the existence of a corporation in a suit involving another issue. If it wishes to challenge the existence of a corporation, the state must attack it *directly* in a separate action.

The reason that a private litigant may not attack the existence of a *de facto* corporation is that if all the parties, including, for example, the creditors and the organizers of the corporation, believe that a *de jure* corporation has been created when they are engaged in their business dealings, no one is prejudiced by the fact that there has been some technical reason that a *de jure* corporation has not been formed. For example, in extending credit, the creditors do so believing that the owners of the corporation are not personally liable for the corporate debts. They have not relied on the personal assets of the corporate owners in extending the credit. Therefore to hold that a *de facto* corporation exists is just to afford the owners the protection that all the parties assumed existed anyway.

There are other reasons for the rule prohibiting collateral attack of the existence of a corporation. Because it is the state's charter and rights that have been granted without technical compliance with the statute, it is up to the state to challenge the validity of the charter. In addition, the courts frequently state that to rule otherwise would lead to the confusion and uncertainty brought about by an attack on the validity of the corporation in every suit in which rights arise as a result of the corporate existence.

The rule against collateral attack has some limitations. For instance, one would not be prevented from proving in court that no corporation, either *de jure* or *de facto,* exists because the defendants merely started calling themselves a corporation without any attempt to form a corporation under the statute.

(1) Corporation by Estoppel. Sometimes the courts will *estop* a party from denying the existence of a corporation even though neither a *de jure* nor a *de facto* corporation has been created. This application of the doctrine of estoppel does not result in the creation of a corporation but merely prevents a party from denying that a corporation exists. The application of the doctrine may come about from agreements, admissions, or conduct of the individuals involved. Thus, if one enters into a contract or signs a promissory note with an apparent corporation, he may be estopped to deny its existence. This does not mean that a corporation is created by application of the doctrine but only that as between the parties the attributes of corporate existence are applied. On the other hand, the doctrine will not be applied by the courts where it would be inequitable, such as where the recognition of the corporate existence is procured as the result of fraud.

CRANSON v. INTERNATIONAL BUSINESS MACHINES CORP.

234 Md. 477, 200 A.2d 33 (1964)

HORNEY, J.

On the theory that the Real Estate Service Bureau was neither a de jure nor a de facto corporation and that Albion C. Cranson, Jr., was a partner in the business conducted by the Bureau and as such was personally liable for its debts, the International Business Machines Corporation brought this action against Cranson for the balance due on electric typewriters purchased by the Bureau. At the same time it moved for summary judgment and supported the motion by affidavit. In due course, Cranson filed a general issue plea and an affidavit in opposition to summary judgment in which he asserted in effect that the Bureau was a de facto corporation and that he was not personally liable for its debts.

The agreed statement of facts shows that in April 1961, Cranson was asked to invest in a new business corporation which was about to be created. Towards this purpose he met with other interested individuals and an attorney and agreed to purchase stock and become an officer and director. Thereafter, upon being advised by the attorney that the corporation had been formed under the laws of Maryland, he paid for and received a stock certificate evidencing ownership of shares in the corporation, and was shown the corporate seal and minute book. The business of the new venture was conducted as if it were a corporation, through corporate bank accounts, with auditors maintaining corporate books and records, and under a lease entered into by the corporation for the office from which it operated its business. Cranson was elected president and all transactions conducted by him for the corporation, including the dealings with I.B.M., were made as an officer of the corporation. At no time did he assume any personal obligation or pledge his individual credit to I.B.M. Due to an oversight on the part of the attorney, of which Cranson was not aware, the certificate of incorporation, which had been signed and acknowledged prior to May 1, 1961, was not filed until November 24, 1961. Between May 17 and November 8, the Bureau purchased eight typewriters from I.B.M., on account of which partial payments were made, leaving a balance due of $4,333.40, for which this suit was brought.

The fundamental question presented by the appeal is whether an officer of a defectively incorporated association may be subjected to personal liability under the circumstances of this case. We think not.

Traditionally, two doctrines have been used by the courts to clothe an officer of a defectively incorporated association with the corporate attribute of limited liability. The first, often referred to as the doctrine of de facto corporations, has been applied in those cases where there are elements showing: (1) the existence of law authorizing incorporation; (2) an effort in good faith to incorporate under the existing law; and (3) actual user or exercise of corporate powers. The second, the doctrine of estoppel to deny the corporate existence, is generally employed where the person seeking to hold the officer personally liable has contracted or otherwise dealt with the association in such a manner as to recognize and in effect admit its existence as a corporate body.

. . . There is, as we see it, a wide difference between creating a corporation by means of the de facto doctrine and estopping a party, due to his conduct in a particular case, from setting up the claim of no incorporation. Although some cases tend to assimilate the doctrines of incorporation de facto and by estoppel, each is a distinct theory and they are not dependent on one another in their application. Where there is a concurrence of the three elements necessary for the application of the de facto corporation doctrine, there exists an entity which is a corporation de jure against all persons but the state. On the other hand, the estoppel theory is applied only to the facts of each particular case and may be invoked even where there is no corporation de facto. Accordingly, even though one or more of the requisites of a de facto corporation are absent, we think that this factor does not preclude the application of the estoppel doctrine in a proper case, such as the one at bar.

I.B.M. contends that the failure of the Bureau to file its certificate of incorporation debarred all corporate existence. But, in spite of the fact that the omission might have prevented the Bureau from being either a corporation de jure or de facto. . . .

We think that I.B.M., having dealt with the Bureau as if it were a corporation and relied on its credit rather than that of Cranson, is estopped to assert that the Bureau was not incorporated at the time the typewriters were purchased. In 1 Clark and Marshall, Private Corporations, § 89, it is stated:

> The doctrine in relation to estoppel is based upon the ground that it would generally be inequitable to permit the corporate existence of an association to be denied by persons who have represented it to be a corporation, or held it out as a corporation, or by any persons who have recognized it as a corporation by dealing with it as such; and by the overwhelming weight of authority, therefore, a person may be estopped to deny the legal incorporation of an association which is not even a corporation de facto.

In cases similar to the one at bar, involving a failure to file articles of incorporation, the courts of other jurisdictions have held that where one has recognized the corporate existence of an association, he is estopped to assert the contrary with respect to a claim arising out of such dealings.

Since I.B.M. is estopped to deny the corporate existence of the Bureau, we hold that

Cranson was not liable for the balance due on account of the typewriters.
(Reversed.)

QUESTIONS FOR DISCUSSION

1. *What is the difference between finding corporate existence* de facto *and preventing a creditor from obtaining a judgment against the officers or stockholders of a defective corporation on the theory of estoppel?*
2. *Did the court find that the Real Estate Service Bureau was a* de facto *corporation in this case? On what primary fact did it base its decision regarding whether the bureau was a* de facto *corporation?*

---

(2) Ignoring the Corporate Entity. There are circumstances under which the courts will ignore the corporate entity and go behind the corporate form in order to hold the owners of the corporation personally liable. The most frequent occasion for "piercing the corporate veil," as this procedure is often called, is the use of the corporate entity in a fraudulent manner. For example, suppose Danny Debtor is heavily indebted to a number of creditors. Danny forms a corporation and transfers most of his property to the corporation. A court would ignore the corporate entity in this situation because it is clearly for the purpose of defrauding Danny's creditors.

It is not necessary that actual fraud be present in order for the courts to set aside the corporate entity. If it is apparent that a manifest injustice would be perpetrated, the court may ignore the corporate entity. On the other hand, the courts do not set aside the corporate form lightly. Generally this occurs only if the corporation is controlled by a single person or very few persons and then only for compelling reasons. Even if it is clear that a corporation has been established to avoid taxes or to prevent creditors from reaching one's personal assets, the court will not ignore the corporate entity in the absence of other compelling reasons.

WHITE v. WINCHESTER LAND DEVELOPMENT CORPORATION
584 S.W.2d 56 (Ky.App. 1979)

Charles T. White and Adelyn White incorporated their card and gift shop as The White House, Inc., on May 23, 1973. On June 6, Charles T. White signed a note to the Winchester Bank in the amount of $20,000 in his capacity as president of the corporation. On July 1, 1974, Adelyn White signed a note to the bank in her capacity as secretary-treasurer of the White House, Inc., in the amount of $19,630.36. After payments, the debt to the bank amounted to $18,350.02. Neither of the Whites signed the notes in their individual names. They also owned all of the stock of The White House, Inc.

The bank filed suit against them individually claiming that they were personally liable for the debt. The lower court found in favor of the bank and the Whites appealed. One of the contentions of the bank was that the corporate veil should be "pierced."

MARTIN, J.

The second issue presented on this appeal deals with appellees' contention that The White House, Inc. is a mere sham designed to defraud its creditors. Thus, it is argued, the corporate veil should be pierced and the Whites found to be personally liable for the obligations of the corporation upon the basis that the corporation is an "alter ego" of the Whites.

Three basic "theories" have been utilized to hold the shareholders of a corporation responsible for corporate liabilities. These have been labeled (1) the instrumentality theory; (2) the alter ego theory; and (3) the equity formulation. Because the bank seeks to rely upon all of these theories—if indeed they are distinct—they bear some examination.

Under the instrumentality theory three elements must be established in order to warrant a piercing of the corporate veil: (1) that the corporation was a mere instrumentality of the shareholder; (2) that the shareholder exercised control over the corporation in such a way as to defraud or to harm the plaintiff; and (3) that a refusal to disregard the corporate entity would subject the plaintiff to unjust loss. The courts adopting this test have been virtually unanimous in requiring that these three elements co-exist before the corporate veil will be pierced.

In this case it cannot be doubted that the Whites exercised a great deal of control over the corporation insofar as they were the sole shareholders. However, mere ownership and control of a corporation by the persons sought to be held liable is not alone a sufficient basis for denial of entity treatment. As to the second element, our courts have placed a great emphasis upon fraudulent organization and have denied entity treatment upon that basis. . . . The record before us is utterly devoid of the essential elements of fraud, *i.e.*, material representation, falsity, scienter, reliance, deception, and injury. A discussion of the third element, that is, a refusal to disregard the corporate entity which would subject the plaintiff to unjust loss, shall be pursued in the next section of the opinion which deals with the problems of renewal and novation; suffice it to say here that the bank's loss was not unjust because it could have secured the loan to The White House, Inc., merely by requiring the Whites to sign those notes in their individual and separate capacities.

As regards the alter ego formulation, the elements thereof have been defined as follows: (1) that the corporation is not only influenced by the owners, but also that there is such unity of ownership and interest that their separateness has ceased; and (2) that the facts are such that an adherence to the normal attributes, *viz*, treatment as a separate entity, of separate corporate existence would sanction a fraud or promote injustice. Again, there was no evidence of fraud adduced, and whatever loss the bank has suffered or may suffer could have been mitigated entirely by securing itself to the extent that the law allows. As to any allegations of failure to adhere to necessary corporate formalities, we again find the resolution (passed by the corporation and forwarded to the bank), . . . instructive. The fact that such a resolution was made and passed evidences that, by observing such a formality, the Whites complied with the strictures of proper corporate existence and thus maintained the separateness of entity so vital to the legitimate corporate function.

While these two formulations are helpful as an analytical framework, issues or "alter ego" do not lend themselves to strict rules and prima facie cases: whether the corporate veil should be pierced depends upon the innumerable equities of each case. Because this issue depends not only upon close-connectedness (which is inevitable in all closely held businesses) but also upon such actions as would indicate unfair dealings, an examination of those other factors is mandated. Indeed, a number of factors are considered in all the cases no matter what "test" is being applied.

Generally speaking, the corporate veil should only be pierced "reluctantly and cautiously" and then only in the presence of a combination of the following factors: (1) undercapitalization; (2) a failure to observe the formalities of corporate existence; (3) nonpayment or overpayment of dividends; (4) a siphoning off of funds by the dominant shareholder(s); and (5) the majority shareholders having guaranteed corporate liabilities in their individual capacities. Because all the factors save undercapitalization are part and parcel of each other, we may dispense with them first.

There is simply no evidence in this case supporting a finding that factors (2) through (5) are true. In fact, there are some elements of this appeal which could be construed as negating those factors. For example, both Exhibit E mentioned *supra* and the form in which the Whites signed the notes indicate that (2) and (5) as set out above are not applicable herein.

The theory of undercapitalization, while somewhat more problematic, is not significant here because Kentucky law does not require any minimum paid-in capital before a corporation begins to do business. More significant are the policy reasons behind the prohibition of undercapitalization—to protect innocent third parties who had no way of knowing that they were dealing with an impecunious entity. . . . In the instant case, the bank cannot be heard to complain that The White House, Inc. was undercapitalized because the bank had knowledge of the financial status of the corporation and could have protected itself.

(Judgment reversed in favor of the Whites.)

QUESTIONS FOR DISCUSSION

1. *What were the three theories for "piercing the corporate veil" that the court mentioned.*
2. *Give an example of what the bank would have to prove in order to have the court hold the Whites personally liable in this case.*

40:4. KINDS OF CORPORATIONS. Corporations may be described in a number of ways. A corporation may be a public, a quasi-public, or a private corporation. Private corporations may be profit or nonprofit corporations. Profit corporations may be public-issue corporations or closed corporations. Corporations may also be classified as foreign or domestic.

(1) The Public Corporation. A public corporation is one formed by the state or the federal government to carry on certain governmental purposes. It is an instrumentality of the government and hence is subject to its control. Towns, counties, and cities, collectively called *municipal corporations,* are examples of public corporations formed by the state in order to perform governmental functions. The Federal Deposit Insurance Corporation and the Tennessee Valley Authority are well-known examples of public corporations formed by the federal government.

(2) The Quasi-Public Corporation. A quasi-public corporation is a private corporation that is chartered to perform a public service. Public utilities, such as electric and gas companies, are examples of quasi-public corporations. Other examples are railroads, bus companies, telephone and telegraph companies, and those organized to maintain or operate bridges or canals or to supply water. Frequently the state will grant a monopoly to a quasi corporation in a certain field. In return, the corporation is subject to heavy regulation in the conduct of its business by the state. The most notable examples are the public utilities, which are generally regulated by public-utilities commissions. These commissions regulate such important matters as the rates that the corporations may charge for their services and the financing of the corporation. Quasi-public corporations are frequently given the power of *eminent domain,* which means that they can take private property upon the payment of reasonable compensation if the property is put to public use. Thus, for example, an electric company can obtain an easement on private land for the erection of poles, and a water company may obtain a similar easement for the laying of pipe to carry water. On the other hand, a corporation that is a purely private undertaking is not a quasi-public corporation if it has not been given the power of eminent domain or other special privilege, even though the nature of the business may be public.

(3) The Private Corporation. A private corporation is one that is formed purely for private purposes and for the benefit of its owners. Perhaps the most important distinction between a public and a private corporation is that the private corporation is not subject to the degree of regulation and control by the state that a public corporation is. In effect, the charter of the private corporation is a contract between the state and the corporation or its organizers. There are, therefore, constitutional limitations on the power of the state to change or control the rights of such corporations.

Private corporations may be *nonprofit* or *nonstock* corporations, or they may be *stock corporations*, that is, organized for profit. A nonprofit corporation is very often organized for a charitable, religious, scientific, or educational purpose. It is not organized for the purpose of making a profit, and hence there is no contemplation of any profits to be distributed to members. Although one's ownership interest in a stock corporation is evidenced by the issuance of shares of stock, under most state statutes one joins the corporation as a "member." How one does this is determined by the bylaws of the organization. The corporation statutes provide for the organization of nonprofit corporations in much the same way that other types of corporations are formed, a subject that will be discussed in Section 40:5. The nonprofit corporation is a convenient organization for allowing various groups to own property and contract without exposing the individual members to liability.

A corporation for profit, or a stock corporation, is best exemplified by the ordinary business corporation. It is organized for the purpose of making a profit, which may be distributed to the owners in the form of dividends. This is the type of corporation that will be the subject of our further study. Such corporations may be either *closed* or *public-issue* corporations. A closed corporation is usually a family corporation or one that is owned by a single or very few individuals. Some states have statutes that allow the owners to elect to be treated as a closed corporation. Under these statutes many of the formalities and mechanics required of other corporations are waived. For example, under Maryland statute the closed corporation may elect to eliminate a board of directors. The public-issue corporation is one whose stock—that is, its ownership interest—is held by members of the public.

(4) The Domestic or Foreign Corporation. Another way to classify corporations is to distinguish them as either domestic or foreign. The distinction is based upon the state of its incorporation or domicile. For example, in Utah a corporation formed in Utah is said to be a *domestic corporation*. A corporation that was formed in Nevada would be a *foreign corporation* in Utah. Corporations generally must file for permission to do business and comply with the applicable state requirements of those states to which it is foreign. A penalty is frequently imposed upon those foreign corporations that fail to comply with state foreign-corporation laws.

40:5. Formation of the Corporation. For the typical small business the state where the corporation is formed is the state in which it will do business. It is most often not worthwhile to incorporate in another state. On the other hand, there are variations between states with regard to their general corporation law, so if one expects the corporation to do business in more than one state, he would do well to analyze the various state statutes in order to incorporate in the state that has the statute most favorable to corporate formation. Consideration should be given to the organization fees and taxes charged by the state; the corporate tax structure of the state; what types, kinds, and classifications of corporate stock may be issued under the laws of the state; and any requirement that may otherwise affect the structure and management of the corporation. Remember also that in those states where the corporation has not been formed but where it does business, the corporation must usually register to do business as a foreign corporation.

Historically, Delaware, New Jersey, and New York have been among the "favorite" states in which to incorporate. In addition to the considerations already mentioned, there are others that may be more subtle. For example, Delaware has long been established as a state that gives corporations favorable treatment. It also has a substantial body of case law established that delineates the rights and duties of corporations, their management, and their stockholders, as well as a judiciary and a bar accustomed to dealing with complex corporate legal issues. Although these considerations are not of great importance to the typical small businessman or businesswoman who is contemplating incorporating his or her business, they are relevant considerations in some cases.

(1) Selecting a Corporate Name. The selection of the name for a business is often very important, and a number of nonlegal considerations should be weighed in such a decision. Whether the name should be descriptive of the business or the product or should contain the names of the owners is among the many factors to be considered. The corporation statutes of most states place some restrictions upon the name that may be selected for the corporation. For example, most state statutes require that the name indicate that the organization is a corporation by including such a term as *incorporated, limited, company, corporation,* or some abbreviation in the name. Among other restrictions found in many corporation statutes are that the name should not imply that the corporation is organized for any purpose not contained in its charter and that the name should not be the same as or confusingly similar to the name of other corporations chartered or licensed to do business in the state.

Most states provide an administrative procedure for reserving a corporate name for a period of time prior to the actual filing for incorporation. One usually inquires of the appropriate state department as to whether a particular name is available. If it is, it may be reserved for a limited time upon the payment of a small fee. One may change the name after incorporation by amending the corporate charter by a procedure discussed in Section 40:6(3). Such a change does not affect the life of the corporation, corporate contracts, or corporate property in any way.

(2) Legal Formation: The Role of Promoters. In order for a corporation to be formed, some individuals obviously have to undertake the steps necessary for corporate formation under the law. The people who do this are often referred to as the *promoters* of the corporation. The promoters are the individuals who undertake to get the business started by obtaining the necessary capital and looking after the other matters—as well as its legal formation—that are required in order for the corporation to begin operation. In so acting, the promoters often enter into contracts and perform other acts prior to the formation of the corporation. Questions frequently arise concerning the role and the duties of the promoters to the corporation that they have formed, as well as the liability of promoters and the corporation on preincorporation contracts. Whether one is a promoter or not is a question of fact. An individual must actually be involved in the formation of the corporation in order to be a promoter. Merely designating someone as a promoter in a contract does not make him or her a promoter.

(3) The Promoter as a Fiduciary. The promoter has a fiduciary duty to the corporation, when it comes into existence, as well as to the subscribers of corporate stock and the original shareholders. Although there are frequently opportunities to do so, the promoters may not make any secret profit at the expense of those to whom they owe a fiduciary duty. They may be made to account for such a profit. On the other hand, the law does not prohibit the promoter from making a profit with the consent of those to whom he owes the duty at the time of the corporate formation. Indeed this may be why the promoter is willing to make the effort required and to incur the risks that may be involved in the formation of the corporation. The fiduciary duty does not extend to those who purchase stock from the original stockholders.

Although the promoters owe a fiduciary duty to those parties already mentioned, the promoters are not agents. The reason is simple. One may not be an agent for a nonexistent principal, and the corporation is not yet in existence. On the other hand, the promoters may become agents of the corporation after it is formed, provided that the corporation agrees to the relationship.

(4) Promoter's Contract Liability. In undertaking to establish a corporation, promoters frequently must enter into contract prior to the actual formation of the corporation. For instance, it may be desirable to lease office space and make employment and sales contracts as well as to incur other obligations. Although these preincorporation agreements are made for the benefit of the corporation, it is the promoter who is liable on them. The reason for this rule is that the promoter is not an agent of the corporation because the corporation is not yet in existence when the contracts are made. Furthermore, in the absence of any further action by the corporation, the corporation is not liable on these contracts when it does come into existence. The promoter remains personally liable on them even though the contracts may have been made in the name of the corporation.

It is possible for the promoter to escape liability on preincorporation contracts by placing a clause in the contract stating that he is not personally liable on the agreement. The other party must, of course, agree to this provision and takes the risk that the corporation may never come into existence. In lieu of such a provision, it is also possible that the other party to the contract may agree to substitute the corporation as a party, after its formation, in place of the promoter. In so doing, the other party agrees to enter into a novation that will discharge the promoter. Such a novation agreement may be either express or implied. In order for the corporation to become liable on preincorporation contracts, it must expressly adopt the contract or knowingly accept the benefits of the contract so that the contract is adopted by implication.

VODOPICH v. COLLIER COUNTY DEVELOPERS, INC.

319 So.2d 43 (Fla.App. 1975)

MOE, A.J.

Plaintiff, Vodopich, a registered real estate broker, participated in a transaction involving the sale of land from Naples Bay Industries, Inc. to the Defendant/Appellee, Mulley, and others. For his services, he was entitled to a 10% commission amounting to $31,500.00. Vodopich subsequently agreed to forego his rights to that commission in consideration for the exclusive right to resell said properties. This agreement was reached orally between Vodopich and Mulley, plus other persons not parties hereto, who were promoting the creation of a corporation which was to become the Co-Defendant, Collier County Development, Inc. At the time this agreement was reached with Mulley, Vodopich was aware of the nonexistence of the contemplated corporate entity.

The corporation was subsequently formed and held title to the properties in question consummating several transactions for the resale of said properties. During this time, Vodopich unsuccessfully demanded the properties be listed as per the agreement for exclusive rights of resale reached with Mulley. Vodopich then brought suit for the resale commissions against Mulley personally and Collier County Developers, Inc.

In the trial court, a jury returned verdicts in favor of Vodopich as against Collier in the amount of $70,710.00, which was the total amount claimed by Vodopich. With regard to the claim against the Defendant, Mulley, however, the jury found him not to be personally liable for Vodopich's commission. It is from this latter decision that Vodopich appeals.

Appellant urges that the trial court erred in giving the following instruction to the jury:

> The rule that one who assumes to act as agent for a principal who has no legal status or existence renders himself individually liable on contracts so made does not apply where the third person has knowledge of a nonexistence or incompetency of the principal.

Appellee contends that this instruction is the law of Florida as stated in *Bryce* v. *Bull,* 1932, 106 Fla. 336, 143 So. 409. We disagree. It is true that similar language appears in *Bryce* as dictum, but the words taken from the factual context of that case have acquired a distorted meaning in the setting of the case at Bar.

We cannot agree with Appellee's assertion that stands as concrete authority for the *Bryce* proposition a corporate promoter will not be liable for contracts entered into on behalf of a nonexistent corporation when the party with whom he is contracting knows of the nonexistence of the contemplated corporation, for even in *Bryce* the Court found the promoter personally liable. In our opinion, before the promoter may escape liability, it must be shown that the parties agreed to bind the corporation alone.

As a general rule, the promoter of a corporation is personally liable on the contracts entered into on behalf of the corporation he is organizing. " 'On the other hand, if the contract is on behalf of the corporation and the person with whom the contract is made agrees to look to the corporation alone for responsibility, the promoter incurs no personal liability.' . . . This exception applies where a promoter contracts in the name of a corporation which is to be formed later and does not intend to be liable on it, and the other party knows that the corporation has not been formed and that the promoter does not intend to be liable." 2 Williston § 306 at p. 426 (1959), quoting from *Frazier* v. *Ash,* . . . 234 F.2d 320, 326–327.

The Supreme Court of Florida, one year after the decision in *Bryce,* heard the case of *Hunt* v. *Adams,* 1933, 111 Fla. 164, 149 So. 24. In referring to the personal liability of an agent contracting for unincorporated societies, the Court, relying on the decision from the previous year in *Bryce,* found the agents not personally liable as a rule "on written contracts made with third persons when executed by them solely in the name of, and on behalf of, the unincorporated charitable . . . society for which they purport to act, where the nature of the written contract entered into, and the particular manner of its execution, is *on its face* such as to *clearly* impute no *intention* on the part of those who sign on behalf of the unincorporated society, *to bind* themselves or their associate *representatives personally.*" (Citations omitted; emphasis added.)

Certainly this indicates that the Supreme

Court in *Bryce* v. *Bull,* supra, did not intend the isolated statement, upon which Defendant relies, to mean that a promoter is relieved on personal liability if the person with whom he contracts merely has knowledge of the nonexistence of the corporate entity for whom the promoter purports to act. If such were true, then, with the corporation's liability not arising until its adoption of the contract, the corporation would have an ingenious method of speculation sanctioned by the law, whereby it could reap all the speculative advantages of a timely adoption without any of the attendant disadvantages.

We feel that the instruction in question is only a partial statement of the law and, in the context of this case, erroneous. The instruction, set forth in the margin[1] which was also given by the trial court, is a more appropriate statement of the law with regard to promoter's liability and is in accord with *Bryce* v. *Bull,* supra.

Accordingly, we hold that there are factual issues in this case that can only be resolved by a properly instructed jury, and the remaining points on appeal are moot in light of this decision.

(Reversed and remanded for new trial.)

QUESTIONS FOR DISCUSSION

1. *What was improper about the original instruction as to the law given the jury by the trial court?*
2. *Under what circumstances may a promoter escape liability on preincorporation contracts?*

---

(5) Corporation Liability on Promoter's Contracts. As a general rule, a corporation is not liable on preincorporation contracts made by its promoters even if the contract is made in its own name and for its own benefit. There are a few states that have statutory provisions to the contrary. The corporation may *adopt* an agreement made by its promoters prior to formation. Although the term *ratification* is also frequently used to describe corporate adoption of preincorporation contracts, this usage is not technically correct. The reason that a *ratification by the principal,* as the term is used in the law of agency, cannot take place in this situation is that the corporation is not in existence when the contract is made by the promoter. Remember that under agency law the ratification of a contract by the principal has the legal effect of relating the legal rights and duties of the principal back to the date the contract was made. Not so with a corporate *adoption* of preincorporation contracts. The rights and obligations of the corporation under the agreement are measured from the time that the contract is adopted rather than from when the contract was made.

Adoption of a preincorporation contract may be made expressly by a formal resolution passed by the board of directors. The adoption may also be made by implication, as when the corporation knowingly accepts the benefits of a preincorporation contract. In order for a preincorporation contract to be adopted, however, it is necessary that the corporation have full knowledge of the facts involved or that it have such notice as would require it to make a further inquiry. Thus merely failing to reply to a demand based upon a claim made under a preincorporation contract would not result in an implied adoption of the agreement. Also the corporation should have an opportunity to decline the benefits of the contract. Thus there is usually no corporate

[1] "The promoter of a corporation is liable on his contract, although made on behalf of the corporation to be formed, unless the person with whom he contracts agrees to look to some other fund for payment. The later formation of the corporation and adoption or ratification of the contract by the corporation may not free the promoter from his liability."

liability for legal services provided in the formation of the corporation because the corporation has no opportunity to decline the benefits of such services.

FORTUNE FURNITURE MANUFACTURING CO., INC. v. MID-SOUTH PLASTIC FABRIC COMPANY, INC.

310 So.2d 725 (Miss. 1975)

Fortune Furniture Manufacturing Company, Inc., received a letter dated July 8, 1968, confirming an oral contract between itself and "Mid-South Plastic Co., Inc.," *and signed* "W. E. Walker, President." *The contract provided that Mid-South was to supply plastic to Fortune for use in its furniture-manufacturing business. Mid-South was organized with a corporate charter dated July 22, 1968, two weeks after the letter signed by Walker.*

A dispute erupted between the parties when Mid-South was unable to supply all of Fortune's needs, and Mid-South sued Fortune. Fortune counterclaimed for an amount $6,066 in excess of Mid-South's claim. The jury found for Fortune, but the trial court sustained a motion made by Mid-South for judgment not withstanding the verdict. Fortune appealed.

GILLESPIE, C.J.

The principal argument of Mid-South is that the letter dated July 8, 1968, written by Walker, purportedly on behalf of Mid-South and as president thereof, is not binding because Mid-South did not come into existence until July 22, 1968, the date of the charter. It is contended by Mid-South that the letter could not have been the act or the creation of Mid-South which was not then a corporate entity and that there was no ratification or adoption of the contract by affirmative action of the company after its incorporation.

The general rule is that a contract made by promoters with a view towards incorporation will be binding upon the corporation if it accepts benefits of the contract with the full knowledge of the terms of the contract.

There are several theories upon which corporations have been held liable, including the theory of adoption and the theory of ratification. In *Pearl Realty Co.* v. *Wells*, 164 Miss. 300, 145 So. 102 (1933), the Court spoke in terms of "ratification" and upheld a contract for services rendered in obtaining a cancellation of a lease to a lot upon which Pearl Realty Company was incorporated after the negotiations between the promoter of the corporation and Wells. In concluding that the corporation had ratified the contract made by the promoters, the Court said:

> It is permissible for promoters to make contracts which, if ratified by corporations after they are organized, will bind the corporations.

An analysis of Pearl Realty Company indicates that the term "ratification" in that opinion does not refer to any formal act of the corporation, but rather that the corporation adopted the contract and received the benefits of it with full knowledge on the part of all parties concerned in the corporation's organization of the manner and conditions under which it had been obtained. The Court held that under such circumstances the contract was binding on the corporation.

The general rule is that adoption may be implied from the acts or acquiescence of a corporation without express acceptance or ratification, and the adopting corporation will be liable if its responsible officers have or are chargeable with knowledge of the facts upon which it acts.

Stillpass and Walker, the only executive officers of Mid-South, initiated and concluded the agreement that was reduced to writing in the letter to Fortune. Stillpass admitted that Mid-South would not have been organized without Fortune's account. The jury was fully justified in finding that these two officers had full knowledge of the terms of the contract and took advantage of it to sell more than $1,000,000 in goods to Fortune as a result of the contract. We are of the opinion that the jury was justified under the evidence in finding that Mid-South was bound by the terms of the letter of July 8, 1968.

(Reversed and remanded.)

Questions for Discussion

1. *Would Walker have been liable in this case had the jury not found in favor of Fortune on the issue of the adoption of the contract by Mid-South? What was Walker's liability, if any, as a result of this decision?*
2. *What acts of the Mid-South officers resulted in the adoption of the contract?*

(6) Promoters, Expenses, and Compensation. Most promoters expect to be compensated in some manner for the effort they put into forming the corporation as well as for the expenses they incur. However, the general rule is that the corporation is not liable to the promoter unless there is a statute to the contrary or unless the corporate charter provides for payment to the promoters. As a practical matter, because the promoters are usually heavily involved in the formation of the corporation, there is usually some sort of agreement with the investors that the promoters will be compensated and that, if appropriate, the required provision will be placed in the corporate charter. Such agreements are not binding on the corporation, however. A very few jurisdictions have held the corporation liable for the expenses of promoters on the theory of estoppel or adoption by implication. The difficulty with this approach, however, as mentioned in Section (5), is that because the corporation is not yet formed, it has no opportunity to decline the services. Of course, if the promoters render services to the corporation after it is formed, with the knowledge and acquiescence of the corporation, then the corporation is liable for the acceptance of such services.

40:6. Articles of Incorporation and Charter. The statutes of most states, including those that have adopted the Model Business Corporation Act, require that a certain number of adults, usually one or three, acting as incorporators sign and acknowledge the articles of incorporation. Copies should then be filed with the appropriate state department, most often the office of the secretary of state. In most states, once the articles have been accepted for filing and the appropriate fees have been paid, the articles are returned to the incorporators as the corporate charter. The issuance of the certificate of incorporation is conclusive evidence of the formation of the corporation except in a proceeding by the state to forfeit the corporate charter. Corporate existence begins under the Model Business Corporation Act once the corporate charter is issued. In some states it does not begin until after an organizational meeting, and in few states not until there is local recording of the certificate of incorporation. Under the Model Business Corporation Act and the statutes of many states, the articles of incorporation shall set forth:

1. The name of the corporation.
2. The period of duration of the corporation, which may be perpetual.
3. The purposes for which the corporation is formed.
4. The name and address of each incorporator.
5. The name and address of the person who is to be a registered agent of the corporation. The registered agent usually must have a residence

in the state in which the corporation is formed. The purpose of a registered agent is to have someone of record who is authorized to receive notices and other matters on behalf of the corporation.

6. The address of the principal office of the corporation.
7. The number and names of those who will serve initially as corporate directors until the first annual meeting of shareholders or until their successors are elected.
8. The aggregate number of shares of stock that the corporation shall have the authority to issue and a statement of the par value of each share or a statement that the shares are without par value.
9. If the shares are to be divided into classes, the number of shares of each class, the designation of each class, and a statement of the preferences, limitations, and relative rights in respect of the shares of each class.
10. A statement of the provisions governing any preferred stock or special class in series if such stock is to be issued.
11. The provisions of any preemptive right to be granted to shareholders.
12. Any provision, not inconsistent with law, which the incorporators elect to set forth for the regulation of the internal affairs of the corporation.

(1) Corporate Bylaws. The corporation may adopt rules for its internal management known as *bylaws.* These rules are, of course, not binding upon parties outside the corporate structure. The power to adopt bylaws resides in the shareholders unless the charter of the corporation or a statute provides for something contrary. The Model Business Corporation Act provides that the power to amend, repeal, or adopt new bylaws is vested in the board of directors unless the articles of incorporation reserve this right to the shareholders. Most statutes also provide that the bylaws may contain any provisions not inconsistent with law or the corporate charter.

The bylaws may govern matters concerning the time and place of the annual meeting of stockholders as well the procedures to be followed in the calling of special meetings. They may also provide for rules governing the selection of directors as well as the organization and meetings of the board of directors. In addition, matters concerning the issue of dividends, the kinds and numbers of corporate officers, and the kinds of books and records to be kept as well as many other subjects may be covered by the bylaws.

(2) Corporate Organization. After the corporate charter is granted, there must be a formal meeting for the purpose of organizing the corporation. In some states an organization meeting of the board of directors is called by the majority of directors named in the articles of incorporation. At this meeting the bylaws are adopted, the officers of the corporation are elected, and other corporate business is transacted. In other jurisdictions the initial organizational meeting is of the shareholders, who may elect a board of directors to complete the organization by electing officers and attending to other matters concerning the beginning of the business of the corporation. The statutes usually provide for notices to be sent out to the applicable people notifying them of the time and the place of the meeting.

(3) Amending the Charter. We have previously made reference to amending the corporate charter. This may be necessary for a number of reasons. It may be desirable to alter the corporate name, to expand the scope of the purpose of the corporation, to increase the number of authorized shares of stock, and so on. Basically an amendment may contain any provision that could lawfully have been included in the original charter. In order to amend the corporate charter, a resolution is usually adopted by the board of directors, approved by the required number of stockholders, and then submitted to the appropriate state office for inclusion with the original corporate charter.

SUMMARY

The corporation is a creature of the state that has a separate legal existence from its creators. It may have perpetual life; may change ownership without affecting its existence; may sue or be sued; and may enter into contracts.

One of the most important characteristics of a corporation is the limited liability of its owners. The management of a corporation may be separate from its owners and, unlike a partnership, the corporation may have perpetual existence. It is typically easier for a corporation to raise capital than it is for a partnership. A corporation may also be treated differently from a partnership for tax purposes.

The authority of the state to grant corporate charters is created by statute. The state may also withdraw a corporate charter under certain circumstances. When a corporation is created in strict compliance with the corporation statute it is said to be a de jure corporation. If there is a bona fide attempt to create a corporation under a state statute and there is actual use or exercise of corporate powers in the belief a corporation is formed, the qualifications for a de facto corporation have been fulfilled. This concept is applied when for some technical reason a de jure corporation has not been formed.

A de facto corporation has the same status as a de jure corporation. The existence may not be collaterally attacked. Under certain conditions the courts may also stop a party from denying the existence of a corporation. If a corporate entity is used in a fraudulent manner the courts may ignore the corporate entity.

Corporations may be classified as public, quasi-public; or private, domestic, or foreign.

A corporation is formed by submitting articles of incorporation to the state. A corporate name must be selected by the promoters who form the corporation. The promoters have a fiduciary duty to the corporation when it comes into existence as well as to subscribers for the stock of the corporation. Promoters are liable on contracts entered into by them prior to the formation of the corporation. They are generally not discharged once the corporation is created nor is the corporation liable on their contracts. The corporation may adopt promoter's contracts.

State statutes normally list the items that must be mentioned in the articles of incorporation. The corporation may also make rules for its internal management known as bylaws.

PROBLEMS

1. Mobile Roofing and Construction Company, Inc., had been incorporated for some twenty-five years when it entered into a contract with Daisy Burrell for the construction of a house. During the term of the contract, Phillip F. Cohen became the sole stockholder, director, and officer of the corporation, although there had been other stockholders, directors, and officers in the past. Burrell then recovered a judgment in the amount of $11,000 against Mobile on a cause of action arising out of the contract. Cohen then "dissolved" the corporation, because of the judgment, by going out of business. For ten or twelve years prior to the trial there had been no stockholders or directors meetings and Cohen had freely withdrawn money from the profits of Mobile. He had personally paid some of the debts of the "dissolved" corporation but not the judgment. Burrell's personal representative then sued Cohen personally to collect the judgment for Burrell's estate contending the court should ignore the corporate entity. Is Cohen personally liable for the judgment? Cohen v. Williams, 318 So. 2d 279 (Ala. 1975).
2. Edward A. Cantor leased some commercial property to Sunshine Greenery, Inc. Sunshine was represented by William J. Brunetti who Cantor knew was just forming the corporation. Brunetti was not asked to personally guarantee the lease. The lease was signed by Brunetti for the corporation on December 16, 1974. On December 3, 1974 a certificate of incorporation was signed by Brunetti as an incorporator and mailed to the secretary of state for the purpose of forming a corporation. For some reason it was not filed officially until December 18, 1974, two days after the lease was signed. Is Brunetti personally liable on the lease? Cantor v. Sunshine Greenery, Inc., 398 A.2d 571 (N.J. App. 1979).
3. On May 18, 1958, Quaker Hill, Inc., contracted to sell a large quantity of nursery stock to Denver Memorial Nursery, Inc. Denver Memorial Nursery was also named as maker of a promissory note for the amount of the sale. The contract and the note were both signed in the name of Denver by "E. D. Parr, Pres." During the course of negotiation of this contract, Parr undertook to organize Denver Memorial Nursery, but it was never incorporated. A different corporation, Mountain View Nurseries, Inc., was organized subsequent to the contract but was never a going concern. Letters to Quaker after the contract was formed used the name Mountain View. When the debt was not paid, Quaker sued Parr, contending that he was personally liable. Parr introduced evidence, including the statement of Quaker's sales-

man, that Quaker knew that the corporation had not been formed at the time the contract was made but insisted that it be signed anyway because the growing season was rapidly passing. Was Parr liable on the contract? Quaker Hill, Inc. v. Parr, 148 Colo. 45, 364 P.2d 1056 (1961).

4. Exum Walker, M.D., contracted to have Joanna M. Knox & Associates, Inc., build a dwelling for $53,870. Joanna M. Knox & Associates, Inc., was not incorporated when the contract was made but did become a corporation some months later. Dr. Walker never paid the final amount of the bill and Joanna M. Knox & Associates, Inc., sued him. One of Walker's grounds of defense was that there was no valid contract with Joanna M. Knox & Associates, Inc., because it was not a corporation at the time the contract was made. Could Walker successfully raise this defense? Exum Walker, M.D., P.C., Pension Trust v. Joanna M. Knox & Associates, Inc., 207 S.E.2d 570 (1974).
5. Harold Baker, Laura Baker, and Theodore Sweetland attempted to organize a corporation in February 1923 and in so doing complied with all statutory requirements except that their attorney failed to record the certificate in the registry for the county in which the corporation was to do business or to file a copy with the secretary of state as was required by the statute. Suit was filed by Bates-Street Shirt Co., alleging that Harold Baker, Laura Baker, and Theodore Sweetland were liable as partners for certain debts because the corporation had never been legally formed. The defendants contended that they were not liable as there was a *de facto* corporation. Were they correct? Baker v. Bates-Street Shirt Co., 6 F.2d 854 (1st Cir. 1925).
6. Walkovszky was injured in New York City when he was run down by a taxicab owned by Seon Cab Corporation. Carlton was claimed by Walkovszky to be the major stockholder of ten corporations, including Seon, each of which had only two cabs registered in its name. The minimum insurance required by law, $10,000, was carried on each cab. In a suit filed by Walkovszky against all ten corporations and Carlton, Walkovszky alleged that the ten corporations were in fact operated as a single enterprise with regard to financing, supplies, repairs, employees, and garaging and that the multiple corporate structure was an attempt to defraud members of the public who might be injured by the cabs. As a result, Walkovszky contended that Carlton should be held personally liable for his injuries. Was Walkovszky correct? Walkovszky v. Carlton, 18 N.Y.2d 414, 223 N.E.2d 6 (1966).
7. Divco-Wayne Corporation sold hearses to a dealer, Martin Vehicle Sales, Inc. These sales were financed by a wholly owned subsidiary of Divco-Wayne, namely, Divco-Wayne Sales Financial Corporation, a separate corporation. Divco Sales Financial Corporation sued Martin to repossess seven hearses. Martin counterclaimed, based on a contract with Divco-Wayne but not with Divco Sales Financial Corporation, the plaintiff. The trial court found in favor of Martin on its counterclaim and entered judgment against

Divco Sales Financial Corporation. Divco Sales Financial Corporation appealed, contending that Martin's contract was with Divco-Wayne and not with Divco Sales Financial Corporation, that it was a separate corporate entity, that Martin did not allege that Divco-Wayne and Divco Sales Financial Corporation were one and the same, and that no fraud was perpetrated by Divco Sales Financial Corporation. Was Divco Sales entitled to a reversal of the trial court's verdict? Divco-Wayne Sales Financial Corp. v. Martin Vehicle Sales, 195 N.E.2d 287 (Ill. 1963).

8. The Railroad Commission of Arkansas ordered the St. Louis & S.F.R. Co. to establish and maintain with another railroad a joint interchange track at Fayetteville. St. Louis failed to comply with the order and was convicted and fined for failure to do so. On appeal the defendant railroad contended that under the applicable statute the commission was empowered to make such an order only after considering petitions signed by at least fifteen "bona fide citizens" residing in the affected territory. St. Louis & S.F.R. Co. contended that because the petition was signed only be eighteen corporations, the statutory requirements had not been met. Was the railroad correct? St. Louis & S.F.R. Co. v. State, 179 S.W. 342 (Ark. 1915).

CHAPTER 41

Corporate Powers

"A corporation . . . being the mere creature of law . . . possesses only those properties which the charter of its creation confers upon it, either expressly or as incidental to its very existence." That quote from the opinion of Chief Justice Marshall in the famous Dartmouth College case[1] perhaps best explains the dimensions of corporate powers, as well as the difference in the powers of the corporation and natural persons. A corporation has only those powers conferred upon it in one way or another by its creator, usually the state. A natural person, on the other hand, has the power to do any act not forbidden by law.

The powers of a corporation, all of which ultimately flow in some manner from the state, may be categorized as express, implied, inherent or incidental, and statutory. The following discussion deals with each of these categories.

41:1. Express Powers. The express powers of a corporation are typically thought of as those that are found in its articles of incorporation or its charter. Frequently many of the express powers of a corporation will be stated in what is known as the *purpose clause* of the articles of incorporation. It is in this clause that the purpose of the corporation is stated. Thus the express powers of individual corporations may vary according to their purpose.

The purpose clause may be quite restricting if one desires to draw it in that manner, but the modern trend is to draw the clause very broadly, conferring many and varied express powers upon the corporation so that it may enter into new lines of business and perform acts in the future, the necessity of which may not be foreseen in the present. Drawing the purpose clause in this manner may eliminate the need for amending the corporate charter at a later date. This trend has proceeded to the point where some states, such as Wisconsin and Nevada, allow the corporate charter simply to state that the corporation may engage in any act not prohibited by law. The Model Business Corporation Act, Section 54(c), provides that the purpose clause in the articles in incorporation may include a statement allowing for "the transaction of any or all lawful business for which corporations may be incorporated under this Act." As a practical matter, however, most purpose clauses at least

[1] *Dartmouth College* v. *Woodward,* 4 Wheat (U.S.) 518, 636; 4 L.Ed. 629 (1819).

state the general line of business in which the corporation is to engage, and some state statutes require at least this much.

The trend in the direction of drafting increasingly broad purpose clauses resulting, in some instances, in almost unlimited express powers has been criticized by some on the ground that the corporate investor should have at least some idea of the type of business in which the corporation is to engage and that that is the objective of the purpose clause. At least in the case of the small, closely held corporation it is probably wise to specify to some degree the line of business in which the corporation will engage. For example, the articles of incorporation may specify that the corporation has the power and is formed for the purpose of engaging in a retail-clothing business or to operate as a general contractor.

41:2. Implied Powers. Obviously corporations engage in many forms of day-to-day activity that may not be expressly spelled out as within its power by the corporate charter. The corporation may, however, have the *implied power* to perform these activities. The implied powers of a corporation are generally defined as those powers that are reasonably necessary and appropriate to carry out the express powers and purposes of the corporation. The implied powers give the modern corporation rather wide latitude as to the type of activities in which it may engage, the express powers in the modern corporate charter typically being drawn very broadly.

You should note, however, that the powers under discussion here are implied from the powers stated expressly in the corporate charter. That is, an implied power must be related or developed in some way from the express powers. In general, in order to be implied, the power must tend directly toward accomplishing the object or purpose for which the corporation is established. It is not enough to show that the activity would be beneficial or profitable to the corporation if it is not consistent with the objectives of the corporation. An example of an implied power would be the power of a corporation formed for the purpose of selling life insurance to hire salespersons. A motel might have the power to construct and operate a swimming pool for its guests.

41:3. Inherent or Incidental Powers. The concept of *inherent* or *incidental* powers of a corporation is often the subject of some confusion. Frequently the terms *incidental or inherent* corporate powers are used interchangeably with the term *implied* powers. This is wrong. They are two different concepts. Inherent or incidental powers are those powers held by the courts to be essential to the existence of *any* corporation. Among those powers would be the power to have a corporate name; to sue and be sued; to enter into contracts; to have a corporate seal; to appoint directors, officers, and agents; and to have perpetual existence, unless some other term is specified, as well as other powers. In other words, all corporations have the same inherent powers unless they are restricted or limited by the corporate charter.

On the other hand, because the implied powers of a corporation flow from the express powers stated in the charter, different corporations have different implied corporate powers.

41:4. STATUTORY POWERS. Today most general corporation statutes list the powers conferred upon any and all corporations unless limited in some manner by the corporate charter. In other words, all corporations are granted these powers without the necessity of listing or mentioning them in the corporate charter. These *statutory powers* generally include those powers that have previously been referred to as incidental or inherent. For example, the Model Business Corporation Act lists a number of statutory powers in Section 4. This section is fairly typical even of those states where the act has not been specifically adopted. These powers are summarized as follows:

1) To have perpetual existence unless limited by the charter,
2) To sue and be sued, complain and defend in its corporate name,
3) To have and use a corporate seal,
4) To purchase, sell, mortgage, lease, exchange, pledge and otherwise dispose of all or part of its assets and property,
5) To lend money,
6) To deal by purchase, subscription or otherwise in the shares or other interests or obligations of corporations or government,
7) To make contracts and incur other liabilities such as by borrowing money,
8) To lend money for its corporate purposes and to invest its funds and hold real and personal property as security,
9) To conduct its business within and without the State,
10) To elect or appoint officers and agents of the corporation and define their duties and compensation,
11) To make and alter by-laws not inconsistent with the corporate charter,
12) To make charitable donations,
13) To transact any business found by the board of directors to be in aid of governmental policy,
14) To pay pensions and establish pension plans as well as incentive plans for employees, officers and directors,
15) To enter into a partnership or joint venture,
16) To have and exercise all powers necessary or convenient to effect its purpose.

The statutes also provide for other powers necessary to the operation of the corporation such as amending its charter, acquiring its own shares, issuing additional stock, declaring and paying dividends, and effecting merger or consolidation with other corporations.

41:5. DEFENSE OF ULTRA VIRES. As previously discussed, a corporation may engage only in acts that it has the power to perform. If it engages in an act without the power to do so, the act is said to be *ultra vires,* that is, "beyond the powers" of the corporation. Care should be taken to distinguish between an act of a corporation that is *ultra vires* and one that is illegal. An illegal act may be *ultra vires,* but not necessarily. That is, the act may be within the power of a corporation but in violation of a statute. Thus a corporation may enter into a contract that is within its power but violates one of the antitrust laws. On the other hand, it is possible that an act of the corporation may be both *ultra vires* and illegal.

A question often arises concerning the status of a contract entered into beyond the powers of the corporation. Under most of the modern statutes and court decisions, the corporation may not raise the defense of *ultra vires* to an executed or partially executed contract. Nor may the other party to a contract raise *ultra vires* in an action involving an executed contract. On the other hand, if the contract is wholly executory, either party may raise *ultra vires* as a defense. For the most part, the defense of *ultra vires* is successfully raised rather infrequently today. The decline of the use may be explained by several factors. In the first place, corporate charters are generally drawn so broadly today that few acts taken by a corporation are in fact *ultra vires.* Also the courts have considerably developed and expanded the doctrine of implied corporate powers. These circumstances coupled with the enactment of modern corporation statutes giving broad powers to corporations and limiting the defense of *ultra vires* have resulted in reduced reliance on the doctrine.

The Model Business Corporation Act, as well as the statutes of a number of states, provides that the defense of *ultra vires* may be raised in the following manner:

1. A stockholder of the corporation may seek an *injunction* to enjoin the corporation from doing an act or transferring or acquiring real or personal property when it is beyond the corporate power to do so.
2. The corporation, acting directly or through a legal representative or through shareholders, may assert *ultra vires* against the incumbent or former officers or directors of the corporation. If the officers of the corporation commit an *ultra vires* act in good faith, as a general rule they will not be held personally liable. On the other hand, if they knowingly disregard the limitations of the charter, they may be held liable for losses sustained by the corporation.
3. The state attorney general may assert *ultra vires* in a proceeding brought to forfeit the corporate charter or to enjoin the corporation from transacting unauthorized business. Remember that because the state grants the corporate charter, it is within its power to revoke it if the corporation is shown to flagrantly ignore its own limitations.

Most of the statutes, as well as the Model Act, also provide that unless the corporation's lack of power or capacity is asserted in a proceeding brought under one of the previously described sections, an act of a corporation is not invalid or unenforceable solely because the corporation lacked the power or capacity to take the action.

CROSS v. THE MIDTOWN CLUB, INC.
365 A.2d 1227 (Conn.App. 1976)

STAPELTON, J.

The following facts are admitted or undisputed: The plaintiff is a member in good standing of the defendant nonstock Connecticut corporation. Each of the individual defendants is a director of the corporation, and together the individual defendants constitute the entire board of directors. The certificate of incorporation sets forth that the sole purpose of the corporation is "to provide facilities for the serving of luncheon or other meals to members." Neither the certificate of incorporation nor the bylaws of the corporation contain any qualifications for membership, nor does either contain any restrictions on the luncheon guests members may bring to the club. The plaintiff sought to bring a female to lunch with him, and both he and his guest were refused seating at the luncheon facility. The plaintiff wrote twice to the president of the corporation to protest the action, but he received no reply to either letter. On three different occasions, the plaintiff submitted applications for membership on behalf of a different female, and only on the third of those occasions did the board process the application, which it then rejected. Shortly after both of the above occurrences, the board of directors conducted two separate pollings of its members, one by mail, the other by a special meeting held to vote on four alternative proposals to amending the bylaws of corporation concerning the admission of women members and guests. None of these proposed amendments to the bylaws received the required number of votes for adoption. Following that balloting, the plaintiff again wrote to the president of the corporation and asked that the directors stop interfering with his rights as a member to bring women guests to the luncheon facility and to propose women for membership. The president's reply was that "the existing bylaws, house rules and customs continue in effect, and therefore [the board] consider[s] the matter closed."

On the basis of all the evidence the court finds: that the corporation had a policy of not accepting women as members or as guests for lunch; that the application of the plaintiff's proposed female candidate for membership was denied because of her sex; and, that the plaintiff has exhausted his effective remedies within the corporation.

The plaintiff's complaint is that the corporation's board of directors has refused to admit the plaintiff's proposed candidate as a member solely because she is a female, and, likewise, that the board has refused to allow the plaintiff to bring female guests to the corporation's luncheon facility. The plaintiff claims that the corporation and its directors, in establishing those policies, have acted ultra vires in that they have exceeded the powers conferred upon them by the certificate of incorporation, the bylaws, and the state statutes regulating corporate powers, and that, in so doing, they have breached the plaintiff's rights as a member of the corporation.

With respect to the defendant's argument that no property right of the plaintiff had been violated, it has been held that equity jurisdiction is not limited to the protection of property rights and that it may be invoked for the protection of personal rights. From early common law Connecticut has recognized that a corporation is empowered to do only those things for which it was created and which are authorized by its charter, and that one member may bring suit against the corporation or its directors for ultra vires acts in violation of the contract between them, as set forth in the corporate charter.

Connecticut has codified the common-law right of a member to proceed against his corporation or its directors in the event of an ultra vires act. In fact, it has been done specifically under the Nonstock Corporation Act. General Statutes § 33–429.

No powers were given to the defendant corporation in its certificate of incorporation, only a purpose, and as a result the only incidental powers which the defendant would have under the common law are those which are necessary to effect its purpose, that being to serve lunch to its members. Since the club was not formed for the purpose of having an exclusively male luncheon club, it cannot be considered necessary to its stated purpose for the club to have the implied power at common law to exclude women members.

Under the Connecticut Nonstock Corporation Act, the corporation could have set forth in its certificate of incorporation that its purpose was to engage in any lawful activity permitted that corporation. General Statutes § 33–427(a). That was not done. Its corporate purposes were very narrowly stated to be solely for providing "facilities for the serving of luncheon or other meals to members." The certificate did not restrict the purpose to the serving of male members. Section 33–428 of the General Statutes provides that the corporate powers of a nonstock corporation are those set forth in the Nonstock Corporation Act, those specifically stated in the certificate of incorporation, neither of which includes the power to exclude women members, and the implied power to "exercise all legal powers necessary or convenient to effect any or all of the purposes stated in its certificate of incorporation. . . ."

We come, thus, to the nub of this controversy and the basic legal question raised by the facts in this case: is it necessary or convenient to the purpose for which this corporation was organized for it to exclude women members? This court concludes that it is not. While a corporation might be organized for the narrower purpose of providing a luncheon club for men only, this one was not so organized. Its stated purpose is broader and this court cannot find that it is either necessary or convenient to that purpose for its membership to be restricted to men. It should be borne in mind that this club is one of the principal luncheon clubs for business and professional people in Stamford. It is a gathering place where a great many of the civic, business, and professional affairs of the Stamford community are discussed in an atmosphere of social intercourse. Given the scope of the entry of women today into the business and professional life of the community and the changing status of women before the law and in society, it would be anomalous indeed for this court to conclude that it is either necessary or convenient to the stated purpose for which it was organized for this club to exclude women as members or guests.

While the bylaws recognize the right of a member to bring guests to the club, the exclusion of women guests is nowhere authorized and would not appear to be any more necessary and convenient to the purpose of the club than the exclusion of women members. The bylaws at present contain no restrictions against female members or guests and even if they could be interpreted as authorizing those restrictions, they would be of no validity in light of the requirement of § 33–459(a) of the General Statutes, that the bylaws must be "reasonable [and] germane to the purposes of the corporation. . . ."

The court therefore concludes that the actions and policies of the defendants in excluding women as members and guests solely on the basis of sex is ultra vires and beyond the power of the corporation and its management under its certificate of incorporation and the Nonstock Corporation Act, and in derogation of the rights of the plaintiff as a member thereof. The plaintiff is entitled to a declaratory judgment to that effect and one may enter accordingly.

Questions for Discussion

1. *Does this decision mean that in Connecticut a club restricted to members of a single sex is illegal? Why did the Court find the defendant corporation in violation of the law?*
2. *What should have been done when the corporation was initially formed if the organizers wanted to restrict membership to one sex?*

MARSILI v. PACIFIC GAS & ELECTRIC COMPANY

51 Cal.App.3d 313 (1975)

Pacific Gas & Electric Company (PG&E) made a $10,000 contribution to Citizens for San Francisco, an association that advocated the defeat of a certain proposition appearing on the ballot in the November 2, 1971, election for the City and County of San Francisco. The proposition, if adopted, would have prohibited the construction in San Francisco of any building more than seventy-two feet in height unless such construction was approved in advance by the voters. The management and directors of PG&E concluded that adoption of the proposition would adversely affect PG&E.

Three stockholders of the PG&E filed an action challenging the contribution on the ground that it was ultra vires *and therefore asking that the individual directors refund the contribution to the corporation. The lower court found against the stockholders, and they appealed.*

KANE, A.J.

Appellants contend that the contribution in question was ultra vires "because neither PG&E's articles of incorporation nor the laws of this state permits PG&E to make political donations. . . ." We disagree.

By definition adopted by appellants themselves, "ultra vires" refers to an act which is beyond the powers conferred upon a corporation by its charter or by the laws of the state of incorporation.

The parties are in agreement that the powers conferred upon a corporation include both express powers, granted by charter or statute, and implied powers to do acts reasonably necessary to carry out the express powers. In California, the express powers which a corporation enjoys include the power to "do any acts incidental to the transaction of its business . . . or expedient for the attainment of its corporate purposes." (Corp. Code, § 802, subd. (b) . . .)

The articles of PG&E are manifestly consistent with this statutory imprimatur. Thus, for example, they authorize all activities and endeavors incidental or useful to the manufacturing, buying, selling and distributing of gas and electric power, including the construction of buildings and other facilities convenient to the achievement of its corporate purposes, and the performance of "all things whatsoever that shall be necessary or proper for the full and complete execution of the purposes for which . . . (the) corporation is formed, and for the exercise and enjoyment of all its powers and franchises."

In addition to the exercise of such express powers, the generally recognized rule is that the management of a corporation, "in the absence of express restrictions, has discretionary authority to enter into contracts and transactions which may be deemed reasonably incidental to its business purposes." (6 Fletcher, Cyclopedia Corporations (1968), § 2486, pp. 314–315.) In short, "a corporation has authority to do what will legitimately tend to effectuate . . . (its) express purpose and objects . . ." (6 Fletcher, ibid., p. 317). California is in accord with this general rule also: " 'Whatever transactions are fairly incidental or auxiliary to the main business of the corporation and necessary or expedient in the protection, care and management of its property may be undertaken by the corporation and be within the scope of its corporated (sic) powers.' " (*David* v. *Pacific Studios Corp.* (1927) 84 Cal.App.611, 615, 258 P. 440, 442.)

No restriction appears in the articles of PG&E which would limit the authority of its board of directors to act upon initiative or referendum proposals affecting the affairs of the company or to engage in activities related to any other legislative or political matter in which the corporation has a legitimate concern. Furthermore, there are no statutory prohibitions in California which preclude a corporation from participating in any type of political activity. In these circumstances, the contribution by PG&E to Citizens for San Francisco was proper if it can fairly be said to fall within the express or implied powers of the corporation.

The crux of the controversy at bench, therefore, is whether a contribution toward the defeat of a local ballot proposition can ever be said to be convenient or expedient to the achievement of legitimate corporate purposes. Appellants take the flat position that in the absence of express statutory authority corporate political contributions are illegal. This contention cannot be sustained. We believe that where, as here, the board of directors reasonably concludes that the adoption of a ballot proposition would have a direct, adverse effect upon the business of the corporation, the board of directors has abundant statutory and charter authority to oppose it.

The law is clear that those to whom the management of the corporation has been entrusted are primarily responsible for judging whether a particular act or transaction is one which is helpful to the conduct of corporate affairs or expedient for the attainment of corporate purposes. Indeed, a court cannot determine that a particular transaction is beyond the powers of a corporation unless it clearly appears to be so as a matter of law.

Neither the court nor minority shareholders can substitute their judgment for that of the corporation "where its board has acted in good faith and used its best business judgment in behalf of the corporation." (*Olson* v. *Basin Oil Co.* (1955) 136 Cal.App.2d 543, 559–560 . . .)

The judgment of the Executive Committee of PG&E was not arbitrary or capricious but was based upon pertinent business considerations that were of direct and immediate concern to the corporation.

(Judgment affirmed.)

QUESTIONS FOR DISCUSSION

1. *Did PG&E have any express power to make the contribution in question?*
2. *Who did the court state had the primary responsibility for judging whether an act was one that was helpful to the conduct of corporate affairs or expedient for the attainment of corporate purposes? Under what circumstances would a court invalidate such a decision?*

41:6. TORTS AND CRIMES. (1) Liability for Torts. Corporations may be liable for torts, just as is any natural person, under the doctrine of *respondeat superior*. The liability results from the acts of its agents committed within the scope of their employment. The corporation is in the role of principal in this instance, and the rules of agency are no different when the corporation is a principal than if the principal were a natural person. The fact that the agent of the corporation has exceeded his authority makes no difference so long as the act is within the scope of his employment. Thus, suppose that Albert Agent is a delivery man for Charm Furniture Company, Inc., and that while making a delivery he negligently ignores a red light at an intersection and is involved in an accident. Although he certainly is not authorized to ignore red lights, he was acting within the scope of his employment and hence Charm Furniture would be liable for the tort.

The courts have also generally held that a corporation may also be liable if the agent injures a third person maliciously or willfully. Such acts are imputable to the corporation, and in a proper case punitive damages may be assessed against it. A fairly common example is the case of the innocent shopper who is arrested for shoplifting by mistake by the store "detective," who in the process subjects the shopper to abuse and even arrest after he should have realized the shopper's innocence.

GILLHAM v. ADMIRAL CORPORATION
523 F.2d 102 (1975)

Admiral Corporation is a manufacturer of television sets. A problem developed in one line of television sets that Admiral manufactured in that the sets had the nasty habit of igniting and catching fire. Admiral received a steady stream of complaints concerning the fires, in some of which home furnishings and dwellings were destroyed and personal injuries were incurred.

Zora Gillham purchased one of the color television sets in question and received severe burns and other serious and crippling injuries caused by a fire originating in her set. She sued Admiral and, in addition to the above facts, established that Admiral could feasibly have reduced or eliminated the fire hazard, that it had continued to market the set after it knew of the problem, and that it had failed to warn prospective purchasers or owners of the danger. The jury awarded Mrs. Gillham compensatory and punitive damages at the end of the trial. The trial court granted Admiral's motion for a judgment not withstanding the verdict and set aside the punitive damages on the ground that Mrs. Gillham did not show that the corporation acted with malice, an essential ingredient for an award of punitive damages under the applicable Ohio law. Mrs. Gillham appealed.

McCREE, J.

Admiral, although apparently conceding that its actions were reprehensible, argues that the district court was correct in refusing to permit the award of punitive damages to stand. It contends that in Ohio punitive damages may not be awarded in the absence of proof that the tortfeasor acted with actual malice, and that this record contains no evidence of malice on Admiral's part. Admiral does not suggest that punitive damages may not, as a matter of law, be awarded in products liability actions, but only that the facts of this case do not permit the implication that Admiral acted maliciously.

In *Columbus Finance, Inc.* v. *Howard,* 42 Ohio St.2d 178, 327 N.E.2d 654 (1975), the Ohio Supreme Court held that Ohio law required a finding of fraud, insult, or malice to sustain punitive damages, but that "actual malice may be inferred from conduct and surrounding circumstances."

Moreover, under Ohio law it is clear that punitive damages may be awarded against a corporation, such as Admiral. *Western Union Telegraph Co.* v. *Smith,* . . . 59 N.E. 890 (1901). In *Western Union,* a suit for wrongful cutting of plaintiff's trees, the court stated that a corporation could be subjected to punitive damages for the tortious acts of its agents within the scope of their employment in any case where a natural person acting for himself would be liable for punitive damages. *Ranells* v. *City of Cleveland,* 41 Ohio St.2d 1, 321 N.E.2d 885 (1975), is not to the contrary. It merely holds punitive damages cannot be assessed against a municipal corporation without specific statutory authorization.

Therefore the question before us is whether no reasonable person, viewing the evidence in the light most favorable to Mrs. Gillham, and drawing all reasonable inferences in her favor, could find that Admiral's conduct was so intentional, reckless, wanton, willful, or gross that an inference of malice could be drawn. We conclude that the evidence, viewed in light of this standard, was sufficient to permit a reasonable person to conclude that Admiral knew that its design posed a grave danger to the lives and property of its customers, and therefore that its failure to redesign the set or warn the public was conduct sufficiently intentional, reckless, wanton, willful, or gross to permit a reasonable inference of malice.

The record supports the conclusion that Admiral knew that its transformers might catch fire because of improper and dangerous design, and that it knew or should have known, even before the set was marketed, that the purported safety device designed to contain the anticipated fires was ineffective. Moreover, the evidence demonstrates that shortly after the set was distributed to the public, and several years before Mrs. Gillham's injury, Admiral was informed that its transformers were causing fires that the safety device was not containing. In spite of this knowledge, and its additional knowledge that its sets were catching fire even when not in use, Admiral neither redesigned the set nor informed purchasers and prospective purchasers of the hazard. It not only continued to distribute the sets, but also misled customers by informing them that the sets were safe.

Accordingly, we conclude that the district court erred in granting the judgment n.o.v. The district court, although concluding that Admiral admittedly was "grossly negligent," believed that its conduct did not show malice. The district court, however, does not sit as trier of fact de novo; its function in deciding this motion for judgment n.o.v. was limited to determining whether a reasonable person could arrive at the conclusion agreed upon by the jury. We hold that the jury could reasonably draw the requisite inference of malice from Admiral's conduct. Therefore its award of punitive damages and attorneys' fees was proper under Ohio law.

(Reversed and remanded with instructions to enter judgment on the verdict.)

QUESTION FOR DISCUSSION

1. *How do you go about showing that a corporation acted with malice toward an individual, as was necessary in this case?*

---

(2) **Liability for Crimes.** A corporation may commit and be charged for certain types of crimes. A corporation obviously cannot be guilty of certain crimes, such as murder, nor can it be imprisoned or executed in the gas chamber. There are many types of crimes, however, for which the punishment is a fine that the legislatures intend to be applicable to corporations. These are usually crimes in which criminal intent is not an element but in which the mere doing of the act or failure to perform the act constitutes the crime. Many federal and state statutes provide that corporations must make reports concerning certain activities. Failure to do so may result in a fine. The types of statutes of which corporations run afoul are the federal and state antitrust laws, workmen's compensation acts, certain tax laws, the Social Security Act, the security acts, and more recently the campaign-contribution laws.

UNITED STATES v. HOUGLAND BARGE LINE, INC.
387 F.Supp. 1110 (W.D. Pa. 1974)

Hougland Barge Line, Inc., was charged by the federal government for violating the Water Pollution Control Act Amendments of 1972 in that it failed to notify the United States Coast Guard of a discharge of oil from its vessels. The statute requires "any person in charge of a vessel" *to notify the Coast Guard of the discharge and provides for a fine of up to $10,000 or imprisonment for up to one year or both for failure to do so. Hougland Barge Line filed a motion with the court to dismiss the charge filed against it by the United States.*

ROSENBERG, D.J.

The defendant argues that a corporation is not includable in the term "person in charge" as such term is used in the pertinent section of the Act because first,

1. In the first sentence of the section, the word "he" is used rather than the word "it."
2. That the second sentence authorizes imprisonment of the "person in charge" and a corporate body cannot be imprisoned.
3. The last sentence authorizes immunity to the person in charge, except in cases of perjury or giving a false statement, and such acts are acts of individuals rather than corporations.
4. That only an individual can be present at the time of a discharge of oil, whereas a corporation cannot be present. . . .

Even if the statute had not defined "person" as including corporations, there is every likelihood that such an interpretation would be made by the courts where the sense and purpose of the statute plainly indicates it. In forbidding monopolies and price fixing, Congress stated in 15 U.S.C. § 1 that "Every person who shall make any contract or engage in any combination or conspiracy declared by sections 1 to 7 of this title to be illegal shall be deemed guilty of a misdemeanor, and, on conviction thereof, shall be punished by fine not exceeding fifty thousand dollars, or by imprisonment not exceeding one year, or by both said punishments in the discretion of the court." This was held time and again to include corporations.

The defendant also argues that as a corporation it cannot be imprisoned, and therefore, this would indicate that it was not intended to apply to corporations. Innumerable federal penal statutes prohibit certain activities, including business entities, and provide penalties for violation of such prohibited acts. Both individuals and corporations are penalized even though a corporation may not be imprisoned. Thus, as illustrated by antitrust cases and Internal Revenue cases, where a statute calls for imprisonment, when imposed against a defendant corporation, only the fine portion of the penalty may be imposed.

As for the defendant's argument that the last sentence of § 1321(b)(5) authorizes immunity for the "person in charge" except in cases of perjury or giving a false statement . . . are acts of individuals rather than corporations, no support is given here to this argument; for it is well known that corporations act through their officers and as such corporations may be guilty of the acts of the individual officers, where such violations have been perpetrated in the corporate processes.

So that where the statute provides protection for a defendant, the protection will be applied on the same basis as imprisonment and fine is translated when applied to a corporation.

In *United States* v. *Knox Coal Company,* 347 F.2d 33, C.A. 3, 1965, the indictment and conviction of a company was upheld for willfully attempting to evade and defeat a large part of taxes due and owing the United States of America. The term "person" in a statute relating to an offense against the United States or to defraud the United States or any agency thereof includes corporations. In addition, a corporation may be convicted of a crime, including a crime of knowledge and willfulness by the conduct of its agents and employees.

The defendant in this case also argues that only an individual can be present at the time of the discharge of oil. There can be no argument with this, but the word "individual" is included in the definition of "person" in § 1321(a)(7) as is the term "corporation." To give credence to the argument of the defendant that corporations cannot be present at the time of the discharge of oil would be as unreasonable as to say that a corporation could not hire individuals, manufacture and sell products, engage in business generally, buy, own and sell real estate or issue bonds and indebtedness obligations. Certainly the corporation is there upon the discharge of oil when any of its duly authorized employees function in its behalf, within the scope of their employment, and perform the duties of the corporation to the only manner in which it is possible for the corporation to act and for which it must bear responsibility.

(Motion of defendant denied.)

QUESTIONS FOR DISCUSSION

1. *How do the courts treat corporations when a statute provides for both fine and imprisonment?*
2. *Does a criminal statute specifically have to mention that it is applicable to corporations in order to be used against them? What major antitrust statute did this court cite in order to illustrate this point?*

UNITED STATES v. SHERPIX, INC.
512 F.2d 1361 (1975)

MacKINNON, J.

In a five-court indictment filed March 23, 1973, appellants, Sherpix, Inc., and its president Louis K. Sher, and seven codefendants, were charged with violating federal and District of Columbia obscenity laws in connection with the distribution and exhibition of the film "Hot Circuit."

A. Conspiracy by Sherpix: The first count of the indictment charges:

Commencing on or about December 1, 1971 and continuing to on or about November 16, 1972, within the District of Columbia and elsewhere, Louis K. Sher, defendant, and Saul Shiffrin, a co-conspirator only, and the original defendants and co-conspirators, Paul Glickler, Richard P. Lerner and Harry Brandt Booking Co., Inc. unlawfully, willfully and knowingly did conspire, combine, confederate and agree together, with each other and others, and through their use of Acorn Films, Inc., Sherpix, Inc. and Trans-Lux Theatre Corporation, to commit offenses against the United States by transporting and causing to be transported in interstate commerce by common carrier and for the purpose of sale, distribution and presentment, an obscene, lewd, lascivious and filthy motion picture. . . .

At trial, Sherpix, Inc. (the corporation) moved to dismiss this count as to it on the ground that the indictment did not assert that Sherpix had conspired with anyone and hence did not charge it with the commission of any crime. However, the trial court denied the motion, and Sherpix was convicted and sentenced under this count.

Both at trial and on appeal, the Government argued that the conspiracy count could be sustained against Sherpix on the grounds that a corporation is criminally responsible for the acts of its officers and thus can be charged with their conspiracies. Since the corporation's president, Sher, was included in the list of conspirators, the Government asserted that this was sufficient to hold the corporation liable. However, this argument is beside the point. The issue here is not whether Sherpix can be found guilty of conspiracy but whether the allegations of the indictment are sufficient to charge Sherpix with an offense.

It is fundamental that the indictment must inform an accused of the nature of the charges against him, and it is a necessary corollary that the indictment must also give the accused notice that he is in fact being accused. The factual allegations in the instant indictment do not measure up to these standards. They fail to charge Sherpix, Inc., with the essential element of a conspiracy, i.e., the agreement with another person to commit a crime. Merely listing the corporation as an instrumentality used by an officer in the conspiracy is not sufficient to establish that the grand jury charged the corporation as a co-conspirator. Individuals or corporations used without their knowledge or consent by others in the commission of offenses may not possess the requisite criminal intent to be guilty of an offense. That the grand jury did not intend to charge Sherpix, Inc. in Count 1 is self-evident from the failure to name Sherpix as a party defendant or to allege that Sherpix "did conspire, combine, cooperate and agree with" any other person to commit offenses against the United States.

We therefore hold that the allegations of the first count are insufficient to charge Sherpix, Inc., with the crime of conspiracy. On a retrial appellant Sherpix, Inc. cannot be tried as a defendant in the conspiracy count as that count is presently charged. We do not express any opinion as to whether a new indictment may be in order.

(Reversed and remanded.)

QUESTIONS FOR DISCUSSION

1. *Did this court hold that a corporation could not be charged and convicted of the crime of conspiracy?*
2. *For what reason did the court hold that the first count of the indictment must be dismissed against Sherpix, Inc.?*

SUMMARY

Corporate powers may be categorized as express, implied, inherent or incidental, and statutory. Express powers are found in the articles of incorporation or corporate charter, frequently in the purpose clause. Implied powers are those powers that are reasonably necessary and appropriate to carry out the express powers and purposes of the corporation. Inherent powers are those powers held by the courts to be essential to the existence of any corporation. Statutory powers are those powers, listed by the corporation statutes, that are conferred upon all corporations unless limited by the corporate charter.

A corporation that engages in an act beyond its corporate powers is said to act *ultra vires.* A corporation generally may not raise the defense of *ultra vires* to an executed or partially executed contract. The defense may be asserted by the state attorney general in a proceeding to forfeit the corporate charter or to obtain an injunction to prevent the corporation from engaging in unauthorized business. A stockholder may use *ultra vires* to prevent the transfer or acquisition of property by the corporation. The corporation may assert *ultra vires* against officers or directors of the corporation.

Corporations may be liable for torts under the doctrine of respondeat superior. A corporation may also be liable for certain crimes, usually those where criminal intent is not an element.

PROBLEMS

1. Free For All Missionary Baptist Church, Inc., leased from Southeastern Beverage and Ice Equipment Company, Inc., certain liquor dispensary equipment for use in an establishment known as "Soul on Top of Peachtree." After making the initial payment, the church defaulted on the monthly rental payments. Southeastern sued the church corporation for breach of the lease. The church defended the suit contending that the lease of the equipment was *ultra vires* and that it was therefore not liable. Is the church correct? Free For All Missionary Baptist Church, Inc. v. Southeastern Beverage & Ice Equipment Company, Inc., 218 S.E.2d 169 (Ga. App. 1975).
2. Rolfe Armored Truck Service, Inc., was charged with violating a Florida criminal statute against maintaining a gambling room, house, apparatus, or other similar place. The penalty for violating the statute was imprisonment for up to three years or a fine not exceeding $5,000. Howard Losey and the corporation sought a writ of prohibition in order to stop the court from trying the case on the ground that the corporation was not a natural person and therefore not subject to the state gambling laws. Should the writ have been granted? State v. Willard, 54 So.2d 183 (Fla. 1951).
3. Bell Houses Ltd. was an English corporation. Its charter stated that it was in the general, civil, and engineering contracting business and that in addition it could acquire real estate and other forms of property. The charter also provided that the corporation could carry on "any other trade or business whatsoever which can, in the opinion of the board of directors be advantageously carried on by the company in connection with . . . the general business of the company." Bell Houses loaned £1 million to City Wall Properties Ltd. for "bridging finance." City Wall agreed to pay a commission of £20,000 for the loan. Later City Wall refused to pay the commission and, when sued by Bell, contended that the loan was *ultra vires* in that Bell was by its charter an engineering and construction company and that this transaction constituted "mortgage brokering." Bell contended that the loan was within the implied powers of the corporation. Was Bell correct? Bell Houses Ltd. v. City Wall Properties Ltd., 2 W.L.R. 1323 (C.A. 1966).
4. San Antonio Public Service Co. was incorporated for the purpose of manufacturing, supplying, and selling electricity and gas for light, heat, and power to the public and municipalities. The company opened a sales room and hired sales personnel to sell gas and electric appliances. The State of Texas filed suit to enjoin San Antonio from such activity and contended that San Antonio's charter did not provide for such activity. Should San Antonio have been enjoined from this activity? State v. San Antonio Public Service Co., 69, S.W.2d 38 (1934).
5. The Silverbrook Cemetery Co. owned a cemetery of some sixty acres on which it had an office building, a garage, and a crematorium. Part of the cemetery was set aside for the use of stone monuments and another part for flat bronze memorials. Silverbrook never sold any stone monuments

but began to sell the bronze markers to customers who purchased burial lots in the cemetery. The Wilmington Memorial Company had been in the monument business for some time. It contended that Silverbrook did not have the power to sell the bronze memorials. Silverbrook's charter did not expressly provide for the sale of markers but did provide that the company could "purchase . . . personal estate of every kind and the same to . . . sell . . ." and "to enter into any and all contracts . . . proper to be in the conduct of its business. . . ." Was Silverbrook entitled to sell the memorials? Wilmington Memorial Co. v. Silverbrook Cemetery Co., 287 A.2d 405 (Del. Ch. 1972).

6. Southern California Edison Co. was a corporation in the business of manufacturing and distributing electricity. J. A. Lighthipe was an employee whose automobile needed repairs. Another employee of the company, Russo, was ordered by his superior to go to Lighthipe's home and bring Lighthipe's car to the company shop for repairs. While Russo was in the process of towing the car, he injured Caleb Chamberlain as the result of his, Russo's, negligence. Chamberlain sued Southern California Edison for his injuries. The company defended on the ground that repairing autos was not one of its purposes and that therefore the repairing of Lighthipe's car was *ultra vires,* hence it should not be liable to Chamberlain. Was Southern Edison correct? Chamberlain v. Southern California Edison Co., 140 P. 25 (1914).

CHAPTER 42

The Financial Structure of Corporations

One of the most important aspects of the corporate form is its financial structure. The methods of raising capital available to the corporation constitute one of the major advantages of the corporate form of organization. The purpose of this chapter is to examine the financial structure of corporations and some of the various methods for financing them. The method of financing most appropriate to an individual corporate situation is beyond the scope of this text and should be covered in any basic finance course. One objective here is to obtain a clear understanding as to the nature of different types of stocks and bonds as well as some of the legal restrictions of their use. The discussion in this chapter concerns corporations in general. Some of the methods of financing may not be appropriate to all corporations. For example, the corporation involved in a small plumbing business may not ever be concerned with the sale of a large issue of bonds in order to finance the operation of its business.

42:1. Corporate Financing. There are two primary ways to finance the corporation: by debt financing and by the sale of ownership interests in the corporation. Debt financing involves the borrowing of money from various financing institutions, such as banks, and the issuance of corporate bonds. The sale of an ownership interest in the corporation involves the sale of corporate stock.

(1) **Bonds.** One method of long-term financing for a corporation is the issuance of bonds in return for loans by the individuals to whom the bonds are issued. This is a form of debt. The typical corporate bond is a written agreement whereby the corporation promises to repay a specific amount of money at a certain date in the future known as the *maturity date.* The agreement also specifies a certain rate of interest to be paid by the corporation until the maturity date. There is also another party involved in this picture known as the *trustee.* Usually the bond may provide for the debt to be secured by certain property of the corporation. A deed of trust is drawn up under which the trustee gets legal title to the property put up as security by the corporation. The lenders or bondholders are the beneficiaries of the deed of trust. Under the deed of trust the property may be sold by the trustee in

the event that the corporation defaults and does not pay the interest on the bond when it comes due or the principal on the maturity date. The trust agreement, which is entered into by the trustee on behalf of the bondholders and the borrowing corporation, is sometimes known as an *indenture.* Usually the provisions of the indenture are printed on the bond itself.

The terms of the agreement may be many and varied, depending to some extent upon the financial strength of the corporation. The bonds may be secured as has been described, or they may be unsecured by any specific property of the corporation, in which case they are known as *debenture* bonds. In this case, the bondholders are unsecured creditors of the corporation. Frequently also the corporation will have a right to *redeem* the bonds, that is, to pay them off prior to the maturity date. Another frequent provision is to allow the bondholder to convert the bond into an ownership interest in the business by obtaining shares of common stock in exchange for the bond. A bond with such a provision is known as a *convertible bond.* Some bonds require the corporate borrower to establish a *sinking fund,* that is, to make periodic cash payments to the trustee toward the retirement of the bonds.

The subject of issuing bonds as a method of financing long-term debt is far too complicated to cover within the scope of this text. Suffice it to say that the terms of the bond are the subject of bargaining between the parties. A bond is a type of debt. The true bondholder has no ownership interest in the corporation. Another method of financing for the corporation is to sell an ownership interest in the corporation, a subject to which we now turn.

(2) **Stock.** Any form of business organization obtains a certain amount of its financing from the owners of the business. In the case of the corporation, this is done by the sale of corporate stock to those who wish to become owners. Those who purchase shares of stock in the corporation are the owners of the corporation are referred to as either *shareholders* or *stockholders.* The shares purchased by the shareholders are evidenced by a *stock certificate.* It is important to note that the stock certificate is only evidence of the shares of the corporation that the certificate represents. The certificate is not the share.

The money or other property given by the shareholders in return for the stock issued by the corporation is frequently referred to as part of the *equity structure* of the corporation. Profits retained by the firm, known as *retained earnings,* constitute the other aspect of the equity structure of the corporation. Sometimes the term *capital* is used to refer to this concept. There are problems involved in the use of the term *capital,* however, for it has many meanings. In some states it refers to the amount paid in originally by the stockholders.

GOLDEN v. OAHE ENTERPRISES, INC.

240 N.W.2d 102 (S.D.) 1976)

Warren Golden, pursuant to an agreement, transferred to Oahe Enterprises, Inc., farming and ranching equipment on January 6, 1967. In exchange Golden was to receive an amount of stock in the corporation. He subsequently became a director and the secretary–treasurer. As a result of a number of disputes as to the number of shares to which Golden was entitled, as well as the value of the stock, a certificate was never issued to Golden. After a period of time and during the dispute between Golden, Oahe, and its president, Donald Emmick, an agreement was reached between Oahe and one Charles Cannon for the purchase by Cannon of real estate owned by Oahe. This real estate was the primary corporate asset. On May 20, 1974, Golden filed an action against Oahe, Emmick, and Cannon (there had been many others) requesting a judgment setting aside the sale of the corporate assets to Cannon, among other things. From an adverse decision, Golden appealed.

WILDS, J.

Respondents have advanced a three-pronged argument in support of their contention that Golden was not a shareholder at the time of the sale of corporate assets to Cannon. They maintain that he was precluded from enjoying shareholder status because the stock certificate was never delivered to him, the stock was not paid for and Golden was never considered to be a stockholder.

It is statutorily provided that ownership of stock is represented by stock certificates. It is well established, however, that such a certificate is mere evidence of ownership. Issuance and delivery of the stock certificate are not essential to ownership. Thus, the fact that Golden did not have possession of a stock certificate is insufficient to defeat his claim to shareholder status.

Once corporate stock is paid for, it is issued regardless of whether a stock certificate is executed and delivered. It is obvious from the corporate records and the findings of then Circuit Court Judge Winans that Warren Golden paid for stock in Oahe Enterprises, Inc., on January 6, 1967, when he executed a bill of sale and transferred equipment to the corporation. The records reflect entry of this same equipment as corporate assets on January 4, 1967.

Respondents' contention that Golden was never considered a stockholder cannot be reconciled with the corporation's bylaws regulating the creation of shareholder status. The relevant bylaws are:

> ARTICLE II
>
> SECTION 6: . . . The original stock transfer books shall be prima facie evidence as to who are the shareholders entitled to examine (the list of stockholders) or transfer books or to vote at any meeting of shareholders.
>
> ARTICLE VI
>
> SECTION 2: . . . The person in whose name shares stand on the books of the corporation shall be deemed by the corporation to be the owner thereof for all purposes.

These bylaws form a binding contract between the corporation and its stockholders. In addition to the corporation and its stockholders, the bylaws are binding upon the directors and third persons who have knowledge of the same and have been brought into privity with them. Certainly the bylaws reflect the purposes and intentions of the incorporators. Application of the above bylaws and principles of law compel the conclusion that Warren Golden was a stockholder at the time of the sale of corporate assets to Cannon. . . .

42:2. Other Forms of Financing. There obviously are forms of financing the corporation other than the sale of stocks or bonds. For instance, a corporation may take out a simple long- or short-term loan with a financing institution. That loan may be secured or unsecured, and the possible terms are limited only by the imagination of the businessman. Financing may also be accomplished by the sale of corporate assets, such as accounts receivable. Finally, the corporation may enter into leasing arrangements rather than the outright purchase of its assets. These various forms of financing may be very complicated and are beyond the scope of our discussion here.

42:3. Nature and Types of Corporate Stock. There are many types of stock that may be issued by the corporation. The stock may be categorized with regard to the stockholders' right to vote, to participate in the division of the assets of the corporation upon dissolution, and to share in dividends. One broad category depends upon whether the stock gives the owner a preference as to the assets of the corporation, which is *preferred stock,* or whether the stock is that of the residual owners of the corporation, or *common stock.* The amount of stock of various types that a corporation may issue depends upon the corporate charter. The charter will state the types and numbers of shares that the corporation may issue. This is known as the *authorized stock.* The corporation may not necessarily sell all the stock that is authorized. These shares are known as *authorized and unissued shares* of stock. Those shares that have been sold are known as *issued* or *outstanding shares.* The corporation may not issue more shares than authorized unless it amends the corporate charter.

If the stock of a corporation has not been sold, it does not constitute an asset of the corporation for the simple reason that the corporation has received nothing in return for the stock. Once the stockholder purchases a share of stock, he is an owner of the corporation, but it is important to note that he does not own any of the corporation's assets. For instance, suppose Sammy Stockholder purchases ten shares of the stock of Fastgrow, Inc. The corporation is the owner of five delivery trucks. Sammy Stockholder has no direct interest in those trucks; he has only an ownership interest in Fastgrow, Inc., which in turn owns the trucks.

(1) **Common Stock.** Common stock carries with it no extraordinary rights to participate in the assets of the corporation upon dissolution. The owners of common stock are the residual owners of the corporation and usually have the largest voice in the management of the corporation. Although the owners of common stock have no special rights or privileges, they also stand to make the greatest gain, if the corporation succeeds, by the distribution of the profits of the corporation through corporate dividends (to be discussed in Chapter 43). On the other hand, if the corporation fails, the common stockholders are the last in line to participate in the dissolution of corporate assets. The various types of creditors and preferred stockholders precede them.

It is possible to have more than one class of common stock. For example, there may be a desire to concentrate control of the corporation in a small number of owners. In order for this to be accomplished, the common stock may be divided into more than one class of common stock: voting and nonvoting stock. There may also be more than one class with regard to participation in dividends paid on the common stock. The small corporation, however, usually has only one class of common stock. In that case, all the stockholders usually have one vote for each share of stock owned and participate equally in dividends paid on the individual shares of stock.

(2) **Preferred Stock.** Sometimes it is necessary for the corporation to grant certain privileges with the sale of a stock in order to induce people to purchase it. Stock may therefore be sold as *preferred stock;* that is, it has claims and rights ahead of the common stock. These preferences may be with regard to the right to participate in the earnings of the corporation or with regard

to the right to participate in the assets of the corporation upon dissolution or a combination of these.

The provisions that may accompany the issue of preferred stock are limited only by the ingenuity of the parties participating in its issue and are largely dictated by economic circumstances. For example, if the stock is preferred with regard to assets and earnings, there may be a provision requiring consent of the preferred stockholders before any debt can be incurred that would have a claim on the corporate assets prior to the claim of the preferred stockholders. On the other hand, preferred stock is frequently *nonvoting* stock, thus leaving control of the corporation to the common stockholders. Another provision for the protection of the preferred stockholders may be to require retained earnings to reach a certain level before any dividends are paid on the common stock.

Other provisions may be included in an issue of preferred stock. The preferred stock may be *participating* or *nonparticipating*. If it is participating preferred stock, then the preferred stock shares in the income with the owners of the common stock. If a dividend is paid to the owners of the common stock, it is also paid to the owners of the preferred stock. For example, the preferred stock may stipulate that a certain cash dividend is to be paid each year, say, $5 per share. There may also be a provision that if the common shareholders receive a dividend in excess of this amount, then the preferred shareholders participate in the distribution of any dividends in excess of $5. There may be many variations on this arrangement, or the preferred stock may not be participating stock at all. Most preferred-stock issues are nonparticipating.

Another provision that may be found in many issues of preferred stock is that it is *cumulative*. This means that if a dividend is not paid as required in a certain year, then that amount is carried forward to the next year. Past and current dividends then must be paid to the preferred shareholders before any dividends can be paid to the common shareholders.

As with bonds, the preferred stock may have a *convertible* feature, which allows the preferred shareholders to convert their shares into common stock. The advantage of this provision is that the preferred stockholders may benefit from an increase in the market price of the common shares if the corporation is successful. This provision has become more popular in recent years.

Preferred-stock issues are often *callable* or *redeemable*. When such a provision is included, it means that the corporation may purchase the stock from the preferred shareholder. The advantage of this type of provision, if the corporation is successful, is that it grants the corporation greater flexibility by eliminating the rigid provisions of the preferred shares. Redeemability must be specifically provided for in order for the corporation to redeem preferred stock.

You should be careful to remember that preferred stock is not a debt of the corporation, although it has some features that make it similar to bonds. Preferred shareholders are not creditors of the corporation unless a dividend has been declared by the board of directors, even if there is a cumulative provision with regard to dividends. Bondholders are creditors, on the other hand, and if the interest is not paid on the bonds when due, bondholders may enforce their rights just as any other creditor may. The sale of preferred stock is an equity form of financing, whereas bonds constitute debt financing.

(3) **Par and No-Par Stock.** The corporate stock may in most cases be set at a specific par value, or it may be no-par stock. *Par value* is an arbitrary value of the stock placed in the corporate charter. It has no relationship to the *market value,* that is, the value at which the stock is actually sold. The original reason for the concept of par value was to protect creditors by assuring that at least a certain value would be received by the corporation in return for the stock. Thus suppose that the corporate charter authorizes 1,000 shares of common stock at $10 par. Each share must be sold for at least $10. We will see in Chapter 43 that if this amount is not actually paid, then the shareholder may be liable to the creditors for the difference. If the stock actually sells for more than the par value, the difference is reflected on the balance sheet. Thus if all the shares are sold at $15 per share, the value of the common stock would be $10,000 with a *paid-in surplus* (the amount paid in excess of par) of $5,000.

There has been some criticism of the use of par value. For one thing some contend that naïve investors may not distinguish between the par value and the market value of stock and thus may be misled. Beginning with New York State in 1912, corporations have been authorized to issue *no-par* stock. When the stock is no-par the board of directors assigns a "stated value" to the stock. This value may be the price at which the stock is sold or some other value. For example, 5,000 shares of stock may be sold for $5 per share. The board of directors may assign a stated value to the stock of $3 per share. Thus the balance sheet of the corporation would reflect the *stated value* of the stock as $15,000 with a paid-in surplus of $10,000.

There are some disadvantages to no-par stock. For instance, some states levy corporate taxes on no-par stock as if it has $100 par value. This assessment can be avoided, of course, by the assignment of an extremely low par value to the stock, such as $1 per share. On the other hand, there are some definite advantages to no-par stock. It avoids the alleged deception of investors already referred to and avoids the problem of deficiency assessments. It also allows the company to market the shares at any price without regard to the artificial par value. Also some states now tax stock transfers at the market value of the stock, thus reducing the advantage of low-par as opposed to no-par stock in those jurisdictions where this is the law.

When the stock of the corporation is sold, it becomes part of the equity structure of the corporation. The amount of money paid by the shareholders at the stated value or at par constitutes a fund that is generally considered to be reserved for creditors and, as we shall see, generally may not be touched by the directors for the payment of dividends. Those amounts paid in excess of the stated or par value constitute what is known as *paid-in surplus.* Over a period of time the corporation may also show a profit that is not returned to the shareholders by way of dividends. This fund is known as *retained earnings*. It is generally out of this fund that dividends may be paid at a later date.

G. LOEWUS & CO. v. HIGHLAND QUEEN PACKING CO.

125 N.J.Eq. 534, 6 A.2d 545 (1939)

BIGELOW, J.

The receiver of the defendant Highland Queen Packing Company, an insolvent New Jersey corporation, prays that its three stockholders be assessed a sum sufficient to pay creditors and administration expenses, on the theory that their stock is not fully paid.

The stock is without par value, issued pursuant to R.S. 14:8–6, N.J.S.A. 14:8–6, "Every corporation organized under this title may issue and may sell its authorized shares without nominal or par value, from time to time, for such consideration as may be prescribed in the certificate of incorporation, or, is so provided in the certificate of incorporation, as from time to time may be fixed by the board of directors. . . .

"Any and all shares without nominal or par value issued as permitted by this article shall be deemed fully paid and nonassessable, and the holder of such shares shall not be liable to the corporation or its creditors in respect thereof."

Defendant's certificate of incorporation, pursuant to the statute, authorizes the board of directors to fix the price of the stock. The directors, at their organization meeting, received and accepted an offer from two of the respondents, Jesse B. Triplett and Boice E. Triplett, to sell to the corporation the business then being conducted by them in consideration of the assumption of debts, and especially a note to the third respondent, Edgar H. Lackey, for $950 and for the further consideration of 300 shares of capital stock to be divided among the three respondents: "It is understood that the said shares of stock shall be issued at the price of $20 per share and representing a total value of $6,000."

The assets and good will of the business turned over by the Tripletts to the corporation were worth only $1,500, so it is alleged. The receiver takes the position that the consideration for the stock fixed by the directors was $20 per share, or 1 total of $6,000, of which only $1,500 or $1,500 plus $1,050, has been paid, and that there is owing by the stockholders the difference or so much thereof as may be necessary to satisfy creditors.

The statutory plan on which stockholders' liability depends is found principally in three sections of our corporation act, which provides that where the capital shall not have been paid in full, and the capital paid shall be insufficient to satisfy debts, each stockholder shall be bound to pay the sum necessary to complete the amount of each share held by him or such proportion thereof as shall be required to satisfy the creditors. In the event less than the full amount of the stock is paid and the company becomes insolvent the stockholders are liable to pay in the balance regardless of any contract or understanding which they had with the corporation for any agreement by the company to accept less than par is void as to creditors. Liability does not depend on a "holding out" to creditors.

Par value stock has a definite value, fixed by the certificate of incorporation, stated in terms of dollars, but it may be issued for money or property or services. Stock without par value is issued for a "consideration" prescribed by the certificate of incorporation or by directors or stockholders. The consideration fixed may be money or property, or anything that constitutes a good and valuable consideration.

Counsel for respondents direct attention to the provision in R.S. 14:8–6, N.J.S.A. 14:8–6, that "shares without nominal or par value issued as permitted by this article shall be deemed fully paid and non-assessable, and the holder of such shares shall not be liable to the corporation or its creditors in respect thereof." A similar provision relating to par value stock issued for property is found in R.S. 14:3–9, N.J.S.A. 14:3–9, and has been part of our statute law many years. "The stock so issued shall be fully paid stock and not liable to any further call." So when stock without par value is issued for less than the prescribed consideration, it is outside the plan of the statute, and the holder thereof, with notice, is liable for the balance of the consideration, or so much thereof as may satisfy creditors.

The question remains whether there was delivered to the corporation in exchange for the stock the full consideration as fixed by the directors. Careful examination of the minutes satisfies me that the only consideration which the Tripletts offered to give and the directors agreed to accept, was the transfer of the business conducted by the Tripletts. The directors, by accepting the offer, fixed the consideration for the stock within the intent of R.S. 14:8–13, N.J.S.A. 14:8–13. The meaning of the statement in the minutes that the stock should be issued at the price of $20 per share, or a total value of $6,000, is not clear. Certainly the parties did not intend that $6,000 should be paid in, additional to the transfer of the business, or even that the difference between $6,000 and the value of the business should be paid in. Probably the sentence has some relation to the deal with Lackey, who was paying $2,000 for a third interest in the enterprise.

The duly fixed consideration for the stock was fully satisfied and the stockholders are not assessable.

QUESTIONS FOR DISCUSSION

1. *What is the difference between par and no-par stock? What type was applicable in this case?*
2. *What value did the court find was given for the stock of Highland Queen Packing Company? Why did the court not assess the three stockholders for the stock they received in this case?*

42:4. ACQUISITION OF CORPORATE OWNERSHIP. There are a number of ways that one may acquire ownership in a corporation: by agreeing to purchase the corporate stock prior to the formation of the corporation; by agreeing to purchase the corporate stock subsequent to the formation of the corporation; and by purchasing the corporate stock from an existing shareholder of the corporation.

(1) Preincorporation Stock Subscriptions. A *stock-subscription agreement* is an agreement to purchase a stated number of shares from a corporation. Such an agreement may be entered into prior to the formation of the corporation or after the corporation is formed. When a corporate formation is contemplated, the promoters of the corporation may attempt to obtain the agreement of individuals to purchase shares of stock in the corporation by entering into a subscription agreement. Once the corporation is formed, the offer to purchase shares may be accepted by the corporation. The subscription is most often viewed as a continuing offer to purchase the stock. There is sometimes uncertainty as to when the offer to purchase stock is accepted by the corporation, however. The board of directors may, of course, pass a resolution accepting the subscription offer. In most states, however, the legal formation of the corporation and the completion of its organization cut off the right of the subscriber to withdraw his offer and thus operate as an acceptance of the offer.

One problem that often occurs is that an individual may be willing to subscribe for a number of shares of stock in a contemplated corporation because of the knowledge that a certain number of others have agreed to purchase shares, and hence the investor is reasonably sure that the success of the corporation will not be hindered by lack of adequate financial support. Should these parties subsequently revoke their subscription offer prior to the formation of the corporation, it might be unfair to other subscribers. One possible solution to this problem is to have the subscribers enter into a contract among themselves to subscribe for a certain number of shares of stock. The subscribers are then bound to each other. However, if the subscriptions are solicited from the public at large, most courts hold that the subscribers are not bound among themselves. Because of this problem a number of states have enacted statutes prohibiting the revocation of an offer to subscribe for shares of stock for a certain period of time after the offer is made. For example, the Model Business Corporation Act provides in Section 17 as follows:

> A subscription for shares of a corporation to be organized shall be irrevocable for a period of six months, unless otherwise provided by the terms of the subscription agreement or unless all of the subscribers consent to the revocation of such subscription.

Because of the problems discussed earlier the political subscriber may wish to have certain conditions included in the subscription agreement. Furthermore the courts generally will imply certain conditions in preincorporation subscription agreements. Among these conditions are that the corporation, when it is formed, will not differ materially from the corporation contemplated by the subscription and that the entire authorized share capital will be subscribed.

On the other hand, the courts generally do not look with favor upon the imposition of numerous conditions in subscription agreements for the reason that they tend to impede corporate formation. The most frequent condition contained in a subscription agreement is that the subscribers will not be bound unless a certain amount of stock is subscribed, and the courts will usually enforce such a condition. Other conditions, however, may be treated by the courts as *special terms* or *conditions subsequent* rather than as conditions precedent to the formation of the subscription agreement. The result of this legal interpretation is that although the subscriber may be entitled to damages for failure to comply with the special term, the subscription agreement is still enforceable and the subscriber must still purchase the stock. Some jurisdictions prohibit conditional subscriptions by statute. We shall see in Chapter 43 that the failure of a stock subscriber to honor the agreement will subject him to liability to the corporation and to creditors of the corporation in the event of corporate insolvency.

BIELINSKI v. MILLER
382 A.2d 357 (N.H. 1978)

Richard J. Bielinski and Richard A. Miller organized a corporation known as Windham Enterprises, Inc. At the first meeting of incorporators, it was voted that 100 shares of stock should be issued to Bielinski, and 100 shares to Richard A. Miller. Each was to pay $1,000.00 in cash.

During the course of the business, Bielinski and Miller applied for a Small Business Administration loan. In the application they stated that each owned 50 percent of the corporation.

On July 2, 1975, at a meeting of the board of directors, Bielinski, due to illness, resigned both as member of the board of directors and president of the corporation. At that meeting the remaining board members subsequently voted to "authorize one (1) share of stock to Joan A. Miller," Richard's wife. The Millers claimed that this vote was confirmed on October 14, 1975, at the annual meeting of stockholders and Directors. Bielinski did not attend and stated the meeting was invalid because he was not given proper notice.

Bielinski then filed suit against the Millers and the corporation. The suit was based upon a New Hampshire statute that provided for relief in the event that there was a deadlock in a closely held corporation having two shareholders each owning 50 percent of the voting stock. The Millers contended Bielinski lacked standing to sue since he was not a shareholder because he had not paid for the corporate stock and no stock certificate had been issued to him.

BOIS, J.

. . . We do find merit, however, in plaintiff's argument that he is entitled to relief under the statute because he is a *subscriber* to fifty percent of the stock of the corporation. "A subscriber is one who subscribes for shares in a corporation, whether before or after incorporation." H. Henn, Law of Corporations § 115, 1 (2d ed. 1970). When plaintiff and Richard Miller each agreed to take, and pay $1,000 for, 100 shares of no par stock, they expressed a clear intent to become subscribers to the corporation and their subsequent actions were consistent with this intent. An agreement to subscribe to a corporation's stock need not be embodied in any particular form or instrument; in fact, the agreement need not even be in writing. Here the subscription agreement was in fact set out in the minutes of the first meeting of the incorporators and was entered into before the corporation formally came into existence. Preincorporation subscriptions are sometimes treated as mere offers, revocable until accepted by the corporation and although "usually formally accepted at the first meeting of the board of directors[,] . . . they can be impliedly accepted." H. Henn, Law of Corporations § 115, 9 (2d ed. 1970). . . . Assuming that the plaintiff's subscription is treated as a mere offer, the corporation's conduct here constituted an implied acceptance of the offer. Additionally, defendants' counsel himself, in argument, agreed that the plaintiff was a subscriber to the stock of Windham Enterprises.

By virtue of a valid and binding subscription, a subscriber is entitled to the rights and privileges of a stockholder even though he has not yet paid for his shares under the subscription agreement. The plaintiff (as well as the defendant) may remain liable to the corporation under the agreement for the payment of the stock.

We therefore hold that, where plaintiff subscribed to fifty percent of the stock that was to be issued by the corporation, he has standing to seek relief under RSA 294:80–a (Supp.1975), provided that he is one of two *equal shareholders* and a "deadlock" exists.

The record contains evidence that one share of stock was authorized to be issued to Joan Miller. If such was done in compliance with the corporate bylaws and all relevant statutory or other requirements, plaintiff may not be a shareholder entitled to relief under the statute. The master made no finding in this regard; we therefore remand for a determination of whether the purported issuance of stock to Joan Miller effectively deprived plaintiff of his fifty percent ownership. If plaintiff is found to be an equal shareholder the master must then determine whether a "deadlock" exists in order to permit granting of the petition.

(Remanded.)

QUESTIONS FOR DISCUSSION

1. *As a result of this decision is Bielinski liable to the corporation for anything?*
2. *Does Bielinski owe the corporation anything if on remand the court finds that Joan Miller owns one share of stock?*

---

(2) Postincorporation Stock Subscriptions. One may enter into a contract for the purchase of shares of stock after the corporation comes into being. The courts of some states distinguish between stock subscriptions in this case and what they term a *purchase and sale.* In those states where the distinction is made, it is the intent of the contract as to when the rights of a stockholder are conveyed to the subscriber that is important. In the South Dakota case of Boroseptic Chemical Co. v. Nelson, 221 N.W. 264 (1928), the court explained the distinction between a purchase and sale and a stock subscription in the following manner:

> If the contract is such as to convey the present rights of a stockholder, it is a subscription in the strict sense. If the contract indicates that it is the intent of the parties that title to the stock and the rights of the stockholder shall not pass until some future time, it is to be construed as an executory contract. . . .
>
> The provision that the stock shall be issued upon completion of payment therefore, at a future date, negatives the intent to have title pass on the making of the contract.

Typically in the purchase and sale the issuance of the shares is conditioned on full payment for them. The purchaser is not a stockholder until the stock certificate is delivered to him. Therefore, in the states that make the distinction between a purchase and sale and a stock subscription, the bankruptcy of the corporation relieves the purchaser of stock of any liability for any balance due for the stock even against creditors' claims because he is not considered a stockholder. If, on the other hand, the agreement is a straight stock-subscription, the stockholder is liable for any unpaid balance due on the stock subscription.

Other states do not make this distinction. In these states any agreement whereby a shareholder agrees to take shares and pay for them should be construed as a stock subscription regardless of whether the agreement is a preincorporation or a postincorporation agreement. In this view the stockholder is liable for any balance due for the stock regardless of whether the certificate has been delivered when the corporation becomes bankrupt or not. A number of legal scholars believe this to be the preferable view on the ground that the shareholder is the equitable owner of the stock regardless of whether he or she has the stock.

SWEENEY v. BRIDAL FAIR, INC.

195 Neb. 166, 237 N.W.2d 138 (1976)

In 1971, as a result of the financial difficulty faced by Bridal Fair, Inc., a special meeting of the stockholders adopted a resolution to increase the capital stock of the corporation from $50,000 to $300,000. The stockholders further authorized 3,850 shares of additional capital stock to be issued and offered to stockholders at $13 per share in the same proportions as their existing stock ownership. The resolution also provided that if a stockholder exercised his or her right to purchase additional stock, he or she was required to pledge the stock as security for a loan to be made to the corporation and to guarantee a portion of the loan. Finally the resolution provided that the stockholders should tender payment in cash for the additional stock within thirty days or forfeit all rights thereto, in which event, the shares were to be issued to the majority stockholder.

The thirty-day period ended on June 10, 1971. Kevin B. Sweeney, John Davis, and Donald Tawzer exercised their rights by June 10, 1971, but the other stockholders failed to do so. The secretary–treasurer of the corporation received a written acceptance of the offer from Burden on June 5, 1971, but did not receive a check from Burden until June 11, 1971, when Burden tried to purchase the remaining stock. At the time of the resolution Burden was the majority stockholder. Sweeney filed suit, contending that Burden's rights were forfeited under the terms of the resolution and that therefore Sweeney should be able to purchase the stock, which would make him the majority stockholder. From an adverse decision, Sweeney appealed.

BOSLAUGH, J.

The plaintiff contends the majority stockholder should be determined as of the time the 30-day period specified in the resolution expired, and that since Burden failed to tender payment in cash within the 30-day period he forfeited his right to purchase the additional shares and the plaintiff then became the majority stockholder. The cross-appellees (Tawzer) contend the forfeiture provision in the resolution conflicted with section 21–2016, R.R.S. 1943, and was invalid.

Section 21–2016, R.R.S. 1943, provides that no penalty working a forfeiture of a subscription to shares of capital stock of a corporation may be declared unless a written demand has been made for the amount due, and the amount due remains unpaid for a period of 20 days after the demand.

A subscription contract is an agreement to purchase original, previously unissued, shares of stock in a corporation. Subscription agreements may be made before or after organization of a corporation. The acceptance of the offer to purchase additional stock to be issued as provided in the resolution constituted a subscription agreement.

The provision in the resolution relating to a forfeiture of the stockholder's rights for failure to pay the amount due within 30 days of May 11, 1971, was contrary to section 21–2016, R.R.S. 1943, and was unenforceable. Since Burden had accepted the offer before June 10, 1971, and paid the amount due on June 11, 1971, he remained the majority stockholder in the corporation.

The offer by its terms expired on June 10, 1971. The stockholders who failed to comply with the resolution by accepting the offer on or before June 10, 1971, lost their rights to purchase additional stock in accordance with the resolution.

(Affirmed.)

QUESTIONS FOR DISCUSSION

1. *Was this a preincorporation or a postincorporation subscription agreement?*
2. *For what practical reasons did Sweeney contest the validity of Burden's acceptance?*

---

(3) **Acquisition of Shares from a Shareholder.** The acquisition of corporate stock from an existing shareholder of the corporation is quite a different thing from what we have been talking about. In the first place, the corporation does not benefit financially from the share of stock transferred between stockholders. The corporation has already received the consideration for the original issue of the stock and has recorded it on its balance sheet. The result is that there is no requirement that full value, or any value for that matter, be given for the stock transferred from an existing shareholder.

The transfer of the stock from an existing shareholder is accomplished by endorsement and delivery of the certificate to the new shareholder and the recording of the transfer on the books of the corporation. In the absence of a restriction in the statutes or the bylaws or charter of a corporation or restriction by contract, corporate shares may be transferred by a stockholder to anyone of his choice. Any restrictions on stock transfers must be reasonable, or they will not be enforced by the courts.

42:5. STATE AND FEDERAL REGULATION OF SECURITY ISSUES. By now it should be clear that there are many opportunities for fraud in the sale of stock to investors and others through such techniques as "watering stock."

Because of many problems in this area, statutes have been passed at both the state and the federal level to regulate the sale of stock as well as other securities. Most states have statutes regulating the sale of securities within the state. The federal regulation of securities is administered by the Securities and Exchange Commission, which regulates not only the sale of securities that come under its jurisdiction but also the various stock exchanges. The SEC has jurisdiction over many types of securities other than stock. Basically any investment agreement whereby the debtor gives the creditor or the investor evidence of the debt or a certificate that indicates participation or financial interest in a profit-sharing agreement may be held to come within the jurisdiction of the SEC. A security may also include investment contracts, voting-trust certificates, certificates for interest in gas or other mineral rights, receipts, or other instruments commonly known as *securities*. The two major acts under which the SEC operates are the Securities Act of 1933 and the Securities Exchange Act of 1934.

(1) The Securities Act of 1933. The Securities Act of 1933 has two basic objectives. One is to provide investors with adequate financial and other information concerning the securities offered for sale. The other is to prohibit various forms of fraud and misrepresentation in the sale of securities. In order to accomplish the first purpose, the law requires that a registration statement be filed with the SEC if certain securities are to be offered for sale to the public. The company issuing the securities or a person in control of such a company must file the statement.

The registration statement requires such information as (1) a description of the registrant's properties and business; (2) a description of the significant provisions of the security offered for sale as well as the security's relationship to the registrant's other capital securities; (3) information about the management of the registrant; and (4) financial statements certified by independent public accountants.

There are certain issues that are exempt from registration. Private offerings to a limited number of people or an offering to an institution that has access to the required information and that does not intend to redistribute the securities is exempt. Also exempt are offerings restricted to the residents of the state in which the issuing company is organized and doing business. The securities of federal, state, municipal, and other governmental instrumentalities are exempt, as are those of charitable institutions, banks, and carriers subject to the Interstate Commerce Act. Another exemption is offerings that are not in excess of a certain amount (which is $1,500,000 in 1982) and that comply with certain regulations promulgated by the SEC. Offerings of "small business investment companies" made in compliance with the rules and regulations of the SEC are exempt. The rules and regulations of the SEC change from time to time, and a careful analysis of them and their applicability is beyond the scope of this text, but you should be aware of their existence.

You should note also that registration with the SEC does not mean that the security is a wise investment or even that the security has any merit.

The only purpose is to assure that there has been adequate disclosure of the company issuing the securities and of the securities themselves. If the facts stated in registration statement and prospectus turn out to be false and misleading, the party responsible is subject to fine or imprisonment or both. If an investor has suffered loss from the purchase of a registered security as a result of a false or misleading statement, the act gives him remedies against the issuing company, its responsible directors, officers, underwriters, controlling interests, sellers of the securities, and others. A *controlling person* is one who controls the issuer of the stock.

The act is rather complicated and its effect rather broad. It may be applied against numerous parties who have contact with the issue of a security and applies to issues of numerous types of securities other than stock. The provisions of the statute should be consulted by any businessman who is involved in the issue of any substantial number of securities to the public.

(2) The Securities Exchange Act of 1934. The Securities Exchange Act of 1934 created the Securities Exchange Commission and attempts to regulate the sale of securities through the various stock exchanges. The Exchange Act has numerous provisions regulating brokers and dealers in securities. It also requires the registration of certain securities before any member, broker, or dealer in securities may effect a transaction on a national exchange. There are some exemptions, primarily for certain obligations of governments. Not only must certain forms be filed with the SEC, but they also must be filed with the appropriate stock exchange.

The Exchange Act also requires the registration of certain over-the-counter securities with the SEC. Such a registration is required when the issuer of the securities has assets totaling more than $1,000,000 and a class of securities held by 500 or more but less than 750 persons. Another requirement that must be met before the act is applicable is that the securities are to be sold by interstate commerce, such as by the use of the mail or telephone calls between jurisdictions.

The Exchange Act also has certain disclosure requirements for "insider transactions." Every person who owns more than 10 percent of a security required to be registered or who is an officer or director of the issuer of such a security must file certain information with the SEC and the exchange when the security is registered, or within ten days after he becomes an officer, director, or owner of more than 10 percent of such securities. The information includes the amount of all the securities of the issuer that he owns and within ten days of the close of any calendar month thereafter any change in the ownership in that month. Furthermore anyone who owns more than 10 percent of the stock listed on a national exchange must file reports with the SEC and the exchange.

The act imposes both criminal and civil liability for failing to comply with its terms or for filing false or misleading statements of fact in the documents filed under the act. Basically this liability is based upon fraud.

HIDELL v. INTERNATIONAL DIVERSIFIED INVESTMENTS
520 F.2d 529 (1975)

In late 1970 and early 1971 Donald and Sally Rosenbloom began to form a real-estate-investment corporation. They began to solicit subscription agreements for the stock of International Diversified Investments. They prepared a prospectus and a form subscription agreement. The agreement provided that it was not to become binding unless subscriptions for 20,000 shares ($500,000) in blocks of 1,000 shares were obtained prior to September 30, 1971.

Thomas and Dorothy Hidell were approached and, after reviewing the prospectus and agreement, signed the subscription agreement. They paid $25,000 between January 21 and September 19, 1971, for the 1,000 shares. When on September 17, 1971, it became clear that the goal of twenty subscriptions would not be reached, the Rosenblooms wrote the six subscribers, asking that the condition be waived. On September 20, 1971, Donald Rosenbloom convinced the Hidells to sign an amendment waiving the condition. He informed them that the company had $190,000 in capital. However, Jack Dougherty, another subscriber, had had misgivings about the investment, and on September 29 Rosenbloom entered into a repurchase agreement for the stock on behalf of International Diversified Investments. Dougherty had required that agreement before he would consent to the proposed amendment of the subscription agreement. The Hidells learned of the repurchase agreement with Dougherty on April 14, 1972, and requested a similar one. Their request was denied.

The Hidells then filed suit against Sally and Donald Rosenbloom and International Diversified Investments, contending that they had violated the Securities Exchange Act of 1934 and the Illinois Securities Act. The district court awarded the Hidells a judgment of $25,000, and the defendents appealed.

PER CURIAM

Defendants argue that the statement in the September 17 letter and any oral statements made to the Hidells on September 20 that an investment of $190,000 had been made in I.D.I. were not false when made because the repurchase agreement was not entered into until nine days after the Hidells consented to the subscription agreement amendment. Alternatively, they contend that the repurchase agreement entered into between Donald Rosenbloom and Dougherty was not a "material fact," as the term is used in Rule 10b–5 and § 12, subd. G of the Illinois Securities Act. We disagree on both grounds.

The record discloses that Donald Rosenbloom was aware of Dougherty's reluctance to sign the subscription agreement amendment before he went to Philadelphia on September 20 to meet with the Hidells. Dougherty testified that he had had conversations with Rosenbloom about a buy-back agreement prior to his receipt of the September 17 letter. And Rosenbloom admitted that, although Dougherty had requested concessions prior to September 17, he never so advised the Hidells. Consequently when Rosenbloom convinced the Hidells to sign the agreement, by stating that an investment of $190,000 had been made, he failed to disclose the fact that one of the subscribers had indicated that he might not so agree without a repurchase agreement. The $190,000 investment statement was, therefore, seriously misleading, even if not wholly false, when made. Whether we view this as a false affirmative statement or as an omission, Rosenbloom himself had reason to believe that it was less than a completely truthful statement of the then current status of the corporation's financing.

Having informed the Hidells that $190,000 had been invested, when he knew of Dougherty's demands and the significant likelihood that a repurchase agreement might be a condition of Dougherty's assent, we believe that Donald Rosenbloom was under an affirmative duty to notify the Hidells of the repurchase agreement on or after September 29 and to afford them the opportunity to cancel their consent to the subscription agreement amendment. Similarly, Sally Rosenbloom, who signed the September 17 resolution approving the amendment to the subscription agreement, and who knew on September 29 of the repurchase agreement, was, as an officer and director of I.D.I., under the same duty of disclosure and is similarly liable under § 10(b), Rule 10b–5, and § 12 of the Illinois Act.

(Judgment affirmed.)

SUMMARY

The sale of bonds is a form of debt financing for a corporation. The debt is payable on the maturity date. If there is a trust agreement securing the bond, it is known as an *indenture.* Debenture bonds are unsecured. A convertible bond may be exchanged for corporate stock. The corporate borrower may also be required to establish a sinking fund.

Corporate stock represents the ownership interest in the corporation. Money or property given by stockholders in return for the stock issued by the corporation becomes part of the equity structure of the corporation. Retained earnings constitute another aspect of the equity structure or capital of the corporation.

The stock of the residual owners of the corporation is known as common stock. Preferred stock is issued to those owners who have a preference to the assets of the corporation. The corporate charter states the amount of authorized stock that the corporation may issue. Preferred stock may have a number of characteristics. It may be voting or nonvoting, participating or nonparticipating, cumulative, convertible, and/or callable.

Corporate stock may or may not have an arbitrary value placed on it called par value. This has no relationship to the market value of the stock. Amounts paid in excess of the stated or par value of the issued stock constitutes paid-in surplus. Profits not returned to shareholders are known as *retained earnings.*

An agreement to purchase a stated number of shares from a corporation is known as a stock-subscription agreement. The Model Business Corporation Act provides that preincorporation subscription agreements are irrevocable for six months unless the agreement provides otherwise.

Some courts distinguish between a postincorporation stock subscription and a purchase and sale. In those states where the distinction is made, the purchaser is not a stockholder until the stock is delivered to him and hence not liable for unpaid stock if the corporation goes bankrupt. Other states do not make this distinction.

Because of problems with the sale of stock and other securities, Congress passed the Securities Act of 1933 and the Securities Exchange Act of 1934. The objective of the Securities Act is to provide investors with adequate information concerning securities and to prohibit fraud and misrepresentation in their sale. The Securities Exchange Act attempts to regulate the sale of securities through the various stock exchanges.

PROBLEMS

1. Certain citizens of Schuyler, Nebraska, prepared articles of incorporation for the formation of a corporation to manufacture chicory. An agreement was signed by one Anton Lednichy, which provided that he would take five shares at $50 each, payment to be made in monthly installments of 4 percent per month. After that, Lednichy made nine payments and then refused to pay any more. Did the corporation have any rights against Lednichy? Nebraska Chicory Co. of Schuyler v. Lednichy, 113 N.W. 245 (1907).

2. The bylaws of the United Envelope Co. provided for the sale of both common and preferred stock. The preferred stock provided for the dividends to be cumulative at the rate of 7 percent per annum. Equal voting power was given to both common and preferred stockholders, and the preferred stockholders had a preference over the common in any distribution of assets upon dissolution. Subsequently a vote was passed to issue additional shares of common stock and to offer it to both the preferred and common stockholders on a proportionate basis at $150 per share, a price materially below its value. Carrie M. Stone, a common stockholder, contended that this preemptive right constituted an additional preference to preferred stockholders because it resulted in a dividend in addition to the 7 percent dividend. Because this extra dividend was not provided for in the issue of preferred stock, Stone contended that it was an illegal preference. Was she correct? Stone v. United Envelope Co., 111 A. 536 (1920).
3. Jordan was a master dental mechanic who conceived an idea to mold prosthetic devices through the use of precision castings that had previously required matching. However, furnace temperatures had to be controlled very precisely. Jordan claimed to know how to do this. Eventually the Metalmold Corporation was formed to engage in this business. Jordan subscribed for 300 shares of $100 par common stock. Other investors paid cash for their stock. Jordan agreed to disclose all his knowledge, skill, and ideas with respect to the casting process. Subsequently a certificate of organization of the corporation was filed, which included a statement that all the stock was paid for. Jordan started to write down his ideas but died before finishing. Everything he wrote could be found in the standard trade journals. The Metalmold Corporation refused to issue stock certificates to Jordan's estate, contending that his process was not new or original and that his representations as to the value of the process were untrue. The corporation contended that Jordan's offer was not "property" that could be accepted in lieu of cash for the stocks and that therefore it was justified in refusing to issue the stock. Was the corporation correct? Trotta v. Metalmold Corp., 96 A.2d 798 (1953).
4. Hart was an accountant who agreed to perform certain CPA and bookkeeping services for United Steel Industries, Inc., for one year in the future in return for the issue to Hart of 5,000 shares of corporate stock. Hart performed some services from January 1965 to May 1965. The stock was issued, however, before any services were rendered. Other stockholders sued to void the issue of stock to Hart and relied on the Texas statute that states, in part, "The consideration paid for the issuance of shares shall consist of money paid, labor done, or property actually received. . . ." Should the issue have been voided? United Steel Industries, Inc. v. Manhart, 405 S.W.2d 231 (1966).

CHAPTER 43

Shareholder Rights and Liabilities

43:1. INTRODUCTION. The owners of the corporation are the shareholders or stockholders. These are the individuals who have invested in the corporation and their ownership interest is usually, but not always, evidenced by certificates indicating the number of shares of the corporation that they own. It is important to remember that the shareholders are not *the* corporation but rather *own* the corporation. The corporation is a separate and distinct legal entity from its shareholders.

The relation between the corporation and the shareholders is basically contractual. The provisions of the contract are determined by the corporate charter and bylaws as well as by the state corporation statutes and relevant judicial decisions. The result of this relationship is that the shareholders have certain rights that cannot be modified or eliminated without their consent. The shareholders also have certain liabilities because of their position. In this chapter we first will examine some of the rights of corporate shareholders and then analyze certain of the liabilities and duties of shareholders.

43:2. WHICH SHAREHOLDERS MAY VOTE. The old common-law rule was that each corporate shareholder had one vote regardless of the number of shares of the corporation that he or she owned. Under the modern corporation statutes, however, each shareholder is entitled to exercise the number of votes equal to the number of shares that he or she owns. This, of course, gives the larger stockholder a greater proportionate voice in corporate affairs. It is possible under the corporation statutes of most states to issue shares of corporate stock that do not have voting rights by issuing different classes of stock, as provided by the corporate charter. Certain classes of stock would have no voting rights, and other classes of stock would be entitled to vote. Such a limitation must be provided for in the corporate charter and cannot be taken away in some other manner, such as by the insertion of a provision in the bylaws. The limitation on the right to vote is most often found in the issue of preferred stock (a class of stock discussed more fully in Chapter 42), which gives its holders preference to corporate assets over other stockholders in the event of corporate dissolution.

Unless the stock is of a nonvoting class, each shareholder is entitled to vote his shares at the regular and special meetings of the stockholders. The stockholders may vote on such matters as the election of corporate directors, merger with another company, and many other issues.

(1) **The Proxy.** Frequently a stockholder may not wish to attend a meeting of the shareholders for one reason or another. This is particularly true in the case of shareholders in large corporations who may not have a personal interest in the business of the corporation other than as an investment. Because the corporation statutes of all jurisdictions provide that the stockholder need not appear personally in order to vote his shares, he or she may vote by *proxy.* He or she accomplishes this by signing a written document granting some individual the power to act as agent in voting his or her shares at the stockholder's meeting. The proxy may limit the authority of the proxy holder to vote the shares in a particular manner, or it may grant the proxy holder the general authority to vote the shares, in which case he may exercise his discretion.

Perhaps the greatest use of proxies is made in the case of large corporations, which may have thousands of stockholders. The management of those corporations frequently solicit proxies from shareholders to vote at annual meetings. Obviously, if one can acquire enough proxies, he can control the corporation although having a rather small ownership interest in it himself. On occasion proxy battles may erupt, with different factions soliciting proxies from the shareholders in an effort to gain control of the corporation. Because of the potential for abuse, proxy solicitation is regulated by the Securities and Exchange Commission under the Securities Exchange Act of 1934. This statute applies to large corporations such as public-utility holding companies and those registered on the national stock exchanges. For example, in the soliciting of proxies, SEC regulations require that sufficient information be provided the stockholder so that he or she knows what is being authorized by the proxy.

(2) **Cumulative Voting.** One of the more important rights of stockholders is to elect the directors of the corporation who manage it. It should be obvious that under normal circumstances the individual or group that controls 51 percent of the corporate stock would be able to elect all the directors if the directors were to be voted upon separately. This would leave the stockholder with a minority interest no representation on the board of directors. Because of the strong views historically in this country that minority interests should be represented, the principal of *cumulative voting* for directors is included in one form or another in most of our corporation statutes. A number of states make cumulative voting mandatory, whereas others make it permissive; that is, cumulative voting may be provided for in the corporate charter. A few states have no provision for cumulative voting.

In *cumulative voting* an individual stockholder has a number of votes equal to the number of shares he or she owns multiplied by the number of directors to be elected. These votes may then be cast in any manner desired. They may be distributed equally or they may all be cast for one director. For example, suppose Sally Stockholder owns 120 shares of stock and six directors are to be elected. She would be able to cast 720 votes for one single director, or she could spread them equally among the six or some lesser number of directors. The strategy, of course, is to cast the votes in such a way as to elect the maximum number of directors. Often a number of minority shareholders may get together and decide how to vote in order to assure representation on the board.

A formula has been developed to determine how many votes are necessary to elect one director of a corporation. First, it is necessary to determine or estimate how many shares are to be voted in the election. Then the following formula may be applied:

$X =$ number of shares needed to elect a given number of directors
$Y =$ the total number of shares voting
$Z =$ the number of directors desired to elect
$N =$ the total number of directors to be elected

$$X = \frac{Y \times Z}{N + 1} + 1$$

In our example, then, assuming that six directors are to be elected and that 400 shares will be voted, in order to determine the number of shares necessary to elect one director:

$$X = \frac{400 \times 1}{6 + 1} + 1$$

$X = 58$ shares to elect one director

It is clear from the example that if Sally casts her votes properly, she can elect two of the six directors and have her interest represented on the board because she owns 120 shares and 58 are necessary to elect a single director.

X **43:4. Voting Agreements and Trusts.** Stockholders will enter into voting agreements, for one reason or another, but most often in order to attempt to control the management of the corporation. These agreements are generally upheld in the absence of fraud or an attempt to take undue advantage of other stockholders. However, some types of voting agreements have been held to be unenforceable. For example, an agreement providing for the sale of shares without the concomitant right to vote has been held to be an unlawful restriction on the right to transfer property. Apart from some problems of public policy, the problem with voting agreements is that if they are violated, the aggrieved shareholder must go to court to enforce the agreement.

Another popular method for gaining control of a corporation is the use of the *voting trust.* Its popularity is due to the fact that the voting trust is perhaps the only sure method for gaining corporate control by binding the shareholders to vote as a unit. Other methods, such as the proxy, may be revoked by the stockholder unless they contain a power coupled with an interest, a subject that was discussed in Chapter 33. A similar problem obtains with regard to shareholder agreements.

A *voting trust* may be established by having the stockholders involved transfer the legal title to their shares to the party serving as *trustee* for the purpose of voting their stock. The shareholders receive *voting-trust certificates* in return and also receive from the trustee any dividends paid on the stock. The trustee, of course, has the power to vote the stock but does not have all the rights of a shareholder. He is limited by the terms of the trust agreement. The voting trust is typically not revocable, but the statues of most states provide a limitation on the length of time of its existence. The Model Business Corporation Act states that the limit of a voting trust is ten years, but some states set it at five years, and at least one sets the limit at twenty-one years.

The voting trust may be established for any legitimate purpose, but perhaps one of the more common reasons is in order for the corporation to obtain financing. Thus as security the creditors may require a majority of the stockholders to agree to a voting trust so that the creditors receive some representation on the board of directors. The trust must be established for a legal purpose or it is invalid. Because of the potential for abuse of voting trusts, the states generally have statutes regulating them. Also the voting trust is a "security" under the Securities Act of 1933 and hence applicable in any cases regulated by the provisions of that act.

WEIL v. BERESTH
220 A.2d 456 (Conn. 1966)

Nathan Weil, Edward Beresth, Gershon Weil, and Raymond Harrison constituted the board of directors of Self Service Sales Corporation and were holders of the majority of its stock. On August 27, 1954, they entered into a written stockholders voting agreement. The agreement provided that as long as each remained a stockholder of Sales, he would vote for the election of the other parties to the agreement to the board of directors; that the bylaws would be amended to reduce the number of directors from five to four; that three would constitute a quorum; and that thereafter none of them would vote to amend the bylaws amended by the agreement unless all the other parties consented.

All went well until April 1965, when Beresth, Gershon Weil, and Harrison, by proxy, voted to adopt new bylaws not consistent with the agreement. The board of directors was expanded from four to five, and Beresth's son was added as a director. Nathan Weil objected to this action and filed suit to enforce the voting agreement. The defendants claimed that the voting agreement violated public policy and was unenforceable. The trial court held that the new bylaws were null and void. The defendants appealed.

KING, C.J.

The defendants correctly concede that a voting agreement is not, per se, invalid under the general, common-law rule. . . . Nor do they claim that the agreement here was entered into for any fraudulent or illegal purpose, even though not apparent on its face, which would deprive it of validity. Thus, the defendants do not claim that this agreement works a fraud on minority stockholders, that it confers any benefit on the plaintiff at the expense of the corporation, or that it contravenes public policy in any similar fashion. They do not claim it is anything other than a voting agreement, and they concede that, at the time of its execution, and thereafter up to and including the present time, it was not controlled by any of the statutes referred to above. Specifically, they disavow any claim that the voting agreement statute is retroactive or in any way controlling on this outstanding voting agreement.

Their sole claim of invalidity is predicated upon an alleged lack of a limit as to the duration of the agreement, which they claim is contrary to the public policy of Connecticut as manifested in the aforementioned proxy, voting trust, and voting agreement statutes. This claim is unsound for at least two reasons.

In the first place, the voting agreement under consideration here is not unlimited in duration. The obligation of each stockholder to vote for the other signatories as directors expressly exists only "so long as he is a stockholder of Sales." Thus, even if there were a public policy forbidding voting agreements of unlimited duration, it cannot be said that that policy has been violated in the instant case.

Although stockholders cannot, by contract, alter their statutory powers as stockholders, they can, as they did here, agree to limit their exercise of those powers. Since the plaintiff concedes that there was no procedural irregularity in the adoption of the new bylaws of 1965, it follows that they are valid and that the trial court was in error in holding any of them null and void. But, to the extent that the vote of April 19, 1965, adopted bylaws conflicting or inconsistent with, or amendatory of, the two bylaws in question, the defendant stockholders are in breach of their contract, and the trial court was not in error insofar as it so concluded.

The plaintiff is entitled to an injunction ordering the defendants to join with him in taking the appropriate steps promptly to call and hold a stockholders' meeting for the following purposes, and at that meeting to vote their stock to carry out those purposes: (1) to reenact the two bylaws which were originally enacted pursuant to paragraphs 2 and 3 of the agreement; (2) to repeal all bylaws or parts of bylaws conflicting or inconsistent therewith; and (3) to withdraw from the board of directors the power to amend those two provisions of the bylaws, either directly or indirectly by the adoption of conflicting or inconsistent bylaws.

(Affirmed in part and reversed in part.)

Questions for Discussion

1. *In this case could the stockholders enter into an agreement limiting the way in which they would cast their vote?*
2. *Could the stockholders agree to alter their exercise of their statutory powers? Why was the court in error in holding the new bylaws null and void?*

43:4. Right to Receive Dividends. One method by which the shareholders of a corporation reap a return on their investment is by the payment to them of dividends on the corporate stock by the corporation. As a general rule, the shareholders have no right to receive dividends from their stock until the board of directors "declares" a dividend. Whether a dividend will be declared is within the discretion of the board of directors. As a general rule, the stockholders have no absolute right to receive dividends just because the corporation is making money. On the other hand, if the shareholders can show that the board of directors has abused its discretion in withholding dividends and has acted in an arbitrary and capricious manner or for noncorporate motives, a court will require the payment of dividends. The mere showing of an accumulation of earnings, however, is not enough to force a distribution of dividends.

Dividends may be paid in cash, stocks, bonds, or other property of the corporation. The most common form of payment, of course, is cash. Stock dividends are in the form of issues of stock to existing shareholders. Because the stock dividend is issued in the same proportion to all shareholders of the same class, it does not affect the relative interests of the shareholders. It merely has the effect of capitalizing the corporate surplus. The Internal Revenue Service does not consider a stock dividend as income to shareholders.

Once the board of directors has declared a cash dividend, the shareholders become creditors of the corporation and may enforce the debt in a legal action against the corporation, just as they could any other debt. Furthermore, once the dividend has formally been declared and made public, the board of directors may not rescind or revoke it. Some cases have held that a stock dividend may be rescinded by the board of directors at any time prior to the issuance of the shares of stock.

Typically the board of directors will declare a dividend payable to the record owners of corporate stock at some date in the near future. The stockholders who own the stock on that date are then the ones who are entitled to receive the dividend. As a general rule, if the stock is transferred after the dividend is declared but prior to the record date, the transferee is entitled to receive the dividend.

We have already stated that the board of directors has the discretion to pay dividends. If the corporation is profitable, they may, for instance, decide not to pay dividends but rather to use the profits for expansion of the business. There are also limitations on when the directors may declare a dividend. The various state statutes govern the conditions under which a dividend may be paid so that creditors will not be defrauded by having all the assets of the corporation paid out to the shareholders in the form of dividends. Most of these statutes fall into three broad categories.

One group of statutes prohibits the payment of dividends if the corporation is "insolvent" or would be "rendered insolvent" by the payment of the dividend. The problem raised by these statutes is concerned with the definition of *insolvent.* Many of the states interpret insolvency in this connection as an inability to pay debts as they mature. California is an example. Under this test a corporation may have assets but be unable to pay their debts because the assets are not liquid. In this circumstance, a dividend could not be paid.

Other states, such as Maine, require that dividends may be paid only out of "net profits." Some statutes speak in terms of "earned surplus," and others mention both. Once again the problem arises with definitions. For instance, if the statute uses the term *net profits,* may a dividend be paid if the corporation earns a profit this year but had a series of losses in the past? In most states the answer appears to be in the affirmative. In some states there is authority to the contrary if the capital of the corporation is impaired. In determining the net profits, many states restrict the payment of dividends only to income. Appreciation of fixed assets may not generally be used in the determination of the amount available for the distribution of dividends.

The third type of statute prohibits the payment of dividends if the effect is to impair the stated capital of the corporation. Under this type of statute it is illegal to pay a dividend if, in doing so, the corporation reduces its net assets below the stated capital. Once again the problem is one of construction. What is meant by the term *capital?* Does it include, for example, what is frequently referred to as *capital surplus,* that is, the excess amount realized by the corporation for the sale of capital stock over its par value?

The Model Business Corporation Act provides that capital surplus is available for distribution to the shareholders provided, along with several other restrictions, that the corporation is solvent. Section 45(a) of the act provides that the directors may pay a dividend only out of the unreserved and unrestricted earned surplus of the corporation. An alternative provides for the payment of dividends out of the unreserved or unrestricted earned surplus of the corporation or out of the "unreserved or unrestricted net earnings of the current fiscal year and the next preceding fiscal year taken as a single period. . . ."

The act, in Section 2, attempts to define *surplus* and *earned surplus*. Many of the states that have not adopted the act have not attempted to define *earned surplus* but leave that task to the courts in the interest of flexibility. The act definitions are as follows:

> (k) "Surplus" means the excess of the net assets of a corporation over its stated capital.
>
> (l) "Earned surplus" means the portion of the surplus of a corporation equal to the balance of its net profits, income, gains and losses from the date of incorporation, or from the latest date when a deficit was eliminated by an application of its capital surplus or stated capital or otherwise, after deducting subsequent distributions to shareholders and transfers to stated capital and capital surplus to the extent such distributions and transfers are made out of earned surplus. . . .
>
> (m) "Capital surplus" means the entire surplus of a corporation other than its earned surplus.

A generalization of the statutes concerning dividends is impossible. Some states prohibit payment of dividends if capital is impaired, and others allow the payment of dividends out of current net profits. Still others prohibit dividends if the effect is to impair capital. In general, however, dividends may not be paid if paying them would render the corporation insolvent, if it is insolvent, or if the payment will impair the capital of the corporation. In a majority of states a dividend may not be paid if the capital of the corporation is impaired because of previous losses incurred by the corporation.

If the directors of a corporation pay out dividends in violation of the state statute, they may be liable to the creditors of the corporation for any losses that result because of the wrongful payment of dividends. The stockholders who receive the wrongfully paid dividends may be forced to disgorge them, although some states require this only if the stockholders have acted in bad faith or with the knowledge that the dividends were illegally paid.

GAY v. GAY'S SUPER MARKETS, INC.
343 A.2d 577 (Me. 1975)

Lawrence E. Gay was a minority shareholder of Gay's Super Markets, Inc. He sued the corporation and its directors, contending that their refusal to declare a dividend for the year 1971 was not done in good faith but for the purpose of driving him out of the ownership structure of the corporation. The directors contended that the refusal was made in good faith because of the need for the funds for the contemplated expansion of the corporation's facilities. The trial court found for the defendants, and plaintiff appealed.

DUFRESNE, J.

Adverting initially to the statute and the bylaws of the defendant corporation, we find that the declaration of dividends is left to the discretion of the board of directors.

Such discretion is, of course, not without limitation:

> As a general rule, the officers of a corporation are the sole judges as to the propriety of declaring dividends, and the courts will not interfere with a proper exercise of their discretion. . . . Yet when the right to a dividend is clear, and there are funds from which it can be properly made, a court of equity will interfere to compel the company to declare it. Directors are not allowed to use their power illegally, wantonly or oppressively. *Belfast and Moosehead Lake Railroad Co.* v. *City of Belfast,* 1885, 77 Me. 445, 454, 1 A. 362.

To justify judicial intervention in cases of this nature, it must, as a general proposition, be shown that the decision not to declare a dividend amounted to fraud, bad faith or an abuse of discretion on the part of the corporate officials authorized to make the determination. *Anderson* v. *Bean,* 1930, 272 Mass. 432, 172 N.E. 647, 652, wherein the Massachusetts Court stated:

> The general principle is that stockholders have no individual interest in the profits of a corporation until a dividend has been declared, that the accumulation of a surplus does not of itself entitle stockholders to a dividend, that the time when a dividend shall be declared and its amount rest in the sound discretion of the corporation or its authorized officers, usually the board of directors, that the action of such officers will not be disturbed if taken in good faith according to law and not in plain violation of the rights of stockholders, and that rational presumptions will be indulged in favor of the honest decision of such officers.

The burden of demonstrating bad faith, fraud, breach of fiduciary duty or abuse of discretion on the part of the directors of a corporation rests on the party seeking judicial mandatory relief respecting the declaration of dividends.

If there are plausible business reasons supportive of the decision of the board of directors, and such reasons can be given credence, a court will not interfere with a corporate board's right to make that decision. It is not our function to referee every corporate squabble or disagreement. It is our duty to redress wrongs, not to settle competitive business interests.

The testimony of director James L. Moody substantially corroborated the evidence given by Carroll Gay. He related that, although Gay's Super Markets, Inc., had approximately $125,000 in cash or its equivalent at the time in question, the directors felt that, due to the expansion commitments of the company, these funds could not be considered as available for dividends. He stated that it was anticipated the new store in Calais would lose about $25,000 in its first year of operation. He further related that it was the board's belief, "the store that does reasonably well from the start normally will lose the first year about the same amount of money as one week's sales. We thought the store would average about $25,000 a week sales in the very early stages." He consistently asserted that the motivating cause for not declaring a dividend was the need for, and the expenses of, expansion.

(Affirmed.)

Questions for Discussion

1. *If Lawrence Gay could have shown that he in fact was forced to sell his stock as a result of the failure of the directors to award a dividend, would this alone have been enough to cause a court to order that dividends be paid?*
2. *If a shareholder establishes that a corporation has been making a profit for a period of years without paying a dividend, would this be sufficient evidence by itself for a court to require that a dividend be paid?*

43:5. Preemptive Rights. When a corporation is formed, the charter will provide for the authorization of a certain number of shares of stock. When a shareholder has purchased a number of shares, he or she has acquired a certain proportional interest in the corporation. If all the stock of the corporation has been sold, the corporation must obtain the permission of the state in order to sell additional shares. The corporation accomplishes this by amending the corporate charter to authorize the issue of more shares of stock. In order that the existing shareholders can maintain their proportionate share of the corporation, they may have what is known as a *preemptive right* to purchase the new issue of stock; that is, each shareholder may have the right to purchase his or her pro rata amount of the new issue before it is offered for sale to others. Suppose, for example, that Hi Fly Corporation has 10,000 authorized shares outstanding, of which Inez Investor owns 3,000 shares. Hi Fly then has its corporate charter amended to allow for the issue of an additional 1,000 shares. Inez has a preemptive right to purchase 300 of those shares in order to maintain her pro rata share of the corporate stock and thus to maintain her relative voting position in the corporation.

Many limitations have developed on the stockholder's preemptive right. For instance, the right may be limited, waived, or modified by the terms of the corporate charter or bylaws. In the majority of states the right is not applicable to treasury stock or to employee stock-purchase plans. Nor does the right exist if the corporation issues shares for services or property given to the corporation or if the shares are used in the case of a corporate consolidation, merger, or reorganization. A question may arise if all the original authorized shares have not been issued and an attempt is made to issue them after a period of time has expired. Do the original stockholders have any preemptive right to these shares? The law is in conflict on this point, but perhaps the majority of states hold that the stockholders have no preemptive right to the shares that have already been authorized but unissued. There is some degree of logic in such a rule, for a shareholder knows the proportion of the total authorized shares he is purchasing, regardless of whether they are all issued. On the other hand, when an entirely new authorization is obtained by an amendment to the charter, this fact would not necessarily have been known to the stockholder when he or she purchased his or her shares.

43:6. Right to Examine Books and Records. The right to inspect the books and records of a corporation is an important one. From such an inspection one may learn information concerning not only the financial condition of the corporation but also the salaries of corporate officers, the names and addresses of other shareholders, and other important matters. The common law gives a stockholder the right to examine the books and records of a corporation at reasonable times as long as the inspection is for a proper purpose. In most states statutes have been enacted that also provide for the right of inspection of corporate books and records by stockholders.

A problem sometimes arises when an individual purchases a very small amount of stock solely in order to inspect the books and records for purposes that may not be in the best interests of the corporation or the other stockholders. At common law the right to inspect was qualified by the condition that the inspection be for a specific purpose and be made in good faith and not merely for the purpose of causing problems for the corporation. The statutes of the various states are not completely consistent in dealing with this problem. Some statutes place no limits on the right of inspection except that it must be at a reasonable time and place. Other statutes seem merely to enlarge the common-law right. For example, the Model Business Corporation Act, Section 52, provides that one who has been a shareholder or holder of a voting-trust certificate for at least six months or who is a shareholder or trust-certificate holder of at least 5 percent of the outstanding shares of the corporation shall have the right to examine the corporate books and records. Those desiring to exercise the right under the act must in addition state the demand in writing, as well as the purpose of making the inspection. Under the act, as well as other statutes, either the shareholder or his attorney or agent may examine the records and make copies of them.

If it appears to the corporation that the purpose of a request to inspect the corporate books and records is not proper, it may refuse to allow the inspection. As a general rule, then, the stockholder's remedy is to go to court seeking an order that the books and records be made available to him. If in fact the request is for an improper purpose, the court will not order the corporation to allow an inspection by the stockholder. In some jurisdictions the burden is on the stockholder to demonstrate that the purpose is proper.

WEIGEL v. O'CONNOR

373 N.E.2d 421 (Ill.App. 1978)

John Weigel was a shareholder in Weigel Broadcasting Company. He filed suit to require the corporation and corporate officers to produce certain corporate records and books for examination. The trial court required the production of some, but not all, of the records.

Weigel appealed.

LINN, J.

Weigel Broadcasting Company, which operates Station WCIU, television Channel 26 in Chicago, was organized by the plaintiff in 1962. The corporation's capital stock consists of 181,669 shares of common stock and 4,588 shares of cumulative preferred stock owned of record by approximately 350 persons. Until 1966 the plaintiff held a controlling interest in the corporation. Currently, he is the owner of record of 16,150 shares, or approximately 9 percent, of the common stock. Defendants O'Connor, Shapiro and Weisberg are chairman of the board, president and secretary of the corporation, respectively. O'Connor and Shapiro hold more than two-thirds of the outstanding common shares. They acquired their controlling interest in 1966 by buying stock from the plaintiff and from the plaintiff's former wife, she having received certain of the stock from the plaintiff as a result of a marital settlement. Defendants O'Connor and Shapiro are also the principal holders of the corporation's cumulative preferred shares, on which dividends first were paid in 1975. No dividend has ever been declared or paid on the common stock held by plaintiff and the other shareholders.

In a letter dated July 22, 1975, plaintiff, through his attorney, made a formal demand to inspect specific corporate books and records. . . .

The demand stated the following purposes for which the plaintiff desired to examine the enumerated books and records: to ascertain the value of his shares and determine the true financial condition of the corporation; to determine the nature and amount of corporate expenses; to determine the source of corporate revenues; to determine the amount and kind of compensation paid to corporate officers and directors; to determine what amount of broadcasting time had been given as trade-outs; to communicate with other minority shareholders; and to allow informed voting by minority shareholders at future meetings.

Section 45 of the Illinois Business Corporation Act gives a stockholder in a corporation, who otherwise qualifies under the provisions of the statute, the right to examine "in person, or by agent or attorney, at any reasonable time or times, for any proper purpose, its books and records of accounts, minutes and record of shareholders and to make extracts therefrom." (Ill. Rev.Stat.1975, ch. 32, par. 157.45.) The Illinois Supreme Court has interpreted the phrase "for any proper purpose" to include "honest motive" and "good faith."

A proper purpose is one which seeks to protect the interests of the corporation as well as the interests of the shareholder seeking the information. A shareholder is legitimately entitled to know anything and everything which the records, books and papers of the company would show so as to protect his interest and so long as he has an honest motive and is not proceeding for vexatious or speculative reasons.

A stockholder must be seeking something more than satisfaction of his curiosity and must not be conducting a general fishing expedition. Proper purpose could be shown by attendance at shareholders' meetings, examination of the books and records of the corporation and from the evidence of an effort on the part of the shareholder to determine the financial situation and the character of management of the corporation. A shareholder's attempts to determine the financial condition of the corporation through an exchange of correspondence with corporate officers evidences proper purpose even when the shareholder does not attend the shareholders' meetings. A desire to learn the reasons for lack of dividends or insubstantial dividends, and suspicion of mismanagement arising from such a dividend policy alone, will constitute a proper purpose.

Although the burden of proving good faith and proper purpose rests with the stockholder, proof of actual mismanagement or wrongdoing is not necessary.

In support of the plaintiff's claim of proper purpose the following evidence was adduced at trial. Weigel was the founder and a longtime shareholder in the corporation in which he retained a substantial investment. Although not always in agreement with current management, the plaintiff actively participated in shareholders' meetings and corresponded with other minority shareholders. Weigel's testimony indicated that, based on information received from corporate insiders, he had reason to fear that the financial security of the corporation was being jeopardized by the personal use of trade-outs by corporate officers, the giving of kick-backs to advertising agents and the uncompensated use of corporate property for the benefit of another corporation. In an attempt to verify the insider information the plaintiff unsuccessfully questioned two of the defendant corporate officers. These factors, coupled with the history of the lack of dividend payments, are sufficient indications of proper purpose to support plaintiff's demand.

A single proper purpose is sufficient to satisfy the statutory requirement. Once that purpose has been established, the shareholder's right of inspection extends to all books and records necessary to make an intelligent and searching investigation. "[T]he right of a stockholder extends to all books, papers, contracts, minutes, or other instruments from which he can derive any information that will enable him to better protect his interests." (5 Fletcher Cyclopedia Corporations, § 2239 at 779 (1976 Revised Volume).) The language "books and records" has been construed to extend to contracts and other papers.

A shareholder need not establish a proper purpose in respect to each document he desires to examine; a proper purpose which would entitle him to the right of inspection generally is sufficient. "[I]t is the general rule that the right of inspection is not qualified by the necessity for it; where the right of inspection exists, refusal of it cannot be justified by proffering him a substitute, or on the ground that he has had the information at the hands of the company, or that it may be obtained from other sources, or that it is not necessary, or that he has recently been allowed inspection." (5 Fletcher Cyclopedia Corporations, § 2249 at 803 (1976 Revised Volume).) Here, the plaintiff made a specific demand for designated documents, not a blanket demand for all books and records of the corporation, although such omnibus demands have been upheld.

(Reversed in favor of plaintiff.)

QUESTIONS FOR DISCUSSION

1. *What is the standard for testing whether a stockholder is entitled to inspect corporate books and records?*
2. *What purposes did the stockholder state as a reason for inspecting the books and records of the corporation?*

43:7. SUITS BY SHAREHOLDERS. On occasion the stockholder may have a complaint against the corporation or its management. The right to vote and thereby control the corporation may be a hollow one if the stockholder does not have sufficient shares and is not capable of obtaining a sufficient number of proxies in order to gain control. An alternative in this instance would be to sell the shares, if possible, to someone else. However, in many cases the shareholder may wish to take his or her cause of action to court. In so doing, it is well to keep in mind that there are basically two types of action that the shareholder may bring: *a direct action* against the corporation and a *derivative* suit.

The shareholder may bring a *direct action* against the corporation to enforce a right that the corporation owes him individually or as a representative of a group of shareholders. For example, the right of a direct action may result from some tort that the corporation has committed against the stockholder or the denial of some contractual right. Such suits may result from a denial of a stockholder's right to vote, to inspect corporate records, to pay dividends, or to attend stockholder meetings. The stockholder brings the direct action in order to gain compensation or a right.

The *derivative action* is of a somewhat different nature. It is brought by the stockholder on behalf of the corporation in order to enforce a right possessed by the corporation, as opposed to a suit for the direct benefit of the individual stockholder. The stockholder, for example, may feel that the corporation has a claim against the corporate management for mismanagement or waste of the corporate assets. Or the stockholder may feel that the corporation has a claim against someone outside the corporation that the corporate management for one reason or another will not enforce. The stockholder cannot enforce these claims directly because they are the claims of the corporation, a separate legal entity. Hence the right to sue on behalf of the corporation, a derivative suit, was developed. The result of a successful derivative suit is that the corporation itself benefits.

WEISS v. NORTHWEST ACCEPTANCE CORPORATION

546 P.2d 1065 (Or. 1976)

DENECKE, J.

Fall Creek Gravel Co., Inc., is a corporation that built logging roads. The plaintiff Weiss is the president and sole stockholder. In the fall of 1969 the defendant, Northwest Acceptance Corporation, financed Fall Creek, including the acquisition of trucks by Fall Creek. Fall Creek acquired the trucks through Automotive Equipment which was Northwest's largest dealer and which had several officers and directors in common with Northwest. Automotive guaranteed the payments due Northwest from Fall Creek on the truck acquisitions. Weiss guaranteed Fall Creek's obligations to Northwest and to other creditors.

In the year or year and a half prior to August 1970, Fall Creek had financial difficulties. The seriousness of its difficulties is in dispute. About August 20, 1970, Northwest began to repossess the equipment. Fall Creek threatened to file bankruptcy. According to plaintiff's evidence, Northwest then agreed with Fall Creek through its president, Weiss, that if Fall Creek voluntarily relinquished the equipment Northwest would allow Fall Creek to use the equipment, finish its jobs and later to liquidate in an advantageous manner.

Plaintiffs' evidence was that in reliance thereon, plaintiffs permitted the equipment to be brought in and sold, but that the sale, and Northwest's subsequent actions, were contrary to what it had promised plaintiffs. At the sale an outsider bid $250 more than Automotive and purchased the equipment for $472,750 which plaintiffs allege was grossly disproportionate to the equipment's true value. Automotive's bid of $472,500 was the total amount Fall Creek owed Northwest. Fall Creek was later adjudicated bankrupt and the plaintiff Auld is trustee in bankruptcy for Fall Creek.

The jury found Northwest had defrauded Fall Creek and Weiss and awarded Weiss general damages and punitive damages and the trustee for Fall Creek, general damages.

Weiss alleged that because of Northwest's acts he was damaged individually in the amount of $1,500,000 because the value of his stock was destroyed and he remains liable for many of Fall Creek's debts.

The jury returned a verdict for Weiss in the amount of $122,164.74, general damages, which the jury found represented the debts and taxes for which Weiss was liable and which Fall Creek would have been able to pay if Northwest had not defrauded Weiss.

Northwest contends assuming the corporation, Fall Creek, had a cause of action against Northwest, that Weiss, as an individual, does not.

The general rule is that a stockholder has no personal right of action against a third party for a wrong to the corporation which lowers the value of the stock. This court has adopted this rule. Conceptually, the rule is based upon the separateness of the corporate entity and its stockholders. Practically, it is based in part upon the rule that corporate creditors are entitled to be paid out of any recovery made to the corporation before the stockholders are entitled to payment.

The plaintiff Weiss acknowledges this to be the general rule, but contends he comes within a recognized and logical exception to the general rule. The exception is that a stockholder may have an individual cause of action against a third party if the third party breaches a duty directly owed to the stockholder as an individual. The stockholder may maintain such a cause of action although it is based upon the conduct of a third party which also creates a cause of action in the corporation.

Weiss contends Northwest breached a special duty owed to him as a personal guarantor of Fall Creek's obligations. These obligations were not obligations of Fall Creek to Northwest but to other creditors for fuel, taxes and a bank loan.

Weiss is in no different position than any other creditor of Fall Creek. Weiss gave credit to Fall Creek in the form of making a guaranty to the bank and to others. His action is similar to that of the bank or others in lending credit to Fall Creek. When, after the credit is extended, the value of the assets of Fall Creek are lessened allegedly by the fraud of Northwest and, therefore, Fall Creek is unable to pay its creditors, the bank and other creditors cannot maintain separate actions against Northwest. The right of recovery is solely that of Fall Creek. That is true of Weiss as well as other creditors.

Weiss cannot recover for his liability on his guaranties of Fall Creek's obligations. That part of the judgment awarding plaintiff Weiss $122,164.74 must be reversed.

Questions for Discussion

1. *Was the action attempted by Weiss in this case a derivative action or a direct action?*
2. *Why did the court conclude that Weiss had no personal right of action against Northwest Acceptance Corporation?*

TAORMINA v. TAORMINA CORP.

78 A.2d 473 (Del.Ch. 1951)

Calogero Taormina was an active stockholder of the Taormina Corporation. In 1939 after a serious illness he returned to Italy and died there in 1946. His widow was named executor in his will. She appointed an attorney to administer Calogero Taormina's estate in Delaware.

When he died, Calogero Taormina was the owner of 498 shares of the corporation and had been since 1935. In 1944 the other stockholders formed a partnership under the name Taormina Company. Each former stockholder had a share of the partnership equal to his proportional interest in the corporation. All the assets of the corporation were transferred to the partnership for little consideration, and the partnership continued the same business formerly performed by the corporation.

Calogero Taormina's estate filed a derivative action against the other shareholders, charging that they unjustly profited themselves at the expense of the corporation. The defendants contended that this was not properly a derivative action and filed a motion to dismiss.

COLCOTT, Ch.

Generally a cause of action belonging to a corporation can be asserted only by the corporation. However, whenever a corporation possesses a cause of action which it either refuses to assert or, by reason of circumstances, is unable to assert, equity will permit a stockholder to sue in his own name for the benefit of the corporation solely for the purpose of preventing injustice when it is apparent that the corporation's rights would not be protected otherwise.

The defendants contend that the allegations of the complaint in the instant case demonstrate that it is in fact an action brought to redress a wrong done to Calogero Taormina as an individual stockholder of the Corporation. They contend that the cause of action sought to be enforced is based upon the withholding by the individual defendants from Calogero Taormina of a share of the profits obtained from the business conducted by the Company. In substance the arguments seek to fasten upon the plaintiffs the contention that Calogero Taormina had a legal right to a share in profits illegally withheld from the Corporation and thus make him an unwilling partner in the alleged conspiracy. The argument is based upon allegations in the complaint to the effect that the actions of the individual defendants deprived Calogero Taormina of his rightful share of corporate profits and gains.

It is true that Calogero Taormina as a minority stockholder suffered financial loss by being deprived of his rightful share as a stockholder of corporate profits, but this is true of most derivative actions. If the argument of the defendants is to be accepted, it would mean that no derivative action could be maintained by the stockholder to redress corporate wrongs if those wrongs had resulted in loss to the value of his stockholdings.

The defendants argue that the only stockholder who can have been injured as a result of the action of the individual defendants is Calogero Taormina for the reason that he is the only stockholder of the corporation who did not participate in the course of action instigated by the individual defendants. They argue that the injury was to him individually and that he must bring suit in his own name for the redress of these wrongs.

The argument ignores the fundamental basis of a derivative stockholder's action which is to enforce a corporate right. . . .

The relief to be obtained in a derivative action is relief to the corporation in which all stockholders, whether guilty or innocent of the wrongs complained of, shall share indirectly. Indeed, it is doubtful whether the result would be different even if the suing stockholder owned all of the stock of the wronged corporation.

I conclude, therefore, that the pending action is a derivative action brought by the plaintiffs to enforce a cause of action owned by the Corporation.

43:8. LIABILITIES OF SHAREHOLDERS. You will recall that we have previously mentioned the fact that corporate shareholders have limited liability. By this we mean that the stockholder's liability is generally limited to the amount of his or her investment in the corporation. If the corporation becomes insolvent, the stockholder generally is not liable to creditors out of his personal assets. There are, however, some special circumstances that warrant further analysis and that involve stockholder liability.

(1) **"Watered Stock."** A corporation may issue its stock for cash, property, labor, or services. When property, labor, or services are given in return for stock, a value must be placed upon them. If the stock is reported as being fully paid but property is accepted as full payment that is in fact worth less than the par value or the stated value of the stock, it is said to be *watered.* Frequently this situation occurs when the property accepted in return for the stock is overvalued.

The question arises as to who determines the value of the property or services received for the company stock. Ordinarily the board of directors places a value on the services or property received in return for payment of the corporation stock. The majority of states follow the Model Business Corporation Act, Section 19, which provides that "In the absence of fraud in the transaction, the judgment of the board of directors or the stockholders, as the case may be, as to the value of consideration receive for shares shall be conclusive." This is sometimes referred to as the *good-faith test* because the judgment of the board is accepted unless it can be shown that there is a lack of "good faith" in the valuation placed upon the property or services received for the company stock. Some jurisdictions do not apply the good-faith test as enunciated in the act but attempt to ascertain the *true value* of the property given in return for the stock. If the true value is less than the par or stated value of the stock then it is said to be *watered.* This test is difficult to apply in practice, and the majority of jurisdictions apply the good-faith test.

If the stock of a corporation is watered, the creditors of the company may be deceived as to its true financial condition. The creditors thus have a right of action against the shareholders who received the watered stock if a creditor is injured as a result of the issuance of stock that is watered. This situation normally occurs if the corporation becomes insolvent. If one purchases watered stock from the original owner of the stock, he or she is not liable unless he or she possesses knowledge that the stock was not fully paid. In most jurisdictions the right of action against the holder of watered stock is the creditor's and not the corporation's. In a minority of states, if the corporation becomes insolvent, a trustee in bankruptcy may sue the holder of watered shares on behalf of the corporation.

The creditors may also recover for stock that was issued to individuals, such as employees, as a bonus without being paid for. If the stock issued as a bonus is stock that has been fully paid and then returned to the corporation, then the holders of such stock are not liable to creditors in the event of corporate insolvency.

The majority of states apply the *fraud* or *misrepresentation* theory of recovery, which requires the creditor to show that he or she relied on the misrepresentation caused by the watered stock. In the other jurisdictions the *statutory-*

obligation theory prevails, which makes the holder of watered stock absolutely liable by statute without the necessity of the creditor's showing any reliance.

BING CROSBY MINUTE MAID CORPORATION v. EATON
297 P.2d 5 (1956)

Wallazz B. Eaton, Sr., formed a corporation to acquire his frozen food business. Forty-five hundred shares at $10 par value were authorized by the commissioner of corporations for sale to Eaton and certain other individuals in consideration of the transfer of the business. The permit required that 1,022 shares be placed in escrow and not transferred without the written consent of the commissioner. The escrowed shares were not to be sold until the shareholders named in the permit waived their rights to dividends or the distribution of assets.

The 1,022 shares were listed on the corporate records as held by Eaton and noted "escrowed." *The remaining 3,478 shares were issued outright to Eaton and later transferred to others. He gave consideration valued at $34,780 for them, but none was ever given for the 1,022 shares. When the corporation became insolvent, one of its creditors, Bing Crosby Minute Maid Corporation, initiated this action to recover the difference between the par value of the stock issued to Eaton and the fair value of the consideration paid for the stock. Eaton contended that he never became owner of the 1,022 shares and that therefore Minute Maid had no action against him for their par value. He also contended that there was no reliance shown by the creditors. The jury awarded a judgment in favor of Minute Maid, but the trial court granted a new trial. Minute Maid and Eaton both appealed.*

SHENK, J.

The plaintiff seeks to base its recovery on the only other exception to the limited liability rule that the record could support, namely, liability for holding watered stock, which is stock issued in return for properties or services worth less than its par value. Accordingly, this case calls for an analysis of the rights of a creditor of an insolvent corporation against a holder of watered stock. Holders of watered stock are generally held liable to the corporation's creditors for the difference between the par value of the stock and the amount paid in.

The defendant's first contention is that because of the escrow he never became an owner of the 1,022 shares and that he therefore never acquired such title to the 1,022 shares as would enable a creditor to proceed against him for their par value. . . . The escrow in the present case permitted the defendant to retain some, but not all, of the incidents of ownership in the 1,022 shares. Although he could not transfer the shares, it appears that despite the escrow he was entitled to count them in determining the extent of his rights to vote and to participate in dividends and asset distributions. The critical feature of the escrow for purposes of the present case is the absence of any restriction on representations that the escrowed shares were outstanding and fully paid. Although the escrow contained provisions designed to protect future stockholders, it afforded no special protection to future creditors of the corporation. Therefore, the escrow did not affect the rights of future creditors and it would appear that despite the escrow the defendant acquired sufficient title to the 1,022 shares to permit the plaintiff to proceed against him for their par value.

The liability of a holder of watered stock has been based on one of two theories: the misrepresentation theory or the statutory obligation theory. The misrepresentation theory is the one accepted in most jurisdictions. The courts view the issue of watered stock as a misrepresentation of the corporation's capital. Creditors who rely on this misrepresentation are entitled to recover the "water" from the holders of the watered shares.

Statutes expressly prohibiting watered stock are commonplace today. In some jurisdictions where they have been enacted, the statutory obligation theory has been applied. Under that theory the holder of watered stock is held responsible to creditors whether or not they have relied on an overvaluation of corporate capital.

In his answer the defendant alleged that in extending credit to the corporation the plaintiff did not rely on the par value of the shares issued, but only on independent investigation and reports as to the corporation's current cash position, its physical assets and its business experience. At the trial the plaintiff's district manager admitted that during the period when the plaintiff extended credit to the corporation, (1) the district manager believed that the original capital of the corporation amounted to only $25,000, and (2) the only financial statement of the corporation that the plaintiff ever saw showed a capital stock account less than $33,000. These admissions would be sufficient to support a finding that the plaintiff did not rely on any misrepresentation arising out of the issuance of watered stock.

It is therefore necessary to determine which theory prevails in this state. The plaintiff concedes that before the enactment of section 1110 of the Corporations Code in 1931, the misrepresentation theory was the only one available to creditors seeking to recover from holders of watered stock. However, he contends that the enactment of that section reflected a legislative intent to impose on the holders of watered stock a statutory obligation to creditors to make good the "water." Section 1110 provides that "The value of the consideration to be received by a corporation for the issue of shares having par value shall be at least equal to the par value thereof, except that: (a) A corporation may issue par value shares, as fully paid up, at less than par, if the board of directors determines that such shares cannot be sold at par. . . ." The statute does not expressly impose an obligation to creditors. Most jurisdictions having similar statutes have applied the misrepresentation theory obviously on the ground that creditors are sufficiently protected against stock watering schemes under that theory. . . . In view of the cases in this state prior to 1931 adopting the misrepresentation theory, it is reasonable to assume that the Legislature would have used clear language expressing an intent to broaden the basis of liability of holders of watered stock had it entertained such an intention. In this state the liability of a holder of watered stock may only be based on the misrepresentation theory.

The plaintiff contends that even under the misrepresentation theory a creditor's reliance on the misrepresentation arising out of the issuance of watered stock should be conclusively presumed. This contention is without substantial merit. If it should prevail, the misrepresentation theory and the statutory obligation theory would be essentially identical. This court has held that under the misrepresentation theory a person who extended credit to a corporation (1) before the watered stock was issued, . . . or (2) with full knowledge that watered stock was outstanding, . . . cannot recover from the holders of the watered stock. These decisions indicate that under the misrepresentation theory reliance by the creditor is a prerequisite to the liability of a holder of watered stock. The trial court was therefore justified in ordering a new trial because of the absence of a finding on that issue.

(The order granting a new trial is affirmed.)

Questions for Discussion

1. *What is the difference between the misrepresentation or fraud theory of watered stock and the statutory-obligation theory?*
2. *What must the creditor prove to recover under the misrepresentation theory that he or she need not prove under the statutory-obligation theory?*

(2) Unpaid Stock Subscriptions. A contract between a corporation and a shareholder whereby the shareholder agrees to pay a certain amount in cash or other property in return for stock is called a *stock subscription.* The stock subscription is a contract between the corporation and the shareholder. If the shareholder does not live up to the terms of the subscription agreement, the corporation has a right of action against the shareholder for the balance due for the stock. Furthermore, stock subscriptions are considered an asset of the corporation. Therefore, in case of corporate insolvency, the creditors of the corporation may recover the amount of the unpaid stock subscriptions from the shareholders. The claims of creditors on unpaid stock subscriptions cannot be defeated by the tactic of having the corporation release its claim against the stockholder on the unpaid subscription agreement. As a general rule, however, the creditor must first proceed against the other assets of the corporation, and when they are exhausted, he then proceeds to recover the unpaid stock subscription. Of course, the corporation itself has a right of action against the stockholder if the stockholder fails to adhere to the subscription agreement.

(3) Illegally Paid Dividends. We have previously discussed some situations in which the corporation may not pay dividends to stockholders. As a general rule, the stockholders who receive illegally paid dividends are liable for their return, although the statutes of a few states provide that stockholders are liable for illegal dividends only if they knew of the illegality or received the dividends in bad faith. Normally the corporate directors are personally liable to the creditors of the corporation, in addition to the stockholders, for dividends that are illegally declared and paid.

SUMMARY

A shareholder may exercise the number of votes equal to the number of shares that he owns. Someone other than the shareholder may vote the shares if he or she has the shareholder's proxy. In order that minority shareholders may be represented on the board of directors, most state statutes provide for cumulative voting for directors.

A voting trust may be established by having stockholders transfer the legal title to their shares to the party serving as trustee for the purpose of voting their stock. The shareholders receive voting trust certificates.

As a general rule, shareholders have no right to receive dividends until the board of directors declares a dividend. The directors have the discretion to declare a dividend that may be paid in either cash, stocks, bonds or other property of the corporation. Some states prohibit the payment of dividends if the corporation is insolvent. Others require that dividends be paid only out of net profits; yet a third group prohibits the payment of dividends if the effect is to impair the stated capital of the corporation. Directors who pay out dividends in violation of a state statute may be liable to creditors for any losses they incur as a result of the improper payment.

In order for existing shareholders to maintain their proportionate share of the corporation, they may have preemptive rights to purchase new issues of stock. There are many limitations to this right.

The right of shareholders to inspect corporate records is defined by statute. The Model Business Corporation Act grants the right of inspection to shareholders of at least 5 percent of the outstanding shares of stock as well as to a few others. A demand for inspection and statement of purpose must also be made. A court will refuse to allow inspection of corporate records for an improper purpose.

A shareholder may bring a direct action against the corporation to enforce a right that the corporation owes him individually or as a representative of a group of shareholders. A derivative action is brought on behalf of the corporation to enforce a right possessed by the corporation.

Shareholders may be liable for watered stock. The value of property or services received for the company stock is determined by the good-faith test under the Model Business Corporation Act. A few jurisdictions apply a true-value test. A majority of states apply the fraud theory of recovery for those claiming to be injured by the issuance of watered stock. Other jurisdictions apply the statutory-obligation theory.

Stockholders may also be liable for unpaid stock subscriptions. The corporation or a creditor of an insolvent corporation has the right of action. Stockholders are also liable for illegally paid dividends.

PROBLEMS

1. Paulette Fownes and others were trustees of a block of voting and nonvoting stock amounting to 22 percent of the outstanding shares of Hubbard Broadcasting, Inc. Hubbard owned several television and radio stations. There came a time when Fownes and the other trustees discussed with the corporation the possibility of redemption of all the common stock held by the trustees at a price of $3.57 per share. The trustees later demanded $7.14 and then $11.29 per share, which was rejected by the corporation as exorbitant.

   On March 14, 1973, the trustees, by letter, demanded the right to examine the share register, books of account, and records of the proceedings of the shareholders and directors. This request was refused by the corporation which instead furnished the trustees with copies of audit reports. The trustees again demanded to see the books and records and were refused by the corporation. Fownes and the other trustees then filed suit to compel the corporation to allow the trustees to inspect the books and records of Hubbard Broadcasting, Inc. Are the trustees entitled to inspect the actual corporate records or does their inspection of a summary of the corporation records meet the corporation's duty? Fownes v. Hubbard Broadcasting, Inc., 225 N.W.2d 534 (Minn. 1975).
2. Honeywell, Inc., manufactured antipersonnel fragmentation bombs during the Vietnam war. One hundred shares of Honeywell stock were purchased by Pillsbury—who was strongly opposed to the war—in order to inspect Honeywell's books and records, including its original shareholder ledger, current shareholder ledger, and all corporate records dealing with weapons and munitions manufacture. Pillsbury's purpose in buying the stock was to try to impress his opinions upon Honeywells' management and other shareholders. Honeywell refused to allow Pillsbury to inspect the books and records. Could Pillsbury compel the corporation to allow him to inspect its books and records? Minnesota Ex Rel. Pillsbury v. Honeywell, Inc., 191 N.W.2d 406 (Minn. 1971).
3. Robert Erdman and three others formed a corporation known as Sol and

Bob, Inc., for the purpose of operating a hairdressing business. Each owned 25 percent of the corporation and was employed by the corporation. In 1970 Erdman terminated his employment with the corporation but retained his stock. Subsequent to Erdman's departure from the business the remaining three employee-stockholders granted themselves large pay increases and bonuses but paid no dividends. Erdman contended that he was entitled to 25 percent of the business and profits because this was really a method of distribution of profits and in fact a dividend. Was he correct? Erdman v. Yolles, 233 N.W.2d 667 (Mich. 1975).

4. M. M. Miles wrote two letters requesting an unlimited inspection of the books and records of the Bank of Heflin. The purpose for the inspection as stated in the letters was to ascertain, among other things, whether any action had been taken contrary to the interest of stockholders, such as misuse of corporate funds. The bank's position was that it owed a duty to its customers not to disclose certain confidential information and refused to allow the inspection. The bank contended that the stockholders were allowed to inspect only certain of the corporate books and records. Was the bank correct? In Re Bank of Heflin v. Miles, 318 So.2d 697 (Ala. 1975).
5. William Stokes owned a substantial amount of the capital stock of Continental Trust Co. of the City of New York. The corporation was organized with 5,000 authorized shares at $100 par value each, or $500,000. Blair & Co. proposed to the corporation that if its shareholders voted to double the authorized shares, they would purchase the additional stock for $450 per share, provided that they could nominate ten of the twenty-one directors. The additional issue was voted by the shareholders, and Stokes demanded the right to subscribe for 221 shares of the new stock at par $100. The corporation refused, and the new stock was eventually sold to Blair & Co. Was Stokes entitled to the 221 shares of the new stock? Stokes v. Continental Trust Co. of the City of New York, 78 N.E. 1090 (N.Y. 1906).
6. McMenomy and others were minority stockholders of Midwest Technical Development Corporation, which was an investment company. In 1965 Midtex, Inc., was organized for the purpose of acquiring the assets of Midwest. All the assets of Midwest were transferred to Midtex as of December 1965. In June of 1962 McMenomy and other minority stockholders had begun a derivative action against eighteen individuals to recover alleged profits made by the individuals in dealing in stock held in the portfolio of Midwest. The defendants contended that the minority stockholders' right of action was defeated because of the transfer of the assets from Midwest to Midtex, which was not a party to this action. Were the defendants correct? McMenomy v. Ryden, 176 N.W.2d 876 (Minn. 1970).
7. August Rueckert entered into a subscription agreement whereby he agreed to purchase ten shares at par value of $100 each in a corporation. The corporation was to be organized, according to the agreement, to buy real estate in the city of Menominee and to "erect thereon and own a building to be occupied or leased for a department store, moving picture theater,

office and other business and recreational purposes." Rueckert refused to purchase the stock, contending that the purpose clause in the charter was different from the subscription agreement. Could Rueckert be compelled to honor the subscription agreement? Menominee Bldg. Co. v. Rueckert, 222 N.W. 162 (Mich. 1928).

8. The Peninsular Telephone Company owned stock in the Florida Telephone Corporation. Peninsular requested that it be allowed to inspect Florida's books for the purpose of learning the names of other shareholders in order to attempt to purchase their stock and thereby gain control of Florida Telephone. Florida refused to allow the inspection, contending that it was for an improper purpose. Was Florida correct? Florida Telephone Corporation v. Florida Ex Rel. Peninsular Telephone Company, 111 So.2d 677 (Fla. 1959).

CHAPTER 44

Management of the Corporation

One of the advantages of the corporate form of organization is the fact that management may be centralized even when there are large numbers of owners. The purpose of this chapter is to examine the role of the stockholders, directors, and officers in the management of the corporation. The law provides for certain functions to be performed by each of these groups.

44:1. The Stockholder's Role in Corporate Management. We have already discussed in large measure the position of the stockholders in the corporation. They are, of course, the residual owners of the company but have no direct role as individuals, or as a body, in the daily operations of the corporation. Nor do the stockholders normally play a role in the policy direction of the corporation. The major role of the stockholders in the corporate management is to elect the board of directors. If the stockholders are dissatisfied with the management and policies set by the board of directors, their remedy is to elect a new board. Under this system a minority of the shareholders may not be able to change the board. The voting rights of stockholders were discussed in Chapter 43. As a practical matter, when large publicly held corporations are involved, a rather small number of shareholders may control the corporation by electing the board of directors through the use of proxies.

(1) **Other Methods of Shareholder Influence.** You should note at this point that although the shareholders are the owners of the corporation and elect the directors, they are not agents of the corporation and have no power to bind the corporation to third parties either based on contract or for torts. On the other hand, you should be aware by this time that the stockholders have other powers in addition to the election of the board of directors and thus do influence the management of the corporation to some extent. For instance, acts that may affect the structure of the corporation or that result in fundamental changes require shareholder approval. Any change such as authorizing additional stock, changing the corporate name, changing the corporate purpose, or selling the assets of the corporation, as well as a merger or consolidation with another corporation, will require the approval of the stockholders. Thus to say that stockholders have no role in the corporate management is not entirely accurate. They do have an indirect role. The influence that the stockholders have in the management of the corporation varies widely from one company to another.

(2) Stockholder Meetings. In order for the stockholders to take action, it is necessary to do so at a regularly or specially called meeting. Most statutes require that there be at least one *regular* meeting of the stockholders a year. In addition there may be any number of *special* meetings for the purpose of transacting various types of business.

In order for a stockholder meeting to be valid, *notice* of the meeting must be given to the stockholders as required by statute. Most statutes require that the notice be given in writing at least ten days, or a similar time period, prior to the meeting. If the meeting is a special meeting, the notice must give the date, time, place, and purpose of the meeting. If the meeting is a regular meeting, the time and place may be stated in the corporate bylaws, and written notice of such a meeting is given to the shareholders. If any unusual action is to be taken at a regular meeting, this must be stated in the notice to stockholders. The purpose of the written notice is to prevent a group of stockholders from voting on significant corporate matters without the opportunity for all stockholders to cast their vote. If notice of a meeting is not given, any action taken during the course of such meeting is usually invalid. However, a stockholder may waive his or her right to receive notice either by signing a formal written waiver or by actually attending the meeting.

Special meetings are usually called by the board of directors of the corporation. However, a number of shareholders may also call such a meeting. The statutes usually require that such a meeting must be called by holders of at least 10 percent of the shares in order to be valid. This requirement, of course, prevents a very small and dissident group of shareholders from calling numerous meetings for no valid purpose.

In order for the business transacted at the stockholder meeting to be valid there must be a quorum present. A quorum is normally a majority of the outstanding shares entitled to vote, but some other percentage may be provided in the corporate charter or bylaws. The Model Business Corporation Act requires that at least one third of the shares eligible to vote must be represented. Once there is a quorum, the vote of a majority of those shares present constitutes an act of the stockholders. Certain fundamental changes in the corporation, to which we have already alluded, may require the vote of a greater number of shareholders.

44:2. Director's Role in Corporate Management. We shall have seen that the stockholders are the owners of the corporation and have to approve a few actions concerning the corporation that are extraordinary. When it comes to the ordinary management and control of the affairs of the corporation, this power is vested in a board of directors elected by the shareholders. The corporation laws of the various states provide for this form of organization. For instance, the Model Business Corporation Act, Section 35, provides that "The business and affairs of a corporation shall be managed by a board of directors except as may be otherwise provided in the articles of incorporation."

The top echelon of the corporate management, then, is the board of directors.

It is from this body that the officers of the corporation and other agents obtain their authority to act on behalf of the corporation. Typically the board will appoint corporate officers and give them authority to carry on the day-to-day operations of the corporation. In addition, executive and other committees of the board may be appointed to deal with certain types of matters. We shall see later that there are certain limitations on the board's powers to delegate its authority. For example, if the board possesses the authority to enact bylaws, this task may not be delegated.

You should note one other important matter in the statute. It says that "the affairs of the corporation shall be managed by a board. . . . An individual director has no authority at all. Only the full board of directors acting as a board has authority. Thus, if Dorothy Director is a member of the board of directors for Fastgrow Corporation, she has no authority to purchase supplies for the corporation. Her only power is to vote as a member of the board to purchase supplies; then, that is an act of the board that may subsequently bind the corporation, but Dorothy acting alone does not have that power.

(1) Qualifications of Directors. The corporate charter and bylaws may state qualifications necessary in order to be a director. There may also be some qualifications stated in the applicable state statute. Perhaps the most frequent requirement is that one be a stockholder in order to serve on the board of directors. The purpose of such a requirement is to minimize the division between ownership and management of the corporation. As a practical matter, however, this requirement can be circumvented by the transfer of a share or two to the person desiring to be a director. The courts have generally upheld such procedures. Another requirement that is sometimes found in varying form in some corporate charters and older corporation statutes is that one be a resident of the state where the corporation is chartered. This would seem to be an unwise requirement for any corporation whose shareholders and business may be widely scattered. New York has a provision requiring one member of the board to be a state resident.

Other qualifications may be stated in the corporate charter or bylaws. Most statutes do not prevent aliens or infants from holding the position of corporate director.

(2) Election, Vacancies, and Removal. The stockholders elect the members of the board of directors. The corporation statutes usually fix the minimum and maximum number of directors of the corporation. Within these limitations the number of directors of the corporation is determined by the corporate charter. The directors may be elected to terms of office that may amount to several years, with several directors standing for election each year. If a vacancy occurs on the board, most statutes provide that the vacancy may be filled by vote of the board for the unexpired term. Some statutes require that the stockholders vote on the appointment at the next annual meeting. Typically, if there is an increase in the number of directors on the board, the new directors must be elected by the stockholders.

A director may resign at any time, or he may be removed from office by the stockholders. Generally a director may be removed for cause, such as neglect of duty or bad faith, without any liability to the corporation. If the shareholders have elected to compensate the directors, the corporation may be liable in damages to a director who is removed without cause prior to the expiration of his term. Furthermore, in the absence of a special provision in the corporate charter, most jurisdictions prohibit the removal of a director by the fellow members of the board.

(3) **Directors' Meetings.** Because the board of directors can act only as a board, it is necessary that they meet. The corporation statutes of the various states have general provisions dealing with the meetings of directors. The purpose of these statutory provisions is to assure that one group of directors, which may constitute a majority, does not act without the knowledge or participation of the other directors. Most statutes provide that no notice of directors meetings is required if the meeting is prescribed in the corporate bylaws. However, if there is a special meeting, a notice of the meeting is generally required. A director may waive his notice of the meeting either by attending the meeting or by signing a written waiver of notice. Some statutes provide that the board of directors may act without a meeting if all members of the board consent to the action taken in a writing that states the action taken.

Most modern corporation statutes eliminate any requirement that the board of directors meet in the state where the company is incorporated. These statutes also usually provide that, in the absence of a contrary bylaw provision, a majority of the directors shall constitute a quorum. Once there is a quorum, an action by a majority of the directors present at a meeting shall constitute an action of the board of directors unless the action requires a greater number than a majority under the charter or bylaws. The method of calling meetings and the notices required for special meetings are usually provided for in the corporate bylaws.

RARE EARTH, INC. v. HOORELBEKE

401 F.Supp.26 (D.C. Mich. 1975)

CANNELLA, D.J.

Rare Earth, Inc. is not as one might surmise, an organization dedicated to environmental activism or the preservation of our natural resources. Rather, it is the corporate entity formed by a group of rock and roll musicians who publicly perform as "RARE EARTH." From this group "comes the dissonant chord" of an intracorporate battle for control resulting from a schism among the band members in July, 1974.

"RARE EARTH" is a rock and roll performing group which has recorded several record albums and which enjoys national prominence in rock music circles. The group was organized in the late 1960s and originally consisted of five performers. A sixth member, Edward Guzman, later joined the band.

The Rare Earth group, consisting of two original members, Hoorelbeke and Bridges, as well as Guzman and the three replacement performers, continued to perform as a unit until July, 1974. However, at that time, the group divided into two factions, each of which now claims control of the corporation and the right to perform as "RARE EARTH." The first faction consists of Bridges, Guzman and two others; the second, of defendants Hoorelbeke and Urso. The Bridges–Guzman faction has presently placed itself in control of the corporation and currently performs as "RARE EARTH." The Hoorelbeke faction challenges this position and claims a majority interest in the corporation and the right to use the "RARE EARTH" mark. The differing views entertained by these individuals has resulted in this litigation.

Prior to the dissension which emerged in July, 1974, between the members of the Rare Earth group, it is undisputed that Peter Hoorelbeke served as a director and president, as well as a 200 share owner of Rare Earth, Inc. However, in mid-July the Bridges faction became aware (through the musical "grapevine") of Hoorelbeke's purported resignation as a band member and as an officer and director of the corporation. Such knowledge did not derive from any written communication from Hoorelbeke to the corporation, but rather resulted from rumor, or, to use a legal term, hearsay. Thereafter, on July 12, a directors' (or shareholders') meeting of Rare Earth, Inc. was convened in Los Angeles. Hoorelbeke was never notified of such meeting and did not attend; the Bridges faction acting upon "their sincere belief that Hoorelbeke had resigned (concluded that) there was no need to notify him of the meeting." At this July 12th meeting, it was decided that Bridges would replace Hoorelbeke as president of Rare Earth, Inc., and that the corporation would commence this action and retain counsel for its prosecution. As we will show infra, the actions taken on July 12th were improper.

Under Michigan law, "A director may resign by written notice to the corporation. The resignation is effective upon its receipt by the corporation or a subsequent time as set forth in the notice of resignation." An identical provision pertains with regard to the resignation of officers.

There is no evidence at bar of a written resignation transmitted by Hoorelbeke to the corporation and thus, as of the July 12th meeting, he remained a director, officer and shareholder of Rare Earth, Inc. This being so, the failure to notice Hoorelbeke of the meeting and his absence therefrom render all actions taken by those present invalid and without effect.

With regard to directors' meetings, Michigan law requires that a "special meeting shall be held upon notice as prescribed in the bylaws" and that a "director is entitled to a notice which will give him ample time to attend the meeting." The statutory requirement that the meeting be convened "upon notice" clearly was not met in the present case as Hoorelbeke received no notice whatsoever. Thus, it is a settled matter of Michigan law that "where a written notice of the meeting of the board of directors is not given although required by either a statute or the corporate bylaws, any action taken by the meeting at which all the directors are not present is void." The import of the foregoing discussion is plain: the failure to notify Director Hoorelbeke of the July 12th meeting and his absence therefrom render all actions taken thereat invalid, including such action as was required to commence this suit either directly or through the appointment of Bridges as President.

If the July 12th meeting is deemed a shareholders' meeting, the actions taken thereat similarly must fail for noncompliance with the notice requirements contained in Mich.Comp. Laws Ann. § 450.1404(1) or with the consent provisions contained in § 450.1407.

A corporation has been defined as a body of individuals united as a single separate entity. When corporate powers are vested in the shareholders or members, they repose in them collectively as a body and not as individuals. That is, individuals have no power to act as or for the corporation except at a corporate meeting called and conducted in accordance with law. 6 Callaghan, Michigan Civil Jurisprudence (1958).

Simply put, as with the case of directors, actions taken at a shareholders' meeting which has not been properly noticed are invalid and without effect.

Thus, we conclude that the meeting of Rare Earth, Inc. which was conducted on July 12, 1974 was a nullity for failure to comply with Michigan law. This being so, the statutory corporate power to commence litigation was not properly exercised. "Suits by corporations must be instituted . . . , as other acts by corporations must be done, by proper authority. The proper persons to authorize the commencement of a suit . . . on behalf of a corporation are primarily the directors, but the power may be vested in the president" as well. 9 W. Fletcher, supra § 4216 at 18–19 (rev. 1964) (footnotes omitted). Here, the actions taken by the directors (or shareholders) toward commencement of this suit are invalid, as are any actions taken by Bridges as president, whose election to that post is similarly tainted. While, "A subsequent ratification is equivalent to original authority for this purpose" (id. at 22), the facts at bar will not support a finding of ratification. There is no proof in the record that a duly constituted meeting or other action by the shareholders, directors or officers of Rare Earth, Inc. subsequently assented to the commencement of this action. The complaint must be dismissed for failure of Rare Earth, Inc., its directors and shareholders to abide by and conform to the dictates and requirements of Michigan Corporate law.

QUESTIONS FOR DISCUSSION

1. *What policy reasons can you give for the requirement that directors and stockholders must be notified of special meetings?*
2. *What was the effect in this case of failing to notify Director Hoorelbeke of the directors' meeting?*

---

(4) Compensation of Directors. Members of the board of directors of a corporation are usually not compensated other than to receive a small honorarium and expenses for traveling to the meetings. Under most state statutes the shareholders may vote to give directors compensation, but the directors may not grant themselves compensation unless authorized by the stockholders or by the corporate charter or bylaws. On the other hand, if an individual who is a director performs some service for the corporation that is beyond the normal duty of a director, he may be entitled compensation for those services.

Frequently the officers of the corporation are also corporate directors. An officer of the corporation, such as the president, is, of course, entitled to compensation for his or her services notwithstanding the fact that that individual also happens to be a director. The general rule is, however, that a director may not vote on his or her compensation; thus the compensation must be voted upon by the other members of the board. There must therefore be a quorum present, excluding the director who is being voted the compensation, in order for the resolution voting the compensation to be effective.

(5) Director's Liability as Fiduciary. Today potential litigants are looking increasingly to the role that corporate officers and directors have in the management of the company. The position of a director is that of a fiduciary. He or she has a fiduciary duty to the corporation and its stockholders. If that fiduciary duty is breached, the director may be liable to the corporation and its shareholders for damages.

Because of his position the corporate director frequently has information that can be used to his own advantage. A director can do nothing that will be either directly or indirectly disadvantageous to the corporation or the shareholders. It should be obvious that a director cannot make a secret profit at the expense of the corporation. For example, he cannot purchase a piece of property that he knows the corporation desires to purchase. He cannot purchase company securities that he knows the corporation is interested in buying. In short, his actions must be to benefit the corporation rather than himself. The director who benefits at the expense of the corporation may be made to account to the corporation for any profits made as the result of such transactions.

Other problem areas for directors involve two possible situations. One problem is whether one may enter into a contract with a corporation of which he is a director. The other situation occurs with regard to contracts between two corporations when an individual sits on the board of directors of each corporation. If a director has an interest in a contract that comes before the board of directors, the safe thing is not to attend the meeting when the matter is voted upon or at least not to cast a vote on the matter. The general rule appears to be that if the director has a personal interest in the contract and his presence is necessary for a quorum and a vote is required to bind the corporation to the contract, then the contract is voidable at the option of the corporation. A minority rule is that such a contract is voidable only upon a showing that it is not beneficial or is unfair to the corporation or that fraud was involved. This rule is in marked contrast to the historical attitude of some courts, most notably in New York, which at one time held that the corporation could not enter into a contract with a director regardless of the benefits to the corporation or the independence of the other directors. Fortunately that rule has been considerably relaxed today.

When a director serves on the board of another corporation, a contract between the two corporations is generally voidable only upon a showing that there was some fraud or that the agreement is unreasonable and unfair. Once again this is a relaxation of the rule absolutely prohibiting such agreements. It is also worthwhile noting at this point that one of the federal antitrust laws, the Clayton Act of 1914, prohibits interlocking directorates when either corporation has capital, surplus, and undivided profits totaling more than $1 million and if an interlocking directorate would tend to result in a substantial lessening of competition. The practical effect is to prohibit an individual from sitting on the boards of directors of two large potential competitors.

A further restriction on directors is imposed by what the courts frequently refer to as the *corporate-opportunity doctrine.* Under this doctrine a director may not, for example, purchase property that is leased to his corporation, purchase property that is needed by the corporation, draw customers away from the corporation to his own or another business in which he has an interest, or take advantage of an offer made to the corporation. If he does any of these things, he violates his fiduciary duty to the corporation by taking advantage of an opportunity presented as a result of his position as corporate director and must account for any profits received as a result of his actions.

Another problem area for a director involves the purchase or sale of the securities of the company of which he or she is a director. May a director who has special knowledge of corporate affairs purchase or sell stock when he knows that corporate circumstances will probably affect the market price of the stock? An example of such information would be the advance knowledge of a dramatic increase or decrease in corporate dividends. Directors and officers in this situation are known as *insiders* because of the inside information they have about the company. The trend of recent court decisions has been to require insiders to reveal information in their possession that might affect the price of the stock. Of course, if disclosure is required, it is facts that must be disclosed and not mere opinions. These court decisions have been influenced to some extent by the federal statutes that regulate this area, most notably the Securities Exchange Act of 1934 discussed in Chapter 42. For example, the Securities and Exchange Commission has adopted Rule 10b–5. This rule requires disclosure by the officer or the director of such facts as may affect the value or the potential value of the stock in a purchase or sale of stock listed on a national stock exchange. This rule also applies to unlisted stock when there has been use of any instrumentality of interstate commerce in the purchase or sale. This rule, of course, extends the duty owed by the officer or the director beyond the stockholders and the corporation to third parties.

PATIENT CARE SERVICES, S.C. v. SEGAL
377 N.E.2d 471 (Ind. 1975)

Marshall B. Segal and David A. Martinez were two physicians who formed Patient Care Services, S.C., a corporation, for the purpose of performing emergency-room services at Little Company of Mary Hospital. Both Segal and Martinez were directors. Segal was president of the corporation and Martinez was secretary–treasurer. After a period of time the relationship between Segal and Martinez grew strained, with some resulting problems with the hospital administration concerning services to be rendered to the emergency room.

On June 21, 1972, Patient Care, through Martinez, presented a plan to the hospital for continuation of the services. This plan was rejected. Emergency-room service was then provided by Medical Services, S.C., a corporation that employed Segal and of which Segal was an incorporator and part owner. Segal contended, among other things, that he had informed Patient Care of his intended action in forming Medical Services. A complaint was filed by Patient Care against Segal. The trial court found for Segal, and this appeal followed.

McNAMARA, J.

It must first be recognized that at least initially the finalization of a written contract to cover the first year of its operation of services at the hospital as well as the preparation and execution of a new one to embrace succeeding years constituted a corporate opportunity for Patient Care. Patient Care was organized for the purpose of providing comprehensive health services to a hospital. For the first year of its incorporation it was engaged in furnishing those services to the hospital, and the hospital was satisfied with those services. Obviously the very nature of its business necessitated a continuation and development of this relationship. When such a corporate opportunity exists, it is inherent in an officer's or director's fiduciary obligations to refrain from purchasing property for himself in which the corporation has an interest, actual or expectant, or which may hinder or defeat the plans and purposes of the corporation in the carrying on or development of the legitimate business for which it was created.

Viewed in this light, this court finds it indisputable that Segal blatantly violated those duties of loyalty and trust which he owed to Patient Care. While an officer and director of that corporation, he helped set up and subsequently took over control of a different corporation organized to perform the very similar, if not identical, services Patient Care was organized to perform.

Segal's response is to point out alleged breaches of contract by Martinez and to argue that any continuation of his relationship with Martinez would have resulted in his peonage to Patient Care. The disagreements with Martinez would still not condone the actions taken by Segal vis-à-vis Patient Care. If Segal felt undercompensated or taken advantage of, he could have resigned from Patient Care. If he felt Martinez breached any contract with him and/or Patient Care, the proper recourse was, and is, for Segal to sue in the courts for breach of contract.

Defendants' third point under this argument is that Segal allegedly acted in good faith. This proposition is predicated on the recognition that Segal gave prior notice to Martinez of his intention to negotiate on his own behalf.

Defendants' case authority holding that a corporate officer or director violates his fiduciary duty to his corporation by failing to inform the corporation of a business opportunity he seized as his own has no applicability to the present case. The cases cannot be inverted to hold that once he gives notice he is ipso facto free to contest with the corporation the business opportunity. In any event, it is clear that the determination of good faith rests upon the existence of many factors of which disclosure is only one. In *Paulman* v. *Kritrer* (1966) 74 Ill. App. 2d 284, . . . this court stated what those factors are. In rejecting a claim that an officer/director had acted in good faith in purchasing on his own behalf property in which the corporation would have been interested, this court stated:

> Whether a corporate officer has seized a corporate opportunity for his own depends not on any single factor nor is it determined by any fixed standard. Numerous factors are to be weighed, including the manner in which the offer was communicated to the officer; the good faith of the officer; the use of corporate assets to acquire the opportunity; the financial ability of the corporation to acquire the opportunity; the degree of disclosure made to the corporation; the action taken by the corporation with reference thereto; and the need or interest of the corporation in the opportunity. These, as well as numerous other factors, are weighed in a given case. The presence or absence of any single factor is not determinative of the issue of corporate opportunity.

(Reversed and remanded.)

QUESTIONS FOR DISCUSSION

1. *What was the "corporate opportunity" that Segal appropriated from Patient Care?*
2. *Do you think that Segal could have rendered services to any other hospital while still involved with Patient Care?*

SMITH v. CITATION MANUFACTURING COMPANY, INC.
587 S.W.2d 39 (Ark. 1979)

Bob R. Smith was elected corporate vice-president by Citation Manufacturing Company, Inc., in January 1975. At the same time he was elected a director of the corporation. He went on the payroll as vice-president about March 1st and was fired from that job in June. He continued as a director until January 1976.

HICKMAN, J.

Citation, an Arkansas corporation located in Siloam Springs, Arkansas, produces and distributes industrial cleaning equipment. Bob Smith had been a distributor of industrial cleaning equipment in California. He was introduced to Citation's Board of Directors as one of the largest distributors in the states. He was hired as vice-president of operations and at the same time elected to Citation's Board of Directors. Elmer Heinrich, the president and chairman of the board, had known Bob Smith, both of them having been in the business for some time, and introduced Smith to the board.

From January, 1975, through June, Citation sold to Smith's California company some $65,000.00 worth of equipment.

Smith sold his company in California about June the 6th. It filed voluntary bankruptcy in October, 1975.

Citation claimed it was unable to collect all that was due for the sales to Smith's company and suffered damages totalling $30,688.88—as a result of Smith's breach of his fiduciary duty.

The chancellor entered these specific findings and conclusions:

From January 27, 1975 through June 1, 1975 Bob Smith had knowledge that Smith Investments and Equipment Sales were having cash flow problems, problems with collecting accounts, problems meeting their obligations as they arose, and other financial problems all of which were not disclosed to Citation. Knowledge of such problems would have been material to Citation's decisions to extend credit and ship goods to Equipment Sales.

In June, 1975, Bob Smith received from Smith Investments two machines that Equipment Sales had purchased from Citation. Smith Investments had not paid Citation for the machines and Bob Smith knew this; he sold them for his own personal gain. . . .

Bob Smith, while serving as an officer and/

or director of Citation, was negligent and breached his implied obligations and fiduciary duties as follows:

> A. By failing to exercise ordinary diligence and good faith in neglecting to provide Citation with the information he had concerning his company's probable inability to pay for goods purchased on credit from Citation.
> B. By failing to warn Citation that his companies would likely be unable to pay for the equipment purchased on credit when diligence and good faith required such warning.
> C. By transferring equipment from Smith Investments and Equipment Sales to himself for satisfaction of his own personal obligations, knowing that Citation had not been paid for such equipment.
> D. By permitting Citation to advance credit to Smith Investments and Equipment Sales in amounts exceeding their ability to pay when he knew or should have known that these companies would not pay for the goods.
> E. By failing to make full and complete disclosure to Citation of the financial condition and affairs of Smith Investments and Equipment Sales as he knew them to be.

As a result of Bob Smith's negligence and breaches of fiduciary duties, Citation has been damaged in the sum of $24,887.17 for which it should have judgment. . . .

Smith was both a director and officer of Citation.

We defined the fiduciary duty of a director in the case of *Sternberg* v. *Blaine,* 179 Ark. 448, 17 S.W.2d 286 (1929) as follows:

> It may therefore be stated as the settled rule in this State that any failure of a director to exercise diligence or good faith which results in loss to a stockholder or creditor, entitles such stockholder or creditor to require the directors whose negligence have caused the loss to pay. In other words, the director whose negligence causes loss is liable for such loss to stockholders and creditors. *Id.* at 453, 17 S.W.2d at 288.

We said in *Bank of Commerce* v. *Goolsby,* 129 Ark. 416, 196 S.W. 803 (1917):

> For all practical purposes they (directors) are trustees when called upon in equity to account for their official conduct.

There is an even greater duty on one who is both a director and an officer in a company.

The evidence in this case clearly supports the chancellor's finding of a breach of the fiduciary relationship. There was testimony Citation did not fully know of the financial condition of Smith's company, and if they had, the goods would not have been shipped.

Smith tried to offer as evidence an unaudited financial statement prepared by an accountant. The chancellor properly excluded it. It was not properly authenticated and no basis was laid for it being admitted as a business record, nor any exception defined in Rule 803(6), Arkansas Rules of Evidence.

(Affirmed in favor of Citation Mfg. Co.)

QUESTION FOR DISCUSSION

1. *What action should Smith have taken to avoid liability in this situation?*

(6) Director's Duty of Care. It is settled law that directors are not insurers of the success of the corporate enterprise. On the other hand, the directors must not act negligently in the management of corporate affairs. It is no defense for a director to claim that he took no active role in the board proceedings and was "ignorant" of any decisions made. In determining director liability for corporate mismanagement, the courts frequently apply the *business-judgment rule.* By this is meant that if the directors exercised reasonable business judgment in their management of the corporation, there is no liability, regardless of the consequences of that judgment. In other words, the courts will not use hindsight or substitute their own judgment for that of the directors so long as the judgment is reasonable. The fact that the judgment turned out wrong, with resulting losses to the corporation, makes no difference. If a mistake is made because of honest business judgment, there is no liability. The rule, however, presupposes that the judgment was reasonably exercised under the circumstances. The director cannot close his eyes to all that is going on around him and escape liability for that reason. On the other hand, most courts hold that the director is entitled to rely on the statements of officers and employees of the corporation as to the activities and condition of the firm, and he is not required to launch an independent investigation into the corporate condition so long as such reliance is reasonable. In other words, the director is not liable for false information supplied to him so long as his reliance on that information is reasonable.

GRAHAM v. ALLIS-CHALMERS MANUFACTURING COMPANY
188 A.2d 125 (Del. 1963)

Allis-Chalmers Manufacturing Company was indicted for violation of the federal antitrust laws. A derivative action was filed against the directors seeking to recover damages alleged to have been suffered by Allis-Chalmers because of those violations. The suit contended that the directors should have learned of the violations by putting into effect a system which would have revealed employee misconduct to them. The trial court found for the defendants, and the plaintiffs appealed.

WOLCOTT, J.

The Board of Directors of fourteen members, four of whom are officers, meets once a month, October excepted, and considers a previously prepared agenda for the meeting. Supplied to the Directors at the meetings are financial and operating data relating to all phases of the company's activities. The Board meetings are customarily of several hours duration in which all the Directors participate actively. Apparently, the Board considers and decides matters concerning the general business policy of the company. By reason of the extent and complexity of the company's operations, it is not practicable for the Board to consider in detail specific problems of the various divisions.

From the Briggs case and others cited by plaintiffs, it appears that directors of a corporation in managing the corporate affairs are bound to use that amount of care which ordinarily careful and prudent men would use in similar circumstances. Their duties are those of control, and whether or not by neglect they have made themselves liable for failure to exercise proper control depends on the circumstances and facts of the particular case.

The precise charge made against these director defendants is that, even though they had no knowledge of any suspicion of wrongdoing on the part of the company's employees, they still should have put into effect a system of watchfulness which would have brought such misconduct to their attention in ample time to have brought it to an end. However, the Briggs case expressly rejects such an idea. On the contrary, it appears that directors are entitled to rely on the honesty and integrity of their subordinates until something occurs to put them on suspicion that something is wrong. If such occurs and goes unheeded, then liability of the directors might well follow, but absent cause for suspicion there is no duty upon the directors to install and operate a corporate system of espionage to ferret out wrongdoing which they have no reason to suspect exists.

The duties of the Allis-Chalmers Directors were fixed by the nature of the enterprise which employed in excess of 30,000 persons, and extended over a large geographical area. By force of necessity, the company's Directors could not know personally all the company's employees. The very magnitude of the enterprise required them to confine their control to the broad policy decisions. That they did this is clear from the record. At the meetings of the Board in which all Directors participated, these questions were considered and decided on the basis of summaries, reports and corporate records. These they were entitled to rely on, not only, we think, under general principles of the common law, but by reason of 8 Del.C. § 141(f) as well, which in terms fully protects a director who relies on such in the performance of his duties.

In the last analysis, the question of whether a corporate director has become liable for losses to the corporation through neglect of duty is determined by the circumstances. If he has recklessly reposed confidence in an obviously untrustworthy employee, has refused or neglected cavalierly to perform his duty as a director, or has ignored either willfully or through inattention obvious danger signs of employee wrongdoing, the law will cast the burden of liability upon him. This is not the case at bar, however, for as soon as it became evident that there were grounds for suspicion, the Board acted promptly to end it and prevent its recurrence.

Plaintiffs say these steps should have been taken long before, even in the absence of suspicion, but we think not, for we know of no rule of law which requires a corporate director to assume, with no justification whatsoever, that all corporate employees are incipient law violators who, but for a tight checkrein, will give free vent to their unlawful propensities.

We therefore affirm the Vice Chancellor's ruling that the individual director defendants are not liable as a matter of law merely because, unknown to them, some employees of Allis-Chalmers violated the anti-trust laws thus subjecting the corporation to loss.

(Affirmed.)

Questions for Discussion

1. *Were the corporate directors in this case entitled to rely on the word of the company employees that no illegal activities were being practiced by the company, or did the directors have to make an independent investigation?*
2. *If there are reasons to be suspicious of the validity of reports filed by employees, what duty does a director have?*

SCHEIN v. CAESAR'S WORLD, INC.
491 F.2d 17 (1974)

Caesar's World, Inc., through its subsidiary, Desert Palace, Inc., entered into an "equipment lease" *with Centronics Data Computer Corporation to install a computerized casino system in the Caesar's Palace Casino. The lease provided that either party had the right to terminate the lease prior to the end of a six-month, rent-free period, which was to begin when* "sufficient equipment and accessories" *were* "installed . . . and operating. . . ."

Simultaneously, and as an inducement to Caesar's to sign the lease, Centronics assigned to Caesar's an option to purchase 510,000 shares of Centronics stock. Centronics retained the right to recapture 50 percent of the stock in the event that the equipment lease was terminated prior to the end of the six-month period.

While Centronics was attempting to install the system, its stock began to rise in value, so Caesar's World exercised its option to obtain the shares. Centronics was never able to place the full system in operation. Caesar's directors agreeably could have obligated themselves to pay $10,000 per month under "Exhibit C" *of the lease and cut off Centronics's right to recapture its stock, which in the meantime had continued to rise substantially in value. A dispute arose between the parties as to Caesar's right to do this, and the directors of Caesar's compromised the dispute by allowing Centronics to recapture 100,000 of the 255,000 shares that it contended it had a right to recapture. Centronics also agreed to register its stock, which would allow Caesar's World to dispose of its holdings more readily.*

Jacob Schein, a stockholder, instituted a derivative action against the officers and directors of Caesar's World, contending that Caesar's should not have entered into the compromise agreement with Centronics. The district court granted the defendant's motion for summary judgment, and Schein appealed. The appellate court adopted the trial court's memorandum as its opinion.

FAY, J.

Initially, it is to be noted that the management of corporate business is vested in the directors of a corporation, the directors having wide discretion in the exercise of business judgment in the performance of their duties. F.S.A., § 608.09. While directors are required to discharge their duties with the diligent care and skill which ordinarily prudent men would exercise under similar circumstances in like positions, they incur no liability to the corporation for issues of business expediency which they resolve through the mere exercise of their business judgment.

It is patently obvious to this Court that in executing the compromise agreement, the directors of Caesar's World fulfilled their corporate duties with the diligent care and skill which ordinarily prudent men would exercise under similar circumstances in like positions. While not passing on the merit of the contention, a persuasive argument could be made that Caesar's right to require execution of "Exhibit C," thus foreclosing Centronics from recapturing one-half of its stock, did not ripen until the commencement of the six-month trial period.

Thus, faced with the prospect of protracted litigation—in which Caesar's World might not prevail—over the contractual rights of the parties to unregistered volatile stock whose value might plummet as spectacularly as it has risen, the defendant directors executed a compromise agreement with Centronics. (While it is not dispositive of the issue at bar, it is noteworthy that an immediate profit of $6,767,000, on a $1,700,000 investment accrued to Caesar's World as a result of the compromise.)

Vested with the management of the business of Caesar's World, it was incumbent upon the corporate directors to evaluate the relative risks of alternative courses of action. Accordingly, the decision of the defendant directors was indisputably a sound exercise of their business judgment which, absent any allegations or showing of bad faith or fraud, is not judicially reviewable. These principles were recently enunciated as follows:

> . . . The management of corporate business is vested in the directors of the corporation, and Courts in Florida have given to directors wide discretion in the exercise of business judgment in the performance of their duties. A Court of equity will not attempt to pass upon questions of mere business expediency or the mere exercise of business judgment, which is vested by law in the governing body of the corporation. *Yarrell Ware & Tr., Inc.* v. *Three Ivory Bros. Mov. Co.*, supra 226 So2d at 890–891. . . .

(Affirmed.)

QUESTIONS FOR DISCUSSION

1. *What standard of behavior for directors and officers was applied by the court in this case?*
2. *Would the decision have been any different had Caesar's lost money on the stock purchase?*

44:3. Officers' Role in Corporate Management. Most of the corporation statutes provide that the officers of a corporation shall be the president, the vice-president, the secretary, and the treasurer. There may be such other officers as the bylaws prescribe. These officers are appointed by the board of directors and are generally given the authority to manage the day-to-day operations of the company. This authority may be conferred by the bylaws and by resolution of the corporate directors. It is important to note that although the officers are agents of the corporation and the general rules of agency apply to their actions, their actual authority is granted only by the board of directors. It is important, therefore, when one is contracting with a corporation through its officers, to make sure that the officer has the authority. This care is particularly necessary when one is contracting on an important matter, such as the purchase of corporate assets, for in such matters if the officer lacks actual authority the courts are reluctant to find any apparent authority to bind the corporation.

The officers of the corporation hold a fiduciary relationship to it just as the directors do. An officer may be held to account to the corporation for any losses it sustains as a result of the breach of that fiduciary duty.

The board of directors has the power to remove an officer at any time that it determines that it is in its best interest to do so. However, if such a removal breaches a contract between the officer and the corporation, then the corporation may be liable in damages to the officer for such a removal.

An officer is, of course, entitled to be compensated for his services to the corporation even if the officer also happens to be a director of the corporation. If the officer is negligent in the performance of such services or commits an illegal or tortious act, he may, of course, be liable in damages to the corporation. He is not liable for mere errors in judgment, however.

SMITH v. DUNLAP

111 So.2d 1 (Ala. 1959)

In June of 1948 the directors of the Alabama Dry Docks and Shipbuilding Company, Inc., adopted a bonus plan providing for additional compensation to officers and other employees based upon a percentage of the company's earnings. A stockholder filed a derivative action questioning the reasonableness of the bonus plan. The trial court sustained the corporation's demurrer and dismissed the suit. The stockholder then appealed.

SIMPSON, J.

In the leading case of *Rogers* v. *Hill,* 289 U.S. 582, 54 S.Ct. 731, 77 L.Ed. 1385, . . . the rule was enunciated that where the amount of a bonus payment to officers of a corporation has no reasonable relation to the value of service for which it is given, it is in reality a gift and the majority stockholders have no power to give away corporate property against the protest of a minority stockholder.

A long line of Alabama cases recognizes the general rule that where officers of a corporation appropriate assets of the corporation to their own use, equity will intervene on behalf of a minority stockholder who is unable to obtain relief within the corporation. (Citation omitted.) The foregoing cases are illustrative of the principle that the receipt of excessive compensation by the officers of a corporation is manifestly an appropriation of corporate assets by said officers to their own use.

The question of whether the compensation is so excessive that it bears no reasonable relation to the value of services rendered is a question of fact to be resolved on final hearing. It was observed in *Gallin* v. *National City Bank of New York,* . . . that "To come within the rule of reason the compensation must be in proportion to the executive's ability, services and time devoted to the company, difficulties involved, responsibilities assumed, success achieved, amounts under jurisdiction, corporate earnings, profits and prosperity, increase in volume or quality of business or both and all other relevant facts and circumstances; nor should it be unfair to stockholders in unduly diminishing dividends properly payable."

After careful study and analysis of the leading cases, text writers, and student comment in law reviews, and after reconciling some of the inconsistencies therein, we conclude that the following principles govern cases of this nature: The amount of compensation to be paid to an officer of a corporation is, in the first instance, within the business discretion of the corporation's board of directors and with this discretion the courts are loath to interfere; generally the decision of the directors as to the amount of such compensation is final; where it appears, however, that the directors have not acted in good faith or that the compensation fixed by them is so excessive that it bears no reasonable relation to the services for which it is given, courts of equity have the power to inquire whether and to what extent payment to the officers constitutes misuse and waste of corporate assets; the power to inquire will, therefore, be exercised by the courts upon a clear showing of excessiveness of compensation or bad faith on the part of the directors; but courts are reluctant and will proceed with great caution in exercising the power to "prune" the payments since it is not intended that a court should be called upon to make a yearly audit and adjust salaries; nor is such an inquiry merely to substitute the court's discretion for the discretion of the directors if that has been honestly and fairly exercised.

We conclude that complainant, a minority stockholder, has sufficiently stated a case for the intervention of equity on behalf of the corporation to inquire as to whether the compensation received by the individual respondents is so excessive that it bears no reasonable relation to the value of the services performed by them. This does not mean, however, that the compensation is per se so excessive; it is a question of fact and the compensation having been regularly fixed by the directors, the burden of proving that it is so excessive is on complainant.

(Reversed and remanded.)

Question for Discussion

1. *What factors did the court say should be considered in a determination of whether the compensation of corporate officers is excessive?*

SUMMARY

Stockholders are the residual owners of the corporation but have no direct role in the daily operation of the company. Most statutes require at least one annual meeting of stockholders and any number of special meetings upon the issuance of a proper notice.

Directors are the top echelon of corporate management and are elected by the stockholders. The individual director has no power as only the full board of directors acting as a board has authority.

Some statutes require a board member also to be a shareholder. Other limitations may be stated in the charter or bylaws. If there is a special meeting of the board, proper notice must be given. Often other than to receive a small honorarium, board members are generally not compensated.

A director is a fiduciary and thus has a high duty of loyalty to the corporation and may be liable to the corporation for breaching that duty. The courts may also impose the corporate-opportunity doctrine against directors. Directors are also considered insiders in their purchase of the corporate stock and thus are subject to the rules of the Securities and Exchange Commission. In determining director liability for corporate management, the courts apply the business-judgment rule.

Officers of the corporation are appointed by the board of directors. They also hold a fiduciary relationship to the corporation and may be removed by the board at any time. An officer is entitled to be compensated for his services.

PROBLEMS

1. Harold F. Stone was chairman of the board of directors of American Lacquer Solvents Company. On December 7, 1967, the board adopted a resolution providing that in consideration of Stone's services American would pay his wife, Rachel, an annual pension of $8,000 if he died before she died. The resolution provided that it could not be revoked without Harold Stone's consent. Several months later Stone and his wife had a marital dispute, and Stone notified the company president that the resolution providing a pension for his wife was to be canceled. A meeting of directors was held,

which five of the seven members attended. They voted unanimously to rescind the resolution. Stone was not notified of the meeting. In November 1, 1968, Harold Stone died. Rachel Stone contended that she should receive the pension. Was she correct? Stone v. American Lacquer Solvents Company, 345 A.2d 174 (Penn. 1975).

2. The directors of the Metals & Chemical Corporation passed a resolution empowering the president to borrow $100,000. The resolution also provided that the proceeds of any funds received from a public sale of the company stock should be applied toward payment of the loan. The loan was made, and subsequently there was a public sale of company stock. The proceeds were not used to pay the loan but were used for other corporate purposes. In response to a lawsuit the directors contended that they were not liable because they had in no way misappropriated or dissipated any funds of the corporation. Were the directors correct? Emmert v. Drake, 224 F.2d 299 (1955).
3. Harry Meth was president of the McClay Company. He had been a co-founder and associated with the company for over fifty years. In 1966–1967 Harry Meth became ill but continued to draw his salary. The corporation had been profitable while Harry Meth was president and remained profitable during this time. A derivative action was filed contending that the payment of the normal salary to Meth during his illness was unreasonable compensation. The trial court agreed. What ruling on appeal? Bermann v. Meth, 258 A.2d 521 (Pa. 1969).
4. Judson B. Glen had been president as well as a director of Bancroft-Whitney Company. The company is a publisher of law books. Glen then agreed to work for Mathew Bender Co., also a publisher of law books. Before he left Bancroft Glen confronted various of its employees in order to induce them to leave Bancroft for Bender Co. He also disclosed the salaries of Bancroft employees to Bender Co. in order to better facilitate Bender's solicitation of them. Ultimately Bender Co. or Glen offered higher salaries to Bancroft's treasurer, three of its four managing editors, two of the four assistant managing editors, three of four indexers, and ten other editors. Bancroft sued Glen for breach of fiduciary duty. Glen contended, among other things, that he had a right to leave Bancroft and to induce others to leave. The trial court found in favor of Glen. Was it correct? Bancroft-Whitney Co. v. Glen, 411, P.2d 921 (1966).
5. Harris Lumber Co. borrowed a sum of money from Merchants & Farmers' Bank. There were three directors of Harris Lumber Co.: D. F. Harris, who was also president; M. A. Nelson, who was also secretary–treasurer; and John Jones. On the day the promissory note was signed, the three directors met at the Merchants & Farmers' Bank and agreed to borrow the money. All agreed that the money should be borrowed. Nelson then executed a promissory note on behalf of the company, payable to the bank. The Harris Lumber Co, refused to pay the note. It contended that the loan was not authorized because there had been no formal meeting of the directors and

no resolution had been passed authorizing the loan. At trial the trial judge refused to allow the introduction of the mortgage based upon Harris Lumber Co.'s contentions, and a judgment was entered in its favor. Was the trial court correct? Merchants & Farmers' Bank v. Harris Lumber Co., 146 S.W. 508 (1912).

6. At a special meeting of the board of directors of El Tejon Oil & Refining Co., a dividend was declared. There was no notice of the meeting and only four of the seven directors attended. No waiver of notice was signed by the remaining three directors. The dividend was paid in cash to all holders of common stock, including the seven directors. The seven directors all immediately returned the cash to the corporation and took promissory notes in return. Meyers, a director, never had his note paid and sued the corporation. The corporation defended on the ground that the dividend was illegal under the circumstances. Meyers contended that the action was ratified by the board of directors. Decide. Meyers v. El Tejon Oil & Refining Co., 174 P.2d 1 (1946).
7. In a celebrated case Socony-Vacuum Oil Co., Inc., was indicted and convicted for violating the federal antitrust laws in the purchase of "distress" gasoline from small refiners in the Midwest in concept with other large oil companies. Socony was fined for this action. A suit seeking to recover the fine from them was filed against the company directors. The directors were well aware of the buying program but contended that the statute under which Socony was convicted was ambiguous and that they were unaware that the company activity violated the law. Should the directors have paid? Simon v. Socony-Vacuum Oil Co., Inc. et al., 38 N.Y.S.2d 270 (1942).

CHAPTER 45

Corporate Dissolution

The existence of a corporation may be terminated by dissolution. There are very definite ways provided by the corporation statutes of the various states by which the corporate existence may come to an end. The mere fact that a corporation ceases doing business or becomes insolvent does not mean that its existence is terminated. The statutory provisions must be adhered to. Sometimes corporate dissolution occurs because of the actions of the stockholders or as the result of a charter provision. Occasionally it is the result of judicial or official action. The purpose of this chapter is to examine the various ways in which the corporate existence may be terminated.

45:1. Dissolution by Stockholder or Pursuant to Charter. Under certain conditions the corporate stockholders have a right to dissolve the corporation. They accomplish this by surrendering the corporate charter to the state. In most jurisdictions, at least a majority of the stockholders must vote to surrender the corporate charter in order for the corporate existence to end. A minority group of shareholders generally does not have the power to dissolve the corporation. On the other hand, there have been instances in which the courts have prevented a majority of the shareholders from dissolving the corporation when such a dissolution would have worked a fraud upon the minority. However, if there is no fraud, the required majority of stockholders may dissolve the corporation by passing a resolution of dissolution at a duly called stockholder's meeting. Articles of dissolution are then prepared according to the applicable statute and filed with the appropriate state official. Note that the fact that the shareholders agree to dissolve the corporation does not terminate it unless the steps required by statute are carried out.

As a general rule, the directors of a corporation do not have the power to dissolve it because the role of the board of directors is to manage the corporation; without a particular statutory provision or a resolution of the stockholders the board has no power to dissolve the corporation. The same is true of preferred shareholders. They have no legal interest in the continuance of the corporation because, as a general rule, the purpose of preferred stock is to give the preferred shareholders a priority as to the corporate assets.

Most commonly, corporate charters provide for the perpetual existence of a corporation. It is possible, however, to provide in the charter for the termination of the corporation upon the happening of some event or condition or upon the expiration of a stated period of time. For instance, a corporation may be formed expressly to accomplish a particular purpose, such as to construct an office building. The corporate charter may provide that the corporation will be dissolved upon the completion of the building. If the corporate existence is terminated as the result of the happening of an event or the expiration of time, the corporation is automatically dissolved. Under some statutes the corporation may continue for a limited period solely for the purpose of winding up its affairs.

45:2. Involuntary Dissolution. There are a number of situations in which a corporation may be involuntarily dissolved. In some of these situations the state, usually through its attorney general, may institute a *quo warranto* proceeding for forfeiture of the corporate charter. The courts will usually grant a forfeiture of the charter only if there is no less drastic remedy available. Among the reasons most often advanced for the forfeiture of the charter are the failure to file the required reports, nonpayment of taxes, and misuser, nonuser, or fraud in the procurement of the corporate charter. *Nonuser* involves the failure of the corporation to organize and carry on operations or a prolonged suspension of operations. *Misuser* means that the corporation has abused or made illegal use of its corporate powers.

The question sometimes arises as to whether a corporation may be dissolved upon the petition of a minority of the stockholders. There are statutes in many states that do in fact provide for the dissolution of a corporation at the behest of minority stockholders under certain circumstances. The circumstances most frequently cited are fraud on the minority by the majority shareholders, gross mismanagement of the corporation, and abandonment of the corporate objectives. If dissensions among the shareholders make the carrying on of the corporate business impossible or unprofitable, most states have statutes allowing for dissolution under these circumstances. If there is no such statute, the modern trend is for the courts to find that they have the power to dissolve the corporation if it is necessary to protect the rights of the minority shareholders or the creditors. Needless to say, this remedy is viewed as rather radical by the courts and is applied only when no less drastic measures are feasible.

PEOPLE BY LEFKOWITZ v. THERAPEUTIC HYPNOSIS
374 N.Y.S.2d 576 (1975)

The people of the state of New York through the state attorney general, Louis J. Lefkowitz, petitioned the court to enjoin James D. McMillen and other individuals from carrying on certain fraudulent business activities. The petitioner also asked that the corporations established by McMillen and his associates be dissolved.

MAHONEY, J.

Subdivision 12 of section 63 of the Executive Law authorizes the Attorney General to apply to the Supreme Court for an order enjoining repeated or fraudulent acts in the carrying on or conduction of any business activity. From the supporting papers and annexed exhibits herein, including the sworn testimony of respondent McMillen taken at investigatory hearings conducted by the Attorney General, it clearly appears that respondent McMillen held himself out to the public, and to his own employees, to be a doctor of psychology, a doctor of theology, a doctor of metaphysical science as well as a college and university graduate, when, in fact, he was none of these nor had he ever attended or graduated from any school above the high school level. Next, respondent McMillen caused, or participated in the causing, numerous newspaper advertisements, brochures, radio and television spot interviews, to be published wherein representations were made that respondent National Institute of Hypnosis Practices, Inc. (NIHPI) was a national corporation overseeing the practice of hypnosis and only licensed those qualified by education and degree to practice that art, when, in fact, NIHPI was a corporation organized by respondent for the sole purpose of issuing certificates to respondent Therapeutic Hypnosis, Inc. (THI) so that corporation could enhance the dignity of its offices by displaying such certificate on its walls.

There is also undisputed evidentiary allegation that respondent McMillen and THI represented to the public that they were qualified and licensed to treat members of the public for physical ailments and, in fact, did so treat some members of the public.

From all of the above, the Court finds that all of the named respondents, corporate and individually, made false, inaccurate and misleading representations to the public that they were professionally qualified to practice hypnosis when, in fact, they were not and, further, that they made false and fraudulent representations to the general public, with the intent to deceive, that they had achieved a high percentage of success in curing people of the habits of smoking and overeating when, in fact, there were not any statistics to justify such representations and such representations were in violation of sections 349 and 350 of the General Business Law.

The Court also finds that respondent NIHPI was incorporated for unlawful purposes, as was THI, all in violation of section 349 of the General Business Law. Further, it is also the finding of the Court that both THI and NIHPI were incorporated for the sole and exclusive purpose of issuing certificates to respondent McMillen and other individuals associated with him to the end that McMillen and others might more successfully hold themselves out to be authorized and degreed practitioners of an implied medical art, all of which was false. . . .

Judgment is granted . . . permanently enjoining the named respondents, individually and corporately, as well as all incorporators, stockholders, officers, directors, agents and employees of the corporate respondents from engaging in any of the fraudulent and illegal practices hereinabove particularized; . . . cancelling certificates of incorporation issue to THI and ordering dissolution of THI, and cancelling certificates of incorporation issue to NIHPI and ordering its dissolution. . . .

45:3. Consequences of Dissolution. The dissolution of a corporation is the legal equivalent of its death. It terminates the corporate existence, and the corporation is no longer able to perform any of the acts normal to a continuing business. The dissolved corporation may not make contracts, hold property, sue or be sued, or perform other corporate acts. Under the modern corporation statutes, however, the corporation may continue a limited existence in order to perform those acts consistent with winding up and liquidating its affairs. The corporation would thus be able to sell its property, collect and pay debts, and do whatever else is necessary to wind up its affairs. Acts that are inconsistent with dissolution, however, are generally void. Thus, if the corporation enters into a contract after dissolution, the contract would be void unless it were of the type authorized by statute in order to wind up the affairs of the corporation.

After the corporation is dissolved, its property and assets belong to the stockholders as residual owners of the corporation. This property is, of course, subject to the claims of creditors, and if the corporation is insolvent upon dissolution, the stockholders will receive nothing. If there are preferred stockholders, they have priority over the common shareholders when the assets of the corporation are distributed. Sometimes a question also arises as to the rights and liabilities of those stockholders who have not fully paid for their stock. If the corporation is solvent, these shareholders participate in the distribution of assets in proportion to the portion of the price of the stock that they have actually paid. On the other hand, if the corporation is insolvent, the shareholder may be assessed for the unpaid portion of his stock. Shareholders may also be liable for assets received prior to the payment of creditors. Those who make such an illegal distribution would also be liable in damages to the creditors.

JOHNSON v. HELICOPTER & AIRPLANE SERVICES CORP.

404 F.Supp. 726 (D.C.Md. 1975)

RAC Corporation was originally incorporated in Delaware. In 1965 it sold all of its assets to another corporation and proceeded to wind up its affairs. A certificate of dissolution was filed on November 19, 1968, with Delaware's secretary of state. After that time RAC continued to prosecute tax claims against the United States, its board of directors continued to meet, it filed tax returns, and it invested its assets in short-term commercial paper. The continued existence of the corporation was authorized under the Delaware corporation statute that gives a corporation three years of existence to wind up its affairs.

RAC Corporation was sued along with several other defendants in a products-liability suit filed by Johnson. The suit was filed more than three years after the certificate of dissolution was issued by Delaware. RAC Corporation filed a motion to dismiss, contending that it lacked capacity to sue or be sued.

YOUNG, J.

At common law, the dissolution of a corporation was its civil death; dissolution abated all pending actions by and against a corporation, thus terminating abruptly its capacity to sue and be sued.

In order to alter the common law and prolong the life of a corporation past dissolution, statutory authority is necessary. . . .

The common law has been supplanted in Delaware, as in all states, by a statute which prolongs the life of a corporation in order to allow the corporation to dispose of its affairs in an orderly fashion.

The continued existence of the corporation is thus strictly limited under Delaware law to a few specific situations:

First, under Section 278, a corporation's existence is continued for three years after dissolution. During that time, the corporation may not conduct the business for which it was originally incorporated. It may conduct only such business as is "incidental and necessary to . . . wind up." . . . The power given to the corporation to dispose of its affairs during this period expressly includes the capacity to sue and be sued; Delaware traditionally held that after dissolution, a corporation could sue and be sued only during this period.

Second, under Section 278, the three-year winding up period may be extended by the Court of Chancery. Such an extension would probably prolong the corporation's capacity to sue and be sued.

Third, automatic prolongation of the life of the corporation beyond the three-year period is provided for by Section 278 for the limited purpose of allowing actions previously commenced by and against the corporation to continue to their expiration. This obviously does not prolong the capacity of the corporation to sue and be sued, but maintains the continuance of the corporation for the benefit of the existing suits.

Fourth, Section 279 of the corporation law provides that the Court of Chancery may at any time appoint receivers or trustees to conduct the business of the corporation.

In summary, then, notwithstanding other indicia of corporate existence continued by the statutes, Section 278 and 279, read jointly, seem to indicate that a corporation has capacity to sue or be sued 1) during the three-year winding-up period; 2) beyond three years if an extension has been procured; or 3) if a receiver or trustee has been appointed by the Court of Chancery.

None of these conditions is present in this case, where suit was brought fully a year after the three-year winding-up period provided in Section 278 had passed. An extension of this period was never procured and a receiver has never been appointed.

RAC is bringing to a close its business, and its activities are consistent with that end. While the corporation is still holding assets, it naturally retains a board of directors, holds meetings, issues reports and invests its assets. To do otherwise might be a breach of the fiduciary duty owed by the directors of the corporation to manage the corporation's affairs. Nevertheless, three years have passed since dissolution, and although RAC is still conducting certain proceedings brought during the winding-up period, it has lost the capacity to sue and be sued in new actions, including the one instituted by the plaintiff here.

(The motion of RAC Corporation to dismiss is granted.)

QUESTIONS FOR DISCUSSION

1. *What is the purpose of the Delaware statute in providing for corporate existence for three years after dissolution?*
2. *May a corporation in Delaware continue its normal business operations after dissolution?*

45:4. CONSOLIDATION AND MERGER. Another action that will result in the termination of corporation existence is the consolidation or merger of the corporation with another corporation.

A *consolidation* occurs when two or more corporations join to form a single new corporation. Both of the original corporations cease to exist, and a brand new corporation is formed, which takes title to all of the assets and property of the former corporation. The new corporation assumes all the debts and liabilities of the former corporation. In order for a consolidation to take place, it is necessary that the boards of directors of the consolidating corporations approve the consolidation as well as at least two thirds of the shareholders of the corporations. If a shareholder does not want to become a shareholder of the newly formed corporation that results from the consolidation, he is entitled to be paid the fair value of his stock. The rationale for this right is that the shareholder should not be forced to become an owner of a new corporation that is different from the one in which he originally invested.

A *merger* of two or more corporations occurs when all the assets and property of the merging corporations is transferred to one of those corporations, which is the survivor. The other corporations terminate their existence so that only one remains, but no new corporation is formed as in a consolidation. Just as in a consolidation, a merger requires the approval of the boards of directors of the merging corporations as well as the vote of, usually, two thirds of the stockholders of the corporations. Shares of stock in the surviving corporation are then issued to the shareholders of the merged corporations. A stockholder who dissents may receive the value of his stock from the surviving corporation.

LAMB v. LEROY CORPORATION

454 P.2d 24 (Nev. 1969)

Nevada Land and Mortgage (N.L.M.) through its wholly owned subsidiary, Nevada Insurance Agency, was indebted to Frank W. Lamb in the amount of $15,087.87. Because of financial reverses N.L.M. contemplated a merger and approached Leroy Corporation with a merger proposal. Leroy Corporation rejected the proposal but agreed to purchase the assets of N.L.M. and a "Sale of Assets Agreement" was signed on May 21, 1965. The assets of N.L.M. were transferred to Leroy in return for Leroy stock. N.L.M. was then dissolved without paying the debt of Lamb. Lamb sued Leroy Corporation, contending that in reality a merger had taken place rather than merely a sale of assets, thus making Leroy liable for the N.L.M. debt. The trial court found in favor of Leroy Corporation and Lamb appealed.

THOMPSON, J.

A consummated agreement of merger or consolidation imposes upon the surviving corporation all liabilities of the constituent corporations so merged or consolidated. The rule is otherwise when the transaction is a bona fide sale of assets. The creditor, Lamb, urges that the transaction between N.L.M. and Leroy was at least a de facto merger.

We do not agree, since the sales agreement did not contemplate a merger or consolidation of the two corporations. Merger is "'. . . a combination whereby one of the constituent companies remains in being—absorbing or merging in itself all the other constituent corporations.'" *Rath* v. *Rath Packing Co.,* 257 Iowa 1277, 136 N.W.2d 410, 415 (1965).

Since such a combination did not here occur, the consequences of merger or consolidation should not be visited upon Leroy. Leroy specifically did not want to combine. The powers of N.L.M. were not transferred to Leroy. Each company maintained a separate identity, and N.L.M. was later dissolved in accordance with statutory procedures therefor.

. . . The transaction was a bona fide sale of assets in exchange for stock. A merger or consolidation of the two companies did not result. The consideration was more than adequate. It was no part of the sales agreement that Leroy was to pay the consideration to the stockholders of N.L.M.; in fact the stock was issued directly to N.L.M. itself. This exchange effectively terminated the relationship of the two corporations and completely executed the terms of the sales agreement. For these reasons this case is controlled by the general rule that when one corporation sells all of its assets to another corporation the purchaser is not liable for the debt of the seller.

(Affirmed.)

Questions for Discussion

1. *Why was the transaction in this case not a merger?*
2. *What is the difference in liability to creditors between the surviving corporation in a merger and the corporation purchasing assets of another corporation?*

SUMMARY

The stockholders may dissolve the corporation by surrendering the corporate charter to the state. As a general rule, a majority of stockholders must vote to do so. The corporate charter may also provide for a limited corporate existence.

Under certain circumstances the state, through its attorney general, may institute quo warranto proceedings to forfeit the corporate charter. Under some conditions a minority of the shareholders may petition to have the corporation dissolved.

Dissolution terminates the corporate existence and its assets belong to the shareholders as residual owners subject to the claims of creditors. After dissolution, the corporation may act to wind up its affairs under the statutes of a number of states.

A corporate consolidation occurs when two or more corporations join to form a single new corporation. A merger of two or more corporations occurs when all the assets and property of the merging corporations is transferred to one of those corporations which is the survivor. For both of these actions most jurisdictions require the approval of the board of directors and two thirds of the stockholders of the corporation.

PROBLEMS

1. A tort claim was filed by the plaintiff against the New York, Chicago & St. Louis Railroad Co. on December 21, 1964. The injury occurred December 22, 1962. The defendant railroad had signed a merger agreement with another railroad on March 1, 1961, and, in fact, was merged with the Norfolk & Western Railway Co. on October 16, 1964. The merger agreement called for the existence of the defendant to cease upon the merger. Defendant filed a motion to set aside the service of summons and to dismiss the suit because of the merger. Should the motion have been granted? Nationwide Insurance Co. v. New York, Chicago & St. Louis Railroad Co., 211 N.E.2d 872 (Ohio App. 1965).
2. The Texas Mutual Life Insurance Company was organized in 1870. It functioned for ten years but ceased activity from 1880 until 1920, when an attempt was made to reorganize it and begin business under the name Bankers' Life Insurance Company. A quo warranto proceeding was initiated by the attorney general of Texas in order to have the corporate charter forfeited. Should the corporation have been allowed to continue or should the court have ordered dissolution? State v. Dilbeck et al., 297 S.W. 1049 (Tex. 1927).
3. Leander Anderson and others were owners of preferred stock of a corporation. The preferred-share contract stated that they should be paid a certain amount upon "the dissolution, liquidation or winding up . . ." of the corporation. The corporation entered into a consolidation agreement with Cliffs Corporation, the new corporation to be called Cleveland-Cliffs Iron Company. The dividends of the preferred shareholders were in arrears, and they filed suit, claiming that the consolidation brought into effect the provisions of the preferred-share agreement requiring them to be paid. Was Anderson correct? Anderson v. Cleveland-Cliffs Iron Co., 87 N.E.2d 384 (Ohio 1948).

SECTION IX

Government Protection of Business and the Consumer

CHAPTER 46

The Antitrust Laws: An Overview

46:1. Introduction. It is virtually impossible for any person actively engaged in business to avoid the impact of government regulation, either state or federal. The larger the business, the more susceptible it is to the many statutes and regulations that have been propagated at the various levels of government. A comprehensive discussion of this field is beyond the scope of a course in business law and indeed is frequently treated in a separate course or two. It is appropriate, however, to mention briefly some of the more significant statutes in the antitrust area, so that you can get some flavor of the degree of control that may be exercised by government. This discussion will omit a multitude of other forms of government regulation that affect business, such as the tax laws, labor laws of all kinds, and a number of other specific statutes.

46:2. The Basis for Government Regulation. As you no doubt know, the United States is a federal form of government. There are two levels, the state and the federal governments. Both of these play a part in the regulation of business, but for large businesses at least, the federal government plays the bigger role. When the United States was formed, the individual states gave up part of their sovereign powers to the federal government. Those powers that they gave up, or that were limited, are contained in the United States Constitution. The federal government, therefore, is a government of limited or enumerated powers. It has only those powers that are granted by the Constitution.

The primary clause that gives Congress the power to regulate business is known as the *commerce clause.* It states that "Congress shall have the power to regulate commerce with foreign nations, among the several states and with the Indian tribes." This clause has been interpreted by the courts to give Congress the power to regulate any activity that has a "close and substantial affect" upon interstate commerce. The courts have even gone a step further and concluded that in a determination of whether an activity has a close and substantial affect upon interstate commerce the impact of all similar activities can be lumped together. This ruling then gives Congress the power to pass statutes that regulate very small businesses if, when lumped together with other similar businesses, they have a close and substantial affect upon interstate commerce. Under this concept there are very few businesses that Congress would not have the power to regulate.

Congress has the exclusive right to regulate foreign commerce. Any state attempt to regulate foreign commerce is unconstitutional. The states, of course, have the power to regulate commerce within their borders. In many cases, the federal government also has the power to regulate that commerce. The state may still regulate, however, if the federal government has not done so or if the state regulation does not conflict with a federal statute or constitute a burden on interstate commerce. If there is a conflict between state and federal legislation or if the federal legislation has preempted the regulation of the activity by the state, then, of course, the state must yield to the federal legislation.

46:3. THE GOALS OF ANTITRUST LEGISLATION. Implicit in the federal antitrust statutes is the presumption that the maintenance of competition between businesses is desirable. Indeed prior to the passage of any of our antitrust statutes the common law made illegal a number of activities that restrained trade. For example, any agreement or conspiracy to restrain trade, with a few exceptions, was illegal under the common law. This philosophy is founded on the premise that if left completely unregulated, monopolies might emerge in certain industries or at the least prices might be maintained at an artificially high level because of agreements between businesses. This statement is really something of an oversimplification, but it is sufficient to make the point.

There are also other reasons behind the passage of some of our so-called antitrust laws other than the maintenance of competition. In fact, it can be argued that some of them limit competition to some degree. For instance, implicit behind the passage of certain statutes, such as the Robinson–Patman Act, is the notion that there is some social good in the preservation of the small business, with the resulting independence and mobility of at least some individuals in our society. This reasoning has been articulated by the Supreme Court in a number of cases. A number of individuals have also argued that there is some danger in the aggregation of great wealth and resources in the hands of too few people. Whatever the reasons, however—and sometimes they are difficult to discern—Congress has passed a number of antitrust statutes, some of which are discussed in the rest of this chapter.

46:4. THE SHERMAN ACT. The first antitrust statute passed by Congress was the Sherman Antitrust Act of 1890. The act is directed at two primary areas. Section 1 of the act makes illegal "Every contract, combination in the form of trust or otherwise, or conspiracy, in restraint of trade or commerce among the several states, or with foreign stations. . . ." The other important section of the act states that "Every person who shall monopolize, or attempt to monopolize, or combine or conspire with any other person or persons to monopolize any part of the trade or commerce . . ." violates Section 2 of the act. These provisions are very broad in scope. The actual activities that result in a violation of the act have been determined by the judiciary since the passage of the act.

There are a number of activities that are proscribed by Section 1. The most well-known offense against Section 1 is price fixing among competitors. It is illegal also to enter into an agreement or a conspiracy to divide up markets, on the basis of either product or geography; to engage in group boycotts; or to agree to limit the production of a product. Agreements to engage in many other activities may also violate the act. The agreement may be implied as a result of the actions of the participants. It is not necessary that there be evidence of an overt agreement.

The types of behavior described previously are what are known as *per se violations* of the Sherman Act. In other words, the mere existence of the agreement itself is enough to establish a violation of the act. It is not a defense to show that the prices fixed are themselves reasonable.

Section 2 outlaws the act of monopolization. It is important to note that *monopolization* is a legal term and not an economic one. Theoretically, at least, the act does not outlaw monopolies but only the act of monopolization. In determining whether a firm has violated this section of the Sherman Act, the courts apply the *rule of reason.* They first examine the actions of the firm in order to determine whether the company has an intent to monopolize. Next they analyze the relevant market in order to determine whether the firm has any monopoly power, that is, the power to affect prices. The percentage that a firm has of the market is relevant to this determination. In theory a large market percentage alone is not enough to put a firm in violation of the act if it has achieved its position by efficiency and fair methods of competition. In practice, however, when a firm obtains a large percentage of the market, such as 90 percent, it takes very little activity on its part in furtherance of the maintenance of this percentage for the court to find the company in violation of the act. There are defenses, of course, such as that the market is so small that it may be able to support only one viable firm.

The Sherman Act has both civil and criminal remedies. If a private litigant successfully prosecutes a civil case under the Sherman Act, it is entitled to recover treble or three times its damages. A new amendment to the act also allows the attorneys general of the various states to file suit for the recovery of damages suffered by citizens of the state as a result of price-fixing violations of the act. The federal government may also prosecute a civil case for violation of the act.

Numerous remedies are available to the court if a firm is found in civil violation of the act, up to dissolution of the firm. This was the remedy applied by the Court in the famous case of *Standard Oil Company* v. *United States,*[1] in which the Court found Standard Oil guilty of violation of the act and dissolved the company.

As a practical matter the criminal penalties of the act are enforced only

[1] 221 U.S. 1 (1911).

for per se violations of the act, although the terms of the statute do not limit the criminal penalties to these types of violations. The penalties for a criminal conviction of the Sherman Act have recently been increased. Violators are no longer misdemeanors but felonies, punishable by up to three years in prison and fines of $100,000 for executives in addition to fines up to $1 million for companies.

UNITED STATES v. CONTAINER CORP. OF AMERICA et al.

393 U.S. 333 (1969)

DOUGLAS, J.

This is a civil antitrust action charging a price-fixing agreement in violation of § 1 of the Sherman Act. . . .

The case as proved is unlike any of other price decisions we have rendered. There was here an exchange of price information but no agreement to adhere to a price schedule as in *Sugar Institute* v. *United States,* 297 U.S. 553, or *United States* v. *Socony-Vacuum Oil Co.,* 310 U.S. 150. There was here an exchange of information concerning specific sales to identified customers, not a statistical report on the average cost to all members, without identifying the parties to specific transactions, as in *Maple Flooring Mfrs. Assns.* v. *United States,* 268 U.S. 563. While there was present here, as in *Cement Mfrs. Protective Assn.* v. *United States,* 268 U.S. 588, an exchange of prices to specific customers, there was absent the controlling circumstance, viz., that cement manufacturers, to protect themselves from delivering to contractors more cement than was needed for a specific job and thus receiving a lower price, exchanged price information as a means of protecting their legal rights from fraudulent inducements to deliver more cement than needed for a specific job.

Here all that was present was a request by each defendant of its competitor for information as to the most recent price charged or quoted, whenever it needed such information and whenever it was not available from another source. Each defendant on receiving that request usually furnished the data with the expectation that it would be furnished reciprocal information when it wanted it. That concerted action is of course sufficient to establish the combination or conspiracy, the initial ingredient of a violation of § 1 of the Sherman Act.

There was of course freedom to withdraw from the agreement. But the fact remains that when a defendant requested and received price information, it was affirming its willingness to furnish such information in return.

There was to be sure an infrequency and irregularity of price exchanges between the defendants; and often the data were available from the records of the defendants or from the customers themselves. Yet the essence of the agreement was to furnish price information whenever requested.

Moreover, although the most recent price charged or quoted was sometimes fragmentary, each defendant had the manuals with which it could compute the price charged by a competitor on a specific order to a specific customer.

Further, the price quoted was the current price which a customer would need to pay in order to obtain products from the defendants furnishing the data.

The defendants account for about 90% of the shipment of corrugated containers from plants in the Southeastern United States. While containers vary as to dimensions, weight, color, and so on, they are substantially identical, no matter who produces them, when made to particular specifications.

The result of this reciprocal exchange of prices was to stabilize prices though at a downward level. Knowledge of a competitor's price usually meant matching that price. The continuation of some price competition is not fatal to the Government's case. The limitation or reduction of price competition brings the case within the ban, for as we held in *United States* v. *Socony-Vacuum Oil Co.,* supra, at 224, n. 59, interference with the setting of price by free market forces is unlawful per se. Price information exchanged in some markets may have no effect on a truly competitive price. But the corrugated container industry is dominated by relatively few sellers. The product is fungible and the competition for sales is price. The demand is inelastic, as buyers place orders only for immediate, short-run needs. The exchange of price data tends toward price uniformity. For a lower price does not mean a larger share of the available business but a sharing of the existing business at a lower return. Stabilizing prices as well as raising them is within the ban of § 1 of the Sherman Act. . . . The inferences are irresistible that the exchange of price information has had an anticompetitive effect in the industry, chilling the vigor of price competition. . . .

Price is too critical, too sensitive a control to allow it to be used even in an informal manner to restrain competition.

(Reversed.)

QUESTIONS FOR DISCUSSION

1. *Describe the nature of the* "agreement." *Did it establish the price to be charged for corrugated containers at a fixed level?*
2. *Would it be a defense to a price-fixing charge to show that the price is reasonable?*

UNITED STATES v. ALUMINUM COMPANY OF AMERICA

148 F.2d 416 (2d Cir. 1945)

L.HAND, J.

"Alcoa" is a corporation, organized under the laws of Pennsylvania on September 18, 1888; its original name, "Pittsburgh Reduction Company," was changed to its present one on January 1, 1907. It has always been engaged in the production and sale of "ingot" aluminum, and since 1895 also in the fabrication of the metal into many finished and semifinished articles.

. . . It is undisputed that throughout this period "Alcoa" continued to be the single producer of "virgin" ingot in the United States; and the plaintiff argues that this without more was enough to make it an unlawful monopoly. . . . "Alcoa's" position is that the fact that it alone continued to make "virgin" ingot in this country did not, and does not, give it a monopoly of the market; that it was always subject to the competition of imported "virgin" ingot, and of what is called "secondary" ingot; and that even if it had not been, its monopoly would not have been retained by unlawful means, but would have been the result of a growth which the Act does not forbid, even when it results in a monopoly.

We conclude therefore that "Alcoa's" control over the ingot market must be reckoned at over ninety percent; that being the proportion which its production bears to imported "virgin" ingot. . . . The producer of so large a proportion of the supply has complete control within certain limits. It is true that, if by raising the price he reduces the amount which can be marketed—as always, or almost always, happens—he may invite the expansion of the small producers who will try to fill the place left open; nevertheless, not only is there an inevitable lag in this, but the large producer is in a strong position to check such competition; and, indeed, if he has retained his old plant and personnel, he can inevitably do so. . . .

Having proved that "Alcoa" had a monopoly of the domestic ingot market, the plaintiff had gone far enough; if it was an excuse, that "Alcoa" had not abused its power, it lay upon "Alcoa" to prove that it had not. But the whole issue is irrelevant anyway, for it is no excuse for "monopolizing" a market that the monopoly has not been used to extract from the consumer more than a "fair" profit. The Act has wider purposes. Indeed, even though we disregarded all but economic considerations, it would by no means follow that such concentration of producing power is to be desired, when it has not been used extortionately. Many people believe that possession of unchallenged economic power deadens initiative, discourages thrift and depresses energy; that immunity from competition is a narcotic, and rivalry is a stimulant, to industrial progress; that the spur of constant stress is necessary to counteract an inevitable disposition to let well enough alone. Such people believe that competitors, versed in the craft as no consumer can be, will be quick to detect opportunities for saving and new shifts in production, and be eager to profit by them. In any event the mere fact that a producer, having command of the domestic market, has not been able to make more than a "fair" profit, is no evidence that a "fair" profit could not have been made at lower prices. . . . True, it might have been thought adequate to condemn only those monopolies which could not show that they had exercised the highest possible ingenuity, had adopted every possible economy, had anticipated every conceivable improvement, stimulated every possible demand. No doubt, that would be one way of dealing with the matter, although it would imply constant scrutiny and constant supervision, such as courts are unable to provide. Be that as it may, that was not the way that Congress chose; it did not condone "good trusts" and condemn "bad" ones; it forbade all. Moreover, in so doing it was not necessarily actuated by economic motives alone. It is possible, because of its indirect social or moral effect, to prefer a system of small producers, each dependent for his success upon his own skill and character, to one in which the great mass of those engaged must accept the direction of a few. These considerations, which we have suggested only as possible purposes of the Act, we think the decisions prove to have been in fact its purposes.

We have been speaking only of the economic reasons which forbid monopoly; but, as we have already implied, there are others, based upon the belief that great industrial consolidations are inherently undesirable, regardless of their economic results. . . . Throughout the history of these statutes it has been constantly assumed that one of their purposes was to perpetuate and preserve, for its own sake and in spite of possible cost, an organization of industry in small units which can effectively compete with each other. We hold that "Alcoa's" monopoly of ingot was of the kind covered by § 2.

It does not follow because "Alcoa" had such a monopoly, that it "monopolized" the ingot market: it may not have achieved monopoly; monopoly may have been thrust upon it. If it had been a combination of existing smelters which united the whole industry and controlled the production of all aluminum ingot, it would certainly have "monopolized" the market. . . .

. . . We may start therefore with the premise that to have combined ninety percent of the producers of ingot would have been to "monopolize" the ingot market; and, so far as concerns the public interest, it can make no difference whether an existing competition is put an end to, or whether prospective competition is prevented. . . . Nevertheless, it is unquestionably true that from the very outset the courts have at least kept in reserve the possibility that the origin of a monopoly may be critical in determining its legality.

. . . (P)ersons may unwittingly find themselves in possession of a monopoly, automatically so to say: that is, without having intended either to put an end to existing competition, or to prevent competition from arising when none had existed; they may become monopolists by force of accident. . . . A market may, for example, be so limited that it is impossible to produce at all and meet the cost of production except by a plant large enough to supply the whole demand. . . . A single producer may be the survivor out of a group of active competitors, merely by virtue of his superior skill, foresight and industry. In such cases a strong argument can be made that, although, the result may expose the public to the evils of monopoly, the Act does not mean to condemn the resultant of those very forces which it is its prime object to foster. . . . The successful competitor, having been urged to compete, must not be turned upon when he wins. . . .

It would completely misconstrue "Alcoa's" position in 1940 to hold that it was the passive beneficiary of a monopoly, following upon an involuntary elimination of competitors by automatically operative economic forces. . . . This continued and undisturbed control did not fall undesigned into "Alcoa's" lap; obviously it could not have done so. It could only have resulted, as it did result, from a persistent determination to maintain the control, with which it found itself vested in 1912. There were at least one or two abortive attempts to enter the industry, but "Alcoa" effectively anticipated and forestalled all competition, and succeeded in holding the field alone. True, it stimulated demand and opened new uses for the metal, but not without making sure that it could supply what it had evoked. There is no dispute as to this; "Alcoa" avows it as evidence of the skill, energy and initiative with which it has always conducted its business; as a reason why, having won its way by fair means, it should be commended, and not dismembered. We need charge it with no moral derelictions after 1912; we may assume that all it claims for itself is true. The only question is whether it falls within the exception established in favor of those who do not seek, but cannot avoid, the control of a market. It seems to us that that question scarcely survives its statement. It was not inevitable that it should always anticipate increases in the demand for ingot and be prepared to supply them. Nothing compelled it to keep doubling and redoubling its capacity before others entered the field. It insists that it never excluded competitors; but we can think of no more effective exclusion than progressively to embrace each new opportunity as it opened, and to face every newcomer with new capacity already geared into a great organization, having the advantage of experience, trade connections and the elite of personnel. Only in case we interpret "exclusion" as limited to manoeuvres not honestly industrial, but actuated solely by a desire to prevent competition, can such a course, indefatigably pursued, be deemed not "exlusionary." So to limit it would in our judgment emasculate the Act; would permit just such consolidations as it was designed to prevent.

In order to fall within § 2, the monopolist must have both the power to monopolize, and the intent to monopolize. To read the passage as demanding any "specific" intent, makes nonsense of it, for no monopolist monopolizes unconscious of what he is doing. So here, "Alcoa" means to keep, and did keep, that complete and exclusive hold upon the ingot market with which it started. That was to "monopolize" that market, however innocently it otherwise proceeded. So far as the judgment held that it was not within § 2, it must be reversed. . . .

(Reversed and remanded.)

Questions for Discussion

1. *Does the Sherman Act outlaw all monopolies?*
2. *Is it a defense to show that despite the presence of monopoly power that power has not been abused and only a "fair" profit has been made by the company?*
3. *What did this court point out are the dangers of monopoly power?*

NOTE: *This case was originally appealed to the Supreme Court but was referred to the Court of Appeals, as four of the Supreme Court Justices disqualified themselves because of previous connections with the case.*

46:5. The Clayton Act. Although the Sherman Act is a significant antitrust tool, it soon became apparent that additional legislation was needed in order to maintain competition. One problem with the Sherman Act is that although it attacks monopolization, a violation often does not occur until a firm already has a significant share of the market. Even if the firm is then found in violation of the act, it is very difficult to fashion a remedy that will restore competition. Furthermore there must be evidence of an intent to monopolize in order to prove a violation of the act. This is often difficult to prove. Therefore in 1914 Congress passed the Clayton Act in an effort to attack many of the activities that lead to reduced competition. These activities are not declared illegal unless their practice may "substantially lessen competition or tend to create a monopoly." The purpose of this standard is to prevent monopolies from occurring and to stop them in their "incipiency." The Clayton Act prohibits four primary types of activities under certain circumstances: (1) price discrimination, (2) tying contracts and exclusive dealings, (3) mergers and acquisitions, and (4) interlocking directorates. The first three activities are prohibited when the incipiency standard is met, and the fourth, the interlocking directorate is prohibited between competing companies over a certain size. Price discrimination will be discussed in Section 46:6.

Section 3 of the Clayton Act prohibits tying contracts or exclusive dealing arrangements if at any time the effect "may be to substantially lessen competition or tend to create a monopoly in any line of commerce." A tying contract involves the sale or the lease of one product on the condition that the purchaser or the lessee agrees to purchase or lease certain other goods. If the seller has some monopoly power over one product, he may use this approach to increase his market share in a product market in which he has no monopoly power. For example, IBM at one time would lease its tabulating machines to customers only if they also agreed to purchase all the tabulating cards from IBM.[2] This arrangement was declared illegal by the Supreme Court.

The original Section 7 of the Clayton Act prohibited the acquisition, by a corporation, of the whole or any part of the stock of another corporation if the effect of the acquisition might substantially lessen competition or tend to create a monopoly. In 1950, Congress passed the Celler–Kefauver Amendment, which also prohibited the acquisition of *assets* that would have the same affect. This provision is designed to reduce the increased concentration that results from mergers and acquisitions. When the incipiency test is met, it prohibits horizontal, vertical, or conglomerate mergers between competitors or potential competitors. A horizontal merger is one in which one company purchases another at the same level of the marketing chain. An example would be a manufacturer of steel's purchasing another manufacturer of steel. A vertical merger would be exemplified by the merger of a manufacturer of shoes and a retailer of shoes. They are at different levels of the chain. A conglomerate merger is one in which the purchased company has no obvious relationship in its product or business to the acquiring firm. Thus the acquisition of a car-rental company by a steel manufacturer would be a conglomerate merger.

Sometimes the courts have also labeled mergers as *product-extension* or *market-extension* mergers. A product-extension merger would occur, for instance, if a soap manufacturer purchased a manufacturer of bleach, a different but related product. A market-extension merger occurs when a company merges with another company doing business in a different geographic area.

In order for the court to find a violation of the act, it need only be shown that there is a probability—not a certainty—that competition will be reduced. In addition, the firms need not be in actual competition in order for the merger to violate the act. The courts have developed the doctrine of *potential competition.* Under this doctrine, if the acquired and acquiring firms are potential competitors, the act may be violated. The doctrine is applicable to firms that are potential, although not actual, competitors in the same geographic area or with respect to the same product.

In the past there has been some problem in the enforcement of the act because the government did not become aware of some mergers until they had actually been consummated for some time or in some cases not at all. As an attempt to deal with this problem the Hart–Scott–Rodino Antitrust Improvements Act of 1976 requires firms of at least $100 million of net assets, or total sales, that plan to acquire firms with net sales or total assets of $10 million to notify the Federal Trade Commission and the Department of Justice

[2] 298 U.S. 131 (1936).

of the intended acquisition. The companies then must wait thirty days before the acquisition can be consummated.

Section 8 of the Clayton Act prohibits a person from serving on the board of directors of two or more competing firms if any one of them has capital surplus and undivided profits aggregating more than $1 million. The presumption is that if the same person is a director of two or more competing corporations, the opportunity for collusion between them would be enhanced.

UNITED STATES v. VON'S GROCERY COMPANY
384 U.S. 270 (1966)

BLACK, J.

On March 25, 1960, the United States brought this action charging that the acquisition by Von's Grocery Company of its direct competitor Shopping Bag Food Stores, both large retail grocery companies in Los Angeles, California, violated § 7 of the Clayton Act . . . as amended in 1950 by the Celler–Kefauver Anti-Merger Act. . . . On March 28, 1960, three days later, the District Court refused to grant the Government's motion for a temporary restraining order and immediately Von's took over all of Shopping Bag's capital stock and assets including 36 grocery stores in the Los Angeles area. After hearing evidence on both sides, the District Court made findings of fact and concluded as a matter of law that there was "not a reasonable probability" that the merger would tend "substantially to lessen competition" or "create a monopoly" in violation of § 7. For this reason the District Court entered judgment for the defendants. . . . The Government appealed directly to this Court. . . . The sole question here is whether the District Court properly concluded on the facts before it that the Government had failed to prove a violation of § 7.

The record shows the following facts relevant to our decision. The market involved here is the retail grocery market in Los Angeles area. In 1958, Von's retail sales ranked third in the area and Shopping Bag's ranked sixth. In 1960 their sales together were 7.5% of the total two and one-half billion dollars of retail groceries sold in the Los Angeles market each year. For many years before the merger both companies had enjoyed great success as rapidly growing companies. From 1948 to 1958 the number of Von's stores in the Los Angeles area practically doubled from 14 to 27, while at the same time the number of Shopping Bag's stores jumped from 15 to 34. During that same decade, Von's sales increased fourfold and its share of the market almost doubled while Shopping Bag's sales multiplied seven times and its share of the market tripled. The merger of these two highly successful, expanding and aggressive competitors created the second largest grocery chain in Los Angeles with sales of almost $172,488,000 annually. In addition the findings of the District Court show that the number of owners operating single stores in the Los Angeles retail grocery market decreased from 5,365 in 1950 to 3,818 in 1961. By 1963, three years after the merger the number of single-store owners had dropped still further to 3,590. During roughly the same period from 1953 to 1962, the number of chains with two or more grocery stores increased from 96 to 150. While the grocery business was being concentrated into the hands of fewer and fewer owners, the small companies were continually being absorbed by the larger firms through mergers. These facts alone are enough to cause us to conclude contrary to the District Court that the Von's–Shopping Bag merger did violate § 7. Accordingly, we reverse. . . .

Like the Sherman Act in 1890 and the Clayton Act in 1914, the basic purpose of the 1950 Celler–Kefauver Act was to prevent economic concentration in the American economy by keeping a large number of small competitors in business. . . . As we said in *Brown Shoe Co.* v. *United States,* 370 U.S. 294, 315, "The dominant theme pervading congressional consideration of the 1950 amendments was a fear of what was considered to be a rising tide of economic concentration in the American economy." To arrest this "rising tide" toward concentration into too few hands and to halt the gradual demise of the small businessman, Congress decided to clamp down with vigor on mergers. . . .

The facts of this case present exactly the threatening trend toward concentration which Congress wanted to halt. The number of small grocery companies in the Los Angeles retail grocery market had been declining rapidly before the merger and continued to decline rapidly afterwards. This rapid decline in the number of grocery store owners moved hand in hand with a significant number of significant absorptions of the small companies by the larger ones. In the midst of this steadfast trend toward concentration, Von's and Shopping Bag, two of the most successful and largest companies in the area, jointly owning 66 grocery stores merged to become the second largest chain in Los Angeles. This merger cannot be defended on the ground that one of the companies was about to fail or that the two had to merge to save themselves from destruction by some larger and more powerful competitor. What we have on the contrary is simply the case of two already powerful companies merging in a way which makes them even more powerful than they were before. If ever such a merger would not violate § 7, certainly it does when it takes place in a market characterized by a long and continuous trend toward fewer and fewer owner–competitors which is exactly the sort of trend which Congress, with power to do so, declared must be arrested.

Appellees' primary argument is that the merger between Von's and Shopping Bag is not prohibited by § 7 because the Los Angeles grocery market was competitive before the merger, has been since, and may continue to be in the future. Even so, § 7 "requires not merely an appraisal of the immediate impact of the merger upon competition, but a prediction of its impact upon competitive conditions in the future; this is what is meant when it is said that the amended § 7 was intended to arrest anticompetitive tendencies in their 'incipiency.' " *U.S.* v. *Philadelphia Nat. Bank,* 374 U.S. 321, 362. It is enough for us that Congress feared that a market marked at the same time by both a continuous decline in the number of small businesses and a large number of mergers would slowly but inevitably gravitate from a market of many small competitors to one dominated by one or a few giants, and competition would thereby be destroyed. Congress passed the Celler–Kefauver Act to prevent such a destruction of competition. Our cases since the passage of that Act have faithfully endeavored to enforce this congressional command. We adhere to them now.

(Reversed and remanded.)

QUESTIONS FOR DISCUSSION

1. *What did the court state is one of the basic purposes of the Celler–Kefauver Act?*
2. *What was the trend in the Los Angeles area regarding concentration in the retail grocery business?*
3. *What argument did the defendants raise to justify the merger? How did the court refute this argument?*
4. *Under what circumstances are mergers prohibited?*

46:6. THE FEDERAL TRADE COMMISSION ACT. In the same year that it passed the Clayton Act (1914) Congress also passed the Federal Trade Commission Act. This act was passed in recognition of the fact that an agency was needed in order to centralize the attack against anticompetitive practices. The result was the creation of the Federal Trade Commission as an independent administrative agency. The commission consists of five commissioners appointed by the president to staggered seven-year terms with the advice and consent of the U.S. Senate. No more than three of the commissioners can be from the same political party. The intent of Congress was to create an independent agency not easily subject—in theory, at least—to political pressures.

In addition to establishing the commission, Section Five of the act declares "unfair methods of competition" in commerce to be illegal. The act was later amended also to make illegal unfair and deceptive practices in commerce. Recently the jurisdiction of the commission has been extended to include acts that *affect* commerce. In addition, the act also gives the commission jurisdiction over Sections 2, 3, 7 and 8 of the Clayton Act.

The types of behavior that constitute an "unfair method of competition" were intentionally left undefined by Congress. The purpose is to allow the commission to deal with new anticompetitive practices that might develop with the passage of time. Generally any action that would constitute a violation of the Sherman or Clayton Act would also be an unfair method of competition. In addition, the courts have also stated that any activity that violates the "spirit" of the Sherman and Clayton acts is also an unfair method of competition. Practices that fall short of violating the Sherman or Clayton Act, therefore, may still result in a violation of the Federal Trade Commission Act. The courts give great weight to the commission's interpretation of what constitutes an unfair method of competition.

The Federal Trade Commission has become very active in the area of consumer protection in recent years, particularly with regard to deceptive advertising, and has gone so far as to require some advertisers whose ads appear deceptive to the commission to run corrective advertising. The commission also has the power under the Magnuson–Moss Warranty Act to regulate the information to be contained in certain warranties. This subject was discussed in some detail in Chapter 17. Finally, the commission has been given many powers in connection with the establishment of trade-regulation rules designed to govern the behavior of business in certain areas.

46:7. The Robinson–Patman Act. Section 2 of the Clayton Act in its original form was directed against discrimination in pricing that injured competitors of the seller. For example, suppose one is the owner of a chain of gas stations. On the opposite corner from one station is a competitor that owns only a single station. By drastically lowering the price at the station across from the competitor and supporting this price reduction by maintaining a higher price at the other stations, one could rather effectively eliminate the competition. Section 2 was directed against this type of injury. There were several loopholes in the statute, however, and there were other types of damage that Section 2 did not reach. For example, it soon became apparent that under the original Section 2 *any* cost savings effected in the sale of a product to one buyer would justify *any* difference in price charged the buyers for the product. As a result, and for other reasons such as the hostility at the time to the growth of chain stores, Congress passed the Robinson–Patman amendment to Section 2 of the Clayton Act in 1936.

Price discrimination occurs when a seller charges a different price for the same product to two different customers. The Robinson–Patman Act prohibits price discrimination that "may be to substantially lessen competition or tend to create a monopoly in any line of commerce." The injury to competition that results from price discrimination may occur on two levels, which for convenience can be called *primary* or *first-line discrimination* and *secondary-line discrimination*. Consider the following example. Suppose Sammy Tireseller sells tires to Joe's Tire Shop in City A at a price of $15 per tire. He sells tires in other cities at a price of $20. Company Competitor also sells tires in City A at a price of $20. Sammy has discriminated in price in his sale of tires to Joe's, which would result in injury to Sammy's competitor at the first or primary level. On the other hand, assume that across the street from Joe's Tire Shop the chain of Roe Wardbuck has a store and that Sammy Tireseller sells the same tire to Wardbuck for $10 that he sells to Joe's for $15. This could result in injury at the secondary or buyer level. The Robinson–Patman Act prohibits price discrimination that may injure competitors at either level.

In addition to direct price discrimination the Robinson–Patman Act also prohibits indirect methods of price discrimination. For example, promotional allowances are prohibited unless offered to all on a reasonably equal basis; brokerage discounts are disallowed; and power is given to the Federal Trade Commission to limit quantity discounts. The act applies to buyers as well as to sellers. It is therefore illegal to knowingly receive goods at a discriminatory price.

There are several situations in which price discrimination is permitted. First, a seller may charge a price differential equal to any cost savings achieved in the sale of the product to one customer versus another. Second, a seller may in good faith lower the price of a product to a buyer in order to meet the equally low price of a competitor. It must be a current customer, however, and the seller may only meet, not beat the competitor's price. Third, a seller does not discriminate in price if the products are different. For example, a seller may sell one model of a product to a customer at one price and another model to another customer at a different price. The differences in the products must be more than superficial. Just calling the same product by a different name or making superficial physical changes would not be sufficient.

The Robinson–Patman Act has been the recipient of a great deal of criticism. In the first place, it is complicated and is not written in a manner that stands as a monument to the English language. More importantly, it has been criticized for its anticompetitive effects. After all, one tool of competition is to discriminate in price, which results in certain cases in added competition rather than less. Those who support the act point out that its purpose is not only to maintain competition but also to protect the smaller business from the overwhelming economic power of the chain stores and other large firms.

UTAH PIE COMPANY v. CONTINENTAL BAKING CO. et al.

386 U.S. 685 (1967)

This suit for damages was brought by Utah Pie Company against Pet Milk Company, Continental Baking Co., and Carnation Company for violation of Section 2(a) of the Clayton Act as amended by the Robinson–Patman Act. The U.S. Court of Appeals reversed a judgment in favor of Utah Pie, who appealed to the U.S. Supreme Court. First, the Court found in favor of Utah Pie against Pet Milk Company and then went on to discuss the case against the other defendants.

WHITE, J.

The product involved is frozen dessert pies—apple, cherry, boysenberry, peach, pumpkin, and mince. The period covered by the suit comprised the years 1958, 1959, and 1960 and the first eight months of 1961. Petitioner is a Utah corporation which for 30 years has been baking pies in its plant in Salt Lake City and selling them in Utah and surrounding States. It entered the frozen pie business in late 1957. It was immediately successful with its new line and built a new plant in Salt Lake City in 1958. The frozen pie market was a rapidly expanding one: 57,060 dozen frozen pies were sold in the Salt Lake City market in 1958, 111,729 dozen in 1959, 184,569 dozen in 1960, and 266,908 dozen in 1961. Utah's share of this market in those years was 66.5%, 34.3%, 45.5%, and 45.3% respectively, its sales volume steadily increasing over the four years. Its financial position also improved. Petitioner is not, however, a large company. At the time of the trial, petitioner operated with only 18 employees, nine of whom were members of the Rigby family, which controlled the business. Its net worth increased from $31,651.98 on October 31, 1957, to $68,802.13 on October 31, 1961. . . .

Each of the respondents is a large company and each of them is a major factor in the frozen pie market in one or more of the regions of the country. Each entered the Salt Lake City frozen pie market before petitioner began freezing dessert pies. None of them had a plant in Utah. . . . The Salt Lake City market was supplied by respondents chiefly from their California operations. They sold primarily on a delivered price basis. . . .

Petitioner's case against Continental is not complicated. Continental was a substantial factor in the market in 1957. But its sales of frozen 22-ounce dessert pies, sold under the "Morton" brand, amounted to only 1.3% of the market in 1958, 2.9% in 1959, and 1.8% in 1960. Its problems were primarily that of cost and in turn that of price, the controlling factor in the market. In late 1960 it worked out a co-packing arrangement in California by which fruit would be processed directly from the trees into the finished pie without large intermediate packing, storing, and shipping expenses. Having improved its position, it attempted to increase its share of the Salt Lake City market by utilizing a local broker and offering short-term price concessions in varying amounts. Its efforts for seven months were not spectacularly successful. Then in June 1961, it took the steps which are the heart of petitioner's complaint against it. Effective for the last two weeks of June it offered its 22-ounce frozen apple pies in the Utah area at $2.85 per dozen. It was then selling the same pies at substantially higher prices in other markets. The Salt Lake City price was less than its direct cost plus an allocation for overhead. Utah's going price at the time for its 24-ounce "Frost 'N' Flame" apple pie sold to Associated Grocers was $3.10 per dozen, and for its "Utah" brand $3.40 per dozen. . . . Utah's response was immediate. It reduced its price on all of its apple pies to $2.75 per dozen. . . . Continental's total sales of frozen pies increased from 3,350 dozen in 1960 to 18,800 dozen in 1961. Its market share increased from 1.8% in 1960 to 8.3% in 1961. The Court of Appeals concluded that Continental's conduct had had only minimal effect, that it had not injured or weakened Utah Pie as a competitor, that it had not substantially lessened competition and that there was no reasonable possibility that it would do so in the future.

We again differ with the Court of Appeals. Its opinion that Utah was not damaged as a competitive force apparently rested on the fact that Utah's sales volume continued to climb in 1961 and on the court's own factual conclusion that Utah was not deprived of any pie business which it otherwise might have had. But this retrospective assessment fails to note that Continental's discriminatory below-cost price caused Utah Pie to reduce its price to $2.75. The jury was entitled to consider the potential impact of Continental's priced reduction absent any responsive price cut by Utah Pie. Price was a major factor in the Salt Lake City market. . . . It could also reasonably conclude that a competitor who is forced to reduce his price to a new all-time low in a market of declining prices will in time feel the financial pinch and will be a less effective competitive force. . . .

. . . We think there was sufficient evidence from which the jury could find a violation of § 2(a) by Continental.

The Carnation Company entered the frozen dessert pie business in 1955 through the acquisition of "Mrs. Lee's Pies" which was then engaged in manufacturing and selling frozen pies in Utah and elsewhere under the "Simple Simon" label. Carnation also quickly found the market extremely sensitive to price. Carnation decided, however, not to enter an economy product in the market, and during the period covered by this suit it offered only its quality "Simple Simon" brand. Its primary method of meeting competition in its markets was to offer a variety of discounts and other reductions, and the technique was not unsuccessful. In 1958, for example, Carnation enjoyed 10.3% of the Salt Lake City market, and although its volume of pies sold in that market increased nearly 100% in the next year, its percentage of the market temporarily slipped to 8.6%. However, 1960 was a turnaround year for Carnation in the Salt Lake City market; it more than doubled its volume of sales over the preceding year and thereby gained 12.1% of the market. And while the price structure in the market deteriorated rapidly in 1961 Carnation's position remained important. . . .

. . . We cannot say that the evidence precluded the jury from finding it reasonably possible that Carnation's conduct would injure competition.

Section 2(a) does not forbid price competition which will probably injure or lessen competition by eliminating competitors, discouraging entry into the market or enhancing the market shares of the dominant sellers. But Congress has established some ground rules for the game. Sellers may not sell like goods to different purchasers at different prices if the result may be to injure competition in either the sellers or the buyers market unless such discriminations are justified as permitted by the Act. This case concerns the sellers market. In this context, the Court of Appeals placed heavy emphasis on the fact that Utah Pie constantly increased its sales volume and continued to make a profit. But we disagree with its apparent view that there is no reasonably possible injury to competition as long as the volume of sales in a particular market is expanding and at least some of the competitors in the market continue to operate at a profit. Nor do we think that the Act only comes into play to regulate the conduct of price discriminators when their discriminatory prices consistently undercut other competitors. It is true that many of the primary line cases that have reached the courts have involved blatant predatory price discriminations employed with the hope of immediate destruction of a particular competitor. On the question of injury to competition such cases present courts with no difficulty, for such pricing is clearly within the heart of the proscription of the Act. . . . We believe that the Act reaches price discrimination that erodes competition as much as it does price discrimination that is intended to have immediate destructive impact. In this case, the evidence shows a drastically declining price structure which the jury could rationally attribute to continued or sporadic price discrimination. The jury was entitled to conclude that "the effect of such discrimination," by each of these respondents, "may be substantially to lessen competition . . . or to injure, destroy, or prevent competition with any person who either grants or knowingly receives the benefit of such discrimination. . . ." The statutory test is one that necessarily looks forward on the basis of proven conduct in the past. Proper application of that standard here requires reversal of the judgment of the Court of Appeals. . . .

It is so ordered.

QUESTIONS FOR DISCUSSION

1. *Was the injury discussed in this case to the first level or second level of competition?*
2. *Did Utah's profits increase during the period of the discrimination? If so, how was it injured?*

46:8. RESALE-PRICE MAINTENANCE. For a period of time an exception to the antitrust laws was allowed for the enactment of so called fair-trade laws in some states. Under these laws a seller who desired to maintain a minimum price for its product could enter into a contract with a retailer who would agree to sell the product at no less than a certain minimum price. In some states a contract with only one retailer would be binding upon all the retailers in that jurisdiction. In 1976 the federal legislation that created this exemption to the Sherman Act was repealed, thus effectively killing off the fair-trade approach to resale-price maintenance.

It is still legal to refuse to sell to a buyer who refuses to sell at a manufacturer's suggested retail price. There are several limitations on this approach, and it is illegal for a manufacturer to try to control the price at which a wholesaler will resell the product to the retailers.

SUMMARY

The basis for federal government regulation of business is found in the United States Constitution, primarily the commerce clause. Congress has the exclusive right to regulate foreign commerce.

One of the major goals of antitrust legislation is to maintain competition. The Sherman Act is directed at restraints of trade and monopolization. The per se rule is applied to certain types of restraints of trade such as price fixing. In determining whether the monopolization section of the Sherman Act has been violated the courts apply the rule of reason.

The Clayton Act makes many of the activities that reduce competition illegal. For instance, price discrimination, tying contracts and exclusive dealings, mergers and acquisitions and interlocking directorates are all made illegal under certain circumstances. The first three activities are tested by the "incipiency" standard.

The Federal Trade Commission Act declares unfair methods of competition to be illegal. The Commission also has the power to establish trade-regulation rules.

The Robinson–Patman Act is an amendment to Section 2 of the Clayton Act. It makes price discrimination illegal if it injures competition at either the primary or secondary level of competition.

PROBLEMS

1. Proctor & Gamble acquired the assets of Clorox Chemical Company. Clorox was a leading manufacturer of liquid bleach. Proctor & Gamble was a diversified manufacturer of household products, including detergents, but did not manufacture bleach. The Federal Trade Commission challenged the merger. Proctor & Gamble contended that because it did not manufacture bleach, competition in this market would not be impaired by the merger. Decide. Federal Trade Commission v. Proctor & Gamble Co., 386 U.S. 568 (1967).
2. Trenton Potteries Company and a number of others entered into an agreement limiting price competition in the sale of sanitary pottery used in bathrooms and lavatories. They were members of a trade organization known as the Sanitary Potters' Association. They were charged with violating Section 1 of the Sherman Act and contended that the prices were reasonable

and had been established for the purpose of eliminating ruinous competition that might cause certain members of the Association to go out of business. The trial court instructed the jury that if it found that there was a price-fixing agreement, it should return a guilty verdict without regard to the reasonableness of the prices fixed. Trenton Potteries appealed an adverse verdict on the ground that the jury instruction was erroneous. Was Trenton Potteries correct? United States v. Trenton Potteries Company et al., 273 U.S. 392 (1927).

3. McLean operated a Sunoco gas station in Jacksonville, Florida, and was one of Sun Oil Company's thirty-eight independent retail dealers in the area. Super Test Oil Company opened a station across the street from McLean and began undercutting him in price. In the subsequent "gas war" McLean began having financial difficulties, so Sun cut the price of the gas that it sold to McLean but not to its other dealers. The Federal Trade Commission charged Sun with price discrimination. Sun contended that it could lower its price to McLean to help McLean meet the price of his competitor. Was Sun correct? Federal Trade Commission v. Sun Oil Co., 371 U.S. 505 (1963).
4. International Salt Co., Inc., possessed patents on two salt-dispensing machines. International leased these machines under leases that required the lessees to purchase all of the unpatented salt and salt tablets consumed in the leased machines from International. International claimed that the requirements were reasonable and necessary to assure proper performance of the machines. Several competitors of International also produced salt of comparable quality. Was the arrangement legal? International Salt Co., Inc. v. United States, 332 U.S. 392 (1947).
5. The Times-Picayune Publishing Co. owned and published a morning and an evening paper in New Orleans, Louisiana. The only competitor was an evening newspaper. The only way one could advertise in the Times-Picayune papers was to place the same ads in both evening and morning papers at the same time. The Justice Department sued Times-Picayune, contending that this arrangement constituted an illegal tying contract. Decide. Times-Picayune Publishing Co., et al. v. United States, 345 U.S. 594 (1953).
6. Colgate-Palmolive Co. set out to prove to the television public that its shaving cream, Rapid Shave, outshaved them all. Its advertising agency prepared three one-minute commercials designed to show that Rapid Shave could soften even the toughness of sandpaper. The ad showed Rapid Shave being applied to a substance that appeared to be sandpaper, and immediately thereafter a razor was shown shaving the substance clean. The announcer in the meantime stated that "To prove Rapid Shave's supermoisturizing power, we put it right from the can onto this tough dry sandpaper. It was apply . . . soak . . . and off in a stroke." In fact, if sandpaper was to be shaved, it had to be soaked in Rapid Shave for approximately eighty minutes. The ad in fact used a mock-up made of Plexiglas to which sand

had been applied. The Federal Trade Commission found the ad deceptive on the theory that it misrepresented the product's moisturizing power. Do you agree? Would it be deceptive to use mashed potatoes as a substitute for ice cream in a commercial extolling the good taste of a particular brand of ice cream? Federal Trade Commission v. Colgate–Palmolive Co., et al., 380 U.S. 374 (1965).

Yes - even though you might not really believe the sandpaper test, it would give the wrong impression.

No - icecream doesn't hold up under bright lights, hence allowed for commercial

CHAPTER 47

Consumer Protection

A great deal of attention has been directed in the last several years to the question of consumer protection. A number of laws have been passed by both the federal and the state governments. Many jurisdictions have established consumer-protection agencies with the power to redress a multitude of consumer complaints. A thorough examination of the entire field of consumer protection would cover such a vast area as to be beyond the scope of this text. In a way, much of the subject matter in this book really is directed toward the protection of consumers. An individual's best protection is an understanding of the consequences of his or her acts as well as the remedies that are available to him or her under the law. This chapter will focus on a number of subject areas that have not been touched upon elsewhere in the text.

47:1. CONSUMER CREDIT. One area that has generated a number of different types of problems for consumers over the years is the extension of consumer credit. A number of statutes have been passed that deal with consumer credit. Perhaps the most significant of these is the Consumer Credit Protection Act.

(1) The Consumer Credit Protection Act. The Consumer Credit Protection Act (CCPA) was passed by Congress in 1968 and has had several amendments since that time. The most significant part of that act is Title I, the Federal Truth in Lending Act. The purpose of this section of the act is to require each creditor to disclose both the finance charge and the annual percentage rate of finance charge in connection with any credit sale or loan to a consumer. One of the objectives of this section of the act is to translate the interest rates charged into a common form and then to require their disclosure so that consumers may compare the rates charged by different lenders. Whether the act has had any significant affect upon interest rates and lending practices, particularly as far as the low-income consumer is concerned, is a subject of considerable debate.

The CCPA applies to transactions that involve the purchase of property on credit as well as the straight borrowing of money. It does not apply to those who loan money in a private transaction, such as to a neighbor or a friend, but rather to those who loan money in the ordinary course of business. It is also applicable to those who arrange the loan, even though they do not actually lend the money. Thus if a seller arranges a loan for a customer with a bank, both the bank and the seller are subject to the act.

Only consumer-credit transactions are covered by the act. Thus only credit extended to a natural person for family, personal, household, or agricultural purposes is regulated under the act. Real-estate financing and transactions involving personal property are both covered under the act. However, transactions other than a real-estate mortgage are excluded if the amount financed exceeds $25,000. Examples of transactions that would be covered under the CCPA include real-estate loans, installment sales and loans, home-improvement loans, and farm loans. Examples of loans not covered would be casual loans between friends; loans exceeding $25,000, except real-estate mortgages; loans to individuals for business purposes; loans to corporations, partnerships, or governments; stockbroker margin accounts; public-utility bills; and commodity-account transactions.

Under the act the Federal Reserve Board is given the power to prescribe regulations to carry out the purpose of the act. The board has issued what is known as Regulation Z, which sets forth in detail the procedures to be followed by lenders in order to comply with the act.

The heart of the act is truth in lending. Note that it is only a disclosure statute and does not establish any ceiling on interest rates or finance charges. The act requires disclosure statements in the extension of credit. The disclosure statement must be made for each transaction covered by the act. A copy of the statement must be given to the borrower, and the lender must retain the original for two years after the transaction or until the debt is paid, whichever is longer.

There are two distinct types of information that must be included in the disclosure statement, the *finance charge* and the *annual percentage rate.* The finance charge is defined by the CCPA as "the sum of all charges, payable directly or indirectly by the person to whom the credit is extended, and imposed directly or indirectly by the creditor as an incident to the extension of credit. . . ." Note that the finance charge includes much more than interest. It includes in addition to interest such charges as any amount that is payable as a discount, points or loan fees, service or carrying charges, credit-report fees, and credit and other types of insurance premiums charged in connection with the loan, as well as other charges. In other words, the finance charge includes all overhead expenses charged beyond the traditional interest charge for the money loaned. Excluded from the finance charge are such things as taxes, recording fees, license fees, default or delinquency charges, and bona fide closing costs in connection with real-estate transactions, such as the costs of title examination.

The finance charge is used in the calculation of the *annual percentage rate.* This rate may then be used by the consumer to shop for credit on a comparative basis, as it reduces all loan costs to a common denominator because of the uniform method of calculation that is required. You should note at this point that the annual percentage rate and the interest rate are not the same thing. The annual percentage rate is based on the finance charge, which includes many more costs than just traditional interest. As a result, the typical disclosure statement will include the finance charges in dollars, the annual percentage rate, and an interest rate. The interest rate will frequently be substantially lower than the annual percentage rate. The disclosure statement must also disclose the amount of each payment and the number of payments.

The disclosures required by the CCPA and Regulation Z apply at several stages of the consumer-credit transaction. The required disclosures must be made when the credit is advertised, when the loan is made, and for periodic billing.

When a sale involves real estate, the act gives the consumer who gives a security interest in real property that is used or expected to be used as his or her principal residence, the right to rescind the transaction. He or she has until midnight of the third business day following the date of consummation of the transaction or date of delivery of the required disclosures, whichever is later. The lender is also required to notify the customer of his or her right to rescind in this situation. Note, however, that the right to rescind does not apply to all loans made under the CCPA.

The act provides for criminal penalties of up to $5,000 or a year in prison or both for anyone who willfully and knowingly fails to comply with the act or gives false or inaccurate information. The act also provides for civil liability. If a creditor fails to disclose information as required, he is liable for twice the amount of the finance charge in connection with the transaction but in any event not less than $100 or more than $1,000. In addition, the successful debtor may collect the costs of enforcing the liability, including reasonable attorney's fees. A creditor may escape liability, however, if within fifteen days of discovering the error and prior to suit by the debtor or receipt of written notice by the debtor, he notifies the debtor of the mistake and makes adjustments so that the debtor is not obligated to pay more than the finance charge or the percentage rate actually disclosed. Nor is a creditor liable if he shows by a preponderance of the evidence that the violation was not intentional and resulted from a bona fide error, despite the existence of procedures to avoid such an error.

JOSEPH v. NORMAN'S HEALTH CLUB, INC., et al.

532 F.2d 86 (1976)

LAY, J.

FACTS

The facts are substantially undisputed. Between 1960 and 1971, Norman R. Saindon owned and operated a chain of health clubs in the St. Louis, Missouri area. "Lifetime memberships" were offered to the public for $360, payable in 24 equal monthly installments of $15 each. Before the Truth in Lending Act became effective, a few memberships were also sold for cash at discounts of 10 to 15 percent off the total installment price. The district court found that it was the intention of the Health Club to sell almost all memberships on the installment plan and to discount all the notes to finance companies. Ninety-eight percent of club members chose to sign installment notes. 386 F. Supp. at 783.

The Club had dealt with seven or more finance companies since 1960, but by the time TILA (Truth in Lending Act) became effective, the Club was tendering all of the notes to only defendants Boston and Consolidated. The Club had negotiated an agreement with each company providing the rate of discount and other terms under which the finance companies would purchase such notes as they determined to be creditworthy. The finance companies were not required to purchase any minimum number of notes under these agreements, but they ultimately rejected only five percent.

The Club's practice was to require customers purchasing a membership on the installment plan to sign a promissory note and to fill out a standard credit application form. The latter was used to provide credit information to the finance companies. Once these forms had been filled out, the Club would often notify one of the finance companies the same day. The finance company would conduct an immediate credit check on the customer and would call the customer to verify that he had signed the membership contract. If the finance company found the new member to be an acceptable credit risk, the Club would assign the note to the finance company without recourse. The finance company would pay the Club the face amount of the note less the amount of the discount provided in the agreements.

Thereafter, the finance company would treat the club member whose note it had accepted just as it treated its own direct loan customers. The finance company would send the club member a payment coupon book as well as instructions that all payments were to be made directly to the finance company and a notice that a late charge would be assessed for late payments. The club member's account was carried on Consolidated's books as a "loan" with the member described as the "customer" and the finance company as the "lender." Defendant Consolidated would notify the club members who had made a certain number of payments that were now preferred customers of Consolidated. Boston similarly designated club accounts as "loans."

The discount, that is, the difference between the face amount of the customer notes and the cash amount which the finance companies would pay the Club upon assignment of each customer note, was substantial. It ranged from $85 to $165 on the $360 notes. The discount rate was the same on all notes at any one time, but it was renegotiated upward from time to time.

AN OVERVIEW OF THE ACT.

The fundamental purpose of the Truth in Lending Act, 15 U.S.C. § 1601 *et seq.*, is to require creditors to disclose the "true" cost of consumer credit, so that consumers can make informed choices among available methods of payment.

The Act was intended to change the practices of the consumer credit industry, and the statute reflects Congress' view that this should be done by imposing disclosure requirements on those who "regularly" extend or offer to extend consumer credit. In interpreting the Act, the Federal Reserve Board and the majority of courts have focused on the substance, rather than the form, of credit transactions, and have looked to the practices of the trade, the course of dealing of the parties, and the intention of the parties in addition to specific contractual obligations.

The statute requires certain information, such as total finance charges and the applicable annual percentage rate, to be disclosed "to each person to whom *consumer* credit is extended and upon whom a finance charge is or may be imposed." 15 U.S.C. § 1631(a). "Credit" includes "the right granted by a creditor to a customer to . . . incur debt and defer its payment." Reg. Z, 12 C.F.R. § 226.2(l). The Act does not apply to loans or credit sales made "for business and commercial purposes," 15 U.S.C. § 1603(1); it covers only extensions of consumer credit, defined in the Act as credit "offered or extended [to] a natural person . . . primarily for personal, family, household, or agricultural purposes." 15 U.S.C. § 1602(h). Failure to comply renders the creditor liable to the consumer in an amount equal to twice the finance charge, but not less than $100 or more than $1000. 15 U.S.C. § 1640(a). In the Act, Congress gave the Federal Reserve Board broad authority to promulgate regulations to ensure compliance. 15 U.S.C. § 1604;. . . Pursuant to the grant of authority, the Board promulgated Regulation Z, 12 C.F.R. § 226.1, *et seq.*

Congress defined "creditor" broadly to include all "who regularly extend, or arrange for the extension of [consumer] credit for which the payment of a finance charge is required, whether in connection with loans, sales of property or services, or otherwise." 15 U.S.C. § 1602(f). The original Act thus contemplated two classes of creditors who were required to make disclosures in appropriate transactions. First are the "extenders" of credit, those who actually provide the funds and carry the risk of the obligation. Second are the "arrangers" of credit, those who negotiate a consumer credit transaction on behalf of a third-party extender. 15 U.S.C. § 1602(f); Reg. Z, 12 C.F.R. § 226.2(f), (m). Only those who "regularly," that is, in the ordinary course of business, arrange or extend credit, are subject to the disclosure requirements.

THE LIABILITY OF THE FINANCE COMPANIES UNDER THE ACT.

In the instant case, the finance companies operated under a definite working arrangement with the Club. The evidence discloses that (1) the finance companies were alerted almost simultaneously with the customer's execution of a note; (2) an immediate credit check was then made by the finance companies; (3) if the customer's note was accepted, the finance companies paid the Club the amount of the note less the discount; (4) the finance companies accepted assignments without recourse to the Club, thus relying solely on the customer for payment; (5) the finance companies often contacted the customer the same day and upon approving him, the companies would send out their payment book describing the manner of payment and notice of late charges (not mentioned in the note assigned); (6) the finance companies carried the note on their books as a "loan" and listed a "finance charge"; and (7) thereafter, the finance companies treated the club member in the same manner as they did their direct consumer loan customers. The situation was no different than if the finance companies had gone to the Club with the prospective member, paid the Club for the membership and then taken the customer's note just as they did take it.

We find, as have all but one of the courts faced with similar facts, that where the third-party financer becomes intimately involved in the relevant credit transactions it may become liable as an extender of credit. Where a finance company becomes an integral part of the seller's financing program, the finance company must bear full responsibility for all disclosures required under the Truth in Lending Act. Here the parties dealt on a prearranged, systematic basis. The fact that the companies did not immediately assume the obligation and could, in their discretion, reject certain notes, assumes little importance in view of the overall course of dealing and the role they actually played in financing 95 percent of the loans. In any event, what is at issue is their liability on the loans they did accept, not on those they rejected. The method of operation used here served to channel the loans to the finance companies, and the Health Club served as a mere conduit between the club members and the actual extenders of credit. In order to find the *arranger* liable, *he* must operate under some "business or other relationship" with the *extender* of credit. However, there is no evidence that the Federal Reserve Board intended these words to mean anything more than pursuing in the regular course of business a practice of channeling credit business to another. This obviously means more than isolated or even periodic discounting of consumer paper between commercial enterprises. Without attempting to limit the definition, we think it clearly encompasses credit transactions which are prearranged and systematically carried out as in this case. It is important to remember, though, that Regulation Z, § 226.2(f) does not discuss the requisite proof of liability as a credit extender; it relates only to the liability of one who arranges credit to be extended by another. The focal issue here, as earlier stated, is whether Congress intended to allow the credit transactions used here to fall outside the Act. As we have discussed, the substance of the transactions is the controlling factor. In that light, we find the arrangement a mere pretext for consumer loans which required disclosure under the Act by both the Club and the finance companies.

THE DISCOUNT AS A FINANCE CHARGE.

Finding liability, the remaining issue is whether the discount constituted a finance charge under the Act.

The district court did not discuss whether the discount on the notes assigned to the finance companies was a finance charge which should have been disclosed by the Club. The district court did note however that unitary price schemes, under which the seller offers goods for the same price whether paid in cash or on the installment plan, can hide a finance charge. 386 F.Supp. at 791.

The Act defines "finance charge" as:

> the sum of *all* charges, payable directly or *indirectly by the person to whom the credit is extended,* and imposed directly or *indirectly* by the creditor as an incident to the extension of credit, including . . . any amount payable under a . . . discount . . . system. 15 U.S.C. § 1605(a) (emphasis added).

The fact that a particular charge may not be included in the definitions of interest found in state usury laws is not controlling, for "finance charge" under TILA was intended to include not only "interest" but many other charges for credit. The Federal Reserve Board has ruled that discounts paid by a seller (such as the Health Club) of consumer accounts receivable must be included in the stated finance charge if and to the extent that they are passed on to the consumer.

In the instant case, it is obvious that all of the discount originally agreed upon by the Club and the finance companies was passed along to the customers as a charge for use of credit. It may be, however, that some portion of the subsequent increases in the discount was not passed along to the customers and was rather absorbed by the Club as a reduction in its profit. On the other hand, adding more members may have permitted the seller so to reduce cost per member that it could provide the same services without increasing the face amount of the note. The district court should explore this matter on remand.

Since the amount of the discounts charged by the finance companies varied from time to time, it is necessary to remand to the district court for computation of the penalty to be assessed under the Act in favor of the plaintiffs and against the finance companies.

(The judgment in favor of the finance companies is vacated and the cause reversed and remanded for further proceedings.)

QUESTIONS FOR DISCUSSION

1. *What does the Court say is the purpose of the Truth in Lending Act?*
2. *If the finance companies did not lend money directly to the customer of the Health Club, why are they required to make disclosures under the Act?*

---

(2) The Uniform Consumer Credit Code. In an attempt to deal with the consumer-credit area in a comprehensive way, the Commissioners on Uniform State Laws in 1968 promulgated the Uniform Consumer Credit Code (UCCC). It has been adopted by nearly a quarter of the states to date and covers many of the same areas as the Federal Consumer Protection Act as well as others. For example, the UCCC deals with such subjects as truth in lending; establishes rate ceilings; and deals with referral sales and door-to-door sales, fine-print clauses in contracts, and such creditor's remedies as deficiency judgments and garnishments, as well as other areas. Its provisions cover credit involving most types of real-estate sales and sales of goods and services by persons who regularly engage in credit selling to persons who use the terms primarily for a personal, family, household, or agricultural purpose. Simple charge accounts that do not involve repayment in installments or interest are not covered.

In most respects the truth-in-lending provisions of the UCCC are the same as those of the CCPA. Both make use of the concept of an annual percentage rate, for example. You should note that the CCPA exempts transactions within the state from federal control as long as there is an adequate provision for enforcement of the state law, and you should also note that the state statutes and regulations are substantially similar to federal law or are stronger. Thus states may now regulate in this area, provided that they enact strong enough provisions and receive an exemption from the Federal Reserve Board.

The UCCC takes a somewhat different approach from the CCPA in that it actually establishes interest-rate ceilings for certain types of loans. These ceilings are set at a rather high level, and lenders are free to set their own interest rates under the ceiling. For instance, the ceiling on most loans is 36 percent of the unpaid balance up to $300; 21 percent for unpaid balances between $300 and $1,000; and 15 percent for the part of the unpaid balance in excess of $1,000. Alternatively the creditor may charge 18 percent across the board. The ceilings are somewhat different for revolving credit transactions. The act is also quite detailed as to what charges are to be included in the rate for the purpose of determining the ceilings. Like the CCPA, the UCCC includes many items in addition to the traditional "interest" charge.

The UCCC also prohibits the seller in a consumer-credit sale or lease from taking a negotiable instrument *other than a check*, and a holder who takes such an instrument does not take the instrument in good faith if he has notice that it is issued in violation of the statute. This provision, of course, severely restricts the concept of a holder in due course in those types of transactions for which the UCCC has been adopted. The holder of a check may still be a holder in due course, as may the holder who takes a negotiable note without notice that the statute has been violated.

Door-to-door solicitation and selling is frequently cited as a practice through which the consumer may be abused. The UCCC deals with this problem in several respects. In order to create a buffer against high-pressure sales tactics in the home, the UCCC provides for a seventy-two-hour "cooling-off" period during which the customer may rescind a credit sale, provided that it was solicited in the customer's home. Note, however, that it is only home-solicitation sales for credit that are subject to the seventy-two-hour period and not other sales. Both the face-to-face solicitation and the signing of the contract must occur in the consumer's home for the provisions of the UCCC to be applicable. Any notice of cancellation given by the consumer must be in writing but need contain no special language. The notice is effective when mailed and must be given before midnight of the third business day after the day on which the contract was signed. The written contract signed by the consumer must inform him of this right, and upon notice the seller must refund any payment made to him by the consumer within ten days.

The UCCC also deals with deficiency judgments. Normally, if a creditor has a security interest in specific goods, he may have the power to repossess those goods and sell them in order to collect the debt. If the sale of the goods does not result in proceeds sufficient to satisfy the debt, the creditor may then sue the debtor for the deficiency. The UCCC prohibits a creditor from seeking a deficiency if he chooses to repossess or accept surrender of goods whose purchase price was $1,000 or less. In other words, the creditor must elect either to sue the debtor for the amount of the debt or to repossess. The right to pursue a deficiency judgment for goods whose purchase price exceeds $1,000 is unaffected.

One potent weapon of the creditor is to garnish a debtor's wages. In other words, through a court action the employer must pay part of the wages earned by the employee directly to the creditor. Most states have limitations on the amount of wages that may be garnished. Usually some percentage of the employee's wages is exempt, or the creditor may be able to garnish wages only above a certain minimum dollar amount. The UCCC exempts from garnishment the greater of 75 percent of an employee's "aggregate disposable earnings" or the amount by which his weekly disposable earnings exceed forty times the federal minimum wage. The CCPA also has a similar provision, except that the exemption applies to thirty times the federal minimum wage. The UCCC protection is limited to garnishments "arising from a consumer credit sale, consumer lease, or consumer loan." Other transactions are exempt. Both acts prohibit an employer from discharging an employee whose wages have been garnished one time. The UCCC prohibition applies regardless of how many garnishment proceedings occur.

(3) **The Fair Credit Reporting Act.** One of the first questions asked by a consumer who is delinquent in paying a bill or who refuses to pay it because of a dispute with the creditor is "How will this affect my credit?" The answer is important because credit information is compiled by various kinds of credit agencies and is then distributed to their customers. This information is gathered from among sources including court records, reports from creditors, interviews with neighbors, and so on. Because of the abuses in credit reporting, Congress passed the Fair Credit Reporting Act in 1971 as an amendment to the CCPA.

Basically two types of reports may be made by credit agencies; *consumer reports* and *investigative reports.* A consumer report is one that is usually made for the purpose of extending credit and contains such information as the consumer's credit background, judgments, tax liens, bankruptcies, and other similar information. The report will also contain information on income and some general personal facts, such as an employment history and any record of arrests, indictments, or convictions. Any investigative report contains much more personal information and is often used in employment decisions and by insurance companies and others. Such a report may also contain information concerning one's personal habits, education, marital status, and many other "facts."

The Fair Credit Reporting Act provides that a consumer must be notified when an investigative report is being prepared. He or she then has the right to request information concerning the nature and scope of the investigation. As to both consumer and investigative reports, the act requires a consumer reporting agency to disclose certain information to the consumer upon his request. Among the matters that must be disclosed are (1) the nature and substance of all information in its files at the time of the request; (2) the sources of the information in its files at the time of the request; and (3) the recipients of any consumer report within certain time limits.

If the consumer disputes information contained in the agency's file, the agency must reinvestigate the information unless the consumer's complaint is frivolous. If the reinvestigation does not resolve the dispute, the consumer may submit a brief statement concerning the dispute to the agency, which must include its substance in subsequent reports. Inaccurate information must be deleted from the agency's file. Furthermore obsolete information must be deleted from the file. In general this includes information that is more than seven years old and bankruptices more than fourteen years old.

The act also limits access to consumer reports. Access is limited to (1) a response to a court order (2) in accordance with written instructions of the consumer to whom it relates and (3) to a person who intends to use the report for employment purposes, for the extension of credit, for the underwriting of insurance, for determination of the consumer's eligibility for a governmental license, or for other legitimate business purposes.

The act also contains a number of provisions governing the procedures to be used by credit-reporting agencies. Obviously, if one is denied credit, he or she should seek the basis for the denial and obtain the information to which he or she is entitled under the act.

LOWRY v. CREDIT BUREAU, INC., OF GEORGIA
444 F. Supp. 541 (1978)

On August 1, 1976, James F. Lowry and Claudia A. Lowry applied for a loan in order to finance a home they had recently constructed. The application was filed with the Decatur Federal Savings and Loan Association. On the application Mr. Lowry listed his name as "James F. Lowry" and his former address as Solana Beach, California.

Decatur Federal maintains a computer terminal which gives it direct access to information stored in the computers of Credit Bureau, Inc. In obtaining the information, the requesting party supplies as much relevant data as possible about the party as to whom they are inquiring. The computer will then supply the names of people for whom it has credit histories and with whom there is a programmed minimum of correspondence between the identifying information of the party for whom information is sought and the parties for whom the computer has stored credit information. The computer will only supply information on those parties where there are at least "fifty points" of correspondence between the subject of the inquiry and the relevant credit records.

The inquiry in the present case resulted in the offering by Credit Bureau of a "James Frank Lowry" of San Francisco, California, whose file showed exactly "fifty points" of correspondence. The file indicated that James Frank Lowry had been adjudicated bankrupt in 1967. Decatur Bank informed its applicant, James F. Lowry of Solana Beach, that there was a problem with his application for a loan.

James F. Lowry then went to the Credit Bureau and after spending the better part of a day there he was told that there was a bankruptcy on his record which he denied. He also denied having ever lived in San Francisco. New reports were then sought from Credit Bureau's computer by Decatur Federal. The reports included the previously noted bankruptcy by a James Frank Lowry. Based upon this, Decatur Federal denied Lowry's request for a loan on September 10, 1976.

On September 13, 1976, James F. Lowry notified Credit Bureau of the denial of his loan and demanded a correction of his credit report. On October 8, 1976, Credit Bureau notified Decatur Federal of its error and eventually Lowry received a loan from Decatur Federal. Lowry was never notified of the correction of the record nor of the outcome of the investigation he requested. Lowry then sued Credit Bureau, Inc. One ground for the suit was for violations of the Fair Credit Reporting Act. The defendants filed a motion for summary judgment.

MURPHY, J.

1. Plaintiffs contend they have been the victim of a violation of defendant Credit Bureau's duty to insure the maximum possible accuracy

of information. The Fair Credit Reporting Act provides in relevant part:

> Whenever a consumer reporting agency prepares a consumer report it shall follow reasonable procedures to assure maximum possible accuracy of the information concerning the individual about whom the report relates. 15 U.S.C. § 1681e(b).

This section imposes an obligation to insure maximum accuracy only in the preparation of a report. The crux of plaintiff's complaint is the potential for confusion of reports inherent in defendant's computer system. Plaintiffs note the potential for confusion was realized in their case and resulted in at least the delay in the grant of a loan.

Plaintiffs' concern is not accuracy in the preparation of credit reports; plaintiffs' concern is the confusion of those reports. Confusion of reports did result, but that does not provide a basis for a federal claim. This district has previously recognized that "in order to pursue a cause of action predicated upon willful or negligent violation of 15 U.S.C. § 1681e(b), the report sought to be attacked must be inaccurate."

The only inaccuracy in the questioned credit reports arises from the presence of plaintiff James Francis Lowry's social security number in the James Frank Lowry file. As noted by both sides, the plaintiff's social security number appeared with the bankrupt James Frank Lowry's file because the computer was programmed to add the information when Decatur Federal's operator accepted the file of James Frank Lowry the first time. The automatic addition of this information may constitute a violation of the 15 U.S.C. § 1681e(b) obligation to provide the maximum possible accuracy in the preparation of a credit report. Preparation may be viewed as a continuing process and the obligation to insure accuracy arises with every addition of information. Plaintiffs may have difficulty demonstrating the existence of damages sustained as a result of this breach [as required under 15 U.S.C. § 1681*o*(1)] but at this stage it cannot be said such proof would be impossible.

2. Plaintiffs contend there has been a breach of defendant Credit Bureau's duty to investigate the accuracy of information "in his file." 15 U.S.C. § 1681i(a). Preliminarily, it must be noted that the circumstances of this case indicate the necessity of reading the language "in his file" to include more than just the computer report which a two month investigation discloses to be the only report relevant to the subject of the inquiry. As encountered in this instance, if a party has credit difficulties because of confusion of two similar computer reports, the subject's "file" must be viewed as the totality of the conflicting information which is causing the credit uncertainty. Under this view, and the language of 15 U.S.C. § 1681i(a), a consumer reporting agency is obligated to reinvestigate the accuracy of information as it relates to the subject of the inquiry. The agency's obligations are not terminated by the fact that the information is accurate about someone else if that information is presented in a manner such as to create inaccurate impressions as to the credit history of a particular individual. To permit the activity encountered here to go uncovered would be contrary to the broadly remedial aims disclosed in 15 U.S.C. § 1681.

The Fair Credit Reporting Act creates an obligation to investigate "within a reasonable period." There is no doubt an investigation was undertaken and changes were made. The only issue here is whether 49 days constitutes a "reasonable period" to determine if one party is the bankrupt referred to in a particular credit report. This element of the complaint is not proper for disposition on summary judgment.

QUESTION FOR DISCUSSION

1. *How may Credit Bureau, Inc., have violated the Fair Credit Reporting Act?*

(4) Credit Cards. Today credit cards are used frequently in consumer sales. They are issued by banks, oil companies, American Express, Carte Blanche, and Diner's Club, as well as many other businesses. Until recently their use presented some difficult legal problems because the laws dealing with commercial transactions were not designed with the credit card in mind.

The distribution of credit cards creates a number of problems, including their unauthorized use. For instance, when a credit card is stolen and is used to purchase merchandise, the loss may be assumed by the credit-card holder, the company issuing the card, or the merchant from whom the goods were purchased. To cut down on the problem, Congress placed a provision in the CCPA prohibiting unsolicited mailings of credit cards. No card may be issued except in response to a request or an application. The CCPA also reduces the exposure of card holders for losses resulting from unauthorized use of the card. Under the act a card holder is liable for unauthorized use of his or her card by a thief only if (1) the liability is not in excess of $50; (2) the card has been "accepted" by the card holder; (3) the issuer gives adequate notice of the card holder's potential liability on the card or on each periodic billing statement; (4) the issuer has provided a method of identification of the card holder by the merchant, either by signature or by photograph; (5) the unauthorized use occurs before the issuer is notified that the card has been lost or stolen; and (6) the issuer provides the card holder with a self-addressed, prestamped notification card.

The truth-in-lending requirements of the CCPA also apply to credit cards. Finance charges must be disclosed in the application for the card and also in the periodic billing through a monthly statement.

(5) Miscellaneous Protections Regarding Credit. There are a multitude of protections for consumers concerning the extension of credit, and they vary widely from state to state. One protection that is afforded by every state is some form of *debtor's exemption.* These exemptions are the assets that a creditor who obtains a judgment against a debtor cannot take in order to satisfy a judgment. In other words, they are exempt from execution. Many states, for instance, allow an exemption for certain types of property, such as a workman's tools, clothing, and so on. On top of that the debtor may exempt property of up to a certain value, such as $1,000. There may also be a *homestead exemption.* This is a sum of money to be given the debtor from the proceeds of the sale of his residence if it is sold by creditors in order to satisfy a debt. The amount of property subject to a debtor's exemption varies widely from one state to another.

An additional protection to the consumer is the limitations placed on the holder-in-due course doctrine. These have come about through court decisions, through state laws, and by action of the Federal Trade Commission. These limitations were discussed in Chapter 24.

47:2 Labeling and Packaging Goods. One of the more difficult problems for consumers is the packaging and labeling of goods in a manner that may be misleading. A substantial number of federal statutes and agencies deal with the problem of labeling for many different types of goods. Examples of some of the more specialized statutes are the Fur Products Labeling and Advertising Act; the Flammable Fabrics Act; the Food, Drug, and Cosmetic Act; and others.

The Federal Trade Commission has jurisdiction under a number of these acts. In general, the standard for determining whether a label is deceptive is the meaning that the label conveys to the ordinary person. The fact that it may be clear to an expert or one otherwise knowledgeable in the field is not enough.

One of the more recent bits of federal legislation concerning consumers is the Fair Packaging and Labeling Act. This act was hailed as a step in the right direction toward eliminating deception in labeling and packaging, although some people have called the act "more packaging than labeling." The act generally requires consumer goods to have a label identifying the product; the manufacturer, packer, or distributor, including its place of business; the net quantity of the contents; and, if the number of servings is stated, the quantity of each serving. One purpose of this information is so that the consumer knows the party against whom a suit may be filed in case he or she has a claim resulting from the use of the product. An additional provision of the act gives the Federal Trade Commission and the Department of Health, Education, and Welfare the authority to promulgate regulations adding requirements to the information contained on the label. For instance, regulations have been passed dealing with *cents-off* or *savings* claims on packages; the use of such terms as *large* or *jumbo;* and the disclosure of information as to the ingredients of nonfoods.

The Magnuson–Moss Warranty Act also deals with the problem of labeling, at least indirectly. One problem for the consumer has been the situation in which he or she purchases a packaged product and the terms of the warranty are placed inside the package so that it cannot be read prior to the purchase. The Federal Trade Commission has passed a regulation under the act that requires the seller to display his warranty so that it may be read at the point of and prior to sale. The warranty may be placed on the outside of the package.

The Federal Trade Commission Act has application in this area, as was discussed in Chapter 46. If a label is deceptive, it may violate the act. One problem, however, is that the act does not have a provision allowing consumers to sue for a violation of the act. One may merely file a complaint with the commission.

SUMMARY

Many of the concepts and laws discussed elsewhere in this text have application to consumer problems. The increasing awareness of consumer problems has resulted in a significant amount of legislation in this area, not only at the federal level but among the states and local jurisdictions as well. Many local jurisdictions have created commissions to handle consumer complaints. In many areas some form of legislation has been passed dealing with such critical consumer services as automobile and appliance sales and repairs.

The Federal Truth in Lending Act is part of the Consumer Credit Protection Act. Only consumer purchases in the ordinary course of business are covered by the act. The act gives the Federal Reserve Board the power to prescribe regulations to carry out the purpose of the act which it has done by issuing Regulation Z.

The CCPA is a disclosure statute. The finance charge and annual percentage rate must be disclosed.

The Uniform Consumer Credit Code has been adopted in a number of states. In addition to disclosure the UCCC also establishes lending rate ceilings and deals with referral and door-to-door sales, fine-print clauses in contracts and certain creditor's remedies.

The Fair Credit Reporting Act was passed because of abuses in credit reporting. Credit agencies may make a consumer report and an investigative report. The consumer must be notified when an investigative report is made. Procedures are established to provide the consumer the opportunity to challenge and clarify any mistakes in the credit agency's file. Access to consumer reports is also limited.

Congress placed a provision in the CCPA prohibiting the unsolicited mailing of credit cards. It also limited the liability of credit card holders for the unauthorized use of the card.

Most states have protections for debtors such as debtor's and homestead exemptions. Limitations have also been placed on the holder in due course doctrine.

The Fair Packaging and Labeling Act requires certain information to be placed on labels. A deceptive label violates the act.

PROBLEMS

1. What are the major objectives of the Federal Truth in Lending Act, and how are they accomplished?
2. Determine whether the following loans are covered by the provisions of the CCPA:
   a. A loan of $500 to X corporation.
   b. A loan in the amount of $30,000 to Jones to purchase a yacht.
   c. A loan in the amount of $30,000 to Jones to purchase a home for his primary residence.
   d. A loan of $500 to XYZ partnership to purchase inventory for a store.
   e. A charge of $100 on Jones's credit card for the purchase of clothes.
3. What is the difference between *interest,* a *finance charge,* and the *annual percentage rate* under the CCPA?
4. Which of the following transactions may be rescinded within three days?
   a. A loan with the debtor's principal residence given as security.
   b. A contract made at a hardware store to purchase a lawnmower on the installment plan.
   c. A contract to purchase a car if the salesman makes the initial contact by an unsolicited visit to Jones at the Jones's home. The contract is then made and signed several hours later at the car dealership.

d. A contract to purchase pots and pans that results from an unsolicited sale in Jones's living room.

5. What is *garnishment?* What limits are placed upon this remedy by the UCCC and the CCPA?
6. Distinguish between an *investigative report* and a *consumer report* prepared by a credit agency.
7. What are the requirements placed upon a credit agency making an investigative report? Who may have access to consumer reports?
8. What rights does a consumer have regarding information contained in a credit agency's file?
9. What liability does a credit-card holder have for charges made on a stolen card?
10. What is a *debtor's exemption?* What is a *homestead exemption?*

CHAPTER 48

Background of the Bankruptcy Code and Chapter 7 Liquidation

48:1. Introduction. The past chapters have focused upon the various means that parties have to adjust relationships between themselves. The debtor–creditor relationship may be governed by various forms of contracts, negotiable instruments, and other legal documents, all of which may state the conditions of repayment of money borrowed by the debtor. These instruments and contracts may also state the remedies available to the creditor in the event that the debtor fails to repay the loan. There are also remedies available to a creditor under state law. Among these are attachment of property before and after judgment, execution against property of the judgment debtor, and garnishment.

Occasionally individuals and businesses find themselves so overwhelmed by debt that there is little or no hope of ever satisfying all the obligations owed by the debtor to various creditors. This may happen because of poor management or because of poor economic conditions. If a debtor finds himself in such a condition, there may be a number of consequences. One of course is the despair that such a condition imposes upon the individual. Second, in the case of a business, is the obvious harm done to its employees who lose their jobs and to customers who on occasion may lose valuable sources of supply. Finally, such a condition throws the various creditors of the debtor into an adversary position among themselves as each creditor attempts to gain some advantage in obtaining repayment of his loan from a debtor whose assets invariably are not sufficient to satisfy all of his debts. Bankruptcy is one method designed to alleviate these problems.

The provisions of a federal law prevail over conflicting state law. However, Congress has not chosen to preempt all state law on the subject of bankruptcy in passing federal bankruptcy laws. In fact some questions in bankruptcy law are resolved by state law. For example, under the Federal Bankruptcy Code an individual may choose to be protected by the debtors' exemptions provided by his state rather than by those stated in the federal law.

The founding fathers were aware of these problems when they wrote our constitution. Article 1, Section 8 of the United States Constitution states that

"The Congress shall have power . . . to establish . . . uniform laws on the subject of bankruptcies throughout the United States." Most states have laws that provide for bankruptcy or other form of relief for debtors. The states also have laws that place certain property of debtors beyond the reach of creditors. These are known as debtor's exemptions and homestead exemptions. The purpose of these exemptions is to protect the debtor against being stripped of every shred of property making it extremely difficult for him to keep even the most rudimentary comforts such as shelter or the tools to make a living.

People react to overwhelming debt in different ways. Debtors may seek to hide their assets from creditors, prefer some creditors over others in the payment of their debts, or give their remaining property to relatives and friends. On the other hand, some creditors may deal with a debtor who is in obvious financial difficulty by grabbing all of the debtor's property that they can get as soon as possible. This approach will probably seal the fate of the business debtor. Other creditors may attempt to "work with" the debtors even to the point of extending further credit in the hope that this will help the debtor improve his financial condition so that all the debts may be paid off.

48:2. History of Federal Bankruptcy Law. Congress enacted the "Bankruptcy Act" in 1898. It later passed a major amendment to the law in 1938 which was known as the Chandler Act. Since that time, no major amendment to the act was made until 1976. Over the years methods of finance as well as economic conditions changed. Courts did not seem to apply the same policy in administering the act, and many of the court decisions handed down were either inconsistent or not all reported. The act had not been changed substantially in the forty years prior to the growth of the consumer credit industry and the general adoption of the Uniform Commercial Code. Because of these problems with the act, Congress enacted the Bankruptcy Code of 1978. This resulted in a complete revision of the bankruptcy law in this country although the basic philosophy of the bankruptcy law remains the same. You should note that the old law was called the Bankruptcy *Act,* whereas the new law is known as the Bankruptcy *Code.* In the following discussion reference will be made on occasion to both the act and the code.

48:3. Overview. The substantive provisions of the Bankruptcy Code of 1978 became effective on October 1, 1979. The code is divided into eight (8) chapters; 1, 3, 5, 7, 9, 11, 13 and 15. The even numbers have been left open to allow for future additions to the code. Chapter 1 (General Provisions), Chapter 3 (Case Administration), and Chapter 5 (Creditors, the Debtor and the Estate) are basically procedural chapters. Chapter 7 (Liquidation) Chapter 9 (Adjustment of Debts of a Municipality), Chapter 11 (Reorganization), and Chapter 13 (Adjustment of Debts of an Individual with Regular Income) are operational chapters. In other words, a case is filed under one of the operational chapters depending upon the qualifications of the debtor and the type of relief sought.

If a debtor seeks to get relief from his debts, he must first decide which chapter of the code he should file a petition under. For instance, if he is an individual, the most common chapter would be Chapter 7, Liquidation. In very general terms this involves turning his assets over to a trustee who will usually liquidate them and pay off the creditors according to their priority. In return, the debtor gets to keep certain property exempted by the law and is discharged from any remaining debts.

An alternative for an individual debtor is to file a petition under Chapter 13. This chapter is designed for the individual debtor who has a regular income to present a plan for paying those debts while under protection of the bankruptcy court. The advantage of this procedure is that the court will protect the debtor from the suits of creditors and other creditor actions in order to give the debtor time to get his affairs in order and pay his bills. The debtor's property need not be liquidated, but of course he must have an income that allows him to pay off the debts under the plan within a limited period of time, usually three years.

If the debtor is a business or corporation an analogous approach to Chapter 13 is Chapter 11, Reorganization. Under this section, a corporation may gain time to reorganize its financial structure and come out of the bankruptcy as a going concern, whereas in a Chapter 7 liquidation the business would essentially be terminated.

48:4. Administrative Structure. Under the Bankruptcy Act of 1898 bankruptcy cases were administered by the United States District Court which acted as a *Court of Bankruptcy.* The same courts also were responsible for adjudicating controversies between the representative of the bankrupt's estate and among the various claimants to the estate. To help the judges, *referees* were appointed to actually handle the case subject to review by the judge. The problem was that the judge wore two hats, one as a bankruptcy court and the other as a regular court adjudicating the controversies among the involved parties. There were bankruptcy rules for the bankruptcy matters, and Federal Rules of Civil Procedure for actions before the District Court.

The Bankruptcy Code of 1978 has attempted to remedy this confusing situation. The new code provides for a transition period until March 31, 1984. Until that date the *Courts of Bankruptcy* under the old act will function as courts of bankruptcy for purposes of the new bankruptcy law and will function as a *separate department* of the District Court. After April 1, 1984, in each judicial district there will be a court of record, known as the *United States Bankruptcy Court,* that will be an adjunct to the district court. Bankruptcy judges will be appointed by the president for fourteen-year terms, subject to confirmation by the Senate. The court will have broad powers over the property of the debtor and over most proceedings arising in or relating to bankruptcy cases. The court may abstain from exercising its jurisdiction if it feels that the dispute can be handled better by another court.

Chapter 15 of the new code sets up a five-year, trial appellate-trustee program in eighteen districts. Under the program the United States Trustee will supervise panels of private trustees. These trustees will remove many of the administrative duties from the bankruptcy judge. This will leave the judge free to resolve disputes without being biased by knowledge obtained in matters involving administration. Thus in this experiment the administrative and judicial functions will be separated. The trustee is the representative of the estate and in that capacity may sue and be sued. He or she has a number of specific powers as well as rights and responsibilities. Basically the trustee's duty is to collect, liquidate, and distribute the estate. The trustee may operate the business of a debtor, except in Chapter 13 cases, and may, subject to numerous limitations, employ people to help him, such as accountants, attorneys, appraisers, and other professional persons.

In a Chapter 7 case, liquidation, an *interim trustee* is appointed promptly and serves until a trustee is elected by the creditors. If the creditors do not elect a trustee, the interim trustee continues to serve.

The situation is somewhat different in a Chapter 11 case, business reorganization. There need not be a trustee and the debtor may remain in possession of his estate. A *debtor in possession* is in the same position as a reorganization trustee. On the other hand, the court may order the appointment of a trustee when a *party in interest* requests it. A party in interest might be a creditor, stockholder, or other similar person. In general, the court may agree to appoint a trustee whenever there has been fraud, dishonesty, or mismanagement or if the appointment would be in the best interests of the creditors, equity security holders, or other interests of the estate.

In Chapter 13 cases, a trustee is required. In a Chapter 9 case there is no trustee.

48:5. Liquidation Under Chapter 7 (Liquidation). Chapter 7 of the Bankruptcy Code, Liquidation, is the chapter under which most consumer bankruptcies are administered. The object of the chapter is to establish a procedure to collect the debtor's nonexempt property, convert it into cash, distribute it according to priorities established under the code, and then discharge the debtor from the debts.

(1) Voluntary Petitions. The first step in a bankruptcy case is the filing of a petition. The proceeding may be either voluntary or involuntary. An involuntary petition is most often instituted by the creditors. In a voluntary proceeding the debtor files the petition. It may be filed jointly by a husband and wife. The term debtor includes a person who resides in the United States or who has a domicile or property in the United States. A municipality may be a debtor. A railroad, bank, insurance company, savings and loan association, or building and loan association may not be a debtor under Chapter 7. Contrary to popular belief, it is not necessary that the debtor be insolvent to file a voluntary petition.

When the petition in a voluntary proceeding is filed, it constitutes *an order for relief.*

(2) **Involuntary Petitions.** Any person may be a debtor in an involuntary case under Chapter 7 except those excluded in voluntary cases, farmers, and charitable corporations. The code defines a farmer as a person who received more than 80 percent of his gross income in the taxable year prior to bankruptcy from a farming operation owned or operated by that person. The term *person* includes individuals, partnerships, and corporations. The term does not apply to governmental units. Farmers are excluded from involuntary cases because of the seasonal nature of their business and the effect this has on the farmer's expenses and debt payments.

The following requirements must be met in order to declare a person an involuntary bankrupt. First, if a debtor has twelve or more creditors, at least three of them who have uncontingent, unsecured claims totaling at least $5,000 must join the petition; or second, if there are fewer than twelve creditors, one or more of the creditors whose claims must total at least $5,000 may file a petition. Third, the debtor must have given the creditors *grounds for relief.* The grounds for relief are either: (a) the debtor must generally not have been paying his bills when due, or (b) within 120 days before the involuntary petition is filed the debtor must have made a general assignment for the benefit of creditors; in other words a custodian was appointed to take possession of the debtor's property. If the debtor does not answer the petition, the court will order the relief requested.

Creditors may be motivated to file a petition to place the debtor in involuntary bankruptcy for several reasons. A major reason is to prevent further deterioration of the debtor's estate in order to assure some payment toward the claims of creditors. An involuntary petition may also be filed by some creditors if they suspect that the debtor is preferring some other creditor(s) in paying his bills.

An obvious reason for additional requirements for an involuntary petition is to protect the debtor from harassment and unnecessary expense. The requirements also serve as a benchmark for creditors to determine when they may take action in the bankruptcy court against an unwilling debtor. If the involuntary petition is not granted, the court may grant court costs to the debtor; the debtor may also obtain damages in the event that a trustee had been appointed who took possession of the debtor's property.

A significant aspect of the code is that the filing of a voluntary or involuntary petition operates as an *automatic stay,* effective against everybody, against a number of activities and actions that might otherwise be commenced against the debtor. In general, the stay applies to the commencement or continuation of a judicial, administrative, or other proceeding against the debtor that arose before bankruptcy and to the commencement or continuation of a proceeding before the U. S. Tax Court concerning the debtor.

There are activities that are excluded from the automatic stay. Among those are criminal proceedings; proceedings to collect alimony or child support; and certain governmental proceedings to enforce regulatory or police powers. The automatic stay is an attempt to maintain the status quo of the debtor's estate at the time of filing of the petition. In return for the automatic-stay provision, the court has the power to provide for the *adequate protection* of parties whose interests may be adversely affected if the stay is continued, such as secured creditors. This means essentially that the secured creditor would receive in value essentially what he bargained for.

(3) **The Debtor's Estate.** (a) *Property.* When a voluntary or involuntary case is begun under the Bankruptcy Code of 1978, an estate is created which consists, in general, of all the debtor's legal or equitable interests in property wherever it is located. Thus the code takes a broad definition of property that may be included in the debtor's estate. Tangible and intangible property is included. Intangible property would include causes of action, interests in patents, patent rights, copyrights, trademarks, contingent remainders, reversions, and other similar interests in real property. Even a worker's vacation pay, accrued but not paid at the time the petition is filed, becomes property of the estate under the new code.

Section 541 of the Bankruptcy Code states the property that constitutes the debtor's estate. It may be summarized as follows:

1. As previously stated, all legal and equitable interests of the debtor in property as of the time of the commencement of the case. Specifically excluded are (a) powers that may be exercised by the debtor solely for the benefit of an entity other than the debtor, such as powers of appointment, and (b) spendthrift trusts that are valid under state law. A spendthrift trust is one that prohibits the beneficiary from obligating the trust estate for his debts. The payments by the trustee to the beneficiary would become part of the debtor's estate however.
2. All interests of the debtor and the debtor's estate provided that the community property is under the sole, equal, or joint management or control of the debtor.
3. Any interest in property recovered by the trustee, or others, for the estate under the provisions of the code.
4. Any interest in property that the debtor inherits or acquires as the result of a divorce decree or property-settlement agreement or property the debtor receives as beneficiary of a life-insurance policy or death-benefit plan becomes part of the estate if two conditions are met. First, the property interest should be acquired *within 180 days after commencement of the estate.* Second, the interest must be an interest that would have been the property of the estate if the interest had been an interest of the debtor on the date of filing the petition.
5. Any interest in property that the estate acquires after commencement of the estate as well as proceeds, product, offspring, rents, and profits from property of the estate. However, earnings from services performed by an *individual* debtor after commencement of the estate are not included.

(b) *Trustee's Powers.* Once the petition is filed, the trustee has the duty to collect the property for the estate. In securing the property for the estate the trustee has a number of powers. These are: (a) the automatic-stay provisions of the code, (b) the turnover provisions of the code, (c) the power to use, sell, or lease property of the estate, (d) the power to obtain unsecured credit and incur unsecured debt in the ordinary course of the debtor's business, (e) the power to assume or reject any executory contract or unexpired lease of the debtor, (f) the power to avoid certain transactions.

We have already discussed the *automatic-stay* provisions of the code. When the petition is filed, other proceedings affecting the debtor are automatically stayed or halted. The purpose of this provision is to allow the trustee to efficiently go about his business of administering and preserving the estate. Not all actions are subject to the automatic stay provisions. Such actions as criminal, alimony, and child-support actions are excluded.

The *turnover provisions* of the code are another important method by which the trustee collects the assets of the estate. These provisions require anyone holding property of the estate on the date when the petition is filed to deliver the property to the trustee. Property of inconsequential value to the estate is exempted unless it has significant use value for the estate. Those who hold property that is to be turned over to the estate are also subject to an accounting. The property must be turned over to the estate even if it may be exempted by the debtor.

Sometimes the trustee in bankruptcy may need to operate the debtor's business or affairs. This will often involve the *use, sale,* or *lease* of the debtor's property during the liquidation of the estate. For the first time, the 1978 code sets forth the rights and responsibilities of the trustee in this process. This was necessary because under prior law the interests of secured creditors of the debtor were often eroded by the imprecision of the law and the fact that trustees were able to easily obtain court orders allowing use of the debtor's property without much regard for the secured creditor's interest in that property. Under the new law, the trustee must give notice of his intent to use, sell, or lease the property before doing so. An opportunity for a hearing must be given to the creditors. On the other hand, no notice or hearing is required if the court has authorized the operation of the debtor's business. If someone has an interest in the property sold or leased, he may request *adequate protection* of that interest from the court.

In general, the bankruptcy trustee has no greater interest in the property than the debtor prior to bankruptcy. Thus, if the property is subject to a security interest when sold by the trustee it would be purchased subject to the security interest. The code does provide for the sale of encumbered property under certain circumstances. For instance, this may occur if the entity possessing the security interest consents to the sale lien free or if the interest is a lien and the price at which the property is to be sold is greater than the value of the interest.

If the trustee operates the business of the debtor, additional borrowing may be required even in Chapter 7 cases. For instance, while liquidating the estate, it may be necessary to maintain insurance coverage or to ship goods. For obvious reasons, lenders may not be overly enthusiastic about lending money to the estate. The trustee may obtain *unsecured* or *secured credit* which is allowable as an administrative expense. If the trustee is unable to obtain credit, the trustee may grant a junior lien on property that is already subject to a security interest. If the trustee is still not able to obtain credit, the court may even authorize a lien on the debtor's property that is superior or equal to an existing lien provided that adequate protection is given to the original lien holder's interest.

When the trustee takes over the debtor's estate, he or she may find some contracts that are beneficial to the estate and others that are detrimental. The Bankruptcy Code of 1978 provides that the trustee may *assume or reject executory contracts and unexpired leases* of the debtor subject to the court's approval. A rejection of course would constitute a breach of the contract and thus give rise to a claim for damages by the other party. A lease or contract that has been assumed by a trustee may be assigned so long as the assignee gives adequate assurance of future performance. Sometimes contracts or leases contain clauses that state that assignment is prohibited or restricted. Such clauses are generally unenforceable against the trustee.

The policy of the bankruptcy law is to prevent the debtor's estate from being dissipated just prior to bankruptcy. There is always the temptation, just prior to bankruptcy, to pay one's favorite creditors, such as a favorite uncle. There also is the temptation, on the part of the debtor, to attempt to keep assets, in excess of the amount permitted, by making bogus transfers to friends or by some other means. The code has several provisions that give the trustee the power to cope with these various activities. These powers are of three types.

The first power is often known as the *strong-arm clause.* This gives the representative of the estate the rights of three types of creditors regardless of whether they actually exist: a creditor with a simple contract with a judicial lien on the property of the debtor as of the time of filing the petition; a creditor with an execution against property of the debtor at the time of filing the petition that is returned unsatisfied; and, a bona fide purchaser of real property of the debtor with such status at the time the petition is filed. The last right is new with the 1978 code. These powers give the trustee the power to avoid any transfer of property by the debtor that could be avoided by these types of creditors. In addition, the trustee may avoid any transaction that is voidable under nonbankruptcy law by a creditor holding an unsecured allowable claim. These powers in effect give the trustee the rights of a judgment creditor and a lien creditor as well as a bona fide purchaser of real estate. The trustee, for example, would have priority against any unsecured creditor. This provision will *not* allow the trustee to avoid perfected security interests.

The second power given to the trustee is to avoid certain statutory liens, that is, liens that arise automatically as a matter of law when certain events happen. The reason for this provision is that if all statutory liens were enforceable in bankruptcy, the state laws could circumvent the federal law and alter the priority of distribution of estate assets established by federal laws. Thus the trustee may avoid a *statutory* lien that first becomes effective upon the happening of a certain event such as the debtor's involvency or commencement of bankruptcy. There are some other situations in which this power may also be invoked. It is important to note that the trustee's power applies to statutory liens and not to those that result from the agreement of the debtor and creditor.

The third power possessed by the trustee allows the avoidance of certain types of transfers, known as *preferential transfers* and *fraudulent transfers.* A preferential transfer is one that enables a creditor to obtain a greater share of the debtor's estate than he would if the transfer had not been made and if he had participated in the distribution of the debtor's estate. Note that these transfers are not illegal under normal circumstances. The reason they are prohibited is that the policy of the code is to distribute the debtor's assets evenly among creditors of the same class. A side benefit of the law is that it discourages creditors from rushing in to grab the debtor's assets through legal actions during his final days prior to bankruptcy. If a transfer is preferential, the creditor must return the property to the debtor's estate. Now let us see which transfers are considered preferential.

(c) *Preferential and fraudulent transfer.* The code sets forth five tests that must be met in order for a transfer to be preferential. If even one of the tests is not met the transfer is not preferential. The trustee may avoid any transfer of property in which:

(a) The transfer was made to or for the benefit of a creditor.
(b) The transfer was made for or on account of an *antecedent* (prior unsecured) *debt* owed by the debtor before the transfer was made.
(c) The transfer was made while the debtor was insolvent.
(d) The transfer was made during the *90 days* immediately preceding the commencement of the case. If the transfer was to an *insider,* the transfer may be avoided if it was made during the period that begins one year before the filing of the petition and ends 90 days before the filing if the insider to whom the transfer was made had reason to know that the debtor was insolvent at the time of the transfer. Examples of insiders are partners of a partnership, relatives of the debtor, or a director or officer of a corporation.
(e) The transfer enables the creditor to whom it was made to receive a greater percentage of his claim than he would receive under the distributive provisions of the code.

Note that the transfer is voidable in most cases even if the transferee had no reason to believe that the debtor was insolvent at the time of the transfer. Note also that under prior law the period was four months; it is now 90 days.

The code also lists six types of transfer that the trustee *cannot* avoid. The first three are protected because they are of the type that do not diminish the size of the estate because the debtor receives value as a result of the transfer. Thus the trustee cannot avoid the following transfers: (1) a transfer that was intended by the debtor and creditor to be a contemporaneous exchange for *new value* given to the debtor and was in fact substantially contemporaneous, such as a check; (2) transfers of a security interest to obtain a purchase-money loan that the debtor used to acquire the property the loan enabled him to purchase after the loan was actually made; (3) transfers to a creditor that are subsequently offset by new value given by the creditor to the debtor.

The next two transfers are protected because of the importance of the type of creditor interest. Thus the following transfers cannot be avoided: (4) transfers to the extent they are in payment of a debt incurred in the ordinary course of business or financial affairs of the debtor and made not later than 45 days after the debt was incurred and made according to ordinary business terms. The purpose of this exception is to allow normal financial relations because it does not detract from the general policy to discourage unusual actions and encourages business with parties in poor financial condition which in turn might help them to regain financial health; (5) the transfer of a perfected security interest in inventory or receivables or the proceeds of either, to the extent that the secured creditor doesn't improve his position during the 90-day period before filing the petition. The final exception is (6) the fixing of a nonavoidable statutory lien.

In summary then, if any one of the five prerequisites is missing or if the transfer is one of the six types protected, the trustee may not set it aside.

The trustee may also avoid *fraudulent transfers.* Under state law a transfer by a debtor to defraud creditors is voidable. A transfer that is voidable as a fraudulent transfer under state law may be set aside by the trustee. In addition, the code has its own provision regarding fraudulent transfers.

The code permits the trustee to void transfers or obligations incurred by the debtor in two situations. First, if within *one year* of filing the petition, the debtor incurred an obligation or made a transfer with the *actual intent* to hinder, delay, or defraud any present or future creditor. Note that this section requires proof of an actual intent to defraud creditors. This is sometimes difficult to do; therefore the code sets forth a number of specific situations in which the trustee may void the transfer or obligation incurred without the necessity of proving actual intent to defraud creditors.

The second situation in which the trustee may void transfers or obligations incurred within one year of the date of filing the petition occurs if the debtor received less than a reasonably equivalent value in exchange for the transfer or obligation and (a) the debtor was insolvent at the time or became insolvent as a result of the transfer, *or* (b) the debtor was engaged in business, or was about to engage in a business or a transaction for which his remaining property was unreasonably small capital, *or* (3) the debtor intended to incur, or believed that he would incur, debts that would be beyond his ability to pay as they matured.

Note that the statute does not avoid transfers made where a necessary equivalent value is given in exchange for the transfer. Value is not the same as consideration. For instance, a promise to perform in the future is not value. Value means either property or the satisfaction of a present or antecedent debt of the debtor. Thus when value is given for the transfer, the estate of the debtor is not depleted because he is receiving something of reasonably equivalent value for what he is transferring.

(d) *Debtor's Exemptions.* One of the basic philosophies underlying bankrupty law is that a person who has become hopelessly overwhelmed with debts should have an opportunity for a fresh start in life without harassment from creditors and the worries and pressures that may result from the burden of debts that the debtor cannot hope to pay. In order to help the debtor begin anew, the law provides that certain basic necessities of life may be claimed by the individual as exempt from creditors' claims.

All states exempt certain property from creditors' claims by statute. In addition, some property is exempt under federal nonbankruptcy law. An example would be social security benefits. The bankruptcy code allows a debtor to claim exemptions provided by state law.

Since the exemptions allowed by the states vary widely, the code provides for exemptions that may be claimed by the debtor *instead* of those allowed under state law. Among the items that the code allows an individual to exempt are: $7,500 of the equity in a residence; up to a $1,200 interest in a motor vehicle; household furnishings, household goods, wearing apparel, animals, crops, and some other household articles for the personal or family use of the debtor or a dependent not to exceed $200 in value for any particular item; $500 for personal or family jewelry; up to $400 cash plus any unused portion of the $7,500 of the home-equity exemption and up to $7,500 in any implements, professional books, or tools of the debtor's trade. There are other exemptions such as up to $4,000 of the loan value of a life-insurance policy owned by the debtor. You should note that a debtor's waiver of his exemptions in favor of an unsecured creditor is unenforceable.

Another provision of the code allows the individual debtor to redeem tangible personal property that is intended primarily for personal, family, or household use and that is subject to a security interest that secures a consumer debt. The debtor must pay the holder of the security interest the amount of the allowed secured claim that is secured by the security intererst.

(e) *Distribution.* We have now arrived at that point where the trustee has collected all of the property of the estate and, usually, has reduced it to cash. The next step is to distribute the property among those creditors who present allowable claims against the estate. The manner of distribution of the estate is determined by a system of *priorities* established by the code. Under this system the claims of creditors are ranked. Higher-ranking claims must all be paid in full before the lower-ranking claims will be paid. If the assets of the estate are not sufficient to satisfy all the claims (which is usually the case) then the assets of the estate are used to satisfy the claims in order of priority until the ranking is reached where the claims exceed the estate assets. At that point the claims of the creditors of that class are satisfied on a pro rata basis. Lower-ranking claims are not paid anything.

Creditors may be either secured creditors, unsecured creditors entitled to a priority, or unsecured creditors that are not priority creditors. Unsecured creditors must file a proof of claim with the court within the time allowed by the code. Secured creditors do not need to file a proof of claim to the extent the secured collateral covers their claim. Proof of claim should be filed within six months after the first date scheduled for the first meeting of creditors unless the court extends the time period.

In distributing the estate, the secured creditors are paid to the extent that the value of their collateral covers the amount of their claim. Secured creditors are unsecured to the extent the amount of their claim exceeds the value of the collateral.

Next to be paid are unsecured claims. The code establishes six priorities for unsecured claims:

1. The first priority is for administrative expenses and certain fees and charges for preserving the estate. The purpose of this priority is to insure those who wind up the estate that they are paid for their efforts. For example, the trustee would be entitled to a fee for his role in administering the estate.
2. The second priority is limited to involuntary bankruptcies and applies to debts incurred after the commencement of the case but before the order for relief or the appointment of a trustee. The purpose of this priority is to give these creditors the same status as if the claim had arisen before the filing of the petition. Since debtors may continue to operate pending a trial on an involuntary petition, creditors must be granted this priority or they would not deal with the debtor. This would make it impossible for the debtor to remain in business in most cases.
3. The third priority is designed to insure that employees will not abandon a failing business for fear that they will not be paid. This increases the chance that the business may be financially rehabilitated. This priority applies to allow unsecured claims for wages, salaries, or commissions, including vacation, severance and sick-leave pay that is earned by an individual within 90 days of the filing of the petition or the date of the cessation of the debtor's business, whichever occurs first. The amount that is given by this priority is limited to $2,000.
4. The fourth priority is established for allowed unsecured claims for contributions to employee-benefit plans. This priority applies only to claims for contributions arising from services rendered within 180 days before the filing of the petition or cessation of the debtor's business, whichever occurs first. It extends only to contributions to the plan to the extent of $2,000 per employee less the amount paid to such employees for wage claims and paid to other employee-benefit plans.
5. The fifth priority extends to unsecured claims up to $900 of the individual

consumer who, before the filing of the petition, deposited money for the purchase of consumer goods or services that were never delivered or provided.

6. The sixth priority is accorded to claims for federal, state, and local taxes, in that order. We shall see in the next section that this priority is also in accord with helping the debtor to get a fresh start since these taxes are also not dischargeable. Thus more of them are paid in the bankruptcy case and not left for the debtor to pay after the case is over.

After the priority claims are paid, any remaining property of the estate is distributed in payment of any allowed unsecured claim, which is not specifically excepted and is filed on time. Next in line are claims filed late because the creditor did not have knowledge or actual notice of the case in time for timely filing, so long as proof of the claim is filed in time for payment. Next paid are claims that are filed late; then, allowed claims for fines, penalties, and the like. If there is still property remaining, then postpetition interest is paid on all paid claims at the legal rate from the date the petition was filed. Finally, any surplus is paid to the debtor.

(4) **Discharge.** The object of bankruptcy for an individual in a Chapter 7 liquidation is a discharge of the individual debtor's indebtedness. You should note that only individuals may be discharged in a Chapter 7 liquidation, not partnerships or corporations. The effect of discharge is to void any present or future judgments against the debtor and stop any legal action concerning dischargeable debts.

(a) *Nondischargeable debts.* The code lists a number of types of debts that are not subject to discharge. Among those debts not subject to discharge are the following:

1. Certain taxes, such as those entitled to a priority that were previously stated in the section on priorities. Also taxes for which a late return or no return at all was filed, as well as taxes for which the debtor made a fraudulent return.
2. Any debt incurred by false pretenses, false representation, or actual fraud *but not* a statement regarding a debtor's financial condition. In addition, discharge of an individual debt may be denied if a materially false statement is made in writing regarding the debtor's financial condition, upon which the creditor reasonably relied, and that the debtor made or published with the intent to deceive.
3. Debts that were not listed or scheduled in the documents required by the code or that were listed or scheduled too late.
4. Debts from fraud or defalcation while acting in a fiduciary capacity, and debts for embezzlement or larceny.
5. Debts to a spouse, former spouse, or child of the debtor for alimony, maintenance, or support unless the debt has been assigned by the spouse

or child to another entity or unless the debt is not actually in the nature of alimony, maintenance, or support.

6. Debts for the willfull and malicious injury by the debtor to another entity or to the property of another entity.
7. Debts for fines, penalties, or forfeiture payable to a governmental unit. This includes tax penalties if the underlying tax is not dischargeable.
8. Debts for an educational loan unless the debt became due more than five years before the date of filing the petition or if excepting the debt would impose an undue hardship on the debtor or the debtor's dependents.
9. Debts from a previous bankruptcy case that survived because the debtor was denied a discharge on grounds other than the fact that the petition was filed within six years of a prior petition or because the debtor waived discharge.

In sum, the types of debts that cannot be discharged are those that arise from taxes, fraudulent behavior, educational loans, and ordered support of a present or former spouse or child. There are obvious policy reasons for denying discharge to these types of debts.

(b) *Grounds for Denying Discharge.* There are also a number of grounds for denying discharge of the debtor. Note that in the previous section we were discussing denial of discharge of *specific debts.* In this section we are dealing with actions of the debtor that may result in discharge being denied for *all* his debts.

The Bankruptcy Code lists a number of grounds for denying a debtor discharge. For instance, if the debtor directly or indirectly transferred, removed, destroyed, mutilated, or concealed his property within one year before the date the petition was filed, discharge will be denied. Similarly, if the debtor concealed, destroyed, or failed to maintain certain business records, discharge might be denied. Discharge would also be denied if the debtor made a false oath or account or used a false claim in the bankruptcy case. Likewise, a failure to explain satisfactorily any loss or deficiency of assets or failure to obey any lawful court order would result in a denial of discharge. Also, discharge will be denied if the debtor has waived discharge after an order for relief is passed and if the waiver is approved by the court.

You should also note that in a Chapter 7 liquidation case discharge will not be granted in a second case commenced within six years of a discharge granted in a prior case. This limitation does not apply to Chapter 13 cases that will be discussed later.

A final point regarding discharge. Under the old Bankruptcy Act discharge would be denied to a debtor who made a materially false written statement regarding his financial condition. An example would be an omission of a significant debt on a credit application. Under the new Bankruptcy Code such a false financial statement would not constitute a ground for denial of discharge.

The reason for the change is that many people might honestly forget a debt in making out the statement. In addition, many creditors put pressure on debtors in filling out the statements which may contribute to omissions in the statement.

(c) *Reaffirmation and Redemption.* One abuse that was frequently practiced by creditors prior to the current Bankruptcy Code was to have debtors reaffirm their debts after discharge. Thus the debtor would once again become obligated to pay a debt, in many states, when it had previously been discharged.

In order to make sure that the debtor understands the rights and consequences of discharge, the code requires a hearing to inform the debtor of these rights and consequences or, if discharge has been denied, to inform the debtor of the reasons for denial.

Under the code, a reaffirmation agreement entered into between the debtor and creditor is subject to a number of limitations. In the first place it is not enforceable unless it satisfies the requirements of nonbankruptcy law. In order to satisfy the requirements of the code, four conditions must be met: (1) the reaffirmation agreement must be made before the granting of discharge; (2) the debtor may rescind the agreement within 30 days after the agreement becomes enforceable; (3) a hearing must be held wherein the court advises the debtor that a reaffirmation agreement is not required, and advises the debtor of the consequences and legal effect of the agreement; and (4) if the debtor is an individual and the debt is a consumer debt not secured by real property of the debtor, the court must approve the agreement.

The test for court approval of the reaffirmation of the agreement is that it does not impose an undue hardship on the debtor or a dependent of the debtor and that it will be in the best interest of the debtor. Another test is that the agreement is entered into in good faith in connection with the redemption of the debtor's property or in settlement of litigation on the question of the debt's nature as dischargeable or nondischargeable.

The code also has a provision concerning the redemption of the property of individual debtors that is intended primarily for personal, family, or household use and that is subject to a lien securing a dischargeable consumer debt. Before the code, a creditor who had a lien on certain property that might not have much market value but would have a much higher replacement cost would be threatened with foreclosure unless the consumer reaffirmed the debt. Thus the secured creditor would realize a greater return than if he merely foreclosed and took the property subject to the lien. The consumer would be inclined to reaffirm since it would cost more to replace the property than to reaffirm and pay off the debt. The new code eliminates this leverage of the secured creditor by, in effect, giving the debtor the right of first refusal on a foreclosure sale of the property involved. The debtor may keep the property by paying the holder of the lien the amount of the allowed secured claim. This allows the debtor to avoid the high replacement costs of a needed item and the creditor gets what he is entitled to without reaffirmation. Also, since under the code the secured creditor only has a claim to the value of the collateral, the debtor only has to pay the value of the collateral should the amount of the claim exceed the value of the collateral.

Our discussion thus far has been limited to Chapter 7 liquidation cases under the code. As listed previously, there are a number of other types of bankruptcies. In the next chapter we will discuss these bankruptcies in light of how they differ from a Chapter 7 bankruptcy and which concepts and facts presented here apply to them.

SUMMARY

One of the objectives of the bankruptcy laws is to allow those who are overwhelmed by debt to get a fresh start. The United States Constitution gives Congress the power to establish a federal bankruptcy law.

Congress enacted the Bankruptcy Act in 1898. This was amended in 1938 by the Chandler Act. In 1976 a new bankruptcy law was passed known as the Bankruptcy Code of 1978, which completely revised the bankruptcy laws. The Bankruptcy Code is divided into eight chapters. The first three are procedural and chapters 7 (Liquidation), 9 (Adjustment of Debts of a Municipality), 11 (Reorganization), and 13 (Adjustment of Debts of an Individual with Regular Income), are operational.

Administratively the code provides for a transition period until March 31, 1984. Until then the courts of bankruptcy will operate as a *separate department* of the United States District Court. After that date there will be courts of record known as the United States Bankruptcy Courts. The code also established a pilot trustee program.

Chapter 7 is the section under which many consumer bankruptcies are administered. The start of the case is the filing of a petition which may be either voluntary or involuntary. In a voluntary case the debtor files the petition which constitutes an *order for relief.*

An involuntary petition is filed by the creditors, but may not be filed against a farmer. If there are twelve or more creditors, at least three with uncontingent, unsecured claims totaling at least $5,000 must join the petition. If there are fewer than twelve creditors, one or more may file the petition if their claims total at least $5,000. In addition, the debtor must have given the creditors *grounds for relief.*

With some exceptions the filing of either a voluntary or involuntary petition operates as an *automatic stay* against a number of activities and actions that might be commenced against the debtor. In return for this action the court has the power to provide for the *adequate protection* of parties whose interests may be adversely affected if the stay is continued.

Once the petition is filed, an estate is created which consists, in general, of all legal or equitable interests of the debtor in property, wherever it is located. The code lists the property that constitutes the debtor's estate.

The trustee has the duty to collect the property of the estate. Among the powers available to the trustee in collecting the estate property are the automatic-stay provisions of the code; the turnover provisions of the code; the power to use, sell, or lease property of the estate; the power to obtain unsecured credit and incur unsecured debt in the ordinary course of the debtor's business; the power to assume or reject any executory contract or unexpired lease of the debtor; and the power to avoid certain transactions.

The trustee has several powers to prevent the debtor's estate from being dissipated by the debtor just prior to bankruptcy. The first, known as the *strong-arm clause*, gives the trustee the rights of certain types of creditors regardless of whether they exist. The second power gives the trustee the right to avoid certain types of statutory liens; a third power gives the trustee the right to void preferential and fraudulent transfers.

A preferential transfer is made for the benefit of a creditor, on account of a prior antecedent debt, while the debtor is insolvent, during the 90 days immediately preceding the commencement of the case; the transfer enables the creditor to receive a greater percentage of his claim than he would have under the distributive provisions of the code. The code also lists a number of types of transfers that cannot be voided. Fraudulent transfers occurring within one year of the commencement of the case may be set aside.

In order to give the debtor the opportunity for a fresh start, the code provides for a number of exemptions to the claims of creditors. As an alternative, the debtor may take advantage of exemptions provided under state law.

The next step in Chapter 7 bankruptcies is to distribute the assets of the estate. Secured creditors are paid to the extent the collateral covers their claim. Next paid are priority unsecured creditors in the order of their priority. Next in line are nonpriority unsecured claims filed on time, followed by those filed late by creditors without actual notice of the case. Next are late claims, then fines and penalties. Postpetition interest is then paid on claims; finally, any surplus is paid to the debtor.

The next step in the bankruptcy process is discharge. Certain debts, such as taxes and alimony, may not be discharged. There are also grounds for denying discharge of the debtor from any debts whatsoever. Among these grounds are concealment of property by the debtor or the making of a false oath or claim in the bankruptcy case.

As a final step, the court requires a hearing to inform the debtor of his rights and the consequences of discharge or, if discharge is denied, to explain the reasons for denial. The code places limitations on the reaffirmation of debts by debtors. The code also allows the debtor to redeem certain personal, household, or family-use property.

PROBLEMS

1. Dudley Debtor owes the local department store $5,000; his family physician, $250; past-due federal income taxes, $8,000; his plumber, $800; the bank, $50,000 for a note and mortgage on his home; and $4,500 in back alimony payments to Wanda Wonderful, his ex-wife. Dudley thinks he should file for bankruptcy. Which debts would be discharged under Chapter 7?

2. Dudley Debtor owns a small retail drug store as a sole proprietor. He has the following creditors: Sam's Stationary Supply, $200; Phillip's Drug Company, $3,000; Art's Ice Cream Supply, $50; Jones Automotive Supply, $800; Jerry's Softdrinks, $120; Ann's Wine Distributors, $2,500; Larry's Cosmetic Supply, $1,200; Wanda's Perfume Wholesaler's, $750; Bruce's Fruit Juices, $225; Sally's Magazine Distributorship, $125; the Gotham News, $78; the Gotham Gazette, $120; Tim's Toy Supply, $320. Dudley has been unable to pay his bills.
   (a) Phillip's and Ann's file an involuntary petition against Dudley but cannot convince the other creditors to join. Would the petition be successful?
   (b) Assume that Gotham News, Tim's, Art's, Sam's, and Larry's file an involuntary petition. Would it be granted?
   (c) Assume Dudley has no debt to Tim's and Sally's. Phillip's and Ann's file an involuntary petition. Would it be granted?
3. Frank Farmer raises turkeys. He has become substantially indebted to the Franford County Co-op for supplies and owes them $10,000. His other bills are to his physician for $75 and to the Town Tractor Company for $450. In August Frank is unable to pay his bills and tells the Co-op that he cannot make further payments until he sells his turkeys over the Thanksgiving and Christmas holidays. The Co-op files an involuntary petition under Chapter 7 in September. Will it be granted?
4. Smith filed a voluntary petition under Chapter 7 on March 15th. At the time she had the following property interests: (a) a bank account containing $1,500; (b) a 1975 Ford Pinto worth $600; (c) accrued vacation pay for two weeks of $500; (d) $5,000 that she received from her grandmother's estate on May 1st, after the petition was filed; (e) a lawsuit filed against another party on February 1st for injuries Smith received as the result of an automobile accident; (f) a power to appoint any person to receive the proceeds of a trust established under the will of Smith's deceased father; and (g) the salary of $250 per week that Smith continued to receive from her employer. To what property is the trustee entitled?
5. Barney was the owner of a small furniture manufacturing company that filed a voluntary petition under Chapter 7. The trustee concluded that the best method of liquidating the business would be to finish the work in process and sell it through Barney's normal marketing channels. Barney has a contract with Sam Supplier. The terms of the contract require Barney to purchase all his glue from Supplier at a price far above the market price. In addition, it is obvious that other supplies must be purchased in order to finish work on the goods in progress, but Barney's estate has no cash. Barney also owns a warehouse that he has used to store his inventory of furniture, but it is empty now and the trustee does not have a need for it in winding up the business. What steps may the trustee take to solve these problems?
6. Dudley Debtor was desperate for cash. He owned some beachfront property worth $125,000 which he sold to his uncle for $75,000 on February 1st in order to pay his creditors. On July 1st when it became obvious that he

could not extricate himself from his financial problems, he sold his aunt a valuable antique grandfather clock for $10. He also sold his car to his brother for the listed bluebook value on August 1st. Since he owed his best friend, Fred, $1,500, he paid this debt in full on September 14th. On September 15th Debtor filed a voluntary petition under Chapter 7. May the trustee set any of these transfers aside and, if so, on what grounds?

7. Johnson owned the following property: a house and furnishings valued at $75,000; a new automobile valued at $5,000; a diamond ring worth $800; certain carpentry tools that he used in his trade; stock in AT&T valued at $6,000; a life-insurance policy with paid-up value of $3,000; clothing valued at $200; and $400 in cash. Johnson files a voluntary petition under Chapter 7. What property may he claim under the federal exemption?
8. Barney Bankrupt purchased a car on credit and financed it through Suburban Bank. The bank had a security interest in the car for the debt in the amount of $4,000. The car was worth $3,000. Barney owed back taxes to the federal government in the amount of $2,000 and back alimony to his ex-wife in the amount of $800. He also had an open account with Earl Clothier, Ltd. in the amount of $700 and an unsecured promissory note with Citizens' Bank in the amount of $2,000. Barney also had two employees in his retail store, who were due wages of $600 each. In addition, Barney owed the following suppliers: (A) $200; (B) $800; (C) $400; and (D) $1,200. A, B, and C filed their claims on time but D was late because the post office lost the notice to creditors. The trustees's costs and fees amounted to $1,200. Barney does not have sufficient assets to pay all the claims. In what order should the trustee pay them?

CHAPTER 49

Reorganization and Adjustment of Debts Under the Bankruptcy Code

Our discussion in the previous chapter focused upon a Chapter 7 bankruptcy. As part of that discussion, a number of subjects have been covered that also apply to bankruptcy proceedings under other Chapters of the code. A debtor may have the option of choosing to proceed under more than one chapter of the code. For instance, depending upon his circumstances, an individual debtor might choose to proceed under either Chapter 7 or to take advantage of Chapter 13, which involves the readjustment of debt for an individual with regular income. It is even possible for an individual to proceed under Chapter 11 although this chapter is most frequently used by businesses that want to reorganize their financial structure without terminating the business. Other debtors are limited with regard to the chapters of the code that they may select. Thus a corporation could not proceed under Chapter 13.

49:1. Chapter 13—Adjustment of Debts for an Individual with Regular Income. Chapter 13 in particular has relevance for individual debtors. Sometimes an individual who has a regular income may find that he has overextended his credit. The debtor may not be able to pay off the debts in an orderly fashion because of harassment by certain creditors, as well as lawsuits and judgments that may be filed against him. Yet, given time, the debtor may be able to eventually pay all the creditors at least a portion of the debt. Chapter 13 is designed to allow the debtor to do this without going the full route of a Chapter 7 liquidation bankruptcy with the attendant need to give up much of his property. This chapter of the code also gives the debtor the protection of the code in formulating and carrying out a repayment plan. Plans outside the code may be upset by the noncooperation of a single creditor.

This chapter of the code replaces old Chapter 13 under the Bankruptcy Act that provided for what were termed *wage-earner's plans.* Only an *individual,* who is not a stockbroker or commodity broker, with regular income may file a petition under Chapter 13. A further limitation is that the individual's liquidated, noncontingent, *unsecured* debts must amount to less than $100,000, and his liquidated, noncontingent, *secured* debts must amount to less than $350,000.

The debtor who wishes to use Chapter 13 begins by filing a petition with the bankruptcy court. If the plan does not work out, the debtor may later convert to Chapter 7 or 11 but may not be compelled to do so by the creditors. As in a Chapter 7 case, the filing acts as an order for relief and operates as an automatic stay of most proceedings filed against the debtor. Usually a plan for repayment is submitted with the petition. The court will also appoint a trustee who has duties that are similar to those in a Chapter 7 case. If the debtor has a business, he may continue to operate it without interference by the trustee.

(1) The Plan. There are certain prerequisites that must be contained in any plan if it is to be approved by the court. The plan must provide for a repayment period of not more than three years unless the court approves a longer period. The court may not approve a period longer than five years. The plan must also provide for the payment of a minimum amount to unsecured creditors. The unsecured creditors must receive at least what they would get under a Chapter 7 liquidation. For that reason, the unsecured creditors do not have to consent to the plan. Unless they agree otherwise, priority claimants, such as administrative claims, wages, or salaries, must be paid in full. The rights of secured and unsecured creditors may be modified, but if the plan classifies claims, it must treat the members of each class equally. The claim of a creditor that is secured only by a security interest in the debtor's principal residence may not be modified.

Secured creditors are treated somewhat differently from unsecured creditors by the code. Holders of an allowed secured claim must consent to the plan unless the plan provides that the holder may retain the lien securing the claim and will receive value in at least the amount of the claim. An alternative is for the debtor to surrender the property that is the subject of the lien to the holder. You will recall that the holder of a secured claim has priority only to the value of the collateral. Thus when the debtor turns over the collateral to the secured creditor, the secured creditor is paid to that extent and there is no reason that he should have the right to approve the plan. If there is any balance due on the secured creditor's claim, after the collateral is returned to him, he becomes an unsecured creditor to the extent of the remaining balance.

The plan must also provide for many other items permitted by the code and the debtor may modify the plan at any time before confirmation by the court. In addition, among other things, the plan may provide for the curing or waiving of any default by the debtor and provide for the assumption or rejection of any executory contract or unexpired lease of the debtor.

Once the creditor has proposed a plan, the court will hold a hearing on the plan. At this time creditors may file objections to the plan. However, if the plan meets the requirements stated in the code and has been proposed in good faith, the court must approve the plan. After the plan is confirmed, the court may order any entity from whom the debtor receives income to pay all or part of any such income to the trustee.

When the court confirms the plan, its provisions bind the debtor and each creditor whether the claim of such creditor is provided for by the plan, and whether such creditor has objected to, has accepted, or has rejected the plan. In addition, unless the plan provides otherwise, the confirmation of the plan vests all of the property of the estate in the debtor, free and clear of any claim or interest of any creditor provided for by the plan. The same priorities apply as in a Chapter 7 case and if there is a trustee, he shares in the first priority.

After the debtor has completed all payments under the plan, the court will grant the debtor discharge, with some exceptions. If the debtor has not completed payments under the plan, the court may grant discharge to the debtor only if the debtor's failure to complete payments is due to circumstances for which he is not accountable, if each unsecured creditor has received not less than the account that would have been paid under Chapter 7 and if modification of the plan is not practicable.

A debtor is generally not limited by the six-year rule as in Chapter 7 bankruptcies. He may file and be discharged under Chapter 13 as frequently as he likes although there are obvious practical limitations. One exception does apply here. The debtor may not receive a second discharge in a case commenced within six years of the date the petition in the first case was filed if the first case involved a discharge under Chapter 13 where the debtor paid less than 100 percent of the allowed unsecured claims or less than 70 percent of those claims and the plan was not the debtor's best effort or proposed in good faith. The purpose of this provision is to prevent the debtor from using the courts to, in effect, defraud creditors by continuously incurring debt and then escaping his obligations by use of the bankruptcy court.

Chapter 13 is extremely useful for the consumer debtor. It allows the debtor to keep much of his property if he can devise an acceptable plan. Many debts may be discharged with part payment and the creditors are no worse off than they would have been under a Chapter 7 liquidation. It does require a debtor with a sufficient income and self-discipline to make the plan work; otherwise it is a waste of time and the debtor might be better advised to go straight through a Chapter 7 bankruptcy.

49:2. Chapter 11—Reorganization. Chapter 11 is another important chapter in the Bankruptcy Code. Although it is concerned primarily with business concerns, an individual may file a petition under Chapter 11. The chapter is most applicable, however, to the business that has encountered financial difficulties, where there is a desire to preserve the business as a going concern and time is needed to get its financial affairs in order. After discharge, the business will remain as a going concern whereas, if the bankruptcy were filed under Chapter 7, the assets of the business would of course be liquidated in order to pay the creditors. Although common-law composition agreements have similar objectives, the disadvantage of that procedure is that it requires agreement of all creditors. Under Chapter 11, nonconsenting creditors may be bound to a confirmed plan and the debtor's financial situation may be substantially restructured.

One of the disadvantages of rehabilitation procedures under the old Bankruptcy Act was the fact that a great deal of delay was involved. Experience showed that the longer the delay, the less chance that the debtor would successfully carry out the rehabilitation plan. Accordingly, the 1978 Bankruptcy Code has combined a number of procedures of the old act into Chapter 11 and streamlined them. Many of the concepts and provisions of Chapter 7 apply to Chapter 11. Our discussion in this section will be limited to those concepts that apply to business reorganizations.

A petition for reorganization may be either a voluntary petition filed by the debtor or an involuntary petition filed under the same conditions as apply to Chapter 7 liquidation cases. Other rules such as those involving preferential and fraudulent transfers and discharge also apply to reorganization cases.

A petition may be filed by anyone who would qualify as a debtor under Chapter 7, as well as by stockbrokers and commodity brokers. A railroad may also be a debtor under Chapter 11 reorganizations.

(1) The Committees. One of the first steps that occurs after the petition is filed and an order for relief passed is the appointment of a committee of creditors by the court. In any case that does not concern a railroad, the court is required to appoint a committee of creditors holding unsecured claims. The purpose of this committee is to represent these creditors in the proceedings and negotiations that follow. Additional committees may also be appointed by the court to represent other creditor interests if it is deemed necessary. For instance, the court may even appoint a committee to represent equity security holders.

The statute provides guidelines for the composition of both the required and permissive committees. Ordinarily the committees will consist of the holders of the seven largest claims among the kinds of claimants represented on the particular committee. Of course, a particular creditor must be willing to serve. If a committee was organized by the creditors before the order for relief was passed, the court will appoint the members of that committee provided that it was fairly chosen and is representative of the different kinds of claims to be represented. The court has the power to change the size of the committees, as well as their membership, if another interested party requests it and establishes that the existing committee is not representative.

The purpose of the various committees is to participate in the negotiations that take place regarding the plan for reorganization of the debtor. The judge does not participate in the meetings, thus freeing him from any bias that might be acquired at the initial meeting as he may have to rule on disputes and legal issues later in the proceedings.

The committee has the power to investigate the acts, conduct, assets, liabilities, and financial condition of the debtor. The committee may also investigate the operation of the debtor's business and the desirability of continuing the business. To aid in these duties the committee may, with the court's approval, employ attorneys, accountants, or other agents to represent or perform services for it. The committee may also question the debtor at its initial meeting. The debtor must answer the questions under oath.

A party in interest may request the court to appoint a trustee who may take possession of the debtor's property and perform the same duties as in a Chapter 7 liquidation case. The trustee may have some additional duties, such as filing a list of creditors and schedule of the assets and liabilities of the debtor. The trustee may also file a plan of reorganization or recommend that the case be converted to a Chapter 7 case.

It is also possible that the court may not appoint a trustee. In that case the debtor will have the duties of a trustee in handling his affairs as a *debtor in possession*. The court may also, on the request of a party in interest, order the appointment of an examiner if it is in the interest of creditors, equity security holders, or if the debtor's fixed, liquidated, unsecured debts exceed $5,000,000. The examiner may investigate the affairs of the debtor, including any allegations of fraud or the incompetence of the debtor.

(2) The Plan. The statute provides that only the debtor may file a plan for the first 120 days after the court passes an order for relief. There are a few exceptions not relevant here. Any party in interest, including the debtor, trustee, creditor, holder, and others may file a plan if and only if: (a) a trustee hasn't been appointed; (b) the debtor has not filed a plan within the 120-day period mentioned earlier; and (c) the debtor has not filed a plan that has been accepted, before 180 days after the date of the order for relief under Chapter 11, by each class, the claims of which are impaired under the plan. The concept of impairment will be further discussed. The time for the debtor to gain acceptance of his plan may be reduced or extended by the court if good cause is shown.

(3) What a Plan Requires. The code states the requirements of any plan as follows:

1. Where the claims and interests are classified, the plan must designate the classes. The plan must specify those claims or interests that are and are not impaired by the plan. A claim is impaired unless the legal, equitable, and contractual rights of the holder of the claim are not altered or unless the holder will receive a cash settlement of the claim. Thus a claim would be impaired if the plan provides for less than full payment or if a debtor's default is not to be cured.
2. The plan must specify the treatment of any impaired class.
3. The plan must provide for the same treatment for each member of a particular class unless the holder of a claim agrees to less favorable treatment.
4. Adequate means to carry out the plan must be provided. For example,

the plan may provide for curing or waiving defaults by the debtor, extending maturity dates on obligations, and selling part of the property of the estate.

5. If the debtor is a corporation, the reorganization plan must provide for inclusion in the corporation charter a provision prohibiting the issuance of nonequity securities, and it must protect any voting rights of security holders.
6. In the selection of any officers, directors, or trustees under the plan, the plan must contain only provisions consistent with the interests of creditors, equity security holders, and public policy.

If the debtor is an individual, the plan may not provide for the sale, use, or lease of exempted property unless the debtor agrees.

(4) What a Plan May Provide. The code lists a number of items that may be but are not required to be included in a plan. The plan may:

1. Impair or leave unimpaired any class of claims, secured or unsecured.
2. Subject to some limitation, provide for the assumption or rejection of any executory contract or unexpired lease.
3. Provide for the settlement or adjustment of claims or interests belonging to the debtor or estate or provide for retention and enforcement of the claim.
4. Provide for the sale and distribution of substantially all of the property of the estate.
5. Include any other appropriate provision not inconsistent with the applicable provisions of the code.

(5) Confirmation of the Plan. We stated earlier that one of the characteristics of Chapter 11 reorganization is that under certain circumstances creditors may be compelled to accept the plan or reorganization. In order for the court to confirm the plan over the objections of certain creditors and classes of creditors, it must contain certain provisions. The court may only confirm the plan if it meets the requirements of the code. As in Chapter 13, confirmation makes the plan binding on the debtor, creditors, equity security holders, and others. The confirmation acts as a discharge of all debts, except for those referred to earlier that would be denied in a liquidation case, such as alimony and certain taxes.

Remember that we have discussed the necessity for the plan to group various individual creditors into classes. This is important for the confirmation step. In order for a plan to be confirmed it is usually necessary for the plan to be accepted by any *class of creditors* whose claims have been impaired. Acceptance by a class is not necessary if the claims of the class have not been impaired. In order for the class of creditors to accept the plan, a majority of the members who have claims totaling two thirds in dollar amount of allowed claims must accept. A class of *ownership* interest accepts the plan when the holders of allowed interests that amount to two thirds of the total ownership interest approve the plan.

Note that once the class accepts the plan it is binding against those individual members of a class that do not accept it. The code protects the interests of these creditors by providing that each holder of a claim or interest who has not accepted the plan must receive from the plan no less than the amount he would have received in a Chapter 7 liquidation case. Of course, if there are insufficient assets to pay all creditors, this sum could be zero.

There are provisions in the code for confirmation of the plan even when an impaired class has not accepted it. These provisions have been referred to as the so called *cram-down rules.* Under these rules, the court may confirm the plan, despite nonacceptance by an impaired class, if the plan treats the class in a "fair and equitable" manner. Among the requirements that a plan must meet in order for it to be considered "fair and equitable" are: (a) with respect to a nonaccepting class of secured claims, the holders of those claims must retain their liens on property of the debtor and if the property is sold their liens will be shifted to the proceeds and (b) with respect to a nonaccepting class of unsecured claims, the claimants must receive the full value of their claim or no claim junior to the claims of the class in question will receive anything.

Once the plan has been accepted, the debtor is discharged from any claims and interests not protected by the plan. If the debtor is unable to comply with the provisions of the plan he may later be subjected to a Chapter 7 liquidation.

49:3. Chapter 9. The remaining substantive section of the code is Chapter 9, which is quite limited in its application and hence only a few words concerning it are appropriate here. This chapter is entitled "Adjustments of Debts of a Municipality" and is limited to municipalities that are insolvent or that cannot pay their debts as they mature. If these bodies are not prohibited from doing so by state law, they may file voluntarily for reorganization under Chapter 9. Many of the provisions of the code that have been considered earlier are appropriate to Chapter 9 proceedings including the creation of a plan of reorganization. Because a governmental unit is involved, the nature of a proceeding involving a municipality is somewhat unique. For instance, a municipality cannot sell its assets in order to satisfy its creditors. Thus the purpose of Chapter 9 is to give the municipality the opportunity to adjust the debts owed creditors with as little loss to the creditors as possible under Federal Court protection and supervision.

SUMMARY

An individual debtor with regular income may enter into a plan to pay his debts under Chapter 13 without entering into a Chapter 7 liquidation of his assets. Chapter 13 is limited to individuals who are not stockbrokers or commodity brokers. Another limitation is that the individual's liquidated, noncontingent, unsecured debts may not exceed $100,000 and the liquidated, noncontingent, secured debts may not exceed $350,000.

The plan must provide for a repayment period not to exceed three years. The court may approve a longer period up to five years. Under the plan, unsecured creditors must receive at least what they would get in a Chapter 7 liquidation case. There are other requirements for the plan. Unsecured creditors need not consent to the plan. Secured creditors must consent, unless they retain their lien securing the claim, and receive value in at least the amount of the claim. Alternatively, the debtor may surrender the secured property to the secured creditor.

Confirmation of the plan by the court binds the debtor, and each creditor and vests all property of the estate in the debtor, free and clear of any claim or interest of any creditor provided for by the plan. After the debtor has completed all payments under the plan, the court will grant the debtor discharge, with some exceptions. The debtor generally is not limited to the Chapter 7 six-year rule. There are some limitations to this based upon the debtor's success in paying the claims stated in the plan.

Chapter 11 is designed to allow business concerns that have financial problems to reorganize under the protection of the bankruptcy court. After discharge, the business remains as a going concern. The petition for reorganization may be either voluntary or involuntary.

Generally, a committee of creditors is established to represent the type of creditors or interest holders from whose ranks the members of the committee are chosen. The code grants a number of powers to the creditor committees, such as investigating the acts, conduct, assets, and liabilities of the debtor. With the approval of the court, the committee may hire various professionals to aid it in its task.

The court may appoint a trustee or leave the debtor in possession of its property, in which case, an examiner may be appointed. The examiner would also have the power to investigate the debtor's affairs.

In Chapter 11 cases, only the debtor may file a plan for the first 120 days after the court passes an order for relief. Other parties may file plans at a later date. The code lists a number of items that *must* be contained in the plan as well as some items that *may* be included.

If the plan meets the requirements of the code, it may be confirmed by the court. Confirmation makes the plan binding on the debtor, creditors, equity security holders, and others. With some exceptions, confirmation acts as a discharge of all debts.

The plan may group creditors into classes. For the plan to be confirmed, it usually must be accepted by a class of creditors whose claims are impaired. If the claims of a class are not impaired, acceptance of the plan is not necessary. If a class member does not approve the plan, he must receive at least the amount he would get in a Chapter 7 liquidation case. The code also has so called *cram-down* rules, which allow the court, under certain circumstances, to confirm the plan without approval by an impaired class.

Chapter 9 is applicable to municipalities. It also involves the submission and confirmation of a plan for adjusting the debts of the municipality.

PROBLEMS

1. Delbert Debtor works for State Company at an annual salary of $40,000.00. He has the following unsecured debts: Friendly Bank, $75,000.00; his wife's mother, $20,000.00; and the credit union at his place of employment, $6,000.00. He also has a mortgage on his home in the amount of $60,000.00. He is contemplating bankruptcy and is considering filing a petition under Chapter 13, so he can keep most of his property. Is Delbert a candidate for Chapter 13?
2. Wally Wagearner has the following unsecured debts: Larry's Department store, $2,000.00; Easy-Credit Finance Co., $5,000.00; Plasticwonder Credit Card, $3,000.00; Sammy Supplier, Inc., $8,000.00; and Credit Union, Inc., $6,000.00. He also has a first mortgage on his house in favor of Citizen's Bank in the amount of $40,000.00 and a second mortgage in favor of Harry's Home Improvement Company in the amount of $10,000.00. Wally files a petition under Chapter 13. He also files a plan under which he agrees to pay the unsecured creditors 50 percent of their claims within 36 months. Wally's only asset is his home, which is valued at $60,000.00. The unsecured creditors object to the plan since it does not provide for the payment of their claims in full. Will the court approve the plan?
3. Wally Wagearner has a large number of unsecured claims. He also has a mortgage on his home in favor of Citizen's Bank in the amount of $50,000.00 and a security interest on his new luxury car, in favor of People's Finance, in the amount of $5,000.00. Wally's only assets are his home and car although his job provides him with a good salary. He files a plan under Chapter 13 under which the unsecured creditors are to be paid 20 percent of their claims and the secured creditors 50 percent. People's Finance objects to the plan. Will the court confirm the plan over People's Finance objections?
4. Disaster Corporation files a plan of reorganization under Chapter 11. Finance Bank, contending that Disaster Corporation is hopelessly in debt, asks the court to convert the reorganization case into a Chapter 7 liquidation case. Will the court grant Finance Bank's request?
5. Danny Debtor filed a plan in 1978 under Chapter 13 of the Bankruptcy Code. The plan provided for payment of 100 percent of the claims of creditors over a three-year period. Although he was able to comply with the plan, severe financial reverses caused him to file another plan one year after he completed performance of the first plan. Under the new plan, Debtor is to pay 70 percent of the creditors' claims over a three-year period. Connie Creditor objects to the plan and opposes any discharge of Debtor

because the new plan has been filed within six years of filing the petition for the first plan. Is Creditor correct?

6. Describe what are referred to as the *cram-down rules* under Chapter 11 of the code. What is the purpose of these rules?
7. (a) What are the advantages to a business of proceeding under Chapter 13 rather than Chapter 7? (b) What is the advantage to a consumer of proceeding under Chapter 13 rather than Chapter 7?
8. Explain the role of committees in a bankruptcy proceeding under Chapter 11.
9. Under what circumstances may the refusal of a class of creditors to accept a plan submitted under Chapter 11 prevent the plan's confirmation?

Appendices

A. Uniform Commercial Code

B. Uniform Partnership Act

C. Uniform Limited Partnership Act

D. Model Business Corporation Act

APPENDIX A

The Uniform Commercial Code*

Author's Note: *The 1972 version of the Uniform Commercial Code is reproduced here. That version substantially amended Article 9 and has been adopted by a majority of the states. Article 8 was substantially amended in 1977 but to date only two states, Minnesota and West Virginia, have adopted it. All fifty states, the District of Columbia and the Virgin Islands have adopted the Uniform Commercial Code, although Louisiana has adopted only Articles 1, 3, 4, 5, 7 and 8.*

1972 OFFICIAL TEXT

ARTICLE 1: GENERAL PROVISIONS

Part 1

Short Title, Construction, Application and Subject Matter of the Act

§ 1–101. Short Title. This Act shall be known and may be cited as Uniform Commercial Code.

§ 1–102. Purposes; Rules of Construction; Variation by Agreement. (1) This Act shall be liberally construed and applied to promote its underlying purposes and policies.

(2) Underlying purposes and policies of this Act are

(a) to simplify, clarify and modernize the law governing commercial transactions;

(b) to permit the continued expansion of commercial practices through custom, usage and agreement of the parties;

(c) to make uniform the law among the various jurisdictions.

(3) The effect of provisions of this Act may be varied by agreement, except as otherwise provided in this Act and except that the obligations of good faith, diligence, reasonableness and care prescribed by this Act may not be disclaimed by agreement but the parties may by agreement determine

the standards by which the performance of such obligations is to be measured if such standards are not manifestly unreasonable.

(4) The presence in certain provisions of this Act of the words "unless otherwise agreed" or words of similar import does not imply that the effect of other provisions may not be varied by agreement under subsection (3).

(5) In this Act unless the context otherwise requires

(a) words in the singular number include the plural, and in the plural include the singular;

(b) words of the masculine gender include the feminine and the neuter, and when the sense so indicates words of the neuter gender may refer to any gender.

§ 1–103. Supplementary General Principles of Law Applicable. Unless displaced by the particular provisions of this Act, the principles of law and equity, including the law merchant and the law relative to capacity to contract, principal and agent, estoppel, fraud, misrepresentation, duress, coercion, mistake, bankruptcy, or other validating or invalidating cause shall supplement its provisions.

§ 1–104. Construction Against Implicit Repeal. This Act being a general act intended as a unified coverage of its subject matter, no part of it shall be deemed to be impliedly repealed by subsequent legislation if such construction can reasonably be avoided.

§ 1–105. Territorial Application of the Act; Parties' Power to Choose Applicable Law. (1) Except as provided hereafter in this section, when a transaction bears a reasonable relation to this state and also to another state or nation the parties may agree that the law either of this state or of such other state or nation shall govern their rights and duties. Failing such agreement this Act applies to transactions bearing an appropriate relation to this state.

(2) Where one of the following provisions of this Act specifies the applicable law, that provision governs and a contrary agreement is effective only to the extent permitted by the law (including the conflict of laws rules) so specified:

Rights of creditors against sold goods. Section 2–402.

Applicability of the Article on Bank Deposits and Collections. Section 4–102.

Bulk transfers subject to the Article on Bulk Transfers. Section 6–102.

Applicability of the Article on Investment Securities. Section 8–106.

Perfection provisions of the Article on Secured Transactions. Section 9–103.

§ 1–106. Remedies to Be Liberally Administered. (1) The remedies provided by this Act shall be liberally administered to the end that the aggrieved party may be put in as good a position as if the other party had fully performed but neither consequential or special nor penal damages may be had except as specifically provided in this Act or by other rule of law.

(2) Any right or obligation declared by this Act is enforceable by action unless the provision declaring it specifies a different and limited effect.

§ 1–107. Waiver or Renunciation of Claim or Right After Breach. Any claim or right arising out of an alleged breach can be discharged in whole or in part without consideration by a written waiver or renunciation signed and delivered by the aggrieved party.

§ 1–108. Severability. If any provision or clause of this Act or application thereof to any person or circumstances is held invalid, such invalidity shall not affect other provisions or applications of the Act which can be given effect without the invalid provision or application, and to this end the provisions of this Act are declared to be severable.

§ 1–109. Section Captions. Section captions are parts of this Act.

Part 2

General Definitions and Principles of Interpretation

§ 1–201. General Definitions. Subject to additional definitions contained in the subsequent Articles of this Act which are applicable to specific Articles or Parts thereof, and unless the context otherwise requires, in this Act:

(1) "Action" in the sense of a judicial proceeding includes recoupment, counterclaim, set-off, suit in equity and any other proceedings in which rights are determined.

(2) "Aggrieved party" means a party entitled to resort to a remedy.

(3) "Agreement" means the bargain of the parties in fact as found in their language or by implication from other circumstances including course of dealing or usage of trade or course of performance as provided in this Act (Sections 1–205 and 2–208). Whether an agreement has legal consequences is determined by the provisions of this Act, if applicable; otherwise by the law of contracts (Section 1–103). (Compare "Contract.")

(4) "Bank" means any person engaged in the business of banking.

(5) "Bearer" means the person in possession of an instrument, document of title, or security payable to bearer or indorsed in blank.

(6) "Bill of lading" means a document evidencing the receipt of goods for shipment issued by a person engaged in the business of transporting or forwarding goods, and includes an airbill. "Airbill" means a document serving for air transportation as a bill of lading does for marine or rail transportation, and includes an air consignment note or air waybill.

(7) "Branch" includes a separately incorporated foreign branch of a bank.

(8) "Burden of establishing" a fact means the burden of persuading the triers of fact that the existence of the fact is more probable than its non-existence.

(9) "Buyer in ordinary course of business" means a person who in good faith and without knowledge that the sale to him is in violation of the ownership rights or security interest of a third party in the goods buys in ordinary course from a person in the business of selling goods of that kind but does not include

a pawnbroker. All persons who sell minerals or the like (including oil and gas) at wellhead or minehead shall be deemed to be persons in the business of selling goods of that kind. "Buying" may be for cash or by exchange of other property or on secured or unsecured credit and includes receiving goods or documents of titled under a pre-existing contract for sale but does not include a transfer in bulk or as security for or in total or partial satisfaction of a money debt.

(10) "Conspicuous": A term or clause is conspicuous when it is so written that a reasonable person against whom it is to operate ought to have noticed it. A printed heading in capitals (as: NON-NEGOTIABLE BILL OF LADING) is conspicuous. Language in the body of a form is "conspicuous" if it is in larger or other contrasting type or color. But in a telegram any stated term is "conspicuous." Whether a term or clause is "conspicuous" or not is for decision by the court.

(11) "Contract" means the total legal obligation which results from the parties' agreement as affected by this Act and any other applicable rules of law. (Compare "Agreement.")

(12) "Creditor" includes a general creditor, a secured creditor, a lien creditor and any representative of creditors, including an assignee for the benefit of creditors, a trustee in bankruptcy, a receiver in equity and an executor or administrator of an insolvent debtor's or assignor's estate.

(13) "Defendant" includes a person in the position of defendant in a cross-action or counterclaim.

(14) "Delivery" with respect to instruments, documents of title, chattel paper or securities means voluntary transfer of possession.

(15) "Document of title" includes bill of lading, dock warrant, dock receipt, warehouse receipt or order for the delivery of goods, and also any other document which in the regular course of business or financing is treated as adequately evidencing that the person in possession of it is entitled to receive, hold and dispose of the document and the goods it covers. To be a document of title a document must purport to be issued by or addressed to a bailee and purport to cover goods in the bailee's possession which are either identified or are fungible portions of an identified mass.

(16) "Fault" means wrongful act, omission or breach.

(17) "Fungible" with respect to goods or securities means goods or securities of which any unit is, by nature or usage of trade, the equivalent of any other like unit. Goods which are not fungible shall be deemed fungible for the purposes of this Act to the extent that under a particular agreement or document unlike units are treated as equivalents.

(18) "Genuine" means free of forgery or counterfeiting.

(19) "Good faith" means honesty in fact in the conduct or transaction concerned.

(20) "Holder" means a person who is in possession of a document of title or an instrument or an investment security drawn, issued or indorsed to him or to his order or to bearer or in blank.

(21) To "honor" is to pay or to accept and pay, or where a credit so engages to purchase or discount a draft complying with the terms of the credit.

(22) "Insolvency proceedings" includes any assignment for the benefit of creditors or other proceedings intended to liquidate or rehabilitate the estate of the person involved.

(23) A person is "insolvent" who either has ceased to pay his debts in the ordinary course of business or cannot pay his debts as they become due or is insolvent within the meaning of the federal bankruptcy law.

(24) "Money" means a medium of exchange authorized or adopted by a domestic or foreign government as a part of its currency.

(25) A person has "notice" of a fact when

(a) he has actual knowledge of it; or

(b) he has received a notice or notification of it; or

(c) from all the facts and circumstances known to him at the time in question he has reason to know that it exists.

A person "knows" or has "knowledge" of a fact when he has actual knowledge of it. "Discover" or "learn" or a word or phrase of similar import refers to knowledge rather than to reason to know. The time and circumstances under which a notice or notification may cease to be effective are not determined by this Act.

(26) A person "notifies" or "gives" a notice or notification to another by taking such steps as may be reasonably required to inform the other in ordinary course whether or not such other actually comes to know of it. A person "receives" a notice or notification when

(a) it comes to his attention; or

(b) it is duly delivered at the place of business through which the contract was made or at any other place held out by him as the place for receipt of such communications.

(27) Notice, knowledge or a notice or notification received by an organization is effective for a particular transaction from the time when it is brought to the attention of the individual conducting that transaction, and in any event from the time when it would have been brought to his attention if the organization had exercised due diligence. An organization exercises due diligence if it maintains reasonable routines for communicating significant information to the person conducting the transaction and there is reasonable compliance with the routines. Due diligence does not require an individual acting for the organization to communicate information unless such communication is part of his regular duties or unless he has reason to know of the transaction and that the transaction would be materially affected by the information.

(28) "Organization" includes a corporation, government or governmental subdivision or agency, business trust, estate, trust, partnership or association, two or more persons having a joint or common interest, or any other legal or commercial entity.

(29) "Party," as distinct from "third party," means a person who has engaged in a transaction or made an agreement within this Act.

(30) "Person" includes an individual or an organization (See Section 1–102).

(31) "Presumption" or "presumed" means that the trier of fact must find the existence of the fact presumed unless and until evidence is introduced which would support a finding of its nonexistence.

(32) "Purchase" includes taking by sale, discount, negotiation, mortgage, pledge, lien, issue or re-issue, gift or any other voluntary transaction creating an interest in property.

(33) "Purchaser" means a person who takes by purchase.

(34) "Remedy" means any remedial right to which an aggrieved party is entitled with or without resort to a tribunal.

(35) "Representative" includes an agent, an officer of a corporation or association, and a trustee, executor or administrator of an estate, or any other person empowered to act for another.

(36) "Rights" includes remedies.

(37) "Security interest" means an interest in personal property or fixtures which secures payment or performance of an obligation. The retention or reservation of title by a seller of goods notwithstanding shipment or delivery to the buyer (Section 2–401) is limited in effect to a reservation of a "security interest." The term also includes any interest of a buyer of accounts or chattel paper which is subject to Article 9. The special property interest of a buyer of goods on identification of such goods to a contract for sale under Section 2–401 is not a "security interest," but a buyer may also acquire a "security interest" by complying with Article 9. Unless a lease or consignment is intended as security, reservation of title thereunder is not a "security interest" but a consignment is in any event subject to the provisions on consignment sales (Section 2–326). Whether a lease is intended as security is to be determined by the facts of each case; however, (a) the inclusion of an option to purchase does not of itself make the lease one intended for security, and (b) an agreement that upon compliance with the terms of the lease the lessee shall become or has the option to become the owner of the property for no additional consideration or for a nominal consideration does make the lease one intended for security.

(38) "Send" in connection with any writing or notice means to deposit in the mail or deliver for transmission by any other usual means of communication with postage or cost of transmission provided for and properly addressed and in the case of an instrument to an address specified thereon or otherwise agreed, or if there be none to any address reasonable under the circumstances. The receipt of any writing or notice within the time at which it would have arrived if properly sent has the effect of a proper sending.

(39) "Signed" includes any symbol executed or adopted by a party with present intention to authenticate a writing.

(40) "Surety" includes guarantor.

(41) "Telegram" includes a message transmitted by radio, teletype, cable, any mechanical method of transmission, or the like.

(42) "Term" means that portion of an agreement which relates to a particular matter.

(43) "Unauthorized" signature or indorsement means one made without actual, implied or apparent authority and includes a forgery.

(44) "Value." Except as otherwise provided with respect to negotiable instruments and bank collections (Sections 3–303, 4–208 and 4–209) a person gives "value" for rights if he acquires them

(a) in return for a binding commitment to extend credit or for the extension of immediately available credit whether or not drawn upon and whether or not a charge-back is provided for in the event of difficulties in collection; or

(b) as security for or in total or partial satisfaction of a pre-existing claim; or

(c) by accepting delivery pursuant to a pre-existing contract for purchase; or

(d) generally, in return for any consideration sufficient to support a simple contract.

(45) "Warehouse receipt" means a receipt issued by a person engaged in the business of storing goods for hire.

(46) "Written" or "writing" includes printing, typewriting or any other intentional reduction to tangible form.

§ 1–202. Prima Facie Evidence by Third Party Documents. A document in due form purporting to be a bill of lading, policy or certificate of insurance, official weigher's or inspector's certificate, consular invoice, or any other document authorized or required by the contract to be issued by a third party shall be prima facie evidence of its own authenticity and genuineness and of the facts stated in the document by the third party.

§ 1–203. Obligation of Good Faith. Every contract or duty within this Act imposes an obligation of good faith in its performance or enforcement.

§ 1–204. Time; Reasonable Time; "Seasonably."

(1) Whenever this Act requires any action to be taken within a reasonable time, any time which is not manifestly unreasonable may be fixed by agreement.

(2) What is a reasonable time for taking any action depends on the nature, purpose and circumstances of such action.

(3) An action is taken "seasonably" when it is taken at or within the time agreed or if no time is agreed at or within a reasonable time.

§ 1–205. Course of Dealing and Usage of Trade. (1) A course of dealing is a sequence of previous conduct between the parties to a particular transaction which is fairly to be regarded as establishing a common basis of understanding for interpreting their expressions and other conduct.

(2) A usage of trade is any practice or method of dealing having such regularity of observance in a place, vocation or trade as to justify an expectation that it will be observed with respect to the transaction in question. The existence and scope of such a usage are to be proved as facts. If it is established

that such a usage is embodied in a written trade code or similar writing the interpretation of the writing is for the court.

(3) A course of dealing between parties and any usage of trade in the vocation or trade in which they are engaged or of which they are or should be aware give particular meaning to and supplement or qualify terms of an agreement.

(4) The express terms of an agreement and an applicable course of dealing or usage of trade shall be construed wherever reasonable as consistent with each other; but when such construction is unreasonable express terms control both course of dealing and usage of trade and course of dealing controls usage of trade.

(5) An applicable usage of trade in the place where any part of performance is to occur shall be used in interpreting the agreement as to that part of the performance.

(6) Evidence of a relevant usage of trade offered by one party is not admissible unless and until he has given the other party such notice as the court finds sufficient to prevent unfair surprise to the latter.

§ 1–206. Statute of Frauds for Kinds of Personal Property Not Otherwise Covered. (1) Except in the cases described in subsection (2) of this section a contract for the sale of personal property is not enforceable by way of action or defense beyond five thousand dollars in amount or value of remedy unless there is some writing which indicates that a contract for sale has been made between the parties at a defined or stated price, reasonably identifies the subject matter, and is signed by the party against whom enforcement is sought or by his authorized agent.

(2) Subsection (1) of this section does not apply to contracts for the sale of goods (Section 2–201) nor of securities (Section 8–319) nor to security agreements (Section 9–203).

§ 1–207. Performance or Acceptance Under Reservation of Rights. A party who will explicit reservation of rights performs or promises performance or assents to performance in a manner demanded or offered by the other party does not thereby prejudice the rights reserved. Such words as "without prejudice," "under protest" or the like are sufficient.

§ 1–208. Option to Accelerate at Will. A term providing that one party or his successor in interest may accelerate payment or performance or require collateral or additional collateral "at will" or "when he deems himself insecure" or in words of similar import shall be construed to mean that he shall have power to do so only if he in good faith believes that the prospect of payment or performance is impaired. The burden of establishing lack of good faith is on the party against whom the power has been exercised.

§ 1–209. Subordinated Obligations. An obligation may be issued as subordinated to payment of another obligation of the person obligated, or a creditor may subordinate his right to payment of an obligation by agreement with either the person obligated or another creditor of the person obligated. Such a subordination does not create a security interest as against either the common debtor or a subordinated creditor. This section shall be construed as declaring the law as it existed prior to the enactment of this section and not as modifying it. Added 1966.

Note: *This new section is proposed as an optional provision to make it clear that a subordination agreement does not create a security interest unless so intended.*

ARTICLE 2: SALES

Part 1

Short Title, Construction and Subject Matter

§ 2–101. Short Title. This Article shall be known and may be cited as Uniform Commercial Code—Sales.

§ 2–102. Scope; Certain Security and Other Transactions Excluded From This Article. Unless the context otherwise requires, this Article applies to transactions in goods; it does not apply to any transaction which although in the form of an unconditional contract to sell or present sale is intended to operate only as a security transaction nor does this Article impair or repeal any statute regulating sales to consumers, farmers or other specified classes of buyers.

§ 2–103. Definitions and Index of Definitions. (1) In this Article unless the context otherwise requires

(a) "Buyer" means a person who buys or contracts to buy goods.

(b) "Good faith" in the case of a merchant means honesty in fact and the observance of reasonable commercial standards of fair dealing in the trade.

(c) "Receipt" of goods means taking physical possession of them.

(d) "Seller" means a person who sells or contracts to sell goods.

(2) Other definitions applying to this Article or to specified Parts thereof, and the sections in which they appear are:

"Acceptance." Section 2–606.
"Banker's credit." Section 2–325.
"Between merchants." Section 2–104.
"Cancellation." Section 2–106(4).
"Commercial unit." Section 2–105.
"Confirmed credit." Section 2–325.
"Conforming to contract." Section 2–106.
"Contract for sale." Section 2–106.
"Cover." Section 2–712.
"Entrusting." Section 2–403.
"Financing agency." Section 2–104.
"Future goods." Section 2–105.
"Goods." Section 2–105.
"Identification." Section 2–501.
"Installment contract." Section 2–612.
"Letters of Credit." Section 2–325.

"Lot." Section 2–105.
"Merchant." Section 2–104.
"Overseas." Section 2–323.
"Person in position of seller." Section 2–707.
"Present sale." Section 2–106.
"Sale." Section 2–106.
"Sale on approval." Section 2–326.
"Sale or return." Section 2–326.
"Termination." Section 2–106.

(3) The following definitions in other Articles apply to this Article:
"Check." Section 3–104.
"Consignee." Section 7–102.
"Consignor." Section 7–102.
"Consumer goods." Section 9–109.
"Dishonor." Section 3–507.
"Draft." Section 3–104.

(4) In addition Article 1 contains general definitions and principles of construction and interpretation applicable throughout this Article.

§ 2–104. Definitions: "Merchant"; "Between Merchants"; "Financing Agency." (1) "Merchant" means a person who deals in goods of the kind or otherwise by his occupation holds himself out as having knowledge or skill peculiar to the practices or goods involved in the transaction or to whom such knowledge or skill may be attributed by his employment of an agent or broker or other intermediary who by his occupation holds himself out as having such knowledge or skill.

(2) "Financing agency" means a bank, finance company or other person who in the ordinary course of business makes advances against goods or documents of title or who by arrangement with either the seller or the buyer intervenes in ordinary course to make or collect payment due or claimed under the contract for sale, as by purchasing or paying the seller's draft or making advances against it or by merely taking it for collection whether or not documents of title accompany the draft. "Financing agency" includes also a bank or other person who similarly intervenes between persons who are in the position of seller and buyer in respect to the goods (Section 2–707).

(3) "Between merchants" means in any transaction with respect to which both parties are chargeable with the knowledge or skill of merchants.

§ 2–105. Definitions: Transferability; "Goods"; "Future" Goods; "Lot"; "Commercial Unit." (1) "Goods" means all things (including specially manufactured goods) which are movable at the time of identification to the contract for sale other than the money in which the price is to be paid, investment securities (Article 8) and things in action. "Goods" also includes the unborn young of animals and growing crops and other identified things attached to realty as described in the section on goods to be severed from realty (Section 2–107).

(2) Goods must be both existing and identified before any interest in them can pass. Goods which are not both existing and identified are "future"

goods. A purported present sale of future goods or of any interest therein operates as a contract to sell.

(3) There may be a sale of a part interest in existing identified goods.

(4) An undivided share in an identified bulk of fungible goods is sufficiently identified to be sold although the quantity of the bulk is not determined. Any agreed proportion of such a bulk or any quantity thereof agreed upon by number, weight or other measure may to the extent of the seller's interest in the bulk be sold to the buyer who then becomes an owner in common.

(5) "Lot" means a parcel or a single article which is the subject matter of a separate sale or delivery, whether or not it is sufficient to perform the contract.

(6) "Commercial unit" means such a unit of goods as by commercial usage is a single whole for purposes of sale and division of which materially impairs its character or value on the market or in use. A commercial unit may be a single article (as a machine) or a set of articles (as a suite of furniture or an assortment of sizes) or a quantity (as a bale, gross, or carload) or any other unit treated in use or in the relevant market as a single whole.

§ 2–106. Definitions: "Contract"; "Agreement"; "Contract for Sale"; "Sale"; "Present Sale"; "Conforming" to Contract; "Termination"; "Cancellation." (1) In this Article unless the context otherwise requires "contract" and "agreement" are limited to those relating to the present or future sale of goods. "Contract for sale" includes both a present sale of goods and a contract to sell goods at a future time. A "sale" consists in the passing of title from the seller to the buyer for a price (Section 2–401). A "present sale" means a sale which is accomplished by the making of the contract.

(2) Goods or conduct including any part of a performance are "conforming" or conform to the contract when they are in accordance with the obligations under the contract.

(3) "Termination" occurs when either party pursuant to a power created by agreement or law puts an end to the contract otherwise than for its breach. On "termination" all obligations which are still executory on both sides are discharged but any right based on prior breach or performance survives.

(4) "Cancellation" occurs when either party puts an end to the contract for breach by the other and its effect is the same as that of "termination" except that the cancelling party also retains any remedy for breach of the whole contract or any unperformed balance.

§ 2–107. Goods to Be Severed From Realty: Recording. (1) A contract for the sale of minerals or the like (including oil and gas) or a structure or its materials to be removed from realty is a contract for the sale of goods within this Article if they are to be severed by the seller but until severance a purported present sale thereof which is not effective as a transfer of an interest in land is effective only as a contract to sell.

(2) A contract for the sale apart from the land of growing crops or other things attached to realty and capable of severance without material

harm thereto but not described in subsection (1) or of timber to be cut is a contract for the sale of goods within this Article whether the subject matter is to be severed by the buyer or by the seller even though it forms part of the realty at the time of contracting, and the parties can by identification effect a present sale before severance.

(3) The provisions of this section are subject to any third party rights provided by the law relating to realty records, and the contract for sale may be executed and recorded as a document transferring an interest in land and shall then constitute notice to third parties of the buyer's rights under the contract for sale.

Part 2

Form, Formation and Readjustment of Contract

§ 2–201. Formal Requirements; Statute of Frauds. (1) Except as otherwise provided in this section a contract for the sale of goods for the price of $500 or more is not enforceable by way of action or defense unless there is some writing sufficient to indicate that a contract for sale has been made between the parties and signed by the party against whom enforcement is sought or by his authorized agent or broker. A writing is not insufficient because it omits or incorrectly states a term agreed upon but the contract is not enforceable under this paragraph beyond the quantity of goods shown in such writing.

(2) Between merchants if within a reasonable time a writing in confirmation of the contract and sufficient against the sender is received and the party receiving it has reason to know its contents, it satisfies the requirements of subsection (1) against such party unless written notice of objection to its contents is given within ten days after it is received.

(3) A contract which does not satisfy the requirements of subsection (1) but which is valid in other respects is enforceable

(a) if the goods are to be specially manufactured for the buyer and are not suitable for sale to others in the ordinary course of the seller's business and the seller, before notice of repudiation is received and under circumstances which reasonably indicate that the goods are for the buyer, has made either a substantial beginning of their manufacture or commitments for their procurement; or

(b) if the party against whom enforcement is sought admits in his pleading, testimony or otherwise in court that a contract for sale was made, but the contract is not enforceable under this provision beyond the quantity of goods admitted; or

(c) with respect to goods for which payment has been made and accepted or which have been received and accepted (Sec. 2–606).

§ 2–202. Final Written Expression; Parol or Extrinsic Evidence. Terms with respect to which the confirmatory memoranda of the parties agree or which are otherwise set forth in a writing intended by the parties as a final expression of their agreement with respect to such terms as are included therein may not be contradicted by evidence of any prior agreement or of a contemporaneous oral agreement but may be explained or supplemented

(a) by course of dealing or usage of trade (Section 1–205) or by course of performance (Section 2–208); and

(b) by evidence of consistent additional terms unless the court finds the writing to have been intended also as a complete and exclusive statement of the terms of the agreement.

§ 2–203. Seals Inoperative. The affixing of a seal to a writing evidencing a contract for sale or an offer to buy or sell goods does not constitute the writing a sealed instrument and the law with respect to sealed instruments does not apply to such a contract or offer.

§ 2–204. Formation in General. (1) A contract for sale of goods may be made in any manner sufficient to show agreement, including conduct by both parties which recognizes the existence of such a contract.

(2) An agreement sufficient to constitute a contract for sale may be found even though the moment of its making is undetermined.

(3) Even though one or more terms are left open a contract for sale does not fail for indefiniteness if the parties have intended to make a contract and there is a reasonably certain basis for giving an appropriate remedy.

§ 2–205. Firm Offers. An offer by a merchant to buy or sell goods in a signed writing which by its terms gives assurance that it will be held open is not revocable, for lack of consideration, during the time stated or if no time is stated for a reasonable time, but in no event may such period of irrevocability exceed three months; but any such term of assurance on a form supplied by the offeree must be separately signed by the offeror.

§ 2–206. Offer and Acceptance in Formation of Contract. (1) Unless otherwise unambiguously indicated by the language or circumstances

(a) an offer to make a contract shall be construed as inviting acceptance in any manner and by any medium reasonable in the circumstances;

(b) an order or other offer to buy goods for prompt or current shipment shall be construed as inviting acceptance either by a prompt promise to ship or by the prompt or current shipment of conforming or nonconforming goods, but such a shipment of non-conforming goods does not constitute an acceptance if the seller seasonably notifies the buyer that the shipment is offered only as an accommodation to the buyer.

(2) Where the beginning of a requested performance is a reasonable mode of acceptance an offeror who is not notified of acceptance within a reasonable time may treat the offer as having lapsed before acceptance.

§ 2–207. Additional Terms in Acceptance or Confirmation. (1) A definite and seasonable expression of acceptance or a written confirmation which is sent within a reasonable time operates as an acceptance even though it states terms additional to or different from those offered or agreed upon, unless acceptance is expressly made conditional on assent to the additional or different terms.

(2) The additional terms are to be construed as proposals for addition to the contract. Between merchants such terms become part of the contract unless:

(a) the offer expressly limits acceptance to the terms of the offer;

(b) they materially alter it; or

(c) notification of objection to them has already been given or is given within a reasonable time after notice of them is received.

(3) Conduct by both parties which recognizes the existence of a contract is sufficient to establish a contract for sale although the writings of the parties do not otherwise establish a contract. In such case the terms of the particular contract consist of those terms on which the writings of the parties agree, together with any supplementary terms incorporated under any other provisions of this Act.

§ 2–208. Course of Performance or Practical Construction. (1) Where the contract for sale involves repeated occasions for performance by either party with knowledge of the nature of the performance and opportunity for objection to it by the other, any course of performance accepted or acquiesced in without objection shall be relevant to determine the meaning of the agreement.

(2) The express terms of the agreement and any such course of performance, as well as any course of dealing and usage of trade, shall be construed whenever reasonable as consistent with each other; but when such construction is unreasonable, express terms shall control course of performance and course of performance shall control both course of dealing and usage of trade (Section 1–205).

(3) Subject to the provisions of the next section on modification and waiver, such course of performance shall be relevant to show a waiver or modification of any term inconsistent with such course of performance.

§ 2–209. Modification, Rescission and Waiver. (1) An agreement modifying a contract within this Article needs no consideration to be binding.

(2) A signed agreement which excludes modification or rescission except by a signed writing cannot be otherwise modified or rescinded, but except as between merchants such a requirement on a form supplied by the merchant must be separately signed by the other party.

(3) The requirements of the statute of frauds section of this Article (Section 2–201) must be satisfied if the contract as modified is within its provisions.

(4) Although an attempt at modification or rescission does not satisfy the requirements of subsection (2) or (3) it can operate as a waiver.

(5) A party who has made a waiver affecting an executory portion of the contract may retract the waiver by reasonable notification received by the other party that strict performance will be required of any term waived, unless the retraction would be unjust in view of a material change of position in reliance on the waiver.

§ 2–210. Delegation of Performance; Assignment of Rights. (1) A party may perform his duty through a delegate unless otherwise agreed or unless the other party has a substantial interest in having his original promisor perform or control the acts required by the contract. No delegation of performance relieves the party delegating of any duty to perform or any liability for breach.

(2) Unless otherwise agreed all rights of either seller or buyer can be assigned except where the assignment would materially change the duty of the other party, or increase materially the burden or risk imposed on him by his contract, or impair materially his chance of obtaining return performance. A right to damages for breach of the whole contract or a right arising out of the assignor's due performance of his entire obligation can be assigned despite agreement otherwise.

(3) Unless the circumstances indicate the contrary a prohibition of assignment of "the contract" is to be construed as barring only the delegation to the assignee of the assignor's performance.

(4) An assignment of "the contract" or of "all my rights under the contract" or an assignment in similar general terms is an assignment of rights and unless the language or the circumstances (as in an assignment for security) indicate the contrary, it is a delegation of performance of the duties of the assignor and its acceptance by the assignee constitutes a promise by him to perform those duties. This promise is enforceable by either the assignor or the other party to the original contract.

(5) The other party may treat any assignment which delegates performance as creating reasonable grounds for insecurity and may without prejudice to his rights against the assignor demand assurances from the assignee (Section 2–609).

Part 3

General Obligation and Construction of Contract

§ 2–301. General Obligations of Parties. The obligation of the seller is to transfer and deliver and that of the buyer is to accept and pay in accordance with the contract.

§ 2–302. Unconscionable Contract or Clause. (1) If the court as a matter of law finds the contract or any clause of the contract to have been unconscionable at the time it was made the court may refuse to enforce the contract, or it may enforce the remainder of the contract without the unconscionable clause, or it may so limit the application of any unconscionable clause as to avoid any unconscionable result.

(2) When it is claimed or appears to the court that the contract or any clause thereof may be unconscionable the parties shall be afforded a reasonable opportunity to present evidence as to its commercial setting, purpose and effect to aid the court in making the determination.

§ 2–303. Allocation or Division of Risks. Where this Article allocates a risk or a burden as between the parties "unless otherwise agreed," the agreement may not only shift the allocation but may also divide the risk or burden.

§ 2–304. Price Payable in Money, Goods, Realty, or Otherwise. (1) The price can be made payable in money or otherwise. If it is payable in whole or in part in goods each party is a seller of the goods which he is to transfer.

(2) Even though all or part of the price is payable in an interest in realty the transfer of the goods and the seller's obligations with reference to them are subject to this Article, but not the transfer of the interest in realty or the transferor's obligations in connection therewith.

§ 2–305. Open Price Term. (1) The parties if they so intend can conclude a contract for sale even though the price is not settled. In such a case the price is a reasonable price at the time for delivery if

(a) nothing is said as to price; or

(b) the price is left to be agreed by the parties and they fail to agree; or

(c) the price is to be fixed in terms of some agreed market or other standard as set or recorded by a third person or agency and it is not so set or recorded.

(2) A price to be fixed by the seller or by the buyer means a price for him to fix in good faith.

(3) When a price left to be fixed otherwise than by agreement of the parties fails to be fixed through fault of one party the other may at his option treat the contract as cancelled or himself fix a reasonable price.

(4) Where, however, the parties intend not to be bound unless the price be fixed or agreed and it is not fixed or agreed there is no contract. In such a case the buyer must return any goods already received or if unable so to do must pay their reasonable value at the time of delivery and the seller must return any portion of the price paid on account.

§ 2–306. Output, Requirements and Exclusive Dealings. (1) A term which measures the quantity by the output of the seller or the requirements of the buyer means such actual output or requirements as may occur in good faith, except that no quantity unreasonably disproportionate to any stated estimate or in the absence of a stated estimate to any normal or otherwise comparable prior output or requirements may be tendered or demanded.

(2) A lawful agreement by either the seller or the buyer for exclusive dealing in the kind of goods concerned imposes unless otherwise agreed an obligation by the seller to use best efforts to supply the goods and by the buyer to use best efforts to promote their sale.

§ 2–307. Delivery in Single Lot or Several Lots. Unless otherwise agreed all goods called for by a contract for sale must be tendered in a single delivery and payment is due only on such tender but where the circumstances give either party the right to make or demand delivery in lots the price if it can be apportioned may be demanded for each lot.

§ 2–308. Absence of Specified Place for Delivery. Unless otherwise agreed

(a) the place for delivery of goods is the seller's place of business or if he has none his residence; but

(b) in a contract for sale of identified goods which to the knowledge of the parties at the time of contracting are in some other place, that place is the place for their delivery; and

(c) documents of title may be delivered through customary banking channels.

§ 2–309. Absence of Specific Time Provisions; Notice of Termination. (1) The time for shipment or delivery or any other action under a contract if not provided in this Article or agreed upon shall be a reasonable time.

(2) Where the contract provides for successive performances but is indefinite in duration it is valid for a reasonable time but unless otherwise agreed may be terminated at any time by either party.

(3) Termination of a contract by one party except on the happening of an agreed event requires that reasonable notification be received by the other party and an agreement dispensing with notification is invalid if its operation would be unconscionable.

§ 2–310. Open Time for Payment or Running of Credit; Authority to Ship Under Reservation. Unless otherwise agreed

(a) payment is due at the time and place at which the buyer is to receive the goods even though the place of shipment is the place of delivery; and

(b) if the seller is authorized to send the goods he may ship them under reservation, and may tender the documents of title, but the buyer may inspect the goods after their arrival before payment is due unless such inspection is inconsistent with the terms of the contract (Section 2–513); and

(c) if delivery is authorized and made by way of documents of title otherwise than by subsection (b) then payment is due at the time and place at which the buyer is to receive the documents regardless of where the goods are to be received; and

(d) where the seller is required or authorized to ship the goods on credit the credit period runs from the time of shipment but post-dating the invoice or delaying its dispatch will correspondingly delay the starting of the credit period.

§ 2–311. Options and Cooperation Respecting Performance. (1) An agreement for sale which is otherwise sufficiently definite (subsection (3) of Section 2–204) to be a contract is not made invalid by the fact that it leaves particulars of performance to be specified by one of the parties. Any such specification must be made in good faith and within limits set by commercial reasonableness.

(2) Unless otherwise agreed specifications relating to assortment of the goods are at the buyer's option and except as otherwise provided in subsections (1) (c) and (3) of Section 2–319 specifications or arrangements relating to shipment are at the seller's option.

(3) Where such specification would materially affect the other party's performance but is not seasonably made or where one party's cooperation is necessary to the agreed performance of the other but is not seasonably forthcoming, the other party in addition to all other remedies

(a) is excused for any resulting delay in his own performance; and

(b) may also either proceed to perform in any reasonable manner or after the time for a material part of his own performance treat the failure to specify or to cooperate as a breach by failure to deliver or accept the goods.

§ 2–312. Warranty of Title and Against Infringement; Buyer's Obligation Against Infringement. (1) Subject to subsection (2) there is in a contract for sale a warranty by the seller that

(a) the title conveyed shall be good, and its transfer rightful; and

(b) the goods shall be delivered free from any security interest or other lien or encumbrance of which the buyer at the time of contracting has no knowledge.

(2) A warranty under subsection (1) will be excluded or modified only by specific language or by circumstances which give the buyer reason to know that the person selling does not claim title in himself or that he is purporting to sell only such right or title as he or a third person may have.

(3) Unless otherwise agreed a seller who is a merchant regularly dealing in goods of the kind warrants that the goods shall be delivered free of the rightful claim of any third person by way of infringement or the like but a buyer who furnishes specifications to the seller must hold the seller harmless against any such claim which arises out of compliance with the specifications.

§ 2–313. Express Warranties by Affirmation, Promise, Description, Sample. (1) Express warranties by the seller are created as follows:

(a) Any affirmation of fact or promise made by the seller to the buyer which relates to the goods and becomes part of the basis of the bargain creates an express warranty that the goods shall conform to the affirmation or promise.

(b) Any description of the goods which is made part of the basis of the bargain creates an express warranty that the goods shall conform to the description.

(c) Any sample or model which is made part of the basis of the bargain creates an express warranty that the whole of the goods shall conform to the sample or model.

(2) It is not necessary to the creation of an express warranty that the seller use formal words such as "warrant" or "guarantee" or that he have a specific intention to make a warranty, but an affirmation merely of the value of the goods or a statement purporting to be merely the seller's opinion or commendation of the goods does not create a warranty.

§ 2–314. Implied Warranty: Merchantability; Usage of Trade. (1) Unless excluded or modified (Section 2–316), a warranty that the goods shall be merchantable is implied in a contract for their sale if the seller is a merchant with respect to goods of that kind. Under this section the serving for value of food or drink to be consumed either on the premises or elsewhere is a sale.

(2) Goods to be merchantable must be at least such as

(a) pass without objection in the trade under the contract description;

and

(b) in the case of fungible goods, are of fair average quality within the description; and

(c) are fit for the ordinary purposes for which such goods are used; and

(d) run, within the variations permitted by the agreement, of even kind, quality and quantity within each unit and among all units involved; and

(e) are adequately contained, packaged, and labeled as the agreement may require; and

(f) conform to the promises or affirmations of fact made on the container or label if any.

(3) Unless excluded or modified (Section 2–316) other implied warranties may arise from course of dealing or usage of trade.

§ 2–315. Implied Warranty: Fitness for Particular Purpose. Where the seller at the time of contracting has reason to know any particular purpose for which the goods are required and that the buyer is relying on the seller's skill or judgment to select or furnish suitable goods, there is unless excluded or modified under the next section an implied warranty that the goods shall be fit for such purpose.

§ 2–316. Exclusion or Modification of Warranties. (1) Words or conduct relevant to the creation of an express warranty and words or conduct tending to negate or limit warranty shall be construed wherever reasonable as consistent with each other; but subject to the provisions of this Article on parol or extrinsic evidence (Section 2–202) negation or limitation is inoperative to the extent that such construction is unreasonable.

(2) Subject to subsection (3), to exclude or modify the implied warranty of merchantability or any part of it the language must mention merchantability and in case of a writing must be conspicuous, and to exclude or modify any implied warranty of fitness the exclusion must be by a writing and conspicuous. Language to exclude all implied warranties of fitness is sufficient if it states, for example, that "There are no warranties which extend beyond the description on the face hereof."

(3) Notwithstanding subsection (2)

(a) unless the circumstances indicate otherwise, all implied warranties are excluded by expressions like "as is," "with all faults" or other language which in common understanding calls the buyer's attention to the exclusion of warranties and makes plain that there is no implied warranty; and

(b) when the buyer before entering into the contract has examined the goods or the sample or model as fully as he desired or has refused to examine the goods there is no implied warranty with regard to defects which an examination ought in the circumstances to have revealed to him; and

(c) an implied warranty can also be excluded or modified by course of dealing or course of performance or usage of trade.

(4) Remedies for breach of warranty can be limited in accordance with the provisions of this Article on liquidation or limitation of damages and on contractual modification of remedy (Sections 2–718 and 2–719).

§ 2–317. Cumulation and Conflict of Warranties Express or Implied. Warranties whether express or implied shall be construed as consistent with each other and as cumulative, but if such construction is unreasonable the intention of the parties shall determine which warranty is dominant. In ascertaining that intention the following rules apply:

(a) Exact or technical specifications displace an inconsistent sample or model or general language of description.

(b) A sample from an existing bulk displaces inconsistent general language of description.

(c) Express warranties displace inconsistent implied warranties other than an implied warranty of fitness for a particular purpose.

§ 2–318. Third Party Beneficiaries of Warranties Express or Implied. Note: *If this Act is introduced in the Congress of the United States this section should be omitted. (States to select one alternative.)*

Alternative A

A seller's warranty whether express or implied extends to any natural person who is in the family or household of his buyer or who is a guest in his home if it is reasonable to expect that such person may use, consume or be affected by the goods and who is injured in person by breach of the warranty. A seller may not exclude or limit the operation of this section.

Alternative B

A seller's warranty whether express or implied extends to any natural person who may reasonably be expected to use, consume or be affected by the goods and who is injured in person by breach of the warranty. A seller may not exclude or limit the operation of this section.

Alternative C

A seller's warranty whether express or implied extends to any person who may reasonably be expected to use, consume or be affected by the goods and who is injured by breach of the warranty. A seller may not exclude or limit the operation of this section with respect to injury to the person of an individual to whom the warranty extends. As amended 1966.

§ 2–319. F.O.B. and F.A.S. Terms. (1) Unless otherwise agreed the term F.O.B. (which means "free on board") at a named place, even though used only in connection with the stated price, is a delivery term under which

(a) when the term is F.O.B. the place of shipment, the seller must at that place ship the goods in the manner provided in this Article (Section 2–504) and bear the expense and risk of putting them into the possession of the carrier; or

(b) when the term is F.O.B. the place of destination, the seller must at his own expense and risk transport the goods to that place and there tender delivery of them in the manner provided in this Article (Section 2–503);

(c) when under either (a) or (b) the term is also F.O. B. vessel, car or other vehicle, the seller must in addition at his own expense and risk load the goods on board. If the term is F.O.B. vessel the buyer must name the vessel and in an appropriate case the seller must comply with the provisions of this Article on the form of bill of lading (Section 2–323).

(2) Unless otherwise agreed the term F.A.S. vessel (which means "free alongside") at a named port, even though used only in connection with the stated price, is a delivery term under which the seller must

(a) at his own expense and risk deliver the goods alongside the vessel in the manner usual in that port or on a dock designated and provided by the buyer; and

(b) obtain and tender a receipt for the goods in exchange for which the carrier is under a duty to issue a bill of lading.

(3) Unless otherwise agreed in any case falling within subsection (1) (a) or (c) or subsection (2) the buyer must seasonably give any needed instructions for making delivery, including when the term is F.A.S. or F.O.B. the loading berth of the vessel and in an appropriate case its name and sailing date. The seller may treat the failure of needed instructions as a failure of cooperation under this Article (Section 2–311). He may also at his option move the goods in any reasonable manner preparatory to delivery or shipment.

(4) Under the term F.O.B. vessel or F.A.S. unless otherwise agreed the buyer must make payment against tender of the required documents and the seller may not tender nor the buyer demand delivery of the goods in substitution for the documents.

§ 2–320. C.I.F. and C. & F. Terms. (1) The term C.I.F. means that the price includes in a lump sum the cost of the goods and the insurance and freight to the named destination. The term C. & F. or C.F. means that the price so includes cost and freight to the named destination.

(2) Unless otherwise agreed and even though used only in connection with the stated price and destination, the term C.I.F. destination or its equivalent requires the seller at his own expense and risk to

(a) put the goods into the possession of a carrier at the port for shipment and obtain a negotiable bill or bills of lading covering the entire transportation to the named destination; and

(b) load the goods and obtain a receipt from the carrier (which may be contained in the bill of lading) showing that the freight has been paid or provided for; and

(c) obtain a policy or certificate of insurance, including any war risk insurance, of a kind and on terms then current at the port of shipment in the usual amount, in the currency of the contract, shown to cover the same goods covered by the bill of lading and providing for payment of loss to the order of the buyer or for the account of whom it may concern; but the seller may add to the price the amount of the premium for any such war risk insurance; and

(d) prepare an invoice of the goods and procure any other documents required to effect shipment or to comply with the contract; and

(e) forward and tender with commercial promptness all the documents in due form and with any indorsement necessary to perfect the buyer's rights.

(3) Unless otherwise agreed the term C. & F. or its equivalent has the same effect and imposes upon the seller the same obligations and risks as a C.I.F. term except the obligation as to insurance.

(4) Under the term C.I.F. or C. & F. unless otherwise agreed the buyer must make payment against tender of the required documents and the seller may not tender nor the buyer demand delivery of the goods in substitution for the documents.

§ 2–321. C.I.F. or C. & F.: "Net Landed Weights"; "Payment on Arrival"; Warranty of Condition on Arrival. Under a contract containing a term C.I.F. or C. & F.

(1) Where the price is based on or is to be adjusted according to "net landed weights," "delivered weights," "out turn" quantity or quality or the like, unless otherwise agreed the seller must reasonably estimate the price. The payment due on tender of the documents called for by the contract is the amount so estimated, but after final adjustment of the price a settlement must be made with commercial promptness.

(2) An agreement described in subsection (1) or any warranty of quality or condition of the goods on arrival places upon the seller the risk of ordinary deterioration, shrinkage and the like in transportation but has no effect on the place or time of identification to the contract for sale or delivery or on the passing of the risk of loss.

(3) Unless otherwise agreed where the contract provides for payment on or after arrival of the goods the seller must before payment allow such preliminary inspection as is feasible; but if the goods are lost delivery of the documents and payment are due when the goods should have arrived.

§ 2–322. Delivery "Ex-Ship." (1) Unless otherwise agreed a term for delivery of goods "ex-ship" (which means from the carrying vessel) or in equivalent language is not restricted to a particular ship and requires delivery from a ship which has reached a place at the named port of destination where goods of the kind are usually discharged.

(2) Under such a term unless otherwise agreed

(a) the seller must discharge all liens arising out of the carriage and furnish the buyer with a direction which puts the carrier under a duty to deliver the goods; and

(b) the risk of loss does not pass to the buyer until the goods leave the ship's tackle or are otherwise properly unloaded.

§ 2–323. Form of Bill of Lading Required in Overseas Shipment; "Overseas." (1) Where the contract contemplates overseas shipment and contains a term C.I.F. or C. & F. or F.O.B. vessel, the seller unless otherwise agreed must obtain a negotiable bill of lading stating that the goods have been loaded on board or, in the case of a term C.I.F. or C. & F., received for shipment.

(2) Where in a case within subsection (1) a bill of lading has been issued in a set of parts, unless otherwise agreed if the documents are not to be sent from abroad the buyer may demand tender of the full set; otherwise only one part of the bill of lading need be tendered. Even if the agreement expressly requires a full set

(a) due tender of a single part is acceptable within the provisions of this Article on cure of improper delivery (subsection (1) of Section 2–508); and

(b) even though the full set is demanded, if the documents are sent from abroad the person tendering an incomplete set may nevertheless require payment upon furnishing an indemnity which the buyer in good faith deems adequate.

(3) A shipment by water or by air or a contract contemplating such shipment is "overseas" insofar as by usage of trade or agreement it is subject to the commercial, financing or shipping practices characteristic of international deep water commerce.

§ 2–324. "No Arrival, No Sale" Term. Under a term "no arrival, no sale" or terms of like meaning, unless otherwise agreed,

(a) the seller must properly ship conforming goods and if they arrive by any means he must tender them on arrival but he assumes no obligation that the goods will arrive unless he has caused the non-arrival; and

(b) where without fault of the seller the goods are in part lost or have so deteriorated as no longer to conform to the contract or arrive after the contract time, the buyer may proceed as if there had been casualty to identified goods (Section 2–613).

§ 2–325. "Letter of Credit" Term; "Confirmed Credit." (1) Failure of the buyer seasonably to furnish an agreed letter of credit is a breach of the contract for sale.

(2) The delivery to seller of a proper letter of credit suspends the buyer's obligation to pay. If the letter of credit is dishonored, the seller may on seasonable notification to the buyer require payment directly from him.

(3) Unless otherwise agreed the term "letter of credit" or "banker's credit" in a contract for sale means an irrevocable credit issued by a financing agency of good repute and, where the shipment is overseas, of good international repute. The term "confirmed credit" means that the credit must also carry the direct obligation of such an agency which does business in the seller's financial market.

§ 2–326. Sale on Approval and Sale or Return; Consignment Sales and Rights of Creditors. (1) Unless otherwise agreed, if delivered goods may be returned by the buyer even though they conform to the contract, the transaction is

(a) a "sale on approval" if the goods are delivered primarily for use, and

(b) a "sale or return" if the goods are delivered primarily for resale.

(2) Except as provided in subsection (3), goods held on approval are not subject to the claims of the buyer's creditors until acceptance; goods held on sale or return are subject to such claims while in the buyer's possession.

(3) Where goods are delivered to a person for sale and such person maintains a place of business at which he deals in goods of the kind involved, under a name other than the name of the person making delivery, then with respect to claims of creditors of the person conducting the business the goods are deemed to be on sale or return. The provisions of this subsection are applicable even though an agreement purports to reserve title to the person making delivery until payment or resale or uses such words as "on consignment" or "on memorandum." However, this subsection is not applicable if the person making delivery

(a) complies with an applicable law providing for a consignor's interest or the like to be evidenced by a sign, or

(b) establishes that the person conducting the business is generally known by his creditors to be substantially engaged in selling the goods of others, or

(c) complies with the filing provisions of the Article on Secured Transactions (Article 9).

(4) Any "or return" term of a contract for sale is to be treated as a separate contract for sale within the statute of frauds section of this Article (Section 2–201) and as contradicting the sale aspect of the contract within the provisions of this Article on parol or extrinsic evidence (Section 2–202).

§ 2–327. Special Incidents of Sale on Approval and Sale or Return. (1) Under a sale on approval unless otherwise agreed

(a) although the goods are identified to the contract the risk of loss and the title do not pass to the buyer until acceptance; and

(b) use of the goods consistent with the purpose of trial is not acceptance but failure seasonably to notify the seller of election to return the goods is acceptance, and if the goods conform to the contract acceptance of any part is acceptance of the whole; and

(c) after due notification of election to return, the return is at the seller's risk and expense but a merchant buyer must follow any reasonable instructions.

(2) Under a sale or return unless otherwise agreed

(a) the option to return extends to the whole or any commercial unit of the goods while in substantially their original condition, but must be exercised seasonably; and

(b) the return is at the buyer's risk and expense.

§ 2–328. Sale by Auction. (1) In a sale by auction if goods are put up in lots each lot is the subject of a separate sale.

(2) A sale by auction is complete when the auctioneer so announces by the fall of the hammer or in other customary manner. Where a bid is made while the hammer is falling in acceptance of a prior bid the auctioneer may in his discretion reopen the bidding or declare the goods sold under the bid on which the hammer was falling.

(3) Such a sale is with reserve unless the goods are in explicit terms put up without reserve. In an auction with reserve the auctioneer may withdraw the goods at any time until he announces completion of the sale. In an auction without reserve, after the auctioneer calls for bids on an article or lot, that article or lot cannot be withdrawn unless no bid is made within a reasonable time. In either case a bidder may retract his bid until the auctioneer's announcement of completion of the sale, but a bidder's retraction does not revive any previous bid.

(4) If the auctioneer knowingly receives a bid on the seller's behalf or the seller makes or procures such a bid, and notice has not been given that liberty for such bidding is reserved, the buyer may at his option avoid the sale or take the goods at the price of the last good faith bid prior to the completion of the sale. This subsection shall not apply to any bid at a forced sale.

Part 4

Title, Creditors and Good Faith Purchasers

§ 2–401. Passing of Title; Reservation for Security; Limited Application of This Section. Each provision of this Article with regard to the rights, obligations and remedies of the seller, the buyer, purchasers or other third parties applies irrespective of title to the goods except where the provision refers to such title. Insofar as situations are not covered by the other provisions of this Article and matters concerning title became material the following rules apply:

(1) Title to goods cannot pass under a contract for sale prior to their identification to the contract (Section 2–501), and unless otherwise explicitly agreed the buyer acquires by their identification a special property as limited by this Act. Any retention or reservation by the seller of the title (property) in goods shipped or delivered to the buyer is limited in effect to a reservation of a security interest. Subject to these provisions and to the provisions of the Article on Secured Transactions (Article 9), title to goods passes from the seller to the buyer in any manner and on any conditions explicitly agreed on by the parties.

(2) Unless otherwise explicitly agreed title passes to the buyer at the time and place at which the seller completes his performance with reference to the physical delivery of the goods, despite any reservation of a security interest and even though a document of title is to be delivered at a different time or place; and in particular and despite any reservation of a security interest by the bill of lading

(a) if the contract requires or authorizes the seller to send the goods to the buyer but does not require him to deliver them at destination, title passes to the buyer at the time and place of shipment; but

(b) if the contract requires delivery at destination, title passes on tender there.

(3) Unless otherwise explicitly agreed where delivery is to be made without moving the goods,

(a) if the seller is to deliver a document of title, title passes at the time when and the place where he delivers such documents; or

(b) if the goods are at the time of contracting already identified and no documents are to be delivered, title passes at the time and place of contracting.

(4) A rejection or other refusal by the buyer to receive or retain the goods, whether or not justified, or a justified revocation of acceptance revests title to the goods in the seller. Such revesting occurs by operation of law and is not a "sale."

§ 2–402. Rights of Seller's Creditors Against Sold Goods. (1) Except as provided in subsections (2) and (3), rights of unsecured creditors of the seller with respect to goods which have been identified to a contract for sale are subject to the buyer's rights to recover the goods under this Article (Sections 2–502 and 2–716).

(2) A creditor of the seller may treat a sale or an identification of goods to a contract for sale as void if as against him a retention of possession by the seller is fraudulent under any rule of law of the state where the goods are situated, except that retention of possession in good faith and current course of trade by a merchant-seller for a commercially reasonable time after a sale or identification is not fraudulent.

(3) Nothing in this Article shall be deemed to impair the rights of creditors of the seller

(a) under the provisions of the Article on Secured Transactions (Article 9); or

(b) where identification to the contract or delivery is made not in current course of trade but in satisfaction of or as security for a pre-existing claim for money, security or the like and is made under circumstances which under any rule of law of the state where the goods are situated would apart from this Article constitute the transaction a fraudulent transfer or voidable preference.

§ 2–403. Power to Transfer; Good Faith Purchase of Goods; "Entrusting." (1) A purchaser of goods acquires all title which his transferor had or had power to transfer except that a purchaser of a limited interest acquires rights only to the extent of the interest purchased. A person with voidable title has power to transfer a good title to a good faith purchaser for value. When goods have been delivered under a transaction of purchase the purchaser has such power even though

(a) the transferor was deceived as to the identity of the purchaser, or

(b) the delivery was in exchange for a check which is later dishonored, or

(c) it was agreed that the transaction was to be a "cash sale," or

(d) the delivery was procured through fraud punishable as larcenous under the criminal law.

(2) Any entrusting of possession of goods to a merchant who deals in

goods of that kind gives him power to transfer all rights of the entruster to a buyer in ordinary course of business.

(3) "Entrusting" includes any delivery and any acquiescence in retention of possession regardless of any condition expressed between the parties to the delivery or acquiescence and regardless of whether the procurement of the entrusting or the possessor's disposition of the goods have been such as to be larcenous under the criminal law.

(4) The rights of other purchasers of goods and of lien creditors are governed by the Articles on Secured Transactions (Article 9), Bulk Transfers (Article 6) and Documents of Title (Article 7).

Part 5

Performance

§ 2–501. Insurable Interest in Goods; Manner of Identification of Goods. (1) The buyer obtains a special property and an insurable interest in goods by identification of existing goods as goods to which the contract refers even though the goods so identified are non-conforming and he has an option to return or reject them. Such identification can be made at any time and in any manner explicitly agreed to by the parties. In the absence of explicit agreement identification occurs

(a) when the contract is made if it is for the sale of goods already existing and identified;

(b) if the contract is for the sale of future goods other than those described in paragraph (c), when goods are shipped, marked or otherwise designated by the seller as goods to which the contract refers;

(c) when the crops are planted or otherwise become growing crops or the young are conceived if the contract is for the sale of unborn young to be born within twelve months after contracting or for the sale of crops to be harvested within twelve months or the next normal harvest season after contracting whichever is longer.

(2) The seller retains an insurable interest in goods so long as title to or any security interest in the goods remains in him and where the identification is by the seller alone he may until default or insolvency or notification to the buyer that the identification is final substitute other goods for those identified.

(3) Nothing in this section impairs any insurable interest recognized under any other statute or rule of law.

§ 2–502. Buyer's Right to Goods on Seller's Insolvency. (1) Subject to subsection (2) and even though the goods have not been shipped a buyer who has paid a part or all of the price of goods in which he has a special property under the provisions of the immediately preceding section may on making and keeping good a tender of any unpaid portion of their price recover them from the seller if the seller becomes insolvent within ten days after receipt of the first installment on their price.

(2) If the identification creating his special property has been made

by the buyer he acquires the right to recover the goods only if they conform to the contract for sale.

§ 2–503. Manner of Seller's Tender of Delivery. (1) Tender of delivery requires that the seller put and hold conforming goods at the buyer's disposition and give the buyer any notification reasonably necessary to enable him to take delivery. The manner, time and place for tender are determined by the agreement and this Article, and in particular

(a) tender must be at a reasonable hour, and if it is of goods they must be kept available for the period reasonably necessary to enable the buyer to take possession; but

(b) unless otherwise agreed the buyer must furnish facilities reasonably suited to the receipt of the goods.

(2) Where the case is within the next section respecting shipment tender requires that the seller comply with its provisions.

(3) Where the seller is required to deliver at a particular destination tender requires that he comply with subsection (1) and also in any appropriate case tender documents as described in subsections (4) and (5) of this section.

(4) Where goods are in the possession of a bailee and are to be delivered without being moved

(a) tender requires that the seller either tender a negotiable document of title covering such goods or procure acknowledgement by the bailee of the buyer's right to possession of the goods; but

(b) tender to the buyer of a non-negotiable document of title or of a written direction to the bailee to deliver is sufficient tender unless the buyer seasonably objects, and receipt by the bailee of notification of the buyer's rights fixes those rights as against the bailee and all third persons; but risk of loss of the goods and of any failure by the bailee to honor the non-negotiable document of title or to obey the direction remains on the seller until the buyer has had a reasonable time to present the document or direction, and a refusal by the bailee to honor the document or to obey the direction defeats the tender.

(5) Where the contract requires the seller to deliver documents

(a) he must tender all such documents in correct form, except as provided in this Article with respect to bills of lading in a set (subsection (2) of Section 2–323); and

(b) tender through customary banking channels is sufficient and dishonor of a draft accompanying the documents constitutes non-acceptance or rejection.

§ 2–504. Shipment by Seller. Where the seller is required or authorized to send the goods to the buyer and the contract does not require him to deliver them at a particular destination, then unless otherwise agreed he must

(a) put the goods in the possession of such a carrier and make such a contract for their transportation as may be reasonable having regard to the nature of the goods and other circumstances of the case; and

(b) obtain and promptly deliver or tender in due form any document necessary to enable the buyer to obtain possession of the goods or otherwise required by the agreement or by usage of trade; and

(c) promptly notify the buyer of the shipment.

Failure to notify the buyer under paragraph (c) or to make a proper contract under paragraph (a) is a ground for rejection only if material delay or loss ensues.

§ 2–505. Seller's Shipment Under Reservation. (1) Where the seller has identified goods to the contract by or before shipment:

(a) his procurement of a negotiable bill of lading to his own order or otherwise reserves in him a security interest in the goods. His procurement of the bill to the order of a financing agency or of the buyer indicates in addition only the seller's expectation of transferring that interest to the person named.

(b) a non-negotiable bill of lading to himself or his nominee reserves possession of the goods as security but except in a case of conditional delivery (subsection (2) of Section 2–507) a non-negotiable bill of lading naming the buyer as consignee reserves no security interest even though the seller retains possession of the bill of lading.

(2) When shipment by the seller with reservation of a security interest is in violation of the contract for sale it constitutes an improper contract for transportation within the preceding section but impairs neither the rights given to the buyer by shipment and identification of the goods to the contract nor the seller's powers as a holder of a negotiable document.

§ 2–506. Rights of Financing Agency. (1) A financing agency by paying or purchasing for value a draft which relates to a shipment of goods acquires to the extent of the payment or purchase and in addition to its own rights under the draft and any document of title securing it any rights of the shipper in the goods including the right to stop delivery and the shipper's right to have the draft honored by the buyer.

(2) The right to reimbursement of a financing agency which has in good faith honored or purchased the draft under commitment to or authority from the buyer is not impaired by subsequent discovery of defects with reference to any relevant document which was apparently regular on its face.

§ 2–507. Effect of Seller's Tender; Delivery on Condition. (1) Tender of delivery is a condition to the buyer's duty to accept the goods and, unless otherwise agreed, to his duty to pay for them. Tender entitles the seller to acceptance of the goods and to payment according to the contract.

(2) Where payment is due and demanded on the delivery to the buyer of goods or documents of title, his right as against the seller to retain or dispose of them is conditional upon his making the payment due.

§ 2–508. Cure by Seller of Improper Tender or Delivery; Replacement. (1) Where any tender or delivery by the seller is rejected because non-conforming and the time for performance has not yet expired, the seller may seasonably notify the buyer of his intention to cure and may then within the contract time make a conforming delivery.

(2) Where the buyer rejects a non-conforming tender which the seller had reasonable grounds to believe would be acceptable with or without money allowance the seller may if he seasonably notifies the buyer have a further reasonable time to substitute a conforming tender.

§ 2–509. Risk of Loss in the Absence of Breach. (1) Where the contract requires or authorizes the seller to ship the goods by carrier

(a) if it does not require him to deliver them at a particular destination, the risk of loss passes to the buyer when the goods are duly delivered to the carrier even though the shipment is under reservation (Section 2–505); but

(b) if it does require him to deliver them at a particular destination and the goods are there duly tendered while in the possession of the carrier, the risk of loss passes to the buyer when the goods are there duly so tendered as to enable the buyer to take delivery.

(2) Where the goods are held by a bailee to be delivered without being moved, the risk of loss passes to the buyer

(a) on his receipt of a negotiable document of title covering the goods; or

(b) on acknowledgment by the bailee of the buyer's right to possession of the goods; or

(c) after his receipt of a non-negotiable document of title or other written direction to deliver, as provided in subsection (4) (b) of Section 2–503.

(3) In any case not within subsection (1) or (2), the risk of loss passes to the buyer on his receipt of the goods if the seller is a merchant; otherwise the risk passes to the buyer on tender of delivery.

(4) The provisions of this section are subject to contrary agreement of the parties and to the provisions of this Article on sale on approval (Section 2–327) and on effect of breach on risk of loss (Section 2–510).

§ 2–510. Effect of Breach on Risk of Loss. (1) Where a tender or delivery of goods so fails to conform to the contract as to give a right of rejection the risk of their loss remains on the seller until cure or acceptance.

(2) Where the buyer rightfully revokes acceptance he may to the extent of any deficiency in his effective insurance coverage treat the risk of loss as having rested on the seller from the beginning.

(3) Where the buyer as to conforming goods already identified to the contract for sale repudiates or is otherwise in breach before risk of their loss has passed to him, the seller may to the extent of any deficiency in his effective insurance coverage treat the risk of loss as resting on the buyer for a commercially reasonable time.

§ 2–511. Tender of Payment by Buyer; Payment by Check. (1) Unless otherwise agreed tender of payment is a condition to the seller's duty to tender and complete any delivery.

(2) Tender of payment is sufficient when made by any means or in any manner current in the ordinary course of business unless the seller demands payment in legal tender and gives any extension of time reasonably necessary to procure it.

(3) Subject to the provisions of this Act on the effect of an instrument on an obligation (Section 3–802), payment by check is conditional and is defeated as between the parties by dishonor of the check on due presentment.

§ 2–512. Payment by Buyer Before Inspection. (1) Where the contract requires payment before inspection non-conformity of the goods does not excuse the buyer from so making payment unless

(a) the non-conformity appears without inspection; or

(b) despite tender of the required documents the circumstances would justify injunction against honor under the provisions of this Act (Section 5–114).

(2) Payment pursuant to subsection (1) does not constitute an acceptance of goods or impair the buyer's right to inspect or any of his remedies.

§ 2–513. Buyer's Right to Inspection of Goods. (1) Unless otherwise agreed and subject to subsection (3), where goods are tendered or delivered or identified to the contract for sale, the buyer has a right before payment or acceptance to inspect them at any reasonable place and time and in any reasonable manner. When the seller is required or authorized to send the goods to the buyer, the inspection may be after their arrival.

(2) Expenses of inspection must be borne by the buyer but may be recovered from the seller if the goods do not conform and are rejected.

(3) Unless otherwise agreed and subject to the provisions of this Article on C.I.F. contracts (subsection (3) of Section 2–321), the buyer is not entitled to inspect the goods before payment of the price when the contract provides

(a) for delivery "C.O.D." or on other like terms; or

(b) for payment against documents of title, except where such payment is due only after the goods are to become available for inspection.

(4) A place or method of inspection fixed by the parties is presumed to be exclusive but unless otherwise expressly agreed it does not postpone identification or shift the place for delivery or for passing the risk of loss. If compliance becomes impossible, inspection shall be as provided in this section unless the place or method fixed was clearly intended as an indispensable condition failure of which avoids the contract.

§ 2–514. When Documents Deliverable on Acceptance; When on Payment. Unless otherwise agreed documents against which a draft is drawn are to be delivered to the drawee on acceptance of the draft if it is payable more than three days after presentment; otherwise, only on payment.

§ 2–515. Preserving Evidence of Goods in Dispute. In furtherance of the adjustment of any claim or dispute

(a) either party on reasonable notification to the other and for the purpose of ascertaining the facts and preserving evidence has the right to inspect, test and sample the goods including such of them as may be in the possession or control of the other; and

(b) the parties may agree to a third party inspection or survey to determine the conformity or condition of the goods and may agree that the findings shall be binding upon them in any subsequent litigation or adjustment.

Part 6

Breach, Repudiation and Excuse

§ 2–601. Buyer's Rights on Improper Delivery. Subject to the provisions of this Article on breach in installment contracts (Section 2–612) and unless otherwise agreed under the sections on contractual limitations of remedy (Sections 2–718 and 2–719), if the goods or the tender of delivery fail in any respect to conform to the contract, the buyer may

(a) reject the whole; or

(b) accept the whole; or

(c) accept any commercial unit or units and reject the rest.

§ 2–602. Manner and Effect of Rightful Rejection. (1) Rejection of goods must be within a reasonable time after their delivery or tender. It is ineffective unless the buyer seasonably notifies the seller.

(2) Subject to the provisions of the two following sections on rejected goods (Sections 2–603 and 2–604),

(a) after rejection any exercise of ownership by the buyer with respect to any commercial unit is wrongful as against the seller; and

(b) if the buyer has before rejection taken physical possession of goods in which he does not have a security interest under the provisions of this Article (subsection 3) of Section 2–711), he is under a duty after rejection to hold them with reasonable care at the seller's disposition for a time sufficient to permit the seller to remove them; but

(c) the buyer has no further obligations with regard to goods rightfully rejected.

(3) The seller's rights with respect to goods wrongfully rejected are governed by the provisions of this Article on Seller's remedies in general (Section 2–703).

§ 2–603. Merchant Buyer's Duties as to Rightfully Rejected Goods. (1) Subject to any security interest in the buyer (subsection (3) of Section 2–711), when the seller has no agent or place of business at the market of rejection a merchant buyer is under a duty after rejection of goods in his possession or control to follow any reasonable instructions received from the seller with respect to the goods and in the absence of such instructions to make reasonable efforts to sell them for the seller's account if they are perishable or threaten to decline in value speedily. Instructions are not reasonable if on demand indemnity for expenses is not forthcoming.

(2) When the buyer sells goods under subsection (1), he is entitled to reimbursement from the seller or out of the proceeds for reasonable expenses of caring for and selling them, and if the expenses include no selling commission then to such commission as is usual in the trade or if there is none to a reasonable sum not exceeding ten percent on the gross proceeds.

(3) In complying with this section the buyer is held only to good faith and good faith conduct hereunder is neither acceptance nor conversion nor the basis of an action for damages.

§ 2–604. Buyer's Options as to Salvage of Rightfully Rejected Goods. Subject to the provisions of the immediately preceding section on perishables if the seller gives no instructions within a reasonable time after notification of rejection the buyer may store the rejected goods for the seller's account or reship them to him or resell them for the seller's account with reimbursement as provided in the preceding section. Such action is not acceptance or conversion.

§ 2–605. Waiver of Buyer's Objections by Failure to Particularize. (1) The buyer's failure to state in connection with rejection a particular defect which is ascertainable by reasonable inspection precludes him from relying on the unstated defect to justify rejection or to establish breach

(a) where the seller could have cured it if stated seasonably; or

(b) between merchants when the seller has after rejection made a request in writing for a full and final written statement of all defects on which the buyer proposes to rely.

(2) Payment against documents made without reservation of rights precludes recovery of the payment for defects apparent on the face of the documents.

§ 2–606. What Constitutes Acceptance of Goods. (1) Acceptance of goods occurs when the buyer

(a) after a reasonable opportunity to inspect the goods signifies to the seller that the goods are conforming or that he will take or retain them in spite of their nonconformity; or

(b) fails to make an effective rejection (subsection (1) of Section 2–602), but such acceptance does not occur until the buyer has had a reasonable opportunity to inspect them; or

(c) does any act inconsistent with the seller's ownership; but if such act is wrongful as against the seller it is an acceptance only if ratified by him.

(2) Acceptance of a part of any commercial unit is acceptance of that entire unit.

§ 2–607. Effect of Acceptance; Notice of Breach; Burden of Establishing Breach After Acceptance; Notice of Claim or Litigation to Person Answerable Over. (1) The buyer must pay at the contract rate for any goods accepted.

(2) Acceptance of goods by the buyer precludes rejection of the goods accepted and if made with knowledge of a non-conformity cannot be revoked because of it unless the acceptance was on the reasonable assumption that the non-conformity would be seasonably cured but acceptance does not of itself impair any other remedy provided by this Article for nonconformity.

(3) Where a tender has been accepted.

(a) the buyer must within a reasonable time after he discovers or should have discovered any breach notify the seller of breach or be barred from any remedy; and

(b) if the claim is one for infringement or the like (subsection (3) of Section 2–312) and the buyer is sued as a result of such a breach he must so notify the seller within a reasonable time after he receives notice of the litigation or be barred from any remedy over for liability established by the litigation.

(4) The burden is on the buyer to establish any breach with respect to the goods accepted.

(5) Where the buyer is sued for breach of a warranty or other obligation for which his seller is answerable over

(a) he may give his seller written notice of the litigation. If the notice states that the seller may come in and defend and that if the seller does not do so he will be bound in any action against him by his buyer by any determination of fact common to the two litigations, then unless the seller after seasonable receipt of the notice does come in and defend he is so bound.

(b) if the claim is one for infringement or the like (subsection (3) of Section 2–312) the original seller may demand in writing that his buyer turn over to him control of the litigation including settlement or else be barred from any remedy over and if he also agrees to bear all expense and to satisfy any adverse judgment, then unless the buyer after seasonable receipt of the demand does turn over control the buyer is so barred.

(6) The provisions of subsections (3), (4) and (5) apply to any obligation of a buyer to hold the seller harmless against infringement or the like (subsection (3) of Section 2–312).

§ 2–608. Revocation of Acceptance in Whole or in Part. (1) The buyer may revoke his acceptance of a lot or commercial unit whose non-conformity substantially impairs its value to him if he has accepted it

(a) on the reasonable assumption that its non-conformity would be cured and it has not been seasonably cured; or

(b) without discovery of such non-conformity if his acceptance was reasonably induced either by the difficulty of discovery before acceptance or by the seller's assurances.

(2) Revocation of acceptance must occur within a reasonable time after the buyer discovers or should have discovered the ground for it and before any substantial change in condition of the goods which is not caused by their own defects. It is not effective until the buyer notifies the seller of it.

(3) A buyer who so revokes has the same rights and duties with regard to the goods involved as if he had rejected them.

§ 2–609. Right to Adequate Assurance of Performance. (1) A contract for sale imposes an obligation on each party that the other's expectation of receiving due performance will not be impaired. When reasonable grounds for insecurity arise with respect to the performance of either party the other may in writing demand adequate assurance of due performance and until he receives such assurance may if commercially reasonable suspend any performance for which he has not already received the agreed return.

(2) Between merchants the reasonableness of grounds for insecurity and the adequacy of any assurance offered shall be determined according to commercial standards.

(3) Acceptance of any improper delivery or payment does not prejudice the aggrieved party's right to demand adequate assurance of future performance.

(4) After receipt of a justified demand failure to provide within a reasonable time not exceeding thirty days such assurance of due performance as is adequate under the circumstances of the particular case is a repudiation of the contract.

§ 2–610. Anticipatory Repudiation. When either party repudiates the contract with respect to a performance not yet due the loss of which will substantially impair the value of the contract to the other, the aggrieved party may

(a) for a commercially reasonable time await performance by the repudiating party; or

(b) resort to any remedy for breach (Section 2–703 or Section 2–711), even though he has notified the repudiating party that he would await the latter's performance and has urged retraction; and

(c) in either case suspend his own performance or proceed in accordance with the provisions of this Article on the seller's right to identify goods to the contract notwithstanding breach or to salvage unfinished goods (Section 2–704).

§ 2–611. Retraction of Anticipatory Repudiation. (1) Until the repudiating party's next performance is due he can retract his repudiation unless the aggrieved party has since the repudiation cancelled or materially changed his position or otherwise indicated that he considers the repudiation final.

(2) Retraction may be by any method which clearly indicates to the aggrieved party that the repudiating party intends to perform, but must include any assurance justifiably demanded under the provisions of this Article (Section 2–609).

(3) Retraction reinstates the repudiating party's rights under the contract with due excuse and allowance to the aggrieved party for any delay occasioned by the repudiation.

§ 2–612. "Installment Contract"; Breach. (1) An "installment contract" is one which requires or authorizes the delivery of goods in separate lots to be separately accepted, even though the contract contains a clause "each delivery is a separate contract" or its equivalent.

(2) The buyer may reject any installment which is non-conforming if the non-conformity substantially impairs the value of that installment and cannot be cured or if the non-conformity is a defect in the required documents; but if the non-conformity does not fall within subsection (3) and the seller gives adequate assurance of its cure the buyer must accept that installment.

(3) Whenever non-conformity or default with respect to one or more installments substantially impairs the value of the whole contract there is a breach of the whole. But the aggrieved party reinstates the contract if he accepts a non-conforming installment without seasonably notifying of cancellation or if he brings an action with respect only to past installments or demands performance as to future installments.

§ 2–613. Casualty to Identified Goods. Where the contract requires for its performance goods identified when the contract is made, and the goods suffer casualty without fault of either party before the risk of loss passes to the buyer, or in a proper case under a "no arrival, no sale" term (Section 2–324) then

(a) if the loss is total the contract is avoided; and

(b) if the loss is partial or the goods have so deteriorated as no longer to conform to the contract the buyer may nevertheless demand inspection and at his option either treat the contract as avoided or accept the goods with due allowance from the contract price for the deterioration or the deficiency in quantity but without further right against the seller.

§ 2–614. Substituted Performance. (1) Where without fault of either party the agreed berthing, loading, or unloading facilities fail or an agreed type of carrier becomes unavailable or the agreed manner of delivery otherwise becomes commercially impracticable but a commercially reasonable substitute is available, such substitute performance must be tendered and accepted.

(2) If the agreed means or manner of payment fails because of domestic or foreign governmental regulation, the seller may withhold or stop delivery unless the buyer provides a means or manner of payment which is commercially a substantial equivalent. If delivery has already been taken, payment by the means or in the manner provided by the regulation discharges the buyer's obligation unless the regulation is discriminatory, oppressive or predatory.

§ 2–615. Excuse by Failure of Presupposed Conditions. Except so far as a seller may have assumed a greater obligation and subject to the preceding section on substituted performance:

(a) Delay in delivery or non-delivery in whole or in part by a seller who complies with paragraphs (b) and (c) is not a breach of his duty under a contract for sale if performance as agreed has been made impracticable by the occurrence of a contingency the non-occurrence of which was a basic assumption on which the contract was made or by compliance in good faith with any applicable foreign or domestic governmental regulation or order whether or not it later proves to be invalid.

(b) Where the causes mentioned in paragraph (a) affect only a part of the seller's capacity to perform, he must allocate production and deliveries among his customers but may at his option include regular customers not then under contract as well as his own requirements for further manufacture. He may so allocate in any manner which is fair and reasonable.

(c) The seller must notify the buyer seasonably that there will be delay or non-delivery and, when allocation is required under paragraph (b), of the estimated quota thus made available for the buyer.

§ 2–616. Procedure on Notice Claiming Excuse. (1) Where the buyer receives notification of a material or indefinite delay or an allocation justified under the preceding section he may by written notification to the seller as to any delivery concerned, and where the prospective deficiency substantially impairs the value of the whole contract under the provisions of this Article relating to breach of installment contracts (Section 2–612), then also as to the whole,

(a) terminate and thereby discharge any unexecuted portion of the contract; or

(b) modify the contract by agreeing to take his available quota in substitution.

(2) If after receipt of such notification from the seller the buyer fails so to modify the contract within a reasonable time not exceeding thirty days the contract lapses with respect to any deliveries affected.

(3) The provisions of this section may not be negated by agreement except in so far as the seller has assumed a greater obligation under the preceding section.

Part 7

Remedies

§ 2–701. Remedies for Breach of Collateral Contracts Not Impaired. Remedies for breach of any obligation or promise collateral or ancillary to a contract for sale are not impaired by the provisions of this Article.

§ 2–702. Seller's Remedies on Discovery of Buyer's Insolvency. (1) Where the seller discovers the buyer to be insolvent he may refuse delivery except for cash including payment for all goods theretofore delivered under the contract, and stop delivery under this Article (Section 2–705).

(2) Where the seller discovers that the buyer has received goods on credit while insolvent he may reclaim the goods upon demand made within ten days after the receipt, but if misrepresentation of solvency has been made to the particular seller in writing within three months before delivery the ten day limitation does not apply. Except as provided in this subsection the seller may not base a right to reclaim goods on the buyer's fraudulent or innocent misrepresentation of solvency or of intent to pay.

(3) The seller's right to reclaim under subsection (2) is subject to the rights of a buyer in ordinary course or other good faith purchaser under this Article (Section 2–403). Successful reclamation of goods excludes all other remedies with respect to them.

§ 2–703. Seller's Remedies in General. Where the buyer wrongfully rejects of revokes acceptance of goods or fails to make a payment due on or before delivery or repudiates with respect to a part or the whole, then with respect to any goods directly affected and, if the breach is of the whole contract (Section 2–612), then also with respect to the whole undelivered balance, the aggrieved seller may

(a) withhold delivery of such goods;

(b) stop delivery by any bailee as hereafter provided (Section 2–705);

(c) proceed under the next section respecting goods still unidentified to the contract;

(d) resell and recover damages as hereafter provided (Section 2–706);

(e) recover damages for non-acceptance (Section 2–708) or in a proper case the price (Section 2–709);

(f) cancel

§ 2–704. Seller's Right to Identify Goods to the Contract Notwithstanding Breach or to Salvage Unfinished Goods. (1) An aggrieved seller under the preceding section may

(a) identify to the contract conforming goods not already identified if at the time he learned of the breach they are in his possession or control;

(b) treat as the subject of resale goods which have demonstrably been intended for the particular contract even though those goods are unfinished.

(2) Where the goods are unfinished an aggrieved seller may in the exercise of reasonable commercial judgment for the purposes of avoiding loss and of effective realization either complete the manufacture and wholly identify the goods to the contract or cease manufacture and resell for scrap or salvage value or proceed in any other reasonable manner.

§ 2–705. Seller's Stoppage of Delivery in Transit or Otherwise. (1) The seller may stop delivery of goods in the possession of a carrier or other bailee when he discovers the buyer to be insolvent (Section 2–702) and may stop delivery of carload, truckload, planeload or larger shipments of express or freight when the buyer repudiates or fails to make a payment due before delivery or if for any other reason the seller has a right to withhold or reclaim the goods.

(2) As against such buyer the seller may stop delivery until

(a) receipt of the goods by the buyer; or

(b) acknowledgment to the buyer by any bailee of the goods except a carrier that the bailee holds the goods for the buyer; or

(c) such acknowledgment to the buyer by a carrier by reshipment or as warehouseman; or

(d) negotiation to the buyer of any negotiable document of title covering the goods.

(3) (a) To stop delivery the seller must so notify as to enable the bailee by reasonable diligence to prevent delivery of the goods.

(b) After such notification the bailee must hold and deliver the goods according to the directions of the seller but the seller is liable to the bailee for any ensuing charges or damages.

(c) If a negotiable document of title has been issued for goods the bailee is not obliged to obey a notification to stop until surrender of the document.

(d) A carrier who has issued a non-negotiable bill of lading is not obliged to obey a notification to stop received from a person other than the consignor.

§ 2–706. Seller's Resale Including Contract for Resale. (1) Under the conditions stated in Section 2–703 on seller's remedies, the seller may resell the goods concerned or the undelivered balance thereof. Where the resale is made in good faith and in a commercially reasonable manner the seller may recover the difference between the resale price and the contract price together with any incidental damages allowed under the provisions of this Article (Section 2–710), but less expenses saved in consequence of the buyer's breach.

(2) Except as otherwise provided in subsection (3) or unless otherwise agreed resale may be at public or private sale including sale by way of one or more contracts to sell or of identification to an existing contract of the seller. Sale may be as a unit or in parcels and at any time and place and on any terms but every aspect of the sale including the method, manner, time,

place and terms must be commercially reasonable. The resale must be reasonably identified as referring to the broken contract, but it is not necessary that the goods be in existence or that any or all of them have been identified to the contract before the breach.

(3) Where the resale is at private sale the seller must give the buyer reasonable notification of his intention to resell.

(4) Where the resale is at public sale

(a) only identified goods can be sold except where there is a recognized market for a public sale of futures in goods of the kind; and

(b) it must be made at a usual place or market for public sale if one is reasonably available and except in the case of goods which are perishable or threaten to decline in value speedily the seller must give the buyer reasonable notice of the time and place of the resale; and

(c) if the goods are not to be within the view of those attending the sale the notification of sale must state the place where the goods are located and provide for their reasonable inspection by prospective bidders; and

(d) the seller must buy.

(5) A purchaser who buys in good faith at a resale takes the goods free of any rights of the original buyer even though the seller fails to comply with one or more of the requirements of this section.

(6) The seller is not accountable to the buyer for any profit made on any resale. A person in the position of a seller (Section 2–707) or a buyer who has rightfully rejected or justifiably revoked acceptance must account for any excess over the amount of his security interest, as hereinafter defined (subsection (3) of Section 2–711).

§ 2–707. "Person in the Position of a Seller." (1) A "person in the position of a seller" includes as against a principal an agent who has paid or become responsible for the price of goods on behalf of his principal or anyone who otherwise holds a security interest or other right in goods similar to that of a seller.

(2) A person in the position of a seller may as provided in this Article withhold or stop delivery (Section 2–705) and resell (Section 2–706) and recover incidental damages (Section 2–710).

§ 2–708. Seller's Damages for Non-acceptance or Repudiation. (1) Subject to subsection (2) and to the provisions of this Article with respect to proof of market price (Section 2–723), the measure of damages for non-acceptance or repudiation by the buyer is the difference between the market price at the time and place for tender and the unpaid contract price together with any incidental damages provided in this Article (Section 2–710), but less expenses saved in consequence of the buyer's breach.

(2) If the measure of damages provided in subsection (1) is inadequate to put the seller in as good a position as performance would have done then the measure of damages is the profit (including reasonable overhead) which the seller would have made from full performance by the buyer, together

with any incidental damages provided in this Article (Section 2–710), due allowance for costs reasonably incurred and due credit for payments or proceeds of resale.

§ 2–709. Action for the Price. (1) When the buyer fails to pay the price as it becomes due the seller may recover, together with any incidental damages under the next section, the price

(a) of goods accepted or of conforming goods lost or damaged within a commercially reasonable time after risk of their loss has passed to the buyer; and

(b) of goods identified to the contract if the seller is unable after reasonable effort to resell them at a reasonable price or the circumstances reasonably indicate that such effort will be unavailing.

(2) Where the seller sues for the price he must hold for the buyer any goods which have been identified to the contract and are still in his control except that if resale becomes possible he may resell them at any time prior to the collection of the judgment. The net proceeds of any such resale must be credited to the buyer and payment of the judgment entitles him to any goods not resold.

(3) After the buyer has wrongfully rejected or revoked acceptance of the goods or has failed to make a payment due or has repudiated (Section 2–610), a seller who is held not entitled to the price under this section shall nevertheless be awarded damages for non-acceptance under the preceding section.

§ 2–710. Seller's Incidental Damages. Incidental damages to an aggrieved seller include any commercially reasonable charges, expenses or commissions incurred in stopping delivery, in the transportation, care and custody of goods after the buyer's breach, in connection with return or resale of the goods or otherwise resulting from the breach.

§ 2–711. Buyer's Remedies in General; Buyer's Security Interest in Rejected Goods. (1) Where the seller fails to make delivery or repudiates or the buyer rightfully rejects or justifiably revokes acceptance then with respect to any goods involved, and with respect to the whole if the breach goes to the whole contract (Section 2–612), the buyer may cancel and whether or not he has done so may in addition to recoverying so much of the price as has been paid

(a) "cover" and have damages under the next section as to all the goods affected whether or not they have been identified to the contract; or

(b) recover damages for non-delivery as provided in this Article (Section 2–713).

(2) Where the seller fails to deliver or repudiates the buyer may also

(a) if the goods have been identified recover them as provided in this Article (Section 2–502); or

(b) in a proper case obtain specific performance or replevy the goods as provided in this Article (Section 2–716).

(3) On rightful rejection or justifiable revocation of acceptance a buyer

has a security interest in goods in his possession or control for any payments made on their price and any expenses reasonably incurred in their inspection, receipt, transportation, care and custody and may hold such goods and resell them in like manner as an aggrieved seller (Section 2–706).

§ 2–712. "Cover"; Buyer's Procurement of Substitute Goods. (1) After a breach within the preceding section the buyer may "cover" by making in good faith and without unreasonable delay any reasonable purchase of or contract to purchase goods in substitution for those due from the seller.

(2) The buyer may recover from the seller as damages the difference between the cost of cover and the contract price together with any incidental or consequential damages as hereinafter defined (Section 2–715), but less expenses saved in consequence of the seller's breach.

(3) Failure of the buyer to effect cover within this section does not bar him from any other remedy.

§ 2–713. Buyer's Damages for Non-Delivery or Repudiation. (1) Subject to the provisions of this Article with respect to proof of market price (Section 2–723); the measure of damages for non-delivery or repudiation by the seller is the difference between the market price at the time when the buyer learned of the breach and the contract price together with any incidental and consequential damages provided in this Article (Section 2–715), but less expenses saved in consequence of the seller's breach.

(2) Market price is to be determined as of the place for tender or, in cases of rejection after arrival or revocation of acceptance, as of the place of arrival.

§ 2–714. Buyer's Damages for Breach in Regard to Accepted Goods. (1) Where the buyer has accepted goods and given notification (subsection (3) of Section 2–607) he may recover as damages for any non-conformity of tender the loss resulting in the ordinary course of events from the seller's breach as determined in any manner which is reasonable.

(2) The measure of damages for breach of warranty is the difference at the time and place of acceptance between the value of the goods accepted and the value they would have had if they had been as warranted, unless special circumstances show proximate damages of a different amount.

(3) In a proper case any incidental and consequential damages under the next section may also be recovered.

§ 2–715. Buyer's Incidental and Consequential Damages. (1) Incidental damages resulting from the seller's breach include expenses reasonably incurred in inspection, receipt, transportation and care and custody of goods rightfully rejected, any commercially reasonable charges, expenses or commissions in connection with effecting cover and any other reasonable expense incident to the delay or other breach.

(2) Consequential damages resulting from the seller's breach include

(a) any loss resulting from general or particular requirements and needs of which the seller at the time of contracting had reason to know and which could not reasonably be prevented by cover or otherwise; and

(b) injury to person or property proximately resulting from any breach of warranty.

§ 2–716. Buyer's Right to Specific Performance or Replevin. (1) Specific performance may be decreed where the goods are unique or in other proper circumstances.

(2) The decree for specific performance may include such terms and conditions as to payment of the price, damages, or other relief as the court may deem just.

(3) The buyer has a right of replevin for goods identified to the contract if after reasonable effort he is unable to effect cover for such goods or the circumstances reasonably indicate that such effort will be unavailing or if the goods have been shipped under reservation and satisfaction of the security interest in them has been made or tendered.

§ 2–717. Deduction of Damages From the Price. The buyer on notifying the seller of his intention to do so may deduct all or any part of the damages resulting from any breach of the contract from any part of the price still due under the same contract.

§ 2–718. Liquidation or Limitation of Damages; Deposits. (1) Damages for breach by either party may be liquidated in the agreement but only at an amount which is reasonable in the light of the anticipated or actual harm caused by the breach, the difficulties of proof of loss, and the inconvenience or nonfeasibility of otherwise obtaining an adequate remedy. A term fixing unreasonably large liquidated damages is void as a penalty.

(2) Where the seller justifiably withholds delivery of goods because of the buyer's breach, the buyer is entitled to restitution of any amount by which the sum of his payments exceeds

(a) the amount to which the seller is entitled by virtue of terms liquidating the seller's damages in accordance with subsection (1), or

(b) in the absence of such terms, twenty per cent of the value of the total performance for which the buyer is obligated under the contract or $500, whichever is smaller.

(3) The buyer's right to restitution under subsection (2) is subject to offset to the extent that the seller establishes

(a) a right to recover damages under the provisions of this Article other than subsection (1), and

(b) the amount or value of any benefits received by the buyer directly or indirectly by reason of the contract.

(4) Where a seller has received payment in goods their reasonable value or the proceeds of their resale shall be treated as payments for the purposes of subsection (2); but if the seller has notice of the buyer's breach before reselling goods received in part performance, his resale is subject to the conditions laid down in this Article on resale by an aggrieved seller (Section 2–706).

§ 2–719. Contractual Modification or Limitation of Remedy. (1) Subject to the provisions of subsections (2) and (3) of this section and of the preceding section on liquidation and limitation of damages,

(a) the agreement may provide for remedies in addition to or in substitution for those provided in this Article and may limit or alter the measure of damages recoverable under this Article, as by limiting the buyer's remedies to return of the goods and repayment of the price or to repair and replacement of non-conforming goods or parts; and

(b) resort to a remedy as provided is optional unless the remedy is expressly agreed to be exclusive, in which case it is the sole remedy.

(2) Where circumstances cause an exclusive or limited remedy to fail of its essential purpose, remedy may be had as provided in this Act.

(3) Consequential damages may be limited or excluded unless the limitation or exclusion is unconscionable. Limitation of consequential damages for injury to the person in the case of consumer goods is prima facie unconscionable but limitation of damages where the loss is commercial is not.

§ 2–720. Effect of "Cancellation" or "Rescission" on Claims for Antecedent Breach. Unless the contrary intention clearly appears, expressions of "cancellation" or "rescission" of the contract or the like shall not be construed as a renunciation or discharge of any claim in damages for an antecedent breach.

§ 2–721. Remedies for Fraud. Remedies for material misrepresentation or fraud include all remedies available under this Article for non-fraudulent breach. Neither rescission or a claim for rescission of the contract for sale nor rejection or return of the goods shall bar or be deemed inconsistent with a claim for damages or other remedy.

§ 2–722. Who Can Sue Third Parties for Injury to Goods. Where a third party so deals with goods which have been identified to a contract for sale as to cause actionable injury to a party to that contract

(a) a right of action against the third party is in either party to the contract for sale who has title to or a security interest or a special property or an insurable interest in the goods; and if the goods have been destroyed or converted a right of action is also in the party who either bore the risk of loss under the contract for sale or has since the injury assumed that risk as against the other;

(b) if at the time of the injury the party plaintiff did not bear the risk of loss as against the other party to the contract for sale and there is no arrangement between them for disposition of the recovery, his suit or settlement is, subject to his own interest, as a fiduciary for the other party to the contract;

(c) either party may with the consent of the other sue for the benefit of whom it may concern.

§ 2–723. Proof of Market Price: Time and Place. (1) If an action based on anticipatory repudiation comes to trial before the time for performance with respect to some or all of the goods, any damages based on market price (Section 2–708 or Section 2–713) shall be determined according to the price of such goods prevailing at the time when the aggrieved party learned of the repudiation.

(2) If evidence of a price prevailing at the times or places described in this Article is not readily available the price prevailing within any reasonable time before or after the time described or at any other place which in commer-

cial judgment or under usage of trade would serve as a reasonable substitute for the one described may be used, making any proper allowance for the cost of transporting the goods to or from such other place.

(3) Evidence of a relevant price prevailing at a time or place other than the one described in this Article offered by one party is not admissible unless and until he has given the other party such notice as the court finds sufficient to prevent unfair surprise.

§ 2–724. **Admissibility of Market Quotations.** Whenever the prevailing price or value of any goods regularly bought and sold in any established commodity market is in issue, reports in official publications or trade journals or in newspapers or periodicals of general circulation published as the reports of such market shall be admissible in evidence. The circumstances of the preparation of such a report may be shown to affect its weight but not its admissibility.

§ 2–725. **Statute of Limitations in Contracts for Sale.** (1) An action for breach of any contract for sale must be commenced within four years after the cause of action has accrued. By the original agreement the parties may reduce the period of limitation to not less than one year but may not extend it.

(2) A cause of action accrues when the breach occurs, regardless of the aggrieved party's lack of knowledge of the breach. A breach of warranty occurs when tender of delivery is made, except that where a warranty explicitly extends to future performance of the goods and discovery of the breach must await the time of such performance the cause of action accrues when the breach is or should have been discovered.

(3) Where an action commenced within the time limited by subsection (1) is so terminated as to leave available a remedy by another action for the same breach such other action may be commenced after the expiration of the time limited and within six months after the termination of the first action unless the termination resulted from voluntary discontinuance or from dismissal for failure or neglect to prosecute.

(4) This section does not alter the law on tolling of the statute of limitations nor does it apply to causes of action which have accrued before this Act becomes effective.

ARTICLE 3: COMMERCIAL PAPER

Part 1

Short-Title, Form and Interpretation

§ 3–101. **Short Title.** This Article shall be known and may be cited as Uniform Commercial Code—Commercial Paper.

§ 3–102. **Definitions and Index of Definitions.** (1) In this Article unless the context otherwise requires

(a) "Issue" means the first delivery of an instrument to a holder or a remitter.

(b) An "order" is a direction to pay and must be more than an authorization or request. It must identify the person to pay with reasonable certainty. It may be addressed to one or more such persons jointly or in the alternative but not in succession.

(c) A "promise" is an undertaking to pay and must be more than an acknowledgment of an obligation.

(d) "Secondary party" means a drawer or indorser.

(e) "Instrument" means a negotiable instrument.

(2) Other definitions applying to this Article and the sections in which they appear are:

"Acceptance." Section 3–410.
"Accommodation party." Section 3–415.
"Alteration." Section 3–407.
"Certificate of deposit." Section 3–104.
"Certification." Section 3–411.
"Check." Section 3–104.
"Definite time." Section 3–109.
"Dishonor." Section 3–507.
"Draft." Section 3–104.
"Holder in due course." Section 3–302.
"Negotiation." Section 3–202.
"Note." Section 3–104.
"Notice of dishonor." Section 3–508.
"On demand." Section 3–108.
"Presentment." Section 3–504.
"Protest." Section 3–509.
"Restrictive Indorsement." Section 3–205.
"Signature." Section 3–401.

(3) The following definitions in other Articles apply to this Article:

"Account." Section 4–104.
"Banking Day." Section 4–104.
"Clearing house." Section 4–104.
"Collecting bank." Section 4–105.
"Customer." Section 4–104.
"Depositary Bank." Section 4–105.
"Documentary Draft." Section 4–104.
"Intermediary Bank." Section 4–105.
"Item." Section 4–104.
"Midnight deadline." Section 4–104.
"Payor bank." Section 4–105.

(4) In addition Article 1 contains general definitions and principles of construction and interpretation applicable throughout this Article.

§ 3–103. Limitations on Scope of Article. (1) This Article does not apply to money, documents of title or investment securities.

(2) The provisions of this Article are subject to the provisions of the

Article on Bank Deposits and Collections (Article 4) and Secured Transactions (Article 9).

§ 3–104. Form of Negotiable Instruments; "Draft"; "Check"; "Certificate of Deposit"; "Note." (1) Any writing to be a negotiable instrument within this Article must

(a) be signed by the maker or drawer; and

(b) contain an unconditional promise or order to pay a sum certain in money and no other promise, order, obligation or power given by the maker or drawer except as authorized by this Article; and

(c) be payable on demand or at a definite time; and

(d) be payable to order or to bearer.

(2) A writing which complies with the requirements of this section is

(a) a "draft" ("bill of exchange") if it is an order;

(b) a "check" if it is a draft drawn on a bank and payable on demand;

(c) a "certificate of deposit" if it is an acknowledgment by a bank of receipt of money with an engagement to repay it;

(d) a "note" if it is a promise other than a certificate of deposit.

(3) As used in other Articles of this Act, and as the context may require, the terms "draft," "check," "certificate of deposit" and "note" may refer to instruments which are not negotiable within this Article as well as to instruments which are so negotiable.

§ 3–105. When Promise or Order Unconditional. (1) A promise or order otherwise unconditional is not made conditional by the fact that the instrument

(a) is subject to implied or constructive conditions; or

(b) states its consideration, whether performed or promised, or the transaction which gave rise to the instrument, or that the promise or order is made or the instrument matures in accordance with or "as per" such transaction; or

(c) refers to or states that it arises out of a separate agreement or refers to a separate agreement for rights as to prepayment or acceleration; or

(d) states that it is drawn under a letter of credit; or

(e) states that it is secured, whether by mortgage, reservation of title or otherwise; or

(f) indicates a particular account to be debited or any other fund or source from which reimbursement is expected; or

(g) is limited to payment out of a particular fund or the proceeds of a particular source, if the instrument is issued by a government or governmental agency or unit; or

(h) is limited to payment out of the entire assets of a partnership, unincorporated association, trust or estate by or on behalf of which the instrument is issued.

(2) A promise or order is not unconditional if the instrument

(a) states that it is subject to or governed by any other agreement; or

(b) states that it is to be paid only out of a particular fund or source except as provided in this section.

§ 3–106. Sum Certain. (1) The sum payable is a sum certain even though it is to be paid

(a) with stated interest or by stated installments; or

(b) with stated different rates of interest before and after default or a specified date; or

(c) with a stated discount or addition if paid before or after the date fixed for payment; or

(d) with exchange or less exchange, whether at a fixed rate or at the current rate; or

(e) with costs of collection or an attorney's fee or both upon default.

(2) Nothing in this section shall validate any term which is otherwise illegal.

§ 3–107. Money. (1) An instrument is payable in money if the medium of exchange in which it is payable is money at the time the instrument is made. An instrument payable in "currency" or "current funds" is payable in money.

(2) A promise or order to pay a sum stated in a foreign currency is for a sum certain in money and, unless a different medium of payment is specified in the instrument, may be satisfied by payment of that number of dollars which the stated foreign currency will purchase at the buying sight rate for that currency on the day on which the instrument is payable or, if payable on demand, on the day of demand. If such an instrument specifies a foreign currency as the medium of payment the instrument is payable in that currency.

§ 3–108. Payable on Demand. Instruments payable on demand include those payable at sight or on presentation and those in which no time for payment is stated.

§ 3–109. Definite Time. (1) An instrument is payable at a definite time if by its terms it is payable

(a) on or before a stated date or at a fixed period after a stated date; or

(b) at a fixed period after sight; or

(c) at a definite time subject to any acceleration; or

(d) at a definite time subject to extension at the option of the holder, or to extension to a further definite time at the option of the maker or acceptor or automatically upon or after a specified act or event.

(2) An instrument which by its terms is otherwise payable only upon an act or event uncertain as to time of occurrence is not payable at a definite time even though the act or event has occurred.

§ 3–110. Payable to Order. (1) An instrument is payable to order when by its terms it is payable to the order or assigns of any person therein specified with reasonable certainty, or to him or his order, or when it is conspicuously designated on its face as "exchange" or the like and names a payee. It may be payable to the order of

(a) the maker or drawer; or

(b) the drawee; or

(c) a payee who is not maker, drawer or drawee; or

(d) two or more payees together or in the alternative; or

(e) an estate, trust or fund, in which case it is payable to the order of the representative of such estate, trust or fund or his successors; or

(f) an office, or an officer by his title as such in which case it is payable to the principal but the incumbent of the office or his successors may act as if he or they were the holder; or

(g) a partnership or unincorporated association, in which case it is payable to the partnership or association and may be indorsed or transferred by any person thereto authorized.

(2) An instrument not payable to order is not made so payable by such words as "payable upon return of this instrument properly indorsed."

(3) An instrument made payable both to order and to bearer is payable to order unless the bearer words are handwritten or typewritten.

§ 3–111. Payable to Bearer. An instrument is payable to bearer when by its terms it is payable to

(a) bearer or the order of bearer; or

(b) a specified person or bearer; or

(c) "cash" or the order of "cash," or any other indication which does not purport to designate a specific payee.

§ 3–112. Terms and Omissions Not Affecting Negotiability. (1) The negotiability of an instrument is not affected by

(a) the omission of a statement of any consideration or of the place where the instrument is drawn or payable; or

(b) a statement that collateral has been given to secure obligations either on the instrument or otherwise of an obligor on the instrument or that in case of default on those obligations the holder may realize on or dispose of the collateral; or

(c) a promise or power to maintain or protect collateral or to give additional collateral; or

(d) a term authorizing a confession of judgment on the instrument if it is not paid when due; or

(e) a term purporting to waive the benefit of any law intended for the advantage or protection of any obligor; or

(f) a term in a draft providing that the payee by indorsing or cashing it acknowledges full satisfaction of an obligation of the drawer; or

(g) a statement in a draft drawn in a set of parts (Section 3–801) to the effect that the order is effective only if no other part has been honored.

(2) Nothing in this section shall validate any term which is otherwise illegal.

§ 3–113. Seal. An instrument otherwise negotiable is within this Article even though it is under a seal.

§ 3–114. Date, Antedating, Postdating. (1) The negotiability of an instrument is not affected by the fact that it is undated, antedated or postdated.

(2) Where an instrument is antedated or postdated the time when it

is payable is determined by the stated date if the instrument is payable on demand or at a fixed period after date.

(3) Where the instrument or any signature thereon is dated, the date is presumed to be correct.

§ 3–115. Incomplete Instruments. (1) When a paper whose contents at the time of signing show that it is intended to become an instrument is signed while still incomplete in any necessary respect it cannot be enforced until completed, but when it is completed in accordance with authority given it is effective as completed.

(2) If the completion is unauthorized the rules as to material alteration apply (Section 3–407), even though the paper was not delivered by the maker or drawer; but the burden of establishing that any completion is unauthorized is on the party so asserting.

§ 3–116. Instruments Payable to Two or More Persons. An instrument payable to the order of two or more persons

(a) if in the alternative is payable to any one of them and may be negotiated, discharged or enforced by any of them who has possession of it;

(b) if not in the alternative is payable to all of them and may be negotiated, discharged or enforced only by all of them.

§ 3–117. Instruments Payable With Words of Description. An instrument made payable to a named person with the addition of words describing him

(a) as agent or officer of a specified person is payable to his principal but the agent or officer may act as if he were the holder;

(b) as any other fiduciary for a specified person or purpose is payable to the payee and may be negotiated, discharged or enforced by him;

(c) in any other manner is payable to the payee unconditionally and the additional words are without effect on subsequent parties.

§ 3–118. Ambiguous Terms and Rules of Construction. The following rules apply to every instrument:

(a) Where there is doubt whether the instrument is a draft or a note the holder may treat it as either. A draft drawn on the drawer is effective as a note.

(b) Handwritten terms control typewritten and printed terms, and typewritten control printed.

(c) Words control figures except that if the words are ambiguous figures control.

(d) Unless otherwise specified a provision for interest means interest at the judgment rate at the place of payment from the date of the instrument, or if it is undated from the date of issue.

(e) Unless the instrument otherwise specifies two or more persons who sign as maker, acceptor or drawer or indorser and as a part of the same transaction are jointly and severally liable even though the instrument contains such words as "I promise to pay."

(f) Unless otherwise specified consent to extension authorizes a single extension for not longer than the original period. A consent to extension, expressed in the instrument, is binding on secondary parties and accommodation makers. A holder may not exercise his option to extend an instrument over the objection of a maker or acceptor or other party who in accordance with Section 3–604 tenders full payment when the instrument is due.

§ 3–119. Other Writings Affecting Instrument. (1) As between the obligor and his immediate obligee or any transferee the terms of an instrument may be modified or affected by any other written agreement executed as a part of the same transaction, except that a holder in due course is not affected by any limitation of his rights arising out of the separate written agreement if he had no notice of the limitation when he took the instrument.

(2) A separate agreement does not affect the negotiability of an instrument.

§ 3–120. Instruments "Payable Through" Bank. An instrument which states that it is "payable through" a bank or the like designates that bank as a collecting bank to make presentment but does not of itself authorize the bank to pay the instrument.

§ 3–121. Instruments Payable at Bank. Note: *If this Act is introduced in the Congress of the United States this section should be omitted.*
(States to select either alternative)

Alternative A—

A note or acceptance which states that it is payable at a bank is the equivalent of a draft drawn on the bank payable when it falls due out of any funds of the maker or acceptor in current account or otherwise available for such payment.

Alternative B—

A note or acceptance which states that it is payable at a bank is not of itself an order or authorization to the bank to pay it.

§ 3–122. Accrual of Cause of Action. (1) A cause of action against a maker or an acceptor accrues

(a) in the case of a time instrument on the day after maturity;

(b) in the case of a demand instrument upon its date or, if no date is stated, on the date of issue.

(2) A cause of action against the obligor of a demand or time certificate of deposit accrues upon demand, but demand on a time certificate may not be made until on or after the date of maturity.

(3) A cause of action against a drawer of a draft of any indorser of an instrument accrues upon demand following dishonor of the instrument. Notice of dishonor is a demand.

(4) Unless an instrument provides otherwise, interest runs at the rate provided by law for a judgment.

(a) in the case of a maker, acceptor or other primary obligor of a demand instrument, from the date of demand;

(b) in all other cases from the date of accrual of the cause of action.

Part 2

Transfer and Negotiation

§ 3–201. Transfer: Right to Indorsement. (1) Transfer of an instrument vests in the transferee such rights as the transferor has therein, except that a transferee who has himself been a party to any fraud or illegality affecting the instrument or who as a prior holder had notice of a defense or claim against it cannot improve his position by taking from a later holder in due course.

(2) A transfer of a security interest in an instrument vests the foregoing rights in the transferee to the extent of the interest transferred.

(3) Unless otherwise agreed any transfer for value of an instrument not then payable to bearer gives the transferee the specifically enforceable right to have the unqualified indorsement of the transferor. Negotiation takes effect only when the indorsement is made and until that time there is no presumption that the transferee is the owner.

§ 3–202. Negotiation. (1) Negotiation is the transfer of an instrument in such form that the transferee becomes a holder. If the instrument is payable to order it is negotiated by delivery with any necessary indorsement; if payable to bearer it is negotiated by delivery.

(2) An indorsement must be written by or on behalf of the holder and on the instrument or on a paper so firmly affixed thereto as to become a part thereof.

(3) An indorsement is effective for negotiation only when it conveys the entire instrument or any unpaid residue. If it purports to be of less it operates only as a partial assignment.

(4) Words of assignment, condition, waiver, guaranty, limitation or disclaimer of liability and the like accompanying an indorsement do not affect its character as an indorsement.

§ 3–203. Wrong or Misspelled Name. Where an instrument is made payable to a person under a misspelled name or one other than his own he may indorse in that name or his own or both; but signature in both names may be required by a person paying or giving value for the instrument.

§ 3–204. Special Indorsement; Blank Indorsement. (1) A special indorsement specifies the person to whom or to whose order it makes the instrument payable. Any instrument specially indorsed becomes payable to the order of the special indorsee and may be further negotiated only by his indorsement.

(2) An indorsement in blank specifies no particular indorsee and may consist of a mere signature. An instrument payable to order and indorsed in blank becomes payable to bearer and may be negotiated by delivery alone until specially indorsed.

(3) The holder may convert a blank indorsement into a special indorsement by writing over the signature of the indorser in blank any contract consistent with the character of the indorsement.

§ 3–205. Restrictive Indorsements. An indorsement is restrictive which either

(a) is conditional; or

(b) purports to prohibit further transfer of the instrument; or

(c) includes the words "for collection," "for deposit," "pay any bank," or like terms signifying a purpose of deposit or collection; or

(d) otherwise states that it is for the benefit or use of the indorser or of another person.

§ 3–206. Effect of Restrictive Indorsement. (1) No restrictive indorsement prevents further transfer or negotiation of the instrument.

(2) An intermediary bank, or a payor bank which is not the depositary bank, is neither given notice nor otherwise affected by a restrictive indorsement of any person except the bank's immediate transferor or the person presenting for payment.

(3) Except for an intermediary bank, any transferee under an indorsement which is conditional or includes the words "for collection," "for deposit," "pay any bank," or like terms (subparagraphs (a) and (c) of Section 3–205) must pay or apply any value given by him for or on the security of the instrument consistently with the indorsement and to the extent that he does so he becomes a holder for value. In addition such transferee is a holder in due course if he otherwise complies with the requirements of Section 3–302 on what constitutes a holder in due course.

(4) The first taker under an indorsement for the benefit of the indorser or another person (subparagraph (d) of Section 3–205) must pay or apply any value given by him for or on the security of the instrument consistently with the indorsement and to the extent that he does so he becomes a holder for value. In addition such taker is a holder in due course if he otherwise complies with the requirements of Section 3–302 on what constitutes a holder in due course. A later holder for value is neither given notice nor otherwise affected by such restrictive indorsement unless he has knowledge that a fiduciary or other person has negotiated the instrument in any transaction for his own benefit or otherwise in breach of duty (subsection (2) of Section 3–304).

§ 3–207. Negotiation Effective Although It May Be Rescinded. (1) Negotiation is effective to transfer the instrument although the negotiation is

(a) made by an infant, a corporation exceeding its powers, or any other person without capacity; or

(b) obtained by fraud, duress or mistake of any kind; or

(c) part of an illegal transaction; or

(d) made in breach of duty.

(2) Except as against a subsequent holder in due course such negotiation is in an appropriate case subject to rescission, the declaration of a constructive trust or any other remedy permitted by law.

§ 3–208. Reacquisition. Where an instrument is returned to or reacquired by a prior party he may cancel any indorsement which is not necessary to his title and reissue or further negotiate the instrument, but any intervening party is discharged as against the reacquiring party and subsequent holders not in due course and if his indorsement has been cancelled is discharged as against subsequent holders in due course as well.

Part 3

Rights of a Holder

§ 3–301. Rights of a Holder. The holder of an instrument whether or not he is the owner may transfer or negotiate it and, except as otherwise provided in Section 3–603 on payment or satisfaction, discharge it or enforce payment in his own name.

§ 3–302. Holder in Due Course. (1) A holder in due course is a holder who takes the instrument

(a) for value; and

(b) in good faith; and

(c) without notice that it is overdue or has been dishonored or of any defense against or claim to it on the part of any person.

(2) A payee may be a holder in due course.

(3) A holder does not become a holder in due course of an instrument:

(a) by purchase of it at judicial sale or by taking it under legal process; or

(b) by acquiring it in taking over an estate; or

(c) by purchasing it as part of a bulk transaction not in regular course of business of the transferor.

(4) A purchaser of a limited interest can be a holder in due course only to the extent of the interest purchased.

§ 3–303. Taking for Value. A holder takes the instrument for value

(a) to the extent that the agreed consideration has been performed or that he acquires a security interest in or a lien on the instrument otherwise than by legal process; or

(b) when he takes the instrument in payment of or as security for an antecedent claim against any person whether or not the claim is due; or

(c) when he gives a negotiable instrument for it or makes an irrevocable commitment to a third person.

§ 3–304. Notice to Purchaser. (1) The purchaser has notice of a claim or defense if

(a) the instrument is so incomplete, bears such visible evidence of forgery or alteration, or is otherwise so irregular as to call into question its validity, terms or ownership or to create an ambiguity as to the party to pay; or

(b) the purchaser has notice that the obligation of any party is voidable in whole or in part, or that all parties have been discharged.

(2) The purchaser has notice of a claim against the instrument when he has knowledge that a fiduciary has negotiated the instrument in payment of or as security for his own debt or in any transaction for his own benefit or otherwise in breach of duty.

(3) The purchaser has notice that an instrument is overdue if he has reason to know

(a) that any part of the principal amount is overdue or that there is an uncured default in payment of another instrument of the same series; or

(b) that acceleration of the instrument has been made; or

(c) that he is taking a demand instrument after demand has been made or more than a reasonable length of time after its issue. A reasonable time for a check drawn and payable within the states and territories of the United States and the District of Columbia is presumed to be thirty days.

(4) Knowledge of the following facts does not of itself give the purchaser notice of a defense or claim

(a) that the instrument is antedated or postdated;

(b) that it was issued or negotiated in return for an executory promise or accompanied by a separate agreement, unless the purchaser has notice that a defense or claim has arisen from the terms thereof;

(c) that any party has signed for accommodation;

(d) that an incomplete instrument has been completed, unless the purchaser has notice of any improper completion;

(e) that any person negotiating the instrument is or was a fiduciary;

(f) that there has been default in payment of interest on the instrument or in payment of any other instrument, except one of the same series.

(5) The filing or recording of a document does not of itself constitute notice within the provisions of this Article to a person who would otherwise be a holder in due course.

(6) To be effective notice must be received at such time and in such manner as to give a reasonable opportunity to act on it.

§ 3–305. Rights of a Holder in Due Course. To the extent that a holder is a holder in due course he takes the instrument free from

(1) all claims to it on the part of any person; and

(2) all defenses of any party to the instrument with whom the holder has not dealt except

(a) infancy, to the extent that it is a defense to a simple contract; and

(b) such other incapacity, or duress, or illegality of the transaction, as renders the obligation of the party a nullity; and

(c) such misrepresentation as has induced the party to sign the instrument with neither knowledge nor reasonable opportunity to obtain knowledge of its character or its essential terms; and

(d) discharge in insolvency proceedings; and

(e) any other discharge of which the holder has notice when he takes the instrument.

§ 3–306. Rights of One Not Holder in Due Course. Unless he has the rights of a holder in due course any person takes the instrument subject to

(a) all valid claims to it on the part of any person; and

(b) all defenses of any party which would be available in an action on a simple contract; and

(c) the defenses of want or failure of consideration, nonperformance of any condition precedent, non-delivery, or delivery for a special purpose (Section 3–408); and

(d) the defense that he or a person through whom he holds the instrument acquired it by theft, or that payment or satisfaction to such holder would be inconsistent with the terms of a restrictive indorsement. The claim of any third person to the instrument is not otherwise available as a defense to any party liable thereon unless the third person himself defends the action for such party.

§ 3–307. Burden of Establishing Signatures, Defenses and Due Course. (1) Unless specifically denied in the pleadings each signature on an instrument is admitted. When the effectiveness of a signature is put in issue

(a) the burden of establishing it is on the party claiming under the signature; but

(b) the signature is presumed to be genuine or authorized except where the action is to enforce the obligation of a purported signer who has died or become incompetent before proof is required.

(2) When signatures are admitted or established, production of the instrument entitles a holder to recover on it unless the defendant establishes a defense.

(3) After it is shown that a defense exists a person claiming the rights of a holder in due course has the burden of establishing that he or some other person under whom he claims is in all respects a holder in due course.

Part 4

Liability of Parties

§ 3–401. Signature. (1) No person is liable on an instrument unless his signature appears thereon.

(2) A signature is made by use of any name, including any trade or assumed name, upon an instrument, or by any word or mark used in lieu of a written signature.

§ 3–402. Signature in Ambiguous Capacity. Unless the instrument clearly indicates that a signature made in some other capacity it is an indorsement.

§ 3–403. Signature by Authorized Representative. (1) A signature may be made by an agent or other representative, and his authority to make it may be established as in other cases of representation. No particular form of appointment is necessary to establish such authority.

(2) An authorized representative who signs his own name to an instrument

(a) is personally obligated if the instrument neither names the person represented nor shows that the representative signed in a representative capacity;

(b) except as otherwise established between the immediate parties, is personally obligated if the instrument names the person represented but does not show that the representative signed in a representative capacity, or if the instrument does not name the person represented but does show that the representative signed in a representative capacity.

(3) Except as otherwise established the name of an organization pre-

ceded or followed by the name and office of an authorized individual is a signature made in a representative capacity.

§ 3–404. Unauthorized Signatures. (1) Any unauthorized signature is wholly inoperative as that of the person whose name is signed unless he ratifies it or is precluded from denying it; but it operates as the signature of the unauthorized signer in favor of any person who in good faith pays the instrument or takes it for value.

(2) Any unauthorized signature may be ratified for all purposes of this Article. Such ratification does not of itself affect any rights of the person ratifying against the actual signer.

§ 3–405. Impostors; Signature in Name of Payee. (1) An indorsement by any person in the name of a named payee is effective if

(a) an impostor by use of the mails or otherwise has induced the maker or drawer to issue the instrument to him or his confederate in the name of the payee; or

(b) a person signing as or on behalf of a maker or drawer intends the payee to have no interest in the instrument; or

(c) an agent or employee of the maker or drawer has supplied him with the name of the payee intending the latter to have no such interest.

(2) Nothing in this section shall affect the criminal or civil liability of the person so indorsing.

§ 3–406. Negligence Contributing to Alteration or Unauthorized Signature. Any person who by his negligence substantially contributes to a material alteration of the instrument or to the making of an unauthorized signature is precluded from asserting the alteration or lack of authority against a holder in due course or against a drawee or other payor who pays the instrument in good faith and in accordance with the reasonable commercial standards of the drawee's or payor's business.

§ 3–407. Alteration. (1) Any alteration of an instrument is material which changes the contract of any party thereto in any respect, including any such change in

(a) the number or relations of the parties; or

(b) an incomplete instrument, by completing it otherwise than as authorized; or

(c) the writing as signed, by adding to it or by removing any part of it.

(2) As against any person other than a subsequent holder in due course

(a) alteration by the holder which is both fraudulent and material discharges any party whose contract is thereby changed unless that party assents or is precluded from asserting the defense;

(b) no other alteration discharges any party and the instrument may be enforced according to its original tenor, or as to incomplete instruments according to the authority given.

(3) A subsequent holder in due course may in all cases enforce the

instrument according to its original tenor, and when an incomplete instrument has been completed, he may enforce it as completed.

§ 3–408. Consideration. Want or failure of consideration is a defense as against any person not having the rights of a holder in due course (Section 3–305), except that no consideration is necessary for an instrument or obligation thereon given in payment of or as security for an antecedent obligation of any kind. Nothing in this section shall be taken to displace any statute outside this Act under which a promise is enforceable notwithstanding lack or failure of consideration. Partial failure of consideration is a defense pro tanto whether or not the failure is in an ascertained or liquidated amount.

§ 3–409. Draft Not an Assignment. (1) A check or other draft does not of itself operate as an assignment of any funds in the hands of the drawee available for its payment, and the drawee is not liable on the instrument until he accepts it.

(2) Nothing in this section shall affect any liability in contract, tort or otherwise arising from any letter of credit or other obligation or representation which is not an acceptance.

§ 3–410. Definition and Operation of Acceptance. (1) Acceptance is the drawee's signed engagement to honor the draft as presented. It must be written on the draft, and may consist of his signature alone. It becomes operative when completed by delivery or notification.

(2) A draft may be accepted although it has not been signed by the drawer or is otherwise incomplete or is overdue or has been dishonored.

(3) Where the draft is payable at a fixed period after sight and the acceptor fails to date his acceptance the holder may complete it by supplying a date in good faith.

§ 3–411. Certification of a Check. (1) Certification of a check is acceptance. Where a holder procures certification the drawer and all prior indorsers are discharged.

(2) Unless otherwise agreed a bank has no obligation to certify a check.

(3) A bank may certify a check before returning it for lack of proper indorsement. If it does so the drawer is discharged.

§ 3–412. Acceptance Varying Draft. (1) Where the drawee's proferred acceptance in any manner varies the draft as presented the holder may refuse the acceptance and treat the draft as dishonored in which case the drawee is entitled to have his acceptance cancelled.

(2) The terms of the draft are not varied by an acceptance to pay at any particular bank or place in the United States, unless the acceptance states that the draft is to be paid only at such bank or place.

(3) Where the holder assents to an acceptance varying the terms of the draft each drawer and indorser who does not affirmatively assent is discharged.

§ 3–413. Contract of Maker, Drawer and Acceptor. (1) The maker or acceptor engages that he will pay the instrument according to its tenor at the time of his engagement or as completed pursuant to Section 3–115 or incomplete instruments.

(2) The drawer engages that upon dishonor of the draft and any necessary notice of dishonor or protest he will pay the amount of the draft to the holder or to any indorser who takes it up. The drawer may disclaim this liability by drawing without recourse.

(3) By making, drawing or accepting the party admits as against all subsequent parties including the drawee the existence of the payee and his then capacity to indorse.

§ 3–414. Contract of Indorser; Order of Liability. (1) Unless the indorsement otherwise specifies (as by such words as "without recourse") every indorser engages that upon dishonor and any necessary notice of dishonor and protest he will pay the instrument according to its tenor at the time of his indorsement to the holder or to any subsequent indorser who takes it up, even though the indorser who takes it up was not obligated to do so.

(2) Unless they otherwise agree indorsers are liable to one another in the order in which they indorse, which is presumed to be the order in which their signatures appear on the instrument.

§ 3–415. Contract of Accommodation Party. (1) An accommodation party is one who signs the instrument in any capacity for the purpose of lending his name to another party to it.

(2) When the instrument has been taken for value before it is due the accommodation party is liable in the capacity in which he has signed even though the taker knows of the accommodation.

(3) As against a holder in due course and without notice of the accommodation oral proof of the accommodation is not admissible to give the accommodation party the benefit of discharges dependent on his character as such. In other cases the accommodation character may be shown by oral proof.

(4) An indorsement which shows that it is not in the chain of title is notice of its accommodation character.

(5) An accommodation party is not liable to the party accommodated, and if he pays the instrument has a right of recourse on the instrument against such party.

§ 3–416. Contract of Guarantor. (1) "Payment guaranteed" or equivalent words added to a signature mean that the signer engages that if the instrument is not paid when due he will pay it according to its tenor without resort by the holder to any other party.

(2) "Collection guaranteed" or equivalent words added to a signature mean that the signer engages that if the instrument is not paid when due he will pay it according to its tenor, but only after the holder has reduced his claim against the maker or acceptor to judgment and execution has been returned unsatisfied, or after the maker or acceptor has become insolvent or it is otherwise apparent that it is useless to proceed against him.

(3) Words of guaranty which do not otherwise specify guarantee payment.

(4) No words of guaranty added to the signature of a sole maker or acceptor affect his liability on the instrument. Such words added to the signature of one of two or more makers or acceptors create a presumption that the signature is for the accommodation of the others.

(5) When words of guaranty are used presentment, notice of dishonor and protest are not necessary to charge the user.

(6) Any guaranty written on the instrument is enforceable notwithstanding any statute of frauds.

§ 3–417. Warranties on Presentment and Transfer. (1) Any person who obtains payment or acceptance and any prior transferor warrants to a person who in good faith pays or accepts that

(a) he has a good title to the instrument or is authorized to obtain payment or acceptance on behalf of one who has a good title; and

(b) he has no knowledge that the signature of the maker or drawer is unauthorized, except that this warranty is not given by a holder in due course acting in good faith

(i) to a maker with respect to the maker's own signature; or

(ii) to a drawer with respect to the drawer's own signature, whether or not the drawer is also the drawee; or

(iii) to an acceptor of a draft if the holder in due course took the draft after the acceptance or obtained the acceptance without knowledge that the drawer's signature was unauthorized; and

(c) the instrument has not been materially altered, except that this warranty is not given by a holder in due course acting in good faith

(i) to the maker of a note; or

(ii) to the drawer of a draft whether or not the drawer is also the drawee; or

(iii) to the acceptor of a draft with respect to an alteration made prior to the acceptance if the holder in due course took the draft after the acceptance, even though the acceptance provided "payable as originally drawn" or equivalent terms; or

(iv) to the acceptor of a draft with respect to an alteration made after the acceptance.

(2) Any person who transfers an instrument and receives consideration warrants to his transferee and if the transfer is by indorsement to any subsequent holder who takes the instrument in good faith that

(a) he has a good title to the instrument or is authorized to obtain payment or acceptance on behalf of one who has a good title and the transfer is otherwise rightful; and

(b) all signatures are genuine or authorized; and

(c) the instrument has not been materially altered; and

(d) no defense of any party is good against him; and

(e) he has no knowledge of any insolvency proceeding instituted with respect to the maker or acceptor or the drawer of an unaccepted instrument.

(3) By transferring "without recourse" the transferor limits the obliga-

tion stated in subsection (2) (d) to a warranty that he has no knowledge of such a defense.

(4) A selling agent or broker who does not disclose the fact that he is acting only as such gives the warranties provided in this section, but if he makes such disclosure warrants only his good faith and authority.

§ 3–418. Finality of Payment or Acceptance. Except for recovery of bank payments as provided in the Article on Bank Deposits and Collections (Article 4) and except for liability for breach of warranty on presentment under the preceding section, payment or acceptance of any instrument is final in favor of a holder in due course, or a person who has in good faith changed his position in reliance on the payment.

§ 3–419. Conversion of Instrument; Innocent Representative. (1) An instrument is converted when

(a) a drawee to whom it is delivered for acceptance refuses to return it on demand; or

(b) any person to whom it is delivered for payment refuses on demand either to pay or to return it; or

(c) it is paid on a forged indorsement.

(2) In an action against a drawee under subsection (1) the measure of the drawee's liability is the face amount of the instrument. In any other action under subsection (1) the measure of liability is presumed to be the face amount of the instrument.

(3) Subject to the provisions of this Act concerning restrictive indorsements a representative, including a depositary or collecting bank, who has in good faith and in accordance with the reasonable commercial standards applicable to the business of such representative dealt with an instrument or its proceeds on behalf of one who was not the true owner is not liable in conversion or otherwise to the true owner beyond the amount of any proceeds remaining in his hands.

(4) An intermediary bank or payor bank which is not a depositary bank is not liable in conversion solely by reason of the fact that proceeds of an item indorsed restrictively (Sections 3–205 and 3–206) are not paid or applied consistently with the restrictive indorsement of an indorser other than its immediate transferor.

Part 5

Presentment, Notice of Dishonor and Protest

§ 3–501. When Presentment, Notice of Dishonor, and Protest Necessary or Permissible. (1) Unless excused (Section 3–511) presentment is necessary to charge secondary parties as follows:

(a) presentment for acceptance is necessary to charge the drawer and indorsers of a draft where the draft so provides, or is payable elsewhere than at the residence or place of business of the drawee, or its date of payment depends upon such presentment. The holder may at his option present for acceptance any other draft payable at a stated date;

(b) presentment for payment is necessary to charge any indorser;

(c) in the case of any drawer, the acceptor of a draft payable at a bank or the maker of a note payable at a bank, presentment for payment is necessary, but failure to make presentment discharges such drawer, acceptor or maker only as stated in Section 3–502(1) (b).

(2) Unless excused (Section 3–511)

(a) notice of any dishonor is necessary to charge any indorser;

(b) in the case of any drawer, the acceptor of a draft payable at a bank or the maker of a note payable at a bank, notice of any dishonor is necessary, but failure to give such notice discharges such drawer, acceptor or maker only as stated in Section 3–502(1) (b).

(3) Unless excused (Section 3–511) protest of any dishonor is necessary to charge the drawer and indorsers of any draft which on its face appears to be drawn or payable outside of the states, territories, dependencies, and possessions of the United States, the District of Columbia and the Commonwealth of Puerto Rico. The holder may at his option make protest of any dishonor of any other instrument and in the case of a foreign draft may on insolvency of the acceptor before maturity make protest for better security.

(4) Notwithstanding any provision of this section, neither presentment nor notice of dishonor nor protest is necessary to charge an indorser who has indorsed an instrument after maturity.

§ 3–502. Unexcused Delay; Discharge. (1) Where without excuse any necessary presentment or notice of dishonor is delayed beyond the time when it is due

(a) any indorser is discharged; and

(b) any drawer or the acceptor of a draft payable at a bank or the maker of a note payable at a bank who because the drawee or payor bank becomes insolvent during the delay is deprived of funds maintained with the drawee or payor bank to cover the instrument may discharge his liability by written assignment to the holder of his rights against the drawee or payor bank in respect of such funds, but such drawer, acceptor or maker is not otherwise discharged.

(2) Where without excuse a necessary protest is delayed beyond the time when it is due any drawer or indorser is discharged.

§ 3–503. Time of Presentment. (1) Unless a different time is expressed in the instrument the time for any presentment is determined as follows:

(a) where an instrument is payable at or a fixed period after a stated date any presentment for acceptance must be made on or before the date it is payable;

(b) where an instrument is payable after sight it must either be presented for acceptance or negotiated within a reasonable time after date or issue whichever is later;

(c) where an instrument shows the date on which it is payable presentment for payment is due on that date;

(d) where an instrument is accelerated presentment for payment is due within a reasonable time after the acceleration;

(e) with respect to the liability of any secondary party presentment for acceptance or payment of any other instrument is due within a reasonable time after such party becomes liable thereon.

(2) A reasonable time for presentment is determined by the nature of the instrument, any usage of banking or trade and the facts of the particular case. In the case of an uncertified check which is drawn and payable within the United States and which is not a draft drawn by a bank the following are presumed to be reasonable periods within which to present for payment or to initiate bank collection:

(a) with respect to the liability of the drawer, thirty days after date or issue whichever is later; and

(b) with respect to the liability of an indorser, seven days after his indorsement.

(3) Where any presentment is due on a day which is not a full business day for either the person making presentment or the party to pay or accept, presentment is due on the next following day which is a full business day for both parties.

(4) Presentment to be sufficient must be made at a reasonable hour, and if at a bank during its banking day.

§ 3–504. How Presentment Made. (1) Presentment is a demand for acceptance or payment made upon the maker, acceptor, drawee or other payor by or on behalf of the holder.

(2) Presentment may be made

(a) by mail, in which event the time of presentment is determined by the time of receipt of the mail; or

(b) through a clearing house; or

(c) at the place of acceptance or payment specified in the instrument or if there be none at the place of business or residence of the party to accept or pay. If neither the party to accept or pay nor anyone authorized to act for him is present or accessible at such place presentment is excused.

(3) It may be made

(a) to any one of two or more makers, acceptors, drawees or other payors; or

(b) to any person who has authority to make or refuse the acceptance or payment.

(4) A draft accepted or a note made payable at a bank in the United States must be presented at such bank.

(5) In the cases described in Section 4–210 presentment may be made in the manner and with the result stated in that section.

§ 3–505. Rights of Party to Whom Presentment Is Made. (1) The party to whom presentment is made may without dishonor require

(a) exhibition of the instrument; and

(b) reasonable identification of the person making presentment and evidence of his authority to make it if made for another; and

(c) that the instrument be produced for acceptance or payment at a place specified in it, or if there be none at any place reasonable in the circumstances; and

(d) a signed receipt on the instrument for any partial or full payment and its surrender upon full payment.

(2) Failure to comply with any such requirement invalidates the presentment but the person presenting has a reasonable time in which to comply and the time for acceptance or payment runs from the time of compliance.

§ 3–506. Time Allowed for Acceptance or Payment. (1) Acceptance may be deferred without dishonor until the close of the next business day following presentment. The holder may also in a good faith effort to obtain acceptance and without either dishonor of the instrument or discharge of secondary parties allow postponement of acceptance for an additional business day.

(2) Except as a longer time is allowed in the case of documentary drafts drawn under a letter of credit, and unless an earlier time is agreed to by the party to pay, payment of an instrument may be deferred without dishonor pending reasonable examination to determine whether it is properly payable, but payment must be made in any event before the close of business on the day of presentment.

§ 3–507. Dishonor; Holder's Right of Recourse; Term Allowing Re-Presentment. (1) An instrument is dishonored when

(a) a necessary or optional presentment is duly made and due acceptance or payment is refused or cannot be obtained within the prescribed time or in case of bank collections the instrument is seasonably returned by the midnight deadline (Section 4–301); or

(b) presentment is excused and the instrument is not duly accepted or paid.

(2) Subject to any necessary notice of dishonor and protest, the holder has upon dishonor an immediate right of recourse against the drawers and indorsers.

(3) Return of an instrument for lack of proper indorsement is not dishonor.

(4) A term in a draft or an indorsement thereof allowing a stated time for re-presentment in the event of any dishonor of the draft by nonacceptance if a time draft or by nonpayment if a sight draft gives the holder as against any secondary party bound by the term an option to waive the dishonor without affecting the liability of the secondary party and he may present again up to the end of the stated time.

§ 3–508. Notice of Dishonor. (1) Notice of dishonor may be given to any person who may be liable on the instrument by or on behalf of the holder or any party who has himself received notice, or any other party who can be compelled to pay the instrument. In addition an agent or bank in whose hands the instrument is dishonored may give notice to his principal or customer or to another agent or bank from which the instrument was received.

(2) Any necessary notice must be given by a bank before its midnight deadline and by any other person before midnight of the third business day after dishonor or receipt of notice of dishonor.

(3) Notice may be given in any reasonable manner. It may be oral or written and in any terms which identify the instrument and state that it has been dishonored. A misdescription which does not mislead the party notified does not vitiate the notice. Sending the instrument bearing a stamp, ticket or writing stating that acceptance or payment has been refused or sending a notice of debit with respect to the instrument is sufficient.

(4) Written notice is given when sent although it is not received.

(5) Notice to one partner is notice to each although the firm has been dissolved.

(6) When any party is in insolvency proceedings instituted after the issue of the instrument notice may be given either to the party or to the representative of his estate.

(7) When any party is dead or incompetent notice may be sent to his last known address or given to his personal representative.

(8) Notice operates for the benefit of all parties who have rights on the instrument against the party notified.

§ 3–509. Protest; Noting for Protest. (1) A protest is a certificate of dishonor made under the hand and seal of a United States consul or vice consul or a notary public or other person authorized to certify dishonor by the law of the place where dishonor occurs. It may be made upon information satisfactory to such person.

(2) The protest must identify the instrument and certify either that due presentment has been made or the reason why it is excused and that the instrument has been dishonored by nonacceptance or nonpayment.

(3) The protest may also certify that notice of dishonor has been given to all parties or to specified parties.

(4) Subject to subsection (5) any necessary protest is due by the time that notice of dishonor is due.

(5) If, before protest is due, an instrument has been noted for protest by the officer to make protest, the protest may be made at any time thereafter as of the date of the noting.

§ 3–510. Evidence of Dishonor and Notice of Dishonor. The following are admissible as evidence and create a presumption of dishonor and of any notice of dishonor therein shown:

(a) a document regular in form as provided in the preceding section which purports to be a protest;

(b) the purported stamp or writing of the drawee, payor bank or presenting bank on the instrument or accompanying it stating that acceptance or payment has been refused for reasons consistent with dishonor;

(c) any book or record of the drawee, payor bank, or any collecting bank kept in the usual course of business which shows dishonor, even though there is no evidence of who made the entry.

§ 3–511. Waived or Excused Presentment, Protest or Notice of Dishonor or Delay Therein. (1) Delay in presentment, protest or notice of dishonor is excused when the party is without notice that it is due or when the delay is caused by circumstances beyond his control and he exercises reasonable diligence after the cause of the delay ceases to operate.

(2) Presentment or notice or protest as the case may be is entirely excused when

(a) the party to be charged has waived it expressly or by implication either before or after it is due; or

(b) such party has himself dishonored the instrument or has countermanded payment or otherwise has no reason to expect or right to require that the instrument be accepted or paid; or

(c) by reasonable diligence the presentment or protest cannot be made or the notice given.

(3) Presentment is also entirely excused when

(a) the maker, acceptor or drawee of any instrument except a documentary draft is dead or in insolvency proceedings instituted after the issue of the instrument; or

(b) acceptance or payment is refused but not for want of proper presentment.

(4) Where a draft has been dishonored by nonacceptance a later presentment for payment and any notice of dishonor and protest for nonpayment are excused unless in the meantime the instrument has been accepted.

(5) A waiver of protest is also a waiver of presentment and of notice of dishonor even though protest is not required.

(6) Where a waiver of presentment or notice or protest is embodied in the instrument itself it is binding upon all parties; but where it is written above the signature of an indorser it binds him only.

Part 6

Discharge

§ 3–601. Discharge of Parties. (1) The extent of the discharge of any party from liability on an instrument is governed by the sections on

(a) payment or satisfaction (Section 3–603); or

(b) tender of payment (Section 3–604); or

(c) cancellation or renunciation (Section 3–605); or

(d) impairment of right of recourse or of collateral (Section 3–606); or

(e) reacquisition of the instrument by a prior party (Section 3–208); or

(f) fraudulent and material alteration (Section 3–407); or

(g) certification of a check (Section 3–411); or

(h) acceptance varying a draft (Section 3–412); or

(i) unexcused delay in presentment or notice of dishonor or protest (Section 3–502).

(2) Any party is also discharged from his liability on an instrument to another party by any other act or agreement with such party which would discharge his simple contract for the payment of money.

(3) The liability of all parties is discharged when any party who has himself no right of action or recourse on the instrument

(a) reacquires the instrument in his own right; or

(b) is discharged under any provision of this Article, except as otherwise provided with respect to discharge for impairment of recourse or of collateral (Section 3–606).

§ 3–602. Effect of Discharge Against Holder in Due Course. No discharge of any party provided by this Article is effective against a subsequent holder in due course unless he has notice thereof when he takes the instrument.

§ 3–603. Payment or Satisfaction. (1) The liability of any party is discharged to the extent of his payment or satisfaction to the holder even though it is made with knowledge of a claim of another person to the instrument unless prior to such payment or satisfaction the person making the claim either supplies indemnity deemed adequate by the party seeking the discharge or enjoins payment or satisfaction by order of a court of competent jurisdiction in an action in which the adverse claimant and the holder are parties. This subsection does not, however, result in the discharge of the liability

(a) of a party who in bad faith pays or satisfies a holder who acquired the instrument by theft or who (unless having the rights of a holder in due course) holds through one who so acquired it; or

(b) of a party (other than an intermediary bank or a payor bank which is not a depositary bank) who pays or satisfies the holder of an instrument which has been restrictively indorsed in a manner not consistent with the terms of such restrictive indorsement.

(2) Payment or satisfaction may be made with the consent of the holder by any person including a stranger to the instrument. Surrender of the instrument to such a person gives him the rights of a transferee (Section 3–201).

§ 3–604. Tender of Payment. (1) Any party making tender of full payment to a holder when or after it is due is discharged to the extent of all subsequent liability for interest, costs and attorney's fees.

(2) The holder's refusal of such tender wholly discharges any party who has a right of recourse against the party making the tender.

(3) Where the maker or acceptor of an instrument payable otherwise than on demand is able and ready to pay at every place of payment specified in the instrument when it is due, it is equivalent to tender.

§ 3–605. Cancellation and Renunciation. (1) The holder of an instrument may even without consideration discharge any party

(a) in any manner apparent on the face of the instrument or the indorsement, as by intentionally cancelling the instrument or the party's signature by destruction or mutilation, or by striking out the party's signature; or

(b) by renouncing his rights by a writing signed and delivered or by surrender of the instrument to the party to be discharged

(2) Neither cancellation nor renunciation without surrender of the instrument affects the title thereto.

§ 3–606. Impairment of Recourse or of Collateral. (1) The holder discharges any party to the instrument to the extent that without such party's consent the holder

(a) without express reservation of rights releases or agrees not to sue any person against whom the party has to the knowledge of the holder a right of recourse or agrees to suspend the right to enforce against such person the instrument or collateral or otherwise discharges such person, except that failure or delay in effecting any required presentment, protest or notice of dishonor with respect to any such person does not discharge any party as to whom presentment, protest or notice of dishonor is effective or unnecessary; or

(b) unjustifiably impairs any collateral for the instrument given by or on behalf of the party or any person against whom he has a right of recourse.

(2) By express reservation of rights against a party with a right of recourse the holder preserves

(a) all his rights against such party as of the time when the instrument was originally due; and

(b) the right of the party to pay the instrument as of that time; and

(c) all rights of such party to recourse against others.

Part 7

Advice of International Sight Draft

§ 3–701. Letter of Advice of International Sight Draft. (1) A "letter of advice" is a drawer's communication to the drawee that a described draft has been drawn.

(2) Unless otherwise agreed when a bank receives from another bank a letter of advice of an international sight draft the drawee bank may immediately debit the drawer's account and stop the running of interest pro tanto. Such a debit and any resulting credit to any account covering outstanding drafts leaves in the drawer full power to stop payment or otherwise dispose of the amount and creates no trust or interest in favor of the holder.

(3) Unless otherwise agreed and except where a draft is drawn under a credit issued by the drawee, the drawee of an international sight draft owes the drawer no duty to pay an unadvised draft but if it does so and the draft is genuine, may appropriately debit the drawer's account.

Part 8

Miscellaneous

§ 3–801. Drafts in a Set. (1) Where a draft is drawn in a set of parts, each of which is numbered and expressed to be an order only if no other part has been honored, the whole of the parts constitutes one draft but a taker of any part may become a holder in due course of the draft.

(2) Any person who negotiates, indorses or accepts a single part of a draft drawn in a set thereby becomes liable to any holder in due course of that part as if it were the whole set, but as between different holders in due course to whom different parts have been negotiated the holder whose title first accrues has all rights to the draft and its proceeds.

(3) As against the drawee the first presented part of a draft drawn in a set is the part entitled to payment, or if a time draft to acceptance and payment. Acceptance of any subsequently presented part renders the drawee liable thereon under subsection (2). With respect both to a holder and to the drawer payment of a subsequently presented part of a draft payable at sight has the same effect as payment of a check notwithstanding an effective stop order (Section 4–407).

(4) Except as otherwise provided in this section, where any part of a draft in a set is discharged by payment or otherwise the whole draft is discharged.

§ 3–802. Effect of Instrument on Obligation for Which It Is Given. (1) Unless otherwise agreed where an instrument is taken for an underlying obligation

(a) the obligation is pro tanto discharged if a bank is drawer, maker or acceptor of the instrument and there is no recourse on the instrument against the underlying obligor; and

(b) in any other case the obligation is suspended pro tanto until the instrument is due or if it is payable on demand until its presentment. If the instrument is dishonored action may be maintained on either the instrument or the obligation; discharge of the underlying obligor on the instrument also discharges him on the obligation.

(2) The taking in good faith of a check which is not postdated does not of itself so extend the time on the original obligation as to discharge a surety.

§ 3–803. Notice to Third Party. Where a defendant is sued for breach of an obligation for which a third person is answerable over under this Article he may give the third person written notice of the litigation, and the person notified may then give similar notice to any other person who is answerable over to him under this Article. If the notice states that the person notified may come in and defend and that if the person notified does not do so he will in any action against him by the person giving the notice be bound by any determination of fact common to the two litigations, then unless after seasonable receipt of the notice the person notified does come in and defend he is so bound.

§ 3–804. Lost, Destroyed or Stolen Instruments. The owner of an instrument which is lost, whether by destruction, theft or otherwise, may maintain an action in his own name and recover from any party liable thereon upon due proof of his ownership, the facts which prevent his production of the instrument and its terms. The court may require security indemnifying the defendant against loss by reason of further claims on the instrument.

§ 3–805. Instruments Not Payable to Order or to Bearer. This Article applies to any instrument whose terms do not preclude transfer and which is otherwise negotiable within this Article but which is not payable to order or to bearer, except that there can be no holder in due course of such an instrument.

ARTICLE 4: BANK DEPOSITS AND COLLECTIONS

Part 1

General Provisions and Definitions

§ 4–101. Short Title. This Article shall be known and may be cited as Uniform Commerical Code—Bank Deposits and Collections.

§ 4–102. Applicability. (1) To the extent that items within this Article are also within the scope of Articles 3 and 8, they are subject to the provisions of those Articles. In the event of conflict the provisions of this Article govern those of Article 3 but the provisions of Article 8 govern those of this Article.

(2) The liability of a bank for action or nonaction with respect to any item handled by it for purposes of presentment, payment or collection is governed by the law of the place where the bank is located. In the case of action or non-action by or at a branch or separate office of a bank, its liability is governed by the law of the place where the branch or separate office is located.

§ 4–103. Variation by Agreement; Measure of Damages; Certain Action Constituting Ordinary Care. (1) The effect of the provisions of this Article may be varied by agreement except that no agreement can disclaim a bank's responsibility for its own lack of good faith or failure to exercise ordinary care or can limit the measure of damages for such lack or failure; but the parties may by agreement determine the standards by which such responsibility is to be measured if such standards are not manifestly unreasonable.

(2) Federal Reserve regulations and operating letters, clearing house rules, and the like, have the effect of agreements under subsection (1), whether or not specifically assented to by all parties interested in items handled.

(3) Action or non-action approved by this Article or pursuant to Federal Reserve regulations or operating letters constitutes the exercise of ordinary care and, in the absence of special instructions, action or non-action consistent with clearing house rules and the like or with a general banking usage not disapproved by this Article, prima facie constitutes the exercise of ordinary care.

(4) The specification or approval of certain procedures by this Article does not constitute disapproval of other procedures which may be reasonable under the circumstances.

(5) The measure of damages for failure to exercise ordinary care in handling an item is the amount of the item reduced by an amount which could not have been realized by the use of ordinary care, and where there is bad faith it includes other damages, if any, suffered by the party as a proximate consequence.

§ 4–104. Definitions and Index of Definitions. (1) In this Article unless the context otherwise requires

(a) "Account" means any account with a bank and includes a checking, time, interest or savings account;

(b) "Afternoon" means the period of a day between noon and midnight;

(c) "Banking day" means that part of any day on which a bank is open to the public for carrying on substantially all of its banking functions;

(d) "Clearing house" means any association of banks or other payors regularly clearing items;

(e) "Customer" means any person having an account with a bank or for whom a bank has agreed to collect items and includes a bank carrying an account with another bank;

(f) "Documentary draft" means any negotiable or nonnegotiable draft with accompanying documents, securities or other papers to be delivered against honor of the draft;

(g) "Item" means any instrument for the payment of money even though it is not negotiable but does not include money;

(h) "Midnight deadline" with respect to a bank is midnight on its next banking day following the banking day on which it receives the relevant item or notice or from which the time for taking action commences to run, whichever is later;

(i) "Properly payable" includes the availability of funds for payment at the time of decision to pay or dishonor;

(j) "Settle" means to pay in cash, by clearing house settlement, in a charge or credit or by remittance, or otherwise as instructed. A settlement may be either provisional or final;

(k) "Suspends payments" with respect to a bank means that it has been closed by order of the supervisory authorities, that a public officer has been appointed to take it over or that it ceases or refuses to make payments in the ordinary course of business.

(2) Other definitions applying to this Article and the sections in which they appear are:

"Collecting bank." Section 4–105.
"Depositary bank." Section 4–105.
"Intermediary bank." Section 4–105.
"Payor bank." Section 4–105.
"Presenting bank." Section 4–105.
"Remitting bank." Section 4–105.

(3) The following definitions in other Articles apply to this Article:

"Acceptance." Section 3–410.
"Certificate of deposit." Section 3–104.
"Certification." Section 3–411.
"Check." Section 3–104.
"Draft." Section 3–104.
"Holder in due course." Section 3–302.

"Notice of dishonor." Section 3–508.
"Presentment." Section 3–504.
"Protest." Section 3–509.
"Secondary party." Section 3–102.

(4) In addition Article 1 contains general definitions and principles of construction and interpretation applicable throughout this Article.

§ 4–105. "Depositary Bank"; "Intermediary Bank"; "Collecting Bank"; "Payor Bank"; "Presenting Bank"; "Remitting Bank." In this Article unless the context otherwise requires:

(a) "Depositary bank" means the first bank to which an item is transferred for collection even though it is also the payor bank;

(b) "Payor bank" means a bank by which an item is payable as drawn or accepted;

(c) "Intermediary bank" means any bank to which an item is transferred in course of collection except the depositary or payor bank;

(d) "Collecting bank" means any bank handling the item for collection except the payor bank;

(e) "Presenting bank" means any bank presenting an item except a payor bank;

(f) "Remitting bank" means any payor or intermediary bank remitting for an item.

§ 4–106. Separate Office of a Bank. A branch or separate office of a bank [maintaining its own deposit ledgers] is a separate bank for the purpose of computing the time within which and determining the place at or to which action may be taken or notices or orders shall be given under this Article and under Article 3.

Note: *The brackets are to make it optional with the several states whether to require a branch to maintain its own deposit ledgers in order to be considered to be a separate bank for certain purposes under Article 4. In some states "maintaining its own deposit ledgers" is a satisfactory test. In others branch banking practices are such that this test would not be suitable.*

§ 4–107. Time of Receipt of Items. (1) For the purpose of allowing time to process items, prove balances and make the necessary entries on its books to determine its position for the day, a bank may fix an afternoon hour of two P.M. or later as a cut-off hour for the handling of money and items and the making of entries on its books.

(2) Any item or deposit of money received on any day after a cut-off hour so fixed or after the close of the banking day may be treated as being received at the opening of the next banking day.

§ 4–108. Delays. (1) Unless otherwise instructed, a collecting bank in a good faith effort to secure payment may, in the case of specific items and with or without the approval of any person involved, waive, modify or extend time limits imposed or permitted by this Act for a period not in excess of an additional banking day without discharge of secondary parties and without liability to its transferor or any prior party.

(2) Delay by a collecting bank or payor bank beyond time limits prescribed or permitted by this Act or by instructions is excused if caused by interruption of communication facilities, suspension of payments by another bank, war, emergency conditions or other circumstances beyond the control of the bank provided it exercises such diligence as the circumstances require.

§ 4–109. Process of Posting. The "process of posting" means the usual procedure followed by a payor bank in determining to pay an item and in recording the payment including one or more of the following or other steps as determined by the bank:

(a) verification of any signature;

(b) ascertaining that sufficient funds are available;

(c) affixing a "paid" or other stamp;

(d) entering a charge or entry to a customer's account;

(e) correcting or reversing an entry or erroneous action with respect to the item.

Part 2

Collection of Items: Depositary and Collecting Banks

§ 4–201. Presumption and Duration of Agency Status of Collecting Banks and Provisional Status of Credits; Applicability of Article; Item Indorsed "Pay Any Bank." (1) Unless a contrary intent clearly appears and prior to the time that a settlement given by a collecting bank for an item is or becomes final (subsection (3) of Section 4–211 and Sections 4–212 and 4–213) the bank is an agent or sub-agent of the owner of the item and any settlement given for the item is provisional. This provision applies regardless of the form of indorsement or lack of indorsement and even though credit given for the item is subject to immediate withdrawal as of right or is in fact withdrawn; but the continuance of ownership of an item by its owner and any rights of the owner to proceeds of the item are subject to rights of a collecting bank such as those resulting from outstanding advances on the item and valid rights of setoff. When an item is handled by banks for purposes of presentment, payment and collection, the relevant provisions of this Article apply even though action of parties clearly establishes that a particular bank has purchased the item and is the owner of it.

(2) After an item has been indorsed with the words "pay any bank" or the like, only a bank may acquire the rights of a holder

(a) until the item has been returned to the customer initiating collection; or

(b) until the item has been specially indorsed by a bank to a person who is not a bank.

§ 4–202. Responsibility for Collection; When Action Seasonable. (1) A collecting bank must use ordinary care in

(a) presenting an item or sending it for presentment; and

(b) sending notice of dishonor or non-payment or returning an item other than a documentary draft to the bank's transferor [or directly to the depositary bank under subsection (2) of Section 4–212] *(see note to Section 4–212)* after learning that the item has not been paid or accepted as the case may be; and

(c) settling for an item when the bank receives final settlement; and

(d) making or providing for any necessary protest; and

(e) notifying its transferor of any loss or delay in transit within a reasonable time after discovery thereof.

(2) A collecting bank taking proper action before its midnight deadline following receipt of an item, notice or payment acts seasonably; taking proper action within a reasonably longer time may be seasonable but the bank has the burden of so establishing.

(3) Subject to subsection (1) (a), a bank is not liable for the insolvency, neglect, misconduct, mistake or default of another bank or person or for loss or destruction of an item in transit or in the possession of others.

§ 4–203. Effect of Instructions. Subject to the provisions of Article 3 concerning conversion of instruments (Section 3–419) and the provisions of both Article 3 and this Article concerning restrictive indorsements only a collecting bank's transferor can give instructions which affect the bank or constitute notice to it and a collecting bank is not liable to prior parties for any action taken pursuant to such instructions or in accordance with any agreement with its transferor.

§ 4–204. Methods of Sending and Presenting; Sending Direct to Payor Bank. (1) A collecting bank must send items by reasonably prompt method taking into consideration any relevant instructions, the nature of the item, the number of such items on hand, and the cost of collection involved and the method generally used by it or others to present such items.

(2) A collecting bank may send

(a) any item direct to the payor bank;

(b) any item to any non-bank payor if authorized by its transferor; and

(c) any item other than documentary drafts to any non-bank payor, if authorized by Federal Reserve regulation or operating letter, clearing house rule or the like.

(3) Presentment may be made by a presenting bank at a place where the payor bank has requested that presentment be made.

§ 4–205. Supplying Missing Indorsement; No Notice from Prior Indorsement. (1) A depositary bank which has taken an item for collection may supply any indorsement of the customer which is necessary to title unless the item contains the words "payee's indorsement required" or the like. In the absence of such a requirement a statement placed on the item by the depositary bank to the effect that the item was deposited by a customer or credited to his account is effective as the customer's indorsement.

(2) An intermediary bank, or payor bank which is not a depositary bank, is neither given notice nor otherwise affected by a restrictive indorsement of any person except the bank's immediate transferor.

§ 4–206. Transfer Between Banks. Any agreed method which identifies the transferor bank is sufficient for the item's further transfer to another bank.

§ 4–207. Warranties of Customer and Collecting Bank on Transfer or Presentment of Items; Time for Claims. (1) Each customer or collecting bank who obtains payment or acceptance of an item and each prior customer and collecting bank warrants to the payor bank or other payor who in good faith pays or accepts the item that

(a) he has a good title to the item or is authorized to obtain payment or acceptance on behalf of one who has a good title; and

(b) he has no knowledge that the signature of the maker or drawer is unauthorized, except that this warranty is not given by any customer or collecting bank that is a holder in due course and acts in good faith

(i) to a maker with respect to the maker's own signature; or

(ii) to a drawer with respect to the drawer's own signature, whether or not the drawer is also the drawee; or

(iii) to an acceptor of an item if the holder in due course took the item after the acceptance or obtained the acceptance without knowledge that the drawer's signature was unauthorized; and

(c) the item has not been materially altered, except that this warranty is not given by any customer or collecting bank that is a holder in due course and acts in good faith

(i) to the maker of a note; or

(ii) to the drawer of a draft whether or not the drawer is also the drawee; or

(iii) to the acceptor of an item with respect to an alteration made prior to the acceptance if the holder in due course took the item after the acceptance, even though the acceptance provided "payable as originally drawn" or equivalent terms; or

(iv) to the acceptor of an item with respect to an alteration made after the acceptance.

(2) Each customer and collecting bank who transfers an item and receives a settlement or other consideration for it warrants to his transferee and to any subsequent collecting bank who takes the item in good faith that

(a) he has a good title to the item or is authorized to obtain payment or acceptance on behalf of one who has a good title and the transfer is otherwise rightful; and

(b) all signatures are genuine or authorized; and

(c) the item has not been materially altered; and

(d) no defense of any party is good against him; and

(e) he has no knowledge of any insolvency proceeding instituted with respect to the maker or acceptor or the drawer of an unaccepted item.

In addition each customer and collecting bank so transferring an item and receiving a settlement or other consideration engages that upon dishonor and any necessary notice of dishonor and protest he will take up the item.

(3) The warranties and the engagement to honor set forth in the two preceding subsections arise notwithstanding the absence of indorsement or words of guaranty or warranty in the transfer or presentment and a collecting bank remains liable for their breach despite remittance to its transferor. Damages for breach of such warranties or engagement to honor shall not exceed the consideration received by the customer or collecting bank responsible plus finance charges and expenses related to the item, if any

(4) Unless a claim for breach of warranty under this section is made within a reasonable time after the person claiming learns of the breach, the person liable is discharged to the extent of any loss caused by the delay in making claim.

§ 4–208. Security Interest of Collecting Bank in Items, Accompanying Documents and Proceeds. (1) A bank has a security interest in an item and any accompanying documents or the proceeds of either

(a) in case of an item deposited in an account to the extent to which credit given for the item has been withdrawn or applied;

(b) in case of an item for which it has given credit available for withdrawal as of right, to the extent of the credit given whether or not the credit is drawn upon and whether or not there is a right of charge-back; or

(c) if it makes an advance on or against the item.

(2) When credit which has been given for several items received at one time or pursuant to a single agreement is withdrawn or applied in part the security interest remains upon all the items, any accompanying documents or the proceeds of either. For the purpose of this section, credits first given are first withdrawn.

(3) Receipt by a collecting bank of a final settlement for an item is a realization on its security interest in the item, accompanying documents and proceeds. To the extent and so long as the bank does not receive final settlement for the item or give up possession of the item or accompanying documents for purposes other than collection, the security interest continues and is subject to the provisions of Article 9 except that

(a) no security agreement is necessary to make the security interest enforceable (subsection (1) (b) of Section 9–203); and

(b) no filing is required to perfect the security interest; and

(c) the security interest has priority over conflicting perfected security interests in the item, accompanying documents or proceeds.

§ 4–209. When Bank Gives Value for Purposes of Holder in Due Course. For purposes of determining its status as a holder in due course, the bank has given value to the extent that it has a security interest in an item provided that the bank otherwise complies with the requirements of Section 3–302 on what constitutes a holder in due course.

§ 4–210. Presentment by Notice of Item Not Payable by, Through or at a Bank; Liability of Secondary Parties. (1) Unless otherwise instructed, a collecting bank may present an item not payable by, through or at a bank by sending to the party to accept or pay a written notice that the bank holds the item for acceptance or payment. The notice must be sent in time to be received on or before the day when presentment is due and the bank must meet any requirement of the party to accept or pay under Section 3–505 by the close of the bank's next banking day after it knows of the requirement.

(2) Where presentment is made by notice and neither honor nor request for compliance with a requirement under Section 3–505 is received by the close of business on the day after maturity or in the case of demand items by the close of business on the third banking day after notice was sent, the presenting bank may treat the item as dishonored and charge any secondary party by sending him notice of the facts.

§ 4–211. Media of Remittance; Provisional and Final Settlement in Remittance Cases. (1) A collecting bank may take in settlement of an item

(a) a check of the remitting bank or of another bank on any bank except the remitting bank; or

(b) a cashier's check or similar primary obligation of a remitting bank which is a member of or clears through a member of the same clearing house or group as the collecting bank; or

(c) appropriate authority to charge an account of the remitting bank or of another bank with the collecting bank; or

(d) if the item is drawn upon or payable by a person other than a bank, a cashier's check, certified check or other bank check or obligation.

(2) If before its midnight deadline the collecting bank properly dishonors a remittance check or authorization to charge on itself or presents or forwards for collection a remittance instrument of or on another bank which is of a kind approved by subsection (1) or has not been authorized by it, the collecting bank is not liable to prior parties in the event of the dishonor of such check, instrument or authorization.

(3) A settlement for an item by means of a remittance instrument or authorization to charge is or becomes a final settlement as to both the person making and the person receiving the settlement.

(a) if the remittance instrument or authorization to charge is of a kind approved by subsection (1) or has not been authorized by the person receiving the settlement and in either case the person receiving the settlement acts seasonably before its midnight deadline in presenting, forwarding for collection or paying the instrument or authorization—at the time the remittance instrument or authorization is finally paid by the payor by which it is payable;

(b) if the person receiving the settlement has authorized remittance by a nonbank check or obligation or by a cashier's check or similar primary obligation of or a check upon the payor or other remitting bank which is not of a kind approved by subsection (1) (b)—at the time of the receipt of such remittance check or obligation; or

(c) if in a case not covered by sub-paragraphs (a) or (b) the person receiving the settlement fails to seasonably present, forward for collection, pay or return a remittance instrument or authorization to it to charge before its midnight deadline—at such midnight deadline.

§ 4–212. Right of Charge-Back or Refund. (1) If a collecting bank has made provisional settlement with its customer for an item and itself fails by reason of dishonor, suspension of payments by a bank or otherwise to receive a settlement for the item which is or becomes final, the bank may revoke the settlement given by it, charge back the amount of any credit given for the item to its customer's account or obtain refund from its customer whether or not it is able to return the items if by its midnight deadline or within a longer reasonable time after it learns the facts it returns the item or sends notification of the facts. These rights to revoke, charge-back and obtain refund terminate if and when a settlement for the item received by the bank is or becomes final (subsection (3) of Section 4–211 and subsections (2) and (3) of Section 4–213).

[(2) Within the time and manner prescribed by this section and Section 4—301, an intermediary or payor bank, as the case may be, may return an unpaid item directly to the depositary bank and may send for collection a draft on the depositary bank and obtain reimbursement. In such case, if the depositary bank has received provisional settlement for the item, it must reimburse the bank drawing the draft and any provisional credits for the item between banks shall become and remain final.]

Note: *Direct returns is recognized as an innovation that is not yet established bank practice, and therefore, Paragraph 2 has been bracketed. Some lawyers have doubts whether it should be included in legislation or left to development by agreement.*

(3) A depositary bank which is also the payor may charge-back the amount of an item to its customer's account or obtain refund in accordance with the section governing return of an item received by a payor bank for credit on its books (Section 4–301).

(4) The right to charge-back is not affected by

(a) prior use of the credit given for the item; or

(b) failure by any bank to exercise ordinary care with respect to the item but any bank so failing remains liable.

(5) A failure to charge-back or claim refund does not affect other rights of the bank against the customer or any other party.

(6) If credit is given in dollars as the equivalent of the value of an item payable in a foreign currency the dollar amount of any charge-back or refund shall be calculated on the basis of the buying sight rate for the foreign currency prevailing on the day when the person entitled to the charge-back or refund learns that it will not receive payment in ordinary course.

§ 4–213. Final Payment of Item by Payor Bank; When Provisional Debits and Credits Become Final; When Certain Credits Become Available for Withdrawal. (1) An item is finally paid by a payor bank when the bank has done any of the following, whichever happens first:

(a) paid the item in cash; or

(b) settled for the item without reserving a right to revoke the settlement and without having such right under statute, clearing house rule or agreement; or

(c) completed the process of posting the item to the indicated account of the drawer, maker or other person to be charged therewith; or

(d) made a provisional settlement for the item and failed to revoke the settlement in the time and manner permitted by statute, clearing house rule or agreement.

Upon a final payment under subparagraphs (b), (c) or (d) the payor bank shall be accountable for the amount of the item.

(2) If provisional settlement for an item between the presenting and payor banks is made through a clearing house or by debits or credits in an account between them, then to the extent that provisional debits or credits for the item are entered in accounts between the presenting and payor banks or between the presenting and successive prior collecting banks seriatim, they become final upon final payment of the item by the payor bank.

(3) If a collecting bank receives a settlement for an item which is or becomes final (subsection (3) of Section 4–211, subsection (2) of Section 4–213) the bank is accountable to its customer for the amount of the item and any provisional credit given for the item in an account with its customer becomes final.

(4) Subject to any right of the bank to apply the credit to an obligation of the customer, credit given by a bank for an item in an account with its customer becomes available for withdrawal as of right

(a) in any case where the bank has received a provisional settlement for the item,—when such settlement becomes final and the bank has had a reasonable time to learn that the settlement is final;

(b) in any case where the bank is both a depositary bank and a payor bank and the item is finally paid,—at the opening of the bank's second banking day following receipt of the item.

(5) A deposit of money in a bank is final when made but, subject to any right of the bank to apply the deposit to an obligation of the customer, the deposit becomes available for withdrawal as of right at the opening of the bank's next banking day following receipt of the deposit.

§ 4–214. Insolvency and Preference. (1) Any item in or coming into the possession of a payor or collecting bank which suspends payment and which item is not finally paid shall be returned by the receiver, trustee or agent in charge of the closed bank to the presenting bank or the closed bank's customer.

(2) If a payor bank finally pays an item and suspends payments without making a settlement for the item with its customer or the presenting bank which settlement is or becomes final, the owner of the item has a preferred claim against the payor bank.

(3) If a payor bank gives or a collecting bank gives or receives a provisional settlement for an item and thereafter suspends payments, the suspension does not prevent or interfere with the settlement becoming final if such finality

occurs automatically upon the lapse of certain time or the happening of certain events (subsection (3) of Section 4–211, subsections (1) (d), (2) and (3) of Section 4–213).

(4) If a collecting bank receives from subsequent parties settlement for an item which settlement is or becomes final and suspends payments without making a settlement for the item with its customer which is or becomes final, the owner of the item has a preferred claim against such collecting bank.

Part 3

Collection of Items: Payor Banks

§ 4–301. Deferred Posting; Recovery of Payment by Return of Items; Time of Dishonor. (1) Where an authorized settlement for a demand item (other than a documentary draft) received by a payor bank otherwise than for immediate payment over the counter has been made before midnight of the banking day of receipt the payor bank may revoke the settlement and recover any payment if before it has made final payment (subsection (1) of Section 4–213) and before its midnight deadline it

(a) returns the item; or

(b) sends written notice of dishonor or nonpayment if the item is held for protest or is otherwise unavailable for return.

(2) If a demand item is received by a payor bank for credit on its books it may return such item or send notice of dishonor and may revoke any credit given or recover the amount thereof withdrawn by its customer, if it acts within the time limit and in the manner specified in the preceding subsection.

(3) Unless previous notice of dishonor has been sent an item is dishonored at the time when for purposes of dishonor it is returned or notice sent in accordance with this section.

(4) An item is returned:

(a) as to an item received through a clearing house, when it is delivered to the presenting or last collecting bank or to the clearing house or is sent or delivered in accordance with its rules; or

(b) in all other cases, when it is sent or delivered to the bank's customer or transferor or pursuant to his instructions.

§ 4–302. Payor Bank's Responsibility for Late Return of Item. In the absence of a valid defense such as breach of a presentment warranty (subsection (1) of Section 4–207), settlement effected or the like, if an item is presented on and received by a payor bank the bank is accountable for the amount of

(a) a demand item other than a documentary draft whether properly payable or not if the bank, in any case where it is not also the depositary bank, retains the item beyond midnight of the banking day of receipt without settling for it or, regardless of whether it is also the depositary bank, does not pay or return the item or send notice of dishonor until after its midnight deadline; or

(b) any other properly payable item unless within the time allowed for acceptance or payment of that item the bank either accepts or pays the item or returns it and accompanying documents.

§ 4–303. When Items Subject to Notice, Stop-Order, Legal Process or Setoff; Order in Which Items May Be Charged or Certified. (1) Any knowledge, notice or stop-order received by, legal process served upon or setoff exercised by a payor bank, whether or not effective under other rules of law to terminate, suspend or modify the bank's right or duty to pay an item or to charge its customer's account for the item, comes too late to so terminate, suspend or modify such right or duty if the knowledge, notice, stop-order or legal process is received or served and a reasonable time for the bank to act thereon expires or the setoff is exercised after the bank has done any of the following:

(a) accepted or certified the item;

(b) paid the item in cash;

(c) settled for the item without reserving a right to revoke the settlement and without having such right under statute, clearing house rule or agreement;

(d) completed the process of posting the item to the indicated account of the drawer, maker or other person to be charged therewith or otherwise has evidenced by examination of such indicated account and by action its decision to pay the item; or

(e) become accountable for the amount of the item under subsection (1) (d) of Section 4–213 and Section 4–302 dealing with the payor bank's responsibility for late return of items.

(2) Subject to the provisions of subsection (1) items may be accepted, paid, certified or charged to the indicated account of its customer in any order convenient to the bank.

Part 4

Relationship Between Payor Bank and Its Customer

§ 4–401. When Bank May Charge Customer's Account. (1) As against its customer, a bank may charge against his account any item which is otherwise properly payable from that account even though the charge creates an overdraft.

(2) A bank which in good faith makes payment to a holder may charge the indicated account of its customer according to

(a) the original tenor of his altered item; or

(b) the tenor of his completed item, even though the bank knows the item has been completed unless the bank has notice that the completion was improper.

§ 4–402. Bank's Liability to Customer for Wrongful Dishonor. A payor bank is liable to its customer for damages proximately caused by the wrongful dishonor of an item. When the dishonor occurs through mistake liability is limited to actual damages proved. If so proximately caused and proved damages may include damages for an arrest or prosecution of the customer or other consequential damages. Whether any consequential damages are proximately caused by the wrongful dishonor is a question of fact to be determined in each case.

§ 4–403. Customer's Right to Stop Payment; Burden of Proof of Loss. (1) A customer may by order to his bank stop payment of any item payable for his account but the order must be received at such time and in such manner as to afford the bank a reasonable opportunity to act on it prior to any action by the bank with resepct to the item described in Section 4–303.

(2) An oral order is binding upon the bank only for fourteen calendar days unless confirmed in writing within that period. A written order is effective for only six months unless renewed in writing.

(3) The burden of establishing the fact and amount of loss resulting from the payment of an item contrary to a binding stop payment order is on the customer.

§ 4–404. Bank Not Obligated to Pay Check More Than Six Months Old. A bank is under no obligation to a customer having a checking account to pay a check, other than a certified check, which is presented more than six months after its date, but it may charge its customer's account for a payment made thereafter in good faith.

§ 4–405. Death or Incompetence of Customer. (1) A payor or collecting bank's authority to accept, pay or collect an item or to account for proceeds of its collection if otherwise effective is not rendered ineffective by incompetence of a customer of either bank existing at the time the item is issued or its collection is undertaken if the bank does not know of an adjudication of incompetence. Neither death nor incompetence of a customer revokes such authority to accept, pay, collect or account until the bank knows of the fact of death or of an adjudication of incompetence and has reasonable opportunity to act on it.

(2) Even with knowledge a bank may for ten days after the date of death pay or certify checks drawn on or prior to that date unless ordered to stop payment by a person claiming an interest in the account.

§ 4–406. Customer's Duty to Discover and Report Unauthorized Signature or Alteration. (1) When a bank sends to its customer a statement of account accompanied by items paid in good faith in support of the debit entries or holds the statement and items pursuant to a request or instructions of its customer or otherwise in a reasonable manner makes the statement and items available to the customer, the customer must exercise reasonable care and promptness to examine the statement and items to discover his unauthorized signature or any alteration on an item and must notify the bank promptly after discovery thereof.

(2) If the bank establishes that the customer failed with respect to an item to comply with the duties imposed on the customer by subsection (1) the customer is precluded from asserting against the bank

(a) his unauthorized signature or any alteration on the item if the bank also establishes that it suffered a loss by reason of such failure; and

(b) an unauthorized signature or alteration by the same wrongdoer on any other item paid in good faith by the bank after the first item and statement was available to the customer for a reasonable period not exceeding fourteen calendar days and before the bank receives notification from the customer of any such unauthorized signature or alteration.

(3) The preclusion under subsection (2) does not apply if the customer establishes lack of ordinary care on the part of the bank in paying the item(s).

(4) Without regard to care or lack of care of either the customer or the bank a customer who does not within one year from the time the statement and items are made available to the customer (subsection (1)) discover and report his unauthorized signature or any alteration on the face or back of the item or does not within three years from that time discover and report any unauthorized indorsement is precluded from asserting against the bank such unauthorized signature or indorsement or such alteration.

(5) If under this section a payor bank has a valid defense against a claim of a customer upon or resulting from payment of an item and waives or fails upon request to assert the defense the bank may not assert against any collecting bank or other prior party presenting or transferring the item a claim based upon the unauthorized signature or alteration giving rise to the customer's claim.

§ 4–407. Payor Bank's Right to Subrogation on Improper Payment. If a payor bank has paid an item over the stop payment order of the drawer or maker or otherwise under circumstances giving a basis for objection by the drawer or maker, to prevent unjust enrichment and only to the extent necessary to prevent loss to the bank by reason of its payment of the item, the payor bank shall be subrogated to the rights

(a) of any holder in due course on the item against the drawer or maker; and

(b) of the payee or any other holder of the item against the drawer or maker either on the item or under the transaction out of which the item arose; and

(c) of the drawer or maker against the payee or any other holder of the item with respect to the transaction out of which the item arose.

Part 5

Collection of Documentary Drafts

§ 4–501. Handling of Documentary Drafts; Duty to Send for Presentment and to Notify Customer of Dishonor. A bank which takes a documentary draft for collection must present or send the draft and accompanying documents for presentment and upon learning that the draft has not been paid or accepted in due course must seasonably notify its customer of such fact even though it may have discounted or bought the draft or extended credit available for withdrawal as of right.

§ 4–502. Presentment of "On Arrival" Drafts. When a draft or the relevant instructions require presentment "on arrival," "when goods arrive" or the like, the collecting bank need not present until in its judgment a reasonable time for arrival of the goods has expired. Refusal to pay or accept because the goods have not arrived is not dishonor; the bank must notify its transferor of such refusal but need not present the draft again until it is instructed to do so or learns of the arrival of the goods.

§ 4–503. Responsibility of Presenting Bank for Documents and Goods; Report of Reasons for Dishonor; Referee in Case of Need. Unless otherwise instructed and except as provided in Article 5 a bank presenting a documentary draft

(a) must deliver the documents to the drawee on acceptance of the draft if it is payable more than three days after presentment; otherwise, only on payment; and

(b) upon dishonor, either in the case of presentment for acceptance or presentment for payment, may seek and follow instructions from any referee in case of need designated in the draft or if the presenting bank does not choose to utilize his services it must use diligence and good faith to ascertain the reason for dishonor, must notify its transferor of the dishonor and of the results of its effort to ascertain the reasons therefor and must request instructions.

But the presenting bank is under no obligation with respect to goods represented by the documents except to follow any reasonable instructions seasonably received; it has a right to reimbursement for any expense incurred in following instructions and to prepayment of or indemnity for such expenses.

§ 4–504. Privilege of Presenting Bank to Deal With Goods; Security Interest for Expenses. (1) A presenting bank which, following the dishonor of a documentary draft, has seasonably requested instructions but does not receive them within a reasonable time may store, sell, or otherwise deal with the goods in any reasonable manner.

(2) For its reasonable expenses incurred by action under subsection (1) the presenting bank has a lien upon the goods or their proceeds, which may be foreclosed in the same manner as an unpaid seller's lien.

ARTICLE 5: LETTERS OF CREDIT

§ 5–101. Short Title. This Article shall be known and may be cited as Uniform Commercial Code—Letters of Credit.

§ 5–102. Scope. (1) This Article applies

(a) to a credit issued by a bank if the credit requires a documentary draft or a documentary demand for payment; and

(b) to a credit issued by a person other than a bank if the credit requires that the draft or demand for payment be accompanied by a document of title; and

(c) to a credit issued by a bank or other person if the credit is not within subparagraphs (a) or (b) but conspicuously states that it is a letter of credit or is conspicuously so entitled.

(2) Unless the engagement meets the requirements of subsection (1), this Article does not apply to engagements to make advances or to honor drafts or demands for payment, to authorities to pay or purchase, to guarantees or to general agreements.

(3) This Article deals with some but not all of the rules and concepts of letters of credit as such rules or concepts have developed prior to this act or may hereafter develop. The fact that this Article states a rule does not by itself require, imply or negate application of the same or a converse rule to a situation not provided for or to a person not specified by this Article.

§ 5–103. Definitions. (1) In this Article unless the context otherwise requires

(a) "Credit" or "letter of credit" means an engagement by a bank or other person made at the request of a customer and of a kind within the scope of this Article (Section 5–102) that the issuer will honor drafts or other demands for payment upon compliance with the conditions specified in the credit. A credit may be either revocable or irrevocable. The engagement may be either an agreement to honor or a statement that the bank or other person is authorized to honor.

(b) A "documentary draft" or a "documentary demand for payment" is one honor of which is conditioned upon the presentation of a document or documents. "Document" means any paper including document of title, security, invoice, certificate, notice of default and the like.

(c) An "issuer" is a bank or other person issuing a credit.

(d) A "beneficiary" of a credit is a person who is entitled under its terms to draw or demand payment.

(e) An "advising bank" is a bank which gives notification of the issuance of a credit by another bank.

(f) A "confirming bank" is a bank which engages either that it will itself honor a credit already issued by another bank or that such a credit will be honored by the issuer or a third bank.

(g) A "customer" is a buyer or other person who causes an issuer to issue a credit. The term also includes a bank which procures issuance or confirmation on behalf of that bank's customer.

(2) Other definitions applying to this Article and the sections in which they appear are:

"Notation of Credit." Section 5–108.

"Presenter." Section 5–112(3).

(3) Definitions in other Articles applying to this Article and the sections in which they appear are:

"Accept" or "Acceptance." Section 3–410.

"Contract for sale." Section 2–106.

"Draft." Section 3–104.

"Holder in due course." Section 3–302.

"Midnight deadline." Section 4–104.

"Security." Section 8–102.

(4) In addition, Article 1 contains general definitions and principles of construction and interpretation applicable throughout this Article.

§ 5–104. Formal Requirements; Signing. (1) Except as otherwise required in subsection (1) (c) of Section 5–102 on scope, no particular form of phrasing is required for a credit. A credit must be in writing and signed by the issuer and a confirmation must be in writing and signed by the confirming bank. A modification of the terms of a credit or confirmation must be signed by the issuer or confirming bank.

(2) A telegram may be a sufficient signed writing if it identifies its sender by an authorized authentication. The authentication may be in code and the authorized naming of the issuer in an advice of credit is a sufficient signing.

§ 5–105. Consideration. No consideration is necessary to establish a credit or to enlarge or otherwise modify its terms.

§ 5–106. Time and Effect of Establishment of Credit. (1) Unless otherwise agreed a credit is established

(a) as regards the customer as soon as a letter of credit is sent to him or the letter of credit or an authorized written advice of its issuance is sent to the beneficiary; and

(b) as regards the beneficiary when he receives a letter of credit or an authorized written advice of its issuance.

(2) Unless otherwise agreed once an irrevocable credit is established as regards the customer it can be modified or revoked only with the consent of the customer and once it is established as regards the beneficiary it can be modified or revoked only with his consent.

(3) Unless otherwise agreed after a revocable credit is established it may be modified or revoked by the issuer without notice to or consent from the customer or beneficiary.

(4) Notwithstanding any modification or revocation of a revocable credit any person authorized to honor or negotiate under the terms of the original credit is entitled to reimbursement for or honor of any draft or demand for payment duly honored or negotiated before receipt of notice of the modification or revocation and the issuer in turn is entitled to reimbursement from its customer.

§ 5–107. Advice of Credit; Confirmation; Error in Statement of Terms. (1) Unless otherwise specified an advising bank by advising a credit issued by another bank does not assume any obligation to honor drafts drawn or demands for payment made under the credit but it does assume obligation for the accuracy of its own statement.

(2) A confirming bank by confirming a credit becomes directly obligated on the credit to the extent of its confirmation as though it were its issuer and acquires the rights of an issuer.

(3) Even though an advising bank incorrectly advises the terms of a credit it has been authorized to advise the credit is established as against the issuer to the extent of its original terms.

(4) Unless otherwise specified the customer bears as against the issuer all risks of transmission and reasonable translation or interpretation of any message relating to a credit.

§ 5–108. "Notation Credit"; Exhaustion of Credit. (1) A credit which specifies that any person purchasing or paying drafts drawn or demands for payment made under it must note the amount of the draft or demand on the letter or advice of credit is a "notation credit."

(2) Under a notation credit

(a) a person paying the beneficiary or purchasing a draft or demand for payment from him acquires a right to honor only if the appropriate notation is made and by transferring or forwarding for honor the documents under the credit such a person warrants to the issuer that the notation has been made; and

(b) unless the credit or a signed statement that an appropriate notation has been made accompanies the draft or demand for payment the issuer may delay honor until evidence of notation has been procured which is satisfactory to it but its obligation and that of its customer continue for a reasonable time not exceeding thirty days to obtain such evidence.

(3) If the credit is not a notation credit

(a) the issuer may honor complying drafts or demands for payment presented to it in the order in which they are presented and is discharged pro tanto by honor of any such draft or demand;

(b) as between competing good faith purchasers of complying drafts or demands the person first purchasing has priority over a subsequent purchaser even though the later purchased draft or demand has been first honored.

§ 5–109. Issuer's Obligation to Its Customer. (1) An issuer's obligation to its customer includes good faith and observance of any general banking usage but unless otherwise agreed does not include liability or responsibility

(a) for performance of the underlying contract for sale or other transaction between the customer and the beneficiary; or

(b) for any act or omission of any person other than itself or its own branch or for loss or destruction of a draft, demand or document in transit or in the possession of others; or

(c) based on knowledge or lack of knowledge of any usage of any particular trade.

(2) An issuer must examine documents with care so as to ascertain that on their face they appear to comply with the terms of the credit but unless otherwise agreed assumes no liability or responsibility for the genuineness, falsification or effect of any document which appears on such examination to be regular on its face.

(3) A non-bank issuer is not bound by any banking usage of which it has no knowledge.

§ 5–110. Availability of Credit in Portions; Presenter's Reservation of Lien or Claim. (1) Unless otherwise specified a credit may be used in portions in the discretion of the beneficiary.

(2) Unless otherwise specified a person by presenting a documentary draft or demand for payment under a credit relinquishes upon its honor all claims to the documents and a person by transferring such draft or demand

or causing such presentment authorizes such relinquishment. An explicit reservation of claim makes the draft or demand non-complying.

§ 5–111. Warranties on Transfer and Presentment. (1) Unless otherwise agreed the beneficiary by transferring or presenting a documentary draft or demand for payment warrants to all interested parties that the necessary conditions of the credit have been complied with. This is in addition to any warranties arising under Articles 3, 4, 7 and 8.

(2) Unless otherwise agreed a negotiating, advising, confirming, collecting or issuing bank presenting or transferring a draft or demand for payment under a credit warrants only the matters warranted by a collecting bank under Article 4 and any such bank transferring a document warrants only the matters warranted by an intermediary under Articles 7 and 8.

§ 5–112. Time Allowed for Honor or Rejection; Withholding Honor or Rejection by Consent; "Presenter." (1) A bank to which a documentary draft or demand for payment is presented under a credit may without dishonor of the draft, demand or credit

(a) defer honor until the close of the third banking day following receipt of the documents; and

(b) further defer honor if the presenter has expressly or impliedly consented thereto.

Failure to honor within the time here specified constitutes dishonor of the draft or demand and of the credit [except as otherwise provided in subsection (4) of Section 5–114 on conditional payment].

Note: *The bracketed language in the last sentence of subsection (1) should be included only if the optional provisions of Section 5–114(4) and (5) are included.*

(2) Upon dishonor the bank may unless otherwise instructed fulfill its duty to return the draft or demand and the documents by holding them at the disposal of the presenter and sending him an advice to that effect.

(3) "Presenter" means any person presenting a draft or demand for payment for honor under a credit even though that person is a confirming bank or other correspondent which is acting under an issuer's authorization.

§ 5–113. Indemnities. (1) A bank seeking to obtain (whether for itself or another) honor, negotiation or reimbursement under a credit may give an indemnity to induce such honor, negotiation or reimbursement.

(2) An indemnity agreement inducing honor, negotiation or reimbursement

(a) unless otherwise explicitly agreed applies to defects in the documents but not in the goods; and

(b) unless a longer time is explicitly agreed expires at the end of ten business days following receipt of the documents by, the ultimate customer unless notice of objection is sent before such expiration date. The ultimate customer may send notice of objection to the person from whom he received the documents and any bank receiving such notice is under a duty to send notice to its transferor before its midnight deadline.

§ 5–114. Issuer's Duty and Privilege to Honor; Right to Reimbursement. (1) An issuer must honor a draft or demand for payment which complies with the terms of the relevant credit regardless of whether the goods or documents conform to the underlying contract for sale or other contract between the customer and the beneficiary. The issuer is not excused from honor of such a draft or demand by reason of an additional general term that all documents must be satisfactory to the issuer, but an issuer may require that specified documents must be satisfactory to it.

(2) Unless otherwise agreed when documents appear on their face to comply with the terms of a credit but a required document does not in fact conform to the warranties made on negotiation or transfer of a document of title (Section 7–507) or of a security (Section 8–306) or is forged or fraudulent or there is fraud in the transaction

(a) the issuer must honor the draft or demand for payment if honor is demanded by a negotiating bank or other holder of the draft or demand which has taken the draft or demand under the credit and under circumstances which would make it a holder in due course (Section 3–302) and in an appropriate case would make it a person to whom a document of title has been duly negotiated (Section 7–502) or a bona fide purchaser of a security (Section 8–302); and

(b) in all other cases as against its customer, an issuer acting in good faith may honor the draft or demand for payment despite notification from the customer of fraud, forgery or other defect not apparent on the face of the documents but a court of appropriate jurisdiction may enjoin such honor.

(3) Unless otherwise agreed an issuer which has duly honored a draft or demand for payment is entitled to immediate reimbursement of any payment made under the credit and to be put in effectively available funds not later than the day before maturity of any acceptance made under the credit.

[(4) When a credit provides for payment by the issuer on receipt of notice that the required documents are in the possession of a correspondent or other agent of the issuer

(a) any payment made on receipt of such notice is conditional; and

(b) the issuer may reject documents which do not comply with the credit if it does so within three banking days following its receipt of the documents; and

(c) in the event of such rejection, the issuer is entitled by charge back or otherwise to return of the payment made.]

[(5) In the case covered by subsection (4) failure to reject documents within the time specified in sub-paragraph (b) constitutes acceptance of the documents and makes the payment final in favor of the beneficiary.]

Note: *Subsections (4) and (5) are bracketed as optional. If they are included the bracketed language in the last sentence of Section 5–112(1) should also be included.*

§ 5–115. Remedy for Improper Dishonor or Anticipatory Repudiation. (1) When an issuer wrongfully dishonors a draft or demand for payment presented under a credit the person entitled to honor has with respect to any documents the rights of a person in the position of a seller (Section 2–707) and may recover from the issuer the face amount of the draft or demand together with incidental damages under Section 2–710 on seller's incidental damages and interest but less any amount realized by resale or other use or disposition of the subject matter of the transaction. In the event no resale or other utilization is made the documents, goods or other subject matter involved in the transaction must be turned over to the issuer on payment of judgment.

(2) When an issuer wrongfully cancels or otherwise repudiates a credit before presentment of a draft or demand for payment drawn under it the beneficiary has the rights of a seller after anticipatory repudiation by the buyer under Section 2–610 if he learns of the repudiation in time reasonably to avoid procurement of the required documents. Otherwise the beneficiary has an immediate right of action for wrongful dishonor.

§ 5–116. Transfer and Assignment. (1) The right to draw under a credit can be transferred or assigned only when the credit is expressly designated as transferable or assignable.

(2) Even though the credit specifically states that it is nontransferable or nonassignable the beneficiary may before performance of the conditions of the credit assign his right to proceeds. Such an assignment is an assignment of an account under Article 9 on Secured Transactions and is governed by that Article except that

(a) the assignment is ineffective until the letter of credit or advice of credit is delivered to the assignee which delivery constitutes perfection of the security interest under Article 9; and

(b) the issuer may honor drafts or demands for payment drawn under the credit until it receives a notification of the assignment signed by the beneficiary which reasonably identifies the credit involved in the assignment and contains a request to pay the assignee; and

(c) after what reasonably appears to be such a notification has been received the issuer may without dishonor refuse to accept or pay even to a person otherwise entitled to honor until the letter of credit or advice of credit is exhibited to the issuer.

(3) Except where the beneficiary has effectively assigned his right to draw or his right to proceeds, nothing in this section limits his right to transfer or negotiate drafts or demands drawn under the credit.

§ 5–117. Insolvency of Bank Holding Funds for Documentary Credit. (1) Where an issuer or an advising or confirming bank or a bank which has for a customer procured issuance of a credit by another bank becomes insolvent before final payment under the credit and the credit is one to which this Article is made applicable by paragraphs (a) or (b) of Section 5–102(1) on scope, the receipt or allocation of funds or collateral to secure or meet obligations under the credit shall have the following results:

(a) to the extent of any funds or collateral turned over after or before the insolvency as indemnity against or specifically for the purpose of payment of drafts or demands for payment drawn under the designated credit, the drafts or demands are entitled to payment in preference over depositors or other general creditors of the issuer or bank; and

(b) on expiration of the credit or surrender of the beneficiary's rights under it unused any person who has given such funds or collateral is similarly entitled to return thereof; and

(c) a charge to a general or current account with a bank if specifically consented to for the purpose of indemnity against or payment of drafts or demands for payment drawn under the designated credit falls under the same rules as if the funds had been drawn out in cash and then turned over with specific instructions.

(2) After honor or reimbursement under this section the customer or other person for whose account the insolvent bank has acted is entitled to receive the documents involved.

ARTICLE 6: BULK TRANSFERS

§ 6–101. Short Title. This Article shall be known and may be cited as Uniform Commercial Code—Bulk Transfers.

§ 6–102. "Bulk Transfer"; Transfers of Equipment; Enterprises Subject to This Article; Bulk Transfers Subject to This Article. (1) A "bulk transfer" is any transfer in bulk and not in the ordinary course of the transferor's business of a major part of the materials, supplies, merchandise or other inventory (Section 9–109) of an enterprise subject to this Article.

(2) A transfer of a substantial part of the equipment (Section 9–109) of such an enterprise is a bulk transfer if it is made in connection with a bulk transfer of inventory, but not otherwise.

(3) The enterprises subject to this Article are all those whose principal business is the sale of merchandise from stock, including those who manufacture what they sell.

(4) Except as limited by the following section all bulk transfers of goods located within this state are subject to this Article.

§ 6–103. Transfers Excepted From This Article. The following transfers are not subject to this Article:

(1) Those made to give security for the performance of an obligation;

(2) General assignments for the benefit of all the creditors of the transferor, and subsequent transfers by the assignee thereunder;

(3) Transfers in settlement or realization of a lien or other security interest;

(4) Sales by executors, administrators, receivers, trustees in bankruptcy, or any public officer under judicial process;

(5) Sales made in the course of judicial or administrative proceedings for the dissolution or reorganization of a corporation and of which notice is sent to the creditors of the corporation pursuant to order of the court or administrative agency;

(6) Transfers to a person maintaining a known place of business in

this State who becomes bound to pay the debts of the transferor in full and gives public notice of that fact, and who is solvent after becoming so bound;

(7) A transfer to a new business enterprise organized to take over and continue the business, if public notice of the transaction is given and the new enterprise assumes the debts of the transferor and he receives nothing from the transaction except an interest in the new enterprise junior to the claims of creditors;

(8) Transfers of property which is exempt from execution.

Public notice under subsection (6) or subsection (7) may be given by publishing once a week for two consecutive weeks in a newspaper of general circulation where the transferor had its principal place of business in this state an advertisement including the names and addresses of the transferor and transferee and the effective date of the transfer.

§ 6–104. Schedule of Property, List of Creditors. (1) Except as provided with respect to auction sales (Section 6–108), a bulk transfer subject to this Article is ineffective against any creditor of the transferor unless:

(a) The transferee requires the transferor to furnish a list of his existing creditors prepared as stated in this section; and

(b) The parties prepare a schedule of the property transferred sufficient to identify it; and

(c) The transferee preserves the list and schedule for six months next following the transfer and permits inspection of either or both and copying therefrom at all reasonable hours by any creditor of the transferor, or files the list and schedule in *(a public office to be here identified).*

(2) The list of creditors must be signed and sworn to or affirmed by the transferor or his agent. It must contain the names and business addresses of all creditors of the transferor, with the amounts when known, and also the names of all persons who are known to the transferor to assert claims against him even though such claims are disputed. If the transferor is the obligor of an outstanding issue of bonds, debentures or the like as to which there is an indenture trustee, the list of creditors need include only the name and address of the indenture trustee and the aggregate outstanding principal amount of the issue.

(3) Responsibility for the completeness and accuracy of the list of creditors rests on the transferor, and the transfer is not rendered ineffective by errors or omissions therein unless the transferee is shown to have had knowledge.

§ 6–105. Notice to Creditors. In addition to the requirements of the preceding section, any bulk transfer subject to this Article except one made by auction sale (Section 6–108) is ineffective against any creditor of the transferor unless at least ten days before he takes possession of the goods or pays for them, whichever happens first, the transferee gives notice of the transfer in the manner and to the persons hereafter provided (Section 6–107).

[**§ 6–106. Application of the Proceeds.** In addition to the requirements of the two preceding sections:

(1) Upon every bulk transfer subject to this Article for which new con-

sideration becomes payable except those made by sale at auction it is the duty of the transferee to assure that such consideration is applied so far as necessary to pay those debts of the transferor which are either shown on the list furnished by the transferor (Section 6–104) or filed in writing in the place stated in the notice (Section 6–107) within thirty days after the mailing of such notice. This duty of the transferee runs to all the holders of such debts, and may be enforced by any of them for the benefit of all.

(2) If any of said debts are in dispute the necessary sum may be withheld from distribution until the dispute is settled or adjudicated.

(3) If the consideration payable is not enough to pay all of the said debts in full distribution shall be made pro rata.]

Note: *This section is bracketed to indicate division of opinion as to whether or not it is a wise provision, and to suggest that this is a point on which State enactments may differ without serious damage to the principle of uniformity.*

In any State where this section is omitted, the following parts of sections, also bracketed in the text, should also be omitted, namely:

Section 6–107(2)(e).
6–108(3)(c).
6–109(2).

In any State where this section is enacted, these other provisions should be also.

Optional Subsection (4)

[(4) The transferee may within ten days after he takes possession of the goods pay the consideration into the (specify court) in the county where the transferor had its principal place of business in this state and thereafter may discharge his duty under this section by giving notice by registered or certified mail to all the persons to whom the duty runs that the consideration has been paid into that court and that they should file their claims there. On motion of any interested party, the court may order the distribution of the consideration to the persons entitled to it.]

Note: *Optional subsection (4) is recommended for those states which do not have a general statute providing for payment of money into court.*

§ 6–107. The Notice. (1) The notice to creditors (Section 6–105) shall state:

(a) that a bulk transfer is about to be made; and

(b) the names and business addresses of the transferor and transferee, and all other business names and addresses used by the transferor within three years last past so far as known to the transferee; and

(c) whether or not all the debts of the transferor are to be paid in full as they fall due as a result of the transaction, and if so, the address to which creditors should send their bills.

(2) If the debts of the transferor are not to be paid in full as they fall due or if the transferee is in doubt on that point then the notice shall state further:

(a) the location and general description of the property to be transferred and the estimated total of the transferor's debts;

(b) the address where the schedule of property and list of creditors (Section 6–104) may be inspected;

(c) whether the transfer is to pay existing debts and if so the amount of such debts and to whom owing;

(d) whether the transfer is for new consideration and if so the amount of such consideration and the time and place of payment; [and]

(e) if for new consideration the time and place where creditors of the transferor are to file their claims.]

(3) The notice in any case shall be delivered personally or sent by registered or certified mail to all the persons shown on the list of creditors furnished by the transferor (Section 6–104) and to all other persons who are known to the transferee to hold or assert claims against the transferor.

§ 6–108. Auction Sales; "Auctioneer." (1) A bulk transfer is subject to this Article even though it is by sale at auction, but only in the manner and with the results stated in this section.

(2) The transferor shall furnish a list of his creditors and assist in the preparation of a schedule of the property to be sold, both prepared as before stated (Section 6–104).

(3) The person or persons other than the transferor who direct, control or are responsible for the auction are collectively called the "auctioneer." The auctioneer shall:

(a) receive and retain the list of creditors and prepare and retain the schedule of property for the period stated in this Article (Section 6–104);

(b) give notice of the auction personally or by registered or certified mail at least ten days before it occurs to all persons shown on the list of creditors and to all other persons who are known to him to hold or assert claims against the transferor; [and]

[(c) assure that the net proceeds of the auction are applied as provided in this Article (Section 6–106).]

(4) Failure of the auctioneer to perform any of these duties does not affect the validity of the sale or the title of the purchasers, but if the auctioneer knows that the auction constitutes a bulk transfer such failure renders the auctioneer liable to the creditors of the transferor as a class for the sums owing to them from the transferor up to but not exceeding the net proceeds of the auction. If the auctioneer consists of several persons their liability is joint and several.

§ 6–109. What Creditors Protected; [Credit for Payment to Particular Creditors]. (1) The creditors of the transferor mentioned in this Article are those holding claims based on transactions or events occurring before the bulk transfer, but creditors who become such after notice to creditors is given (Sections 6–105 and 6–107) are not entitled to notice.

[(2) Against the aggregate obligation imposed by the provisions of this Article concerning the application of the proceeds (Section 6–106 and subsection (3) (c) of 6–108) the transferee or auctioneer is entitled to credit for sums paid to particular creditors of the transferor, not exceeding the sums believed in good faith at the time of the payment to be properly payable to such creditors.]

§ 6–110. Subsequent Transfers. When the title of a transferee to property is subject to a defect by reason of his non-compliance with the requirements of this Article, then:

(1) a purchaser of any of such property from such transferee who pays no value or who takes with notice of such non-compliance takes subject to such defect, but

(2) a purchaser for value in good faith and without such notice takes free of such defect.

§ 6–111. Limitation of Actions and Levies. No action under this Article shall be brought nor levy made more than six months after the date on which the transferee took possession of the goods unless the transfer has been concealed. If the transfer has been concealed, actions may be brought or levies made within six months after its discovery.

Note to Article 6: *Section 6–106 is bracketed to indicate division of opinion as to whether or not it is a wise provision, and to suggest that this is a point on which State enactments may differ without serious damage to the principle of uniformity.*

In any State where Section 6–106 is not enacted, the following parts of sections, also bracketed in the text, should also be omitted, namely:

Sec. 6–107(2)(e).
6–108(3)(c).
6–109(2).

In any State where Section 6–106 is enacted, these other provisions should be also.

ARTICLE 7: WAREHOUSE RECEIPTS, BILLS OF LADING AND OTHER DOCUMENTS OF TITLE

Part 1

General

§ 7–101. Short Title. This Article shall be known and may be cited as Uniform Commercial Code—Documents of Title.

§ 7–102. Definitions and Index of Definitions. (1) In this Article, unless the context otherwise requires:

(a) "Bailee" means the person who by a warehouse receipt, bill of lading or other document of title acknowledges possession of goods and contracts to deliver them.

(b) "Consignee" means the person named in a bill to whom or to whose order the bill promises delivery.

(c) "Consignor" means the person named in a bill as the person from whom the goods have been received for shipment.

(d) "Delivery order" means a written order to deliver goods directed to a warehouseman, carrier or other person who in the ordinary course of business issues warehouse receipts or bills of lading.

(e) "Document" means document of title as defined in the general definitions in Article 1 (Section 1–201).

(f) "Goods" means all things which are treated as movable for the purposes of a contract of storage or transportation.

(g) "Issuer" means a bailee who issues a document except that in relation to an unaccepted delivery order it means the person who orders the possessor of goods to deliver. Issuer includes any person for whom an agent or employee purports to act in issuing a document if the agent or employee has real or apparent authority to issue documents, notwithstanding that the issuer received no goods or that the goods were misdescribed or that in any other respect the agent or employee violated his instructions.

(h) "Warehouseman" is a person engaged in the business of storing goods for hire.

(2) Other definitions applying to this Article or to specified Parts thereof, and the sections in which they appear are:

"Duly negotiate." Section 7–501.

"Person entitled under the document." Section 7–403(4).

(3) Definitions in other Articles applying to this Article and the sections in which they appear are:

"Contract for sale." Section 2–106.

"Overseas." Section 2–323.

"Receipt" of goods. Section 2–103.

(4) In addition Article 1 contains general definitions and principles of construction and interpretation applicable throughout this Article.

§ 7–103. Relation of Article to Treaty, Statute, Tariff, Classification or Regulation. To the extent that any treaty or statute of the United States, regulatory statute of this State or tariff, classification or regulation filed or issued pursuant thereto is applicable, the provisions of this Article are subject thereto.

§ 7–104. Negotiable and Non-Negotiable Warehouse Receipt, Bill of Lading or Other Document of Title. (1) A warehouse receipt, bill of lading or other document of title is negotiable

(a) if by its terms the goods are to be delivered to bearer or to the order of a named person; or

(b) where recognized in overseas trade, if it runs to a named person or assigns.

(2) Any other document is non-negotiable. A bill of lading in which it is stated that the goods are consigned to a named person is not made negotiable by a provision that the goods are to be delivered only against a written order signed by the same or another named person.

§ 7–105. Construction Against Negative Implication. The omission from either Part 2 or Part 3 of this Article of a provision corresponding to a provision made in the other Part does not imply that a corresponding rule of law is not applicable.

Part 2

Warehouse Receipts: Special Provisions

§ 7–201. Who May Issue a Warehouse Receipt; Storage Under Government Bond. (1) A warehouse receipt may be issued by any warehouseman.

(2) Where goods including distilled spirits and agricultural commodities are stored under a statute requiring a bond against withdrawal or a license for the issuance of receipts in the nature of warehouse receipts, a receipt issued for the goods has like effect as a warehouse receipt even though issued by a person who is the owner of the goods and is not a warehouseman.

§ 7–202. Form of Warehouse Receipt; Essential Terms; Optional Terms. (1) A warehouse receipt need not be in any particular form.

(2) Unless a warehouse receipt embodies within its written or printed terms each of the following, the warehouseman is liable for damages caused by the omission to a person injured thereby:

(a) the location of the warehouse where the goods are stored;

(b) the date of issue of the receipt;

(c) the consecutive number of the receipt;

(d) a statement whether the goods received will be delivered to the bearer, to a specified person, or to a specified person or his order;

(e) the rate of storage and handling charges, except that where goods are stored under a field warehousing arrangement a statement of that fact is sufficient on a non-negotiable receipt;

(f) a description of the goods or of the packages containing them;

(g) the signature of the warehouseman, which may be made by his authorized agent;

(h) if the receipt is issued for goods of which the warehouseman is owner, either solely or jointly or in common with others, the fact of such ownership; and

(i) a statement of the amount of advances made and of liabilities incurred for which the warehouseman claims a lien or security interest (Section 7–209). If the precise amount of such advances made or of such liabilities incurred is, at the time of the issue of the receipt, unknown to the warehouseman or to his agent who issues it, a statement of the fact that advances have been made or liabilities incurred and the prupose thereof is sufficient.

(3) A warehouseman may insert in his receipt any other terms which are not contrary to the provisions of this Act and do not impair his obligation of delivery (Section 7–403) or his duty of care (Section 7–204). Any contrary provisions shall be ineffective.

§ 7–203. Liability of Non-Receipt or Misdescription. A party to or purchaser for value in good faith of a document of title other than a bill of lading relying in either case upon the description therein of the goods may recover from the issuer damages caused by the non-receipt or misdescription of the goods, except to the extent that the document conspicuously indicates that the issuer does not know whether any part or all of the goods in fact were received or conform to the description, as where the description is in terms of marks or labels or kind, quantity, or condition, or the receipt or description is qualified by "contents, condition and quality unknown," "said to contain" or the like, if such indication be true, or the party or purchaser otherwise has notice.

§ 7–204. Duty of Care; Contractual Limitation of Warehouseman's Liability. (1) A warehouseman is liable for damages for loss of or injury to the goods caused by his failure to exercise such care in regard to them as a reasonably careful man would exercise under like circumstances but unless otherwise agreed he is not liable for damages which could not have been avoided by the exercise of such care.

(2) Damages may be limited by a term in the warehouse receipt or storage agreement limiting the amount of liability in case of loss or damage, and setting forth a specific liability per article or item, or value per unit of weight, beyond which the warehouseman shall not be liable; provided, however, that such liability may on written request of the bailor at the time of signing such storage agreement or within a reasonable time after receipt of the warehouse receipt be increased on part or all of the goods thereunder, in which event increased rates may be charged based on such increased valuation, but that no such increase shall be permitted contrary to a lawful limitation of liability contained in the warehouseman's tariff, if any. No such limitation is effective with respect to the warehouseman's liability for conversion to his own use.

(3) Reasonable provisions as to the time and manner of presenting claims and instituting actions based on the bailment may be included in the warehouse receipt or tariff.

(4) This section does not impair or repeal . . .

Note: *Insert in subsection (4) a reference to any statute which imposes a higher responsibility upon the warehouseman or invalidates contractual limitations which would be permissible under this Article.*

§ 7–205. Title Under Warehouse Receipt Defeated in Certain Cases. A buyer in the ordinary course of business of fungible goods sold and delivered by a warehouseman who is also in the business of buying and selling such goods takes free of any claim under a warehouse receipt even though it has been duly negotiated.

§ 7–206. Termination of Storage at Warehouseman's Option. (1) A warehouseman may on notifying the person on whose account the goods are held and any other person known to claim an interest in the goods require payment of any charges and removal of the goods from the warehouse at the termination of the period of storage fixed by the document, or, if no period is fixed, within a stated period not less than thirty days after the notification. If the goods are not removed before the date specified in the notification, the warehouseman may sell them in accordance with the provisions of the section on enforcement of a warehouseman's lien (Section 7–210).

(2) If a warehouseman in good faith believes that the goods are about to deteriorate or decline in value to less than the amount of his lien within the time prescribed in subsection (1) for notification, advertisement and sale, the warehouseman may specify in the notification any reasonable shorter time for removal of the goods and in case the goods are not removed, may sell them at public sale held not less than one week after a single advertisement or posting.

(3) If as a result of a quality or condition of the goods of which the warehouseman had no notice at the time of deposit the goods are a hazard to other property or to the warehouse or to persons, the warehouseman may sell the goods at public or private sale without advertisement on reasonable notification to all persons known to claim an interest in the goods. If the warehouseman after a reasonable effort is unable to sell the goods he may dispose of them in any lawful manner and shall incur no liability by reason of such disposition.

(4) The warehouseman must deliver the goods to any person entitled to them under this Article upon due demand made at any time prior to sale or other disposition under this section.

(5) The warehouseman may satisfy his lien from the proceeds of any sale or disposition under this section but must hold the balance for delivery on the demand of any person to whom he would have been bound to deliver the goods.

§ 7–207. Goods Must Be Kept Separate; Fungible Goods. (1) Unless the warehouse receipt otherwise provides, a warehouseman must keep separate the goods covered by each receipt so as to permit at all times identification and delivery of those goods except that different lots of fungible goods may be commingled.

(2) Fungible goods so commingled are owned in common by the persons entitled thereto and the warehouseman is severally liable to each owner for that owner's share. Where because of overissue a mass of fungible goods is insufficient to meet all the receipts which the warehouseman has issued against it, the persons entitled include all holders to whom overissued receipts have been duly negotiated.

§ 7–208. Altered Warehouse Receipts. Where a blank in a negotiable warehouse receipt has been filled in without authority, a purchaser for value and without notice of the want of authority may treat the insertion as authorized. Any other unauthorized alteration leaves any receipt enforceable against the issuer according to its original tenor.

§ 7–209. Lien of Warehouseman. (1) A warehouseman has a lien against the bailor on the goods covered by a warehouse receipt or on the proceeds thereof in his possession for charges for storage or transportation (including demurrage and terminal charges), insurance, labor, or charges present or future in relation to the goods, and for expenses necessary for preservation of the goods or reasonably incurred in their sale pursuant to law. If the person on whose account the goods are held is liable for like charges or expenses in relation to other goods whenever deposited and it is stated in the receipt that a lien is claimed for charges and expenses in relation to other goods, the warehouseman also has a lien against him for such charges and expenses whether or not the other goods have been delivered by the warehouseman. But against a person to whom a negotiable warehouse receipt is duly negotiated a warehouseman's lien is limited to charges in an amount or at a rate specified on the receipt or if no charges are so specified then to a reasonable charge for storage of the goods covered by the receipt subsequent to the date of the receipt.

(2) The warehouseman may also reserve a security interest against the bailor for a maximum amount specified on the receipt for charges other than those specified in subsection (1), such as for money advanced and interest. Such a security interest is governed by the Article on Secured Transactions (Article 9).

(3) (a) A warehouseman's lien for charges and expenses under subsection (1) or a security interest under subsection (2) is also effective against any person who so entrusted the bailor with possession of the goods that a pledge of them by him to a good faith purchaser for value would have been valid but is not effective against a person as to whom the document confers no right in the goods covered by it under Section 7–503.

(b) A warehouseman's lien on household goods for charges and expenses in relation to the goods under subsection (1) is also effective against all persons if the depositor was the legal possessor of the goods at the time of deposit. "Household goods" means furniture, furnishings and personal effects used by the depositor in a dwelling.

(4) A warehouseman loses his lien on any goods which he voluntarily delivers or which he unjustifiably refuses to deliver.

§ 7–210. Enforcement of Warehouseman's Lien. (1) Except as provided in subsection (2), a warehouseman's lien may be enforced by public or private sale of the goods in bloc or in parcels, at any time or place and on any terms which are commercially reasonable, after notifying all persons known to claim an interest in the goods. Such notification must include a statement of the amount due, the nature of the proposed sale and the time and place of any public sale. The fact that a better price could have been obtained by a sale at a different time or in a different method from that selected by the warehouseman is not of itself sufficient to establish that the sale was not made in a commercially reasonable manner. If the warehouseman either sells the goods in the usual manner in any recognized market therefor, or if he sells at the price current in such market at the time of his sale, or if he has otherwise sold in conformity with commercially reasonable practices among dealers in the type of goods sold, he has sold in a commercially reasonable manner. A sale of more goods than apparently necessary to be offered to insure satisfaction of the obligation is not commercially reasonable except in cases covered by the preceding sentence.

(2) A warehouseman's lien on goods other than goods stored by a merchant in the course of his business may be enforced only as follows:

(a) All persons known to claim an interest in the goods must be notified.

(b) The notification must be delivered in person or sent by registered or certified letter to the last known address of any person to be notified.

(c) The notification must include an itemized statement of the claim, a description of the goods subject to the lien, a demand for payment within a specified time not less than ten days after receipt of the notification, and a conspicuous statement that unless the claim is paid within the time the goods will be advertised for sale and sold by auction at a specified time and place.

(d) The sale must conform to the terms of the notification.

(e) The sale must be held at the nearest suitable place to that where the goods are held or stored.

(f) After the expiration of the time given in the notification, an advertisement of the sale must be published once a week for two weeks consecutively in a newspaper of general circulation where the sale is to be held. The advertisement must include a description of the goods, the name of the person on whose account they are being held, and the time and place of the sale. The sale must take place at least fifteen days after the first publication. If there is no newspaper of general circulation where the sale is to be held, the advertisement must be posted at least ten days before the sale in not less than six conspicuous places in the neighborhood of the proposed sale.

(3) Before any sale pursuant to this section any person claiming a right in the goods may pay the amount necessary to satisfy the lien and the reasonable expenses incurred under this section. In that event the goods must not be sold, but must be retained by the warehouseman subject to the terms of the receipt and this Article.

(4) The warehouseman may buy at any public sale pursuant to this section.

(5) A purchaser in good faith of goods sold to enforce a warehouseman's lien takes the goods free of any rights of persons against whom the lien was valid, despite noncompliance by the warehouseman with the requirements of this section.

(6) The warehouseman may satisfy his lien from the proceeds of any sale pursuant to this section but must hold the balance, if any, for delivery on demand to any person to whom he would have been bound to deliver the goods.

(7) The rights provided by this section shall be in addition to all other rights allowed by law to a creditor against his debtor.

(8) Where a lien is on goods stored by a merchant in the course of his business the lien may be enforced in accordance with either subsection (1) or (2).

(9) The warehouseman is liable for damages caused by failure to comply with the requirements for sale under this section and in case of willful violation is liable for conversion.

Part 3

Bills of Lading: Special Provisions

§ 7–301. Liability for Non-Receipt or Misdescription; "Said to Contain"; "Shipper's Load and Count"; Improper Handling. (1) A consignee of a non-negotiable bill who has given value in good faith or a holder to whom a negotiable bill has been duly negotiated relying in either case upon the description therein of the goods, or upon the date therein shown, may recover from the issuer damages caused by the misdating of the bill or the nonreceipt or misdescription of the goods, except to the extent that the document indicates that the issuer does not know whether any part or all of the goods in fact were received or conform to the description, as where the description is in terms of marks or labels or kind, quantity, or condition or the receipt or description is qualified by "contents or condition of contents of packages unknown," "said to contain," "shipper's weight, load and count" or the like, if such indication be true.

(2) When goods are loaded by an issuer who is a common carrier, the issuer must count the packages of goods if package freight and ascertain the kind and quantity if bulk freight. In such cases "shipper's weight, load and count" or other words indicating that the description was made by the shipper are ineffective except as to freight concealed by packages.

(3) When bulk freight is loaded by a shipper who makes available to the issuer adequate facilities for weighing such freight, an issuer who is a common carrier must ascertain the kind and quantity within a reasonable time after receiving the written request of the shipper to do so. In such cases "shipper's weight" or other words of like purport are ineffective.

(4) The issuer may by inserting in the bill the words "shipper's weight, load and count" or other words of like purport indicate that the goods were loaded by the shipper; and if such statement be true the issuer shall not be liable for damages caused by the improper loading. But their omission does not imply liability for such damages.

(5) The shipper shall be deemed to have guaranteed to the issuer the accuracy at the time of shipment of the description, marks, labels, number, kind, quantity, condition and weight, as furnished by him; and the shipper shall indemnify the issuer against damage caused by inaccuracies in such particulars. The right of the issuer to such indemnity shall in no way limit his responsibility and liability under the contract of carriage to any person other than the shipper.

§ 7–302. Through Bills of Lading and Similar Documents. (1) The issuer of a through bill of lading or other document embodying an undertaking to be performed in part by persons acting as its agents or by connecting carriers is liable to anyone entitled to recover on the document for any breach by such other persons or by a connecting carrier of its obligation under the document but to the extent that the bill covers an undertaking to be performed overseas or in territory not contiguous to the continental United States or an undertaking including matters other than transportation this liability may be varied by agreement of the parties.

(2) Where goods covered by a through bill of lading or other document embodying an undertaking to be performed in part by persons other than the issuer are received by any such person, he is subject with respect to his own performance while the goods are in his possession to the obligation of the issuer. His obligation is discharged by delivery of the goods to another such person pursuant to the document, and does not include liability for breach by any other such persons or by the issuer.

(3) The issuer of such through bill of lading or other document shall be entitled to recover from the connecting carrier or such other person in possession of the goods when the breach of the obligation under the document occurred, the amount it may be required to pay to anyone entitled to recover on the document therefor, as may be evidenced by any receipt, judgment, or transcript thereof, and the amount of any expense reasonably incurred by it in defending any action brought by anyone entitled to recover on the document therefor.

§ 7–303. Diversion; Reconsignment; Change of Instructions. (1) Unless the bill of lading otherwise provides, the carrier may deliver the goods to a person or destination other than that stated in the bill or may otherwise dispose of the goods on instructions from

(a) the holder of a negotiable bill; or

(b) the consignor on a non-negotiable bill notwithstanding contrary instructions from the consignee; or

(c) the consignee on a non-negotiable bill in the absence of contrary instructions from the consignor, if the goods have arrived at the billed destination or if the consignee is in possession of the bill; or

(d) the consignee on a non-negotiable bill if he is entitled as against the consignor to dispose of them.

(2) Unless such instructions are noted on a negotiable bill of lading, a person to whom the bill is duly negotiated can hold the bailee according to the original terms.

§ 7–304. Bills of Lading in a Set. (1) Except where customary in overseas transportation, a bill of lading must not be issued in a set of parts. The issuer is liable for damages caused by violation of this subsection.

(2) Where a bill of lading is lawfully drawn in a set of parts, each of which is numbered and expressed to be valid only if the goods have not been delivered against any other part, the whole of the parts constitute one bill.

(3) Where a bill of lading is lawfully issued in a set of parts and different parts are negotiated to different persons, the title of the holder to whom

the first due negotiation is made prevails as to both the document and the goods even though any later holder may have received the goods from the carrier in good faith and discharged the carrier's obligation by surrender of his part.

(4) Any person who negotiates or transfers a single part of a bill of lading drawn in a set is liable to holders of that part as if it were the whole set.

(5) The bailee is obliged to deliver in accordance with Part 4 of this Article against the first presented part of a bill of lading lawfully drawn in a set. Such delivery discharges the bailee's obligation on the whole bill.

§ 7–305. Destination Bills. (1) Instead of issuing a bill of lading to the consignor at the place of shipment a carrier may at the request of the consignor procure the bill to be issued at destination or at any other place designated in the request.

(2) Upon request of anyone entitled as against the carrier to control the goods while in transit and on surrender of any outstanding bill of lading or other receipt covering such goods, the issuer may procure a substitute bill to be issued at any place designated in the request.

§ 7–306. Altered Bills of Lading. An unauthorized alteration or filling in of a blank in a bill of lading leaves the bill enforceable according to its original tenor.

§ 7–307. Lien of Carrier. (1) A carrier has a lien on the goods covered by a bill of lading for charges subsequent to the date of its receipt of the goods for storage or transportation (including demurrage and terminal charges) and for expenses necessary for preservation of the goods incident to their transportation or reasonably incurred in their sale pursuant to law. But against a purchaser for value of a negotiable bill of lading a carrier's lien is limited to charges stated in the bill or the applicable tariffs, or if no charges are stated then to a reasonable charge.

(2) A lien for charges and expenses under subsection (1) on goods which the carrier was required by law to receive for transportation is effective against the consignor or any person entitled to the goods unless the carrier had notice that the consignor lacked authority to subject the goods to such charges and expenses. Any other lien under subsection (1) is effective against the consignor and any person who permitted the bailor to have control or possession of the goods unless the carrier had notice that the bailor lacked such authority.

(3) A carrier loses his lien on any goods which he voluntarily delivers or which he unjustifiably refuses to deliver.

§ 7–308. Enforcement of Carrier's Lien. (1) A carrier's lien may be enforced by public or private sale of the goods, in bloc or in parcels, at any time or place and on any terms which are commercially reasonable, after notifying all persons known to claim an interest in the goods. Such notification must include a statement of the amount due, the nature of the proposed sale and the time and place of any public sale. The fact that a better price could have been obtained by a sale at a different time or in a different method from that selected by the carrier is not of itself sufficient to establish that the sale was not made in a commercially reasonable manner. If the carrier either sells the goods in the usual manner in any recognized market therefor or if he sells at the price current in such market at the time of his sale or if he has otherwise sold in conformity with commercially reasonable practices among dealers in the type of goods sold he has sold in a commercially reasonable manner. A sale of more goods than apparently necessary to be offered to ensure satisfaction of the obligation is not commercially reasonable except in cases covered by the preceding sentence.

(2) Before any sale pursuant to this section any person claiming a right in the goods may pay the amount necessary to satisfy the lien and the reasonable expenses incurred under this section. In that event the goods must not be sold, but must be retained by the carrier subject to the terms of the bill and this Article.

(3) The carrier may buy at any public sale pursuant to this section.

(4) A purchaser in good faith of goods sold to enforce a carrier's lien takes the goods free of any rights of persons against whom the lien was valid, despite noncompliance by the carrier with the requirements of this section.

(5) The carrier may satisfy his lien from the proceeds of any sale pursuant to this section but must hold the balance, if any, for delivery on demand to any person to whom he would have been bound to deliver the goods.

(6) The rights provided by this section shall be in addition to all other rights allowed by law to a creditor against his debtor.

(7) A carrier's lien may be enforced in accordance with either subsection (1) or the procedure set forth in subsection (2) of Section 7–210.

(8) The carrier is liable for damages caused by failure to comply with the requirements for sale under this section and in case of willful violation is liable for conversion.

§ 7–309. Duty of Care; Contractual Limitation of Carrier's Liability. (1) A carrier who issues a bill of lading whether negotiable or non-negotiable must exercise the degree of care in relation to the goods which a reasonably careful man would exercise under like circumstances. This subsection does not repeal or change any law or rule of law which imposes liability upon a common carrier for damages not caused by its negligence.

(2) Damages may be limited by a provision that the carrier's liability shall not exceed a value stated in the document if the carrier's rates are dependent upon value and the consignor by the carrier's tariff is afforded an opportunity to declare a higher value or a value as lawfully provided in the tariff, or where no tariff is filed he is otherwise advised of such opportunity; but no such limitation is effective with respect to the carrier's liability for conversion to its own use.

(3) Reasonable provisions as to the time and manner of presenting claims and instituting actions based on the shipment may be included in a bill of lading or tariff.

Part 4

Warehouse Receipts and Bills of Lading: General Obligations

§ 7–401. Irregularities in Issue of Receipt or Bill or Conduct of Issuer. The obligations imposed by this Article on an issuer apply to a document of title regardless of the fact that

(a) the document may not comply with the requirements of this Article or of any other law or regulation regarding its issue, form or content; or

(b) the issuer may have violated laws regulating the conduct of his business; or

(c) the goods covered by the document were owned by the bailee at the time the document was issued; or

(d) the person issuing the document does not come within the definition of warehouseman if it purports to be a warehouse receipt.

§ 7–402. Duplicate Receipt or Bill; Overissue. Neither a duplicate nor any other document of title purporting to cover goods already represented by an outstanding document of the same issuer confers any right in the goods, except as provided in the case of bills in a set, overissue of documents for fungible goods and substitutes for lost, stolen or destroyed documents. But the issuer is liable for damages caused by his overissue or failure to identify a duplicate document as such by conspicuous notation on its face.

§ 7–403. Obligation of Warehouseman or Carrier to Deliver; Excuse. (1) The bailee must deliver the goods to a person entitled under the document who complies with subsections (2) and (3), unless and to the extent that the bailee establishes any of the following:

(a) delivery of the goods to a person whose receipt was rightful as against the claimant;

(b) damage to or delay, loss or destruction of the goods for which the bailee is not liable [, but the burden of establishing negligence in such cases is on the person entitled under the document];

Note: *The brackets in (1)(b) indicate that State enactments may differ on this point without serious damage to the principle of uniformity.*

(c) previous sale or other disposition of the goods in lawful enforcement of a lien or on warehouseman's lawful termination of storage;

(d) the exercise by a seller of his right to stop delivery pursuant to the provisions of the Article on Sales (Section 2–705);

(e) a diversion, reconsignment or other disposition pursuant to the provisions of this Article (Section 7–303) or tariff regulating such right;

(f) release, satisfaction or any other fact affording a personal defense against the claimant;

(g) any other lawful excuse.

(2) A person claiming goods covered by a document of title must satisfy the bailee's lien where the bailee so requests or where the bailee is prohibited by law from delivering the goods until the charges are paid.

(3) Unless the person claiming is one against whom the document con-

fers no right under Sec. 7–503(1), he must surrender for cancellation or notation of partial deliveries any outstanding negotiable document covering the goods, and the bailee must cancel the document or conspicuously note the partial delivery thereon or be liable to any person to whom the document is duly negotiated.

(4) "Person entitled under the document" means holder in the case of a negotiable document, or the person to whom delivery is to be made by the terms of or pursuant to written instructions under a non-negotiable document.

§ 7–404. No Liability for Good Faith Delivery Pursuant to Receipt or Bill. A bailee who in good faith including observance of reasonable commercial standards has received goods and delivered or otherwise disposed of them according to the terms of the document of title or pursuant to this Article is not liable therefore. This rule applies even though the person from whom he received the goods had no authority to procure the document or to dispose of the goods and even though the person to whom he delivered the goods had no authority to receive them.

Part 5

Warehouse Receipts and Bills of Lading: Negotiation and Transfer

§ 7–501. Form of Negotiation and Requirements of "Due Negotiation." (1) A negotiable document of title running to the order of a named person is negotiated by his indorsement and delivery. After his indorsement in blank or to bearer any person can negotiate it by delivery alone.

(2) (a) A negotiable document of title is also negotiated by delivery alone when by its original terms it runs to bearer.

(b) When a document running to the order of a named person is delivered to him the effect is the same as if the document had been negotiated.

(3) Negotiation of a negotiable document of title after it has been indorsed to a specified person requires indorsement by the special indorsee as well as delivery.

(4) A negotiable document of title is "duly negotiated" when it is negotiated in the manner stated in this section to a holder who purchases it in good faith without notice of any defense against or claim to it on the part of any person and for value, unless it is established that the negotiation is not in the regular course of business or financing or involves receiving the document in settlement or payment of a money obligation.

(5) Indorsement of a non-negotiable document neither makes it negotiable nor adds to the transferee's rights.

(6) The naming in a negotiable bill of a person to be notified of the arrival of the goods does not limit the negotiability of the bill nor constitute notice to a purchaser thereof of any interest of such person in the goods.

§ 7–502. Rights Acquired by Due Negotiation. (1) Subject to the following section and to the provisions of Section 7–205 on fungible goods, a holder to whom a negotiable document of title has been duly negotiated acquires thereby:

(a) title to the document;

(b) title to the goods;

(c) all rights accruing under the law of agency or estoppel, including rights to goods delivered to the bailee after the document was issued; and

(d) the direct obligation of the issuer to hold or deliver the goods according to the terms of the document free of any defense or claim to him except those arising under the terms of the document or under this Article. In the case of a delivery order the bailee's obligation accrues only upon acceptance and the obligation acquired by the holder is that the issuer and any indorser will procure the acceptance of the bailee.

(2) Subject to the following section, title and rights so acquired are not defeated by any stoppage of the goods represented by the document or by surrender of such goods by the bailee, and are not impaired even though the negotiation or any prior negotiation constituted a breach of duty or even though any person has been deprived of possession of the document by misrepresentation, fraud, accident, mistake, duress, loss, theft or conversion, or even though a previous sale or other transfer of the goods or document has been made to a third person.

§ 7–503. Document of Title to Goods Defeated in Certain Cases. (1) A document of title confers no right in goods against a person who before issuance of the document had a legal interest or a perfected security interest in them and who neither

(a) delivered or entrusted them or any document of title covering them to the bailor or his nominee with actual or apparent authority to ship, store or sell or with power to obtain delivery under this Article (Section 7–403) or with power of disposition under this Act (Sections 2–403 and 9–307) or other statute or rule of law; nor

(b) acquiesced in the procurement by the bailor or his nominee of any document of title.

(2) Title to goods based upon an unaccepted delivery order is subject to the rights of anyone to whom a negotiable warehouse receipt or bill of lading covering the goods has been duly negotiated. Such a title may be defeated under the next section to the same extent as the rights of the issuer or a transferee from the issuer.

(3) Title to goods based upon a bill of lading issued to a freight forwarder is subject to the rights of anyone to whom a bill issued by the freight forwarder is duly negotiated; but delivery by the carrier in accordance with Part 4 of this Article pursuant to its own bill of lading discharges the carrier's obligation to deliver.

§ 7–504. Rights Acquired in the Absence of Due Negotiation; Effect of Diversion; Seller's Stoppage of Delivery. (1) A transferee of a document, whether negotiable or non-negotiable, to whom the document has been delivered but not duly negotiated, acquires the title and rights which his transferor had or had actual authority to convey.

(2) In the case of a non-negotiable document, until but not after the bailee receives notification of the transfer, the rights of the transferee may be defeated

(a) by those creditors of the transferor who could treat the sale as void under Section 2–402; or

(b) by a buyer from the transferor in ordinary course of business if the bailee has delivered the goods to the buyer or received notification of his rights; or

(c) as against the bailee by good faith dealings of the bailee with the transferor.

(3) A diversion or other change of shipping instructions by the consignor in a non-negotiable bill of lading which causes the bailee not the deliver to the consignee defeats the consignee's title to the goods if they have been delivered to a buyer in ordinary course of business and in any event defeats the consignee's rights against the bailee.

(4) Delivery pursuant to a non-negotiable document may be stopped by a seller under Section 2–705, and subject to the requirement of due notification there provided. A bailee honoring the seller's instructions is entitled to be indemnified by the seller against any resulting loss or expense.

§ 7–505. Indorser Not a Guarantor for Other Parties. The indorsement of a document of title issued by a bailee does not make the indorser liable for any default by the bailee or by previous indorsers.

§ 7–506. Delivery Without Indorsement: Right to Compel Indorsement. The transferee of a negotiable document of title has a specifically enforceable right to have his transferor supply any necessary indorsement but the transfer becomes a negotiation only as of the time the indorsement is supplied.

§ 7–507. Warranties on Negotiation or Transfer of Receipt or Bill. Where a person negotiates or transfers a document of title for value otherwise than as a mere intermediary under the next following section, then unless otherwise agreed he warrants to his immediate purchaser only in addition to any warranty made in selling the goods

(a) that the document is genuine; and

(b) that he has no knowledge of any fact which would impair its validity or worth; and

(c) that his negotiation or transfer is rightful and fully effective with respect to the title to the document and the goods it represents.

§ 7–508. Warranties of Collecting Bank as to Documents. A collecting bank or other intermediary known to be entrusted with documents on behalf of another or with collection of a draft or other claim against delivery of documents warrants by such delivery of the documents only its own good faith and authority. This rule applies even though the intermediary has purchased or made advances against the claim or draft to be collected.

§ 7–509. Receipt or Bill: When Adequate Compliance With Commercial Contract. The question whether a document is adequate to fulfill the obligations of a contract for sale or the conditions of a credit is governed by the Articles on Sales (Article 2) and on Letters of Credit (Article 5).

Part 6

Warehouse Receipts and Bills of Lading: Miscellaneous Provisions

§ 7–601. Lost and Missing Documents. (1) If a document has been lost, stolen or destroyed, a court may order delivery of the goods or issuance of a substitute document and the bailee may without liability to any person comply with such order. If the document was negotiable the claimant must post security approved by the court to indemnify any person who may suffer loss as a result of non-surrender of the document. If the document was not negotiable, such security may be required at the discretion of the court. The court may also in its discretion order payment of the bailee's reasonable costs and counsel fees.

(2) A bailee who without court order delivers goods to a person claiming under a missing negotiable document is liable to any person injured thereby, and if the delivery is not in good faith becomes liable for conversion. Delivery in good faith is not conversion if made in accordance with a filed classification or tariff or, where no classification or tariff is filed, if the claimant posts security with the bailee in an amount at least double the value of the goods at the time of posting to indemnify any person injured by the delivery who files a notice of claim within one year after the delivery.

§ 7–602. Attachment of Goods Covered by a Negotiable Document. Except where the document was originally issued upon delivery of the goods by a person who had no power to dispose of them, no lien attaches by virtue of any judicial process to goods in the possession of a bailee for which a negotiable document of title is outstanding unless the document be first surrendered to the bailee or its negotiation enjoined, and the bailee shall not be compelled to deliver the goods pursuant to process until the document is surrendered to him or impounded by the court. One who purchases the document for value without notice of the process or injunction takes free of the lien imposed by judicial process.

§ 7–603. Conflicting Claims; Interpleader. If more than one person claims title or possession of the goods, the bailee is excused from delivery until he has had a reasonable time to ascertain the validity of the adverse claims or to bring an action to compel all claimants to interplead and may compel such interpleader, either in defending an action for nondelivery of the goods, or by original action, whichever is appropriate.

ARTICLE 8: INVESTMENT SECURITIES

Part 1

Short Title and General Matters

§ 8–101. Short Title. This Article shall be known and may be cited as Uniform Commercial Code—Investment Securities.

§ 8–102. Definitions and Index of Definitions. (1) In this Article unless the context otherwise requires

(a) A "security" is an instrument which

(i) is issued in bearer or registered form; and

(ii) is of a type commonly dealt in upon securities exchanges or markets or commonly recognized in any area in which it is issued or dealt in as a medium for investment; and

(iii) is either one of a class or series or by its terms is divisible into a class or series of instruments; and

(iv) evidences a share, participation or other interest in property or in an enterprise or evidences an obligation of the issuer.

(b) A writing which is a security is governed by this Article and not by Uniform Commercial Code—Commercial Paper even though it also meets the requirements of that Article. This Article does not apply to money.

(c) A security is in "registered form" when it specifies a person entitled to the security or to the rights it evidences and when its transfer may be registered upon books maintained for that purpose by or on behalf of an issuer or the security so states.

(d) A security is in "bearer form" when it runs to bearer according to its terms and not by reason of any indorsement.

(2) A "subsequent purchaser" is a person who takes other than by original issue.

(3) A "clearing corporation" is a corporation all of the capital stock of which is held by or for a national securities exchange or association registered under a statute of the United States such as the Securities Exchange Act of 1934.

(4) A "custodian bank" is any bank or trust company which is supervised and examined by state or federal authority having supervision over banks and which is acting as custodian for a clearing corporation.

(5) Other definitions applying to this Article or to specified Parts thereof and the sections in which they appear are:

"Adverse claim." Section 8–301.

"Bona fide purchaser." Section 8–302.

"Broker." Section 8–303.

"Guarantee of the signature." Section 8–402.

"Intermediary Bank." Section 4–105.

"Issuer." Section 8–201.

"Overissue." Section 8–104.

(6) In addition Article 1 contains general definitions and principles of construction and interpretation applicable throughout this Article.

§ 8–103. Issuer's Lien. A lien upon a security in favor of an issuer thereof is valid against a purchaser only if the right of the issuer to such lien is noted conspicuously on the security.

§ 8–104. Effect of Overissue; "Overissue." (1) The provisions of this Article which validate a security or compel its issue or reissue do not apply to the extent that validation, issue or reissue would result in overissue; but

(a) if an identical security which does not constitute an overissue is reasonably available for purchase, the person entitled to issue or validation may compel the issuer to purchase and deliver such a security to him against surrender of the security, if any, which he holds; or

(b) if a security is not so available for purchase, the person entitled to issue or validation may recover from the issuer the price he or the last purchaser for value paid for it with interest from the date of his demand.

(2) "Overissue" means the issue of securities in excess of the amount which the issuer has corporate power to issue.

§ 8–105. Securities Negotiable; Presumptions. (1) Securities governed by this Article are negotiable instruments.

(2) In any action on a security

(a) unless specifically denied in the pleadings, each signature on the security or in a necessary indorsement is admitted;

(b) when the effectiveness of a signature is put in issue the burden of establishing it is on the party claiming under the signature but the signature is presumed to be genuine or authorized;

(c) when signatures are admitted or established production of the instrument entitles a holder to recover on it unless the defendant establishes a defense or a defect going to the validity of the security; and

(d) after it is shown that a defense or defect exists the plaintiff has the burden of establishing that he or some person under whom he claims is a person against whom the defense or defect is ineffective (Section 8–202).

§ 8–106. Applicability. The validity of a security and the rights and duties of the issuer with respect to registration of transfer are governed by the law (including the conflict of laws rules) of the jurisdiction of organization of the issuer.

§ 8–107. Securities Deliverable; Action for Price. (1) Unless otherwise agreed and subject to any applicable law or regulation respecting short sales, a person obligated to deliver securities may deliver any security of the specified issue in bearer form or registered in the name of the transferee or indorsed to him or in blank.

(2) When the buyer fails to pay the price as it comes due under the contract of sale the seller may recover the price

(a) of securities accepted by the buyer; and

(b) of other securities if efforts at their resale would be unduly burdensome or if there is no readily available market for their resale.

Part 2

Issue–Issuer

§ 8–201. "Issuer." (1) With respect to obligations on or defenses to a security "issuer" includes a person who

(a) places or authorizes the placing of his name on a security (otherwise than as authenticating trustee, registrar, transfer agent or the like) to evidence that it represents a share, participation or other interest in his property or in an enterprise or to evidence his duty to perform an obligation evidenced by the security; or

(b) directly or indirecting creates fractional interests in his rights or property which fractional interests are evidenced by securities; or

(c) becomes responsible for or in place of any other person described as an issuer in this section.

(2) With respect to obligations on or defenses to a security a guarantor is an issuer to the extent of his guaranty whether or not his obligation is noted on the security.

(3) With respect to registration of transfer (Part 4 of this Article) "issuer" means a person on whose behalf transfer books are maintained.

§ 8–202. Issuer's Responsibility and Defenses; Notice of Defect or Defense. (1) Even against a purchaser for value and without notice, the terms of a security include those stated on the security and those made part of the security by reference to another instrument, indenture or document or to a constitution, statute, ordinance, rule, regulation, order or the like to the extent that the terms so referred to do not conflict with the stated terms. Such a reference does not of itself charge a purchaser for value with notice of a defect going to the validity of the security even though the security expressly states that a person accepting it admits such notice.

(2) (a) A security other than one issued by a government or governmental agency or unit even though issued with a defect going to its validity is valid in the hands of a purchaser for value and without notice of the particular defect unless the defect involves a violation of constitutional provisions in which case the security is valid in the hands of a subsequent purchaser for value and without notice of the defect.

(b) The rule of subparagraph (a) applies to an issuer which is a government or governmental agency or unit only if either there has been substantial compliance with the legal requirements governing the issue or the issuer has received a substantial consideration for the issue as a whole or for the particular security and a stated purpose of the issue is one for which the issuer has power to borrow money or issue the security.

(3) Except as otherwise provided in the case of certain unauthorized signatures on issue (Section 8–205), lack of genuineness of a security is a complete defense even against a purchaser for value and without notice.

(4) All other defenses of the issuer including nondelivery and conditional delivery of the security are ineffective against a purchaser for value who has taken without notice of the particular defense.

(5) Nothing in this section shall be construed to affect the right of a party to a "when, as and if issued" or a "when distributed" contract to cancel the contract in the event of a material change in the character of the security which is the subject of the contract or in the plan or arrangement pursuant to which such security is to be issued or distributed.

§ 8–203. Staleness as Notice of Defects or Defenses. (1) After an act or event which creates a right to immediate performance of the principal obligation evidenced by the security or which sets a date on or after which the security is to be presented or surrendered for redemption or exchange, a purchaser is charged with notice of any defect in its issue or defense of the issuer

(a) if the act or event is one requiring the payment of money or the delivery of securities or both on presentation or surrender of the security and such funds or securities are available on the date set for payment or exchange and he takes the security more than one year after that date; and

(b) if the act or event is not covered by paragraph (a) and he takes the security more than two years after the date set for surrender or presentation or the date on which such performance became due.

(2) A call which has been revoked is not within subsection (1).

§ 8–204. Effect of Issuer's Restrictions on Transfer. Unless noted conspicuously on the security a restriction on transfer imposed by the issuer even though otherwise lawful is ineffective except against a person with actual knowledge of it.

§ 8–205. Effect of Unauthorized Signature on Issue. An unauthorized signature placed on a security prior to or in the course of issue is ineffective except that the signature is effective in favor of a purchaser for value and without notice of the lack of authority if the signing has been done by

(a) an authenticating trustee, registrar, transfer agent or other person entrusted by the issuer with the signing of the security or of similar securities or their immediate preparation for signing; or

(b) an employee of the issuer or of any of the foregoing entrusted with responsible handling of the security.

§ 8–206. Completion or Alteration of Instrument. (1) Where a security contains the signatures necessary to its issue or transfer but is incomplete in any other respect.

(a) any person may complete it by filling in the blanks as authorized; and

(b) even though the blanks are incorrectly filled in, the security as completed is enforceable by a purchaser who took it for value and without notice of such incorrectness.

(2) A complete security which has been improperly altered even though fraudulently remains enforceable but only according to its original terms.

§ 8–207. Rights of Issuer With Respect to Registered Owners. (1) Prior to due presentment for registration of transfer of a security in registered form the issuer or indenture trustee may treat the registered owner as the person exclusively entitled to vote, to receive notifications and otherwise to exercise all the rights and powers of an owner.

(2) Nothing in this Article shall be construed to affect the liability of the registered owner of a security for calls, assessments or the like.

§ 8–208. Effect of Signature of Authenticating Trustee, Registrar or Transfer Agent. (1) A person placing his signature upon a security as authenticating trustee, registrar, transfer agent or the like warrants to a purchaser for value without notice of the particular defect that

(a) the security is genuine; and

(b) his own participation in the issue of the security is within his capacity and within the scope of the authorization received by him from the issuer; and

(c) he has reasonable grounds to believe that the security is in the form and within the amount the issuer is authorized to issue.

(2) Unless otherwise agreed, a person by so placing his signature does not assume responsibility for the validity of the security in other respects.

Part 3

Purchase

§ 8–301. Rights Acquired by Purchaser; "Adverse Claim"; Title Acquired by Bona Fide Purchaser. (1) Upon delivery of a security the purchaser acquires the rights in the security which his transferor had or had actual authority to convey except that a purchaser who has himself been a party to any fraud or illegality affecting the security or who as a prior holder had notice of an adverse claim cannot improve his position by taking from a later bona fide purchaser. "Adverse claim" includes a claim that a transfer was or would be wrongful or that a particular adverse person is the owner of or has an interest in the security.

(2) A bona fide purchaser in addition to acquiring the rights of a purchaser also acquires the security free of any adverse claim.

(3) A purchaser of a limited interest acquires rights only to the extent of the interest purchased.

§ 8–302. "Bona Fide Purchaser." A "bona fide purchaser" is a purchaser for value in good faith and without notice of any adverse claim who takes delivery of a security in bearer form or of one in registered form issued to him or indorsed to him or in blank.

§ 8–303. "Broker." "Broker" means a person engaged for all or part of his time in the business of buying and selling securities, who in the transaction concerned acts for, or buys a security from or sells a security to a customer. Nothing in this Article determines the capacity in which a person acts for purposes of any other statute or rule to which such person is subject.

§ 8–304. Notice to Purchaser of Adverse Claims. (1) A purchaser (including a broker for the seller or buyer but excluding an intermediary bank) of a security is charged with notice of adverse claims if

(a) the security whether in bearer or registered form has been indorsed "for collection" or "for surrender" or for some other purpose not involving transfer; or

(b) the security is in bearer form and has on it an unambiguous statement that it is the property of a person other than the transferor. The mere writing of a name on a security is not such a statement.

(2) The fact that the purchaser (including a broker for the seller or buyer) has notice that the security is held for a third person or is registered

in the name of or indorsed by a fiduciary does not create a duty of inquiry into the rightfulness of the transfer or constitute notice of adverse claims. If, however, the purchaser (excluding an intermediary bank) has knowledge that the proceeds are being used or that the transaction is for the individual benefit of the fiduciary or otherwise in breach of duty, the purchaser is charged with notice of adverse claims.

§ 8–305. Staleness as Notice of Adverse Claims. An act or event which creates a right to immediate performance of the principal obligation evidenced by the security or which sets a date on or after which the security is to be presented or surrendered for redemption or exchange does not of itself constitute any notice of adverse claims except in the case of a purchase

(a) after one year from any date set for such presentment or surrender for redemption or exchange; or

(b) after six months from any date set for payment of money against presentation or surrender of the security if funds are available for payment on that date.

§ 8–306. Warranties on Presentment and Transfer. (1) A person who presents a security for registration of transfer or for payment or exchange warrants to the issuer that he is entitled to the registration, payment or exchange. But a purchaser for value without notice of adverse claims who receives a new, reissued or re-registered security on registration of transfer warrants only that he has no knowledge of any unauthorized signature (Section 8–311) in a necessary indorsement.

(2) A person by transferring a security to a purchaser for value warrants only that

(a) his transfer is effective and rightful; and

(b) the security is genuine and has not been materially altered; and

(c) he knows no fact which might impair the validity of the security.

(3) Where a security is delivered by an intermediary known to be entrusted with delivery of the security on behalf of another or with collection of a draft or other claim against such delivery, the intermediary by such delivery warrants only his own good faith and authority even though he has purchased or made advances against the claim to be collected against the delivery.

(4) A pledgee or other holder for security who redelivers the security received, or after payment and on order of the debtor delivers that security to a third person makes only the warranties of an intermediary under subsection (3).

(5) A broker gives to his customer and to the issuer and a purchaser the warranties provided in this section and has the rights and privileges of a purchaser under this section. The warranties of and in favor of the broker acting as an agent are in addition to applicable warranties given by and in favor of his customer.

§ 8–307. Effect of Delivery Without Indorsement; Right to Compel Indorsement. Where a security in registered form has been delivered to a purchaser without a necessary indorsement he may become a bona fide purchaser only as of the time the indorsement is supplied, but against the transferor the transfer is complete upon delivery and the purchaser has a specifically enforceable right to have any necessary indorsement supplied.

§ 8–308. Indorsement, How Made; Special Indorsement; Indorser Not a Guarantor; Partial Assignment. (1) An indorsement of a security in registered form is made when an appropriate person signs on it or on a separate document an assignment or transfer of the security or a power to assign or transfer it or when the signature of such person is written without more upon the back of the security.

(2) An indorsement may be in blank or special. An indorsement in blank includes an indorsement to bearer. A special indorsement specifies the person to whom the security is to be transferred, or who has power to transfer it. A holder may convert a blank indorsement into a special indorsement.

(3) "An appropriate person" in subsection (1) means

(a) the person specified by the security or by special indorsement to be entitled to the security; or

(b) where the person so specified is described as a fiduciary but is no longer serving in the described capacity,—either that person or his successor; or

(c) where the security or indorsement so specifies more than one person as fiduciaries and one or more are no longer serving in the described capacity,—the remaining fiduciary or fiduciaries, whether or not a successor has been appointed or qualified; or

(d) where the person so specified is an individual and is without capacity to act by virtue of death, incompetence, infancy or otherwise,—his executor, administrator, guardian or like fiduciary; or

(e) where the security or indorsement so specifies more than one person as tenants by the entirety or with right of survivorship and by reason of death all cannot sign,—the survivor or survivors; or

(f) a person having power to sign under applicable law or controlling instrument; or

(g) to the extent that any of the foregoing persons may act through an agent,—his authorized agent.

(4) Unless otherwise agreed the indorser by his indorsement assumes no obligation that the security will be honored by the issuer.

(5) An indorsement purporting to be only of part of a security representing units intended by the issuer to be separately transferable is effective to the extent of the indorsement.

(6) Whether the person signing is appropriate is determined as of the date of signing and an indorsement by such a person does not become unauthorized for the purposes of this Article by virtue of any subsequent change of circumstances.

(7) Failure of a fiduciary to comply with a controlling instrument or with the law of the state having jurisdiction of the fiduciary relationship, including any law requiring the fiduciary to obtain court approval of the transfer,

does not render his indorsement unauthorized for the purposes of this Article.

§ 8–309. Effect of Indorsement Without Delivery. An indorsement of a security whether special or in blank does not constitute a transfer until delivery of the security on which it appears or if the indorsement is on a separate document until delivery of both the document and the security.

§ 8–310. Indorsement of Security in Bearer Form. An indorsement of a security in bearer form may give notice of adverse claims (Section 8–304) but does not otherwise affect any right to registration the holder may possess.

§ 8–311. Effect of Unauthorized Indorsement. Unless the owner has ratified an unauthorized indorsement or is otherwise precluded from asserting its ineffectiveness

(a) he may assert its ineffectiveness against the issuer or any purchaser other than a purchaser for value and without notice of adverse claims who has in good faith received a new, reissued or re-registered security on registration of transfer; and

(b) an issuer who registers the transfer of a security upon the unauthorized indorsement is subject to liability for improper registration (Section 8–404).

§ 8–312. Effect of Guaranteeing Signature or Indorsement. (1) Any person guaranteeing a signature of an indorser of a security warrants that at the time of signing

(a) the signature was genuine; and

(b) the signer was an appropriate person to indorse (Section 8–308); and

(c) the signer had legal capacity to sign. But the guarantor does not otherwise warrant the rightfulness of the particular transfer.

(2) Any person may guarantee an indorsement of a security and by so doing warrants not only the signature (subsection 1) but also the rightfulness of the particular transfer in all respects. But no issuer may require a guarantee of indorsement as a condition to registration of transfer.

(3) The foregoing warranties are made to any person taking or dealing with the security in reliance on the guarantee and the guarantor is liable to such person for any loss resulting from breach of the warranties.

§ 8–313. When Delivery to the Purchaser Occurs; Purchaser's Broker as Holder. (1) Delivery to a purchaser occurs when

(a) he or a person designated by him acquires possession of a security; or

(b) his broker acquires possession of a security specially indorsed to or issued in the name of the purchaser; or

(c) his broker sends him confirmation of the purchase and also by book entry or otherwise identifies a specific security in the broker's possession as belonging to the purchaser; or

(d) with respect to an identified security to be delivered while still in the possession of a third person when that person acknowledges that he holds for the purchaser.

(e) appropriate entries on the books of a clearing corporation are made under Section 8–320.

(2) The purchaser is the owner of a security held for him by his broker, but is not the holder except as specified in subparagraphs (b), (c) and (e) of subsection (1). Where a security is part of a fungible bulk the purchaser is the owner of a proportionate property interest in the fungible bulk.

(3) Notice of an adverse claim received by the broker or by the purchaser after the broker takes delivery as a holder for value is not effective either as to the broker or as to the purchaser. However, as between the broker and the purchaser the purchaser may demand delivery of an equivalent security as to which no notice of an adverse claim has been received.

§ 8–314. Duty to Deliver, When Completed. (1) Unless otherwise agreed where a sale of a security is made on an exchange or otherwise through brokers

(a) the selling customer fulfills his duty to deliver when he places such a security in the possession of the selling broker or of a person designated by the broker or if requested causes an acknowledgment to be made to the selling broker that it is held for him; and

(b) the selling broker including a correspondent broker acting for a selling customer fulfills his duty to deliver by placing the security or a like security in the possession of the buying broker or a person designated by him or by effecting clearance of the sale in accordance with the rules of the exchange on which the transaction took place.

(2) Except as otherwise provided in this section and unless otherwise agreed, a transferor's duty to deliver a security under a contract of purchase is not fulfilled until he places the security in form to be negotiated by the purchaser in the possession of the purchaser or of a person designated by him or at the purchaser's request causes an acknowledgment to be made to the purchaser that it is held for him. Unless made on an exchange a sale to a broker purchasing for his own account is within this subsection and not within subsection (1).

§ 8–315. Action Against Purchaser Based Upon Wrongful Transfer. (1) Any person against whom the transfer of a security is wrongful for any reason, including his incapacity, may against any one except a bona fide purchaser reclaim possession of the security or obtain possession of any new security evidencing all or part of the same rights or have damages.

(2) If the transfer is wrongful because of an unauthorized indorsement, the owner may also reclaim or obtain possession of the security or new security even from a bona fide purchaser if the ineffectiveness of the purported indorsement can be asserted against him under the provisions of this Article on unauthorized indorsements (Section 8–311).

(3) The right to obtain or reclaim possession of a security may be specifically enforced and its transfer enjoined and the security impounded pending the litigation.

§ 8–316. Purchaser's Right to Requisites for Registration of Transfer on Books. Unless otherwise agreed the transferor must on due demand supply his purchaser with any proof of his authority to transfer or with any other requisite which may be necessary to obtain registration of the transfer of the security but if the transfer is not for value a transferor need not do so unless the purchaser furnishes the necessary expenses. Failure to comply with a demand made within a reasonable time gives the purchaser the right to reject or rescind the transfer.

§ 8–317. Attachment or Levy Upon Security. (1) No attachment or levy upon a security or any share or other interest evidenced thereby which is outstanding shall be valid until the security is actually seized by the officer making the attachment or levy but a security which has been surrendered to the issuer may be attached or levied upon at the source.

(2) A creditor whose debtor is the owner of a security shall be entitled to such aid from courts of appropriate jurisdiction, by injunction or otherwise, in reaching such security or in satisfying the claim by means thereof as is allowed at law or in equity in regard to property which cannot readily be attached or levied upon by ordinary legal process.

§ 8–318. No Conversion by Good Faith Delivery. An agent or bailee who in good faith (including observance of reasonable commercial standards if he is in the business of buying, selling or otherwise dealing with securities) has received securities and sold, pledged or delivered them according to the instructions of his principal is not liable for conversion or for participation in breach of fiduciary duty although the principal had no right to dispose of them.

§ 8–319. Statute of Frauds. A contract for the sale of securities is not enforceable by way of action or defense unless

(a) there is some writing signed by the party against whom enforcement is sought or by his authorized agent or broker sufficient to indicate that a contract has been made for sale of a stated quantity of described securities at a defined or stated price; or

(b) delivery of the security has been accepted or payment has been made but the contract is enforceable under this provision only to the extent of such delivery or payment; or

(c) within a reasonable time a writing in confirmation of the sale or purchase and sufficient against the sender under paragraph (a) has been received by the party against whom enforcement is sought and he has failed to send written objection to its contents within ten days after its receipt; or

(d) the party against whom enforcement is sought admits in his pleading, testimony or otherwise in court that a contract was made for sale of a stated quantity of described securities at a defined or stated price.

§ 8–320. Transfer or Pledge within a Central Depository System. (1) If a security

(a) is in the custody of a clearing corporation or of a custodian bank or a nominee of either subject to the instructions of the clearing corporation; and

(b) is in bearer form or indorsed in blank by an appropriate person or registered in the name of the clearing corporation or custodian bank or a nominee of either; and

(c) is shown on the account of a transferor or pledgor on the books of the clearing corporation;

then, in addition to other methods, a transfer or pledge of the security or any interest therein may be effected by the making of appropriate entries on the books of the clearing corporation reducing the account of the transferor or pledgor and increasing the account of the transferee or pledgee by the amount of the obligation or the number of shares or rights transferred or pledged.

(2) Under this section entries may be with respect to like securities or interests therein as a part of a fungible bulk and may refer merely to a quantity of a particular security without reference to the name of the registered owner, certificate or bond number or the like and, in appropriate cases, may be on a net basis taking into account other transfers or pledges of the same security.

(3) A transfer or pledge under this section has the effect of a delivery of a security in bearer form or duly indorsed in blank (Section 8–301) representing the amount of the obligation or the number of shares or rights transferred or pledged. If a pledge or the creation of a security interest is intended, the making of entries has the effect of a taking of delivery by the pledgee or a secured party (Sections 9–304 and 9–305). A transferee or pledgee under this section is a holder.

(4) A transfer or pledge under this section does not constitute a registration of transfer under Part 4 of this Article.

(5) That entries made on the books of the clearing corporation as provided in subsection (1) are not appropriate does not affect the validity or effect of the entries nor the liabilities or obligations of the clearing corporation to any person adversely affected thereby.

Part 4

Registration

§ 8–401. Duty of Issuer to Register Transfer. (1) Where a security in registered form is presented to the issuer with a request to register transfer, the issuer is under a duty to register the transfer as requested if

(a) the security is indorsed by the appropriate person or persons (Section 8–308); and

(b) reasonable assurance is given that those indorsements are genuine and effective (Section 8–402); and

(c) the issuer has no duty to inquire into adverse claims or has discharged any such duty (Section 8–403); and

(d) any applicable law relating to the collection of taxes has been complied with; and

(e) the transfer is in fact rightful or is to a bona fide purchaser.

(2) Where an issuer is under a duty to register a transfer of a security the issuer is also liable to the person presenting it for registration or his principal for loss resulting from any unreasonable delay in registration or from failure or refusal to register the transfer.

§ 8–402. Assurance that Indorsements Are Effective. (1) The issuer may require the following assurance that each necessary indorsement (Section 8–308) is genuine and effective

(a) in all cases, a guarantee of the signature (subsection (1) of Section 8–312) of the person indorsing; and

(b) where the indorsement is by an agent, appropriate assurance of authority to sign;

(c) where the indorsement is by a fiduciary, appropriate evidence of appointment or incumbency;

(d) where there is more than one fiduciary, reasonable assurance that all who are required to sign have done so;

(e) where the indorsement is by a person not covered by any of the foregoing, assurance appropriate to the case corresponding as nearly as may be to the foregoing.

(2) A "guarantee of the signature" in subsection (1) means a guarantee signed by or on behalf of a person reasonably believed by the issuer to be responsible. The issuer may adopt standards with respect to responsibility provided such standards are not manifestly unreasonable.

(3) "Appropriate evidence of appointment or incumbency" in subsection (1) means

(a) in the case of a fiduciary appointed or qualified by a court, a certificate issued by or under the direction or supervision of that court or an officer thereof and dated within sixty days before the date of presentation for transfer; or

(b) in any other case, a copy of a document showing the appointment or a certificate issued by or on behalf of a person reasonably believed by the issuer to be responsible or, in the absence of such a document or certificate, other evidence reasonably deemed by the issuer to be appropriate. The issuer may adopt standards with respect to such evidence provided such standards are not manifestly unreasonable. The issuer is not charged with notice of the contents of any document obtained pursuant to this paragraph (b) except to the extent that the contents relate directly to the appointment or incumbency.

(4) The issuer may elect to require reasonable assurance beyond that specified in this section but if it does so and for a purpose other than that specified in subsection 3(b) both requires and obtains a copy of a will, trust, indenture, articles of co-partnership, by-laws, or other controlling instrument it is charged with notice of all matters contained therein affecting the transfer.

§ 8–403. Limited Duty of Inquiry. (1) An issuer to whom a security is presented for registration is under a duty to inquire into adverse claims if

(a) a written notification of an adverse claim is received at a time and in a manner which affords the issuer a reasonable opportunity to act on it prior to the issuance of a new, reissued or re-registered security and the notification identifies the claimant, the registered owner and the issue of which the security is a part and provides an address for communications directed to the claimant; or

(b) the issuer is charged with notice of an adverse claim from a controlling instrument which it has elected to require under subsection (4) of Section 8–402.

(2) The issuer may discharge any duty of inquiry by any reasonable means, including notifying an adverse claimant by registered or certified mail at the address furnished by him or if there be no such address at his residence or regular place of business that the security has been presented for registration of transfer by a named person, and that the transfer will be registered unless within thirty days from the date of mailing the notification, either

(a) an appropriate restraining order, injunction or other process issues from a court of competent jurisdiction; or

(b) an indemnity bond sufficient in the issuer's judgment to protect the issuer and any transfer agent, registrar or other agent of the issuer involved, from any loss which it or they may suffer by complying with the adverse claim is filed with the issuer.

(3) Unless an issuer is charged with notice of an adverse claim from a controlling instrument which it has elected to require under subsection (4) of Section 8–402 or receives notification of an adverse claim under subsection (1) of this section, where a security presented for registration is indorsed by the appropriate person or persons the issuer is under no duty to inquire into adverse claims. In particular

(a) an issuer registering a security in the name of a person who is a fiduciary or who is described as a fiduciary is not bound to inquire into the existence, extent, or correct description of the fiduciary relationship and thereafter the issuer may assume without inquiry that the newly registered owner continues to be the fiduciary until the issuer receives written notice that the fiduciary is no longer acting as such with respect to the particular security;

(b) an issuer registering transfer on an indorsement by a fiduciary is not bound to inquire whether the transfer is made in compliance with a controlling instrument or with the law of the state having jurisdiction of the fiduciary relationship, including any law requiring the fiduciary to obtain court approval of the transfer; and

(c) the issuer is not charged with notice of the contents of any court record or file or other recorded or unrecorded document is in its possession and even though the transfer is made on the indorsement of a fiduciary to the fiduciary himself or to his nominee.

§ 8–404. Liability and Non-Liability for Registration. (1) Except as otherwise provided in any law relating to the collection of taxes, the issuer is not liable to the owner or any other person suffering loss as a result of the registration of a transfer of a security if

(a) there were on or with the security the necessary indorsements (Section 8–308); and

(b) the issuer has no duty to inquire into adverse claims or has discharged any such duty (Section 8–403).

(2) Where an issuer has registered a transfer of a security to a person not entitled to it the issuer on demand must deliver a like security to the true owner unless

(a) the registration was pursuant to subsection (1); or

(b) the owner is precluded from asserting any claim for registering the transfer under subsection (1) of the following section; or

(c) such delivery would result in overissue, in which case the issuer's liability is governed by Section 8–104.

§ 8–405. Lost, Destroyed and Stolen Securities. (1) Where a security has been lost, apparently destroyed or wrongfully taken and the owner fails to notify the issuer of that fact within a reasonable time after he has notice of it and the issuer registers a transfer of the security before receiving such a notification, the owner is precluded from asserting against the issuer any claim for registering the transfer under the preceding section or any claim to a new security under this section.

(2) Where the owner of a security claims that the security has been lost, destroyed or wrongfully taken, the issuer must issue a new security in place of the original security if the owner

(a) so requests before the issuer has notice that the security has been acquired by a bona fide purchaser; and

(b) files with the issuer a sufficient indemnity bond; and

(c) satisfies any other reasonable requirements imposed by the issuer.

(3) If, after the issue of the new security, a bona fide purchaser of the original security presents it for registration of transfer, the insurer must register the transfer unless registration would result in overissue, in which event the issuer's liability is governed by Section 8–104. In addition to any rights on the indemnity bond, the issuer may recover the new security from the person to whom it was issued or any person taking under him except a bona fide purchaser.

§ 8–406. Duty of Authenticating Trustee, Transfer Agent or Registrar. (1) Where a person acts as authenticating trustee, transfer agent, registrar, or other agent for an issuer in the registration of transfers of its securities or in the issue of new securities or in the cancellation of surrendered securities

(a) he is under a duty to the issuer to exercise good faith and due diligence in performing his functions; and

(b) he has with regard to the particular functions he performs the same obligation to the holder or owner of the security and has the same rights and privileges as the issuer has in regard to those functions.

(2) Notice to an authenticating trustee, transfer agent, registrar or other such agent is notice to the issuer with respect to the functions performed by the agent.

ARTICLE 9: SECURED TRANSACTIONS; SALES OF ACCOUNTS AND CHATTEL PAPER

Part 1

Short Title, Applicability and Definitions

§ 9–101. Short Title. This Article shall be known and may be cited as Uniform Commercial Code—Secured Transactions.

§ 9–102. Policy and Subject Matter of Article. (1) Except as otherwise provided in Section 9–104 on excluded transactions, this Article applies

(a) to any transaction (regardless of its form) which is intended to create a security interest in personal property or fixtures including goods, documents, instruments, general intangibles, chattel paper or accounts; and also

(b) to any sale of accounts or chattel paper.

(2) This Article applies to security interests created by contract including pledge, assignment, chattel mortgage, chattel trust, trust deed, factor's lien, equipment trust, conditional sale, trust receipt, other lien or title retention contract and lease or consignment intended as security. This Article does not apply to statutory liens except as provided in Section 9–310.

(3) The application of this Article to a security interest in a secured obligation is not affected by the fact that the obligation is itself secured by a transaction or interest to which this Article does not apply. Amended in 1972.

Note: *The adoption of this Article should be accompanied by the repeal of existing statutes dealing with conditional sales, trust receipts, factor's liens where the factor is given a non-possessory lien, chattel mortgages, crop mortgages, mortgages on railroad equipment, assignment of accounts and generally statutes regulating security interests in personal property.*

Where the state has a retail installment selling act or small loan act, that legislation should be carefully examined to determine what changes in those acts are needed to conform them to this Article. This Article primarily sets out rules defining rights of a secured party against persons dealing with the debtor; it does not prescribe regulations and controls which may be necessary to curb abuses arising in the small loan business or in the financing of consumer purchases on credit. Accordingly there is no intention to repeal existing regulatory acts in those fields by enactment or re-enactment of Article 9. See Section 9–203(4) and the Note thereto.

§ 9–103. Perfection of Security Interests in Multiple State Transactions. (1) Documents, instruments and ordinary goods.

(a) This subsection applies to documents and instruments and to goods other than those covered by a certificate of title described in subsection (2), mobile goods described in subsection (3), and minerals described in subsection (5).

(b) Except as otherwise provided in this subsection, perfection and the effect of perfection or non-perfection of a security interest in collateral are governed by the law of the jurisdiction where the collateral is when the last event occurs on which is based the assertion that the security interest is perfected or unperfected.

(c) If the parties to a transaction creating a purchase money security interest in goods in one jurisdiction understand at the time that the security interest attaches that the goods will be kept in another jurisdiction, then the law of the other jurisdiction governs the perfection and the effect of perfection or non-perfection of the security interest from the time it attaches until thirty days after the debtor receives possession of the goods and thereafter if the goods are taken to the other jurisdiction before the end of the thirty-day period.

(d) When collateral is brought into and kept in this state while subject to a security interest perfected under the law of the jurisdiction from which the collateral was removed, the security interest remains perfected, but if action is required by Part 3 of this Article to perfect the security interest,

(i) if the action is not taken before the expiration of the period of perfection in the other jurisdiction or the end of four months after the collateral is brought into this state, whichever period first expires, the security interest becomes unperfected at the end of that period and is thereafter deemed to have been unperfected as against a person who became a purchaser after removal;

(ii) if the action is taken before the expiration of the period specified in subparagraph (i) the security interest continues perfected thereafter;

(iii) for the purpose of priority over a buyer of consumer goods (subsection (2) of Section 9–307), the period of the effectiveness of a filing in the jurisdiction from which the collateral is removed is governed by the rules with respect to perfection in subparagraphs (i) and (ii).

(2) Certificate of title.

(a) This subsection applies to goods covered by a certificate of title issued under a statute of this state or of another jurisdiction under the law of which indication of a security interest on the certificate is required as a condition of perfection.

(b) Except as otherwise provided in this subsection, perfection and the effect of perfection or non-perfection of the security interest are governed by the law (including the conflict of laws rules) of the jurisdiction issuing the certificate until four months after the goods are removed from that jurisdiction and thereafter until the goods are registered in another jurisdiction, but in any event not beyond surrender of the certificate. After the expiration of that period, the goods are not covered by the certificate of title within the meaning of this section.

(c) Except with respect to the rights of a buyer described in the next paragraph, a security interest, perfected in another jurisdiction otherwise than by notation on a certificate of title, in goods brought into this state and thereafter covered by a certificate of title issued by this state is subject to the rules stated in paragraph (d) of subsection (1).

(d) If goods are brought into this state while a security interest therein is perfected in any manner under the law of the jurisdiction from which the goods are removed and a certificate of title is issued by this state and the certificate does not show that the goods are subject to the security interest or that they may be subject to security interests not shown on the certificate, the security interest is subordinate to the rights of a buyer of the goods who is not in the business of selling goods of that kind to the extent that he gives value and receives delivery of the goods after issuance of the certificate and without knowledge of the security interest.

(3) Accounts, general intangibles and mobile goods.

(a) This subsection applies to accounts (other than an account described in subsection (5) on minerals) and general intangibles and to goods which are mobile and which are of a type normally used in more than one jurisdiction, such as motor vehicles, trailers, rolling stock, airplanes, shipping containers, road building and construction machinery and commercial harvesting machinery and the like, if the goods are equipment or are inventory leased or held for lease by the debtor to others, and are not covered by a certificate of title described in subsection (2).

(b) The law (including the conflict of laws rules) of the jurisdiction in which the debtor is located governs the perfection and the effect of perfection or nonperfection of the security interest.

(c) If, however, the debtor is located in a jurisdiction which is not a part of the United States, and which does not provide for perfection of the security interest by filing or recording in that jurisdiction, the law of the jurisdiction in the United States in which the debtor has its major executive office in the United States governs the perfection and the effect of perfection or non-perfection of the security interest through filing. In the alternative, if the debtor is located in a jurisdiction which is not a part of the United States or Canada and the collateral is accounts or general intangibles for money due or to become due, the security interest may be perfected by notification to the account debtor. As used in this paragraph, "United States" includes its territories and possessions and the Commonwealth of Puerto Rico.

(d) A debtor shall be deemed located at his place of business if he has one, at his chief executive office if he has more than one place of business, otherwise at his residence. If, however, the debtor is a foreign air carrier under the Federal Aviation Act of 1958, as amended, it shall be deemed located at the designated office of the agent upon whom service of process may be made on behalf of the foreign air carrier.

(e) A security interest perfected under the law of the jurisdiction of the location of the debtor is perfected until the expiration of four months after a change of the debtor's location to another jurisdiction, or until perfection would have ceased by the law of the first jurisdiction, whichever period first expires. Unless perfected in the new jurisdiction before the end of that period, it becomes unperfected thereafter and is deemed to have been unperfected as against a person who became a purchaser after the change.

(4) Chattel paper.

The rules stated for goods in subsection (1) apply to a possessory security interest in chattel paper. The rules stated for accounts in subsection (3) apply to a non-possessory security interest in chattel paper, but the security interest may not be perfected by notification to the account debtor.

(5) Minerals.

Perfection and the effect of perfection or non-perfection of a security interest which is created by a debtor who has an interest in minerals or the like (including oil and gas) before extraction and which attaches thereto as extracted, or which attaches to an account resulting from the sale thereof at the wellhead or minehead are governed by the law (including the conflict of laws rules) of the jurisdiction wherein the wellhead or minehead is located. Amended in 1972.

§ 9–104. Transactions Excluded From Article. This Article does not apply

(a) to a security interest subject to any statute of the United States, to the extent that such statute governs the rights of parties to and third parties affected by transactions in particular types of property; or

(b) to a landlord's lien; or

(c) to a lien given by statute or other rule of law for services or materials except as provided in Section 9–310 on priority of such liens; or

(d) to a transfer of a claim for wages, salary or other compensation of an employee; or

(e) to a transfer by a government or governmental subdivision or agency; or

(f) to a sale of accounts or chattel paper as part of a sale of the business out of which they arose, or an assignment of accounts or chattel paper which is for the purpose of collection only, or a transfer of a right to payment under a contract to an assignee who is also to do the performance under the contract or a transfer of a single account to an assignee in whole or partial satisfaction of a preexisting indebtedness; or

(g) to a transfer of an interest in or claim in or under any policy of insurance, except as provided with respect to proceeds (Section 9–306) and priorities in proceeds (Section 9–312); or

(h) to a right represented by a judgment (other than a judgment taken on a right to payment which was collateral); or

(i) to any right of set-off; or

(j) except to the extent that provision is made for fixtures in Section 9–313, to the creation or transfer of an interest in or lien on real estate, including a lease or rents thereunder; or

(k) to a transfer in whole or in part of any claim arising out of tort; or

(l) to a transfer of an interest in any deposit account (subsection (1) of Section 9–105), except as provided with respect to proceeds (Section 9–306) and priorities in proceeds (Section 9–312).

Amended in 1972.

§ 9–105. Definitions and Index of Definitions. (1) In this Article unless the context otherwise requires:

(a) "Account debtor" means the person who is obligated on an account, chattel paper or general intangible;

(b) "Chattel paper" means a writing or writings which evidence both a monetary obligation and a security interest in or a lease of specific goods, but a charter or other contract involving the use or hire of a vessel is not chattel paper. When a transaction is evidenced both by such a security agreement or a lease and by an instrument or a series of instruments, the group of writings taken together constitutes chattel paper;

(c) "Collateral" means the property subject to a security interest, and includes accounts and chattel paper which have been sold;

(d) "Debtor" means the person who owes payment or other performance of the obligation secured, whether or not he owns or has rights in the collateral, and includes the seller of accounts or chattel paper. Where the debtor and the owner of the collateral are not the same person, the term "debtor" means the owner of the collateral in any provision of the Article dealing with the collateral, the obligor in any provision dealing with the obligation, and may include both where the context so requires;

(e) "Deposit account" means a demand, time, savings, passbook or like account maintained with a bank, savings and loan association, credit union or like organization, other than an account evidenced by a certificate of deposit;

(f) "Document" means document of title as defined in the general definitions of Article 1 (Section 1–201), and a receipt of the kind described in subsection (2) of Section 7–201;

(g) "Encumbrance" includes real estate mortgages and other liens on real estate and all other rights in real estate that are not ownership interests;

(h) "Goods" includes all things which are movable at the time the security interest attaches or which are fixtures (Section 9–313), but does not include money, documents, instruments, accounts, chattel paper, general intangibles, or minerals or the like (including oil and gas) before extraction. "Goods" also includes standing timber which is to be cut and removed under a conveyance or contract for sale, the unborn young of animals, and growing crops;

(i) "Instrument" means a negotiable instrument (defined in Section 3–104), or a security (defined in Section 8–102) or any other writing which evidences a right to the payment of money and is not itself a security agreement or lease and is of a type which is in ordinary course of business transferred by delivery with any necessary indorsement or assignment;

(j) "Mortgage" means a consensual interest created by a real estate mortgage, a trust deed on real estate, or the like;

(k) An advance is made "pursuant to commitment" if the secured party has bound himself to make it, whether or not a subsequent event of default or other event not within his control has relieved or may relieve him from his obligation;

(l) "Security agreement" means an agreement which creates or provides for a security interest;

(m) "Secured party" means a lender, seller or other person in whose favor there is a security interest, including a person to whom accounts or chattel paper have been sold. When the holders of obligations issued under an indenture of trust, equipment trust agreement or the like are represented by a trustee or other person, the representative is the secured party;

(n) "Transmitting utility" means any person primarily engaged in the railroad, street railway or trolley bus business, the electric or electronics communications transmission business, the transmission of goods by pipeline, or the transmission or the production and transmission of electricity, steam, gas or water, or the provision of sewer service.

(2) Other definitions applying to this Article and the sections in which they appear are:

"Account." Section 9–106.
"Attach." Section 9–203.
"Construction mortgage." Section 9–313(1).
"Consumer goods." Section 9–109(1).
"Equipment." Section 9–109(2).
"Farm products." Section 9–109(3).
"Fixture." Section 9–313(1).
"Fixture filing." Section 9–313(1).
"General intangibles." Section 9–106.
"Inventory." Section 9–109(4).
"Lien creditor." Section 9–301(3).
"Proceeds." Section 9–306(1).
"Purchase money security interest." Section 9–107.
"United States." Section 9–103.

(3) The following definitions in other Articles apply to this Article:

"Check." Section 3–104.
"Contract for sale." Section 2–106.
"Holder in due course." Section 3–302.
"Note." Section 3–104.
"Sale." Section 2–106.

(4) In addition Article 1 contains general definitions and principles of construction and interpretation applicable throughout this Article. Amended in 1966, 1972.

§ 9–106. Definitions: "Account"; "General Intangibles." "Account" means any right to payment for goods sold or leased or for services rendered which is not evidenced by an instrument or chattel paper, whether or not it has been earned by performance. "General intangibles" means any personal property (including things in action) other than goods, accounts, chattel paper, documents, instruments, and money. All rights to payment earned or unearned under a charter or other contract involving the use or hire of a vessel and all rights incident to the charter or contract are accounts. Amended in 1966, 1972.

§ 9–107. Definitions: "Purchase Money Security Interest." A security interest is a "purchase money security interest" to the extent that it is

(a) taken or retained by the seller of the collateral to secure all or part of its price; or

(b) taken by a person who by making advances or incurring an obligation gives value to enable the debtor to acquire rights in or the use of collateral if such value is in fact so used.

§ 9–108. When After-Acquired Collateral Not Security for Antecedent Debt. Where a secured party makes an advance, incurs an obligation, releases a perfected security interest, or otherwise gives new value which is to be secured in whole or in part by after-acquired property his security interest in the after-acquired collateral shall be deemed to be taken for new value and not as security for an antecedent debt if the debtor acquires his rights in such collateral either in the ordinary course of his business or under a contract of purchase made pursuant to the security agreement within a reasonable time after new value is given.

§ 9–109. Classification of Goods; "Consumer Goods"; "Equipment"; "Farm Products"; "Inventory." Goods are:

(1) "consumer goods" if they are used or bought for use primarily for personal, family or household purposes;

(2) "equipment" if they are used or bought for use primarily in business (including farming or a profession) or by a debtor who is a non-profit organization or a governmental subdivision or agency or if the goods are not included in the definitions of inventory, farm products or consumer goods;

(3) "farm products" if they are crops or livestock or supplies used or produced in farming operations or if they are products of crops or livestock in their unmanufactured states (such as ginned cotton, wool-clip, maple syrup, milk and eggs), and if they are in the possession of a debtor engaged in raising, fattening, grazing or other farming operations. If goods are farm products they are neither equipment nor inventory;

(4) "inventory" if they are held by a person who holds them for sale or lease or to be furnished under contracts of service or if he has so furnished them, or if they are raw materials, work in process or materials used or consumed in a business. Inventory of a person is not to be classified as his equipment.

§ 9–110. Sufficiency of Description. For the purposes of this Article any description of personal property or real estate is sufficient whether or not it is specific if it reasonably identifies what is described.

§ 9–111. Applicability of Bulk Transfer Laws. The creation of a security interest is not a bulk transfer under Article 6 (see Section 6–103).

§ 9–112. Where Collateral Is Not Owned by Debtor. Unless otherwise agreed, when a secured party knows that collateral is owned by a person who is not the debtor, the owner of the collateral is entitled to receive from the secured party any surplus under Section 9–502(2) or under Section 9–504(1), and is not liable for the debt or for any deficiency after resale, and he has the same right as the debtor

(a) to receive statements under Section 9–208;

(b) to receive notice of and to object to a secured party's proposal to retain the collateral in satisfaction of the indebtedness under Section 9–505;

(c) to redeem the collateral under Section 9–506;

(d) to obtain injunctive or other relief under Section 9–507(1); and

(e) to recover losses caused to him under Section 9–208(2).

§ 9–113. Security Interests Arising Under Article on Sales. A security interest arising solely under the Article on Sales (Article 2) is subject to the provisions of this Article except that to the extent that and so long as the debtor does not have or does not lawfully obtain possession of the goods

(a) no security agreement is necessary to make the security interest enforceable; and

(b) no filing is required to perfect the security interest; and

(c) the rights of the secured party on default by the debtor are governed by the Article on Sales (Article 2).

§ 9–114. Consignment. (1) A person who delivers goods under a consignment which is not a security interest and who would be required to file under this Article by paragraph (3) (c) of Section 2–326 has priority over a secured party who is or becomes a creditor of the consignee and who would have a perfected security interest in the goods if they were the property of the consignee, and also has priority with respect to identifiable cash proceeds received on or before delivery of the goods to a buyer, if

(a) the consignor complies with the filing provision of the Article on Sales with respect to consignments (paragraph (3) (c) of Section 2–326) before the consignee receives possession of the goods; and

(b) the consignor gives notification in writing to the holder of the security interest if the holder has filed a financing statement covering the same types of goods before the date of the filing made by the consignor; and

(c) the holder of the security interest receives the notification within five years before the consignee receives possession of the goods; and

(d) the notification states that the consignor expects to deliver goods on consignment to the consignee, describing the goods by item or type.

(2) In the case of a consignment which is not a security interest and in which the requirements of the preceding subsection have not been met, a person who delivers goods to another is subordinate to a person who would have a perfected security interest in the goods if they were the property of the debtor. Added in 1972.

Part 2

Validity of Security Agreement and Rights of Parties Thereto

§ 9–201. General Validity of Security Agreement. Except as otherwise provided by this Act a security agreement is effective according to its terms between the parties, against purchasers of the collateral and against creditors. Nothing in this Article validates any charge or practice illegal under any statute or regulation thereunder governing usury, small loans, retail installment sales, or the like, or extends the application of any such statute or regulation to any transaction not otherwise subject thereto.

§ 9–202. Title to Collateral Immaterial. Each provision of this Article with regard to rights, obligations and remedies applies whether title to collateral is in the secured party or in the debtor.

§ 9–203. Attachment and Enforceability of Security Interest; Proceeds; Formal Requisites. (1) Subject to the provisions of Section 4–208 on the security interest of a collecting bank and Section 9–113 on a security interest arising under the Article on Sales, a security interest is not enforceable against the debtor of third parties with respect to the collateral and does not attach unless

(a) the collateral is in the possession of the secured party pursuant to agreement, or the debtor has signed a security agreement which contains a description of the collateral and in addition, when the security interest covers crops growing or to be grown or timber to be cut, a description of the land concerned; and

(b) value has been given; and

(c) the debtor has rights in the collateral.

(2) A security interest attaches when it becomes enforceable against the debtor with respect to the collateral. Attachment occurs as soon as all of the events specified in subsection (1) have taken place unless explicit agreement postpones the time of attaching.

(3) Unless otherwise agreed a security agreement gives the secured party the rights to proceeds provided by Section 9–306.

(4) A transaction, although subject to this Article, is also subject to ....*, and in the case of conflict between the provisions of this Article and any such statute, the provisions of such statute control. Failure to comply with any applicable statute has only the effect which is specified therein. Amended in 1972.

Note: *At * in subsection (4) insert reference to any local statute regulating small loans, retail installment sales and the like.*

The foregoing subsection (4) is designed to make it clear that certain transactions, although subject to this Article, must also comply with other applicable legislation.

This Article is designed to regulate all the "security" aspects of transactions within its scope. There is, however, much regulatory legislation, particularly in the consumer field, which supplements this Article and should not be repealed by its enactment. Examples are small loan acts, retail installment selling acts and the like. Such acts may provide for licensing and rate regulation and may prescribe particular forms of contract. Such provisions should remain in force despite the enactment of this Article. On the other hand if a retail installment selling act contains provisions on filing, rights on default, etc., such provisions should be repealed as inconsistent provisions as to deficiencies, penalties, etc., in the Uniform Consumer Credit Code and other recent related legislation should remain because those statutes were drafted after the substantial enactment of the Article and with the intention of modifying certain provisions of this Article as to consumer credit.

§ 9–204. After-Acquired Property; Future Advances. (1) Except as provided in subsection (2), a security agreement may provide that any or all obligations covered by the security agreement are to be secured by after-acquired collateral.

(2) No security interest attaches under an after-acquired property clause to consumer goods other than accessions (Section 9–314) when given as additional security unless the debtor acquires rights in them within ten days after the secured party gives value.

(3) Obligations covered by a security agreement may include future advances or other value whether or not the advances or value are given pursuant to commitment (subsection (1) of Section 9–105). Amended in 1972.

§ 9–205. Use or Disposition of Collateral Without Accounting Permissible. A security interest is not invalid or fraudulent against creditors by reason of liberty in the debtor to use, commingle or dispose of all or part of the collateral (including returned or repossessed goods) or to collect or compromise accounts or chattel paper, or to accept the return of goods or make repossessions, or to use, commingle or dispose of proceeds, or by reason of the failure of the secured party to require the debtor to account for proceeds or replace collateral. This section does not relax the requirements of possession where perfection of a security interest depends upon possession of the collateral by the secured party or by a bailee. Amended in 1972.

§ 9–206. Agreement Not to Assert Defenses Against Assignee; Modification of Sales Warranties Where Security Agreement Exists. (1) Subject to any statute or decision which establishes a different rule for buyers or lessees of consumer goods, an agreement by a buyer or lessee that he will not assert against an assignee any claim or defense which he may have against the seller or lessor is enforceable by an assignee who takes his assignment for value, in good faith and without notice of a claim or defense, except as to defenses of a type which may be asserted against a holder in due course of a negotiable instrument under the Article on Commercial Paper (Article 3). A buyer who as part of one transaction signs both a negotiable instrument and a security agreement makes such an agreement.

(2) When a seller retains a purchase money security interest in goods the Article on Sales (Article 2) governs the sale and any disclaimer, limitation or modification of the seller's warranties. Amended in 1962.

§ 9–207. Rights and Duties When Collateral Is in Secured Party's Possession. (1) A secured party must use reasonable care in the custody and preservation of collateral in his possession. In the case of an instrument or chattel paper reasonable care includes taking necessary steps to preserve rights against prior parties unless otherwise agreed.

(2) Unless otherwise agreed, when collateral is in the secured party's possession

(a) reasonable expenses (including the cost of any insurance and payment of taxes or other charges) incurred in the custody, preservation, use or operation of the collateral are chargeable to the debtor and are secured by the collateral;

(b) the risk of accidental loss or damage is on the debtor to the extent of any deficiency in any effective insurance coverage;

(c) the secured party may hold as additional security any increase or profits (except money) received from the collateral, but money so received, unless remitted to the debtor, shall be applied in reduction of the secured obligation;

(d) the secured party must keep the collateral identifiable but fungible collateral may be commingled;

(e) the secured party may repledge the collateral upon terms which do not impair the debtor's right to redeem it.

(3) A secured party is liable for any loss caused by his failure to meet any obligation imposed by the preceding subsections but does not lose his security interest.

(4) A secured party may use or operate the collateral for the purpose of preserving the collateral or its value or pursuant to the order of a court of appropriate jurisdiction or, except in the case of consumer goods, in the manner and to the extent provided in the security agreement.

§ 9–208. Request for Statement of Account or List of Collateral. (1) A debtor may sign a statement indicating what he believes to be the aggregate amount of unpaid indebtedness as of a specified date and may send it to the secured party with a request that the statement be approved or corrected and returned to the debtor. When the security agreement or any other record kept by the secured party identifies the collateral a debtor may similarly request the secured party to approve or correct a list of the collateral.

(2) The secured party must comply with such a request within two weeks after receipt by sending a written correction or approval. If the secured party claims a security interest in all of a particular type of collateral owned by the debtor he may indicate that fact in his reply and need not approve or correct an itemized list of such collateral. If the secured party without reasonable excuse fails to comply he is liable for any loss caused to the debtor thereby; and if the debtor has properly included in his request a good faith statement of the obligation or a list of the collateral or both the secured party may claim a security interest only as shown in the statement against persons misled by his failure to comply. If he no longer has an interest in the obligation or collateral at the time the request is received he must disclose the name and address of any successor in interest known to him and he is liable for any loss caused to the debtor as a result of failure to disclose. A successor in interest is not subject to this section until a request is received by him.

(3) A debtor is entitled to such a statement once every six months without charge. The secured party may require payment of a charge not exceeding $10 for each additional statement furnished.

Part 3

Rights of Third Parties; Perfected and Unperfected Security Interests; Rules of Priority

§ 9–301. Persons Who Take Priority Over Unperfected Security Interests; Rights of "Lien Creditor." (1) Except as otherwise provided in subsection (2), an unperfected security interest is subordinate to the rights of

(a) persons entitled to priority under Section 9–312;

(b) a person who becomes a lien creditor before the security interest is perfected;

(c) in the case of goods, instruments, documents, and chattel paper, a person who is not a secured party and who is a transferee in bulk or other buyer not in ordinary course of business or is a buyer of farm products in ordinary course of business, to the extent that he gives value and receives delivery of the collateral without knowledge of the security interest and before it is perfected;

(d) in the case of accounts and general intangibles, a person who is not a secured party and who is a transferee to the extent that he gives value without knowledge of the security interest and before it is perfected.

(2) If the secured party files with respect to a purchase money security interest before or within ten days after the debtor receives possession of the collateral, he takes priority over the rights of a transferee in bulk or of a lien creditor which arise between the time the security interest attaches and the time of filing.

(3) A "lien creditor" means a creditor who has acquired a lien on the property involved by attachment, levy or the like and includes an assignee for benefit of creditors from the time of assignment, and a trustee in bankruptcy from the date of the filing of the petition or a receiver in equity from the time of appointment.

(4) A person who becomes a lien creditor while a security interest is perfected takes subject to the security interest only to the extent that it secures advances made before he becomes a lien creditor or within 45 days thereafter or made without knowledge of the lien or pursuant to a commitment entered into without knowledge of the lien. Amended in 1972.

§ 9–302. When Filing Is Required to Perfect Security Interest; Security Interests to Which Filing Provisions of This Article Do Not Apply. (1) A financing statement must be filed to perfect all security interests except the following:

(a) a security interest in collateral in possession of the secured party under Section 9–305;

(b) a security interest temporarily perfected in instruments or documents without delivery under Section 9–304 or in proceeds for a 10-day period under Section 9–306;

(c) a security interest created by an assignment of a beneficial interest in a trust or a decedent's estate;

(d) a purchase money security interest in consumer goods; but filing is required for a motor vehicle required to be registered; and fixture filing is required for priority over conflicting interests in fixtures to the extent provided in Section 9–313;

(e) an assignment of accounts which does not alone or in conjunction with other assignments to the same assignee transfer a significant part of the outstanding accounts of the assignor;

(f) a security interest of a collecting bank (Section 4–208) or arising under the Article on Sales (see Section 9–113) or covered in subsection (3) of this section;

(g) an assignment for the benefit of all the creditors of the transferor, and subsequent transfers by the assignee thereunder.

(2) If a secured party assigns a perfected security interest, no filing under this Article is required in order to continue the perfected status of the security interest against creditors of and transferees from the original debtor.

(3) The filing of a financing statement otherwise required by this Article is not necessary or effective to perfect a security interest in property subject to

(a) a statute or treaty of the United States which provides for a national or international registration or a national or international certificate of title or which specifies a place of filing different from that specified in this Article for filing of the security interest; or

(b) the following statutes of this state; [list any certificate of title statute covering automobiles, trailers, mobile homes, boats, farm tractors, or the like, and any central filing statute*.]; but during any period in which collateral is inventory held for sale by a person who is in the business of selling goods of that kind, the filing provisions of this Article (Part 4) apply to a security interest in that collateral created by him as debtor; or

(c) a certificate of title statute of another jurisdiction under the law of which indication of a security interest on the certificate is required as a condition of perfection (subsection (2) of Section 9–103).

(4) Compliance with a statute or treaty described in subsection (3) is equivalent to the filing of a financing statement under this Article, and a security interest in property subject to the statute or treaty can be perfected only by compliance therewith except as provided in Section 9–103 on multiple state transactions. Duration and renewal of perfection of a security interest perfected by compliance with the statute or treaty are governed by the provisions of the statute or treaty; in other respects the security interest is subject to this Article. Amended in 1972.

*__Note:__ *It is recommended that the provisions of certificate of title acts for perfection of security interests by notation on the certificates should be amended to exclude coverage of inventory held for sale.*

§ 9–303. When Security Interest Is Perfected; Continuity of Perfection. (1) A security interest is perfected when it has attached and when all of the applicable steps required for perfection have been taken. Such steps are specified in Sections 9–302, 9–304, 9–305 and 9–306. If such steps are taken before the security interest attaches, it is perfected at the time when it attaches.

(2) If a security interest is originally perfected in any way permitted under this Article and is subsequently perfected in some other way under this Article, without an intermediate period when it was unperfected, the security interest shall be deemed to be perfected continuously for the purposes of this Article.

§ 9–304. Perfection of Security Interest in Instruments, Documents, and Goods Covered by Documents; Perfection by Permissive Filing; Temporary Perfection Without Filing or Transfer of Possession. (1) A security interest in chattel paper or negotiable documents may be perfected by filing. A security interest in money or instruments (other than instruments which constitute part of chattel paper) can be perfected only by the secured party's taking possession, except as provided in subsections (4) and (5) of this section and subsections (2) and (3) of Section 9–306 on proceeds.

(2) During the period that goods are in the possession of the issuer of a negotiable document therefor, a security interest in the goods is perfected by perfecting a security interest in the document, and any security interest in the goods otherwise perfected during such period is subject thereto.

(3) A security interest in goods in the possession of a bailee other than one who has issued a negotiable document therefor is perfected by issuance of a document in the name of the secured party or by the bailee's receipt of notification of the secured party's interest or by filing as to the goods.

(4) A security interest in instruments or negotiable documents is perfected without filing or the taking of possession for a period of 21 days from the time it attaches to the extent that it arises for new value given under a written security agreement.

(5) A security interest remains perfected for a period of 21 days without filing where a secured party having a perfected security interest in an instrument, a negotiable document or goods in possession of a bailee other than one who has issued a negotiable document therefor.

(a) makes available to the debtor the goods or documents representing the goods for the purpose of ultimate sale or exchange or for the purpose of loading, unloading, storing, shipping, transshipping, manufacturing, processing or otherwise dealing with them in a manner preliminary to their sale or exchange, but priority between conflicting security interests in the goods is subject to subsection (3) of Section 9–312; or

(b) delivers the instrument to the debtor for the purpose of ultimate sale or exchange or of presentation, collection, renewal or registration of transfer.

(6) After the 21-day period in subsections (4) and (5) perfection depends upon compliance with applicable provisions of this Article. Amended in 1972.

§ 9–305. When Possession by Secured Party Perfects Security Interest Without Filing. A security interest in letters of credit and advices of credit (subsection (2) (a) of Section 5–116), goods, instruments, money, negotiable documents or chattel paper may be perfected by the secured party's taking possession of the collateral. If such collateral other than goods covered by a negotiable document is held by a bailee, the secured party is deemed to have possession from the time the bailee receives notification of the secured party's interest. A security interest is perfected by possession from the time possession is taken without relation back and continues only so long as possession is retained, unless otherwise specified in this Article. The security interest may be otherwise perfected as provided in this Article before or after the period of possession by the secured party. Amended in 1972.

§ 9–306. "Proceeds"; Secured Party's Rights on Disposition of Collateral. (1) "Proceeds" includes whatever is received upon the sale, exchange, collection or other disposition of collateral or proceeds. Insurance payable by reason of loss or damage to the collateral is proceeds, except to the extent that it is payable to a person other than a party to the security agreement. Money, checks, deposit accounts, and the like are "cash proceeds." All other proceeds are "non-cash proceeds."

(2) Except where this Article otherwise provides, a security interest continues in collateral notwithstanding sale, exchange or other disposition thereof unless the disposition was authorized by the secured party in the security agreement or otherwise, and also continues in any identifiable proceeds including collections received by the debtor.

(3) The security interest in proceeds is a continuously perfected security interest if the interest in the original collateral was perfected but it ceases to be a perfected security interest and becomes unperfected ten days after receipt of the proceeds by the debtor unless

(a) a filed financing statement covers the original collateral and the proceeds are collateral in which a security interest may be perfected by filing in the office or offices where the financing statement has been filed and, if the proceeds are acquired with cash proceeds, the description of collateral in the financing statement indicates the types of property constituting the proceeds; or

(b) a filed financing statement covers the original collateral and the proceeds are identifiable cash proceeds; or

(c) the security interest in the proceeds is perfected before the expiration of the ten day period.

Except as provided in this section, a security interest in proceeds can be perfected only by the methods or under the circumstances permitted in this Article for original collateral of the same type.

(4) In the event of insolvency proceedings instituted by or against a debtor, a secured party with a perfected security interest in proceeds has a perfected security interest only in the following proceeds;

(a) in identifiable non-cash proceeds and in separate deposit accounts containing only proceeds;

(b) in identifiable cash proceeds in the form of money which is neither commingled with other money nor deposited in a deposit account prior to the insolvency proceedings;

(c) in identifiable cash proceeds in the form of checks and the like which are not deposited in a deposit account prior to the insolvency proceedings; and

(d) in all cash and deposit accounts of the debtor in which proceeds have been commingled with other funds, but the perfected security interest under this paragraph (d) is

(i) subject to any right to set-off; and

(ii) limited to an amount not greater than the amount of any cash proceeds received by the debtor within ten days before the institution of the insolvency proceedings less the sum of (I) the payments to the secured party on account of cash proceeds received by the debtor during such period and (II) the cash proceeds received by the debtor during such period to which the secured party is entitled under paragraphs (a) through (c) of this subsection (4).

(5) If a sale of goods results in an account or chattel paper which is transferred by the seller to a secured party, and if the goods are returned to or are repossessed by the seller or the secured party, the following rules determine priorities:

(a) If the goods were collateral at the time of sale, for an indebtedness of the seller which is still unpaid, the original security interest attaches again to the goods and continues as a perfected security interest if it was perfected at the time when the goods were sold. If the security interest was originally perfected by a filing which is still effective, nothing further is required to continue the perfected status; in any other case, the secured party must take possession of the returned or repossessed goods or must file.

(b) An unpaid transferee of the chattel paper has a security interest in the goods against the transferor. Such security interest is prior to a security interest asserted under paragraph (a) to the extent that the transferee of the chattel paper was entitled to priority under Section 9–308.

(c) An unpaid transferee of the account has a security interest in the goods against the transferor. Such security interest is subordinate to a security interest asserted under paragraph (a).

(d) A security interest of an unpaid transferee asserted under paragraph (b) or (c) must be perfected for protection against creditors of the transferor and purchasers of the returned or repossessed goods.

Amended in 1972.

§ 9–307. Protection of Buyers of Goods. (1) A buyer in ordinary course of business (subsection (9) of Section 1–201) other than a person buying farm products from a person engaged in farming operations takes free of a security interest created by his seller even though the security interest is perfected and even though the buyer knows of its existence.

(2) In the case of consumer goods, a buyer takes free of a security interest even though perfected if he buys without knowledge of the security interest, for value and for his own personal, family or household purposes unless prior to the purchase the secured party has filed a financing statement covering such goods.

(3) A buyer other than a buyer in ordinary course of business (subsection (1) of this section) takes free of a security interest to the extent that it secures future advances made after the secured party acquires knowledge of the purchase, or more than 45 days after the purchase, whichever first occurs, unless made pursuant to a commitment entered into without knowledge of the purchase and before the expiration of the 45 day period. Amended in 1972.

§ 9–308. Purchase of Chattel Paper and Instruments. A purchaser of chattel paper or an instrument who gives new value and takes possession of it in the ordinary course of his business has priority over a security interest in the chattel paper or instrument.

(a) which is perfected under Section 9–304 (permissive filing and temporary perfection) or under Section 9–306 (perfection as to proceeds) if he acts without knowledge that the specific paper or instrument is subject to a security interest; or

(b) which is claimed merely as proceeds of inventory subject to a security interest (Section 9–306) even though he knows that the specific paper or instrument is subject to the security interest.

Amended in 1972.

§ 9–309. Protection of Purchasers of Instruments and Documents. Nothing in this Article limits the rights of a holder in due course of a negotiable instrument (Section 3–302) or a holder to whom a negotiable document of title has been duly negotiated (Section 7–501) or a bona fide purchaser of a security (Section 8–301) and such holders or purchasers take priority over an earlier security interest even though perfected. Filing under this Article does not constitute notice of the security interest to such holders or purchasers.

§ 9–310. Priority of Certain Liens Arising by Operation of Law. When a person in the ordinary course of his business furnishes services or materials with respect to goods subject to a security interest, a lien upon goods in the possession of such person gives a statute or rule of law for such materials or services takes priority over a perfected security interest unless the lien is statutory and the statute expressly provides otherwise.

§ 9–311. Alienability of Debtor's Rights: Judicial Process. The debtor's rights in collateral may be voluntarily or involuntarily transferred (by way of sale, creation of a security interest, attachment, levy, garnishment or other judicial process) notwithstanding a provision in the security agreement prohibiting any transfer or making the transfer constitute a default.

§ 9–312. Priorities Among Conflicting Security Interests in the Same Collateral. (1) The rules of priority stated in other sections of this Part and in the following sections shall govern when applicable: Section 4–208 with respect to the security interests of collecting banks in items being collected, accompanying documents and proceeds; Section 9–103 on security interests related to other jurisdictions; Section 9–114 on consignments.

(2) A perfected security interest in crops for new value given to enable the debtor to produce the crops during the production season and given not more than three months before the crops become growing crops by planting

or otherwise takes priority over an earlier perfected security interest to the extent that such earlier interest secures obligations due more than six months before the crops become growing crops by planting or otherwise, even though the person giving new value had knowledge of the earlier security interest.

(3) A perfected purchase money security interest in inventory has priority over a conflicting security interest in the same inventory and also has priority in identifiable cash proceeds received on or before the delivery of the inventory to a buyer if

(a) the purchase money security interest is perfected at the time the debtor receives possession of the inventory; and

(b) the purchase money secured party gives notification in writing to the holder of the conflicting security interest if the holder had filed a financing statement covering the same types of inventory (i) before the date of the filing made by the purchase money secured party, or (ii) before the beginning of the 21-day period where the purchase money security interest is temporarily perfected without filing or possession (subsection (5) of Section 9–304); and

(c) the holder of the conflicting security interest receives the notification within five years before the debtor receives possession of the inventory; and

(d) the notification states that the person giving the notice has or expects to acquire a purchase money security interest in inventory of the debtor, describing such inventory by item or type.

(4) A purchase money security interest in collateral other than inventory has priority over a conflicting security interest in the same collateral or its proceeds if the purchase money security interest is perfected at the time the debtor receives possession of the collateral or within ten days thereafter.

(5) In all cases not governed by other rules stated in this section (including cases of purchase money security interests which do not quality for the special priorities set forth in subsections (3) and (4) of this section), priority between conflicting security interests in the same collateral shall be determined according to the following rules:

(a) Conflicting security interests rank according to priority in time of filing or perfection. Priority dates from the time a filing is first made covering the collateral or the time the security interest is first perfected, whichever is earlier, provided that there is no period thereafter when there is neither filing nor perfection.

(b) So long as conflicting security interests are unperfected, the first to attach has priority.

(6) For the purposes of subsection (5) a date of filing or perfection as to collateral is also a date of filing or perfection as to proceeds.

(7) If future advances are made while a security interest is perfected by filing or the taking of possession, the security interest has the same priority for the purposes of subsection (5) with respect to the future advances as it does with respect to the first advance. If a commitment is made before or while the security interest is so perfected, the security interest has the same

priority with respect to advances made pursuant thereto. In other cases a perfected security interest has priority from the date the advance is made. Amended in 1972.

§ 9–313. Priority of Security Interests in Fixtures. (1) In this section and in the provisions of Part 4 of this Article referring to fixture filing, unless the context otherwise requires

(a) goods are "fixtures" when they become so related to particular real estate that an interest in them arises under real estate law

(b) a "fixture filing" is the filing in the office where a mortgage on the real estate would be filed or recorded of a financing statement covering goods which are or are to become fixtures and conforming to the requirements of subsection (5) of Section 9–402

(c) a mortgage is a "construction mortgage" to the extent that it secures an obligation incurred for the construction of an improvement on land including the acquisition cost of the land, if the recorded writing so indicates.

(2) A security interest under this Article may be created in goods which are fixtures or may continue in goods which become fixtures, but no security interest exists under this Article in ordinary building materials incorporated into an improvement on land.

(3) This Article does not prevent creation of an encumbrance upon fixtures pursuant to real estate law.

(4) A perfected security interest in fixtures has priority over the conflicting interest of an encumbrancer or owner of the real estate where

(a) the security interest is a purchase money security interest, the interest of the encumbrancer or owner arises before the goods become fixtures, the security interest is perfected by a fixture filing before the goods become fixtures or within ten days thereafter, and the debtor has an interest of record in the real estate or is in possession of the real estate; or

(b) the security interest is perfected by a fixture filing before the interest of the encumbrancer or owner is of record, the security interest has priority over any conflicting interest of a predecessor in title of the encumbrancer or owner, and the debtor has an interest of record in the real estate or is in possession of the real estate; or

(c) the fixtures are readily removable factory or office machines or readily removable replacements of domestic appliances which are consumer goods, and before the goods become fixtures the security interest is perfected by any method permitted by this Article; or

(d) the conflicting interest is a lien on the real estate obtained by legal or equitable proceedings after the security interest was perfected by any method permitted by this Article.

(5) A security interest in fixtures, whether or not perfected, has priority over the conflicting interest of an encumbrancer or owner of the real estate where

(a) the encumbrancer or owner has consented in writing to the security interest or has disclaimed an interest in the goods as fixtures; or

(b) the debtor has a right to remove the goods as against the encumbrancer or owner. If the debtor's right terminates, the priority of the security interest continues for a reasonable time.

(6) Notwithstanding paragraph (a) of subsection (4) but otherwise subject to subsections (4) and (5), a security interest in fixtures is subordinate to a construction mortgage recorded before the goods become fixtures if the goods become fixtures before the completion of the construction. To the extent that it is given to refinance a construction mortgage, a mortgage has this priority to the same extent as the construction mortgage.

(7) In cases not within the preceding subsections, a security interest in fixtures is subordinate to the conflicting interest of an encumbrancer or owner of the related real estate who is not the debtor.

(8) When the secured party has priority over all owners and encumbrancers of the real estate, he may, on default, subject to the provisions of Part 5, remove his collateral from the real estate but he must reimburse any encumbrancer or owner of the real estate who is not the debtor and who has not otherwise agreed for the cost of repair of any physical injury, but not for any diminution in value of the real estate caused by the absence of the goods removed or by any necessity of replacing them. A person entitled to reimbursement may refuse permission to remove until the secured party gives adequate security for the performance of this obligation. Amended in 1972.

§ 9–314. Accessions. (1) A security interest in goods which attaches before they are installed in or affixed to other goods takes priority as to the goods installed or affixed (called in this section "accessions") over the claims of all persons to the whole except as stated in subsection (3) and subject to Section 9–315(1).

(2) A security interest which attaches to goods after they become part of a whole is valid against all persons subsequently acquiring interests in the whole except as stated in subsection (3) but is invalid against any person with an interest in the whole at the time the security interest attaches to the goods who has not in writing consented to the security interest or disclaimed an interest in the goods as part of the whole.

(3) The security interests described in subsections (1) and (2) do not take priority over

(a) a subsequent purchaser for value of any interest in the whole; or

(b) a creditor with a lien on the whole subsequently obtained by judicial proceedings; or

(c) a creditor with a prior perfected security interest in the whole to the extent that he makes subsequent advances.

If the subsequent purchase is made, the lien by judicial proceedings obtained or the subsequent advance under the prior perfected security interest is made or contracted for without knowledge of the security interest and before it is perfected. A purchaser of the whole at a foreclosure sale other than the holder of a perfected security interest purchasing at his own foreclosure sale is a subsequent purchaser within this section.

(4) When under subsections (1) or (2) and (3) a secured party has an

interest in accessions which has priority over the claims of all persons who have interests in the whole, he may on default subject to the provisions of Part 5 remove his collateral from the whole but he must reimburse any encumbrancer or owner of the whole who is not the debtor and who has not otherwise agreed for the cost of repair of any physical injury but not for any diminution in value of the whole caused by the absence of the goods removed or by any necessity for replacing them. A person entitled to reimbursement may refuse permission to remove until the secured party gives adequate security for the performance of this obligation.

§ 9–315. Priority When Goods Are Commingled or Processed. (1) If a security interest in goods was perfected and subsequently the goods or a part thereof have become part of a product or mass, the security interest continues in the product or mass if

(a) the goods are so manufactured, processed, assembled or commingled that their identity is lost in the product or mass; or

(b) a financing statement covering the original goods also covers the product into which the goods have been manufactured, processed or assembled.

In a case to which paragraph (b) applies, no separate security interest in that part of the original goods which has been manufactured, processed or assembled into the product may be claimed under Section 9–314.

(2) When under subsection (1) more than one security interest attaches to the product or mass, they rank equally according to the ratio that the cost of the goods to which each interest originally attached bears to the cost of the total product or mass.

§ 9–316. Priority Subject to Subordination. Nothing in this Article prevents subordination by agreement by any person entitled to priority.

§ 9–317. Secured Party Not Obligated on Contract of Debtor. The mere existence of a security interest or authority given to the debtor to dispose of or use collateral does not impose contract or tort liability upon the secured party for the debtor's acts or omissions.

§ 9–318. Defenses Against Assignee; Modification of Contract After Notification of Assignment; Term Prohibiting Assignment Ineffective; Identification and Proof of Assignment. (1) Unless an account debtor has made an enforceable agreement not to assert defenses or claims arising out of a sale as provided in Section 9–206 the rights of an assignee are subject to

(a) all the terms of the contract between the account debtor and assignor and any defense or claim arising therefrom; and

(b) any other defense or claim of the account debtor against the assignor which accrues before the account debtor receives notification of the assignment.

(2) So far as the right to payment or a part thereof under an assigned contract has not been fully earned by performance, and notwithstanding notification of the assignment, any modification of or substitution for the contract made in good faith and in accordance with reasonable commercial standards

is effective against an assignee unless the account debtor has otherwise agreed but the assignee acquires corresponding rights under the modified or substituted contract. The assignment may provide that such modification or substitution is a breach by the assignor.

(3) The account debtor is authorized to pay the assignor until the account debtor receives notification that the amount due or to become due has been assigned and that payment is to be made to the assignee. A notification which does not reasonably identify the rights assigned is ineffective. If requested by the account debtor, the assignee must seasonably furnish reasonable proof that the assignment has been made and unless he does so the account debtor may pay the assignor.

(4) A term in any contract between an account debtor and an assignor is ineffective if it prohibits assignment of an account or prohibits creation of a security interest in a general intangible for money due or to become due or requires the account debtor's consent to such assignment or security interest. Amended in 1972.

Part 4

Filing

§ 9–401. Place of Filing; Erroneous Filing; Removal of Collateral.

First Alternative Subsection (1)

(1) The proper place to file in order to perfect a security interest is as follows:

(a) when the collateral is timber to be cut or is minerals or the like (including oil and gas) or accounts subject to subsection (5) of Section 9–103, or when the financing statement is filed as a fixture filing (Section 9–313) and the collateral is goods which are or are to become fixtures, then in the office where a mortgage on the real estate would be filed or recorded;

(b) in all other cases, in the office of the [Secretary of State].

Second Alternative Subsection (1)

(1) The proper place to file in order to perfect a security interest is as follows:

(a) when the collateral is equipment used in farming operations, or farm products, or accounts or general intangibles arising from or relating to the sale of farm products by a farmer, or consumer goods, then in the office of the .... in the county of the debtor's residence or if the debtor is not a resident of this state then in the office of the .... in the county where the goods are kept, and in addition when the collateral is crops growing or to be grown in the office of the .... in the county where the land is located;

(b) when the collateral is timber to be cut or is minerals or the like (including oil and gas) or accounts subject to subsection (5) of Section 9–103, or when the financing statement is filed as a fixture filing (Section 9–313) and the collateral is goods which are or are to become fixtures, then in the office where a mortgage on the real estate would be filed or recorded;

(c) in all other cases, in the office of the [Secretary of State].

Third Alternative Subsection (1)

(1) The proper place to file in order to perfect a security interest is as follows:

(a) when the collateral is equipment used in farming operations, or farm products, or accounts or general intangibles arising from or relating to the sale of farm products by a farmer, or consumer goods, then in the office of the .... in the county of the debtor's residence or if the debtor is not a resident of this state then in the office of the .... in the county where the goods are kept, and in addition when the collateral is crops growing or to be grown in the office of the .... in the county where the land is located;

(b) when the collateral is timber to be cut or is minerals or the like (including oil and gas) or accounts subject to subsection (5) of Section 9–103, or when the financing statement is filed as a fixture filing (Section 9–313) and the collateral is goods which are or are to become fixtures, then in the office where a mortgage on the real estate would be filed or recorded;

(c) in all other cases, in the office of the [Secretary of State] and in addition, if the debtor has a place of business in only one county of this state, also in the office of .... of such county, or, if the debtor has no place of business in this state, but resides in the state, also in the office of .... of the county in which he resides.

Note: *One of the three alternatives should be selected as subsection (1).*

(2) A filing which is made in good faith in an improper place or not in all of the places required by this section is nevertheless effective with regard to any collateral as to which the filing complied with the requirements of this Article and is also effective with regard to collateral covered by the financing statement against any person who has knowledge of the contents of such financing statement.

(3) A filing which is made in the proper place in this state continues effective even though the debtor's residence or place of business or the location of the collateral or its use, whichever controlled the original filing, is thereafter changed.

Alternative Subsection (3)

[(3) A filing which is made in the proper county continues effective for four months after a change to another county of the debtor's residence or place of business or the location of the collateral, whichever controlled the original filing. It becomes ineffective thereafter unless a copy of the financing statement signed by the secured party is filed in the new county within said period. The security interest may also be perfected in the new county after the expiration of the four-month period; in such case perfection dates from the time of perfection in the new county. A change in the use of the collateral does not impair the effectiveness of the original filing.]

(4) The rules stated in Section 9–103 determine whether filing is necessary in this state.

(5) Notwithstanding the preceding subsections, and subject to subsection (3) of Section 9–302, the proper place to file in order to perfect a security interest in collateral, including fixtures, of a transmitting utility is the office of the [Secretary of State]. This filing constitutes a fixture filing (Section 9–313) as to the collateral described therein which is or is to become fixtures.

(6) For the purposes of this section, the residence of an organization is its place of business if it has one or its chief executive office if it has more than one place of business. Amended in 1962 and 1972.

Note: *Subsection (6) should be used only if the state chooses the Second or Third Alternative Subsection (1).*

§ 9–402. Formal Requisites of Financing Statement; Amendments; Mortgage as Financing Statement. (1) A financing statement is sufficient if it gives the name of the debtor and the secured party, is signed by the debtor, gives an address of the secured party from which information concerning the security interest may be obtained, gives a mailing address of the debtor and contains a statement indicating the types, or describing the items, of collateral. A financing statement may be filed before a security agreement is made or a security interest otherwise attaches. When the financing statement covers crops growing or to be grown, the statement must also contain a description of the real estate concerned. When the financing statement covers timber to be cut or covers minerals or the like (including oil and gas) or accounts subject to subsection (5) of Section 9–103, or when the financing statement is filed as a fixture filing (Section 9–313) and the collateral is goods which are or are to become fixtures, the statement must also comply with subsection (5). A copy of the security agreement is sufficient as a financing statement if it contains the above information and is signed by the debtor. A carbon, photographic or other reproduction of a security agreement or a financing statement is sufficient as a financing statement if the security agreement so provides or if the original has been filed in this state.

(2) A financing statement which otherwise complies with subsection (1) is sufficient when it is signed by the secured party instead of the debtor if it is filed to perfect a security interest in

(a) collateral already subject to a security interest in another jurisdiction when it is brought into this state, or when the debtor's location is changed to this state. Such a financing statement must state that the collateral was brought into this state or that the debtor's location was changed to this state under such circumstances; or

(b) proceeds under Section 9–306 if the security interest in the original collateral was perfected. Such a financing statement must describe the original collateral; or

(c) collateral as to which the filing has lapsed; or

(d) collateral acquired after a change of name, identity or corporate structure of the debtor (subsection (7)).

(3) A form substantially as follows is sufficient to comply with subsection (1):

Name of debtor (or assignor) ....
Address ....
Name of secured party (or assignee) ....
Address ....

1. This financing statement covers the following types (or items) of property:
   (Describe) ....
2. (If collateral is crops) The above described crops are growing or are to be grown on:
   (Describe Real Estate) ....
3. (If applicable) The above goods are to become fixtures on*

   * Where appropriate substitute either "The above timber is standing on. . . ." or "The above minerals or the like (including oil and gas) or accounts will be financed at the wellhead or minehead of the well or mine located on. . . ."

   (Describe Real Estate) ....
   and this financing statement is to be filed [for record] in the real estate records. (If the debtor does not have an interest of record) The name of a record owner is ....
4. (If products of collateral are claimed) Products of the collateral are also covered.

(use whichever is applicable)	.... Signature of Debtor (or Assignor) .... Signature of Secured Party (or Assignee)

(4) A financing statement may be amended by filing a writing signed by both the debtor and the secured party. An amendment does not extend the period of effectiveness of a financing statement. If any amendment adds collateral, it is effective as to the added collateral only from the filing date of the amendment. In this Article, unless the context otherwise requires, the term "financing statement" means the original financing statement and any amendments.

(5) A financing statement covering timber to be cut or covering minerals or the like (including oil and gas) or accounts subject to subsection (5) of Section 9–103, or a financing statement filed as a fixture filing (Section 9–313) where the debtor is not a transmitting utility, must show that it covers this type of collateral, must recite that it is to be filed [for record] in the real estate records, and the financing statement must contain a description

of the real estate [sufficient if it were contained in a mortgage of the real estate to give constructive notice of the mortgage under the law of this state]. If the debtor does not have an interest of record in the real estate, the financing statement must show the name of a record owner.

(6) A mortgage is effective as a financing statement filed as a fixture filing from the date of its recording if

(a) the goods are described in the mortgage by item or type; and

(b) the goods are or are to become fixtures related to the real estate described in the mortgage; and

(c) the mortgage complies with the requirements for a financing statement in this section other than a recital that it is to be filed in the real estate records; and

(d) the mortgage is duly recorded.

No fee with reference to the financing statement is required other than the regular recording and satisfaction fees with respect to the mortgage.

(7) A financing statement sufficiently shows the name of the debtor if it gives the individual, partnership or corporate name of the debtor, whether or not it adds other trade names or names of partners. Where the debtor so changes his name or in the case of an organization its name, identity or corporate structure that a filed financing statement becomes seriously misleading, the filing is not effective to perfect a security interest in collateral acquired by the debtor more than four months after the change, unless a new appropriate financing statement is filed before the expiration of that time. A filed financing statement remains effective with respect to collateral transferred by the debtor even though the secured party knows of or consents to the transfer.

(8) A financing statement substantially complying with the requirements of this section is effective even though it contains minor errors which are not seriously misleading. Amended in 1972.

Note: *Language in brackets is optional.*

Note: *Where the state has any special recording system for real estate other than the usual grantor-grantee index (as, for instance, a tract system or a title registration or Torrens system) local adaptations of subsection (5) and Section 9–403(7) may be necessary. See Mass.Gen.Laws Chapter 106, Section 9–409.*

§ 9–403. What Constitutes Filing; Duration of Filing; Effect of Lapsed Filing; Duties of Filing Officer. (1) Presentation for filing of a financing statement and tender of the filing fee or acceptance of the statement by the filing officer constitutes filing under this Article.

(2) Except as provided in subsection (6) a filed financing statement is effective for a period of five years from the date of filing. The effectiveness of a filed financing statement lapses on the expiration of the five year period unless a continuation statement is filed prior to the lapse. If a security interest perfected by filing exists at the time insolvency proceedings are commenced by or against the debtor, the security interest remains perfected until termina-

tion of the insolvency proceedings and thereafter for a period of sixty days or until expiration for a period of sixty days or until expiration of the five year period, whichever occurs later. Upon lapse the security interest becomes unperfected, unless it is perfected without filing. If the security interest becomes unperfected upon lapse, it is deemed to have been unperfected as against a person who became a purchaser or lien creditor before lapse.

(3) A continuation statement may be filed by the secured party within six months prior to the expiration of the five year period specified in subsection (2). Any such continuation statement must be signed by the secured party, identify the original statement by file number and state that the original statement is still effective. A continuation statement signed by a person other than the secured party of record must be accompanied by a separate written statement of assignment signed by the secured party of record and complying with subsection (2) of Section 9–405, including payment of the required fee. Upon timely filing of the continuation statement, the effectiveness of the original statement is continued for five years after the last date to which the filing was effective whereupon it lapses in the same manner as provided in subsection (2) unless another continuation statement is filed prior to such lapse. Succeeding continuation statements may be filed in the same manner to continue the effectiveness of the original statement. Unless a statute on disposition of public records provides otherwise, the filing officer may remove a lapsed statement from the files and destroy it immediately if he has retained a microfilm or other photographic record, or in other cases after one year after the lapse. The filing officer shall so arrange matters by physical annexation of financing statements to continuation statements or other related filings, or by other means, that if he physically destroys the financing statements of a period more than five years past, those which have been continued by a continuation statement or which are still effective under subsection (6) shall be retained.

(4) Except as provided in subsection (7) a filing officer shall mark each statement with a file number and with the date and hour of filing and shall hold the statement or a microfilm or other photographic copy thereof for public inspection. In addition the filing officer shall index the statement according to the name of the debtor and shall note in the index the file number and the address of the debtor given in the statement.

(5) The uniform fee for filing and indexing and for stamping a copy furnished by the secured party to show the date and place of filing for an original financing statement or for a continuation statement shall be $.......... if the statement is in the standard form prescribed by the [Secretary of State] and otherwise shall be $.........., plus in each case, if the financing statement is subject to subsection (5) of Section 9–402, $.......... The uniform fee for each name more than one required to be indexed shall be $.......... The secured party may at his option show a trade name for any person and an extra uniform indexing fee of $.......... shall be paid with respect thereto.

(6) If the debtor is a transmitting utility (subsection (5) of Section 9–

401) and a filed financing statement so states, it is effective until a termination statement is filed. A real estate mortgage which is effective as a fixture filing under subsection (6) of Section 9–402 remains effective as a fixture filing until the mortgage is released or satisfied of record or its effectiveness otherwise terminates as to the real estate.

(7) When a financing statement covers timber to be cut or covers minerals or the like (including oil and gas) or accounts subject to subsection (5) of Section 9–103, or is filed as a fixture filing, [it shall be filed for record and] the filing officer shall index it under the names of the debtor and any owner of record shown on the financing statement in the same fashion as if they were the mortgagors in a mortgage of the real estate described, and, to the extent that the law of this state provides for indexing of mortgages under the name of the mortgagee, under the name of the secured party as if he were the mortgagee thereunder, or where indexing is by description in the same fashion as if the financing statement were a mortgage of the real estate described. Amended in 1972.

Note: *In states in which writings will not appear in the real estate records and indices unless actually recorded the bracketed language in subsection (7) should be used.*

§ 9–404. Termination Statement. (1) If a financing statement covering consumer goods is filed on or after ...., then within one month or within ten days following written demand by the debtor after there is no outstanding secured obligation and no commitment to make advances, incur obligations or otherwise give value, the secured party must file with each filing officer with whom the financing statement was filed, a termination statement to the effect that he no longer claims a security interest under the financing statement, which shall be identified by file number. In other cases whenever there is no outstanding secured obligation and no commitment to make advances, incur obligations or otherwise give value, the secured party must on written demand by the debtor send the debtor, for each filing officer with whom the financing statement was filed, a termination statement to the effect that he no longer claims a security interest under the financing statement, which shall be identified by file number. A termination statement signed by a person other than the secured party of record must be accompanied by a separate written statement of assignment signed by the secured party of record complying with subsection (2) of Section 9–405, including payment of the required fee. If the affected secured party fails to file such a termination statement as required by this subsection, or to send such a termination statement within ten days after proper demand therefor, he shall be liable to the debtor for one hundred dollars, and in addition for any loss caused to the debtor by such failure.

(2) On presentation to the filing officer of such a termination statement he must note it in the index. If he has received the termination statement in duplicate, he shall return one copy of the termination statement to the secured party stamped to show the time of receipt thereof. If the filing officer

has a microfilm or other photographic record of the financing statement, and of any related continuation statement, statement of assignment and statement of release, he may remove the originals from the files at any time after receipt of the termination statement, or if he has no such record, he may remove them from the files at any time after one year after receipt of the termination statement.

(3) If the termination statement is in the standard form prescribed by the [Secretary of State], the uniform fee for filing and indexing the termination statement shall be $......, and otherwise shall be $......, plus in each case an additional fee of $ ...... for each name more than one against which the termination statement is required to be indexed. Amended in 1972.

Note: *The date to be inserted should be the effective date of the revised Article 9.*

§ 9–405. Assignment of Security Interest; Duties of Filing Officer; Fees. (1) A financing statement may disclose an assignment of a security interest in the collateral described in the financing statement by indication in the financing statement of the name and address of the assignee or by an assignment itself or a copy thereof on the face or back of the statement. On presentation to the filing officer of such a financing statement the filing officer shall mark the same as provided in Section 9–403(4). The uniform fee for filing, indexing and furnishing filing data for a financing statement so indicating an assignment shall be $...... if the statement is in the standard form prescribed by the [Secretary of State] and otherwise shall be $......., plus in each case an additional fee of $ ...... for each name more than one against which the financing statement is required to be indexed.

(2) A secured party may assign of record all or part of his rights under a financing statement by the filing in the place where the original financing statement was filed of a separate written statement of assignment signed by the secured party of record and setting forth the name of the secured party of record and the debtor, the file number and the date of filing of the financing statement and the name and address of the assignee and containing a description of the collateral assigned. A copy of the assignment is sufficient as a separate statement if it complies with the preceding sentence. On presentation to the filing officer of such a separate statement, the filing officer shall mark such separate statement with the date and hour of the filing. He shall note the assignment on the index of the financing statement, or in the case of a fixture filing, or a filing covering timber to be cut, or covering minerals or the like (including oil and gas) or accounts subject to subsection (5) of Section 9–103, he shall index the assignment under the name of the assignor as grantor and, to the extent that the law of this state provides for indexing the assignment of a mortgage under the name of the assignee, he shall index the assignment of the financing statement under the name of the assignee. The uniform fee for filing, indexing and furnishing filing data about such a separate statement of assignment shall be $...... if the statement is in the standard form prescribed by the [Secretary of State] and otherwise shall be $......, plus in

each case an additional fee of $...... for each name more than one against which the statement of assignment is required to be indexed. Notwithstanding the provisions of this subsection, an assignment of record of a security interest in a fixture contained in a mortgage effective as a fixture filing (subsection (6) of Section 9–402) may be made only by an assignment of the mortgage in the manner provided by the law of this state other than this Act.

(3) After the disclosure or filing of an assignment under this section, the assignee is the secured party of record. Amended in 1972.

§ 9–406. Release of Collateral; Duties of Filing Officer; Fees. A secured party of record may by his signed statement release all or a part of any collateral described in a filed financing statement. The statement of release is sufficient if it contains a description of the collateral being released, the name and address of the debtor, the name and address of the secured party, and the file number of the financing statement. A statement of release signed by a person other than the secured party of record must be accompanied by a separate written statement of assignment signed by the secured party of record and complying with subsection (2) of Section 9–405, including payment of the required fee. Upon presentation of such a statement of release to the filing officer he shall mark the statement with the hour and date of filing and shall note the same upon the margin of the index of the filing of the financing statement. The uniform fee for filing and noting such a statement of release shall be $...... if the statement is in the standard form prescribed by the [Secretary of State] and otherwise shall be $......, plus in each case an additional fee of $...... for each name more than one against which the statement of release is required to be indexed. Amended in 1972.

[§ 9–407. Information From Filing Officer]. [(1) If the person filing any financing statement, termination statement, statement of assignment, or statement of release, furnishes the filing officer a copy thereof, the filing officer shall upon request note upon the copy the file number and date and hour of the filing of the original and deliver or send the copy to such person.]

[(2) Upon request of any person, the filing officer shall issue his certificate showing whether there is on file on the date and hour stated therein, any presently effective financing statement naming a particular debtor and any statement of assignment thereof and if there is, giving the date and hour of filing of each such statement and the names and addresses of each secured party therein. The uniform fee for such a certificate shall be $ ...... if the request for the certificate is in the standard form prescribed by the [Secretary of State] and otherwise shall be $ .... Upon request the filing officer shall furnish a copy of any filed financing statement or statement of assignment for a uniform fee of $ ...... per page.] Amended in 1972.

Note: *This section is proposed as an optional provision to require filing officers to furnish certificates. Local law and practices should be consulted with regard to the advisability of adoption.*

§ 9–408. Financing Statements Covering Consigned or Leased Goods. A consignor or lessor of goods may file a financing statement using the terms "consignor," "consignee," "lessor," "lessee" or the like instead of the terms specified in Section 9–402. The provisions of this Part shall apply as appropriate to such a financing statement but its filing shall not of itself be a factor in determining whether or not the consignment or lease is intended as security (Section 1–201 (37)). However, if it is determined for other reasons that the consignment or lease is so intended, a security interest of the consignor or lessor which attaches to the consigned or leased goods is perfected by such filing. Added in 1972.

Part 5

Default

§ 9–501. Default; Procedure When Security Agreement Covers Both Real and Personal Property. (1) When a debtor is in default under a security agreement, a secured party has the rights and remedies provided in this Part and except as limited by subsection (3) those provided in the security agreement. He may reduce his claim to judgment, foreclose or otherwise enforce the security interest by any available judicial procedure. If the collateral is documents the secured party may proceed either as to the documents or as to the goods covered thereby. A secured party in possession has the rights, remedies and duties provided in Section 9–207. The rights and remedies referred to in this subsection are cumulative.

(2) After default, the debtor has the rights and remedies provided in this Part, those provided in the security agreement and those provided in Section 9–207.

(3) To the extent that they give rights to the debtor and impose duties on the secured party, the rules stated in the subsections referred to below may not be waived or varied except as provided with respect to compulsory disposition of collateral (subsection (3) of Section 9–504 and Section 9–505) and with respect to redemption of collateral (Section 9–506) but the parties may by agreement determine the standards by which the fulfillment of these rights and duties is to be measured if such standards are not manifestly unreasonable:

(a) subsection (2) of Section 9–502 and subsection (2) of Section 9–504 insofar as they require accounting for surplus proceeds of collateral;

(b) subsection (3) of Section 9–504 and subsection (1) of Section 9–505 which deal with disposition of collateral;

(c) subsection (2) of Section 9–505 which deals with acceptance of collateral as discharge of obligation;

(d) Section 9–506 which deals with redemption of collateral; and

(e) subsection (1) of Section 9–507 which deals with the secured party's liability for failure to comply with this Part.

(4) If the security agreement covers both real and personal property, the secured party may proceed under this Part as to the personal property or he may proceed as to both the real and the personal property in accordance

with his rights and remedies in respect of the real property in which case the provisions of this Part do not apply.

(5) When a secured party has reduced his claim to judgment the lien of any levy which may be made upon his collateral by virtue of any execution based upon the judgment shall relate back to the date of the perfection of the security interest in such collateral. A judicial sale, pursuant to such execution, is a foreclosure of the security interest by judicial procedure within the meaning of this section, and the secured party may purchase at the sale and thereafter hold the collateral free of any other requirements of this Article. Amended in 1972.

§ 9–502. Collection Rights of Secured Party. (1) When so agreed and in any event on default the secured party is entitled to notify an account debtor or the obligor on an instrument to make payment to him whether or not the assignor was theretofore making collections on the collateral, and also to take control of any proceeds to which he is entitled under Section 9–306.

(2) A secured party who by agreement is entitled to charge back uncollected collateral or otherwise to full or limited recourse against the debtor and who undertakes to collect from the account debtors or obligors must proceed in a commercially reasonable manner and may conduct his reasonable expenses of realization from the collections. If the security agreement secures an indebtedness, the secured party must account to the debtor for any surplus, and unless otherwise agreed, the debtor is liable for any deficiency. But, if the underlying transaction was a sale of accounts or chattel paper, the debtor is entitled to any surplus or is liable for any deficiency only if the security agreement so provides. Amended in 1972.

§ 9–503. Secured Party's Right to Take Possession After Default. Unless otherwise agreed a secured party has on default the right to take possession of the collateral. In taking possession a secured party may proceed without judicial process if this can be done without breach of the peace or may proceed by action. If the security agreement so provides the secured party may require the debtor to assemble the collateral and make it available to the secured party at a place to be designated by the secured party which is reasonably convenient to both parties. Without removal a secured party may render equipment unusable, and may dispose of collateral on the debtor's premises under Section 9–504.

§ 9–504. Secured Party's Right to Dispose of Collateral After Default; Effect of Disposition. (1) A secured party after default may sell, lease or otherwise dispose of any or all of the collateral in its then condition or following any commercially reasonable preparation or processing. Any sale of goods is subject to the Article on Sales (Article 2). The proceeds of disposition shall be applied in the order following to

(a) the reasonable expenses of retaking, holding, preparing for sale or lease, selling, leasing and the like and, to the extent provided for in the agreement and not prohibited by law, the reasonable attorneys' fees and legal expenses incurred by the secured party;

(b) the satisfaction of indebtedness secured by the security interest under which the disposition is made;

(c) the satisfaction of indebtedness secured by any subordinate security interest in the collateral if written notification of demand therefor is received before distribution of the proceeds is completed. If requested by the secured party, the holder of a subordinate security interest must seasonably furnish reasonable proof of his interest, and unless he does so, the secured party need not comply with his demand.

(2) If the security interest secures an indebtedness, the secured party must account to the debtor for any surplus, and, unless otherwise agreed, the debtor is liable for any deficiency. But if the underlying transaction was a sale of accounts or chattel paper, the debtor is entitled to any surplus or is liable for any deficiency only if the security agreement so provides.

(3) Disposition of the collateral may be by public or private proceedings and may be made by way of one or more contracts. Sale or other disposition may be as a unit or in parcels and at any time and place and on any terms but every aspect of the disposition including the method, manner, time, place and terms must be commercially reasonable. Unless collateral is perishable or threatens to decline speedily in value or is of a type customarily sold on a recognized market, reasonable notification of the time and place of any public sale or reasonable notification of the time after which any private sale or other intended disposition is to be made shall be sent by the secured party to the debtor, if he has not signed after default a statement renouncing or modifying his right to notification of sale. In the case of consumer goods no other notification need be sent. In other cases notification shall be sent to any other secured party from whom the secured party has received (before sending his notification to the debtor or before the debtor's renunciation of his rights) written notice of a claim of an interest in the collateral. The secured party may buy at any public sale and if the collateral is of a type customarily sold in a recognized market or is of a type which is the subject of widely distributed standard price quotations he may buy at private sale.

(4) When collateral is disposed of by a secured party after default, the disposition transfers to a purchaser for value all of the debtor's rights therein, discharges the security interest under which it is made and any security interest or lien subordinate thereto. The purchaser takes free of all such rights and interests even though the secured party fails to comply with the requirements of this Part or of any judicial proceedings

(a) in the case of a public sale, if the purchaser has no knowledge of any defects in the sale and if he does not buy in collusion with the secured party, other bidders or the person conducting the sale; or

(b) in any other case, if the purchaser acts in good faith.

(5) A person who is liable to a secured party under a guaranty, indorsement, repurchase agreement or the like and who receives a transfer of collateral from the secured party or is subrogated to his rights has thereafter the rights and duties of the secured party. Such a transfer of collateral is not a sale or disposition of the collateral under this Article. Amended in 1972.

§ 9–505. Compulsory Disposition of Collateral; Acceptance of the Collateral as Discharge of Obligation. (1) If the debtor has paid sixty percent of the cash price in the case of a purchase money security interest in consumer goods or sixty percent of the loan in the case of another security interest in consumer goods, and has not signed after default a statement renouncing or modifying his rights under this Part a secured party who has taken possession of collateral must dispose of it under Section 9–504 and if he fails to do so within ninety days after he takes possession the debtor at his option may recover in conversion or under Section 9–507(1) on secured party's liability.

(2) In any other case involving consumer goods or any other collateral a secured party in possession may, after default, propose to retain the collateral in satisfaction of the obligation. Written notice of such proposal shall be sent to the debtor if he has not signed after default a statement renouncing or modifying his rights under this subsection. In the case of consumer goods no other notice need be given. In other cases notice shall be sent to any other secured party from whom the secured party has received (before sending his notice to the debtor or before the debtor's renunciation of his rights) written notice of a claim of an interest in the collateral. If the secured party receives objection in writing from a person entitled to receive notification within twenty-one days after the notice was sent, the secured party must dispose of the collateral under Section 9–504. In the absence of such written objection the secured party may retain the collateral in satisfaction of the debtor's obligation. Amended in 1972.

§ 9–506. Debtor's Right to Redeem Collateral. At any time before the secured party has disposed of collateral or entered into a contract for its disposition under Section 9–504 or before the obligation has been discharged under Section 9–505(2) the debtor or any other secured party may unless otherwise agreed in writing after default redeem the collateral by tendering fulfillment of all obligations secured by the collateral as well as the expenses reasonably incurred by the secured party in retaking, holding and preparing the collateral for disposition, in arranging for the sale, and to the extent provided in the agreement and not prohibited by law, his reasonable attorneys' fees and legal expenses.

§ 9–507. Secured Party's Liability for Failure to Comply With This Part. (1) If it is established that the secured party is not proceeding in accordance with the provisions of this Part disposition may be ordered or restrained on appropriate terms and conditions. If the disposition has occurred the debtor or any person entitled to notification or whose security interest has been made known to the secured party prior to the disposition has a right to recover from the secured party any loss caused by a failure to comply with the provisions of this Part. If the collateral is consumer goods, the debtor has a right to recover in any event an amount not less than the credit service charge plus ten percent of the principal amount of the debt or the time price differential plus 10 percent of the cash price.

(2) The fact that a better price could have been obtained by a sale at a different time or in a different method from that selected by the secured

party is not of itself sufficient to establish that the sale was not made in a commercially reasonable manner. If the secured party either sells the collateral in the usual manner in any recognized market therefor or if he sells at the price current in such market at the time of his sale or if he has otherwise sold in conformity with reasonable commercial practices among dealers in the type of property sold he has sold in a commercially reasonable manner. The principles stated in the two preceding sentences with respect to sales also apply as may be appropriate to other types of disposition. A disposition which has been approved in any judicial proceeding or by any bona fide creditors' committee or representative of creditors shall conclusively be deemed to be commercially reasonable, but this sentence does not indicate that any such approval must be obtained in any case nor does it indicate that any disposition not so approved is not commercially reasonable.

ARTICLE 10: EFFECTIVE DATE AND REPEALER

§ 10–101. Effective Date. This Act shall become effective at midnight on December 31st following its enactment. It applies to transactions entered into and events occurring after that date.

§ 10–102. Specific Repealer; Provision for Transition. (1) The following acts and all other acts and parts of acts inconsistent herewith are hereby repealed:

(Here should follow the acts to be specifically repealed including the following:

Uniform Negotiable Instruments Act
Uniform Warehouse Receipts Act
Uniform Sales Act
Uniform Bills of Lading Act
Uniform Stock Transfer Act
Uniform Conditional Sales Act
Uniform Trust Receipts Act

Also any acts regulating:

Bank collections
Bulk sales
Chattel mortgages
Conditional sales
Factor's lien acts
Farm storage of grain and similar acts
Assignment of accounts receivable)

(2) Transactions validly entered into before the effective date specified in Section 10–101 and the rights, duties and interests flowing from them remain valid thereafter and may be terminated, completed, consummated or enforced as required or permitted by any statute or other law amended or repealed by this Act as though such repeal or amendment had not occurred.

Note: *Subsection (1) should be separately prepared for each state. The foregoing is a list of statutes to be checked.*

§ 10–103. General Repealer. Except as provided in the following section, all acts and parts of acts inconsistent with this Act are hereby repealed.

§ 10–104. Laws Not Repealed. (1) The Article on Documents of Title (Article 7) does not repeal or modify any laws prescribing the form or contents of documents of title or the services or facilities to be afforded by bailees, or otherwise regulating bailees' businesses in respects not specifically dealt with herein; but the fact that such laws are violated does not affect the status of a document of title which otherwise complies with the definition of a document of title (Section 1–201).

[(2) This Act does not repeal ....*, cited as the Uniform Act for the Simplification of Fiduciary Security Transfers, and if in any respect there is any inconsistency between that Act and the Article of this Act on investment securities (Article 8) the provisions of the former Act shall control.]

Note: *At * in subsection (2) insert the statutory reference to the Uniform Act for the Simplication of Fiduciary Security Transfers if such Act has previously been enacted. If it has not been enacted, omit subsection (2).*

ARTICLE 11: (REPORTERS' DRAFT)

Effective Date and Transition Provisions

This material has been numbered Article 11 to distinguish it from Article 10, the transition provision of the 1962 Code, which may still remain in effect in some states to cover transition problems from pre-Code law to the original Uniform Commercial Code. Adaptation may be necessary in particular states. The terms "[old Code]" and "[new Code]" and "[old U.C.C.]" and "[new U.C.C.]" are used herein, and should be suitably changed in each state.

This draft was prepared by the Reporters and has not been passed upon by the Review Committee, the Permanent Editorial Board, the American Law Institute, or the National Conference of Commissioners on Uniform State Laws. It is submitted as a working draft which may be adapted as appropriate in each state. The "Discussions" in the Appendix were written by the Reporters to assist in understanding the purpose of the drafts.

§ 11–101. Effective Date. This Act shall become effective at 12:01 A.M. on ______, 19__.

§ 11–102. Preservation of Old Transition Provision. The provisions of [here insert reference to the original transition provision in the particular state] shall continue to apply to [the new U.C.C.] and for this purpose the [old U.C.C. and new U.C.C.] shall be considered one continuous statute.

§ 11–103. Transition to [New Code]—General Rule. Transactions validly entered into after [effective date of old U.C.C.] and before [effective date of new U.C.C.], and which were subject to the provisions of [old U.C.C.] and which would be subject to this Act as amended if they had been entered into after the effective date of [new U.C.C.] and the rights, duties and interests flowing from such transactions remain valid after the latter date and may be terminated, completed, consummated or enforced as required or permitted by the [new U.C.C.]. Security interests arising out of such transactions which are perfected when [new U.C.C.] becomes effective shall remain perfected until they lapse as provided in [new U.C.C.], and may be continued as permitted by [new U.C.C.], except as stated in Section 11–105.

§ 11–104. Transition Provision on Change of Requirement of Filing. A security interest for the perfection of which filing or the taking of possession was required under [old U.C.C.] and which attached prior to the effective date of [new U.C.C.] but was not perfected shall be deemed perfected on the effective date of [new U.C.C.] if [new U.C.C.] permits perfection without filing or authorizes filing in the office or offices where a prior ineffective filing was made.

§ 11–105. Transition Provision on Change of Place of Filing. (1) A financing statement or continuation statement filed prior to [effective date of new U.C.C.] which shall not have lapsed prior to [the effective date of new U.C.C.] shall remain effective for the period provided in the [old Code], but not less than five years after the filing.

(2) With respect to any collateral acquired by the debtor subsequent to the effective date of [new U.C.C.], any effective financing statement or continuation statement described in this section shall apply only if the filing or filings are in the office or offices that would be appropriate to perfect the security interests in the new collateral under [new U.C.C.].

(3) The effectiveness of any financing statement or continuation statement filed prior to [effective date of new U.C.C.] may be continued by a continuation statement as permitted by [new U.C.C.], except that if [new U.C.C.] requires a filing in an office where there was no previous financing statement, a new financing statement conforming to Section 11–106 shall be filed in that office.

(4) If the record of a mortgage of real estate would have been effective as a fixture filing of goods described therein if [new U.C.C.] had been in effect on the date of recording the mortgage, the mortgage shall be deemed effective as a fixture filing as to such goods under subsection (6) of Section 9–402 of the [new U.C.C.] on the effective date of [new U.C.C.].

§ 11–106. Required Refilings. (1) If a security interest is perfected or has priority when this Act takes effect as to all persons or as to certain persons without any filing or recording, and if the filing of a financing statement would be required for the perfection or priority of the security interest against those persons under [new U.C.C.], the perfection and priority rights of the security interest continue until 3 years after the effective date of [new U.C.C.]. The perfection will then lapse unless a financing statement is filed as provided in subsection (4) or unless the security interest is perfected otherwise than by filing.

(2) If a security interest is perfected when [new U.C.C.] takes effect under a law other than [U.C.C.] which requires no further filing, refiling or recording to continue its perfection, perfection continues until and will lapse 3 years after [new U.C.C.] takes effect, unless a financing statement is filed as provided in subsection (4) or unless the security interest is perfected otherwise than by filing, or unless under subsection (3) of Section 9–302 the other law continues to govern filing.

(3) If a security interest is perfected by a filing, refiling or recording under a law repealed by this Act which required further filing, refiling or recording to continue its perfection, perfection continues and will lapse on the date provided by the law so repealed for such further filing, refiling or recording unless a financing statement is filed as provided in subsection (4) or unless the security interest is perfected otherwise than by filing.

(4) A financing statement may be filed within six months before the perfection of a security interest would otherwise lapse. Any such financing statement may be signed by either the debtor or the secured party. It must identify the security agreement, statement or notice (however denominated in any statute or other law repealed or modified by this Act), state the office where and the date when the last filing, refiling or recording, if any, was made with respect thereto, and the filing number, if any, or book and page, if any, of recording and further state that the security agreement, statement or notice, however denominated, in another filing office under the [U.C.C.] or under any statute or other law repealed or modified by this Act is still effective. Section 9–401 and Section 9–103 determine the proper place to file such a financing statement. Except as specified in this subsection, the provisions of Section 9–403(3) for continuation statements apply to such a financing statement.

§ 11–107. Transition Provisions as to Priorities. Except as otherwise provided in [Article 11], [old U.C.C.] shall apply to any questions of priority if the positions of the parties were fixed prior to the effective date of [new U.C.C.]. In other cases questions of priority shall be determined by [new U.C.C.].

§ 11–108. Presumption that Rule of Law Continues Unchanged. Unless a change in law has clearly been made, the provisions of [new U.C.C.] shall be deemed declaratory of the meaning of the [old U.C.C.].

APPENDIX B

Uniform Partnership Act

Part I

Preliminary Provisions

§ **1. Name of Act.** This act may be cited as Uniform Partnership Act.
§ **2. Definition of Terms.** In this act, "Court" includes every court and judge having jurisdiction in the case.

"Business" includes every trade, occupation, or profession.

"Person" includes individuals, partnerships, corporations, and other associations.

"Bankrupt" includes bankrupt under the Federal Bankruptcy Act or insolvent under any state insolvent act.

"Conveyance" includes every assignment, lease, mortgage, or encumbrance.

"Real property" includes land and any interest or estate in land.
§ **3. Interpretation of Knowledge and Notice.** (1) A person has "knowledge" of a fact within the meaning of this act not only when he has actual knowledge thereof, but also when he has knowledge of such other facts as in the circumstances shows bad faith.

(2) A person has "notice" of a fact within the meaning of this act when the person who claims the benefit of the notice

(a) States the fact to such person, or

(b) Delivers through the mail, or by other means of communication, a written statement of the fact to such person or to a proper person at his place of business or residence.
§ **4. Rules of Construction.** (1) The rule that statutes in derogation of the common law are to be strictly construed shall have no application to this act.

(2) The law of estoppel shall apply under this act.

(3) The law of agency shall apply under this act.

(4) This act shall be so interpreted and construed as to effect its general purpose to make uniform the law of those states which enact it.

(5) This act shall not be construed so as to impair the obligations of any contract existing when the act goes into effect, nor to affect any action or proceedings begun or right accrued before this act takes effect.

§ 5. Rules for Cases Not Provided for in this Act. In any case not provided for in this act the rules of law and equity, including the law merchant, shall govern.

Part II

Nature of Partnership

§ 6. Partnership Defined. (1) A partnership is an association of two or more persons to carry on as co-owners a business for profit.

(2) But any association formed under any other statute of this state, or any statute adopted by authority, other than the authority of this state, is not a partnership under this act, unless such association would have been a partnership in this state prior to the adoption of this act; but this act shall apply to limited partnerships except in so far as the statutes relating to such partnerships are inconsistent herewith.

§ 7. Rules for Determining the Existence of a Partnership. In determining whether a partnership exists, these rules shall apply:

(1) Except as provided by Section 16 persons who are not partners as to each other are not partners as to third persons.

(2) Joint tenancy, tenancy in common, tenancy by the entireties, joint property, common property, or part ownership does not of itself establish a partnership, whether such co-owners do or do not share any profits made by the use of the property.

(3) The sharing of gross returns does not of itself establish a partnership, whether or not the persons sharing them have a joint or common right or interest in any property from which the returns are derived.

(4) The receipt by a person of a share of the profits of a business is prima facie evidence that he is a partner in the business, but no such inference shall be drawn if such profits were received in payment:

(a) As a debt by installments or otherwise,

(b) As wages of an employee or rent to a landlord,

(c) As an annuity to a widow or representative of a deceased partner,

(d) As interest on a loan, though the amount of payment vary with the profits of the business.

(e) As the consideration for the sale of a good-will of a business or other property by installments or otherwise.

§ 8. Partnership Property. (1) All property originally brought into the partnership stock or subsequently acquired by purchase or otherwise, on account of the partnership, is partnership property.

(2) Unless the contrary intention appears, property acquired with partnership funds is partnership property.

(3) Any estate in real property may be acquired in the partnership name. Title so acquired can be conveyed only in the partnership name.

(4) A conveyance to a partnership in the partnership name, though

without words of inheritance, passes the entire estate of the grantor unless a contrary intent appears.

Part III

Relations of Partners to Persons Dealing with the Partnership

§ 9. Partner Agent of Partnership as to Partnership Business. (1) Every partner is an agent of the partnership for the purpose of its business, and the act of every partner, including the execution in the partnership name of any instrument, for apparently carrying on in the usual way the business of the partnership of which he is a member binds the partnership, unless the partner so acting has in fact no authority to act for the partnership in the particular matter, and the person with whom he is dealing has knowledge of the fact that he has no such authority.

(2) An act of a partner which is not apparently for the carrying on of the business of the partnership in the usual way does not bind the partnership unless authorized by the other partners.

(3) Unless authorized by the other partners or unless they have abandoned the business, one or more but less than all the partners have no authority to:

(a) Assign the partnership property in trust for creditors or on the assignee's promise to pay the debts of the partnership,

(b) Dispose of the good-will of the business,

(c) Do any other act which would make it impossible to carry on the ordinary business of a partnership,

(d) Confess a judgment,

(e) Submit a partnership claim or liability to arbitration or reference.

(4) No act of a partner in contravention of a restriction on authority shall bind the partnership to persons having knowledge of the restriction.

§ 10. Conveyance of Real Property of the Partnership. (1) Where title to real property is in the partnership name, any partner may convey title to such property by a conveyance executed in the partnership name; but the partnership may recover such property unless the partner's act binds the partnership under the provisions of paragraph (1) of section 9 or unless such property has been conveyed by the grantee or a person claiming through such grantee to a holder for value without knowledge that the partner, in making the conveyance, has exceeded his authority.

(2) Where title to real property is in the name of the partnership, a conveyance executed by a partner, in his own name, passes the equitable interest of the partnership, provided the act is one within the authority of the partner under the provisions of paragraph (1) of section 9.

(3) Where title to real property is in the name of one or more but not all the partners, and the record does not disclose the right of the partner-

ship, the partners in whose name the title stands may convey title to such property, but the partnership may recover such property if the partners' act does not bind the partnership under the provisions of paragraph (1) of section 9, unless the purchaser or his assignee, is a holder for value, without knowledge.

(4) Where the title to real property is in the name of one or more or all the partners, or in a third person in trust for the partnership, a conveyance executed by a partner in the partnership name, or in his own name, passes the equitable interest of the partnership, provided the act is one within the authority of the partner under the provisions of paragraph (1) of section 9.

(5) Where the title to real property is in the names of all the partners a conveyance executed by all the partners passes all their rights in such property.

§ 11. Partnership Bound by Admission of Partner. An admission or representation made by any partner concerning partnership affairs within the scope of his authority as conferred by this act is evidence against the partnership.

§ 12. Partnership Charged with Knowledge of or Notice to Partner. Notice to any partner of any matter relating to partnership affairs, and the knowledge of the partner acting in the particular matter, acquired while a partner or then present to his mind, and the knowledge of any other partner who reasonably could and should have communicated it to the acting partner, operate as notice to or knowledge of the partnership, except in the case of a fraud on the partnership committed by or with the consent of that partner.

§ 13. Partnership Bound by Partner's Wrongful Act. Where, by any wrongful act or omission of any partner acting in the ordinary course of the business of the partnership or with the authority of his co-partners, loss or injury is caused to any person, not being a partner in the partnership, or any penalty is incurred, the partnership is liable therefor to the same extent as the partner so acting or omitting to act.

§ 14. Partnership Bound by Partner's Breach of Trust. The partnership is bound to make good the loss:

(a) Where one partner acting within the scope of his apparent authority receives money or property of a third person and misapplies it; and

(b) Where the partnership in the course of its business receives money or property of a third person and the money or property so received is misapplied by any partner while it is in the custody of the partnership.

§ 15. Nature of Partner's Liability. All partners are liable

(a) Jointly and severally for everything chargeable to the partnership under sections 13 and 14.

(b) Jointly for all other debts and obligations of the partnership; but any partner may enter into a separate obligation to perform a partnership contract.

§ 16. Partner by Estoppel. (1) When a person, by words spoken or written or by conduct, represents himself, or consents to another representing him to any one, as a partner in an existing partnership or with one or more persons not actual partners, he is liable to any such person to whom such representation has been made, who has, on the faith of such representation, given credit to the actual or apparent partnership, and if he has made such representation or consented to its being made in a public manner he is liable to such person, whether the representation has or has not been made or communicated to such person so giving credit by or with the knowledge of the apparent partner making the representation or consenting to its being made.

(a) When a partnership liability results, he is liable as though he were an actual member of the partnership.

(b) When no partnership liability results, he is liable jointly with the other persons, if any, so consenting to the contract or representation as to incur liability, otherwise separately.

(2) When a person has been thus represented to be a partner in an existing partnership, or with one or more persons not actual partners, he is an agent of the persons consenting to such representation to bind them to the same extent and in the same manner as though he were a partner in fact, with respect to persons who rely upon the representation. Where all the members of the existing partnership consent to the representation, a partnership act or obligation results; but in all other cases it is the joint act or obligation of the person acting and the person consenting to the representation.

§ 17. Liability of Incoming Partner. A person admitted as a partner into an existing partnership is liable for all the obligations of the partnership arising before his admission as though he had been a partner when such obligations were incurred, except that this liability shall be satisfied only out of partnership property.

Part IV

Relations of Partners to One Another

§ 18. Rules Determining Rights and Duties of Partners. The rights and duties of the partners in relation to the partnership shall be determined, subject to any agreement between them, by the following rules:

(a) Each partner shall be repaid his contributions, whether by way of capital or advances to the partnership property and share equally in the profits and surplus remaining after all liabilities, including those to partners, are satisfied; and must contribute towards the losses, whether of capital or otherwise, sustained by the partnership according to his share in the profits.

(b) The partnership must indemnify every partner in respect of payments made and personal liabilities reasonably incurred by him in the ordinary and proper conduct of its business, or for the preservation of its business or property.

(c) A partner, who in aid of the partnership makes any payment or advance beyond the amount of capital which he agreed to contribute, shall be paid interest from the date of the payment or advance.

(d) A partner shall receive interest on the capital contributed by him only from the date when repayment should be made.

(e) All partners have equal rights in the management and conduct of the partnership business.

(f) No partner is entitled to remuneration for acting in the partnership business, except that a surviving partner is entitled to reasonable compensation for his services in winding up the partnership affairs.

(g) No person can become a member of a partnership without the consent of all the partners.

(h) Any difference arising as to ordinary matters connected with the partnership business may be decided by a majority of the partners; but no act in contravention of any agreement between the partners may be done rightfully without the consent of all the partners.

§ 19. Partnership Books. The partnership books shall be kept, subject to any agreement between the partners, at the principal place of business of the partnership, and every partner shall at all times have access to and may inspect and copy any of them.

§ 20. Duty of Partners to Render Information. Partners shall render on demand true and full information of all things affecting the partnership to any partner or the legal representative of any deceased partner or partner under legal disability.

§ 21. Partner Accountable as a Fiduciary. (1) Every partner must account to the partnership for any benefit, and hold as trustee for it any profits derived by him without the consent of the other partners from any transaction connected with the formation, conduct, or liquidation of the partnership or from any use by him of its property.

(2) This section applies also to the representatives of a deceased partner engaged in the liquidation of the affairs of the partnership as the personal representatives of the last surviving partner.

§ 22. Right to an Account. Any partner shall have the right to a formal account as to partnership affairs:

(a) If he is wrongfully excluded from the partnership business or possession of its property by his co-partners.

(b) If the right exists under the terms of any agreement,

(c) As provided by section 21,

(d) Whenever other circumstances render it just and reasonable.

§ 23. Continuation of Partnership Beyond Fixed Term. (1) When a partnership for a fixed term or particular undertaking is continued after the termination of such term or particular undertaking without any express agreement, the rights and duties of the partners remain the same as they were at such termination, so far as is consistent with a partnership at will.

(2) A continuation of the business by the partners or such of them as habitually acted therein during the term, without any settlement or liquidation of the partnership affairs, is prima facie evidence of a continuation of the partnership.

Part V

Property Rights of a Partner

§ 24. Extent of Property Rights of a Partner. The property rights of a partner are (1) his rights in specific partnership property, (2) his interest in the partnership, and (3) his right to participate in the management.

§ 25. Nature of a Partner's Right in Specific Partnership Property. (1) A partner is co-owner with his partners of specific partnership property holding as a tenant in partnership.

(2) The incidents of this tenancy are such that:

(a) A partner, subject to the provisions of this act and to any agreement between the partners, has an equal right with his partners to possess specific partnership property for partnership purposes; but he has no right to possess such property for any other purpose without the consent of his partners.

(b) A partner's right in specific partnership property is not assignable except in connection with the assignment of rights of all the partners in the same property.

(c) A partner's right in specific partnership property is not subject to attachment or execution, except on a claim against the partnership. When partnership property is attached for a partnership debt the partners, or any of them, or the representatives of a deceased partner, cannot claim any right under the homestead or exemption laws.

(d) On the death of a partner his right in specific partnership property vests in the surviving partner or partners, except where the deceased was the last surviving partner, when his right in such property vests in his legal representative. Such surviving partner or partners, or the legal representative of the last surviving partner, has no right to possess the partnership property for any but a partnership purpose.

(e) A partner's right in specific partnership property is not subject to dower, curtesy, or allowances to widows, heirs, or next of kin.

§ 26. Nature of Partner's Interest in the Partnership. A partner's interest in the partnership is his share of the profits and surplus, and the same is personal property.

§ 27. Assignment of Partner's Interest. (1) A conveyance by a partner of his interest in the partnership does not of itself dissolve the partnership, nor, as against the other partners in the absence of agreement, entitle the assignee, during the continuance of the partnership to interfere in the management or administration of the partnership business or affairs, or to require any information or account of partnership transactions, or to inspect the partnership books; but it merely entitles the assignee to receive in accordance with his contract the profits to which the assigning partner would otherwise be entitled.

(2) In case of a dissolution of the partnership, the assignee is entitled to receive his assignor's interest and may require an account from the date only of the last account agreed to by all the partners.

§ 28. Partner's Interest Subject to Charging Order. (1) On due application to a competent court by any judgment creditor of a partner, the court which entered the judgment, order, or decree, or any other court, may charge the interest of the debtor partner with payment of the unsatisfied amount of such judgment debt with interest thereon; and may then or later appoint a receiver of his share of the profits, and of any other money due or to fall due to him in respect of the partnership, and make all other orders, directions, accounts and inquiries which the debtor partner might have made, or which the circumstances of the case may require.

(2) The interest charged may be redeemed at any time before foreclosure, or in case of a sale being directed by the court may be purchased without thereby causing a dissolution:

(a) With separate property, by any one or more of the partners, or

(b) With partnership property, by any one or more of the partners with the consent of all the partners whose interests are not so charged or sold.

(3) Nothing in this act shall be held to deprive a partner of his right, if any, under the exemption laws, as regards his interest in the partnership.

Part VI

Dissolution and Winding up

§ 29. **Dissolution Defined.** The dissolution of a partnership is the change in the relation of the partners caused by any partner ceasing to be associated in the carrying on as distinguished from the winding up of the business.

§ 30. **Partnership Not Terminated by Dissolution.** On dissolution the partnership is not terminated, but continues until the winding up of partnership affairs is completed.

§ 31. **Causes of Dissolution.** Dissolution is caused: (1) Without violation of the agreement between the partners,

(a) By the termination of the definite term or particular undertaking specified in the agreement,

(b) By the express will of any partner when no definite term or particular undertaking is specified,

(c) By the express will of all the partners who have not assigned their interests or suffered them to be charged for their separate debts, either before or after the termination of any specified term or particular undertaking.

(d) By the expulsion of any partner from the business bona fide in accordance with such a power conferred by the agreement between the partners;

(2) In contravention of the agreement between the partners, where the circumstances do not permit a dissolution under any other provision of this section, by the express will of any partner at any time;

(3) By any event which makes it unlawful for the business of the partnership to be carried on or for the members to carry it on in partnership;

(4) By the death of any partner;

(5) By the bankruptcy of any partner or the partnership;

(6) By decree of court under section 32.

§ 32. Dissolution by Decree of Court. (1) On application by or for a partner the court shall decree a dissolution whenever:

(a) A partner has been declared a lunatic in any judicial proceeding or is shown to be of unsound mind,

(b) A partner becomes in any other way incapable of performing his part of the partnership contract.

(c) A partner has been guilty of such conduct as tends to affect prejudicially the carrying on of the business,

(d) A partner wilfully or persistently commits a breach of the partnership agreement, or otherwise so conducts himself in matters relating to the partnership business that it is not reasonably practicable to carry on the business in partnership with him,

(e) The business of the partnership can only be carried on at a loss,

(f) Other circumstances render a dissolution equitable.

(2) On the application of the purchaser of a partner's interest under sections 27 or 28:

(a) After the termination of the specified term or particular undertaking,

(b) At any time if the partnership was a partnership at will when the interest was assigned or when the charging order was issued.

§ 33. General Effect of Dissolution on Authority of Partner. Except so far as may be necessary to wind up partnership affairs or to complete transactions begun but not then finished, dissolution terminates all authority of any partner to act for the partnership,

(1) With respect to the partners,

(a) When the dissolution is not by the act, bankruptcy or death of a partner; or

(b) When the dissolution is by such act, bankruptcy or death of a partner, in cases where section 34 so requires.

(2) With respect to persons not partners, as declared in section 35.

§ 34. Right of Partner to Contribution from Copartners After Dissolution. Where the dissolution is caused by the act, death or bankruptcy of a partner, each partner is liable to his copartners for his share of any liability created by any partner acting for the partnership as if the partnership had not been dissolved unless

(a) The dissolution being by act of any partner, the partner acting for the partnership had knowledge of the dissolution, or

(b) The dissolution being by the death or bankruptcy of a partner, the partner acting for the partnership had knowledge or notice of the death or bankruptcy.

§ 35. Power of Partner to Bind Partnership to Third Persons After Dissolution. (1) After dissolution a partner can bind the partnership except as provided in Paragraph (3)

(a) By any act appropriate for winding up partnership affairs or completing transactions unfinished at dissolution;

(b) By any transaction which would bind the partnership if dissolution had not taken place, provided the other party to the transaction

(I) Had extended credit to the partnership prior to dissolution and had no knowledge or notice of the dissolution; or

(II) Though he had not so extended credit, had nevertheless known of the partnership prior to dissolution, and, having no knowledge or notice of dissolution, the fact of dissolution had not been advertised in a newspaper of general circulation in the place (or in each place if more than one) at which the partnership business was regularly carried on.

(2) The liability of a partner under paragraph (1b) shall be satisfied out of partnership assets alone when such partner had been prior to dissolution.

(a) Unknown as a partner to the person with whom the contract is made; and

(b) So far unknown and inactive in partnership affairs that the business reputation of the partnership could not be said to have been in any degree due to his connection with it.

(3) The partnership is in no case bound by any act of a partner after dissolution

(a) Where the partnership is dissolved because it is unlawful to carry on the business, unless the act is appropriate for winding up partnership affairs; or

(b) Where the partner has become bankrupt; or

(c) Where the partner has no authority to wind up partnership affairs; except by a transaction with one who

(I) Had extended credit to the partnership prior to dissolution and had no knowledge or notice of his want of authority; or

(II) Had not extended credit to the partnership prior to dissolution, and, having no knowledge or notice of his want of authority, the fact of his want of authority has not been advertised in the manner provided for advertising the fact of dissolution in paragraph (1bII).

(4) Nothing in this section shall affect the liability under section 16 of any person who after dissolution represents himself or consents to another representing him as a partner in a partnership engaged in carrying on business.

§ 36. Effect of Dissolution on Partner's Existing Liability. (1) The dissolution of the partnership does not of itself discharge the existing liability of any partner.

(2) A partner is discharged from any existing liability upon dissolution of the partnership by an agreement to that effect between himself, the partnership creditor and the person or partnership continuing the business; and such agreement may be inferred from the course of dealing between the creditor having knowledge of the dissolution and the person or partnership continuing the business.

(3) Where a person agrees to assume the existing obligations of a dissolved partnership, the partners whose obligations have been assumed shall be discharged from any liability to any creditor of the partnership who, know-

ing of the agreement, consents to a material alteration in the nature or time of payment of such obligations.

(4) The individual property of a deceased partner shall be liable for all obligations of the partnership incurred while he was a partner but subject to the prior payment of his separate debts.

§ 37. Right to Wind Up. Unless otherwise agreed the partners who have not wrongfully dissolved the partnership or the legal representative of the last surviving partner, not bankrupt, has the right to wind up the partnership affairs; provided, however, that any partner, his legal representative or his assignee, upon cause shown, may obtain winding up by the court.

§ 38. Rights of Partners to Application of Partnership Property. (1) When dissolution is caused in any way, except in contravention of the partnership agreement, each partner as against his co-partners and all persons claiming through them in respect of their interests in the partnership, unless otherwise agreed, may have the partnership property applied to discharge its liabilities, and the surplus applied to pay in cash the net amount owing to the respective partners. But if dissolution is caused by expulsion of a partner, bona fide under the partnership agreement and if the expelled partner is discharged from all partnership liabilities, either by payment or agreement under section 36(2), he shall receive in cash only the net amount due him from the partnership.

(2) When dissolution is caused in contravention of the partnership agreement the rights of the partners shall be as follows:

(a) Each partner who has not caused dissolution wrongfully shall have,

(I) All the rights specified in paragraph (1) of this section, and

(II) The right, as against each partner who has caused the dissolution wrongfully, to damages for breach of the agreement.

(b) The partners who have not caused the dissolution wrongfully, if they all desire to continue the business in the same name, either by themselves or jointly with others, may do so, during the agreed term for the partnership and for that purpose may possess the partnership property, provided they secure the payment by bond approved by the court, or pay to any partner who has caused the dissolution wrongfully, the value of his interest in the partnership at the dissolution, less any damages recoverable under clause (2aII) of the section, and in like manner indemnify him against present or future partnership liabilities.

(c) A partner who has caused the dissolution wrongfully shall have:

(I) If the business is not continued under the provisions of paragraph (2b) all the rights of a partner under paragraph (1), subject to clause (2aII), of this section,

(II) If the business is continued under paragraph (2b) of this section the right as against his co-partners and all claiming through them in respect of their interests in the partnership, to have the value of his interest in the partnership, less any damages caused to his co-partners by the dissolution, ascertained and paid to him in cash, or the payment secured by bond approved by the court, and to be released from all existing liabilities of the partnership;

but in ascertaining the value of the partner's interest the value of the good-will of the business shall not be considered.

§ 39. Rights Where Partnership Is Dissolved for Fraud or Misrepresentation. Where a partnership contract is rescinded on the ground of the fraud or misrepresentation of one of the parties thereto, the party entitled to rescind is, without prejudice to any other right, entitled.

(a) To a lien on, or right of retention of, the surplus of the partnership property after satisfying the partnership liabilities to third persons for any sum of money paid by him for the purchase of an interest in the partnership and for any capital or advances contributed by him; and

(b) To stand, after all liabilities to third persons have been satisfied, in the place of the creditors of the partnership for any payments made by him in respect of the partnership liabilities; and

(c) To be indemnified by the person guilty of the fraud or making the representation against all debts and liabilities of the partnership.

§ 40. Rules for Distribution. In settling accounts between the partners after dissolution, the following rules shall be observed, subject to any agreement to the contrary:

(a) The assets of the partnership are;

(I) The partnership property,

(II) The contributions of the partners necessary for the payment of all the liabilities specified in clause (b) of this paragraph.

(b) The liabilities of the partnership shall rank in order of payment, as follows:

(I) Those owing to creditors other than partners,

(II) Those owning to partners other than for capital and profits,

(III) Those owing to partners in respect of capital,

(IV) Those owing to partners in respect of profits.

(c) The assets shall be applied in the order of their declaration in clause (a) of this paragraph to the satisfaction of the liabilities.

(d) The partners shall contribute, as provided by section 18(a) the amount necessary to satisfy the liabilities; but if any, but not all, of the partners are insolvent, or, not being subject to process, refuse to contribute, the other parties shall contribute their share of the liabilities, and, in the relative proportions in which they share the profits, the additional amount necessary to pay the liabilities.

(e) An assignee for the benefit of creditors or any person appointed by the court shall have the right to enforce the contributions specified in clause (d) of this paragraph.

(f) Any partner or his legal representative shall have the right to enforce the contributions specified in clause (d) of this paragraph, to the extent of the amount which he has paid in excess of his share of the liability.

(g) The individual property of a deceased partner shall be liable for the contributions specified in clause (d) of this paragraph.

(h) When partnership property and the individual properties of the partners are in possession of a court for distribution, partnership creditors shall have priority on partnership property and separate creditors on individual property, saving the rights of lien or secured creditors as heretofore.

(i) Where a partner has become bankrupt or his estate is insolvent the claims against his separate property shall rank in the following order:

(I) Those owing to separate creditors,

(II) Those owing to partnership creditors,

(III) Those owing to partners by way of contribution.

§ 41. Liability of Persons Continuing the Business in Certain Cases. (1) When any new partner is admitted into an existing partnership, or when any partner retires and assigns (or the representative of the deceased partner assigns) his rights in partnership property to two or more of the partners, or to one or more of the partners and one or more third persons, if the business is continued without liquidation of the partnership affairs, creditors of the first or dissolved partnership are also creditors of the partnership so continuing the business.

(2) When all but one partner retire and assign (or the representative of a deceased partner assigns) their rights in partnership property to the remaining partner, who continues the business without liquidation of partnership affairs, either alone or with others, creditors of the dissolved partnership are also creditors of the person or partnership so continuing the business.

(3) When any partner retires or dies and the business of the dissolved partnership is continued as set forth in paragraphs (1) and (2) of this section, with the consent of the retired partners or the representative of the deceased partner, but without any assignment of his right in partnership property, rights of creditors of the dissolved partnership and of the creditors of the person or partnership continuing the business shall be as if such assignment had been made.

(4) When all the partners or their representatives assign their rights in partnership property to one or more third persons who promise to pay the debts and who continue the business of the dissolved partnership, creditors of the dissolved partnership are also creditors of the person or partnership continuing the business.

(5) When any partner wrongfully causes a dissolution and the remaining partners continue the business under the provisions of section 38(2b), either alone or with others, and without liquidation of the partnership affairs, creditors of the dissolved partner are also creditors of the person or partnership continuing the business.

(6) When a partner is expelled and the remaining partners continue the business either alone or with others, without liquidation of the partnership affairs, creditors of the dissolved partnership are also creditors of the person or partnership continuing the business.

(7) The liability of a third person becoming a partner in the partnership continuing the business, under this section, to the creditors of the dissolved partnership shall be satisfied out of partnership property only.

(8) When the business of a partnership after dissolution is continued

under any conditions set forth in this section the creditors of the dissolved partnership, as against the separate creditors of the retiring or deceased partner or the representative of the deceased partner, have a prior right to any claim of the retired partner or the representative of the deceased partner against the person or partnership continuing the business, on account of the retired or deceased partner's interest in the dissolved partnership or on account of any consideration promised for such interest or for his right in partnership property.

(9) Nothing in this section shall be held to modify any right of creditors to set aside any assignment on the ground of fraud.

(10) The use by the person or partnership continuing the business of the partnership name, or the name of a deceased partner as part thereof, shall not of itself make the individual property of the deceased partner liable for any debts contracted by such person or partnership.

§ 42. Rights of Retiring or Estate of Deceased Partner When the Business Is Continued. When any partner retires or dies, and the business is continued under any of the conditions set forth in section 41(1, 2, 3, 5, 6), or section 38(2b), without any settlement of accounts as between him or his estate and the person or partnership continuing the business, unless otherwise agreed, he or his legal representative as against such persons or partnership may have the value of his interest at the date of dissolution ascertained, and shall receive as an ordinary creditor an amount equal to the value of his interest in the dissolved partnership with interest, or, at his option or at the option of his legal representative, in lieu of interest, the profits attributable to the use of his right in the property of the dissolved partnership; provided that the creditors of the dissolved partnership as against the separate creditors, or the representative of the retired or deceased partner, shall have priority on any claim arising under this section, as provided by section 41(8) of this act.

§ 43. Accrual of Actions. The right to an account of his interest shall accrue to any partner, or his legal representative, as against the winding up partners or the surviving partners or the person or partnership continuing the business, at the date of dissolution, in the absence of any agreement to the contrary.

Part VII

Miscellaneous Provisions

§ 44. When Act Takes Effect. This act shall take effect on the _______day of _______one thousand nine hundred and _______.

§ 45. Legislation Repealed. All acts or parts of acts inconsistent with this act are hereby repealed.

APPENDIX C

Uniform Limited Partnership Act

Author's note: *The Uniform Limited Partnership Act was initially written in 1916. A revised version was written in 1976 but only 2 states, Connecticut and Wyoming, have adopted it to date. 46 states have adopted the 1916 version reproduced here. Minnesota adopted the 1916 version but has enacted a statute to shift to the 1976 version.*

§ 1. Limited Partnership Defined. A limited partnership is a partnership formed by two or more persons under the provisions of Section 2, having as members one or more general partners and one or more limited partners. The limited partners as such shall not be bound by the obligations of the partnership.

§ 2. Formation. (1) Two or more persons desiring to form a limited partnership shall

(a) Sign and swear to a certificate, which shall state

I. The name of the partnership,

II. The character of the business,

III. The location of the principal place of business,

IV. The name and place of residence of each member; general and limited partners being respectively designated.

V. The term for which the partnership is to exist,

VI. The amount of cash and a description of and the agreed value of the other property contributed by each limited partner,

VII. The additional contributions, if any, agreed to be made by each limited partner and the times at which or events on the happening of which they shall be made.

VIII. The time, if agreed upon, when the contribution of each limited partner is to be returned.

IX. The share of the profits or the other compensation by way of income which each limited partner shall receive by reason of his contribution,

X. The right, if given, of a limited partner to substitute an assignee as contributor in his place, and the terms and conditions of the substitution,

XI. The right, if given, of the partners to admit additional limited partners,

XII. The right, if given, of one or more of the limited partners to priority over other limited partners, as to contributions or as to compensation by way of income, and the nature of such priority.

XIII. The right, if given, of the remaining general partner or partners to continue the business on the death, retirement or insanity of a general partner, and

XIV. The right, if given, of a limited partner to demand and receive property other than cash in return for his contribution.

(b) File for record the certificate in the office of [here designate the proper office].

(2) A limited partnership is formed if there has been substantial compliance in good faith with the requirements of paragraph (1).

§ 3. Business Which May Be Carried On. A limited partnership may carry on any business which a partnership without limited partners may carry on, except [here designate the business to be prohibited].

§ 4. Character of Limited Partner's Contribution. The contributions of a limited partner may be cash or other property, but not services.

§ 5. A Name Not to Contain Surname of Limited Partner; Exceptions. (1) The surname of a limited partner shall not appear in the partnership name, unless

(a) It is also the surname of a general partner, or

(b) Prior to the time when the limited partner became such the business had been carried on under a name in which his surname appeared.

(2) A limited partner whose name appears in a partnership name contrary to the provisions of paragraph (1) is liable as a general partner to partnership creditors who extend credit to the partnership without actual knowledge that he is not a general partner.

§ 6. Liability for False Statements in Certificate. If the certificate contains a false statement, one who suffers loss by reliance on such statement may hold liable any party to the certificate who knew the statements to be false.

(a) At the time he signed the certificate, or

(b) Subsequently, but within a sufficient time before the statement was relied upon to enable him to cancel or amend the certificate, or to file a petition for its cancellation or amendment as provided in Section 25(3).

§ 7. Limited Partner Not Liable to Creditors. A limited partner shall not become liable as a general partner unless, in addition to the exercise of his rights and powers as a limited partner, he takes part in the control of the business.

§ 8. Admission of Additional Limited Partners. After the formation of a limited partnership, additional limited partners may be admitted upon filing an amendment to the original certificate in accordance with the requirements of Section 25.

§ 9. Rights, Powers and Liabilities of a General Partner. (1) A general partner shall have all the rights and powers and be subject to all the restrictions and liabilities of a partner in a partnership without limited partners, except that without the written consent or ratification of the specific act by all the limited partners, a general partner or all of the general partners have no authority to

(a) Do any act in contravention of the certificate,

(b) Do any act which would make it impossible to carry on the ordinary business of the partnership,

(c) Confess a judgment against the partnership,

(d) Possess partnership property, or assign their rights in specific partnership property, for other than a partnership purpose,

(e) Admit a person as a general partner,

(f) Admit a person as a limited partner, unless the right so to do is given in the certificate,

(g) Continue the business with partnership property on the death, retirement or insanity of a general partner, unless the right so to do is given in the certificate.

§ 10. Rights of a Limited Partner. (1) A limited partner shall have the same rights as a general partner to

(a) Have the partnership books kept at the principal place of business of the partnership, and at all times to inspect and copy any of them,

(b) Have on demand true and full information of all things affecting the partnership, and a formal account of partnership affairs, whenever circumstances render it just and reasonable, and

(c) Have dissolution and winding up by decree of court.

(2) A limited partner shall have the right to receive a share of the profits or other compensation by way of income, and to the return of his contribution as provided in Sections 15 and 16.

§ 11. Status of Person Erroneously Believing Himself a Limited Partner. A person who has contributed to the capital of a business conducted by a person or partnership erroneously believing that he has become a limited partner in a limited partnership, is not, by reason of his exercise of the rights of a limited partner, a general partner with the person or in the partnership carrying on the business, or bound by the obligations of such person or partnership; provided that on ascertaining the mistake he promptly renounces his interest in the profits of the business, or other compensation by way of income.

§ 12. One Person Both General and Limited Partner. (1) A person may be a general partner and a limited partner in the same partnership at the same time.

(2) A person who is a general, and also at the same time a limited partner, shall have all the rights and powers and be subject to all the restrictions of a general partner; except that, in respect to his contribution, he shall have the rights against the other members which he would have had if he were not also a general partner.

§ 13. Loans and Other Business Transactions with Limited Partner. (1) A limited partner also may loan money to and transact other business with the partnership, and, unless he is also a general partner, receive on account of resulting claims against the partnership, with general creditors, a pro rata share of the assets. No limited partner shall in respect to any such claim

(a) Receive or hold as collateral security any partnership property, or

(b) Receive from a general partner or the partnership any payment, conveyance, or release from liability, if at the time the assets of the partnership are not sufficient to discharge partnership liabilities to persons not claiming as general or limited partners,

(2) The receiving of collateral security, or a payment, conveyance, or release in violation of the provisions of paragraph (1) is a fraud on the creditors of the partnership.

§ 14. Relation of Limited Partners Inter Se. Where there are several limited partners the members may agree that one or more of the limited partners shall have a priority over other limited partners as to the return of their contributions, as to their compensation by way of income, or as to any other matter. If such an agreement is made it shall be stated in the certificate, and in the absence of such a statement all the limited partners shall stand upon equal footing.

§ 15. Compensation of Limited Partner. A limited partner may receive from the partnership the share of the profits or the compensation by way of income stipulated for in the certificate; provided, that after such payment is made, whether from the property of the partnership or that of a general partner, the partnership assets are in excess of all liabilities of the partnership except liabilities to limited partners on account of their contributions and to general partners.

§ 16. Withdrawal or Reduction of Limited Partner's Contribution. (1) A limited partner shall not receive from a general partner or out of partnership property any part of his contribution until

(a) All liabilities of the partnership, except liabilities to general partners and to limited partners on account of their contributions, have been paid or there remains property of the partnership sufficient to pay them.

(b) The consent of all members is had, unless the return of the contribution may be rightfully demanded under the provisions of paragraph (2), and

(c) The certificate is cancelled or so amended as to set forth the withdrawal or reduction.

(2) Subject to the provisions of paragraph (1) a limited partner may rightfully demand the return of his contribution

(a) On the dissolution of a partnership, or

(b) When the date specified in the certificate for its return has arrived, or

(c) After he has given six months' notice in writing to all other members, if no time is specified in the certificate either for the return of the contribution or for the dissolution of the partnership,

(3) In the absence of any statement in the certificate to the contrary or the consent of all members, a limited partner, irrespective of the nature

of his contribution, has only the right to demand and receive cash in return for his contribution.

(4) A limited partner may have the partnership dissolved and its affairs wound up when

(a) He rightfully but unsuccessfully demands the return of his contribution, or

(b) The other liabilities of the partnership have not been paid, or the partnership property is insufficient for their payment as required by paragraph (1a) and the limited partner would otherwise be entitled to the return of his contribution.

§ 17. Liability of Limited Partner to Partnership. (1) A limited partner is liable to the partnership

(a) For the difference between his contribution as actually made and that stated in the certificate as having been made, and

(b) For any unpaid contribution which he agreed in the certificate to make in the future at the time and on the conditions stated in the certificate.

(2) A limited partner holds as trustee for the partnership

(a) Specific property stated in the certificate as contributed by him, but which was not contributed or which has been wrongfully returned, and

(b) Money or other property wrongfully paid or conveyed to him on account of his contribution.

(3) The liabilities of a limited partner as set forth in this section can be waived or compromised only by the consent of all members; but a waiver or compromise shall not affect the right of a creditor of a partnership, who extended credit or whose claim arose after the filing and before a cancellation or amendment of the certificate, to enforce such liabilities.

(4) When a contributor has rightfully received the return in whole or in part of the capital of his contribution, he is nevertheless liable to the partnership for any sum, not in excess of such return with interest, necessary to discharge its liabilities to all creditors who extended credit or whose claims arose before such return.

§ 18. Nature of Limited Partner's Interest in Partnership. A limited partner's interest in the partnership is personal property.

§ 19. Assignment of Limited Partner's Interest. A limited partner's interest is assignable.

(2) A substituted limited partner is a person admitted to all the rights of a limited partner who has died or has assigned his interest in a partnership.

(3) An assignee, who does not become a substituted limited partner, has no right to require any information or account of the partnership transactions or to inspect the partnership books; he is only entitled to receive the share of the profits or other compensation by way of income, or the return of his contribution, to which his assignor would otherwise be entitled.

(4) An assignee shall have the right to become a substituted limited partner if all the members (except the assignor) consent thereto or if the

assignor, being thereunto empowered by the certificate, gives the assignee that right.

(5) An assignee becomes a substituted limited partner when the certificate is appropriately amended in accordance with Section 25.

(6) The substituted limited partner has all the rights and powers, and is subject to all the restrictions and liabilities of his assignor, except those liabilities of which he was ignorant at the time he became a limited partner and which could not be ascertained from the certificate.

(7) The substitution of the assignee as a limited partner does not release the assignor from liability to the partnership under Section 6 and 17.

§ 20. Effect of Retirement, Death or Insanity of a General Partner. The retirement, death or insanity of a general partner dissolves the partnership, unless the business is continued by the remaining general partners

(a) Under a right so to do stated in the certificate, or

(b) With the consent of all members.

§ 21. Death of Limited Partner. (1) On the death of a limited partner his executor or administrator shall have all the rights of a limited partner for the purpose of settling his estate, and such power as the deceased had to constitute his assignee a substituted limited partner.

(2) The estate of a deceased limited partner shall be liable for all his liabilities as a limited partner.

§ 22. Rights of Creditors of Limited Partner. (1) On due application to a court of competent jurisdiction by any judgment creditor of a limited partner, the court may charge the interest of the indebted limited partner with payment of the unsatisfied amount of the judgment debt; and may appoint a receiver, and make all other orders, directions, and inquiries which the circumstances of the case may require.

In those states where a creditor on beginning an action can attach debts due the defendant before he has obtained a judgment against the defendant it is recommended that paragraph (1) of this section read as follows:

On due application to a court of competent jurisdiction by any creditor of a limited partner, the court may charge the interest of the indebted limited partner with payment of the unsatisfied amount of such claim; and may appoint a receiver, and make all other orders, directions, and inquiries which the circumstances of the case may require.

(2) The interest may be redeemed with the separate property of any general partner, but may not be redeemed with partnership property.

(3) The remedies conferred by paragraph (1) shall not be deemed exclusive of others which may exist.

(4) Nothing in this act shall be held to deprive a limited partner of his statutory exemption.

§ 23. Distribution of Assets. (1) In settling accounts after dissolution the liabilities of the partnership shall be entitled to payment in the following order:

(a) Those to creditors, in the order of priority as provided by law, except those to limited partners on account of their contributions, and to general partner,

(b) Those to limited partners in respect to their share of the profits and other compensation by way of income on their contributions,

(c) Those to limited partners in respect to the capital of their contributions,

(d) Those to general partners other than for capital and profits,

(e) Those to general partners in respect to profits,

(f) Those to general partners in respect to capital.

(2) Subject to any statement in the certificate or to subsequent agreement, limited partners share in the partnership assets in respect to their claims for capital, and in respect to their claims for profits or for compensation by way of income on their contributions respectively, in proportion to the respective amounts of such claims.

§ 24. When Certificate Shall Be Cancelled or Amended. (1) The certificate shall be cancelled when the partnership is dissolved or all limited partners cease to be such.

(2) A certificate shall be amended when

(a) There is a change in the name of the partnership or in the amount or character of the contribution of any limited partner,

(b) A person is substituted as a limited partner,

(c) An additional limited partner is admitted,

(d) A person is admitted as a general partner,

(e) A general partner retires, dies or becomes insane, and the business is continued under Section 20.

(f) There is a change in the character of the business of the partnership,

(g) There is a false or erroneous statement in the certificate,

(h) There is a change in the time as stated in the certificate for the dissolution of the partnership or for the return of a contribution,

(i) A time is fixed for the dissolution of the partnership, or the return of a contribution, no time having been specified in the certificate, or

(j) The members desire to make a change in any other statement in the certificate in order that it shall accurately represent the agreement between them.

§ 25. Requirements for Amendment and for Cancellation of Certificate. (1) The writing to amend a certificate shall

(a) Conform to the requirements of Section 2(1a) as far as necessary to set forth clearly the change in the certificate which it is desired to make, and

(b) Be signed and sworn to by all members, and an amendment substituting a limited partner or adding a limited or general partner shall be signed also by the member to be substituted or added, and when a limited partner is to be substituted, the amendment shall also be signed by the assigning limited partner.

(2) The writing to cancel a certificate shall be signed by all members.

(3) A person desiring the cancellation or amendment of a certificate, if any person designated in paragraphs (1) and (2) as a person who must execute the writing refuses to do so, may petition the [here designate the proper court] to direct a cancellation or amendment thereof.

(4) If the court finds that the petitioner has a right to have the writing executed by a person who refuses to do so, it shall order the [here designate the responsible official in office designated in Section 2] in the office where the certificate is recorded to record the cancellation or amendment of the certificate; and where the certificate is to be amended, the court shall also cause to be filed for record in said office a certified copy of its decree setting forth the amendment.

(5) A certificate is amended or cancelled when there is filed for record in the office [here designate the office designated in Section 2] where the certificate is recorded

(a) A writing in accordance with the provisions of paragraph (1), or (2) or

(b) A certified copy of the order of court in accordance with the provisions of paragraph (4).

(6) After the certificate is duly amended in accordance with this section, the amended certificate shall thereafter be for all purposes the certificate provided for by this act.

§ 26. Parties to Actions. A contributor, unless he is a general partner, is not a proper party to proceedings by or against a partnership, except where the object is to enforce a limited partner's right against or liability to the partnership.

§ 27. Name of Act. This act may be cited as The Uniform Limited Partnership Act.

§ 28. Rules of Construction. (1) The rule that statutes in derogation of the common law are to be strictly construed shall have no application to this act.

(2) This act shall be so interpreted and construed as to effect its general purpose to make uniform the law of those states which enact it.

(3) This act shall not be so construed as to impair the obligations of any contract existing when the act goes into effect, nor to affect any action on proceedings begun or right accrued before this act takes effect.

§ 29. Rules for Cases Not Provided for in this Act. In any case not provided for in this act the rules of law and equity, including the law merchant, shall govern.

§ 30.[1] Provisions for Existing Limited Partnerships. (1) A limited partnership formed under any statute of this state prior to the adoption of this act, may become a limited partnership under this act by complying with the provisions of Section 2; provided the certificate sets forth

(a) The amount of the original contribution of each limited partner, and the time when the contribution was made, and

(b) That the property of the partnership exceeds the amount sufficient to discharge its liabilities to persons not claiming as general or limited partners by an amount greater than the sum of the contributions of its limited partners.

(2) A limited partnership formed under any statute of this state prior to the adoption of this act, until or unless it becomes a limited partnership under this act, shall continue to be governed by the provisions of [here insert proper reference to the existing limited partnership act or acts], except that such partnership shall not be renewed unless so provided in the original agreement.

§ 31.[1] **Act [Acts] Repealed.** Except as affecting existing limited partnerships to the extent set forth in Section 30, the act (acts) of [here designate the existing limited partnership act or acts] is (are) hereby repealed.

[1] Sections 30, 31, will be omitted in any state which has not a limited partnership act.

APPENDIX D

Model Business Corporation Act

(1969 Text as Amended to 1979.)

§ **1. Short Title.** This Act shall be known and may be cited as the "[supply name of state] . . . Business Corporation Act."
§ **2. Definitions.** As used in this Act, unless the context otherwise requires, the term:

(a) "Corporation" or "domestic corporation" means a corporation for profit subject to the provisions of this Act, except a foreign corporation.

(b) "Foreign corporation" means a corporation for profit organized under laws other than the laws of this State for a purpose or purposes for which a corporation may be organized under this Act.

(c) "Articles of incorporation" means the original or restated articles of incorporation or articles of consolidation and all amendments thereto including articles of merger.

(d) "Shares" means the units into which the proprietary interests in a corporation are divided.

(e) "Subscriber" means one who subscribes for shares in a corporation, whether before or after incorporation.

(f) "Shareholder" means one who is a holder of record of shares in a corporation. If the articles of incorporation or the by-laws so provide, the board of directors may adopt by resolution a procedure whereby a shareholder of the corporation may certify in writing to the corporation that all or a portion of the shares registered in the name of such shareholder are held for the account of a specified person or persons. The resolution shall set forth (1) the classification of shareholder who may certify, (2) the purpose or purposes for which the certification may be made, (3) the form of certification and information to be contained therein, (4) if the certification is with respect to a record date or closing of the stock transfer books within which the certification must be received by the corporation and (5) such other provisions with respect to the procedure as are deemed necessary or desirable. Upon receipt by the corporation of a certification complying with the procedure, the persons specified in the certification shall be deemed, for the purpose or purposes set forth in the certification, to be the holders of record of the number of shares specified in place of the shareholder making the certification.

(g) "Authorized shares" means the shares of all classes which the corporation is authorized to issue.

(h) "Treasury shares" means shares of a corporation which have been issued, have been subsequently acquired by and belong to the corporation, and have not, either by reason of the acquisition or thereafter, been cancelled or restored to the status of authorized but unissued shares. Treasury shares shall be deemed to be "issued" shares, but not "outstanding" shares.

(i) "Net assets" means the amount by which the total assets of a corporation exceed the total debts of the corporation.

(j) "Stated capital" means, at any particular time, the sum of (1) the par value of all shares of the corporation having a par value that have been issued, (2) the amount of consideration received by the corporation for all shares of the corporation without par value that have been issued, except such part of the consideration therefor as may have been allocated to capital surplus in a manner permitted by law, and (3) such amounts not included in clauses (1) and (2) of this paragraph as have been transferred to stated capital of the corporation, whether upon the issue of shares as a share dividend or otherwise, minus all reductions from such sum as have been effected in a manner permitted by law. Irrespective of the manner of designation thereof by the laws under which a foreign corporation is organized, the stated capital of a foreign corporation shall be determined on the same basis and in the same manner as the stated capital of a domestic corporation, for the purpose of computing fees, franchise taxes and other charges imposed by this Act.

(k) "Surplus" means the excess of the net assets of a corporation over its stated capital.

(l) "Earned surplus" means the portion of the surplus of a corporation equal to the balance of its net profits, income, gains and losses from the date of incorporation, or from the latest date when a deficit was eliminated by an application of its capital surplus or stated capital or otherwise, after deducting subsequent distributions to shareholders and transfers to stated capital and capital surplus to the extent such distributions and transfers are made out of earned surplus. Earned surplus shall include also any portion of surplus allocated to earned surplus in mergers, consolidations or acquisitions of all or substantially all of the outstanding shares or of the property and assets of another corporation, domestic or foreign.

(m) "Capital surplus" means the entire surplus of a corporation other than its earned surplus.

(n) "Insolvent" means inability of a corporation to pay its debts as they become due in the usual course of its business.

(o) "Employee" includes officers but not directors. A director may accept duties which make him also an employee.

§ 3. Purposes. Corporations may be organized under this Act for any lawful purpose or purposes, except for the purpose of banking or insurance.

§ 4. General Powers. Each corporation shall have power.

(a) To have perpetual succession by its corporate name unless a limited period of duration is stated in its articles of incorporation.

(b) To sue and be sued, complain and defend, in its corporate name.

(c) To have a corporate seal which may be altered at pleasure, and to use the same by causing it, or a facsimile thereof, to be impressed or affixed or in any other manner reproduced.

(d) To purchase, take, receive, lease, or otherwise acquire, own, hold, improve, use and otherwise deal in and with, real or personal property, or any interest therein, wherever situated.

(e) To sell, convey, mortgage, pledge, lease, exchange, transfer and otherwise dispose of all or any part of its property and assets.

(f) To lend money and use its credit to assist its employees.

(g) To purchase, take, receive, subscribe for, or otherwise acquire, own, hold, vote, use, employ, sell, mortgage, lend, pledge, or otherwise dispose of, and otherwise use and deal in and with, shares or other interests in, or obligations of, other domestic or foreign corporations, associations, partnerships or individuals, or direct or indirect obligations of the United States or of any other government, state, territory, governmental district or municipality or of any instrumentality thereof.

(h) To make contracts and guarantees and incur liabilities, borrow money at such rates of interest as the corporation may determine, issue its notes, bonds, and other obligations, and secure any of its obligations by mortgage or pledge of all or any of its property, franchises and income.

(i) To lend money for its corporate purposes, invest and reinvest its funds, and take and hold real and personal property as security for the payment of funds so loaned or invested.

(j) To conduct its business, carry on its operations and have offices and exercise the powers granted by this Act, within or without this State.

(k) To elect or appoint officers and agents of the corporation, and define their duties and fix their compensation.

(l) To make and alter by-laws, not inconsistent with its articles of incorporation or with the laws of this State, for the administration and regulation of the affairs of the corporation.

(m) To make donations for the public welfare or for charitable, scientific or educational purposes.

(n) To transact any lawful business which the board of directors shall find will be in aid of governmental policy.

(o) To pay pensions and establish pension plans, pension trusts, profit sharing plans, stock bonus plans, stock option plans and other incentive plans for any or all of its directors, officers and employees.

(p) To be a promoter, partner, member, associate, or manager of any partnership, joint venture, trust or other enterprise.

(q) To have and exercise all powers necessary or convenient to effect its purposes.

§ 5. Indemnification of Officers, Directors, Employees and Agents.

(a) A corporation shall have power to indemnify any person who was or is a party or is threatened to be made a party to any threatened, pending or completed action, suit or proceeding, whether civil, criminal, administrative or investigative (other than an action by or in the right of the corporation) by reason of the fact that he is or was a director, officer, employee or agent of the corporation, or is or was serving at the request of the corporation as a director, officer, employee or agent of another corporation, partnership, joint venture, trust or other enterprise, against expenses (including attorneys' fees), judgments, fines and amounts paid in settlement actually and reasonably incurred by him in connection with such action, suit or proceeding if he acted in good faith and in a manner he reasonably believed to be in or not opposed to the best interests of the corporation, and, with respect to any criminal action or proceeding, had no reasonable cause to believe his conduct was unlawful. The termination of any action, suit or proceeding by judgment, order, settlement, conviction, or upon a plea of nolo contendere or its equivalent, shall not, of itself, create a presumption that the person did not act in good faith and in a manner which he reasonably believed to be in or not opposed to the best interests of the corporation, and, with respect to any criminal action or proceeding, had reasonable cause to believe that his conduct was unlawful.

(b) A corporation shall have power to indemnify any person who was or is a party or is threatened to be made a party to any threatened, pending or completed action or suit by or in the right of the corporation to procure a judgment in its favor by reason of the fact that he is or was a director, officer, employee or agent of the corporation, or is or was serving at the request of the corporation as a director, officer, employee or agent of another corporation, partnership, joint venture, trust or other enterprise against expenses (including attorneys' fees) actually and reasonably incurred by him in connection with the defense or settlement of such action or suit if he acted in good faith and in a manner he reasonably believed to be in or not opposed to the best interests of the corporation and except that no indemnification shall be made in respect of any claim, issue or matter as to which such person shall have been adjudged to be liable for negligence or misconduct in the performance of his duty to the corporation unless and only to the extent that the court in which such action or suit was brought shall determine upon application that, despite the adjudication of liability but in view of all circumstances of the case, such person is fairly and reasonably entitled to indemnity for such expenses which such court shall deem proper.

(c) To the extent that a director, officer, employee or agent of a corporation has been successful on the merits or otherwise in defense of any action, suit or proceeding referred to in subsections (a) or (b), or in defense of any claim, issue or matter therein, he shall be indemnified against expenses (including attorneys' fees) actually and reasonably incurred by him in connection therewith.

(d) Any indemnification under subsections (a) or (b) (unless ordered by a court) shall be made by the corporation only as authorized in the specific case upon a determination that indemnification of the director, officer, employee or agent is proper in the circumstances because he has met the applicable standard of conduct set forth in subsections (a) or (b). Such determination shall be made (1) by the board of directors by a majority vote of a quorum consisting of directors who were not parties to such action, suit or proceeding, or (2) if such a quorum is not obtainable, or, even if obtainable a quorum of distinterested directors so directs, by independent legal counsel in a written opinion, or (3) by the shareholders.

(e) Expenses (including attorneys' fees) incurred in defending a civil or criminal action, suit or proceeding may be paid by the corporation in advance of the final disposition of such action, suit or proceeding as authorized in the manner provided in subsection (d) upon receipt of an undertaking by or on behalf of the director, officer, employee or agent to repay such amount unless it shall ultimately be determined that he is entitled to be indemnified by the corporation as authorized in this section.

(f) The indemnification provided by this section shall not be deemed exclusive of any other rights to which those indemnified may be entitled under any by-law, agreement, vote of shareholders or disinterested directors or otherwise, both as to action in his official capacity and as to action in another capacity while holding such office, and shall continue as to a person who has ceased to be a director, officer, employee or agent and shall inure to the benefit of the heirs, executors and administrators of such a person.

(g) A corporation shall have power to purchase and maintain insurance on behalf of any person who is or was a director, officer, employee or agent of the corporation, or is or was serving at the request of the corporation as a director, officer, employee or agent of another corporation, partnership, joint venture, trust or other enterprise against any liability asserted against him and incurred by him in any such capacity or arising out of his status as such, whether or not the corporation would have the power to indemnify him against such liability under the provisions of this section.

§ 6. Right of Corporation to Acquire and Dispose of its Own Shares. A corporation shall have the right to purchase, take, receive or otherwise acquire, hold, own, pledge, transfer or otherwise dispose of its own shares, but purchases of its own shares, whether direct or indirect, shall be made only to the extent of unreserved and unrestricted earned surplus available therefor, and, if the articles of incorporation so permit or with the affirmative vote of the holders of a majority of all shares entitled to vote thereon, to the extent of unreserved and unrestricted capital surplus available therefor.

To the extent that earned surplus or capital surplus is used as the measure of the corporation's right to purchase its own shares, such surplus shall be restricted so long as such shares are held as treasury shares, and upon the disposition or cancellation of any such shares the restriction shall be removed pro tanto.

Notwithstanding the foregoing limitation, a corporation may purchase or otherwise acquire its own shares for the purpose of:

(a) Eliminating fractional shares.

(b) Collecting or compromising indebtedness to the corporation.

(c) Paying dissenting shareholders entitled to payment for their shares under the provisions of this Act.

(d) Effecting, subject to the other provisions of this Act, the retirement of its redeemable shares by redemption or by purchase at not to exceed the redemption price.

No purchase of or payment for its own shares shall be made at a time when the corporation is insolvent or when such purchase or payment would make it insolvent.

§ 7. Defense of Ultra Vires. No act of a corporation and no conveyance or transfer of real or personal property to or by a corporation shall be invalid by reason of the fact that the corporation was without capacity or power to do such act or to make or receive such conveyance or transfer, but such lack of capacity or power may be asserted:

(a) In a proceeding by a shareholder against the corporation to enjoin the doing of any act or the transfer of real or personal property by or to the corporation. If the unauthorized act or transfer sought to be enjoined is being, or is to be, performed or made pursuant to a contract to which the corporation is a party, the court may, if all of the parties to the contract are parties to the proceeding and if it deems the same to be equitable, set aside and enjoin the performance of such contract, and in so doing may allow to the corporation or to the other parties to the contract, as the case may be, compensation for the loss or damage sustained by either of them which may result from the action of the court in setting aside and enjoining the performance of such contract, but anticipated profits to be derived from the performance of the contract shall not be awarded by the court as a loss or damage sustained.

(b) In a proceeding by the corporation, whether acting directly or through a receiver, trustee, or other legal representative, or through shareholders in a representative unit, against the incumbent or former officers or directors of the corporation.

(c) In a proceeding by the Attorney General, as provided in this Act, to dissolve the corporation, or in a proceeding by the Attorney General to enjoin the corporation from the transaction of unauthorized business.

§ 8. Corporate Name. The corporate name:

(a) Shall contain the word "corporation," "company," "incorporated" or "limited," or shall contain an abbreviation of one of such words.

(b) Shall not contain any word or phrase which indicates or implies that it is organized for any purpose other than one or more of the purposes contained in its articles of incorporation.

(c) Shall not be the same as, or deceptively similar to, the name of any domestic corporation existing under the laws of this State or any foreign corporation authorized to transact business in this State, or a name the exclusive right to which is, at the time, reserved in the manner provided in this Act, or the name of a corporation which has in effect a registration of its corporate name as provided in this Act, except that this provision shall not apply if the applicant files with the Secretary of State either of the following: (1) the written consent of such other corporation or holder of a reserved or registered name to use the same or deceptively similar name and one or more words are added to make such name distinguishable from such other name, or (2) a certified copy of a final decree of a court of competent jurisdiction establishing the prior right of the applicant to the use of such name in this State.

A corporation with which another corporation, domestic or foreign, is merged, or which is formed by the reorganization or consolidation of one or more domestic or foreign corporations or upon a sale, lease or other disposition to or exchange with, a domestic corporation of all or substantially all the assets of another corporation, domestic or foreign, including its name, may have the same name as that used in this State by any of such corporations if such other corporation was organized under the laws of, or is authorized to transact business in, this State.

§ 9. **Reserved Name.** The exclusive right to the use of a corporate name may be reserved by:

(a) Any person intending to organize a corporation under this Act.

(b) Any domestic corporation intending to change its name.

(c) Any foreign corporation intending to make application for a certificate of authority to transact business in this State.

(d) Any foreign corporation authorized to transact business in this State and intending to change its name.

(e) Any person intending to organize a foreign corporation and intending to have such corporation make application for a certificate of authority to transact business in this State.

The reservation shall be made by filing with the Secretary of State an application to reserve a specified corporate name, executed by the applicant. If the Secretary of State finds that the name is available for corporate use, he shall reserve the same for the exclusive use of the applicant for a period of one hundred and twenty days.

The right to the exclusive use of a specified corporate name so reserved may be transferred to any other person or corporation by filing in the office of the Secretary of State a notice of such transfer, executed by the applicant for whom the name was reserved, and specifying the name and address of the transferee.

§ 10. **Registered Name.** Any corporation organized and existing under the laws of any state or territory of the United States may register its corporate name under this Act, provided its corporate name is not the same as, or deceptively similar to, the name of any domestic corporation existing under the laws of this State, or the name of any foreign corporation authorized to transact business in this State, or any corporate name reserved or registered under this Act. *(Remainder of text omitted.)*

§ 11. **Renewal of Registered Name** *(Text omitted).*

§ 12. **Registered Office and Registered Agent.** Each corporation shall have and continuously maintain in this State:

(a) A registered office which may be, but need not be, the same as its place of business.

(b) A registered agent, which agent may be either an individual resident in this State whose business office is identical with such registered office, or a domestic corporation, or a foreign corporation authorized to transact business in this State, having a business office identical with such registered office.

§ 13. Change of Registered Office or Registered Agent *(Text omitted).*

§ 14. Service of Process on Corporation. The registered agent so appointed by a corporation shall be an agent of such corporation upon whom any process, notice or demand required or permitted by law to be served upon the corporation may be served.

Whenever a corporation shall fail to appoint or maintain a registered agent in this State, or whenever its registered agent cannot with reasonable diligence be found at the registered office, then the Secretary of State shall be an agent of such corporation upon whom any such process, notice or demand may be served. Service on the Secretary of State of any such process, notice, or demand shall be made by delivering to and leaving with him, or with any clerk having charge of the corporation department of his office, duplicate copies of such process, notice or demand. In the event any such process, notice or demand is served on the Secretary of State, he shall immediately cause one of the copies thereof to be forwarded by registered mail, addressed to the corporation at its registered office. Any service so had on the Secretary of State shall be returnable in not less than thirty days. *(Remainder of text omitted.)*

§ 15. Authorized Shares. Each corporation shall have power to create and issue the number of shares stated in its articles of incorporation. Such shares may be divided into one or more classes, any or all of which classes may consist of shares with par value or shares without par value, with such designations, preferences, limitations, and relative rights as shall be stated in the articles of incorporation. The articles of incorporation may limit or deny the voting rights of or provide special voting rights for the shares of any class to the extent not inconsistent with the provisions of this Act.

Without limiting the authority herein contained, a corporation, when so provided in its articles of incorporation, may issue shares of preferred or special classes:

(a) Subject to the right of the corporation to redeem any of such shares at the price fixed by the articles of incorporation for the redemption thereof.

(b) Entitling the holders thereof to cumulative, noncumulative or partially cumulative dividends.

(c) Having preference over any other class or classes of shares as to the payment of dividends.

(d) Having preference in the assets of the corporation over any other class or classes of shares upon the voluntary or involuntary liquidation of the corporation.

(e) Convertible into shares of any other class or into shares of any series of the same or any other class, except a class having prior or superior rights and preferences as to dividends or distribution of assets upon liquidation, but shares without par value shall not be converted into shares with par value unless that part of the stated capital of the corporation represented by such shares without par value is, at the time of conversion, at least equal to the aggregate par value of the shares into which the shares without par value are to be converted or the amount of any such deficiency is transferred from surplus to stated capital.

§ 16. Issuance of Shares of Preferred or Special Classes in Series. If the articles of incorporation so provide, the shares of any preferred or special class may be divided into and issued in series. If the shares of any such class are to be issued in series, then each series shall be so designated as to distinguish the shares thereof from the shares of all other series and classes. Any or all of the series of any such class and the variations in the relative rights and preferences as between different series may be fixed and determined by the articles of incorporation, but all shares of the same class shall be identical except as to the following relative rights and preferences, as to which there may be variations between different series:

(a) The rate of dividend.

(b) Whether shares may be redeemed and, if so, the redemption price and the term and conditions of redemption.

(c) The amount payable upon shares in event of voluntary and involuntary liquidation.

(d) Sinking fund provisions, if any, for the redemption or purchase of shares.

(e) The terms and conditions, if any, on which shares may be converted.

(f) Voting rights, if any.

If the articles of incorporation shall expressly vest authority in the board of directors, then, to the extent that the articles of incorporation shall not have established series and fixed and determined the variations in the relative rights and preferences as between series, the board of directors shall have authority to divide any or all of such classes into series and, within the limitations set forth in this section and in the articles of incorporation, fix and determine the relative rights and preferences of the shares of any series so established.

In order for the board of directors to establish a series, where authority so to do is contained in the articles of incorporation, the board of directors shall adopt a resolution setting forth the designation of the series and fixing and determining the relative rights and preferences thereof, or so much thereof as shall not be fixed and determined by the articles of incorporation.

Prior to the issue of any shares of a series established by resolution adopted by the board of directors, the corporation shall file in the office of the Secretary of State a statement setting forth:

(a) The name of the corporation.

(b) A copy of the resolution establishing and designating the series, and fixing and determining the relative rights and preferences thereof.

(c) The date of adoption of such resolution.

(d) That such resolution was duly adopted by the board of directors.

Such statement shall be executed in duplicate by the corporation by its president or a vice president and by its secretary or an assistant secretary, and verified by one of the officers signing such statement, and shall be delivered to the Secretary of State. If the Secretary of State finds that such statement conforms to law, he shall, when all franchise taxes and fees have been paid as in this Act prescribed:

(1) Endorse on each of such duplicate originals the word "Filed," and the month, day, and year of the filing thereof.

(2) File one of such duplicate originals in his office.

(3) Return the other duplicate original to the corporation or its representative.

Upon the filing of such statement by the Secretary of State, the resolution establishing and designating the series and fixing and determining the relative rights and preferences thereof shall become effective and shall constitute an amendment of the articles of incorporation.

§ 17. Subscriptions for Shares. A subscription for shares of a corporation to be organized shall be irrevocable for a period of six months, unless otherwise provided by the terms of the subscription agreement or unless all of the subscribers consent to the revocation of such subscription.

Unless otherwise provided in the subscription agreement, subscriptions for shares, whether made before or after the organization of a corporation, shall be paid in full at such time, or in such installments and at such times, as shall be determined by the board of directors. Any call made by the board of directors for payment on subscriptions shall be uniform as to all shares of the same class or as to all shares of the same series, as the case may be. In case of default in the payment of any installment or call when such payment is due, the corporation may proceed to collect the amount due in the same manner as any debt due the corporation. The by-laws may prescribe other penalties for failure to pay installments or calls that may become due, but no penalty working a forfeiture of a subscription, or of the amounts paid thereon, shall be declared as against any subscriber unless the amount due thereon shall remain unpaid for a period of twenty days after written demand has been made therefor. If mailed, such written demand shall be deemed to be made when deposited in the United States mail in a sealed envelope addressed to the subscriber at his last post-office address known to the corporation, with postage thereon prepaid. In the event of the sale of any shares by reason of any forfeiture, the excess of proceeds realized over the amount due and unpaid on such shares shall be paid to the delinquent subscriber or to his legal representative.

§ 18. Consideration for Shares. Shares having a par value may be issued for such consideration expressed in dollars, not less than the par value thereof, as shall be fixed from time to time by the board of directors.

Shares without par value may be issued for such consideration expressed in dollars as may be fixed from time to time by the board of directors unless the articles of incorporation reserve to the shareholders the right to fix the consideration. In the event that such right be reserved as to any shares, the shareholders shall, prior to the issuance of such shares, fix the consideration to be received for such shares, by a vote of the holders of a majority of all shares entitled to vote thereon.

Treasury shares may be disposed of by the corporation for such consideration expressed in dollars as may be fixed from time to time by the board of directors.

That part of the surplus of a corporation which is transferred to stated capital upon the issuance of shares as a share dividend shall be deemed to be the consideration for the issuance of such shares.

In the event of the issuance of shares upon the conversion or exchange of indebtedness or shares, the consideration for the shares so issued shall be (1) the principal sum of, and accrued interest on, the indebtedness so exchanged or converted, or the stated capital then represented by the shares so exchanged or converted, and (2) that part of surplus, if any, transferred to stated capital upon the issuance of shares for the shares so exchanged or converted, and (3) any additional consideration paid to the corporation upon the issuance of shares for the indebtedness or shares so exchanged or converted.

§ 19. Payment for Shares. The consideration for the issuance of shares may be paid, in whole or in part, in cash, in other property, tangible or intangible, or in labor or services actually performed for the corporation. When payment of the consideration for which shares are to be issued shall have been received by the corporation, such shares shall be deemed to be fully paid and nonassessable.

Neither promissory notes nor future services shall constitute payment or part payment for the issuance of shares of a corporation.

In the absence of fraud in the transaction, the judgment of the board of directors or the shareholders, as the case may be, as to the value of the consideration received for shares shall be conclusive.

§ 20. Stock Rights and Options. Subject to any provisions in respect thereof set forth in its articles of incorporation, a corporation may create and issue, whether or not in connection with the issuance and sale of any of its shares or other securities, rights or options entitling the holders thereof to purchase from the corporation shares of any class or classes. Such rights or options shall be evidenced in such manner as the board of directors shall approve and, subject to the provisions of the articles of incorporation, shall set forth the terms upon which, the time or times within which and the price or prices at which such shares may be purchased from the corporation upon the exercise of any such right or option. If such rights or options are to be issued to directors, officers or employees as such of the corporation or of any subsidiary thereof, and not to the shareholders generally, their issuance shall be approved by the affirmative vote of the holders of a majority of the shares entitled to vote thereon or shall be authorized by and consistent with a plan approved or ratified by such a vote of shareholders. In the absence of fraud in the transaction, the judgment of the board of directors as to the adequacy of the consideration received for such rights or options shall be conclusive. The price or prices to be received for any shares having a par value, other than treasury shares to be issued upon the exercise of such rights or options, shall not be less than the par value thereof.

§ 21. Determination of Amount of Stated Capital. In case of the issuance by a corporation of shares having a par value, the consideration received therefor shall constitute stated capital to the extent of the par value of such shares, and the excess, if any, of such consideration shall constitute capital surplus.

In case of the issuance by a corporation of shares without par value, the entire consideration received therefor shall constitute stated capital unless the corporation shall determine as provided in this section that only a part thereof shall be stated capital. Within a period of sixty days after the issuance of any shares without par value, the board of directors may allocate to capital surplus any portion of the consideration received for the issuance of such shares. No such allocation shall be made of any portion of the consideration received for shares without par value having a preference in the assets of the corporation in the event of involuntary liquidation except the amount, if any, of such consideration in excess of such preference.

If shares have been or shall be issued by a corporation in merger or consolidation or in acquisition of all or substantially all of the outstanding shares or of the property and assets of another corporation, whether domestic or foreign, any amount that would otherwise constitute capital surplus under the foregoing provisions of this section may instead be allocated to earned surplus by the board of directors of the issuing corporation except that its aggregate earned surplus shall not exceed the sum of the earned surpluses as defined in this Act of the issuing corporation and of all other corporations, domestic or foreign, that were merged or consolidated or of which the shares or assets were acquired.

The stated capital of a corporation may be increased from time to time by resolution of the board of directors directing that all or a part of the surplus of the corporation be transferred to stated capital. The board of directors may direct that the amount of the surplus so transferred shall be deemed to be stated capital in respect of any designated class of shares.

§ 22. Expenses of Organization, Reorganization and Financing. The reasonable charges and expenses of organization or reorganization of a corporation, and the reasonable expenses of and compensation for the sale or underwriting of its shares, may be paid or allowed by such corporation out of the consideration received by it in payment for its shares without thereby rendering such shares not fully paid or assessable.

§ 23. Shares Represented by Certificates and Uncertificated Shares. The shares of a corporation shall be represented by certificates or shall be uncertificated shares. Certificates shall be signed by the chairman or vice chairman of the board of directors or the president or a vice president and by the treasurer or an assistant treasurer or the secretary or an assistant secretary of the corporation, and may be sealed with the seal of the corporation or a facsimile thereof.

Any of or all the signatures upon a certificate may be a facsimile. In case any officer, transfer agent or registrar who has signed or whose facsimile signature has been placed upon such certificate shall have ceased to be such officer, transfer agent or registrar before such certificate is issued, it may be issued by the corporation with the same effect as if he were such officer, transfer agent or registrar at the date of its issue.

Every certificate representing shares issued by a corporation which is authorized to issue shares of more than one class shall set forth upon the face or back of the certificate, or shall state that the corporation will furnish to any shareholder upon request and without charge, a full statement of the designations, preferences, limitations, and relative rights of the shares of each class authorized to be issued, and if the corporation is authorized to issue any preferred or special class in series, the variations in the relative rights and preferences between the shares of each such series so far as the same have been fixed and determined and the authority of the board of directors to fix and determine the relative rights and preferences of subsequent series.

Each certificate representing shares shall state upon the face thereof:

(a) That the corporation is organized under the laws of this State.

(b) The name of the person to whom issued.

(c) The number and class of shares, and the designation of the series, if any, which such certificate represents.

(d) The par value of each share represented by such certificate, or a statement that the shares are without par value.

No certificate shall be issued for any share until such share is fully paid.

Unless otherwise provided by the articles of incorporation or by-laws, the board of directors of a corporation may provide by resolution that some or all of any or all classes and series of its shares shall be uncertificated shares, provided that such resolution shall not apply to shares represented by a certificate until such certificate is surrendered to the corporation. Within a reasonable time after the issuance or transfer of uncertificated shares, the corporation shall send to the registered owner thereof a written notice containing the information required to be set forth or stated on certificates pursuant to the second and third paragraphs of this section. Except as otherwise expressly provided by law, the rights and obligations of the holders of uncertificated shares and the rights and obligations of the holders of certificates representing shares of the same class and series shall be identical.

§ 24. Fractional Shares. A corporation may (1) issue fractions of a share, either represented by a certificate or uncertificated, (2) arrange for the disposition of fractional interests by those entitled thereto, (3) pay in cash the fair value of fractions of a share as of a time when those entitled to receive such fractions are determined, or (4) issue scrip in registered or bearer form which shall entitle the holder to receive a certificate for a full share or an uncertificated full share upon the surrender of such scrip aggregating a full share. A certificate for a fractional share or an uncertificated fractional share shall, but scrip shall not unless otherwise provided therein, entitle the holder to exercise voting rights, to receive dividends thereon, and to participate in any of the assets of the corporation in the event of liquidation. The board of directors may cause scrip to be issued subject to the condition that it shall become void if not exchanged for certificates representing full shares or uncertificated full shares before a specified date, or subject to the condition that the shares for which scrip is exchangeable may be sold by the corporation and the proceeds thereof distributed to the holders of scrip, or subject to any other conditions which the board of directors may deem advisable.

§ 25. Liability of Subscribers and Shareholders. A holder of or subscriber to shares of a corporation shall be under no obligation to the corporation or its creditors with respect to such shares other than the obligation to pay to the corporation the full consideration for which such shares were issued or to be issued.

Any person becoming an assignee or transferee of shares or of a subscription for shares in good faith and without knowledge or notice that the full consideration therefor has not been paid shall not be personally liable to the corporation or its creditors for any unpaid portion of such consideration.

An executor, administrator, conservator, guardian, trustee, assignee for the benefit of creditors, or receiver shall not be personally liable to the corporation as a holder of or subscriber to shares of a corporation but the estate and funds in his hands shall be so liable.

No pledgee or other holder of shares as collateral security shall be personally liable as a shareholder.

§ 26. Shareholders' Preemptive Rights. The shareholders of a corporation shall have no preemptive right to acquire unissued or treasury shares of the corporation, or securities of the corporation convertible into or carrying a right to subscribe to or acquire shares, except to the extent, if any, that such right is provided in the articles of incorporation.

§ 26A. Shareholders' Preemptive Rights [Alternative]. Except to the extent limited or denied by this section or by the articles of incorporation, shareholders shall have a preemptive right to acquire unissued or treasury shares or securities convertible into such shares or carrying a right to subscribe to or acquire shares.

Unless otherwise provided in the articles of incorporation,

(a) No preemptive right shall exist

(1) to acquire any shares issued to directors, officers or employees pursuant to approval by the affirmative vote of the holders of a majority of the shares entitled to vote thereon or when authorized by and consistent with a plan theretofore approved by such a vote of shareholders; or

(2) to acquire any shares sold otherwise than for cash.

(b) Holders of shares of any class that is preferred or limited as to dividends or assets shall not be entitled to any preemptive right.

(c) Holders of shares of common stock shall not be entitled to any preemptive right to shares of any class that is preferred or limited as to dividends or assets or to any obligations, unless convertible into shares of common stock or carrying a right to subscribe to or acquire shares of common stock.

(d) Holders of common stock without voting power shall have no preemptive right to shares of common stock with voting power.

(e) The preemptive right shall be only an opportunity to acquire shares or other securities under such terms and conditions as the board of directors may fix for the purpose of providing a fair and reasonable opportunity for the exercise of such right.

§ 27. By-laws. The initial by-laws of a corporation shall be adopted by its board of directors. The power to alter, amend or repeal the by-laws or adopt new by-laws, subject to repeal or change by action of the shareholders, shall be vested in the board of directors unless reserved to the shareholders by the articles of incorporation. The by-laws may contain any provisions for the regulation and management of the affairs of the corporation not inconsistent with law or the articles of incorporation.

§27A. By-laws and Other Powers in Emergency [Optional]. The board of directors of any corporation may adopt emergency by-laws, subject to repeal or change by action of the shareholders, which shall, notwithstanding any different provision elsewhere in this Act or in the articles of incorporation or by-laws, be operative during any emergency in the conduct of the business of the corporation resulting from an attack on the United States or any nuclear or atomic disaster. The emergency by-laws may make any provision that may be practical and necessary for the circumstances of the emergency, including provisions that:

(a) A meeting of the board of directors may be called by any officer or director in such manner and under such conditions as shall be prescribed in the emergency by-laws;

(c) The director or directors in attendance at the meeting, or any greater number fixed by the emergency by-laws, shall constitute a quorum; and

(c) The officers or other persons designated on a list approved by the board of directors before the emergency, all in such order of priority and subject to such conditions, and for such period of time (not longer than reasonably necessary after the termination of the emergency) as may be provided in the emergency by-laws or in the resolution approving the list shall, to the extent required to provide a quorum at any meeting of the board of directors, be deemed directors for such meeting.

The board of directors, either before or during any such emergency, may provide, and from time to time modify, lines of succession in the event that during such an emergency any or all officers or agents of the corporation shall for any reason be rendered incapable of discharging their duties.

The board of directors, either before or during any such emergency, may, effective in the emergency, change the head office or designate several alternative head offices or regional offices, or authorize the officers so to do.

To the extent not inconsistent with any emergency by-laws so adopted, the by-laws of the corporation shall remain in effect during any such emergency and upon its termination the emergency by-laws shall cease to be operative.

Unless otherwise provided in emergency by-laws, notice of any meeting of the board of directors during any such emergency may be given only to such of the directors as it may be feasible to reach at the time and by such means as may be feasible at the time, including publication or radio.

To the extent required to constitute a quorum at any meeting of the board of directors during any such emergency, the officers of the corporation who are present shall, unless otherwise provided in emergency by-laws, be deemed, in order of rank and within the same rank in order of seniority, directors for such meeting.

No officer, director or employee acting in accordance with any emergency by-laws shall be liable except for willful misconduct. No officer, director or employee shall be liable for any action taken by him in good faith in such an emergency in furtherance of the ordinary business affairs of the corporation even though not authorized by the by-laws then in effect.

§28. Meetings of Shareholders. Meetings of shareholders may be held at such place within or without this State as may be stated in or fixed in accordance with the by-laws. If no other place is stated or so fixed, meetings shall be held at the registered office of the corporation.

An annual meeting of the shareholders shall be held at such time as may be stated in or fixed in accordance with the by-laws. If the annual meeting is not held within any thirteen-month period the Court of. . . . . may, on the application of any shareholder, summarily order a meeting to be held.

Special meetings of the shareholders may be called by the board of directors, the holders of not less than one-tenth of all the shares entitled to vote at the meeting, or such other persons as may be authorized in the articles of incorporation or the by-laws.

§29. Notice of Shareholders' Meetings. Written notice stating the place, day and hour of the meeting and, in case of a special meeting, the purpose or purposes for which the meeting is called, shall be delivered not less than ten nor more than fifty days before the date of the meeting, either personally or by mail, by or at the direction of the president, the secretary, or the officer or persons calling the meeting, to each shareholder of record entitled to vote at such meeting. If mailed, such notice shall be deemed to be delivered when deposited in the United States mail addressed to the shareholder at his address as it appears on the stock transfer books of the corporation, with postage thereon prepaid.

§30. Closing of Transfer Books and Fixing Record Date. For the purpose of determining shareholders entitled to notice of or to vote at any meeting of shareholders or any adjournment thereof, or entitled to receive payment of any dividend, or in order to make a determination of shareholders for any other proper purpose, the board of directors of a corporation may provide that the stock transfer books shall be closed for a stated period but not to exceed, in any case, fifty days. If the stock transfer books shall be closed for the purpose of determining shareholders entitled to notice of or to vote at a meeting of shareholders, such books shall be closed for at least ten days immediately preceding such meeting. In lieu of closing the stock transfer books, the by-laws, or in the absence of an applicable by-law the board of directors, may fix in advance a date as the record date for any such determination of shareholders, such date in any case to be not more than fifty days and, in case of a meeting of shareholders, not less than ten days prior to the date on which the particular action, requiring such determination of shareholders, is to be taken. If the stock transfer books are not closed and no record date is fixed for the determination of shareholders entitled to notice of or to vote at a meeting of shareholders, or shareholders entitled to receive payment of a dividend, the date on which notice of the meeting is mailed or the date on which the resolution of the board of directors declaring such dividend is adopted, as the case may be, shall be the record date for such determination of shareholders. When a determination of shareholders entitled to vote at any meeting of shareholders has been made as provided in this section, such determination shall apply to any adjournment thereof.

§31. Voting Record. The officer or agent having charge of the stock transfer books for shares of a corporation shall make a complete record of the shareholders entitled to vote at such meeting or any adjournment thereof, arranged in alphabetical order, with the address of and the number of shares held by each. Such record shall be produced and kept open at the time and place of the meeting and shall be subject to the inspection of any shareholder during the whole time of the meeting for the purposes thereof.

Failure to comply with the requirements of this section shall not affect the validity of any action taken at such meeting.

An officer or agent having charge of the stock transfer books who shall fail to prepare the record of shareholders, or produce and keep it open for inspection at the meeting, as provided in this section, shall be liable to any shareholder suffering damage on account of such failure, to the extent of such damage.

§32. Quorum of Shareholders. Unless otherwise provided in the articles of incorporation, a majority of the shares entitled to vote, represented in person or by proxy, shall constitute a quorum at a meeting of shareholders, but in no event shall a quorum consist of less than one-third of the shares entitled to vote at the meeting. If a quorum is present, the affirmative vote of the majority of the shares represented at the meeting and entitled to vote on the subject matter shall be the act of the shareholders, unless the vote of a greater number or voting by classes is required by this Act or the articles of incorporation or by-laws.

§33. Voting of Shares. Each outstanding share, regardless of class, shall be entitled to one vote on each matter submitted to a vote at a meeting of shareholders, except as may be otherwise provided in the articles of incorporation. If the articles of incorporation provide for more or less than one vote for any share, on any matter, every reference in this Act to a majority or other proportion of shares shall refer to such a majority or other proportion of votes entitled to be cast.

Neither treasury shares, nor shares held by another corporation if a majority of the shares entitled to vote for the election of directors of such other corporation is held by the corporation, shall be voted at any meeting or counted in determining the total number of outstanding shares at any given time.

A shareholder may vote either in person or by proxy executed in writing by the shareholder or by his duly authorized attorney-in-fact. No proxy shall be valid after eleven months from the date of its execution, unless otherwise provided in the proxy.

[Either of the following prefatory phrases may be inserted here: "The articles of incorporation may provide that" or "Unless the articles of incorporation otherwise provide"] . . . at each election for directors every shareholder entitled to vote at such election shall have the right to vote, in person or by proxy, the number of shares owned by him for as many persons as there are directors to be elected and for whose election he has a right to vote, or to cumulate his votes by giving one candidate as many votes as the number of such directors multiplied by the number of his shares shall equal, or by distributing such votes on the same principle among any number of such candidates.

Shares standing in the name of another corporation, domestic or foreign, may be voted by such officer, agent or proxy as the by-laws of such other corporation may prescribe, or, in the absence of such provision, as the board of directors of such other corporation may determine.

Shares held by an administrator, executor, guardian or conservator may be voted by him, either in person or by proxy, without a transfer of such shares into his name. Shares standing in the name of a trustee may be voted by him, either in person or by proxy, but no trustee shall be entitled to vote shares held by him without a transfer of such shares into his name.

Shares standing in the name of a receiver may be voted by such receiver, and shares held by or under the control of a receiver may be voted by such receiver without the transfer thereof into his name if authority to do so be contained in an appropriate order of the court by which such receiver was appointed.

A shareholder whose shares are pledged shall be entitled to vote such shares until the shares have been transferred into the name of the pledgee, and thereafter the pledgee shall be entitled to vote the shares so transferred.

On and after the date on which written notice of redemption of redeemable shares has been mailed to the holders thereof and a sum sufficient to redeem such shares has been deposited with a bank or trust company with irrevocable instruction and authority to pay the redemption price to the holders thereof upon surrender of certificates therefor, such shares shall not be entitled to vote on any matter and shall not be deemed to be outstanding shares.

§34. Voting Trusts and Agreements among Shareholders. Any number of shareholders of a corporation may create a voting trust for the purpose of conferring upon a trustee or trustees the right to vote or otherwise represent their shares, for a period of not to exceed ten years, by entering into a written voting trust agreement specifying the terms and conditions of the voting trust, by depositing a counterpart of the agreement with the corporation at its registered office, and by transferring their shares to such trustee or trustees for the purposes of the agreement. Such trustee or trustees shall keep a record of the holders of voting trust certificates evidencing a beneficial interest in the voting trust, giving the names and addresses of all such holders and the number and class of the shares in respect of which the voting trust certificates held by each are issued, and shall deposit a copy of such record with the corporation at its registered office. The counterpart of the voting trust agreement and the copy of such record so deposited with the corporation shall be subject to the same right of examination by a shareholder of the corporation, in person or by agent or attorney, as are the books and records of the corporation, and such counterpart and such copy of such record shall be subject to examination by any holder of record of voting trust certificates, either in person or by agent or attorney, at any reasonable time for any proper purpose.

Agreements among shareholders regarding the voting of their shares shall be valid and enforceable in accordance with their terms. Such agreements shall not be subject to the provisions of this section regarding voting trusts.

§35. Board of Directors. All corporate powers shall be exercised by or under authority of, and the business and affairs of a corporation shall be managed under the direction of, a board of directors except as may be otherwise provided in this Act or the articles of incorporation. If any such provision is made in the articles of incorporation, the powers and duties conferred or imposed upon the board of directors by this Act shall be exercised or performed to such extent and by such person or persons as shall be provided in the articles of incorporation. Directors need not be residents of this State or shareholders of the corporation unless the articles of incorporation or by-laws so require. The articles of incorporation or by-laws may prescribe other qualifications for directors. The board of directors shall have authority to fix the compensation of directors unless otherwise provided in the articles of incorporation.

A director shall perform his duties as a director, including his duties as a member of any committee of the board upon which he may serve, in good faith, in a manner he reasonably believes to be in the best interests of the corporation, and with such care as an ordinarily prudent person in a like position would use under similar circumstances. In performing his duties, a director shall be entitled to rely on information, opinions, reports or statements, including financial statements and other financial data, in each case prepared or presented by:

(a) one or more officers or employees of the corporation whom the director reasonably believes to be reliable and competent in the matters presented,

(b) counsel, public accountants or other persons as to matters which the director reasonably believes to be within such person's professional or expert competence, or

(c) a committee of the board upon which he does not serve, duly designated in accordance with a provision of the articles of incorporation or the by-laws, as to matters within its designated authority, which committee the director reasonably believes to merit confidence, but he shall not be considered to be acting in good faith if he has knowledge concerning the matter in question that would cause such reliance to be unwarranted. A person who so performs his duties shall have no liability by reason of being or having been a director of the corporation.

A director of a corporation who is present at a meeting of its board of directors at which action on any corporate matter is taken shall be presumed to have assented to the action taken unless his dissent shall be entered in the minutes of the meeting or unless he shall file his written dissent to such action with the secretary of the meeting before the adjournment thereof or shall forward such dissent by registered mail to the secretary of the corporation immediately after the adjournment of the meeting. Such right to dissent shall not apply to a director who voted in favor of such action.

§ 36. Number and Election of Directors. The board of directors of a corporation shall consist of one or more members. The number of directors shall be fixed by, or in the manner provided in, the articles of incorporation or the by-laws, except as to the number constituting the initial board of directors, which number shall be fixed by the articles of incorporation. The number of directors may be increased or decreased from time to time by amendment to, or in the manner provided in, the articles of incorporation or the bylaws, but no decrease shall have the effect of shortening the term of any incumbent director. In the absence of a by-law providing for the number of directors, the number shall be the same as that provided for in the articles of incorporation. The names and addresses of the members of the first board of directors shall be stated in the articles of incorporation. Such persons shall hold office until the first annual meeting of shareholders, and until their successors shall have been elected and qualified. At the first annual meeting of shareholders and at each annual meeting thereafter the shareholders shall elect directors to hold office until the next succeeding annual meeting, except in case of the classification of directors as permitted by this Act. Each director shall hold office for the term for which he is elected and until his successor shall have been elected and qualified.

§ 37. Classification of Directors. When the board of directors shall consist of nine or more members, in lieu of electing the whole number of directors annually, the articles of incorporation may provide that the directors be divided into either two or three classes, each class to be as nearly equal in number as possible, the term of office of directors of the first class to expire at the first annual meeting of shareholders after their election, that of the second class to expire at the second annual meeting after their election, and that of the third class, if any, to expire at the third annual meeting after their election. At each annual meeting after such classification the number of directors equal to the number of the class whose term expires at the time of such meeting shall be elected to hold office until the second succeeding annual meeting, if there be two classes, or until the third succeeding annual meeting, if there be three classes. No classification of directors shall be effective prior to the first annual meeting of shareholders.

§ 38. Vacancies. Any vacancy occurring in the board of directors may be filled by the affirmative vote of a majority of the remaining directors though less than a quorum of the board of directors. A director elected to fill a vacancy shall be elected for the unexpired term of his predecessor in office. Any directorship to be filled by reason of an increase in the number of directors may be filled by the board of directors for a term of office continuing only until the next election of directors by the shareholders.

§ 39. Removal of Directors. At a meeting of shareholders called expressly for that purpose, directors may be removed in the manner provided in this section. Any director or the entire board of directors may be removed, with or without cause, by a vote of the holders of a majority of the shares then entitled to vote at an election of directors.

In the case of a corporation having cumulative voting, if less than the entire board is to be removed, no one of the directors may be removed if the votes cast against his removal would be sufficient to elect him if then cumulatively voted at an election of the entire board of directors, or, if there be classes of directors, at an election of the class of directors of which he is a part.

Whenever the holders of the shares of any class are entitled to elect one or more directors by the provisions of the articles of incorporation, the provisions of this section shall apply, in respect to the removal of a director or directors so elected, to the vote of the holders of the outstanding shares of that class and not to the vote of the outstanding shares as a whole.

§ 40. Quorum of Directors. A majority of the number of directors fixed by or in the manner provided in the by-laws or in the absence of a by-law fixing or providing for the number of directors, then of the number stated in the articles of incorporation, shall constitute a quorum for the transaction of business unless a greater number is required by the articles of incorporation or the by-laws. The act of the majority of the directors present at a meeting at which a quorum is present shall be the act of the board of directors, unless the act of a greater number is required by the articles of incorporation or the by-laws.

§ 41. Director Conflicts of Interest. No contract or other transaction between a corporation and one or more of its directors or any other corporation, firm, association or entity in which one or more of its directors are directors or officers or are financially interested, shall be either void or voidable because of such relationship or interest or because such director or directors are present at the meeting of the board of directors or a committee thereof which authorizes, approves or ratifies such contract or transaction or because his or their votes are counted for such purpose, if:

(a) the fact of such relationship or interest is disclosed or known to the board of directors or committee which authorizes, approves or ratifies the contract or transaction by a vote or consent sufficient for the purpose without counting the votes or consents of such interested directors; or

(b) the fact that such relationship or interest is disclosed or known to the shareholders entitled to vote and they authorize, approve or ratify such contract or transaction by vote or written consent; or

(c) the contract or transaction is fair and reasonable to the corporation.

Common or interested directors may be counted in determining the presence of a quorum at a meeting of the board of directors or a committee thereof which authorizes, approves or ratifies such a contract or transaction.

§ 42. Executive and Other Committees. If the articles of incorporation or the by-laws so provide, the board of directors, by resolution adopted by a majority of the full board of directors, may designate from among its members an executive committee and one or more other committees each of which, to the extent provided in such resolution or in the articles of incorporation or the by-laws of the corporation, shall have and may exercise all the authority of the board of directors, except that no such committee shall have authority to (i) declare dividends or distributions, (ii) approve or recommend to shareholders actions or proposals required by this Act to be approved by shareholders, (iii) designate candidates for the office of director, for purposes of proxy solicitation or otherwise, or fill vacancies on the board of directors or any committee thereof, (iv) amend the by-laws, (v) approve a plan of merger not requiring shareholder approval, (vi) reduce earned or capital surplus, (vii) authorize or approve the reacquisition of shares unless pursuant to a general formula or method specified by the board of directors, or (viii) authorize or approve the issuance or sale of, or any contract to issue or sell, shares or designate the terms of a series of a class of shares, provided that the board of directors, having acted regarding general authorization for the issuance or sale of shares, or any contract therefor, and, in the case of a series, the designation thereof, may, pursuant to a general formula or method specified by the board by resolution or by adoption of a stock option or other plan, authorize a committee to fix the terms upon which such shares may be issued or sold, including, without limitation, the price, the dividend rate, provisions for redemption, sinking fund, conversion, voting or preferential rights, and provisions for other features of a class of shares, or a series of a class of shares, with full power in such committee to adopt any final resolution setting forth all terms thereof and to authorize the statement of the terms of a series for filing with the Secretary of State under this Act.

Neither the designation of any such committee, the delegation thereto of authority, nor action by such committee pursuant to such authority shall alone constitute compliance by any member of the board of directors, not a member of the committee in question, with his responsibility to act in good faith, in a manner he reasonably believes to be in the best interests of the corporation, and with such care as an ordinarily prudent person in a like position would use under similar circumstances.

§ 43. Place and Notice of Directors' Meetings; Committee Meetings. Meetings of the board of directors, regular or special, may be held either within or without this State.

Regular meetings of the board of directors or any committee designated thereby may be held with or without notice as prescribed in the by-laws. Special meetings of the board of directors or any committee designated thereby shall be held upon such notice as is prescribed in the by-laws. Attendance of a director at a meeting shall constitute a waiver of notice of such meeting, except where a director attends a meeting for the express purpose of objecting to the transaction of any business because the meeting is not lawfully called or convened. Neither the business to be transacted at, nor the purpose of, any regular or special meeting of the board of directors or any committee designated thereby need be specified in the notice or waiver of notice of such meeting unless required by the by-laws.

Except as may be otherwise restricted by the articles of incorporation or by-laws, members of the board of directors or any committee designated thereby may participate in a meeting of such board or committee by means of a conference telephone or similar communications equipment by means of which all persons participating in the meeting can hear each other at the same time and participation by such means shall constitute presence in person at a meeting.

§ 44. Action by Directors without a Meeting. Unless otherwise provided by the articles of incorporation or by-laws, any action required by this Act to be taken at a meeting of the directors of a corporation, or any action which may be taken at a meeting of the directors or of a committee, may be taken without a meeting if a consent in writing, setting forth the action so taken, shall be signed by all of the directors, or all of the members of the committee, as the case may be. Such consent shall have the same effect as a unanimous vote.

§ 45. Dividends. The board of directors of a corporation may, from time to time, declare and the corporation may pay dividends in cash, property, or its own shares, except when the corporation is insolvent or when the payment thereof would render the corporation insolvent or when the declaration or payment thereof would be contrary to any restriction contained in the articles of incorporation, subject to the following provisions:

(a) Dividends may be declared and paid in cash or property only out of the unreserved and unrestricted earned surplus of the corporation, except as otherwise provided in this section.

[Alternative] (a) Dividends may be declared and paid in cash or property only out of the unreserved and unrestricted earned surplus of the corporation, or out of the unreserved and unrestricted net earnings of the current fiscal year and the next preceding fiscal year taken as a single period, except as otherwise provided in this section.

(b) If the articles of incorporation of a corporation engaged in the business of exploiting natural resources so provide, dividends may be declared and paid in cash out of the depletion reserves, but each such dividend shall be identified as a distribution of such reserves and the amount per share paid from such reserves shall be disclosed to the shareholders receiving the same concurrently with the distribution thereof.

(c) Dividends may be declared and paid in its own treasury shares.

(d) Dividends may be declared and paid in its own authorized but unissued shares out of any unreserved and unrestricted surplus of the corporation upon the following conditions:

(1) If a dividend is payable in its own shares having a par value, such shares shall be issued at not less than the par value thereof and there shall be transferred to stated capital at the time such dividend is paid an amount of surplus equal to the aggregate par value of the shares to be issued as a dividend.

(2) If a dividend is payable in its own shares without par value, such shares shall be issued at such stated value as shall be fixed by the board of directors by resolution adopted at the time such dividend is declared, and there shall be transferred to stated capital at the time such dividend is paid an amount of surplus equal to the aggregate stated value so fixed in respect of such shares; and the amount per share so transferred to stated capital shall be disclosed to the shareholders receiving such dividend concurrently with the payment thereof.

(e) No dividend payable in shares of any class shall be paid to the holders of shares of any other class unless the articles of incorporation so provide or such payment is authorized by the affirmative vote or the written consent of the holders of at least a majority of the outstanding shares of the class in which the payment is to be made.

A split-up or division of the issued shares of any class into a greater number of shares of the same class without increasing the stated capital of the corporation shall not be construed to be a share dividend within the meaning of this section.

§ 46. Distributions from Capital Surplus. The board of directors of a corporation may, from time to time, distribute to its shareholders out of capital surplus of the corporation a portion of its assets, in cash or property, subject to the following provisions:

(a) No such distribution shall be made at a time when the corporation is insolvent or when such distribution would render the corporation insolvent.

(b) No such distribution shall be made unless the articles of incorporation so provide or such distribution is authorized by the affirmative vote of the holders of a majority of the outstanding shares of each class whether or not entitled to vote thereon by the provisions of the articles of incorporation of the corporation.

(c) No such distribution shall be made to the holders of any class of shares unless all cumulative dividends accrued on all preferred or special classes of shares entitled to preferential dividends shall have been fully paid.

(d) No such distribution shall be made to the holders of any class of shares which would reduce the remaining net assets of the corporation below the aggregate preferential amount payable in event of involuntary liquidation to the holders of shares having preferential rights to the assets of the corporation in the event of liquidation.

(e) Each such distribution, when made, shall be identified as a distribution from capital surplus and the amount per share disclosed to the shareholders receiving the same concurrently with the distribution thereof.

The board of directors of a corporation may also, from time to time, distribute to the holders of its outstanding shares having a cumulative preferential right to receive dividends, in discharge of their cumulative dividend rights, dividends payable in cash out of the capital surplus of the corporation, if at the time the corporation has no earned surplus and is not insolvent and would not thereby be rendered insolvent. Each such distribution when made, shall be identified as a payment of cumulative dividends out of capital surplus.

§ 47. Loans to Employees and Directors. A corporation shall not lend money to or use its credit to assist its directors without authorization in the particular case by its shareholders, but may lend money to and use its credit to assist any employee of the corporation or of a subsidiary, including any such employee who is a director of the corporation, if the board of directors decides that such loan or assistance may benefit the corporation.

§ 48. Liabilities of Directors in Certain Cases. In addition to any other liabilities, a director shall be liable in the following circumstances unless he complies with the standard provided in this Act for the performance of the duties of directors:

(a) A director who votes for or assents to the declaration of any dividend or other distribution of the assets of a corporation to its shareholders contrary to the provisions of this Act or contrary to any restrictions contained in the articles of incorporation, shall be liable to the corporation, jointly and severally with all other directors so voting or assenting, for the amount of such dividend which is paid or the value of such assets which are distributed in excess of the amount of such dividend or distribution which could have been paid or distributed without a violation of the provisions of this Act or the restrictions in the articles of incorporation.

(b) A director who votes for or assents to the purchase of the corporation's own shares contrary to the provisions of this Act shall be liable to the corporation, jointly and severally with all other directors so voting or assenting, for the amount of consideration paid for such shares which is in excess of the maximum amount which could have been paid therefor without a violation of the provisions of this Act.

(c) A director who votes for or assents to any distribution of assets of a corporation to its shareholders during the liquidation of the corporation without the payment and discharge of, or making adequate provision for, all known debts, obligations, and liabilities of the corporation shall be liable to the corporation, jointly and severally with all other directors so voting or assenting, for the value of such assets which are distributed, to the extent that such debts, obligations and liabilities of the corporation are not thereafter paid and discharged.

Any director against whom a claim shall be asserted under or pursuant to this section for the payment of a dividend or other distribution of assets of a corporation and who shall be held liable thereon, shall be entitled to contribution from the shareholders who accepted or received any such dividend or assets knowing such dividend or distribution to have been made in violation of this Act, in proportion to the amounts received by them.

Any director against whom a claim shall be asserted under or pursuant to this section shall be entitled to contribution from the other directors who voted for or assented to the action upon which the claim is asserted.

§ 49. Provisions Relating to Actions by Shareholders. No action shall be brought in this State by a shareholder in the right of a domestic or foreign corporation unless the plaintiff was a holder of record of shares or of voting trust certificates therefor at the time of the transaction of which he complains, or his shares or voting trust certificates thereafter devolved upon him by operation of law from a person who was a holder of record at such time.

In any action hereafter instituted in the right of any domestic or foreign corporation by the holder or holders of record of shares of such corporation or of voting trust certificates therefor, the court having jurisdiction, upon final judgment and a finding that the action was brought without reasonable cause, may require the plaintiff or plaintiffs to pay to the parties named as defendant the reasonable expenses, including fees of attorneys, incurred by them in the defense of such action.

In any action now pending or hereafter instituted or maintained in the right of any domestic or foreign corporation by the holder or holders of record of less than five percent of the outstanding shares of any class of such corporation or of voting trust certificates therefor, unless the shares or voting trust certificates so held have a market value in excess of twenty-five thousand dollars, the corporation in whose right such action is brought shall be entitled at any time before final judgment to require the plaintiff or plaintiffs to give security for the reasonable expenses, including fees of attorneys, that may be incurred by it in connection with such action or may be incurred by other parties named as defendant for which it may become legally liable. Market value shall be determined as of the date that the plaintiff institutes the action or, in the case of an intervenor, as of the date that he becomes a party to the action. The amount of such security may from time to time be increased or decreased, in the discretion of the court, upon showing that the security provided has or may become inadequate or is excessive. The corporation shall have recourse to such security in such amount as the court having jurisdiction shall determine upon the termination of such action, whether or not the court finds the action was brought without reasonable cause.

§ 50. Officers. The officers of a corporation shall consist of a president, one or more vice presidents as may be prescribed by the by-laws, a secretary, and a treasurer, each of whom shall be elected by the board of directors at such time and in such manner as may be prescribed by the by-laws. Such other officers and assistant officers and agents as may be deemed necessary may be elected or appointed by the board of directors or chosen in such other manner as may be prescribed by the by-laws. Any two or more offices may be held by the same person, except the offices of president and secretary.

All officers and agents of the corporation, as between themselves and the corporation, shall have such authority and perform such duties in the management of the corporation as may be provided in the by-laws, or as may be determined by resolution of the board of directors not inconsistent with the by-laws.

§ 51. Removal of Officers. Any officer or agent may be removed by the board of directors whenever in its judgment the best interests of the corporation will be served thereby, but such removal shall be without prejudice to the contract rights, if any, of the person so removed. Election or appointment of an officer or agent shall not of itself create contract rights.

§ 52. Books and Records: Financial Reports to Shareholders; Examination of Records. Each corporation shall keep correct and complete books and records of account and shall keep minutes of the proceedings of its shareholders and board of directors and shall keep at its registered office or principal place of business, or at the office of its transfer agent or registrar, a record of its shareholders, giving the names and addresses of all shareholders and the number and class of the shares held by each. Any books, records and minutes may be in written form or in any other form capable of being converted into written form with a reasonable time.

Any person who shall have been a holder of record of shares or of voting trust certificates therefor at least six months immediately preceding his demand or shall be the holder of record of, or the holder of record of voting trust certificates for, at least five per cent of all the outstanding shares of the corporation, upon written demand stating the purpose thereof, shall have the right to examine, in person, or by agent or attorney, at any reasonable time or times, for any proper purpose its relevant books and records of accounts, minutes, and record of shareholders and to make extracts therefrom.

Any officer or agent who, or a corporation which, shall refuse to allow any such shareholder or holder of voting trust certificates, or his agent or attorney, so to examine and make extracts from its books and records of account, minutes, and record of shareholders, for any proper purpose, shall be liable to such shareholder or holder of voting trust certificates in a penalty of ten percent of the value of the shares owned by such shareholder, or in respect of which such voting trust certificates are issued, in addition to any other damages or remedy afforded him by law. It shall be a defense to any action for penalties under this section that the person suing therefor has within two years sold or offered for sale any list of shareholders or of holders of voting trust certificates for shares of such corporation or any other corporation or has aided or abetted any person in procuring any list of shareholders or of holders of voting trust certificates for any such purpose, or has improperly used any information secured through any prior examination of the books and records of account, or minutes, or record of shareholders or of holders of voting trust certificates for shares of such corporation or any other corporation, or was not acting in good faith or for a proper purpose in making his demand.

Nothing herein contained shall impair the power of any court of competent jurisdiction, upon proof by a shareholder or holder of voting trust certificates of proper purpose, irrespective of the period of time during which such shareholder or holder of voting trust certificates shall have been a shareholder of record or a holder of record of voting trust certificates, and irrespective of the number of shares held by him or represented by voting trust certificates held by him, to compel the production for examination by such shareholder or holder of voting trust certificates of the books and records of account, minutes and record of shareholders of a corporation.

Each corporation shall furnish to its shareholders annual financial statements, including at least a balance sheet as of the end of each fiscal year and a statement of income for such fiscal year, which shall be prepared on the basis of generally accepted accounting principles, if the corporation prepares financial statements for such fiscal year on that basis for any purpose, and may be consolidated statements of the corporation and one or more of its subsidiaries. The financial statements shall be mailed by the corporation to each of its shareholders within 120 days after the close of each fiscal year and, after such mailing and upon written request, shall be mailed by the corporation to any shareholder (or holder of a voting trust certificate for its shares) to whom a copy of the most recent annual financial statements has not previously been mailed. In the case of statements audited by a public accountant, each copy shall be accompanied by a report setting forth his opinion thereon; in other cases, each copy shall be accompanied by a statement of the president or the person in charge of the corporation's financial accounting records (1) stating his reasonable belief as to whether or not the financial statements were prepared in accordance with generally accepted accounting principles and, if not, describing the basis of presentation, and (2) describing any respects in which the financial statements were not prepared on a basis consistent with those prepared for the previous year.

§ 53. Incorporators. One or more persons, or a domestic or foreign corporation, may act as incorporator or incorporators of a corporation by signing and delivering in duplicate to the Secretary of State articles of incorporation for such corporation.

§ 54. Articles of Incorporation. The articles of incorporation shall set forth:

(a) The name of the corporation.

(b) The period of duration, which may be perpetual.

(c) The purpose or purposes for which the corporation is organized which may be stated to be, or to include, the transaction of any or all lawful business for which corporations may be incorporated under this Act.

(d) The aggregate number of shares which the corporation shall have authority to issue; if such shares are to consist of one class only, the par value of each of such shares, or a statement that all of such shares are without par value; or, if such shares are to be divided into classes, the number of shares of each class, and a statement of the par value of the shares of each such class or that such shares are to be without par value.

(e) If the shares are to be divided into classes, the designation of each class and a statement of the preferences, limitations and relative rights in respect of the shares of each class.

(f) If the corporation is to issue the shares of any preferred or special class in series, then the designation of each series and a statement of the variations in the relative rights and preferences as between series insofar as the same are to be fixed in the articles of incorporation, and a statement of any authority to be vested in the board of directors to establish series and fix and determine the variations in the relative rights and preferences as between series.

(g) If any preemptive right is to be granted to shareholders, the provisions therefor.

(h) Any provision, not inconsistent with law, which the incorporators elect to set forth in the articles of incorporation for the regulation of the internal affairs of the corporation, including any provision restricting the transfer of shares and any provision which under this Act is required or permitted to be set forth in the by-laws.

(i) The address of its initial registered office, and the name of its initial registered agent at such address.

(j) The number of directors constituting the initial board of directors and the names and addresses of the persons who are to serve as directors until the first annual meeting of shareholders or until their successors be elected and qualify.

(k) The name and address of each incorporator.

It shall not be necessary to set forth in the articles of incorporation any of the corporate powers enumerated in this Act.

§ 55. Filing of Articles of Incorporation. Duplicate originals of the articles of incorporation shall be delivered to the Secretary of State. If the Secretary of State finds that the articles of incorporation conform to law, he shall, when all fees have been paid as in this Act prescribed:

(a) Endorse on each of such duplicate originals the word "Filed," and the month, day and year of the filing thereof.

(b) File one of such duplicate originals in his office.

(c) Issue a certificate of incorporation to which he shall affix the other duplicate original.

The certificate of incorporation, together with the duplicate original of the articles of incorporation affixed thereto by the Secretary of State, shall be returned to the incorporators or their representative.

§ 56. Effect of Issuance of Certificate of Incorporation. Upon the issuance of the certificate of incorporation, the corporate existence shall begin, and such certificate of incorporation shall be conclusive evidence that all conditions precedent required to be performed by the incorporators have been compiled with and that the corporation has been incorporated under this Act, except as against this State in a proceeding to cancel or revoke the certificate of incorporation or for involuntary dissolution of the corporation.

§ 57. Organization Meeting of Directors. After the issuance of the certificate of incorporation an organization meeting of the board of directors named in the articles of incorporation shall be held, either within or without this State, at the call of a majority of the directors named in the articles of incorporation, for the purpose of adopting by-laws, electing officers and transacting such other business as may come before the meeting. The directors calling the meeting shall give at least three days' notice thereof by mail to each director so named, stating the time and place of the meeting.

§ 58. Right to Amend Articles of Incorporation. A corporation may amend its articles of incorporation from time to time, in any and as many respects as may be desired, so long as its articles of incorporation as amended contain only such provisions as might be lawfully contained in original articles of incorporation at the time of making such amendment, and, if a change in shares or the rights of shareholders, or an exchange, reclassification or cancellation of shares or rights of shareholders is to be made, such provisions as may be necessary to effect such change, exchange, reclassification or cancellation.

In particular, and without limitation upon such general power of amendment, a corporation may amend its articles of incorporation, from time to time, so as:

(a) To change its corporate name.

(b) To change its period of duration.

(c) To change, enlarge or diminish its corporate purposes.

(d) To increase or decrease the aggregate number of shares, or shares of any class, which the corporation has authority to issue.

(e) To increase or decrease the par value of the authorized shares of any class having a par value, whether issued or unissued.

(f) To exchange, classify, reclassify or cancel all or any part of its shares, whether issued or unissued.

(g) To change the designation of all or any part of its shares, whether issued or unissued, and to change the preferences, limitations, and the relative rights in respect of all or any part of its shares, whether issued or unissued.

(h) To change shares having a par value, whether issued or unissued, into the same or a different number of shares without par value, and to change shares without par value, whether issued or unissued, into the same or a different number of shares having a par value.

(i) To change the shares of any class, whether issued or unissued, and whether with or without par value, into a different number of shares of the same class or into the same or a different number of shares, either with or without par value, of other classes.

(j) To create new classes of shares having rights and preferences either prior and superior or subordinate and inferior to the shares of any class then authorized, whether issued or unissued.

(k) To cancel or otherwise affect the right of the holders of the shares of any class to receive dividends which have accrued but have not been declared.

(l) To divide any preferred or special class of shares, whether issued or unissued, into series and fix and determine the designations of such series and the variations in the relative rights and preferences as between the shares of such series.

(m) To authorize the board of directors to establish, out of authorized but unissued shares, series of any preferred or special class of shares and fix and determine the relative rights and preferences of the shares of any series so established.

(n) To authorize the board of directors to fix and determine the relative rights and preferences of the authorized but unissued shares of series theretofore established in respect of which either the relative rights and preferences have not been fixed and determined or the relative rights and preferences theretofore fixed and determined are to be changed.

(o) To revoke, diminish, or enlarge the authority of the board of directors to establish series out of authorized but unissued shares of any preferred or special class and fix and determine the relative rights and preferences of the shares of any series so established.

(p) To limit, deny or grant to shareholders of any class the preemptive right to acquire additional or treasury shares of the corporation, whether then or thereafter authorized.

§ 59. Procedure to Amend Articles of Incorporation. Amendments to the articles of incorporation shall be made in the following manner:

(a) The board of directors shall adopt a resolution setting forth the proposed amendment and, if shares have been issued, directing that it be submitted to a vote at a meeting of shareholders, which may be either the annual or a special meeting. If no shares have been issued, the amendment shall be adopted by resolution of the board of directors and the provisions for adoption by shareholders shall not apply. The resolution may incorporate the proposed amendment in restated articles of incorporation which contain a statement that except for the designated amendment the restated articles of incorporation correctly set forth without change the corresponding provisions of the articles of incorporation as theretofore amended, and that the restated articles of incorporation together with the designated amendment supersede the original articles of incorporation and all amendments thereto.

(b) Written notice setting forth the proposed amendment or a summary of the changes to be affected thereby shall be given to each shareholder of record entitled to vote thereon within the time and in the manner provided in this Act for the giving of notice of meetings of shareholders. If the meeting be an annual meeting, the proposed amendment of such summary may be included in the notice of such annual meeting.

(c) At such meeting a vote of the shareholders entitled to vote thereon shall be taken on the proposed amendment. The proposed amendment shall be adopted upon receiving the affirmative vote of the holders of a majority of the shares entitled to vote thereon, unless any class of shares is entitled vote thereon shall be taken on the proposed amendment. The proposed amendment shall be adopted upon receiving vote of the holders of a majority of the shares of each class of shares entitled to vote thereon as a class and of the total shares entitled to vote thereon.

Any number of amendments may be submitted to the shareholders, and voted upon by them, at one meeting.

§ 60. Class Voting on Amendments. The holders of the outstanding shares of a class shall be entitled to vote as a class upon a proposed amendment, whether or not entitled to vote thereon by the provisions of the articles of incorporation, if the amendment would:

(a) Increase or decrease the aggregate number of authorized shares of such class.

(b) Increase or decrease the par value of the shares of such class.

(c) Effect an exchange, reclassification or cancellation of all or part of the shares of such class.

(d) Effect an exchange, or create a right of exchange, of all or any part of the shares of another class into the shares of such class.

(e) Change the designations, preferences, limitations or relative rights of the shares of such class.

(f) Change the shares of such class, whether with or without par value, into the same or a different number of shares, either with or without par value, of the same class or another class or classes.

(g) Create a new class of shares having rights and preferences prior and superior to the shares of such class, or increase the rights and preferences or the number of authorized shares, of any class having rights and preferences prior or superior to the shares of such class.

(h) In the case of a preferred or special class of shares, divide the shares of such class into series and fix and determine the designation of such series and the variations in the relative rights and preferences between the shares of such series, or authorize the board of directors to do so.

(i) Limit or deny any existing preemptive rights of the shares of such class.

(j) Cancel or otherwise affect dividends on the shares of such class which have accrued but have not been declared.

§ 61. Articles of Amendment. The articles of amendment shall be executed in duplicate by the corporation by its president or a vice president and by its secretary or an assistant secretary, and verified by one of the officers signing such articles, and shall set forth:

(a) The name of the corporation.

(b) The amendments so adopted.

(c) The date of the adoption of the amendment by the shareholders, or by the board of directors where no shares have been issued.

(d) The number of shares outstanding, and the number of shares entitled to vote thereon, and if the shares of any class are entitled to vote thereon as a class, the designation and number of outstanding shares entitled to vote thereon of each such class.

(e) The number of shares voted for and against such amendment, respectively, and, if the shares of any class are entitled to vote thereon as a class, the number of shares of each such class voted for and against such amendment, respectively, or if no shares have been issued, a statement to that effect.

(f) If such amendment provides for an exchange, reclassification or cancellation of issued shares, and if the manner in which the same shall be effected is not set forth in the amendment, then a statement of the manner in which the same shall be effected.

(g) If such amendment effects a change in the amount of stated capital, then a statement of the manner in which the same is effected and a statement, expressed in dollars, of the amount of stated capital as changed by such amendment.

§ 62. Filing of Articles of Amendment. Duplicate originals of the articles of amendment shall be delivered to the Secretary of State. If the Secretary of State finds that the articles of amendment conform to law, he shall, when all fees and franchise taxes have been paid as in this Act prescribed:

(a) Endorse on each of such duplicate originals the word "Filed," and the month, day and year of the filing thereof.

(b) File one of such duplicate originals in his office.

(c) Issue a certificate of amendment to which he shall affix the other duplicate original.

The certificate of amendment, together with the duplicate original of the articles of amendment affixed thereto by the Secretary of State, shall be returned to the corporation or its representative.

§ 63. Effect of Certificate of Amendment. The amendment shall become effective upon the issuance of the certificate of amendment by the Secretary of State, or on such later date, not more than thirty days subsequent to the filing thereof with the Secretary of State, as shall be provided for in the articles of amendment.

No amendment shall affect any existing cause of action in favor of or against such corporation, or any pending suit to which such corporation shall be a party, or the existing rights of persons other than shareholders; and, in the event the corporate name shall be changed by amendment, no suit brought by or against such corporation under its former name shall abate for that reason.

§ 64. Restated Articles of Incorporation. A domestic corporation may at any time restate its articles of incorporation as theretofore amended, by a resolution adopted by the board of directors.

Upon the adoption of such resolution, restated articles of incorporation shall be executed in duplicate by the corporation by its president or a vice president and by its secretary or assistant secretary and verified by one of the officers signing such articles and shall set forth all of the operative provisions of the articles of incorporation as theretofore amended together with a statement that the restated articles of incorporation correctly set forth without change the corresponding provisions of the articles of incorporation as theretofore amended and that the restated articles of incorporation supersede the original articles of incorporation and all amendments thereto.

Duplicate originals of the restated articles of incorporation shall be delivered to the Secretary of State. If the Secretary of State finds that such restated articles of incorporation conform to law, he shall, when all fees and franchise taxes have been paid as in this Act prescribed:

(1) Endorse on each of such duplicate originals the word "Filed," and the month, day and year of the filing thereof.

(2) File one of such duplicate originals in his office.

(3) Issue a restated certificate of incorporation, to which he shall affix the other duplicate original.

The restated certificate of incorporation, together with the duplicate original of the restated articles of incorporation affixed thereto by the Secretary of State, shall be returned to the corporation or its representative.

Upon the issuance of the restated certificate of incorporation by the Secretary of State, the restated articles of incorporation shall become effective and shall supersede the original articles of incorporation and all amendments thereto.

§ 65. Amendment of Articles of Incorporation in Reorganization Proceedings. Whenever a plan of reorganization of a corporation has been confirmed by decree or order of a court of competent jurisdiction in proceedings for the reorganization of such corporation, pursuant to the provisions of any applicable statute of the United States relating to reorganizations of corporations, the articles of incorporation of the corporation may be amended, in the manner provided in this section, in as many respects as may be necessary to carry out the plan and put it into effect, so long as the articles of incorporation as amended contain only such provisions as might be lawfully contained in original articles of incorporation at the time of making such amendment.

In particular and without limitation upon such general power of amendment, the articles of incorporation may be amended for such purpose so as to:

(A) Change the corporate name, period of duration or corporate purposes of the corporation;

(B) Repeal, alter or amend the by-laws of the corporation;

(C) Change the aggregate number of shares or shares of any class, which the corporation has authority to issue;

(D) Change the preferences, limitations and relative rights in respect of all or any part of the shares of the corporation, and classify, reclassify or cancel all or any part thereof, whether issued or unissued;

(E) Authorize the issuance of bonds, debentures or other obligations of the corporation, whether or not convertible into shares of any class or bearing warrants or other evidences of optional rights to purchase or subscribe for shares of any class, and fix the terms and conditions thereof; and

(F) Constitute or reconstitute and classify or reclassify the board of directors of the corporation, and appoint directors and officers in place of or in addition to all or any of the directors or officers then in office.

Amendments to the articles of incorporation pursuant to this section shall be made in the following manner:

(a) Articles of amendment approved by decree or order of such court shall be executed and verified in duplicate by such person or persons as the court shall designate or appoint for the purpose, and shall set forth the name of the corporation, the amendments of the articles of incorporation approved by the court, the date of the decree or order approving the articles of amendment, the title of the proceedings in which the decree or order was entered, and a statement that such decree or order was entered by a court having jurisdiction of the proceedings for the reorganization of the corporation pursuant to the provisions of an applicable statute of the United States.

(b) Duplicate originals of the articles of amendment shall be delivered to the Secretary of State. If the Secretary of State finds that the articles of amendment conform to law, he shall, when all fees and franchise taxes have been paid as in his Act prescribed:

(1) Endorse on each of such duplicate originals the word "Filed," and the month, day and year of the filing thereof.

(2) File one of such duplicate originals in his office.

(3) Issue a certificate of amendment to which he shall affix the other duplicate original.

The certificate of amendment, together with the duplicate original of the articles of amendment affixed thereto by the Secretary of State, shall be returned to the corporation or its representative.

The amendment shall become effective upon the issuance of the certificate of amendment by the Secretary of State, or on such later date, not more than thirty days subsequent to the filing thereof with the Secretary of State, as shall be provided for in the articles of amendment without any action thereon by the directors or shareholders of the corporation and with the same effect as if the amendments had been adopted by unanimous action of the directors and shareholders of the corporation.

§ 66. Restriction on Redemption or Purchase of Redeemable Shares. No redemption or purchase of redeemable shares shall be made by a corporation when it is insolvent or when such redemption or purchase would render it insolvent, or which would reduce the net assets below the aggregate amount payable to the holders of shares having prior or equal rights to the assets of the corporation upon involuntary dissolution.

§ 67. Cancellation of Redeemable Shares by Redemption or Purchase. When redeemable shares of a corporation are redeemed or purchased by the corporation, the redemption or purchase shall effect a cancellation of such shares, and a statement of cancellation shall be filed as provided in this section. Thereupon such shares shall be restored to the status of authorized but unissued shares, unless the articles of incorporation provide that such shares when redeemed or purchased shall not be reissued, in which case the filing of the statement of cancellation shall constitute an amendment to the articles of incorporation and shall reduce the number of shares of the class so cancelled which the corporation is authorized to issue by the number of shares so cancelled.

The statement of cancellation shall be executed in duplicate by the corporation by its president or a vice president and by its secretary or an assistant secretary, and verified by one of the officers signing such statement, and shall set forth:

(a) The name of the corporation.

(b) The number of redeemable shares cancelled through redemption or purchase, itemized by classes and series.

(c) The aggregate number of issued shares, itemized by classes and series, after giving effect to such cancellation.

(d) The amount, expressed in dollars, of the stated capital of the corporation after giving effect to such cancellation.

(e) If the articles of incorporation provide that the cancelled shares shall not be reissued, the number of shares which the corporation will have authority to issue itemized by classes and series, after giving effect to such cancellation.

Duplicate originals of such statement shall be delivered to the Secretary of State. If the Secretary of State finds that such statement conforms to law, he shall, when all fees and franchise taxes have been paid as in this Act prescribed:

(1) Endorse on each of such duplicate originals the word "Filed," and the month, day and year of the filing thereof.

(2) File one of such duplicate originals in his office.

(3) Return the other duplicate original to the corporation or its representative.

Upon the filing of such statement of cancellation, the stated capital of the corporation shall be deemed to be reduced by that part of the stated capital which was, at the time of such cancellation, represented by the shares so cancelled.

Nothing contained in this section shall be construed to forbid a cancellation of shares or a reduction of stated capital in any other manner permitted by this Act.

§ 68. Cancellation of Other Reacquired Shares. A corporation may at any time, by resolution of its board of directors, cancel all or any part of the shares of the corporation of any class reacquired by it, other than redeemable shares redeemed or purchased, and in such event a statement of cancellation shall be filed as provided in this section. *(Remainder of text omitted.)*

§ 69. Reduction of Stated Capital in Certain Cases. A reduction of the stated capital of a corporation, where such reduction is not accompanied by any action requiring an amendment of the articles of incorporation and not accompanied by a cancellation of shares, may be made in the following manner: *(Remainder of text omitted.)*

§ 70. Special Provisions Relating to Surplus and Reserves. The surplus, if any, created by or arising out of a reduction of the stated capital of a corporation shall be capital surplus.

The capital surplus of a corporation may be increased from time to time by resolution of the board of directors directing that all or a part of the earned surplus of the corporation be transferred to capital surplus.

A corporation may, by resolution of its board of directors, apply any part or all of its capital surplus to the reduction or elimination of any deficit arising from losses, however incurred, but only after first eliminating the earned surplus, if any, of the corporation by applying such losses against earned surplus and only to the extent that such losses exceed the earned surplus, if any. Each such application of capital surplus shall, to the extent thereof, effect a reduction of capital surplus.

A corporation may, by resolution of its board of directors, create a reserve or reserves out of its earned surplus for any proper purpose or purposes, and may abolish any such reserve in the same manner. Earned surplus of the corporation to the extent so reserved shall not be available for the payment of dividends or other distributions by the corporation except as expressly permitted by this Act.

§ 71. Procedure for Merger. Any two or more domestic corporations may merge into one of such corporations pursuant to a plan of merger approved in the manner provided in this Act.

The board of directors of each corporation shall, by resolution adopted by each such board, approve a plan of merger setting forth:

(a) The names of the corporations proposing to merge, and the name of the corporation into which they propose to merge, which is hereinafter designated as the surviving corporation.

(b) The terms and conditions of the proposed merger.

(c) The manner and basis of converting the shares of each corporation into shares, obligations or other securities of the surviving corporation or of any other corporation or, in whole or in part, into cash or other property.

(d) A statement of any changes in the articles of incorporation of the surviving corporation to be effected by such merger.

(e) Such other provisions with respect to the proposed merger as are deemed necessary or desirable.

§ 72. Procedure for Consolidation. Any two or more domestic corporations may consolidate into a new corporation pursuant to a plan of consolidation approved in the manner provided in this Act.

The board of directors of each corporation shall, by a resolution adopted by each such board, approve a plan of consolidation setting forth:

(a) The names of the corporations proposing to consolidate, and the name of the new corporation into which they propose to consolidate, which is hereinafter designated as the new corporation.

(b) The terms and conditions of the proposed consolidation.

(c) The manner and basis of converting the shares of each corporation into shares, obligations or other securities of the new corporation or of any other corporation or, in whole or in part, into cash or other property.

(d) With respect to the new corporation, all of the statements required to be set forth in articles of incorporation for corporations organized under this Act.

(e) Such other provisions with respect to the proposed consolidation as are deemed necessary or desirable.

§ 72-A. Procedure for Share Exchange. All the issued or all the outstanding shares of one or more classes of any domestic corporation may be acquired through the exchange of all such shares of such class or classes by another domestic or foreign corporation pursuant to a plan of exchange approved in the manner provided in this Act.

The board of directors of each corporation shall, by resolution adopted by each such board, approve a plan of exchange setting forth:

(a) The name of the corporation the shares of which are proposed to be acquired by exchange and the name of the corporation to acquire the shares of such corporation in the exchange, which is hereinafter designated as the acquiring corporation.

(b) The terms and conditions of the proposed exchange.

(c) The manner and basis of exchanging the shares to be acquired for shares, obligations or other securities of the acquiring corporation or any other corporation, or, in whole or in part, for cash or other property.

(d) Such other provisions with respect to the proposed exchange as are deemed necessary or desirable.

The procedure authorized by this section shall not be deemed to limit the power of a corporation to acquire all or part of the shares of any class or classes of a corporation through a voluntary exchange or otherwise by agreement with the shareholders.

§ 73. Approval by Shareholders.

(a) The board of directors of each corporation in the case of a merger or consolidation, and the board of directors of the corporation the shares of which are to be acquired in the case of an exchange, upon approving such plan of merger, consolidation or exchange, shall, by resolution, direct that the plan be submitted to a vote at a meeting of its shareholders, which may be either an annual or a special meeting. Written notice shall be given to each shareholder of record, whether or not entitled to vote at such meeting, not less than twenty days before such meeting, in the manner provided in this Act for the giving of notice of meetings of shareholders, and, whether the meeting be an annual or a special meeting, shall state that the purpose or one of the purposes is to consider the proposed plan of merger, consolidation or exchange. A copy or a summary of the plan of merger, consolidation or exchange, as the case may be, shall be included in or enclosed with such notice.

(b) At each such meeting, a vote of the shareholders shall be taken on the proposed plan. The plan shall be approved upon receiving the affirmative vote of the holders of a majority of the shares entitled to vote thereon of each such corporation, unless any class of shares of any such corporation is entitled to vote thereon as a class, in which event, as to such corporation, the plan shall be approved upon receiving the affirmative vote of the holders of a majority of the shares of each class of shares entitled to vote thereon as a class and of the total shares entitled to vote thereon. Any class of shares of any such corporation shall be entitled to vote as a class if any such plan contains any provision which, if contained in a proposed amendment to articles of incorporation, would entitle such class of shares to vote as a class and, in the case of an exchange, if the class is included in the exchange.

(c) After such approval by a vote of the shareholders of each such corporation, and at any time prior to the filing of the articles of merger, consolidation or exchange, the merger, consolidation or exchange may be abandoned pursuant to provisions therefor, if any, set forth in the plan.

(d) (1) Notwithstanding the provisions of subsections (a) and (b), submission of a plan of merger to a vote at a meeting of shareholders of a surviving corporation shall not be required if:

(i) the articles of incorporation of the surviving corporation do not differ except in name from those of the corporation before the merger,

(ii) each holder of shares of the surviving corporation which were outstanding immediately before the effective date of the merger is to hold the same number of shares with identical rights immediately after,

(iii) the number of voting shares outstanding immediately after the merger, plus the number of voting shares issuable on conversion of other securities issued by virtue of the terms of the merger and on exercise of rights and warrants so issued, will not exceed by more than 20 percent the number of voting shares outstanding immediately before the merger, and

(iv) the number of participating shares outstanding immediately after the merger, plus the number of participating shares issuable on conversion of other securities issued by virtue of the terms of the merger and on exercise of rights and warrants so issued, will not exceed by more than 20 percent the number of participating shares outstanding immediately before the merger.

(2) As used in this subsection:

(i) "voting shares" means shares which entitle their holders to vote unconditionally in elections of directors;

(ii) "participating shares" means shares which entitle their holders to participate without limitation in distribution of earnings or surplus.

§ 74. Articles of Merger, Consolidation or Exchange.

(a) Upon receiving the approvals required by Sections 71, 72 and 73, articles of merger or articles of consolidation shall be executed in duplicate by each corporation by its president or a vice president and by its secretary or an assistant secretary, and verified by one of the officers of each corporation signing such articles, and shall set forth:

(1) The plan of merger or the plan of consolidation;

(2) As to each corporation, either (i) the number of shares outstanding, and, if the shares of any class are entitled to vote as a class, the designation and number of outstanding shares of each such class, or (ii) a statement that the vote of shareholders is not required by virtue of subsection 73(d);

(3) As to each corporation the approval of whose shareholders is required, the number of shares voted for and against such plan, respectively, and, if the shares of any class are entitled to vote as a class, the number of shares of each such class voted for and against such plan, respectively.

(b) Duplicate originals of the articles of merger, consolidation or exchange shall be delivered to the Secretary of State. If the Secretary of State finds that such articles conform to law, he shall, when all fees and franchise taxes have been paid as in this Act prescribed:

(1) Endorse on each of such duplicate originals the word "Filed," and the month, day and year of the filing thereof.

(2) File one of such duplicate originals in his office.

(3) Issue a certificate of merger, consolidation or exchange to which he shall affix the other duplicate original.

(c) The certificate of merger, consolidation or exchange together with the duplicate original of the articles affixed thereto by the Secretary of State, shall be returned to the surviving, new or acquiring corporation, as the case may be, or its representative.

§ 75. Merger of Subsidiary Corporation. Any corporation owning at least ninety percent of the outstanding shares of each class of another corporation may merge such other corporation into itself without approval by a vote of the shareholders of either corporation. Its board of directors shall, by resolution, approve a plan of merger setting forth:

(A) The name of the subsidiary corporation and the name of the corporation owning at least ninety percent of its shares, which is hereinafter designated as the surviving corporation.

(B) The manner and basis of converting the shares of the subsidiary corporation into shares, obligations or other securities of the surviving corporation or of any other corporation or, in whole or in part, into cash or other property.

A copy of such plan of merger shall be mailed to each shareholder of record of the subsidiary corporation.

Articles of merger shall be executed in duplicate by the surviving corporation by its president or a vice president and by its secretary or an assistant secretary, and verified by one of its officers signing such articles, and shall set forth:

(a) The plan of merger;

(b) The number of outstanding shares of each class of the subsidiary corporation and the number of such shares of each class owned by the surviving corporation; and

(c) The date of the mailing to shareholders of the subsidiary corporation of a copy of the plan of merger.

On and after the thirtieth day after the mailing of a copy of the plan of merger to shareholders of the subsidiary corporation or upon the waiver thereof by the holders of all outstanding shares duplicate originals of the articles of merger shall be delivered to the Secretary of State. If the Secretary of State finds that such articles conform to law, he shall, when all fees and franchise taxes have been paid as in this Act prescribed:

(1) Endorse on each of such duplicate originals the word "Filed," and the month, day and year of the filing thereof,

(2) File one of such duplicate originals in his office, and

(3) Issue a certificate of merger to which he shall affix the other duplicate original.

The certificate of merger, together with the duplicate original of the articles of merger affixed thereto by the Secretary of State, shall be returned to the surviving corporation or its representative.

§ 76. Effect of Merger, Consolidation or Exchange. A merger, consolidation or exchange shall become effective upon the issuance of a certificate of merger, consolidation or exchange by the Secretary of State, or on such later date, not more than thirty days subsequent to the filing thereof with the Secretary of State, as shall be provided for in the plan.

When a merger of consolidation has become effective:

(a) The several corporations parties to the plan of merger or consolidation shall be a single corporation, which, in the case of a merger, shall be that corporation designated in the plan of merger as the surviving corporation, and, in the case of a consolidation, shall be the new corporation provided for in the plan of consolidation.

(b) The separate existence of all corporations parties to the plan of merger or consolidation, except the surviving or new corporation, shall cease.

(c) Such surviving or new corporation shall have all the rights, privileges, immunities and powers and shall be subject to all the duties and liabilities of a corporation organized under this Act.

(d) Such surviving or new corporation shall thereupon and thereafter possess all the rights, privileges, immunities, and franchises, of a public as well as of a private nature, of each of the merging or consolidating corporations;

and all property, real, personal, and mixed, and all debts due on whatever account, including subscriptions to shares, and all other choses in action, and all and every other interest of or belonging to or due to each of the corporations so merged or consolidated, shall be taken and deemed to be transferred to and vested in such single corporation without further act or deed; and the title to any real estate, or any interest therein, vested in any of such corporations shall not revert or be in any way impaired by reason of such merger or consolidation.

(e) Such surviving or new corporation shall thenceforth be responsible and liable for all the liabilities and obligations of each of the corporations so merged or consolidated; and any claim existing or action or proceeding pending by or against any of such corporations may be prosecuted as if such merger or consolidation had not taken place, or such surviving or new corporation may be substituted in its place. Neither the rights of creditors nor any liens upon the property of any such corporation shall be impaired by such merger or consolidation.

(f) In the case of a merger, the articles of incorporation of the surviving corporation shall be deemed to be amended to the extent, if any, that changes in its articles of incorporation are stated in the plan of merger; and, in the case of a consolidation, the statements set forth in the articles of consolidation and which are required or permitted to be set forth in the articles of incorporation of corporations organized under this Act shall be deemed to be the original articles of incorporation of the new corporation.

When a merger, consolidation or exchange has become effective, the shares of the corporation or corporations party to the plan that are, under the terms of the plan, to be converted or exchanged, shall cease to exist, in the case of a merger or consolidation, or be deemed to be exchanged in the case of an exchange, and the holders of such shares shall thereafter be entitled only to the shares, obligations, other securities, cash or other property into which they shall have been converted or for which they shall have been exchanged, in accordance with the plan, subject to any rights under Section 80 of this Act.

§ 77. Merger, Consolidation or Exchange of Shares between Domestic and Foreign Corporations. One or more foreign corporations and one or more domestic corporations may be merged or consolidated, or participate in an exchange, in the following manner, if such merger, consolidation or exchange is permitted by the laws of the state under which each such foreign corporation is organized:

(a) Each domestic corporation shall comply with the provisions of this Act with respect to the merger, consolidation or exchange, as the case may be, of domestic corporations and each foreign corporation shall comply with the applicable provisions of the laws of the state under which it is organized.

(b) If the surviving or new corporation in a merger or consolidation is to be governed by the laws of any state other than this State, it shall comply with the provisions of this Act with respect to foreign corporations if it is to transact business in this State, and in every case it shall file with the Secretary of State of this State:

(1) An agreement that it may be served with process in this State in any proceeding for the enforcement of any obligation of any domestic corporation which is a party to such merger or consolidation and in any proceeding for the enforcement of the rights of a dissenting shareholder of any such domestic corporation against the surviving or new corporation;

(2) An irrevocable appointment of the Secretary of State of this State as its agent to accept service of process in any such proceeding; and

(3) An agreement that it will promptly pay to the dissenting shareholders of any such domestic corporation, the amount, if any, to which they shall be entitled under provisions of this Act with respect to the rights of dissenting shareholders.

§ 78. Sale of Assets in Regular Course of Business and Mortgage or Pledge of Assets. The sale, lease, exchange, or other disposition of all, or substantially all, the property and assets of a corporation in the usual and regular course of its business and the mortgage or pledge of any or all property and assets of a corporation whether or not in the usual and regular course of business may be made upon such terms and conditions and for such consideration, which may consist in whole or in part of cash or other property, including shares, obligations or other securities of any other corporation, domestic or foreign, as shall be authorized by its board of directors; and in any such case no authorization or consent of the shareholders shall be required.

§ 79. Sale of Assets Other Than in Regular Course of Business. A sale, lease, exchange, or other disposition of all, or substantially all, the property and assets, with or without the good will, of a corporation; if not in the usual and regular course of its business, may be made upon such terms and conditions and for such consideration, which may consist in whole or in part of cash or other property, including shares, obligations or other securities of any other corporation, domestic or foreign, as may be authorized in the following manner:

(a) The board of directors shall adopt a resolution recommending such sale, lease, exchange, or other disposition and directing the submission thereof to a vote at a meeting of shareholders, which may be either an annual or a special meeting.

(b) Written notice shall be given to each shareholder of record, whether or not entitled to vote at such meeting, not less than twenty days before such meeting, in the manner provided in this Act for the giving of notice of meetings of shareholders, and, whether the meeting be an annual or a special meeting, shall state that the purpose, or one of the purposes is to consider the proposed sale, lease, exchange, or other disposition.

(c) At such meeting the shareholders may authorize such sale, lease, exchange, or other disposition and may fix, or may authorize the board of directors to fix, any or all of the terms and conditions thereof and the consideration to be received by the corporation therefor. Such authorization shall require the affirmative vote of the holders of a majority of the shares of the corporation entitled to vote thereon, unless any class of shares is entitled to vote thereon as a class, in which event such authorization shall require the affirmative vote of the holders of a majority of the shares of each class of shares entitled to vote as a class thereon and of the total shares entitled to vote thereon.

(d) After such authorization by a vote of shareholders, the board of directors nevertheless, in its discretion, may abandon such sale, lease, exchange, or other disposition of assets, subject to the rights of third parties under any contracts relating thereto, without further action or approval by shareholders.

§ 80. Right of Shareholders to Dissent and Obtain Payment for Shares.

(a) Any shareholder of a corporation shall have the right to dissent from, and to obtain payment for his shares in the event of, any of the following corporate actions.

(1) Any plan of merger or consolidation to which the corporation is a party, except as provided in subsection (c);

(2) Any sale or exchange of all or substantially all of the property and assets of the corporation not made in the usual or regular course of its business, including a sale in dissolution, but not including a sale pursuant to an order of a court having jurisdiction in the premises or a sale for cash on terms requiring that all or substantially all of the net proceeds of sale be distributed to the shareholders in accordance with their respective interests within one year after the date of sale;

(3) Any plan of exchange to which the corporation is a party as the corporation the shares of which are to be acquired;

(4) Any amendment of the articles of incorporation which materially and adversely affects the rights appurtenant to the share of the dissenting shareholders in that it:

(i) alters or abolishes a preferential right of such shares;

(ii) creates, alters or abolishes a right in respect of the redemption of such shares, including a provision respecting a sinking fund for the redemption or repurchase of such shares;

(iii) alters or abolishes a preemptive right of the holder of such shares to acquire shares or other securities;

(iv) excludes or limits the right of the holder of such shares to vote on any matter, or to cumulate his votes, except as such right may be limited by dilution through the issuance of shares or other securities with similar voting rights; or

(5) Any other corporate action taken pursuant to a shareholder vote with respect to which the articles of incorporation, the bylaws, or a resolution of the board of directors directs that dissenting shareholders shall have a right to obtain payment for their shares.

(b) (1) A record holder of shares may assert dissenters' rights as to less than all of the shares registered in his name only if he dissents with respect to all the shares beneficially owned by any one person, and discloses the name and address of the person or persons on whose behalf he dissents. In that event, his rights shall be determined as if the shares as to which he has dissented and his other shares were registered in the names of different shareholders.

(2) A beneficial owner of shares who is not the record holder may assert dissenters' rights with respect to shares held on his behalf, and shall be treated as a dissenting shareholder under the terms of this section and Section 31 if he submits to the corporation at the time of or before the assertion of these rights a written consent of the record holder.

(c) The right to obtain payment under this section shall not apply to the shareholders of the surviving corporation in a merger if a vote of the shareholders of such corporation is not necessary to authorize such merger.

(d) A shareholder of a corporation who has a right under this section to obtain payment for his shares shall have no right at law or in equity to attack the validity of the corporate action that gives rise to his right to obtain payment, nor to have the action set aside or rescinded, except when the corporate action is unlawful or fraudulent with regard to the complaining shareholder or to the corporation.

§ 81. Procedures for Protection of Dissenters' Rights.

(a) As used in this section:

(1) "Dissenter" means a shareholder or beneficial owner who is entitled to and does assert dissenters' rights under Section 80, and who has performed every act required up to the time involved for the assertion of such rights.

(2) "Corporation" means the issuer of the shares held by the dissenter before the corporate action, or the successor by merger or consolidation of that issuer.

(3) "Fair value" of shares means their value immediately before the effectuation of the corporate action to which the dissenter objects, excluding any appreciation or depreciation in anticipation of such corporate action unless such exclusion would be inequitable.

(4) "Interest" means interest from the effective date of the corporate action until the date of payment, at the average rate currently paid by the corporation on its principal bank loans, or, if none, at such rate as is fair and equitable under all the circumstances.

(b) If a proposed corporate action which would give rise to dissenters' rights under Section 80(a) is submitted to a vote at a meeting of shareholders, the notice of meeting shall notify all shareholders that they have or may have a right to dissent and obtain payment for their shares by complying with the terms of this section, and shall be accompanied by a copy of Sections 80 and 81 of this Act.

(c) If the proposed corporate action is submitted to a vote at a meeting of shareholders, any shareholder who wishes to dissent and obtain payment for his shares must file with the corporation, prior to the vote, a written notice of intention to demand that he be paid fair compensation for his shares if the proposed action is effectuated, and shall refrain from voting his shares in approval of such action. A shareholder who fails in either respect shall acquire no right to payment for his shares under this section or Section 80.

(d) If the proposed corporate action is approved by the required vote at a meeting of shareholders, the corporation shall mail a further notice to all shareholders who gave due notice of intention to demand payment and who refrained from voting in favor of the proposed action. If the proposed corporate action is to be taken without a vote of shareholders, the corporation shall send to all shareholders who are entitled to dissent and demand payment for their shares a notice of the adoption of the plan of corporate action. The notice shall (1) state where and when a demand for payment must be sent and certificates of certificated shares must be deposited in order to obtain payment, (2) inform holders of uncertificated shares to what extent transfer of shares will be restricted from the time that demand for payment is received, (3) supply a form for demanding payment which includes a request for certification of the date on which the shareholder, or the person on whose behalf the shareholder dissents, acquired beneficial ownership of the shares, and (4) be accompanied by a copy of Sections 80 and 81 of this Act. The time set for the demand and deposit shall be not less than 30 days from the mailing of the notice.

(e) A shareholder who fails to demand payment, or fails (in the case of certificated shares) to deposit certificates, as required by a notice pursuant to subsection (d) shall have no right under this section or Section 80 to receive payment for his shares. If the shares are not represented by certificates, the corporation may restrict their transfer from the time of receipt of demand for payment until effectuation of the proposed corporate action, or the release of restrictions under the terms of subsection (f). The dissenter shall retain all other rights of a shareholder until these rights are modified by effectuation of the proposed corporate action.

(f) (1) Within 60 days after the date set for demanding payment and depositing certificates, if the corporation has not effectuated the proposed corporate action and remitted payment for shares pursuant to paragraph (3), it shall return any certificates that have been deposited, and release uncertificated shares from any transfer restrictions imposed by reason of the demand for payment.

(2) When uncertificated shares have been released from transfer restrictions, and deposited certificates have been returned, the corporation may at any later time send a new notice conforming to the requirements of subsection (d), with like effect.

(3) Immediately upon effectuation of the proposed corporate action, or upon receipt of demand for payment if the corporate action has already been effectuated, the corporation shall remit to dissenters who have made demand and (if their shares are certificated) have deposited their certificates the amount which the corporation estimates to be the fair value of the shares, with interest if any has accrued. The remittance shall be accompanied by:

(i) the corporation's closing balance sheet and statement of income for

a fiscal year ending not more than 16 months before the date of remittance, together with the latest available interim financial statements;
(ii) a statement of the corporation's estimate of fair value of the shares; and
(iii) a notice of the dissenter's right to demand supplemental payment, accompanied by a copy of Sections 80 and 81 of this Act.

(g) (1) If the corporation fails to remit as required by subsection (f), or if the dissenter believes that the amount remitted is less than the fair value of his shares, or that the interest is not correctly determined, he may send the corporation his own estimate of the value of the shares or of the interest, and demand payment of the deficiency.

(2) If the dissenter does not file such an estimate within 30 days after the corporation's mailing of its remittance, he shall be entitled to no more than the amount remitted.

(h) (1) Within 60 days after receiving a demand for payment pursuant to subsection (g), if any such demands for payment remain unsettled, the corporation shall file in an appropriate court a petition requesting that the fair value of the shares and interest thereon be determined by the court.

(2) An appropriate court shall be a court of competent jurisdiction in the county of this state where the registered office of the corporation is located. If, in the case of a merger or consolidation or exchange of shares, the corporation is a foreign corporation without a registered office in this state, the petition shall be filed in the county where the registered office of the domestic corporation was last located.

(3) All dissenters, wherever residing, whose demands have not been settled shall be made parties to the proceeding as in an action against their shares. A copy of the petition shall be served on each such dissenter; if a dissenter is a nonresident, the copy may be served on him by registered or certified mail or by publication as provided by law.

(4) The jurisdiction of the court shall be plenary and exclusive. The court may appoint one or more persons as appraisers to receive evidence and recommend a decision on the question of fair value. The appraisers shall have such power and authority as shall be specified in the order of their appointment or in any amendment thereof. The dissenters shall be entitled to discovery in the same manner as parties in other civil suits.

(5) All dissenters who are made parties shall be entitled to judgment for the amount by which the fair value of their shares is found to exceed the amount previously remitted, with interest.

(6) If the corporation fails to file a petition as provided in paragraph (1) of this subsection, each dissenter who made a demand and who has not already settled his claim against the corporation shall be paid by the corporation the amount demanded by him, with interest, and may sue therefor in an appropriate court.

(i) (1) The costs and expenses of any proceeding under subsection (h), including the reasonable compensation and expenses of appraisers appointed by the court, shall be determined by the court and assessed against the corporation, except that any part of the costs and expenses may be apportioned and assessed as the court may deem equitable against all or some of the dissenters who are parties and whose action in demanding supplemental payment the court finds to be arbitrary, vexatious, or not in good faith.

(2) Fees and expenses of counsel and of experts for the respective parties may be assessed as the court may deem equitable against the corporation and in favor of any or all dissenters if the corporation failed to comply substantially with the requirements of this section, and may be assessed against either the corporation or a dissenter, in favor of any other party, if the court finds that the party against whom the fees and expenses are assessed acted arbitrarily, vexatiously, or not in good faith in respect to the rights provided by this section and Section 80.

(3) If the court finds that the services of counsel for any dissenter were of substantial benefit to other dissenters similarly situated, and should not be assessed against the corporation, it may award to these counsel reasonable fees to be paid out of the amounts awarded to the dissenters who were benefited.

(j) (1) Notwithstanding the foregoing provisions of this section, the corporation may elect to withhold the remittance required by subsection (f) from any dissenter with respect to shares of which the dissenter (or the person on whose behalf the dissenter acts) was not the beneficial owner of the date of the first announcement to news media or to shareholders of the terms of the proposed corporate action. With respect to such shares, the corporation shall, upon effectuating the corporate action, state to each dissenter its estimate of the fair value of the shares, state the rate of interest to be used (explaining the basis thereof), and offer to pay the resulting amounts on receiving the dissenter's agreement to accept them in full satisfaction.

(2) If the dissenter believes that the amount offered is less than the fair value of the shares and interest determined according to this section, he may within 30 days after the date of mailing of the corporation's offer, mail the corporation his own estimate of fair value and interest, and demand their payment. If the dissenter fails to do so, he shall be entitled to no more than the corporation's offer.

(3) If the dissenter makes a demand as provided in paragraph (2), the provisions of subsections (h) and (i) shall apply to further proceedings on the dissenter's demand.

§ 82. Voluntary Dissolution by Incorporators. A corporation which has not commenced business and which has not issued any shares, may be voluntarily dissolved by its incorporators at any time in the following manner:

(a) Articles of dissolution shall be executed in duplicate by a majority of the incorporators, and verified by them, and shall set forth:

(1) The name of the corporation.

(2) The date of issuance of its certificate of incorporation.

(3) That none of its shares has been issued.

(4) That the corporation has not commenced business.

(5) That the amount, if any, actually paid in on subscriptions for its shares, less any part thereof disbursed for necessary expenses, has been returned to those entitled thereto.

(6) That no debts of the corporation remain unpaid.

(7) That a majority of the incorporators elect that the corporation be dissolved.

(b) Duplicate originals of the articles of dissolution shall be delivered to the Secretary of State. If the Secretary of State finds that the articles of dissolution conform to law, he shall, when all fees and franchise taxes have been paid as in this Act prescribed:

(1) Endorse on each of such duplicate originals the word "Filed," and the month, day and year of the filing thereof.

(2) File one of such duplicate originals in his office.

(3) Issue a certificate of dissolution to which he shall affix the other duplicate original.

The certificate of dissolution, together with the duplicate original of the articles of dissolution affixed thereto by the Secretary of State, shall be returned to the incorporators or their representative. Upon the issuance of such certificate of dissolution by the Secretary of State, the existence of the corporation shall cease.

§ 83. Voluntary Dissolution by Consent of Shareholders. A corporation may be voluntarily dissolved by the written consent of all of its shareholders.

Upon the execution of such written consent, a statement of intent to dissolve shall be executed in duplicate by the corporation by its president or a vice president and by its secretary or an assistant secretary, and verified by one of the officers signing such statement, which statement shall set forth:

(a) The name of the corporation.

(b) The names and respective addresses of its officers.

(c) The names and respective addresses of its directors.

(d) A copy of the written consent signed by all shareholders of the corporation.

(e) A statement that such written consent has been signed by all shareholders of the corporation or signed in their names by their attorneys thereunto duly authorized.

§ 84. Voluntary Dissolution by Act of Corporation. A corporation may be dissolved by the act of the corporation, when authorized in the following manner:

(a) The board of directors shall adopt a resolution recommending that the corporation be dissolved, and directing that the question of such dissolution be submitted to a vote at a meeting of shareholders, which may be either an annual or a special meeting.

(b) Written notice shall be given to each shareholder of record entitled to vote at such meeting within the time and in the manner provided in this Act for the giving of notice of meetings of shareholders, and, whether the meeting be an annual or special meeting, shall state that the purpose, or one of the purposes, of such meeting is to consider the advisability of dissolving the corporation.

(c) At such meeting a vote of shareholders entitled to vote thereat shall be taken on a resolution to dissolve the corporation. Such resolution shall be adopted upon receiving the affirmative vote of the holders of a majority of the shares of the corporation entitled to vote thereon, unless any class of shares is entitled to vote thereon as a class, in which event the resolution shall be adopted upon receiving the affirmative vote of the holders of a majority of the shares of each class of shares entitled to vote thereon as a class and of the total shares entitled to vote thereon.

(d) Upon the adoption of such resolution, a statement of intent to dissolve shall be executed in duplicate by the corporation by its president or a vice president and by its secretary or an assistant secretary, and verified by one of the officers signing such statement, which statement shall set forth:

(1) The name of the corporation.

(2) The names and respective addresses of its officers.

(3) The names and respective addresses of its directors.

(4) A copy of the resolution adopted by the shareholders authorizing the dissolution of the corporation.

(5) The number of shares outstanding, and, if the shares of any class are entitled to vote as a class, the designation and number of outstanding shares of each such class.

(6) The number of shares voted for and against the resolution, respectively, and, if the shares of any class are entitled to vote as a class, the number of shares of each such class voted for and against the resolution, respectively.

§ 85. Filing of Statement of Intent to Dissolve. Duplicate originals of the statement of intent to dissolve, whether by consent of shareholders or by act of the corporation, shall be delivered to the Secretary of State. If the Secretary of State finds that such statement conforms to law, he shall, when all fees and franchise taxes have been paid as in this Act prescribed:

(a) Endorse on each of such duplicate originals the word "Filed," and the month, day and year of the filing thereof.

(b) File one of such duplicate originals in his office.

(c) Return the other duplicate original to the corporation or its representative.

§ 86. Effect of Statement of Intent to Dissolve. Upon the filing by the Secretary of State of a statement of intent to dissolve, whether by consent of shareholders or by act of the corporation, the corporation shall cease to carry on its business, except insofar as may be necessary for the winding up thereof, but its corporate existence shall continue until a certificate of dissolution has been issued by the Secretary of State or until a decree dissolving the corporation has been entered by a court of competent jurisdiction as in this Act provided.

§ 87. Procedure after Filing of Statement of Intent to Dissolve. After the filing by the Secretary of State of a statement of intent to dissolve:

(a) The corporation shall immediately cause notice thereof to be mailed to each known creditor of the corporation.

(b) The corporation shall proceed to collect its assets, convey and dispose of such of its properties as are not to be distributed in kind to its shareholders, pay, satisfy and discharge its liabilities and obligations and do all other acts required to liquidate its business and affairs, and, after paying or adequately providing for the payment of all its obligations, distribute the remainder of its assets, either in cash or in kind, among its shareholders according to their respective rights and interests.

(c) The corporation, at any time during the liquidation of its business and affairs, may make application to a court of competent jurisdiction within the state and judicial subdivision in which the registered office or principal place of business of the corporation is situated, to have the liquidation continued under the supervision of the courts as provided in this Act.

§ 88. Revocation of Voluntary Dissolution Proceedings by Consent of Shareholders *(Text omitted).*

§ 89. Revocation of Voluntary Dissolution Proceedings by Act of Corporation *(Text omitted).*

§ 90. Filing of Statement of Revocation of Voluntary Dissolution Proceedings *(Text omitted).*

§ 91. Effect of Statement of Revocation of Voluntary Dissolution Proceedings. Upon the filing by the Secretary of State of a statement of revocation of voluntary dissolution proceedings, whether by consent of shareholders or by act of the corporation, the revocation of the voluntary dissolution proceedings shall become effective and the corporation may again carry on its business.

§ 92. Articles of Dissolution. If voluntary dissolution proceedings have not been revoked, then when all debts, liabilities and obligations of the corporation have been paid and discharged, or adequate provision has been made therefor, and all of the remaining property and assets of the corporation have been distributed to its shareholders, articles of dissolution shall be executed in duplicate by the corporation by its president or a vice president and by its secretary or an assistant secretary, and verified by one of the officers signing such statement, which statement shall set forth:

(a) The name of the corporation.

(b) That the Secretary of State has theretofore filed a statement of intent to dissolve the corporation, and the date on which such statement was filed.

(c) That all debts, obligations and liabilities of the corporation have been paid and discharged or that adequate provision has been made therefor.

(d) That all the remaining property and assets of the corporation have been distributed among its shareholders in accordance with their respective rights and interests.

(e) That there are no suits pending against the corporation in any court, or that adequate provision has been made for the satisfaction of any judgment, order or decree which may be entered against it in any pending suit.

§ 93. Filing of Articles of Dissolution. Duplicate originals of such articles of dissolution shall be delivered to the Secretary of State. If the Secretary of State finds that such articles of dissolution conform to law, he shall, when all fees and franchise taxes have been paid as in this Act prescribed:

(a) Endorse on each of such duplicate originals the word "Filed," and the month, day and year of the filing thereof.

(b) File one of such duplicate originals in his office.

(c) Issue a certificate of dissolution to which he shall affix the other duplicate original.

The certificate of dissolution, together with the duplicate original of the articles of dissolution affixed thereto by the Secretary of State, shall be returned to the representative of the dissolved corporation. Upon the issuance of such certificate of dissolution the existence of the corporation shall cease, except for the purpose of suits, other proceedings and appropriate corporate action by shareholders, directors and officers as provided in this Act.

§ 94. Involuntary Dissolution. A corporation may be dissolved involuntarily by a decree of the .... court in an action filed by the Attorney General when it is established that:

(a) The corporation has failed to file its annual report within the time required by this Act, or has failed to pay its franchise tax on or before the first day of August of the year in which such franchise tax becomes due and payable; or

(b) The corporation procured its articles of incorporation through fraud; or

(c) The corporation has continued to exceed or abuse the authority conferred upon it by law; or

(d) The corporation has failed for thirty days to appoint and maintain a registered agent in this State; or

(e) The corporation has failed for thirty days after change of its registered office or registered agent to file in the office of the Secretary of State a statement of such change.

§ 95. Notification to Attorney General. The Secretary of State, on or before the last day of December of each year, shall certify to the Attorney General the names of all corporations which have failed to file their annual reports or to pay franchise taxes in accordance with the provisions of this Act, together with the facts pertinent thereto. He shall also certify, from time to time, the names of all corporations which have given other cause for dissolution as provided in this Act, together with the facts pertinent thereto. Whenever the Secretary of State shall certify the name of a corporation to the Attorney General as having given any cause for dissolution, the Secretary of State shall concurrently mail to the corporation at its registered office a notice that such certification has been made. Upon the receipt of such certification, the Attorney General shall file an action in the name of the State against such corporation for its dissolution. Every such certificate from the Secretary of State to the Attorney General pertaining to the failure of a corporation to file an annual report or pay a franchise tax shall be taken and received in all courts as prima facie evidence of the facts therein stated. If, before action is filed, the corporation shall file its annual report or pay its franchise tax, together with all penalties thereon, or shall appoint or maintain a registered agent as provided in this Act, or shall file with the Secretary of State the required statement of change of registered office or registered agent, such fact shall be forthwith certified by the Secretary of State to the Attorney General and he shall not file an action against such corporation for such cause. If, after action is filed, the corporation shall file its annual report or pay its franchise tax, together with all penalties thereon, or shall appoint or maintain a registered agent as provided in this Act, or shall file with the Secretary of State the required statement of change of registered office or registered agent, and shall pay the costs of such action, the action for such cause shall abate.

§ 96. Venue and Process *(Text omitted).*

§ 97. Jurisdiction of Court to Liquidate Assets and Business of Corporation. The .... courts shall have full power to liquidate the assets and business of a corporation:

(a) In an action by a shareholder when it is established:

(1) That the directors are deadlocked in the management of the corporate affairs and the shareholders are unable to break the deadlock, and that irreparable injury to the corporation is being suffered or is threatened by reason thereof; or

(2) That the acts of the directors or those in control of the corporation are illegal, oppressive or fraudulent; or

(3) That the shareholders are deadlocked in voting power, and have failed, for a period which includes at least two consecutive annual meeting dates, to elect successors to directors whose terms have expired or would have expired upon the election of their successors; or

(4) That the corporate assets are being misapplied or wasted.

(b) In an action by a creditor:

(1) When the claim of the creditor has been reduced to judgment and an execution thereon returned unsatisfied and it is established that the corporation is insolvent; or

(2) When the corporation has admitted in writing that the claim of the creditor is due and owing and it is established that the corporation is insolvent.

(c) Upon application by a corporation which has filed a statement of intent to dissolve, as provided in this Act, to have its liquidation continued under the supervision of the court.

(d) When an action has been filed by the Attorney General to dissolve a corporation and it is established that liquidation of its business and affairs should precede the entry of a decree of dissolution.

Proceedings under clause (a), (b) or (c) of this section shall be brought in the county in which the registered office or the principal office of the corporation is situated.

It shall not be necessary to make shareholders parties to any such action or proceeding unless relief is sought against them personally.

§ 98. Procedure in Liquidation of Corporation by Court. In proceedings to liquidate the assets and business of a corporation the court shall have power to issue injunctions, to appoint a receiver or receivers pendente lite, with such powers and duties as the court, from time to time, may direct, and to take such other proceedings as may be requisite to preserve the corporate assets wherever situated, and carry on the business of the corporation until a full hearing can be had.

After a hearing had upon such notice as the court may direct to be given to all parties to the proceedings and to any other parties in interest designated by the court, the court may appoint a liquidating receiver or receivers with authority to collect the assets of the corporation, including all amounts owing to the corporation by subscribers on account of any unpaid portion of the consideration for the issuance of shares. Such liquidating receiver or receivers shall have authority, subject to the order of the court, to sell, convey and dispose of all or any part of the assets of the corporation wherever situated, either at public or private sale. The assets of the corporation or the proceeds resulting from a sale, conveyance or other disposition thereof shall be applied to the expenses of such liquidation and to the payment of the liabilities and obligations of the corporation, and any remaining assets or proceeds shall be distributed among its shareholders according to their respective rights and interests. The order appointing such liquidating receiver or receivers shall state their powers and duties. Such powers and duties may be increased or diminished at any time during the proceedings.

The court shall have power to allow from time to time as expenses of the liquidation compensation to the receiver or receivers and to attorneys in the proceeding, and to direct the payment thereof out of the assets of the corporation or the proceeds of any sale or disposition of such assets.

A receiver of a corporation appointed under the provisions of this section shall have authority to sue and defend in all courts in his own name as receiver of such corporation. The court appointing such receiver shall have exclusive jurisdiction of the corporation and its property, wherever situated.

§ 99. Qualifications of Receivers. A receiver shall in all cases be a natural person or a corporation authorized to act as receiver, which corporation may be a domestic corporation or a foreign corporation authorized to transact business in this State, and shall in all cases give such bond as the court may direct with such sureties as the court may require.

§ 100. Filing of Claims in Liquidation Proceedings. In proceedings to liquidate the assets and business of a corporation the court may require all creditors of the corporation to file with the clerk of the court or with the receiver, in such form as the court may prescribe, proofs under oath of their respective claims. If the court requires the filing of claims it shall fix a date, which shall be not less than four months from the date of the order, as the last day for the filing of claims, and shall prescribe the notice that shall be given to creditors and claimants of the date so fixed. Prior to the date so fixed the court may extend the time for the filing of claims. Creditors and claimants failing to file proofs of claim on or before the date so fixed may be barred, by order of court, from participating in the distribution of the assets of the corporation.

§ 101. Discontinuance of Liquidation Proceedings. The liquidation of the assets and business of a corporation may be discontinued at any time during the liquidation proceedings when it is established that cause for liquidation no longer exists. In such event the court shall dismiss the proceedings and direct the receiver to redeliver to the corporation all its remaining property and assets.

§ 102. Decree of Involuntary Dissolution. In proceedings to liquidate the assets and business of a corporation, when the costs and expenses of such proceedings and all debts, obligations and liabilities of the corporation shall have been paid and discharged and all of its remaining property and assets distributed to its shareholders, or in case its property and assets are not sufficient to satisfy and discharge such costs, expenses, debts and obligations, all the property and assets have been applied so far as they will go to their payment, the court shall enter a decree dissolving the corporation, whereupon the existence of the corporation shall cease.

§ 103. Filing of Decree of Dissolution. In case the court shall enter a decree dissolving a corporation, it shall be the duty of the clerk of such court to cause a certified copy of the decree to be filed with the Secretary of State. No fee shall be charged by the Secretary of State for the filing thereof.

§ 104. Deposit with State Treasurer of Amount Due Certain Shareholders. Upon the voluntary or involuntary dissolution of a corporation, the portion of the assets distributable to a creditor or shareholder who is unknown or cannot be found, or who is under disability and there is no person legally competent to receive such distributive portion, shall be reduced to cash and deposited with the State Treasurer and shall be paid over to such creditor or shareholder or to his legal representative upon proof satisfactory to the State Treasurer of his right thereto.

§ 105. Survival of Remedy After Dissolution. The dissolution of a corporation either (1) by the issuance of a certificate of dissolution by the Secretary of State, or (2) by a decree of court when the court has not liquidated the assets and business of the corporation as provided in this Act, or (3) by expiration of its period of duration, shall not take away or impair any remedy available to or against such corporation, its directors, officers, or shareholders, for any right or claim existing, or any liability incurred, prior to such dissolution if action or other proceeding thereon is commenced within two years after the date of such dissolution. Any such action or proceeding by or against the corporation may be prosecuted or defended by the corporation in its corporate name. The shareholders, directors and officers shall have power to take such corporate or other action as shall be appropriate to protect such remedy, right or claim. If such corporation was dissolved by the expiration of its period of duration, such corporation may amend its articles of incorporation at any time during such period of two years so as to extend its period of duration.

§ 106. Admission of Foreign Corporation. No foreign corporation shall have the right to transact business in this State until it shall have procured a certificate of authority so to do from the Secretary of State. No foreign corporation shall be entitled to procure a certificate of authority under this Act to transact in this State any business which a corporation organized under this Act is not permitted to transact. A foreign corporation shall not be denied a certificate of authority by reason of the fact that the laws of the state or country under which such corporation is organized governing its organization and internal affairs differ from the laws of this State, and nothing in this Act contained shall be construed to authorize this State to regulate the organization or the internal affairs of such corporation.

Without excluding other activities which may not constitute transacting business in this State, a foreign corporation shall not be considered to be transacting business in this State, for the purposes of this Act, by reason of carrying on in this State any one or more of the following activities:

(a) Maintaining or defending any action or suit or any administrative or arbitration proceeding, or effecting the settlement thereof or the settlement of claims or disputes.

(b) Holding meetings of its directors or shareholders or carrying on other activities concerning its internal affairs.

(c) Maintaining bank accounts.

(d) Maintaining offices or agencies for the transfer, exchange and registration of its securities, or appointing and maintaining trustees or depositaries with relation to its securities.

(e) Effecting sales through independent contractors.

(f) Soliciting or procuring orders, whether by mail or through employees or agents or otherwise, where such orders require acceptance without this State before becoming binding contracts.

(g) Creating as borrower or lender, or acquiring, indebtedness or mortgages or other security interests in real or personal property.

(h) Securing or collecting debts or enforcing any rights in property securing the same.

(i) Transacting any business in interstate commerce.

(j) Conducting an isolated transaction completed within a period of thirty days and not in the course of a number of repeated transactions of like nature.

§ 107. Powers of Foreign Corporation. A foreign corporation which shall have received a certificate of authority under this Act shall, until a certificate of revocation or of withdrawal shall have been issued as provided in this Act, enjoy the same, but no greater, rights and privileges as a domestic corporation organized for the purposes set forth in the application pursuant to which such certificate of authority is issued; and, except as in this Act otherwise provided, shall be subject to the same duties, restrictions, penalties and liabilities now or hereafter imposed upon a domestic corporation of like character.

§ 108. Corporate Name of Foreign Corporation. No certificate of authority shall be issued to a foreign corporation unless the corporate name of such corporation:

(a) Shall contain the word "corporation," "company," "incorporated," or "limited," or shall contain an abbreviation of one of such words, or such corporation shall, for use in this State, add at the end of its name one of such words or an abbreviation thereof.

(b) Shall not contain any word or phrase which indicates or implies that it is organized for any purpose other than one or more of the purposes contained in its articles of incorporation or that it is authorized or empowered to conduct the business of banking or insurance.

(c) Shall not be the same as, or deceptively similar to, the name of any domestic corporation existing under the laws of this State or any foreign corporation authorized to transact business in this State, or a name the exclusive right to which is, at the time, reserved in the manner provided in this Act, or the name of a corporation which has in effect a registration of its name as provided in this Act except that this provision shall not apply if the foreign corporation applying for a certificate of authority files with the Secretary of State any one of the following:

(1) a resolution of its board of directors adopting a fictitious name for use in transacting business in this State which fictitious name is not deceptively similar to the name of any domestic corporation or of any foreign corporation authorized to transact business in this State or to any name reserved or registered as provided in this Act, or

(2) the written consent of such other corporation or holder of a reserved or registered name to use the same or deceptively similar name and one or more words are added to make such name distinguishable from such other name, or

(3) a certified copy of a final decree of a court of competent jurisdiction establishing the prior right of such foreign corporation to the use of such name in this State.

§ 109. Change of Name by Foreign Corporation. Whenever a foreign corporation which is authorized to transact business in this State shall change its name to one under which a certificate of authority would not be granted to it on application therefor, the certificate of authority of such corporation shall be suspended and it shall not thereafter transact any business in this State until it has changed its name to a name which is available to it under the laws of this State or has otherwise complied with the provisions of this Act.

§ 110. Application for Certificate of Authority. A foreign corporation, in order to procure a certificate of authority to transact business in this State, shall make application therefore to the Secretary of State, which application shall set forth:

(a) The name of the corporation and the state or country under the laws of which it is incorporated.

(b) If the name of the corporation does not contain the word "corporation," "company," "incorporated," or "limited," or does not contain an abbreviation of one of such words, then the name of the corporation with the word or abbreviation which it elects to add thereto for use in this State.

(c) The date of incorporation and the period of duration of the corporation.

(d) The address of the principal office of the corporation in the state or country under the laws of which it is incorporated.

(e) The address of the proposed registered office of the corporation in this State, and the name of its proposed registered agent in this State at such address.

(f) The purpose or purposes of the corporation which it proposes to pursue in the transaction of business in this State.

(g) The names and respective addresses of the directors and officers of the corporation.

(h) A statement of the aggregate number of shares which the corporation has authority to issue, itemized by classes, par value of shares, shares without par value, and series, if any, within a class.

(i) A statement of the aggregate number of issued shares itemized by classes, par value of shares, shares without par value, and series, if any, within a class.

(j) A statement, expressed in dollar, of the amount of stated capital of the corporation, as defined in this Act.

(k) An estimate, expressed in dollars, of the value of all property to be owned by the corporation for the following year, wherever located, and an estimate of the value of the property of the corporation to be located within this State during such year, and an estimate, expressed in dollars, of the gross amount of business which will be transacted by the corporation during such year, and an estimate of the gross amount thereof which will be transacted by the corporation at or from places of business in this State during such year.

(l) Such additional information as may be necessary or appropriate in order to enable the Secretary of State to determine whether such corporation is entitled to a certificate of authority to transact business in this State and to determine and assess the fees and franchise taxes payable as in this Act prescribed.

Such application shall be made on forms prescribed and furnished by the Secretary of State and shall be executed in duplicate by the corporation by its president or a vice president and by its secretary or an assistant secretary, and verified by one of the officers signing such application.

§ 111. Filing of Application for Certificate of Authority. Duplicate originals of the application of the corporation for a certificate of authority shall be delivered to the Secretary of State, together with a copy of its articles of incorporation and all amendments thereto, duly authenticated by the proper officer of the state or country under the laws of which it is incorporated.

If the Secretary of State finds that such application conforms to law, he shall, when all fees and franchise taxes have been paid as in this Act prescribed:

(a) Endorse on each of such documents the word "Filed," and the month, day and year of the filing thereof.

(b) File in his office one of such duplicate originals of the application and the copy of the articles of incorporation and amendments thereto.

(c) Issue a certificate of authority to transact business in this State to which he shall affix the other duplicate original application.

The certificate of authority, together with the duplicate original of the application affixed thereto by the Secretary of State, shall be returned to the corporation or its representative.

§ 112. Effect of Certificate of Authority. Upon the issuance of a certificate of authority by the Secretary of State, the corporation shall be authorized to transact business in this State for those purposes set forth in its application, subject, however, to the right of this State to suspend or to revoke such authority as provided in this Act.

§ 113. Registered Office and Registered Agent of Foreign Corporation. Each foreign corporation authorized to transact business in this State shall have and continuously maintain in this State:

(a) A registered office which may be, but need not be, the same as its place of business in this State.

(b) A registered agent, which agent may be either an individual resident in this State whose business office is identical with such registered office, or a domestic corporation, or a foreign corporation authorized to transact business in this State, having a business office identical with such registered office.

§ 114. Change of Registered Office or Registered Agent of Foreign Corporation *(Text omitted).*

§ 115. Service of Process on Foreign Corporation. The registered agent so appointed by a foreign corporation authorized to transact business in this State shall be an agent of such corporation upon whom any process, notice or demand required or permitted by law to be served upon the corporation may be served.

Whenever a foreign corporation authorized to transact business in this State shall fail to appoint or maintain a registered agent in this State, or whenever any such registered agent cannot with reasonable diligence be found at the registered office, or whenever the certificate of authority of a foreign corporation shall be suspended or revoked, then the Secretary of State shall be an agent of such corporation upon whom any such process, notice, or demand may be served. Service on the Secretary of State of any such process, notice or demand shall be made by delivering to and leaving with him, or with any clerk having charge of the corporation department of his office, duplicate copies of such process, notice or demand. In the event any such process, notice or demand is served on the Secretary of State, he shall immediately cause one of such copies thereof to be forwarded by registered mail, addressed to the corporation at its principal office in the state or country under the laws of which it is incorporated. Any service so had on the Secretary of State shall be returnable in not less than thirty days.

The Secretary of State shall keep a record of all processes, notices and demands served upon him under this section, and shall record therein the time of such service and his action with reference thereto.

Nothing herein contained shall limit or affect the right to serve any process, notice or demand, required or permitted by law to be served upon a foreign corporation in any other manner now or hereafter permitted by law.

§ 116. Amendment to Articles of Incorporation of Foreign Corporation *(Text omitted).*

§ 117. Merger of Foreign Corporation Authorized to Transact Business in This State *(Text omitted).*

§ 118. Amended Certificate of Authority *(Text omitted).*

§ 119. Withdrawal of Foreign Corporation *(Text omitted).*

§ 120. Filing of Application for Withdrawal *(Text omitted).*

§ 121. Revocation of Certificate of Authority *(Text omitted).*

§ 122. Issuance of Certificate of Revocation *(Text omitted).*

§ 123. Application to Corporations Heretofore Authorized to Transact Business in This State *(Text omitted).*

§ 124. Transacting Business without Certificate of Authority. No foreign corporation transacting business in this State without a certificate of authority shall be permitted to maintain any action, suit or proceeding in any court of this State, until such corporation shall have obtained a certificate of authority. Nor shall any action, suit or proceeding be maintained in any court of this State by any successor or assignee of such corporation on any right, claim or demand arising out of the transaction of business by such corporation in this State, until a certificate of authority shall have been obtained by such corporation or by a corporation which has acquired all or substantially all of its assets.

The failure of a foreign corporation to obtain a certificate of authority to transact business in this State shall not impair the validity of any contract or act of such corporation, and shall not prevent such corporation from defending any action, suit or proceeding in any court of this State.

A foreign corporation which transacts business in this State without a certificate of authority shall be liable to this State, for the years or parts thereof during which it transacted business in this State without a certificate of authority, in an amount equal to all fees and franchise taxes which would have been imposed by this Act upon such corporation had it duly applied for and received a certificate of authority to transact business in this State as required by this Act and thereafter failed all reports required by this Act, plus all penalties imposed by this Act for failure to pay such fees and franchise taxes. The Attorney General shall bring proceedings to recover all amounts due this State under the provisions of this Section.

§ 125. Annual Report of Domestic and Foreign Corporations. Each domestic corporation, and each foreign corporation authorized to transact business in this State, shall file, within the time prescribed by this Act, an annual report setting forth:

(a) The name of the corporation and the state or country under the laws of which it is incorporated.

(b) The address of the registered office of the corporation in this State, and the name of its registered agent in this State at such address, and, in case of a foreign corporation, the address of its principal office in the state or country under the laws of which it is incorporated.

(c) A brief statement of the character of the business in which the corporation is actually engaged in this State.

(d) The names and respective addresses of the directors and officers of the corporation.

(e) A statement of the aggregate number of shares which the corporation has authority to issue, itemized by classes, par value of shares, shares without par value, and series, if any, within a class.

(f) A statement of the aggregate number of issued shares, itemized by classes, par value of shares, shares without par value, and series, if any, within a class.

(g) A statement, expressed in dollars, of the amount of stated capital of the corporation, as defined in this Act.

(h) A statement, expressed in dollars, of the value of all the property owned by the corporation, wherever located, and the value of the property of the corporation located within this State, and a statement, expressed in dollars, of the gross amount of business transacted by the corporation for the twelve months ended on the thirty-first day of December preceding the date herein provided for the filing of such report and the gross amount thereof transacted by the corporation at or from places of business in this State. If, on the thirty-first day of December preceding the time herein provided for the filing of such report, the corporation had not been in existence for a period of twelve months, or in the case of a foreign corporation had not been authorized to transact business in this State for a period of twelve months, the statement with respect to business transacted shall be furnished for the period between the date of incorporation or the date of its authorization to transact business in this State, as the case may be, and such thirty-first day of December. If all the property of the corporation is located in this State and all of its business is transacted at or from places of business in this State, or if the corporation elects to pay the annual franchise tax on the basis of its entire stated capital, then the information required by this subparagraph need not be set forth in such report.

(i) Such additional information as may be necessary or appropriate in order to enable the Secretary of State to determine and assess the proper amount of franchise taxes payable by such corporation.

Such annual report shall be made on forms prescribed and furnished by the Secretary of State, and the information therein contained shall be given as of the date of the execution of the report, except as to the information required by subparagraphs (g), (h) and (i) which shall be given as of the close of business on the thirty-first day of December next preceding the date herein provided for the filing of such report. It shall be executed by the corporation by its president, a vice president, secretary, an assistant secretary, or treasurer, and verified by the officer executing the report, or, if the corporation is in the hands of a receiver or trustee, it shall be executed on behalf of the corporation and verified by such receiver or trustee.

§ 126. Filing of Annual Report of Domestic and Foreign Corporations *(Text omitted).*

§ 127. Fees, Franchise Taxes and Charges to Be Collected by Secretary of State *(Text omitted).*

§ 128. Fees for Filing Documents and Issuing Certificates *(Text omitted).*

§ 129. Miscellaneous Charges *(Text omitted).*

§ 130. License Fees Payable by Domestic Corporations *(Text omitted).*

§ 131. License Fees Payable by Foreign Corporations *(Text omitted).*

§ 132. Franchise Taxes Payable by Domestic Corporations *(Text omitted).*

§ 133. Franchise Taxes Payable by Foreign Corporations *(Text omitted).*

§ 134. Assessment and Collection of Annual Franchise Taxes *(Text omitted).*

§ 135. Penalties Imposed upon Corporations *(Text omitted).*

§ 136. Penalties Imposed upon Officers and Directors. Each officer and director of a corporation, domestic or foreign, who fails or refuses within the time prescribed by this Act to answer truthfully and fully interrogatories propounded to him by the Secretary of State in accordance with the provisions of this Act, or who signs any articles, statement, report, application or other document filed with the Secretary of State which is known to such officer or director to be false in any material respect, shall be deemed to be guilty of a misdemeanor, and upon conviction thereof may be fined in any amount not exceeding .... dollars.

§ 137. Interrogatories by Secretary of State. The Secretary of State may propound to any corporation, domestic or foreign, subject to the provisions of this Act, and to any officer or director thereof, such interrogatories as may be reasonably necessary and proper to enable him to ascertain whether such corporation has complied with all the provisions of this Act applicable to such corporation. Such interrogatories shall be answered within thirty days after the mailing thereof, or within such additional time as shall be fixed by the Secretary of State, and the answers thereto shall be full and complete and shall be made in writing and under oath. If such interrogatories be directed to an individual they shall be answered by him, and if directed to a corporation they shall be answered by the president, vice president, secretary or assistant secretary thereof. The Secretary of State need not file any document to which such interrogatories relate until such interrogatories be answered as herein provided, and not then if the answers thereto disclose that such document is not in conformity with the provisions of this Act. The Secretary of State shall certify to the Attorney General, for such action as the Attorney General may deem appropriate, all interrogatories and answers thereto which disclose a violation of any of the provisions of this Act.

§ 138. Information Disclosed by Interrogatories. Interrogatories propounded by the Secretary of State and the answers thereto shall not be open to public inspection nor shall the Secretary of State disclose any facts or information obtained therefrom except insofar as his official duty may require the same to be made public or in the event such interrogatories or the answers thereto are required for evidence in any criminal proceedings or in any other action by this State.

§ 139. Power of Secretary of State. The Secretary of State shall have the power and authority reasonably necessary to enable him to administer this Act efficiently and to perform the duties therein imposed upon him.

§ 140. Appeal from Secretary of State *(Text omitted).*

§ 141. Certificates and Certified Copies to Be Received in Evidence *(Text omitted).*

§ 142. Forms to Be Furnished by Secretary of State *(Text omitted).*

§ 143. Greater Voting Requirements. Whenever, with respect to any action to be taken by the shareholders of a corporation, the articles of incorporation require the vote or concurrence of the holders of a greater proportion of the shares, or of any class or series thereof, than required by this Act with respect to such action, the provisions of the articles of incorporation shall control.

§ 144. Waiver of Notice. Whenever any notice is required to be given to any shareholder or director of a corporation under the provisions of this Act or under the provisions of the articles of incorporation or by-laws of the corporation, a waiver thereof in writing signed by the person or persons entitled to such notice, whether before or after the time stated therein, shall be equivalent to the giving of such notice.

§ 145. Action by Shareholders without a Meeting. Any action required by this Act to be taken at a meeting of the shareholders of a corporation, or any action which may be taken at a meeting of the shareholders, may be taken without a meeting if a consent in writing, setting forth the action so taken, shall be signed by all of the shareholders entitled to vote with respect to the subject matter thereof.

Such consent shall have the same effect as a unanimous vote of shareholders, and may be stated as such in any articles or document filed with the Secretary of State under this Act.

§ 146. Unauthorized Assumption of Corporate Powers. All persons who assume to act as a corporation without authority so to do shall be jointly and severally liable for all debts and liabilities incurred or arising as a result thereof.

§ 147. Application to Existing Corporations *(Text omitted).*

§ 148. Application to Foreign and Interstate Commerce *(Text omitted).*

§ 149. Reservation of Power *(Text omitted).*

§ 150. Effect of Repeal of Prior Acts *(Text omitted).*

§ 151. Effect of Invalidity of Part of This Act *(Text omitted).*

§ 152. Repeal of Prior Acts *(Text omitted).*

GLOSSARY

Ab initio. Literally, from the beginning, as in "void *ab initio.*"

Abandonment of Property. The voluntary release of property by the owner without vesting it in anyone else.

Abate. To reduce, remove, or diminish, as in "to *abate* a nuisance."

Acceleration Clause. A term in an instrument, such as a promissory note, providing that when a payment is missed the entire balance comes due immediately.

Acceptance. Performance of acts that are within the powers of the agreement offered, implying confirmation. Compliance by the offeree with the terms and conditions of an offer.

Accession. An exception to the rule of property conveyancing. Accession is the right to possess one's own property and any accessory that naturally or artifically arises out of it (such as crops from a farm).

Accommodation Party. One who signs a negotiable instrument as the acceptor, drawer, or maker for purposes of turning that negotiable instrument over to another. He is a surety.

Accord and Satisfaction. A settlement of a claim in which the creditor agrees to accept in payment something different from what might be legally enforced.

Account. The claim of one person against another.

Accretion. A gradual increase of an existing piece of property that is not apparent when it is proceeding, but noticeable only after a period of time.

Action in Rem. An action against a thing.

Administrator. In general, a representative given authority by a court of law to be an overseer or trustee; one appointed to administer the estate of one who dies intestate.

Adverse Possession. The acquiring of land that belongs to another. The claimant may acquire the title to that land not by right, but because the true owner has not exercised his right of ejectment.

Affidavit. A statement of fact sworn to be the truth before an authorized officer.

Agency Coupled with an Interest. An agent who has authority given to him by a principal but who also has an interest in the subject matter of the agency.

Agent. A person appointed by a principal to be the latter's representative. The principal is responsible for the agent's acts if agent is acting within the scope of his authority.

Alien. A person who is not a citizen of the country in which he is residing.

Alienation. The transfer of real property and possession of lands from one person to another.

Allonge. A piece of paper that is permanently affixed to an instrument that specifies further indorsement and becomes a part of the instrument.

Alteration. A change in a contract that alters the elements of the contract but leaves its general intent intact.

Ancillary. Subordinate or auxiliary to the principal; an agreement may be *ancillary* to the primary contract.

Answer. A statement by the defendant of defense or denial of the plaintiff's demands.

Antedate. To make a contract take effect prior to the delivery of the subject matter.

Appellant. The party who pursues the appeal.

Appellee. The party against whom an appeal is taken.

Arbitration. Way of settling differences through examination by an impartial party selected jointly by disagreeing parties.

Assault. An attempt to commit a violent physical act, placing one in reasonable apprehension of physical harm.

Assignment. The intentional transference of rights or ownerships from one party to another party.

Assignee. A person to whom rights or ownerships have been transferred.

Assignor. The party who assigns his rights or ownership over to another party.

Assumpsit. A common-law form of action for the recovery of damages for the breach of a simple contract.

Attachment. A proceeding to take a defendant's property into legal custody, to be used to pay a debt defendant owes to the plaintiff.

Attestation. The witnessing of the executing of a written instrument.

Attorney at Law. An officer of the court who may act professionally in legal formalities, negotiations, or proceedings for his client.

Attorney in Fact. An attorney given limited but legal authority to act for a principal in matters not of legal character.

Attractive Nuisance. Doctrine that one has a duty to protect young children when he maintains on his property a condition that is dangerous to young children because it is likely to attract them to the premises and because of their inability to appreciate the danger.

Bailment. Possession and duty to account for something that belongs to someone else.

Bankrupt. A person who is insolvent; one who has by formal decree of a court declared to be subject to be proceeded against under the bankruptcy laws, or who by his own voluntary action is able to take benefit of their protection.

Battery. An unpermitted touching.

Bearer. Person in possession of a bill or note.

Beneficiary. A receiver of another's benefits. One may be the *beneficiary* of a trust or insurance policy.

Bequest. Testamentary gift of personal property.

Bilateral Contract. A contract where each party promises some performance.

Bill of Exchange. An order from one person to another to pay a certain sum of money to a third person, such as a check or draft.

Bill of Lading. A written acknowledgment of the receipt of goods and specifications as to how goods will be transported. It also specifies the limits of liability to the carrier.

Binder. A preliminary written agreement that sets forth the most important items of a contract and gives temporary protection to the holder before the formal contract is drawn up.

Blank Indorsement. An indorsement of a bill or note by merely signing the name of the indorser, which makes the note payable to bearer.

Blue Sky Laws. Laws that make regulation and supervision of investment companies mandatory in order to prevent unregistered stock dealers from selling fraudulent stocks.

Bona fide. Without fraud; in good faith.

Bond. A contract or obligation, such as a bail bond or performance bond.

Broker. A person who brings parties together to buy and sell property.

Bulk Transfer. Any transfer, not in the ordinary course of the transferor's business, of a major part of the materials, supplies, merchandise, or other inventory of an enterprise.

Business Trust. An organizational form of business where trustees are given freedom to conduct business affairs without interference from stockholders. Property is placed in the hands of trustees who manage and deal with it for the use and benefit of beneficiaries.

Bylaws. The rules designated by a corporation for the conduct and duties of its shareholders.

Cancellation. Destroying or otherwise making unusable a written instrument.

Capital. Amount of wealth a person or organization has at one point in time; assets of a permanent or fixed nature or employed in carrying on a business.

Capital Surplus. The difference between what new issues of stock are sold for and their par value.

Cashier's Check. A bill of exchange drawn by a bank upon itself. It is a primary obligation of the bank.

Caveat Emptor. "Let the buyer beware." The doctrine that the buyer purchases at his own risk.

Certificate of Deposit. A negotiable promissory note upon which a bank promises to pay interest.

Certiorari. A method of bringing the record of an inferior tribunal before a higher court to determine whether the lower court acted within the limits of its authority.

Charitable Trust. Includes all gifts held in trust for religious and charitable organizations.

Charter. An instrument that grants power, such as a corporate charter. Also, an instrument to hire or lease a vehicle or vessel.

Chattel Paper. Paper that represents both a monetary obligation and a security interest in or a lease of specific goods.

Chattel Personal. An article of personal property; a movable thing.

Chattel Real. A right to real property that does not amount to a freehold interest, such as a leasehold estate.

Chose in Action. Means "rights of action." The

rights of action involved are breach of contract or injury of property, but not torts, which could not be enforced without action.

Chose in Possession. A thing which the person has not only the right to enjoy but also the actual enjoyment thereof.

Civil Law. The legal system established in the countries of continental Europe, and others, based on codes as opposed to the common-law system of England. Also, that body of law dealing with suits between private parties as opposed to criminal law.

Codicil. An addition to a will that republishes the whole will, so far as it is not revoked or altered by the codicil.

Cognovit. A "statement of confession," often times referred to as a "power of attorney" or simply as a "power." It is the written authority of a debtor to enter judgment against that debtor.

Collateral. Related to, but not strictly a part of, the main thing or matter under consideration; tending to support the main result. Also, that property in which a creditor may be given a security interest by a debtor.

Common Carrier. One who undertakes for hire or reward to transport from place to place goods of all who choose to employ him.

Common Law. A law made up of the opinions and actions of judges. Its rules arise from the application of reason to changing conditions of society. Common law is the legal system of England, the United States, and many other countries.

Common Stock. Proprietary interest in a corporation. The owners of common stock are the residual owners of the corporation.

Community Property. Property acquired by married couples together; property acquired during marriage when not acquired as separate property by either husband or wife.

Complaint. A statement of the substantive facts upon which the plaintiff's claim to relief is founded.

Composition of Creditors. An agreement made between a debtor and his creditors whereby the creditors, in return for some payment in the present or future, agree to accept a payment less than the full amount of their claim.

Conditional Sale. A transaction in which a seller transfers the possession of goods to a buyer, with the understanding that title to the goods shall not pass from seller to buyer until buyer has paid the price.

Conditions Concurrent. Conditions in a contract that are mutually dependent and that are to be performed at the same time.

Condition Subsequent. A condition in a contract, the happening of which will defeat an obligation pursuant thereto.

Confession of Judgment. Voluntary submission to court jurisdiction, giving by consent what could otherwise be obtained only by summons and complaint.

Confusion of Goods. Such an intermixture of goods owned by different persons that the property can no longer be distinguished.

Consequential Damages. Those that flow naturally, but indirectly, from a wrong act as, for example, lost profits.

Conservator. A temporary guardian having the duty of conserving the property of an incompetent until a guardian is appointed.

Consideration. A necessary element for a valid contract. Something given in return, other than a promise, or a forbearance, or the creation, modification, or destruction of a legal relation, or a return promise, bargained for and given in exchange for the promise.

Consignee. One to whom something is consigned or shipped.

Consolidation. To unite two things, without destroying each one's individual characteristics; for example, a *consolidation* of two or more different corporations.

Constructive Notice. A legal substitute for actual notice.

Constructive Trust. A trust employed by equity to prevent a person from retaining property to which he is not in good conscience entitled.

Contract. A promise or a set of promises for the breach of which the law gives a remedy, or the performance of which the law in some way recognizes as a duty.

Contributory Negligence. Negligence of an injured party which contributes in the slightest degree to his harm.

Conversion. Any unauthorized act depriving a person of his property or right to possession of his property permanently or for an indefinite time.

Conveyance. A transfer of the title of land from one person or class of persons to another.

Corporation. An artificial person or being created by laws of a state or nation and endorsed by law, with the capacity of perpetual succession, whose powers are defined by the certificate of incorporation.

Counterclaim. A claim asserted by a defendant against a plaintiff.

Covenant. An agreement between two or more persons to do or permit the doing of a particular act.

Cover. As used in the Uniform Commercial Code, the purchase of substitute goods by a buyer where the seller has breached the contract for the sale of goods.

Curtesy. The freehold estate in possession which a husband has in his wife's land, upon her death, for the remainder of his own life.

***Cy Pres* Doctrine.** (*Cy pres*, "as near as possible.") As used in trust law, the doctrine that the intention of the party is to be carried out as nearly as may be when it would be illegal or impossible to give the instrument literal effect.

Damages. A monetary award that may be recovered in the courts by a person who has suffered loss, injury, or detriment to his person, rights, or property as a result of another's acts or omissions. Damages may be nominal, punitive, or compensatory.

Debenture. An instrument that serves as evidence of indebtedness.

Debt. That created when one person binds himself to pay money to another.

Deceit. Any trick, collusion, contrivance, false representation, or underhanded practice used to defraud another.

Declaration. The plea by which a plaintiff in a suit at law sets out his cause of action.

Decree. The judgment of an equity court.

Deed. An instrument that on its face assumes to convey the property described.

De facto. Defined as in fact; actually; indeed; as in "a *de facto* corporation."

Defamation. False and malicious words, imputing conduct, which injuriously affect a person's reputation or tend to degrade a person in society.

Defendant. The one against whom a suit is brought.

De jure. Of right, legitimate, lawful, or by right and just title, as in "a *de jure* corporation."

Demurrer. A formal allegation that the facts as stated in a pleading, even if admitted, are not sufficient to put the demurrant to the necessity of answering or proceeding further with the cause.

Deposition. A method of discovery used in a lawsuit whereby a party is called by another party to answer, under oath, questions concerning the case.

Derivative Suit. A suit by the stockholders of a corporation, against other parties, to recover sums due on behalf of the corporation.

Descent. The act of succeeding to the ownership of an estate by inheritance or any act of law; title acquired by one party from another as his heir at law.

Detinue. Wrongful detention of personal property.

Devise. The direction of a testator of sound mind as to the disposition of his property after death.

Dictum. An opinion expressed by a court, but lacking the force of adjudication because it is not necessarily involved in the case.

Discharge. To relieve, as of a debt, responsibility, or accusation; to absolve; to clear; or to set free.

Disclaimer. The repudiation or denial of another's claim or right.

Dishonor. To refuse to accept or pay; used in a commercial sense.

Disparagement. Matter that is intended by its publisher to be understood, or that is reasonably understood, to cast doubt upon the existence or extent of another's property, chattels, or intangible things.

Dissolution. A final dissolving or termination, which takes place either by a judgment of a court of competent jurisdiction, or by legislative repeal, such as of a charter of a corporation.

Divestiture. Any cutting short of an interest prior to its otherwise normal termination. To give up; for example, to *divest* oneself of certain property.

Dividend. A declared distribution of earnings to stockholders by a corporation.

Documentary Draft. As used in the Uniform Commercial Code, any negotiable or non-negotiable draft, with accompanying documents, securities, or other papers as collateral security for payment.

Domestic Corporation. A corporation created

by or under the laws of the state in which the corporation's domicile is in question.

Dominant Tenement. In an easement appurtenant to various lots within a restricted area, the existence of two estates is required; the *dominant tenement* is the estate deriving the benefit of the easement.

Dormant Partner. A partner who is unknown as such to those doing business with the partnership.

Dower. A common-law concept granting to a wife a life interest in any realty owned by the husband during the course of the marriage, which right is enforceable at his death.

Draft. A bill of exchange. A three-party instrument, which may be negotiable or nonnegotiable.

Duress. Unlawful constraint whereby one is forced to do some act against one's will.

Easement. The right one person has to use the land of another for a specific purpose. It is distinct from the ownership of the soil itself.

Easement Appurtenant. An easement incapable of separate existence apart from the land to which it is annexed. *See* Dominant tenement; Servient tenement.

Easement in Gross. Easement that does not require a dominant tenement. It is an interest in or right to use the land of another for some purpose.

Ecclesiastical Law. The laws of the established church; at one time, that body of jurisprudence administered by the ecclesiastical courts of England.

Ejectment. Action for trial of title to land.

Eleemosynary. Constituted for the perpetual distribution of alms or bounty of the founders; charitable; as in "an *eleemosynary* institution."

Eminent Domain. The right of the people or government to take private property for public use.

Encumbrance. A claim or lien on real property. It may also denote a lien or charge on personal property.

Entity. A real being; existence; for example, "a corporation is a separate legal *entity.*"

Equity. To do what is right in matters of property and maintenance of civil rights. A system of jurisprudence developed in England that administers justice according to a particular practice and procedure. Distinguished from the law courts, where different practice and procedures are followed.

Equitable Conversion. Under this doctrine, as soon as a valid realty sales contract is made equity considers the buyer as the beneficial owner of the realty and the seller as trustee for him.

Escrow. A written instrument that by its terms imports a legal obligation and which is deposited by a grantor with a third party to be kept by him until the performance of a condition or happening of a certain event, and then to be delivered to grantor.

Estate. The condition or circumstance in which an owner stands with reference to his property. The type of interest that one has in property.

Estate by Entirety. Co-ownership of property by husband and wife in which there is unity of estate, possession, control, and time. Upon the death of one, the survivor takes the entire estate.

Estoppel. A legal bar stopping or prohibiting one as a matter of law from saying anything contrary to his own previous word or deed.

Eviction. A turning out of possession, or placing the party in such a situation that, his expulsion being inevitable, he voluntarily surrenders the possession to some expulsion.

Ex contractu. Arising from or out of a contract.

Ex delicto. Arising as the result of a wrong or tort.

Executor, Executrix. A person or corporation empowered to discharge the duties of a fiduciary and appointed as such by a testator in his will.

Executory. That which has not yet been performed, as in "that contract is still *executory.*"

Exemplary Damages. Damages that are punitive in nature over and above what is necessary to compensate the plaintiff for his loss as a result of the wrongful act of the defendant. Generally given in a civil case only upon the showing that the defendant acted willfully and with malice.

Exoneration. Right that a person has who has been compelled to pay what another should be forced to pay in full.

Factor. An agent to whom property is consigned for sale.

False Imprisonment. The unlawful detention of a person for any length of time whereby he is deprived of his liberty.

Fee Simple. The largest estate or ownership interest that a person can have; absolute ownership of land, with no qualification or conditions.

Felony. A crime that is declared to be so by statutory or common-law jurisdiction. Generally a crime of a more serious nature than those designated as misdemeanors.

Fiduciary. A person who has a duty to act for the benefit of another person; also, having to do with trust and confidence. An attorney has a *fiduciary* relationship to his or her client.

Fixture. An article that was a chattel but which by being physically annexed or affixed to realty becomes an accessory thereto and part thereof.

Forbearance. Refraining from acting; for example, waiting for payment of a debt after it becomes due.

Foreclosure. The process of terminating the rights of a mortgagor to redeem his estate or property. In general the process involves enforcing payment of the debt secured by a mortgage by taking and selling the mortgaged property.

Foreign Corporation. A corporation existing under laws of a state other than the one under which it was organized.

Forfeiture. That which is lost or the right to which is alienated by a crime, offense, neglect of duty, or breach of contract.

Fraud. A tort that consists of a false representation of a material fact which is intended to deceive another so that he acts in reliance on the misrepresentation to his injury. It implies perjury, concealment, falsification, or misrepresentation.

Freehold. An estate in land of uncertain duration; generally a life estate or fee.

Fungible. Of such a nature that different batches or samples are interchangeable or capable of mutual substitution, such as coal or wheat.

Garnishment. A proceeding or process whereby the property, money, or credits of one person, generally called the debtor, in possession of one owing by another, generally designated the garnishee, are applied to payment of the debt of the debtor by means of a process against the debtor and garnishee.

Gift. A voluntary transfer of property by one to another without consideration or compensation therefor.

Gift *causa mortis*. A gift made in the expectation of imminent death on condition that the donor die. Upon the happening of that condition, the title to the property vests in the surviving donee.

Gift *inter vivos*. Literally, a gift between the living, which becomes absolute during the life of the donor and donee.

Good Faith. Freedom from knowledge or circumstances that would put a party on notice to make further inquiry. Honesty in fact in the conduct or transaction concerned.

Goods. As used in the Uniform Commercial Code, goods means all things (including specially manufactured goods) which are movable at the time of identification to the contract for sale other than the money in which the price is to be paid, investment securities (subtitle 8), and things in action. "Goods" also includes the unborn young of animals, growing crops, and other identified things attached to realty as described in the section on goods to be severed from realty.

Goodwill. The value of a business over and above the value of its physical property, which may result from favorable consideration of customers.

Grand Jury. A separate tribunal whose proceedings are secret. The primary function of the grand jury is to determine whether an individual ought to be indicted for the commission of a felony.

Grantee. One to whom some valuable interest or right in land is conveyed.

Grantor. The party who conveys any trust property or interest in land to a grantee by deed.

Gratuitous. Without valuable or legal consideration.

Guarantor. One who agrees to see the engagement of another performed. One who undertakes collaterally to answer for the payment of another's debt.

Guardian. A person who legally has care of the person or property, or both, of another who is incompetent to act for himself.

Habeas corpus. Literally, "you have the body." The name given to a number of writs requiring one who has custody of an individual to bring him before the court.

Heir. One to whom the property of a deceased person goes either by descent or by will.

Holographic Will. A will that is entirely written, dated, and signed by the hand of the testator himself.

Holder in Due Course. As used in the Uniform Commercial Code, one who takes an instrument (a) for value; (b) in good faith; and (c) without notice that it is overdue or has been dishonored, or knowledge of any defense against or claim to it on the part of any person.

Homestead. An estate created, not only for protection of the family as a whole but for units of the family. Under the laws of many states a homestead exemption is created which consists of property protected from levy and execution.

Illusory. Deceiving, or tending to deceive, or fallacious.

Implied. Where the intention of a party is not manifest in expressed or explicit words but is gathered from that party's action and the surrounding circumstances.

Incapacity. Legal inability to act.

Inchoate. Partial or incomplete.

Incidental Damages. As used in the Uniform Commercial Code, the term implies damages available to a buyer or seller of goods that are in the nature of expenses reasonably incurred by a party as a result of the other party's breach and which are the natural and probable result of the breach of contract.

Indemnity. The obligation resting on one person to make good any loss or damage another has incurred or may incur by acting at his request or for his benefit.

Indenture. A deed between two parties in which they enter into reciprocal and corresponding grants or obligations toward each other.

Indictment. A finding by a grand jury that reasonable grounds exist to believe a crime has been committed.

Indorsement. Act by a payee of a negotiable note or other document of signing his name on the back, with or without further qualifying words, which operates to transfer the note or other document.

Infant. A person who lacks capacity because he or she has not reached the age of majority, most commonly 18 or 21 years of age.

Information. A formal charge of a crime designed to specifically inform the defendant of what he is charged with in order that he may intelligently plead thereto. Generally used to charge one with a misdemeanor.

Injunction. An order of a court requiring a party to perform some act or refrain from performing some act.

In personam. An action against a particular person or persons, as opposed to things.

Insanity. That degree or quantity of mental disorder that relieves one of criminal responsibility for his actions or of civil responsibility. Lack of mental capacity.

Insolvent. A situation where a person or business has not sufficient property to pay all his debts. Sometimes used also to refer to a situation where a party's liabilities exceed his or her assets.

In status quo. Being placed in the same position in which the party was at the time of the inception of a contract which is sought to be rescinded.

Intangible Property. Property that has no intrinsic or marketable value of itself but rather is representative of or evidence of value, such as a stock certificate.

Inter Alia. Among other things.

Interlocutory. Not final; temporary or provisional.

Interrogatories. Written questions propounded by one party and served on an adversary, who must serve written answers thereto under oath. A tool for gaining discovery of another party's case.

Intestate. A person is said to *die intestate* when he dies without making a will.

Inventory. An itemized list or enumeration of property, article by article.

Invitees. One who is at a place at the invitation of another and for the benefit of the person giving the invitation. A customer of a store is an invitee.

Joint and Several. When a creditor may sue one of two or more parties for the full amount owed either individually or all together, the liability is said to be *joint and several.*

Joint Stock Company. An association of individuals for the purpose of profit, possessing a common capital contributed by members composing it, such capital being divided into transferable stocks.

Joint Tenancy. A type of ownership in which the rights to the entire property go to the survivor among the joint tenants. A joint tenancy requires unity of interest, title, time, and possession.

Joint Venture. An undertaking by two or more persons jointly to carry out a simple business enterprise for profit.

Jurisdiction. The power to hear and determine the subject matter in controversy between parties, or to adjudicate or exercise any judicial power over them.

Jury. A body or group of people, finally sworn to try issues in a particular case.

Laches. In equity, a failure to do something that a party should have done; negligence.

Landlord. One entitled to rent for use and occupancy of any housing accommodation. One who owns an estate in land who is entitled to be paid rent from one to whom he has leased the premises.

Last Clear Chance. A chance that arises to escape peril that one has negligently placed oneself in. A doctrine in tort law whereby a plaintiff may overcome a defense of contributory negligence.

Latent. Hidden; concealed; dormant; not appearing on the face of the thing.

Law Merchant. That body of law customs and regulations developed by merchants to regulate their transactions, and which has become part of the common law of England.

Lease. Any agreement that gives rise to a relationship of landlord and tenant. Also used in renting the use of personal as well as real property.

Legacy. A disposition of personal property by will.

Letters Testamentary. The authentic evidence that a will has been provided, and that permission has been given to the executor, by the appropriate court, to exercise powers conferred by the testator in his will.

Levy. To raise or collect money, as by assessment, execution, or other legal process.

Libel. A defamatory statement made by writing. A false written accusation, which is published, intended to injure a person's reputation or expose a person to public contempt or ridicule.

License. The grant of permission to do a particular thing, exercise a certain privilege, or carry on a particular business.

Lien. Security for a debt, duty, or other obligation.

Life Estate. An estate whose duration is limited to the life of the person holding it or some other person.

Limited Partnership. A partnership created pursuant to statute that requires at least one general partner with unlimited liability, and one or more partners whose liability is limited to their investment in the partnership and who have no say in the management of the partnership.

Liquidated. Specifying exactly what and how much is involved, as in the terms *liquidated damages* or *liquidated amount.*

Long Arm Statute. A statute that gives the courts jurisdiction over one who resides in another state by virtue of some action of the individual which brings him into contact with the state asserting jurisdiction.

Malice. Committing an unlawful act without just cause or lawful excuse.

Mandamus. A writ issued by a court requiring an officer or body to perform a nondiscretionary act required by virtue of its position or duties.

Marshaling of Assets. The ranking of assets in order so as to apply the assets to the satisfaction of various claims in a specific order.

Maturity Date. The day when a note becomes due and payable, or time when an action could be maintained thereon to enforce payment.

Mechanic's Lien. A lien against a building or construction site in favor of persons furnishing labor or materials.

Mediation. The act of a third person who intervenes in a dispute between two contending parties to reconcile them or persuade them to adjust or settle their dispute.

Merchant. As used in the Uniform Commercial Code, "Merchant" means a person who deals in goods of kind or otherwise by his occupation holds himself out as having knowledge or skill peculiar to the practice or goods involved in the transaction, or to whom such knowledge or skill may be attributed by his employment of an agent or broker or other intermediary who by his occupation holds himself out as having such knowledge or skill.

Merchantable. In the Uniform Commercial Code, to be *merchantable* goods must be at least such as pass without objection in the trade; of fair average quality within the description if fungible, or fit for the ordinary purposes for which such goods are used; run, within the variations permitted by the agreement of even kind, quality and quantity within each unit and among all units involved; are adequately contained, packaged and labeled as the agreement may require; and conform to the promises or affirmations of fact made on the container or label, if any.

Merger. Uniting of a lesser and greater estate in the same person. Also, a doctrine whereby a document of a lower order is united with a document of a higher order, as in the case of a contract for the sale of land merging in the deed.

Midnight Deadline. As used in the Uniform Commercial Code, with respect to a bank the midnight deadline is midnight on its next banking day following the banking day on which it receives the relevant item or notice, or from which the time for taking action commences to run, whichever is later.

Misdemeanor. In general, as designated by the legislature or under the common law, a crime or offense that is less in nature than a felony and for the conviction of which the consequences are generally less.

Misfeasance. The performance of an act that might be lawfully done, but in an improper manner, by which another person receives an injury.

Mitigation of Damages. The duty of one who has been injured to take action to prevent damages from increasing or to limit his or her damages.

Monopoly. Ownership or control that dominates the means of production or the product in a business or occupation. In its pure sense, a market that is characterized by a single seller.

Mortgage. A type of security interest in personal or real property. Under the common law, an estate to secure payment of a debt that becomes void if the obligation is performed.

Motion. An application for a rule or order made orally or in writing to a court or judge. A request that the court do something.

Necessaries. Those items that are required for one to maintain his or her station in life.

Negligence. A failure to use ordinary care. Failure to use such care as persons of ordinary prudence are accustomed to exercise under the same or similar circumstances.

Negotiable. Capable of being transferred. Under the Uniform Commercial Code, transferable by delivery or indorsement so as to vest in the transferor the rights of the transferee.

Non obstanto veredicto. Literally, "notwithstanding the verdict," for example, a judgment may be granted by a court notwithstanding the verdict of the jury.

Novation. Substitution of a new contract between new or different parties.

Nuisance. A tort. That which annoys or disturbs one in the enjoyment of possession of his property which arises from the unreasonable, unwarranted, or unlawful use by another person of his property or other improper conduct.

Nuncupative Will. A will that is not written but is declared orally by the testator in his last illness, in the manner specified by statute.

Obligor. The person owing a duty, debt, or other obligation to another.

Offeree. In a contractual situation, the person to whom an offer is made.

Offeror. In a contractual situation, the person making the offer.

Option. An agreement to hold an offer open for a specified time given in return for the payment or promise of consideration.

Order Instrument. An instrument which by its terms is payable to the order or assigns of any person therein specified with reasonable certainty.

Output Contract. Contract in which a seller agrees to sell his entire output to a buyer.

Par Value. The face value of the stock as opposed to its market value.

Pari delicto. Literally, in equal fault or guilt.

Parol Evidence. Oral evidence. The parol-evidence rule prohibits parol evidence that contradicts the clear, unambiguous terms of a written document.

Partition. The division of property or its proceeds in case it is sold.

Penal Damages. Exemplary or punitive damages.

Pendente lite. Pending a suit, during the actual progress of the suit; during litigation.

Peremptory Challenge. The challenging of a potential juror on an arbitrary basis, without the need of showing cause as to why he or she should not sit on the jury.

Per curiam. By the entire court. Appellate opinions that are not attributed to a specific judge are handed down *per curiam.*

Perfection. As used in the Uniform Commercial Code, the status of a security interest where it is effective in protecting the creditor's security interest in a debtor's property against the competing interests of other third parties in that property.

Per se. Literally, itself, as in "words that amount to slander *per se.*"

Personal Property. All property other than real property.

Petit Jury. As opposed to the grand jury, the petit jury sits in the trial of civil and criminal cases, decides issues of fact, applies the law to those facts, and awards a verdict.

Plaintiff. The party in a lawsuit who sues another party, who is called the *defendant.* The plaintiff begins the lawsuit by asserting a claim.

Pledge. A bailment of personal property as a security for some debt.

Postdate. To date an instrument for some date in the future.

Precatory Words. Words of entreaty, request, or seeking permission as opposed to words of command.

Preemptive Right. The right of stockholders to purchase their pro rata amount of a new issue of stock offered by the corporation.

Presentment. As used in the Uniform Commercial Code regarding negotiable instruments, presentment is a demand for acceptance or payment made upon the maker, acceptor, drawee, or other payee by or on behalf of the holder.

Prima facie. Literally, on first sight. Presumed to be true unless disproved by some evidence to the contrary. A *prima facie case* is said to be established when a party has submitted sufficient evidence to require his or her opponent to answer it.

Private Carrier. One who holds himself out as ready to furnish transportation for hire only to those with whom he chooses to deal in accordance with such contracts as he makes with them.

Privilege. An advantage, benefit, protection, or immunity afforded a particular group of individuals.

Privity. A mutual relationship to the same rights of property. Privity of contract is the connection that exists between two or more contracting parties.

Probate. The act of proving a will. Also, a general term used to describe all matters over which probate courts have jurisdiction.

Proceeds. As used in the Uniform Commercial Code concerning secured transactions, *proceeds* includes whatever is received when collateral or proceeds are sold, exchanged, collected, or otherwise disposed of. The term also includes the account arising when the right to payment is earned under a contract right.

Profit a prendre. The right to take soil, gravel, minerals, and the like from another's land.

Promissory Estoppel. Relates primarily to those promises, primarily to charitable institutions, which lack consideration but where, because of surrounding facts, injustice can only be avoided by enforcing the promise.

Promissory Note. Evidence of an indebtedness that exists independent of the note.

Promoter. One who promotes, urges on, encourages, invites, or advances. As used regarding the formation of corporations, one who acts to form a corporation.

Proxy. An authority or power to do a certain thing on behalf of another.

Public Policy. The policy of society as indicated by its statutes, ordinances, legislature records, and court decisions.

Pur autre Vie. During the life of another, as in "a life estate *pur autre vie.*"

Quantum meruit. Literally, "as much as he deserves." A class of obligations imposed by law without regard to the intention or assent of the parties bound.

Quasi Contract. An obligation imposed by law in order that justice may be done and to avoid unjust enrichment of one party at the expense of another.

Quiet Title. A proceeding to establish the

plaintiff's title to land by bringing an adverse party into court and requiring him to prove his claim or to forever be estopped from asserting it.

Quitclaim Deed. A deed passing any title that the grantor may have in property, but not professing that such title is valid and containing no warranty for title.

Quo warranto. A writ intended to prevent the exercise of powers not conferred by law.

Ratification. The approval, by act, word, or conduct, of that which was attempted.

Recourse. Resort to a person who is secondarily liable after the default of the person who is primarily liable. Used with reference to negotiable instruments.

Redeem. To purchase back, as mortgaged property, by paying what is due on the mortgage debt, or to receive back by paying the obligation.

Reformation. An equitable remedy by which a written instrument is made or construed to express the real intention of the parties when some error or mistake has been committed in the instrument.

Reimbursement. Refund or payment back.

Rejection. Refusal after consideration. Used with regard to the law of contracts as where an offeree rejects the offer made by an offeror.

Remainder. An estate in land limited to take effect after another estate is determined. In wills, the remainder of an estate is what remains after all other provisions of the will have been satisfied.

Replevin. A possessory action that is appropriate to determine which of the two contending persons is the owner of specific personal property. One of the old common-law writs to recover possession of personal property.

Repudiate. To deny.

Requirements Contract. A contract in which buyer agrees to buy all his required commodities from one seller.

Res judicata. Literally, "the thing is decided." The doctrine under which in a civil case the same parties cannot relitigate the same issues once a final decision has been rendered.

Rescind. To avoid, withdraw, or cancel a contract.

Respondeat Superior. A doctrine of implied agency by reason of relationship of master and servant.

Restraint of Trade. An action whose performance would limit competition in business or restrict the promissor in the exercise of gainful occupation.

Restrictive Indorsement. As used in the Uniform Commerical Code, an indorsement is restrictive that either is conditional; or supports to prohibit further transfer of the instrument; or includes the words "for collection," "for deposit," "pay any bank," or like terms signifying a purpose of deposit or collection or otherwise states that it is for the benefit or use of the indorser or of another person.

Resulting Trust. A trust raised by implication of law and presumed to have been contemplated by the parties.

Retained Earnings. That portion of profits earned that is not paid out in the form of dividends, which has the purpose of expanding or maintaining the business.

Reversion. The estate left in the grantor during the continuance of a particular estate. Also, the residue left in the grantor or his heirs after termination of a particular estate.

Revocation. The recall of a power or authority conferred on the vacating of an instrument previously made. As used in contract law, when the offeror withdraws or revokes his offer.

Riparian Rights. Rights of an owner of the land adjoining the banks of waterways such as rivers and streams.

Scienter. Knowledge making a person responsible for the consequences of acts. It is an integral element of making false promises or committing a breach of trust.

Seal. As used in the common law, a particular sign made to attest in the most formal manner the execution of an instrument. Under the old common law a contract under seal was a formal contract and required no consideration. This use has been abolished today, and the seal is less important.

Secondary Party. As used in the Uniform Commercial Code regarding negotiable instruments, a drawer or indorser; a party whose liability is secondary to that of a primary party.

Security Interest. As used in the Uniform Commercial Code, an interest in personal property or fixtures that secures payment or performance of an obligation.

Seisin. The right of possession of real property.

Servient Tenement. An estate that owes service to a dominant tenement. Used frequently to describe the estate that gives the benefit to another in an easement appurtenant.

Setoff. A counter action against the plaintiff that grows out of a matter independent of the plaintiff's cause of action.

Settlor. A person who creates a trust, the trustor.

Several Liability. Liability applied to a number of persons; usually implying that each one is liable separately.

Severalty. Sale and exclusive possession. An estate in severalty is held by one person in his own right only.

Shareholder. One who holds or owns a share or shares in a joint stock or incorporated company in a common fund or in some property. He is not a corporation creditor, and has no rights as such.

Sine qua non. Literally, that without which the thing cannot be.

Slander. Oral defamation. Words falsely spoken, which are published and injurious to the reputation of another.

Special Warranty Deed. A deed warranting title only against those claiming by, through, or under the grantor.

Specific Performance. An equitable remedy, which compels such substantial performance of the contract as will do justice between the parties.

Stale Check. As used in the Uniform Commercial Code, a check that is not presented for payment within six months of issue.

Stare decisis. The principle that states that courts are obliged to adhere to results of decided cases and to refrain from disturbing general principles that have been established by judicial determination.

Statute of Descent. Statute that determines how the property of an estate shall pass when one dies without leaving a valid will.

Statute of Frauds. The Statute passed originally in England in 1677 and adopted in one form or another in most jurisdictions within the United States. It sets forth those types of contractual situations in which the contract must be in writing or a written memorandum thereof signed by the party to be charged in order to be enforceable.

Statute of Limitation. The Statute of the state determining that after a lapse of a specified time, a claim shall not be enforced in a judicial proceeding.

Stockholder. The owner or holder of shares in a corporation having capital stock represented by shares.

Stock Subscription. A contract by which a subscriber is bound to pay the company certain amounts in return for its stock.

Strict Liability. A tort doctrine pursuant to which under certain circumstances a party may be held liable without fault for injuries befalling another.

Subinfeudation. A system in feudal England by which tenants were granted smaller estates out of those which they held of their lords, to be held of themselves as inferior lords.

Sublease. Agreement by which a lessee underlets the premises or a part thereof to a third person for a period less than the lessee's term.

Subpoena. An order to a person to appear as a witness in court.

Subrogation. The putting of one to whom a particular right does not legally belong in the position of legal owner of the right.

Summary Judgment. A motion to determine if there is sufficient evidence to justify a trial upon the issues made by the pleadings.

Summons. A call by authority, or by command of a superior, to appear at a place named or to attend to some duty.

Surety. One who binds himself for the payment of money or the performances of an act for another already bound.

Surplus. As defined in the Model Corporation Act, the excess of net assets of a corporation over its stated capital.

Surrender. To yield to power or possession of another; to give or deliver up possession of anything on compulsion or demand.

Tacit. Done or made in silence; implied or indicated, but not actually expressed.

Tacking. Successive, uninterrupted possessions of persons between whom privity exists. Used in connection with the concept of adverse possession where the consecutive periods of possession of property by two or more people can be tacked or cumulated to equal the required period.

Tangible. Having physical existence; able to be felt or touched.

Tenant. One who holds or possesses lands or tenements by any kind of title, either in fee, for life, for years, or at will.

Tenant-in-chief. Under feudal law, a tenant who held his land directly from the king.

Tenant in Common. A type of ownership by two or more people who have unity of possession.

Tender. An offer to perform an act which the party offering is bound to perform.

Testate. Having left a will.

Testator. One who makes a will.

Third Party Beneficiary. One to whom the promisee owes or is believed to owe a duty which is discharged by the promissor' performances. The contractual right the third party beneficiary acquires is to enforce a promise made for his benefit which he otherwise would not be able to enforce.

Tort. A private or civil wrong that is not a breach of contract; the violation of some duty owed another that causes his injury.

Trade Acceptance. A draft or bill of exchange drawn by the seller on the purchaser of goods sold and accepted by such purchaser.

Trade Fixture. A fixture placed in a building rented by the tenant to carry on his trade or business.

Trading Partnership. A partnership that engages in the business of buying and selling.

Trespass. At common law, an unlawful act that causes injury to the person, property, or relative rights of another.

Trespass *de bonis asportatis.* A common-law action brought by an owner of goods to recover damages for taking and carrying them away.

Trespass on the Case. A common-law form of action to recover damages for consequential injury arising either in tort or in contract.

Trespass *quare clausum fregit.* The breaking and entering by force and arms of the plaintiff's property.

Trespass *vi et armis.* The action for an injury to a person that is the immediate result of an unlawful act.

Trial Court. The court that determines the facts of the case, applies the law to those facts, and arrives at a verdict.

Trover. A remedy to recover personal chattels wrongfully converted by another to his own use.

Trust. An obligation on a person arising out of confidence reposed in him, to apply property faithfully and according to such confidence. A type of arrangement where the legal title to property is given to a trustee and the equitable title or interest to a beneficiary.

Trust Corpus. Property that is the subject of a trust.

Trustee. The party to whom the trust property is given who holds it for the benefit of another.

Ultra vires. An act beyond the powers of the corporation.

Unconscionable. Conduct is *unconscionable* when it is so harsh as to be shocking to the conscience.

Undue Influence. Influence that destroys free agency and brings about results that are not the product of the free will of the person influenced.

Unilateral Contract. A contract in which a promise is given in exchange for an act or forbearance.

Usury. The taking of a greater premium or interest for the use of money loaned than the law allows.

Valid. Good or sufficient in point of law; sustainable and effective in law.

Venue. The place or territory within which either party to a suit may require the case to be tried.

Verdict. The determination of the jury upon the facts in issue in a case.

Vested. Fixed, not subject to defeat, giving the rights of absolute ownership.

Void. Having no legal force or existence.

Void *ab initio.* Having never come into effect; void from the beginning.

Voidable. Capable of being avoided or declared invalid.

Voir Dire. Preliminary examination that the court may make of one presented as a witness or juror where his competency, interest, etc., are questioned. Generally used in regard to the questioning of prospective members of a jury.

Voting Trust. A device by which stockholders divorce voting rights from ownership by retaining the ownership and transferring voting rights to trustees in whom voting rights of all depositors in the trust are pooled.

Waiver. A giving up, relinquishing, or surrender of some known right; takes place where a person dispenses with the performance of something that he has a right to exact.

Ward. A person placed under the care of a guardian, most often an infant.

Warehouseman. A person lawfully engaged in the business of storing goods for profit.

Warrant. Authorization; command; for example, an arrest warrant or search warrant may be issued by a member of the judicial branch.

Warranty. An engagement or undertaking, express or implied, that a certain fact regarding the subject of the contract is or shall be as it is expressly or impliedly declared or promised to be.

Warranty Deed. A deed with covenants of a general warranty whereby the grantor of real property makes certain warranties of title to the grantee.

Waste. Every wrongful act of mismanagement of the property rights, or interests of an estate, causing a loss or damage.

Watered Stock. Stock that purports to be paid in full, but that in fact has not been fully paid for.

Will. An expression of a person's desires as to the particular dispositions to be made of his or her property.

Winding up. Administration of assets for the purpose of terminating business and discharging the obligations of the business to its members.

Writ. A mandatory precept issuing from a court of justice.

INDEX